SOME FUNDAMENTAL PHYSICAL CONSTANTS

Constant	Symbol	Value
speed of light in a vacuum	c	3.00×10^8 m/s
gravitational constant	G	6.67×10^{-11} m^3/kg·s^2
Avogadro's number	N_A	6.02×10^{23} mol^{-1}
universal gas constant	R	8.31 J/mol·K
Boltzmann's constant	k	1.38×10^{-23} J/K
elementary charge	e	1.60×10^{-19} C
permittivity of free space	ϵ_0	8.85×10^{-12} C^2/N·m^2
	$1/(4\pi\epsilon_0)$	8.99×10^9 kg·m^3·s^{-2}·C^{-2}
permeability of free space	μ_0	$4\pi \times 10^{-7}$ T·m/A
electron mass	m_e	9.11×10^{-31} kg
proton mass	m_p	1.67×10^{-27} kg
neutron mass	m_n	1.67×10^{-27} kg
Planck's constant	h	6.63×10^{-34} J·s
$h/2\pi$	$\hbar$	1.05×10^{-34} J·s
		$= 6.58 \times 10^{-22}$ MeV·s
	$\hbar c$	197 MeV·fm
electron charge-to-mass ratio	$-e/m_e$	-1.76×10^{11} C/kg
proton-electron mass ratio	m_p/m_e	1840
molar volume of ideal gas at STP		22400 cm^3/mol
Bohr magneton	μ_B	9.27×10^{-24} J/T
magnetic flux quantum	$\Phi_0 = h/2e$	2.07×10^{-15} Wb
Bohr radius	a_0	0.529×10^{-10} m
Rydberg constant	R_∞	1.10×10^7 m^{-1}

SOME ASTRONOMICAL CONSTANTS

Constant	Symbol	Value
standard gravity at Earth's surface	g	9.80665 m/s^2
equatorial radius of Earth	R_e	6.374×10^6 m
mass of Earth	M_e	5.976×10^{24} kg
mass of Moon		7.350×10^{22} kg $= 0.0123\ M_e$
mean radius of Moon's orbit around Earth		3.844×10^8 m
mass of Sun	$M_\odot$	1.989×10^{30} kg
mean radius of Earth's orbit around Sun	AU	1.496×10^{11} m
period of Earth's orbit around Sun	yr	3.156×10^7 s
diameter of our galaxy		7.5×10^{20} m
mass of our galaxy		2.7×10^{41} kg $= (1.4 \times 10^{11})\ M_\odot$
Hubble parameter	H	2.1×10^{-18} s^{-1}

SOME SI BASE UNITS

Physical Quantity	Name of Unit	Symbol
length	meter	m
mass	kilogram	kg
time	second	s
electric current	ampere	A
thermodynamic temperature	kelvin	K
amount of substance	mole	mol

SOME SI DERIVED UNITS

Physical Quantity	Name of Unit	Symbol	SI Unit
frequency	hertz	Hz	s^{-1}
energy	joule	J	$kg \cdot m^2/s^2$
force	newton	N	$kg \cdot m/s^2$
pressure	pascal	Pa	$kg/m \cdot s^2$
power	watt	W	$kg \cdot m^2/s^3$
electric charge	coulomb	C	$A \cdot s$
electric potential	volt	V	$kg \cdot m^2/A \cdot s^3$
electric resistance	ohm	Ω	$kg \cdot m^2/A^2 \cdot s^3$
capacitance	farad	F	$A^2 \cdot s^4/kg \cdot m^2$
inductance	henry	H	$kg \cdot m^2/A^2 \cdot s^2$
magnetic flux	weber	Wb	$kg \cdot m^2/A \cdot s^2$
magnetic flux density	tesla	T	$kg/A \cdot s^2$

SI UNITS OF SOME OTHER PHYSICAL QUANTITIES

Physical Quantity	SI Unit
speed	m/s
acceleration	m/s^2
angular speed	rad/s
angular acceleration	rad/s^2
torque	$kg \cdot m^2/s^2$, or $N \cdot m$
heat flow	J, or $kg \cdot m^2/s^2$, or $N \cdot m$
entropy	J/K, or $kg \cdot m^2/K \cdot s^2$, or $N \cdot m/K$
thermal conductivity	$W/m \cdot K$

SOME CONVERSIONS OF NON-SI UNITS TO SI UNITS

Energy:

1 electron-volt (eV) $= 1.6022 \times 10^{-19}$ J
1 erg $= 10^{-7}$ J
1 British thermal unit (BTU) $= 1055$ J
1 calorie (cal) $= 4.186$ J
1 kilowatt-hour (kWh) $= 3.6 \times 10^6$ J

Mass:

1 gram (g) $= 10^{-3}$ kg
1 atomic mass unit (u) $= 931.5$ MeV/c^2
$\qquad\qquad = 1.661 \times 10^{-27}$ kg
1 MeV/$c^2 = 1.783 \times 10^{-30}$ kg

Force:

1 dyne $= 10^{-5}$ N
1 pound (lb or #) $= 4.448$ N

Length:

1 centimeter (cm) $= 10^{-2}$ m
1 kilometer (km) $= 10^3$ m
1 fermi $= 10^{-15}$ m
1 Angstrom (Å) $= 10^{-10}$ m
1 inch (in or ") $= 0.0254$ m
1 foot (ft) $= 0.3048$ m
1 mile (mi) $= 1609.3$ m
1 astronomical unit (AU) $= 1.496 \times 10^{11}$ m
1 light-year (ly) $= 9.46 \times 10^{15}$ m
1 parsec (ps) $= 3.09 \times 10^{16}$ m

Angle:

1 degree (°) $= 1.745 \times 10^{-2}$ rad
1 min (') $= 2.909 \times 10^{-4}$ rad
1 second (") $= 4.848 \times 10^{-6}$ rad

Volume:

1 liter (L) $= 10^{-3}$ m^3

Power:

1 kilowatt (kW) $= 10^3$ W
1 horsepower (hp) $= 745.7$ W

Pressure:

1 bar $= 10^5$ Pa
1 atmosphere (atm) $= 1.013 \times 10^5$ Pa
1 pound per square inch (lb/in^2) $= 6.895 \times 10^3$ Pa

Time:

1 year (yr) $= 3.156 \times 10^7$ s
1 day (d) $= 8.640 \times 10^4$ s
1 hour (h) $= 3600$ s
1 minute (min) $= 60$ s

Speed:

1 mile per hour (mi/h) $= 0.447$ m/s

Magnetic field:

1 gauss $= 10^{-4}$ T

Volume I

PHYSICS

FOR SCIENTISTS AND ENGINEERS

Volume I

PHYSICS

FOR SCIENTISTS AND ENGINEERS

THIRD EDITION

Paul M. Fishbane
University of Virginia

Stephen G. Gasiorowicz
University of Minnesota

Stephen T. Thornton
University of Virginia

PEARSON

Prentice Hall

Upper Saddle River, New Jersey 07458

Library of Congress Cataloging-in-Publication Data
Fishbane, Paul M.
    Physics for scientists and engineers / Paul M. Fishbane, Stephen G. Gasiorowicz,
Stephen T. Thornton.—3rd ed.
        p.   cm.
    Includes indexes.
    ISBN 0-13-142094-1—ISBN 0-13-141883-1
ISBN 0-13-141881-5—ISBN 0-13-141882-3
    1. Physics. I. Gasiorowicz, Stephen. II. Thornton, Stephen T. III. Title.

QC23.2.F58   2005
530—dc22

2003058209

Senior Editor: Erik Fahlgren
Editor in Chief, Science: John Challice
Development Editor: Catherine Flack
Executive Project Manager: Ann Heath
Editor in Chief, Development: Carol Trueheart
Executive Managing Editor: Kathleen Schiaparelli
Assistant Managing Editor: Beth Sweeten
Vice President ESM Production and Manufacturing: David W. Riccardi
Manufacturing Manager: Trudy Pisciotti
Manufacturing Buyer: Alan Fischer
Creative Director: Carole Anson
Art Director: John Christiana
Director of Creative Services: Paul Belfanti
Assistant Managing Editor, AV Production & Management: Patty Burns
AV Art Editor: Adam Velthaus
Artwork: Imagineering
Media Editor: Paul Draper
Assistant Managing Editor, Science Media: Nicole Bush
Assistant Managing Editor, Science Supplements: Becca Richter
Executive Marketing Manager: Mark Pfaltzgraff
Associate Editor: Christian Botting
Editorial Assistant: Andrew Sobel
Interior Design: Dina Curro
Cover Design: Kiwi Design
Cover Photo: Antonio Mo/Getty Images, Inc. -Taxi
Photo Research: Yvonne Gerin
Photo Editor: Nancy Seise
Production Services/Composition: Preparé Inc.

© 2005, 1996, 1993 Pearson Education Inc.
Pearson Prentice Hall
Pearson Education, Inc.
Upper Saddle River, New Jersey 07458

Printed in the United States of America
10 9 8 7 6 5 4 3 2 1

ISBN 0-13-141883-1

Pearson Education Ltd., *London*
Pearson Education Australia Pty. *Limited, Sydney*
Pearson Education *Singapore Pte. Ltd.*
Pearson Education North Asia Ltd., *Hong Kong*
Pearson Education Canada, Ltd., *Toronto*
Pearson Educación de Mexico, *S.A. de C.V*
Pearson Education—*Japan, Tokyo*
Pearson Education *Malaysia, Pte. Ltd.*

Dedication

To our students, the most important element in the making of this book

Brief Contents

1 Tooling Up 1

2 Straight-Line Motion 28

3 Motion in Two and Three Dimensions 60

4 Newton's Laws 87

5 Applications of Newton's Laws 119

6 Work and Kinetic Energy 151

7 Potential Energy
and Conservation of Energy 183

8 Linear Momentum, Collisions,
and the Center of Mass 209

9 Rotations of Rigid Bodies 246

10 More on Angular Momentum and Torque 280

11 Statics 313

12 Gravitation 338

13 Oscillatory Motion 366

14 Waves 397

15 Superposition and Interference
of Waves 435

16 Properties of Fluids 462

17 Temperature and Ideal Gases 491

18 Heat Flow and the First Law
of Thermodynamics 515

19 The Molecular Basis of Thermal Physics 548

20 The Second Law of Thermodynamics . . . 573

21 Electric Charge

22 Electric Field

23 Gauss' Law

24 Electric Potential

25 Capacitors and Dielectrics

26 Currents in Materials

27 Direct-Current Circuits

28 The Effects of Magnetic Fields

29 The Production and Properties
of Magnetic Fields

30 Faraday's Law

31 Magnetism and Matter

32 Inductance and Circuit Oscillations

33 Alternating Currents

34 Maxwell's Equations and Electromagnetic
Waves

35 Light

36 Mirrors and Lenses and Their Uses

37 Interference

38 Diffraction

39 Special Relativity

40 Quantum Physics

41 Quantization of Angular Momentum and
of Energy Values

42 Quantum Effects in Large Systems of
Fermions and Bosons

43 Quantum Engineering

44 Nuclear Physics

45 Particles and Cosmology

Appendices A-1

Answers to "What Do You Think?" . . . Q-1

Answers to Understanding
the Concepts Questions C-1

Answers to Odd-Numbered
Problems O-1

Photo Credits P-1

Index I-1

Contents

PREFACE   xvii

1 TOOLING UP   1

1-1 A Little Background 2

1-2 Fundamental Physical Quantities and Their Units 4

1-3 Accuracy and Significant Figures 9

1-4 Dimensional Analysis 12

1-5 Estimates: How a Little Reasoning Goes a Long Way 14

1-6 Scalars and Vectors 15

Summary, 22   Understanding the Concepts, 22
Problems, 23

2 STRAIGHT-LINE MOTION   28

2-1 Displacement 28

2-2 Speed and Velocity 31

2-3 Acceleration 36

2-4 Motion with Constant Acceleration 40

2-5 Freely Falling Objects 44

*2-6 Integration and Motion in One Dimension 48

Summary, 51   Understanding the Concepts, 52
Problems, 53

3 MOTION IN TWO AND THREE DIMENSIONS   60

3-1 Position and Displacement 60

3-2 Velocity and Acceleration 62

3-3 Motion with Constant Acceleration 66

3-4 Projectile Motion 68

3-5 Uniform Circular Motion 74

3-6 Relative Motion 77

Summary, 80   Understanding the Concepts, 81
Problems, 82

4 NEWTON'S LAWS   87

4-1 Forces and Newton's First Law 88

4-2 Newton's Second Law of Motion 92

4-3 Newton's Third Law of Motion 97

*4-4 Noninertial Frames 99

4-5 Using Newton's Laws: Identifying Forces and Free-Body Diagrams 103

4-6 Using Newton's Laws: Finding the Motion 109

Summary, 110   Understanding the Concepts, 111
Problems, 112

Note: Sections marked with a * can be considered optional.

5 APPLICATIONS OF NEWTON'S LAWS   119

5-1 Common Forces Revisited 119

5-2 Friction 127

5-3 Drag Forces 133

5-4 Forces and Circular Motion 136

5-5 Fundamental Forces 141

Summary, 142   Understanding the Concepts, 143 Problems, 144

6 WORK AND KINETIC ENERGY   151

6-1 Kinetic Energy, Work, and the Work–Energy Theorem 152

6-2 Constant Forces in More Than One Dimension 159

6-3 Forces That Vary with Position 162

6-4 Conservative and Nonconservative Forces 169

6-5 Power 172

*6-6 Kinetic Energy at Very High Speeds 174

Summary, 175   Understanding the Concepts, 176 Problems, 177

7 POTENTIAL ENERGY AND CONSERVATION OF ENERGY   183

7-1 Potential Energy and Conservative Forces 184

7-2 Energy Conservation and Allowed Motion 191

7-3 Motion in Two or Three Dimensions 195

7-4 Is Energy Conservation a General Principle? 198

Summary, 201   Understanding the Concepts, 201 Problems, 202

8 LINEAR MOMENTUM, COLLISIONS, AND THE CENTER OF MASS   209

8-1 Momentum and Its Conservation 210

8-2 Collisions and Impulse 212

8-3 Perfectly Inelastic Collisions; Explosions 216

8-4 Elastic Two-Body Collisions in One Dimension 220

8-5 Elastic Collisions in Two and Three Dimensions 223

8-6 Center of Mass 225

*8-7 Rocket Motion 234

8-8 Momentum Transfer at High Energies 237

Summary, 238   Understanding the Concepts, 239 Problems, 240

9 ROTATIONS OF RIGID BODIES   246

9-1 Simple Rotations of a Rigid Body 247

9-2 Rotational Kinetic Energy 251

9-3 Evaluation of Rotational Inertia 253

9-4 Torque 259

9-5 Angular Momentum and Its Conservation 265

9-6 Rolling 267

Summary, 272   Understanding the Concepts, 273 Problems, 274

10 MORE ON ANGULAR MOMENTUM AND TORQUE 280

10-1 Generalization of Angular Momentum 280

10-2 Generalization of Torque 285

10-3 The Dynamics of Rotation 288

10-4 Conservation of Angular Momentum 293

10-5 Work and Energy in Angular Motion 298

10-6 Collecting Parallels Between Rotational and Linear Motion 300

*10-7 Quantization of Angular Momentum 301

*10-8 Precession 302

Summary, 305   Understanding the Concepts, 306   Problems, 307

12-5 A Closer Look at Gravitation 355

*12-6 Einstein's Theory of Gravitation 357

Summary, 359   Understanding the Concepts, 360   Problems, 361

13 OSCILLATORY MOTION 366

13-1 The Kinematics of Simple Harmonic Motion 367

13-2 A Connection to Circular Motion 370

13-3 Springs and Simple Harmonic Motion 371

13-4 Energy and Simple Harmonic Motion 374

13-5 The Simple Pendulum 378

13-6 More About Pendulums 381

13-7 Damped Harmonic Motion 383

13-8 Driven Harmonic Motion 385

Summary, 388   Understanding the Concepts, 389   Problems, 390

11 STATICS 313

11-1 Static Conditions for Rigid Bodies 313

11-2 Gravity and Rigid Bodies 316

11-3 Applications of Statics 318

11-4 Solids and How They Respond to Forces 323

Summary, 328   Understanding the Concepts, 329   Problems, 330

12 GRAVITATION 338

12-1 Early Observations of Planetary Motion 339

12-2 Newton's Inverse-Square Law 340

12-3 Planets and Satellites 345

12-4 Gravitation and Extended Objects 349

14 WAVES 397

14-1 Types of Waves 398
14-2 The Wave Equation 401
14-3 Periodic Waves 403
14-4 Traveling Waves 405
14-5 Energy and Power in Waves 412
14-6 Standing Waves 414
14-7 More About Sound 418
14-8 The Doppler Effect 423
*14-9 Shock Waves 426
Summary, 428   Understanding the Concepts, 429
Problems, 430

15 SUPERPOSITION AND INTERFERENCE OF WAVES 435

15-1 The Superposition Principle 436
15-2 Standing Waves Through Interference 437
15-3 Beats 439
15-4 Spatial Interference Phenomena 443
15-5 Pulses 446
*15-6 Fourier Decomposition of Waves 452
*15-7 Pulses and the Uncertainty Principle 454
Summary, 456   Understanding the Concepts, 457
Problems, 457

16 PROPERTIES OF FLUIDS 462

16-1 States of Matter 462
16-2 Density and Pressure 464
16-3 Pressure in a Fluid at Rest 466
16-4 Buoyancy and Archimedes' Principle 469
16-5 Fluids in Motion 474
16-6 The Equation of Continuity 475
16-7 Bernoulli's Equation 476
16-8 Applications of Bernoulli's Equation 478
*16-9 Real Fluids 481
Summary, 483   Understanding the Concepts, 484
Problems, 485

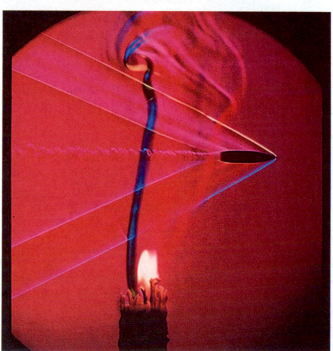

17 TEMPERATURE AND IDEAL GASES 491

17-1 Temperature and Thermal Equilibrium 492
17-2 Ideal Gases and Absolute Temperature 494
17-3 Thermal Expansion 497
17-4 The Equation of State of Gases 499
*17-5 Blackbody Radiation 503
Summary, 507   Understanding the Concepts, 507
Problems, 508

18 HEAT FLOW AND THE FIRST LAW OF THERMODYNAMICS 515

18-1 Changes in Thermal Systems 516
18-2 Heat Flow 518
18-3 Heat Flow in Materials 523
18-4 The Mechanical Equivalent of Heat 526
18-5 Work Done by Thermal Systems 528
18-6 The First Law of Thermodynamics 531
18-7 Internal Energy of Ideal Gases 532
18-8 More Applications for Ideal Gases 534
Summary, 538   Understanding the Concepts, 539
Problems, 540

19 THE MOLECULAR BASIS OF THERMAL PHYSICS 548

19-1 A Microscopic View of Gases 549
19-2 Pressure and Molecular Motion 550
19-3 The Meaning of Temperature 552
19-4 Probability Distributions 554
19-5 The Velocity Distribution of Gases 557
19-6 The Maxwell–Boltzmann Distribution 559
*19-7 Collisions and Transport Phenomena 562
Summary, 566   Understanding the Concepts, 567
Problems, 567

20 THE SECOND LAW OF THERMODYNAMICS 573

20-1 Beyond Energy Conservation 574
20-2 The Second Law of Thermodynamics 580
20-3 The Carnot Cycle 581
20-4 Other Types of Engines 586
20-5 Entropy and the Second Law 590
20-6 Entropy and Ideal Gases 594
20-7 The Meaning of Entropy 596
Summary, 600   Understanding the Concepts, 601
Problems, 601

APPENDIX I
THE SYSTÈME INTERNATIONALE (SI) OF UNITS A-1

APPENDIX II
SOME FUNDAMENTAL PHYSICAL CONSTANTS A-2

APPENDIX III
OTHER PHYSICAL QUANTITIES A-3

APPENDIX IV
MATHEMATICS A-5

APPENDIX V
PERIODIC TABLE OF THE ELEMENTS A-8

APPENDIX VI
SIGNIFICANT DATES IN THE DEVELOPMENT OF PHYSICS A-9

ANSWERS TO "WHAT DO YOU THINK?" Q-1

ANSWERS TO UNDERSTANDING THE CONCEPTS QUESTIONS C-1

ANSWERS TO ODD-NUMBERED PROBLEMS O-1

PHOTO CREDITS P-1

INDEX I-1

Applications in the Text

Note: This list includes applied topics that receive significant discussion in the chapter text, Think About This sections, Conceptual Examples and worked Examples. Many other applied topics are introduced briefly or appear in the end-of-chapter Understanding the Concepts Questions and Problems. Topics that appear in a Think About This section appear in blue.

Chapter 1

What does the global positioning system measure? 6
What governs the proportions of things? 14
Estimating the size of a molecule in the human body 15

Chapter 2

Filming a 100-m dash 31
Determining a runner's acceleration 37
What is an accelerometer and what are its uses? 39
Calculating takeoff speed and runway length for a T-83 training jet 43

Chapter 3

Plotting the trajectory of a car 62
Is it possible to experience free fall for long periods? 73
"Lateral acceleration" air bags in cars 77
Crossing a river 79
Relative motion and the jet stream 79

Chapter 4

How does an air bag protect you in a crash? 97
How does a rocket work? 99
Skateboard jumping 101
Is Earth's surface an inertial frame? 103

Chapter 5

The strength of fishing line 122
Friction acts backward—how can it allow us to walk forward? 132
Accelerating a car without spinning the wheels 132
Where does friction come from? 133
How do several skydivers land at the same location? 134
Using a bathroom scale in space 137
Turning on a icy road 138

Chapter 6

What work is done when we lift an object from the floor? 154
How can we estimate the energy of a snow avalanche? 155
Moving a sofa across a carpeted floor 155
How did the Egyptians build the pyramids? 159
Marketing James Watt's steam engine 173

Chapter 7

Calculating the force experienced by the Moon 197
How does our society produce and use energy? 200

Chapter 8

Comparing the rates of momentum change in Ping-Pong and bowling balls 210
Velocity in billiard ball collisions 211
How are the foundations for tall buildings made? 214
Calculating the speed of decay in different nuclei 219
How does a jet engine work? 220
Finding the speed of a Saturn V rocket after combustion ceases 236
How do we know that atoms have a nuclear structure? 237

Chapter 9

Can you run a car using rotational kinetic energy? 252
Can angular momentum be transferred in collisions? 267

Chapter 10

What is the role of internal forces in the torque on a system? 288
How does a submarine navigate 'silently'? 294
Finding the angular momentum of a comet 294
Making a sci-fi movie about an asteroid that collides with Earth 303

Chapter 11

What forces keep a rock climber in equilibrium? 315
Is cement an indispensable element of construction? 317
Finding the tension in the cable of a construction crane 319
Why don't construction cranes tip over? 322
What makes a solid strong? 327

Chapter 12

Retrograde motion of Mars 340
Calculating the height above Earth of a satellite 343
How do we know Earth's mass? 345
What determines how big mountains can get? 355

Chapter 13

Why were pendulum clocks so useful? 379

Examining damped harmonic motion in a gong struck by a hammer 385

How can we reduce the wind-driven sway of skyscrapers? 386

Chapter 14

Waves moving into a harbor 406

Why are longitudinal waves in solids called sound? 409

How do we know what's inside Earth? 411

Why does a distant shout sound fainter than a nearby one? 414

Exciting standing waves on a guitar string 415

What do standing waves look like in two and three dimensions? 417

Waves in a pipe organ 421

Chapter 15

The frequency of tuning forks 442

How do noise-canceling headphones work? 447

How are reflected and transmitted waves of use to us? 452

Does the reciprocal relation have practical consequences? 456

Chapter 16

The density of a neutron star 465

Lifting a car with a hydraulic press 468

Finding the height of a column of mercury 468

Water in a garden hose 476

How is fluid speed in a closed pipe measured? 478

Why does a giraffe need a strong heart? 479

Why do racing cars have wings? 481

Of what use are the dimples on a golf ball? 483

Chapter 17

Is it easy to make a pot of tea in a space station? 493

Why can a jar lid be loosened by hot water? 498

Allowing for linear expansion in a steel bridge 498

Temperature changes in a stored can of gasoline 499

Why shouldn't you pressurize your tires maximally in hot climates? 500

How do we get the fizz into dispensed drinks? 502

Estimating the temperature at the Sun's surface 505

Chapter 18

What are the origins of the terms calorie, calorimetry, caloric? 521

Why are latent heats of vaporization significantly larger than latent heats of fusion? 522

Melting a tin soldier 523

Keeping an igloo cozy 524

Chapter 19

Comparing neon and argon atoms 553

What does absolute zero mean at the molecular level? 553

How likely is it that random guessing will produce a perfect score? 556

Why is there so little hydrogen in Earth's atmosphere? 559

How does the molecular picture explain thermal energy conduction in materials? 565

Chapter 20

How do perpetual motion machines work? 580

A Carnot engine with ideal gas 583

Running a Carnot engine in reverse 584

The stirling engine 586

The Diesel, Otto and Brayton cycles 587

Why is it so hard to make automobile engines that don't pollute? 588

Heat pumps and refrigerators 588

Preface

This text is designed for a calculus-based physics course at the beginning university and college level. It is written with the expectation that students either have taken or are currently taking a beginning course in calculus. Students taking a physics course based on this book should leave with a solid conceptual understanding of the fundamental physical laws, how these laws can be applied to solve many problems, and how physics is relevant to the world around them.

"Understanding" encompasses our three overriding goals for this book:

Doing Things Correctly.   First, we want a book that is fundamentally *right*. This is sometimes taken to mean "rigorous," but we want to emphasize that we do not feel "rigorous" is a synonym for "difficult." Rather, we associate "right" with showing all the evidence, with using the evidence correctly to support the point being made, and with making the point in a way that will allow the student to say, "I see where that comes from." We try to avoid the phrase "It can be shown that . . ." and the attitude that "It is true because we say it is."

Conceptual Emphasis.   Second, we want students to understand the material at as deep a conceptual level as possible. We are aware that there is a large gap between being able to get the right answer to a physics problem and comprehending the physical concepts that lie behind it. We want students to be able to answer the *why* as well as the *how*. The student who has a conceptual understanding can not only do problems for which he or she has models, but can also approach a new problem with confidence. Many of the changes to the third edition are designed to address this issue, including the addition of Conceptual Examples, a "What Do You Think" step at the end of most Examples, and a substantial increase in the number of end-of-chapter qualitative questions.

Modern Physics Integrated.   Third, without sacrificing the essential aspects of classical physics, we have included modern notions throughout the book. Classical topics have lost none of their importance and must form the basis of any first course. However, what is traditionally called "modern" physics—the topics centered on relativity and quantum physics—began about a century ago. It hardly seems possible to ignore these topics in view of their importance for today's technology and, more critically, for understanding today's world. Many of the ideas of modern physics are not mathematically difficult. However, they can be nonintuitive, and we think it is important that students begin to develop intuition about this material as early as possible. Although much of this material appears in optional sections (marked with an asterisk), in many cases it is intertwined with the classical material. The uncertainty principle and its role in both classical and quantum physics, information on atomic structure and spectra, information on band structure or on blackbody radiation, and the nature and role of fundamental forces are a few of the topics that are included in this way. We conclude the text in what has become the traditional way, with chapters on modern physics. We think the preparation we laid down for this material in earlier chapters will make it more easily assimilated.

A few words about mathematics: The idea of getting it *right* applies to mathematical derivation as well as to qualitative aspects. Our approach is to introduce the mathematics that students need to know the first time they need to know it, in the context of the physics being presented. We try to make that material self-contained so that the student can understand the material without having to go elsewhere for mathematical help. In this way, the mathematics appears in progressive degrees of difficulty. We believe that this approach fosters better understanding and less reliance on formula memorization. We also feel that the ability to make quantitative estimates is one of the most important skills that a scientist or engineer can have. We have made the development of that ability an important part of our approach, both in the text and in the problems. Finally, as in real-world problems, we vary the number of significant figures in examples and in problems. We feel students should maintain an awareness of significant figures and not end up thinking all problems involve the same number of significant figures.

The Third Edition

With the help of reviewers and users of the second edition, we have made a thorough review of content and organization, with some material moved both within and among chapters to enhance the logical progression and structure of the material. We have rewritten much of the material with clarity in mind.

Organizational Changes.

Changes to the third edition include:

- Redistribution of the material in the 2nd-edition chapter "Properties of Solids"—for example, the material on heat conduction in solids now appears in the appropriate chapter on thermal physics, while material on stress and strain now appears in the chapter on statics.

- Redistribution of the material on waves between Chapters 14 and 15 to create a more logical division between single waves, and the superposition of several.

- Consolidation of some material that we feel does not affect the basic understanding of the subject—for example, both the "physical optics" and "magnetism in matter" chapters are more compact.

Conceptual Examples. We have added a new type of Example, designed to help the student think about the material in a way that emphasizes conceptual understanding of the content. These may have some modest algebraic content, such as a simple estimate or reasoning involving inequalities. There are two or more of these per chapter.

CONCEPTUAL EXAMPLE 17–4 We learn by experience that we can loosen a metal lid that is stuck on a glass jar by pouring hot water over the lid. Why does this work?

Answer As the lid's temperature rises, it expands. Movement occurs where the glass and metal are stuck and the lid releases (Fig. 17–11). In fact, you could dip the entire system (lid and glass container) in hot water, and the different expansions of the metal and the glass will lead to the same result. Note that the same difference in the thermal expansion is used in making a good seal: lids are placed on jars when the contents are hot.

▶ **FIGURE 17–11** By pouring hot water over this lid, you can take advantage of thermal expansion to free it from the jar.

Think About This. The primary purpose of these sections is to pose and answer questions about a new idea or the application of the material discussed. When writing these sections, we ask the kinds of questions a good student might be asking on his or her own, and which the majority of students will find intriguing.

Worked Examples. We have introduced a new structure into the Worked Examples to serve as a model and to build problem-solving skills. The goal of this new structure is to emphasize visualization and the identification of the knowns, unknowns, and concepts to be used. We avoid using a stiff and uncompromising framework, but most examples are broken down into a series of steps:

- **Setting It Up.** In most cases this step begins with a sketch and, for mechanics problems, the preparation of a free-body diagram. The figures accompanying most Examples are in a student sketch style that the student can realistically be expected to emulate. To reinforce this first step, all the figures in the end-of-chapter Problems are also drawn in this style. Because a sketch represents a first step in problem solving, the sketches in the end-of-chapter problems sometimes provide a crucial hint to the student. This step also includes some reasoning on how to determine what is being asked when it may not be completely obvious.

- **Strategy.** Here the concepts used to solve the problem are outlined and applied to the situation; it is where we "talk it through." This is the heart of the solution, the part where most students will succeed or fail.

- **Working It Out.** The strategy outlined in the preceding part is carried through in a series of well-defined steps. This part should be straightforward if the previous steps have been done properly.

- **What Do You Think?** The final step in solving a problem is to confirm that the answer makes sense. This section of the solution reinforces the example by asking the student a thought-provoking conceptual question associated with the problem just solved. These questions should confirm that the student has understood the concepts or send them looking for checks to the answers. Answers to "What Do You Think?" questions are provided at the end of the book.

Vectors. These are now represented with an arrow over the letter rather than in bold-face (for example, $\vec{F} = m\vec{a}$) to be more consistent with the way professors write them in lecture and the way students write them in homework and exams.

Questions. The end-of-chapter material includes qualitative questions under the heading "Understanding the Concepts." We have increased the number of these conceptual questions by nearly 50 percent.

HOW DOES AN AIR BAG PROTECT YOU IN A CRASH?

Large forces imply large accelerations. A car accident or a fall from a great height may be deadly because of the rapid deceleration, the result of large forces that your body may not be equipped to withstand. For protection it is necessary to find a way to bring you to a stop by providing a smaller deceleration over a larger time. Air bags in automobiles work on this principle; when a collision stops a car very suddenly, a passenger would suffer a very sudden deceleration in a subsequent collision with the steering wheel or the windshield. This is mitigated by the very rapid release of an air bag, which is deep enough and "soft" enough to allow the passenger to slow down over a longer period of time. Firefighters similarly use large elastic safety nets to catch people who have to jump from burning buildings. When the deceleration is for fun, the same principle applies. Bungee cords are made of a very elastic material, and there are no bungee *chains*, which would have the unfortunate effect of stopping you "on a dime." Still another application is provided by airplane ejection seats, which in the past were powered by explosives beneath the seat. The rapid acceleration of these mechanisms often led to serious damage to the pilot. Today ejection seats are powered by small rockets that can supply a smaller acceleration over a longer period of time, rather than a large acceleration over a very short period of time and hence a safer ejection (Fig. 4–12). One other example comes to mind: You may have seen drawings in which Superman catches Lois Lane just before she hits the ground. That very action would imply a rapid deceleration that would be just as bad for Lois as hitting the ground. Superman would do better not to wait for the last instant and instead slow Lois down over a longer period of time.

▲ **FIGURE 4–12** Test ejection of a pilot from an AMX jet fighter. The jet plumes below the seat are due to the ejection rocket.

EXAMPLE 4–3 You need to deliver a box of bowling balls to a bowling alley. The balls will be placed in a box that is initially at rest but that you want to push into the bowling alley. The box itself has a mass that is very small compared to even one bowling ball. You start with one ball in the box, exert a given force of given strength upon the box for a time period Δt, and at the end of that time the box moves at a speed of 3.2 m/s. You then repeat the procedure with more bowling balls in the box; you exert the same amount of force on the box for the same period of time (Δt) and find the box to have a final speed of 0.4 m/s. How many balls are in the box now?

Setting It Up The two cases are shown in Figs. 4–11a and b at the particular time, after an interval Δt, when the speeds are v_1 and v_2, respectively. You know that an identical force of constant magnitude F acts on two different masses m_1 and m_2 for identical time periods Δt, where $m_2 = nm_1$. Here m_1 is the mass of one bowling ball and n is the number of balls in the box, which is the quantity we want to find. The resulting speeds after time Δt are v_1 and v_2, respectively, and are given.

Strategy In the two cases described the box containing the bowling ball(s) is subject to the same force. Moreover, we can ignore the mass of the box. Using Newton's second law, we can find the accelerations a_1 and a_2 during the period Δt when the force operates. These accelerations are

$$a_1 = \frac{F}{m_1} \quad \text{and} \quad a_2 = \frac{F}{m_2} = \frac{F}{nm_1}.$$

Although we do not know the numerical values of the two accelerations, we do know the speeds v_1 and v_2 after a fixed period of acceleration. Further, we learned in Chapter 2 that an object that starts at rest and undergoes a fixed acceleration $\vec{a}$ for a given period of time Δt has the velocity $\Delta \vec{v} = \vec{v} = \vec{a} \, \Delta t$. In our one-dimensional case, then, we have

$$v_1 = a_1 \Delta t = \frac{F \, \Delta t}{m_1} \quad \text{and} \quad v_2 = a_2 \Delta t = \frac{F \, \Delta t}{nm_1}.$$

We now have enough information to solve for the unknown, n.

Working It Out We can solve for the ratio F/m in terms of v_1 and Δt and substitute it into the equation for v_2, which we can then solve for n. Alternatively, we can simply take the ratio of the two speeds:

$$\frac{v_1}{v_2} = \frac{(F \, \Delta t / m_1)}{(F \, \Delta t / nm_1)} = n.$$

Numerical substitution gives $n = (3.2 \text{ m/s})/(0.4 \text{ m/s}) = 8$ bowling balls.

What Do You Think? Suppose the mass of the box is not negligible. What would be the effect?

(a)

(b)

▲ **FIGURE 4–11** Delivering bowling balls to an alley. In (a) the box has one ball, and in (b) there is an unknown number of balls.

Problems. Approximately 40 percent of the end-of-chapter Problems are new or revised. We have separated the Problems into two categories. The first group of Problems is organized by section. The second group, called General Problems, resembles the situations that are met in real-life science and engineering. They also help to develop the student's appreciation of the links that exist throughout physics as well as how to approach problems for which the clues may be more obscure. All Problems are labeled I, II, or III. Level I Problems are the least difficult. These Problems develop student recognition of particular physics concepts and build confidence. Level II Problems typically have more than one step and require an increased understanding of the material; the General Problems carry this requirement a step further. Level III Problems are especially challenging, in some cases demanding significant synthesis of concepts in the text. The gradations in problem range and difficulty allow you to tailor the Problems you assign to the capabilities of your class and to the subjects that interest you the most.

Versions of the Text

The third edition is available in two hardback versions, Extended with Modern Physics (Chapters 1–45) and Standard (Chapters 1–40), as well as in three softcover volumes: Volume I (Chapters 1–20), Volume II (Chapters 21–38), and Volume III (Chapters 39–45).

Supplements

Instructor Supplements

Instructor's Resource CD-ROM Package (0-13-039150-6) This CD-ROM set includes virtually every electronic asset you'll need in and out of the classroom. Though you can navigate freely through the CDs to find the resources you want, the included software allows you to browse and search through the catalog of assets. The CD-ROMs are organized by chapter and include all text illustrations and tables from *Physics for Scientists and Engineers, Third Edition*, in JPEG, Microsoft PowerPoint, and Adobe PDF formats. Instructors can preview, sequence, and play back images, as well as perform keyword searches, add lecture notes, and incorporate their own digital resources. IRCD1 also contains TestGen, a powerful dual-platform, fully networkable software program for creating tests ranging from short quizzes to long exams. Questions from the third edition *Test Item File*, including algorithmic versions, are supplied, and professors can use the Question Editor to modify existing questions or create new questions. The CD-ROMs also contain additional Powerpoint Presentations, electronic versions of the *Test Item File*, the *Instructor's Solutions Manual*, the *Instructor's Resource Manual*, and the end-of-chapter Understanding the Concepts questions and problems from *Physics for Scientists and Engineers, Third Edition*.

Instructor's Solutions Manual (Vol. I: 0-13-039157-3; Vol. II: 0-13-144741-6) Authored by *Jerry Shi (Pasadena City College)*, the ISM contains detailed, worked solutions to every problem from the text, as well as answers to the even-numbered "Understanding the Concepts" questions.

Instructor's Resource Manual (0-13-141738-X) This IRM contains lecture outlines, notes, demonstration suggestions, and other teaching resources.

Test Item File (0-13-039158-1) This test bank contains approximately 3000 multiple choice, short answer, and true/false questions, many conceptual in nature. All are referenced to the corresponding text section and ranked by level of difficulty.

Transparency Pack (0-13-039166-2) Includes approximately 400 full-color transparencies of images from the text.

PH GradeAssist Instructor's Quick Start Guide (0-13-141740-1) This guide will help adopting instructors register for and use PH GradeAssist, Prentice Hall's own online homework system. It also contains the access code necessary to create their accounts and access the course material.

Peer Instruction: A User's Manual (0-13-565441-6)
Eric Mazur (Harvard University)
A manual with ready-to-use resources for an innovative new approach to teaching introductory physics, developed by a well-known physicist and leader in physics education.

Just-in-Time Teaching: Blending Active Learning with Web Technology (0-13-085034-9)
Gregor Novak (Air Force Academy), Andrew Gavrin (IUPUI), Wolfgang Christian (Davidson College), and Evelyn Patterson (Air Force Academy)
In this resource book for educators, the four authors fully discuss just what *Just-in-Time Teaching* is. *Just-in-Time Teaching (JiTT)* is a teaching and learning methodology designed to engage students by using feedback from pre-class Web assignments to adjust classroom lessons. This allows students to receive rapid response to the specific questions and problems they are having instead of more generic lectures that may or may not address what students actually need help with. Many teachers have found that this process also encourages students to take more control of the learning process and become active and interested learners.

Physlets®: Teaching Physics with Interactive Curricular Material (0-13-029341-5)
Wolfgang Christian and Mario Belloni (Davidson College)
This manual/CD package shows physics instructors—both Web novices and Java-savvy programmers alike—how to author their own interactive curricular material using Physlets—Java applets written for physics pedagogy that can be embedded directly into HTML documents and that can interact with the user. It demonstrates the use of Physlets in conjunction with JavaScript to deliver a wide variety of Web-based interactive physics activities. It also provides examples of Physlets created for classroom demonstrations, traditional and Just-in-Time Teaching homework problems, pre- and post-laboratory exercises, and Interactive Engagement activities. More than just a technical how-to book, the manual gives instructors some ideas about the new possibilities that Physlets offer and is designed to make the transition to using Physlets quick and easy.

Student Supplements

Student Study Guide with Selected Solutions
(Vol. I: 0-13-100070-5; Vol. II: 0-13-146500-7)
David Reid (Eastern Michigan University)
The print study guide provides the following for each chapter:

- Objectives
- Chapter Review with Examples and integrated quizzes
- Reference Tools & Resources (equation summaries, important tips, and tools)
- Practice Problems by Carl Adler (East Carolina University)
- Selected Solutions for several end-of-chapter problems

PH GradeAssist Student Quick Start Guide (0-13-141741-X)   This guide contains information on how to register and use PH GradeAssist. It also contains the access code necessary for students to create their accounts and access the course.

Ranking Task Exercises in Physics: Student Edition (0-13-144851-X)
Thomas L. O'Kuma (Lee College), David P. Maloney (Indiana University–Purdue University at Fort Wayne), and Curtis J. Hieggelke (Joliet Junior College)
Ranking Task Exercises are a unique resource for physics instructors who are looking for tools to incorporate more conceptual analysis in their course. This supplement contains approximately 220 Ranking Task Exercises that cover all classical physics topics. Ranking Tasks are an innovative type of conceptual exercise that ask students to make comparative judgments about a set of variations on a particular physical situation.

Physlet® Physics (0-13-101969-4)
Wolfgang Christian and Mario Belloni (Davidson College)
This text and CD-ROM package provides ready-to-run interactive Physlet-based curricular material for both teachers and students. Physlets, award-winning Java applets written by Christian and Belloni, have been widely adopted by the physics teaching community. This book provides the first class-tested collection of ready-to-run Physlet-based material that is easy to assign (like an end-of-chapter problem of a textbook) and easy to use (material is on a CD and in the book). Neither a Web server nor an Internet connection is required.

Interactive Physics Workbook, Second Edition (0-13-067108-8)
Cindy Schwarz (Vassar College), John Ertel (Naval Academy), MSC Software
This interactive workbook and hybrid CD-ROM package is designed to help students visualize and work with specific physics problems through simulations created with Interactive Physics files. Forty problems of varying degrees of difficulty require students to make predictions, change variables, run, and visualize motion on the computer. The workbook/study guide provides instructions, physics review, hints, and questions. The accompanying hybrid CD-ROM contains everything students need to run the simulations.

Tutorials in Introductory Physics and Homework Package (0-13-097069-7)
Lillian C. McDermott, Peter S. Shaffer, and the Physics Education Group
(all of the University of Washington)
This landmark book presents a series of physics tutorials designed by a leading physics education research group. Emphasizing the development of concepts and scientific reasoning skills, the tutorials focus on the specific conceptual and reasoning difficulties that students tend to encounter. The tutorials cover a range of topics in Mechanics, Electricity and Magnetism, and Waves & Optics.

The Portable TA: A Physics Problem-Solving Guide, Second Edition
(Vol. I: 0-13-231713-3; Vol. II: 0-13-231721-4)
Andrew Elby
This two-volume set contains a collection of problems with carefully detailed strategies and solutions that provide students with additional problem-solving techniques.

MCAT Physics Study Guide (0-13-627951-1)
Joseph Boone (California Polytechnic State University-San Luis Obispo)
Since most MCAT questions require more thought and reasoning than simply plugging numbers into an equation, this study guide is designed to refresh students' memories about the topics they've covered in class. Additional review, practice problems, and review questions are included.

Science on the Internet: A Student's Guide (0-13-028253-7)
Andrew Stull and Carl Adler
This useful resource gives clear steps to access Prentice Hall's regularly updated science resources, along with an overview of general World Wide Web navigation strategies. Available free for students when packaged with the text.

Media Resources

Course Management Options

Course management systems offer you a robust architecture for communicating with your students and letting them communicate with each other, allowing you and your students to post course-related documents, managing your roster and gradebook, and providing on-line assessment. For schools with a local WebCT or Blackboard license, we offer a complete downloadable content cartridge that will give you a rapid start to your on-line course materials. Adapt and customize our materials to your needs, using the tools of these systems. For instructors without the benefit of a local course management system, we also offer Course Compass, Pearson Education's own nationally hosted course management system, powered by Blackboard. Course Compass combines the power of a full-featured course management system with a quick, easy-to-use interface.

The content cartridge for *Physics for Scientists and Engineers, Third Edition*, includes all the material from the Companion Website, selected resources from the Instructor's Resource CD-ROMs, and all of the questions from the TestGenerator test bank, plus additional materials designed specifically to work in concert with innovative teaching methods. The latter include several activities for *Just-in-Time Teaching*, by *Gregor Novak* and *Andrew Gavrin*; and conceptual, quantitative, and MCAT practice problem sets for on-line assessment.

On-line Assessment Options

You may not need the full capabilities of a course management system (such as posting documents or managing a bulletin board) but may prefer to use an on-line assessment system. On-line assessment provides students instantaneous feedback and repeated practice, and it offers instructors relief from hours of grading and managing a gradebook. On-line assessment systems feature algorithmically-generated questions and quizzes, allow you to create and modify assignments and question pools, and provide tools for analyzing your gradebook entries. Prentice Hall has a powerful new entry in the on-line assessment space: PH GradeAssist. Ask your PH representative about getting a PHGA demonstration and an Instructor's Quick Start Guide. In addition, Prentice Hall partners with other systems, such as WebAssign, to provide an ample selection of book-specific questions to include in your assignments.

PH GradeAssist

This nationally hosted system includes assessment banks associated with Just-in-Time Teaching materials, and other conceptual and quantitative questions. In addition, most of the even-numbered end-of-chapter Problems and questions for *Physics for Scientists and Engineers*, *Third Edition*, have been converted for use in PHGA, and the majority of them have an algorithmically-generated variant. You select which questions to assign, and you may edit them or create new questions. You also control various important parameters, such as how much questions are worth and when a student can take a quiz.

WebAssign

WebAssign's nationally hosted homework delivery service harnesses the power of the Internet and puts it to work for you by collecting and grading homework. You can create assignments from a database of end-of-chapter questions and problems from the third edition of *Physics for Scientists and Engineers*, or write and customize your own. You have complete control over the homework your students receive, including due date, content, feedback, and question formats.

Companion Website (*http://physics.prenhall.com/fishbane*)

The Companion Website is a quick, interactive resource that allows students to check their understanding with practice quizzes and to explore the material of each chapter further by mining the World Wide Web.

- Reference Tools and Resources by David Reid (Eastern Michigan University)
- Practice Questions by Carl Adler (East Carolina University)
- Algorithmically generated Practice Problems by Carl Adler
- On-line Destinations (links to related sites) by Carl Adler
- Applications, with links to related Internet sites, by Gregor Novak and Andrew Gavrin (Indiana University–Purdue University, Indianapolis)

All quiz modules are scored by the computer; results can be automatically e-mailed to the student's professor or teaching assistant.

Acknowledgments

Three authors alone cannot accomplish a project of this magnitude. We are grateful to the many people who have contributed to making the third edition a better text. A special thanks goes to Professor Jerry Shi (Pasadena City College). In addition to writing the *Instructor's Solutions Manual*, he was an invaluable resource in checking and refining the problem sets and answers and coordinating the independent critique and feedback of the problems and answers from Jenny Quan (Pasadena City College) and David Curott (emeritus, University of North Alabama).

We would like to acknowledge and thank all the instructors who provided valuable feedback for the third edition:

H. Biritz
Georgia Institute of Technology

Tim Bolton
Kansas State University

David Boness
Seattle University

Brian Borovsky
Grinnell College

Chih-Yung Chien
Johns Hopkins University

Krishna M. Chowdary
Bucknell University

Stephane Coutu
Pennsylvania State University

Michael Dennin
University of California–Irvine

N. John DiNardo
Drexel University

J. Finkelstein
San Jose State University

Francis M. Gasparini
SUNY University at Buffalo

Michael Grady
SUNY College at Fredonia

Benjamin Grinstein
University of California–San Diego

John B. Gruber
San Jose State University

Randy Harris
University of California–Davis

Kevin Kimberlin
Bradley University

Amitabh Lath
Rutgers University

Ronald S. MacTaylor
Salem State College

Daniel Marlow
Princeton University

William W. McNairy
Duke University

M. Howard Miles
Washington State University (emeritus)

John Milsom
University of Arizona

Gary A. Morris
Rice University

Elena Murino Di Ventra
Virginia Polytechnic Institute and State University

Michael G. Nichols
Creighton University

Michael B. Ottinger
Missouri Western State College

Kevin T. Pitts
University of Illinois

Andrea Raspini
SUNY College at Fredonia

Dennis Rioux
University of Wisconsin-Oshkosh

Joseph Rothberg
University of Washington

Andrew Scherbakov
Georgia Institute of Technology

Bartlett M. Sheinberg
Houston Community College

Marllin L. Simon
Auburn University

Ross L. Spencer
Brigham Young University

Mark W. Sprague
East Carolina University

Michael G. Strauss
University of Oklahoma

Laszlo Takacs
University of Maryland–Baltimore County

Michael J. Tammaro
University of Rhode Island

Anatoli Vankov
Rochester Institute of Technology

T.S. Venkataraman
Drexel University

Walter D. Wales
University of Pennsylvania

Scott A. Yost
University of Tennessee

R. K. P. Zia
Virginia Polytechnic Institute and State University

We would also like to acknowledge reviewers of the previous editions:

Maris Abolins
Michigan State University

V. K. Agarwal
Moorhead State University

Ricardo Alarcon
Arizona State University

Bradley Antanaitis
Lafayette College

Thomas Armstrong
University of Kansas

Philip S. Baringer
University of Kansas

John E. Bartelt
Vanderbilt University

William Bassichis
Texas A & M University

Benjamin F. Bayman
University of Minnesota

Carl Bender
Washington University

Hans-Uno Bengtsson
University of California, Los Angeles

Robert Bowden
Virginia Polytechnic Institute and State University

Bennet Brabson
Indiana University

Michael Browne
University of Idaho

Timothy Burns
Leeward Community College

Alice Chance
Western Connecticut State University

Edward Chang
University of Massachusetts

Robert Clark
Texas A & M University

Albert Claus
Loyola University

Robert Coakley
University of Southern Maine

Lucien Cremaldi
University of Mississippi

W. Lawrence Croft
Mississippi State University

Chris L. Davis
University of Louisville

Robin Davis
University of Kansas

Jack Denson
Mississippi State University

James Dicello
Clarkson University

P. E. Eastman
University of Waterloo

Robert J. Endorf
University of Cincinnati

Arnold Feldman
University of Hawaii at Manoa

A. L. Ford
Texas A & M University

Gabor Forgacs
Clarkson University

William Fickinger
Case Western Reserve University

Rex Gandy
Auburn University

Alexander B. Gardner
Howard University

Simon George
California State University, Long Beach

James Gerhart
University of Washington

Robert E. Gibbs
Eastern Washington University

Wallace L. Glab
Texas Tech University

James R. Goff
Pima Community College

Alan I. Goldman
Iowa State University

Phillip Gutierrez
University of Oklahoma

Frank Hagelberg
SUNY–Albany

Robert F. Harder
George Fox College

Bruce Harmon
Iowa State University

Warren W. Hein
South Dakota State University

Joseph Hemsky
Wright State University

Jerome Hosken
City College of San Francisco

Joey Houston
Michigan State University

Francis L. Howell
University of North Dakota

Alvin Jenkins
North Carolina State University

Karen Johnston
North Carolina State University

Evan W. Jones
Sierra College

Garth Jones
University of British Columbia

Leonard Kahn
University of Rhode Island

Alain E. Kaloyeros
SUNY–Albany

Charles Kaufman
University of Rhode Island

Robert J. Kearney
University of Idaho

Thomas Keil
Worcester Polytechnic University

Carl Kocher
Oregon State University

Arthur Z. Kovacs
Rochester Institute of Technology

Claude Laird
University of Kansas

Vance Gordon Lind
Utah State University

A. Eugene Livingston
University of Notre Dame

B. A. Logan
University of Ottawa

Karl Ludwig
Boston University

Robert Marande
Pennsylvania State University

David Markowitz
University of Connecticut

Erwin Marquit
University of Minnesota

Marvin L. Marshak
University of Minnesota

Charles R. McKenzie
Salisbury State University

Norman McNeal
Sauk Valley Community College

Forrest Meiere
Indiana University–Purdue University

Roy Middleton
University of Pennsylvania

Irvin A. Miller
Drexel University

George Miner
University of Dayton

Thomas Muller
University of California, Los Angeles

Richard Murphy
University of Missouri–Kansas City

Lorenzo Narducci
Drexel University

Peter Nemethy
New York University

David Ober
Ball State University

Gottlieb S. Oehrlein
SUNY–Albany

Jay Orear
Cornell University

Micheal J. O'Shea
Kansas State University

Dan Overcash
Auburn University

Patrick Papin
San Diego State University

Kwangjai Park
University of Oregon

Robert A. Pelcovits
Brown University

R. J. Peterson
University of Colorado

Frank Pinski
University of Cincinnati

Lawrence Pinsky
University of Houston

Stephen Pinsky
Ohio State University

Richard Plano
Rutgers University

Hans Plendl
Florida State University

Shafigur Rahman
Allegheny College

Don D. Reeder
University of Wisconsin–Madison

Peter Riley
University of Texas, Austin

John Lewis Robinson
University of British Columbia

L. David Roper
Virginia Polytechnic Institute and State University

Ernest Rost
University of Colorado

Richard Roth
Eastern Michigan University

Carl Rotter
West Virginia University

Mendel Sachs
SUNY–Buffalo

Francesca Sammarruca
University of Idaho

Charles Scherr
University of Texas

Eric Sheldon
University of Lowell

Charles Shirkey
Bowling Green State University

Robert Simpson
University of New Hampshire

James Smith
University of Illinois

J. C. Sprott
University of Wisconsin

Malcolm Steuer
University of Georgia

Thor Stromberg
New Mexico State University

William G. Sturrus
Youngstown State University

Richard E. Swanson
Sandhills Community College

Leo H. Takahashi
Pennsylvania State University

Smio Tani
Marquette University

Robert Tribble
Texas A & M University

Rod Varley
Hunter College

Gianfranco Vidali
Syracuse University

John Wahr
University of Colorado

William Walker
University of California, Santa Barbara

Fa-chung Wang
Prairie View A & M University

Gail S. Welsh
Salisbury State University

George Williams
University of Utah

Finally, we would like to thank the publishing team at Prentice Hall who have helped us to carry this project through. Our editor, Erik Fahlgren, who directed the project, has been a constant source of ideas, encouragement, and material help. He has been a most constructive listener and has played the devil's advocate very well when it was necessary. Catherine Flack, our development editor, reminded us that this book is meant for students first and that a failure to communicate at the appropriate level is the worst kind of failure. She too has been willing to listen, and her role as the surrogate student has been immensely helpful. Her help has been crucial to the project. Thanks to Christian Botting, who coordinated the supplements program, and Paul Draper, who managed the media, including the Web-based material. Each of them have been crucial. Special thanks to Andrew Sobel, who worked behind the scenes and handled a massive string of entangled cords like Horowitz playing Chopin. It would be difficult to see how this work could have been completed without his unerring attention to detail. Last but not least, the production of a book such as this one is an enormous task demanding the most careful attention. We want to thank John Christiana for creating a design that is appealing and economical. We also want to thank Beth Sweeten and Fran Daniele for the elaborate management job that has led this book to press. To all the individuals listed above, and to the many others at Prentice Hall who have worked to make this book a success, we extend our heartfelt thanks.

The cumulative and accelerating nature of science and technology make it more imperative than ever that our emerging scientists and engineers understand how few and how solid are the pillars of the enterprise. From this view, the distinctions between "science" and "engineering" and between "classical physics" and "modern physics" melt. We want this book to make evident the pillars of physics as well as the highly interconnected structure that has been erected on those pillars.

Paul Fishbane
pmf2r@virginia.edu

Stephen Gasiorowicz
gasior@umn.edu

Stephen Thornton
stt@virginia.edu

About the Authors

Paul M. Fishbane

Paul Fishbane has been teaching undergraduate courses at the University of Virginia, where he is Professor of Physics, for more than 30 years. In addition to this text, he is the co-author of a modern physics text also published by Prentice Hall. He received his doctoral degree from Princeton University in 1967 and has published over 100 papers in his field, theoretical high-energy physics. He has held visiting appointments at the State University of New York at Stony Brook, Los Alamos Scientific Lab, CERN Laboratory in Switzerland, Amsterdam's NIKHEF laboratory, France's Institut de Physique Nucleaire, l'Université Paris-Sud, and l'École Polytechnique. He has been active for many years at the Aspen Center for Physics, where current issues in physics are discussed with an international group of participants. Among his many other interests, we'll mention in alphabetical order antiques, biking, the kitchen, and music. The rest of his time is spent trying to keep up with his family, including his grandchildren, Ruby and Ivy.

Stephen G. Gasiorowicz

Stephen Gasiorowicz was born in Poland and received his Ph.D. in physics at the University of California, Los Angeles, in 1952. After spending eight years at the Lawrence Radiation Laboratory in Berkeley, California, he joined the faculty of the University of Minnesota, where he is now Professor Emeritus. His field of research was theoretical high-energy physics. As a visiting professor, he has spent extended periods of time at the Niels Bohr Institute, NORDITA in Copenhagen, the Max Planck Institute for Physics and Astrophysics in Munich, DESY in Hamburg, Fermilab in Batavia, and the Universities of Marseille and Tokyo. He has been a frequent visitor to and officer of the Aspen Center of Physics. He has written books on elementary particle physics and quantum physics, and co-authored a modern physics text also published by Prentice Hall. He enjoys biking, canoeing, and skiing, and he is constantly amazed watching his grandchildren Hannah, Becca, Kyle, and Eliza grow up.

Stephen T. Thornton

Stephen Thornton completed his doctoral research at Oak Ridge National Laboratory while completing his Ph.D. at the University of Tennessee. He joined the faculty at the University of Virginia in 1968 and became the first Director of its Institute of Nuclear and Particle Physics in 1984. He has held a Max Planck fellowship and two Fulbright fellowships and has performed nuclear physics experiments at accelerators throughout the United States and Europe. His recent interests include teaching and developing distance-learning courses in physics and physical science for K–12 teachers, and he helped establish the Center for Science Education at the University of Virginia. He has had over 120 research papers published and co-authored three textbooks. He has two grown sons and is married to former NASA astronaut Dr. Kathryn Thornton, with whom he has three daughters. His interests include keeping up with his children, snow skiing, scuba diving, and traveling.

Volume I

PHYSICS

FOR SCIENTISTS AND ENGINEERS

◀ A wide range of physical activity is implied in this set of tools. They exert forces and make things move; they allow one to measure and act with precision; they suggest control of a set of actions. The design and construction of these tools is itself a technological exploit. Some tools are designed with the aid of computers, which themselves are a formidable tool in the exploration of the world, and the materials from which they are constructed reveal a mastery of the properties of metals and plastics.

Tooling Up

W hy physics? What is it that makes an understanding of the fundamentals of science so essential in today's world? Unlike our distant ancestors, we no longer rely on our innate senses but employ high technology to learn about the world around us. We no longer rely just on the strength of our own bodies to build or move but also on machines. We no longer gather food or building materials only from our local environment but also move raw materials around the globe on a daily basis. We strive to improve our lives by accessing information in ever greater amounts and at ever greater rates. Our society is heavily reliant on technology and all technology is rooted in the basic sciences. To gain an understanding of the workings of our society, to actively contribute to it as scientists or engineers, or to make use of it as consumers, we need to understand some of the basics of physics.

In this chapter we will gather a few of the essential ingredients, both mathematical and physical, that will help us to gain an understanding of the physical laws described in the rest of this book. We will discuss the different regimes of the physical world. We will describe the essentials of measurement and the meaning of accuracy in measurement. We will explain the role of the fundamental quantities of our physical world, namely space, time, and mass. We will describe the system of units that allows us to communicate the results of our observations in a universal way. Finally, in our discussion of vectors, we will introduce a significant descriptive tool for the real world.

1–1 A Little Background

To see forward, first look backward. The earliest steps toward the sciences arose out of the recognition of patterns of regularity. Many of these patterns—the sequence of day and night, the seasons, and the regular motion of the heavenly bodies—raised questions about their causes. But it was less the proposed answers to these questions than *the attention given to observation* that began to move us toward science. Ancient observations of the position and timing of objects and events in the sky were surprisingly accurate. The Babylonians could predict the motion of planets, and the description of the appearance of a brilliant "new star" in A.D. 1054—now known to be the Crab supernova—was recorded by Chinese astronomers in enough day-by-day detail to confirm today's nuclear physics calculations of the star's brightness.

Blind reverence for authority can cripple scientific progress. For example, based on the authority of the ancient Greek scientists, the idea that gravity makes heavier objects fall more quickly than lighter ones persisted from Greek times all the way to the Renaissance. An important component of the Renaissance was a reintroduction of the role of experiment in natural science. Among those who must be given credit for challenging authority and helping to reestablish experiment to its primary level are William Gilbert, for magnetism, and Galileo Galilei, for motion. Galileo had a more immediate impact on the rise of physics because of his influence on Isaac Newton, whose laws of motion of 1665 form the underpinning of much of the physics that we will study.

The give and take between experimental observation, the mathematical formulation of descriptive and predictive theories, and further experimental tests of predictions of these theories form the *scientific method*. Figure 1–1 evokes the operation of the scientific method. One does not start from scratch every time one encounters a new set of data; instead one builds on the body of known knowledge. That is why there is a balance in the figure; only in the most extreme circumstances is one forced to formulate hypotheses that break with what we already know, and even then any new hypothesis must be consistent with all that has gone before. We can cite quantum mechanics as a true scientific revolution; nevertheless it was a revolution that had to take into account the enormous success of Newtonian physics. In the scientific method, human imagination is subject to the checks and balances of experiment and of a long history of scientific development.

Progress in our understanding of the physical world has moved along two fronts. On one side, new concepts are built—sometimes in small steps, sometimes in large ones—on earlier concepts. These concepts summarize an ever-increasing body of experimental information and permit an extrapolation of our ideas into areas where patterns of regularity had not previously been seen or even suspected. We speak of opening up new fields. Thus, for example, our current deeper understanding of weather patterns is built on the existence of larger computers and on ever-improving techniques for modeling the behavior of fluids and gases. On the other side, progress is driven by an improvement of experimental techniques that has allowed scientists to probe nature in domains hitherto inaccessible to experimentation. For example, much modern science and technology can be traced to the development of pumps efficient enough to allow the creation of nearly

▶ **FIGURE 1–1** The scientific method is represented here as an interplay between observation and the formal structure of ideas and mathematical techniques that we know as a theory. Observation sometimes leads to new theory, always with the requirement that any new theory must remain consistent with the large body of existing knowledge, both experimental and theoretical, and theory suggests further experiments that can confirm or deny the acceptability of the theory.

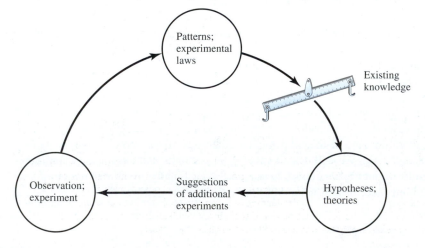

Patterns; experimental laws

Existing knowledge

Observation; experiment

Suggestions of additional experiments

Hypotheses; theories

perfect vacuums in sealed vessels. Among other things, these vacuums have allowed us to make the beams of particles that are used to probe nuclei and their constituents—pure science—as well as to use similar beams to help construct tiny integrated circuits.

Scales of the Observable World

Physics underwent a series of revolutionary developments in the period 1900 to 1925, and the field is sometimes divided into classical (pre-1900) and modern (post-1900) physics. Another way to think of physics is in terms of *scale* (Fig. 1–2). Newton's laws of motion were thought for a long time to be universally applicable. However, starting in the late nineteenth century observations began to reveal that there are scales at which these laws must be replaced by a different picture of motion. We need quantum mechanics to describe the behavior of matter at distances on the atomic scale and below; we need special relativity to describe motion at speeds on the scale of the speed of light; and we need general relativity when large masses (compared with, say, that of the Sun) are involved. The behavior of atoms and their constituents cannot be understood without quantum mechanics and special relativity, and the behavior of the universe as a whole cannot be understood without general relativity. Because matter is made up of atoms, and because some space-based technology demands that we take general relativity into account, the importance of understanding these laws of physics is obvious. Nevertheless, for objects that are large on an atomic scale or move slowly in comparison with the speed of light, these laws reduce to the simpler laws of motion set out by Newton. Thus for the description of most of the world, and for many—but by no means all—engineering applications, we can use Newton's laws without having to think about the fact that they are actually approximations to another, deeper set of laws.

The regime of nature that can be described in the context of Newton's laws includes the great subjects of classical physics—mechanics, waves, thermodynamics, and electricity and magnetism—and forms the content of most of this book. In this regime, the impact of laws other than Newton's laws reveal themselves only in the form of properties and constants that can be taken from experiment. Physical properties such as the ability to conduct electricity or to change the direction of light propagation can be calculated in terms of the underlying theory, quantum mechanics, only with the greatest difficulty, if at all. But this is unimportant as far as how they are used in classical physics, where these properties can be described by empirical quantities such as electrical conductivity or the index of refraction of light whose numerical values can be taken from experiment.

The realm of the very small (quantum physics) and the realm of the very fast (special relativity) form what we might call the frontier of physics, where we practice one of the permanent goals of science—to build our understanding. Where we can, we will consistently

(a)

(b)

(c)

▲ **FIGURE 1–2** (a) A scanning tunneling microscope image of a material surface, with individual atoms visible. This realm is the realm of quantum mechanics, a world very different from that in which we live. (b) An image from our daily lives, a regime with which we are all familiar, where Newton's laws can be used to understand physical phenomena. (c) We know from historical record that the Crab nebula is the result of a supernova, a great stellar explosion. At its heart is a very compact remnant—a neutron star—of the original star. Astronomical objects such as these serve as laboratories that allow us to test our understanding of many aspects of the natural world. A supernova brings in both very large and very small scales: it is explicable with an understanding of the gravitational and electromagnetic forces as well as of nuclear phenomena.

try to relate classical physics to the underlying frontier physics. We must do this because technology is pushing engineering practice into these frontier areas, and quantum physics is increasingly essential to modern engineering. It took very little time for the laser to move from the physics research laboratory into the CD player and grocery scanner, and the everyday medical technique of magnetic resonance imaging relies on a deep understanding of the behavior of nuclei in magnetic fields. In the near future few engineers will be able to function without some understanding of quantum mechanics and relativity.

Students sometimes feel that physics is somehow separate from engineering, but in fact it is only the goals of the two disciplines that differ. The aim of much of engineering is to use knowledge of the basic structure and functioning of materials to advance technology, and it is this knowledge that is the domain of physics. Both the advancement of technology and the attainment of the necessary basic knowledge are extremely important in the twenty-first century. That is why a knowledge of physics is so essential to future engineers.

There is one last point to make, and not the least. There are a lot of pages in this book, but there are only a few laws of physics. Beneath a lot of complexity, there is simplicity. The very same laws of physics that explain the structure of atomic nuclei explain the behavior of the Crab nebula. The behavior of an automobile engine is explained by the same laws of physics that describe the cooling of a container of hot water. Keep this in mind. Physics is not the accumulation of detail; it is the ability to understand the detail with a unifying overview that reveals the essential.

1–2 Fundamental Physical Quantities and Their Units

Scientific Notation

The range of numbers that appear in the physical world is truly enormous. For example, the mass of Earth is about 5,980,000,000,000,000,000,000,000 kilograms (kg), and the diameter of a proton is about 0.000000000000001 meter (m). This many zeros are inconvenient, and we employ a shorthand method of writing very large and very small numbers. By using powers of 10, Earth's mass is more easily written as 5.98×10^{24} kg, and the diameter of a proton is written as 10^{-15} m. In this notation, 10^3 represents 1000 and 10^{-4} means 0.0001. We shall use this standard *scientific notation* throughout this book.

A considerable advantage of scientific notation is that multiplication and division are easily performed by adding and subtracting exponents of 10. Thus the product $100 \times 100 = 10,000$ can be written as $10^2 \times 10^2 = 10^{2+2} = 10^4$. The awkward multiplication $0.00000055 \times 24,000$ can be done more easily as $(5.5 \times 10^{-7}) \times (2.4 \times 10^4) = (5.5 \times 2.4) \times 10^{-7+4} = 13 \times 10^{-3} = 1.3 \times 10^1 \times 10^{-3} = 1.3 \times 10^{-2}$. Where division is involved, we simply change the sign of an exponent and use the multiplication rules. For example,

$$\frac{7.5 \times 10^{-3}}{2.5 \times 10^{-4}} = \frac{7.5}{2.5} \times 10^{-3} \times 10^{+4} = 3.0 \times 10 = 30.$$

Fundamental Physical Quantities

The three quantities *distance*, *time*, and *mass* play a fundamental role in our exploration of the physical world (Fig. 1–3). As Tables 1–1, 1–2, and 1–3 show, these quantities cover an enormous range of values in our universe. These three quantities are already intuitively familiar, as is the idea that we measure them in a certain set of units. The purpose of units is to provide a common set of standards. The importance of standardization is evident. You may wear a size 7 shoe in the United States, but this unit is not of much use in Europe, where a different system is used; there, your shoe size would be 38.

Hundreds of years ago, people used what was readily available as standards for measurement. Length measurements such as the foot came into use in this manner. Over time, measurement systems have become both more precise and more universal. For an early example, French scientists established the forerunner of the International System of measurements in 1791. They defined the meter, the second, and the kilogram: The meter—roughly one yard—was defined as one ten-millionth (10^{-7}) of the distance along Earth's surface between the equator and the North Pole, the second as 1/86,400 of a mean solar day, and the kilogram as the mass of a certain volume of water. In 1889,

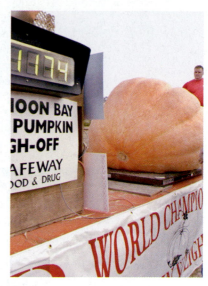

(a)   (b)   (c)

▲ **FIGURE 1–3** (a) Measuring the position of a racecar driver using a simple ruler. Length measurements have different instruments for different needs. (b) Clocks and watches that display time in a digital fashion are commonplace. (c) The entrants in the pumpkin contest demand an objective, repeatable way to determine mass.

an international organization called the General Conference on Weights and Measures was formed to meet periodically to refine these units of measure. In 1960, this organization decided to name a system of units based on the meter, second, and kilogram. These form the International System, with the abbreviation **SI** (for the French words **Système International**). This system is also known as the *metric*, or *mks, system* (after *m*eter, *k*ilogram, and *s*econd). As the tables suggest, we employ the SI throughout.

Length: The definition of the meter has changed several times. In 1889, one meter was defined as the distance between two finely engraved marks on a bar of platinum–iridium that was kept in a vault outside Paris. Even though several copies of this bar were distributed throughout the world, such a standard of length had many shortcomings. For instance, with progress in optical techniques, the scratches on the bar were seen to be fuzzy and imprecise. In 1960, the standard of length was changed to depend upon an atomic constant—the wavelength of a particular orange-red light emitted by an isotope of krypton $\left(^{86}\text{Kr}\right)$ gas. Because our ability (and need) to measure length has led us to require even greater accuracy, this standard also became insufficiently precise. Therefore, in 1983, the Seventeenth General Conference on Weights and Measures established a standard of length based on the speed of light in vacuum (denoted by the letter c). A **meter** (m) is now defined as the distance light travels in vacuum during $1/299,792,458$ second. Some orders of magnitude for lengths are given in Table 1–1.

TABLE 1–1 • Orders of Magnitude for Length

Parameter	Length (m)	Parameter	Length (m)
Proton	10^{-15}	Earth–Moon distance	10^{9}
Hydrogen atom	10^{-10}	Earth–Sun distance	10^{11}
Flu virus	10^{-7}	Diameter of solar system	10^{13}
One bit on a DVD	10^{-6}	Distance to nearest star (Proxima Centauri)	10^{17}
Raindrop	10^{-3}	Diameter of our galaxy (Milky Way)	10^{21}
Height of person	10^{0}	Distance to nearest galaxy	10^{22}
One mile	10^{3}	Distance to edge of observable universe	10^{26}
Diameter of Earth	10^{7}		

Time: The second was originally defined as 1/86,400 of the mean solar day, which is the time interval, averaged over a year, from noon of one day to noon of the next. This definition is insufficient because Earth's rotation is both slightly irregular and gradually slowing down from year to year. Therefore, in 1967, a definition of the second was adopted that depends on an atomic standard. The **second** (s) is now defined as the duration of 9,192,631,770 periods of a particular vibration of a cesium atom isotope $\left(^{133}\text{Cs}\right)$. Clocks based on this standard are, in effect, identical because all atoms of ^{133}Cs are indistinguishable and because frequency can be measured in the laboratory to an accuracy of about 4 parts in 10^{13}. Some orders of magnitude for time are given in Table 1–2.

TABLE 1–2 • Orders of Magnitude for Time

Parameter	Time (s)	Parameter	Time (s)
Time for light to cross proton	10^{-23}	Class lecture	10^3
Time for light to cross atom	10^{-19}	One Earth day	10^5
Period of visible light wave	10^{-15}	One Earth year	10^7
Period of vibration for standard cesium clock	10^{-10}	Age of Greek antiquities	10^{11}
Time required for one operation in a personal computer	10^{-9}	Age of first humanoids	10^{14}
Half-life of muon	10^{-6}	Age of Earth	10^{17}
Period of highest audible sound	10^{-4}	Age of universe	10^{18}
Period of human heartbeat	10^0		

Mass: The kilogram was originally defined as the mass of one liter of water under certain conditions of temperature and pressure. In 1901, the standard **kilogram** (kg) was defined as the mass of a particular cylinder of platinum–iridium alloy kept at the International Bureau of Weights and Measures in France. Duplicate copies of the cylinder made of this particularly stable alloy are kept in laboratories such as the National Institute of Standards and Technology in Maryland. Although the standards of time and length can be reproduced to precisions of 1 part in 10^{12}, the standard of mass can be reproduced only to perhaps 1 part in 10^8 or 10^9. This standard of mass leaves much to be desired. We would like to find an atomic or natural standard for mass, but even though we know that all atoms of the same type have the same mass, nobody knows how to count atoms with the required accuracy. Some orders of magnitude for mass are given in Table 1–3.

TABLE 1–3 • Orders of Magnitude for Mass

Parameter	Mass (kg)	Parameter	Mass (kg)
Electron	10^{-30}	Battleship	10^8
Hydrogen atom	10^{-27}	Moon	10^{23}
Uranium atom	10^{-24}	Earth	10^{25}
Dust particle	10^{-13}	Sun	10^{30}
Raindrop	10^{-6}	Our galaxy (Milky Way)	10^{41}
Piece of paper	10^{-2}	Observable universe	10^{52}
Human	10^2		

THINK ABOUT THIS . . .

WHAT DOES THE GLOBAL POSITIONING SYSTEM MEASURE?

Many of you have used the Global Positioning System (or GPS for short) to keep track of where you are on Earth (Fig. 1–4). This system consists of a network of 24 satellites orbiting Earth at an altitude of 20,000 km. The satellites carry very accurate atomic clocks—accurate to about 4 parts in 10^{13}—and emit regular signals. Moreover, the satellite positions are closely tracked and known very accurately. The emitted signals arrive at the speed of light at a receiver that you carry, and because six of the satellites are in your line of sight at any point on Earth, your receiver can compare the times at which signals from different satellites arrive. In that way what is measured is in fact time. These time measurements can be translated to position as your apparatus recognizes, through the time differences for signal arrival, that you are closer to one satellite than to another and in this way triangulates your position. You can find your position to within a meter with the best receiver available, and with the type of receiver many hikers carry today, accuracy of a couple of hundred meters or less is commonplace.

The accuracy with which you learn your position depends on knowing accurately your distance from each satellite. This distance is determined by the travel time and the speed of light. But the speed of light is not so simple to determine, in part because the speed of light in air differs from its speed in vacuum. The accuracy of your position measurement is also limited by your receiver's capacity to measure the tiny difference in arrival times of signals from different satellites. It is interesting to know that some elements of Einstein's general theory of relativity, one of the most advanced and complex fields of research, are essential to the design of the GPS. The theory takes into account how Earth's mass affects the frequency of any ticking clock, and the correction is important. The utility of the GPS is a reply to anyone who tells you that the general theory of relativity is irrelevant to the real world.

▲ **FIGURE 1–4** With the GPS, travelers—from airplane pilots to hikers—can locate their positions with remarkable accuracy.

Other Systems of Units

The SI is by far the most important and widely accepted system of units in the world today. Two other systems, however, are still in common use.

cgs: The *cgs system* is based on the *centimeter, gram,* and *second* and is a metric system derived directly from SI. Different systems may be used sometimes because they are easy to use; for example, density (mass per unit volume) is still normally quoted in grams per cubic centimeter (g/cm^3) because most densities are nearer unity in this system than in SI. For example, the density of water is 1 g/cm^3 in cgs but is 10^3 kg/m^3 in SI. The definition of the units of the cgs system is based on those of SI:

$$1 \text{ cm} \equiv 0.01 \text{ m} \quad \text{and} \quad 1 \text{ g} \equiv 0.001 \text{ kg},$$

where we have used the symbol $\equiv$ to indicate a definition.

British Engineering System: The *British engineering system,* or British system, is based on units of the inch, pound, and second. This system is used only in the United States and in parts of the British Commonwealth, where it is in the process of being replaced by SI. Even in the United States, scientists seldom use the British system, but existing technology based on this system may require its use in engineering applications. The desirability of international trade suggests that even this limited use will eventually disappear.

The British system of units is now *defined* in terms of SI units. The unit of length, the *inch* (in.), is defined as 1 in. $\equiv$ 2.54 cm (Fig. 1–5). The unit of mass in the British system is called a *slug,* but it is seldom used. A slug is equal to 14.5939 kg. (You are perhaps used to thinking of the unit of mass as the pound. In fact, the pound is not a unit of mass at all, but rather one of a force, here the particular force known as the *weight.* See Chapter 4 for a discussion of this issue.) Finally, the unit of time—the *second*—is the same in the British system as it is in SI.

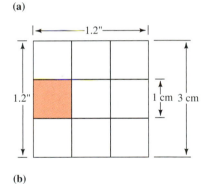

(a)

(b)

▲ **FIGURE 1–5** (a) One inch is exactly 2.54 cm. (b) A square 3.0 cm × 3.0 cm in size has an area 9.0 cm^2, or $(3.0/2.54)^2$ in^2 = 1.4 $in.^2$

Unit Prefixes

A useful set of prefixes in SI replace given powers of 10, as shown in Table 1–4. For example, the distance between New York City and Los Angeles is 4,483,000 m,

TABLE 1–4 • Unit Prefixes for Powers of 10												
Prefix	**Symbol**	**Multiple**	**Prefix**	**Symbol**	**Multiple**	**Prefix**	**Symbol**	**Multiple**	**Prefix**	**Symbol**	**Multiple**
Exa†	E	10^{18}	Mega	M	10^6	Deci†	d	10^{-1}	Nano	n	10^{-9}
Peta†	P	10^{15}	Kilo	k	10^3	Centi	c	10^{-2}	Pico	p	10^{-12}
Tera	T	10^{12}	Hecto†	h	10^2	Milli	m	10^{-3}	Femto†	f	10^{-15}
Giga	G	10^9	Deka†	da	10^1	Micro	μ	10^{-6}	Atto†	a	10^{-18}

†Except for centi, the prefixes near 1 (10^0) are rarely used in the United States. Similarly, the very small and very large multiples (powers of ±15, ±18) are not normally used. You should become familiar with all others.

or 4.483×10^6 m. Neither of these quantities is convenient for everyday use. However, we can use the prefix *kilo-*, which stands for 10^3, to state this distance as 4483 kilometers (km). A finger's width, for example, would not normally be written as 0.015 m or as 1.5×10^{-2} m, but rather as 1.5 centimeters, or 1.5 cm. The standard of mass, the kilogram (kg), is often more convenient to use than the gram (g). In most countries of the world, produce prices are quoted per kilogram because that is a convenient mass unit. You will soon appreciate the simplicity of SI and obtain a feel for its units of length and mass.

Many special units, both within and outside SI, are present because of historical precedent or simply because they are easy to use. For example, the quantity 10^{-6} m is equivalent to a micrometer, or μm, in SI, but is also known as a micron. Another example is the angstrom, Å, a unit often used in atomic physics for the wavelength of light. One angstrom has the value 10^{-10} m, or 0.1 nm [1 nanometer (nm) $= 10^{-9}$ m]. The range of visible light is about 3000 to 7000 Å, or 300 to 700 nm. Both the nanometer and the angstrom are used, although the angstrom is being used less and less frequently.

At the other extreme of the length scale, astronomers and astrophysicists find the meter to be too small for practical use, and they employ three other distance measures: (1) the *astronomical unit* (AU), which is the mean distance between Earth and the Sun, or 1.496×10^{11} m, is useful for distances within our solar system; (2) the *light-year* (ly), the distance that light travels in one year, roughly 0.95×10^{16} m, useful for dealing with interstellar distances; and (3) the *parsec* (pc), which is 3.0857×10^{16} m, equal to about 3.26 ly or 2.063×10^5 AU, also useful for interstellar distance measurements.

Units and Unit Conversions

The angstrom and the astronomical unit are examples of *derived units*, units defined in terms of the basic set (the meter, the kilogram, and the second). These units are simple changes in scale of the basic units. A more complicated type of derived unit involves combinations of the fundamental units. Sometimes these combinations are given names, sometimes not. Speed is measured in meters per second (m/s), but its unit is not given a separate name. When we study power, however, we will find it to be measured in kilogram-meters squared per second cubed $(kg \cdot m^2/s^3)$; this more complicated unit is given a name, the watt (W), and is a *derived* unit in SI.

There are still other units that had their own names from the start and were not obviously combinations of length, time, and mass when they first came into use. They measure quantities that only later were understood to be expressible in terms of length, time, and mass, so that the units could be expressed in SI in terms of the meter, second, and kilogram. These units are "adoptions" of the SI. One example is the unit of pressure, the pascal, (Pa).

We need to be aware of units when we work with equations involving physical quantities. If all the quantities in such an equation are expressed within a *single* system of units—for example, if all the units are SI units—then the units on both sides of such relations must match. This provides us with a very useful tool for checking these relations: If units on both sides of an equation cannot be made to match, there must be an error somewhere within the logic that led to the equation. Further, if a situation arises involving more than one unit for a particular quantity—such as both meters and inches for length—then we must find a simple way to *convert* between different systems of units so that only one unit appears. As we will now see, we can change units conveniently and systematically with the primary conversion equations that relate one set of units to another. Examples of such primary equations are the expressions

$$1 \text{ in} = 2.54 \text{ cm} \quad \text{and} \quad 100 \text{ cm} = 1 \text{ m.}$$

The first of these can be rewritten as

$$1 = \frac{2.54 \text{ cm}}{1 \text{ in}} \quad \text{or} \quad 1 = \frac{1 \text{ in}}{2.54 \text{ cm}}.$$

Any equation can be multiplied by the pure number 1 without change; by judicious choice of the factor 1, units can be canceled and replaced by others. For example,

$$15 \text{ in} = (15 \text{ in})(1) = (15 \text{ in})\left(\frac{2.54 \text{ cm}}{1 \text{ in}}\right).$$

The inch unit cancels, and

$$15 \text{ in} = (15)(2.54 \text{ cm}) = 38 \text{ cm}.$$

Some other examples of conversions of this type are

$$1 \text{ yd} = 1 \text{ yd}\left(\frac{36 \text{ in}}{1 \text{ yd}}\right)\left(\frac{2.54 \text{ cm}}{1 \text{ in}}\right)\left(\frac{1 \text{ m}}{100 \text{ cm}}\right) = 0.9144 \text{ m}$$

and

$$1 \text{ mi} = 1 \text{ mi}\left(\frac{5280 \text{ ft}}{1 \text{ mi}}\right)\left(\frac{12 \text{ in}}{1 \text{ ft}}\right)\left(\frac{2.54 \text{ cm}}{1 \text{ in}}\right)\left(\frac{1 \text{ m}}{100 \text{ cm}}\right)\left(\frac{1 \text{ km}}{1000 \text{ m}}\right) = 1.609 \text{ km}.$$

The technique described here is applicable to more complicated examples.

EXAMPLE 1–1 Given that the speed of light is 2.998×10^8 m/s, what is the distance in meters that light travels in 1 yr? [This distance is the *light-year* (ly).]

Strategy This is a problem involving the conversion of units. The fact that the speed of light c is 2.998×10^8 m/s tells us that light travels 2.998×10^8 m in 1 s. Thus we are interested in expressing c in meters per year to find the distance, in meters, that light travels in 1 yr.

Working It Out We want a conversion for time. We proceed by writing unity as a series of ratios; for example, we know 60 s = 1 min, so that

$$1 = \frac{60 \text{ s}}{1 \text{ min}}.$$

We want to go all the way from seconds to years. Thus we need

$$1 = \frac{60 \text{ s}}{1 \text{ min}}, \quad 1 = \frac{60 \text{ min}}{1 \text{ h}}, \quad 1 = \frac{24 \text{ h}}{1 \text{ d}}, \quad 1 = \frac{365.25 \text{ d}}{1 \text{ yr}}.$$

(For more precision, we use 365.25 days (d) as an average year instead of just 365 d.) The speed of light is then

$$c = (2.998 \times 10^8 \text{ m/s})\left(\frac{60 \text{ s}}{1 \text{ min}}\right)\left(\frac{60 \text{ min}}{1 \text{ h}}\right)\left(\frac{24 \text{ h}}{1 \text{ d}}\right)\left(\frac{365.25 \text{ d}}{1 \text{ yr}}\right)$$

$$= 9.461 \times 10^{15} \text{ m/yr}.$$

We have calculated that 1 ly = 9.461×10^{15} m.

What Do You Think? We went to a lot of trouble to find the distance light travels in a year. What is so special about the unit 1 yr, and if there is nothing special, why do we bother with such quantities? *Answers to* **What Do You Think?** *questions are given in the back of the book.*

1–3 Accuracy and Significant Figures

Uncertainty in Measurement

Physics rests on experiment, and experiment requires measurement. But measurements are, at best, only approximate, more or less so depending on the instrument doing the measurement. Although you might be happy to know the distance of the route you took between New York and Los Angeles to within the tenth of a mile possible on your car's odometer, you would use a tape measure rather than the car's odometer to measure the length of your driveway. An **uncertainty** is an indication of the accuracy of a measurement. The uncertainty depends on the accuracy and calibration of the instrument that is making the measurement and on how well the instrument can be read. We can best illustrate the meaning of uncertainty with an example. If the width of a page of paper is measured with a ruler to be 21.6 cm with an uncertainty of 1 mm (or 0.1 cm), which is about the best you could manage with a typical ruler, it would be correct to say that the width is 21.6 cm ± 0.1 cm, or 21.6 ± 0.1 cm. (The ± is read as "plus or minus.") Here, 21.6 cm is called the *central value* and 0.1 cm the *uncertainty* around that central value. In this case, the basis of the uncertainty lies in how well our eyes can read the ruler and on the precision with which the ruler was made. We often use the term *percentage uncertainty* as a measure of the ratio of the uncertainty of a quantity to its central value. The percentage uncertainty is found by multiplying this ratio by 100. The percentage uncertainty of our paper measurement is thus

$$(100\%)\left(\frac{0.1 \text{ cm}}{21.6 \text{ cm}}\right) = 0.5\%.$$

We also say that a quantity is known to one part in some total. For example, in saying that the paper width was 21.6 cm ± 0.1 cm, we could equally well say that the width is known to 1 part in 216.

We can find the area of the paper by measuring the length and multiplying by the width. Suppose that we measure the paper's length to be 27.9 ± 0.1 cm. The percentage uncertainty for the length is 0.4 percent. We find the area by multiplying 21.6 cm by 27.9 cm, which equals 603 cm^2. Because the measurements of width and length both contain uncertainties, there will also be an uncertainty in the paper's area. But what is this uncertainty? A correct way to proceed is to use a process called quadrature: If P_1 and P_2 are the percentage uncertainties of two quantities being multiplied (or divided), the net uncertainty is $\sqrt{P_1^2 + P_2^2}$. If P_1 and P_2 are about the same size, we can more simply get an *approximate* idea of the percentage uncertainty of the product by adding the percentage uncertainties. In the case of the paper's area, the percentage uncertainty is approximately 0.5% + 0.4% = 0.9%. This means an uncertainty of $(0.009)(603 \text{ cm}^2) = 5 \text{ cm}^2$, and so the area of the paper is 603 ± 5 cm^2. If P_1 and P_2 are rather different from one another, then the quadrature process shows that it is the larger of the two that dominates in the product (or division) of the two. Thus, for example, if you used a ruler with a percentage uncertainty of 1 percent for measuring the length of your piece of paper but a ruler with a percentage uncertainty of 0.1 percent for the width measurement, the area percentage uncertainty will be approximately 1 percent.

Significant Figures

Physical quantities are never known with certainty unless they are merely definitions. The degree of uncertainty in a quantity is implied in the number of digits assigned to its numerical value. Thus, when we say that an object is 2.00 m long, we mean that it is between 1.995 and 2.005 m long. If we want to say that the length is somewhere between 1.9995 and 2.0005 m, we say that the length is 2.000 m. In the first case, three significant figures are used to describe the object's length; in the second case, the number of significant figures is 4. When we say that a sheet of paper has an area of 603 cm^2, we are using three significant figures, meaning that there is a good probability that the area lies between 602.5 and 603.5 cm^2.

Zeros that are used only to set a decimal point are not part of our count of significant figures. Thus 0.00035 has 2 significant figures, not 6. To take a more extreme example, we mentioned in Section 1–1 that the mass of Earth is 5,980,000,000,000,000,000,000,000 kg. Surely we do not know Earth's mass to 25 significant figures! Scientific notation provides a way to avoid this ambiguity. When we write the mass of Earth as 5.98×10^{24} kg, we indicate unambiguously that we know the mass to 3 significant figures; if we knew the mass to only 2 significant figures, we would write 6.0×10^{24} kg.

In many cases a quantity is made up of the product of several other quantities each known to a differing number of significant figures. Our discussion of percentage uncertainties above for P_1 and P_2 when they are very different from one another (one significant figure difference is a factor of 10 in the uncertainty) shows that the number of significant figures in the product is generally that of the least well-known factor. When sums are involved, we can use the same rule, but when there are differences and significant cancellations occur, one has to be more careful (see Example 1–4).

Keep in mind that definitions assign exact values. Thus the inch is *defined* as 2.54 cm, meaning that 1 in = 2.54000 ... cm, or *exactly* 2.54 cm. In effect, there are an infinite number of significant figures in the number of centimeters in an inch.

EXAMPLE 1–2   How many centimeters are there in 1 mi?

Strategy   The number of centimeters in an inch is given by definition, as is the number of inches in a foot and the number of feet in a mile. We can thus apply a string of unit conversions to connect a centimeter to a mile.

Working It Out   There are 5280 ft in 1 mi by definition, there are 12 in in 1 ft by definition, and there are 2.54 cm in 1 in by definition. Thus the number of centimeters in a mile is

$$1 \text{ mi} \times \left(\frac{5280 \text{ ft}}{1 \text{ mi}} \right) \times \left(\frac{12 \text{ in}}{1 \text{ ft}} \right) \times \left(\frac{2.54 \text{ cm}}{1 \text{ in}} \right) = 160{,}934.4 \text{ cm.}$$

What Do You Think?   There are 160,934.4 cm in a mile by definition. Are you obliged to keep all these figures when you have a unit conversion from miles to centimeters? *Answers to **What Do You Think?** questions are given in the back of the book.*

EXAMPLE 1–3   Suppose that you have an apparatus that can measure a length in millimeters to two significant figures and a time in seconds to five significant figures and that you measure speed by dividing a length by a time. A fellow student claims that one insect has been measured walking at 0.523 m/min and another at 0.516 m/min. Can the student really tell the difference between these two speeds?

Setting It Up   In Figure 1–6 we indicate the tools necessary for this measurement. The question is properly interpreted as: Given the ability of your apparatus to measure speed, how many significant figures are there in a speed measurement and are the two numbers claimed different to within the number of significant figures you expect?

Strategy   You are directly dividing a measured distance by the time measured to walk that distance: speed = distance/time. As distance is measured to two significant figures and time to five significant figures, the number of significant figures in the numerator is fewer than the number of significant figures in the denominator. It is the least accurate of the quantities in a calculation that dictates the number of significant figures in the answer, so there are two significant figures in the distance–time ratio.

Working It Out   According to the reasoning above, the speed of 0.523 m/min should be rounded to 0.52 m/min and 0.516 m/min should similarly be rounded to 0.52 m/min, and there is no evidence for a difference in the speed of the two insects. To within the accuracy of the measurement, the insects are moving at the same speed.

What Do You Think?   In working through the solution, we retained two significant figures when we converted millimeters to meters. Why is that allowed? *Answers to **What Do You Think?** questions are given in the back of the book.*

▲ **FIGURE 1–6**   To measure the speed of a walking insect, both a ruler and a watch are necessary, each with its own precision.

It is tempting when using a hand calculator to keep many digits even when the problem calls for fewer. In fact, it is a good strategy to keep many digits through a calculation, because it is sometimes necessary to do this to get the right answer to a sufficient number of significant figures, particularly when subtractions are involved (see Example 1–4). At the end of a calculation, however, the result should be whittled down to only as many figures as are significant, and this number is generally dictated by the input parameter with the fewest significant figures.

EXAMPLE 1–4   An angle θ in radians (see Section 3–5) is known to one significant figure, $\theta = 0.005$ rad. Calculate the deviation from the small angle approximation

$$\frac{1}{\sin \theta} - \frac{1}{\theta}.$$

Working it out   If we keep only one significant figure in each term, we would use $\sin(0.005) \cong 0.005$ and find

$$\frac{1}{\sin \theta} - \frac{1}{\theta} = \frac{1}{\sin(0.005)} - \frac{1}{0.005} = \frac{1}{0.005} - \frac{1}{0.005} = 0.$$

But if we keep more digits in our result for $\sin \theta$, $\sin(0.005) \cong 0.00499998$, then we find

$$\frac{1}{\sin \theta} - \frac{1}{\theta} \cong \frac{1}{0.00499998} - \frac{1}{0.005} \cong 0.0008.$$

This result has been rounded off to include one significant figure.

What Do You Think?   When you try to use your hand calculator to check the result of the last example, you might find that it does not have enough digits available to do so. Does that mean you should toss your calculator in the wastebasket? *Answers to **What Do You Think?** questions are given in the back of the book.*

1-4 Dimensional Analysis

Dimensions

Three basic elements enter the description of any physical quantity: its spatial properties, how much matter is involved, and its temporal elements. All descriptions of matter, relationships, and events involve combinations of these three basic characteristics, and all measurements can be reduced ultimately to the measurement of three fundamental physical quantities: length, time, and mass. Any physical quantity, no matter how complex, can be expressed as an algebraic combination of these three basic quantities. Speed, for example, is a length per time.

Length, time, and mass therefore have significance far beyond that of providing the basis of a system of units. They specify the three **primary dimensions**. We use the abbreviations $[L]$, $[T]$, and $[M]$ for these primary dimensions (the square brackets here and below indicate that we are dealing with dimensions). The **dimension** of a physical quantity is the algebraic combination of $[L]$, $[T]$, and $[M]$ from which the quantity is formed. The speed v provides an example. The dimension of v is

$$[v] = [L/T], \quad \text{or} \quad [v] = [LT^{-1}].$$

Do not confuse the dimension of a quantity with the units in which it is measured. A speed may have units of meters per second, miles per hour, or, for that matter, light-years per century. All of these different choices of units are consistent with the dimension $[LT^{-1}]$.

Any physical quantity has dimensions that are algebraic combinations $[L^q T^r M^s]$ of the primary dimensions, where the superscripts q, r, and s refer to the order (or power) of the dimension. Thus, for example, an area has dimension $[L^2]$. If all of the exponents q, r, and s are zero, the combination will be dimensionless. The number π, which is the ratio of two quantities each of which have dimension of length (the circumference and the radius of a circle), is an example of a dimensionless quantity. The exponents q, r, and s can be positive integers, negative integers, or even fractional powers.

Matching Dimensions

Study of the dimensions of an equation—*dimensional analysis*—is an important exercise with several different roles in science and engineering. Any equation that relates physical quantities must have consistent dimensions; that is, *the dimensions on one side of an equation must be the same as those on the other side.* This provides a valuable check for any calculation. Dimensional analysis can also reveal how changes in one quantity in a physical situation leads to changes in dependent quantities. Finally, when there is reason to believe that only certain physical quantities can enter into a physical situation, dimensional analysis can provide us with powerful and confirming insights.

Let's look at some examples of dimensional analysis. In Chapter 7, we derive a relation between the height h of a dropped object and the speed of that object. This relation involves the *acceleration of gravity*, g, a quantity whose dimension is $[g] = [LT^{-2}]$. The relation reads

$$gh = \tfrac{1}{2}v^2.$$

Let's compare the dimensions on each side of this equation. The dimension of h is $[L]$, so the left-hand side has dimensions $[LT^{-2}][L] = [L^2 T^{-2}]$. The right-hand side has dimensions of speed squared, $[LT^{-1}]^2 = [L^2 T^{-2}]$. Thus the dimensions match. If, through error, we had written a relation $gh^2 = v^2/2$, then this check would have revealed the error. Note that dimensional analysis does not help us understand or calculate numerical factors, in this case $\tfrac{1}{2}$.

CONCEPTUAL EXAMPLE 1–5 A discussion among five friends results in five suggestions for formulas for the time t it will take you to walk home based on the distance d you need to walk and the average walking speed v that you can muster, namely

(a) $t = d/(2v)$,　(b) $t = d/v^2$,　(c) $t = v/d$,

(d) $t = d/v$,　(e) $t = d \times v$.

Based on your knowledge that speed can be measured in miles per hour, which of the formulas above could be correct?

Answer We can use dimensional analysis here. Based on the fact that speed can have units of miles per hour, we deduce that the dimensions of speed are given by

$$[v] = [\text{miles}]/[\text{hours}] = [L/T].$$

The dimension of a correct formula for t must be $[T]$, and as the dimension of d is $[L]$, we can eliminate (b) {dimensions $[L/(L/T)^2] = [T^2]$}, (c) {dimensions $[(L/T)/L] = [T^{-1}]$}, and (e) {dimensions $[L \times (L/T)] = [L^2/T]$}. Both (a) and (d) are dimensionally correct and are possible formulas. Only (d) is in fact correct, but to show that you would have to go beyond dimensional analysis.

EXAMPLE 1–6 Newton's law of universal gravitation gives the magnitude of the force between two objects of mass m_1 and m_2 separated by a distance r as

$$F = G\frac{m_1 m_2}{r^2}.$$

The SI unit of the force is the newton, equivalent to $1\ \text{kg} \cdot \text{m/s}^2$. Find the dimensions of the gravitational constant, G.

Strategy By solving the equation above for the quantity G, we can use the fact that the dimensions of both sides of the resulting equation match, and if we know the dimensions of each quantity in the expression for G, we can find the dimensions of the combination by algebra. From the information given on the unit of force, the dimensions of force must be $[MLT^{-2}]$. We now know the dimensions of every quantity in the expression for G.

Working It Out We have $G = Fr^2/m_1 m_2$. Writing the dimensions for both sides gives

$$[G] = \frac{[F][L^2]}{[M^2]} = \frac{[MLT^{-2}][L^2]}{[M^2]}$$
$$= [MLT^{-2}] \times [M^{-2}L^2] = [M^{-1}L^3T^{-2}].$$

Note that the individual dimensions can be consolidated inside the square brackets or left within their own brackets—whichever is easiest.

What Do You Think? Given the dimensions of G, are its units uniquely chosen? *Answers to **What Do You Think?** questions are given in the back of the book.*

To see how dimensional analysis can allow us to derive relations between physical quantities, consider the simple pendulum (Fig. 1–7). This system consists of a small bob of mass m on the end of a light string of length ℓ. A pendulum swings. When it is displaced away from the vertical direction, gravity pulls it back down. It overshoots the minimum, swings to the other side, swings all the way back to the starting point, then repeats its motion. One full cycle of this motion takes a time τ called the *period*. But how does the period depend on the mass of the pendulum bob?

To answer this, we must gather a list of physical parameters on which the period might depend. We would not expect air resistance to play a large role, because if you set your pendulum up within a vacuum, the motion will not be very different. That leaves us with a list of candidates consisting of the bob mass m with dimension $[M]$, the length ℓ of the string with dimension $[L]$, and the acceleration of gravity, g, discussed above. The latter quantity has dimension $[g] = [LT^{-2}]$. There are no other dimensional quantities on which the period of the pendulum should depend.

The dimension of the period is time $[T]$. We now look for an algebraic combination of m, ℓ, and g that has the dimension of τ. We want to find q, r, and s so that

$$[\tau] = [m^q][\ell^r][g^s],$$

or, in terms of the dimensions,

$$[T] = [M^q][L^r][L^sT^{-2s}].$$

There are no powers of $[M]$ on the left-hand side, so $q = 0$. From dimensional analysis alone, we have learned a remarkable fact: *The mass does not enter at all into the period.* In Problem 1–75 at the end of the chapter, we will continue this treatment and show that, in fact, the only combination of the parameters with the same dimension as that of the period is

$$\sqrt{\frac{\ell}{g}}.$$

▲ **FIGURE 1–7** The simple pendulum consists of a light string from which a bob—a small dense object—is suspended.

The pendulum period τ must then be proportional to this quantity. We do not learn the dimensionless numerical coefficient by which the square root is multiplied to give an equality for τ rather than a proportionality. But we do learn the dependence of τ on ℓ and g, and we learn that the mass does not enter into the result. One measurement of the period of a pendulum with known length would determine the unknown numerical coefficient.

Warning: This example also illustrates the limits of dimensional analysis. If for example our system contained two masses, say m_1 and m_2, then the ratio m_1/m_2 is dimensionless and could appear in any way in any formula derived from dimensional analysis.

▲ **FIGURE 1–8** The scaling of bones. (From Galileo's *Dialogue Concerning Two New Sciences*, 1638.)

There is a basic scale within objects of all sizes, from a planet to a microbe: the atom, which is a building block from which our world is built up. Moreover, the strengths of materials depend on interactions between atoms, and as we do not know how to derive these interactions independently, we must take these interactions as given. In other words, we must take the strength of materials as given. Given this strength, Galileo, in a process closely related to dimensional analysis, studied the question of what sort of bones are required to support animals of different sizes (see his *Dialogue Concerning Two New Sciences,* 1638). He was able to understand that the bones necessary to support an elephant of several tons must be proportionally thicker than the bones of a human weighing around 150 lb (Fig. 1–8), and the basic morphology of an elephant—its proportions—accordingly differs from that of a human. This type of analysis, often referred to as *scaling analysis*, shows that the type of insect exoskeleton that supports an ant 1 cm long would be completely inadequate for a hypothetical ant 1 m in length. More importantly, scaling analysis helps structural engineers determine what size of beam is necessary to support a large structure. It tells aeronautical engineers how to use small models to simulate the behavior of large aircraft and filmmakers how to film ripples in a bathtub yet convince us that they are mighty waves on the ocean. ■

1–5 Estimates: How a Little Reasoning Goes a Long Way

We may sometimes want to make a quick calculation. We may want to check a complicated numerical calculation to see if the answer is reasonable. We may want to make a cost estimate or an estimate of the amount of materials needed for a project. Or we may not have access to all the data needed, so that only an estimate, or a very rough approximation, is possible. In these cases, we perform an *order-of-magnitude* calculation, in which variables are rounded off to the nearest power of 10 or to some other easily handled number. The final result of a calculation with variables so dramatically rounded off is accurate only to within an order of magnitude, but such an estimate can often be extremely useful.

For example, the sight of a Christmas tree lot may make us think about how many natural trees are sold at Christmas time. There are roughly 100 million families in the United States, but perhaps only half of them—50 million—have a Christmas tree. It is reasonable to assume that about half of those families with a tree buy a natural tree. We thus arrive at an estimate of about 25 million natural trees sold.

Now, let's say that we have a friend who owns a 200-acre Christmas tree farm (Fig. 1–9), and we wonder how many trees he can plant. We reasonably suppose that trees are planted about 6 ft apart, so that each tree takes up about 36 ft^2. Considering space occupied by roads, buildings, or other uses, we might change this estimate to one tree per 50 ft^2. But how big is 200 acres? Most of us do not remember the precise size of an acre, but we might hazard a guess that a typical suburban house lot is about a quarter-acre and that it is perhaps 100 ft across by 100 ft deep, or 10^4 ft^2. This indicates that an acre is about 40,000 ft^2. (In fact, 1 acre = 43,560 ft^2, so our guess is not far off.) If each tree requires about 50 ft^2, then our friend can plant 40,000/50 = 800 trees—let's take 1000 for simplicity—per acre, for a total of 200,000 trees. This is a reasonable order-of-magnitude calculation.

▲ **FIGURE 1–9** Images such as this one allow us to make useful estimates—in this case, a set of questions revolving around the number of Christmas trees that are planted.

EXAMPLE 1–7 Estimate the average area available to each person in the United States and then to each person on Earth.

Setting It Up In a problem such as this it is sometimes just as important to know what is *not* given and must be found from other sources—prior knowledge, an appendix to this book, an encyclopedia, the Internet, and so forth. We shall need to know the number of people in the United States and the world and the surface area of each in order to make our estimate.

Strategy The average area per person means simply the total area divided by the number of persons. We'll need to calculate the ratio twice, once for the United States and once for the world.

Working It Out There are some 300 million people in the United States. The United States is roughly a rectangle about 3000 mi (5000 km) from east to west and about 2000 mi (3000 km) from north to south, giving an approximate total area of 15×10^6 km^2. With $(1 \text{ km})^2 = (10^3 \text{ m})^2 = 10^6 \text{ m}^2$, this translates to 15×10^{12} m^2.

Dividing 15×10^{12} m^2 by 3×10^8 people, we obtain 5×10^4 m^2 per person. This is a square roughly 200 m on a side.

For Earth, we might estimate a total of about 6 billion people. We should remember from geography that the circumference of Earth is 25,000 mi, or about 40,000 km. We divide 40,000 km by 6 (circumference $= 2\pi r$, and we will approximate 2π as 6) to obtain a radius of 7000 km. The total surface area of Earth is $4\pi r^2 = 12 \times 7 \times 7 \times 10^6$ km^2, or roughly 5×10^8 km^2. However, only about one-third of Earth's surface area is land—roughly 2×10^8 km^2, or 2×10^{14} m^2. Dividing 2×10^{14} m^2 by 6×10^9 peo-ple, we estimate an area on Earth of about 3×10^4 m^2 per person. The United States has roughly the same population density as the land masses of Earth as a whole.

What Do You Think? The estimate made in this example required you to find by one means or another the radius of Earth. Suppose your number for the radius were larger than the number used here. Would your estimate for the result of this example be larger or smaller? *Answers to* **What Do You Think?** *questions are given in the back of the book.*

EXAMPLE 1–8 To a good approximation, a human body consists mainly of water. One mole (mol) of water, which consists of about 6×10^{23} molecules, has a mass of 18 g. Assuming that the molecules of water in your body are closely packed together, make a rough estimate of the size of a molecule.

Strategy To start, if we know the volume V_1 of one water molecule, we can make an estimate of the linear size d of the molecule by taking the cube root. With the assumption that the molecules are "closely packed," we are approximating the volume of one molecule to be the average volume taken up by each molecule. And to estimate the volume taken up by one molecule under the conditions stated, we can estimate the volume of your body and divide by the estimated number N of molecules. This leaves us with the task of estimating the number of molecules N in your body.

The number N is determined by finding out how many moles n you contain. (A mole is about 6×10^{23} molecules. It is the quantity of a substance whose weight in grams equals the substance's molecular weight.) If you have a mass M in grams, the number of moles you contain is $n = M/(\text{mass of 1 mole})$, where for water the mass of 1 mol is 18 g. Then $N = (6 \times 10^{23})n$. With this information we can proceed to perform the calculation.

Working It Out For your mass, let's use 60 kg, which is equivalent to a weight of about 132 lb. Because $60 \text{ kg} = 60 \times 10^3 \text{ g} = 6.0 \times 10^4$ g, the number of moles n is

$$n = \frac{\text{total mass}}{\text{mass of 1 mol}} = \frac{6.0 \times 10^4 \text{ g}}{18 \text{ g}} = 3.3 \times 10^3 \text{ mol}.$$

In turn, the total number of molecules N is

$$N = (6 \times 10^{23} \text{ molecules/mol})(3.3 \times 10^3 \text{ mol})$$
$$= 2 \times 10^{27} \text{ molecules}.$$

To find the volume taken up by one molecule, we divide the total volume V of your body by N. We might estimate that your body forms a solid with height 2 m, width 0.5 m, and depth 0.3 m, so that

$$V = (2 \text{ m})(0.5 \text{ m})(0.3 \text{ m}) = 0.3 \text{ m}^3.$$

Thus the volume V_1 taken up by each molecule is

$$V_1 = \frac{V}{N} = \frac{0.3 \text{ m}^3}{2 \times 10^{27} \text{ molecules}} = 0.15 \times 10^{-27} \text{ m}^3/\text{molecule},$$

and with the close-packing assumption this is the same as the volume of the molecule itself. Finally, if a water molecule is approximated by a little cube with sides of length d,

$$d = (V_1)^{1/3} = (0.15 \times 10^{-27} \text{ m}^3)^{1/3} \cong 0.5 \times 10^{-9} \text{ m},$$

or about 5×10^{-10} m. Because 1 Å is 10^{-10} m, the size of a water molecule is estimated to be several angstroms. In fact, the separation between the hydrogen atoms in a water molecule is about 2 Å.

What Do You Think? Suppose that instead of being closely packed the molecules in your body are on the average separated by a distance 10 times greater than their typical radius. Is the typical radius of a molecule then (a) 10 times less than the answer above, (b) 1000 times less than the answer above, (c) 10 times greater than the answer above, or (d) 1000 times greater than the answer above? *Answers to* **What Do You Think?** *questions are given in the back of the book.*

1–6 Scalars and Vectors

The mathematical descriptions of physical systems in this book deal with two types of quantities. One type is an ordinary algebraic quantity called a **scalar**. A scalar has no direction associated with it. The statement that the mass of a ball is $\frac{1}{4}$ kg specifies all we need to know about its mass. The same is true of the time it takes the ball to travel a certain distance—the statement of a number for the time says it all. Other examples of scalar quantities include temperature, the energy of a moving body, and electric charge. Some scalar quantities, such as mass, are always positive, whereas others, such as electric charge, can be positive or negative.

There are physical quantities, however, that cannot be described by scalars; a direction is needed for a complete description of these quantities. Specifying the velocity of a ball requires specifying not only the speed (how fast it is going) but also the direction in which it is traveling. Quantities that must be described by both a magnitude (always positive) and a direction are called **vectors**. Vectors describe displacement, velocity, acceleration, force, electric fields, and numerous other quantities. They play an important role in physics, and this section summarizes some of their properties.

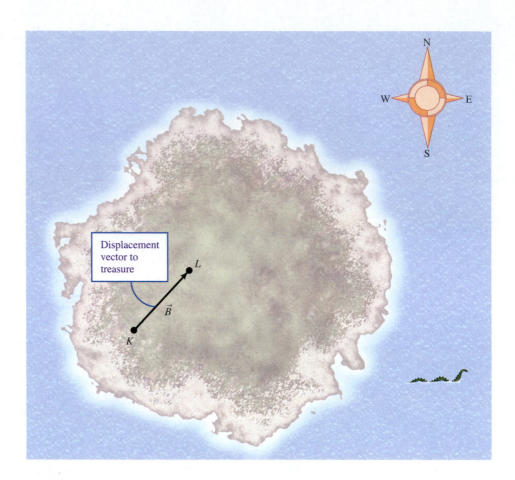

▶ **FIGURE 1–10** The displacement vector $\vec{B}$ from some point K to a second point L on the map for a treasure hunt.

The Displacement Vector

Displacement, which is the difference between two positions of an object, is an important vector quantity. We shall use displacement to describe many of the properties of vectors. To help understand displacement, imagine a treasure hunt in which we must proceed from some point K to a second point L that is 30 paces northeast of K. The displacement from K to L may be drawn as an arrow on a map (Fig. 1–10), and that arrow is the pictorial representation of the displacement vector from K to L.

We can give the vector a name—$\vec{B}$ in Fig. 1–10. The vector $\vec{B}$ has two attributes: a length, or **magnitude** (30 paces), and a direction (to the northeast). We refer to the point where the vector starts as the tail and the point where it ends as the tip, or head, of the vector. Although the length of the displacement vector and the direction in which it points are fixed, the position of the tail (and the tip) of the vector is not. We can shift a vector by moving it to another location in such a way that the vector retains its original direction and length. Thus vector $\vec{B}$ represents the displacement from *any* starting point to the point that is 30 paces away from and to the northeast of the starting point (Fig. 1–11).

In the text, we denote all vectors with an overarrow letter, such as $\vec{B}$. The magnitude of $\vec{B}$ is sometimes denoted by $|\vec{B}|$ but more usually by the unadorned symbol B.

▶ **FIGURE 1–11** The vector $\vec{B}$ represents the displacement 30 paces to the northeast from *any* point. Vectors can be shifted about without changing their characteristics as long as their magnitude (length) and direction (orientation) are unchanged. Thus each of the vectors drawn is the same.

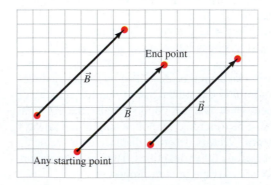

Addition and Subtraction of Vectors

Addition: The result of two successive displacements is also a displacement, which we call a *net displacement*. For example, on an organized day hike the path heads from Base to Coyote Point and then on to Sunset Rock (Fig. 1–12). The initial displacement from Base to Coyote Point is 3.5 km due east of Base. We call this displacement $\vec{A}$ in Figure 1–12. Our second displacement is specified by vector $\vec{B}$; it starts at Coyote Point and proceeds 3.0 km northeast to Sunset Rock. The net displacement takes us from Base to Sunset Rock, and we denote this by the vector $\vec{R}$. The vector $\vec{R}$ is the *sum of the two vectors* $\vec{A}$ and $\vec{B}$:

$$\vec{R} = \vec{A} + \vec{B} \tag{1–1}$$

The sum of the two vectors, which is known as the **resultant vector**, is formed as follows. Draw vector $\vec{A}$, then place the tail of vector $\vec{B}$ on the tip of vector $\vec{A}$. The line from the tail of $\vec{A}$ to the tip of $\vec{B}$ is vector $\vec{R}$. The addition of vectors is *commutative*, that is, the order of the vectors does not matter, so

$$\vec{A} + \vec{B} = \vec{B} + \vec{A}. \tag{1–2}$$

It is easy to see this in Figure 1–13 when we use the method of placing the tail of the second vector at the tip of the first. The figure shows the sum in both orders, and the result is the same. Note that the magnitude R is not the sum of the magnitudes A and B! A glance at Figure 1–12 shows you that Base and Sunset Rock are not separated by the sum of the lengths of the vectors $\vec{A}$ and $\vec{B}$; only if those vectors were aligned would that be the case.

Figure 1–14 shows the result of adding one more displacement vector to the series of displacements in Figure 1–12: The vector $\vec{C}$ takes us from Sunset Rock to Joe's Bar and Grill on the figure. Vector $\vec{C}$ has magnitude 1.0 km and points in a southerly direction. When $\vec{C}$ is added to $\vec{R} = \vec{A} + \vec{B}$, we obtain a new vector, $\vec{S}$, which represents the net displacement from Base to Joe's Bar and Grill:

$$\vec{S} = \vec{A} + \vec{B} + \vec{C}. \tag{1–3}$$

How To Draw

1. Draw first displacement

2. Draw second displacement

3. Draw resultant vector

▲ **FIGURE 1–12** A day hike involves first a walk from Base to Coyote Point, then a walk from Coyote Point to Sunset Rock. This sequence of two displacement vectors leads to a net displacement. Here, we use a series of steps to outline how to add vectors to arrive at a resultant vector.

▲ **FIGURE 1–13** (a) We can find the sum $\vec{R}$ of $\vec{A}$ and $\vec{B}$ by placing $\vec{A}$ and $\vec{B}$ tail to head *in either order.* (b) By combining the two graphs in a parallelogram, the equality of the sum in either order is apparent.

◀ **FIGURE 1–14** One more displacement vector, $\vec{C}$, is added to the displacements in Fig. 1–12. The net displacement is $\vec{S}$.

▶ **FIGURE 1–15** (a) The net displacement $\vec{S}$ is found by first adding $\vec{A}$ to $\vec{B}$, then adding the result to $\vec{C}$. (b) The net displacement $\vec{S}$ is found by first adding $\vec{B}$ to $\vec{C}$, then adding the result to $\vec{A}$. The net displacement is the same in both cases, which shows that vector addition is associative.

Figure 1–15 uses this vector to show that vector addition is *associative*. This term means that we can group the addition in any way we find convenient:

$$\vec{S} = (\vec{A} + \vec{B}) + \vec{C} = \vec{A} + (\vec{B} + \vec{C}). \qquad (1\text{–}4)$$

Subtraction: The *null vector* $\vec{0}$ is a special vector with zero magnitude. It has the property that $\vec{A} + \vec{0} = \vec{A}$. With the help of this vector, we can define the negative, $-\vec{B}$, of a vector $\vec{B}$: When $-\vec{B}$ is added to the vector $\vec{B}$, the sum is the null vector:

$$\vec{0} = \vec{B} + (-\vec{B}). \qquad (1\text{–}5)$$

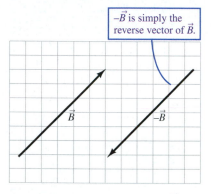

▲ **FIGURE 1–16** The vector $-\vec{B}$ is obtained from $\vec{B}$ by reversing the direction of $\vec{B}$ while leaving its magnitude unchanged.

This means, as shown in Figure 1–16, that $-\vec{B}$ is a vector that has the same magnitude as $\vec{B}$, but $-\vec{B}$ points in the opposite direction. The subtraction $\vec{A} - \vec{B}$ of two vectors is simply the addition of $\vec{A}$ and $-\vec{B}$:

$$\vec{T} = \vec{A} - \vec{B} = \vec{A} + (-\vec{B}). \qquad (1\text{–}6)$$

This vector is shown in Figure 1–17.

Scalar Multiplication and Unit Vectors

When vectors are multiplied by scalars, the result is a vector. Twice the displacement $\vec{B}$ is just a displacement with twice the magnitude of $\vec{B}$, but $\vec{B}$ continues to point in the same direction. More generally, $b\vec{B}$ has a length b times that of the vector $\vec{B}$ (Fig. 1–18). The vector $4\vec{B}$ is formed by the *scalar multiplication* of 4 and $\vec{B}$.

This allows us to define *unit vectors* as follows (Fig. 1–19): Any vector $\vec{U}$ is written as $\vec{U} = U\hat{u}$, where U is the magnitude of $\vec{U}$ and $\hat{u}$ is a **unit vector** that points in the direction of $\vec{U}$ and has a magnitude of 1. We consistently denote unit vectors by the use of the caret over them. We include all the units (meter, meter per second, etc.) of our vector with the magnitude U, so that the unit vector $\hat{u}$ is *dimensionless*. In other words, the unit vector simply specifies a direction. Other notations for the unit vector will be introduced as needed.

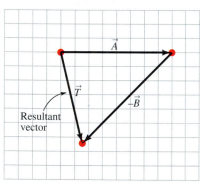

▲ **FIGURE 1–17** Vector difference $\vec{T} = \vec{A} - \vec{B}$.

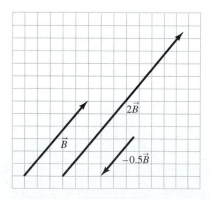

▲ **FIGURE 1–18** Scalar multiplication of a vector $\vec{B}$ by a scalar quantity b is a vector with the same (or opposite) direction as $\vec{B}$ but a magnitude or length scaled by the factor b.

▲ **FIGURE 1–19** Any vector $\vec{U}$ can be broken into a scalar magnitude U times a unit vector (here labeled $\hat{u}$) that points in the same direction as $\vec{U}$ ($\vec{U} = U\hat{u}$). All units are placed into U so that $\hat{u}$ is dimensionless.

Components

It is convenient to use coordinate axes to describe the location of points. In two dimensions, such as the surface of a table, we shall often specify two perpendicular axes, the x-axis and the y-axis (Fig. 1–20a). With these axes, we can specify a point P by giving its coordinates (x_1, y_1), that is, how far it is along the x-direction and the y-direction from the origin O, rather than giving a direction and a distance. The point P may also be described with a *position vector* $\vec{D}$ that extends from the origin to P. We use the term **position vector** to denote the displacement *as measured from the origin of a particular coordinate frame.* The position vector differs from the displacements used above in that it is tied to a particular point.

Let us now use our axes as another way to think about a displacement. Using our knowledge of vector addition and scalar multiplication, we arrive at this displacement in two steps: We first make a displacement of magnitude x_1 along the x-axis, then we follow it with a displacement of magnitude y_1 along the y-axis (Fig. 1–20b). To write this in vectorial form, we employ a unit vector $\hat{i}$ that points in the x-direction and another unit vector $\hat{j}$ that points in the y-direction. (We shall use this notation throughout.) Then $x_1\hat{i}$ is a vector pointing in the x-direction whose magnitude is the absolute value of x_1, whereas $y_1\hat{j}$ is a vector pointing in the y-direction whose magnitude is the absolute value of y_1. Thus

$$\vec{D} = x_1\hat{i} + y_1\hat{j}. \tag{1–7}$$

Figure 1–21 applies this reasoning to a more general vector $\vec{V}$. We draw the vector in the xy-plane, and as above

$$\vec{V} = V_x\hat{i} + V_y\hat{j}. \tag{1–8}$$

The **component vectors** of $\vec{V}$ are the vectors $V_x\hat{i}$ and $V_y\hat{j}$, while we refer to V_x and V_y as the **components** of the vector $\vec{V}$. The coordinate frame is specified in the figure, *as it always must be if we refer to particular components.* Figure 1–21 also shows another way to describe the vector $\vec{V}$. It follows from Pythagoras' theorem that the length of $\vec{V}$ is

$$V = \sqrt{V_x^2 + V_y^2}. \tag{1–9}$$

We see from the figure that $\vec{V}$ makes an angle θ with the +x-direction. The angle θ is given by

$$\tan\theta = \frac{V_y}{V_x} \quad \text{or} \quad \theta = \tan^{-1}\left(\frac{V_y}{V_x}\right). \tag{1–10}$$

Equivalently, the vector components are described in terms of V and θ by

$$V_x = V\cos\theta \quad \text{and} \quad V_y = V\sin\theta. \tag{1–11}$$

We have therefore shown that:

In two dimensions, a vector may be described either with a magnitude V and an angle θ measured from the x-axis or with components V_x and V_y.

We saw in our earlier discussion how to add vectors graphically, placing them in sequence tip to tail. But addition (and subtraction) are especially simple to handle in terms of components. The component vectors add independently, so the components do too. Thus, for example, if $\vec{V}$ and $\vec{W}$ are two vectors with components (V_x, V_y) and (W_x, W_y), then

$$\vec{V} + \vec{W} = (V_x + W_x)\hat{i} + (V_y + W_y)\hat{j}. \tag{1–12}$$

(a)

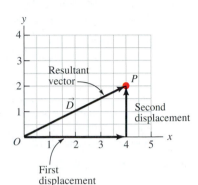

(b)

▲ **FIGURE 1–20** (a) A point P can be located in a two-dimensional coordinate system by specifying the two coordinates $x_1 = 4$ and $y_1 = 2$ along their x- and y-axes, respectively. The position of point P with respect to the origin, O, is denoted by the vector $\vec{D}$. (b) The displacement $\vec{D}$ from O to P can be produced with two successive displacements taken in either order: a displacement in the x-direction and a displacement in the y-direction.

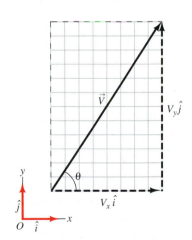

▲ **FIGURE 1–21** A vector $\vec{V}$ in two dimensions is described either by a magnitude V and an angle θ or by the components V_x and V_y.

EXAMPLE 1–9 A chess piece has been moved on a chess board from point A, whose (x, y) coordinates (measured in centimeters) are $(2.0, 3.0)$, to point B, whose (x, y) coordinates, again in centimeters, are $(5.0, 4.0)$. The corresponding displacement vector is labeled $\vec{D}$. (a) Describe $\vec{D}$ in terms of the position vectors $\vec{A}$ and $\vec{B}$ of the points A and B, respectively. (b) Calculate the length of $\vec{D}$ and the angle it makes with the x-axis.

Setting It Up Figure 1–22 describes the situation, including the specified coordinate system as well as the two points A and B, the position vectors $\vec{A}$ and $\vec{B}$, and the displacement vector $\vec{D}$. When we say that we want to "describe" the vector $\vec{D}$, we mean that we want to specify it, and this can be done by specifying its components.

Strategy (a) A glance at our sketch tells us immediately that $\vec{A} + \vec{D} = \vec{B}$. We can now solve this equation for $\vec{D}$:

$$\vec{D} = \vec{B} - \vec{A}.$$

Since the coordinates of both $\vec{A}$ and $\vec{B}$ are specified, we merely need to carry out the required subtraction of coordinates. (b) Since we will have solved for the coordinates of $\vec{D}$ in part (a), this part is a simple application of Eqs. (1–9) and (1–10).

Working It Out (a) We have

$$\vec{A} = (2.0\text{ cm})\hat{i} + (3.0\text{ cm})\hat{j} \quad\text{and}\quad \vec{B} = (5.0\text{ cm})\hat{i} + (4.0\text{ cm})\hat{j}.$$

Then, from Eq. (1–12),

$$\vec{D} = (5.0\text{ cm} - 2.0\text{ cm})\hat{i} + (4.0\text{ cm} - 3.0\text{ cm})\hat{j}$$

$$= (3.0\text{ cm})\hat{i} + (1.0\text{ cm})\hat{j}.$$

The minus signs appear because, in $\vec{B} - \vec{A}$, the components of $\vec{A}$ are *subtracted* from those of $\vec{B}$.

(b) We have, from Eq. (1–9),

$$D = \sqrt{D_x^2 + D_y^2} = \sqrt{(3.0\text{ cm})^2 + (1.0\text{ cm})^2} = 3.2\text{ cm}$$

and, from Eq. (1–10),

$$\tan\theta = \frac{D_y}{D_x} = \frac{1.0\text{ cm}}{3.0\text{ cm}} = 0.33;$$

that is, $\theta = 18°$.

What Do You Think? Suppose that you had changed the placement of the origin by moving it 2 cm in the negative x-direction. [Thus the (x, y) coordinates of point A (measured in centimeters) would be $(4.0, 3.0)$ whereas those of point B would be $(7.0, 4.0)$.] Without doing any calculations, what can you say about the displacement vector $\vec{D}$? *Answers to **What Do You Think?** questions are given in the back of the book.*

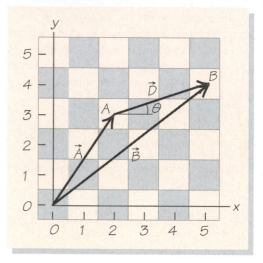

▲ **FIGURE 1–22** The two points A and B have position vectors $\vec{A}$ and $\vec{B}$, respectively. The vector $\vec{D}$ is the displacement from point A to point B.

EXAMPLE 1–10 Figure 1–23a specifies a set of points J, K, L, and M that refer to the day hike described earlier: J is Base, K is Coyote Point, L is Sunset Rock, and M is Joe's Bar and Grill. In the figure, distances are measured in tenths of kilometers. Express the vectors $\vec{A}$, $\vec{B}$, and $\vec{C}$ by describing the displacements from J to K, from K to L, and from L to M, respectively. Describe these displacements in component form relative to a set of axes in which the x-axis points to the east and the y-axis points north. Use your description to calculate the sum of the three vectors. Give the length and inclination with respect to the x-axis of the resultant vector.

Setting It Up Figure 1–23a includes the set of x- and y-axes specified in the problem.

Strategy The breakdown of the displacements as drawn into components is a matter of geometry and can be read off the sketch. After that, vector addition is carried out by adding components.

Working It Out The vector $\vec{A}$ lies along the x-direction; thus, it has no y-component and is given by

$$\vec{A} = (3.5\text{ km})\hat{i}.$$

The vector $\vec{B}$ points in the northeast direction; that is, it makes an angle $\phi = 45°$ with the x-axis. Thus $\sin\phi = \cos\phi = 1/\sqrt{2} \cong 0.71$. From Eqs. (1–11) we have $B_x = B\cos\phi = (3.0\text{ km})(0.71) = 2.1\text{ km}$ and B_y has the same value. Thus

$$\vec{B} = B_x\hat{i} + B_y\hat{j} = (2.1\text{ km})\hat{i} + (2.1\text{ km})\hat{j}.$$

(a)

(b)

◀ **FIGURE 1–23**
(a) A set of coordinate axes has been added to the map of Fig. 1–12.
(b) The vector $\vec{S}$ is constructed by adding the components of $\vec{A}$, $\vec{B}$, and $\vec{C}$.

Finally, the vector $\vec{C}$ has a y-component only. Because it points to the $-y$-direction, it has a component that is -1.0 km:

$$\vec{C} = (-1.0 \text{ km})\hat{j}.$$

Now that we have the three vectors, we can find their sum $\vec{S}$ by adding the components:

$$\begin{aligned}
\vec{S} = \vec{A} + \vec{B} + \vec{C} &= (3.5 \text{ km})\hat{i} + (2.1 \text{ km})\hat{i} + (2.1 \text{ km})\hat{j} \\
&\quad + (-1.0 \text{ km})\hat{j} \\
&= (3.5 \text{ km} + 2.1 \text{ km})\hat{i} + (2.1 \text{ km} - 1.0 \text{ km})\hat{j} \\
&= (5.6 \text{ km})\hat{i} + (1.1 \text{ km})\hat{j}.
\end{aligned}$$

To find the length and angle of $\vec{S}$, given its components, we use Eqs. (1–9) and (1–10):

$$S = \sqrt{(5.6 \text{ km})^2 + (1.1 \text{ km})^2} = 5.7 \text{ km},$$

$$\tan\theta = \frac{1.1 \text{ km}}{5.6 \text{ km}} = 0.20,$$

or $\theta = 11°$. This is the angle the resultant vector makes with the x-axis.

What Do You Think? If you had chosen your x-axis tilted with respect to the one used in the example, you would have found a resultant vector whose inclination with respect to the x-axis would have differed from that found in the example. But we have argued that displacements are independent of coordinate systems. Is this a real conflict? *Answers to **What Do You Think?** questions are given in the back of the book.*

Vectors in Three-Dimensional Space

The vectors we have considered thus far are two dimensional. Vectors can also represent quantities in three dimensions. A vector in three dimensions can be specified with a *Cartesian* or *Euclidean* set of axes x, y, and z. (The two-dimensional x- and y-axes are also called Cartesian.) Figure 1–24 illustrates such a three-dimensional system, showing the conventional orientation of the three axes. The orientation of the axes is best described using a *right-hand rule*, as this figure shows. Start with the usual x- and y-axes (x to the east, say, and y to the north). To find the direction of the z-axis:

Point the fingers of your right hand along the x-axis. Now curl these fingers in the direction of the y-axis. Your thumb will point along the z-axis.

This right-hand rule has become a well-established convention and will appear in many places in this book. Figure 1–25 shows how a vector $\vec{V}$ is decomposed into three components along the three axes.

The three unit vectors for the three axes are denoted by $\hat{i}$, $\hat{j}$, and $\hat{k}$; the unit vector $\hat{k}$ points in the z-direction (Fig. 1–26). A point P in three-dimensional space is now assigned the coordinates (x, y, z) in a given frame, and its displacement vector from the origin—its position vector—is

$$\vec{D} = x\hat{i} + y\hat{j} + z\hat{k}. \tag{1–13}$$

Similarly, any vector $\vec{V}$ has the components (V_x, V_y, V_z) such that

$$\vec{V} = V_x\hat{i} + V_y\hat{j} + V_z\hat{k}. \tag{1–14}$$

Pythagoras' theorem again tells us the magnitude, or length, of $\vec{V}$:

$$V = \sqrt{V_x^2 + V_y^2 + V_z^2}. \tag{1–15}$$

Vector Equations

Vector equations are equations in which vector quantities appear on both sides of the equal sign. For example, we shall learn in Chapter 4 that a force, which is a vector quantity, causes an acceleration, which is also a vector quantity, proportional to the force. The equality force = a constant × acceleration is a vector equation. The equality in such an equation means either that the magnitudes and direction of the vector on each side are the same or, equivalently, that the respective components of the vector equation are equal. In two dimensions, a vector equation is really shorthand for two separate equations for the components; in three dimensions, a vector equation represents three separate equations.

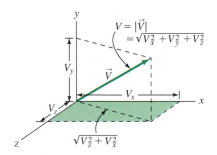

▲ **FIGURE 1–24** Cartesian coordinate system for three dimensions. A right-hand rule specifies the orientation of the three axes. The z-axis points out of the plane of the paper. The thumb points along the z-axis; the fingers curl from the x-axis toward the y-axis.

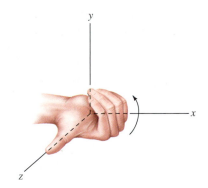

▲ **FIGURE 1–25** Vector $\vec{V}$ decomposed into three components in three-dimensional space.

▲ **FIGURE 1–26** The unit vectors $\hat{i}$, $\hat{j}$, and $\hat{k}$ are aligned along the three Cartesian axes x, y, and z, respectively. Each has unit magnitude.

Summary

In this chapter we have given you a set of tools and methods that you can use to tackle the ideas about the physical world that are the real subject of this book. What is covered here is an indispensable minimum and we will add more tools as we go along. We have also touched on a number of ideas—that length, time, and mass are fundamental "dimensions"; the nature of the scientific method; and how useful estimates can be. You won't need to think about these consciously all the time, but we'll be using them throughout.

The range of quantities relevant to our understanding of the physical world is so large that it is useful to employ scientific notation. In this notation, any number can be represented by a decimal number from 1 to 10 multiplied by a power of 10. The quantities that appear in physics and engineering have units as well as sizes. The International System of units, or SI, provides reproducible and precise definitions of mass, length, and time. The SI units are the kilogram (kg), meter (m), and second (s). Some quantities are used so often that their units are given a special name (e.g., force is measured in newtons, N, in SI), but these units are derived units: They can always be expressed in terms of the three primary units (e.g., $1 \, \text{N} = 1 \, \text{kg} \cdot \text{m/s}^2$). Units that appear in equations can be manipulated algebraically, and after conversion to a single unit system, the units on both sides of any correct equation will match.

Mass, length, and time are the quantities with the three primary dimensions, abbreviated $[M]$, $[L]$, $[T]$. Dimensions should not be confused with units, which refer to a particular choice of unit system. Any physical quantity has dimensions that are rational combinations of the primary dimensions. Dimensions can be manipulated algebraically, and both sides of any correct equation will have the same dimensions. When we analyze the dimensions of an equation, we are performing a dimensional analysis. Dimensional analysis is useful for checking the answers to problems, for learning scaling laws, and for discovering relations between physical quantities.

Numbers that represent physical quantities can be measured only to a certain accuracy. An explicit way to indicate this accuracy is to write a physical quantity x as a central value $\pm$ an uncertainty. Calculations involving physical quantities are meaningful only to within the known accuracy of those quantities. When several numbers of different accuracies are involved in a calculation, the least accurate quantity primarily determines the accuracy of the result. A second way to indicate the known accuracy of a physical quantity is through the use of significant figures, that is, the number of digits in the value between 1 and 10, which are then multiplied by the power of 10 in scientific notation.

The ability to estimate is one that should be cultivated. An educated first guess is a valuable start to the solution of any problem. An order-of-magnitude calculation is such a guess. Similarly, when you arrive at an answer to a physical problem, it is always wise to ask yourself whether it makes sense.

Some, but not all, physical quantities include directional information. Displacement and velocity, for example, include directional information, and such quantities are represented by vectors, in contrast to others, such as mass or time, that are scalars. The vectors $\vec{A}$ and $\vec{B}$ are mathematical objects with both magnitude and direction. They obey the rule

$$\vec{A} + \vec{B} = \vec{B} + \vec{A}. \tag{1–2}$$

Vectors can be expressed in graphical form; we draw them as arrows of length equal to their magnitude within a particular coordinate system. The simplest such system is the Cartesian system, with mutually perpendicular x-, y-, and z-axes for three-dimensional space. We can express any vector $\vec{V}$ in terms of the unit vectors—vectors of unit length—for a given coordinate system. Thus the unit vectors $\hat{i}$, $\hat{j}$, and $\hat{k}$ point along the x-, y-, and z-axes, respectively. Then

$$\vec{V} = V_x \hat{i} + V_y \hat{j} + V_z \hat{k}. \tag{1–14}$$

The quantities V_x, V_y, and V_z are the components of $\vec{V}$. The magnitude of $\vec{V}$ is

$$V = \sqrt{V_x^2 + V_y^2 + V_z^2}. \tag{1–15}$$

Vector equations are relations that equate different vectors. In such an equation, the components of the vectors are equal on both sides, so that, in three dimensions, a vector equation is an elegant and compact way to state three separate equations.

Understanding the Concepts

1. If space were somehow four dimensional instead of three dimensional, would the concept of displacement vectors still make sense? In what way could the idea of vector displacement be generalized?

2. How might the measurement of distance in centimeters, meters, and kilometers rather than in inches, feet, yards, and miles be more convenient? How might it be less convenient?

3. What is your height in centimeters and meters?

4. What is your mass (notice that we do not say *weight*) in kilograms?

5. Angles can be measured in units of degrees (1° = 1/360th of a full circle) or of radians (1 rad = 1/2π of a full circle). What are the *dimensions* of these units?

6. The human pulse and the swing of a pendulum are possible time units. Are they ideal ones?

7. How can length be defined in terms of the speeds of light and time?

8. Three vectors all have the same magnitude. Is it possible to add them together to obtain a null vector?

9. Three vectors have different magnitudes given by 4, 5, and 7 m. Is it possible for the three vectors to add to zero? Prove your result by drawing a figure.

10. It has been said that a recent measurement of the magnetization of an electron has been carried to an accuracy comparable to determining the distance between Los Angeles and New York to within the width of a human hair. What level of accuracy is that in parts per thousand, or million, or billion?

11. Is the vector sum of the two unit vectors $\hat{i} + \hat{j}$ also a unit vector? Under what conditions is it possible for two unit vectors to add to a resultant vector that also has unit magnitude?

12. Suppose we were in radio contact with some distant civilization. How would we communicate how large we are? What assumptions are you making in providing your suggestions?

13. A useful estimate is that there are $\pi \times 10^7$ s in a year. How accurate is this?

14. How does Pythagoras' theorem enter in the determination of the length of the vector $\vec{v} = a\hat{i} + b\hat{j}$?

15. The next time you weigh yourself, consider the uncertainty of the numbers you read. What would you estimate the percentage uncertainty in this measurement of your weight to be?

16. A small worm absorbs the oxygen it needs through its surface. Assume that the oxygen needed by an animal is proportional to its mass and estimate how much the absorption of oxygen per unit area would have to increase if the worm were to increase each of its dimensions by a factor of 10. You may find it interesting to know that the lungs of a human being have about 100 m² of surface available for the absorption of oxygen.

17. A mouse eats the equivalent of about one-quarter of its mass in food every day. You do not. Why?

18. The estimate made in Example 1–7 required you to find by one means or another the radius of Earth. Suppose your number for the radius were 10 percent larger than the number used here. How much would this change your estimate for the result of this example?

19. A 1-kg meteorite adds its mass to that of Earth when it comes crashing in from outer space. Is it necessary to revise all previous calculations involving the effect of Earth on other objects?

20. True or false: Two vectors that have exactly the same length will give a vector of zero length when one is subtracted from the other.

21. What is the minimum number of vectors needed to describe the position of an object in three-dimensional space? How many different sets of vectors can do this?

22. A river flows from west to east at 6 mph. A swimmer enters the river on the south bank and swims straight across at 3 mph relative to the water. Draw a vector diagram for the velocity of the swimmer relative to the bank. If the river were flowing twice as fast, what would change about the cross-river trip?

Problems

1–1 Scientific Notation

1. (I) Twenty thousand jelly beans are in a jar, and 15 percent of them are green. Express in scientific notation the number of green jelly beans.

2. (I) What is the product of 10^5 and 10^{-4}? The ratio $10^{-4}/10^5$?

3. (I) Your bank account contains $356.00. How many cents is that, in scientific notation?

4. (II) Express the number e^{84} in standard scientific notation (powers of 10). The mathematical constant e has a value 2.718....

5. (II) Calculate the cube root of the number 10^{21} as well as the square of the resulting number.

1–2 Length, Time, and Mass

6. (I) The Empire State Building is 1472 ft high. Express this height in both meters and centimeters.

7. (I) What is your height in atomic diameters (see Table 1–1)?

8. (I) A 5-ft, 5-in-tall skier should use skis 5 cm longer than her height. How long should her skis be? Skis are made in 5-cm intervals (150 cm, 155 cm, etc.). What length skis should she buy if she rounds off to the nearest 5 cm?

9. (I) At one time, grapes sold in Italy for 1.25 euros per kilogram. If the conversion rate between dollars and euros was then 1 euro = $0.94, what was the price of grapes in dollars per kilogram?

10. (II) Gasoline is heavily taxed in Europe, with a recent cost of 1.20 euros per liter. What was this price in dollars per gallon if the currency conversion is 1.04 euros = $1 US? One gallon (gal) = 3.8 liters (L).

11. (II) The acceleration due to gravity, g, is 9.80 m/s² in SI. Convert this to the British system, where length is measured in feet rather than meters.

12. (II) The gravitational constant G is 6.67×10^{-11} m³/s² · kg. What is G in units of cm³/s² · g?

13. (II) The radius of the Moon (assumed spherical) is 1.74×10^3 km, and its mass is 7.35×10^{22} kg. What is the density of the Moon in grams per cubic centimeter?

14. (II) A neutron star has a radius of 15 km and a mass of 1.4×10^{31} kg. What is the density of the neutron star in metric tons per cubic centimeter?

15. (II) Gasoline consumption in Europe is measured in liters (L) per 100 km. For example, a small Opel uses 7.0 L/100 km, while the gasoline consumption of a large Mercedes is 23 L/100 km. Convert these to miles per gallon.

1–3 Accuracy and Significant Figures

16. (I) A student wishes to make a measurement of the road distance from his dormitory to the physics building of his university. He uses his car's trip odometer, which measures distance only in units of a tenth of a mile. (a) He makes one trip and the odometer reads 0.3 mi. What can he say is the distance and, in particular, to how many significant figures? (b) One day he has nothing better to do with his time and, adding trips both to and from, he makes 100 trips; his odometer measures 27.2 mi. What can he now say is the distance? How is this result consistent with the result of part (a)? (c) A friend challenges the student on his measurement and, upon reflection, the student is not sure whether he made 99, 100, or 101 trips. How should he modify his statement of the distance?

17. (II) A well-known approximation to π is $\pi \cong 22/7$. What percentage error does this result have? How much better is the approximation $355/113$?

18. (II) The net force F on a mass m moving at speed v in a circular path of radius r has magnitude $F = mv^2/r$. The mass is measured to be 0.00535 kg, the radius is 0.3 m, and the speed is 1.1 m/s. Give the magnitude of the net force. Pay attention to the number of significant figures.

19. (II) A rectangular box is stated to have width 0.75 ± 0.02 m, length 0.5 ± 0.1 m, and height 0.582 ± 0.058 m. What is the volume, stated to the appropriate number of significant figures with an uncertainty?

20. (II) You wish to determine your density (mass per unit volume) by two measurements: by weighing yourself on a digital scale and by submerging yourself in a tank of water with vertical sides and noting the rise in water level on a scale marked in centimeters (Fig. 1–27). The surface area of water in the tank is 1.5 m². The digital scale gives the weight in pounds at 0.5-lb intervals, and you cannot make a reading of the water level to better than 0.5-cm accuracy. Suppose that the weight reading is 213 lb and the water level changes from 152 to 158.5 cm. What is your density? Express your result as a central value with a percentage error.

▲ **FIGURE 1–27** Problem 20.

21. (III) If you want to know the area of a circle to 10 percent accuracy, how accurately should you measure the diameter of the circle?

22. (III) Consider the infinite series $\sum_{n=0}^{\infty} (x^n/n!)$. [The symbol $n!$ (n factorial) means the product $1 \times 2 \times 3 \times \cdots \times n$. By definition, $0! = 1$.] If $x = 0.100000$, how many terms in the series suffice to give a result correct to six significant figures?

1–4 Dimensional Analysis

23. (I) The kinetic energy of a baseball is denoted by $mv^2/2 = p^2/2m$, where m is the baseball's mass and v is its speed. This relation can be used to define p, the baseball's momentum. Use dimensional analysis to find the dimensions of momentum.

24. (I) One of Einstein's most famous results is contained in the formula $E = mc^2$, where E is the energy content of the mass m and c is the speed of light. What are the dimensions of E?

25. (I) A length L that appears in atomic physics is given by the formula $L = h/m_ec$, where m_e is the mass of an electron, c is the speed of light, and h is a constant known as Planck's constant. What are the dimensions of h?

26. (II) What are the dimensions of h^2/m^3G, where h is a constant called Planck's constant, m is a mass, and G is the gravitational constant? The dimensions of the constants in this formula can be found in the list of physical constants given in Appendix II.

27. (II) A fellow student has proposed that an object accelerates under the influence of the gravity of the Sun according to the formula

$$a = \frac{MG}{r},$$

where a is the magnitude of the acceleration measured in feet per seconds squared, M is the mass of the Sun, the physical constant G is 6.67×10^{-11} m³/s²·kg, and r is the distance to the Sun's center. Can this formula be correct?

28. (II) You are told that the speed of sound in a metal depends only on the density ρ ($[ML^{-3}]$) and on the bulk modulus of the metal, B, which has dimensions $[ML^{-1}T^{-2}]$. Express the sound speed in terms of ρ and B.

29. (III) A force F acting on a body of mass m a distance r from some origin has magnitude $F = Ame^{-\alpha r}/r^4$, where A and α are both constants. The constant $e = 2.718\cdots$. Given that the force has dimensions kilogram-meter per seconds squared, what are the dimensions of (a) the constant α and (b) the constant A?

30. (II) In quantum mechanics, the fundamental constant called Planck's constant, h, has dimensions of $[ML^2T^{-1}]$. Construct a quantity with the dimensions of length using h, a mass m, and c, the speed of light.

31. (II) It is known that the quantity Kq^2/hc is dimensionless (K is a numerical constant; h and c are as discussed in Problem 30; q is an unknown quantity). (a) What are the dimensions of q? (b) What are the dimensions of q^2/R, where R is a length?

1–5 How a Little Reasoning Goes a Long Way

32. (I) How many times does an average person's heart beat in a lifetime? Estimate the number of times an automobile tire rotates in a trip across the United States.

33. (II) A criminal posing as a tourist wants to smuggle $25 million in gold across the U.S. border in his station wagon. Is he likely to make it? Use estimates and explain your reasoning.

34. (II) Estimate the volume of concrete used for the construction of the tunnel under the English Channel. The system consists of two railway tunnels and a service tunnel and is 30 km long.

35. (II) Estimate the area used for the storage of one bit on a 1.44-MB 3 1/2-in diameter floppy disk (1 byte = 8 bits).

36. (II) What is the weight of water in a full 5-gal can? About what volume of water can a typical person carry?

37. (II) Make separate estimates of the number of automobile mechanics in the United States based on (a) your total bill for automobile repairs compared with a reasonable average salary for a mechanic, (b) how many hours it takes to repair a car, and (c) the number of people you know who are automobile mechanics compared with the total number of people you know.

38. (II) The Sun is 93 million miles from Earth. What is the diameter of the Sun? [A dime held at arm's length will just about cover the surface of the Sun (Fig. 1–28).]

▲ **FIGURE 1–28** Problem 38.

39. (II) A typical cloud contains droplets of water with an average radius of 0.5×10^{-4} m. How many droplets are needed for a cloud that provides a rainfall of 0.5 cm in your city?

40. (II) Pine trees can only be purchased from the local state nursery in bundles of 100. You want to plant trees over 1 acre (0.405 hectares, or 4050 m^2) in 2-m intervals. Your plot of land is square. How many bundles do you order?

41. (II) Earth's radius is approximately 6400 km, and about two-thirds of the surface is water. Make a reasonable guess as to the average depth of the oceans and seas and calculate the volume of all the water. Express your answer in cubic centimeters. If 1 liter (L) of some medicine were poured into the oceans and stirred sufficiently to be *uniformly* distributed throughout all the oceans, what is the degree of dilution in powers of 10?

42. (II) A strong radioactive point source emits 10^8 gamma particles per second uniformly in all directions. What fraction of these particles will hit a circular detector of diameter 4 cm that is 2 m away from the source?

43. (II) There are some 200 million vehicles in the United States, and each one is driven about 15,000 mi/yr. Estimate the number of automobiles on the road at any one moment.

44. (II) Suppose that all the eighteen-wheeler trucks in the United States line up bumper to bumper on I-80. Can they form a continuous line from New York to San Francisco?

45. (II) Suppose that the circumference of Earth is a perfect circle of exactly 25,000 mi. Somebody prepares a wire that is supposed to go around the equator completely but makes it 2 m too long by mistake. If this 2-m-too-long wire were placed around the equator in a perfect circle with the ends of the wire just touching each other, by how much would the wire be off the ground?

46. (II) Make a rough estimate of the number of apples produced each year in a 10-acre orchard.

47. (II) The mass of one atom of hydrogen is 1.7×10^{-27} kg. Given that the mass of the Sun is 2.0×10^{30} kg, how many atoms of hydrogen would it contain if it consisted purely of hydrogen? Actually, it consists of 70 percent hydrogen and 30 percent helium by mass, and the mass of a helium atom is 6.6×10^{-27} kg. How many hydrogen atoms does the Sun have?

48. (II) Suppose that oil consists of molecules of CH_2. This means that 1 mol of oil, consisting of 6.02×10^{23} molecules, has a mass of 14 g. One milliliter (mL) of oil has a mass of 0.95 g. When this much oil is poured on water, it does not spread forever but spreads until it makes a circular film of area 1.5×10^7 cm^2. How does the fact that the oil does not spread forever support the idea of atoms and molecules? Assuming that the oil slick is one molecule thick (a *monomolecular layer*), with the molecules touching, and that the molecules are spheres, estimate the size of an oil molecule.

49. (II) By estimating the size of Manhattan Island and the average amount of living space that a person uses and by supposing that 2 million people live in Manhattan, make an estimate of the average building height there.

1–6 Scalars and Vectors

50. (I) Draw the vector $3\hat{i} + 4\hat{j}$ by first drawing the x-component vector, then the y-component vector, then adding them graphically. Multiply the vector by a factor of 2 and repeat the exercise.

51. (I) A girl runs around a circular lake. Devise a simple coordinate system to describe her position and direction of travel at any time assuming that she begins at the south end of the lake and runs clockwise at a speed of 3 m/s.

52. (I) What is the resultant vector when the vectors $\vec{A} = 6\hat{i} - 5\hat{j}$ and $\vec{B} = 8\hat{i} + 3\hat{j}$ are added together? When $\vec{B}$ is subtracted from $\vec{A}$?

53. (I) A drunken sailor stumbles four paces north, six paces northeast, two paces east, and five paces west. Describe the final location from the initial position by a single displacement vector.

54. (II) The access point to a septic system buried in a back yard is described by 50 ft northwest and 24 ft northeast from the corner of the house. Using a coordinate system with north in the y-direction and east in the x-direction, write out the two vectors with the magnitude in meters. Use unit vectors.

55. (II) A football player catches the kickoff on the 10-yd line and runs straight up the field for 15 yd, turns left for 15 yd, goes straight up the field for 10 yd, turns right for 20 yd, reverses his field (makes a 180° turn) for 5 yd, and then streaks straight up the field for a touchdown. Define a coordinate system and draw the entire path in vector form.

56. (II) Draw a vector $\vec{V}$ that points in the northwesterly direction, making an angle α with the northerly direction, as in Figure 1–29. If north is chosen as the y-direction and east as the x-direction, what is the x-component of $\vec{V}$?

▲ **FIGURE 1–29**   Problem 56.

57. (II) Suppose that in Problem 56 you choose north as the x-direction and west as the y-direction. What is the x-component of $\vec{V}$ in this case?

58. (II) Refer to the situation outlined in Problems 56 and 57. Choose the x-axis as the line that makes an angle of 45° with the northerly direction and is inclined to the east and the y-axis as the line that makes a 45° angle with the westerly direction and is inclined to the north. What is the x-component of $\vec{V}$ in this case?

59. (II) In computer-aided drafting programs, lines can be specified in either rectangular or polar coordinates (Fig. 1–30). In such programs, if the coordinates (x, y) are given, the cursor draws a line from its current position a distance x to the right and a distance y up. If polar coordinates (r, θ) are given, the line is drawn

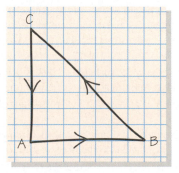

▲ **FIGURE 1–30**   Problem 59.

from its current position through an angle θ in the counterclockwise direction from the positive x-axis through a distance r. Give the instructions both in terms of (x, y) and (r, θ) for the drafting of the triangle shown in the figure starting at point A.

60. (II) Consider the following vectors: $\vec{A} = -2\hat{i} - 3\hat{j}$, $\vec{B} = \hat{i} + 2\hat{j} + 3\hat{k}$, $\vec{C} = 3\hat{j} + 3\hat{k}$, and $\vec{D} = -2\hat{i} - \hat{k}$. Find (a) $\vec{A} + \vec{B} + \vec{C} + \vec{D}$, (b) $\vec{A} - \vec{D}$, (c) $\vec{A} + \vec{D} - \vec{B}$, and (d) $|\vec{A} - \vec{C}|$.

61. (II) Vectors $\vec{A}, \vec{B}, \vec{C}$, and $\vec{D}$ are shown in Fig. 1–31. (a) Give the vectors in component form. (b) Determine the following quantities both algebraically and graphically: $2\vec{A} + \vec{C} - \vec{D}$, $\vec{B} + \vec{C}/2$, and $|\vec{D} - \vec{B}|$.

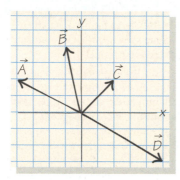

▲ **FIGURE 1–31** Problem 61.

62. (II) Suppose that you have three vectors, $\vec{A} = 3\hat{i} + 4\hat{j}$, $\vec{B} = 2\hat{i} - 2\hat{j} + 4\hat{k}$, and $\vec{C} = -\hat{i} + 5\hat{j} - 3\hat{k}$. Show that the sum of these three vectors can alternatively be computed by first summing $\vec{A}$ and $\vec{B}$ and then summing the resultant with $\vec{C}$ or by first summing $\vec{B}$ and $\vec{C}$ and then summing the resultant with $\vec{A}$.

63. (II) In all of our sketches of vectors in a plane, we used *orthogonal* axes, the horizontal x-axis, and the vertical y-axis. Consider a pair of nonorthogonal axes, for example a horizontal x-axis and a y-axis that makes an angle of 60° with the horizontal. Show what the *components* of an arbitrary vector look like along these axes and express these components in terms of the length of the vector and the angle that it makes with the x-axis.

General Problems

64. (I) A human hair has a diameter of about 10^{-4} m. Given that all of the hair on a head can be gathered into a lock of diameter 4 cm, estimate the fraction of the human scalp area from which hairs spring and the number of hairs on the head.

65. (I) There are 100 cm in 1 m, 1000 kg in 1 metric ton, and 1000 g in 1 kg. (a) How many metric tons are in 1 g? (b) How many cubic meters are in 1 cm³?

66. (I) Measure the height, width, and thickness of this book in centimeters. Estimate its mass. Calculate its density. An object will float in water if its density is less than that of water. Without doing the experiment, would you expect this book to float in water?

67. (I) Given that the speed of light is 2.998×10^8 m/s and the distance from Mars to Earth at some particular time is 1.5×10^8 mi, how long does it take light to travel from Mars to Earth at that time?

68. (I) Eighteen grams of water is known to contain 6.02×10^{23} molecules of H_2O. What is the mass in kilograms of one molecule of H_2O?

69. (II) The density of a human body is approximately 1 g/cm³, which is also the density of water. Use the result of the calculation in Problem 68 to estimate the number of molecules of water that a typical human body would contain if it were made up entirely of water.

70. (II) A silver nucleus consists of 108 closely packed nucleons (protons and neutrons) and has a radius of 5×10^{-15} m. A neutron star is basically an overgrown nucleus, with neutrons only, closely packed in the same way as nucleons are in the silver nucleus. If the radius of a neutron star is 12 km, how many neutrons does it contain?

71. (II) Imagine that molecules of H_2O are stacked up in a cubic array, like a large number of cubical boxes, with a water molecule at the center of each cube. Let the side of each cube be L. Given that the density of water is 1 g/cm³, estimate the distance L from the data given in Problem 68.

72. (II) The gasoline usage rate required to propel an automobile is very roughly proportional to the mass of the automobile. Assuming that the proportions and types of materials of an automobile do not change, calculate the percentage of gasoline savings that would be realized if cars were reduced by 20 percent in each of their three space dimensions.

73. (II) A typical star has a mass of about 2×10^{30} kg and there are about 10^{11} stars in a galaxy. What is the mass of this typical galaxy? Assume that stars are made primarily of hydrogen; the mass of a hydrogen atom is 1.67×10^{-27} kg. How many hydrogen atoms are there in a galaxy?

74. (II) Determine the thickness of a page of this book to an accuracy better than 10 percent. Explain your method and give your uncertainty.

75. (II) We have seen in the text that the period of a simple pendulum is independent of the mass of the pendulum bob. Further, we have seen that the dimensional relation between the period τ, the pendulum length ℓ, and the acceleration of gravity g takes the form

$$[\tau] = [\ell^r][g^s].$$

Use the fact that the dimension of τ is $[T]$, that of ℓ is $[L]$, and that of g is $[L/T^2]$ to show that

$$\tau \propto \sqrt{\frac{\ell}{g}}.$$

76. (II) You need to know the area of a square bedroom to 5 percent accuracy in order to purchase carpet. How accurately do you need to know the length of the side?

77. (II) The water supply of Pittsburgh is contaminated by an oil spill. Make some reasonable assumptions in order to estimate how many trucks per day are needed to bring in a minimum supply of water. What if each person were allowed to take a bath every three days?

78. (II) Assume that houses are set on quarter-acre lots and each house receives four pieces of mail a day. Estimate how far a postal carrier walks in one day and how much mail he or she carries. What is the mass of all this mail?

79. (II) According to Kepler's third law of planetary motion, the square of the period of a planet is proportional to the cube of its mean distance from the Sun. Given that Earth, whose period is 1 yr, is 1.5×10^8 km away from the Sun, calculate the distance from the Sun to Venus, whose period is 0.61 yr, and the period of Saturn, which is 14×10^8 km away from the Sun.

80. (II) When two parallel plates of area A separated by a distance y move with relative speed v with respect to each other in a fluid (Fig. 1–32), there is a frictional force (the *viscosity*) given by the formula

$$F = \eta \frac{vA}{y}.$$

What are the dimensions of the coefficient of viscosity η?

▲ FIGURE 1–32 Problem 80.

81. (II) The mass of a vertical cylinder of atmosphere of cross section 1 cm² is approximately the same as the mass of a cylinder of the same area of water that is 30 ft high. Use this to estimate the number of "molecules of air" in the atmosphere given that, on average, one molecule of air is 1.6 times as massive as one molecule of water. (See Example 1–8.)

82. (II) Show that $(\vec{A} + \vec{B}) + \vec{C} = \vec{A} + (\vec{B} + \vec{C})$.

83. (II) A vector $\vec{u}$ in the xy-plane has x- and y-components $u\cos\theta$ and $u\sin\theta$, respectively, where θ is the angle that $\vec{u}$ makes with the $+x$-axis. A second vector $\vec{v}$, also lies in the xy-plane, and it is perpendicular to $\vec{u}$. (a) Draw a figure to show that there are two possibilities for the vector $\vec{v}$: Its x- and y-components are either $-v\sin\theta$ and $v\cos\theta$ or $v\sin\theta$ and $-v\cos\theta$. (b) Check that $v_x u_x + v_y u_y = 0$ independent of θ.

84. (II) For objects that move in a circle about an origin O, it can be convenient to use the mutually perpendicular unit vectors $\hat{i}_r$ and $\hat{i}_t$, defined as in Figure 1–33. If a Cartesian coordinate system has its origin at O, with an x-axis chosen so that the angle between it and the line OP is θ, then (a) show that $\hat{i}_r = \hat{i}\cos\theta + \hat{j}\sin\theta$, (b) calculate the y-component of $\hat{i}_r$, and (c) express $\hat{i}_t$ as a combination of $\hat{i}$ and $\hat{j}$.

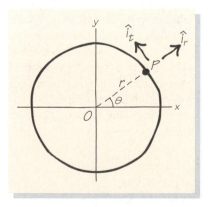

▲ FIGURE 1–33 Problem 84.

85. (II) A vector $\vec{r}$ has length r and points in the direction shown in Figure 1–34. The angles θ and ϕ are drawn in the figure. (a) Show that the x-component of $\vec{r}$ is $r\sin\theta\cos\phi$. (b) Show that the z-component of $\vec{r}$ is $r\cos\theta$. (c) Find the y-component of $\vec{r}$ in terms of r, θ, and ϕ.

▲ FIGURE 1–34 Problem 85.

86. (II) The number of molecules in 22.4 L of air at the surface of Earth is around 6.0×10^{23}. A liter is 10^{-3} m³. Let us say that the volume of air you take in and then expel with each breath is 2.5 L. Let us also say that the air around Earth is a uniform blanket 8 km thick. (Actually, it drops as height increases, but our assumption is not far off.) Finally, Michelangelo Buonarotti lived for 91 yr and died sufficiently long ago that all the "molecules of air" he ever breathed are thoroughly dispersed throughout the atmosphere. Under all these conditions, how many molecules of air breathed by Michelangelo are in your lungs right now?

87. (III) A stretched wire has three physical attributes: the density λ, or mass per unit length; the total length ℓ; and the tension τ. The latter is related to how hard the wire is being pulled to keep it stretched and has dimensions of $[MLT^{-2}]$. Show by dimensional analysis that if the time t_0 of one back-and-forth vibration of the wire in a direction perpendicular to its length depends only on these three quantities, then t_0 has the form $t_0 = $ (a constant)$\ell\sqrt{\lambda/\tau}$.

▶ These runners leave the blocks with the idea of making a total displacement from the starting to the finishing line. The runner who makes this displacement in the least time is the winner. A full description of the displacement as a function of time tells us all we can know about the motion of the runners.

Straight-Line Motion

Understanding motion is one of our key goals. Motion occurs over all the scales of our world, from the heavenly bodies to the motion of particles in atoms and nuclei. We need to understand the motion of airplanes and automobiles when we design airport runways and interstate off-ramps. Motion is at the heart of physical science, whether we are interested in the rotation of the blades of a steam turbine, the pulse of electricity that results from the punching of a computer keyboard, or the takeoff of a high jumper. That is why we begin with a study of **mechanics**, the science of motion and its causes. The *description* of motion is a subset of mechanics known as **kinematics**, which is the subject of this chapter and the next. Once we know how to describe motion, we will explore the *causes* of motion. In this chapter, we study the motion of an object in a straight line; that is, one-dimensional motion. This simpler situation will prepare us for motion in two and three dimensions—motion on a plane and motion in space—the topic of Chapter 3.

▶ 2–1 Displacement

Consider the motion of athletes running a 100-m dash. Some runners are able to attain a tremendous advantage right at the beginning of the race—they leap ahead of the others. Other runners have a late kick that allows them to take the lead at the end of the race (Fig. 2–1). To understand this motion, we want to break it down into variables that we can measure: distance and time. Suppose that we set up an electronic timing system along the path of the 100-m dash that records the times of the runners every 5 m. The timing system will give us the distance each runner travels as a function of time.

Let us first consider the distance traveled by the athletes. To do this, we construct a coordinate system with an *x*-axis using the variable *x* to indicate the distance traveled. As in our discussion of vectors in Chapter 1, the choice of origin of the *x*-axis is up to us, but it is convenient to take the origin at the starting line, which we therefore set at $x = 0$ m. We also choose the direction that the runners take to be the $+x$-direction. Figure 2–2a shows the data points that represent the distance (the *x*-value) traveled by one runner—actually, the distance traveled by a point on the runner's chest (we do this so that we do not have to worry about the complicated motion of the runner's entire body)—plotted against the time it takes the runner to reach that distance. We choose the starting time to be $t = 0$ s, again for convenience.

Figure 2–2b shows a curve that interpolates the data points of Fig. 2–2a to reasonable accuracy. It also contains an extension (the dashed line) that shows the runner slowing down and coming to rest after a short time interval. The figure can be used to read the time taken to reach any distance *x* within the 100 m, even those values not given directly by the data. The curve, which we call $x(t)$, represents the distance from the starting point, *x*, as a function of time *t*. We use the notion of *displacement*, a concept we met in Chapter 1, to study this curve further. The displacement is the *change* in the position of an object. If we denote the first position at time t_1 by x_1 and the second position at time t_2 by x_2, then the displacement is the final position minus the initial position; this is defined mathematically by

$$\Delta x \equiv x_2 - x_1. \tag{2–1}$$

We use the capital Greek letter Δ (delta) to indicate a change (or difference) in a variable from one value to another. The time interval is, similarly,

$$\Delta t = t_2 - t_1. \tag{2–2}$$

There is an important distinction between Δx and Δt: The displacement Δx can have a sign, whereas the time interval Δt is normally positive. In our example of the runner, *x* increases steadily with time, and the possibility that Δx can have two signs does not play an important role. But if one of the runners were to set off in the opposite direction, his or her displacement in a given (positive) time interval would be negative—that very fact would tell us that the runner was moving in the $-x$-direction. If the runner who took off in the wrong direction realized there was a problem and turned around, then we would have a negative displacement during some time interval and a positive one during some other time interval.

▲ **FIGURE 2–1** The motion of runners in the 100-m dash provides us with a test case to examine motion in detail and allows us to refine our ideas of motion.

(a)

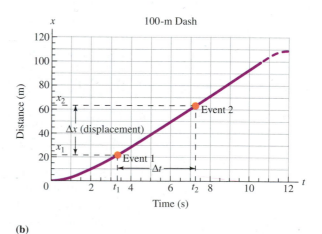

(b)

▲ **FIGURE 2–2** (a) Distance plotted against time at 5-m intervals for one athlete in the 100-m dash. (b) The curve is an interpolation of the data shown in part (a), with the dashed part an extrapolation. The runner is at position x_1 at time t_1 (event 1) and at position x_2 at time t_2 (event 2). In the time interval $\Delta t = t_2 - t_1$ between event 1 and event 2, the displacement of the runner is $\Delta x = x_2 - x_1$.

(a) The position vectors $\vec{x}_1$ and $\vec{x}_2$ run from the origin to the positions x_1 and x_2, respectively, of the runner. The displacement vector $\Delta\vec{x} = \vec{x}_2 - \vec{x}_1$ is also drawn. (b) The information contained in part (a) can be plotted as a curve of position versus time.

(a)

(b)

While the position depends on the choice of origin, the displacement in Eq. (2–1) does not. Suppose that the origin were placed 10 m back of the starting blocks. Then the position at time t_1 would be given by $x_1 + 10$ m, and the position at time t_2 would be given by $x_2 + 10$ m. The displacement would still be given by

$$\Delta x = (x_2 + 10\text{ m}) - (x_1 + 10\text{ m}) = x_2 - x_1. \qquad (2\text{–}3)$$

This relationship is shown in Fig. 2–3.

Finally, let us mention the *net displacement* over some definite time interval. If a few minutes after the race is over the runners return to the starting blocks, then their net displacement is zero for that interval; we simply take the difference between the final and initial positions, and these positions are the same. Of course, there are many ways to have a net displacement of zero, or any other value, over a long time interval. The net displacement, or the displacement over a long time interval, contains less information about the details of the motion than data from a series of displacements over short time intervals.

EXAMPLE 2–1 Use Fig. 2–2 to find the runner's position for the dash at times 2 and 5 s. What is the displacement between these two times?

Setting It Up We label the positions x_1 at $t_1 = 2$ s and x_2 at $t_2 = 5$ s. These are given as data on the graph.

Strategy The displacement is simply $\Delta x = x_2 - x_1$. We can use the diagram in Fig. 2–2 to read off the distances traveled, x_1 and x_2.

Working It Out With the origin at the beginning of the race, the position at 2 s is $x_1 = 10$ m and that at 5 s is $x_2 = 40$ m. Then

$$\Delta x = x_2 - x_1 = (40\text{ m}) - (10\text{ m}) = 30\text{ m}.$$

This displacement is positive, indicating movement in the $+x$-direction.

Displacement Is a Vector

For one-dimensional motion, keeping track of the sign of x as well as its magnitude provides all possible information for both x and the displacement. The sign of x is the manifestation in one dimension of the fact that the displacement is a vector. For motion in two or three dimensions the displacement has both a magnitude and a *direction*, and the full vectorial description is important. A two- or three-dimensional displacement is not simply forward or backward along a line (positive or negative) but can take any possible direction. We can say, for example, that an ant had a displacement of 2 cm in a northeast direction or to the south. In anticipation of the fact that we will eventually have to deal with the vectorial nature of displacement, we define the displacement as

$$\Delta\vec{x} \equiv \vec{x}_2 - \vec{x}_1, \qquad (2\text{–}4)$$

where $\vec{x}_2$ is the *position vector* at time t_2 (in the example of our runner, the vector from the starting line to the position of the runner at time t_2) and $\vec{x}_1$ is the position vector at t_1. In one dimension use of the full vector notation is unnecessary, although we could write the vector $\vec{x}_1$ as the product of the distance from the origin x_1 (including the sign) and the unit vector $\hat{i}$ in the positive x-direction, but we don't gain much by this. Instead, we can work with components like x_1, always recalling that the sign is important. This is how we proceed in this chapter; in later chapters, where more than one-dimensional motion occurs, we'll employ a more specific vector notation.

2–2 Speed and Velocity

Speed

The quantities *speed* and *velocity* both describe how fast the position of an object changes. The speed of an automobile or airplane is a familiar concept that we will refine here. The **average speed** for an object in motion is defined as the total distance traveled divided by the time taken to travel that distance:

$$\text{average speed} \equiv \frac{\text{total distance traveled}}{\text{time interval}}. \qquad (2\text{–}5)$$

Speed is a scalar quantity and is always positive. As the time interval over which the object moves changes, the average speed may also change. For example, from the graph of Fig. 2–2, during the first 5.1 s of the 100-m dash, the runner has traveled 40 m. The average speed over this interval is

$$\text{for 0 to 5.1 s: average speed} = \frac{40 \text{ m}}{5.1 \text{ s}} = 7.8 \text{ m/s}.$$

For the last 5.4 s, however, the runner progresses 60 m, and the average speed is

$$\text{for 5.1 to 10.5 s: average speed} = \frac{60 \text{ m}}{5.4 \text{ s}} = 11 \text{ m/s}.$$

The runner completes the 100-m dash in 10.5 s. Over the entire 100-m dash, then, the runner's average speed is (100 m)/(10.5 s), or 9.5 m/s.

We can imagine taking the time interval to be a very small one. Then we would be calculating the *instantaneous speed*, or just speed for short. We'll learn more about this procedure in the discussion of velocity below. For example, if we find that (distance traveled in 1 ms), divided by (1 ms) at a clock time $t = 1.5$ s, we get the speed *at* $t = 1.5$ s. (Strictly speaking, we should take the time interval so small that there is no ambiguity about whether we were "at" or merely "around" $t = 1.5$ s.) This speed, which is what your car's speedometer measures, can vary from moment to moment.

EXAMPLE 2–2 A photographer is assigned the task of filming the 100-m dash of the runner whose data are given in Fig. 2–2. The photographer is stationary at a point 20 m in front of the starting blocks. When the runner reaches the 10-m mark, the photographer's cart starts traveling at 10 m/s in the same direction as the runner until the cart reaches the 100-m mark, at which point it stops. Superimpose on Fig. 2–2b the drawing of a continuous line representing the position of the photographer's cart from the beginning to the end of the race.

Setting It Up We must recognize here that, starting from the time the runner reaches the 10-m mark, the photographer's cart moves at a constant speed v, here 10 m/s.

Strategy Constant speed means that position changes *linearly* with time, and this "curve" is a straight line, one whose slope on the plot of x versus t is the value of the speed.

Working It Out We get the particular value of the slope from the value of the speed: 10 m/s. Thus the straight line must have a change of 10 units of length for every unit of time. The straight line that represents this motion is drawn in Fig. 2–4.

What Do You Think? Draw several straight lines through the origin on the position-versus-time graph to represent motions of different carts like the one the photographer used. Draw a line that

shows you the order in which the carts arrive at a certain place; draw another line that shows you how far different carts have traveled in a fixed time. Which slope describes the fastest cart? *Answers to **What Do You Think?** questions are given in the back of the book.*

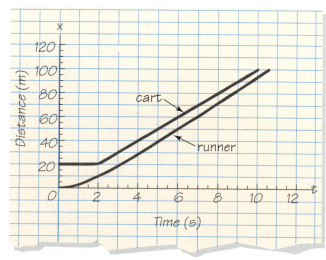

▲ **FIGURE 2–4** The motion of the photographer's cart is the straight line adjacent to the plot of the runner's motion.

Problem-Solving Techniques

Problem solving is important. The techniques we present here represent the basis for a plan of action on *any* problem. As we proceed, other techniques will fit into the general scheme.

1. Read the problem, then read it again. Failure to read the problem carefully is perhaps the source of more false starts and wrong answers than any other cause.

2. Draw a sketch or diagram of the problem to help you to visualize the situation presented by the problem. We illustrate this aspect of the technique in Fig. 2B1–1.

3. Write down the given and known quantities.

4. Make sure you understand which quantities are to be found.

5. There are generally only a few principles applicable to the solution of a problem. Think about which principles link the quantities to be determined to those that are known.

6. Use the principles that apply to the situation to guide you toward equation(s) that relate the quantities in the problem. Take care—at times, certain equations apply to a given situation and others do not. The rest is mathematics. Several equations may need to be manipulated together at times. Count the number of equations available to see if there are enough equations to determine the unknowns. There should be as many equations as there are unknowns.

7. When you solve for an unknown in terms of the known quantities, use symbols, not numbers. Wait until the end to replace symbols with numbers and units. It is important to include units; the answer may require them, and the proper cancellation of units provides a valuable accuracy check.

8. When you arrive at a number, think about it. Does it make sense? If you find that it takes 3 min to drive from New York City to Los Angeles, you have probably made a mistake!

9. Use *any* checks you can find for your result.

▶ **FIGURE 2B1–1** Some suggestions for illustrations.

How To Draw Sketches

1. **Problem Statement:** A girl starts from rest at $x = -10$ m (time $= 0$), walks to $x = 20$ m at $t = 25$ s, turns around, and walks back to $x = -10$ m, where she stops at 45 s. Sketch a plot of position versus time.

2. **Thinking Process:** Let's first make a sketch of the motion along the x-axis and mark the start, turn around, and stop positions as 1, 2, and 3.

(a)

Note that the positions 1, 2, and 3 are "at rest." For these three positions, the speed is zero and their slope on the x-versus t-plot must be zero.

3. **Make a Sketch:**

(b)

4. **Check Sketch:** The curve goes through all three known positions. However, the slope (remember the speed) at position 3 is not zero so we must fix that.

5. **Redraw Sketch:**

(c)

Velocity

Although the notion of speed may be useful for sports, it is of limited utility in mechanics, where a different variable, velocity, plays an important role. Average speed refers to the total distance traveled; **velocity** refers to how fast the *displacement* changes. Velocity, like speed, is measured over a certain time interval. If a car has a displacement Δx in a particular time interval Δt, then the car's **average velocity**, v_{av}, over that time interval is defined (in one dimension) by

$$v_{av} \equiv \frac{\text{displacement}}{\text{time interval}} \tag{2–6}$$

$$= \frac{x_2 - x_1}{t_2 - t_1} \tag{2–7}$$

$$= \frac{\Delta x}{\Delta t}. \tag{2–8}$$

In one-dimensional motion, since the displacement Δx has a sign, *so does the velocity v* (or v_{av}). If v_{av} is positive, there is motion in the $+x$-direction, and if it is negative, there is motion in the $-x$-direction. The dimensions of velocity are $[LT^{-1}]$, with SI units of meters per second.

In our discussion of displacement, we pointed out that when we go from one-dimensional motion to two- or three-dimensional motion we will need to use the fact that the displacement is in fact a vector. In two- and three-dimensional motion, the vector nature of the velocity will likewise be crucial.

Warning: We caution the reader that it is *only* in this chapter—for one-dimensional motion—that we attach a sign to the symbol v to indicate whether the velocity is positive or negative. Everywhere else—for two- and three-dimensional motion—the symbol v will stand for the *magnitude* of the vector describing the velocity, and the magnitude is always positive. Similar remarks apply for displacement and acceleration.

Average Velocity: The average velocity $\Delta x/\Delta t$ provides us with only limited information of the details of the motion. Consider, for example, a car being driven along a straight road for 1 h (so that $\Delta t = 1$ h). If the displacement of the car in that time is $\Delta x = 30$ mi along the road, then the above expression will read

$$v_{av} = 30 \text{ mi/h}.$$

Since the displacement is in the positive direction, the average velocity is positive. Its magnitude only depends on the net displacement, so that any changes from the $+x$-direction to the $-x$-direction and any slowing down or speeding up are not reflected in that number. The car could have traveled at 60 mi/h to a point 45 mi down the road, then turned around and traveled 15 mi back to the finishing point at 60 mi/h. It could have traveled a steady 30 mi/h or it could have traveled the 30 mi at 120 mi/h, arriving in 15 min, and simply parked at the finish point until the hour was up. All that the definition in Eq. (2–8) has given us is the *average velocity* over the time interval in question. It is perfectly consistent to have a zero average velocity over a finite time interval even though a considerable distance may have been covered. On any round trip the net displacement is zero. Of course, an Olympic swimmer might resent being told that, whatever her efforts, her average velocity will be zero in a race that finishes at the same point as it starts. It is because of the limited information content of the average velocity that we turn to the concept of *instantaneous velocity.*

Instantaneous Velocity: The definition of the average velocity includes a time interval. We learn more about the details of the motion when smaller time intervals are used. For example, if we had asked about the average velocity during the 15-min intervals in the 1-h trip just discussed, we would already have a more detailed picture of the car's motion. Dividing the hour into sixty 1-min intervals and finding the average velocity during each minute would give us still more information.

It is possible to make the time interval Δt progressively smaller, finding the average velocity for each interval. Suppose that we look at the particular time interval from some time t to a time $t + \Delta t$. In this time interval, the displacement $\Delta x = x(t + \Delta t) - x(t)$ occurs. The average velocity during this interval is the ratio of Δx to Δt. Now, if Δt becomes very small, so does Δx, *but their ratio remains finite.* (For example, if a car travels at a constant velocity of 30 mi/h, you could learn that value by measuring the distance traveled over 1 h or one hundredth of a second—in each case the ratio displacement divided by the time interval is exactly the same.) We say that we are taking the *limit* as Δt approaches zero, symbolized by $\Delta t \rightarrow 0$. This limit refers to a particular time t and gives us the average velocity over a shorter and shorter time interval around that time. The **instantaneous velocity** at time t is the velocity of an object at that given instant of time, and it is defined as the limit of the average velocity as $\Delta t \rightarrow 0$:

$$v(t) = \lim_{\Delta t \to 0} \frac{x(t + \Delta t) - x(t)}{\Delta t} = \lim_{\Delta t \to 0} \frac{\Delta x}{\Delta t}. \tag{2–9}$$

When we use the term *velocity*, we mean instantaneous velocity, unless we state otherwise, and we refer to the instantaneous velocity without subscript in Eq. (2–9). The right-hand side of Eq. (2–9) is the *definition* of a derivative in calculus. Velocity is the derivative of displacement with respect to the time:

$$v(t) = \lim_{\Delta t \to 0} \frac{\Delta x}{\Delta t} = \frac{dx}{dt} \qquad (2\text{–}10)$$

INSTANTANEOUS VELOCITY

The instantaneous speed v is the magnitude of the velocity:

$$v \equiv |\vec{v}|, \qquad (2\text{–}11)$$

that is, the velocity stripped of any directional information. In one-dimensional motion, the speed is just the absolute value of the velocity. In more than one dimension, where the velocity is explicitly a vector, the speed is the length of that vector. We shall limit the use of the word *speed* in our discussion of mechanics to the definition Eq. (2–11).

Graphing the Motion: Figure 2–5 graphs the motion of an automobile as it moves along a straight road during a ride that begins at the time t_1 at point A and ends with the car parked at point Z at time t_2. The vertical axis shows the position, in kilometers, and the horizontal axis the time, in minutes. The curve allows us to study the average and instantaneous velocities of the car.

First of all we see that, before the start time t_1 and after the arrival time t_2, x does not change, so that $\Delta x = 0$ during these periods and therefore the velocities are zero. A flat line in the x-versus-t graph means zero derivative and hence zero velocity—the car is not moving. Next, we see that the total displacement A to Z is $+5.0$ km and the time of travel is 6.0 min, that is, 0.1 h. Thus the average velocity is $\Delta x/\Delta t = 5.0 \text{ km}/0.1 \text{ h} = 50$ km/h. We can learn more detail about the motion by looking at shorter time intervals. For example, the displacement in the last minute of travel (from W to Z) is 1.15 km, and this allows us to calculate the average velocity in the last minute as

$$\frac{\Delta x}{\Delta t} = \frac{1.15 \text{ km}}{1.0 \text{ min}} = \frac{1.15 \text{ km}}{1.0 \text{ min}} \times \frac{60 \text{ min}}{1 \text{ h}} = 69 \text{ km/h}.$$

We note here an important point: The ratio $\Delta x/\Delta t$—the average velocity—is the *slope of the straight line that joins the initial and final points*.

As we shorten the time interval, we get closer to the instantaneous velocity. We can get close to the instantaneous velocity at 3.0 min (point C) by looking at the *short* time

▶ FIGURE 2–5 As two end points defining a time interval move closer together, the time interval moves toward zero, and the average velocity for that interval then becomes the slope of the curve of displacement versus time at that point. The slope of a line (in green) tangent to the curve (see points BCD) of displacement versus time gives the instantaneous velocity at a time corresponding to the tangent point. Here we see the line that determines the velocity at point C. At the end of the curve, the line tangent to the curve is flat (has zero slope), so that the instantaneous velocity there is zero. The automobile represented here has come to a stop.

interval of 24 s from 2 min 48 s (point B) to 3 min 12 s (point D). During that time the displacement is from 3.28 km to 3.48 km. Thus,

$$\frac{\Delta x}{\Delta t} = \frac{0.20\ \text{km}}{24\ \text{s}} = \frac{0.20\ \text{km}}{24\ \text{s}} \times \frac{60\ \text{s}}{1\ \text{min}} \times \frac{60\ \text{min}}{1\ \text{h}} = 30\ \text{km/h}.$$

As the interval Δt becomes smaller, the line joining the two points and whose slope represents the average velocity between the two points becomes closer and closer to a tangent to the curve. In the limit of very tiny Δt, the slope of this line is the instantaneous velocity. As the slope of the tangent line changes from point to point on the curve, so does the instantaneous velocity change from moment to moment as you move along the curve. If you move a point on a transparent ruler along the curve so that the ruler's edge is always tangential to the curve, you can visualize the changes in instantaneous velocity during the trip.

All this is completely consistent with what we know from calculus: The derivative dx/dt at any given time t is the slope of the function $x(t)$ at that time, and the slope of the function $x(t)$ at time t is the slope of the tangent to the curve at that time. Thus *the slope of the tangent to our curve of x versus t at any time is the instantaneous velocity of the automobile at that time.*

Finally, let us write an explicit form for the position as a function of time when velocity has the constant value $v = v_0$. We know that for constant velocity the average velocity is the same over any interval, $v_{av} = v_0$. We could then, for example, solve for $x(t)$ in Eq. (2–7) written in the form $v_{av} = v_0 = [x(t) - x_0]/(t - 0)$. We find immediately

$$x(t) = v_0 t + x_0. \tag{2–12}$$

This should not be a surprising result. Starting at an initial point x_0, the position changes linearly with t, where v_0, the coefficient of t, is the slope of the x- versus t-curve, that is, the time derivative of x with respect to t.

EXAMPLE 2–3 We want to analyze a runner's motion around the 30.0-m mark in the 100-m dash using the data in Table 2–1. First calculate the average velocity of a runner over two time intervals $\Delta t = 3.91$ s and $\Delta t = 8.20$ s, with the time intervals centered around $x = 30.0$ m. Also calculate the average velocity over the smallest time interval available from the table near $x = 30.0$ m.

Setting It Up Calculation of an average velocity requires knowing displacements and the time intervals over which those displacements take place. This information is precisely what is given in Table 2–1.

TABLE 2–1 • Times for a 100-m Dash			
Distance (m)	Time (s)	Distance (m)	Time (s)
0	0	55	6.37
5	1.36	60	6.83
10	2.01	65	7.28
15	2.57	70	7.74
20	3.09	75	8.20
25	3.60	80	8.65
30	4.09	85	9.11
35	4.55	90	9.57
40	5.01	95	10.04
45	5.47	100	10.50
50	5.92		

Strategy The average velocity can be calculated using Eqs. (2–6) through (2–8), with the sign of the average velocity determined by the sign of the displacement. This will require us to read off the displacements and their corresponding time intervals from the table.

Working It Out According to Table 2–1, the time points on either side of 30.0 m that give a time interval Δt of 3.91 s are $t_1 = 2.01$ s and $t_2 = 5.92$ s; for $\Delta t = 8.20$ s, the times are $t_1 = 0$ s and $t_2 = 8.20$ s. We obtain the distances that correspond to each of these time points from Table 2–1. The average velocities are then

$$\text{for } \Delta t = 3.91\ \text{s: } v_{av} = \frac{x_2 - x_1}{t_2 - t_1} = \frac{50\ \text{m} - 10\ \text{m}}{5.92\ \text{s} - 2.01\ \text{s}} = 10\ \text{m/s};$$

$$\text{for } \Delta t = 8.20\ \text{s: } v_{av} = \frac{x_2 - x_1}{t_2 - t_1} = \frac{75\ \text{m} - 0\ \text{m}}{8.20\ \text{s} - 0\ \text{s}} = 9.2\ \text{m/s}.$$

The smallest available time interval around 30.0 m in Table 2–1 is the 0.95-s interval from 25 m (3.60 s) to 35 m (4.55 s), and

$$\text{for } \Delta t = 0.95\ \text{s: } v_{av} = \frac{x_2 - x_1}{t_2 - t_1} = \frac{35\ \text{m} - 25\ \text{m}}{4.55\ \text{s} - 3.60\ \text{s}} = 11\ \text{m/s}.$$

What Do You Think? With the data in the table, we found the smallest time interval listed around a particular time (or place) and used it to calculate the average velocity for that interval. Could you use graphical techniques to find the velocity at a particular time, that is, the instantaneous velocity at that time? If so, how would you proceed? Is this more precise than using only the data in the table? *Answers to **What Do You Think?** questions are given in the back of the book.*

CONCEPTUAL EXAMPLE 2–4 Consider the one-dimensional motion described in parts (a) to (d) of Fig. 2–6. Match the graphs to the following descriptions: (1) a rock dropped off the roof of a building, (2) fooling around with a time machine, (3) a rock thrown vertically upward, and (4) a baton in a relay race being passed from one runner to the next.

Answer 1. If x marks the height, the graph (b) describes a rock dropped off the roof of a building. You would start at a large value of x, which would decrease starting from the moment the rock is dropped, then stop, as the rock then stays at ground level.

2. You might suppose that a time machine (no, don't buy stock in the company selling them!) would take you back in time. You could start somewhere at some time, say, $t = t_0$, move, then somehow go back in time, move to a different position, and when the clock reached $t = t_0$ again, you would be in two different positions at the same time! This two-places-at-one-time property is visible in graph (d), as can be seen by following the vertical line that corresponds to a given time.

3. For vertical motion x again measures the height. Starting from when the rock is thrown, x increases, reaches a maximum at a later time, then decreases as the rock falls. That describes graph (a).

4. A straight line on a position–time plot represents constant velocity. Two runners will generally run at different speeds. Thus graph (c) represents a succession of two runners, the first slower than the second.

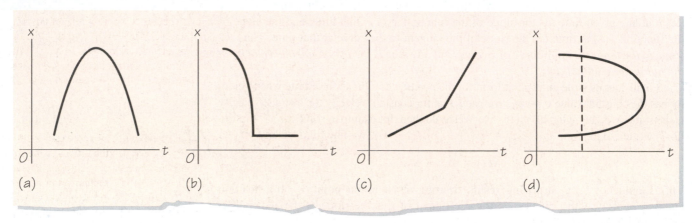

▲ **FIGURE 2–6** A series of curves on a graph of position versus time.

2–3 Acceleration

Just as velocity is defined as the rate of change of position, an object's acceleration is defined as the rate of change of its velocity with respect to time. Velocity can change because of a change in speed or because of a change in direction or both. An airplane taking off undergoes acceleration as its speed increases; an automobile going around a sharp curve undergoes acceleration because the direction of its motion changes. Since we are dealing with one-dimensional motion in this chapter, it is mainly changes in the magnitude of the velocity that count, along with what happens at "turn-around" places.

In our discussion of velocity, there is a particularly simple case: uniform, or constant, velocity, in which a position change is proportional to the time interval. There is also a particularly simple case for acceleration, in which the velocity change is proportional to the time interval; we refer to this case as uniform, or constant, acceleration.

Average Acceleration

The runner in the 100-m dash of Fig. 2–2 starts off with $v = 0$ m/s at $t = 0$ s. Two seconds later, he is moving with $v = 8$ m/s. His velocity has changed by $\Delta v = 8$ m/s over a time interval $\Delta t = 2$ s. We define the **average acceleration** over a time period, a_{av}, in terms of velocity v_1 at time t_1 and velocity v_2 at time t_2:

$$a_{av} \equiv \frac{v_2 - v_1}{t_2 - t_1} = \frac{\Delta v}{\Delta t}. \tag{2–13}$$

In one-dimensional motion the velocity change may well be negative, even though both the initial and final velocities are positive—the change in velocity is the *difference* between two velocities. This would occur, for example, when the object is slowing down. More generally, the acceleration in one-dimensional motion can have a sign opposite to that of the velocity. The dimensions of acceleration are $[LT^{-2}]$, with SI units of meters per second squared.

EXAMPLE 2–5 A runner in the 100-m dash accelerates to 10 m/s at 4 s and maintains this velocity for the next 4 s. She then realizes that she is going to win and slows over the next 4.7 s to reach a velocity of 8 m/s at the end of the race. She has run the 100-m dash in 12.7 s. What is the runner's average acceleration over the time periods 0 to 4 s, 4 to 8 s, and 8 to 12.7 s?

Setting It Up We plot the velocity versus the time in Fig. 2–7 to match the description, assuming $t = 0$ at the start. The sections with linearly changing velocity correspond to the periods of uniform (constant) acceleration. All required velocities and times are given in the problem except for the additional magnitude $v = 0$ m/s at $t = 0$ s. The runner is moving to the right always, so that the velocities are always positive.

▲ **FIGURE 2–7** The dashed line could represent the actual velocity, which is not specified in the problem statement; the segmented straight lines are the means of finding the average acceleration.

Strategy Equation (2–13) is used to find the average acceleration in terms of the given times and velocities. We must be careful to keep track of signs.

Working It Out

$$\text{for 0 to 4 s: } a_{av} = \frac{v_2 - v_1}{t_2 - t_1} = \frac{10 \text{ m/s} - 0 \text{ m/s}}{4 \text{ s} - 0 \text{ s}} = 2.5 \text{ m/s}^2;$$

$$\text{for 4 to 8 s: } a_{av} = \frac{v_2 - v_1}{t_2 - t_1} = \frac{10 \text{ m/s} - 10 \text{ m/s}}{8 \text{ s} - 4 \text{ s}} = 0 \text{ m/s}^2;$$

$$\text{for 8 to 12.7 s: } a_{av} = \frac{v_2 - v_1}{t_2 - t_1} = \frac{8 \text{ m/s} - 10 \text{ m/s}}{12.7 \text{ s} - 8 \text{ s}} = -0.42 \text{ m/s}^2.$$

The greatest average acceleration occurs at the beginning of the race, when the runner is attempting to reach her greatest speed. Although she is running at her highest velocity during the middle part of the race, her average acceleration during this period is zero. During the time interval when she is slowing down at the end of the race, her average acceleration is negative.

What Do You Think? Consider the (one-dimensional) motion described in parts (a) to (d) of Fig. 2–8. Match the graphs to the following descriptions: (1) a somewhat idealized ping-pong match, (2) the baton being carried by one and then another relay runner, (3) a ball being thrown into the air, and (4) driving along a road with badly synchronized red lights. *Answers to* **What Do You Think?** *questions are given in the back of the book.*

▲ **FIGURE 2–8** Series of curves on graph of velocity versus time.

In Example 2–5, the fact that the runner's average acceleration at the end of the race was negative means that the magnitude of the velocity is decreasing rather than increasing. You will sometimes see the term **deceleration** to describe situations in which the *magnitude* of the velocity decreases. For one-dimensional motion the sign of the acceleration determines its direction, just as it does for the velocity.

Instantaneous Acceleration

When we considered velocity, it was the instantaneous velocity that provided the most detailed and useful information about an object's motion, and for the same reasons we want to work with the **instantaneous acceleration**. In fact, when we use the term *acceleration*, we shall be referring to the instantaneous acceleration, unless otherwise

100-m Dash

(a)

Slope gives instantaneous acceleration

$a = \dfrac{5.9 \text{ m/s}}{2.7 \text{ s}}$

$= 2.2 \text{ m/s}^2$ at 2 s

$\Delta v = 5.9$ m/s

$\Delta t = 2.7$ s

(b)

(c)

stated. We define the instantaneous acceleration as the limit of $\Delta v / \Delta t$ as the time interval Δt goes to zero. This is again a derivative:

$$a \equiv \lim_{\Delta t \to 0} \frac{\Delta v}{\Delta t} = \frac{dv}{dt}. \tag{2–14}$$

INSTANTANEOUS ACCELERATION

We have discovered that we can find the velocity of a runner from a plot of position versus time; analogously, we can determine the acceleration from a plot of velocity versus time. Consider again our 100-m-dash runner (Fig. 2–9a). We obtain the acceleration at any time t by finding the slope of the tangent to the curve of v as a function of t at the particular time t. In Fig. 2–9b, we show an enlarged view of the region around times $t = 2$ s and $t = 4$ s, with a drawing of the tangent to the curve of v versus t at $t = 2.0$ s. We can determine the acceleration of our runner either by finding the slopes of tangents to the velocity curve or by taking the algebraic time derivative of the function $v(t)$. You can see, for example, that the tangent at $t = 2$ s is positive (positive acceleration), and as it goes to $t = 4$ s, the slope remains positive but decreases in value. Thus the acceleration decreases in value. This is reflected in Fig. 2–9c, which plots the acceleration of the runner versus time. This curve was derived from the velocity curve of Fig. 2–9b by reading off the slope of the tangent line at each point. Notice that the acceleration is initially very high as the runner gains speed, but by 4 s, when the runner moves at only a very slowly increasing speed, the acceleration drops nearly to zero.

◀ **FIGURE 2–9** (a) Velocity of runner whose position is shown in Fig. 2–2. This curve can be determined, for example, by finding the slope at each point along the position-versus-time curve and plotting it. Figure 2–5 illustrates the procedure. (b) The instantaneous acceleration of the runner is found by measuring the slope of the tangents to the curve of velocity versus time. Two such tangents are drawn in blue in this enlarged view; their slope gives the acceleration at $t = 2$ s and $t = 4$ s. The slope of the tangent at $t = 2$ s is calculated here. (c) Acceleration of runner as a function of time. This curve can be found by plotting the slope of the tangents to the velocity-versus-time curve as a function of time.

CONCEPTUAL EXAMPLE 2–6 Figure 2–10a shows a graph of the velocity of a runner who hears the starting gun at $t = 0$. He starts his run at $t = 0.1$ s—that is his reaction time. Is the acceleration infinitely large at that point?

Answer If we take Figure 2–10a seriously, with a truly instantaneous jump at $t = 0.1$ s, then it is indeed infinite. The slope of the v-versus-t curve is perfectly vertical, and that means an infinite slope. This idealized situation corresponds to the runner making the transition from at rest to full speed instantaneously. Of course, if we were to look carefully at the velocity as a function of time near 0.1 s, it would look more like Fig. 2–10b, which shows a more gradual rise to full speed, with a slope that is never infinite. A good runner's acceleration can be large, but it is limited.

(a)

(b)

▶ **FIGURE 2–10** In (a) we see an idealized version of the acceleration that in reality looks more like that plotted in (b).

By using our discussion about how velocity depends on position and how acceleration depends on velocity, we can make a link here from acceleration to position. If we examine the derivatives presented in Eqs. (2–10) and (2–14), we find that

$$a = \frac{dv}{dt} = \frac{d}{dt}\left(\frac{dx}{dt}\right) = \frac{d^2x}{dt^2}.$$ (2–15)

The acceleration is the time derivative of the velocity; equivalently, the acceleration is *the second time derivative of the displacement.*

THINK ABOUT THIS . . .

WHAT IS AN ACCELEROMETER AND WHAT ARE ITS USES?

An accelerometer is a device that measures acceleration; it does so in its simplest form by using a spring to translate acceleration into a measurement of the position of a mass. Figure 2–11 shows the basic arrangement for a one-dimensional motion. As we shall see in Chapter 5, a spring has the characteristic that it will stretch (or compress if a is negative) by an amount $x_f = (\text{const})a$, where the constant is characteristic of the mass and the particular spring. The length x_f is measurable and can be used to read the acceleration.

Accelerometers have two basic uses. First, they can be used as navigation devices because the measurement of acceleration can be turned into a measurement of position, as we shall see below. Second, they can be used to test vibrations and shock, which are characterized by accelerations within the test system. This would include crash testing for passenger vehicles (cars and planes) as well as monitoring how buildings and other structures respond to events such as earthquakes or explosions. Air bags are triggered by accelerometers.

▲ **FIGURE 2–11** Schematic view of an accelerometer for one-dimensional motion. The amount by which the mass moves is a measure of the acceleration.

EXAMPLE 2–7 The position x of an experimental rocket moving along a long rail is measured to be $x(t) = (5\ \text{m/s})t + (8\ \text{m/s}^2)t^2 + (4\ \text{m/s}^3)t^3 - (0.25\ \text{m/s}^4)t^4$ over the first 10 s of its motion, where t is in seconds and x is in meters. Find the velocity and acceleration of the rocket over the first 10 s and display the results graphically.

Setting It Up A plot of the position x as a function of time is shown in Fig. 2–12a.

Strategy The velocity and acceleration of the rocket can be determined by taking the time derivatives in Eqs. (2–10) and (2–15), respectively. This is a straightforward exercise in calculus. In particular, we use the rule $(d/dt)At^n = nAt^{n-1}$.

Working It Out We have

$$v = \frac{dx}{dt} = (5\ \text{m/s}) + (16\ \text{m/s}^2)t + (12\ \text{m/s}^3)t^2 - (1\ \text{m/s}^4)t^3;$$

$$a = \frac{dv}{dt} = (16\ \text{m/s}^2) + (24\ \text{m/s}^3)t - (3\ \text{m/s}^4)t^2.$$

Velocity and acceleration are measured in meters per second and meters per second squared, respectively, when time is measured in seconds. We plot these results over the time period 0 to 10 s in Figs. 2–12b and c, respectively. Note that although the position x is zero at time $t = 0$, neither the velocity nor the acceleration is zero at this time.

What Do You Think? The statement of this example specifies that $x(t)$ describes the motion over the first 10 s of motion. Suppose you wanted to use this formula to describe the motion for times beyond 10 s. What's to stop you from doing this? [*Hint*: In a real situation will the rocket keep accelerating? What are the consequences for its motion if it doesn't continue to accelerate?] Answers to **What Do You Think?** questions are given in the back of the book.

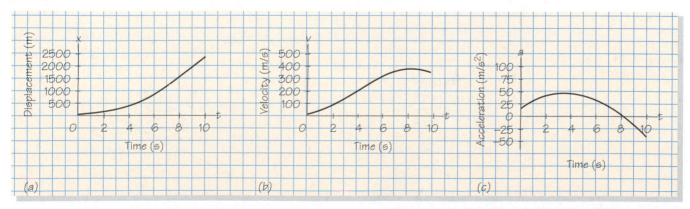

(a)   (b)   (c)

▲ **FIGURE 2–12** We have plotted the displacement, the instantaneous velocity, and the instantaneous acceleration as functions of time.

If we wished, we could continue and look at the rate of change of acceleration, da/dt, or even higher derivatives. But we don't need to do this because it is acceleration, not its changes, that plays a primary role in the laws of motion (see Chapter 4).

2–4 Motion with Constant Acceleration

The simplest example of acceleration is constant acceleration. It is a physically important case because the motion of freely falling objects near Earth's surface is motion with constant acceleration.

A constant acceleration implies that the value of the instantaneous acceleration equals the value of the average acceleration:

$$\text{for constant acceleration: } a_{av} = \frac{v(t_2) - v(t_1)}{t_2 - t_1} = a, \qquad (2\text{–}16)$$

where a is the constant acceleration and t_2 and t_1 are any values of time that we take for a velocity measurement. For convenience, we'll give the last time t_2 the more general label t and choose the first time t_1 to be 0. We also relabel $v(t_1) = v(0) = v_0$, and with this relabeling, we can rearrange the above equation to read

$$v = at + v_0. \qquad (2\text{–}17)$$

VELOCITY IN ONE DIMENSION UNDER CONSTANT ACCELERATION

The velocity's *linear* dependence on time is characteristic of constant acceleration (see also Section 2–6). The sign of the change in the velocity, whether it is increasing or decreasing, is the sign of a.

Let us now turn to the question of how the position changes with time under constant acceleration. We have already seen that when the velocity is constant the displacement changes linearly with time. Here, the velocity changes linearly with time; if velocity is increasing, then we would expect the displacement to increase even more rapidly with time. We analyze this in two steps: First we establish that the formula $x = v_0t + x_0$ that applies for constant velocity [Eq. (2–12)] also applies for nonconstant velocity if v_0 is replaced by v_{av}, that is, the speed averaged over the time interval from 0 to t. Second, we find v_{av} for constant acceleration and insert it into the formula for the position.

For the first step, we write Eq. (2–7) as

$$v_{av} = \frac{x(t) - x(0)}{t - 0} = \frac{x(t) - x_0}{t}.$$

This equation is solved for x to yield

$$x = v_{av}t + x_0. \qquad (2\text{–}18)$$

For the second step, we find v_{av} for constant acceleration. As you can see from Fig. 2–13, the average velocity in this case is simply the average of the initial and final

▶ **FIGURE 2–13** When acceleration is constant, the velocity changes linearly with time.

velocities over the total time period t. The initial velocity is v_0 and the final velocity at time t is $v = v(t)$, giving

$$v_{av} = \tfrac{1}{2}(v_0 + v). \tag{2-19}$$

AVERAGE VELOCITY IN ONE DIMENSION UNDER CONSTANT ACCELERATION

For this case, v_{av} is just the velocity at time $t/2$. When we substitute this result for v_{av} into Eq. (2-18), we find that for constant acceleration

$$x = \tfrac{1}{2}(v_0 + v)t + x_0 = \tfrac{1}{2}vt + \tfrac{1}{2}v_0 t + x_0. \tag{2-20}$$

We now substitute v from Eq. (2-17) into this equation and determine the position as a function of time:

$$x = \tfrac{1}{2}at^2 + v_0 t + x_0. \tag{2-21}$$

POSITION IN ONE DIMENSION UNDER CONSTANT ACCELERATION

Part of this result, $v_0 t + x_0$, describes the position of the object if it moved with a constant velocity v_0 starting from the position x_0. The additional quadratic term in t is due to the constant acceleration a. As we had expected, the position varies more rapidly than linearly with time.

We should discuss one more useful relation. The labeled equations (2-17) and (2-21) above, which contain the variable time, allow us to deduce a relationship between displacement, speed, and acceleration that does not involve time. To do so, we first solve for the time t from Eq. (2-17):

$$\text{for constant acceleration: } t = \frac{v - v_0}{a}. \tag{2-22}$$

If we substitute this expression for time into the first part of Eq. (2-20), we have

$$x = \left(\frac{v_0 + v}{2}\right)\left(\frac{v - v_0}{a}\right) + x_0$$

for constant acceleration, or

$$x - x_0 = \frac{v^2 - v_0^2}{2a}. \tag{2-23}$$

This equation can be written as

$$v^2 = v_0^2 + 2a(x - x_0). \tag{2-24}$$

SPEED AS A FUNCTION OF POSITION UNDER CONSTANT ACCELERATION

This result gives us the speed at any position x in terms of the constant acceleration a and the initial speed and position. Time does not enter into this result. Note the use of the word *speed* here, as only the magnitude of the velocity appears.

The labeled equations of this section are particularly useful, but you need not memorize them. You can deduce all the information in these equations from the simple statement that the acceleration is constant; alternatively, simply remember that the position changes quadratically with time and has a linear term in t as well [Eq. (2-21)]. The series of examples that follow illustrates some of the uses of these results.

EXAMPLE 2–8 An amateur bowler releases a ball with an initial velocity of 3.0 m/s; the ball slows down with a constant negative acceleration of -0.20 m/s^2. How far does the ball roll before stopping, and how long does it take to stop?

Setting It Up Choose a coordinate system with $x = 0$ m at the point where the ball leaves the bowler's hand (Fig. 2–14a). The x-axis is along the direction of the ball's motion. In Fig. 2–14b, we have sketched the constant acceleration as well as the linearly decreasing velocity and the quadratically varying position that correspond to the acceleration. We start the clock at $t = 0$ s when the ball leaves the bowler's hand. The initial conditions are then $x_0 = 0$ m,

▲ **FIGURE 2–14** (a) The origin O marks the spot ($x = 0$ m) where the ball leaves the bowler's hand. (b) Sketches of acceleration, velocity, and displacement as functions of time.

$t_0 = 0$ s, and $v_0 = 3.0$ m/s. The constant acceleration a_0 is negative in this coordinate system.

Strategy The crucial point is that the acceleration is constant here, so we can use our derived results. The first part of the question asks us to find the connection between speed and position, for which we can use Eq. (2–24) (an equation for which all is known except x, which is the displacement of the ball). Thus we can solve Eq. (2–24) for the x-value for which the final velocity v has become zero. For the second part, we can use the connection between velocity and time [Eq. (2–22)] to find the time when the velocity is zero.

Working It Out From Eq. (2–24), the solution for x is

$$x = x_0 + \frac{v^2 - v_0^2}{2a}.$$

All the quantities on the right-hand side are known, giving the numerical value

$$x = (0 \text{ m}) + \frac{(0 \text{ m/s})^2 - (3.0 \text{ m/s})^2}{2(-0.20 \text{ m/s}^2)} = 22.5 \text{ m}.$$

Use Eq. (2–22) to determine the time of motion,

$$t = \frac{(0 \text{ m/s}) - (3.0 \text{ m/s})}{-0.20 \text{ m/s}^2} = 15 \text{ s}.$$

Any checks you can find for your answer are helpful. In particular, here use Eq. (2–21) with $t = 15$ s to determine the displacement once more:

$$x = \frac{(-0.20 \text{ m/s}^2)(15 \text{ s})^2}{2} + (3.0 \text{ m/s})(15 \text{ s}) + (0 \text{ m}) = 22.5 \text{ m},$$

which confirms the earlier result. The equations for constant acceleration are consistent with one another.

What Do You Think? A bowling alley is usually polished very smoothly, as is a bowling ball. Why does a ball slow down after it leaves the bowler's hand? *Answers to* **What Do You Think?** *questions are given in the back of the book.*

EXAMPLE 2–9 A runner bursts out of the starting block 0.10 s after the gun signals the start of a race. She runs at constant acceleration for the next 1.9 s of the race. If she has gone 8.0 m after 2.0 s, what are her acceleration and velocity at this time?

Setting It Up We draw a coordinate system in Fig. 2–15, which also contains a sketch of the acceleration as a function of time. In this case the only thing we know about the acceleration is that it is constant and positive—we don't know its value a. We do, however, know that $x(t_f) = 8.0$ m, where t_f is the final time, 2.0 s. We also know the initial position x_0 and the initial velocity v_0.

Strategy The runner is not moving during the first 0.10 s of the race, and thus her acceleration is zero during this period of time. During the next 1.9 s, she has an acceleration that is not zero. We can use the results of this section *only if the acceleration is constant* during the entire time, so a correct strategy is to break up the motion into constant-acceleration pieces. In the first piece ($t = 0$ s to $t = 0.10$ s, nothing at all happens. For the second piece, consider a time

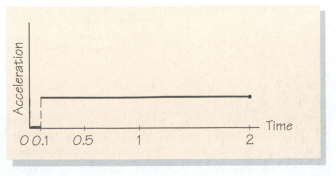

▲ **FIGURE 2–15** The acceleration of the runner is zero in the first one-tenth second and a positive constant thereafter.

$t' = t - 0.10$ s. (This is equivalent to starting your clock at 0.1 s rather than at 0 s.) For times t of 0.10 and 2.0 s, t' is 0 and 1.9 s, respectively. Over the time period $t' = 0$ s to $t' = 1.9$ s, we can use the results of this section, including that the initial time t'_0 is 0. The initial conditions of the problem are therefore $t'_0 = 0$ s, $x_0 = 0$ m, and $v_0 = 0$ m/s, and we want the acceleration and velocity at $t'_f = t_f - 0.1$ s = 1.9 s and $x = 8.0$ m. An examination of the labeled equations shows that there are two unknowns (v and a) in Eqs. (2--17) and (2--24). Equation (2--19) does not allow us to determine either v or a. Equation (2--21), however, allows us to determine the acceleration. Knowing the acceleration, we can then use Eq. (2--17) to determine the velocity at any time, in particular for the final time.

Working It Out We insert the known values (with the primed values of time) into Eq. (2--21) and solve for the acceleration:

$$8.0 \text{ m} = \frac{a(1.9 \text{ s})^2}{2} + (0 \text{ m/s})(0 \text{ s}) + (0 \text{ m}) = 10 \text{ m};$$

$$a = \frac{16 \text{ m}}{3.6 \text{ s}^2} = 4.4 \text{ m/s}^2.$$

Now at $t' = 1.9$ s ($t = 2.0$ s) after the runner starts,

$$v = at' + v_0 = (4.4 \text{ m/s}^2)(1.9 \text{ s}) + 0 \text{ m/s} = 8.4 \text{ m/s}.$$

What happens if the initial time is not $t = 0$? There is nothing very special about the time $t = 0$ in motion under constant acceleration; in fact, Example 2--9 illustrated the difficulty with Eqs. (2--17) through (2--24) if the initial time must always be zero. That is why it is useful to allow t_0 to be an arbitrary time in the equations relating acceleration, velocity, and displacement. In this case, the important equations of this section are as follows:

$$v = a(t - t_0) + v_0, \tag{2-25a}$$

FOR CONSTANT ACCELERATION a:

$$v_{av} = \tfrac{1}{2}(v_0 + v) \quad \text{(unchanged)}, \tag{2-25b}$$

$$x = \tfrac{1}{2}a(t - t_0)^2 + v_0(t - t_0) + x_0, \tag{2-25c}$$

$$v^2 = v_0^2 + 2a(x - x_0) \quad \text{(unchanged)}. \tag{2-25d}$$

The only difference between the four labeled equations (2--17), (2--19), (2--21), and (2--24) and Eqs. (2--25a) to (2--25d) is that t is replaced by $t - t_0$ in the latter equations. Remember, Eqs. (2--25a) to (2--25d) are valid only when the acceleration is constant between the times t_0 and t. When $t_0 = 0$ in Eqs. (2--25a) to (2--25d), we obtain the four labeled equations. By letting $t_0 = 0.10$ s and $t = 2.0$ s, Example 2--9 can now be worked more easily, as the acceleration is constant during this time interval.

EXAMPLE 2-10 A T-38 training jet (Fig. 2--16a) has an acceleration of 3.6 m/s^2 that lasts 5.0 s during the initial phase of takeoff. The afterburner engines are then turned up to full power for an acceleration of 5.1 m/s^2. The speed needed for takeoff is 164 knots (1 m/s = 1.94 knots). Calculate the length of runway needed and the total time of takeoff.

Setting It Up We draw the acceleration as a function of time in Fig. 2--16b, labeling the two constant values of acceleration that enter, a_1 (3.6 m/s^2) and a_2 (5.1 m/s^2). We label the point where acceleration changes, $t_1 = 5.0$ s.

Strategy There are two different constant accelerations in this example; therefore, we need to divide the problem into two parts. For the period from $t = 0$ to t_1, we have the values $x_0 = 0$ m and $v_0 = 0$ m/s at $t_0 = 0$ s. We then use Eqs. (2--25a) and (2--25d) to find the velocity and distance, respectively, at t_1. We then repeat the exercise for the second period of acceleration, with initial values of position and time coming from the final values of the first period. For this period we first find the time for which the velocity takes a final value $v_f = 164$ knots. We then use this value of time to find the dis-

tance traveled in the second period and add it to the distance traveled in the first period to obtain the runway length.

Working It Out For the first period, Eq. (2--25a) gives the velocity at 5.0 s:

for $t = 5.0$ s: $v = (3.6 \text{ m/s}^2)(5.0 \text{ s} - 0 \text{ s}) + (0 \text{ m/s}) = 18$ m/s.

Next we use Eq. (2--25c) to determine the distance the jet has traveled:

for $t = 5.0$ s:
$x = \tfrac{1}{2}(3.6 \text{ m/s}^2)(5.0 \text{ s} - 0 \text{ s})^2 + (0 \text{ m/s})(5.0 \text{ s} - 0 \text{ s})$
$\quad + (0 \text{ m}) = 45$ m.

We now move to the second phase of takeoff, where full power is applied. We have a new set of initial conditions beginning with the time $t_0 = 5.0$ s; namely, $x_0 = 45$ m and $v_0 = 18$ m/s. We want to find the time and distance corresponding to a final velocity of 164 knots. Let us first convert this value into SI units. The final velocity of 164 knots is

$$(164 \text{ knots})\frac{1 \text{ m/s}}{1.94 \text{ knots}} = 84.4 \text{ m/s}.$$

(continues on next page)

We use Eq. (2–25a) to find the time t at takeoff; the acceleration is now 5.1 m/s^2. From Eq. (2–25a) we have

$$a(t - t_0) = v - v_0,$$

$$t = \frac{v - v_0}{a} + t_0 = \frac{84.4 \text{ m/s} - 18 \text{ m/s}}{5.1 \text{ m/s}^2} + 5.0 \text{ s} = 18 \text{ s}.$$

Note that we waited until we had solved for the variable t before inserting the numerical values (with units) for v, v_0, a, and t_0. This technique also serves as a check as the cancellation of units gives the expected result, seconds in this case.

Equation (2–25c) can be used directly to determine the takeoff distance because all the variables for the second phase, except x, are now known:

$$x = \frac{(5.1 \text{ m/s}^2)(18 \text{ s} - 5.0 \text{ s})^2}{2} + (18 \text{ m/s})(18 \text{ m/s} - 5.0 \text{ s})$$

$$+ 45 \text{ m} = 7.1 \times 10^2 \text{ m}$$

Because we have included $x_0 = 45$ m from the first phase, 710 m (or 2330 ft) is the total amount of runway used.

What Do You Think? Figure 2–17 shows a sequence of constant accelerations undergone by a body in one-dimensional motion. Ignoring the sharp points of rapid change in the acceleration, sketch the velocity of a particle that starts at $t = 0$ with velocity $v = 0$. *Answers to **What Do You Think?** questions are given in the back of the book.*

(a)

(b)

▲ **FIGURE 2–16** (a) A T-38 jet accelerating during takeoff. In part (b) we have made a graph of acceleration versus time.

▶ **FIGURE 2–17**

2–5 Freely Falling Objects

In Section 2–4, we mentioned an important example of constant acceleration: gravity. The **acceleration due to gravity** is given the symbol $\vec{g}$, and its magnitude is approximately 9.80 m/s^2.[†] Ignoring the effects of air resistance, any object dropped in the vicinity of Earth's surface will move with constant acceleration $\vec{g}$. We call this motion **free fall**. The direction of $\vec{g}$ is down, toward Earth's center, a direction easily found using a plumb bob (a string with a mass at its end).

Galileo Galilei, who can be considered the first modern physicist, systematically investigated the motions of falling objects. Centuries earlier, Aristotle had suggested (incorrectly) that the speed of a falling object depends on the weight of the object and that this speed is proportional to the distance fallen. Galileo questioned this wisdom; his approach was to test by experiment, and he performed precise measurements whose results he described in mathematical language. He determined that the distance that objects fall

[†] There are variations of the order of 1 percent in the magnitude and direction of $\vec{g}$ over Earth's surface.

after starting from rest is proportional to the square of the time; equivalently, the speed of a falling object is proportional to the square root of the distance fallen. As Eqs. (2–21) and (2–24) show, this type of motion is characteristic of constant acceleration.

If we drop a hard rubber ball and a sheet of paper simultaneously from the same height, we observe that the paper floats down and the ball reaches the floor first (Figure 2–18a). The ball experiences a greater acceleration and a larger final velocity than does the sheet of paper. This is because air resistance affects the sheet of paper to a much greater degree than it affects the ball. If we wad the paper up and repeat the experiment (Fig. 2–18b), the effect of air resistance on the paper is decreased and the falling times for the paper and the ball in this experiment are more nearly equal. If the same experiment is done in a vacuum, the falling times are, as best as we can measure, the same for all objects. We shall assume for now that we are dealing with small, heavy objects ("particles") with negligible air resistance, although we must remember that in many situations air resistance can be an important effect. A falling particle near Earth's surface undergoes a constant acceleration that is the same for *all* particles, independent of their composition or their mass. We should emphasize that this fact applies to all free-fall motions, including motion that is initially *upward*. For such objects a constant acceleration downward will eventually produce a downward velocity—an initially upward velocity is only a question of the initial value. A tennis ball that is tossed up moves with the same constant acceleration downward as a dropped rock.

To study the effects of gravity quantitatively, it is easiest to set up a coordinate system with a direction perpendicular to Earth's surface. Let us align the y-axis with the vertical direction. We have two choices for this direction. If we choose the positive direction toward the center of Earth, then the acceleration of gravity is $+g$. If we choose the positive direction of the y-axis away from Earth's center, then the acceleration of gravity is $-g$. We could work problems by using either choice of axes, but here we choose the latter because it will facilitate our future discussion of motion in two and three dimensions. The vertical motion of any freely moving object for which air resistance can be ignored is then summarized by the four labeled equations of Section 2–4 for constant acceleration with $a = -g$. Thus for a freely falling body,

$$v = -gt + v_0, \tag{2–26a}$$

$$v_{av} = \tfrac{1}{2}(v_0 + v) \quad \text{(unchanged)}, \tag{2–26b}$$

$$y = \tfrac{1}{2}(-gt^2) + v_0 t + y_0, \tag{2–26c}$$

$$v^2 = v_0^2 - 2g(y - y_0). \tag{2–26d}$$

Similarly, g can be inserted into Eqs. (2–25a) to (2–25d). Remember that it is only the near constancy of the acceleration of gravity near Earth's surface that makes these equations applicable. Figure 2–19 represents the up-and-down motion of an object with the corresponding velocity and acceleration vectors.

Air resistance important for paper

(a)

(b)

▲ **FIGURE 2–18** Photos taken under a flashing strobe light allow us to see the position of a moving object at equal time intervals. (a) The rubber ball falls directly to the floor, whereas the sheet of paper floats down more slowly. (b) After the sheet of paper is wadded up and the experiment is repeated, the ball and the paper fall almost together.

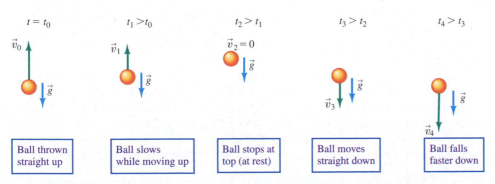

▲ **FIGURE 2–19** Details of the motion of a ball tossed in the air. It slows, stops, and falls back. Although position and velocity change with time, the acceleration is constant.

EXAMPLE 2–11 How much time elapses before a ball hits the ground after it has been dropped from rest from a height of 100 m (Fig. 2–20)? What is the ball's velocity just before it hits the ground?

Setting It Up In Fig. 2–20 we indicate that the ground is at a level $y = 0$ m, whereas $y = 100$ m is the position from which the ball is dropped. The given initial conditions at $t_0 = 0$ s are $y_0 = 100$ m and $v_0 = 0$ m/s.

▶ **FIGURE 2–20** The ball is released at 100 m from the ground at $t = 0$ s.

Strategy To find the time of flight t knowing the displacement and the initial conditions, we use Eq. (2–26c). Knowing this time, we can then use Eq. (2–26a) to calculate the velocity.

Working It Out From Eq. (2–26c) we have

$$0 \text{ m} = \frac{1}{2}(-9.80 \text{ m/s}^2)t^2 + (0 \text{ m/s})t + 100 \text{ m};$$

$$t^2 = -\frac{200 \text{ m}}{-9.80 \text{ m/s}^2} = 20.4 \text{ s}^2, \quad t = 4.52 \text{ s}.$$

Then Eq. (2–26c) gives the velocity:

$$v = -(9.80 \text{ m/s}^2)(4.52 \text{ s}) + 0 \text{ m/s} = -(44.3 \text{ m/s}).$$

Notice that the final velocity has a large magnitude and is *negative;* that is, it is in a downward direction. The negative sign appears because we chose the direction of positive displacement (along the y-axis) to be up.

What Do You Think? Rather than simply being dropped, the ball is tossed upward. Immediately after leaving the hand, the ball's acceleration is (a) upward, at least until reaching the top of its trajectory; (b) downward with magnitude g; or (c) zero, for a brief moment before gravity starts to act. *Answers to* **What Do You Think?** *questions are given in the back of the book.*

In Example 2–11 we calculated the velocity of an object falling freely from rest. Assume that an object starts at $y_0 = h$ with $v_0 = 0$ at $t_0 = 0$. What is the magnitude of the velocity when the object hits the ground ($y = 0$)? We use Eq. (2–26d) to determine

$$v^2 = 0 - 2g(0 - h) = 2gh;$$

for constant acceleration: $\quad v = \sqrt{2gh}.$ (2–27)

The positive sign of the square root is appropriate because it refers to the *magnitude* of the velocity. If the y-axis is up, the velocity $v = -\sqrt{2gh}$.

EXAMPLE 2–12 Calculate the time elapsed for the ball of Example 2–11 to drop from 100 to 75 m and from 75 to 50 m.

Setting It Up We are using the same data as in Example 2–11.

Strategy We calculate the times t_1 and t_2 at 75 and 50 m, respectively, given initial values $y_0 = 100$ m at $t_0 = 0$ s. We can use exactly the same procedure as was used in Example 2–11 to find the time t_1 to get to 75 m. We can also calculate directly the time elapsed $(t_2 - t_1)$ in falling from 75 to 50 m.

Working It Out Using Eq. (2–26c),

$$y = \frac{-gt^2}{2} + y_0, \quad t^2 = \frac{2(y_0 - y)}{g};$$

fall from 100 to 75 m:

$$t_1^2 = \frac{2(100 \text{ m} - 75 \text{ m})}{9.80 \text{ m/s}^2} = 5.10 \text{ s}^2 \quad \text{or} \quad t_1 = 2.26 \text{ s};$$

fall from 100 to 50 m:

$$t_2^2 = \frac{2(100 \text{ m} - 50 \text{ m})}{9.80 \text{ m/s}^2} = 10.2 \text{ s}^2 \quad \text{or} \quad t_2 = 3.19 \text{ s}.$$

Thus the times of flight are

from 100 to 75 m: $t = 2.26 \text{ s} - 0 \text{ s} = 2.26 \text{ s};$

from 75 to 50 m: $t = 3.19 \text{ s} - 2.26 \text{ s} = 0.93 \text{ s}.$

The ball's velocity increases as it falls, and it takes less time to drop a given distance as time progresses (Fig. 2–21).

▲ **FIGURE 2–21** The time for a ball to drop a given distance decreases as the ball gains velocity.

EXAMPLE 2–13 Calculate the position of the ball at $t = 1$ s and $t = 2$ s under the conditions stated in Example 2–11.

Setting It Up We again use the data from Example 2–11.

Strategy The initial conditions are the same as before. In this case we can use Eq. (2–26c) to find y as a function of time and then at $t = 1$ s and $t = 2$ s.

Working It Out With the given initial conditions,

$$y = -\tfrac{1}{2}gt^2 + 100 \text{ m};$$

◀ **FIGURE 2–22** The distance traveled in a given time interval increases as the square of the total time elapsed for an object in free fall.

at $t = 1$ s: $\quad y = \dfrac{(-9.80 \text{ m/s}^2)(1 \text{ s})^2}{2} + 100 \text{ m} = 95.1 \text{ m};$

at $t = 2$ s: $\quad y = \dfrac{(-9.80 \text{ m/s}^2)(2 \text{ s})^2}{2} + 100 \text{ m} = 80.4 \text{ m}.$

We show these distances in Fig. 2–22 for the 1-s equal time intervals. During the first time interval ($t = 0$ to $t = 1$ s), the ball travels $100 \text{ m} - 95.1 \text{ m} = 4.9 \text{ m}$, but during the second interval ($t = 1$ s to $t = 2$ s), the ball travels $95.1 \text{ m} - 80.4 \text{ m} = 14.7 \text{ m}$.

What Do You Think? Figure 2–23 illustrates the increasing distance a freely falling object travels in each time interval. What measurements could you make on this picture to verify that it corresponds to constant acceleration? *Answers to **What Do You Think?** questions are given in the back of the book.*

▶ **FIGURE 2–23** This multiple image of a ball falling under the influence of gravity is made by flashing a stroboscope 11 times at equal time intervals over 0.5 s. It shows that the ball falls greater distances in the same time interval for later times.

EXAMPLE 2–14 A ball is thrown straight up with a speed of 10.0 m/s from a third-floor window that is located 15.0 m above the ground. Calculate the maximum height of the ball, the ball's velocity when it hits the ground, and the total time it takes to reach the ground.

Setting It Up See Fig. 2–24, which also labels the y-axis, the initial ($t = 0$) height y_0, and the initial (upward) velocity v_0.

Strategy A ball thrown upward reaches its maximum height when its velocity is zero. At this point, the ball's direction of motion changes. We use this fact in Eq. (2–26a); we set $v = 0$ in that equation and solve for the time, call it t_{top}. Then we can find the maximum height above the ground, y_{max}, by substituting $t = t_{\text{top}}$ in Eq. (2–26c).

We can work the remainder of this example in two possible ways. We can either take a new initial condition to be at the maximum height of the ball with an initial velocity of zero or we can keep the original initial conditions. We'll work the example using the second strategy, so the initial conditions are as already given. We can use the connection between speed and distance, Eq. (2–26d), to find the final speed when the ball hits the ground ($y = 0$). It is not a problem that we have the speed and not the velocity because we know in this case that the ball is heading downward when it hits the ground. Finally, we can solve Eq. (2–26a) to find the flight time to the ground.

▲ **FIGURE 2–24** A ball is initially thrown up.

(continues on next page)

Working It Out   To determine the time when the ball reaches its maximum height, use Eq. (2–26a):

$$v = 0 \text{ m/s} = gt_{top} + v_0 = -(9.80 \text{ m/s}^2)t_{top} + 10.0 \text{ m/s};$$

$$t_{top} = \frac{10.0 \ \cancel{\text{m/s}}}{9.80 \ \cancel{\text{m/s}^2}} = 1.02 \text{ s}.$$

The maximum height is found by using this value of t in Eq. (2–26c):

$$y_{max} = \tfrac{1}{2}gt_{top}^2 + v_0 t_{top} + y_0$$
$$= \tfrac{1}{2}(-9.80 \text{ m/s}^2)(1.02 \text{ s})^2 + (10.0 \text{ m/s})(1.02 \text{ s})$$
$$+ 15.0 \text{ m} = 20.1 \text{ m}.$$

To calculate the final velocity v, we use Eq. (2–26d) with $y = 0$:

$$v^2 = v_0^2 + 2g(y - y_0)$$
$$= (10.0 \text{ m/s})^2 + 2(-9.80 \text{ m/s}^2)(0 \text{ m} - 15.0 \text{ m})$$
$$= 100 \text{ m}^2/\text{s}^2 + 294 \text{ m}^2/\text{s}^2 = 394 \text{ m}^2/\text{s}^2;$$

$$v = -19.9 \text{ m/s}.$$

In the last step we chose the negative root of v^2 based on our knowledge that the ball is moving downward. (We can choose either the positive or negative solution of the square root of v^2, whichever is physically sensible.)

Finally, from Eq. (2–26a) the flight time is as follows:

$$v = -19.9 \text{ m/s} = gt + v_0 = (-9.80 \text{ m/s}^2)t + 10.0 \text{ m/s};$$

$$t = \frac{19.9 \ \cancel{\text{m/s}} + 10.0 \ \cancel{\text{m/s}}}{9.80 \ \cancel{\text{m/s}^2}} = 3.05 \text{ s}.$$

What Do You Think?   Make sketches of position as a function of time and velocity as a function of time for the motion described in this example. What is the acceleration at the top of the ball's motion? *Answers to **What Do You Think?** questions are given in the back of the book.*

*2–6 Integration and Motion in One Dimension

We have learned how to find the velocity of an object if its displacement $x(t)$ is known ($v = dx/dt$); we can also determine the acceleration of an object if its velocity $v(t)$ is known ($a = dv/dt$). What about the inverse? Can we determine the displacement if the velocity is known or the velocity if the acceleration is known? What happens if the acceleration is not constant? You will learn in your calculus course (if you have not already) that the techniques of **integration** allow us to find the displacement of an object if we are given its velocity and to find its velocity if we are given the acceleration.

Displacement as an Integral of Velocity over Time

Let us start with the formula for the instantaneous velocity given in Eq. (2–10), $v(t) = dx(t)/dt$. Let us first consider $v(t)$ to be constant. Under these circumstances, the displacement is the time elapsed multiplied by the velocity, that is, $\Delta x = v \, \Delta t$. [This is just a rearrangement of the constant-velocity version of Eq. (2–10), namely $v = \Delta x/\Delta t$.] If we graph the relation $\Delta x = v \, \Delta t$ on a v–t plot, as in Fig. 2–25a, we see that the displacement Δx is *equal to the area under the curve of v as a function of time* (area = height × width). If the velocity changes from v_0 during the time interval $t = 0$ to $t = t_1$ to v_1 from the time interval $t = t_1$ to $t = T$ (Fig. 2–25b), as might be the case, for example, in a relay race in which two successive runners do not have the same velocity, then the total displacement is the sum of two displacements:

$$\Delta x = \Delta x_1 + \Delta x_2 = v_0 \times (t_1 - 0) + v_1 \times (T - t_1).$$

Figure 2–25b shows that this is again the sum of the areas under the plot of velocity versus time, that is, the total area under the v–t curve. The generalization of this process is straightforward. We may have a velocity that changes continuously with time (Fig. 2–25c). We then divide the total time interval from an initial time t_i to a final time t_f into many small equal time intervals Δt. These new time intervals are so short that the velocity cannot change very much over their duration; in other words, the velocity can be taken as a constant in these intervals. Let's concentrate on the tiny time interval from t_k to $t_k + \Delta t$ in Fig. 2–25c, where t_k is some intermediate time between t_i and t_f. The velocity $v(t_k)$ at the beginning of the time interval must be similar to the velocity $v(t_k + \Delta t)$ at the end; further, it should be nearly the same as the average value in the time interval. We therefore write this average value as just $v(t_k)$ itself. For this interval, then,

$$\Delta x_k = v(t_k) \, \Delta t.$$

This typical interval is shown as the darker area of the shaded curve in Fig. 2–25c.

Note: Sections marked with a * can be considered optional.

The total displacement $x_f - x_i$ over the total time interval t_i to t_f is then a sum over the small displacements $\Delta x_k = v_k \, \Delta t$. There must be N equal, tiny time intervals Δt to span the time interval from t_i to t_f:

$$N \, \Delta t = t_f - t_i;$$

$$x_f - x_i = \sum_{k=1}^{N} \Delta x_k = \sum_{k=1}^{N} v_k \, \Delta t. \qquad (2\text{–}28)$$

This equation approximates the full area under the curve in Fig. 2–25c; we have now seen that it also represents the displacement $x_f - x_i$ over the period from t_i to t_f. If we are concerned that our treatment of approximating the exact area by a sum over the areas of skinny rectangles isn't accurate enough because v changes significantly over the time interval Δt, we can simply make Δt smaller and smaller. In the limit that each time interval Δt goes to zero, we have an exact result:

$$x_f - x_i = \lim_{\Delta t \to 0} \sum_{k=1}^{N} v_k \, \Delta t. \qquad (2\text{–}29)$$

Integration

The right-hand side of Eq. (2–29) is the definition of the integral of $v(t)$ over time:

$$x_f - x_i = \int_{t_i}^{t_f} v(t) \, dt. \qquad (2\text{–}30)$$

This expression is called a **definite integral** because the limits of integration—here t_f and t_i—are specified. The symbol $\int$ replaces the summation sign Σ when we sum over an infinite number of infinitesimal intervals of Δt. The area under the curve of velocity versus time between t_i and t_f in Fig. 2–25c is exactly the displacement $x_f - x_i$.

If the integration limits are not specified, we have

$$x = \int v(t) \, dt + C, \qquad (2\text{–}31)$$

where C is a constant of integration determined from the initial conditions discussed earlier (i.e., x_0 and v_0 at t_0). Equation (2–31) is called an **indefinite integral**. You will find a table of some indefinite integrals in Table IV–8 in the Appendix.

In Eqs. (2–30) and (2–31), we speak of taking the integral of $v(t)$ over the time. The function $v(t)$ in this case is called the *integrand*. The integral of a function between two limits is the area under the curve that represents a graph of the function versus its variable. For example, we have shown that the integral of $v(t)$ over t is the total area under the curve $v(t)$ versus t.

Velocity as an Integral of Acceleration over Time

The analysis of how velocity varies with time when the acceleration is constant shows that on a graph of acceleration versus time the change in velocity over a certain time interval is equal to the area under the a–t curve (Fig. 2–26). Just as displacement is the integral of the velocity over time, we find that the velocity is an integral of the acceleration over time:

$$v_f - v_i = \int_{t_i}^{t_f} a(t) \, dt, \qquad (2\text{–}32)$$

$$v = \int a(t) \, dt + C. \qquad (2\text{–}33)$$

We have again written both the definite and indefinite integral forms.

(a)

(b)

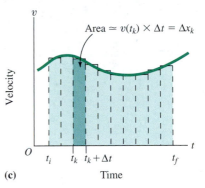

(c)

▲ **FIGURE 2–25** (a) The displacement over the dark area is $\Delta x_k = v \, \Delta t_k$, where v is the (constant) velocity in this interval. (b) If we have a sequence of two constant velocities, we can sum the two areas to find the displacement. (c) For varying velocity, we divide up the time into a series of intervals over which the velocity is (approximately) constant. When the areas of all the intervals are summed, we find the net displacement between initial time t_i and final time t_f, which is the shaded area under the curve of velocity versus time. This area is the integral

$$x_f - x_i = \int_{t_i}^{t_f} v \, dt.$$

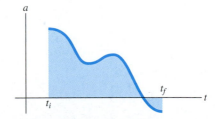

▲ **FIGURE 2–26** The change in velocity is found by integrating the acceleration over time.

CONCEPTUAL EXAMPLE 2–15 The acceleration a of a mass attached to the end of a spring and moving in a groove aligned with the x-axis is sinusoidal in time, $a(t) = A \sin(\omega t)$. The quantity ω, which has dimensions $[1/T]$, is a constant characteristic of the mass and the spring. (See Chapter 13 for a more complete discussion of this system.) How would you go about using this information to find the position of the mass as a function of time? How would you check that you made your calculation correctly? You can assume that you have a table of integrals and derivatives to help you with the mathematical details.

Answer We can put together the steps we have separately described in this section: The velocity is the integral over time of the acceleration, and the position is the integral over time of the velocity.

Thus we want to carry out two integrals in succession to find the position of the mass as a function of time. Since the integral of a sine function is proportional to a cosine function (see Table IV–9 in the Appendix), that is the form of the velocity. And the integral of a cosine function is proportional to a sine function (see the same table), so the position of the mass—the integral of the velocity—is a sine function. Both the original acceleration and the position are sine functions. The mass moves back and forth with time, and its position is proportional to the original acceleration. As a good check, we could take a derivative with respect to time of the position to find the velocity, then a derivative of the velocity to find the acceleration (equivalent to taking two time derivatives of the position to find the acceleration).

More on Integration

Integration is the inverse of differentiation. We can see this for ourselves as follows: An integral of the form

$$I(T) \equiv \int_{t_0}^{T} f(t)\, dt$$

measures the area under the graph of $f(t)$ as a function of t extending from t_0 to T. (In fact, we have used this in finding, for example, a displacement for a time-varying velocity.) The derivative of our integral is defined by

$$\frac{dI(T)}{dT} = \lim_{\Delta T \to 0} \frac{I(T + \Delta T) - I(T)}{\Delta T}.$$

The first term in the numerator is the area under the curve of $f(t)$ extending from t_0 to $T + \Delta T$; the second is the area extending only to T. The numerator is the difference of these, and this difference is the area under the curve extending from T to $T + \Delta T$. As ΔT becomes smaller and smaller, the area is just the width, ΔT, multiplied by the height, which as $\Delta T \to 0$ is just the height, $f(T)$. Thus we clearly see from our graphical interpretation that $dI(T)/dT = f(T)$. Since T is arbitrary, the formula holds everywhere; in other words, we can just write it in terms of the time t:

$$\frac{dI(t)}{dt} = f(t).$$

The derivative of an integral with respect to the upper limit of the integration is the integrand.

The role of the lower limit on the integral is just to set a starting point. Suppose we were to shift the lower limit on the integral from t_0 to t_1. We choose $t_1 > t_0$ for convenience. We then have

(area from t_0 to T) = (area from t_0 to t_1) + (area from t_1 to T),

or, equivalently,

$$\int_{t_0}^{T} f(t)\, dt = \int_{t_0}^{t_1} f(t)\, dt + \int_{t_1}^{T} f(t)\, dt.$$

As far as the dependence on T is concerned, the first integral on the right is just a number, independent of T. Thus when we differentiate our expression, this term plays no role—the derivative of a constant is zero. The significance of this is the following: If we know that some function of t may be written in the form $f(t) = dI(t)/dt$, it then follows that we can write $I(t)$ as an integral over $f(t)$. However, because of the independence of this result on the lower limit on the integral (a constant), we can only conclude that

$$I(T) = \int^{T} f(t)\, dt + C,$$

where C is an arbitrary constant. The integral does not tell us where the area calculation starts from (that is why the lower limit has been left off), and this ambiguity is represented by C. Once we decide on a lower limit on the integral, say t_0, then we can immediately fix C. Its value is $I(t_0)$, since the integral for which the upper and lower limits are equal is zero.

We can easily write some integrals based on some known derivatives and the inverse nature of integration and differentiation:

(i) $\quad\quad \dfrac{dt^n}{dt} = nt^{n-1} \quad\quad$ leading to $\quad\quad \displaystyle\int^T t^{n-1}\, dt = \dfrac{T^n}{n} + C; \quad\quad$ (2–34)

(ii) $\quad\quad \dfrac{d\sin t}{dt} = \cos t \quad\quad$ leading to $\quad\quad \displaystyle\int^T (\cos t)\, dt = (\sin T) + C; \quad\quad$ (2–35)

(iii) $\quad\quad \dfrac{de^{at}}{dt} = ae^{at} \quad\quad$ leading to $\quad\quad \displaystyle\int^T e^{at}\, dt = \dfrac{1}{a} e^{aT} + C. \quad\quad$ (2–36)

Special Case of Constant Acceleration

The value of Eqs. (2–30) and (2–32) is that they can be used to find the position and the velocity of an object whether its acceleration is constant or not. We can demonstrate the consistency of these integration techniques by using them to derive once more the equations that describe the constant-acceleration situation. Suppose that the magnitude of the constant acceleration is a. Beginning with Eq. (2–32), with the initial conditions x_0 and v_0 for $t = t_0$, we have

$$ v - v_0 = \int_{t_0}^t a\, dt. $$

The integral of a constant C_1 is, according to Eq. (2–35), $C_1 t$, so

$$ \text{for constant acceleration: } v - v_0 = at \Big|_{t_0}^t = a(t - t_0), $$

where the vertical line indicates that the total function is to be evaluated at the two limits of integration (t_0 and t) and the results subtracted. Rearranging this result gives Eq. (2–25a),

$$ \text{for constant acceleration: } v = v_0 + a(t - t_0). $$

We integrate this result once more to determine x:

for constant acceleration:

$$ x - x_0 = \int_{t_0}^t v(t)\, dt = \int_{t_0}^t (v_0 + at - at_0)\, dt $$

$$ = \left(v_0 t + \tfrac{1}{2}at^2 - at_0 t \right) \Big|_{t_0}^t $$

$$ = v_0(t - t_0) + \tfrac{1}{2}a(t^2 - t_0^2) - at_0(t - t_0) $$

$$ = v_0(t - t_0) + \tfrac{1}{2}at^2 - at_0 t + \tfrac{1}{2}at_0^2 = v_0(t - t_0) + \tfrac{1}{2}a(t - t_0)^2, $$

$$ x = \tfrac{1}{2}a(t - t_0)^2 + v_0(t - t_0) + x_0, $$

which is Eq. (2–25c). Integration techniques allow us to obtain these results directly.

Summary

Kinematics is the mathematical description of motion. Quantities of particular physical significance are displacement, instantaneous velocity, and instantaneous acceleration. These are all vector quantities, expressible in one dimension as magnitudes with signs. The displacement Δx measures the change in an object's position; when an object moves from position x_1 to position x_2, its displacement is for one dimension,

$$ \Delta x \equiv x_2 - x_1. \quad\quad (2–1) $$

The instantaneous velocity at any given time is defined as the limit of the average velocity as $\Delta t \to 0$:

$$v(t) = \lim_{\Delta t \to 0} \frac{x(t + \Delta t) - x(t)}{\Delta t}. \qquad (2\text{--}9)$$

It is thus the rate of change, or derivative, of the displacement at that time:

$$v(t) = \lim_{\Delta t \to 0} \frac{\Delta x}{\Delta t} = \frac{dx}{dt}. \qquad (2\text{--}10)$$

The instantaneous acceleration is defined as the rate of change of velocity:

$$a \equiv \lim_{\Delta t \to 0} \frac{\Delta v}{\Delta t} = \frac{dv}{dt}. \qquad (2\text{--}14)$$

We can also define the average speed and average velocity, but these concepts are used less frequently.

Useful relations between displacement, velocity, and acceleration that are valid for constant acceleration in one dimension are

for constant acceleration a:

$$v = a(t - t_0) + v_0, \qquad (2\text{--}25a)$$

$$v_{av} = \tfrac{1}{2}(v_0 + v), \qquad (2\text{--}19, 2\text{--}25b)$$

$$x = \tfrac{1}{2}a(t - t_0)^2 + v_0(t - t_0) + x_0, \qquad (2\text{--}25c)$$

$$v^2 = v_0^2 + 2a(x - x_0). \qquad (2\text{--}24, 2\text{--}25d)$$

Graphical techniques are useful in determining velocity. Graphically, velocity is the slope of a curve of displacement versus time. Similar analysis can determine acceleration from a curve of velocity versus time.

The acceleration due to gravity, which has magnitude g, is an important example of a constant acceleration. In this case, the acceleration points to Earth's center. (This result is accurate to the extent that there is no air resistance or friction in the motion.) A particle falling a distance h from rest has the final velocity

for constant acceleration: $v = \sqrt{2gh}. \qquad (2\text{--}27)$

By using integration techniques, the displacement and velocity can be determined from the velocity and acceleration, respectively:

$$x_f - x_i = \int_{t_i}^{t_f} v(t)\, dt, \qquad (2\text{--}30)$$

$$v_f - v_i = \int_{t_i}^{t_f} a(t)\, dt. \qquad (2\text{--}32)$$

These last two expressions are valid even if the acceleration is not constant.

Understanding the Concepts

1. Why is it a good idea to increase the space between your car and the car in front of you when the speed of the cars increases?
2. A piece of chalk is thrown straight up; at some point it reaches a maximum height and begins to drop. What is the velocity at the maximum height? Can there be a nonzero acceleration at this point even though the velocity is zero?
3. In a series of thought experiments, an object is dropped from rest from a given height on a variety of planets. Each of these planets

has a different acceleration due to gravity, g_x. Describe how the time of fall varies with g_x. How does the speed of the object at the end of the fall vary with g_x?

4. You are in the unfortunate position of being in an elevator with 20 bowling balls when the elevator cable breaks, causing both you and the elevator to fall under the acceleration of gravity. The emergency brake has not yet cut in. What is happening inside the elevator?

5. A falling object moves faster and faster the farther it falls. Does this mean that an object dropped from an arbitrarily large height will hit the ground moving with an arbitrarily high speed? If you answered yes, how do you explain the fact that meteors, which can come from very far away, don't hit the ground with a nearly infinite speed?

6. If an object that is restricted to moving along a straight line has a positive initial velocity and if the acceleration is always negative, can the velocity remain positive?

7. An astronaut in full gear can jump up 0.8 m on Earth. What arguments would you use to estimate the height of a jump on the moon, where the acceleration of gravity has magnitude 1.6 m/s^2?

8. What is the role of an air bag placed where a falling object is expected to land? How can an air bag prevent injury to someone who jumps from a height?

9. An object moving along the x-axis on a straight horizontal rail starts by moving rapidly to the right, slows, and comes to a stop, then starts moving more and more rapidly to the left. True or false: This description is consistent with motion with constant acceleration (even around the region where the object reverses the direction of its motion).

10. For the data shown in Table 2–1, will there be any difference between the average speed over some interval and the magnitude of the average velocity over the same interval? How would you answer the same question if the motion were not on a straight track?

11. What should the velocity of the runner in Fig. 2–2 be at $t = 0$ s? Do the data justify your conclusion? Explain.

12. "Zeno's paradox" comes to us from ancient Greece. It concerns the difficulty that a runner might have in catching a tortoise near the finish line of a race if the tortoise is ahead of the runner at one point, as follows. At some time, the tortoise is a distance L in front of the runner. After a time interval Δt, the runner is $L/2$ behind the tortoise. After a later time $\Delta t/2$, the runner is $L/4$ behind the tortoise. After a time $\Delta t/4$, the runner is $L/8$ behind the tortoise. The runner always appears to be behind! Where did the Greeks go wrong? By the way, the correct answer to this question was given only in Newton's time and lies behind the crucial concepts of calculus.

13. The velocity of an object moving in one dimension is measured at equal distance intervals. It is found that the magnitude of the velocity is proportional to the square root of the distance traveled. What can you say about the motion?

14. If the velocity of an object is positive, is its acceleration necessarily positive? Is there any connection between the sign of the velocity and the sign of the acceleration?

15. True or false: A freely falling body is moving with four times the speed when it has fallen twice as far.

16. A juggler tosses a beanbag straight up with initial speed v_0 under the influence of gravity, lets a second beanbag drop from rest, and tosses a third straight down with initial speed v_0. Compare the subsequent accelerations of the three beanbags.

17. In what order do the beanbags of the previous question hit the floor? What are the relative speeds of the three bags when they hit the floor?

18. A beanbag is tossed straight up. It rises, reaches a maximum height, then falls back down. What is the acceleration of the beanbag at its maximum height?

19. You measure the velocity of a bicycle moving in a straight line. How would you determine the rotations per second made by the wheel; in particular, what additional measurements are needed?

20. Given a stopwatch and a measuring rod, how would you determine the average acceleration experienced by someone jumping on a trampoline?

21. Describe some situations, other than free fall under the influence of gravity, in which an object could be undergoing constant acceleration.

22. You are given a measuring rod and a movie camera with a rather precisely known speed of the motion of the film. How would you use this to determine (to some degree of accuracy) the instantaneous velocity of a person jumping up and down on a trampoline? How would you use your apparatus to measure the instantaneous acceleration?

23. Consider a super-ball that drops from a certain height onto a rigid surface. It starts off with zero velocity at time $t = 0$. Sketch the velocity as a function of time from the time that the ball is dropped to just before it hits the floor at time $t = 10$ s. Assuming that the starting velocity on the rebound has the same magnitude (though opposite direction) as the ball when it reaches the ground, sketch the velocity as a function of time after the rebound. Do all this on the same graph. When will the ball have zero velocity again? What will its position be at that time?

24. If you consider the fact that in nature we can never have perfectly sharp angles in the velocity directions, round off the sketch in the question above at the point where the ball reverses direction. Describe in words what happens to the acceleration during the turn-around motion of the ball. We have restricted ourselves in sketching the motion as a function of time, with t along the horizontal axis and x along the vertical axis. Suppose we were to reverse these, so that x is along the horizontal axis and t along the vertical one. Sketch the following motions with your exchanged axes: (a) a car is at rest; (b) a car is moving slowly; (c) a car is moving rapidly. What is the main disadvantage of using this set of axes?

Problems

2–1 Displacement

1. (I) A grasshopper jumps along a groove aligned with the x-axis. Starting at the origin, the grasshopper's first jump has a displacement $+32$ cm, the second jump has a displacement -27 cm, the third a displacement -23 cm, and the fourth a displacement $+39$ cm. What is the net displacement? At what position is the grasshopper after all four jumps?

2. (I) Using the data in Table 2–1, draw position vectors to the runner for 40 and 80 m. Write and draw the displacement from 40 to 80 m.

3. (I) A gym teacher organizes a series of indoor races in the gym, which is 42 m in length. The students run from one end to the other and back again. After three round trips, what is the distance traveled by each student and what is the displacement vector? Draw a graph of the magnitude of the displacement vector as a function of time if it takes 7 s to run each 42-m leg. Assume that the speed is constant.

2–2 Speed and Velocity

4. (I) In 1991, Carl Lewis edged out Leroy Burrell in the World Championships in Tokyo to set a new world record in the 100-m dash. Their times at 10-m intervals are given here. Calculate the average velocity for Lewis's world record for 0 to 50 m, 50 to 100 m, and 0 to 100 m.

	Time (s)	
Distance (m)	Lewis	Burrell
10	1.88	1.83
20	2.96	2.89
30	3.88	3.79
40	4.77	4.68
50	5.61	5.55
60	6.46	6.41
70	7.30	7.28
80	8.13	8.12
90	9.00	9.01
100	9.86	9.88

5. (I) (a) Plot the path of an automobile that travels from a starting point to a point 15 km along a straight road at 75 km/h. It stops for 25 min, then continues on the same straight road for 40 km at 100 km/h. After a 5-min stop, it returns to its starting point at 60 km/h. Draw your position axis as horizontal and your time axis as vertical. (b) On the same plot, draw the path of an automobile that starts from the same spot 25 min after the first one and travels at 74 km/h in the original direction of the first automobile. Where and how often will the two cars meet?

6. (I) A car moving at 65 mi/h passes a pickup truck moving at 50 mi/h. The car goes on for 30 mi, then stops at a rest stop for 20 min. The car resumes its journey, again at 65 mi/h. Assuming the truck did not stop and maintained its speed, did it pass the car while it was stopped? If so, how long after the initial passing does it take for the car to catch the truck again? Solve this problem by graphical means.

7. (I) Redraw the same paths for the two vehicles of Problem 6 on a plot in which the horizontal axis is the time axis and the position, x, is along the vertical axis. Suppose that you took this new plot and simply relabeled the axes, so that the vertical direction represents time and the horizontal direction represents position. Interpret the paths of the two vehicles.

8. (I) An automobile travels north, covering a distance of 30 mi in 35 min, stops for 20 min, and then continues north for 20 mi, taking 25 min. Assume that the car moves uniformly during each segment of the trip. Calculate the average velocity of the total trip. Calculate the average velocity for the first half (by time) and the last half of the trip.

9. (II) An automobile driver travels north for 2 min at 30 mi/h, then stops at a red light for 30 s before proceeding again for 3 min at 45 mi/h. He then stops at a stop sign for 3 s, drives forward at 30 mi/h for 2 min, and finally stops for gas. (a) How far does the automobile travel? (b) What is the average velocity? Use units of miles and minutes.

10. (II) The position of a falling particle is given by $x = x_0 + v_0 t - \frac{1}{2} g t^2$. What is the velocity of the particle as a function of time? Calculate the average velocity during the time intervals t of 0 to 1 s, 1 to 2 s, and, more generally, t to $t + 1$ seconds.

11. (II) Use the velocity of a particle as a function of time, tabulated below, to calculate the position of the particle at each of the times; assume that at $t = 0$ s the particle was at the origin and at rest. [*Hint*: A graph is simplest.]

Time (s)	Velocity (m/s)
0.5	0.75
1.5	1.75
2.5	8.75
3.5	21.75
4.5	39.25
5.5	62.75
6.5	90.75
7.5	122.75

12. (II) The height of a bungee jumper above ground level is given as a function of time t by

$$y = (25 \text{ m}) \cos[\pi t/(6 \text{ s})] + (38 \text{ m}).$$

(a) Sketch the function $y(t)$ from $t = 0$ s to $t = 5$ s. (b) Calculate the average velocity of the jumper between $t = 2$ s and $t = 3$ s and between $t = 3$ s and $t = 4$ s. (c) What is the instantaneous velocity of the jumper when he is closest to the ground?

13. (II) An old brain teaser reads as follows: Two trains leave different stations 80 km apart and travel toward each other on a straight track. One train has a speed of 80 km/h and the other has a speed of 160 km/h. A very fast insect leaves the slower train and heads toward the faster train at a speed of 240 km/h. Upon encountering the second train, it turns around and just as rapidly returns to the first train. It continues these maneuvers until it is squashed between the two trains when they collide. Graph what has happened and use your graph to determine how far the insect will have traveled. Can you think of a way to estimate your result or to calculate it rapidly?

14. (II) Traffic signals are placed along a straight road at positions $x = 0$ m, $x = 600$ m, and $x = 1200$ m (Fig. 2–27). The time intervals during which the signals are green are shown by the thick lines in the figure. (a) Draw the displacement-versus-time curves (fastest and slowest) for a car that passes through all the lights when the car moves with constant speed. (b) Draw a similar set of lines for a car traveling in the opposite direction. (c) Assuming that the lights are timed such that a car passes through all lights in the middle of the time interval, what is the speed for which the lights are timed? (d) What is the fastest constant speed of a car that makes it through all the signals, assuming it arrives at the first light at the optimal moment?

▲ **FIGURE 2–27** Problem 14.

15. (III) The distance an ant moves in a straight-line motion is given by $x = 0.010t^3 - 0.050t^2 + 1.5t$ centimeters, where t is in seconds. Calculate the velocities for t values of 1, 5, and 10 s. What is the average velocity for the first 10 s? Why is the formula unrealistic for long times?

16. (III) The displacement of a particle as a function of t is described by the equation $x = \sqrt{(2.0\,\text{m}^2/\text{s})(t + 1.0\,\text{s})}$, where t is measured in seconds. (a) Plot $x(t)$ between $t = 0$ s and $t = 5.0$ s. (b) Calculate the average velocity between $t = 1.0$ s and $t = 5.0$ s, between $t = 2.0$ s and $t = 4.0$ s, and between $t = 2.8$ s and 3.2 s. (c) Compare these results with the instantaneous velocity at $t = 3.0$ s.

2–3 Acceleration

17. (I) An automobile badly in need of repairs is able to accelerate at a constant value of $0.40\,\text{m/s}^2$. How long does it take the automobile to get to 35 mi/h?

18. (I) A bicyclist is pedaling at a constant speed of 10 m/s when she decides to slow down. She stops pedaling and sits up, and the combined effects of wind resistance and road friction cause a negative acceleration of $-0.3\,\text{m/s}^2$. If this acceleration does not change, how long would it take her to slow to 5 m/s?

19. (I) A car is said to go from rest to 60 mi/h in 9.0 s. Assuming that the acceleration is uniform, what is its value in units of g?

20. (II) Car A leaves a city and travels along a straight road for 1.5 min at 60 km/h. It then accelerates uniformly for 0.25 min until it reaches a speed of 80 km/h. It proceeds at that speed for 2.0 min, then decelerates uniformly for 0.50 min until it comes to rest. Car B leaves the same city along the same road and accelerates uniformly for 1.6 min until it reaches a speed of 120 km/h. It then decelerates uniformly until it comes to rest again after 1.6 min. (a) Plot the curve of the cars' motions on a graph in which the vertical axis is the speed v and the horizontal axis is the time t. (b) Plot the motions of the cars on a graph in which the vertical axis is the distance x from the city and the horizontal axis is time t. (c) How far will the two cars have traveled during the different stages?

21. (II) An automobile starting from rest at $t = 0$ s undergoes constant acceleration on a straight line (Fig. 2–28). It is observed to pass two marks separated by 64 m, the first at $t = 8$ s and the second at $t = 12$ s. What is the value of the acceleration?

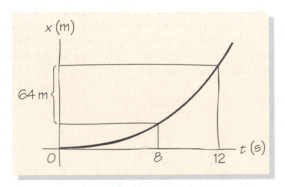

▲ **FIGURE 2–28** Problem 21.

22. (II) Inclined planes are convenient tools to study motion under a constant acceleration. The time of passage of a ball rolling on an inclined plane is measured by three light gates positioned 60 cm apart. The ball passes the light gates at 0.30, 1.15, and 1.70 s. Find the acceleration of the ball.

23. (II) Suppose the position of a particle is described by $x = A\sin(\omega t)$. Calculate the velocity and acceleration of the particle as a function of time.

24. (II) Consider the motion of the particle whose velocity is tabulated in Problem 11. Use these data to make a table of approximate values of the acceleration for $t = 1$ s, 2 s, ..., 7 s.

25. (II) The position x of a block attached to a spring as a function of time is given by the formula $x = A\sin(\pi t/12)$, as shown in Fig. 2–29. Describe in words the motion of the block.

▲ **FIGURE 2–29** Problem 25.

26. (II) The position of a particle is given by $x = At^2 + Be^{\alpha t}$. The particle is initially ($t = 0$) at $x = -1.5$ cm with $v = 0.25$ cm/s. At 0.10 s, the velocity is observed to be 0.045 cm/s. What is the acceleration at 1.0 s?

27. (II) Consider an object whose acceleration is determined by its velocity, as in the equation $a = A - (v/t_0)$. Here t_0 and A are constants with the dimensions of time and acceleration, respectively. Assume that the object starts out at $t = 0$ s with an initial velocity v_0. Sketch the behavior of the acceleration and of the velocity. Describe the motion after a long time.

2–4 Motion with Constant Acceleration

28. (I) A drag racer reaches 128 mi/h in a $\frac{1}{4}$-mi race. Assuming a constant acceleration, what was the elapsed time?

29. (I) A car traveling 25 mi/h must reach a minimum of 50 mi/h within a 1000-ft access lane. What must the car's constant acceleration be?

30. (I) An airplane starting from rest reached its takeoff velocity of 212 mi/h over a runway of 6000 ft. How long did this take if the plane rolled with a constant acceleration?

31. (I) A ball rolling straight down a ramp undergoes a constant acceleration of $0.50\,\text{m/s}^2$. What is the average velocity over the period 1 to 2 s assuming the ball started from rest at $t = 0$ s?

32. (I) A rocket accelerates uniformly from rest to a speed of 4.2×10^3 mi/h in 125 s. Over what distance does the rocket accelerate?

33. (I) A soccer ball rolls with an initial velocity of 8.0 m/s in an easterly direction across a flat field. Friction slows the ball down at the rate of $0.50\,\text{m/s}^2$. (a) Express the ball's velocity as a function of time. (b) Where is the ball 5.0 s after it starts to roll?

34. (I) A bowling ball is rolled down the alley with an initial velocity of $+10$ m/s. There is a small amount of friction, and it produces an acceleration of $-0.2\,\text{m/s}^2$. (a) What is the velocity at 1 and 2 s? (b) What is the average velocity over the first 2 s?

35. (II) A lead weight falls from a height of 6 m onto a muddy surface. It comes to rest after penetrating 0.4 cm into the surface. What was the magnitude of the average acceleration during the impact? How long did it take to stop?

36. (II) The speed of a landing airplane is 80 m/s. After touching ground it rolls a distance of 400 m on the runway at a constant velocity. It then decelerates at 3.0 m/s² until it stops. (a) Sketch the displacement–time and velocity–time curves. (b) Calculate the distance traveled on the ground and the time interval between touch-down and full stop.

37. (II) A car travels at a constant velocity of 20 m/s toward an intersection. When the car is 80 m from the intersection, the traffic light turns yellow. The driver continues with constant velocity for 1.2 s and then applies the brakes with a constant acceleration such that the car stops just at the intersection. (a) Sketch the displacement–time and velocity–time curves for the motion. (b) Determine the acceleration of the car during the braking period.

38. (II) A car accelerates from rest at 3.0 m/s² for 4 s, travels at a constant speed for 7 s, accelerates at 1.0 m/s² for 15 s, and then decelerates to rest at 2.5 m/s² (Fig. 2–30, not to scale). How far has the car traveled?

▲ **FIGURE 2–30**   Problem 38.

39. (II) A child is in an open-cage elevator facing out on a hotel lobby. The elevator is descending at a constant speed of 1 m/s. The child lets a penny drop from his hand when the elevator is 20 m above the floor of the lobby. How much time does the penny spend in the air? Ignore air resistance.

40. (II) Suppose that a runner were capable of a constant acceleration of 2.8 m/s² for the entire length of a 100-m dash. (a) How long would it take the runner to run the first 10 m? (b) How long for the first 50 m? (c) For the second 50 m? (d) For the entire 100 m? (e) Compare to the times of Problem 4.

41. (II) In 1979, the Japanese tested a magnetically levitated train. The train is both suspended and propelled by magnetic forces. The train traveled on a straight 7000-m-long track starting from rest; it reached a peak speed of 144 m/s before it came to rest again. Both the acceleration and deceleration were constant and of the same magnitude. The entire length of the track was used. (a) What was the magnitude of the acceleration (and deceleration)? (b) How much time was spent on the trip from one end of the track to the other?

42. (II) Your bus is leaving the stop, accelerating at a constant rate of 0.6 m/s². You turn the corner to see the bus pulling out of the stop 30 m ahead of you. What is the minimum steady speed with which you must run to catch the bus? Olympic sprinters can run at 10 m/s.

43. (II) A speeder is traveling along a straight road at 75 mi/h. He passes a standing police car, which starts to chase him. The police car accelerates from 0 to 85 mi/h in 13 s and travels at 85 mi/h thereafter. (a) Sketch the positions of both cars on the same x-versus-t graph. (b) How far from its starting point does the police car overtake the speeder? (c) What is the elapsed time?

44. (II) A car is moving at 35 mi/h when the driver sees a light turn red. She hits the brake pedal when she is 90 ft from the light and the deceleration of the automobile has magnitude 3.0 m/s². Does the car stop before it arrives at the light? How far does it travel before stopping?

45. (II) In Problem 44, how long does it take to stop from the moment the brakes are applied?

46. (II) Two automobiles are geared quite differently and are to be used for a drag race over a distance of 400 m. Car A accelerates at a constant value of 5.0 m/s² for the first 200 m, then at a constant value of 2.5 m/s² for the remaining 200 m. Car B accelerates at a constant rate of 4.5 m/s² for the first 200 m but at 3.0 m/s² for the remaining distance. (a) Give the value of the speed of each automobile at the 200-m mark and the time it took each to get there. (b) What are the finishing times for the race and the values of the respective speeds at the end of the race?

47. (II) An electron in the picture tube of a TV set traveling in a straight line accelerates uniformly from speed 3×10^4 to 5×10^6 m/s along a length of 2 cm. (a) How much time does the electron spend in this 2-cm region? (b) What is the magnitude of the electron's acceleration?

48. (II) You have an old, heavy automobile that does not accelerate very rapidly but can maintain acceleration for a long period. Suppose that your car has a maximum acceleration that takes it from 0 to 50 mi/h in 18 s. What would the speed be if this average acceleration were maintained for 36 s? How far does your car go during the first 20 s, and how far would it travel in 36 s under the above conditions?

49. (II) An elevator accelerates from the ground with a uniform acceleration a. After 3 s, an object is dropped out of an opening in the floor of the elevator and that object hits the ground 3.5 s later. How large is the acceleration? How high was the elevator when the object was dropped?

50. (II) Two small objects A and B are suspended from the ends of a rope thrown over a pulley (Fig. 2–31). Object A is 1.2 m above B when the system is released from rest. Object A descends with a downward acceleration of 0.3 m/s² and, because of the rope, B accelerates upward at the same rate. How much time elapses before the objects bump into each other?

▲ **FIGURE 2–31**   Problem 50.

51. (II) A car moving at 60 mi/h can be brought to rest in 4 s. Assuming that the deceleration is uniform, how far will the car travel between the time the brakes are applied and the time the car stops?

52. (II) A mountain climber is attached to a rope. She slips, and after she has fallen straight down 8 m, the rope starts to decelerate her. If the constant deceleration is $5g$ (five times the acceleration due to gravity), how much will the rope have to stretch? (In reality, the deceleration depends on the stretching of the rope.)

53. (II) A bullet traveling at 600 m/s penetrates a block of wood and comes to rest with a constant deceleration after traveling 20 cm (Fig. 2–32). What is the magnitude of the deceleration? How long does it take the bullet to stop?

▲ FIGURE 2–32 Problem 53.

2–5 Freely Falling Objects

54. (I) The tower of Pisa is 54.5 m tall. Assuming that Galileo dropped his object from rest from the top of the tower and that the effects of air resistance were negligible, how long would it have taken the object to fall?

55. (I) A story claims that someone who fell off New York City's Empire State Building (which has approximately 100 floors) was overheard to say "so far, so good" as he passed a third-floor window. Make some estimates to see if this is possible.

56. (I) A string is to have a series of lead sinkers tied to it. The first is tied at the bottom, and the second is tied 10 cm up from the bottom. The string can be held at its top and dropped from a height onto the top of a drum on which the first sinker already rests; each time a sinker hits the drum, a tap is heard. How far above the bottom sinker must the third, fourth, and fifth sinkers be tied so that the series of four taps is spaced by equal time intervals when the string is dropped?

57. (I) The acceleration due to gravity on the surface of the Moon is only about one-sixth the acceleration due to gravity on Earth's surface. In the celebrated experiment of the dropped feather performed by an astronaut on the Moon, how long did it take for the feather to drop 1 m to the surface if it started from rest?

58. (I) The acceleration due to gravity on the surface of Jupiter is 25.9 m/s². If it were possible to perform the experiment, how long would it take for an object that is initially at rest to fall a distance of 10 m on Jupiter's surface? Ignore any effects due to "air" resistance.

59. (II) A rock is thrown nearly straight upward from the edge of the top of a building at an initial speed of 22 m/s. Its trajectory will take it just past the edge, so that it lands on the ground. How much later must a second rock be dropped from rest at the same initial height of 10 m so that the two rocks hit the ground at the same time?

60. (II) An astronaut shipwrecked on a distant planet with unknown characteristics is on top of a cliff, which he wishes to descend. He does not know the acceleration due to gravity on the planet, and he has only a good watch with which to make measurements. He wants to learn the height of the cliff, and to do this, he makes two measurements (Fig. 2–33). First, he lets a rock fall from rest off the cliff edge; he finds that the rock takes 4.15 s to reach the distant ground. Second, he releases the rock from the same spot but tosses it upward so that it rises a height of what he estimates to be 2 m before it falls to the ground below. This time the rock takes 6.30 s to reach the ground. What is the height of the cliff?

▲ FIGURE 2–33 Problem 60.

61. (II) A ball is dropped from the roof of a 25-m-tall building. What is the velocity of the object when it touches the ground? Suppose that the ball is a perfect golf ball and it bounces such that the velocity as it leaves the ground has the same magnitude but the opposite direction as the velocity with which it reached the ground. How high will the ball bounce? Now suppose, instead, that the ball bounces back to a height of 20 m. What was the velocity with which it left the ground?

62. (II) A ball is thrown upward from the ground. It passes a window 10 m above the ground and is seen to descend past the window 2.2 s after it went by on its way up (Fig. 2–34). It reaches the ground 3.6 s after it was thrown. Use this information to calculate the acceleration due to gravity, g.

▲ FIGURE 2–34 Problem 62.

*2–6 Integration and Motion in One Dimension

63. (II) An object moves with an acceleration that depends on time and has the form $a = -4\sqrt{t}$ meters per seconds squared, where t is measured in seconds. Its velocity at $t = 0$ s is 15 m/s. How far will the object travel before it comes to a stop?

64. (II) A horizontally moving rocket has a chemical burn rate that produces an acceleration given by $a = 10 - 2t$, where the units are in SI. The rocket starts from rest at $x = 0$ and $t = 0$ and the fuel burns out at $t = 5$ s. (a) What is the velocity of the rocket as a function of time? (b) What is the position of the rocket at 5 s?

65. (II) The velocity of an object moving in a viscous fluid is given by the expression

$$v(t) = (4.0 \text{ m/s}) + (8.0 \text{ m/s})e^{-0.5t},$$

where t is measured in seconds. Calculate the acceleration of the object as a function of time. Using this result and the expression for $v(t)$, write the acceleration as a function of the velocity.

66. (I) A machine causes an object's speed to increase exponentially, $v(t) = v_0 e^{at}$, where $a = 0.5$ s^{-1} and $v_0 = 1$ m/s. If the object starts from the origin, how far has it traveled after 2 s?

67. (II) The height of a mass suspended from a ceiling by a spring at different times t is given by the formula $y = (0.1 \text{ m}) \sin(3 \pi t)$, where t is measured in seconds. (a) Plot the height as a function of time for times $t = 0$ s to $t = 1.0$ s. (b) Use your graph to determine the instantaneous velocity at time $t = 0.15$ s. (c) A measurement of the slopes at different times shows that the instantaneous velocity can be represented by the formula $\vec{v} = A[\cos(3\pi t)]\hat{j}$, where again t is measured in seconds. Use your measurement from part (b) to determine A. (Do not forget the units of A.) (d) Plot v as a function of t and use that graph to determine the instantaneous acceleration at time $t = 0.15$ s.

68. (II) The acceleration of gravity g is a constant only for a limited range of height differences. A better approximation, one that might hold over a larger range of height differences, is that g decreases linearly with height, $g = g_0 - hg'$, where h is the height measured from the ground surface and g' is a (small) constant of the appropriate dimensions. (a) Find the speed of a dropped object as a function of height assuming it was dropped starting from rest from a height h_0. (b) Find the speed of a dropped object as a function of time assuming it was dropped starting from rest from a height h_0.

69. (III) A powerful rocket moves for a short time with an acceleration that grows with time according to the formula $a = \alpha t^2$. If the rocket is to accelerate from rest in this way until it reaches a speed v_f, how long must the acceleration be maintained?

70. (III) When the effect of air resistance is taken into account, the acceleration of a falling object is described by the equation $a = ge^{-bt}$, where g is the acceleration due to gravity, $b = 0.5$ s^{-1}, and t is the time measured from the moment of release. (a) Calculate the velocity and displacement (from the place of release) of the object as a function of time. (b) Sketch these curves. (c) How long does it take for the object to fall 50 m? (d) Compare this time with the time it would take to fall 50 m without air resistance, that is, with $b = 0$.

General Problems

71. (I) The simplest juggling act involves two objects, one of which is transferred from one hand to the other when the second object is tossed upward (Fig. 2–35). Perform an experiment that will tell you how fast you can transfer an object from one hand to the other (e.g., by transferring something back and forth 20 times while a friend times you). Estimate from this how high you would have to toss the second object to perform the juggling act. How high would you have to toss the objects if you wanted to juggle three of them?

▲ FIGURE 2–35 Problem 71.

72. (I) A test of your reaction time is to catch a 12-in ruler held vertically by another person. Put your thumb and one finger near the bottom of the ruler, and as soon as the other person releases the ruler, squeeze your thumb and finger together to prevent the ruler from falling. Suppose that in such a test the ruler is grabbed after 5 in of it has passed your hand. What is the time interval between the visual detection of movement and the squeezing together of the fingers?

73. (I) A high jumper can jump 2 m on Earth. All other things being equal, how high could the same jumper jump on the surface of the Moon, where the acceleration of gravity is one-sixth that of Earth?

74. (I) You could probably jump off an 8-ft wall without hurting yourself (but do not try it!). Estimate what your deceleration would be when you hit the ground.

75. (II) A Moon rock is thrown upward with velocity 7 m/s. After 7 s, it has a downward velocity of 4 m/s. What is the acceleration due to gravity on the Moon? How high above the starting point did the rock go before it began to fall?

76. (II) Two long-separated friends, June and Bill, spot each other in an airport terminal from a distance of 20 m. They start to run toward each other (Fig. 2–36). Bill accelerates at a constant rate of 0.9 m/s^2 and June at a constant rate of 1.0 m/s^2. How far from June's initial position do they meet?

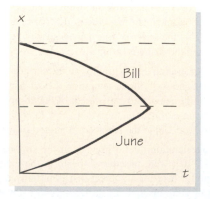

▲ FIGURE 2–36 Problem 76.

77. (II) An object moves in one dimension (described by an x-value) with a constant value J of the *third* derivative of position with respect to time. Write an equation for the position as a function of time given an initial position x_0 and an initial velocity v_0 (the sign of v_0 indicates the initial direction of motion).

78. (II) The velocity of an object moving along a straight axis is given as a function of time by $v = (4 \text{ m/s}^4)t^3 - (1 \text{ m/s}^2)t$, where v is measured in meters per second and t is measured in seconds. How far did the object move in the period from 0.5 to 1.5 s? Find the average velocity in this period and compare it to the maximum and minimum values of the velocity in the same period. You may want to make a graph to help.

79. (II) A tennis ball is dropped from a height of 10 m. It falls onto an electrical switch and bounces back to a height of 9 m. The switch is connected to an electronic device that shows that the time of contact between the ball and the switch was 0.002 s. Calculate (a) the velocity with which the ball hit the switch, (b) the velocity with which the ball left the switch, and (c) the average acceleration during the time of contact with the switch.

80. (II) A water balloon is dropped from the top of a tower, 200 m off the ground. An alert archer at the base of the tower sees the balloon and shoots an arrow straight up toward the balloon 5 s after the balloon is dropped. The arrow's initial velocity is 40 m/s. Where does the arrow intercept the balloon?

81. (II) There are several known cases of paratroopers whose chutes did not open as they fell but who survived by falling into brush or snow or onto a steep hillside. It is possible to survive a fall when the deceleration on impact is some 500 m/s^2, equivalent to about 50g. What is the distance traveled within a snowbank while a paratrooper comes to a stop if the deceleration is constant with a magnitude of 50g? The speed with which a para-trooper would enter the snowbank is about 40 m/s. Over how much time does the paratrooper decelerate?

82. (III) One test for the effects of the acceleration of gravity is to tie a set of weights to a string, with the second lowest separated from the lowest by L_0, the third lowest from the second lowest by L_1, and so forth, and to then drop the string (Fig. 2–37). Supposing that free fall corresponds to motion with constant acceleration, how should the separations $L_1, L_2, \ldots, L_n$ (where n is the number of weights) be related to L_0 if the sounds made by the weights as they land form a steady beat? The lowest weight starts at the surface onto which the other weights fall.

▲ FIGURE 2–37   Problem 82.

► The water droplets that fly up from this beautiful fountain in Monaco follow parabolic paths. Such paths are a consequence of the constant vertical acceleration associated with local gravity.

Motion in Two and Three Dimensions

I n Chapter 2 we considered the description of motion in one dimension—linear motion. We will now extend this description to motion in space. We observe such motion in the curved path of a thrown ball, in the swing of the pendulum of a grandfather clock, or in the orbits of the planets around the Sun. While for one-dimensional motion the directional aspect is encapsulated in signs, for motion in two or three dimensions we must use vectors to describe the directional aspect properly. With the help of the mathematical apparatus provided by vectors, we'll find that it is straightforward to describe the motion of objects in a plane or in space in a manner that builds on our earlier work with one-dimensional motion.

3–1 Position and Displacement

The motion of a planet orbiting the Sun traces out a path in space. Similarly, a rock thrown off a cliff follows a certain path, or **trajectory**, as does any pointlike object as it moves through space. For motion in a plane, think of a skater on a lake whose skates leave marks that specify the trajectory of the motion. Figure 3–1 depicts a particle, for example the skater, moving in a two-dimensional plane. We label the plane as the *xy*-plane and introduce a Cartesian coordinate system that contains an origin and *x*- and

(a)   (b)

Position vector describes position at given time.

◀ **FIGURE 3–1** (a) The ice skater is gliding over the path indicated by the blue trajectory line. (b) Position vectors $\vec{r}_P$ and $\vec{r}_Q$ point from the origin to the positions P and Q at the two times t_1 and t_2, respectively, along the skater's path of motion. The displacement vector between these times is $\Delta\vec{r} \equiv \vec{r}_Q - \vec{r}_P$.

y-axes. The particle is at the position P at time t_1; this position is described by the position vector $\vec{r}_P$, which points from the origin to the point P. At a later time t_2 the particle is located at position Q and is described by the position vector $\vec{r}_Q$. The change in the particle's position between times t_1 and t_2—the final position minus the initial position—can be described by the displacement vector $\Delta\vec{r}$; this vector is defined by

$$\Delta\vec{r} \equiv \vec{r}_Q - \vec{r}_P. \tag{3–1}$$

The vector $\Delta\vec{r}$ points from the tip of vector $\vec{r}_P$ to the tip of vector $\vec{r}_Q$ and describes the direction of the displacement as well as its magnitude. Whereas the position vectors $\vec{r}_Q$ and $\vec{r}_P$ depend on the choice of origin, the displacement vector $\Delta\vec{r}$ is independent of the choice of origin. To see this clearly, let's imagine that there is a new origin O' such that the vector from O' to O is the fixed vector $\vec{b}$ (Fig. 3–2). The position vector of point P in the new coordinate system is $\vec{r}_P + \vec{b}$ and that of point Q in the new system is $\vec{r}_Q + \vec{b}$. If we calculate the displacement, which is the difference $(\vec{r}_Q + \vec{b}) - (\vec{r}_P + \vec{b})$, the vector $\vec{b}$ cancels. In other words, we have again arrived at the displacement vector defined in Eq. (3–1). This can also be seen in the graphical representation in Fig. 3–2. The displacement is independent of our choice of origin.

As a particle moves, the components of its position vector (with respect to the Cartesian coordinate axes) change with time:

$$\vec{r}(t) = x(t)\hat{i} + y(t)\hat{j}. \tag{3–2}$$

For three-dimensional motion, we would proceed exactly as in Section 1–6: We set up three axes, define three mutually perpendicular unit vectors $\hat{i}$, $\hat{j}$, and $\hat{k}$, and write a position vector in the form

$$\vec{r}(t) = x(t)\hat{i} + y(t)\hat{j} + z(t)\hat{k}. \tag{3–3}$$

The fact that there is more than one vector component to the motion is the only difference between one-dimensional motion and two- or three-dimensional motion.

Displacement vector $\Delta\vec{r}$ is independent of origin.

$\vec{r}_P' = \vec{r}_P + \vec{b}$
$\vec{r}_Q' = \vec{r}_Q + \vec{b}$

◀ **FIGURE 3–2** A displacement vector $\Delta\vec{r}$ is independent of the origin. Here O and O' are two origins, and although the initial and final position vectors to points P and Q do depend on the origins, the difference between these position vectors does not.

EXAMPLE 3–1 The position of a bumper car in an amusement park ride (Fig. 3–3a) is described as a function of time by the coordinates $x = c_1t^2 + c_2t + c_3$ and $y = d_1t^2 + d_2t + d_3$, where $c_1 = 0.20$ m/s^2, $c_2 = 5.0$ m/s, $c_3 = 0.50$ m, $d_1 = -1.0$ m/s^2, $d_2 = 10.0$ m/s, and $d_3 = 2.0$ m. Find the position vectors of the car at $t = 3.0$ s and $t = 6.0$ s and the displacement vector between these times. Plot the trajectory, that is, a curve of y versus x that traces the path of the car on the floor.

Setting It Up We plot the locations of the car along the x-axis and the y-axis as a function of time in Figs. 3–3b and 3–3c, respectively. In the first instance we are given the position as a function of time and are asked to find the position at a particular time; in other words, we are actually given what we must find! We'll label the desired displacement vector $\Delta\vec{r}$.

Strategy The first part is a straightforward numerical substitution. For the displacement, we calculate the difference between the position vectors (components x and y) at 6.0 and 3.0 s. This difference is the displacement vector. As for plotting the trajectories, the simplest way to proceed is to start with an xy-plane. We can then mark the x- and y-values at a given time as a point on this graph. A half-second later, say, there is another point that can be marked, and so forth. By connecting those consecutive points starting from the one at the earliest time, we mark out the trajectory.

Working It Out We insert the two values of time (3.0 and 6.0 s) into the equations for x and y:

for $t = 3.0$ s:
$$x(t) = (0.20 \text{ m/s}^2)(3.0 \text{ s})^2 + (5.0 \text{ m/s})(3.0 \text{ s}) + 0.50 \text{ m} = 17 \text{ m},$$
$$y(t) = (-1.0 \text{ m/s}^2)(3.0 \text{ s})^2 + (10.0 \text{ m/s})(3.0 \text{ s}) + 2.0 \text{ m} = 23 \text{ m};$$

for $t = 6.0$ s:
$$x(t) = (0.20 \text{ m/s}^2)(6.0 \text{ s})^2 + (5.0 \text{ m/s})(6.0 \text{ s}) + 0.50 \text{ m} = 38 \text{ m},$$
$$y(t) = (-1.0 \text{ m/s}^2)(6.0 \text{ s})^2 + (10.0 \text{ m/s})(6.0 \text{ s}) + 2.0 \text{ m} = 26 \text{ m}.$$

With these components, Eq. (3–2) gives us the position vectors of the car at the two times:

for $t = 3.0$ s: $\vec{r}(t) = (17\hat{i} + 23\hat{j})$ m;
for $t = 6.0$ s: $\vec{r}(t) = (38\hat{i} + 26\hat{j})$ m.

Thus the displacement vector of the car between 3.0 and 6.0 s is [Eq. (3–1)]

$$\Delta\vec{r} = \vec{r}(t = 6.0 \text{ s}) - \vec{r}(t = 3.0 \text{ s})$$
$$= (38\hat{i} + 26\hat{j}) \text{ m} - (17\hat{i} + 23\hat{j}) \text{ m} = (21\hat{i} + 3\hat{j}) \text{ m}.$$

Finally, we plot y versus x, moment by moment, in Fig. 3–3d. This curve is the trajectory of the car.

What Do You Think? Why does the trajectory curve look so similar to the curve of y versus time? There are other vector descriptions of motion in a plane. Can you think of another such set? *Answers to **What Do You Think?** questions are given in the back of the book.*

(a)

(b)

(c)

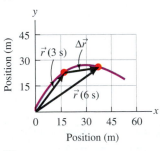
(d)

▲ **FIGURE 3–3** (a) An xy-coordinate system can be laid out on the floor. (b) Location along x-axis of car's motion from 0 to 8 s. (c) Car's location along y-axis. (d) By combining the results of part (b) and part (c), we can plot the trajectory, a graph of the y-position versus the x-position.

3–2 Velocity and Acceleration

Velocity

As for the one-dimensional motion described in Chapter 2, the velocity of a particle describes the rate of change of the position of the particle as it moves on its trajectory. Generally we will consider two-dimensional motion as we work through the chapter, as it is simpler than considering three-dimensional motion, but the approach applies perfectly well to three dimensions. Using Eq. (3–1) for the particle's displacement, the average velocity $\vec{v}_{av}$ over the finite time interval from t to $t + \Delta t$ is accordingly defined by

$$\vec{v}_{av} \equiv \frac{\vec{r}(t + \Delta t) - \vec{r}(t)}{\Delta t} = \frac{\Delta \vec{r}}{\Delta t}. \tag{3–4}$$

Equation (3–4) shows that the direction of $\vec{v}_{av}$ is the same as the direction of the displacement vector $\Delta \vec{r}$.

As the time Δt tends towards zero, the displacement over that interval becomes smaller and smaller, and as we'll describe in more detail later, the displacement vector $\Delta \vec{r}$ becomes tangent to the particle's trajectory at the location of the moving particle. Then, as in Eq. (2–11), the *instantaneous velocity* $\vec{v}(t)$ is obtained by letting Δt become infinitesimally small:

$$\vec{v}(t) \equiv \lim_{\Delta t \to 0} \frac{\vec{r}(t + \Delta t) - \vec{r}(t)}{\Delta t} = \frac{d\vec{r}}{dt}. \tag{3–5}$$

We have recognized that in the limit $\Delta t \to 0$ we arrive at the time derivative of the position vector. The instantaneous velocity can change from moment to moment. The direction of $\vec{v}$ at time t is tangent to the trajectory curve at that time (Fig. 3–4). Of course, we already know that its magnitude is by definition the particle's speed.

We can write the velocity vector in terms of components by using Eqs. (3–2) and (3–5):

$$\vec{v} = \frac{d}{dt}\vec{r}(t) = \frac{d}{dt}[x(t)\hat{i} + y(t)\hat{j}] \tag{3–6}$$

$$= \frac{dx}{dt}\hat{i} + \frac{dy}{dt}\hat{j}. \tag{3–7}$$

(The unit vectors $\hat{i}$ and $\hat{j}$ are constant in magnitude and direction, so their derivatives are zero.) We write Eq. (3–7) in the form

$$\vec{v} = v_x\hat{i} + v_y\hat{j} \tag{3–8}$$

$$= \vec{v}_x + \vec{v}_y, \tag{3–9}$$

where

$$v_x = \frac{dx}{dt}, \tag{3–10a}$$

$$v_y = \frac{dy}{dt}, \tag{3–10b}$$

and the component vectors are

$$\vec{v}_x = \frac{dx}{dt}\hat{i}, \tag{3–11a}$$

$$\vec{v}_y = \frac{dy}{dt}\hat{j}. \tag{3–11b}$$

The component vectors $\vec{v}_x$ and $\vec{v}_y$ of the velocity vector $\vec{v}$ are drawn in Fig. 3–4. The magnitude of the velocity $\vec{v}$ can be written in terms of the components of $\vec{v}$:

$$v = |\vec{v}| = \sqrt{v_x^2 + v_y^2}. \tag{3–12}$$

The angle θ that the velocity vector $\vec{v}$ makes with the x-axis is determined in terms of the components of the velocity by

$$\tan \theta = \frac{v_y}{v_x}. \tag{3–13}$$

▲ FIGURE 3–4 The velocity vector $\vec{v}$ at point P is tangent to the particle's trajectory at that point. The component vectors $\vec{v}_x$ and $\vec{v}_y$ of the velocity vector $\vec{v}$ at that point are also included.

EXAMPLE 3–2 Use the data presented in Example 3–1 to find the bumper car's average velocity over the period from 3.0 to 6.0 s and the car's instantaneous velocity at $t = 3.0$ s.

Setting It Up The data of Example 3–1 give the position of the car as a function of time.

Strategy The average velocity is given by Eq. (3–4), and this requires us to know the displacement for a given time interval. That information is available from Example 3–1. For the instantaneous velocity we use Eq. (3–7) and evaluate the derivatives of $x(t)$ and $y(t)$.

Working It Out Given the result of Example 3–1, that the displacement vector of the bumper car between $t = 3.0$ s and $t = 6.0$ s is $\Delta \vec{r} = (21\hat{i} + 3.0\hat{j})$ m, we have

$$\vec{v}_{av} = \frac{\Delta \vec{r}}{\Delta t} = \frac{(21\hat{i} + 3.0\hat{j}) \text{ m}}{6.0 \text{ s} - 3.0 \text{ s}} = (7.0\hat{i} + 1.0\hat{j}) \text{ m/s}.$$

As for the instantaneous velocity, we require

$$\frac{dx}{dt} = \frac{d}{dt}(c_1 t^2 + c_2 t + c_3) = 2c_1 t + c_2,$$

$$\frac{dy}{dt} = \frac{d}{dt}(d_1 t^2 + d_2 t + d_3) = 2d_1 t + d_2.$$

Substituting the numerical values of $c_1, c_2, d_1,$ and d_2 at $t = 3.0$ s from Example 3–1, we have

$$\frac{dx}{dt} = 2(0.20 \text{ m/s}^2)(3.0 \text{ s}) + (5.0 \text{ m/s}) = 6.2 \text{ m/s},$$

$$\frac{dy}{dt} = 2(-1.0 \text{ m/s}^2)(3.0 \text{ s}) + (10.0 \text{ m/s}) = 4.0 \text{ m/s}.$$

Thus the velocity at $t = 3.0$ s is

$$\vec{v} = \frac{dx}{dt}\hat{i} + \frac{dy}{dt}\hat{j} = (6.2 \text{ m/s})\hat{i} + (4.0 \text{ m/s})\hat{j}.$$

This velocity vector is shown in Fig. 3–5.

What Do You Think? Could you have used the graph of the trajectory in Fig. 3–3d to read off the velocity of the car?

▶ **FIGURE 3–5** The trajectory of the car's path in Example 3–1 is plotted for times up to 10 s; also shown are the position vector $\vec{r}$, the velocity $\vec{v}$, and the acceleration $\vec{a}$ at $t = 3.0$ s.

Acceleration

Acceleration describes how rapidly velocity changes with time. This "change" could be in the magnitude (the speed) or the speed could remain the same while the *direction* of the velocity vector changes or both magnitude and direction may change. As for motion in one dimension, acceleration is found from velocity in the same way that velocity is found from displacement. For a finite time interval Δt, the average acceleration is defined as

$$\vec{a}_{av} \equiv \frac{\vec{v}(t + \Delta t) - \vec{v}(t)}{\Delta t} = \frac{\Delta \vec{v}}{\Delta t}. \tag{3–14}$$

The instantaneous acceleration at time t is the limit of the average acceleration as Δt approaches zero, which is a derivative:

$$\vec{a} \equiv \lim_{\Delta t \to 0} \frac{\vec{v}(t + \Delta t) - \vec{v}(t)}{\Delta t} = \frac{d\vec{v}}{dt}. \tag{3–15}$$

The instantaneous acceleration is in principle a function of time, meaning that its three components are generally functions of time. As for velocity, we can express acceleration in terms of its components; for two dimensions (again for economy) we have

$$\vec{a} = \frac{dv_x}{dt}\hat{i} + \frac{dv_y}{dt}\hat{j} \qquad (3\text{–}16)$$

$$= a_x\hat{i} + a_y\hat{j}. \qquad (3\text{–}17)$$

Here, the components of the acceleration vector are

$$a_x = \frac{dv_x}{dt} = \frac{d^2x}{dt^2}, \qquad (3\text{–}18a)$$

$$a_y = \frac{dv_y}{dt} = \frac{d^2y}{dt^2}. \qquad (3\text{–}18b)$$

EXAMPLE 3–3 Calculate the instantaneous acceleration, magnitude and direction, of the bumper car in Example 3–1 at $t = 1.0$ s and $t = 3.0$ s.

Setting It Up We will want to use the known velocity vector of the car calculated in Example 3–2 using data for the position vectors from Example 3–1.

Strategy The acceleration vector is the time derivative of the known velocity vector, Eqs. (3–18a) and (3–18b). The acceleration is a function of time, into which we will then substitute particular values of time.

Working It Out From Eqs. (3–18),

$$a_x = \frac{dv_x}{dt} = \frac{d}{dt}(2c_1t + c_2) = 2c_1$$

and

$$a_y = \frac{dv_y}{dt} = \frac{d}{dt}(2d_1t + d_2) = 2d_1.$$

Thus [Eq. (3–17)]

$$\vec{a} = 2c_1\hat{i} + 2d_1\hat{j}.$$

In this case, the car's acceleration is a constant—it is independent of time—and so is exactly the same for $t = 1.0$ s and for $t = 3.0$ s.

Given the values of c_1 and d_1 from Example 3–1 (0.20 and -1.0 m/s^2, respectively), the numerical value of the acceleration is

$$\vec{a} = (0.40\hat{i} - 2.0\hat{j}) \text{ m/s}^2.$$

The magnitude of the acceleration is

$$a = |\vec{a}| = \sqrt{a_x^2 + a_y^2} = \sqrt{(0.40)^2 + (-2.0)^2} \text{ m/s}^2$$
$$= \sqrt{4.2} \text{ m/s}^2 = 2.0 \text{ m/s}^2.$$

The acceleration vector makes an angle θ with the x-axis, which is shown in Fig. 3–5; the angle θ is derived from

$$\tan\theta = \frac{a_y}{a_x} = \frac{-2.0 \text{ m/s}^2}{0.40 \text{ m/s}^2} = -5.0;$$

so $\theta = -79°$; that is, the direction of the acceleration is at $-79°$ to the horizontal, almost directly toward the $-y$-direction.

What Do You Think? We started this example by taking the time derivative of the known velocity vector. Could we instead have started with the position vector as a function of time (given in Example 3–1)?

Representing Trajectories

Look again at the trajectory of the bumper car discussed in Examples 3–1, 3–2, and 3–3 (Fig. 3–5); this trajectory is a curve representing the car's position on the floor of the amusement park ride (its x-position versus its y-position). In Fig. 3–5, we show a position vector $\vec{r}$, a velocity vector $\vec{v}$, and an acceleration vector $\vec{a}$ at $t = 3$ s. Although the figure shows $\vec{a}$ at the point corresponding to $t = 3.0$ s, we in fact saw that the acceleration of the car is independent of time and hence would be drawn as the same vector everywhere along the curve.

We can create a graphical representation like that in Fig. 3–5 for any motion. As an object moves, its trajectory will be traced out by the tip of the position vector $\vec{r}$ as $\vec{r}$ changes with time. The velocity vector $\vec{v}$ at any time t is a vector of magnitude $|\vec{v}|$ that is *tangential* to the trajectory at time t. This is quite intuitive. However, it is not quite so obvious how to think about the acceleration.

Since the acceleration is to the velocity as the velocity is to the displacement, one thing we could do is to repeat the procedure of the preceding paragraph with velocity and acceleration. A plot of the tip of the vector $\vec{v}$ can be drawn; it is the curve of the points whose *horizontal coordinate* at time t is v_x and whose *vertical coordinate* at that time is v_y. We might call this the "velocity trajectory." The acceleration at time t is given by a vector whose magnitude is $|\vec{a}|$ and whose direction is tangential to the velocity trajectory at time t. *The acceleration vector $\vec{a}$ is tangent to the velocity trajectory but not to the trajectory itself.*

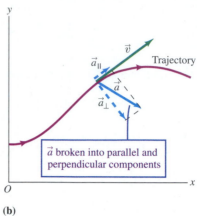

(a)

(b)

▲ **FIGURE 3–6** Velocity $\vec{v}$ and acceleration $\vec{a}$ of a particle following some trajectory. (a) The acceleration of the particle is separated into x- and y-components. (b) The acceleration of the particle is separated into components parallel and perpendicular to the path.

We can illustrate the consequences of these facts using Fig. 3–6, which shows the path of an object with $\vec{v}$ and $\vec{a}$ indicated at one time along the path. In Fig. 3–6a, the acceleration is separated into its a_x and a_y components. Alternatively, we can separate the acceleration $\vec{a}$ into components that are parallel (tangential) and perpendicular (normal) to the velocity vector (Fig. 3–6b). We label these components $a_\parallel$ and $a_\perp$, respectively. The component $a_\parallel$ of $\vec{a}$ that is parallel to $\vec{v}$ affects the magnitude but not the direction of $\vec{v}$. Similarly, the $a_\perp$ component changes the direction but not the magnitude of $\vec{v}$. It is useful to refer separately to the parallel and perpendicular components of an object's acceleration because they affect the velocity differently.

CONCEPTUAL EXAMPLE 3–4 The motion of bumper cars is extremely erratic: You are colliding with other cars or you are trying to use evasive techniques. Figure 3–7 shows the path of a bumper car. Consider the points A, B, C, D, and E. (a) At which point did a collision most likely take place? (b) At which point did evasive action most likely take place? (c) Can you determine where the magnitude of the velocity is the greatest? (d) Can you determine where the magnitude of the acceleration is the greatest?

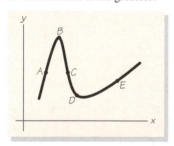

▲ **FIGURE 3–7** The motion of a bumper car can be quite erratic as it slams into other cars and takes evasive action to avoid collisions. Here we have the trajectory, or path, of such a car.

Answer (a) A collision most likely took place at point B because there is an abrupt change in direction. The driver could not change the direction so quickly without some outside effect.

(b) Evasive action probably took place at point D because there is a rapid but smooth change in direction as the driver turned quickly. The motion change is typically more abrupt at collisions.

(c, d) We can't tell where the magnitudes of the velocity and acceleration are the greatest from the trajectory alone because we do not know the times associated with points along the trajectory. We might guess that the acceleration was a maximum during the collision at point B because the velocity would change dramatically during the collision. On the other hand, if the bumper car were traveling very slowly at the time of a collision at point B then the collision might not be a very violent one and the acceleration would not necessarily be very large. You simply do not have enough information on a trajectory to tell. A plot such as Fig. 3–7 does not contain *all* the information about the motion.

3–3 Motion with Constant Acceleration

When an object moves with *constant acceleration*—meaning constant in both magnitude and direction—it can move only in a straight line (one dimension) or a plane (two dimensions). The plane of motion is formed by the initial velocity vector and the acceleration vector $\vec{a}$. The motion remains in this plane because, as Fig. 3–8 illustrates, the initial velocity vector has no component $v_{0\perp}$ perpendicular to the specified plane, and since the acceleration is in the plane, $v_\perp$ can never change and become nonzero. Motion near

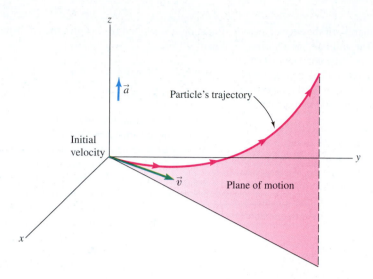

◀ FIGURE 3–8 Projectile motion lies in a plane, the plane formed by the initial velocity vector and the acceleration vector.

Earth's surface, that is, motion under the sole influence of gravity, with air resistance ne-glected, provides an everyday example. If we throw a rock, it moves in a plane defined by the initial direction of the motion and the constant (vertical) acceleration of gravity.

At this point we'll simplify our notation by defining the plane of the motion as the xy-plane. We'll suppose for the moment that the initial velocity can have both x- and y-components, as can the (constant) acceleration. We can then use the results for one di-mension from Chapter 2 to write *independently* the x- and y-components of position $\vec{r}$ and velocity $\vec{v}$ in terms of the constant-acceleration components. In other words, we can think of the x- and y-motions as separate from each other, governed only by their own separate constant accelerations. We use Eqs. (2–17) and (2–21) to find

x-component of $\vec{r}$: $x = x_0 + v_{0x}t + \frac{1}{2}a_x t^2,$ (3–19)

x-component of $\vec{v}$: $v_x = v_{0x} + a_x t;$ (3–20)

y-component of $\vec{r}$: $y = y_0 + v_{0y}t + \frac{1}{2}a_y t^2,$ (3–21)

y-component of $\vec{v}$: $v_y = v_{0x} + a_x t.$ (3–22)

Here, x_0 and y_0 are the components of $\vec{r} = \vec{r}_0$ at an initial time $t = 0$ and v_{0x} and v_{0y} are the components of $\vec{v} = \vec{v}_0$ at time $t = 0$. Together these quantities are the given **initial conditions**. In vectorial form, the initial conditions are

$$\vec{r}_0 = x_0\hat{i} + y_0\hat{j}$$ (3–23)

and

$$\vec{v}_0 = v_{0x}\hat{i} + v_{0y}\hat{j}$$ (3–24)

at $t = 0$.

Equations (3–19) through (3–22), which give position and velocity for motion with constant acceleration $\vec{a}$, can be written more compactly in vector form:

$$\vec{r} = \vec{r}_0 + \vec{v}_0 t + \frac{1}{2}\vec{a}t^2$$ (3–25)

MOTION WITH CONSTANT ACCELERATION

$$\vec{v} = \vec{v}_0 + \vec{a}t$$ (3–26)

This form of the kinematic equations has the additional benefit that it does not refer to any particular set of axes. Remember that these important and useful results are valid *only* when $\vec{a}$ is constant. We can easily see the important features of these compact equations. In particular, you can see that for any direction for which the acceleration component is zero the position (or, equivalently, the displacement) component changes *linearly* with time, corresponding to a *constant*-velocity component. For any direction for which the acceleration component is not zero, the position component changes *quadratically* in time, corresponding to a *linearly* changing velocity component.

EXAMPLE 3–5 A wayward golf ball rolls off the edge of a vertical cliff overlooking the Pacific Ocean. The golf ball has a horizontal velocity component of 10 m/s and no vertical component when it leaves the cliff. Describe the subsequent motion. (The golf ball provides us with our first glimpse of projectile motion. In the following section we will look at this important type of motion in more detail.)

Setting It Up The displacement, velocity, and acceleration all lie in the same plane, which we assign to be the xy-plane. In Fig. 3–9a we include a coordinate system, placing the origin at the point where the ball leaves the cliff and with the y-direction pointing up. We are given initial values of velocity. A "description" of the motion consists of writing the position as a function of time. Given this, further quantities, such as velocity as a function of time, can be found by differentiation.

Strategy This is a case of motion under constant acceleration. In this case the ball's constant acceleration is that of gravity, and thus $\vec{a} = \vec{g}$. The vector $\vec{g}$ points toward Earth's center—vertically downward—and has magnitude 9.8 m/s². Because we have constant acceleration, we can use Eqs. (3–19) through (3–22) to describe the motion, for which we know the initial values (at $t = 0$, the moment when the ball rolls off the cliff). As emphasized above, we can say that because there is no component of acceleration in the x-direction, the horizontal velocity component is constant and will remain at its initial value.

Working It Out We start with initial values: The golf ball's initial position and velocity in our chosen coordinate system are $x_0 = 0$ m, $v_{0x} = 10$ m/s, $y_0 = 0$ m, and $v_{0y} = 0$ m/s. Next we specify that in our coordinate system the acceleration has components $a_x = 0$ m/s² and $a_y = -9.8$ m/s². We determine the velocity components as a function of time from Eqs. (3–20) and (3–22):

$$v_x = 10 \text{ m/s}$$

and

$$v_y = 0 \text{ m/s} + (-9.8 \text{ m/s}^2)t = (-9.8 \text{ m/s}^2)t. \quad (3\text{–}27)$$

Equations (3–19) and (3–21) give the ball's position as a function of time:

$$x = 0 \text{ m} + (10 \text{ m/s})t + \tfrac{1}{2}(0 \text{ m/s}^2)t^2 = (10 \text{ m/s})t,$$
$$y = 0 \text{ m} + (0 \text{ m/s})t + \tfrac{1}{2}(-9.8 \text{ m/s}^2)t^2 = (-4.9 \text{ m/s}^2)t^2. \quad (3\text{–}28)$$

Figure 3–9a shows the trajectory of the golf ball. (We'll discuss trajectories under constant acceleration in more detail in Section 3–4.) It also shows the velocity vector and its components at 0.5-s intervals for the first 2 s of the motion. The horizontal component of the velocity stays constant, whereas the vertical component changes linearly with time. Further, the total velocity vector is a tangent to the ball's path of motion at each point along its trajectory.

Figure 3–9b shows the position vector $\vec{r}$, velocity $\vec{v}$, and acceleration $\vec{a}$ at $t = 1$ s and $t = 2$ s. Whereas $\vec{a}$ remains constant, $\vec{r}$ and $\vec{v}$ change with time. The three vectors $\vec{r}, \vec{v},$ and $\vec{a}$ do not generally point in the same direction at a given time during the golf ball's motion. The directions of $\vec{r}$ and $\vec{v}$ are specified by angles θ and θ', respectively, with respect to the x-axis. These angles are

$$\tan \theta = \frac{y}{x} = \frac{(-4.9 \text{ m/s}^2)t^2}{(10 \text{ m/s})t} = (-0.49 \text{ s}^{-1})t$$

and

$$\tan \theta' = \frac{v_y}{v_x} = \frac{(-9.8 \text{ m/s}^2)t}{(10 \text{ m/s})} = (-0.98 \text{ s}^{-1})t,$$

respectively. Both angles vary with time.

What Do You Think? According to Fig. 3–9, the ball appears to drop into the ocean about 25 m from the cliff. In this problem what determines how far from the base of the cliff the ball enters the water?

▶ **FIGURE 3–9** (a) The velocity vector $\vec{v}$ and components $\vec{v}_x$ and $\vec{v}_y$ of the golf ball are shown at 0.5-s intervals up to 2.0 s. (b) Position $\vec{r}$, velocity $\vec{v}$, and acceleration $\vec{a}$ of the golf ball for $t_1 = 1$ s and $t_2 = 2$ s.

(a)

(b)

3–4 Projectile Motion

A golf ball in motion is an example of a *projectile* that moves under the effect of gravity. In the absence of air resistance, what is the trajectory of a projectile? The motion is that of constant acceleration due to gravity, and this constant acceleration $\vec{g}$ has only a vertical component; we can use all the constant-acceleration results of the previous section to find the trajectory. The ball's motion is best described by separating it into horizontal and vertical components—as we have already emphasized, the horizontal motion is *independent* of the vertical motion—and then applying the kinematic equations for constant acceleration.

Usually it is easiest to place the origin at the starting point, assigning the y-direction vertically and the x-direction along the horizontal (Fig. 3–10), as we did in Example 3–5. The initial position of the ball is $x_0 = y_0 = 0$; the initial velocity at $t = 0$ is $\vec{v}_0$. The flight of the golf ball starts at an initial angle to the horizontal that we call the *elevation angle* θ_0. Then $\vec{v}_0$ has components

$$v_{0x} = v_0 \cos \theta_0 \quad \text{and} \quad v_{0y} = v_0 \sin \theta_0. \tag{3–29}$$

The components of the acceleration are the constants

$$a_x = 0 \quad \text{and} \quad a_y = -g. \tag{3–30}$$

Using Eqs. (3–19) through (3–22), the components of $\vec{r}$ and $\vec{v}$ (the position and velocity of the ball, respectively) are

$$x = 0 + (v_0 \cos \theta_0)t + \tfrac{1}{2}(0)t^2 = (v_0 \cos \theta_0)t, \tag{3–31}$$

$$y = 0 + (v_0 \sin \theta_0)t + \tfrac{1}{2}(-g)t^2 = (v_0 \sin \theta_0)t - \tfrac{1}{2}gt^2, \tag{3–32}$$

and

$$v_x = v_0 \cos \theta_0 + (0)t = v_0 \cos \theta_0, \tag{3–33}$$

$$v_y = v_0 \sin \theta_0 - gt. \tag{3–34}$$

The Trajectory

We can find the trajectory of the golf ball by plotting its height y versus its x-position. We know both x and y as functions of time, and we can eliminate the time dependence by using Eq. (3–31) to find the time t as a function of x. We then insert the result for t into Eq. (3–32) to find the trajectory, that is, the height y as a function of x, with the time dependence eliminated:

$$t = \frac{x}{v_0 \cos \theta_0}; \tag{3–35}$$

$$y = (v_0 \sin \theta_0)\frac{x}{v_0 \cos \theta_0} - \frac{1}{2}g\left(\frac{x}{v_0 \cos \theta_0}\right)^2$$

$$= (\tan \theta_0)x - \left(\frac{g}{2v_0^2 \cos^2 \theta_0}\right)x^2. \tag{3–36}$$

The coefficients of x and x^2 in Eq. (3–36) are both constants, so the trajectory has the form

$$y = C_1 x - C_2 x^2. \tag{3–37}$$

This is the equation of a parabola passing through the origin with its axis parallel to the y-axis. *The trajectory of all objects moving with constant acceleration is parabolic.* Parabolic motion is illustrated in the chapter-opening photograph and Fig. 3–10 as well as in Fig. 3–11, which shows the position of a ball at equal time intervals.

The trajectory and the time dependence of the components of displacement have some simple characteristics that can be useful in our study of projectile motion—range, flight time, and maximum height. These are easily extracted from the motion, and we discuss them further below.

Range: We define the **range** R of a projectile launched from the ground ($y = 0$) to be the horizontal distance that the projectile travels over level ground; that is, it lands at the same height from which it started. The quantity R is the value of x when the projectile has returned to the ground, that is, when y again equals zero. If we insert $y = 0$ into Eq. (3–37), we have

$$y = 0 = R(C_1 - C_2 R), \tag{3–38}$$

(a)

(b)

▲ **FIGURE 3–10** (a) A golf ball leaves a tee with an initial velocity of magnitude v_0 at an elevation angle θ_0. (b) The side view of the motion shows a parabolic trajectory.

▲ **FIGURE 3–11** Motion of a ball bouncing along the floor and moving under the influence of gravity. In the air, the ball moves with constant acceleration, which in this case is directed downward due to gravity. The velocity vector changes throughout the motion, although its horizontal component does not. The velocity's vertical component changes linearly with time. The resulting trajectory forms a series of parabolas.

▶ **FIGURE 3–12** For a fixed initial speed and if air resistance is ignored, a projectile's trajectory will have a maximum range for an elevation angle of 45°. The range is the horizontal distance the projectile travels to reach the same height from which it started.

where we have set $x = R$. To find R, set the factor $C_1 - C_2 R = 0$ in Eq. (3–38), or $R = C_1/C_2$. Inserting the values of C_1 and C_2 from Eq. (3–36) yields

$$R = \frac{C_1}{C_2} = \frac{\tan\theta_0(2v_0^2\cos^2\theta_0)}{g} = \frac{2v_0^2}{g}\left(\frac{\sin\theta_0}{\cos\theta_0}\right)\cos^2\theta_0 = \frac{v_0^2}{g}\,2\sin\theta_0\cos\theta_0.$$

From trigonometry, $\sin(2\theta_0) = 2\sin\theta_0\cos\theta_0$, and we find

$$R = \frac{v_0^2}{g}\sin 2\theta_0. \qquad (3\text{–}39)$$

The range R depends on the initial speed v_0 and the elevation angle (the initial angle) of the projectile. As θ_0 increases progressively from 0° to 45° and then to 90°, the range $R\,[\propto \sin(2\theta_0)]$ starts out at zero, increases to a maximum at $\theta_0 = 45°$ [i.e., $\sin(2\theta_0) = 1$], then decreases back down to zero at $\theta_0 = 90°$. So, to throw or kick a ball over level ground as far as you can, send it upward at a 45° angle. For this case, which gives the maximum range, we have

$$R_{\text{max}} = \frac{v_0^2}{g}. \qquad (3\text{–}40)$$

If the projectile is launched at an angle higher or lower than 45°, the range is shorter (Fig. 3–12). Note that according to Eq. (3–39) there are *two* initial angles for which a projectile has the same range for a given initial speed (Fig. 3–12). For example, in softball a pop fly at 75° and a line drive at 15° can both be caught by the shortstop (compare the two trajectories in Fig. 3–12).

Flight Time: Let T be the total flight time of a ball. Figure 3–13 shows that the ball reaches its maximum height exactly halfway through its trajectory, at time $t = T/2$. At this point, its motion is horizontal and the vertical component of velocity is zero. We can find $T/2$ by setting $v_y = 0$ in Eq. (3–34), $0 = v_0\sin\theta_0 - g(T/2)$. We solve for T to find that

$$T = \frac{2v_0}{g}\sin\theta_0. \qquad (3\text{–}41)$$

▶ **FIGURE 3–13** A projectile (a ball) moving under the force of gravity is at its maximum height when $v_y = 0$. At that moment, the ball is traveling horizontally. We have marked the velocity at this point as $\vec{v}_2$.

This time the simple factor $\sin \theta_0$ enters. Look again at the motion at 75° and 15° in Fig. 3–12 and you will understand that in softball the fly ball's flight time is greater than that of the line drive.

Maximum Height: The maximum height $y_{max} = h$ is reached at time $T/2$. From Eq. (3–32), we find the height at this time,

$$h = (v_0 \sin \theta_0)\frac{2v_0}{2g}\sin \theta_0 - \frac{1}{2}g\left(\frac{2v_0}{2g}\sin \theta_0\right)^2 = v_0^2\frac{\sin^2 \theta_0}{g} - gv_0^2\frac{\sin^2 \theta_0}{2g^2}$$

$$= v_0^2\frac{\sin^2 \theta_0}{2g}. \tag{3–42}$$

We use Eqs. (3–36), (3–39), (3–41), and (3–42) to determine a projectile's trajectory, range, flight time, and maximum height, respectively. The range and flight time refer to the special case where the ball returns to its original height. These equations need not be memorized; instead, it is important to understand how they were obtained. We apply these methods again in Examples 3–6 through 3–10.

EXAMPLE 3–6 To win a bet that he can drive a golf ball a horizontal distance of 250 m, an amateur golfer goes to a cliff overlooking the ocean. The cliff is 52 m above the ocean. The golfer strikes the golf ball so that the ball's initial speed is 48 m/s and the elevation angle (from the horizontal) is 36°. Does he win his bet? What is the horizontal distance actually covered by the ball?

Setting It Up Figure 3–14 shows the situation. We place the origin of our coordinate system at the tee where the ball's motion starts, letting y extend upward. We know the initial conditions ($t = 0$ when the ball is struck), which with our coordinate system are $x_0 = 0$ m, $y_0 = 0$ m, $v_0 = 48$ m/s, and $\theta_0 = 36°$. We want to determine the distance R' from the tee to the point at which the golf ball reaches the ocean ($y = -52$ m).

▲ **FIGURE 3–14** A golf ball is driven off a cliff into the ocean.

Strategy We use the trajectory equation to find the value of x at which the golf ball reaches the ocean surface. Note that we cannot use Eq. (3–39) to calculate the range because that result applies only to level ground; we don't want the horizontal distance when the ball returns to $y = 0$ m. However, we can still use Eq. (3–36) to find the value of x when $y = -52$ m.

Working It Out Equation (3–36) reads in our case

$$y = -52 \text{ m} = (\tan \theta_0)R' - \left(\frac{g}{2v_0^2 \cos^2 \theta_0}\right)R'^2.$$

Rearranging this equation yields

$$R'^2 - \frac{2v_0^2 \cos^2 \theta_0 \tan \theta_0}{g}R' + \frac{2yv_0^2 \cos^2 \theta_0}{g} = 0,$$

$$R'^2 + bR' + c = 0,$$

where

$$b = -\frac{2v_0^2 \cos^2 \theta_0 \tan \theta_0}{g} \quad \text{and} \quad c = \frac{2yv_0^2 \cos^2 \theta_0}{g}.$$

Solving this quadratic equation to find R' gives

$$R' = \frac{-b \pm \sqrt{b^2 - 4c}}{2}.$$

Inserting the values of b and c, we obtain

$$R' = \frac{v_0^2 \cos^2 \theta_0 \tan \theta_0}{g} \pm \frac{1}{2}\sqrt{\frac{4v_0^4 \cos^4 \theta_0 \tan^2 \theta_0}{g^2} - \frac{8yv_0^2 \cos^2 \theta_0}{g}}.$$

Now inserting $y = -52$ m and the initial values to determine R' yields

$$R' = \frac{(48 \text{ m/s})^2 \cos^2 36° \tan 36°}{9.8 \text{ m/s}^2}$$

$$\pm \frac{1}{2}\sqrt{\frac{4(48 \text{ m/s})^4 \cos^4 36° \tan^2 36°}{(9.8 \text{ m/s}^2)^2} - \frac{8(-52 \text{ m})(48 \text{ m/s})^2 \cos^2 36°}{9.8 \text{ m/s}^2}}$$

$$= 281 \text{ m or} -57 \text{ m}.$$

Now, did the golfer drive the ball a distance of 281 m or -57 m? The positive value must be correct. The golfer wins his bet.

What Do You Think? We stated the positive solution (281 m) must be the correct solution to the problem, but the negative solution (-57 m) also is a solution. What is the physical meaning of the negative solution?

EXAMPLE 3–7 What was the maximum height above the ocean of the golf ball in Example 3–6, and how long was the golf ball in flight?

Setting It Up We can again refer to Fig. 3–14. We denote the maximum height above the ground by h. In this case we want the maximum height *above the ocean*. We also want to find the total time T of the trip.

Strategy The maximum height of the golf ball occurs when the vertical component of the velocity is zero, and Eq. (3–42) will give h. The value we seek is $h + 52$ m. As for T, we can find it by using Eq. (3–31), together with the knowledge that the total horizontal distance traveled is 281 m.

Working It Out From Eq. (3–42),

$$h = \frac{v_0^2 \sin^2 \theta_0}{2g} = \frac{(48 \text{ m/s})^2 \sin^2 36°}{2(9.8 \text{ m/s}^2)} = 41 \text{ m}.$$

The answer is therefore 41 m + 52 m = 93 m.

From Eq. (3–31) with a horizontal distance traveled of 281 m,

$$281 \text{ m} = (48 \text{ m/s})(\cos 36°)T;$$

$$T = \frac{281 \text{ m}}{(48 \text{ m/s}) \cos 36°} = 7.2 \text{ s}.$$

What Do You Think? On level ground the horizontal distance covered by the golf ball depends on $\sin 2\theta_0$ [see Eq. (3–39)], where θ_0 is the initial elevation angle, and the range is a maximum for $\theta_0 = 45°$. However, in this example we do not have level ground. Will the maximum horizontal travel distance still occur for $\theta_0 = 45°$?

CONCEPTUAL EXAMPLE 3–8 A major league pitcher and you, the student, compete in throwing a baseball as far as possible in an initially horizontal direction. Assume that you each throw so that each ball leaves the hand at exactly the same height. Whose ball will go further and why?

Answer The time it takes the ball to hit the ground is determined by the height from which the ball starts. Since the ball leaves the hand horizontally, the initial vertical velocity component is the same for both of you, namely zero, and hence the time it takes to hit the ground is the same for both of you. But the initial horizontal velocity component of the ball is expected to be larger for the professional pitcher, so that in the same time it covers a larger distance. The independence of the two components of the motion is again key here. Figure 3–15 illustrates the equal fall time for two projectiles that fit this description; in this photograph one of the projectiles has an initial speed of zero.

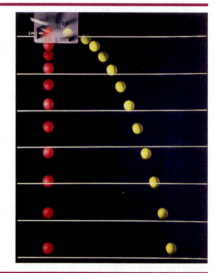

▶ **FIGURE 3–15** Two balls released simultaneously have two different trajectories, but in a given time each moves the same vertical distance. The difference in their motions is the magnitude of their (constant) x-components of velocity.

EXAMPLE 3–9 A group of engineering students constructs a slingshot device that lobs water balloons. The device is constructed so that the angle of the lob can be adjusted, and it has a launch speed (the balloon's initial speed) of 12 m/s. There is a target 14 m away at the same elevation. How should they adjust the initial angle so that they reach the target?

Setting It Up The slingshot setup is shown in Fig. 3–16. The students must find a value of launch angle θ_0 that will produce a given range R for a given initial speed v_0.

▲ **FIGURE 3–16** The students can orient their slingshot in two ways to get the same range for the same initial speed—just one is shown here.

Strategy In this case the range equation for horizontal ground, Eq. (3–39), can be used, and we can solve it for the launch angle.

Working It Out With $R = 14$ m and $v_0 = 12$ m/s, Eq. (3–39) gives

$$R = 14 \text{ m} = \frac{(12 \text{ m/s})^2 \sin 2\theta_0}{9.8 \text{ m/s}^2},$$

or

$$\sin 2\theta_0 = 0.95.$$

This equation has *two* solutions, $2\theta_0 = 72°$ and $2\theta_0 = 108°$, or θ_0 is 36° and 54°. These are the two possible initial angles that result in a given range, as in Fig. 3–12. We have drawn one of these trajectories (Fig. 3–16). *A reminder*: There will always be two initial angles that generate the same range, except for maximum range, which is produced only by the limiting angle 45°.

What Do You Think? From the standpoint of surprise, which of the two solutions, 36° or 54°, might be best for the students to use if the target were human?

EXAMPLE 3–10 A boy would rather shoot coconuts down from a tree than climb the tree or wait for the coconuts to drop. The boy aims his slingshot directly at a coconut, but at the same moment that his rock leaves the slingshot, the coconut falls from the tree. Show that the rock will hit the coconut.

Setting It Up We establish the launch point of the rock as the origin of a suitable xy-coordinate system in Fig. 3–17. The coconut is at the point (x_0, y_0). We are asked if the two objects moving under the influence of gravity will be at the same spot at the same time. For one of the trajectories, the initial velocity is given by an angle that would take it to the coconut if there were no gravity.

▲ **FIGURE 3–17** If the coconut falls at the same time the rock leaves the slingshot, both the coconut and rock fall the same distance.

Strategy We must compare a trajectory that includes both a horizontal and a vertical component (the rock) versus one that has only a vertical component (the coconut). For that reason it will be useful to think in terms of these components. It is useful to first consider what would happen if there were no gravity, then to see how the presence of gravity modifies the positions of both the rock and coconut.

Working It Out The rock has an initial velocity (v_{x0}, v_{y0}). If there were no gravity acting, the rock would follow a straight-line path that would place it at the point $(x_0 = v_{x0}t, y_0 = v_{y0}t)$ after a time t. This is the time necessary for the rock to reach the coconut (which is still at the tree since gravity has been ignored so far). Now let's include the effect of gravity. First, consider what happens to the coconut. During the time t that the rock travels toward the coconut, the coconut falls the distance $gt^2/2$ (Fig. 3–17). In other words, the height of the coconut after time t is [Eq. (3–21)]

$$y = y_0 - \frac{1}{2}gt^2.$$

Next, consider what happens to the rock when we include gravity. The rock's horizontal velocity component remains constant at v_{x0}. However, the vertical velocity component of the rock is changing under the effect of gravity and, after time t, Eq. (3–21) shows us that the rock's height is not $v_{y0}t$ but rather

$$y = v_{y0}t - \frac{1}{2}gt^2.$$

The rock is a height $gt^2/2$ below the height it would have if it followed a straight-line motion, which is precisely the distance the coconut falls (Fig. 3–17). Thus the rock will hit the coconut at the common point $\frac{1}{2}gt^2$ below the coconut's starting point. In effect, the parabolic path of the rock "tracks" the falling coconut.

What Do You Think? The real world is usually somewhat different than the idealized case discussed in textbooks. What are some reasons why the rock may not hit the coconut?

THINK ABOUT THIS...

IS IT POSSIBLE TO EXPERIENCE FREE FALL FOR LONG PERIODS?

Every jump puts you in free fall. Some of you may have done bungee jumping, where you can be in free fall for a couple of seconds until the cord starts to pull. What would it be like to be in free fall for longer periods? NASA has equipped a KC-135 airplane that allows training astronauts and others to experience longer periods of free fall. The plane is equipped to coast following a parabolic trajectory identical to that of a projectile. For the 25 s of the dive, the occupants are in free fall along with the airplane (see Fig. 3–18). Much of the film *Apollo 13*, which recounts the dramatic story of a mission to the Moon that barely made it back to Earth, was shot within the NASA plane. The best place to experience free fall is the International Space Station, which is orbiting around Earth in a free fall in which Earth's curvature allows the surface to "fall away" from the projectile's path as the projectile proceeds. The personnel inhabiting the station may be in free fall for months.

◀ **FIGURE 3–18** The interior of the airplane used by NASA for a free-fall environment during that part of the flight where the plane follows the same parabolic path taken by a projectile in free fall. In this photograph, astronauts in training are experiencing some of the same effects they will feel during a stay in the International Space Station.

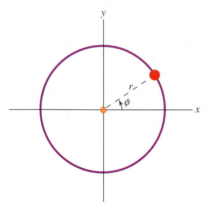

▲ FIGURE 3–19 This turntable is rotating with two orange capital letters "E". They are blurred due to the motion.

▲ FIGURE 3–20 Radial variable r and angular variable ϕ of plane polar coordinates. The symbol ρ is sometimes used instead of r.

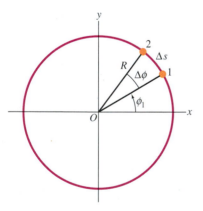

▲ FIGURE 3–21 During the time interval Δt a particle in circular motion traverses an arc length Δs and an angle $\Delta \phi$.

3–5 Uniform Circular Motion

We are quite familiar with objects that move in circles at constant speeds, for example a carnival merry-go-round, the platter of a microwave oven, or a phonograph turntable. We are in fact being carried in a circular motion with Earth as it rotates each day. An object undergoes **uniform circular motion** when it travels at a constant speed along a circular path (see Fig. 3–19). Circular motion is a special case of motion on a plane. We can best describe uniform circular motion by specifying a radius r and an angular coordinate ϕ, as shown in Fig. 3–20. The origin of the coordinate system is at the center of the circle and the angle ϕ is measured from the $+x$-axis, increasing in the counterclockwise direction. We measure ϕ, which is dimensionless, in units of *radians* (rad) (an entire rotation around one circle is 2π radians or $360°$). The coordinates r and ϕ are referred to as *plane polar coordinates*. For circular motion, the radius is fixed, so we can let the radius $r = R$, a constant, and in this case we only need one variable, the angle ϕ, to describe the position of an object (Fig. 3–20).

Although uniform circular motion is motion at constant speed, it is motion with *acceleration*, and this is a crucial point to remember. In this acceleration the *direction* of the velocity but not its magnitude changes constantly. (For nonuniform circular motion both the magnitude and the direction change.) As we shall see, the acceleration in uniform circular motion is a vector that always points to the center of the circular path.

Consider a point object moving in uniform circular motion at a radius R from point 1 to point 2, as shown in Fig. 3–21. During the time interval Δt needed to travel between these points, the object moves through an angle $\Delta \phi$. This distance along the circle is the *arc length* and we denote it by Δs. We have

$$\text{arc length} = \Delta s = R \, \Delta \phi. \tag{3–43}$$

The average speed of motion of the object moving from point 1 to point 2 is in turn

$$v = \frac{\Delta s}{\Delta t} = \frac{R \, \Delta \phi}{\Delta t} = R \frac{\Delta \phi}{\Delta t}. \tag{3–44}$$

If we take the limit of Δt, we arrive at the instantaneous speed

$$v = \frac{ds}{dt} = R \frac{d\phi}{dt}. \tag{3–45}$$

Because we are dealing with *uniform* circular motion here, the average speed and the instantaneous speed will be the same.

It is useful to define the **angular speed** ω of the object moving along the circle as the rate of change of the angle ϕ; in terms of the derivative this is

$$\omega \equiv \frac{d\phi}{dt}. \tag{3–46}$$

The speed v is simply related to ω,

$$v = \omega R. \tag{3–47}$$

In uniform circular motion, v is a constant, and hence, from Eq. (3–47), ω *is a constant in uniform circular motion.* This is simply understood by realizing that in uniform circular motion the angle is swept out at a steady rate. The angular speed ω is measured in radians per second (rad/s) in SI and has dimensions $[1/T]$.

The **period** T is the time an object takes to make one complete revolution, for example, the time it takes the friend you're watching on the carousel to come around again. The distance traveled in one revolution is $2\pi R$, and for uniform circular motion, the speed is a constant v. We can then determine the period T,

$$\text{distance traveled during one revolution} = 2\pi R = vT; \tag{3–48}$$

$$T = \frac{2\pi R}{v} = \frac{2\pi R}{\omega R} = \frac{2\pi}{\omega}. \tag{3–49}$$

The **frequency** f is the number of revolutions that the object makes per unit time; for example, Earth makes 1.156×10^{-5} revolutions per second (or 1 revolution every 86,400 s, i.e., a day). Frequency is the inverse of the period T:

$$f \equiv \frac{1}{T}. \tag{3-50}$$

This relation is easy to understand: If the period is 2 s, the frequency is $\frac{1}{2}$ revolution per second, or if the period is 0.1 s, the frequency is 10 revolutions per second. Using the last two equations, we see that the relation between the angular speed and frequency is

$$\omega = 2\pi f. \tag{3-51}$$

The SI unit of frequency f is the hertz (Hz), defined as one cycle (or revolution) per second (cps).

Acceleration in Uniform Circular Motion: Let's look in more detail at acceleration in uniform circular motion by considering how direction changes as an object moves in a circle at constant speed. We aim to show that the acceleration always points to the center of the circle and find its magnitude.

Figure 3–22a shows a particle located at point A, determined by angle ϕ_1 at time t_1. At a later time t_2 the particle is at point B and the particle's angular position is given by ϕ_2. The *direction* of the velocity vector is always tangential to the circle in the direction of motion and therefore changes continuously with time. Figure 3–22a includes the velocity vectors $\vec{v}_1$ at time t_1 and $\vec{v}_2$ at time t_2. These two vectors have the same magnitude but different directions. The *change* in the velocity vector over the time period Δt (where $\Delta t = t_2 - t_1$) is given by $\Delta \vec{v} = \vec{v}_2 - \vec{v}_1$. We have drawn this vector difference in Fig. 3–22b midway between points A and B. We see that $\Delta \vec{v}$, and hence $\vec{a}_{av} = \Delta \vec{v}/\Delta t$, points toward the center of the circle. To see this more clearly for $\Delta \vec{v}$, just mentally shift the position of $\Delta \vec{v}$ so that it acts at the midway point in Fig. 3–22b, which puts its direction toward $-y$ (down). If we make the time interval Δt smaller, as in Fig. 3–23a, then $\Delta \phi$ also gets smaller, and $\Delta \vec{v}$—and consequently $\vec{a}_{av}$—points ever more closely to the center of the circle (Fig. 3–23b). In the limit in which Δt goes to zero, the ratio $\Delta \vec{v}/\Delta t$ gives us the instantaneous acceleration. *The instantaneous acceleration points precisely to the center of the circle.* We accordingly call this a **centripetal acceleration**, from the Latin meaning "center seeking."

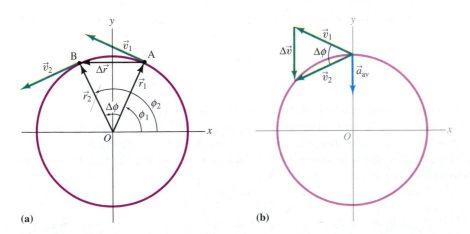

(a) (b)

▲ **FIGURE 3–22** (a) During time Δt from t_1 to t_2 a particle in motion has changed its position vector from $\vec{r}_1$ to $\vec{r}_2$ and its velocity vector from $\vec{v}_1$ to $\vec{v}_2$. For uniform circular motion, the magnitude of $\vec{v}$ is constant; however, the direction is always perpendicular to $\vec{r}$ and is therefore changing continuously. (b) To form the vector difference $\Delta v = \vec{v}_2 - \vec{v}_1$, we translate the vector $\vec{v}_2$ so that its tail meets the tail of the vector $\vec{v}_1$. We do this at the midway point in the particle's path from time t_1 to time t_2. In part (a), the angle between $\vec{r}_1$ and $\vec{r}_2$ is $\Delta\phi$, the same angle as that between $\vec{v}_1$ and $\vec{v}_2$. We see from the figure that the average change in velocity, which is proportional to $\vec{a}_{av}$, points toward the center of the circle.

▶ **FIGURE 3–23** (a) The time difference Δt is smaller than in Figs. 3–21a and b; so too are $\Delta\phi$, $\Delta\vec{r}$, and $\Delta\vec{v}$. (b) The resulting $\vec{a}_{av}$ points more precisely to the center of the circle. In the limit that $\Delta t \to 0$, $\vec{a}_{av}$ becomes the instantaneous acceleration $\vec{a}(t)$ and points exactly to the center of the circle. **(a)**

(b)

Note that $\Delta\vec{v}$ and $\vec{a}_{av}$ both point down.

To obtain the magnitude of the acceleration, notice from Figs. 3–23a and b that the angle $\Delta\phi$ between $\vec{v}_1$ and $\vec{v}_2$ is the same as the angle between $\vec{r}_1$ and $\vec{r}_2$, so they form similar isosceles triangles (see also Fig. 3–22). Such triangles obey the relation

$$\frac{\Delta v}{v} = \frac{\Delta r}{r}. \tag{3–52}$$

(Note that here Δr is *not* a radius change; it is more closely related to the arc length covered in the time interval.) We write this in the form $\Delta v = (v/r)\,\Delta r$, and, after dividing by Δt, we obtain

$$\frac{\Delta v}{\Delta t} = \frac{v}{r}\frac{\Delta r}{\Delta t}.$$

By taking the limit as Δt goes to zero, we are led to the derivative form for the instantaneous acceleration:

$$a = \lim_{\Delta t \to 0} \frac{\Delta v}{\Delta t} = \frac{v}{r}\lim_{\Delta t \to 0}\frac{\Delta r}{\Delta t} = \frac{v}{r}v = \frac{v^2}{r}. \tag{3–53}$$

An alternative form for the acceleration is expressed in terms of the angular speed ω rather than v:

$$a = \frac{v^2}{r} = \frac{(\omega r)^2}{r} = r\omega^2. \tag{3–54}$$

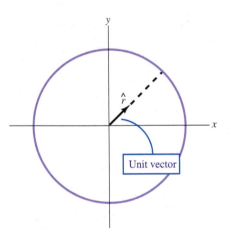

▲ **FIGURE 3–24** The unit vector $\hat{r}$ has length 1 and points away from the origin.

Unit vector

Equations (3–53) and (3–54) express the magnitude of the acceleration.

Since the instantaneous acceleration of an object undergoing uniform circular motion is centripetal, we use a unit vector in the radial direction to help us express this. We define this unit vector by

$$\hat{r} \equiv \frac{\vec{r}}{r}, \tag{3–55}$$

where the caret over the unit vector $\hat{r}$ distinguishes this vector from the position vector $\vec{r}$. In contrast to the Cartesian unit vectors $\hat{i}$ and $\hat{j}$, which are fixed in space, the unit vector in the radial direction varies with the angle ϕ. Notice that the unit vector $\hat{r}$ points outward from the origin (Fig. 3–24). In terms of this vector, the centripetal acceleration is

$$\vec{a} = -\frac{v^2}{r}\hat{r}. \tag{3–56}$$

THE ACCELERATION IN UNIFORM CIRCULAR MOTION

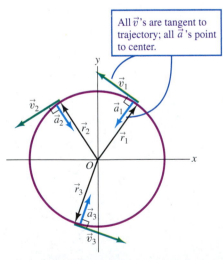

▲ **FIGURE 3–25** The instantaneous position $\vec{r}$, instantaneous velocity $\vec{v}$, and instantaneous acceleration $\vec{a}$ are shown for several positions around the path of a particle in uniform circular motion.

All $\vec{v}$'s are tangent to trajectory; all $\vec{a}$'s point to center.

We emphasize the negative sign here: Since the unit vector $\hat{r}$ points away from the origin, the acceleration points *toward* the origin.

The relationship of $\vec{r}$, $\vec{v}$, and $\vec{a}$ is shown in Fig. 3–25 for various times during a particle's circular orbit. Note that $\vec{a}$ is always perpendicular to $\vec{v}$. All the vectors are constant in magnitude but *vary continuously in direction*.

EXAMPLE 3–11 Automobiles include automatic protection systems such as side airbags and roll bars. One well-known German automobile manufacturer has a system that deploys when the "lateral acceleration" exceeds $3g$ in magnitude. Let's assume that this lateral (sideways) acceleration occurs when the car is rounding a curve. What is the relationship between the radius of the car's circular motion and its speed in this situation? How fast could the car go around a curve of radius 20 m, which might be a typical city street corner, before the automatic system deploys?

Setting It Up Figure 3–26 labels the unknown speed v and known radius R for the circular motion. These together imply a centripetal acceleration a (the "lateral acceleration") whose magnitude is given.

Strategy This is a straightforward use of the relation between a, R, and v. The relation is Eq. (3–54), and we can use it to solve for v.

Working It Out If $a = v^2/R = 3g$, then $v^2 = 3gR$. For $R = 20$ m, we have

$$v^2 = 3gR = 3(9.8 \text{ m/s}^2)(20 \text{ m}) = 590 \text{ m}^2/\text{s}^2,$$

$$v = 24 \text{ m/s} = 54 \text{ mph}.$$

Our experience tells us that this is indeed too high a speed with which to round a city street corner.

What Do You Think? Would it be a good idea for the manufacturer simply to set the speed as a criterion for deployment of their airbag?

▲ **FIGURE 3–26** Car turning in circular motion.

EXAMPLE 3–12 A child's top spins uniformly at 4.0 Hz. What is the centripetal acceleration on the outside surface if the radius of the top is 3.0 cm?

Setting It Up Here we are given the rotation frequency f and the radius r of the circular motion and want the magnitude of the centripetal acceleration.

Strategy We can find ω immediately given f using $\omega = 2\pi f$. We can then use Eq. (3–54), which gives a in terms of ω and r.

Working It Out The angular speed ω is

$$\omega = (2\pi \text{ rad})(4.0 \text{ Hz}) = 25 \text{ rad/s}.$$

The radius in question is 3.0 cm $= 3.0 \times 10^{-2}$ m, so the acceleration has magnitude

$$a = \omega^2 r = (25 \text{ rad/s})^2 (3.0 \times 10^{-2} \text{ m}) = 19 \text{ m/s}^2.$$

Remember that the unit *radian* is a dimensionless quantity. It is a measure of angle and we use it in the measurement of angular speed. However, the unit is dropped in this example during the calculation of the acceleration, as it would not be appropriate to write the answer for acceleration as 19 $\text{rad}^2 \cdot \text{m/s}^2$.

What Do You Think? Since the angular velocity is constant for all parts of the top, the expression $v = \omega r$ shows that the outermost parts of the top move the fastest. Which parts of the top have the largest acceleration?

3–6 Relative Motion

Consider, as in Fig. 3–27a, a boy in a bus moving with constant velocity. The boy is tossing a ball and catching it. He is observed by a woman standing on the sidewalk who sees the bus moving past her. The boy and the woman both observe the ball, and we will call them *observers* of the motion of the objects around them. Each observer sees the situation (or, more formally, measures the motion) from his or her **frame of reference**. Let us think about how the description of the ball's motion can vary according to the frame of reference of the observers and how these descriptions are related. This is an important exercise as some descriptions of physical systems are more easily performed in one frame of reference than in another.

Our two observers are said to have a **relative motion**. Let us place the origin of a frame of reference B on the boy and the origin of a second frame of reference W on the woman. From frame W, the bus, along with the boy in it, moves to the right with velocity $\vec{u}$. On the other hand, the boy, or any observer in frame B, observes the woman to be moving with velocity $-\vec{u}$ (the same speed but opposite direction).

(a)

Path of ball according to boy

(b)

Path of ball according to woman

(c)

▲ **FIGURE 3–27** A boy and a woman are in relative motion. (a) The woman observes the bus and the boy on the bus tossing a ball to travel with constant velocity. (b) The boy observes the trajectory of his ball going directly up and down. (c) The woman sees the trajectory of the boy's ball as a parabola.

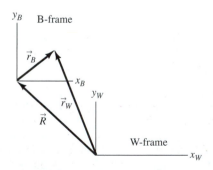

▲ **FIGURE 3–28** The position of an object (the ball) can be measured by an observer in frame W (whose origin is centered on the woman) or by an observer in frame B (whose origin is centered on the boy). The positions in the two frames can be related once the relative position of the two frames is known.

Let us now consider the motion of the ball according to our two observers (Figs. 3–27b and c). In Fig. 3–28 we set up position vectors in each system to denote the position of the ball. The position vector of the ball as measured by the boy is $\vec{r}_B$, and the position vector according to the woman is $\vec{r}_W$. In addition, we let position vector $\vec{R}$ describe the position of the boy according to the woman, that is, the position of the origin of frame B as measured from frame W. We have

$$\vec{r}_W = \vec{R} + \vec{r}_B. \qquad (3\text{–}57)$$

To see how the observations of the ball compare, we need one more ingredient. We recognize that from frame W the position of the boy moves with speed u or, more precisely, that

$$\vec{u} = \frac{d\vec{R}}{dt}. \qquad (3\text{–}58)$$

If we now take the derivatives of Eq. (3–57), we obtain the relation between the velocity $\vec{v}_W$ of the ball as measured by the woman and the velocity $\vec{v}_B$ that the boy measures:

$$\frac{d\vec{r}_W}{dt} = \frac{d\vec{R}}{dt} + \frac{d\vec{r}_B}{dt} = \vec{u} + \frac{d\vec{r}_B}{dt}, \qquad (3\text{–}59)$$

or

$$\vec{v}_W = \vec{u} + \vec{v}_B. \qquad (3\text{–}60)$$

The woman sees the ball moving at the velocity of the bus plus the velocity that the boy measures it to have.

While the two observers disagree on the velocity of the ball, we can show that they agree on the ball's *acceleration*. To do so, take the derivative of the velocities in Eq. (3–60) to obtain the accelerations:

$$\vec{a}_W = \frac{d\vec{v}_W}{dt} = \frac{d\vec{u}}{dt} + \frac{d\vec{v}_B}{dt} = \frac{d\vec{u}}{dt} + \vec{a}_B \qquad (3\text{–}61)$$

or, since the speed of the bus is constant, $d\vec{u}/dt = 0$,

$$\vec{a}_W = \vec{a}_B. \qquad (3\text{–}62)$$

So we find that if the relative speed of the two observers is a constant, then the two observers will agree as to the acceleration of an object, in this case the ball.

How will the two observers describe the ball's motion, assuming the boy tosses the ball vertically with respect to himself? The boy sees the ball go up and down, which is just the projectile motion described in Chapter 2 (or the full projectile motion of this chapter with no horizontal initial velocity). The woman, however, sees the ball in parabolic motion with an initial (and constant) horizontal velocity $\vec{u}$.

One more interesting case: If the bus accelerates, $\vec{u}$ is no longer constant. The boy will accelerate with the bus, but the ball in midair won't. As soon as the ball leaves the boy's hand, it only experiences the vertical acceleration of gravity, and its horizontal motion is determined at the instant that it left the boy's hand. While it is in the air, the boy's velocity will have changed, and when the ball comes down, the boy may no longer be underneath it. From the point of view of the woman, if she keeps her eyes on the ball, she will see exactly what she saw before because from her point of view the ball is only subject to a vertical acceleration and a certain initial horizontal velocity, that of the bus when the ball left the boy's hand.

In this situation $d\vec{u}/dt$ is no longer zero. Equations (3–57), (3–60), and (3–61) are still valid, but Eq. (3–62) becomes

$$\vec{a}_W = \vec{A} + \vec{a}_B, \qquad (3\text{–}63)$$

where $\vec{A}$, the acceleration of the bus, is

$$\vec{A} = \frac{d\vec{u}}{dt} = \frac{d^2\vec{R}}{dt^2}.$$

EXAMPLE 3–13 A boat must cross a river that is 150 m wide. The river has a current of 3 km/h, and the boat can be rowed through the water with a uniform speed of 4 km/h with respect to the water. Set up two coordinate systems to describe the displacement of the boat: one fixed on the bank and the other fixed to a spot moving with the current of the river. Using these coordinate systems, express the position vector of the boat at time t; assume that the boat leaves the dock at an angle θ to the $-x$-axis with respect to a point moving with the water. Calculate θ such that the boat lands at a point exactly opposite the starting point. How long will the trip take?

Setting It Up The preparation of appropriate sketches is quite useful here. We draw two figures, one that includes the origin O and axes of a frame (the "dock frame") fixed on the dock on the river bank (Fig. 3–29a) and a second that includes the origin O' and axes of a frame (the "raft frame") fixed to a hypothetical spot in the water—a raft that drifts with the current (Fig. 3–29b). The river is assumed to move with a velocity $\vec{u}$ in the $+x$-direction in the dock frame.

Strategy The origin O' (the raft) is seen by the observer on the fixed dock to move with the velocity $\vec{u} = u\hat{i}$, where u is the speed of the current (3 km/h).

If the boat can move at speed $v' = 4$ km/h with respect to the water, then, as observed in the frame moving with the water (the raft frame), the boat moves with velocity

$$\vec{v}_b' = (-v'\cos\theta)\hat{i} + (v'\sin\theta)\hat{j},$$

where we do not distinguish between the unit vectors of the two frames. Here we have set the boat off at the angle specified in Fig. 3–29. The primes indicate that we are describing the boat in the raft frame because the speed of the raft is specified in this frame. As long as the rower does not look at the bank and fixes his vision on a spot in the water, he sees himself moving with speed v' at the angle θ through the water. An observer on the dock, however, sees the boat move with velocity

$$\vec{v}_b = \vec{u} + \vec{v}_b' = (u - v'\cos\theta)\hat{i} + (v'\sin\theta)\hat{j},$$

consistent with Eq. (3–57). At this point our problem is solved if angle θ is chosen such that the dockside observer sees no horizontal velocity (x-component of velocity); with this choice of θ, he will see the boat moving straight across the water to the point on the opposite shore, as in Fig. 3–29a.

Working It Out We choose θ so that the x-component of $\vec{v}_b$ vanishes,

$$\cos\theta = \frac{u}{v'} = \frac{3 \text{ km/h}}{4 \text{ km/h}} = 0.75;$$

$$\theta = 41°.$$

The dockside observer sees the boat move straight across the river with speed $v'\sin\theta$, and the trip takes a time

$$t = \frac{y}{v'\sin\theta} = \frac{(0.15 \text{ km})(60 \text{ min/h})}{(4 \text{ km/h})(\sin 41°)} \cong 3 \text{ min.}$$

What Do You Think? If there were no river current, would a trip straight across take (a) more time, (b) less time, or (c) the same time as worked out in the example?

(a)

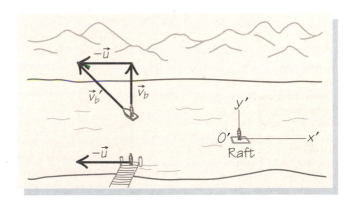

(b)

▲ **FIGURE 3–29** (a) As seen by an observer O standing on the dock, the boat moves straight across the river. (b) As seen by an observer O' moving with the water, the boat moves at an angle θ.

CONCEPTUAL EXAMPLE 3–14 A high-elevation wind called the jet stream flows on average from west to east. A U.S. coast-to-coast airplane flight takes about 5 h west to east but about 30 min longer east to west. Explain this in terms of the relative motion discussed in this section.

Answer Whatever the direction, the airplane flies at the same speed relative to the air. However, the air is not at rest with respect to Earth's surface. If we use the terminology of this section, the jet stream moves with velocity $\vec{u}$ with respect to an observer fixed on Earth with direction from west to east. The airplane has a velocity $\vec{v}_A$ with respect to the atmosphere, but it is positive in one direc-

tion (say, from Los Angeles to Washington) and negative for the opposite direction. The velocity $\vec{v}_B$ of the airplane with respect to an observer fixed on Earth is the sum $\vec{v}_A + \vec{u}$. When $\vec{v}_A$ and $\vec{u}$ have the same direction (west to east), their sum has a larger magnitude than when they point in opposite directions (east to west), so the speed is larger west to east than east to west. The jet stream can flow at over 100 mph in places, and it can significantly affect flight times.

What Do You Think? Suppose a plane encountered the west-to-east jet stream as it was flying north to south. Would this lengthen the flight time?

Summary

A particle moving in space follows a trajectory, or path. In three dimensions, the position vector of such a particle is

$$\vec{r} = x(t)\hat{i} + y(t)\hat{j} + z(t)\hat{k}. \tag{3–3}$$

The displacement vector $\Delta\vec{r}$ between times t and $t + \Delta t$ is the difference between the position vectors at these times. If P and Q label the particle's position at t and $t + \Delta t$, respectively,

$$\Delta\vec{r} \equiv \vec{r}_Q - \vec{r}_P. \tag{3–1}$$

The instantaneous velocity $\vec{v}$ is found from the displacement vector $\Delta\vec{r}$ over small time intervals:

$$\vec{v}(t) \equiv \lim_{\Delta t \to 0} \frac{\vec{r}(t + \Delta t) - \vec{r}(t)}{\Delta t} = \frac{d\vec{r}}{dt}. \tag{3–5}$$

In two dimensions, which we describe by x- and y-coordinates, the velocity is expressed in terms of its component vectors as

$$\vec{v} = \vec{v}_x + \vec{v}_y, \tag{3–9}$$

where, in terms of unit vectors,

$$\vec{v}_x = \frac{dx}{dt}\hat{i} \quad \text{and} \quad \vec{v}_y = \frac{dy}{dt}\hat{j}. \tag{3–11a,b}$$

In terms of its components, the magnitude of the velocity vector is

$$v = |\vec{v}| = \sqrt{v_x^2 + v_y^2}. \tag{3–12}$$

Further, the angle θ that $\vec{v}$ makes with the x-axis is given by

$$\tan\theta = \frac{v_y}{v_x}. \tag{3–13}$$

The instantaneous acceleration is a derivative of the velocity:

$$\vec{a} \equiv \lim_{\Delta t \to 0} \frac{\vec{v}(t + \Delta t) - \vec{v}(t)}{\Delta t} = \frac{d\vec{v}}{dt}. \tag{3–15}$$

Like velocity, acceleration can also be expressed in vector components. The acceleration has components a_x and a_y, which are derivatives of the x- and y-components of the velocity.

For constant acceleration, we have

$$\vec{r} = \vec{r}_0 + \vec{v}_0 t + \tfrac{1}{2}\vec{a}t^2, \tag{3–25}$$

$$\vec{v} = \vec{v}_0 + \vec{a}t \tag{3–26}$$

These vector equations represent a set of three component equations. Constant acceleration means each component of the acceleration is constant, and the kinematic equations above show that each component of the motion is independent of the other components. With constant acceleration, the motion is restricted to the plane formed by the initial velocity and the acceleration, so it is really two-dimensional motion.

In the absence of air resistance, a projectile moves under the influence of gravity in the vicinity of Earth's surface with a constant-acceleration vector $\vec{a} = \vec{g}$. The trajectory of such a projectile is a parabola. The range (the horizontal distance a projectile launched from the ground travels over level ground), the maximum height of the trajectory, and the projectile's flight time can all be calculated and follow directly from the equations for motion under constant acceleration.

In addition to motion under constant acceleration, we also looked at uniform circular motion. This motion is most simply described by means of plane polar coordinates, with r as the radial coordinate and ϕ as the angular coordinate. Angles are measured in radians (which are dimensionless), so that the arc length formed by the angle $\Delta\phi$ in a circle of radius R is

$$\text{arc length} = R\,\Delta\phi. \tag{3–43}$$

The angular speed of an object in circular motion is

$$\omega \equiv \frac{d\phi}{dt}. \tag{3–46}$$

The speed of a particle moving in circular motion with angular speed ω is

$$v = \omega R. \tag{3–47}$$

The period T is the time to make one complete revolution, and the frequency f is the inverse of T. The relation between angular speed and frequency is

$$\omega = 2\pi f. \tag{3–51}$$

The velocity of a particle in uniform circular motion is tangent to the circle. The acceleration of this particle has magnitude

$$a = \frac{v^2}{r} = r\omega^2. \qquad (3\text{-}53, \ 3\text{-}54)$$

The direction of the particle's acceleration is centripetal; that is, it points inward along a radius toward the center of the circle.

Finally, we looked at relative motion. Suppose that two observers, A and B, are in relative motion. These observers are said to be in different frames. Observer A measures the velocity of observer B to be $\vec{u}$, the relative velocity between A and B. Then, if the velocities of an object as measured by observers A and B are $\vec{v}_A$ and $\vec{v}_B$, respectively, these velocities are related by

$$\vec{v}_A = \vec{u} + \vec{v}_B. \qquad (3\text{-}60)$$

If $\vec{u}$ is a constant, then the two observers agree on the acceleration of the observed object.

Understanding the Concepts

1. A boy wants to knock down a coconut with a rock and a slingshot. He knows it is unlikely that the coconut will fall while he is shooting. Does he aim directly at the coconut or does he aim a little higher?

2. What factors determine how far an athlete travels in the long jump? Which factor do you think must be the most important?

3. Why is it important to have limits on wind speed for world-record track and field events?

4. Under what conditions is it possible to have a constant speed yet a nonzero acceleration?

5. You wish to row a boat to a point directly opposite your starting point on the bank of a swiftly moving river. The best speed you can make through the water—meaning with respect to the water—is 2 mi/h. The downstream current of the river is also 2 mi/h. Can you reach your desired goal in a finite time without doing some walking? Explain your answer.

6. There are two initial angles for which the range of a projectile is the same (except in the instance where the initial angle is 45°). What are the differences in the two trajectories that might lead a football quarterback to choose one of them over the other in throwing a pass? If you are familiar with football, you will know that the quarterback usually chooses one of these two trajectories.

7. Two cannons side by side have the same angle of inclination. One of them can shoot a shell with double the initial velocity of the other one. The ratio of the ranges turns out to be 4 : 1. Why is that?

8. A ball is thrown straight up in an elevator moving up with uniform velocity. The ball comes straight down. What is the value of its velocity relative to the ground at the top of its flight?

9. Your lecture demonstrations may include a car that, even while moving, shoots a ball vertically off a platform on the car. If the car is moving with constant speed on a horizontal surface, the ball ends up at the same position on the car that it left. What would happen if the car were decelerating as it released the ball?

10. The Earth–Sun distance is approximately 400 times the Earth–Moon distance. Make a rough estimate of the ratio of the centripetal acceleration of Earth around the Sun to that of the Moon around Earth.

11. Suppose that you measure the period for successive rotations of a phonograph turntable as being unchanging. Is this enough to assure you that the turntable has uniform circular motion?

12. With a given initial speed, there are two initial firing angles for which a projectile has the same range. How many initial velocities give the same maximum height? The same time of flight?

13. You are running in the rain. At what angle should you hold your umbrella for the most protection? Assume that no wind is blowing, so that the raindrops fall straight down. You may need to assume some value for the raindrops' speed as well as your own.

14. You are on the rotating platform of a merry-go-round and throw a ball straight up. Describe the motion of the ball according to a person standing on the ground beside the platform. If it is thrown up at the outside edge of the platform, where will it land?

15. Suppose you are standing on a moving merry-go-round and you throw a ball straight up in the air. Where will it land if you throw it (a) very high and (b) just above your head? Assume that you are standing about midway between the center and the edge of the rotating circular platform.

16. Earth is a rotating platform because it turns on an axis. If a projectile initially aimed at a particular point on the equator is shot from the North Pole, will it arrive at that point, assuming that it could travel such a great distance? Describe the motion of the projectile as seen by someone on Earth.

17. Suppose a projectile is fired from Washington, D.C., toward the equator. How does this situation differ from the one discussed in Question 16?

18. A motorcyclist rides on a large, rotating platform in an amusement park. The cyclist starts at the center of the rotating platform and travels in a radial direction with a constant speed. Is it correct to say that when he or she has traveled some distance from the platform center the acceleration due to the circular motion (which is directed in the inward radial direction) will stop the motorcycle? Analyze the motion.

19. The claim is made that the horizontal and vertical motions of a projectile are independent. Which of the following statements support this claim? (a) If a bullet is dropped from the hand while one is fired horizontally from the same height, they both hit the ground at the same time (ignoring air resistance). (b) A ball thrown horizontally will hit the ground earlier than one thrown at an upward angle with the same initial speed. (c) A thrown ball covers the maximum horizontal distance for a given initial speed when it is thrown at a 45° upward angle.

20. Any change in an object's velocity, whether the change is in magnitude or the direction, means that there has been a (nonzero) acceleration. True or false?

21. Does the parabolic motion of projectiles apply to a missile traveling from a launch pad to a target area 5000 mi downrange?

22. You are playing a table tennis match, with the table placed centrally on a rotating platform. What sort of adjustments to your play do you think you will have to make due to the rotation?

Problems

3–1 Position and Displacement

1. (I) A car travels 21 km to the northeast, then 15 km to the east, before it travels 28 km to the north. Express the position vector from where the car starts to the point at which each turn occurs. What is the car's total displacement?

2. (I) A particle is located by the position described by the vector $\vec{r} = (c_1 - c_2 t)\hat{i} + (d_1 + d_2 t + d_3 t^2)\hat{j}$, where $c_1 = 11$ m, $c_2 = 1.5$ m/s, $d_1 = -12$ m, $d_2 = -2.0$ m/s, and $d_3 = 0.85$ m/s^2. At what time(s) does the particle pass through the position $x = 0$ m? At what time(s), and where, does the particle cross the line $x = y$? Sketch the particle's trajectory from $t = -10$ s to $t = +10$ s.

3. (I) A gym teacher organizes a series of indoor races that follow along the walls of the gym; the race starts from corner A, continues to corner B, and so forth. The gym is a rectangle with the distance $AB = 25$ m and the distance $BC = 35$ m. Suppose that the origin of a coordinate system is at point A, leg AB is in the $+x$-direction, and leg BC is in the $+y$-direction. Express the position vector of a running student at each of the four corners.

4. (I) A runner races with a uniform speed of 27.0 km/h around a circular track of radius 172 m. Draw the track and the runner's displacement vector after 20 s, 40 s, 60 s, and 2 min assuming that at $t = 0$ s the runner is at the three o'clock position. Assume counterclockwise motion.

5. (I) A treasure map locates the site of a treasure by reference to two starting points, A and B. Point A is chosen to be the origin, and B is at the point $(2.5$ km$)\hat{i}$. The instructions state that the treasure lies at the intersection of two lines. One line starts at A and passes through the point $(2.0$ km$)\hat{i} + (4.0$ km$)\hat{j}$; the other line, starting at B, passes through the point $(6.0$ km$)\hat{i} - (8.0$ km$)\hat{j}$. Sketch the instructions and find the location of the intersection graphically as well as algebraically. Express the vector that gives the displacement of the intersection point from point C, whose location relative to the coordinate system specified by A and B is given by $(1.2$ km$)\hat{i} - (2.2$ km$)\hat{j}$.

6. (II) The position of a particle in a given coordinate system is $\vec{r}(t) = (-6 + 4t^2)\hat{i} + (-4 + 3t)\hat{j}$, where the distances are in meters when t is in seconds. At what time will the particle cross the y-axis? At what time will it cross the x-axis? Can you find an equation that relates the y-coordinate to the x-coordinate and therefore gives the trajectory in the xy-plane? Where would the x- and y-axes have to be moved so that at $t = 0$ s the trajectory passes through the origin?

7. (II) Figure 3–30 shows the trajectory of a bicyclist traveling with a constant speed of 25.2 km/h. Express the position vector of the bicyclist as a function of time.

▲ FIGURE 3–30   Problem 7.

8. (II) The x- and y-coordinates of the position of a land surveyor are shown on Fig. 3–31 as a function of time. (a) Plot the trajectory of the surveyor. (b) Calculate the area of the plot that she surveyed.

▲ FIGURE 3–31   Problem 8.

9. (II) The position of a particle is given in a certain coordinate system by the vector $\vec{r}(t) = (4$ m$) \cos(\pi t/T)\hat{i} - (4$ m$) \sin(\pi t/T)\hat{j}$. Find the displacement vector at times $t = T/3$, $t = T/2$, and $t = 2T$. In each case, compute the distance to the origin. What is the angle that the position vector makes with the $+x$-axis for arbitrary t? (The angles are in radians).

10. (III) In a "shoot the coconut" lecture demonstration, the position of the coconut is given by the vector $(h_0 - \frac{1}{2}gt^2)\hat{j}$, whereas that of the projectile, aimed at the coconut at time $t = 0$, is given by $(-L + ut)\hat{i} - [(h_0 ut/L) - (\frac{1}{2}gt^2)]\hat{j}$. Show that the two will always collide and find the time at which this takes place (Fig. 3–32). Express the displacement vector of the coconut relative to the projectile.

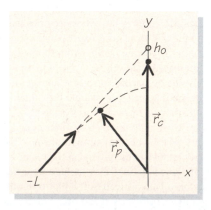

▲ FIGURE 3–32   Problem 10.

3–2 Velocity and Acceleration

11. (I) A tennis ball leaves a racquet in some direction with an initial speed of 45 m/s. After 2.0 s, it has slowed down to 37 m/s. Assume that there is no bouncing on the ground involved. What is the magnitude of the average acceleration? What would you need to know to determine the direction of the acceleration?

12. (I) A particle is observed to move with the coordinates $x(t) = (1.5$ m/s$)t + (-0.5$ m/s$^2)t^2$ and $y(t) = 6$ m $+ (-3$ m/s$)t + (1.5$ m/s$^2)t^2$. What are the particle's position, velocity, and acceleration? At what time(s) are the velocity's horizontal and vertical components equal?

13. (I) At a given moment, a fly moving through the air has a velocity vector that changes with time according to $v_x = 2.2$ m/s, $v_y = (3.7 \text{ m/s}^2)t$, and $v_z = (-1.2 \text{ m/s}^3)t^2 + 3.3$ m/s, where t is measured in seconds. What is the fly's acceleration?

14. (I) A particle moves in such a way that its coordinates are

$$x(t) = A \cos \omega t, \qquad y(t) = A \sin \omega t.$$

Calculate the x- and y-components of the velocity and the acceleration of the particle.

15. (II) A whale traveling southwest at 70 km/h is spotted 5.0 km to the northwest off the coast of Malibu. Photographers jump into a boat that can move at 30 km/h. With what velocity will the photographers intercept the whale assuming that their boat travels a straight-line path? What is the position vector of the whale from the original point on the coast when the photographers reach the whale?

16. (II) A lifeguard standing on a tower throws a buoy to a swimmer 5 m from the tower (Fig. 3–33). The lifeguard, positioned 3 m above the water, pulls in the rope at a speed of 1 m/s. How fast is the swimmer coming to the shore when he is (a) 4 m and (b) 3 m from the water's edge?

▲ **FIGURE 3–33** Problem 16.

17. (II) An engineering student holds her open compass perpendicular to the drafting board, touching the board with both tips of the compass (Fig. 3–34). She slowly closes the compass so that the tips move toward each other with a speed of $2v_0 = 0.060$ m/s. Initially, the angle between the arms of the compass is $2\theta = 60°$. The arms of the compass are $L = 15$ cm long. (a) Calculate the velocity of the joint of the compass as a function of time. (b) Give the velocity of the joint at the moment that she starts to close the compass and at the moment that the two tips reach each other.

▲ **FIGURE 3–34** Problem 17.

18. (II) Calculate the velocity vectors for the coconut and projectile in Problem 10. Express the difference between the two velocity vectors (the relative velocity). What is the magnitude of the relative velocity? What is the angle that its direction makes with the x-axis?

19. (II) Calculate the velocity vector for the particle described by the position vector given in Problem 9. Calculate the angle that the velocity vector makes with the x-axis and show that the velocity vector is always perpendicular to the position vector.

20. (II) A particle moving with an initial velocity $\vec{v} = (30 \text{ m/s})\hat{j}$ undergoes an acceleration $\vec{a} = [3.5 \text{ m/s}^2 + (0.7 \text{ m/s}^5)t^3]\hat{i} + [2 \text{ m/s}^2 - (0.3 \text{ m/s}^4)t^2]\hat{j}$. What are the particle's position and velocity after 30 s assuming that it starts at the origin?

21. (II) The motion of a planet about a star is described by the vector $\vec{r}_1 = R \cos(2\pi \, ft)\hat{i} + R \sin(2\pi \, ft)\hat{j}$. The position vector of another planet about the same star is $\vec{r}_2 = 4R \cos(\pi \, ft/4)\hat{i} + 4R \sin(\pi \, ft/4)\hat{j}$. Show that both of these describe circular motion with constant speed. Calculate the position vector of the second planet relative to the first planet. Sketch the path by noting that the speed of one of the planets in its circular motion is eight times larger than the speed of the other. This is like the motion of a planet whose year is 8 Earth years.

22. (II) Calculate the acceleration vectors for the two planets in Problem 21. Express the vector that describes the acceleration of the second planet relative to the first one.

23. (II) A bag is dropped from a hot-air balloon. Its height is given by the formula $h = H - ut - (u/B)e^{-Bt}$. What are the dimensions of B? What is the initial velocity? What is the velocity as $t \rightarrow \infty$? Calculate the accelerations at $t = 0$ and at $t = \infty$.

24. (II) A car and a truck start from a common spot and travel in straight lines at respective speeds of 30 and 40 km/h. Exactly 1 h later they telephone each other and find that they are separated by exactly 50 km. At what relative directions did they travel?

3–3 Motion with Constant Acceleration

25. (I) A launching mechanism accelerates a baseball horizontally at 24 m/s² for 0.5 s. The baseball's initial velocity is zero. With what velocity will it leave the launcher? The baseball leaves the launcher at the same speed when the launcher is turned in the vertical direction. How high will the baseball go?

26. (I) An airplane is flying due south on a level course at a speed of 600 km/h. At an altitude of 7.50×10^3 m directly above a mountain top, the airplane meets severe turbulence and descends with a vertical acceleration of 4.00 m/s² for 15 s. (a) What is the total displacement of the plane in the 15 s? (b) What is the plane's velocity at the end of the 15 s? (c) What is the plane's position vector with respect to the mountain top at the end of the 15 s?

27. (II) A boy shoots a rock with an initial velocity of 21 m/s straight up from his slingshot. He quickly reloads and shoots another rock in the same way 3.0 s later. (a) At what time and (b) at what height do the rocks meet? (c) What is the velocity of each rock when they meet?

28. (II) A gymnast works out on a trampoline. At the instant that she leaves the trampoline, a point on her waist is 2.3 m above the floor and at the center of the trampoline. At that instant, the point has an upward velocity of 7.8 m/s and a horizontal velocity of 3.0 m/s. Write equations that describe the subsequent motion of that point and find its maximum height.

29. (II) A softball pitcher pitches a ball at a speed of 65 mph (roughly constant over the pitch distance). The distance from the pitcher's mound to home plate is 43 ft, but the pitcher releases the ball about 41 ft from home plate. (a) How long does it take for the ball to reach home plate? (b) A good pitcher can make the ball curve to the left about 6 in over the pitch distance. If this is due to a constant acceleration, what is its magnitude and in which direction does it act?

30. (II) A man in the crow's nest of a sailing ship moving through smooth seas at a steady 8 km/h accidentally lets a cannonball drop from his station, which is 6.5 m above the deck at the top of the main mast. (a) Assuming that he dropped the ball from a position immediately adjacent to the vertical mast, where does the ball land with respect to the mast? (b) How long does it take for the ball to fall to the deck? (c) In the time it takes the ball to fall, how far has the ball moved with respect to an observer fixed on the shore?

3–4 Projectile Motion

31. (I) A runner attempting a broad jump leaves the ground with a horizontal velocity of magnitude 9.0 m/s. Assuming the horizontal component of velocity is unaffected, what vertical component of velocity must the runner acquire to jump 9.5 m?

32. (I) An engineering student wants to throw a ball out of a third-story dormitory window (10 m off the ground) onto a target on the ground placed 8.0 m away from the building. (a) If the student throws the ball horizontally, with what velocity must it be thrown? (b) What must the velocity of the ball be if it is thrown up at an elevation angle of 29°? (c) What is the ball's time of flight in case (b)?

33. (I) A projectile is shot at an angle of 34° to the horizontal with an initial speed of 225 m/s. What is the speed at the maximum height of the trajectory?

34. (I) At what points in a projectile's trajectory above level ground is the magnitude of the velocity a maximum and a minimum? What are these velocities in terms of the initial speed v_0 and elevation angle θ_0?

35. (I) A projectile is shot at an angle of 25° to the horizontal over level ground. Assuming air resistance plays no role, what angle does the projectile make with the horizontal when it lands?

36. (I) Find the initial angle if the range of a projectile is twice its maximum height.

37. (II) (a) Show that the range R can be expressed in terms of the maximum height h and in particular that $R = 4h \cot \theta_0$. (b) Show that when the range is a maximum $h = R/4$.

38. (II) A grasshopper can jump a maximum horizontal distance of 65 cm. Assuming that it spends effectively no time on the ground, that it is tireless, and that it moves in a straight line through a succession of jumps, what is the distance that it can cover in an hour?

39. (II) At what angle must a projectile be fired so that its maximum height and its range are equal? What other angle will give the same range?

40. (II) A boy wants to knock down a coconut with a rock and his slingshot. He observes that the coconut is about 3.0 m above his slingshot and the tree is 4.0 m away along the ground. He knows from experience that the release speed of his rock is 20 m/s. How far above the coconut should he aim?

41. (II) Potato cannons are popular devices made out of PVC pipe that utilize hair spray as an explosive mechanism. One group built a cannon that shoots small potatoes over 100 m. Let's ignore friction in our analysis. (a) What will be the muzzle velocity for a range of 120 m? (b) What height is reached for this maximum horizontal distance? (c) The students want to use the device to get rid of pigeons roosting on tall buildings. How high can they reach?

42. (II) Galileo throws a rock from the top of the Leaning Tower of Pisa at an upward angle of 60° with speed v_0. The rock is in flight for 6.5 s and hits the ground 15 m from the base of the building. Ignore air resistance and ignore the fact that the tower tilts a bit. (a) What is the speed v_0? (b) How high off the ground is the top of the tower? (c) What is the speed of the rock just before it hits the ground?

43. (II) In the Battle of Hastings in A.D. 1066, during which the Normans of France defeated the Saxons in England, an important role was played by Norman archers who shot arrows over a wall of shields erected by the Saxons. If the Norman bows had a maximum range of 350 m and the arrows were shot at an elevation of 55°, how close were the Normans to the Saxons? Assume that the arrows reached their target.

44. (II) A punter kicks a football during a critical football game. The ball leaves his foot from ground level with a speed of 28 m/s at an angle of 50° to the horizontal. At the very top of its flight, the ball hits a wandering seagull. The ball and the seagull each stop dead and fall vertically from the point of collision (Fig. 3–35). In the following, ignore air resistance. (a) With what speed is the ball moving when it strikes the seagull? (b) How high was the unfortunate seagull when it met the ball? (c) What is the speed of the seagull when it hits the ground?

▲ **FIGURE 3–35** Problem 44.

45. (II) A place kicker attempts a field goal, giving the ball an initial velocity of 30 m/s at an angle of 32° with the level field. The uprights are 35 m from the point at which the ball is kicked and the horizontal bar is 4.0 m from the ground. (a) At what time after the kick will the ball pass the goal posts? (b) Is the kick successful, and by how many meters does the ball clear or pass beneath the bar?

46. (II) Astronaut Alan Shepard had the good fortune to play golf on the Moon. The acceleration of gravity on the surface of the Moon is only about one-sixth of that on the surface of Earth. Assuming that Shepard was not noticeably hampered by his space suit and that his best drive on Earth (unhampered by air resistance) was some 210 yd, how far could he drive a ball on the Moon? Derive a general answer to this question for planet X in terms of f, where g_x, the acceleration of gravity on that planet, is fg.

47. (II) A projectile is launched over flat ground and the effects of air resistance are minimal. At what angles with respect to the ground should the launcher be oriented so that the projectile's range is half its maximum range? Why are there two possible angles? What are the angles so that the range is zero, that is, the projectile lands at the foot of the launcher?

48. (II) You must throw a baseball to hit a target on the ground 50 m from the base of a building that is 20 m in height. You are standing at a point on the edge of the roof nearest the target. (a) With what velocity must you throw the baseball if it is to leave the hand horizontally? (b) With what velocity must you throw the baseball if it is to leave the hand at an angle of 45° up from the horizontal? (c) What is the horizontal component of the initial value of the velocity in case (b)?

3–5 Circular Motion

49. (I) The space shuttle is in a circular orbit 220 km above Earth's surface and completes an Earth revolution every 89 min. (a) What is the shuttle's speed? (b) Its acceleration?

50. (I) The Moon circles Earth at a distance of 3.84×10^5 km. The period is approximately 28 d. What is the magnitude of the moon's acceleration, in units of g, as the Moon orbits Earth?

51. (I) A runner in the 200-m dash must make part of the dash around a curve that forms the arc of a circle. This arc has a radius of curvature of 25 m. Assuming that she runs at a steady speed and completes the 200 m in 25.5 s, what is her centripetal acceleration while she is running the curve?

52. (I) A rock placed in a plastic bag is tied to a rope 1.2 m long. The rock is whirled in a horizontal circle. (a) What is the rock's centripetal acceleration if the period of motion is 1.8 s? (b) The plastic bag will break if the radial acceleration exceeds 56 m/s^2. With what speed must the rock be whirled if the plastic bag is to be broken?

53. (I) A passenger on the outer edge of a merry-go-round, 7.5 m from the central pivot, learns that when the merry-go-round is in steady motion his centripetal acceleration is 3.3 m/s^2. How long does it take to make one revolution?

54. (II) The shaft of the engine of a car rotates at 4000 rev/min. A flywheel 20 cm in diameter rotates with the shaft. Calculate the centripetal acceleration of a point on the rim of the flywheel and express it in units of g, the acceleration of gravity.

55. (II) Safety requires that the centripetal acceleration of cars traveling along highway curves may not exceed one-tenth of the acceleration of gravity. How small can the radius of curvature of a curve be for a road where the posted speed limit is 65 mph?

56. (II) Suppose that a point object is in uniform circular motion, moving steadily at a distance R from some central point. The time for one revolution is T. Use dimensional analysis to find the dependence of the centripetal acceleration on T and R. Compare this result to the acceleration derived from a detailed analysis of uniform circular motion performed in Section 3–6.

57. (II) A mass is tethered to a post and moves in a circular path of radius $r = 0.35$ m on an air table—friction free—at a constant speed $v = 18$ m/s. We employ the coordinate system shown in Fig. 3–36. (a) If at $t = 0$ s the mass is at $\theta = 0°$, what are the coordinates (x, y) of the mass at $t = 0.1$ s? (b) What is the acceleration vector of the mass at $t = 0$ s? (c) What is the acceleration vector of the mass when $\theta = 90°$?

▲ **FIGURE 3–36** Problem 57.

58. (II) The space shuttle is moving in a circular orbit with a speed of 7.8 km/s and a period of 87 min. In order to return to Earth, the shuttle fires its retro engines opposite to its direction of motion. The engines provide a deceleration of 6 m/s^2 that is constant in magnitude and direction. What is the total acceleration of the shuttle?

59. (II) An electron in a research apparatus follows a circular path. On the electron's first circuit of the apparatus, its speed is v_0 and the radius of its circular path is R. Each time it makes one circuit, it passes a short region where it receives a "kick" and gains an additional speed of $v_0/100$. The electron follows a circular path such that the magnitude of its acceleration is always the same. What is the radius of the circular path after the electron has received 10 kicks?

60. (II) An automobile moves on a circular track of radius 1.00 km. It starts from rest from the point $(x, y) = (1.00$ km, 0 km$)$ and moves counterclockwise with a steady *tangential* acceleration such that it returns to the starting point with a speed of 30.0 m/s after one lap. (The origin of the Cartesian coordinate system is at the center of the circular track.) What is the car's velocity (magnitude and direction) when it is one-eighth of the way around the track? Express the position and velocity at this point in terms of the unit vectors along the x- and y-axes.

3–6 Relative Motion

61. (I) A sailor wants to travel due east from Miami at a velocity of 15 km/h with respect to a coordinate system fixed on land. The sailor must contend with the Gulf Stream, which moves north at 5 km/h. With what velocity with respect to the water should the sailboat proceed under sail?

62. (I) During an uphill portion of a bicycle race, a cyclist reads a message on a board informing him that the leader is 30 s ahead and that the leader is traveling at 24 km/h. The cyclist's speedometer informs him that he is traveling at 21 km/h. (a) What is the speed of the leader with respect to the cyclist? (b) How far in front of the cyclist is the leader assuming that the speeds have not changed in the last half minute or so?

63. (I) Rain is falling steadily but there is no wind. You are in an automobile that moves at 80 km/h, and you see from the drops on a side window that the rain makes streaks at a 58° angle with respect to the vertical. What is the vertical velocity of the raindrops?

64. (II) A cyclist's top speed on a flat road is v. This is an "air speed" because the limiting speed for a cyclist is determined by the wind resistance. In other words, this is her top speed with respect to the air. She cycles a flat course straight north for a distance L, turns around, and cycles straight south for the same distance. In the following, ignore the time it takes her to turn around and assume that the cyclist can maintain her top speed with respect to the air. (a) Write a formula for the total course time t_0 in terms of L and v. (b) There is a north wind blowing at speed v_w. Write a new formula for the course time t_1 of the cyclist, including the effect of v_w. (c) Show that for $v_w \ll v$ the course time can be approximated by $t_1 = t_0 [1 + (v_w^2/v^2)]$. To show this result, you may want to use the approximation $(1 - x)^{-1} \cong 1 + x$ for $x \ll 1$. (d) Plot the time t_1 as a function of v_w and show that it is always greater than t_0. What happens at $v_w = v$ and why?

65. (II) An athlete can jump vertically a distance of 40 in starting from a standing position. (a) Find the speed with which the athlete left the ground. (b) Assuming that he can leave the ground with the same starting speed as in part (a), find the jump height when the athlete jumps at a 45° angle. What horizontal distance is covered in this jump? (Assume flat ground.)

66. (II) Consider Problem 24. What is the relative velocity of the car and the truck in this problem? More precisely, what is the velocity of the truck as seen by an observer stationed in the car?

67. (II) An airplane flies due south with respect to the ground at an air speed of 900 km/h for 2.0 h before turning and moving southwest with respect to the ground for 3.0 h. During the entire trip, a wind blows in the easterly direction at 120 km/h. (a) What is the plane's average speed with respect to the ground? (b) What is the plane's average velocity with respect to the ground? (c) What is the final position vector?

68. (II) An airplane is to fly due north from New Orleans to St. Louis, a distance of 673 mi. On that day and at the altitude of the flight, a wind blows from the west at a steady speed of 85 mi/h. The airplane can maintain an air speed of 320 mi/h. Ignore the periods of takeoff and landing. (a) In what direction must the airplane fly in order to arrive at St. Louis without changing direction? Draw a diagram and label this direction with an angle. Would this calculation change if the distance between the cities were twice as great? (b) What is the flying time for this flight? (c) Recalculate the flying time if the airplane heads due north until it reaches the latitude of St. Louis and then flies due west into the wind to reach the city.

69. (III) Earth has a radius of 6.4×10^6 m, and its orbit around the Sun has a radius of some 1.5×10^{11} m. Earth simultaneously rotates about its own axis and moves around the Sun (Fig. 3–37). Assume a circular orbit and that Earth's axis of rotation is perpendicular to its orbital plane. (a) What is the speed with respect to the Sun of the point on Earth's equator nearest the Sun? (b) Of the point on Earth's equator farthest from the Sun? (c) Of the two points on Earth's equator, midway between the points in parts (a) and (b)?

Earth rotates about its own axis

Sun

▲ **FIGURE 3–37** Problem 69.

General Problems

70. (I) Neutron stars are the very dense remnants of partially collapsed bright stars. One such (hypothetical) star has a radius of 40 km and rotates with a period of 26 s. What are the velocity and acceleration of a particle on the surface due to the rotation? Compare this acceleration with the acceleration of gravity on Earth.

71. (II) A golfer wants to land a golf ball on the green located 155 m away horizontally but 4.0 m higher. The golfer chooses an eight iron that he knows will result in the ball leaving the tee at an elevation angle of 65°. (a) With what velocity should the ball leave the tee? (b) What is the maximum height of the ball above the green?

72. (II) A softball player hits the ball when it is 0.9 m above home plate. The ball leaves the bat at an elevation angle of 30°. What initial speed must the ball have to clear a fence 1.8 m high located 65 m away in left field?

73. (II) A boat is required to traverse a river that is 150 m wide. The current in the river moves with a speed of 6 km/h. The boat can be rowed on still water with a speed of 10 km/h. Set up a convenient coordinate system in which to describe the various displacements. Using this coordinate system, write down the position vector of the boat at time t assuming that the boat moves with uniform speed and that it leaves one side with the velocity vector making an angle θ with the direction of the river. Calculate θ such that the boat lands at a point exactly opposite the starting point. How long will the trip take?

74. (II) A possible way to measure g is to send a mass traveling horizontally with a known speed off the edge of a vertical drop, such as a table. The table height can be measured, as can the distance from the base of the table where the mass strikes the floor. (a) If the speed of the mass just as it leaves the table is 2.50 m/s and the distance from the table base to the point at which the mass strikes the floor is 108 cm, how long was the mass in flight? (b) The table height is 86 cm. What is the value of g obtained from this data? (c) Use $g = 9.80$ m/s² and neglect the air resistance to calculate the speed of the mass when it arrives at the floor.

75. (II) A sailor on top of a mast 26 m high drops a hammer. The ship is rolling with a maximum angle away from the vertical of no more than 15°. At the moment that the hammer is dropped, the mast is exactly vertical while the top of the mast is moving laterally at a speed of 3.6 m/s. Will the hammer fall into the sea or onto the deck given that the ship is 19 m wide?

76. (II) Someone standing on a down-tilting inclined plane making an angle of 30° with the horizontal throws a ball horizontally with a velocity of 10 m/s. How far down the plane will the ball strike the plane?

77. (III) A juggler is able to handle four balls simultaneously. He takes 0.3 s to cycle each ball through his hands, throw the ball, and be ready to catch the next ball. (a) With what velocity must he throw each ball vertically? (b) What is the position of the other three balls when he has just caught one of the balls? (c) How high must he throw the balls if he is to juggle five balls?

78. (III) A cannon can project a cannonball from its barrel with a certain muzzle speed v_0. Ignoring the effects of air resistance, what formula expresses the distance the cannonball travels before it reaches the ground as a function of the angle θ that the barrel makes with the ground? Unlike the projectile treated in Section 3–4, the cannonball is fired from the edge of a cliff of height h_0 above the level plain at which it is aimed. Show that the angle θ that gives the largest horizontal range is given by $\sin^2 \theta = v_0^2/2(v_0^2 + 2gh_0)$. [*Hint*: In calculus, you learn that a function of a variable such as θ has a maximum (or minimum) at an angle θ_0, the angle for which the derivative of the function with respect to θ is zero.]

79. (III) A wheel 72 cm in diameter rolls along a road, with the center moving in a straight line at a uniform speed of 18 km/h. What are the position vector, the velocity vector, and the acceleration vector of a fixed point on the rim of the wheel relative to a fixed point on the straight line followed by the wheel on the road?

◄ The idea that an apple falling from the tree might have something in common with the Moon rotating around the Earth originated with Isaac Newton, who showed that the same force governs the falling of an apple and the Moon's movement around the Earth. This observation, popularly linked with Newton watching an apple fall, is one of the truly significant discoveries in science.

Newton's Laws

In the previous chapters, we explored quantities that *describe* motion, such as velocity and acceleration. This subject—kinematics—tells us such things as how, given an object's acceleration, we can find its velocity, or displacement. What it does not tell us is how it got that acceleration in the first place. What is it that makes an object fall with a constant acceleration? What governs the way a baseball's velocity changes when it connects with a baseball bat? The study of the *cause* of changes in motion is called **dynamics**, and we can only understand dynamics by understanding the concept of forces, "pushes" or "pulls" that act on an object and influence its motion.

The three laws that describe how forces govern motion were described by Isaac Newton in 1687. Now known as Newton's laws, they are based on careful and extensive observation. These laws provide an extraordinarily accurate description of the motions of all material objects—small or large, simple or complicated—on the everyday scale. It took physicists more than two centuries to discover that some motions cannot be treated using these laws. We must go beyond Newton's laws to describe motion within systems as small as the atom and for motion at speeds near the speed of light (3.0×10^8 m/s); we'll look at these questions more thoroughly in the latter chapters of this book. Newton's laws represent a tremendous achievement in their simplicity and breadth. In this book we will apply Newton's laws to the motion of the heavenly bodies, fluids, springs, projectiles, electric charges, and many other systems. In this chapter and the next we study the concept of forces, the basic three laws, and how these laws apply to some simple situations.

▶ **FIGURE 4–1** The woman exerts a force $\vec{F}$ on the wagon. Other forces must also act on the wagon if it travels with constant velocity as with a constant velocity the net force has to be zero.

4–1 Forces and Newton's First Law

For most of us, everyday experience implies that the "natural" state of motion of a body, when left alone, is the state of being at rest. A sled comes to rest at the bottom of a hill, a piece of furniture will not move unless it is pushed, a car will ultimately come to rest if the engine is turned off. In ancient times, Aristotle observed the motions around him and argued that a body would move only when subjected to a **force**; otherwise, in the absence of these forces, it would come to rest.

We have long since understood that Aristotle's ideas on this subject are wrong, but forces do play a central role in motion. In simplest terms, a force is something that acts to push or pull on an object. For example, Fig. 4–1 shows an adult pulling a child in a wagon. The adult exerts a force on the wagon by pulling it. The compression or extension of a spring represents another situation in which forces act (Fig. 4–2). If the spring is extended, it acts to pull in any object attached to its end; if the spring is compressed, it acts to push out the object at its end. You can on your own think of two or three forces you have experienced or observed today. Forces have two important characteristics (based on experiments that test these ideas):

a. A force is a **vector**, so that it has a direction as well as a magnitude. In the case of the wagon, the adult's pulling force acts along the handle rod (as shown in Fig. 4–1), and the direction depends on the angle that the handle rod makes.

b. Forces are additive, which means that when several forces act together, the subsequent motion of the object acted on is the same as if a single force equal to the vector sum of the individual forces were acting. That single force is the **net force**.

▲ **FIGURE 4–2** (a) A mass attached to the end of a relaxed spring is at rest. (b) The spring has been compressed to the left and pushes the mass to the right. What will happen next?

Figure 4–3 shows a sled being pulled by two ropes. If the forces exerted by the ropes are denoted by $\vec{F}_1$ and $\vec{F}_2$, respectively, then the net force is the vector sum of these, so that

$$\vec{F}_{net} = \vec{F}_1 + \vec{F}_2. \qquad (4\text{–}1)$$

More generally, for any number of forces acting, we have

$$\vec{F}_{net} = \sum_i \vec{F}_i. \qquad (4\text{–}2)$$

▶ **FIGURE 4–3** Ignoring gravity, two forces, $\vec{F}_1$ and $\vec{F}_2$, act on a sled, resulting in the net force $\vec{F}_{net}$.

EXAMPLE 4–1 Three small children each tug at the same teddy bear as in Fig. 4–4a. (All the forces are in the horizontal plane.) The three forces on the teddy bear are $\vec{F}_1 = (-5 \text{ units})\hat{k}$, $\vec{F}_2 = (5 \text{ units})\hat{i}$, and $\vec{F}_3 = (-5 \text{ units})\hat{i} + (5 \text{ units})\hat{k}$. What is the net force on the teddy bear? Ignore the force of gravity on it.

Strategy The unknown net force is a vectorial sum of three given individual forces. This is a straightforward application of vector addition; the net force on the teddy bear is the vector sum of the individual forces (Fig. 4–4b). This is most easily accomplished by adding the components separately.

Working It Out We have

$$\vec{F}_{\text{net}} = \vec{F}_1 + \vec{F}_2 + \vec{F}_3 = -5\hat{k} + 5\hat{i} + (-5\hat{i} + \hat{k})$$
$$= (-5 + 5)\hat{k} + (5 - 5)\hat{i} = 0.$$

The three forces add vectorially to zero.

We have not specified the units of force in this example. When the net force is zero, the units are irrelevant. We shall return to the question of dimensions and units in the next section.

What Do You Think? A large piece of machinery is being maneuvered into place by seven workers. Five workers pull on the machine with rods that have fixed direction (not all in the same plane) and are pulled with constant given forces, whereas two workers pull on ropes with a fixed pull strength (fixed magnitude), but the direction in which they act can be changed. Is it always possible to make the net force on the machine zero? *Answers to* **What Do You Think?** *questions are given in the back of the book.*

(a)

(b)

▲ **FIGURE 4–4** (a) Three children pull on a teddy bear in the xz-plane (horizontal). (b) Diagram for forces on the teddy bear. We have ignored the force of gravity.

The Aristotelian interpretation of experience was that to maintain motion a net force must act on a moving object. Galileo challenged this view in the first half of the seventeenth century. By carrying out experiments with objects sliding and rolling on a succession of ever smoother surfaces, so that friction became less and less important, he was led to the conclusion that moving objects possess an *inertia*, that is, they persist in their motion unless *something acts on them to make them change their motion*. Think of a hockey puck sliding across the ice, a situation in which friction is indeed small. The puck would go a long way if the surface were perfectly level.

Isaac Newton refined this within his **first law of motion**, also known as the **law of inertia**, which states:

> **When there is no net force acting on an object, that object maintains its motion with a constant velocity.**

We must also include the converse statement:

> **When an object moves with constant velocity, the net force acting on it must be zero.**

Constant velocity means that both the magnitude *and the direction* of the velocity do not change.

Recall the discussion of Section 3–6, where observers in different frames were moving at constant velocity with respect to each other. With this perspective there is another way to phrase the first law. In that situation, if one of these observers sees an object moving at constant velocity, then the other observers will all agree that the object moves at constant velocity, although they each will have a different value for that velocity. One of these observers, the one that is moving with the object, will say that the object is at rest. But that observer is not singled out in any significant way—zero velocity is a special case of constant velocity. From this perspective, the first law can be rephrased as:

When there is no net force on an object, it will be at rest in a reference frame that is one of a set of reference frames moving at constant velocity to one another.

When there is no net force on an object, there exists a reference frame in which the object will be at rest. It, and all other reference frames that move with a constant velocity relative to this one, forms the set of inertial reference frames. The name "inertial frame" comes from the fact that observers in inertial frames all agree that the first law, the law of inertia, holds. We see here that the first law is about the existence of inertial frames. In fact, the *converse* statement of the first law of motion, stated on p.89 above, is only true if the object has a constant velocity as seen in an inertial reference frame. (Later we'll talk about noninertial reference frames.)

CONCEPTUAL EXAMPLE 4–2 You are in an airplane flying at 30,000 ft with a constant velocity of magnitude 600 mph. You pick up a pretzel stick and let it drop from your hand. Are you traveling at 600 mph toward the pretzel stick?

Answer Once you let the pretzel stick go, the only force acting on it is that of gravity, which is aligned directly down. Both before and after it is dropped there is no net *horizontal* force acting on it, so the pretzel stick maintains its original horizontal motion of 600 mph relative to the ground and stays in the same horizontal position relative to you. Note the use of the qualifier "relative to the ground"—in fact, Newton's first law tells us that such qualifiers are necessary.

Having stated two versions of the law of inertia, let us now look in more detail at some issues raised by it. Mathematically we can express constant velocity by saying that

$$\frac{d\vec{v}}{dt} = 0,$$

or equivalently,

$$\vec{v} = \vec{v}_0, \tag{4–3}$$

where $\vec{v}_0$ is a fixed vector. It is important to realize that Eq. (4–3) is a vector equation, which means that each vectorial *component* of the velocity is a constant. An object at rest is just a special case, with $\vec{v}_0 = 0$, of Eq. (4–3).

Since acceleration is the rate of change of velocity, this also means that in the absence of a net force an object does not accelerate,

$$\vec{a} = 0. \tag{4–4}$$

Finally, we know from Chapter 3 that constant velocity means that an object's position vector $\vec{r}$ changes linearly with time; that is, $d\vec{r}/dt =$ a constant. In other words, all three position coordinates change linearly with time:

$$\vec{r} = \vec{r}_0 + \vec{v}_0 t. \tag{4–5}$$

At this point we might think about how this applies to the situation described in Example 4–1. What is the motion of the teddy bear as a result of the children tugging on it? We found that there are three forces on the toy, but the **net force** on it is zero. (Keep in mind that this means that the x-, y-, and z-components of the net force are all zero.) According to the first law, then, the bear's velocity is unchanging. If it is at rest when the three children first apply their forces, it remains at rest.

How does Newton's first law square with our experience? The answer to that question depends on how familiar you are with the action of friction. To understand the first law, you have to be able to visualize how friction acts and how it can be limited. Consider the horizontal motion depicted in Fig. 4–1. When we look at this figure, it may appear as though there is just one force acting upon the wagon, the pull of the woman's arm, and this would suggest the first law is wrong: The wagon is moving at constant speed, and you think that there is only a single force acting on it. However, if you consider the motion again, you may recognize the presence of friction (acting especially in the mechanical part of the axle), the force of gravity, and the force exerted by the ground that keeps the wagon from sinking. These must all add up to give a net force of zero. We can get a sense of a nearly friction-free world at a curling rink (Fig. 4–5), where the stone is subject to very little friction, and it is more apparent that the first law applies. We will look at the forces of friction and air resistance in more detail in Chapter 5.

▲ **FIGURE 4–5** In the game of curling, the broom melts a layer of ice and reduces the friction for the sliding rock. Even the reduced friction eventually brings the rock to rest.

A First Look at Some Common Forces and Their Properties

In our discussion of the first law, we have dealt with forces in a qualitative way. Now we'll look in a little more detail. Our qualitative notions about pushes and pulls arise to a large extent from *contact forces* that are associated with physical contact. Let's consider the wagon shown in Fig. 4–1. The child's mother pulls on a handle, and the handle pulls on the wagon to which it is attached. The force of friction, which we argued had to cancel the horizontal force acting on the wagon through the handle, is also associated with contact, either within the wheel system or in the contact of the wheel and the ground.

Some forces act without physical contact, and we say that they *act at a distance*. Consider the following application of the first law: A person leans over a railing and holds a short rope with a bowling ball attached to it (Fig. 4–6). There is no question that the person has to exert a force on the rope quite similar to the force she would have to exert when pulling a wagon. The rope in turn acts on the ball. The force that a taut rope can exert *as a pull* on whatever it is attached to is the force of **tension**. Yet the bowling ball is at rest, so it must be subject to a zero net force. There must be a force that just cancels the pull on the rope, and that is the force due to **gravity**, also known as the ball's **weight**. This force is due to the presence of Earth, and as the ball is not touching Earth, we must conclude that gravity acts at a distance. Now suppose the rope is cut (Fig. 4–6b). The ball no longer remains at rest. It falls with a changing velocity, as we know very well from our discussion of falling objects or projectile motion in Chapters 2 and 3. Thus, according to the first law, the net force on the ball is no longer zero. In fact, it is just the force of gravity, acting alone, without tension or any other force to cancel it and bring the net force to zero. If we watched the ball a little more, we would see it land and come to rest on the ground (Fig. 4–6c). The ball has arrived at another situation in which no net force acts on it. Yet gravity continues to act, so now there is a new force that cancels gravity: another contact force called the **normal force**. This force is associated with the material strength of the solids that make up the ground or any other solid surface. It is what keeps one solid object from penetrating the surface of another solid surface.

Is gravity the only force that acts at a distance? All the forces of electricity and magnetism also act at a distance, and these forces are the ones responsible for the properties of materials. This means that on a fundamental level these forces govern tension, friction, the normal force, and any other contact force. Is this a contradiction? Not at all: The distance involved in the contact forces we have discussed is of atomic size. The atoms of the mother's hand and those of the wagon handle interact—forces act between them—over imperceptibly small distances. Contact forces refer to forces that act over distances too short to be visible to the eye.

Relative Motion and Reference Frames

In Section 3–6 we described the uniform (constant-velocity) motion of a bus. We can say now that no net force acts on the bus, and we can use this situation to look more deeply at the form of Newton's first law that describes the existence of inertial frames. We'll vary the situation of Section 3–6 a little. We continue to have an observer A standing at the curb, measuring the bus to move with constant velocity $\vec{v}_A$, but this time we'll place the second observer, we'll call her B, walking parallel to and at the same speed as the bus. Observer A measures observer B to have the same velocity $\vec{v}_1$ as the bus. Figure 4–7a shows the situation from the point of view of observer A, whereas Fig. 4–7b depicts observer B's rather different point of view. Observer B keeps up with the bus, so that she sees the bus at rest; that is, to observer B, the bus has velocity $\vec{v}_B = 0$. The coordinate system, or frame of reference, centered on observer B is the one referred to in the statement of Newton's first law in which an object with no net force acting on it is at rest.

Which of these observers sees the "true" situation? It is true that observer B sees observer A move with velocity $-\vec{v}_1$; observer B even sees the street moving with that same velocity. But so what? Our "commonsense" preference is for the point of view of observer A because his reference frame is more familiar. But if you have ever had the experience of sitting in a very slowly moving train leaving a station on a track adjacent to another train, you will recall a disorientation as to whether it is your own train or the other train that is moving—we really cannot distinguish who moves and who doesn't.

(a)

(b)

(c)

▲ **FIGURE 4–6** Bowling ball held suspended by means of a rope. (a) The tension in the rope is a force equal and opposite to the weight of the bowling ball yielding a zero net force. (b) After the rope is cut, the net force becomes the weight, and the bowling ball accelerates downward where (c) it comes to rest, and there is once more no net force acting on it.

(a) According to person A

(b) According to Person B

▲ **FIGURE 4–7** Observer A stands watching a bus moving down a street, and observer B moves at the same speed as the moving bus but on the sidewalk. In (a) we see things from the point of view of observer A's frame; in (b) we see things from the point of view of observer B's (and C's) frame.

In other words, there is no "true" situation—A's point of view is no better or worse than that of B. As each of our observers measures the bus to have a constant velocity, each observer concludes that there is no net force acting on the bus, and Newton's law holds in the same way for each observer.

We have concluded that the first law provides no clue as to who is at rest. When observers A and B try to determine which one of them is at rest or exactly what his or her absolute velocity is by looking at the bus, they cannot do so. Another way to say this is that among a set of inertial reference frames there is no way, at least based on Newton's first law, to choose one frame as "special" or "preferred." While an object may be at rest only in a particular inertial frame—here the bus is at rest in the frame centered on observer B—it is not at rest in an absolute sense. We shall see in Section 4–2 that, *even in the presence of net forces*, there is no fundamental way to distinguish between different *inertial* frames—so there is no way to say if any particular inertial frame is at rest in an absolute sense.

(a)

4–2 Newton's Second Law of Motion

When the net force acting on an object is zero, the object maintains constant velocity. What happens to this object when a (nonzero) net force acts on it? Newton's second law answers this question both qualitatively and quantitatively. We can start by saying:

An object acted upon by a net force accelerates.

Further, the object will accelerate, or change its velocity, in the same direction as the net force.

Consider a sled on an ice surface slick enough to be friction free (Fig. 4–8). If the sled were at rest under these conditions and you could push it with a given force—say by bracing your back against a rock and pushing with your feet—the sled would start to move, that is, it would accelerate starting from zero velocity to some velocity that depends on the duration and strength of the push. In fact, the acceleration of the sled is proportional to the strength, or magnitude, of the push; if the sled is pushed twice as hard (and we'll show what "twice" means below), its acceleration will double.

Which is easiest to push in this way, an empty sled or one loaded with kids (Figs. 4–8a and b)? Experience tells us that the specific response of an object to a given net force also depends on the mass of the object. Mass measures an object's resistance to a change in its motion, the property that we earlier called *inertia*, and to reinforce this point, the mass in this context is sometimes called the **inertial mass**. If the object is already moving, then the object's resistance to change can be its resistance either to speeding up or to slowing down; if it is at rest, we are dealing with its resistance to starting to move. The greater the mass of an object, the greater its resistance to a change in motion, that is, the greater its inertia. It is easier to set the empty sled in motion than the full one.

To state all this in a more quantitative fashion, if the mass of the sled in our example were doubled and then pushed with the same force you used the first time, the sled would move with half of the initial acceleration. For a given force the less the object's mass, the

(b)

▲ **FIGURE 4–8** You exert a fixed force on a sled. The resulting acceleration of the sled is smaller in (b) than in (a), because the sled has more mass. As a result, the empty sled aquires a much greater speed than does the full sled.

greater the acceleration (Fig. 4–8a); the greater the mass, the less the acceleration (Fig. 4–8b). As we'll see in more detail in the next subsection, we can even build a mass scale in this way. Experiments with a set of masses will then exhibit an *inverse* relation between mass and the rate of change of velocity (acceleration) for a given force.

This basic relation between the net force on an object, $\vec{F}_{\text{net}}$, the mass of the object, m, and the object's acceleration, $\vec{a}$, was discovered by Newton and has come to be known as **Newton's second law of motion**. It is stated quantitatively as $\vec{a} = \vec{F}_{\text{net}}/m$ or, in more standard form,

$$\vec{F}_{\text{net}} = m\vec{a}. \tag{4–6}$$

NEWTON'S SECOND LAW

Acceleration is the rate of change of velocity so $\vec{a} = d\vec{v}/dt$, and in terms of the velocity, Newton's second law becomes

$$\vec{F}_{\text{net}} = m\frac{d\vec{v}}{dt}. \tag{4–7}$$

It is important to remember that the force appearing here is the net force; that is, the vector sum of all the forces acting on the object. As with any vectorial expression, Newton's second law is equivalent to three scalar equations for the components:

$$F_{x,\text{net}} = ma_x = m\frac{dv_x}{dt}; \tag{4–8a}$$

$$F_{y,\text{net}} = ma_y = m\frac{dv_y}{dt}; \tag{4–8b}$$

$$F_{z,\text{net}} = ma_z = m\frac{dv_z}{dt}. \tag{4–8c}$$

We shall refer interchangeably to Newton's second law in its vector form, Eq. (4–6) or (4–7), or in its equivalent component form, Eqs. (4–8).

There are some very important properties of the inertial mass (or just mass for simplicity) to take into account. Mass is a scalar quantity; it has no direction associated with it. It is always positive. As a consequence, the acceleration of an object is in the same direction as the force exerted on it. Furthermore, masses are additive; that is, if an object contains two parts with masses m_1 and m_2, the total mass of that object is[†]

$$m = m_1 + m_2. \tag{4–9}$$

Is the Second Law Just a Definition of Force?

In Chapter 5 and beyond, we will explore many applications of Newton's second law. At this point, however, we must consider what the second law really means. Does Eq. (4–6) merely define the force? We can measure an object's acceleration with a meter stick and a watch. Do we use that measurement merely to define the quantity $\vec{F}/m$? The power of the second law lies in the fact that it goes beyond this mere definition. Here we shall explain how forces and masses are defined and how Newton's second law allows us to make predictions of the motion of objects.

We'll begin with the calibration of a set of masses. Suppose that we have a large supply of lumps of gold. We will label one of them A and call it our standard mass, $m_A = 1$ kg. We also have available a spring and a very smooth (frictionless) table (Fig. 4–9a). As you know from experience, a relaxed spring (one that is in an "equilibrium" position) neither pulls nor pushes, but a compressed spring exerts a force that will push away an object in contact with it. The force exerted by a given spring depends only on how much the spring is compressed. We compress the spring by 1 cm, place lump A against it, release the spring, and measure the instantaneous acceleration of the lump for the particular compression of the spring. (Actually a hand is needed to compress and then release the spring, although we have not included the hand in the figure.) Suppose that when

[†]When the object has a mass that changes with time (as in rocket motion), we need to proceed differently, but the argument will still remain valid.

Spring in lane 1 is in equilibrium position. Gold nuggets are placed in front of springs in lanes 2 and 3, and springs compressed by 1 cm.

(a)

Springs in lanes 2 and 3 are released and accelerate nuggets A and B.

(b)

Springs are compressed by more than 1 cm.

(c)

Acceleration of nuggets increases over that in (b).

(d)

▲ **FIGURE 4–9** (a, b) A series of experiments with a spring and a set of masses allows us to construct a mass scale and to learn about the forces exerted by the spring. (c, d) Once we have established masses and know the force law, we can predict the motion of a mass.

the spring is compressed by 1 cm and then released, lump A has an acceleration of, say, $a_A = 2 \text{ m/s}^2$, as in Fig. 4–9b. (Notice that this measurement must be made instantaneously, at the moment that the spring is released. Although this measurement is correlated with the measurement of the speed with which the mass leaves the spring, it is not the same measurement!) At this point, Newton's second law provides a precise definition of the force exerted by this spring when it is compressed by 1 cm. Because $F = m_A a_A$, the force is $F = (1 \text{ kg})(2 \text{ m/s}^2) = 2 \text{ kg} \cdot \text{m/s}^2$. Every time the spring is compressed by 1 cm, it will exert this force.

We can now measure the masses of our other lumps of gold. Place a second lump, B, in front of the spring and compress it by the same 1 cm (Fig. 4–9a). When the spring is released (Fig. 4–9b), the initial instantaneous acceleration of lump B is measured to be, say, $a_B = 3 \text{ m/s}^2$. By Newton's second law, the mass m_B is

$$m_B = \frac{F}{a_B} = \frac{m_A a_A}{a_B} = m_A \frac{2 \text{ m/s}^2}{3 \text{ m/s}^2} = \frac{2}{3} m_A = \frac{2}{3} \text{ kg}.$$

In this way, we can determine the mass of each of our lumps of gold—or of any object for that matter.

The force of the spring is always the same whenever it is compressed by 1 cm. Let's now compress the spring by a different amount, place lump A in front of it (Fig. 4–9c), then measure the acceleration of lump A when the spring is released (Fig. 4–9d). This time, we measure, say, $a'_A = 5 \text{ m/s}^2$. We can again use Newton's second law to define the force that the spring exerts when it is compressed by the new amount; it is $F' = m_A a'_A = (1 \text{ kg})(5 \text{ m/s}^2) = 5 \text{ kg} \cdot \text{m/s}^2$. We can continue this process to define the force exerted by the spring for different compressions. In the same way we can figure out the pulling force exerted by the spring when it is stretched. We have found a *force law* for the spring, which in this case is simply a catalog of how much force the particular spring exerts for any given compression or extension. A force law describes the force due to a particular source and how that force depends on variables such as position or time. Throughout this book, we shall study the force laws describing many phenomena, including friction, gravity, springs, and electric charges.

At this point, Newton's second law enables us to make predictions about the motion of an object for the first time. If we compress the spring to the position shown in

Fig. 4–9c and place lump B against it, we can predict lump B's acceleration. We have already determined that the force exerted by the spring in this second position has magnitude $F' = 5 \text{ kg} \cdot \text{m/s}^2$. When the spring is released, as in Fig. 4–9d, the second law *predicts* the instantaneous acceleration to be

$$a'_B = \frac{F'}{m_B} = \frac{5 \text{ kg} \cdot \text{m/s}^2}{2/3 \text{ kg}} = 7.5 \text{ m/s}^2.$$

Once we know how much force is exerted by the spring for a given compression or stretch, we can use the spring to learn about other forces. Suppose we have an unknown force X. Rather than studying the motions that force X causes, we act with it on a mass—any mass—which is also attached to our standard spring. The spring length is allowed to adjust so that the mass does not accelerate. According to the first law, the spring force is exactly canceling force X, and because we know the spring force, we now know force X. As an example, if we turn our spring to the vertical and suspend a block of known mass from it so that the block is motionless (Fig. 4–10), the force of gravity acting on the block must exactly balance the upward force of the stretched spring. By observing how much the spring is stretched, we have measured the force of gravity on the block. This is done in a way that involves no motion! We could now predict how the block would accelerate if there were no spring and only gravity acted on it.

▲ **FIGURE 4–10** By using Newton's first law and a known force, such as that exerted by the stretched spring, we can measure other forces—in this case, the force of gravity acting on the (stationary) block.

EXAMPLE 4–3 You need to deliver a box of bowling balls to a bowling alley. The balls will be placed in a box that is initially at rest but that you want to push into the bowling alley. The box itself has a mass that is very small compared to even one bowling ball. You start with one ball in the box, exert a given force of given strength upon the box for a time period Δt, and at the end of that time the box moves at a speed of 3.2 m/s. You then repeat the procedure with more bowling balls in the box; you exert the same amount of force on the box for the same period of time (Δt) and find the box to have a final speed of 0.4 m/s. How many balls are in the box now?

Setting It Up The two cases are shown in Figs. 4–11a and b at the particular time, after an interval Δt, when the speeds are v_1 and v_2, respectively. You know that an identical force of constant magnitude F acts on two different masses m_1 and m_2 for identical time periods Δt, where $m_2 = nm_1$. Here m_1 is the mass of one bowling ball and n is the number of balls in the box, which is the quantity we want to find. The resulting speeds after time Δt are v_1 and v_2, respectively, and are given.

Strategy In the two cases described the box containing the bowling ball(s) is subject to the same force. Moreover, we can ignore the mass of the box. Using Newton's second law, we can find the accelerations a_1 and a_2 during the period Δt when the force operates. These accelerations are

$$a_1 = \frac{F}{m_1} \quad \text{and} \quad a_2 = \frac{F}{m_2} = \frac{F}{nm_1}.$$

Although we do not know the numerical values of the two accelerations, we do know the speeds v_1 and v_2 after a fixed period of acceleration. Further, we learned in Chapter 2 that an object that starts at rest and undergoes a fixed acceleration $\vec{a}$ for a given period of time Δt has the velocity $\Delta \vec{v} = \vec{v} = \vec{a} \, \Delta t$. In our one-dimensional case, then, we have

$$v_1 = a_1 \Delta t = \frac{F \, \Delta t}{m_1} \quad \text{and} \quad v_2 = a_2 \Delta t = \frac{F \, \Delta t}{nm_1}.$$

We now have enough information to solve for the unknown, n.

Working It Out We can solve for the ratio F/m in terms of v_1 and Δt and substitute it into the equation for v_2, which we can then solve for n. Alternatively, we can simply take the ratio of the two speeds:

$$\frac{v_1}{v_2} = \frac{(F \, \Delta t / m_1)}{(F \, \Delta t / nm_1)} = n.$$

Numerical substitution gives $n = (3.2 \text{ m/s})/(0.4 \text{ m/s}) = 8$ bowling balls.

What Do You Think? Suppose the mass of the box is not negligible. What would be the effect?

(a)

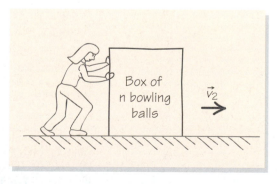

(b)

▲ **FIGURE 4–11** Delivering bowling balls to an alley. In (a) the box has one ball, and in (b) there is an unknown number of balls.

Units: We first discussed the dimensions and units of mass, length, and time in Chapter 1. Because acceleration has dimensions of $[LT^{-2}]$ and units of meters per seconds squared in SI, force has dimensions of $[M \cdot LT^{-2}]$ and, in SI, units of kilogram-meters per seconds squared, or **newtons** (N):

$$1\ \text{N} \equiv 1\ \text{kg} \cdot \text{m/s}^2. \qquad (4\text{--}10)$$

In other words, a force of 1 N exerted upon an object with a mass of 1 kg will produce an acceleration of 1 m/s^2.

In the system of centimeters, grams, and seconds (the cgs system), the force $1\ \text{g} \cdot \text{cm/s}^2$ is called the *dyne*:

$$1\ \text{dyne} \equiv 1\ \text{g} \cdot \text{cm/s}^2 = 10^{-5}\ \text{N}. \qquad (4\text{--}11)$$

Here, again, a force of 1 dyne acting on a mass of 1 g causes an acceleration of 1 cm/s^2.

Another force unit in everyday use is the pound (lb); 1 lb = 4.448 N. The pound is used in the British engineering system, in which mass is measured in slugs.

How Do Observers in Different Inertial Frames See Newton's Second Law?

Let's revisit the two streetside observers A and B of Section 4–1 (Fig. 4–7). They are in different inertial frames moving with fixed velocity with respect to one another. In Section 4–1, B was moving with the constant velocity of the bus, but here we are going to let the bus change its velocity, so we introduce a new label $\vec{u}$ for the fixed velocity of B with respect to A. These observers will generally measure the bus to have different velocities, $\vec{v}_A$ and $\vec{v}_B$, whether the bus is accelerating or not.

Now suppose there is a net force on the bus—the bus is accelerating. Do each of the observers see the same acceleration? If they do, then they will agree on the force acting. We have already seen in Eq. (3–60) how the two observers see the velocity of a bus: $\vec{v}_A = \vec{v}_B + \vec{u}$, or

$$\vec{v}_B = \vec{v}_A - \vec{u}.$$

Let's now see how observers A and B measure the *rate of change* in the bus's velocity—this is the important measurement for the second law. The rate of change is a derivative with respect to time, so we take the derivative with respect to time of both sides of the equation:

$$\frac{d\vec{v}_B}{dt} = \frac{d\vec{v}_A}{dt} - \frac{d\vec{u}}{dt}. \qquad (4\text{--}12)$$

The observers are in inertial frames, meaning that their relative velocity $\vec{u}$ is constant and its derivative is zero. Therefore

$$\frac{d\vec{v}_B}{dt} = \frac{d\vec{v}_A}{dt}. \qquad (4\text{--}13)$$

Our observers *agree* on the acceleration of the bus, a result we first encountered in Chapter 3. Assuming now that the mass m of the bus does not change with time,[†] we multiply both sides of Eq. (4–13) by m to find that

$$m\frac{d\vec{v}_B}{dt} = m\frac{d\vec{v}_A}{dt},$$

or, according to the second law,

$$\vec{F}_B = \vec{F}_A, \qquad (4\text{--}14)$$

where $\vec{F}_B$ is the force on an accelerating object, here the bus, as measured by observer B and $\vec{F}_A$ is the force measured by observer A. We conclude that *observers in different inertial frames agree on the net force acting on an object.*

We saw in Section 4–1 that observers in different inertial frames could not use the first law as a way of deciding who was moving and who was standing still; here we see

---

[†]When the forces holding the system together are strong, special relativity implies a modification. In most of the situations we meet, the effect is very small.

that as they also agree on the net force acting, the second law cannot help us to answer this question either. **Observers in different inertial frames cannot by experiment tell which of them is moving and which of them is at rest.** The statement of the equivalence of all inertial frames is sometimes called the *relativity principle*. This result takes us a long way from Aristotle: not only is the state of being at rest not special, there is no fundamental way to distinguish that state from any state of steady motion. Once Einstein understood that the relativity principle applied to any physical situation, not just those involving the mechanical motion of objects under the influence of mechanical forces, he was led to the theory of relativity (see Chapter 39).

THINK ABOUT THIS . . .

HOW DOES AN AIR BAG PROTECT YOU IN A CRASH?

Large forces imply large accelerations. A car accident or a fall from a great height may be deadly because of the rapid deceleration, the result of large forces that your body may not be equipped to withstand. For protection it is necessary to find a way to bring you to a stop by providing a smaller deceleration over a larger time. Air bags in automobiles work on this principle; when a collision stops a car very suddenly, a passenger would suffer a very sudden deceleration in a subsequent collision with the steering wheel or the windshield. This is mitigated by the very rapid release of an air bag, which is deep enough and "soft" enough to allow the passenger to slow down over a longer period of time. Firefighters similarly use large elastic safety nets to catch people who have to jump from burning buildings. When the deceleration is for fun, the same principle applies. Bungee cords are made of a very elastic material, and there are no bungee *chains*, which would have the unfortunate effect of stopping you "on a dime." Still another application is provided by airplane ejection seats, which in the past were powered by explosives beneath the seat. The rapid acceleration of these mechanisms often led to serious damage to the pilot. Today ejection seats are powered by small rockets that can supply a smaller acceleration over a longer period of time, rather than a large

acceleration over a very short period of time and hence a safer ejection (Fig. 4–12). One other example comes to mind: You may have seen drawings in which Superman catches Lois Lane just before she hits the ground. That very action would imply a rapid deceleration that would be just as bad for Lois as hitting the ground. Superman would do better not to wait for the last instant and instead slow Lois down over a longer period of time.

Pilot in seat

▲ **FIGURE 4–12** Test ejection of a pilot from an AMX jet fighter. The jet plumes below the seat are due to the ejection rocket.

4–3 Newton's Third Law of Motion

The first and second laws involve forces acting *on* objects, and such forces have a specific source. The push on the sled that we described earlier did not come from nowhere; it came from a person. Newton noticed, however, that forces always come in pairs. When you push on the sled, the sled pushes back on you. *When a force due to object B acts on object A, then a force due to object A also acts on object B.* It is obvious that when you push on a wall the wall pushes back on you—that is why you may have to dig in your feet and why you would move backward if you pushed on a wall while you were standing on ice. It is less obvious that when Earth tugs on an apple, causing it to fall (the force is gravity), the apple also tugs on Earth, causing Earth to accelerate toward the apple. Earth and the apple exert gravitational forces *on each other*. We can refine the notion we adopted when discussing the first two laws of motion—that forces act upon objects—to say that forces do not simply act *on* objects; rather, forces act *between* two objects or between an object and its surroundings. Objects are said to *interact* when forces act between them.

Newton extended this statement to a quantitative form in the **third law of motion**:

When a force due to object B acts on object A, then an equal and opposite force due to object A acts on object B.

According to Newton's third law, the force on Earth due to the apple is equal in magnitude but opposite in direction to the force on the apple due to Earth.

A mathematical statement of the third law is the following: Let the force on object A due to object B be $\vec{F}_{AB}$. Then an equal and opposite force $\vec{F}_{BA}$ acts on object B due to object A:

$$\vec{F}_{BA} = -\vec{F}_{AB} \qquad (4\text{-}15)$$

NEWTON'S THIRD LAW

Notice that on the left we have a force acting on object A and on the right a force acting on object B. These forces act on different objects—this is not a pair of canceling forces acting on the same object! The third law is sometimes called *the law of equal action and reaction*. Do not let this particular phrasing mislead you into believing that the *accelerations* of the two objects are the same. Each object accelerates according to the second law, which means that the acceleration depends on the mass of the object. When the apple accelerates toward Earth, Earth accelerates towards the apple, but the Earth's acceleration is far far less than that of the apple's.

The third law is illustrated in Examples 4–4 and 4–5. In these examples, we imagine an outer space environment where we can think of astronauts and satellites as isolated from all forces except the forces they exert on one another. Such a situation is more difficult to arrange on Earth.

EXAMPLE 4–4 An astronaut and a satellite are in an environment where they can be considered to form an isolated system with no external forces acting on that system (Fig. 4–13a). The astronaut tugs on the satellite with a force of 10.0 N toward herself (to the right and down in the figure). What is the force on the astronaut?

Setting It Up The figure labels the the two objects involved as the astronaut A and the satellite S. We are given the force on the satellite due to the astronaut, $\vec{F}_{SA} = 10.0$ N to the right and down.

Strategy This is an application of the third law: If we know the force on the satellite due to the astronaut, then the force on the astronaut due to the satellite, $\vec{F}_{AS}$, is equal in magnitude and oppositely directed.

Working It Out We have

$$\vec{F}_{AS} = -\vec{F}_{SA} = 10.0 \text{ N to the left and up in the figure}$$

This force is drawn in Fig. 4–13b.

What Do You Think? An astronaut is at work repairing the Hubble Telescope in outer space. Having just finished a successful maneuver, she slaps the telescope-carrying space ship in a gesture of satisfaction. She is loosely tethered to the space ship. What will be the effect of the slap?

(a) (b) (c)

▲ **FIGURE 4–13** (a) An astronaut A is interacting with a satellite S. (b) According to Newton's third law, the force the astronaut exerts on the satellite is equal and opposite to the force the satellite exerts on the astronaut. (c) Even though the force on the astronaut has the same magnitude as the force on the satellite, the accelerations of astronaut and satellite are quite different as they have unequal masses.

EXAMPLE 4–5 Assume that the mass of the astronaut in Example 4–4 is 75.5 kg and that of the satellite is 755 kg. What is the acceleration of each?

Setting It Up We can refer to Figs. 4–13a and b. We label the known masses of astronaut and satellite by m_A and m_S, respectively. Here we want to know their accelerations given the forces on them and their masses.

Strategy In the previous example we assessed the forces acting on each of the two objects using the third law. In this example we must apply the second law twice to find the accelerations of the objects.

Working It Out The forces, and thus the accelerations, are all directed along the line between the astronaut and satellite, and we can therefore drop the vector notation. According to the second law, the satellite has an acceleration of magnitude

$$a_S = \frac{F_{SA}}{m_S} = \frac{10.0 \text{ N}}{755 \text{ kg}} = 0.0132 \text{ m/s}^2.$$

This acceleration of the satellite is directed to the right (and down). For the astronaut, the acceleration has magnitude

$$a_A = \frac{F_{AS}}{m_A} = \frac{10.0 \text{ N}}{75.5 \text{ kg}} = 0.132 \text{ m/s}^2.$$

This acceleration is directed to the left (and up). Thus, as Fig. 4–13c shows, the astronaut experiences an acceleration whose magnitude is 10 times larger than that of the satellite. The forces may be equal, but the result of those forces is very different. This difference follows from the difference in the masses of the satellite and the astronaut.

What Do You Think? Consider a sled containing a passenger on flat, slick ice. The sled is given a push by a person. Does the passenger exert a force on the sled when the push is given?

Examples 4–4 and 4–5 demonstrate why the force that the apple exerts on Earth is not directly observable. With a small apple of mass $m_a = 0.1$ kg, the force on the apple due to Earth ($\vec{F}_{aE}$) is approximately 1 N, directed downward. This leads to an acceleration of magnitude $F_{aE}/m_a = 10$ m/s² when the apple falls from the tree. According to Newton's third law, the upward force $\vec{F}_{Ea}$ that the apple exerts on Earth also has magnitude 1 N. But because Earth's mass m_E is approximately 6×10^{24} kg, its upward acceleration has magnitude $F_{Ea}/m_E \cong 2 \times 10^{-25}$ m/s², much too small to be observable. The evidence that led Newton to the third law involved the forces of the Earth–Moon system, where the law's effects are observable.

THINK ABOUT THIS...

HOW DOES A ROCKET WORK?

In Fig. 4–14, we see a very entertaining demonstration that works on the same principle as a rocket's propulsion system. When the valve of the fire extinguisher is opened, a force is exerted on a mass of CO_2 gas, causing it to be accelerated from the fire extinguisher. (The origin of this force is the pressure in the cannister, but that is not what is important here; we only need *some* force to act on a mass and cause it to be expelled.) By Newton's third law, the expelled gas exerts an equal and opposite force on the extinguisher. If the extinguisher is held rigidly by the man on the wagon, the force due to the expelled gas acts on the entire system, which in the figure is accelerated to the left. This is the principle of the rocket: The rocket accelerates matter in one direction and by Newton's third law it is itself accelerated in the opposite direction.

▲ **FIGURE 4–14** Action and reaction go together thanks to the third law. In this case, the release of CO_2 from a fire extinguisher causes the initially stationary cart to be propelled in a direction opposite that of the released gas.

*4–4 Noninertial Frames

In the discussion of reference frames to this point, we focused on **inertial frames**, that is, frames moving with a uniform velocity. For such frames there is no net force on an observer that is at rest with respect to the coordinates that define the frame. We now consider accelerating or **noninertial frames of reference**. For example a set of coordinate axes centered on an airplane accelerating down a runway or on a car going around

Man in falling elevator is trying to determine why the woman in the other falling elevator experiences no apparent force.

▶ **FIGURE 4–15** If the person in the right-hand elevator observing the person in the left-hand elevator does not know that each elevator is in free fall, then he will have to have recourse to a fictitious force to explain his observations.

a corner define noninertial frames. Consider another example: Two adjacent glass elevators start at rest, at the same height and time and drop in free fall. A person inside the first elevator moves under the influence of the force of gravity and falls with acceleration $\vec{g}$. An observer in the second elevator, who is also in free fall, looks at the first elevator (Fig. 4–15). When he does so, he has the choice of two explanations of his observations:

1. We are each in free fall, subject to the same gravitational force, and as a consequence we are both falling with acceleration $\vec{g}$. Since we started out together, with the same initial velocity, we are still together, at rest relative to each other. We are both in noninertial frames.

2. The person in the first elevator is at rest relative to me. She is however subject to a gravitational force and ought to be accelerating downward with acceleration $\vec{g}$. Since she is still at rest relative to me, it must be that there is some force that just counters gravity, so that she is *not* accelerating. I wonder what the source of that force is.

In the first interpretation, the observer in the second elevator knows somehow that he is in free fall and interprets the motion of the first elevator accordingly. In the second interpretation, the observer in the second elevator doesn't know he is accelerating under the influence of gravity and assumes that he is in an inertial frame—Newton's laws of motion are operative—and he must therefore invent a *fictitious force* (sometimes called a *pseudoforce* or a *noninertial force*) to make sense of his observations. The force is fictitious because the observer cannot find any source for it. More generally, it is necessary to invent such fictitious forces to make sense of observations made in in any noninertial frame of reference.

You may wonder if this is realistic. After all, wouldn't the observer "feel" that he is falling? But suppose the "observer" is just a camera or some other scientific instrument looking at the person in the first elevator and that in the film of that person you could see Earth below and thereby know that gravity was acting on him or her. Then you really couldn't decide from the film which interpretation was correct.

In the example above, the observer in the noninertial frame had to invent a fictitious force to "maintain" the first law. Similarly, fictitious forces may be needed to "maintain" the second law for an observer in a noninertial frame. Consider a wheeled food cart sitting in the aisle of an airplane taxiing along a runway with a constant velocity.

In a reference frame centered on the plane the cart is at rest, and in a reference frame fixed to Earth both the cart and the plane are moving with a constant velocity. In both cases the first law—the law of inertia—holds. But suppose now the airplane accelerates down the runway for takeoff. What happens to the wheeled cart? If you are inside the airplane, you will see the cart start to accelerate backward down the aisle, even though no visible identifiable force has acted upon it. You would have to invent a fictitious force to account for its motion. (From the reference frame fixed to Earth, this would pose no problem: The cart would be moving forward with the same constant velocity it had before the plane accelerated.)

To take a last example, consider an observer turning around on her heels (Fig. 4–16). She sees a parked automobile 3 m away, and as she turns—you can certainly try this yourself—the car appears to move in a circle about her: she sees the automobile as having a velocity that is constantly changing direction. According to the observer, it is accelerating (as in Chapter 3—all objects in circular motion are accelerating as the direction of their motion is always changing). If there is acceleration, then according to Newton's laws, there is a force acting. But what is the force that accelerates the automobile? The observer again has to invent a fictitious force to explain the phenomenon. This invented force would have to be a bizarre one: A second automobile 30 m away will have an acceleration 10 times as large as the car 3 m away ($a = \omega^2 r$, and the angular speed ω is set by the observer's rotation), so that the fictitious force that acts on that farthest automobile is, if we use the second law, 10 times as large! The conclusion, and the main point to be made here, is that Newton's laws do not apply to motions observed from noninertial frames of reference. A certain amount of care is necessary to determine whether all possible sources of force have been identified and whether an observer is in an inertial or noninertial frame. We shall normally deal with real forces in inertial frames and give ample warning when accelerating (noninertial) frames are involved.

▲ **FIGURE 4–16** A person pivots in the street. From her point of view the near car is accelerating in circular motion, and the far car is similarly accelerating with even greater magnitude.

CONCEPTUAL EXAMPLE 4–6 Consider a skateboarder who jumps a barrier while the skateboard moves under the barrier. We'll assume here that the boarder has practiced on a stationary board and always jumps straight up off the board. If the board with boarder moves with constant velocity along a flat surface, then by jumping straight up with respect to the board, both boarder and board will continue to move with their original horizontal component of motion, and after the barrier is passed, the boarder falls neatly back onto the board (Fig. 4–17). Consider now a variant: The same board has a little rocket engine triggered by the jump, and it accelerates as the boarder is jumping. What happens? Describe the process from the point of view of (a) the other boarders watching the stunt and (b) an observer *on* the board—for example, a small camera that sends a video to someone watching a monitor.

Answer The skateboarder will land behind the board. (a) From the point of view of the audience, this is easy to understand. The skateboarder launches when the board has a certain horizontal speed. Because after leaving the board there are no further horizontal forces acting on the boarder, his horizontal velocity component remains fixed. But during the time the boarder moves in the air, the board speeds up, so the audience sees the boarder fall behind.

(b) The observer on the board will also see the skateboarder fall behind. In fact, he sees the skateboarder accelerating to the rear with a magnitude equal to the board's acceleration as the audience sees it. Being a believer in Newton's second law, he ascribes the boarder's acceleration as due to a force, but since he is in a noninertial frame, that force will have no apparent source. He might use that fact to decide he is indeed a noninertial observer.

▲ **FIGURE 4–17** As long as the skateboard is in uniform motion and the skateboarder jumps vertically off it, the boarder will end up right back on the board.

EXAMPLE 4–7 Professor A is standing on a train platform; her friend, professor B, is leaving the station in a train with acceleration α in the $+x$-direction (Fig. 4–18). Professor A considers herself to be at rest and states that there is no net force acting on her. What does professor B observe professor A's motion to be, and how might he interpret that motion?

Setting It Up Figure 4–18 includes the x-axis. We call the known velocity of the train as measured from the platform $\vec{u}$ (actually only its change is known, and that is all that matters here), and we call $\vec{v}_A$ the velocity of professor A as measured from a frame of reference at rest with respect to professor B, the motion we want to find.

Strategy We can first simplify matters by recognizing that the motion is in one dimension and drop the explicit vector notation. We also recognize that the velocity of the platform as seen by professor B is $-u$ and that since professor A is at rest with respect to the platform, this is also the velocity that professor B would ascribe to professor A. We still need to interpret this result, which we can do by translating it into the fact that professor B will see professor A accelerating.

Working It Out We have found that professor B measures professor A to have velocity

$$v_A = -u.$$

By taking a derivative of this, we find that professor B measures professor A to have acceleration

$$\frac{dv_A}{dt} = -\frac{du}{dt} = -\alpha.$$

Professor B sees professor A accelerating *backward* (minus sign) with an acceleration of magnitude α. He would then say that, according to the second law, there must be a force on professor A of $-m\alpha$ that is responsible for giving her this motion. He would not, however, be able to find an identifiable agent for this force and might in this way decide that he is in a noninertial frame.

What Do You Think? Suppose each professor drops a tennis ball on the floor and each observes their own ball bounce. What would each conclude?

(a) According to Professor A   (b) According to Professor B

▲ **FIGURE 4–18** (a) Professor A's view. (b) Because professor B is in a noninertial frame, Newton's second law does not apply to his measurements of professor A's motion, and he sees professor A accelerating.

Suppose an observer sees an object (including perhaps himself) accelerating. How does the observer know if the object accelerates in response to a force or if instead the effect is associated with the observer being in a noninertial frame? The answer to this question is that many experiments have been performed with forces in many different situations, and as a consequence we know a good deal about the sources of force. We know, for example, that magnets and electric charges are associated with forces that act at a distance, and we know how to detect contact or friction forces of various kinds. Once we know the sources of forces, we can check to see if any such sources are present and then check whether they are acting in our given situation. If there are none, then the observed effect is due to measurements having been made in an accelerating frame.

IS EARTH'S SURFACE AN INERTIAL FRAME?

The answer to this question is no. Earth rotates about the Sun, and it rotates about its own axis. In addition, there are minor wobbles associated with tidal effects and even the effects of other planets. These effects are detectable with precision equipment and are significant when motion on the planetary scale is concerned—Earth's rotation has a dominant effect on the weather. For most of the situations we encounter, however, the fact that a frame fixed to Earth's surface accelerates is a very minor effect, and we can treat a coordinate system fixed to Earth as an inertial frame. For example, at the surface Earth's rotation has a 1 percent effect on the value of *g*. We have to conclude that an inertial frame is an idealization that we can get quite close to in a laboratory but not actually achieve. ∎

4–5 Using Newton's Laws: Identifying Forces and Free-Body Diagrams

We have seen that Newton's second law allows us to predict motion when we know the forces that are acting or, using the first law, allows us to adjust forces so that there will *not* be motion. Much of this text is devoted to exploring these issues, and in this section and the next we set out a framework for the study of motion in the presence of forces. In this section, we will discuss the first step in this approach, in which we identify all the forces that act on an object and find the net force. In the next section, we show how to use the second law to find the subsequent motion of the object.

Newton's second law, $\vec{F}_{net} = m\vec{a}$, relates the mass of an object, the net force acting on it, and its acceleration. To use the second law, *we must know exactly what object we are talking about*. This step may be less obvious than it first seems. For example, if we want to analyze the motion of a wagon being pulled by a child we must consider only the forces acting *on the wagon*, not the forces that act on the child. It is only the forces on the wagon and the mass of the wagon itself that will determine its motion according to Newton's second law. We must accordingly be able to isolate the wagon. Or we could isolate the system of wagon and child, find all the forces acting on that system, and determine the system's motion. After all, the wagon itself consists of several parts. So in this context, the word *system* refers to whatever is being isolated.

We can best isolate an object in a sketch that we shall refer to as a **free-body diagram** for the object, in this case the wagon. The free-body diagram starts with the isolated wagon and indicates each individual force, magnitude and direction, that acts on it. Figure 4–19 shows how to set up a free-body diagram. In addition to the forces, the free-body diagram should indicate a set of coordinate axes so that we can use this diagram to help us separate the vectors into their vector components. To avoid the effects of noninertial frames, place the axes in an inertial frame—usually attached to some fixed, stationary point—rather than attached to an accelerating point. In Fig. 4–19, we have placed the origin of the coordinate system at a spot on the ground adjacent to the wagon. It is often convenient to use a set of Cartesian axes: (x, y) for planar figures or (x, y, z) for three dimensions. Newton's second law breaks down into separate equations for the vector components along these mutually perpendicular axes. Any orientation of the axes is acceptable, but certain choices will be easier to use. For example, it is often convenient to orient the *y*-axis vertically in the study of falling objects so that the force of gravity has only a *y*-component.

While it is frequently convenient to draw the resulting acceleration of the object in question on the free-body diagram, you should take the utmost care *not to confuse the acceleration with a force*. The acceleration is the object's *response* to the net force acting on it; *it is not a force itself*. In figures in this book, we single out the acceleration vector in blue when we include it in the free-body diagram, whereas force vectors are always drawn in pink.

How to Prepare a Free-body Diagram

1. Make a simple sketch of the system.

2. Choose the body to be isolated (wagon).

3. Add convenient coordinate system.

4. Identify forces that act on wagon. Label them on diagram. Identify labels if necessary.

5. Draw forces acting on single point, usually at center of object.

$\vec{F}_g$ = Gravitational force on cart from Earth
$\vec{F}_H$ = Hand force on cart handle from boy
$\vec{f}$ = Frictional force on cart wheels
$\vec{F}_N$ = Normal force on cart from ground

▲ **FIGURE 4–19** How to draw a free-body diagram.

EXAMPLE 4–8 Block 1 is glued to the top of block 2 (Fig. 4–20a). The masses of the blocks are m_1 and m_2, respectively. A rope is attached to block 2, pulling it horizontally to the right with a force of constant magnitude T along a perfectly smooth horizontal surface (this is the tension force identified in Section 4–1). What equations govern the motion of block 1? Solve this problem in two ways: (a) Consider the system of the two blocks glued together and (b) consider block 1 isolated.

(a)

(b) Isolate m_1 and m_2 together

(c) Isolate only m_1

(d) Isolate only m_2

▲ **FIGURE 4–20** (a) Two blocks that are glued together are pulled by a rope. Free-body diagrams for (b) the two-block system, (c) block 1 alone, and (d) block 2 alone.

Setting It Up When we want to find the equation describing the motion of, say, block 1, we mean that we want to write Newton's second law for it.

Strategy This problem starts with the preparation of free-body diagrams. These diagrams involve first identifying the object or system, then identifying all the *external* forces on that object or system.

Once this is done, we can immediately write the second law. Here we apply the method to two objects: (a) the system of the two blocks glued together and (b) block 1 alone. There is an additional piece of information for us: The acceleration of block 1 and of the system of two blocks will be the same, since they are glued together. That means that if in part (a) we find an expression for the acceleration of the system we have found an expression for the acceleration of block 1. We also note that whatever forces act in the vertical direction (e.g., gravity) they must cancel out entirely because there is no motion of either block 1 or the block 1–block 2 system in the vertical direction. Thus we can ignore all vertical forces.

For part (a), in which we consider the system of two blocks glued together, we find a direct result for the acceleration of the system and therefore the acceleration of block 1 alone. For part (b), in which we consider a free-body diagram for block 1 alone, we are going to discover as we work it through that we are going to have to consider a separate free-body diagram for block 2 as well. Method (b) turns out to be more complicated, but the results are the same.

Working It Out (a) Figure 4–20b is a free-body diagram for the block 1–block 2 system. The net force on the system is the force $\vec{T}$ due to the rope as all vertical forces cancel. The mass of the system is $m_1 + m_2$. Thus Newton's second law for the system is

$$\vec{T} = (m_1 + m_2)\vec{a}_{\text{sys}}.$$

This equation gives us $\vec{a}_{\text{sys}} = \vec{T}/(m_1 + m_2)$ and $\vec{a}_1 = \vec{a}_{\text{sys}}$.

(b) Figure 4–20c is a free-body diagram for block 1 alone. This time, the net (horizontal) force on block 1 is exclusively a contact force due to block 2. This contact force, which is at this point unknown, is written as $\vec{F}_{12}$, where the subscript specifies that we have a force *on* block 1 *due to* block 2. Thus we have

$$m_1\vec{a}_1 = \vec{F}_{12}.$$

This certainly does not appear to be equivalent to our result in part (a), at least superficially. To go further, we must also look at the free-body diagram for block 2 (Fig. 4–20d). From that diagram, we find

$$m_2\vec{a}_2 = \vec{T} + \vec{F}_{21}.$$

Here, $\vec{F}_{21}$ is the force on block 2 due to block 1. According to Newton's third law, however, $\vec{F}_{21} = -\vec{F}_{12}$; so, the expression for Newton's second law applied to block 2 is

$$m_2\vec{a}_2 = \vec{T} - \vec{F}_{12}.$$

We then solve for the unknown force $\vec{F}_{12}$, with the result $\vec{F}_{12} = \vec{T} - m_2\vec{a}_2$. We substitute this back into our equation for block 1:

$$m_1\vec{a}_1 = \vec{F}_{12} = \vec{T} - m_2\vec{a}_2.$$

Finally, the blocks move together and thus $\vec{a}_2 = \vec{a}_1 = \vec{a}_{\text{sys}}$, so that

$$m_1\vec{a}_1 = \vec{T} - m_2\vec{a}_1,$$

or $(m_1 + m_2)\vec{a}_{\text{sys}} = \vec{T}$. We can now recognize the equation we found in part (a) and discover that the two methods give the same answer.

What Do You Think? Did the presence of the glue have any important role to play in the solution to this problem? In other words, if there were no glue and no friction between the two blocks, would the motion of block 1 be different?

External and Internal Forces

Example 4–8 illustrates the important difference between *external* and *internal* forces. Internal forces are those that act within the system we are isolating. These forces may act to hold the system rigidly together or they may simply act between different parts of the system. If we consider the two blocks in Example 4–8 as a single, isolated system, the internal forces would be associated with the glue holding the two masses together. We can examine these internal forces, $\vec{F}_{12}$ and $\vec{F}_{21}$, in Fig. 4–21a. The fact that these forces are equal and opposite (Newton's third law) is, as we have seen in the example, an essential component of our analysis.

External forces, in contrast, are forces that act on the system from outside. In Example 4–8, the net force $\vec{T}$ is an external force acting on the two-block system (Fig. 4–21b). Once a system has been isolated, only the external forces acting on it influence its overall motion. As we shall see in more detail in Chapter 8, internal forces do not enter into the second law because, by Newton's third law, they cancel in pairs and so do not contribute to the net force. For example, we saw that the internal forces $\vec{F}_{12}$ and $\vec{F}_{21}$ of the two-block system canceled each other and did not enter into the solution of the problem. In the statement of the second law, *the net force is the net external force.* All of the objects we deal with are, in fact, complicated systems composed of many atoms held together by internal forces. But even though these internal forces act within the object, they cannot influence its overall acceleration.

If a given system is broken up into separate pieces, the internal forces for the original system may become external forces for the pieces. The motion of the smaller pieces is then governed by these external forces. In Example 4–8, the isolation of block 1 compelled us to treat $\vec{F}_{12}$ as an external force on block 1. In that same example, we made two different choices of the objects to be isolated; hence, we made different choices as to which forces to treat as external and which ones as internal. Each time, we found that only the external forces affected the motion of the blocks. We saw that a judicious choice of which object or system of objects to isolate can simplify the calculation of the object's motion considerably. Take a careful look at the free-body diagrams in Fig. 4–20 to learn how to isolate a given object and recognize the forces on it.

(a) Internal forces

(b) Net external force

▲ **FIGURE 4–21** (a) Internal forces within system of two blocks studied in Example 4–8. (b) Net external force on that same system. Some vertical forces act (gravity and contact forces), but these do not contribute to the net external forces because they cancel.

CONCEPTUAL EXAMPLE 4–9 As a result of the tug, the astronaut and the satellite in Example 4–5 (Fig. 4–13) are accelerated toward each other. What happens when they collide, assuming that the astronaut sticks to the satellite?

Answer If we look at the total system, that is, the satellite and astronaut together, we see that there are no external forces acting on

the system, and therefore the system will maintain its initial velocity. An observer who sees the system at rest at the beginning will see it at rest at the end. The tug is a force internal to the system that results in some rearrangement of the overall system but not in movement of the system as a whole. We will revisit this sort of question in more detail in Chapter 8.

A Second Look at Some Common Forces

In Section 4–1 we first mentioned some common forces that appear in many physical situations: gravity, normal forces, tension, and friction. We will now look again at these forces, this time within the context of free-body diagrams, both as a more complete illustration of the preparation of free-body diagrams and as a way to understand more about the forces themselves. We start with a familiar example: a sled of mass m moving down a snow-covered hill. What is the free-body diagram for the sled?

Three forces act on the sled, one of which acts at a distance. The force that acts at a distance is the force that pulls the sled vertically downward—the force of gravity $\vec{F}_g$, drawn downward in the diagram of this situation in Fig. 4–22a and in the free-body diagram of Fig. 4–22b. We draw this force vector with its tail at some point in the sled.

Figure 4–22 shows a second force, this time a contact force. It is the normal force $\vec{F}_N$, and it is exerted by the hill on the sled. As we stated earlier, this force must be present because in its absence the force of gravity would cause the sled to accelerate down *into* the surface of the hill. The direction of the normal force is *normal*, or perpendicular, to the surface of the hill because that is the direction with no component of motion of the sled or any object on any solid surface. The normal force is the reason this book remains upon the surface of the table and the reason you can sit in a chair. The magnitude of the

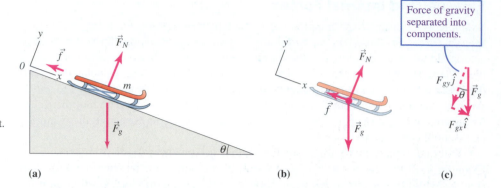

► FIGURE 4–22 (a) A sled on an inclined plane, with the forces acting on it. (b) Free-body diagram for the sled. (c) The force of gravity is decomposed into components perpendicular and parallel to the plane.

normal force will always adjust itself to a situation (to a break point), and the adjustment is such as to make the normal force cancel any forces that might make the sled penetrate the hill. The normal force would be different if a child sat on the sled than for the sled alone because the force of gravity is different for the sled alone than it is for the sled with a child on it. The normal force is simple to use in practice—just as simple as gravity. In contrast to gravity, though, which is a direct and fundamental force between two massive objects (here, Earth and the sled), a normal force results from the addition of many complicated intermolecular forces within and between the materials making up the object and the surface on which it rests. Fortunately, we do not have to worry about these complications, as we are looking at a larger system that includes these forces as internal forces. We draw the normal force with its tail at the point chosen for the force of gravity.

Finally, there is a third force on the sled, included in Fig. 4–22 in the same way as the other two forces, and that is the friction force, which we label $\vec{f}$. Friction, like the normal force, is an approximation to a complicated interaction between the sled rails and the snow, and friction is also a type of contact force. Its direction is opposite to the motion at the surface, here parallel to the hill and uphill. If the sled were moving uphill, the force of friction would act downhill. Friction is always parallel to the surface and is therefore always perpendicular to any normal force at the surface.

The second law as applied to the sled, $\vec{F}_{\text{net}} = m\vec{a}$, becomes

$$\vec{F}_g + \vec{F}_N + \vec{f} = m\vec{a}. \tag{4–16}$$

A vector equation such as this one stands for three equations for the three components, so we must now break this equation into component form. To do so, we must first choose a set of axes. These axes should be placed in an inertial (nonaccelerating) frame to avoid the complications associated with noninertial frames. A convenient choice is one in which the decomposition of the force vectors is simplest. Another convenient choice is one in which the acceleration is parallel to one axis. When there are several forces and they cannot *all* point along an axis, as in this example, it is best to align the axes with as many forces as possible. A good choice for this example is shown in Fig. 4–22, where the y-axis is perpendicular to the hill and the x-axis points downhill. Then two of the three forces—$\vec{F}_N$ and $\vec{f}$—are along these axes. By including the axes in the free-body diagram (Fig. 4–22b), it becomes simpler to read off the components of the second law.

In terms of our coordinate axes, the forces $\vec{F}_N$ and $\vec{f}$ are simple:

$$\vec{F}_N = F_N\hat{j}, \qquad \vec{f} = -f\hat{i}.$$

In contrast, the force of gravity has components in both the x- and y-directions. For the breakdown of $\vec{F}_g$ into components, we have drawn a useful diagram in Fig. 4–22c. From this figure we find that

$$\vec{F}_g = (F_g \sin \theta)\hat{i} + (-F_g \cos \theta)\hat{j},$$

where F_g stands for the (positive) magnitude of the force of gravity. Because all three forces have only x- or y-components, any motion due to the forces—any acceleration, in other words—occurs entirely in the xy-plane. Any component of motion in the z-direction (perpendicular to the page) is a constant-velocity component.

Equation (4–16) can now be written as

$$(F_g \sin \theta)\hat{i} + (-F_g \cos \theta)\hat{j} - f\hat{i} + F_N\hat{j} = ma_x\hat{i} + ma_y\hat{j},$$

or in component form,

$$\text{in the } x\text{-direction: } F_g \sin \theta - f = ma_x; \qquad (4\text{--}17)$$

$$\text{in the } y\text{-direction: } -F_g \cos \theta + F_N = ma_y. \qquad (4\text{--}18a)$$

We can immediately simplify Eq. (4–18a) by setting $a_y = 0$ as we know that the sled does not leave the hill surface. It is this constraint that determines the magnitude of $\vec{F}_N$. With $a_y = 0$, Eq. (4–18a) becomes

$$\text{in the } y\text{-direction: } -F_g \cos \theta + F_N = 0. \qquad (4\text{--}18b)$$

Equation (4–17) is the equation that describes the unknown sled acceleration a_x. It depends on the magnitude of the friction force, f. As we shall see in Chapter 5, f depends on the value of F_N, which, in turn, we can find from Eq. (4–18b).

Problem-Solving Techniques

The preparation of a free-body diagram is an important skill. We can lay out some general guidelines using Fig. 4–19 to help us identify steps:

1. Identify and isolate the object in question. Make a sketch with the object clearly labeled.

2. Identify all the forces acting on the isolated object. Draw each force on the free-body diagram as a labeled arrow; include rough approximations of the direction and magnitude of each force. These arrows should all start from a single point somewhere in the "middle" of the object—remember that until we begin to worry about the rotations of extended objects we are treating our objects as points.

3. Draw a set of coordinate axes with the origin at a fixed point of the diagram; that is, the origin should not be attached to the object itself. Choose these axes so that you can easily pick out the components of the various forces along them.

4. Remember that force equations are vector equations and they stand for a set of equations for the components. Depending on the problem, you may have to separate the forces into their components with respect to the coordinate axes you have chosen.

5. If you include an arrow that represents your guess as to the acceleration of the object, distinguish it clearly from the arrows representing forces. Remember, the acceleration is the response to a force, not a force itself. Figure 4B1–1 summarizes these steps and gives some further hints for finding the net force.

How to Draw

More on Drawing Free-Body Diagrams

(a) Draw free-body diagram
(see Fig. 4–19)

Add
$\vec{F}_N + \vec{F}_H + \vec{F}_g + \vec{f}$

Sum $= \vec{F}_{net}$

(b) Add forces, tail to head

$\vec{F}_{net} = F_{net, x}\,\hat{i} + F_{net, y}\,\hat{j}$

$= F_{net, x}\,\hat{i}$

$= \rightarrow$

(c) Add components

▲ **FIGURE 4B1–1** More hints on drawing free-body diagrams. Parts (b) and (c) show two ways to find the net force, the first a graphical method and the second a method in which components are used.

CONCEPTUAL EXAMPLE 4–10 Two rock climbers—climber 1 and climber 2—are loosely tied together by a rope that can be assumed to be *almost* inextensible. Since the rope is loose, there is no tension. Climber 1 falls but is saved because the upper climber is at a belay point and when the rope pulls taut the fall is arrested. What are the forces that act on the climbers at various times? Which forces on the two-climber system are internal and which are external at the end of the action, when climber 1 has been saved and all is motionless?

Answer In Fig. 4–23 we sketch the situation before and after the fall. Before the fall, the downward force on each climber is the force of gravity, that is, down with magnitudes $m_1 g$ and $m_2 g$, respectively. In each case the normal force due to the rocks acts on the hands and feet of the climbers and just cancels these. While climber 1 is falling, the only force acting on him is the force of gravity; climber 2 is unaffected by the events. Since the rope is almost inextensible, the falling climber is almost instantaneously arrested once the rope has been drawn straight. During the brief period of arrest, he experiences a large deceleration and therefore a very strong upward force. This is due to a rapid rise in the rope tension. This large tension also acts downward on climber 2. It obliges him to adjust his already secure hold in such a way that he can maintain the increased upward contact forces from the rock that must act if he is to remain stationary. Helped by the belay, we assume that he can indeed do this, and then the forces acting on him are the downward force of gravity and downward force of the final tension of the rope, magnitude T. Because this final tension keeps climber 1 from accelerating, it is equal in magnitude to the force of gravity on climber 1, $T = m_1 g$. In turn, climber 2 must be able to maintain an upward contact force from the rocks that is equal and opposite to the net downward force on him, magnitude $T + m_2 g = (m_1 + m_2)g$. He does this with an appropriate "stiffness" in his hands and feet, which is where his muscles and balance come in.

The most important internal force for the two-climber system at the end of the action when the rope is taut and nothing is moving is the tension. There are also many complicated internal forces within climber 2 that are enabling him to hold on. None of these would appear in a free-body diagram for the entire system. The *net* force of gravity, $(m_1 + m_2)g$, is an external force for the system, as is the normal force from the rocks that climber 2 is holding.

▲ **FIGURE 4–23** The falling climber is saved from falling by a second climber with the aid of the rope.

EXAMPLE 4–11 A swinging golf club strikes a golf ball (Fig. 4–24a). During a period of contact, the force of the club (magnitude F_c) on the ball makes an upward angle θ with respect to the horizontal. You may assume that the ball is no longer in contact with the tee but that the club contact continues. Draw a free-body diagram for the golf ball and specify the forces in the coordinate system that you choose.

Strategy As usual, in the preparation of the free-body diagram, we must isolate the object whose motion we want to analyze (here the object is the golf ball), then identify each force acting on it, including both magnitude and direction. A judicious choice of coordinate system will help simplify the expression of Newton's second law for the object.

Working It Out The ball in Fig. 4–24a has left the tee, so there is no contact force from it. The only forces acting on the ball are the force due to the club, $\vec{F}_c$ (a contact force), and the force of gravity $\vec{F}_g$. Figure 4–24b isolates the ball and shows the forces that act on it. It also contains a coordinate system for which the forces are decomposed as

$$\vec{F}_c = (F_c \cos \theta)\,\hat{i} + (F_c \sin \theta)\,\hat{j}$$

and

$$\vec{F}_g = -mg\,\hat{j},$$

where m is the mass of the golf ball.

(a)   **(b)**

◀ **FIGURE 4–24** (a) Golf club hitting a golf ball. (b) Free-body diagram for the golf ball. The ball has already left the tee, so there is no contact force from the tee, but the ball is still in contact with the golf club at this point.

What Do You Think? In this example a net force with a horizontal component and, if $F_c \sin\theta > mg$, with an upward vertical component acts on the ball. Does this mean that during the period that the contact force acts the ball will accelerate in the horizontal direction? Furthermore, will the ball accelerate upward during the contact period?

4–6 Using Newton's Laws: Finding the Motion

We have learned how to use free-body diagrams to help us write Newton's second law for an object that is acted on by forces. This is a very important step along the way to finding the motion of that object. The object's motion is described by *solving* the equations expressing Newton's second law. It is for this reason that we call those equations the *equations of motion*. We can solve them only if we know the particular *force law*; that is, the force expressed as a function of variables such as position, velocity, time, or any other parameters of the problem. Force laws are determined in experiments that measure these forces in controlled conditions. Many of the most common forces (e.g., the force of a spring) depend only on the position of the object—and the simplest case of all occurs when the force is constant, that is, independent of the position of the object (e.g., gravity). We'll see how to solve the equations of motion for a constant force in Example 4–12. We will need to use the kinematic equations that we developed for constant acceleration in Chapters 2 and 3.

EXAMPLE 4–12 The constant net force $\vec{F}$, shown in Fig. 4–25, acts on a nugget of gold whose mass is m. (You could imagine that the nugget is in midair, on a tabletop, or whatever you like; we only ask you to assume that the vector sum of all the forces acting is constant.) What is the subsequent motion of the nugget given that its initial velocity is $\vec{v}_0$?

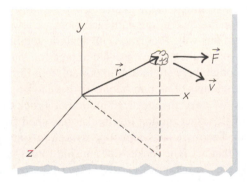

▲ **FIGURE 4–25** A gold nugget is located at a displacement $\vec{r}$ from the origin of a coordinate system and moves with velocity $\vec{v}$.

Strategy Finding the motion in a situation where Newton's second law can be applied means finding the acceleration using that law, then solving the kinematic equations of Chapters 2 and 3 that determine position and velocity given that acceleration.

Working It Out Because $\vec{F}$ is a constant vector, it points in a fixed direction. For convenience, we choose the x-axis to point in the same direction as the force, so that

$$\vec{F} = F\hat{\imath}. \tag{4–19}$$

The force has no components in the y- and z-directions so the nugget of gold will not have a component of acceleration in these directions. The nugget can nevertheless have motion with constant-velocity components in these directions, and these components are the com-

ponents of $\vec{v}_0$. Motion of constant velocity in the y- and z-directions means that these components of the nugget's position vector change linearly with time:

$$y(t) = y_0 + v_{0y}t, \tag{4–20a}$$

$$z(t) = z_0 + v_{0z}t. \tag{4–20b}$$

The quantities v_{0y} and v_{0z} are the constant components of the velocity vector in the y- and z-directions, and y_0 and z_0 are the values of the y- and z-components of the position at time $t = 0$.

For the x-direction things are different. The x-component of Newton's second law is

$$F_x = F = ma_x, \tag{4–21}$$

or, equivalently,

$$F = m\frac{d^2x}{dt^2}. \tag{4–22}$$

We want to find the function $x(t)$ that satisfies this equation. Equivalently, Eq. (4–21) tells us that a_x has a constant value F_x/m, and we want to find the position of the object given this constant acceleration. We have already encountered this situation in Section 2–4, and we can turn to that section for the answer. Once the *initial conditions* are stated, the answer is given by Eq. (2–21), and we restate the result in Eq. (4–23). The initial conditions are as follows: We place the location of the object at $t = 0$ to be $x = x_0$, and the x-component of the velocity at that time is $dx/dt = v_{0x}$. The x-component of the position of the nugget is then

$$x(t) = x_0 + v_{0x}t + \frac{F}{2m}t^2. \tag{4–23}$$

This result obeys both the equation of motion [Eq. (4–22)] and the initial conditions at $t = 0$. We can verify this by direct differentiation and direct substitution.

What Do You Think? If the nugget is twice as massive, the acceleration will have half the magnitude it has here. Does this mean that the nugget goes half as far in time t?

▲ **FIGURE 4-26**   (a) an unstretched spring; (b) a 1 kg mass hangs from the spring stretching it out; (c) with a 2 kg mass, the spring stretches more. The amount of stretch is proportional to the weight.

The motion described in Example 4–12 is really only a review of our study of kinematics with constant acceleration from Chapters 2 and 3. What is new here is the idea that constant acceleration is associated with a constant force. We will be returning to this example in Chapter 5. It is an important example because it is applicable to the force law of gravity near Earth's surface.

When the force on an object is not constant but depends explicitly on the object's position, the solutions to the equations of motion become more complex. In fact, there are only a few cases of the force law for which we can find simple expressions for this type of motion. A spring is one example, and we shall see later that the force law for a spring that has been compressed or stretched by an amount x from a relaxed position is proportional to x, always acting in a direction opposite to the compression or stretching (i.e., if x is positive, the force is negative, and if x is negative, the force is positive). The larger the compression or stretch, the stronger the spring force (Fig. 4–26). Mathematically, the proportionality is expressed as

$$\text{spring force law: } F \propto -x.$$

We also have Newton's second law,

$$F = m\frac{d^2x}{dt^2}.$$

The equation of motion for a spring therefore has the form

$$-x \propto \frac{d^2x}{dt^2}.$$

The equation shows proportionality rather than equality because we have left off all the constants. When all the constants are properly included, we are left with a *differential equation* for x. The meaning of the equation in this case is that x must be a function of time whose second derivative is proportional to the negative of the function itself. Only certain functions will satisfy such an equation.[†]

When the equations of motion have simple mathematical solutions, as they indeed have for the spring, we say we have an *analytic solution*. In fact, there are few force and motion problems that we can solve in this way. Problems with analytic solutions populate textbooks such as this one, and you might get the mistaken impression that all problems can be solved in this way. Although there are indeed important, real-life problems that can be solved analytically, the equations of motion frequently have to be solved numerically with the aid of computers. Whereas the idealized motion of a rock thrown up in the air is simple, the problem of the exact motion of a rocket launched from Cape Canaveral is quite a different matter. The force laws cannot be written so simply in this case because of the range of forces that act. The rocket's propulsion forces need to be adjusted to compensate for varying forces, including air resistance, the acceleration of gravity, and wind forces, which all vary with altitude. Sophisticated computer calculations are needed to figure out these adjustments so that we know just how a rocket will behave as it rises into orbit.

Summary

Newton's three laws express the dynamics of motion by showing how forces acting between objects determine the subsequent motion of those objects. The first law states what happens to an object—moving or at rest—when no net force acts on it:

$\vec{F}_{net}$ is the vector sum of any individual forces that act on an object. When $\vec{F}_{net}$ is zero, the object moves with constant velocity.

Forces, which are vectorial quantities, act on objects and cause them to accelerate. For a given force, this acceleration is inversely proportional to the mass m of the object in question. This is expressed in Newton's second law,

$$\vec{F}_{net} = m\vec{a}. \tag{4–6}$$

In SI, the force is measured in newtons, abbreviated N, where $1 \text{ N} \equiv 1 \text{ kg} \cdot \text{m/s}^2$.

[†]We shall see later that these solutions are sinusoidal: the object moves back and forth.

Forces act *between* objects. If objects A and B interact, that is, if there are forces acting between them, then Newton's third law states that the force on object A due to object B, $\vec{F}_{AB}$, is equal and opposite to the force on object B due to object A, $\vec{F}_{BA}$:

$$\vec{F}_{BA} = -\vec{F}_{AB}. \qquad (4\text{--}15)$$

Observers in reference frames moving with respect to one another observe the motion of a given object differently. An observer who verifies that Newton's second law holds, with known or identifiable sources of forces, is said to be in an inertial frame. If a second observer moves with constant velocity relative to the first, the second observer is also in an inertial frame; if there is nonuniform relative motion, the second observer is in a noninertial frame. Observers in inertial frames agree on the forces they see acting on an object. There is no experiment they can perform to decide who is moving in an absolute sense. Observers in noninertial frames disagree on the forces that act on an object and, in effect, Newton's second law does not hold from the point of view of an accelerating observer.

If we know the nature of the forces that act on an object, then Newton's laws can help us determine the motion of the object. Conversely, the laws allow us to measure the forces acting on an object by measuring the object's motion. If we want to determine the motion of an object, we must know the forces that act on it, and knowing a force means knowing how the force due to a particular source varies with position, time, or other variables.

Newton's third law explains why forces that act within an object or within a system of objects have no effect on the motion of the object. These forces are called internal forces. Only forces external to the object or system determine an object's motion—we refer to these as the external forces.

To best use Newton's laws, we draw free-body diagrams that conceptually isolate an object or system. We include all the forces acting on it, keeping in mind that a force is a vector. Once a free-body diagram has been prepared, we choose a convenient set of axes to write the three components of the second law. These equations can be solved, either analytically or numerically, to find the object's motion.

Understanding the Concepts

1. A small but dense mass is swinging freely at the end of a light string. A very sharp knife cuts the string when the mass is at the bottom of its swing; the knife does not disturb anything else. What is the subsequent flight of the mass?

2. A baseball is hit out of the park. Sketch a trajectory and show the forces (use arrows) that act on the ball at various points along its path.

3. Someone pushes on a wall. What experiment can you propose to determine the force with which the person pushes?

4. If you were in a freely falling elevator, the contact force on you due to the floor would drop to zero. If the elevator were to accelerate rapidly upward, the contact force between you and the floor would increase. Why?

5. You are standing in an elevator that is at rest. The elevator starts moving up. In due course you reach your desired floor, and the elevator slows down and stops. Assume that you are standing on a scale during the whole trip. Describe and explain how the pointer on the scale will move from beginning to end.

6. An astronaut is working while in orbit. When the astronaut assembles a piece of equipment, will he or she notice a difference between working with components of large mass as opposed to components of small mass?

7. Two masses on a smooth flat surface are successively pushed against a compressed spring. The spring is compressed by the same amount each time. One mass is twice as heavy as the other, as measured by a vertical spring scale. True or false: The spring is released, and when the speed each mass attains is measured, they are found to be identical, just as the speeds of falling objects of different mass are the same.

8. When a satellite travels around Earth in a circular orbit, it moves at a constant speed. Does Newton's first law apply in this situation? Is the velocity constant? Is there a force present?

9. There is a well-known parlor trick in which a tablecloth is pulled sharply from beneath a dinner setting, leaving the setting in place. Why does this work?

10. Is a spaceship heading from Earth to the Moon traveling in a force-free environment? Explain. What about a spaceship traveling from Earth to Mars that is currently in a region far away from either planet?

11. An adult sits on a child's table. The table is about to break. Is it correct to say that it is the weight of the adult that is causing the table to break?

12. If you have ever ridden a bicycle with friends, you know that, even without pedaling, different cyclists descend the same hill at different speeds. Isn't this in conflict with the claim that all falling objects accelerate at the same rate under the influence of gravity?

13. From your experience with forces, which of these are contact forces: friction, the force of gravity, the normal force on an object on the floor, the force due to a magnet, the tension in a rope used in a pulley?

14. A box is placed on a table. The box's weight and the normal force on it are equal and opposite. Is this an example of Newton's third law?

15. Shortly after jumping from an airplane, a parachutist will descend with constant velocity. Why is this?

16. Using Newton's second law, devise a system (other than the compressed spring described in Section 4–3) that could be used to measure masses.

17. Determining the mass of objects in space, where the motion is that of free fall, can be difficult. How do you think it can be done?

18. Newton stated that a reference frame at rest with respect to the distant stars would be a good inertial system. Comment on this. Suggest systems that would be good inertial systems for experiments conducted (a) in a physics lab, (b) on the space shuttle, (c) on a ship at sea, and (d) on Mars.

19. A rubber ball and a golf ball have the same mass but the rubber ball has a larger radius than the golf ball. If they are identically accelerated with the same initial force, why might the golf ball go farther in the atmosphere?

20. A fellow student has drawn the free-body diagram in Fig. 4–27 for a block on a smooth ramp, with the arrows representing different forces. What, if anything, is wrong with this diagram?

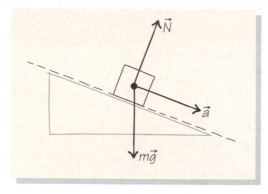

▲ **FIGURE 4–27**   Question 20.

21. Describe the forces that are responsible for the acceleration of an automobile. In particular, in what way is friction between the tires and the road responsible for the acceleration? What, then, is the role of the engine?

22. If you tie a rock to one end of a piece of string and swing it in a circle, you experience a force on your hand. Why?

23. A girl stands on a merry-go-round halfway along a line from the center to the rim. She jumps straight up into the air and lands back on the merry-go-round. Analyze where she lands and how this depends on the rotation speed of the platform.

24. When a baseball hits a bat, a force must be exerted on the ball to change its direction of motion. Describe the forces between the ball and the bat. What effect does the ball have on the bat? Does the batter feel the effects of any of these forces?

25. Give three examples of noninertial forces.

26. You put your hand on a table, place a brick on top of the hand, and give the brick a smart blow with a hammer. Your experience is quite different than what it would be if the brick were not present. Why?

27. A marble is placed on the side of a bowl. It rolls down into the bowl, then continues to roll up and down the bowl's side until it finally comes to rest at the bottom. Explain this in terms of the first and second laws.

28. Describe qualitatively how you might go about constructing an accelerometer, which measures the acceleration of a ship.

[*Hint*: Think about how a spring placed between your back and the back of a seat would react when an automobile in which you are sitting accelerates forward. Such devices are of great importance for "blind" navigation, which might be required for a submarine. They allow us to follow the path when we know the acceleration as a function of time and then reconstruct the velocity and the position as functions of time. (This procedure is described in Chapter 2.) It is, in fact, difficult to construct an accurate version of such a device without knowing about rotational motion.]

29. You are standing on a scale in an elevator. The elevator suddenly starts to move upward. What happens to the reading of the scale and why?

30. A diver jumps from a high platform and experiences a feeling of weightlessness. Is the force of gravity no longer acting on the diver?

31. An apple hangs from a tree, firmly attached to the branch. Does Earth exert a force on the apple even though the apple is at rest? If so, then by Newton's third law, the apple exerts a force on Earth. What keeps Earth from accelerating toward the apple in this situation?

32. A 5-kg mass is placed on a table. A professor states that the normal force on the mass due to the table is (5 kg) (9.8 m/s^2) and that this is a consequence of Newton's third law. Is this a sound analysis?

33. In Example 4–7, we discussed professor A, who is at rest on a train platform, and professor B, in the accelerating train, to illustrate the consequences of the noninertial nature of professor B's reference frame. How do we know that professor A's frame is inertial?

34. Consider a horse that pulls on a cart. By Newton's third law, the cart pulls on the horse with a force of equal magnitude but in the opposite direction. How can there be any motion?

35. What makes a car go forward when the engine is turned on and the transmission is engaged?

36. You are standing in a stationary bus and suddenly find yourself thrown backward. What does Newton's second law say about that?

37. A car is stationary on a flat parking lot. The force of gravity acts downward and an equal and opposite normal force acts upward. Is it correct to say that these forces are equal and opposite because of Newton's third law?

38. In a tug of war, one side is stronger than the other. Assuming that both sides exert themselves to the maximum, discuss the motion of a handkerchief tied to the rope being pulled.

39. A fellow student states that forces cause an object to move. Criticize this statement.

40. How can we be sure that an object moving with a constant velocity has no force on it? Perhaps we are observing the object from a noninertial frame of reference.

41. Baron Munchausen claimed that it is possible for a very strong man to pull himself off the ground and rise into the air by pulling on his bootstraps. Discuss this mode of liftoff in the cold light of Newton's laws.

Problems

4–1   Forces and Newton's First Law

1. (I) In applying Newton's laws, we must identify the forces acting on an object. Are there any forces acting on the following objects? If so, list them: (a) the space shuttle in Earth orbit; (b) an ice skater coasting on ice; (c) the *Voyager I* spacecraft far past the orbit of the planet Pluto.

2. A skater is gliding at what is very close to constant speed on a frozen lake. (There is negligible friction between blade and ice.)

A stiff but very steady wind is blowing. What are the forces acting on the skater? Do they balance?

3. (I) In a tug of war, a red ribbon tied around a point on the rope between the two teams moves with a uniform velocity of 0.1 m/s in the *y*-direction. One team exerts a force on the rope of 600 N in the *y*-direction. What force does the other team exert on the rope?

4. (I) A boat sailing in the northeasterly direction with constant speed experiences a wind force of magnitude 3 × 10^3 N from the

south. What is the force on the sailboat due to the resistance of the water to motion through it?

5. (I) In a classic demonstration, Otto von Guericke used 16 horses—8 on each side—to try to pull apart two hemispheres forming a sphere from which air had been evacuated. Could he as well have used only 8 horses on one side, with the other side tied to a sturdy tree?

6. (II) Three nonzero forces act on a particle at the origin of a coordinate system: $\vec{F}_1$ is in the z-direction, whereas $\vec{F}_2$ and $\vec{F}_3$ lie in the xy-plane. Can you arrange the magnitudes and directions of $\vec{F}_2$ and $\vec{F}_3$ (keeping them always in the xy-plane) so that the particle does not accelerate?

7. (II) (a) A spider is suspended from a single vertical thread; the spider has a mass $m = 30$ mg. The spider is acted upon by the force of gravity, which is directed downward and has magnitude 3.0×10^{-4} N, and by the *tension T* in the thread, a common type of contact force that always acts in the direction of the thread and is directed away from the point at which it is attached. In this case, the tension acts in the upward direction. What is the magnitude of the tension? (b) The spider is now attached to two threads of equal length that make a 120° angle with each other, as in Fig. 4–28. The spider is motionless, waiting for a victim. What is the tension in each thread? [*Hint:* The tension is a vector directed away from the attachment point. Newton's first law must be satisfied in its vector form.]

▲ **FIGURE 4–28** Problem 7.

8. (II) A very large—consider it to be infinitely large—mesh of stiff wires makes a horizontal plane and large electric charges are placed at the intersections of the wires. A mass with another charge is placed below one of the charges on the mesh. Electric charges exert forces on one another that are proportional to the strength of the charges. Each charge on the mesh repels the lone charge with a force that varies with the distance between the charges. The forces are directed along the line between the charges. Show that the net force on the lone charge is a repulsion directed straight downward.

4–2 Newton's Second Law of Motion

9. (I) A car coasts along a road with initial velocity $\vec{v}_0$. It inevitably slows down and finally comes to rest. (a) Describe why this is so. (b) An observer traveling with uniform velocity $\vec{v}_0$ starts out at rest relative to the car. What does she see with the passage of time? How does she explain what happens?

10. (I) The force of gravity on an apple (mass 0.15 kg) has a magnitude of about 1.5 N. What is the acceleration of the apple as it falls toward Earth? How large is the force of gravity on a falling SUV, mass 2500 kg, if it has the same acceleration as the apple? In each case, assume that only the force of gravity acts.

11. (I) The force of gravitation attracts two masses m_1 and m_2 to each other. If the masses are separated by a distance d, the magnitude of the force on each mass is

$$F = Gm_1m_2/d^2,$$

where G is a constant. Suppose $m_2 = 3m_1$. Make a sketch of the two masses with vectors that indicate the direction of the forces on the two bodies. Draw the lengths of the vectors to correspond to the magnitude of the two forces. Repeat the sketch but replace the force vectors with acceleration vectors.

12. (I) The forces acting on an airplane are the following:

gravity: $F_g = 6.0 \times 10^5$ N down;

engine thrust: $F_E = 2.0 \times 10^5$ N forward;

lift: $F_L = 6.0 \times 10^5$ N up;

air drag: $F_D = 1.5 \times 10^4$ N backward.

What is the net force on the airplane? In what direction, if any, is it accelerating?

13. (II) A forensic expert wants to examine the striations on a bullet fired by a gun. A bullet of mass 2.0 g is fired from the gun with muzzle speed 400 m/s into a special resistive material. The bullet is stopped in a distance of 14 cm. If we assume the negative acceleration is constant, what is the acceleration of the bullet inside the material and what force is exerted on the bullet as it accelerates?

14. (II) A spring exerts a force when it is compressed or extended. The force is proportional to the distance x by which it is compressed or stretched away from its equilibrium position. The direction of the spring's force is toward its equilibrium position. Draw diagrams with the spring compressed and extended. Choose a direction for x. Write mathematical expressions for the force in both situations.

15. (II) Five forces, all of the same magnitude F, act on an object of mass m at the origin of the coordinate system shown in Fig. 4–29. Two of the forces are aligned along the x-axis; one is oriented in the $+x$-direction and one in the $-x$-direction. Two other forces are similarly aligned with the z-axis. The fifth force points in the $+y$-direction. What are the direction and magnitude of the acceleration of the object? If we had specified the forces in a different order, would the answer have been different?

▲ **FIGURE 4–29** Problem 15.

16. (II) An object can move along a line between two points A and B separated by a distance $2L$. The object is attracted to each point by a force that is a constant c times the distance between the object and the point. Find an expression for the net force on the object. The net force tends to pull the object back to a particular point. Where is that point?

17. (II) Two forces act on an object of mass 2.5 kg: force $\vec{F}_1$ that is directed along the $+x$-direction and has magnitude 0.50 N and force $\vec{F}_2$ that points at a 45° angle in the $+y$ and $-x$ quadrant and has magnitude 2.0 N. Find the additional force, if any, such that the object will accelerate in the $+y$-direction with magnitude 1.5 m/s².

18. (II) A sports car of mass 720 kg accelerates from 0 to 60 mi/h in 6.7 s. What is the average force, in newtons, that the road exerts on the wheels of the car? A huge SUV of mass 2400 kg can accelerate from 0 to 60 mi/h in 9.7 s. What force does the road exert on the SUV? What would the SUV's acceleration be if it were acted on by the same external force as the sports car?

19. (II) A car of mass 1150 kg accelerates from rest to 100 km/h in 11 s. With additional streamlining, the same car undergoes acceleration to the same speed in 9.0 s. What is the difference in the force exerted by the air (the *drag force*) on the car in the two cases? For this problem, assume the drag force is constant. (This assumption is a poor one in practice.)

20. (II) A common type of contact force is that provided by a taut rope. Suppose that a taut rope is attached to an object. In what direction can the rope apply a force to the object? This force is the tension.
(a) A cart being pulled horizontally by a light rope (the word "light" in this context means that you can ignore the mass of the rope in all your considerations) has a mass of 25 kg and accelerates in the horizontal direction at 2.40 m/s². What is the tension in the rope? If the same cart is pulled so that it accelerates at 0.65 m/s², what is the tension? Assume that the only force acting on the cart is the tension.
(b) Suppose that the rope passes over a fixed pulley (Fig. 4–30) and the cart accelerates horizontally at 1.4 m/s². What is the upward force on the pulley?

▲ **FIGURE 4–30** Problem 20.

21. (II) A father pulls his identical twins on identical sleds tied one after the other (Fig. 4–31). He exerts a force $\vec{F}$ that makes an angle of 30° with the horizontal and that leads to an acceleration of the two sleds. What is the tension in the rope that he is pulling? What is the tension in the rope that connects the front sled to the rear sled? The mass of each sled plus its twin is m. Assume that there is no friction between the sleds' runners and the snow surface.

▲ **FIGURE 4–31** Problem 21.

22. (II) Electrons, mass about 10^{-30} kg, are constituents of atoms and respond to electrical forces. Such forces, of varying strengths, can be generated in the laboratory. Suppose that a constant electrical force of 5×10^{-14} N acts on an electron. What is the speed of the electron after 10^{-10} s? After 10^{-9} s?

4–3 Newton's Third Law of Motion

23. (I) A hook is screwed into a ceiling; one end of a string is tied to the hook and the other end has an 8-kg mass attached to it. Assuming that the force of gravity acting on a mass m has magnitude mg, where $g = 9.8$ m/s², what force does the hook exert on the string?

24. (I) A 3000-kg pickup truck pulls a 1200-kg boat on a trailer, and they are accelerating together at 1.2 m/s². What is the horizontal force that the truck and boat trailer exert on the road?

25. (I) A falling automobile, mass 950 kg, has an acceleration of magnitude 9.8 m/s² when only the force of gravity acts on it. What is the magnitude of the upward acceleration of Earth? Take the mass of Earth to be 6.0×10^{24} kg.

26. (I) The force exerted on a satellite by an astronaut is $(6.5 \text{ N})\hat{i} + (3.7 \text{ N})\hat{j} + (-4.7 \text{ N})\hat{k}$. What is the force exerted by the satellite on the astronaut? What is the magnitude of this force?

27. (II) In Problem 1, you were asked to find the forces acting on certain objects. By Newton's third law, these original objects are the sources for forces that act on other objects. What are these other objects for each of the following original objects: (a) the space shuttle in an Earth orbit; (b) an ice skater coasting on ice; (c) the *Voyager I* spacecraft far past the orbit of the planet Pluto?

28. (II) A force of magnitude 8.0 N pushes on a horizontally stacked set of blocks on a frictionless surface (Fig. 4–32) with masses $m_1 = 2.0$ kg, $m_2 = 3.0$ kg, and $m_3 = 4.0$ kg. (a) What is the acceleration of the stack? (b) What are the forces on block 1 as well as the net force on this block? (c) Repeat part (b) for block 2. (d) Repeat part (b) for block 3.

▲ **FIGURE 4–32** Problem 28.

29. (II) Two blocks connected by a spring are placed on a frictionless flat table. The blocks are pulled apart and then released. A measurement of their accelerations relative to the table shows that the acceleration of one block is twice that of the other. (a) What can you say about the masses of the blocks? (b) After the blocks compress the spring enough, they are pushed apart again. Suppose that while this is happening the spring is cut, so that the blocks are no longer connected. What can you say about the velocities of the blocks after this happens?

30. (II) Three blocks of equal mass are stacked vertically. Block 1 is on top and rests on block 2, which in turn rests on block 3, which rests on a table. There is a downward force on the table of 3 N. (a) What are the forces acting on block 3? (b) What are the forces acting on block 2? (c) What are the forces acting on block 1?

31. (II) The engine of a train pulls five cars each of mass $m = 21,000$ kg. During a period of acceleration, the force between the engine and the first car is 15,000 N. (a) What is the acceleration of the train? (b) Sketch the forces acting on the first car. (c) How large is the tension in the hook between the first and second cars?

32. (II) Repeat Problem 28, this time with the blocks stacked in the reverse order, that is, block 3 to the left and block 1 to the right.

33. (II) Three charges move through space with no forces acting on them except the electric forces that they exert on each other. In an appropriate coordinate system, some of the forces can be broken down as follows: The force that charge 1 exerts on charge 2 is $\vec{F}_{21} = (2\,\text{N})\hat{i} + (-3\,\text{N})\hat{j} + (1\,\text{N})\hat{k}$; the force that charge 1 exerts on charge 3 is $\vec{F}_{31} = (-3\,\text{N})\hat{i} + (2\,\text{N})\hat{j} + (-3\,\text{N})\hat{k}$. What is the total, or net, force on charge 1?

*4–4 Noninertial Frames

34. (I) An observer inside an elevator that is in free fall will see any object that was initially at rest in midair inside the elevator remain in that position. How does he explain this fact, assuming that he knows about the existence of gravity?

35. (I) Consider the situation described in Example 4–7, with $v_A = 0.0$ and $v_B = 3.0$ m/s. The train is accelerating in the x-direction at 0.70 m/s², according to professor A. Can professor B tell that he is accelerating? Does he think that professor A is accelerating? Do our physical senses help us solve these questions? What is the acceleration of professor A according to professor B? Is this a real acceleration?

36. (II) An observer sits in a cylindrical bathtub that can rotate about a vertical axis. At first, the tub is at rest—relative to a neighboring tree for example. A little later, the tub is rotating at high speed around its vertical axis. The observer is still at rest relative to the tub, but the water level is no longer flat; it runs up on the sides of the tub. How does the observer explain this?

37. (II) A simple demonstration of the effect of a noninertial frame can be made by dangling a watch from a chain while you are taking a trip in a jet. What is the angle of the chain with respect to a vertical window edge when (a) the jet is parked at the gate, (b) the jet is accelerating with some acceleration a' along the runway just before takeoff, and (c) the jet is at cruising speed at 30,000 ft? You can try this type of experiment in an accelerating automobile.

38. (II) A mass of 2 kg lies on a horizontal table that is placed in the back of a truck (Fig. 4–33a). The mass is held in position by a string and a force of 6 N acts on it in the $-y$-direction. The truck then accelerates in the $+x$-direction. As a result, an observer in the truck sees that the string holding the mass on the table makes an angle of 20° with respect to its initial orientation, as shown in Fig. 4–33b.

Truck at constant speed $\vec{a} = 0$

(a)

Truck accelerates $\vec{a}$

(b)

▲ **FIGURE 4–33** Problem 38.

What is the acceleration of the truck? Describe what the observer in the truck sees and how he might interpret his observations.

4–5 Using Newton's Laws I: Free-Body Diagrams

39. (I) A car is accelerating straight ahead on a flat road. Draw the free-body diagram for the car.

40. (I) A child slides down a water slide at an amusement park. What forces are present? Draw a free-body diagram for the child.

41. (I) Consider an astronaut sitting in a rocket just after blastoff from Cape Canaveral. Draw a free-body diagram showing the forces present on the astronaut sitting in her seat.

42. (I) Consider the arrangement shown in Fig. 4–34. The whole system is at rest. Draw all the forces, draw the free-body diagram for each mass, and give the equations of motion for each mass.

▲ **FIGURE 4–34** Problem 42.

43. (I) A person throws a medicine ball into the air. Draw separate free-body diagrams for the ball and for the person (a) for the time just before the ball leaves the hands and (b) after the ball has been thrown.

44. (I) A piano sits within a large freight elevator. (a) Draw a free-body diagram for the piano alone. (b) Draw a free-body diagram for the elevator alone.

45. (II) Using a rope, a horse pulls a wagon that is initially at rest. (a) Draw a free-body diagram for the horse. Include friction, Earth's gravity, the influence of the wagon, and any other forces that may be operating. Draw the direction of the forces as accurately as possible. (b) Repeat part (a) for the wagon. (c) Repeat part (a) for Earth, including only the horse and wagon as influences. (d) Why doesn't the horse accelerate backward toward the wagon? In answering, think about the different forces acting on the horse and wagon.

46. (II) A rope spans a gap by being attached at two points on either side of the gap. The two points are at the same height, and the rope is attached in such a way that the tension in it is 50 N. (See Problem 20 for a discussion of tension.) (a) Can the rope be perfectly horizontal? You can assume here that the rope is so light that you can neglect any mass it has. (b) A mass of 2.0 kg is hung by another (light) rope tied to the gap-spanning rope at its midway point. Can the gap-spanning rope remain horizontal? (c) If your answer to part (a) is no, what is the angle the appropriate parts of the gap-spanning rope make with the horizontal?

47. (II) An engine pulls three identical railroad cars along a track. The train's acceleration is a. Draw a free-body diagram for the train as a whole, assigning masses to the cars and engine and putting in all the forces, including those between cars. Which forces are external and which are internal?

48. (II) A block of mass M sits on a rough horizontal surface. A rope inclined upward at an angle θ with the horizontal exerts a force of magnitude T on the block. The block remains stationary. Draw a free-body diagram and express all the forces on the block in terms of M, g, θ, and T.

49. (II) Consider a brick sliding down an inclined plane of 21°. Draw a free-body diagram for the brick that includes the gravity, friction, and normal forces. Choose a coordinate system and give the force components in equation form.

50. (II) Draw two free-body diagrams for a tug of war in which the strengths of the two sides are unbalanced. (The two objects in the diagram will be the two groups of contestants.) Be sure to include all forces!

51. (II) A heavy man stands on a ladder that is leaning against a rough-surfaced wall and rests on the ground (Fig. 4–35). Draw the free-body diagram for the system of man and ladder. Ignore the mass of the ladder. If the ground is very smooth, the ladder may start to slip. Why?

▲ **FIGURE 4–35** Problem 51.

52. (II) Two astronauts, Joe and Moe, are taking a space walk. They are connected to the ship only by a slack rope. They have a light rope between them and decide to have an outer space tug of war. In their space suits, Joe has a mass of 100 kg, whereas Moe has a mass of 110 kg. The loser of the tug of war is the first astronaut to cross an imaginary line midway between them. (a) Draw a free-body diagram for each astronaut. What is the constraint of Newton's third law? (b) Show that Moe wins the tug of war. (c) Miffed at his loss, Joe decides to challenge Moe to a tug of war with their feet firmly planted on Earth. Draw the free-body diagram for each astronaut. (d) Joe wins the tug of war this time. How is this possible?

53. (II) An electrically charged mass, when placed between two charged parallel plates, experiences a constant force F_e perpendicular to the plates. Suppose that such a mass moving vertically with some speed v_y enters the region between two such plates, which are oriented vertically. Draw a free-body diagram for the mass while it is between the plates. Include the force of gravity. Note that there are two possibilities for the direction of F_e; choose one or the other.

54. (II) Objects made of iron are attracted by an electromagnet when a switch is thrown activating it. Suppose that a small piece of iron is suspended by a thread in the vicinity of an electromagnet, as shown in Fig. 4–36. Draw a free-body diagram for the iron object (a) immediately after the switch is closed and the electromagnet is activated and (b) when the switch has been closed for a long time.

▲ **FIGURE 4–36** Problem 54.

55. (II) A person is standing in an elevator that is moving upward with a constant speed. (a) What are the forces acting on the person? (b) Suppose that the elevator is accelerating upward with an acceleration $\vec{g}$ that is equal in magnitude to the acceleration that the person would have if the only force acting on him were gravity. What are the forces acting on the person and how large are they? (c) Suppose that the cable breaks and the elevator, together with its unfortunate passenger, is falling freely. What forces are acting on the person?

4–6 Using Newton's Laws II: Finding the Motion

56. (I) An object of mass 43 g that can move only in the xy-plane is at rest at the origin of the xy-coordinate system at time $t = 0$ s. Constant forces $\vec{F}_1 = (0.071, 0, 0)$ N and $\vec{F}_2 = (0, 0.081, 0)$ N act on the object. (a) Draw the free-body diagram for the object. (b) State the initial conditions, that is, the position and velocity vectors at $t = 0$ s. (c) What are the position and velocity of the object at time $t = 1.200$ s? (d) At $t = 3.600$ s?

57. (II) An object of mass M is subject to two constant forces: $\vec{F}_1$, pointing in the x-direction, and $\vec{F}_2$, pointing in the y-direction. Show that the motion is one of constant acceleration. Find the magnitude and direction of this uniform acceleration.

58. (II) A third constant force $\vec{F}_3 = (F_{3x}, F_{3y}, 0)$ acts on the object in Problem 56 at $t = 0$ s in addition to the forces $\vec{F}_1$ and $\vec{F}_2$. (a) What must F_{3x} and F_{3y} be so that the object does not accelerate? (b) What must F_{3x} and F_{3y} be so that at $t = 5.000$ s the object is at position $\vec{r} = (1.000, 1.000, 0)$ m?

59. (III) At some time t, the displacement of an object moving in one dimension of $m = 2.0$ kg is measured to be $x = At^{3/4}$, where $A = 0.030$ m/s$^{3/4}$. Determine the net force acting on this object. Note that the force will depend on the time.

60. (III) When a small sphere starting from rest falls through a liquid, it experiences a so-called linear drag force $\vec{F}_D$, in addition to the force of gravity mg directed downward. The force $\vec{F}_D$ is directed upward and has a magnitude proportional to the velocity, $F_D = bv$, where b is a constant. The equation of motion—Newton's second law—for motion in the vertical direction z (measured with the $+$-direction downward) reads $m\, dv/dt = mg - bv$. (a) Show that this equation is solved when the velocity takes the form $v(t) = mg(1 - e^{-bt/m})/b$. (b) Sketch the speed as a function of time. (c) Sketch the position as a function of time.

61. (III) The force on a mass m that can move along the x-axis is given by $F = -kx$. (This force is characteristic of the spring; k is a property of this particular spring and is called the spring constant.) (a) Show that, with an appropriate choice of the constant c, $x = A\cos(ct) + B\sin(ct)$ is a solution of the equation of motion. What is the value of c? (b) If the object is at $x = 0$ at $t = 0$ and if its velocity at that time is v_0, what are A and B?

General Problems

62. (I) The gravitational force that the Sun exerts on Earth has a magnitude of 3.5×10^{22} N and is directed toward the Sun. What force does Earth exert on the Sun?

63. (II) A load of mass 53.2 kg is to be lifted by two ropes each going over pulleys. The ropes are separated in such a way that both ropes initially make an angle of 65° with the horizontal (Fig. 4–37). The magnitude of the forces exerted by each of the ropes is 333 N. (a) Draw a free-body diagram for the load. The only forces acting on it are the forces of the ropes, directed along the ropes, and the force of gravity, which points down and has a magnitude given by the mass of the load times the constant acceleration $g = 9.80$ m/s². (b) What is the net force on the load? (c) What is the initial acceleration (magnitude and direction) of the load?

▲ **FIGURE 4–37** Problem 63.

64. (II) An object of weight $F_g = 20$ N is suspended as shown in Fig. 4–38. The rope and pulleys are very light and friction is negligible. Determine the weight of the object that maintains the equilibrium given that the angle between the rope and the horizontal is 35° on both sides of the movable pulley.

▲ **FIGURE 4–38** Problem 64.

65. (II) A window washer sits on a board supported by two ropes that go over pulleys attached to a scaffold. He holds the other ends of the ropes in his hands (Fig. 4–39). He pulls on them in such a way that he moves upward at constant speed. Draw a free-body diagram for the window washer only and give all the equations; assume that the tension in a given rope is the same everywhere in the rope, even after it wraps around a pulley. Suppose that the effect of the window washer's efforts leads to an upward acceleration of both the window washer and the board. How are the equations modified?

▲ **FIGURE 4–39** Problem 65.

66. (II) A dead tennis ball is dropped onto a granite slab and it bounces back to about 20 percent of the height from which it was dropped. What forces acted on the tennis ball for it to exhibit this behavior? In what direction do these forces point? Estimate their magnitude.

67. (II) A train consists of an engine and three cars tied closely together. The mass of the whole train is 1.7×10^5 kg; the engine alone has a mass of 80,000 kg and each car has a mass of 30,000 kg. Through its wheels, the engine can exert a horizontal force of 3.0×10^4 N on Earth. (a) When the engineer wishes to accelerate forward, in which direction should the engine exert its force on Earth? What is the magnitude of the force exerted by Earth on the engine then? (b) When the engineer acts to accelerate the train, what is the force on the set of three cars and what is the subsequent acceleration of these cars? (Treat the three cars as a unified whole.) (c) When the engineer acts to accelerate the train, what are all the forces on, and subsequent acceleration of, the second car of the train?

68. (II) A boat is being pulled up the middle of a canal at a steady speed by two horses, one on each side of the canal (Fig. 4–40). The ropes tying the boat to the horses each make an angle of 30° with the lengthwise direction of the canal. (a) Draw a free-body diagram for the boat. Include the friction force due to the boat sliding through the water. (b) If the force exerted on each horse by its rope is 2000 N, what is the force of friction that the boat experiences due to the water?

▲ **FIGURE 4–40** Problem 68.

69. (II) A passenger in a stationary elevator is acted on both by the force of gravity, which is directed downward, and by a contact force directed upward by the floor. The contact force from the floor cancels the force of gravity, and in this way there is no net force on the passenger, which explains why she measures herself to be at rest with respect to the walls. (a) What is the free-body diagram on the passenger when the elevator descends at constant velocity? (b) The elevator cable breaks. (Do not be alarmed—the emergency brake will act if you finish this problem in time.)

As seen from the outside, the passenger and elevator fall together and accelerate under the force of gravity. A free-body diagram for the passenger would show the single force of gravity acting. What does the passenger observe and how would she draw a free-body diagram for herself?

70. (II) A parachutist has jumped from a plane. (a) What are the forces acting on the parachutist when the parachute opens? Suppose that the force due to air resistance opposes the parachutist's motion and is proportional to the speed of the system; that is, $\vec{F}_{\text{air resistance}} = -A\vec{v}$. (b) What are the dimensions of the constant A? (c) Sketch the downward acceleration of the parachutist as a function of velocity. Show that the velocity of the system becomes constant at some point. This constant velocity is called the *terminal velocity*. (d) If the force of gravity on the system is 1000 N and the terminal velocity is 6.0 m/s, what is the value of the constant A?

71. (II) A parachutist experiences two forces: One is the force of gravity, which is of the form $\vec{F}_g = -mg\hat{j}$. Here, m is the mass of the parachutist and g is the acceleration due to gravity. The other force is a drag force, and, in contrast to Problem 70, it has the form $\vec{F}_d = Av^2\hat{j}$ in this problem, where the velocity of the parachutist is given by $-v\hat{j}$. (a) What are the dimensions (and units) of A? (b) Complete the equation $dv/dt = \ldots$. (c) At some point, the parachutist reaches a terminal velocity that is constant. What is it?

72. (II) An egg of mass 80 g is in a stiff box of mass 200 g surrounded on all sides by padding that keeps the egg in place. A force $\vec{F} = (1.2\ \text{N})\hat{i} + (-0.08\ \text{N})\hat{j}$ acts on the outside of the box, in addition to the force of gravity (which is oriented in the $-y$-direction). (a) Draw free-body diagrams for the egg–box system and for the egg. (b) Express Newton's second law as it applies to the egg. (c) What are the acceleration of the egg–box system and net force on the egg?

73. (II) A bicyclist rides a flat course at constant speed v with respect to the ground into a stiff but steady headwind of speed v_w, also with respect to the ground. (a) Draw a free-body diagram for the cyclist–cycle system. List the forces acting on this system, including the wind resistance, which pushes in his face, and the frictional forces with the ground, which propel him forward. (b) If you are observing the cyclist from a reference frame at rest with respect to the ground, what is the vector sum of all the forces? (c) If you are observing the cyclist from a frame at rest with respect to the bicycle, what are the speed of the cyclist, the speed of the wind, and the vector sum of all the forces? (d) If you are observing the cyclist from a frame at rest with respect to the air, what are the speed of the cyclist, the speed of the wind, and the vector sum of all the forces?

74. (II) An electron is placed midway between two charged parallel plates oriented vertically. The electron experiences a constant force perpendicular to the plates. Suppose that the electron, whose mass is 9.0×10^{-31} kg, is moving vertically with $v_y = 2.0 \times 10^6$ m/s and enters the region between the two plates (Fig. 4–41). This region has a vertical height of 1 cm, and while in it, the electron experiences a force of 3.0×10^{-18} N. (a) How long does the electron stay in the region? (b) After the electron has passed through the region, what is its horizontal velocity component?

▲ **FIGURE 4–41** Problem 74.

75. (II) The air stream from a leaf blower exerts a force F_0 on the low setting and a force $2F_0$ on the high setting. When on the low setting the leaf blower accelerates a soda can at 0.20 m/s² and an empty box at 0.10 m/s². (a) With what acceleration will each object be accelerated by the leaf blower on the high setting? (b) With what acceleration will the leaf blower accelerate the box if the can is placed inside the box for both the low and high settings? (c) The soda can's mass is measured and found to have the value 4.3 g. What is the mass of the empty box?

76. (III) The acceleration of an object dropped from a height is smaller at the equator than at the poles, even when the force of gravity is assumed to be the same everywhere on the surface of Earth. (a) Assuming that the difference has something to do with the rotation of Earth, obtain an expression for the difference between the accelerations in terms of the radius of Earth and its rotation frequency. (b) Given that the radius of Earth is $R = 6.38 \times 10^6$ m and that the acceleration at the poles is $g_0 = 9.80$ m/s², calculate the acceleration of a falling object at the equator. (c) Calculate the acceleration of a falling object as a function of the angle of latitude in terms of g_0. (d) If experiment shows a deviation from your prediction, what effects could be responsible for that deviation?

77. (III) As shown in Fig. 4–42, a plumb bob suspended from a frame settles to a steady position as the frame slides down a frictionless inclined plane. What is the angle the plumb bob makes with the vertical during the slide?

▲ **FIGURE 4–42** Problem 77.

◀ It is a contact force exerted by the starting blocks against the foot of a sprinter that accelerates him forward. An equal and opposite force is exerted on the starting block, and, because the block is attached to Earth, on Earth itself.

Applications of Newton's Laws

I n this chapter we apply Newton's laws to a variety of situations in which forces act. We will consider several types of forces, including gravity, tension, normal forces, friction, and drag forces, all of which act on us and the objects around us. We shall also look at the role of forces in circular motion as well as the features of motion in rotating frames. Finally, we will look at how the forces are ultimately described in terms of more fundamental forces—in particular, ones that act at a microscopic level.

5–1 Common Forces Revisited

The forces that we encountered in Chapter 4 demand further study, as we have not yet explored all of their significant features. These forces include the force of *gravity* $\vec{F}_g$, here treated as a constant force that acts on every object, the tension $\vec{T}$ exerted by a taut rope, and the normal force $\vec{F}_N$ that keeps you from falling through your chair. We'll study friction in Section 5–2.

Gravity

The parabolic form of a projectile's trajectory near Earth (Fig. 5–1) is due to the force of gravity. As we saw in Chapter 3, projectiles in the vicinity of Earth's surface travel with a

▲ **FIGURE 5–1** The horse follows a parabolic path as it jumps the rail.

constant acceleration $\vec{a}$ that points downward and has magnitude g. With Newton's second law, we can say that the force of gravity causes the projectile's constant acceleration and that the force has constant magnitude near Earth's surface and is always directed down toward the center of Earth. More precisely, the acceleration $\vec{a}$ of our projectile is given in terms of the force of gravity $\vec{F}_g$ on the projectile and the inertial mass m of the projectile as

$$\vec{a} = \frac{\vec{F}_g}{m}. \tag{5–1}$$

A very special characteristic of the force of gravity is that, at any given location, *it causes all objects to accelerate in the same way—no matter what their mass.* In other words, the right-hand side of Eq. (5–1) is independent of the mass m. The only way this can happen is for the *force itself to be proportional to the mass,* so that the factor m in the denominator cancels with another such factor in the force itself.

Let's look more carefully at the meaning of this remarkable fact. That an object has constant acceleration under the influence of gravity means only that this force takes the form

$$\vec{F}_g = m_g\vec{g}, \tag{5–2}$$

where $\vec{g}$ is a constant vector with dimensions of acceleration and points to the center of Earth.[†] The "gravitational mass" m_g is the property of the object that determines the strength of the gravitational force acting on it. However, the acceleration of *all* objects under the influence of gravity is *precisely the same*; mathematically, this statement means

$$m_g = m. \tag{5–3}$$

This relation between gravitational mass and inertial mass is remarkable because, at least until Einstein's theory of general relativity, not developed until about 1915, we had no reason to think that the force of gravity has anything special to do with the inertial mass. (See Chapter 12 for further discussion.) We recall from Chapter 4 that the inertial mass determines the *response* to a force, and tests to determine the inertial mass are possible without using gravity at all—say, by seeing how much the object accelerates when *any* known force, even one that has absolutely nothing to do with gravity, acts on it. Conversely, tests to determine the force of gravity on an object can be performed without using motion at all—say, by observing the equilibrium stretch of a spring while the object hangs from it under the influence of gravity. Nevertheless, the equality of Eq. (5–3) has been experimentally verified to a very high degree of accuracy.[‡] Thus the force of gravity has the simple form

$$\vec{F}_g = m_g\vec{g} = m\vec{g}. \tag{5–4}$$

FORCE OF GRAVITY

The acceleration $\vec{a}$ of any object at Earth's surface under the influence of gravity alone is then given by the constant vector $\vec{g}$:

$$\vec{a} = \frac{\vec{F}_g}{m} = \vec{g}. \tag{5–5}$$

As described in Chapter 3, the magnitude of $\vec{g}$ at Earth's surface is roughly

$$g = 9.80 \ \text{m/s}^2. \tag{5–6}$$

Experiments show that this value varies by about 1 percent over Earth's surface, with the higher values occurring at the poles. This variation is due to irregularities in the shape and density of Earth and to Earth's rotation.

[†]The force near any astronomical object has the same form and direction, but the magnitude of g is different. We'll learn more about this in Chapter 12.
[‡]The so-called Eötvös experiments verify that the inertial and gravitational masses are the same to 1 part in 10^{12}.

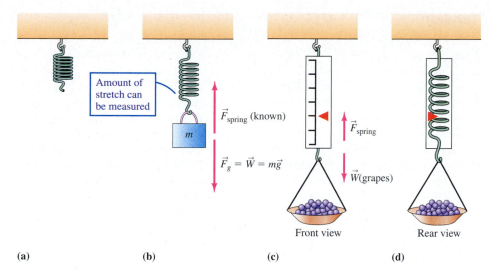

(a) (b) (c) (d)

▲ **FIGURE 5–2** The weight of an object is found by attaching it to a spring, for which the amount of stretch corresponds to a given force. In (a) the spring is unstretched. (b) Calibration (the force corresponding to a given amount of stretch) is mapped out by hanging a series of known masses. (c, d) Once the calibration is done, the weight of any other object can be found.

We noted in Chapter 4 that the force of gravity on an object is commonly called the object's weight, $\vec{W}$,

$$\vec{W} = \vec{F}_g = m\vec{g}. \tag{5–7}$$

WEIGHT

We can experimentally determine the weight of an object by balancing the force of gravity against a second calibrated force such as that exerted by the stretched spring shown in its unstretched state in Fig. 5–2a and stretched, with a mass attached, in Fig. 5–2b. Once we know how much force is exerted for a given stretch, we can find the weight of any object by suspending it from the spring and observing the amount of stretch (Figs. 5–2c and d). In fact, when you step on a bathroom scale, you are indirectly measuring the force of gravity on your body in a similar way; the second force is the calibrated bathroom scale, which is in essence a compressible spring. While we might in everyday conversation sometimes use the words *weight* and *mass* interchangeably, this is not correct. Be careful not to confuse the weight of an object, measured in newtons or pounds, with its mass. We know from Chapter 4 that the mass of an object is the quantity of matter it contains, which is measured by its inertia—its resistance to any change in motion. The mass and weight of an object, though, are numerically proportional to each other through g—the weight of an object equals the object's mass times $\vec{g}$. In other words, the mass is intrinsic to the object, but its weight depends on where it is: If you went to the Moon, where a falling object falls with a different acceleration $\vec{g}_{Moon}$, a 1-kg mass would still be a 1-kg mass, but its weight would have magnitude mg_{Moon} rather than mg_{Earth}.

CONCEPTUAL EXAMPLE 5–1 Let us imagine that the inertial mass of any object is exactly 5 percent larger than its gravitational mass. How would this affect our treatment of the projectile motion?

Answer The force of gravity $\vec{F}_g = m_g\vec{g}$ can be determined by weighing the object. However the two factors cannot be determined without knowing more about gravity (this is the subject of Chapter 12). The acceleration due to the force of gravity is given by

$$\vec{a} = \frac{\vec{F}_g}{m} = \frac{m_g}{m}\vec{g} = \frac{1}{1.05}\vec{g}.$$

It is this quantity that is determined by dropping an object from a given height, and it is this acceleration that has the value 9.81 m/s² on Earth. Furthermore it is this quantity that enters into projectile motion. Thus nothing would be changed, *provided* that the ratio of m to m_g (here 1.05) is a universal constant, and does not change from object to object.

(a)

(b)

Rope tension is transmitted around corners.

(c)

Tension

We know from experience that it is possible to pull objects using ropes. The tension $\vec{T}$ is a force exerted on an object by a flexible rope (or wire or cable or string) that is directed *along* the direction of the rope away from the object. By *flexible*, we mean that the rope goes around corners (by means of frictionless idealized pulleys, a way to avoid the effects of friction), not that it stretches. Tension always pulls, never pushes (Figs. 5–3a, b, and c). For light (negligible-mass) ropes, the *magnitude of the tension is the same everywhere along the rope*. (A rope of negligible mass—the kind of idealized rope we employ in these chapters—is a rope whose mass is small compared to other masses in the problem. When the mass of the rope is not negligible, then the tension will vary along the rope.)

The tension will adjust itself to different values according to the situation. For example, a rope from which a bucket is suspended (Figs. 5–3b and c) has less tension than does the same rope with a piano suspended from it because in the two cases the tension cancels very different weights. If we were able to look inside a taut rope, we would find that tension arises from the molecular forces that hold the rope together and give it flexibility. We do not need to know the microscopic details to be able to use tension in everyday problems.

◀ **FIGURE 5–3** (a) The tension force pulls but cannot push. (b, c) A rope of negligible mass with a given tension can maintain the magnitude of the tension even if the direction of the rope is changed. Tension can be transmitted around corners without change of magnitude because we are assuming an ideal pulley.

EXAMPLE 5–2 Consider Fig. 5–4a, which shows a fishing line hanging from a hook attached to the ceiling. This fishing line is rated as 10-lb-test line, which means that it should hold as long as the tension within it does not exceed 10 lb. A box of mass 2.0 kg is attached to the line. Find the tension in the line. Will the line hold? What will happen if a box of mass 5.0 kg is attached instead?

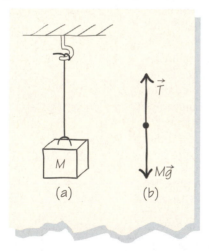

▲ **FIGURE 5–4** (a) Mass hangs from hook using 10-lb-test line. (b) Free-body diagram for a 2.0-kg package suspended from a line.

Setting It Up We are given the mass of the box, which we label as M. If we can calculate the unknown line tension's magnitude T, we can decide whether the line will break. We draw a sketch of the situation (Fig. 5–4a).

Strategy An important first step is to adapt the sketch and make it into a free-body diagram for the box (Fig. 5–4b). With the aid of this diagram we can write Newton's law for the box and use the condition that the mass is stationary—that is, that there is no acceleration, so that the net force is zero—to find T.

Working It Out The free-body diagram shows us that the tension and the force of gravity are the only forces acting on the box. Thus the net force on it is zero:

$$\vec{F}_{\text{net}} = \vec{T} - M\vec{g} = 0.$$

We solve this for T:

$$T = Mg.$$

The sign is positive because we want only the magnitude of the tension force. With $M = 2.0$ kg,

$$T = Mg = (2.0 \text{ kg})(9.8 \text{ m/s}^2) = 19.6 \text{ N}$$

$$= (19.6 \text{ N})\left(\frac{1 \text{ lb}}{4.45 \text{ N}}\right) = 4.4 \text{ lb}.$$

The line will hold. But with $M = 5.0$ kg,

$$T = (5.0 \text{ kg})(9.8 \text{ m/s}^2) = 49 \text{ N} = (49 \text{ N})\left(\frac{1 \text{ lb}}{4.45 \text{ N}}\right) = 11 \text{ lb}.$$

The line may not hold with the heavier mass suspended from it as the tension within it exceeds its breaking tension.

What Do You Think? Why do fishermen use fishing line that is rated much higher than the weight of any fish they are likely to catch? *Answers to **What Do You Think?** questions are given in the back of the book.*

How can we measure the tension within a rope at a given point? One way would be to cut the rope at the given point and insert a calibrated spring scale. The scale will stretch by an amount corresponding to the tension in the rope. This experiment could be performed at different points along the rope to show that the tension is the same throughout.

(a)

(b)

(c)

◀ **FIGURE 5–5** Some normal forces. These forces are always perpendicular to the contact surface, keeping the object on which the force acts from entering the surface.

▼ **FIGURE 5–6** (a) A scale can be used to measure the normal force on an apple. (b) We can draw the forces acting on the apple.

(a)

(b) How to Draw Free-Body Diagrams

1) Draw gravitational force

$\vec{F}_{g,a}$ gravitational force on apple

2) Add normal force

$\vec{F}_N$ normal force on apple from scale

3) Draw free-body diagram

$\vec{F}_N$

$\vec{F}_{g,a}$

4) Apple doesn't accelerate

Therefore, $\vec{F}_{net} = 0$

$\vec{F}_{net} = \vec{F}_{g,a} + \vec{F}_N = 0$

(b)

Normal Force

Let's look at the forces acting on an apple placed on a table (Fig. 5–5). In addition to the force of gravity, a *normal force* $\vec{F}_N$ acts on the apple. In Chapter 4, we were introduced to this force and acknowledged that such a normal force must exist—without it, gravity would cause the apple to accelerate into the surface of the table. The normal force is the result of the complex interaction between the molecules of the table. We call a material that can give rise to a normal force a *solid*. The normal force on an object acts only when the object is in contact with the table. The normal forces on the objects shown in Fig. 5–5 adjust themselves to cancel the components of forces perpendicular to the surface. If the weight is too large, however, the supporting surface will collapse. The normal force pushes but never pulls. It acts perpendicular to and away from the surface at the object's point of contact. If this were not true, an apple on a table would accelerate to one side of the table or the other. In fact, whatever the tilt of the table, the apple does not move into its surface because *the normal force always acts perpendicular to and away from the surface*. Because of this property, the normal force cannot counteract any forces parallel to the surface. For example, the normal force cannot oppose the frictional force parallel to the table's surface in the case of the tilted table in Fig. 5–5b and cannot keep the apple from rolling down the table. All the normal force can do is keep the apple from penetrating the table.

We can see how the magnitude of $\vec{F}_N$ is determined in a situation by the use of Newton's first law. An apple of 0.25 kg sitting on a table must experience an upwardly directed normal force that is equal and opposite to the downward force of gravity; the magnitude of this force is $mg = (0.25 \text{ kg})(9.8 \text{ m/s}^2) = 2.5 \text{ N}$. If a pumpkin of mass 2.5 kg sits on the table, the normal force is 25 N. As we'll see in the examples below, the normal force on the apple will differ in magnitude from the apple's weight if we tilt the table.

An ordinary bathroom scale gives us a way to read directly the magnitude of the normal force. The spring in the scale opposes the force of gravity in exactly the same way as does the normal force and therefore gives the magnitude of the normal force via a calibrated scale. In Fig. 5–6, a scale measures the normal force of 2.5 N that acts on the apple (although the force is often expressed in units other than newtons). The scale could equally well be placed on other surfaces that exert normal forces. For example, the normal force exerted on a hand by a wall can be measured directly if a scale is inserted between the wall and the hand.

There are other forces similar to the tension and the normal force. Among these we could include the support forces supplied by hooks, connection points, bearings, and so forth. The direction and magnitude of such forces are determined by the requirement that the attachment point in question does not accelerate. For example, a hook from which a 20-lb weight is suspended must supply an upward force of 20 lb if the mass is not to fall.

Applying Newton's Laws with Constant Forces

In many everyday situations, the tension and the normal force are constant. Examples involving constant tensions, constant normal forces, and gravity provide us with solvable equations of motion. We shall investigate several such examples here.

It may be helpful first to recall the motion that results when *any* constant net force $\vec{F}$ acts on an object of mass m. The object's velocity and position as a function of time are solutions to Newton's second law, $\vec{F} = m\vec{a}$, and also involve the initial conditions of the motion. In one dimension, where the acceleration, the velocity, and the position have only x-components, these solutions are

$$\text{for constant force: } x = x_0 + v_0 t + \frac{1}{2}at^2 = x_0 + v_0 t + \frac{1}{2}\frac{F}{m}t^2; \quad (5\text{–}8)$$

$$\text{for constant force: } v = v_0 + at = v_0 + \frac{F}{m}t. \quad (5\text{–}9)$$

The constants x_0 and v_0 refer here to the values of position and velocity at time $t = 0$ and are often given or known quantities. For a force with three constant components, we must apply the solution to each component separately. Remember that these formulas apply only when $\vec{F}$ is constant.

EXAMPLE 5–3 A coffee cup of mass 75 g is placed on a slippery (frictionless) ramp tilted at an angle 20° to the horizontal. The coffee cup starts from rest and slides down the ramp. How far down the ramp has the cup moved after 2.0 s?

Setting It Up Figure 5–7a sketches the situation. We label the given mass of the cup as m and the ramp angle as θ. We want the distance x that the cup has moved as a function of time t.

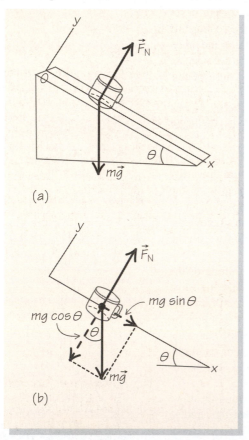

(a)

(b)

▲ **FIGURE 5–7** (a) A coffee cup on a frictionless, tilted ramp. (b) Free-body diagram for the coffee cup, including the vector decomposition of the forces.

Strategy We first prepare a free-body diagram, choose a convenient set of axes, and decompose the forces into their components along these axes (Fig. 5–7b). We then write Newton's second law, which will determine acceleration. At that point kinematic equations can be used to find the displacement. Because there is no friction, the only forces acting on the cup are gravity $\vec{F}_g$ and the normal force $\vec{F}_N$. Note that $\vec{F}_N$ is perpendicular to the surface of the inclined plane and is *not* oriented in the vertical direction.

The origin is placed at the starting point of the coffee cup. We choose the axes shown in Fig. 5–7b because we know from experience the acceleration $\vec{a}$ of the cup will be along the ramp, with no component perpendicular to the ramp. Accordingly, the x-direction points down along the ramp and the acceleration will have only an x-component a_x, $\vec{a} = a_x\hat{i}$. We must then decompose the forces acting along this set of axes and express Newton's second law. While only the component of forces along x will act to accelerate the cup, the constraint that there is *no* acceleration perpendicular to the surface may still be helpful, and so we will write Newton's law for that direction as well. (Although we may not use this in this example, writing equations for all vector components is a good habit to get into and will be necessary when friction becomes an issue.)

Finally, it is always useful to have a way to check our result. Here we can use the fact that we expect no movement if the plane is horizontal.

Working It Out With our choice of axes, the normal force has only a y-component:

$$\vec{F}_N = F_N\hat{j}.$$

The magnitude F_N is as yet unknown, but we will find it below from the requirement that the acceleration has no y-component. The force of gravity, however, has both x- and y-components (Fig. 5–7b):

$$\vec{F}_g = (mg\sin\theta)\hat{i} - (mg\cos\theta)\hat{j}.$$

Newton's second law has the vector form $\vec{F}_g + \vec{F}_N = m\vec{a}$, or

$$(mg\sin\theta)\hat{i} - (mg\cos\theta)\hat{j} + F_N\hat{j} = ma_x\hat{i}.$$

Each component of this equation is a separate equation of motion:

$$\text{for the } x\text{-component: } mg\sin\theta = ma_x; \quad (5\text{–}10)$$

$$\text{for the } y\text{-component: } -mg\cos\theta + F_N = 0. \quad (5\text{–}11)$$

Now we solve the equations of motion. Equation (5–10) gives

$$a_x = g \sin \theta. \qquad (5\text{–}12)$$

Equation (5–11) shows that $\vec{F}_N$ has magnitude $mg \cos \theta$.

To find how the position of the coffee cup changes with time, we use the kinematic equation for constant acceleration in one dimension: Eq. (5–8). In this equation, the cup starts at the origin, so $x_0 = 0$; moreover, $v_0 = 0$ because the cup is initially at rest. We then insert Eq. (5–12) into Eq. (5–8):

$$x = \tfrac{1}{2} a_x t^2 = \tfrac{1}{2}(g \sin \theta) t^2.$$

At $t = 2.0$ s,

$$x = (0.5)(9.8 \text{ m/s}^2)(\sin 20°)(2.0 \text{ s})^2 = 6.7 \text{ m}.$$

We find the cup has moved 6.7 m after 2.0 s. Does this seem far to you? The lack of friction is an important effect! As is typical for problems involving gravity, the answer is independent of the mass of the cup. The angle $\theta = 0°$ presents a special limit; in this case, the plane is horizontal, and there should be no acceleration whatsoever. This is ensured by the $\sin \theta$ factor in the acceleration, which is zero when $\theta = 0°$. In this limit the normal force has magnitude mg, as expected. A second limit is the case $\theta = 90°$, when $\sin \theta = 1$. Here, $a_x = g$ and $F_N = 0$, also as expected.

What Do You Think? It seems unwise to put a coffee cup on an inclined frictionless ramp. What would change in this example if we put other objects—a golf ball or a car—on the ramp?

EXAMPLE 5–4 Because of a wager, a woman wishes to lift a professional football player off his feet. The player is a large interior lineman (a tackle) with a mass of 149 kg. (He weighs 328 lb.) The woman has devised a system for the task, which is shown in Fig. 5–8a. We will assume that all pulleys, ropes, and miscellaneous gear in the apparatus have negligible mass and are frictionless. What is the magnitude of the downward force the woman must exert on the end of the rope in order to lift the lineman?

Setting It Up The given mass of the lineman is M; the mass of the woman is denoted by m. The downward force the woman exerts at the point that the lineman is lifted is just the tension T in the rope, and this is the quantity we want to find.

Strategy Even a smooth lift at constant velocity is sufficient—the lineman does not need to accelerate continuously to be lifted. This is a condition that the net force of the ropes on the lineman balances his weight. The tension in the rope, which is the force that must be supplied by the woman pulling on the rope, is the same throughout the rope.

If we examine Fig. 5–8a, we see that all four rope segments, 1 through 4, pull upward on the lineman. Figure 5–8b illustrates the external forces on the isolated system more clearly. Since it is a single massless rope passing around all the pulleys, the tension in each one is the same, namely T. The free-body diagram for the system is Fig. 5–8c. When the woman begins to pull on the rope, it acquires a tension of magnitude T, and the sum of the tensions $T_1 + T_2 + T_3 + T_4 = 4T$ increases from zero. As long as this sum is less than the weight of the lineman, Mg, he will remain on the ground. But when this sum becomes equal to Mg, there is no net force on the lineman, and any additional tension—no matter how small—will start him accelerating upward. Thus the condition to lift the lineman is

$$4T = Mg.$$

We need only plug in the numbers.

Working It Out We have

$$T = \tfrac{1}{4} Mg = \tfrac{1}{4}(149 \text{ kg})(9.80 \text{ m/s}^2) = 365 \text{ N}.$$

This is the magnitude of the smallest downward force that the woman must apply to her end of the rope. Provided that she weighs more than 365 N (82.1 lb), which corresponds to a mass $m = W/g = 37.3$ kg, she can apply this force just by hanging on the rope. (If she weighs more than 82.1 lb, then she can "partially" hang from the rope, meaning that her feet don't completely lose contact with the ground.)

The arrangement described here is called a *block and tackle* and is used, for example, to enable a single person to lift an engine from a car.

What Do You Think? Why did we make the assumption that all pulleys, ropes, and miscellaneous gear in the apparatus have negligible mass and are frictionless?

(a)

(b)

(c)

◀ **FIGURE 5–8** (a) A lineman lifted by a rope via a system of pulleys. (b) The lineman is clearly isolated, with the external forces shown. (c) Free-body diagram for the lineman.

In Examples 5–5 and 5–6 we investigate further all three forces discussed so far in this section—tension, the normal force, and gravity. These examples also introduce something new, in that two masses are involved. Each mass requires an identification of the forces acting on it, a free-body diagram, and an equation of motion.

EXAMPLE 5–5 Masses $m_1 = 1.1$ kg and $m_2 = 2.3$ kg are attached to opposite ends of a massless rope draped over a pulley (Fig. 5–9a). (This device is called an *Atwood machine*.) Mass m_2 rests on a scale that measures the normal force exerted on m_2. There is no motion. What is the reading on the scale, and what is the tension in the rope?

Setting It Up We want the rope tension, magnitude T, and the normal force, magnitude F_N, that the scale exerts on m_2.

Strategy We start with a free-body diagram for each mass (Fig. 5–9b), including a coordinate, here y, and find the equations of motion for each mass. Because there is no acceleration, these equations express the fact that net force on each mass is zero. All three forces—gravity, tension, and normal force—are vertical, so we have only one component (the y-component) to consider and we can drop the vector notation. There are two masses, hence two equations, and these should be enough to determine the two unknown forces.

Working It Out Take the upward direction as positive. The forces acting on m_1 are gravity $(-m_1 g)$ and the rope tension (T), and the net force on m_1 is

$$F_{\text{net}, 1} = -m_1 g + T = 0. \qquad (5\text{–}13)$$

Three forces act on mass m_2: $-m_2 g$, T, and F_N. Notice that the same value of T acts on each mass, pulling (acting upward) in each case. The net force on m_2 is

$$F_{\text{net}, 2} = -m_2 g + T + F_N = 0. \qquad (5\text{–}14)$$

We then can solve Eqs. (5–13) and (5–14) for the tension and the normal force (the scale reading). Solving Eq. (5–13),

$$T = m_1 g = (1.1 \text{ kg})(9.8 \text{ m/s}^2) = 11 \text{ N}.$$

If we insert this value of T into Eq. (5–14), we can solve for the normal force and hence the scale reading:

$$F_N = m_2 g - T = m_2 g - m_1 g = (m_2 - m_1)g$$
$$= (2.3 \text{ kg} - 1.1 \text{ kg})(9.8 \text{ m/s}^2)$$
$$= (1.2 \text{ kg})(9.8 \text{ m/s}^2) \cong 12 \text{ N}.$$

For comparison, the weight of m_2 alone is $(2.3 \text{ kg})(9.8 \text{ m/s}^2) = 23$ N.

(a)

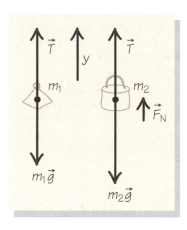

(b)

▲ **FIGURE 5–9** (a) An Atwood machine. (b) Free-body diagrams for the masses.

What Do You Think? What would the relation between the masses have to be for the scale to read zero?

EXAMPLE 5–6 Consider Fig. 5–10: Two masses $m_1 = 1.00$ kg and $m_2 = 2.00$ kg are connected by a rope that passes over an ideal pulley. Mass m_1 hangs straight down, while m_2 slides without friction on a ramp inclined at an angle θ. At $t = 0$, the system is started from rest in the position shown in Fig. 5–10a. Describe the motion of the two masses for $\theta = 25.0°$. Find the angle θ' for which the system remains motionless.

Setting It Up We are given two masses with forces acting on them. One of those forces, the rope tension, is unknown. We want to find the motion of the masses, that is, their acceleration, from which we can work out position. We also want to describe a geometry for which there is no acceleration.

Strategy In this problem, we again have two masses for which we must write separate equations of motion: $\vec{F}_{\text{net on 1}} = m_1 \vec{a}_1$ and $\vec{F}_{\text{net on 2}} = m_2 \vec{a}_2$. As they are connected by a rope, the two masses "move together" (the rope is assumed to be inextensible), and in particular the *magnitudes of the accelerations* are the same, $a_1 = a_2 = a$. We begin by drawing a free-body diagram for each mass (Fig. 5–10b). A force of common magnitude acts on each mass; this force is the tension $\vec{T}$. The tension and gravity $m_1 \vec{g}$ act on m_1. Three forces act on m_2: the normal force $\vec{F}_N$, tension, and gravity $m_2 \vec{g}$.

Once the free-body diagrams are drawn, we choose coordinate systems and write Newton's second law. Remember that the coordinate systems we choose are for our convenience and do not affect the result, and we *do not have to choose the same coordinate system*

(a)

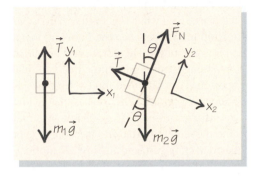

(b)

▲ **FIGURE 5–10** (a) Two masses connected by a rope via a pulley. (b) Free-body diagrams for masses m_1 and m_2.

for the two masses. In fact, choosing different systems makes this problem easier to solve (Fig. 5–10b). We will have to count equations and determine if there are enough to solve for the unknowns, in particular a.

As for finding the angle for which $a = 0$, if we had found enough equations to allow us to solve for a in the first part, we could see whether there is an angle for which $a = 0$. This is much simpler if we leave our expressions in algebraic form, a very good general rule.

Working It Out We choose the origin of each coordinate system to be at the location of the respective mass at $t = 0$. Because m_1 moves only in the vertical direction, labeled y_1, its acceleration $\vec{a}_1$ is aligned with y_1. For m_1, then, we need look only at the component of Newton's second law along y_1,

$$T - m_1 g = m_1 a_1. \quad (5\text{–}15)$$

Although the mass m_2 has forces acting in two directions, it moves only in the direction labeled x_2 in Fig. 5–10b. Thus $\vec{a}_2$ is aligned with x_2. We must now decompose the forces acting on m_2 into components in this coordinate system. The normal force $\vec{F}_N$ is in the $+y_2$-direction and $\vec{T}$ is in the $-x_2$-direction. The third force, gravity, has two components; we can determine these components by recalling from geometry that the angle θ indicated in Fig. 5–10b is the same as the ramp angle θ in Fig. 5–10a. Then the force of gravity $m_2\vec{g}$ has x_2-component $m_2 g \sin \theta$ and y_2-component $-m_2 g \cos \theta$. Thus Newton's second law for mass m_2, $\vec{F}_{\text{net on 2}} = \vec{F}_N + \vec{T} + m_2\vec{g} = m_2\vec{a}_2$, breaks down into two component equations:

for the x_2-component: $-T + m_2 g \sin \theta = m_2 a_2$; (5–16)

for the y_2-component: $F_N - m_2 g \cos \theta = 0$. (5–17)

The three equations (5–15), (5–16), and (5–17) are not enough to solve for the four unknowns T, F_N, a_1, and a_2. The needed fourth equation is the expression that states that the the two masses have accelerations of the same magnitudes,

$$a_1 = a_2 \equiv a. \quad (5\text{–}18)$$

The single acceleration magnitude a is then substituted in Eqs. (5–15), (5–16), and (5–17), which become three equations for the three unknowns a, T, and F_N. Equation (5–17) gives the normal force,

$$F_N = m_2 g \cos \theta.$$

The tension cancels in the sum of Eqs. (5–15) and (5–16):

$$T - m_1 g - T + m_2 g \sin \theta = m_1 a + m_2 a.$$

We are left with an equation for a with solution

$$a = \frac{m_2 \sin \theta - m_1}{m_1 + m_2} g. \quad (5\text{–}19)$$

Now that we have found the accelerations we can find positions as a function of time by using the constant-acceleration equations (5–8). Each mass starts from rest at the origin of its respective coordinate system, so $v_0 = 0$ for both masses, and both y_{10} ($\equiv y_1$ at $t = 0$) and x_{20} ($\equiv x_2$ at $t = 0$) are zero. Thus Eq. (5–8) gives

for m_1: $y_1 = \frac{1}{2} at^2$;

for m_2: $x_2 = \frac{1}{2} at^2$.

The masses move only in these directions. Together with Eq. (5–19), these equations describe the motion fully.

For the case $\theta = 25.0°$, we use $\sin 25.0° = 0.423$, so

$$a = \frac{(2.00 \text{ kg})(0.423) - 1.00 \text{ kg}}{1.00 \text{ kg} + 2.00 \text{ kg}}(9.80 \text{ m/s}^2) = -0.503 \text{ m/s}^2.$$

Thus $y_1 = (-0.503 \text{ m/s}^2)t^2/2 = x_2$. Note the minus sign in the acceleration: The sign indicates that m_1 drops and m_2 moves up the ramp. Equation (5–19) shows that the acceleration—including its sign—depends on the masses and the ramp angle.

Finally, to find the angle for which the acceleration is zero, we note that, according to Eq. (5–19), the acceleration is zero at an angle θ' for which

$$m_2 \sin \theta' - m_1 = 0, \quad (5\text{–}20)$$

or $\sin \theta' = m_1/m_2$. For this particular problem the forces will balance and acceleration will be zero for $\sin \theta' = (1.00 \text{ kg})/(2.00 \text{ kg})$, or $\theta' = 30.0°$.

What Do You Think? Consider the system at the angle $\theta' = 30.0°$. (a) What happens for masses $m_1 = 3.0$ kg and $m_2 = 6.0$ kg? (b) What happens for masses $m_1 = 4.0$ kg and $m_2 = 6.0$ kg? (c) What happens for masses $m_1 = 3.0$ kg and $m_2 = 5.0$ kg?

5–2 Friction

Friction is a familiar concept. It is a contact force that impedes sliding, and we experience it in all aspects of our lives. Sometimes friction is useful to us: It is friction that holds nails and screws in place (Fig. 5–11); if there were no friction between our feet and the ground, we could not walk, and if there were no friction between the wheels of a car and the road, the engine would cause the wheels to spin but there would be no forward or backward motion. In other situations friction is not a desirable phenomenon and we do our best to minimize it. Even with the oil added to a car's motor to reduce frictional forces in the engine, as much as 20 percent of gasoline consumption goes to overcome friction in the engine. Lubrication reduces friction and therefore also reduces surface wear—automobile engines now tend to last longer because internal friction has been reduced through more precise manufacturing and more effective lubrication. We have to take into account the forces of friction if we are to understand any realistic mechanics problem.

Static and Kinetic Friction

Suppose you want to slide a crate full of books from one place to another. You push on it with a small horizontal force but nothing happens. Even when you push as hard as you can, the crate does not move. Why not? **Static friction** acts between the floor and the crate in the absence of motion in such a way as to *prevent* motion. This force must be variable because it balances each of your own different-strength pushes. Suppose that you finally get the crate moving with the help of another person. The combined force overcomes the static friction because *static friction has a maximum magnitude*. Friction has to do with the interlocking of microscopically rough surfaces (for more see p. 133), and the external force that overcomes the interlocking and gets the movement started has to be large enough to bend or break the tiny protuberances on the contacting surfaces that impede motion.

▲ **FIGURE 5–11** Nails are held in place by the force of friction, which can be quite substantial.

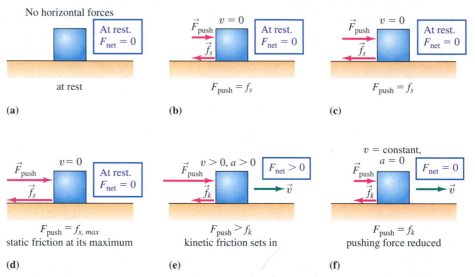

(a) at rest $F_{push} = f_s$ (b) $F_{push} = f_s$ (c)

▶ **FIGURE 5–12** Sequence in which an increasing pushing force is opposed by an equal and opposite static friction force. Static friction can increase only to a certain point, after which the crate accelerates. Kinetic friction, which applies when the crate actually moves, is smaller than the maximum size of static friction, so it takes less pushing force to move the crate once its motion has begun.

$F_{push} = f_{s, max}$ static friction at its maximum (d)

$F_{push} > f_k$ kinetic friction sets in (e)

$F_{push} = f_k$ pushing force reduced (f)

Once the crate is moving, it is easier to keep it moving at a constant speed. There is still friction opposing your push, but it is now **kinetic** (or **sliding**) **friction**, that is, friction associated with motion. Experiment shows that the magnitude of kinetic friction is smaller than the maximum value of static friction. The entire sequence of getting the crate started and keeping it moving is illustrated in Fig. 5–12. Static friction acts in a direction *opposite* to the component of an applied force *along* the surface; sliding friction acts *opposite* to the direction of the velocity of a sliding object at its point or points of contact. In each case the friction force is parallel to the surface (Fig. 5–13a). Figure 5–13b illustrates the magnitude of friction when an object such as the crate starts from rest and is pushed with a steadily increasing external force. The magnitude of static friction increases as the external force increases until the "break point" arrives, at which point kinetic friction enters. This force does not vary with the external pushing force. When the applied force decreases to zero, kinetic friction continues to act until the body stops. At that point we revert to the static situation.

▶ **FIGURE 5–13** (a) Friction opposes the motion of the crate. (b) With a steadily increasing push, static friction will increase in magnitude until the crate starts to move, at which point kinetic friction, whose magnitude does not vary with the pushing force, takes over.

(a)

(b)

CONCEPTUAL EXAMPLE 5–7 In the discussion above we stated that a large enough force has to be applied to overcome static friction, and the applied force must be large enough to bend or break the small protuberances that keep a surface from being perfectly flat. Given the fact that measurements of the coefficient of static friction between a block of wood and a table give the same answer after many experiments, would you argue in favor of bending or breaking the surface imperfections?

Answer If static friction were due primarily to obstacles that have to be broken in order to facilitate motion, then after a few experiments all the obstacles would have been broken off, and static friction would decrease significantly. Because this does not happen, the bending of deformations must be the primary source of friction. We should add that if you actively smooth a surface by, for example, sanding it, then you are presumably doing some breaking of the protuberances on it.

Quantitative Properties of Friction

Quantitative tests on friction were made by Leonardo da Vinci some 200 years before Newton's work on dynamics. Leonardo experimented with a set of blocks of varying sizes sliding on table tops and discovered some surprising facts. He found that both static and kinetic friction are independent of the surface area of the blocks in contact with the table top. Moreover, both *static and kinetic friction are proportional to the magnitude F_N of the normal force exerted by the table top on the blocks*. The experiments that led Leonardo to his conclusions are quite simple: Take a given block and turn it so that faces of different areas are in contact with the table top. The friction force on the block is the same no matter what face is down.

The proportionality constant that relates the friction force and the normal force is the **coefficient of friction** μ. This (positive dimensionless) constant is determined experimentally. As the maximum value of static friction is generally not equal to the force of kinetic friction, we distinguish two coefficients: μ_s for static friction and μ_k for kinetic friction.

If we write the force of static friction as $\vec{f}_s$ and that of kinetic friction as $\vec{f}_k$, their magnitudes are given by

$$0 \le f_s \le \mu_s F_N; \tag{5–21}$$

STATIC FRICTION

$$f_k = \mu_k F_N. \tag{5–22}$$

KINETIC FRICTION

Equation (5–21) expresses a range because, as we have described above, static friction takes a value that depends on the external conditions. The experimental fact that the maximum value of static friction exceeds kinetic friction implies the inequality

$$\mu_s > \mu_k. \tag{5–23}$$

You experience this when you have to exert a greater force to get a crate of books moving than you have to exert to keep it moving. We also make the assumption, reasonably well satisfied by experiment, that μ_k *is independent of the relative speed of the two surfaces.*

The coefficients of friction depend on the two surfaces involved. We know from everyday experience that a basketball shoe on a basketball court involves a larger coefficient of friction than does the blade of an ice skate on a frozen lake. A lubricating material—such as sweat—between the basketball shoe and the court will drastically reduce the coefficient of friction. In ice skating, the lubricating material is a layer of liquid water between blade and ice (Fig. 5–14a); an ice skater can make static friction take over by adroit use of the blade—in Fig. 5–14b the skater is pushing off or making a turn by digging into the ice rather than gliding over it. Table 5–1 shows some typical values of coefficients of static friction. The materials listed are generally unlubricated ("dry"). Coefficients of kinetic friction can be anywhere from roughly 25 percent to 100 percent of the corresponding coefficients of static friction.

The values in Table 5–1 are meant only to be indicative—the coefficients of friction are sharply dependent on such things as the cleanliness of the surfaces, their roughness, and so forth. Two very rough objects may have a large coefficient of friction that can be reduced once the objects are smoothed. But if two objects of the same material

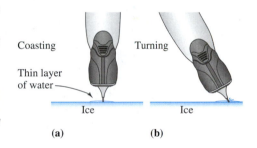

▲ **FIGURE 5–14** (a) A layer of water lubricates and decreases the action of kinetic friction for an ice skater. (b) By manipulating the blade of the skate, a skater makes static friction act in the sideways direction and can make a turn.

TABLE 5–1 • Some Coefficients of Static Friction	
Materials	μ_s
Automobile brake shoes on a brake drum	1.2
Dry tire on dry asphalt	1.0
Hard steel on hard steel	0.8
Oak on oak, parallel to the grain	0.6
Book on a table	0.3
Wet tire on wet asphalt	0.2
Ice on wood	0.05
Teflon on steel	0.04

are smoothed too much and are free of dirt and oxidation as well, the coefficient of friction may rise virtually to infinity because the surfaces weld together! In effect, the molecules at the surface between the two objects can interact just as they do at the interior, and the two objects become one.

Example 5–8 illustrates one method of measuring the coefficient of static friction.

EXAMPLE 5–8 A box of mass m is set at rest on a horizontal surface; there is friction between the box and the surface. The surface is slowly raised at one side so that it becomes a ramp, making an angle θ with the horizontal. By analyzing the forces on the box, find the critical ramp angle θ_c at which the box will start to slide.

Setting It Up Figure 5–15a illustrates the situation. We include in the figure a coordinate system that simplifies the solution: The $+x$-direction is oriented down along the ramp and the y-direction is perpendicular to the ramp. The origin is placed at the initial position of the mass.

(a)

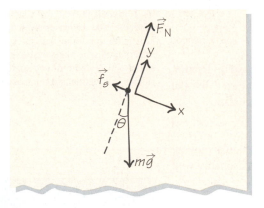

(b)

▲ **FIGURE 5–15** (a) Box of mass m experiencing a friction force on a ramp. The ramp angle θ at which the mass starts to slide yields the coefficient of static friction. (b) Free-body diagram for the box.

Strategy The free-body diagram is Fig. 5–15b. The three forces acting on the box before it starts to move are the force of gravity (vertically down), the normal force $\vec{F}_N$ (perpendicular to the surface), and the force of static friction, $\vec{f}_s$ (up the ramp). The friction force opposes the motion that would take place if there were no friction. The

magnitude of the force of static friction is determined by the fact that the mass remains motionless for a sufficiently shallow ramp angle. The maximum value of static friction depends on F_N [see Eq. (5–21)], so we must also find its value. We can use the fact that there is no acceleration perpendicular to the ramp to find F_N.

As the ramp angle increases, the components of the normal and gravity forces along and perpendicular to the ramp change. As long as the box is not slipping (there is no net force—we refer to the box as being in *static equilibrium*), the friction force (along the ramp) is static friction. Once the value of f_s reaches its maximum value $\mu_s F_N$, however, the box will slip, and this determines the critical value of the ramp angle θ_c.

Working It Out We start with the box stationary, so that Newton's second law with zero acceleration reads $\vec{F}_g + \vec{F}_N + \vec{F}_s = 0$, or

$$(mg \sin \theta)\hat{i} - (mg \cos \theta)\hat{j} + F_N\hat{j} - f_s\hat{i} = 0.$$

In component form, the equations of motion are

for the x-component: $mg \sin \theta - f_s = 0;$ (5–24a)

for the y-component: $F_N - mg \cos \theta = 0.$ (5–24b)

The x-component equation relates the force of static friction to the force of gravity,

$$f_s = mg \sin \theta. \quad (5–25)$$

The y-component equation determines F_N as a function of θ:

$$F_N = mg \cos \theta. \quad (5–26)$$

As θ (and $\sin \theta$) increases, the force of friction from Eq. (5–25) that is needed to hold the box in static equilibrium also increases. Eventually, static friction reaches its maximum value, $\mu_s F_N$. Beyond that point, the box will start to slide. Setting static friction to its maximum value, $f_s = \mu_s F_N$, Eq. (5–25) determines the critical angle θ_c at which the box starts to slide. Equation (5–25) then becomes

$$\mu_s F_N = mg \sin \theta_c,$$

or, from Eq. (5–26),

$$\mu_s mg \cos \theta_c = mg \sin \theta_c.$$

Cancel the factor mg from this equation to find θ_c:

$$\frac{\sin \theta_c}{\cos \theta_c} = \tan \theta_c = \mu_s. \quad (5–27)$$

This equation tells us that if we measure the angle at which the box begins to slip, we measure μ_s, and this is in fact a useful way to determine coefficients of static friction between surfaces.

What Do You Think? Imagine that a person pushes down on the box in a direction opposite to the normal force. How does this change the frictional force? What happens to the critical angle?

Example 5–8 suggests a related experiment that will measure the coefficient of kinetic friction. Once the box begins to slip, kinetic friction acts, given by $f_k = \mu_k F_N = \mu_k mg \cos \theta$. If we use Newton's second law and apply it along the x-direction, then

instead of Eq. (5–24a), we find

$$mg \sin \theta - \mu_k mg \cos \theta = ma_x. \qquad (5\text{–}28)$$

Because μ_k is smaller than μ_s, the ramp can be lowered back down, decreasing θ while the mass is still sliding, and the box will continue to slide. There is a second critical value of θ—call it θ_c'—for which the forces in Eq. (5–28) cancel and the object no longer accelerates. Instead, the object slides at constant velocity. This critical angle is given by

$$\frac{\sin \theta_c'}{\cos \theta_c'} = \tan \theta_c' = \mu_k. \qquad (5\text{–}29)$$

Thus θ_c' measures the coefficient of kinetic friction.

EXAMPLE 5–9

A professor with a light eraser (assume massless) in her hand leans against a blackboard. Her straight arm makes an angle of 60° with the horizontal, and the force $\vec{F}_{\text{prof}}$ exerted by her arm on the eraser has magnitude $F_{\text{prof}} = 50$ N. The coefficient of static friction between the eraser and the blackboard is $\mu_s = 0.15$. Does the eraser slip?

Setting It Up We sketch the situation in Fig. 5–16b, and we can use this figure as a free-body diagram for the eraser. We choose a coordinate system with x into the board and y vertically upward. We are interested in finding the force of friction between the board and the eraser. If that value exceeds the maximum value of static friction, the eraser will slip.

Strategy After drawing the free-body diagram to allow us to pick out the forces acting, we apply Newton's law of motion to the eraser under the assumption that it is not slipping. The components of that equation should allow us to solve for the magnitude of the friction force to find out if it exceeds its maximum value.

Working It Out If there were no friction (a perfectly slippery board), the eraser would slide up. Therefore, the static friction force must be down, in the $-y$-direction. As long as there's no acceleration, Newton's first law applies to the eraser, $\vec{F}_N + \vec{f}_s + \vec{F}_{\text{prof}} = 0$, and we have

$$-F_N \hat{i} - f_s \hat{j} + (F_{\text{prof}} \cos \theta) \hat{i} + (F_{\text{prof}} \sin \theta) \hat{j} = 0.$$

The two component equations are (θ is the angle with respect to x, here 60°)

for the x-component: $-F_N + F_{\text{prof}} \cos \theta = 0$;

for the y-component: $-f_s + F_{\text{prof}} \sin \theta = 0$.

The x-component equation determines F_N from the requirement that it balances the horizontal component of the professor's force $\vec{F}_{\text{prof}}$. Once we have found that $F_N = F_{\text{prof}} \cos \theta$, we can determine the *maximum* value of static friction, $f_{s,\text{max}} = \mu_s F_N = \mu_s F_{\text{prof}} \cos \theta$. When this maximum value is exceeded, the eraser begins to slip. Thus, when we substitute the maximum value of static friction into the y-component equation, we find a condition for the critical angle θ_c for which the eraser begins to slip:

$$-\mu_s F_{\text{prof}} \cos \theta_c + F_{\text{prof}} \sin \theta_c = 0;$$

$$\frac{\sin \theta_c}{\cos \theta_c} = \tan \theta_c = \mu_s.$$

Note the striking feature that the critical angle at which the eraser starts to slip is independent of the force the professor exerts! Numerical substitution yields $\tan \theta_c = 0.15$, or $\theta_c = 8.5°$. This angle is less than the 60° angle made by the arm, so the eraser indeed slips.

What Do You Think? Suppose that the eraser were not massless. Is the critical angle θ_c still independent of the professor's force?

(a)

(b)

▲ **FIGURE 5–16** (a) Professor erases the blackboard. (b) Sketch of forces and free-body diagram for the eraser.

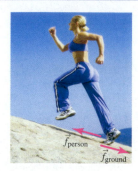

▲ **FIGURE 5–17** A runner uses the force of static friction between shoe and ground to accelerate forward.

We have stated that friction acts against motion, but at the start of the chapter we also stated that the presence of friction allows us to walk and cars to accelerate. Friction can act to produce a positive acceleration on us. How is this possible? The answer lies in the fact that people are extended and flexible systems and for such systems static friction can act *forward on the system as a whole*, even though it will always act to *impede motion at the point of contact*. Walking or running is perhaps the most familiar example of this phenomenon (Fig. 5–17). The runner is exercising a muscle that pushes the contact foot backward, and without friction the motion of the foot at the point of contact would therefore be backward. (You can easily test this by trying to walk on a frozen puddle.) But friction opposes the backward motion at the point of contact and therefore is a force that acts in the *forward* direction on the runner's foot. With sufficient stiffness in the runner, this forward friction force acts on the entire system to move her forward. By Newton's third law, there is a corresponding backward-directed force on Earth. An automobile similarly moves forward because of static friction, as we describe more quantitatively in Example 5–10.

EXAMPLE 5–10

An automobile with four-wheel drive and a powerful engine has a mass of 1000 kg. Its weight is evenly distributed on its four wheels, whose coefficient of static friction with the dry road is 0.8. If the car starts from rest on a horizontal surface, what is the greatest forward acceleration that it can attain without spinning its wheels?

Setting It Up The situation is illustrated in Fig. 5–18a. The car mass is m and the coefficient of static friction between the wheels and the road is μ_s. The unknowns are the car's acceleration, magnitude a, and in particular maximum magnitude a_{max}.

Strategy It is static friction, magnitude f_s, that accelerates the car forward, and the automobile has its greatest forward acceleration when f_s is at its maximum. The engine of the car creates the rotational motion shown in Figure 5–18a in the wheels. If there were no friction between the wheels and the road, the wheels would simply spin. When there is no slipping, it is *static* friction that acts, and μ_s, not μ_k, enters the problem. The motion of the tires at the point of contact with the road is to the rear in the absence of friction. The force of friction opposes this rearward motion, so the direction of $\vec{F}_s$ is toward the front of the car. The forward frictional force is the *only* external horizontal force acting on the car and hence determines its forward acceleration.

Figure 5–18b shows the forces acting on the car, and Fig. 5–18c is the free-body diagram. Using it, we can write Newton's second law:

$$\vec{F}_N + m\vec{g} + \vec{f}_s = m\vec{a}.$$

(There is actually a separate normal force and friction force at each wheel, but because the weight is distributed evenly over each wheel, all the normal and friction forces at each wheel are equal.) The acceleration is maximum when f_s is a maximum. To find it, we'll need to break the Newton's law equation into components and find the magnitude of the normal force.

Working It Out With the coordinate system in the figure, $\vec{F}_N = F_N\hat{j}$, $m\vec{g} = -mg\hat{j}$, $\vec{f}_s = f_s\hat{i}$, and the acceleration is forward, so $\vec{a} = a\hat{i}$. Newton's second law is then

$$F_N\hat{j} - mg\hat{j} + f_s\hat{i} = ma\hat{i},$$

and the component equations are

for the x-component: $f_s = ma$;

for the y-component: $N - mg = 0$.

Force of static friction on rear and front tires. The same force acts on the left side tires.

(a)

(b)

(c)

▲ **FIGURE 5–18** (a) A car accelerates forward under the influence of static friction. (b) Forces acting. (c) Free-body diagram for the car.

We find a_{max} when static friction is at its maximum value $\mu_s F_N = \mu_s mg$, where we have used $F_N = mg$ from the y-component equation. Thus

$$\mu_s mg = ma_{max}.$$

The mass of the car cancels out of this expression, leaving

$$a_{max} = \mu_s g = (0.8)(9.8 \text{ m/s}^2) \cong 8 \text{ m/s}^2.$$

This is quite a significant forward acceleration. Note that as μ_s decreases, the maximum acceleration decreases; in other words, when μ_s is zero, the automobile can only spin its wheels.

What Do You Think? Cars with front-wheel drive perform better in snow than those with rear-wheel drive. Why do you suppose this is?

THINK ABOUT THIS . . .

WHERE DOES FRICTION COME FROM?

To understand the origins of friction, let's look closely at two surfaces that rub against one another. Figure 5–19 shows a microscopic view of two such surfaces. Because of the hills and valleys present on any rough surface—and all surfaces are rough when viewed closely enough—the area on two surfaces that actually touch together is a small fraction of the area that appears to be in contact. Friction forces are due to three major effects: the interlocking of surface irregularities, the attraction between the contact points due to forces between the molecules of the two objects (the objects "adhere"), and the "plowing out" of softer materials by harder ones. The coefficient of static friction can be greater than that of kinetic friction because the materials have a longer time to "settle in" together.

This description helps us understand why the friction force is independent of the apparent surface area that is in contact while it is dependent on the normal force. The normal force is a measure of how strongly the two surfaces are pressed together; when the normal force is large, the two surfaces are pressed strongly together. The rough surfaces shown in Fig. 5–19 mesh more closely when pressed strongly together, and the *actual* surface area in contact increases. In fact, the normal force is a good measure of the actual surface contact area. Whether we place the broad side or the narrow end of a brick on a table, approximately the same surface areas are in actual contact, even if the apparent contact areas are vastly different. So the friction force comes from the interaction of the two surfaces at the atomic level and is proportional to the real area that is in contact at this microscopic level.

The study of friction, wear, and lubrication is called *tribology*, a subject of obvious importance. Despite much effort, a truly fundamental understanding of friction remains elusive. The discovery of Teflon™—a very slippery coating material that you have likely seen in frying pans—was a happy accident, not the result of a planned development program. The discoverer has stated that he was lucky he was not blown up in the process.

(a)

(b)

▲ **FIGURE 5–19** If we use both of the figures chosen: (a) The two surfaces appear smooth, (b) but a microscopic view of the contact region reveals rough surfaces.

5–3 Drag Forces

A spoon dropping through molasses, an automobile moving at highway speeds, and the space shuttle using a parachute to slow down during landing are all subject to a substantial *drag force*, which is a resistive force somewhat like friction. Drag forces act like sliding friction in that *they act in a direction opposite to that of the motion*, but they differ from sliding friction in two ways: They depend on the *speed v* of the object that is moving through the medium and there is no equivalent to a normal force to set their magnitude.

In many everyday situations, an automobile moving on a highway, for example, the drag force $\vec{F}_D$ is found by experiment to have magnitude

$$F_D = \tfrac{1}{2}\rho A C_D v^2, \tag{5–30}$$

where ρ is the mass density (the mass per unit volume) of the medium through which the object moves, A is the maximum cross-sectional area presented by the moving object, and C_D is the *drag coefficient*. The drag coefficient is dimensionless and depends on the object's shape. A highly streamlined object might have a drag coefficient as small as 0.1, whereas a particularly awkward shape will have a drag coefficient greater than 1. The most streamlined automobiles have drag coefficients around 0.25.

Terminal Speed

The fact that the drag force on an object increases with the speed has an important consequence. Consider a parachutist falling through the air (Fig. 5–20) and acted upon by both gravity and a drag force, as described by Eq. (5–30). When the parachutist first starts to fall,

▲ **FIGURE 5–20** When the drag force and the force of gravity acting on a parachutist are equal and opposite, the parachutist has reached his terminal speed.

his speed is slow, so the drag force is small and there is an acceleration $\vec{g}$ due to gravity. As the parachutist's speed increases, so does the drag; at some point the drag will be equal and opposite to the force of gravity. Then the parachutist no longer accelerates and his speed remains constant at a maximum value. The maximum speed v_t is called the *terminal speed.* Some typical values are 5 m/s for a parachutist, 8 m/s for a falling raindrop, 40 m/s for a baseball, and 60 m/s for a skydiver (parachutist with unopened parachute).

Let's calculate the terminal velocity in the case of falling objects, where only the force of gravity and the drag force [Eq. (5–30)] act. The terminal speed is found from the condition that the forces of gravity and drag are equal and opposite: $\vec{F}_g = -\vec{F}_D$, or in terms of magnitudes

$$mg = \tfrac{1}{2}\rho A C_D v_t^2 .$$

We solve for the terminal speed:

$$v_t = \sqrt{\frac{2mg}{\rho A C_D}}. \tag{5–31}$$

CONCEPTUAL EXAMPLE 5–11 Consider a skydiver falling through the atmosphere. How can he change his terminal speed? When several skydivers jump out of the same airplane and want to connect together, how do they all get to the same location (Fig. 5–21)?

Answer The terminal speed is attained when the magnitude of the drag force equals that of gravity. The skydiver cannot change the force of gravity on himself, but he can change the drag force and subsequent terminal speed by varying the area A and the drag coefficient C_D [see Eq. (5–31)]. Thus skydivers can make themselves compact to increase their speed through the air or spread their arms and legs to slow down. Skydivers can even change direction slightly by varying their drag characteristics, and their clothes may be loose so that they can be "deployed." For many skydivers to meet up together and form a chain or circle, some skydivers have to increase their terminal speeds and others must decrease theirs.

▲ **FIGURE 5–21** These skydivers had to change their terminal speeds and directions in order to arrive at the same location at the same time and link up.

EXAMPLE 5–12 The maximum force with which a certain automobile engine can accelerate a car is 3200 N. The density of air is about 1.2 kg/m^3, the cross-sectional area of the automobile is 3.4 m^2, and the drag coefficient is 0.50. Assuming that the drag force of Eq. (5–30) is the only force resisting the motion, what is the maximum speed of the automobile?

Setting It Up Figure 5–22a shows the forces, with Fig. 5–22b the free-body diagram for the automobile. The term v_t labels the terminal speed and that is what we wish to find.

Strategy The forces on the car are $\vec{F}_{\text{engine}}$ and $\vec{F}_{\text{drag}}$. (Friction from the road actually propels the car forward, but this occurs because of forces provided by the engine, so we will refer to this force as $\vec{F}_{\text{engine}}$.) These must be equal and opposite at terminal speed, when there is no further acceleration. Since the two forces point in opposite directions, this is in fact a condition that the force magnitudes be equal. Everything is known in the condition except for v_t, and we can solve for it.

Working It Out Our condition reads $F_{\text{engine}} = F_{\text{drag}}$, or, using Eq. (5–30) for the drag force,

$$F_{\text{engine}} = \tfrac{1}{2}\rho A C_D v_t^2 .$$

We solve this equation for the terminal speed squared:

$$v_t^2 = \frac{2F_{\text{engine}}}{\rho A C_D} = \frac{2(3200 \text{ N})}{(1.2 \text{ kg/m}^3)(3.4 \text{ m}^2)(0.50)} = 3.1 \times 10^3 \text{ m}^2/\text{s}^2 .$$

The terminal speed of the automobile is then

$$v_t = 56 \text{ m/s}, \qquad \text{or } 125 \text{ mi/h}.$$

What Do You Think? In what ways would reducing the drag coefficient of automobiles be useful? Have drag coefficients for cars changed over the years?

(a)

▲ **FIGURE 5–22** (a) Car moving on road. (b) Free-body diagram for an automobile that moves under the influence of the force provided by the engine and the drag of the air through which it moves.

In Example 5–12, air drag was used to determine the maximum speed of an automobile. Road friction, sometimes referred to as rolling friction, also plays an important role in opposing the motion of an automobile. Road friction is fairly constant over a large range of speeds and, as a rule of thumb, road friction has an effect equal to that of drag at about 40 mi/h. As the speed increases, the relative size of drag compared to road friction increases.

A Better Approximation to Drag Force

Precise experiments reveal that the force law for the drag force depends in a more complicated way on the medium through which an object moves and on the shape and size of the moving object than Eq. (5–30) suggests. In fact, at low speeds the drag force is proportional to v, not v^2. It is only after the speed has increased enough to produce turbulence in the medium that the v^2 term comes in. Thus, a better approximation to the drag force magnitude than Eq. (5–30) is

$$F_D = bv + cv^2. \qquad (5\text{–}32)$$

The coefficients b and c contain information on the shape of the moving object as well as on the medium in which it moves. The first term always dominates for sufficiently low speeds, while the second term dominates for higher speeds.

EXAMPLE 5–13 A marble of mass 5.0 g falls into a jar of oil. The drag force on the marble is given by Eq. (5–32) with $b = 0.20$ kg/s and $c = 0.10$ kg/m. (These values are typical for real fluids such as oil and for an object the size and shape of a marble.) Find the value of the speed for which the two terms in the drag force are equal. Which of the two terms is dominant when the drag force is comparable to the force of gravity on the marble? Will the two terms ever be equal as the marble falls through the oil?

Setting It Up Figure 5–23 shows the progress of the marble as it falls through the fluid.

Strategy We need only evaluate terms in Eq. (5–32) for this situation, and this equation will also show which drag force is dominant. As before, we can recognize that the terminal speed is reached when the overall drag force matches the force of gravity. This speed may be less than or greater than the speed for which the two terms in the drag force equation are comparable. If the speed is less, then the b term always dominates in the drag; if it is greater, then the c term will dominate near the terminal speed.

Working It Out Let the value of the speed for which the two terms in the drag force are equal be v'. Then v' is determined by

$$bv' = cv'^2.$$

This equation is solved by

$$v' = \frac{b}{c} = \frac{0.20 \text{ kg/s}}{0.10 \text{ kg/m}} = 2.0 \text{ m/s}.$$

At terminal speed, the magnitudes of the force of gravity and the drag force are equal; that is, $F_D = mg$. For now, let's utilize only the first term in the drag, bv, $bv = mg$:

$$v = \frac{mg}{b} = \frac{(5.0 \times 10^{-3} \text{ kg})(9.8 \text{ m/s}^2)}{0.20 \text{ kg/s}} = 0.25 \text{ m/s}.$$

When the values of the two terms in the drag force are equal—when $bv = cv^2$—the speed of the marble is 2.0 m/s. However, when the gravity and drag are the same size, the b term in the drag force dominates; this yields a speed of only 0.25 m/s. The marble will never reach speeds at which the c term has increased to where it is comparable to the b terms.

What Do You Think? The speeds in this example are fairly high and perhaps represent a marble falling in oil at a relatively high temperature. What happens to oil at low temperature?

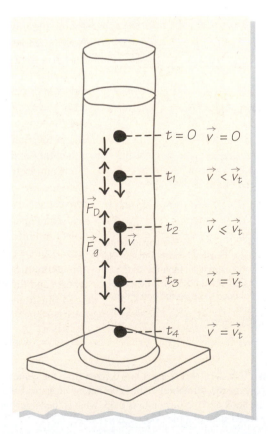

▲ **FIGURE 5–23** The marble is captured at various stages of its progress through the oil. When the marble has reached the terminal speed, the magnitudes of the drag force and gravity are equal.

(a)

Rope breaks, $\vec{T} = 0$. Book flies off in a direction tangent to circle along instantaneous $\vec{v}$.

(b)

(c)

▲ **FIGURE 5–24** (a) The tension in the rope holding a whirling book points toward the center of the circle and maintains the circular motion of the book. (b) If the rope breaks, there is no force on the book, which then moves in a straight line tangent to the circle. (c) Top view of (b).

5–4 Forces and Circular Motion

If you were to tie this book to the end of a rope and swing it smoothly over your head in a nearly horizontal circle, keeping your hand as close to the center of the circle as possible (Fig. 5–24a), you would feel the rope become taut. (Because of gravity, the rope could not be perfectly horizontal, but we'll ignore gravity here.) This tautness is tension in the rope. As we discussed in Section 5–1, a rope under tension pulls on whatever it is connected to—in this case your hand at one end and the book at the other. In this situation the pull of the rope is the only horizontal force on the book, so there is a net horizontal force on the book, and the book will accelerate in the horizontal plane. This isn't surprising, though; we saw in Chapter 3 that an object in circular motion is accelerating even if it has constant speed (*uniform circular motion*). There is acceleration because the *direction* of the velocity vector is changing. If the constant speed of the uniform circular motion of the book is v and the radius of its circular motion is r, then Eq. (3–56) tells us that the acceleration vector $\vec{a}$ of the book is

$$\vec{a} = -\frac{v^2}{r}\hat{r}. \qquad (5\text{–}33)$$

The unit vector $\hat{r}$ points outward from the center of the circular motion. The minus sign in Eq. (5–33) indicates that the acceleration is *toward* the center of the circle.

Newton's second law, $\vec{F} = m\vec{a}$, tells us that the force $\vec{F}$ required to keep a mass m in uniform circular motion is

$$\vec{F} = -\frac{mv^2}{r}\hat{r}. \qquad (5\text{–}34)$$

NEWTON'S SECOND LAW FOR UNIFORM CIRCULAR MOTION

Again, the minus sign reminds us that the force vector points to the center of the circle. Any force that points in toward the center of a circle is called a *centripetal force*; in this case it is the rope tension that is a centripetal force. Similarly, any acceleration toward the center of a circle is a *centripetal acceleration*. Equation (5–34) tells us the magnitude and direction of the rope tension. Because the rope leads from the book to the center of the circle, the direction of the tension is indeed toward the center. If the rope were cut, the tension that provided the centripetal force would no longer be present. According to the first law, then, the book would move without acceleration in the horizontal plane; that is, in a straight line (Figs. 5–24b and c).

Centripetal forces, whether they are realized by a rope tension or any other source of force, *are not constant forces*. They are constantly changing in direction and therefore cause an acceleration that is constantly varying.

The tension of a rope is not the only force that can be centripetal and maintain an object in uniform circular motion, and we'll look at other forces in the next few examples. A planet in circular orbit about the Sun or a satellite in circular orbit about Earth are maintained in their circular motion by gravitational forces. Friction between the wheels and the road is the force that takes an automobile around a curve or, as Example 5–14 illustrates, that holds an object on a rotating turntable in its circular path.

EXAMPLE 5–14 A book sits 0.75 m from the center of a rotating playground turntable. If the coefficient of static friction between the book and the turntable is 0.55, what is the largest possible angular speed such that the book will not slide off?

Setting It Up Figure 5–25a shows the book on the turntable. We are not given the mass m of the book, so we leave it in symbolic form. (It will in fact turn out to be unimportant.)

Strategy Figure 5–25b is the free-body diagram for the book. The force maintaining the book in uniform circular motion is static friction $\vec{f}_s$ between the book and the turntable. The largest rotation frequency that is possible without the book slipping occurs when the friction force reaches its maximum value, $f_s = \mu_s F_N$.

Newton's second law, $\vec{F}_{net} = \vec{F}_N + m\vec{g} + \vec{F}_s = m\vec{a}$, must be decomposed into equations for the vertical direction and for the horizontal plane. For uniform circular motion, the horizontal motion equations are represented by a single radial component equation, Eq. (5–34). There is no acceleration component in the vertical direction; that is, the direction in which $\vec{F}_N$ and $\vec{F}_g = m\vec{g}$ act. The condition that the vertical acceleration is zero will give us F_N and hence the maximum value of friction. Once we have this value, we can use it in the equation for the horizontal component of Newton's law. We can then use this equation to determine the maximum speed, that is, the speed for which the radial acceleration just matches the maximum force of friction. Finally, we will have to translate the maximum speed to a maximum angular speed.

(a)

(b)

◀ **FIGURE 5–25**
(a) Book rotates on playground turntable. (b) Free-body diagram for the book.

Working It Out The vertical component of Newton's second law is

$$F_N - mg = 0.$$

This gives $F_N = mg$. In turn, the maximum value of static friction is

$$f_s = \mu_s F_N = \mu_s mg.$$

The maximum radial acceleration $\vec{a}_{max}$ occurs for the maximum value of static friction. The radial component of Newton's second law, the coefficient of $\hat{r}$ in Eq. (5–34), is then

$$\mu_s mg = m\frac{v_{max}^2}{r}.$$

Note that the mass of the book cancels. We want to express this result in terms of an angular speed. From Chapter 3, we know that if the turntable is rotating with angular speed ω, then a point on the turntable that is a distance r from the center moves with speed $v = \omega r$. Thus our radial equation of motion becomes

$$\mu_s g = \frac{v_{max}^2}{r} = \frac{(\omega_{max} r)^2}{r} = \omega_{max}^2 r.$$

We solve for ω_{max}:

$$\omega_{max} = \sqrt{\frac{\mu_s g}{r}} = \sqrt{\frac{(0.55)(9.8 \text{ m/s}^2)}{0.75 \text{ m}}} = 2.7 \text{ rad/s}.$$

What Do You Think? If the angular velocity in this example exceeds the value calculated, what will happen to the book? How can you explain that based on the free-body diagram of Fig. 5–25b?

Normal forces can also act centripetally, as in Example 5–15.

EXAMPLE 5–15 A space station is an environment where the astronauts are in permanent free fall along with the station and so are in "weightless" conditions—a scale beneath their feet would register no contact force. One way to avoid weightlessness would be to give the station some rotation. Accordingly, consider a station that consists of a hollow circular tube that is rotating around its central axis, with the astronauts moving at the outermost radius (Fig. 5–26). If that radial distance is 50 m, what must be the speed v of a point on the outer wall such that a bathroom scale will read the same as it would on Earth?

Setting It Up We note that the reading of a bathroom scale on Earth is the weight mg, where m is the astronaut's mass. We are not given m.

Strategy Rather than setting up Newton's second law in what might be a rather complicated coordinate system, let's try to think through this problem to gain more insight. An astronaut of mass m who stands on a scale on the outermost wall within the space station moves in a circular path of radius R at constant speed v. He therefore accelerates toward the center of the circle (the axis) with magnitude v^2/R from Eq. (5–33). The force that causes this acceleration is the normal force $\vec{F}_N$, of magnitude mv^2/R, which is supplied by the tube's outer wall (Fig. 5–26b). The bathroom scale indicates this force, just as it would indicate the magnitude of the (upwardly directed) normal force on the astronaut if he were standing on the scale on Earth. Because the normal force on Earth balances gravity, it would have magnitude mg. Thus the scale on the space station reads as it would read on Earth provided that

$$mv^2/R = mg.$$

This can be solved for v.

(a)

(b)

◀ **FIGURE 5–26**
(a) Space station from the movie *2001: A Space Odyssey*. (b) A normal force $\vec{F}_N$ pointing to the circle's center acts on each astronaut. The presence of this force gives the astronauts the illusion of weight.

(continues on next page)

Working It Out The mass m cancels out from this equation and we can solve for the speed v:

$$v = \sqrt{gR} = \sqrt{(9.8 \text{ m/s}^2)(50 \text{ m})} = 22 \text{ m/s}.$$

This is a fairly considerable speed—about 50 mi/h. Motion such as this involves internal forces that the station must be constructed to withstand. These internal forces are proportional to v^2. The space station would have to be extremely well constructed, and hence extremely expensive, to have a large enough rotation to give the astro-

nauts the illusion of being on Earth. Actual space stations will have to make do with much less of a normal force—if any at all. Each piece of material that they are constructed from has to be transported from Earth and so must be very light.

What Do You Think? What would happen if the astronaut tried to walk on the inside surface of the innermost wall of the rotating station? Or on the "side walls," the ones parallel to the plane of rotation?

We have now discussed how tension, a friction force and a normal force can each act as the centripetal force responsible for circular motion. In particular, static friction is the centripetal force that causes an automobile to make a turn (the tires holding to the road rather than skidding). But static friction is limited to a maximum value. Above this value sliding friction takes over and the automobile skids. Thus the speed and/or radius with which an automobile can make a turn is limited. It is therefore useful to "bank" a curve. In a banked curve, the road is tilted so that the normal force, perpendicular to the road surface, has a component pointing to the center of the circle of the curve. In this way, the normal force acts together with static friction to make up the centripetal force. The combination accelerates the automobile through the curve at a greater speed than could be achieved without skidding if the road were flat. Example 5–16 illustrates this point in the extreme case where there is no friction.

EXAMPLE 5–16 It is a dark and stormy night, and a driver advancing along an icy road must negotiate a turn (Fig. 5–27a). The road traces out an arc of a circle with radius 320 m and is banked at

(a)

(b)

▲ **FIGURE 5–27** (a) Car making turn on icy, banked road. (b) Free-body diagram for the automobile, seen face on. The center of the circle described by the road curve is to the right.

an angle of 5.1°. Assume that the friction between the tires and the road is zero. At what speed should the driver take the curve to avoid sliding off the road?

Setting It Up The translation of the question to a quantitative setup is "What is the speed for which the car will not skid?" We denote the given radius of the circular path by r; other quantities are specified in the figure.

Strategy Figure 5–27b is a free-body diagram for the car as seen from the front together with a coordinate system. Note that here we have chosen axes that are vertical and horizontal rather than along and perpendicular to the plane. The center of the circle is to the $+x$-direction. In the absence of friction, only the normal force and gravity act on the car, and we want to write Newton's law and then break it into horizontal and vertical components. The vertical component has no acceleration, as the car stays on a horizontal path if there is no slipping. The horizontal component will have the acceleration for the circular motion of the car. These two equations should be enough to determine uniquely the speed.

Working It Out Newton's second law is $\vec{F}_{\text{net}} = \vec{F}_N + \vec{F}_g = m\vec{a}$, or in our coordinate system

$$(F_N \cos \theta)\hat{j} + (F_N \sin \theta)\hat{i} - (mg)\hat{j} = \frac{mv^2}{r}\hat{i}.$$

In component form, we have

$$\text{for the } x\text{-component: } F_N \sin \theta = mv^2/r; \tag{5–35}$$

$$\text{for the } y\text{-component: } F_N \cos \theta - mg = 0. \tag{5–36}$$

Equation (5–36) determines F_N:

$$F_N = \frac{mg}{\cos \theta}.$$

Note that, unlike other situations that we have studied, in this case the normal force is larger than the force of gravity. This result is inserted into Eq. (5–35) and solved for v:

$$\frac{mg}{\cos\theta}\sin\theta = \frac{mv^2}{r};$$

$$v = \sqrt{gr\tan\theta}.$$

With $r = 320$ m and $\theta = 5.1°$ ($\tan 5.1° = 0.089$),

$$v = \sqrt{(9.8 \text{ m/s}^2)(320 \text{ m})(0.089)} = 17 \text{ m/s} \cong 60 \text{ km/h}.$$

Only the x-component of F_N acts as the centripetal force to carry the car around the curve. When the automobile moves without sliding, the vertical component of F_N just balances the force of gravity, and the horizontal component provides the centripetal force, as in Eq. (5–34).

Note that this system is balanced quite delicately. For a given banking, curvature, and speed, Eq. (5–35) uniquely determines F_N.

If F_N is too small—if the speed is too slow—the product $F_N \cos\theta$ cannot balance gravity, and the car slips *down* the slope and off the road. If F_N is too large—if the speed is too fast—the vertical component of the normal force will exceed mg, and the car will skid *up* the slope and off the road. Although you might have been tempted to think that this question involves the *maximum* speed with which the driver can take the curve, there is in fact only one possible speed.

What Do You Think? We treated this example under the assumption that the road was perfectly slick. Suppose instead that there was a nonzero coefficient of static friction. Without doing any calculations, how does this new element affect our arguments about how delicately balanced the system is?

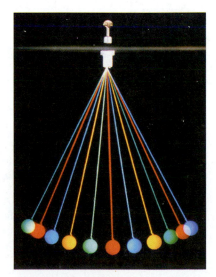

◀ **FIGURE 5–28** (a) Simple pendulum of mass m and length ℓ. The path followed by the bob traces a circular arc with the speed along the path varying with angle θ. (b) Free-body diagram for the bob of the simple pendulum. (c) Components of the force of gravity on the pendulum bob.

Circular Motion with Changing Speed

What happens in circular motion with changing angular speed? Figure 5–28a illustrates a *simple pendulum*, a point mass m (the bob) suspended from a string of negligible mass that has a length ℓ. The pendulum swings in a plane under the influence of gravity. During this swing, the mass on the end follows a circular path determined by the length of the string, but it is *not* uniform circular motion because the speed is not constant. To understand this, we draw a free-body diagram (Fig. 5–28b) for the bob, which is located at angle θ. Two forces act on it: gravity and tension. The bob accelerates according to Newton's second law, $\vec{F}_{net} = \vec{F}_g + \vec{T} = m\vec{a}$. The forces can be broken into components perpendicular to the path (radial forces) and components along the path (tangential forces). The tension is purely radial and inwardly directed (negative), whereas the force of gravity has an outward (positive) radial component $mg\cos\theta$ and a tangential component $mg\sin\theta$ (Fig. 5–28c). Thus the second law expressed in component form is

for the radial component: $-T + mg\cos\theta = -\dfrac{mv^2}{r}$;

for the tangential component: $mg\sin\theta = ma_{tan} = m\dfrac{dv}{dt}$.

Here the tangential component of the acceleration, a_{tan}, is the rate of change of speed.

The speed of the pendulum changes because a tangential force is present. The radial forces simply guide the mass on a circular path and *do not change its speed*. Figure 5–29 enables us to see that the pendulum changes its speed.

For *any* curving path there will generally be forces acting that have components perpendicular to the path and tangential to the path. Force components perpendicular to the path change the direction of the object's motion but not its speed. The tangential components of the net force change the magnitude of the velocity but not its direction.

▲ **FIGURE 5–29** As the pendulum bob moves under the influence of gravity and the string tension, it speeds up to a maximum at the bottom of the swing and moves most slowly at the top. This is revealed here by an open-shutter photograph of the swing with light flashes at equal time intervals of $\frac{1}{10}$ s.

CONCEPTUAL EXAMPLE 5–17 At the beginning of this section we referred to attaching this book to a rope and swinging it in a horizontal circle. Let's consider this situation again, and again ignore the gravitational force and consider only the forces that make the book describe a circle. We know that when the book is in uniform circular motion, the force is *central*. In other words, the rope must lead back to the center of the circular motion. What forces were required to bring the book to this state?

Answer In order to start the book moving and bring its speed up to its final value, there must be a tangential force component exerted by the rope on the book through the rope tension. Thus the tension cannot be purely radial but must in addition have a tangential component. This is managed by moving the hand responsible for swinging the book away from the center of the circle—the hand itself will follow a smaller circle about the motion center. The larger the tangential component of the tension and the longer it operates, the faster the book swings. As the book swings faster, the radial tension force also increases, but the hand holding the rope must take this increasing radial component into account with a firmer hold. In reality there are forces acting such as gravity and air resistance that will slow the motion down unless a continuous effort is made to keep it up, as you would quickly realize if you were asked to keep the book swinging for a significant period of time.

EXAMPLE 5–18 A skater of mass 52 kg is swinging in a circle of radius 2.0 m at a speed 3.0 m/s. He maintains this motion by holding onto a rope attached through a frictionless bearing to a fixed pole at the center of the circle (Fig. 5–30). The tension of the rope provides the centripetal force for his circular motion, and we can regard the ice as frictionless. As he swings around the pole, however, he runs into a rough patch of ice 48 cm long, where the coefficient of friction is 0.10. (a) What is the tension T in the rope before he hits the rough patch? (b) What is his speed just after he passes the rough patch? (c) How must the tension change to keep him on the same circular path after he has passed the rough patch?

▲ **FIGURE 5–30** Ice skater rotating about pole by means of a rope.

Setting It Up The given mass, radius of motion, and speed are labeled *m, R,* and *v,* respectively. We label the coefficient of kinetic friction in the rough patch as μ_k. The unknown tension has magnitude T.

Strategy For part (a), we simply use the fact that tension supplies the centripetal force required for the motion. For part (b), we note that the friction of the rough patch is a constant tangential force, which slows the skater down. The patch is small enough for us to as-

sume that the motion of the skater through the patch is a straight line. The frictional force yields the tangential acceleration (negative) through Newton's second law. We then use our kinematic formulas for one-dimensional motion to find the speed change. For part (c), we redo the exercise of part (a) but with the reduced speed.

Working It Out (a) Newton's second law for the circular motion reads

$$T = \frac{mv^2}{r} = \frac{(52\ \text{kg})(3.0\ \text{m/s})^2}{2.0\ \text{m}} = 2.3 \times 10^2\ \text{N}.$$

Compare this to the skater's weight, $mg = (52\ \text{kg})(9.8\ \text{m/s}^2) = 5.1 \times 10^2\ \text{N}.$

(b) The magnitude of the friction force is

$$f_k = \mu_k F_N = \mu_k mg = (0.10)(510\ \text{N}) = 51\ \text{N}.$$

Here f_k/m is the magnitude of the tangential acceleration, pointing oppositely to the motion (friction will slow the skater down). We then use the one-dimensional formula relating speed change to distance traveled for a constant force, Eq. (2–24):

$$v^2 = v_0^2 + 2a\ \Delta x = v_0^2 - 2\frac{f_k}{m}\Delta x$$

$$= (3.0\ \text{m/s})^2 - 2\frac{51\ \text{N}}{52\ \text{kg}}(48\ \text{cm})\frac{1\ \text{m}}{100\ \text{cm}} = 8.1\ \text{m}^2/\text{s}^2.$$

After passing the rough patch, the skater has slowed to

$$v = \sqrt{8.1\ \text{m}^2/\text{s}^2} = 2.8\ \text{m/s}.$$

(c) We simply follow the first procedure and compute the tension with the new speed:

$$T = \frac{mv^2}{r} = \frac{(52\ \text{kg})(2.8\ \text{m/s})^2}{2.0\ \text{m}} = 2.1 \times 10^2\ \text{N}.$$

This is 20 N less than the tension in the rope before the rough patch is encountered.

What Do You Think? You are a fairly good skater and decide to try the motion described in this example. You find a suitable, safe place. What should you be concerned about when performing the maneuver?

Circular Motion and Noninertial Frames

The doughnut-shaped space station in Example 5–15 illustrates how a normal force can be the centripetal force that moves an object in a circle. This system also nicely illustrates the effects of noninertial frames, first mentioned in Section 4–4. An astronaut in the space station is an observer in a noninertial frame. Seen from outside, the astronaut is in uniform circular motion and is therefore accelerating. The centripetal force responsible for this acceleration is the normal force, which can be read by a scale at the astronaut's feet. But what does the astronaut see? As he perceives it, everything around him within the station is at rest. The scale reads the usual kind of normal force, directed toward the center of the circle.

The astronaut's analysis of the situation is that he is at rest and there is no identifiable source for the force that keeps him on the scale—this force is therefore a fictitious force. It is directed outward (*centrifugal*, meaning "directed away from the center") because it balances the inward-directed normal force that the scale reads.

While the notion of a centrifugal force is a commonplace one, we should keep in mind that *it is not a force in the same sense as the forces that appear in Newton's laws.* It is the consequence of an observer being in a rotating (and hence accelerating) reference frame. We use the language of the centrifugal force to interpret the invisible hand that forces us against the door when an automobile makes a high-speed turn. Seen from outside the automobile, a passenger is accelerating centripetally. A centripetal normal force from the door of the automobile pushing on the passenger is the force responsible for moving the passenger in his curving path. The automobile passenger views things differently. The passenger perceives herself to be at rest with respect to the car, yet she feels the normal force from the door. The passenger then invents a "centrifugal force" to balance the normal force from the door.

5–5 Fundamental Forces

The forces that we have dealt with so far include the force of gravity (weight), normal forces, friction, tension, drag, and spring forces. With the exception of weight, these forces are what we can call secondary forces, the manifestation on a macroscopic level of something happening at the microscopic level. At the microscopic level, the constituents of matter interact through the fundamental forces of electricity and magnetism. The forces themselves are well understood, but there are so many atoms and molecules in any piece of matter or in any object, and they interact in such a complex way, that even with the largest computers at our disposal we would be unable to derive a spring force, say, from the electromagnetic interactions between the molecules of the spring. Thus we must be content to use the spring force—a secondary force—as a simple effective force to solve problems relevant to the everyday scale. The force of gravity is a different case. It corresponds to a more general law—the fundamental force of *universal gravitation*, first understood by Newton—but used over a limited range.

What we now refer to as the fundamental forces, the ones that underlie the secondary forces, are not necessarily the forces that were thought of as the fundamental forces 200, 100, or even 25 years ago. Our progress on the understanding of what is truly fundamental has been steady, marked especially with a series of *unifications*. Thus what were once thought to be distinct fundamental forces are now viewed as different aspects of the same fundamental force.

We now believe that there are only three fundamental forces in nature. These are the force of **universal gravitation**, the **electroweak** force, and the **strong** force (often called the nuclear force). The expression of the force of universal gravitation (see Chapter 12) was one of Newton's greatest discoveries. Albert Einstein replaced Newton's expression in his law of general relativity, making some new predictions of tiny experimental effects that are not predicted by Newtonian gravitation. Einstein also made persistent but unsuccessful efforts to unify gravitation with the other fundamental forces. The electroweak force is a relatively recent discovery, dating from a theory proposed by Sheldon L. Glashow, Abdus Salam, and Steven Weinberg in the 1960s and 1970s, that was verified experimentally in the early 1980s. This work unified two forces formerly thought to be independent and fundamental: the *weak* force, responsible mainly for some types of radioactive processes in nuclei and important in the evolution of the universe, and the forces of *electromagnetism* (Chapters 21 to 34). On the scale appropriate to the secondary forces that we deal with every day, the electromagnetism aspect of the electroweak force is dominant, and it is often convenient to refer simply to the electromagnetic force. Electromagnetism is itself the result of a nineteenth-century unification of the forces of *electricity* and *magnetism*; these two forces were previously thought to be different kinds of forces. The third fundamental force, the strong force, is responsible for holding together the nuclei of atoms. Both the strong force and the electroweak force have been the object of more recent attempts at unification, which have yet to bear fruit. Since the 1980s, there has also been a new effort to bring gravitation into the unification program. Unification continues to be a fascinating beacon that guides to a simpler description of the universe.

On the everyday scale of the secondary forces that we have discussed in this chapter, only the force of gravity is a direct aspect of a fundamental force: universal gravitation. All the other forces that we have investigated are ultimately due to the electromagnetic force, which binds atoms and molecules together into ordinary matter; for example, contact forces result from forces exerted at the atomic scale. We do not directly see the strong force on our everyday scale; it holds nuclei together so tightly that, for most practical purposes, we can think of them as indivisible lumps of matter. Only when we discuss nuclear energy or the composition of stars do strong forces come into practical play.

Within the nucleus, the typical strong force is about 100 times larger than the electromagnetic forces. The gravitational force between two protons in the nucleus is many, many orders of magnitude smaller than either the strong or the electroweak force. But as the strong force acts over such a limited range, we experience the electroweak and gravitational forces as the dominant forces at the everyday scale. The gravitational force dominates on the astronomical scale because the atoms of matter are electrically neutral, and so electromagnetic forces cancel very nearly to zero.

Summary

This chapter is devoted to describing various forces that occur in nature and to applying the problem-solving techniques developed in Chapter 4. These techniques are ones that we use throughout the book.

The force of gravity (or weight), tension, and the normal force are common forces for which it is often possible to solve the equations of motion given by Newton's second law. The force of gravity is expressed as

$$\vec{F}_g = m\vec{g}, \tag{5-4}$$

where $\vec{g}$ is a vector of constant magnitude that points down toward Earth's center.

The tension $\vec{T}$ is a variable force that is determined according to the circumstance and exerted by ropes (or wires or cables or strings). Tension always pulls on a mass in the direction of the rope; it is transmitted everywhere along the rope, taking a single constant magnitude throughout the rope when the mass of the rope is negligible.

The normal force $\vec{F}_N$ is also variable. It is directed perpendicularly to a surface and acts to cancel the effects of any other forces that might make a mass accelerate into the surface.

The friction force acts when two surfaces slide or attempt to slide across one another. Static friction $\vec{f}_s$ is variable and acts in a direction that would oppose any sliding motion that would occur if there were no friction. Its magnitude can increase up to a maximum value proportional to the magnitude of the normal force:

$$\text{static friction: } 0 \leq f_s \leq \mu_s F_N. \tag{5-21}$$

Kinetic friction $\vec{f}_k$ acts when sliding actually occurs and is also proportional to F_N:

$$\text{kinetic friction: } f_k = \mu_k F_N. \tag{5-22}$$

The constants μ_s and μ_k are the coefficients of friction, and generally

$$\mu_s > \mu_k. \tag{5-23}$$

Another type of friction, the drag force, occurs when objects move through fluids or gases. The drag most frequently encountered varies with the speed squared:

$$F_D = \tfrac{1}{2}\rho A C_D v^2. \tag{5-30}$$

Any drag force that increases with speed leads to a situation in which objects accelerating within media can be accelerated only up to a terminal speed.

Forces are responsible for accelerating objects moving in a curved path. When the motion is uniform circular motion, then Newton's second law takes the form

$$\vec{F} = -\frac{mv^2}{r}\hat{r}. \tag{5-34}$$

The vector $-\hat{r}$ is directed to the center of the circle and the force is said to be centripetal. Any of the forces we mentioned above can act as the centripetal force that keeps an object in uniform circular motion. When both the direction and the magnitude of the velocity change, then $\vec{F}$ has a component tangential to the motion. This component causes the magnitude of the velocity to change, while a force component perpendicular to the motion causes the direction of the motion to change.

All the forces of nature are ultimately described in terms of three fundamental forces: the force of universal gravitation, the electroweak force, and the strong force. The electromagnetic force, which is part of the electroweak force, is responsible for most of the secondary forces, including tension, friction, drag, normal forces, and spring forces.

Understanding the Concepts

1. A tightrope walker moves to the center of a thin wire that was initially stretched taut to a horizontal position. Why is it that the wire cannot remain horizontal?

2. We have said that it is actually the force of friction that is responsible for both the acceleration of automobiles and our ability to walk. What is the role of the engine or of the muscles in these processes?

3. An observer sees a mass hanging motionless from a vertical string. Under what circumstances is the tension in the string greater than or less than the weight of the mass?

4. What are some factors that could limit how fast a hot rod can go in a $\frac{1}{4}$-mi race?

5. How does the fact that a rope has mass complicate solving a problem about lifting a load with a pulley? What is the effect of friction in the pulley?

6. Why is it helpful for an automobile with an engine in the front to have front-wheel drive? Why is it useful to put sand in the trunk of your car in winter if your car has rear-wheel drive?

7. Why do bicyclists or motorcyclists "lean into" a curve? In explaining why, make use of the fact that we are able to balance ourselves best when we feel that the net force on us is coming from directly beneath our feet. Also think about the forces that friction must oppose.

8. How might the result of Example 5–10 vary if the weight was not evenly distributed over the wheels and/or the car was equipped with rear-wheel drive instead of four-wheel drive?

9. Suppose that a rope has tension because a mass is suspended from its end. Let's say that the rope is now cut, a spring is inserted at the cut, and we observe the stretch of the spring as a measure of the rope's tension. If the mass of the rope is negligible, does the observed tension of the rope depend on where the cut is made along the string? Does your answer change if the rope cannot be considered massless?

10. A bowl of water with floating ice cubes is placed on a scale. The ice cubes melt. Does the reading of the scale change? Why or why not?

11. What is the role of the keel, which runs along the center of the bottom of a sailboat? Some sailboats have centerboards rather than keels. These are simply large boards that can be lowered or raised in the position of a keel. What is the role of a centerboard? Why might you prefer a keel to a centerboard or vice versa?

12. Suppose the woman in Example 5–4 had just the limiting mass of 37.3 kg and the suspended lineman ate a cheeseburger, adding 500 g to his mass. Assuming the woman doesn't let go of the rope, describe qualitatively the resulting motion.

13. The block and tackle illustrated in Example 5–4 employed the same rope four times to allow a mass m_1 to lift a mass $m_2 = 4m_1$. Could you have increased the number of pulleys in that example so that the same rope is employed 100 times and a mere baby could have lifted the football lineman? What are the practical limitations in this sort of process?

14. We have referred to massless ropes and the tension in them. What physical considerations allow you to think of a rope as massless?

15. The speed of the boats (shells) used in scull racing is, to a good approximation, independent of the number of people rowing (provided that the number is larger than three or four). At first, this might appear strange: The more people there are, the larger the forward propelling force available to overcome the drag of the water. Can you explain this seeming contradiction?

16. Describe a series of experiments that would have allowed Leonardo da Vinci to decide that the force of friction depends only on the normal force on an object.

17. True or false:
(a) Sliding friction can accelerate an object.
(b) Sliding friction can be used to increase the speed of an object.
(c) Static friction can accelerate an object.
(d) Static friction can be used to increase the speed of an object.

18. At the moment a car in a loop-the-loop roller coaster is at the top of the loop (directly below the track), can there be a normal force on it? Such a force would point straight down, in the same direction as that of gravity.

19. The riders on the loop-the-loop of Question 18 experience a feeling of near weightlessness close to the top of the loop. Why? Would the coins in their pockets fall out?

20. Suppose a block of wood sits on the floor of an elevator with the elevator moving upward with a constant velocity. Will the normal force differ from what it would be when the elevator is stationary?

21. Suppose a block of wood sits on the floor of an elevator with the elevator moving upward with a positive acceleration. Will the normal force differ from what it would be when the elevator is stationary?

22. Suppose that it were possible for a ship to sail all the way around the world along a great circle. Is a centripetal force necessary to keep the ship moving in this circle? What force or forces would act centripetally?

23. Look at the pendulum photo in Fig. 5–29. Is the net force on the pendulum bob constant throughout its motion? Is there a point at which the net force equals zero? Describe the forces acting on the pendulum bob when it just stops and turns around on the right side.

24. When you sit in the passenger seat of a car that makes a tight turn to the left, you could be thrown out of the car if the door should open (and you are not belted in). Why?

25. At the beginning of Section 5–4, we spoke of swinging a book on the end of a rope in a nearly horizontal circle. Could the motion form a perfectly horizontal plane?

26. Imagine that you and your partner are on skates on a perfectly frictionless ice surface. You hold on to opposite ends of a rope and pull toward each other. What happens if you each have exactly the same mass? If you have half your partner's mass?

27. Why is it hard to run when the ground is icy?

28. You could attach the end of a stick to a mass and use the stick to pull the mass, just as with a rope. Is the associated force a tension force? If not, in what ways does it differ from tension?

29. Tarzan swings from tree to tree on a vine. At which point in his swing is the vine most likely to break and why?

30. A person on a rapidly moving Ferris wheel feels that she is about to fly off the seat when her seat reaches the top of the circular path. Why is that?

31. A hemispherical bowl is placed open end up on a table and rotates around its own vertical axis. A die is allowed to slide down from the edge into the bowl. Describe the motion of the die as seen by a hypothetical observer at the bottom of the bowl, assuming that there is no friction between the bowl and the die.

32. The person in Fig. 5–17 is moving along briskly at a constant speed. What is the net force on her? Describe the various external forces acting on her.

33. Consider Eq. (5–32) for the drag force, which has two terms. Which term, if any, dominates at low speeds and which, if any, dominates at high speeds? How would you decide the size of the domains for which one term is much larger than the other?

34. You sit on a comfy chair sinking into the seat cushion. Is a normal force acting as you do? Describe how the normal force is acting in this case.

Problems

5–1   Some Simple Constant Forces

1. (I) An 8-g bullet that travels at 500 m/s is fired into a rigidly fixed block of wood. The bullet is found 7 cm into the wood. What was the average force exerted by the wood opposing the bullet's motion? Assume that the deceleration was uniform.

2. (I) In Example 5–4, a woman lifts a football lineman (mass 149 kg) by pulling down on one end of a rope with 365 N of force. Suppose that the woman has a mass of 50 kg. If a bathroom scale were beneath her feet as she lifted the lineman, what would the scale read in pounds? Analyze the free-body diagram for the woman and remember that a scale of this type reads the upward normal force that the floor exerts on her.

3. (I) A helium balloon just manages to lift 100 kg (including the mass of the balloon, the helium it contains, and its payload) off the ground and then hovers 1 m off the ground. The upward force that maintains the balloon is *buoyancy* (we shall treat this force in Chapter 16). What is the magnitude of the buoyancy in this case?

4. (I) A woman of mass 61 kg sits in a racing car. When she depresses the accelerator, the car accelerates in a straight line to 210 km/h in 7.3 s. What are the direction and magnitude of the force she experiences? Where is the force applied?

5. (I) A hockey puck of mass 0.10 kg slides without friction on ice. In an appropriate coordinate system, its velocity $\vec{v}_1 = (1.4 \text{ m/s})\hat{i} + (3.0 \text{ m/s})\hat{j}$. A constant force $\vec{F} = (4.0 \text{ N})\hat{i}$ is then applied to the puck. After how many seconds will the puck have a speed of 6.0 m/s?

6. (II) A man of mass 80 kg is escaping a burning building using a rope that will break if the tension exceeds 600 N. (a) With what acceleration must he slide down the rope if it is not to break? (b) How far down the rope is he and what is his velocity after 5.0 s assuming he drops with the minimum acceleration of part (a)?

7. (II) A metal rod of mass 5.6 kg and length 3.5 m is suspended from the ceiling. (a) What is the tension in the rod at a distance of 2.0 m from the top? (b) What is it at a distance of 3.0 m from the top?

8. (II) A brick hangs from a string attached to the ceiling. When a horizontal force of 12 N is applied to the brick, the string makes an angle of 25° with the vertical (Fig. 5–31). What is the mass of the brick?

9. (II) Figure 5–32 shows a person applying a horizontal force in trying to push a 25-kg block up a frictionless plane inclined at an angle of 15°. (a) Calculate the force needed just to keep the block in equilibrium. (b) Suppose that she applies three times that force. What will be the acceleration of the block?

▲ **FIGURE 5–32**   Problem 9.

10. (II) Two blocks of mass m_1 and m_2 are placed in contact on a smooth surface with the more massive one (m_1) on the left. A force of magnitude F pointing to the right is applied to the block on the left. (a) What is the acceleration of the system? (b) What force acts on the block on the right?

11. (II) An automobile of mass 1200 kg pulls another automobile of mass 1400 kg with a tow rope. (a) In order to pull out onto a highway, the automobile must accelerate to 55 mi/h in an access lane that is only 120 m long. What must its acceleration be? (b) What is the tension in the tow rope?

12. (II) An Atwood machine consists of a massless string connecting two masses over a massless, frictionless pulley (Fig. 5–33). In this case, the masses are 1.70 and 1.65 kg. The system is released from rest with the 1.7-kg mass 2.15 m above the floor and the 1.65-kg mass on the floor. (a) What is the acceleration of the 1.7-kg mass? Of the 1.65-kg mass? (b) What is the speed of the 1.7-kg mass just before it hits the floor? (c) How long does it take the 1.7-kg mass to reach the floor?

▲ **FIGURE 5–31**   Problem 8.

▲ **FIGURE 5–33**   Problem 12.

13. (II) Two blocks of masses M and m are connected by a light rope that passes over a frictionless pulley. Mass M sits on an inclined plane with an angle of inclination of $\theta = 30°$ (Fig. 5–34). The coefficient of static friction between mass M and the inclined plane is 0.20, while $m = 3.0$ kg. Determine the largest and smallest possible values of M for which the system remains in equilibrium. Calculate the force of static friction on the block of mass M if $M = 6.0$ kg.

▲ **FIGURE 5–34**   Problem 13.

14. (II) Consider a variation on an Atwood machine in which the masses are each on a (frictionless) incline (Fig. 5–35). The mass sliding on incline 1, m_1, is 1.50 kg, and the angle of this incline is $\theta_1 = 62°$. If the mass on the second incline, m_2, is 2.50 kg, what is the angle θ_2 so that the system does not accelerate?

▲ **FIGURE 5–35**   Problem 14.

15. (II) Suppose the woman in Example 5–4 had just the limiting mass of 37.3 kg and that the suspended lineman ate a cheeseburger, adding 500 g to his mass. Assuming the woman doesn't let go of the rope, describe quantitatively the resulting motion.

16. (II) In the text, we stated that the tension is the same everywhere in a massless rope. Here we ask you to find the tension as a function of position for a stationary hanging rope with constant mass density $\lambda = $ mass/unit length. [*Hint*: Consider a particular point along the rope and find how much mass lies below that point. The tension at that point must be enough to support that mass.]

17. (III) A compound Atwood machine is constructed by replacing one of the masses of a conventional Atwood machine (see Problem 12) with the pulley of another Atwood machine (Fig. 5–36). Altogether there are three masses, two ropes, and two pulleys; the ropes and pulleys are to be considered massless and friction free. Describe the method by which the motions of the masses of this machine can be analyzed. [*Hint*: There are two independent tensions and two independent accelerations, which must be found by analyzing the equations of motion for the three masses simultaneously.] Solve these equations for arbitrary masses. What happens if $m_2 = m_3 \neq m_1$? [*Hint*: If m_1 accelerates, so does the pulley, so m_2 and m_3 share an additional acceleration.]

▲ **FIGURE 5–36**   Problem 17.

18. (III) A double-pulley system—like that described in Problem 17—has masses $m_1 = 2.00$ kg, $m_2 = 1.20$ kg, and $m_3 = 0.800$ kg. (a) What are the accelerations of all the masses? (b) What are the tensions in all the ropes?

19. (III) Consider the three-pulley arrangement shown in Fig. 5–37. The three masses m_1, m_2, and m_3 have the values 4.00, 10.00, and 6.00 kg, respectively. All the pulleys are frictionless and the strings are massless. What are the tensions in all the strings and what are the accelerations of the masses?

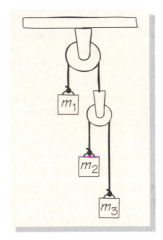

▲ **FIGURE 5–37**   Problem 19.

5–2   Friction

20. (I) A boy of mass 42 kg runs into an ice-skating rink in street shoes and begins a smooth slide. He starts with a speed of 5.2 m/s and comes to a stop after sliding for 4.7 s. What is the coefficient of kinetic friction of his shoes on ice?

21. (I) A truck of mass 5000 kg accelerates on a straight road at 1.2 m/s². Assuming that air resistance is negligible, what is the minimum coefficient of static friction between the road and the tires? (Don't worry about differences in normal forces on each wheel and assume four-wheel drive.)

22. (I) The coefficient of static friction between a worker's shoes and the floor is 0.81, while the coefficient of static friction between the floor and a crate is 0.43. The worker, mass 80 kg, pushes the crate, which has a mass of 140 kg. What is the friction force on the worker due to friction between the shoes and the floor when the crate starts to slide?

23. (I) A car of mass 1200 kg is moving at 25 m/s. The driver suddenly sees a dog crossing the road, slams on the brakes, and manages to stop the car in 4.2 s. What is the minimum coefficient of static friction between the tires and the road? (Assume that the acceleration is constant and that there is no skidding.)

24. (I) A rope is connected to a 25-kg cement block placed on a board leaning against a wall at an angle of 25° with respect to the horizontal (Fig. 5–38). The coefficient of kinetic friction between the cement block and board is $\mu_k = 0.4$. (a) What is the tension in the rope if it is pulled at constant speed straight up the board? (b) What is the tension if the rope is pulled up at constant speed at an angle 40° from the horizontal?

▲ **FIGURE 5–38**   Problem 24.

25. (I) A worker must push a 58-kg crate across a floor. The coefficient of kinetic friction between the crate and the floor is $\mu_k = 0.63$. What is the minimum force that the worker must exert to keep the crate moving once the crate starts moving?

26. (II) A man wants to push a package of shingles of total mass 15 kg up a roof being built at an angle of 27°. The coefficient of kinetic friction between the package and the roofing paper already in place is $\mu_k = 0.55$. (a) How much force does the man have to exert on the package directly along the slope of the roof to cause the package to accelerate at 0.15 m/s²? (b) If the coefficient of static friction is 0.58, will the package remain on the roof?

27. (II) A person learning to snow ski will use the snowplow position; it is a rudimentary way of keeping one's skiing speed under control. Let's imagine that a beginning skier finds herself on an icy slope of 22°. Only by setting her skis in a good snowplow position, with the tips of both skis pointed inward and the inner edges dug in, is she able to keep from accelerating. Effectively, what coefficient of sliding friction is created by the snowplow?

28. (II) A crate of mass 250 kg is loaded on the back of a truck. The coefficient of static friction between the crate and the truck bed is μ_s. The truck decelerates such that it comes to a stop from a speed of 60 mi/h (26.7 m/s) in a distance of 140 m. How large must μ_s be so that the crate does not slide forward on the truck bed?

29. (II) The coefficient of static friction between a car of mass 1500 kg and an asphalt road is $\mu_s = 0.70$. (a) What is the shortest distance over which the car can accelerate from rest to a speed of 96 km/h? (b) How long will this take? (Because we neglect drag and rolling friction, the distance and time will be unusually short.)

30. (II) A block of mass $m = 2.0$ kg is placed on a horizontal surface. The coefficient of static friction between the block and the surface is $\mu_s = 0.40$. A light rope is tied to the block and thrown over a frictionless pulley (Fig. 5–39). The free end of the rope is pulled with a slowly increasing force T. At what value of T will the block start to move if the angle that the rope makes with the horizontal is 50°?

▲ **FIGURE 5–39**   Problem 30.

31. (II) A 500-g box is placed on a board at a 35° incline and accelerates from rest down the board at 0.5 m/s². (a) How long does the box take to travel down the board of length 0.8 m? (b) What frictional force opposes the motion of the box? (c) What is the coefficient of friction between box and board?

32. (II) A block of mass 0.70 kg rests on top of another block of mass 0.90 kg, which rests on a frictionless surface. The coefficient of static friction between the blocks is $\mu_s = 0.45$. What is the maximum horizontal force that can be applied to the upper block so that the blocks accelerate together without the upper block sliding on the lower one? If the horizontal force is applied to the lower block instead, what is the maximum force that will give rise to the same motion?

33. (II) A 50-kg box rests on a rough horizontal surface with which the box has a large coefficient of static friction, $\mu_s = 0.75$. The box is pulled by means of a light rope with a force of magnitude F, making an angle θ with the horizontal. (a) Find the magnitude of the force F that will just start the box moving horizontally as a function of θ. (b) Show that there is some angle θ for which F takes a minimum value. What is this value for our case and what is the force F corresponding to this value? Explain why, physically, there is such a minimum value.

34. (II) A pile of snow at the crest of a roof with a slope of 40° from the horizontal starts to slide off. The distance from the crest to the edge of the roof is 8 m, and the coefficient of kinetic friction for the snow on the roof is 0.1. (a) What is the speed of the pile of snow when it reaches the edge of the roof? (b) Assuming that it is 6 m from the edge of the roof to the ground, how far out from the base of the building does the snow land?

35. (II) Consider again the professor of Example 5–9. Her mass is 55 kg and the coefficient of static friction between her and the floor is μ_1. What is the minimum value of μ_1 for which she will not slip on the floor?

36. (III) A mass m_1 rests on top of another mass, m_2, which in turn rests on a frictionless horizontal surface (Fig. 5–40). A light cord is attached to m_2, which is used to pull on it with a force F. (a) Find the acceleration of each object when the surface between the two objects is frictionless. (b) Find the acceleration of each object when the surface between the two objects is rough

▲ **FIGURE 5–40**   Problem 36.

enough to ensure that m_1 does not slide on m_2. (c) What are the magnitude and direction of the contact forces exerted by the lower object on the upper one assuming that the upper object is sliding on the lower object with a nonzero coefficient of kinetic friction μ_k? (d) Find the acceleration of each object if the surface between the two objects is such that the upper object is sliding on the lower one under the influence of kinetic friction, with a coefficient of kinetic friction μ_k.

5–3 Drag Forces

37. (I) Estimate the drag force on an automobile cruising at 65 mi/h. Assume that the drag coefficient C_D is 0.45 and that the car's cross-sectional area is 4 m². Take air to have a density of 1.25 kg/m³.

38. (I) A parachute is rigged so that a parachutist of total mass 116 kg with full gear reaches the ground at a terminal speed of 4.9 m/s. Assuming that the drag force on the parachutist, moving with speed v, has a magnitude equal to kv^2, what is the value of k?

39. (I) A ball of mass 500 g is observed to reach its terminal speed of 18 m/s after being dropped from the top of a tall building. Assume that the density of air is 1.25 kg/m³ and the drag coefficient C_D is 0.40. What is the effective cross-sectional area of the ball? You can use Eq. (5–30) here.

40. (I) A race car of mass 800 kg has a maximum acceleration from rest of 4.8 m/s². Assume that the car's engine is such that the force on the tires is constant and that the car's effective cross-sectional area into the air is 1.8 m². If the car's top speed is observed to be 90 m/s, what is its drag coefficient C_D? Take the density of air as 1.25 kg/m³.

41. (II) A barge moving at uniform speed is pulled by two horses moving on opposite sides of a canal in which the barge floats. The ropes connecting the horses to the barge make an angle of 32° with the line of motion of the barge. If the resistance to the motion of the barge is characterized by a frictional force given by $F = -(220 \text{ N} \cdot \text{s/m})v$, where v is the speed of the barge in meters per second and the tension in each of the ropes is 74 N, what is the speed of the barge?

42. (II) A sphere of radius r_1 and mass m_1 falls through the air and is found to have a terminal speed v_1. Equation (5–30) applies. (a) What is the terminal speed v_2 for a sphere with twice the radius and the same mass density as the first sphere? (b) Generalize the result of part (a) to find the terminal speed for a sphere with a radius z times the first radius but the same mass density.

43. (II) The terminal speed of a skydiver of mass 75 kg can be controlled by the orientation of her body and can range from 40 to 60 m/s. Assume that she can change the area presented to the ground by a factor of 1.5 in going from the minimum to the maximum terminal speed and that the larger area presented to the ground slows the skydiver down. Now, how does the drag coefficient change? (Express your answer as a fraction.)

44. (II) Assume that the drag force on a parachute is given by Eq. (5–30) in the text. The effective area of the parachute is 30 m² and the density of air is 1.25 kg/m³. If a 90-kg parachutist finds that his terminal speed is 6.0 m/s, what is the drag coefficient?

45. (II) A parachutist of 60-kg mass uses a parachute similar to the one used by the parachutist of the previous problem. Determine the terminal speed of this lighter person.

46. (II) A marble of mass m falls through a fluid and is subject to the drag force $\vec{F}_D = -A\vec{v}$, where $\vec{v}$ is the velocity of the marble. The marble will reach a terminal speed given by $\vec{v}_t = m\vec{g}/A$. Use dimensional analysis to estimate how long it will take to reach the terminal speed. [*Hint*: A characteristic "time" can be constructed from A, g, and m.]

5–4 Forces and Circular Motion

47. (I) An airplane of mass 2×10^4 kg executes a banked turn of radius 30 km while flying at 200 m/s. What acceleration will the passengers have as a result of the turn?

48. (I) A rock swings in a nearly horizontal circle at the end of a string whose breaking tension is 26 N. The circular path is 0.35 m in radius, and the rock's mass is 220 g. What is the maximum speed the rock can have before the string breaks?

49. (I) A man of mass 65 kg stands at the edge of a merry-go-round of radius 5.3 m. The merry-go-round turns at 6.0 rev/min. What are the magnitude and direction of the net force on the man?

50. (I) An accelerometer shows that an airplane flying at 650 km/h undergoes a vertical acceleration of 0.30 g's (1 g = 9.8 m/s²) at a certain moment. What is the radius of curvature of the airplane's (horizontal) path at that point?

51. (II) An automobile makes a turn whose radius is 150 m (Fig. 5–41). The road is banked at an angle of 18°, and the coefficient of friction between the wheels and road is 0.3. Find the maximum and minimum speeds for the car to stay on the road without skidding up or down the banked road.

▲ **FIGURE 5–41** Problem 51.

52. (II) A merry-go-round has a circular platform that is 1 m from the central axis at its inner edge and is 5 m from the central axis at its outer edge. The ride turns at a rate of one full rotation every 10 s. A passenger holds himself to the surface with a pair of very sticky shoes and is most comfortable when he orients his body length along the line of the net force on him. Determine the angle his body makes to the vertical (a) 1 m from the axis, (b) 3 m from the axis, and (c) 5 m from the axis.

53. (II) A spring is 0.6 m long. One end is permanently attached to a pivot on a horizontal table top. The force that it exerts on a body attached to the other end is $(4.2 \text{ N/m})x$, where x is the distance that the spring is stretched beyond its normal length. Suppose a mass of 400 g is attached to the other end of the spring and the whole system is set in circular motion. (The entire motion takes place on the horizontal table top.) How far will the spring be stretched if the mass rotates with an angular velocity of 8 rad/s?

54. (II) A student carrying an accelerometer on a large roller coaster measures a value of 0.93 m/s² perpendicular to the track when going over the top of a circular portion of the track. At another curve the student finds the accelerometer to read 2.20 m/s² at the top of the curve. The student has no way of knowing how fast she is going in both situations. When she gets off the roller coaster, she estimates that the radius of curvature for the second curve is about one-half the value of the other. What was her relative speed through the second curve with respect to the first?

55. (II) The coefficient of static friction between a small stone and a horizontal turntable is measured in the following way. The stone is placed on the turntable at a distance R from the axis, and the speed of rotation is slowly increased to 33 rev/min. When the experiment is repeated for several different values of R, it is found that the stone remains on the turntable if $R < 21$ cm and that it slides off with increasing speed of rotation if $R > 21$ cm. Determine μ_s from these data.

56. (II) A Ferris wheel in an amusement park has a radius of 30 m and makes one complete turn every 75 s. Calculate the normal force that a passenger of mass 60 kg experiences through the seat of the pants (the seat bottom is parallel to the ground) when the passenger is (a) at the bottom of the path, nearest the ground, and (b) at the maximum height of the path.

57. (II) Assume that the acceleration of the Moon due to Earth's gravity is 0.0027 m/s^2. What is the velocity of the Moon with respect to Earth if the period of the Moon's motion around Earth is 28 days? Do not look up the distance between Earth and the Moon; instead, calculate it. Compare this result with the distance given in Appendix III–1.1.

58. (II) A fighter pilot makes a dive almost vertically down and pulls up while traveling at 1500 km/h in a turn of radius 1.75 km. How many g's will the fighter pilot feel at the bottom of the dive? Because the pilot will black out if the number of g's is greater than 11, is this a safe maneuver? (The number of g's is the acceleration in units of $g = 9.80$ m/s^2.)

59. (II) A mass of 1.00 kg hangs from a rope placed through a hole in a smooth, frictionless table. At the other end of the rope is attached a puck of mass 400 g, 80 cm from the hole in the table. The puck swings in a circular orbit around the hole (Fig. 5–42). With what speed must the puck rotate to keep it 80 cm from the hole?

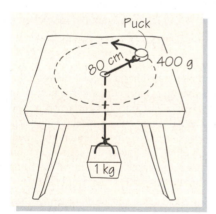

▲ **FIGURE 5–42** Problem 59.

60. (II) A mass m at the end of a string of length r moves at constant speed v in a circle on a frictionless table. (a) Find the tension in the string in terms of m, r, and v. (b) A second mass identical to the first is attached at the midpoint of the string and the two are whirled; the speed of the outer mass is again v. Draw free-body diagrams for the two masses and calculate the tensions in terms of m, r, and v.

61. (II) Consider the *conical pendulum*, a mass on the end of a massless string, with the other end of the string fixed on a ceiling. Given the proper push, this pendulum can swing in a circle at a given angle θ, maintaining the same height h throughout its swing, as shown in Fig. 5–43. (a) What is the free-body diagram for such a pendulum? (b) If the mass of the pendulum is 0.2 kg, the length of the pendulum is 0.5 m, and the angle at which it swings is $\theta = 10°$, what is the speed of the mass as it swings?

▲ **FIGURE 5–43** Problems 61 and 63.

62. (II) An object moves along a trajectory that is parallel to the unit vector $\hat{r}$ of some fixed coordinate system. Are any of the statements below true? Which ones and why or why not?
(a) Any force acting on the object is perpendicular to $\hat{r}$.
(b) If the magnitude of the velocity of the object is unchanging, the force acting on it is zero.
(c) Any force acting on the object must be aligned with $\hat{r}$.

63. (II) A heavy bob is attached to one end of a string whose other end is attached to a hook on the ceiling. The system acts as a conical pendulum (see Fig. 5–43), with the string making an angle of 30° with the vertical and the bob traveling in the horizontal plane at an angular velocity of $\pi/2$ rad/s. How long is the string?

64. (II) A small mass slides without friction in a horizontal circular path around the sides of a circular bowl. The bottom of the bowl may be described as a parabola, with the height h above the bottom varying quadratically with the distance r from the axis: $h = br^2$. The mass is observed to move in its circular path with a speed v. What is the height of the path?

65. (II) A small block slides in a horizontal circle on the inside of a conical surface, with the cone making an angle of 44° with the vertical (Fig. 5–44). Assuming that there is no friction between the block and the surface and the block slides with an angular speed of 3.8 rad/s, at what vertical height above the apex of the cone does the block slide?

▲ **FIGURE 5–44** Problem 65.

66. (II) Two light strings 1.0 m in length are attached to a vertical support 1.0 m apart, and a mass of 5.0 kg at the end of the two strings is whirled about the vertical z-axis (Fig. 5–45, see next page). Both strings are taut, so that they and the vertical support form an equilateral triangle. The tension in the upper string is measured to be 150 N. (a) What is the tension in the lower string? (b) How much time does it take for the apparatus to make one complete circuit around the vertical support?

▲ **FIGURE 5–45** Problem 66.

67. (II) The road in Example 5–16 is now in better condition, and the coefficient of static friction between the tires and the road is 0.8. With what maximum constant speed can the motorist now negotiate the curve? Friction acts here both to provide a centripetal force and to help oppose the vertical forces. Is there one and only one speed with which the curve can be taken, as in Example 5–16? Why or why not?

68. (II) A train of mass 1.5×10^5 kg is traveling horizontally at 80 km/h and rounding a bend whose radius of curvature is 2 km. At the same time it is decelerating at a rate of 0.2 m/s^2. The length of the train is negligible compared with the size of the bend, and the train can be treated as a point. What net force does the track exert on the train? Give an approximate answer in which the net change in speed is small compared to the speed itself. Such questions help engineers decide how "robustly" a track must be constructed or how much to bank it.

69. (II) A small puck of mass $m = 0.1$ kg moves in a circle of radius 0.3 m on a table top; the puck is tied with a massless string to a tether at the origin. The coefficient of kinetic friction between the puck and the table top is $\mu_k = 0.25$. At $t = 0$ s, the puck has a velocity in the $+y$-direction of magnitude 8.0 m/s. (a) What is the tension in the string at $t = 0$ s? (b) What is the tension in the string at the end of one revolution?

70. (II) A pendulum hangs at rest from a hook in a ceiling of a building. The building is located at a latitude such that the radius vector from the center of Earth to the building makes an angle θ with Earth's axis of rotation. Assume that the force of gravity is the same everywhere and points directly to the center of Earth. What is the angle that the pendulum makes with the vertical because of Earth's rotation? (By definition, Earth rotates about its axis once a day.)

71. (III) A satellite of mass 3000 kg travels in a circular orbit 180 km above Earth, where the acceleration due to gravity is 5 percent smaller than on Earth's surface. Assume that, in a year, the satellite loses 5.0 km in altitude because of the drag of the extremely thin atmosphere at that altitude. What would you estimate the density of air to be at that altitude given that the effective area of the satellite is 6.0 m^2 and the drag coefficient C_D in Eq. (5–30) is 1.0?

72. (III) Consider a ball thrown outward from the center of a platform that rotates counterclockwise with uniform angular velocity ω (Fig. 5–46). An observer standing off the platform (in an inertial reference frame) will describe the ball as moving with uniform velocity in a straight line. (Ignore the effect of gravity; imagine looking down on the platform from above, so that you do not see the up-and-down motion of the ball.) The inertial observer will see that the ball reaches a horizontal distance r from the center in time $t = r/v$, where v is the speed of the ball. (a) Show that, in time t, a point at a radius r on the platform will

have moved a distance $d = v\omega t^2$. (b) An observer moving with the platform will see the ball curve away to the right (as seen from the center of the platform). Show that the perceived acceleration is perpendicular to the velocity vector and that its magnitude is $2\omega v$. (c) What is the direction of the perceived acceleration of a ball thrown by a person on the platform from the rim toward the center? It is not surprising that an observer in an accelerated frame sees force-free motion that nevertheless undergoes acceleration. Under certain circumstances—for example, in the study of global wind motion on a rotating Earth—it is convenient to study motion as seen in an accelerating frame. The frame-imposed acceleration is then attributed to a purely fictitious force, which, for rotating systems like the one treated in this problem, is called the *Coriolis force*.

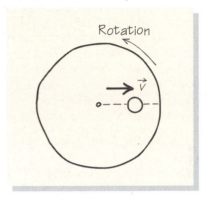

▲ **FIGURE 5–46** Problem 72.

General Problems

73. (I) A 5.0-kg sphere at the end of a 1.2-m cable swings in a horizontal circle on a frictionless surface at the rate of one revolution every 1.4 s. What is the tension in the cable?

74. (II) A small die that is placed 4 in from the center of a turntable begins to slide. The turntable is rotating at 45 rev/min. What is the coefficient of static friction between the die and the turntable?

75. (II) Three masses (from left to right: 0.3, 0.4, and 0.2 kg) are connected by light cords to make a "train" sliding on a frictionless surface. They are accelerated by a constant horizontal force $F = 1.5$ N that pulls the rightmost mass to the right. What is the tension T in the cord (a) between the 0.3- and 0.4-kg masses and (b) between the 0.4- and 0.2-kg masses?

76. (II) A string 6.95 m long is strung between two pegs (4.96 m apart) on a ceiling. A mass of 3.88 kg is attached to a point 2.96 m along the string. What are the tensions in the two segments of the string?

77. (II) Consider a system of masses connected by light ropes that pass over massless and frictionless pulleys (Fig. 5–47). (a) When

▲ **FIGURE 5–47** Problem 77.

m_1 is displaced vertically by Δx_1, what is the displacement Δx_2 of m_2? (b) For $m_1 = 1.2$ kg and $m_2 = 1.8$ kg, calculate the respective accelerations of the two masses. (c) What is the tension in the string for the masses given in part (b)?

78. (II) Masses $m_1 = 0.80$ kg and $m_2 = 1.10$ kg are connected by a taut rope. Mass m_2 is just over the edge of a ramp inclined at an angle of $\theta = 25°$, as in Fig. 5–48, and the masses have a coefficient of kinetic friction $\mu_k = 0.25$ with the surface. At $t = 0$ s, the system is given an initial speed of $v_0 = 1.2$ m/s, which starts mass m_2 down the ramp. (a) Draw the free-body diagram for each mass. (b) Solve the equations of motion to predict the motion of the system with time. Assume that the rope is long enough so that mass m_1 does not hit the pulley.

▲ FIGURE 5–48 Problem 78.

79. (II) Two masses of 5 and 10 kg are tied together by a massless spring. A force of 12 N acts on the 10-kg mass directly away from the 5-kg mass. At a particular instant the 5-kg mass has an acceleration of 3 m/s² in the direction of the larger mass. What is the acceleration of the 10-kg mass?

80. (II) A light cord attached to a ball will break if the tension in the string exceeds 40 N. The ball has a mass of 150 g. A student swings the ball in a circle in the vertical plane. (a) You bet that the cord will break when the ball is at the top of the circle. Will you win the bet? (b) If the length of the cord is 1.0 m, how fast will the ball be moving when the cord breaks?

81. (II) One of the entertainments at the carnival is a rotating cylinder. The participants step in and place themselves against the interior wall. The cylinder starts to rotate more and more rapidly, and at some point the floor falls away, leaving the customers stuck like so many flies to the wall. If the cylinder were to slow down without the floor coming back up, the participants would begin to slip down. In terms of the relevant parameters, express the rotational speed ω at which this happens.

82. (II) A motorcycle moves in a horizontal circular path on the inside surface of a vertical cylinder of radius 8 m. Assuming that the coefficient of static friction between the wheels of the motorcycle and the wall is 0.9, how fast must the motorcycle move so that it stays in the horizontal path?

83. (II) Consider the conical pendulum described in Problem 61. Express the angular velocity in terms of the string angle θ and the string length ℓ.

84. (II) A tractor of mass 800 kg is pulling a sled loaded with 1450 kg of hay bales. The coefficient of kinetic friction between the sled and the ground is 0.68. (a) What horizontal force must the tractor exert to move at constant speed? (b) What is the tension in the rope between the tractor and sled? (c) If the tractor stops, how much horizontal force must it exert to get the sled moving again if $\mu_s = 0.70$?

85. (II) A bicyclist traveling at 10 m/s rides around an unbanked curve. If the coefficient of friction between the tires and the road is $\mu = 0.4$, what is the shortest turn the bicyclist can safely make? Is the coefficient of friction here static or kinetic?

86. (II) Two cars are traveling at 60 mi/h, one behind the other. The driver of the second car reacts by braking 0.8 s after she observes the sudden braking of the car ahead of her. The front car has a mass of 1200 kg, and the coefficient of friction between the tires and the road with the brakes applied is 0.8. The second car has mass 1600 kg, and the coefficient of friction with the brakes applied is 0.7. How far behind must the second car have been in order to avoid hitting the first car?

87. (II) A stunt motorcyclist rides with uniform speed on the inside rim of a vertical circular ramp of radius 8 m (Fig. 5–49). How fast must the motorcyclist travel to avoid leaving the surface at the top of the loop?

▲ FIGURE 5–49 Problem 87.

88. (II) A pith ball is a typically small ball that is very light and can be given electrical charge—electrical charges exert forces on one another. If two such balls, labeled 1 and 2, are given charges q_1 and q_2, respectively, then the force on one due to the other is repulsive (directed along the line between them and tending to push them apart), with magnitude roughly $(9 \times 10^9) \times q_1 \times q_2 \times d^{-2}$ N, where d is their separation, measured in meters, and the unit of charge is the coulomb (C). Two tiny pith balls, mass 0.5 g, are hung from nearly the same point on the ceiling by identical strings, 60 cm long, and given identical electrical charges q. At equilibrium, the balls separate, and the strings are found to make an angle of 60° with one another. Find q.

89. (III) The new white belt of a long horizontal conveyor is moving with a constant speed $v = 3.0$ m/s. A small block of carbon is placed on the belt with zero initial velocity relative to the ground. The block will slip a bit before moving with the belt, leaving a black mark on the belt (Fig. 5–50). How long is that mark if the coefficient of kinetic friction between the carbon block and the belt is 0.20 and the coefficient of static friction is 0.30?

▲ FIGURE 5–50 Problem 89.

As the elevator moves up and down, the force of gravity and the contact forces from the floor do work on the people inside. This work can be linked to changes in energy, and eventually to the principle of the conservation of energy.

Work and Kinetic Energy

A pitcher throws a baseball—can we predict the motion of the ball? In the previous chapters we've seen that as long as we have enough information about initial positions, velocities, and the forces that act on the ball, we can, in principle, use Newton's laws to predict everything that is predictable about the ball's motion. Nevertheless, the ball has a complex shape and spin, and we would have a hard time being certain that we had accurate information on the ball's interaction with the air. Even if we did have all the correct information, the equations would be complex and unlikely to have analytic solutions. Nowadays we can resort to numerical computation on ever more powerful computers when details of a complex motion are needed. For example, the motion of a rocket bound for Pluto is complex because it is affected by the gravitational forces due to Earth, the Sun, and other planets along the way. These forces are all calculable, but Newton's second law must be solved numerically, and such a solution has to be carried out with high precision to get a suitable trajectory.

Scientists of the eighteenth and nineteenth centuries did not have computers and instead developed powerful analytical methods to deal with certain aspects of the motion of objects. Their research led to a much deeper understanding of the dynamical properties of systems than the "black box" of computer programs can possibly provide. One of their most important discoveries was the concept of *energy*. Energy is a quantity that an object or a physical system possesses. As we shall learn in future chapters, there are many forms of energy—kinetic energy, potential energy, thermal energy, electrical energy, chemical energy, nuclear energy, and mass energy. The names of these different forms of energy were due to the circumstances in which they were studied, but the forms of energy are closely linked and the names are unimportant. When considering energy, what is important is to keep in mind the deep physical law, known as the **principle of the conservation of energy**, which states that energy cannot be created or destroyed, only transformed from one form to another. In other words, all the forms of energy are interchangeable, fundamentally one and the same. The law of the conservation of energy is one of the most fundamental of all physical laws, and we will be occupied with it in this chapter and the next.

In this chapter, we are concerned with only one form of energy: energy of motion, or more technically, *kinetic energy*. We start our study of kinetic energy through consideration of a precisely defined quantity called *work*. A force acting on an object can do work on the object when the object moves. Thus all the forces acting on a baseball moving through the air do work on it. The hand that throws the ball, gravity, and the friction, or drag, forces due to the air all do work on the baseball and in doing so cause the kinetic energy of the ball to change. The close relation between work and energy is a practical tool, useful for learning about the aspects of the baseball's motion that don't depend directly on time—for example, how fast the baseball is moving when it is at a certain location. In Chapter 7, we'll change our focus away from work and toward seeing how we can view the connection between work and energy as a conservation law[†] for energy. In doing so, we'll begin to expand our ideas about energy to forms other than kinetic energy.

6–1 Kinetic Energy, Work, and the Work–Energy Theorem

We begin by rethinking some simple features of motion in one dimension. Consider, for example, a bobsled with mass m being pushed by some bobsledders on flat ground (Fig. 6–1) over an icy surface. (We'll assume there is no friction between the sled and the surface.) The bobsled's motion is described by a position x and a velocity v—as the motion is strictly one dimensional, we can dispense with vector notation. When the bobsled is at rest, we assign it a value of **kinetic energy** $K = 0$. When the bobsled moves, its kinetic energy will have a nonzero value. In other words, its kinetic energy has something to do with its motion. This energy is supplied by the forces applied by the three men pushing the sled. We have already learned that if an object such as the bobsled changes its velocity from $v = 0$ to a nonzero value, it must, by definition, accelerate. For our initial analyses we'll suppose this to be a *constant acceleration a*. After moving a distance Δx, subject to this acceleration and starting from rest, the bobsled will acquire a speed given by [see, e.g., Eq. (2–24)]

$$2a \, \Delta x = v^2. \tag{6–1}$$

▲ **FIGURE 6–1** As the bobsled is pushed, it gains kinetic energy.

We also know that a net force F_{net} leads to an acceleration, and here this force is supplied by the men pushing the sled. For constant acceleration the net force will be *constant*. Thus our result may be rewritten in a different form. Multiplying by $\frac{1}{2}m$ and replacing the combination ma by F_{net} gives

$$F_{net} \, \Delta x = \tfrac{1}{2}mv^2. \tag{6–2}$$

This equation is the root of all that will follow in this chapter. On the right we have an expression that is associated with the motion of our bobsled, which we identify with the bobsled's **kinetic energy**. On the left are the quantities that produce this kinetic energy: a net force acting on the bobsled and the sled's displacement. This side of our relation is the **work** W_{net} that the net force does on the bobsled. Equation (6–2) is the starting point for all that follows in this chapter in that we will find ways to generalize work to situations where the force can vary and where the motion is more than one dimensional. Similarly, we will find ways to generalize the kinetic energy to cases where the motion is not one dimensional and the object is not treated as a point. The main point to re-

---

[†]It is important to realize from the outset that when we talk about a conservation law, we mean that there is something that stays constant in time, not the more colloquial meaning of not wasting it!

member throughout is that *when a net force acts to displace an object, it transmits (kinetic) energy to it.* The energy transmitted will be W_{net}, which is *the net work done by the net force on the accelerating object,* or briefly, the *work.* In the case of our constant force in one dimension,

$$W_{net} \equiv F_{net}\,\Delta x. \qquad (6\text{–}3)$$

CONSTANT NET FORCE IN ONE DIMENSION

The object's resulting *kinetic energy K* is given by

$$K \equiv \tfrac{1}{2}mv^2. \qquad (6\text{–}4)$$

KINETIC ENERGY

If the object is not initially at rest but has initial velocity v_0, then the results of Chapter 2 for constant acceleration show us that Eq. (6–1) is replaced by

$$v^2 - v_0^2 = 2a\,\Delta x \qquad (6\text{–}5)$$

and Eq. (6–2) is replaced by

$$W_{net} = \tfrac{1}{2}mv^2 - \tfrac{1}{2}mv_0^2 = \Delta K, \qquad (6\text{–}6)$$

WORK–ENERGY THEOREM

where for a constant force, $W_{net} = F_{net}\,\Delta x$. This equation is known as the **work–energy theorem**. It cements our interpretation that the left side of Eq. (6–2), the net work, is the kinetic energy transmitted. The work done by the net external force to accelerate the object from its initial velocity v_0 to its final velocity v equals the *change* in the object's kinetic energy from its initial value $K_0 = \tfrac{1}{2}mv_0^2$ to the final value $K = \tfrac{1}{2}mv^2$. The kinetic energy change is $\Delta K = K - K_0$. If we write Eq. (6–6) as

$$K_0 + W_{net} = K, \qquad (6\text{–}7)$$

then we can view the work–energy theorem as saying that if we start with an object with kinetic energy K_0 and do net work on it, the sum of these is the final kinetic energy. Again, the net work is the amount of additional kinetic energy that our object has been given. We will find more general versions of the work–energy theorem for motion in three dimensions and nonconstant forces.

Remember, work is something done to an object by a force as the object is displaced (Fig. 6–2); kinetic energy is something that the object possesses by virtue of having work done on it.

The *sign* of the work can be either positive or negative, and accordingly the kinetic energy can either increase or decrease. We see from the definition of work in Eq. (6–3) that the sign of W_{net} is determined by the relative sign of F_{net} and of the displacement Δx. If the net force is directed in the same direction as the displacement (they have the same sign), then the net work done by that force is positive, and it increases the kinetic energy of the object to which the force is applied. If the force acts in a direction *opposite* to the displacement (they have opposite signs), the work done is negative, and the object's kinetic energy decreases. Consider our bobsled. Suppose that the bobsledders have acted to give the sled a forward velocity and decide that they want to bring it to a halt, always on level ground. They therefore run to the front of the sled and apply a force opposite to its motion, doing work on the sled even as the sled continues to move forward (Fig. 6–3). The displacement is to the right while the pushing force is to the left. The directions of the force and the displacement are opposite, so their product (the work done) is negative, and the change in kinetic energy is negative—the sled slows down.

A cautionary note: *Work* and *energy* have technical meanings which we have been careful to spell out, but these terms also have colloquial meanings that may differ from our definitions. For example, holding a bag of groceries while waiting for a bus would ordinarily be described as work. In our definition this cannot be work because work involves displacement and the bag of groceries does not move. In a technical sense the

▲ **FIGURE 6–2** As he raises the barbell from a position near the floor to a position at his shoulder, this weightlifter does positive work. However, the weightlifter does no work in holding the barbell stationary at his shoulder.

> Bobsledders do work on sled to slow it down.

▶ **FIGURE 6–3** Bobsledders bring the sled to a stop by applying a force opposite to the displacement of the sled.

force exerted by our muscles does no work on the bag. The word *energy* also has a variety of colloquial meanings, but in physics the meaning of the word is very tightly prescribed, even if we deal with energy in many different contexts.

THINK ABOUT THIS . . .

WHAT WORK IS DONE WHEN WE LIFT AN OBJECT FROM THE FLOOR?

We remarked above that no net work is done on a bag of groceries when it is held steady—the fact that the person holding the bag may feel tired as a result is irrelevant in the definition of work, which requires a displacement. That same person may also have lifted the bag from the floor. Is there any net work involved in that case? We can break up the motion into three parts: (i) starting at floor level the bag is brought up to lifting speed in a short time (and moves a short distance in that time); (ii) the bag is raised smoothly, at a constant speed, to the desired height; and (iii) in a short period of time the bag is brought to a halt. During stage (i) the person lifting must exert an upward force exceeding the force of gravity on the bag. There is a net force, there is a displacement in the same direction, and there is positive net

work done on the bag. The bag accelerates, gaining kinetic energy. During stage (ii) there is a displacement but no net force—the contact force of your hand is equal and opposite to the force of gravity—and hence no net work. The kinetic energy of the bag remains unchanged. Finally, during stage (iii) there is a net force, a (small) displacement, and net work is done. The kinetic energy changes, but the change is negative. The net work is also negative and is due primarily to gravity, which points down, even as the displacement continues upward. Over all three stages we can say that no net work is done. The positive net work done in stage (i) is canceled by the negative net work done in stage (iii): The bag started at rest and it ends at rest.

Units of Work and Energy

Work and kinetic energy must have the same dimensions. Their units correspond to these dimensions. Work is measured in $(\text{newtons})(\text{meters}) = (\text{kg} \cdot \text{m/s}^2)(\text{m}) = (\text{kg})(\text{m/s})^2$, and kinetic energy is measured in $(\text{kg})(\text{m/s})^2$; these are indeed the same. The unit of work and energy is so important that it has a special name in SI, the **joule** (J). One joule is the work done by the force of one newton pushing an object a distance of one meter along the direction of the force. Thus $1 \text{ J} = 1 \text{ N} \cdot \text{m}$.

In the cgs system, the unit of energy is the **erg**, which is the work done by a force of one dyne in moving an object a distance of one centimeter. The conversion is the following:

$$1 \text{ erg} = (1 \text{ dyne})(1 \text{ cm}) = (1 \text{ g} \cdot \text{cm/s}^2)(1 \text{ cm})$$

$$= (1 \text{ g} \cdot \text{cm}^2/\text{s}^2)(1 \text{ kg}/10^3 \text{ g})(1 \text{ m}^2/10^4 \text{ cm}^2)$$

$$= 10^{-7} \text{ kg} \cdot \text{m}^2/\text{s}^2 = 10^{-7} \text{ J}.$$

EXAMPLE 6–1   Consider the kinetic energies given in Table 6–1.

(a) Given that the electron mass is 0.9×10^{-30} kg, how fast is it moving in its orbit around the nucleus?

(b) If the sprinter runs 100 m in 10 s, what is her mass?

(c) Given that Earth's mass is 6×10^{24} kg, how fast is it moving around the Sun?

Strategy   In each of the cases we can use Table 6–1 and the relation $K = \frac{1}{2}mv^2$ [Eq. (6–4)] and solve for the unknown.

Working It Out (a) Here the unknown is the speed v. It follows from $K = \frac{1}{2}mv^2$ and from the given values of K and m that the electron speed is

$$v = (2K/m)^{1/2} \approx 10^6 \text{ m/s}.$$

(b) In this case we know $v = 10$ m/s; hence we use $m = 2K/v^2 = 20$ kg. This doesn't mean that we are dealing with a five-year-old prodigy sprinter! Rather, Table 6–1 only gives values to the nearest power of 10.

(c) With Earth's mass m, we find its orbital speed from $v = (2K/m)^{1/2} = 2 \times 10^4$ m/s.

TABLE 6–1 • Some Orders of Magnitude for Kinetic Energies

System	Kinetic Energy (J)
Electron in orbit around a nucleus	10^{-18}
Molecule of air at room temperature	10^{-17}
Electron in a TV tube	10^{-15}
Walking ant	10^{-8}
Falling raindrop	10^{-3}
Running human	10^3
Automobile on a highway	10^5
Cruising airplane	10^{11}
Large earthquake	10^{17}
Earth in orbital motion around the Sun	10^{33}

THINK ABOUT THIS . . .

HOW CAN WE ESTIMATE THE ENERGY OF A SNOW AVALANCHE?

Estimating a quantity always involves having some knowledge of the physical system involved. In this case, you may know that a snow avalanche typically starts in a steep mountain gulley and that it is a "shelf" of snow forming a layer over a previously frozen surface which slides down the gulley. This layer will likely have formed as the result of a snowfall on the earlier surface, so a reasonable value for its thickness is 15 cm (about 6 in), and you could guess that a plaque perhaps 50 m^2 in area breaks off and slides. The volume is thus $50 \times 50 \times 0.15$ m$^3 = 375$ m^3. You might also know that a foot of snow is equivalent to about an inch of rain, suggesting that snow is roughly 10 percent as dense as liquid water, which has a mass density of 1 g/cm$^3 = 10^3$ kg/m^3. We therefore give our snow shelf a density of 10^2 kg/m^3, and the mass of the moving snow is the product of volume and density, around 4×10^4 kg. At this point we can use the work–energy theorem. We might take a height $h = 500$ m for our mountain valley, and we'll assume that the avalanche moves approximately vertically. The work done by gravity uses the fact that the force has magnitude $mg = (4 \times 10^4 \text{ kg})(10 \text{ m/s}^2) = 4 \times 10^5$ kg · m/s^2. We then estimate the work done by gravity as $W = F \times h = mg \times h = (4 \times 10^5)(500)$ J $= 2 \times 10^8$ J. This is our estimate for the kinetic energy. It is interesting to convert this estimate to a speed for the mass of snow: We have $v^2 = 2K/m = 2(mgh)/m = 2gh$. (Not surprisingly, this is the speed squared we would find from the kinematics of an object of any mass falling under constant acceleration g.) Numerically, $v^2 = 2(10)(500)(\text{m/s})^2$, or $v = 100$ m/s. This is nearly 200 mi/h, and even though it does not take into account drag due to air, it is not far off of measured values.

CONCEPTUAL EXAMPLE 6–2

A furniture mover pushes a sofa across a carpeted floor. The mover applies a horizontal force of magnitude F to the sofa, and he and the sofa move with uniform velocity. Discuss the notion of work and energy for this situation. What is different if the sofa is pushed with the same force on the frictionless surface of a slick floor (but the mover has nonslip soles on his shoes)?

Answer The sofa moves with uniform velocity, that is, it does not accelerate. Therefore there is no *net force* acting on it. The force applied by the mover has magnitude F, and therefore the force of kinetic friction between sofa and floor has the same magnitude and points in the opposite direction. If the net force is zero, no work is done on the sofa; equivalently, the sofa's kinetic energy is unchanging. (There is no motion in the vertical direction, and we can treat the entire problem as a one-dimensional one.) The mover does work $W = F \Delta x$ in displacing the sofa by Δx, but this work alone does not go into increasing the kinetic energy of the sofa; only the net work does that. Is there nevertheless a transfer of energy associated with the work done by the mover? Yes, the work done by the mover *must* involve a transfer of energy, but it does not go into the kinetic energy of an object. Instead the work is just the right amount to overcome the work done by friction in response to the sofa's motion. The work done by the mover goes into heating the sofa's feet and the floor and into abrasion. We will learn about this in more detail later.

In the case of the slippery floor (no kinetic friction), the *net* force is just the force exerted by the worker. Net work is done as the sofa moves, so the kinetic energy of the sofa changes, that is, there is an acceleration. This increased kinetic energy is given by the work–energy theorem [Eq. (6–6)].

What Do You Think? We speak of the mover's energy being used to push the sofa. What energy are we talking about? *Answers to **What Do You Think?** questions are given in the back of the book.*

EXAMPLE 6–3 In a movie stunt, a car of mass 1200 kg falls a vertical distance of 24 m starting from rest (Fig. 6–4a). What is the work done by the force of gravity on the car? Use the work–energy theorem to find the final velocity of the car just before it hits the water. (Treat the car as a pointlike object.)

Setting It Up We draw a sketch of the car falling in Fig. 6–4b. We only need the vertical y-axis, which points upward. The car, of given mass m, has a downward displacement $\Delta y = y_f - y_i = 0\text{ m} - 24\text{ m} = -24\text{ m}$ and an initial speed $v_0 = 0$. We want to find the work done by gravity on the car and the car's final speed v_f.

Strategy The force of gravity is the only force acting, so it makes the only contribution to the net work. We can calculate the work done by gravity and then use the work–energy theorem to find the final speed.

Working It Out Gravity has magnitude mg and is oriented downward, so the y-component of the net force is $F_{\text{net}} = -mg$. Therefore the net work is

$$W_{\text{net}} = F_{\text{net}}\,\Delta y = (-mg)\,\Delta y$$
$$= (1200\text{ kg})(-9.8\text{ m/s}^2)(-24\text{ m}) = 2.8 \times 10^5\text{ J}.$$

Note that this work is positive. We can now use the work–energy theorem, Eq. (6–6), to find v_f. The initial kinetic energy, K_i, is zero because the car starts from rest. Thus we have

$$\Delta K = K_f = W_{\text{net}} = mg\,\Delta y, \qquad \tfrac{1}{2}mv_f^2 = mg\,\Delta y.$$

The mass cancels and we find that

$$v_f = \sqrt{2g\,\Delta y} = \sqrt{2(-9.8\text{ m/s}^2)(-24\text{ m})}$$
$$= 22\text{ m/s}, \quad \text{about 79 km/h (nearly 50 mi/h)}.$$

(We could have solved this problem in Chapter 2; we consider it here to bring in the concepts of work and kinetic energy.)

What Do You Think? If the mass of the car is doubled, is (i) the work done by gravity and (ii) the final velocity of the car (a) doubled, (b) halved, or (c) the same? (Answer this without doing any calculations.)

Gravity does work on car.

▶ **FIGURE 6–4** (a) Gravity pulls the car down. (b) The car leaves the draw bridge at $y = 24$ m.

Work Done by Individual Forces

It is possible to calculate the work done by each force acting on an object. For example, if each force is constant and all the forces act along the x-axis, then the work done on an object by a particular force $\vec{F} = F\hat{i}$ as the object moves through a displacement Δx is given by

$$W = F\,\Delta x.$$

We anticipated being able to think about the work done by individual forces in Conceptual Example 6–2 when we calculated the work done by the mover as distinct from the net work. Remember, however, that the work–energy theorem, Eq. (6–6), specifies the *net* work, that is, the work done by the *net* force. To apply it, we have to consider *all* the forces acting on an object. When a piano is slowly lowered at constant speed by a rope, the force of gravity is canceled by the rope tension and there is no acceleration. Here, the net force is zero, so the net work is also zero. This is consistent with the fact that there is no change in kinetic energy and no change in speed.

Because the net force is the sum of the individual forces acting, the net work can be decomposed into a sum of the work done by each of the individual forces. *This is a result that will hold for the most general definition of work.* It is sometimes simpler to find the work W_n done by individual forces F_n and then take the algebraic sum $\sum W_n$ to find the net work than to find the net force and calculate the work done by it. In the case of the piano lowered by a rope at constant speed, positive work is done by gravity as the piano is lowered. The tension of the rope, however, which points upward, does the same magnitude of *negative* work on the piano. The forces cancel and the net work is zero in this case.

CONCEPTUAL EXAMPLE 6–4 You are standing in a stationary elevator. Suddenly the elevator accelerates upward for some distance and then stops very suddenly. What do you expect will happen to *you*? Explain this in terms of work and energy.

Answer In a stationary elevator, I know I am subject to the force of gravity (downward), but since I am not falling, there must be an upward normal force due to the presence of the solid floor under my feet to compensate and cancel gravity. When the elevator accelerates, I accelerate with it. The forces that act on *me* are still the force of gravity and the normal force. Now, however, the normal force exceeds that of gravity, so that there is a net upward force that accelerates me. My speed increases with that of the elevator, and therefore my kinetic energy increases. This energy is supplied by the work done on me by the net force, in accordance with the work–energy theorem. When the elevator suddenly stops, I have a certain amount of kinetic energy, associated with an upward velocity. If the elevator stops quickly enough, I would leave the floor. In that case I would be subject only to the downward force of gravity. There is negative work done on me by that force as I continue to rise, and the work–

energy theorem implies that my kinetic energy has to decrease as I move upward. At some point my kinetic energy goes to zero. As I fall back to the floor of the elevator, my kinetic energy increases again because the work done by the force of gravity is positive. When I hit the floor of the elevator, I experience an upward normal force larger than the force of gravity, so that the net work done is negative—the normal force points upward while my displacement as I come to rest is in the downward direction. This brings me to rest again, removing the kinetic energy that I had in coming down. (In a real situation, I would reduce the acceleration by bending my knees.) Once both the elevator and I are stationary again, the force of gravity and the normal force cancel, so that there is no further net work done on me by external forces. My kinetic energy is zero.

What Do You Think? Describe in terms of work and energy what happens to the elevator together with its passenger seen as a single system—in other words, imagine you were looking at the elevator from the outside, without being able to see the various actions within. [*Hint*: What is it that moves and stops the elevator?]

EXAMPLE 6–5 A box of books of mass 100 kg is pushed with constant speed in a straight line over a rough floor with a coefficient of kinetic friction 0.2. Find the work done by the force that pushes the box if the box is moved 3 m across the floor.

Setting It Up We draw a free-body diagram in Fig. 6–5, including an *x*–*y* coordinate system and the forces. We know the mass *m* of the box of books, the coefficient of kinetic friction μ_k between the box and the floor, and the distance *d* that the box moves along the *x*-axis. The box moves with constant speed. We want to find the work *W* done by the force *F* that pushes the box.

Strategy Because the box moves with a constant speed (no acceleration), there is no *net* force on the box. Therefore, we can set the net forces in both *x*- and *y*-directions equal to zero. This allows us to determine the pushing force $\vec{F}$, and we can then find the work it does on the box.

Working It Out Because the box moves with a constant velocity, the *net* horizontal force must vanish. Thus the pushing force $\vec{F}$, must be equal in magnitude but opposite in direction to the force of friction $\vec{f}$, whose magnitude is given by $f = \mu_k F_N$. The forces in the vertical direction must cancel and so $F_N = mg$, and friction has magnitude $\mu_k mg$. Hence the magnitude of $\vec{F}$ is also $\mu_k mg$. The force $\vec{F}$ is oriented along the same direction as the displacement, so that the work done by the pushing force is positive; this work is

$$W = Fd = \mu_k mgd$$
$$= (0.2)(100 \text{ kg})(9.8 \text{ m/s}^2)(3 \text{ m}) = 6 \times 10^2 \text{ J}.$$

What Do You Think? Suppose the force is unchanged but the coefficient of friction gradually increases as the box moves into a re-

gion of rougher floor. Use the work–energy theorem to explain qualitatively what happens.

(a)

▶ **FIGURE 6–5**
(a) A variety of forces act on the box of books in motion. This picture of the situation includes a coordinate system.
(b) Free-body diagram for a box of books.

(b)

EXAMPLE 6–6

A crate of mass 96 kg is pushed across a horizontal floor by a force $\vec{F}$. The coefficient of kinetic friction between the crate and the floor is $\mu_k = 0.27$. The crate moves with uniform velocity. What is the magnitude of $\vec{F}$? Suppose that at some point the crate passes onto a new section of floor, where $\mu'_k = 0.085$. The pushing force on the crate is unchanged. After 1.25 m on the new section of the floor, the crate moves with a speed $v_f = 2.3$ m/s. What was the original speed v_i of the crate?

Setting It Up   We illustrate the situation in Fig. 6–6a, including an x–y coordinate system. We know the crate mass m, the coefficients of kinetic friction for the two floor sections, the final speed of the crate after being pushed a distance $\Delta x = 1.25$ m in the second floor section, and the fact that the speed was constant in the first floor section. We want to find the magnitude of the pushing force and the speed of the crate in the first floor section.

Strategy   We draw a free-body diagram in Fig. 6–6b and 6c and denote all the forces acting on the crate: gravity, normal force, pushing force, and friction. With a zero net force (constant speed) on the first section, we have enough information to find each of the forces there, including the pushing force. This pushing force acts in the second floor section, where we can use the work–energy theorem in a situation where the friction force has changed magnitude. This will tell us by how much the crate's kinetic energy has changed, and from that we can find the original speed.

Working It Out   The free-body diagram permits us to find the pushing force, which, because the crate moves with constant speed on the first floor section, must balance the force of friction. This friction force has magnitude

$$f = \mu_k mg = (0.27)(96 \text{ kg})(9.8 \text{ m/s}^2) = 2.5 \times 10^2 \text{ N},$$

and the pushing force $\vec{F}$ must have just this magnitude: $F = 2.5 \times 10^2$ N. We have used the cancellation of the vertical forces to find the magnitude of the normal force and hence the magnitude of the friction force.

On the new floor section, the force of friction is less, whereas the pushing force remains the same. Thus there is a (constant) net force in the direction of motion and the crate accelerates uniformly. On the new section, the force of friction has magnitude $f' = \mu'_k mg$. Therefore the net force on the crate has magnitude $F_{net} = F - f' = (\mu_k - \mu'_k)mg$ and acts in the direction of motion of the crate. The net work done on the crate as it moves a distance Δx on the new section of floor is then

$$W_{net} = F_{net}\,\Delta x = (\mu_k - \mu'_k)mg\,\Delta x.$$

According to the work–energy theorem, this is the increase in kinetic energy of the crate as it moves over the new section of floor:

$$W_{net} = K_f - K_i = \tfrac{1}{2}mv_f^2 - \tfrac{1}{2}mv_i^2.$$

We solve this equation for the initial speed:

$$v_i^2 = v_f^2 - \frac{2W_{net}}{m} = v_f^2 - \frac{2(\mu_k - \mu'_k)mg\,\Delta x}{m}$$
$$= (2.3 \text{ m/s})^2 - 2(0.27 - 0.085)(9.8 \text{ m/s}^2)(1.25 \text{ m})$$
$$= 0.76 \text{ m}^2/\text{s}^2,$$
$$v_i = 0.87 \text{ m/s}.$$

What Do You Think?   Suppose that instead of hitting a smoother section of the floor, the coefficient of friction does not change but the floor begins to slope downward at the $x = 0$ m point. Use the work–energy theorem to explain what happens.

▶ **FIGURE 6–6**   (a) The floor changes composition to a more slippery surface at $x = 0$ m. (b) Forces acting on the crate. (c) Free-body diagram for the crate.

HOW DID THE EGYPTIANS BUILD THE PYRAMIDS?

While we don't know the fine details, we can be sure that the pyramids (Fig. 1–10), a truly immense project, were constructed with the aid of at least some of the classic "simple machines"—ramp, lever, block and tackle (see Example 5–4), screw thread, and so forth (Fig. 6–7). The expression for work contains both the displacement of the object being moved and the magnitude of the force. We can do the same amount of work by applying a larger force over a smaller distance or a smaller force over a larger distance. The simple machines are devices that do a given amount of work by employing a smaller force over a longer distance; in a sense they amplify your ability to do work by allowing you to exchange a large force applied over a short distance for a smaller force applied over a larger distance. Hoisting a several-ton block to its place in the pyramids requires an immense amount of work. It would not have been possible for the Egyptians to lift such a block vertically, but they could get it to the top of the pyramid by applying a smaller force than the weight of the block over a longer distance, thereby doing the same work on the block. This would still require many workers but is achievable by pushing the block up a sloping ramp built along the side of the pyramid, rather than over a purely vertical displacement. (We'll describe the work done as an object moves along a ramp in Section 6–2.) To help transport the stone blocks from their source, the Egyptians likely used another simple machine, the lever, in the form of a huge crowbar, to raise the blocks high enough for logs to be slid underneath. The stone blocks could then be pushed with less force than if the blocks were pushed along the ground because rolling involves less friction than sliding. The Egyptians made use of these elegant and practical force "amplifiers" thousands of years ago, and we still use them today. It is convenient to use a ramp to put a refrigerator in a truck, to use a lever in the form of a screwdriver to pry open the lid of a paint can, or to employ a block and tackle as in Example 5–4. In each case, we say we are using a "mechanical advantage," or that we are using "leverage."

▲ **FIGURE 6–7** A simple machine is a kind of force amplifier. It does the same work that a large force would do over a small distance by applying a small force over a large distance. Thus a lever lifts a large mass a height *h* by applying a force much less than the mass weight over a distance much greater than *h*. Here, we can see a lever, a ramp, a wedge, and various screw-driven devices.

6–2 Constant Forces in More Than One Dimension

We have so far discussed the concepts of kinetic energy, work, and the work–energy theorem as they apply to objects that move only in one dimension. We now want to extend these concepts to two and three dimensions. We continue to assume that the net force $\vec{F}$ is a constant, meaning that it is a vector that does not change with time and has the same magnitude and direction at every point in space. Throughout, we'll take care to emphasize that force is a vector, as are displacement, velocity, and acceleration. (For simplicity, we have dropped the subscript "net" on the net force).

The application of energy and work to two and three dimensions is a straightforward one; we know that Newton's laws include the information that motion and the effects of forces on motion are *independently applicable to the different Cartesian directions*. Let's first apply these concepts to two dimensions for simplicity. Newton's second law, $\vec{F} = m\vec{a}$, encompasses two separate equations, one for the x-direction and one for the y-direction. If we label the components of all vector quantities with the appropriate axes, then we can derive an equation like the work–energy theorem—Eq. (6–2)—for each direction:

$$F_x \, \Delta x = \tfrac{1}{2}mv_x^2 - \tfrac{1}{2}mv_{0x}^2, \tag{6–8a}$$

$$F_y \, \Delta y = \tfrac{1}{2}mv_y^2 - \tfrac{1}{2}mv_{0y}^2. \tag{6–8b}$$

Here, the displacement vector is $\Delta \vec{r} = \Delta x \, \hat{i} + \Delta y \, \hat{j}$. [Note that the quantity $\tfrac{1}{2}mv_x^2$ in Eq. (6–8a), say, is not the "x-component" of kinetic energy; kinetic energy involves the speed squared, and the speed involves all the components of velocity.] The velocity $\vec{v}$ and the initial velocity $\vec{v}_0$ have each been separated into their components, as has the net force $\vec{F}$.

Let's now take the sum of these two equations. The square of the *magnitude* of the velocity—speed squared (v^2)—is the sum of the x- and y-components of the velocity squared:

$$v^2 = v_x^2 + v_y^2.$$

The summed equations then form a generalization of the work–energy theorem:

$$F_x\,\Delta x + F_y\,\Delta y = \tfrac{1}{2}mv^2 - \tfrac{1}{2}mv_0^2. \tag{6-9}$$

Defining kinetic energy just as we have before—Eq. (6–5), $K = \tfrac{1}{2}mv^2$— the right-hand side of this equation is once again the change in kinetic energy.

The left-hand side of Eq. (6–9) generalizes the definition of the work done on an object in two dimensions. The work done easily generalizes to three dimensions:

$$\text{for constant force: } W = F_x\,\Delta x + F_y\,\Delta y + F_z\,\Delta z. \tag{6-10}$$

This definition of the work done contains the components of the two vectors $\vec{F}$ and $\Delta\vec{r}$. One takes each component of the force and multiplies it by the corresponding component of the displacement, with the resulting terms added together. This combination of two vectors occurs in situations other than this one, and because it appears so frequently, it is given a name: the **scalar product** (or **dot product**) $\vec{A}\cdot\vec{B}$ of two vectors $\vec{A}$ and $\vec{B}$. The work, then, is the scalar product of $\vec{F}$ and $\Delta\vec{r}$, namely $\vec{F}\cdot\Delta\vec{r}$. A scalar product, although it is the product of two vectors, is itself a scalar quantity. If you are not already familiar with this concept, refer to the box The Scalar Product.

From Eq. (6–10) we can now see that the work W done by a constant force $\vec{F}$ acting on an object that moves through a displacement $\Delta\vec{r}$ is

$$W = \vec{F}\cdot\Delta\vec{r} \tag{6-11}$$

Using this expression for work, the work–energy theorem takes exactly the same form as before—$W_{\text{net}} = \Delta K$. Keep in mind that Eq. (6–11) applies only as long as the net force is a constant vector. (We'll see later how to deal with a varying force.)

We can also write the work differently than Eq. (6–10). Using the general properties of the scalar product (see the box), we have

$$W = \vec{F}\cdot\Delta\vec{r} = F\,\Delta r\cos\theta, \tag{6-12}$$

where θ is the angle between the two vectors (Fig. 6–8). Equivalently, the work done is the simple product of the magnitude of force *in the direction of the displacement* and the magnitude of the displacement. Put another way, *only the component of the force along the direction of the displacement* (here $F\cos\theta$) *counts in the work*. From this fact we can conclude that a force perpendicular to the motion of an object does no work on the object. A frequently occurring example of this is the normal force $\vec{F}_N$, which is perpendicular to the surface on which an object moves; $\vec{F}_N$ does no work on that object. Another important example of a force that does no work in this way is the centripetal force responsible for uniform circular motion (see Section 6–4). Note also that the quantity $F\,\Delta r\cos\theta$ can be positive or negative, depending on whether the displacement is "with" or "against" the force.

Remember that, although the scalar product is formed from two vectors, it is itself a scalar. Work is always a scalar quantity.

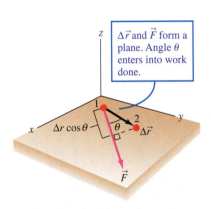

▲ **FIGURE 6–8** The two nonparallel vectors $\Delta\vec{r}$ and $\vec{F}$ always form a plane. We have labeled that plane as the xy-plane. The angle between these vectors determines the work done by the constant force on an object undergoing displacement $\Delta\vec{r}$, namely $W = F\,\Delta r\cos\theta$. We have indicated the component of $\Delta\vec{r}$ along the force, namely $\Delta r\cos\theta$.

Problem-Solving Techniques

The Scalar Product

In Chapter 1, we discussed the definition of vectors and the multiplication of vectors by scalars. The product $b\vec{A}$ of a scalar b and a vector $\vec{A}$ is a vector. It points in the same direction as $\vec{A}$ and has magnitude $|b|A$, where A is the magnitude of $\vec{A}$. One way to multiply two vectors—$\vec{A}$ and $\vec{B}$, for example—is the **scalar product** $\vec{A} \cdot \vec{B}$. The scalar product is a scalar quantity whose value is

$$\vec{A} \cdot \vec{B} \equiv AB \cos \theta. \quad (B1\text{–}1)$$

Here, θ is the angle between the directions of the two vectors (Fig. 6B1–1a). The scalar product has the properties that

$$\vec{A} \cdot \vec{B} \equiv \vec{B} \cdot \vec{A} \quad (B1\text{–}2)$$

and

$$\vec{A} \cdot (\vec{B} + \vec{C}) = (\vec{A} \cdot \vec{B}) + (\vec{A} \cdot \vec{C}). \quad (B1\text{–}3)$$

If two vectors are perpendicular (*orthogonal*) to each other, then $\theta = 90°$ and $\cos \theta = 0$, and their scalar product is zero. If the vectors are parallel to each other, then the scalar product takes on its maximum value, that is, the product of the magnitudes of the two vectors. The scalar product of a vector with itself is the square of its magnitude, $\vec{A} \cdot \vec{A} = A^2$. The unit vectors $\hat{i}$, $\hat{j}$ and $\hat{k}$ along some set of orthogonal axes x, y, and z have the property that

$$\hat{i} \cdot \hat{i} = \hat{j} \cdot \hat{j} = \hat{k} \cdot \hat{k} = 1. \quad (B1\text{–}4)$$

Because they are orthogonal to each other,

$$\hat{i} \cdot \hat{j} = \hat{j} \cdot \hat{k} = \hat{i} \cdot \hat{k} = 0. \quad (B1\text{–}5)$$

Two vectors $\vec{A}$ and $\vec{B}$ can be decomposed into their vector components: $\vec{A} = A_x\hat{i} + A_y\hat{j} + A_z\hat{k}$ and $\vec{B} = B_x\hat{i} + B_y\hat{j} + B_z\hat{k}$. The rules in Eqs. (B1–4) and (B1–5) allow us to write the scalar product of $\vec{A}$ and $\vec{B}$ as

$$\begin{aligned}\vec{A} \cdot \vec{B} &= (A_x\hat{i} + A_y\hat{j} + A_z\hat{k}) \\ &\quad \cdot (B_x\hat{i} + B_y\hat{j} + B_z\hat{k}) \quad (B1\text{–}6) \\ &= A_xB_x + A_yB_y + A_zB_z.\end{aligned}$$

Thus the scalar product of two vectors is the sum of the product of the components of the two vectors.

The scalar product is a scalar quantity, so it remains the same even if the axes of our coordinate system are rotated. If we consider two vectors $\vec{A}$ and $\vec{B}$, we may choose our coordinate frame in such a way that $\vec{A}$ lies along the x-axis, $\vec{A} = A\hat{i}$. The other axes can be arranged so that the vector $\vec{B}$ has only x- and y-components, $\vec{B} = B_x\hat{i} + B_y\hat{j}$. These vectors are shown in Fig. 6B1–1b, which is a view looking down on the plane formed by $\vec{A}$ and $\vec{B}$. According to Eq. (B1–6), the scalar product is then given by

$$\vec{A} \cdot \vec{B} = AB_x.$$

Thus the scalar product of two vectors may be described as the *product of the length of one vector and the projection of the other vector along the direction of the first one.* [Because $B_x = B \cos \theta$, we recover here our original definition: Eq. (B1–1).] This way of looking at things makes it clear that the orientation of coordinate axes is irrelevant to the value of the scalar product.

Scalar products are useful in many mathematical and physical manipulations. Consider, for example, the following result from analytic geometry: For a triangle whose sides have lengths a, b, c and where the angle between known sides a and b is θ,

$$c^2 = a^2 + b^2 - 2ab \cos \theta. \quad (B1\text{–}7)$$

In vector notation, and as in Fig. 6B1–2, if two sides of a triangle are denoted by the vectors $\vec{a}$ and $\vec{b}$, then the third side is given by $\vec{c} = \vec{a} - \vec{b}$. Squaring both sides ("squaring" a vector means taking the scalar product of the vector with itself) gives

$$\vec{c}^2 = (\vec{a} - \vec{b})^2 = \vec{a}^2 + \vec{b}^2 - 2\vec{a} \cdot \vec{b}, \quad (B1\text{–}8)$$

which implies Eq. (B1–7). In this book, in addition to all the applications associated with energy, we'll meet the scalar product again in fluid motion, electricity and magnetism, and other places.

(a)

(b)

▲ **FIGURE 6B1–2** Illustrating the rule for the length of the third side of a triangle if the length of two sides and the angle between them are known.

▲ **FIGURE 6B1–1** (a) Two nonparallel vectors $\vec{A}$ and $\vec{B}$ have been displaced so that their tails meet at the same point. They are oriented in space with the angle θ between them. (b) The x-axis of our coordinate system has been redefined so that $\vec{A}$ lies along the $+x$-direction. The scalar product is independent of the orientation of the axes.

EXAMPLE 6–7 A mover has to place a box of books of mass m in a truck, and it is too heavy to lift directly. He therefore uses a ramp that makes an angle θ with the horizontal and pushes horizontally, applying a force $\vec{F}$ to the box. Find the magnitude of $\vec{F}$ such that the box moves up the plane with acceleration $\vec{a}$. What is the work done by $\vec{F}$? (Assume that the ramp is rough, with coefficient of kinetic friction μ_k, and that the distance the box moves along the ramp is d.)

Setting it Up We illustrate the situation in Fig. 6–9a, including a suitable coordinate system.

Strategy We first draw a free-body diagram and use it to find the components of the forces along the axes. Newton's second law then allows us to find the force necessary to push the box up the ramp with acceleration $\vec{a}$. We know the displacement, so we can then find the work done by this force.

Working It Out   Figure 6–9b is the free-body diagram for the box. We then separate the forces into their components in Fig. 6–9c. The component of the force along the ramp is $F \cos \theta$, so the work done by this force on the box as it moves a distance d is $W = Fd \cos \theta$.

Now we find the magnitude F. The y-components of the forces must add up to zero because there is no acceleration in the direction perpendicular to the ramp. We thus have

$$F_N - F \sin \theta - mg \cos \theta = 0.$$

From this equation, $F_N = F \sin \theta + mg \cos \theta$. There is also an acceleration of the box up the ramp, which is determined by Newton's second law applied to the x-direction:

$$ma = F \cos \theta - mg \sin \theta - \mu_k F_N$$
$$= F \cos \theta - mg \sin \theta - \mu_k (F \sin \theta + mg \cos \theta).$$

This equation can be solved for F:

$$F = \frac{ma + mg \sin \theta + \mu_k mg \cos \theta}{\cos \theta - \mu_k \sin \theta}.$$

The work done by the force in accelerating the box is then

$$W = Fd \cos \theta = md \frac{a + g \sin \theta + \mu_k g \cos \theta}{1 - \mu_k \tan \theta}.$$

Notice that this is not the net work, as that involves the pushing force, friction and gravity.

What Do You Think?   What happens if the ramp angle is larger? Give a qualitative answer.

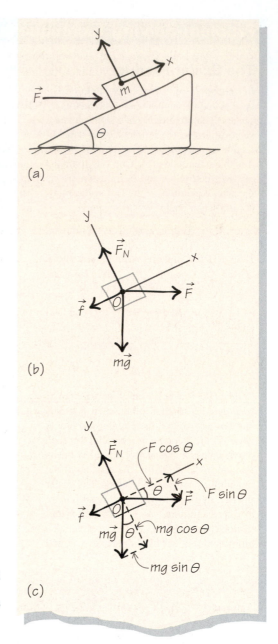

▶ **FIGURE 6–9**   (a) Box being pushed up a ramp. (b) Free-body diagram for the block on the inclined plane. (c) The forces are decomposed into components along the plane surface and perpendicular to it.

6–3 Forces That Vary with Position

In our study of the work–energy theorem so far the force acting on an object has been constant. Many forces in nature and in engineering, however, such as the gravitational force or the forces exerted by springs, vary with position. It is possible to have a force whose magnitude varies from point to point, one whose direction varies from point to point, or one for which both the magnitude and direction vary from point to point. We want to generalize the definition of work to include these cases so that the work–energy theorem continues to hold.

Variable Forces in One Dimension

Let's first go back to motion in one dimension. The only change from the constant-force situation of Section 6–1 is that the force magnitude F depends on position x, so that $F = F(x)$. Figure 6–10a shows a force that varies with position and is one possible form of $F(x)$. We may adapt what we have learned so far to this situation by approximating the curve in Fig. 6–10a by a series of "steps"—their width being some interval in x—that come close to matching $F(x)$. An examination of Fig. 6–10b shows that the width of these steps is made to vary: If $F(x)$ is fairly flat in a certain region, the width of the step (its Δx) can be relatively large. If $F(x)$ changes rapidly, we must make the width of the steps in that range very small. Even if $F(x)$ in any interval differs a little from the step height across the interval, the error in treating F as a constant (e.g., taking its value in the middle of the width of the interval) is small as far as the work calculation is concerned, since if the interval Δx is small, $F \Delta x$ will be small.

To rephrase this more mathematically, divide the total displacement, $x_f - x_0$, into a series of small intervals across each of which, to within a small error, the force is constant. When the force is changing only slowly with position, it will remain roughly constant over a relatively large interval (see the parts of Fig. 6–10a marked "1"), whereas if the force varies considerably with a small change of position (see the part of Fig. 6–10a marked "2"), it will remain roughly constant over only a very small interval. Figure 6–10b shows how the width of the intervals would have to vary so that $F(x)$ is approximately constant over each interval. For mathematical simplicity we now take the width of the intervals to be all the same, and to ensure our assumption that the force is almost constant over this interval, we set our common interval width to the width of the smallest interval in Fig. 6–10b, as drawn in Fig. 6–10c. We call this width Δx. We have now approximated our variable force as a series of constant forces, each varying only very slightly from its value in a neighboring interval.

Let us now denote the (average) value of the force in the interval from x_0 to x_1 by F_1, the (average) value of the force in the interval from x_1 to x_2 by F_2, and so on, as in Fig. 6–11. Let us also denote the velocities at the edges of the intervals as follows: at x_0 the velocity is v_0, at x_1 it is v_1, at x_2 it is v_2, and so on. There are altogether $N = (x_f - x_0)/\Delta x$ intervals. The work–energy theorem, Eq. (6–2), applied in succession to each of these N intervals gives

$$F_1 \Delta x = \tfrac{1}{2}mv_1^2 - \tfrac{1}{2}mv_0^2,$$
$$F_2 \Delta x = \tfrac{1}{2}mv_2^2 - \tfrac{1}{2}mv_1^2,$$
$$F_3 \Delta x = \tfrac{1}{2}mv_3^2 - \tfrac{1}{2}mv_2^2,$$
$$\vdots$$
$$F_N \Delta x = \tfrac{1}{2}mv_f^2 - \tfrac{1}{2}mv_{N-1}^2.$$

Here, v_N is equivalent to the final speed, so we denote it v_f. If we add all these equations, we see that all the intermediate kinetic energies cancel, leaving

$$F_1 \Delta x + F_2 \Delta x + F_3 \Delta x + \cdots + F_N \Delta x = \sum_{i=1}^{N} F_i \Delta x = \tfrac{1}{2}mv_f^2 - \tfrac{1}{2}mv_0^2.$$

The right side is the change in kinetic energy between the initial point x_0 and the final point x_f. The left side of this equation is the work done in going from x_0 to x_1 plus the

(a)

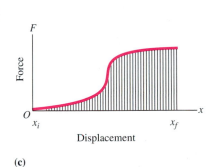

(b)

(c)

▲ **FIGURE 6–10** (a) A smoothly varying force. (b) If there is a region where the force changes more rapidly, the intervals in that region can always be made smaller so that the force can be thought of as constant in each interval. (c) Take the width of each interval to be the width of the smallest interval in (b).

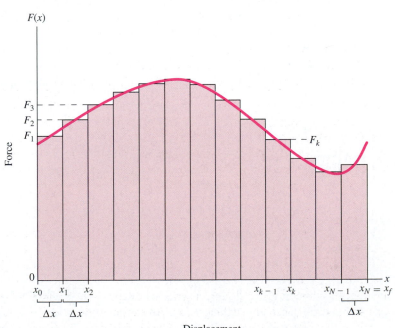

◄ **FIGURE 6–11** A variable force $F(x)$ has an approximately constant value $F(x_k) = F_j$ within the jth small interval, of width Δx, in the region between x_0 and x_N. The work done by the force on an object as the object moves between these limits is the sum of the work W_k done in each interval, for which we can use the constant-force formula for work, $W_k = F_k \Delta x$. This sum, W, is the integral of $F(x)$ over x, and it is equal to the area under the curve of F versus x.

work done in going from x_1 to x_2, and so on. Each of these individual terms can be interpreted as the area of the rectangle formed by the interval width and the function height in that interval. The sum of these terms is approximately the area under the curve of F versus x. As the number of terms becomes infinite—as we make the width of our intervals infinitesimally small—the approximation becomes exact. The work done over the entire distance from x_0 to x_f is the *total area under the curve of force versus position*. In other words, it is the integral of the function $F(x)$ over the interval starting at point x_0 and ending at point x_f. (For a review of integration refer to the box Integration, a Quick Review.) We may therefore write the above equation as

$$\int_{x_0}^{x_f} F(x) \, dx = K_f - K_0 = \Delta K. \qquad (6\text{--}13)$$

In terms of the work–energy theorem—that is, the change in kinetic energy of a system is the net work done on the system—we can identify the left side as the net work done,

$$W = \int_{x_0}^{x_f} F(x) \, dx. \qquad (6\text{--}14)$$

(Again, we have not bothered with the subscript "net.") We can also say that the work done by any one force that varies with position in one dimension has this form. In the cases that we deal with in this chapter the integral can be done explicitly, or at least numerically. In the simple case of a constant force of magnitude F_0, this form reduces to

$$W = (F_0) \int_{x_0}^{x_f} dx = (F_0)(x_f - x_0) = F_0 \, \Delta x,$$

which gives us back the result of Eq. (6–3).

Work Done by a Spring

One of the most important examples of a one-dimensional variable force is the force exerted on a mass by a spring attached to it (Fig. 6–12). The force takes the form

$$F = -kx. \qquad (6\text{--}15)$$

Here, x measures the displacement of the mass from an equilibrium position; k is a constant characteristic of the particular spring, known as the **spring constant.** This force law is known as **Hooke's law,** after its seventeenth-century discoverer, Robert Hooke. Note the sign: The spring force always acts to bring the mass back to $x = 0$. When x is positive, the mass is on the right side in Fig. 6–13a and the force points to the left, and when x is negative, the mass is on the left side of the origin and the force points to the right. The universal importance of this force is that it applies to virtually any system that has an equilibrium point in which a small movement away from that point brings in forces that tend to bring you back to the point; this is called a stable equilibrium point.

▲ **FIGURE 6–12** A spring exerts a force on an object that tends to bring the object back to the equilibrium position. Here the spring is stretched past the equilibrium position.

▶ **FIGURE 6–13** (a) The force exerted by a spring changes as the spring stretches (or compresses). Twice the stretch corresponds to twice the force. (b) The area under the curve of force versus x is the work the spring force performs on a mass attached to its end. The work done as the spring stretches from $x = 0$ to $x = L$ is the area shaded in red.

Examples include pendulums and the forces between the atoms that make up solids. This will be discussed in more detail in the next chapter.

Let's now calculate the work done on the mass by the spring force when the mass moves from the equilibrium position (where the spring force is zero) to the position $x = L$ (Fig. 6–13a). Note that if the mass is at rest, it cannot move away from $x = 0$ under the influence of the spring force alone; the spring force is zero there. But the mass may be acted on by other forces or may have started its motion at a different point than $x = 0$. If other forces are involved, then the work done on the mass by the spring force alone is not the *net* work. No matter; we can always calculate the work done by the spring force alone as the mass is displaced. To calculate the work done by the spring, we must integrate the (nonconstant) force over the displacement, as in Eq. (6–14):

$$W = \int_0^L (-kx)\, dx.$$

We use the general integration formula for powers from Appendix IV–8, with $p = 1$, to find that

$$W = -k \int_0^L x\, dx = -\tfrac{1}{2} k x^2 \Big|_0^L = -\tfrac{1}{2} k L^2.$$

The work done by the spring on the mass varies with the square of the distance moved. What does the negative sign mean? The spring force is in the direction opposite to the displacement of the mass, so the work done by the spring is negative. This is sensible given that the force acts to bring the mass back to $x = 0$, and if the spring force were the only force acting, the mass would slow down as it goes from zero to L. Figure 6–13b illustrates how the work done is equal to the area under the curve of F versus x, here shaded in red. The area of the triangle is indeed $\tfrac{1}{2} k L^2$, and it is negative since F itself is negative.

Problem–Solving Techniques

Integration, a Quick Review

Here we want to remind you of some important features of the process of integration. (A more complete discussion of integration is contained in Section 2–6.) Integration of a function is the inverse of differentiation of that function. In other words, if we integrate some function $f(x)$ and then differentiate the result, we get the function $f(x)$ back again. More precisely, suppose that the functions $f(x)$ and $g(x)$ are related by

$$g(x) = \int_{x_0}^x f(x')\, dx'. \quad \text{(B2–1)}$$

Then

$$\frac{dg(x)}{dx} = f(x). \quad \text{(B2–2)}$$

As we saw in the subsection Variable Forces in One Dimension, we can interpret work as the area under the curve of F versus x. If the function F is negative and the displacements

are positive, as in Fig. 6B2–1, the area and hence the work are negative. Or, if the displacement is negative and the force positive, the work will again be negative. In the language of integrals, this occurs because the integral changes sign when we reverse the limits on it:

$$\int_{x_f}^{x_i} f(x)\, dx = -\int_{x_i}^{x_f} f(x)\, dx. \quad \text{(B2–3)}$$

If the integral on the right-hand side is positive, then the integral on the left-hand side—which represents the work done in a displacement from x_f to x_i—is negative.

(a)

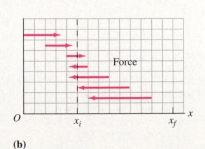

(b)

▶ **FIGURE 6B2–1** (a) In the region between x_i and x_f, the shaded area under the curve is negative, (b) corresponding to negative values for the force $F(x)$.

CONCEPTUAL EXAMPLE 6–8 What is the amount of work done by the spring force in moving from $x = 0$ to $x = L$ compared to the work done moving from $x = 0$ to $x = 2L$?

Answer We can use the graph in Fig. 6–13b for this question. The work done by the spring force as the mass moves from one value of x to another is the area under the curve of F versus x, between the two values of x. We consider the shaded triangle under the force curve (here, a straight line with slope k) from $x = 0$ to $x = L$ and a second shaded triangle under the force curve from $x = 0$ to $x = 2L$. These triangles are similar, but the second one has twice the base length and twice the height. The second triangle has four times the area of the first, so the spring force does four times as much work when the mass moves twice as far.

EXAMPLE 6–9 A worker pushes a 20-kg crate straight across a 1-m-long section of horizontal floor with a constant force of 29 N. This section of floor has the peculiarity that it becomes rougher from beginning to end, and the crate is moving at 1.2 m/s when it arrives at the start of this section. The coefficient of kinetic friction is 0.15 at the start and 0.25 at the finish, varying linearly with distance in between. What is the speed of the crate at the end of the section?

Setting It Up Figure 6–14 indicates a coordinate system in which the starting point for the section is at $x = 0$ and the end point is at $x_f = 1$ m. We let m be the given mass of the crate; v_i the given (positive) initial velocity; v_f the unknown final velocity; F, which acts along the $+x$-direction, the given pushing force; and f_k the force of kinetic friction, taking the form $f_k = -\mu_k F_N$. The minus sign in f_k shows it acts in the $-x$-direction. Finally, we know the initial and final values of μ_k, namely μ_k^i and μ_k^f, as well as the fact that μ_k varies linearly with distance over the range $x = 0$ to $x = x_f = 1$ m. This translates into a coefficient of friction that takes the algebraic form $\mu_k = \mu_k^i + (\mu_k^f - \mu_k^i)(x/x_f)$.

Strategy This type of question is tailor made for the work–energy theorem. We use it to find the final energy, hence the final speed, in terms of the calculable initial kinetic energy and the calculable net work. The net work involves the net force in the x-direction, and the net force is composed of both F and f_k. The free-body diagram in Fig. 6–6b suffices and allows us to see that $f_k = \mu_k mg$.

Working It Out Begin with the net work:
$F_{net} = F + f_k = F - [\mu_k^i + (\mu_k^f - \mu_k^i)(x/x_f)]mg$. Therefore

$$W_{net} = \int_0^{x_f} F_{net}\, dx = \int_0^{x_f} \left[F - \mu_k^i mg - (\mu_k^f - \mu_k^i)\left(\frac{x}{x_f}\right)mg \right] dx$$

$$= [F - \mu_k^i mg]x_f - [(\mu_k^f - \mu_k^i)mg]\frac{x_f^2}{2x_f}$$

$$= [F - \mu_k^i mg]x_f - [(\mu_k^f - \mu_k^i)mg]\frac{x_f}{2}.$$

Here we have used the result that the integral of unity is x and the integral of x is $\frac{1}{2}x^2$; in each case we evaluate at the upper and lower limit and take the difference, and we have included the multiplicative constants as well. Numerically,

$$W_{net} = [29\text{ N} - (0.15)(20\text{ kg})(9.8\text{ m/s}^2)](1\text{ m})$$
$$- [(0.25 - 0.15)(20\text{ kg})(9.8\text{ m/s}^2)]\frac{1\text{ m}}{2} = -10\text{ J}.$$

The sign is negative because the friction force dominates. You can see that the first term in square brackets is very nearly zero. It is composed of a term involving the pushing force and a canceling term involving the initial value of kinetic friction, suggesting that the crate was moving at constant velocity over a section of floor with coefficient of kinetic friction 0.15 before it arrived at the section treated here. In any case, with negative net work, we expect the speed to decrease.

The initial kinetic energy is $K_i = \frac{1}{2}mv_i^2 = \frac{1}{2}(20\text{ kg})(1.2\text{ m/s})^2 = 14\text{ J}$. We then use the work–energy theorem in the form of Eq. (6–7),

$$K_f = W_{net} + K_i = -10\text{ J} + 14\text{ J} = 4\text{ J}.$$

This gives us a final speed according to

$$v_f^2 = (2/m)K_f = (2/20\text{ kg})(4\text{ J}) = 0.4\text{ (m/s)}^2, \text{ or } v_f = 0.6\text{ m/s}.$$

What Do You Think? Does this example have anything to do with the spring force?

▲ **FIGURE 6–14** Coordinate system for forces acting on crate.

Forces That Vary in Both Magnitude and Direction

For many common forms of motion forces appear that vary in magnitude and/or direction: When an object moves under the influence of a force, either the force may change direction as the object moves or the object may change its direction as it moves. Think of driving your car along a steep and winding mountain road. Gravity is constant in both direction and magnitude, but you are constantly changing your direction. Thus the work done on you by gravity near one location—a scalar product of the form $m\vec{g} \cdot \Delta\vec{r}$, where $\Delta\vec{r}$ is a small local displacement at that first location—may be different in another location, in this case because the direction of the displacement changes.

Therefore we want to generalize the net work done, the quantity that appears in the work–energy theorem, to encompass the possibility either that the force varies its direc-

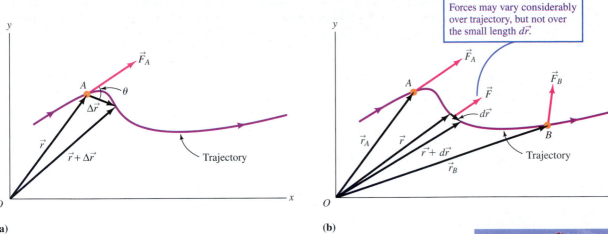

Forces may vary considerably over trajectory, but not over the small length $d\vec{r}$.

(a)   (b)

▲ **FIGURE 6–15** (a) An object moves from position $\vec{r}$ at time t to position $\vec{r} + \Delta\vec{r}$ at time $t + \Delta t$ while a force $\vec{F}_A$ acts on it. The work done on the object by the force over the displacement $\Delta\vec{r}$ is $\vec{F}_A \cdot \Delta\vec{r}$. (b) In a finite time, there is a net displacement from point A to point B.

tion from point to point or that the net displacement $\vec{r}_f - \vec{r}_i$ of the moving object results from a rather complicated *path* in space (Fig. 6–15) or both. (Here we have drawn the path in a plane, but the path could move through three-dimensional space.) The path is described with a changing position vector $\vec{r}$. At the point A, which is a distance r from the origin, the force on the object is $\vec{F}_A$. This force does not change very much as the object moves through a small displacement $\Delta\vec{r}$ from the point labeled by position vector $\vec{r}$ to an adjacent point labeled by position vector $\vec{r} + \Delta\vec{r}$ (Fig. 6–15a). As $\Delta\vec{r} \rightarrow 0$, this displacement is *tangential* to the curve at the tip of $\vec{r}$, whereas the force vector, which can be regarded as constant over a sufficiently small displacement, points in some other direction. The work done on the object in moving through the small interval is then $\vec{F}_A \cdot \Delta\vec{r}$. We find the total work done for a displacement from point A with position vector $\vec{r}_A$ to point B with position vector $\vec{r}_B$ (Fig. 6–15b) by summing the small contributions from each small interval along the path. In the limit that the small displacements go to zero, the sum takes the form of an integral:

$$W = \lim_{\Delta\vec{r} \rightarrow 0} \sum \vec{F} \cdot \Delta\vec{r} = \int_{\vec{r}_A}^{\vec{r}_B} \vec{F} \cdot d\vec{r}. \qquad (6-16)$$

This formula is the general definition of work and is consistent with all our earlier definitions. The integral that appears here is called a **line integral** because it depends not only on the beginning and ending points A and B but also, in general, on the path, or line, taken to move between these points. We shall explore the properties of this integral in Section 6–4.

With the definition of work of Eq. (6–16), *the work–energy theorem in the form given by Eq. (6–6), $W_{net} = \Delta K$, applies for the most general case.*

No Work Is Done in Uniform Circular Motion

We conclude this section with an important observation: *No net work is done on a particle that undergoes uniform circular motion* (Fig. 6–16). This holds for *any* part of a circular trajectory at constant speed. Recall from Section 3–5 or 5–4 that an object undergoing uniform circular motion experiences an acceleration that is directed along the radius toward the center of the circle. Thus the force is directed in the (negative) radial direction and is always *perpendicular* to the direction of motion, which is tangential to the circular trajectory. If the infinitesimal displacement along an arc is $d\vec{s}$, as in Fig. 6–17, then the scalar product $\vec{F} \cdot d\vec{s}$ is zero because the force has no component in the direction of the displacement. Thus no work is done.

If the motion is circular with varying speed, a tangential force $\vec{F}_t$ must be present (one that is parallel to the direction of the displacement). Work is done in this case with

▲ **FIGURE 6–16** As long as a Ferris wheel passenger is moving in uniform circular motion, the net work done on the passenger along any segment of the arc of the circle is zero. The net force that uniformly moves the passenger in a circle is made up of a combination of gravity and normal forces. If the Ferris wheel moves with a uniform velocity, why does it need a motor?

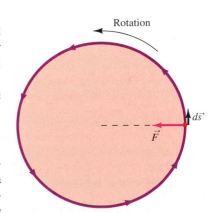

▲ **FIGURE 6–17** In the uniform circular motion of an object, the infinitesimal displacement is tangent to the circle. The force responsible for this motion is directed toward the center of the circle and is perpendicular to the displacement. The work done by this force on the object is therefore zero.

a consequent change in energy. Because the force is directed along (or opposite to) the displacement $d\vec{s}$, the scalar product $\vec{F}_t \cdot d\vec{s}$ for the infinitesimal work done has nonzero magnitude $F_t \, ds$. We integrate this quantity to find the total work.

EXAMPLE 6–10 A ball with a mass of 8.0 kg is attached to the end of a rod 1.5 m long of negligible mass. The rod is attached to a vertical shaft in such a way that the rod is held perpendicular to the shaft (Fig. 6–18). The attachment point of the rod and shaft is a frictionless pivot point that allows the rod to rotate about the shaft. A tangential force of constant magnitude F is applied to the ball for one-quarter turn. As a result, the ball and rod rotate about the shaft with an angular speed of 1.2 rev/s. If the ball is initially at rest, what is F?

Setting It Up We indicate the tangential force F and the velocity v in the figure. The rotation is about the z-axis, which is vertical. We know the mass m of the ball, the radius r of the circular motion of the ball (r being rod length), and the ball's angular velocity ω after starting from rest and rotating a quarter turn.

Strategy If we find the ball's final speed from the given final angular speed, we can then find the ball's final kinetic energy. Because the initial kinetic energy is zero, we know the change in kinetic energy and can use the work–energy theorem to find the corresponding net work. Given the distance over which the force acts, this allows us to find the net force itself.

Working It Out The angular speed after the force has acted is 1.2 rev/s = $(1.2 \text{ rev/s})(2\pi \text{ rad/rev})$ = 7.5 rad/s, so the speed of the ball at the end of the push is $v = \omega r = (7.5 \text{ s}^{-1})(1.5 \text{ m})$ = 11 m/s. Thus

$$K_f - K_i = \tfrac{1}{2}mv_f^2 - \tfrac{1}{2}mv_i^2$$

$$= \tfrac{1}{2}(8.0 \text{ kg})(11 \text{ m/s})^2 = 4.8 \times 10^2 \text{ J}.$$

The work–energy theorem states that this is the net work, and this work is done only by the applied force. This force, applied for a quarter turn, acts over a distance $\frac{1}{4}(2\pi r) = \frac{1}{4}\pi(3.0 \text{ m})$ = 2.4 m. Thus the force has magnitude

$$F = \frac{K_f - K_i}{\text{distance}} = \frac{4.8 \times 10^2 \text{ J}}{2.4 \text{ m}} = 2.0 \times 10^2 \text{ N}.$$

What Do You Think? Is the tangential force referred to here the only force acting on the ball?

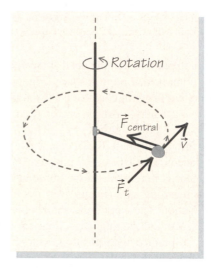

▲ **FIGURE 6–18** Forces acting on ball.

CONCEPTUAL EXAMPLE 6–11 Consider a stone twirled at the end of a length of rope. Due to the centripetal tension of the rope, the stone is undergoing nearly circular motion in a horizontal plane. The rope is slowly shortened by pulling it in, and as you may have experienced in a similar situation, the stone then moves more rapidly. How is this consistent with the fact that the force due to the rope is perpendicular to the path of the stone? [*Hint*: Sketch the path of the stone and reexamine the statement of the question.]

Answer If the stone were really undergoing circular motion with the rope tension the only (horizontal) force acting on it, this force would be perpendicular to the motion and could do no work; by the work–energy theorem, its kinetic energy and hence its speed could not change. But if the rope is being drawn in, the path of the stone is actually a spiral, and in Fig. 6–19 we see that the line from the center of rotation O to the stone is not quite perpendicular to the spiral path. Thus there is a small component of the force vector (the tension of the rope) along the in-going part of the spiral. This component of the force speeds up the stone as it moves along its path. You can see this in another way if you think of a rope shortening as it wraps around a pole. The pole must have a nonzero radius, and this means that the motion is not quite circular about the center of the pole. We'll see in Chapter 10 that we can view this as a matter of conserving angular

momentum and that the conservation of angular momentum implies an increase in speed as the radius of the motion decreases.

▲ **FIGURE 6–19** In spiral motion the direction from the center of the motion is not quite perpendicular to the direction of the motion, as it is in circular motion. The work done is therefore not zero.

6–4 Conservative and Nonconservative Forces

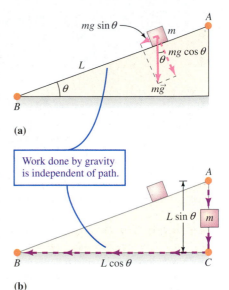

One of the most important questions that arises when an object undergoes a certain displacement concerns *path dependence*: How does the work done by a force depend on the path taken by the object during its displacement? We are not necessarily considering net force here; certainly when the work–energy theorem is applied, we must do so. But the question of path dependence is one that is best addressed in the context of individual forces. We illustrate what is involved by considering, in order, gravity, friction, and the spring force.

Gravity: Let us consider the work that gravity does when a box of books of mass m moves from the top (point A) of a ramp of length L to the bottom (point B). Figure 6–20a shows the geometry of the ramp, which makes an angle θ with the horizontal. The force of gravity acts vertically downward on the box with magnitude mg. This force can be decomposed into a normal component $mg \cos \theta$ perpendicular to the ramp and a parallel component $mg \sin \theta$ pointing along the ramp toward its bottom. When the box moves down the ramp, as in Fig. 6–20a, the motion is perpendicular to the normal component of gravity, and the work done by that component of the force is therefore zero. The work done by the force of gravity to move the object from A to B (a total length L) along the direction of motion thus involves only the component $mg \sin \theta$ of gravity *along* the ramp,

$$W = (mg \sin \theta)L. \qquad (6\text{–}17)$$

Consider now a second path from the top of the ramp to the bottom. This second displacement is achieved by moving the box vertically off the back of the ramp (from A to C in Fig. 6–20b) and then horizontally to the previous end point (from C to B in Fig. 6–20b). For the first leg, the component of the force along the motion is mg, and the distance through which the force acts is $L \sin \theta$. Thus the work done by gravity on the box over this first leg is $(mg)L \sin \theta$. The box then moves horizontally; gravity does no work on the box during this portion of the displacement because the force of gravity is perpendicular to the motion. Thus the total work done by gravity on the box for this second path is $mgL \sin \theta$, the same as that calculated in Eq. (6–17). We would find the same result for other paths: The work done by gravity depends only on the difference between the final and initial heights of the box, in this case $L \sin \theta$; it does not depend on the path by which this height difference is reached. In fact, if the height decreases by an amount h, the work done in moving through any path is mgh (see Example 6–3). In this particular case, $h = L \sin \theta$.

Friction: Let us next consider the work done by sliding friction on a coffee cup of mass m that is pushed across a rough horizontal surface. (While there are other forces besides friction acting on the cup, it is only the work done by friction that interests us here.) The friction force on the cup has magnitude $\mu_k mg$ and is directed *opposite* to the motion. Suppose that the cup in Fig. 6–21 is moved (with the aid of some external force) along the $+x$-axis from $x = 0$ to $x = L$ and then back again. For the first half of the motion, the displacement of the cup is $+L\hat{\imath}$, whereas the force of friction is directed to the $-x$-direction:

▲ **FIGURE 6–20** (a) An object of mass m moves from point A to point B on an inclined plane under the influence of gravity. Gravity does positive (or negative) work on the object as it moves down (or up) the plane. (b) The object now moves from point A to point B by a different path: a vertical motion from point A to point C followed by a horizontal movement from C to B. The work done by gravity is exactly the same as in part (a).

◀ **FIGURE 6–21** A cup moves along a horizontal surface from $x = 0$ to $x = L$ and back again. One of the forces acting is friction, and the work done by friction in the back-and-forth motion is not zero.

$\vec{f} = -\mu_k mg\,\hat{i}$. In the motion back to the origin, the displacement is $-L\hat{i}$, whereas the force of friction is directed to the $+x$-direction: $\vec{F} = +\mu_k mg\,\hat{i}$. Then the work done by friction is

$$W = W_{\text{out}} + W_{\text{back}} = (-\mu_k mg)(L) + (\mu_k mg)(-L) = -2\mu_k mgL.$$

Rather than canceling, friction does the same negative work coming back as going out, and there is work done even though the net displacement is zero. Moreover, the amount of work depends on L; that is, it depends on the path taken by the cup. The least amount of work is associated with the shortest path. In this case the starting and finishing points are the same and the shortest path is $L = 0$, that is, there is no motion at all. You can guess that if the starting and finishing points were different, the path over which friction does the least amount of work is the straight line between the points.

Spring Force: Equation (6–14), $W = \displaystyle\int_{x_0}^{x_f} F(x)\, dx$, is an expression for the work done that applies to any one-dimensional force that depends only on the position of the object. It is a general property of integration that one-dimensional integrals such as these depend only on the end points of the integration, not on any intermediate points. Thus the work done by such is independent of the path. This is illustrated with the spring force in the following example.

EXAMPLE 6–12 A mass m is attached to the end of a spring with spring constant k. The equilibrium position of the mass is at $x = 0$ (Fig. 6–22a). Consider two paths by which the mass can move from the point $x = L$ to the point $x = 0$: Path 1 is the direct motion (Fig. 6–22b) and path 2 is the displacement from $x = L$ to $x = a\,(a < 0)$ followed by a displacement to $x = b\,(b > L)$ and completed with a final displacement to $x = 0$ (Fig. 6–22c). Find the work done on the mass by the spring for each path.

Strategy Because we know the spring force $F = -kx$, we can find the work by using $W = \int F\, dx$ between suitable limits of integration.

Working It Out In the case of path 1, we have

$$W = \int_L^0 (-kx)\, dx = -\frac{kx^2}{2}\Big|_L^0 = -\left(0 - \frac{kL^2}{2}\right) = \frac{kL^2}{2}.$$

The work done by the spring on the mass is positive.

In the case of path 2, there are three contributions to W—corresponding to the three displacement steps described—and their sum is

$$W = \int_L^a (-kx)\, dx + \int_a^b (-kx)\, dx + \int_b^0 (-kx)\, dx$$

$$= -\frac{kx^2}{2}\Big|_L^a - \frac{kx^2}{2}\Big|_a^b - \frac{kx^2}{2}\Big|_b^0$$

$$= -\frac{ka^2}{2} + \frac{kL^2}{2} - \frac{kb^2}{2} + \frac{ka^2}{2} - \frac{k \times 0^2}{2} + \frac{kb^2}{2} = \frac{kL^2}{2}.$$

The spring does the same work over path 2 as over path 1. You can convince yourself rather quickly, given the general way the terms canceled in the work done over path 2, that the equality of the work done over the two paths would have been the same whatever the form of $F(x)$.

What Do You Think? Suppose that the spring constant k were not really constant but varied slightly with x so that $k = k_0(1 + bx)$. Give an argument, based only on general properties of integrals, that

the work done by this force as the object under its influence moves from one point to another is independent of the path between the points.

▲ **FIGURE 6–22** Equilibrium position of the mass attached to the spring. (b) The spring is pulled to $x = L$ and then moves back to $x = 0$. (c) A more complicated path for motion from $x = L$ to $x = 0$.

Conservative and Nonconservative Forces

We have seen that when an object moves from one position to another as forces act on it, the work done by gravity or the spring force is independent of the path taken, whereas the work done is dependent on the path taken if the force of friction acts on the object. We must be aware that the work, the integral of Eq. (6–16), may or may not depend on the path. We can categorize forces by whether the work they do is independent or not of the path taken. Forces for which the work is independent of the path are called **conservative forces**; those for which the work depends on the path are called **nonconservative forces**. Gravity and the spring force are conservative, whereas sliding friction is nonconservative.

> **The work done by a conservative force in moving an object from one position to another is independent of the path taken by the object.**

We can reformulate this statement in terms of closed paths (paths that end at the same point at which they start):

> **The work done by a conservative force in moving an object along any closed path is zero.**

Here is why these two statements are equivalent. One possible closed path is the path for which the object does not move at all, in which case W is obviously zero. But the statement that the work done is path independent means that the same amount of work is done along this path as along *any* path for which the object starts and finishes at the same point. Thus a conservative force does no work on an object moving along any closed path, and indeed this is a definition of a conservative force.

As an example, consider a bag of concrete. If we move it from the back of a truck to the ground, the force of gravity does positive work. If we then lift the bag from the ground back to the truck, gravity does negative work. The total work done by gravity in this process must be zero because the bag of concrete is back where it started—more precisely it is back at the same height at which it started.

We can now restate our conclusions about forces that act in one dimension, which were earlier stated in terms of path independence. *Any* force that can be written in the form $F = F(x)$ is a conservative force. This includes constant forces as a special case.

Conservative forces matter for two reasons. First, we shall see in Chapter 7 that we can dispense with work for these forces and instead use the notion of a *potential energy*. Second, *most of the forces we deal with are conservative*. This is true of all the fundamental forces in nature. In particular, *central forces*—those for which the force is directed along a line from a fixed center and whose magnitude depends only on the distance from the center—are conservative. The gravitational force between the Sun and the planets is an important example of a central force.

THINK ABOUT THIS . . .

FRICTION APPEARS TO BE A CONSTANT FORCE; WHY ISN'T IT CONSERVATIVE?

You might think from this discussion that kinetic (sliding) friction, which we have explicitly shown is not conservative, provides a counterexample to the idea that forces with constant magnitude are always conservative. After all, isn't friction on a horizontal surface a constant? This apparent difficulty is resolved by noting that friction is *not*, in fact, constant. Its *magnitude* is constant if the coefficient of sliding friction is constant. On a horizontal surface the magnitude of the friction force is $\mu_k F_N$, and F_N is the magnitude of gravity in simple situations. However, the *direction* of the friction force is not constant. The direction depends on the direction of motion of the ob-

ject on which the force acts. When the object moves to the right, friction acts to the left; when the object moves to the left, friction acts to the right. Mathematically,

$$\vec{f} = \mu_k F_N \text{ in the direction opposite to } \vec{v},$$

which means that, strictly speaking, $\vec{f}$ depends on velocity. The work done by friction when an object moves out cannot possibly cancel the work done by friction as the object comes back along a reversed path, as was the case for the spring, because along each leg the friction does negative work. We can safely conclude that *sliding friction is nonconservative*.

The fact that we won't be able to associate a "potential energy" with nonconservative forces does not put us into any fundamental difficulty. We will simply have to deal with conservative and nonconservative forces differently. The work–energy theorem works for conservative and nonconservative forces alike, and we can always separate the work done by the two types of forces. To understand why this might be useful, consider a physical situation where there is an unknown frictional or drag force present, for example, an accelerating automobile subject to an unknown drag force from the air. We may then use the work–energy theorem to calculate the work done by this unknown force. The theorem in this case reads

$$\left(\begin{array}{c}\text{initial kinetic}\\ \text{energy}\end{array}\right) + \left(\begin{array}{c}\text{work done by}\\ \text{known force}\end{array}\right) + \left(\begin{array}{c}\text{work done by}\\ \text{unknown drag force}\end{array}\right) = \left(\begin{array}{c}\text{final kinetic}\\ \text{energy}\end{array}\right).$$

The term *work done by unknown drag force* can be regarded as the energy lost, or **dissipated**. Rearranging, we find the work done by the unknown drag force.

$$\left(\begin{array}{c}\text{work done by}\\ \text{unknown drag force}\end{array}\right) = \left(\begin{array}{c}\text{final kinetic}\\ \text{energy}\end{array}\right) - \left(\begin{array}{c}\text{initial kinetic}\\ \text{energy}\end{array}\right) - \left(\begin{array}{c}\text{work done by}\\ \text{known force}\end{array}\right).$$

The work done by the known force can be either positive or negative.

EXAMPLE 6–13 A mass of 50 g is placed on the end of a spring with a spring constant of 6.0 N/m. The mass is released from rest at $x = 10$ cm and moves to $x = 0$ (Fig. 6–23). Opposing the force of the spring is an unknown drag force. Find the work done on the mass by the drag force if the velocity of the mass at the equilibrium position $x = 0$ is 0.85 m/s.

Setting It Up The motion is all along one direction that we take as the x-axis. Figure 6–23 shows the mass both at the equilibrium position $x = 0$ and at the position released from rest, $x_0 = 10$ cm. We know the values of the mass m at the end of a spring and the spring constant k as well as the position of initial release from rest x_0 and the speed v_1 when the mass passes back through the equilibrium position.

Strategy We must put several elements in place to be able to use the work–energy theorem and find the work done by the drag force. The net work is the difference between the final kinetic energy K_f and the initial kinetic energy K_i; we know both K_i (which is zero because the mass starts from rest) and $K_f \left(=\frac{1}{2}mv_1^2\right)$. The net work, however, is also the algebraic sum of the work done by the spring, W_s, and the work done by the drag force, W_D. Thus, if we can find W_s, we can solve for W_D.

Working It Out The spring force $F = -kx$, so the work done by the spring on the mass as it moves from point x_0 to zero is

$$W_s = \int_{x_0}^{0} (-kx)\,dx = -\int_{0}^{x_0} (-kx)\,dx = \frac{1}{2}kx^2\Big|_{0}^{x_0} = \frac{1}{2}kx_0^2.$$

We now apply the work–energy theorem:

$$W_{\text{net}} = W_s + W_D = K_f - K_i,$$

or

$$W_D = K_f - K_i - W_s = \frac{1}{2}mv_1^2 - 0 - \frac{1}{2}kx_0^2$$
$$= (0.5)(50 \times 10^{-3}\,\text{kg})(0.85\,\text{m/s})^2$$
$$- (0.5)(6.0\,\text{N/m})(0.10\,\text{m})^2 = -0.012\,\text{J},$$

where W_D, the work done by the unknown drag force, is negative, showing that energy is dissipated. Drag forces always dissipate energy, that is, they never add energy. This is because drag forces always act in a direction opposite to the direction of the displacement.

What Do You Think? Drag forces may arise in different ways. Give some ideas of how energy is dissipated if the mass at the end of the spring slides on a rough surface. Do the same if the spring and masses are immersed in molasses.

▲ **FIGURE 6–23** Coordinate system for mass on spring.

6–5 Power

Up to this point, we have said nothing about how *rapidly* work is done. When you go up a flight of steps, you are doing a certain amount of work, essentially mgh, where h is the vertical distance. But you can go up the steps very slowly or you can run up as fast as you can. In the first case you are doing the given amount of work over a long period. In the second case the rate at which you do the work is larger. **Power** P is *the rate at which work is done:*

$$P \equiv \frac{dW}{dt}. \qquad (6\text{–}18)$$

DEFINITION OF POWER

It is straightforward to calculate the power for a constant force. In one dimension, we have Eq. (6–3), $W = F\,\Delta x$, where F is the constant force and Δx is the displacement from some fixed starting point. If we divide by Δt and take the limit of small Δt (so that Δx is also small), then we can recognize $\Delta x/\Delta t \to v$; on the left side we have $W/\Delta t$, that is, the work done per unit time, or power. We have shown that

$$P = Fv. \qquad (6\text{–}19)$$

In two or three dimensions, the work done by a constant force is $W = \vec{F} \cdot \Delta \vec{r}$ [Eq. (6–11)]. It is easy to generalize our result for one dimension to

$$P = \vec{F} \cdot \frac{d\,\Delta\vec{r}}{dt} = \vec{F} \cdot \vec{v}. \qquad (6\text{–}20)$$

Since the quantities on the right side of this expression can change with time, the power P in Eq. (6–20) is more appropriately called the **instantaneous power**.

The SI units of power are joules per second (J/s), and the unit has been given its own name, the **watt** (W for short; do not confuse this unit with the algebraic symbol W that we use for work). One watt is the power generated when a force of one newton displaces an object moving with a speed of one meter per second. Another commonly used unit of power is the **horsepower** (hp):

$$1 \text{ hp} = 550 \text{ ft. lb/s} = 746 \text{ W}.$$

Horsepower is sometimes used today as a measure of the power of automobile engines, but cars could just as well be rated in watts. You undoubtedly know that the watt is commonly used to rate the power output of lightbulbs (we study *electrical energy* in Chapter 26). A useful measure of electrical energy is based on the watt or, more conveniently, the kilowatt (kW). The **kilowatt-hour** (kWh) is the amount of work done when one kilowatt of power is generated for one hour.[†] Because there are 3600 s in 1 h,

$$1 \text{ kWh} = 3.6 \times 10^6 \text{ J}.$$

EXAMPLE 6–14 Early in the nineteenth century, James Watt wanted to market his newly developed steam engine to a society that until then had relied heavily on horses for mechanical work. So Watt invented a unit that made it clear how useful a steam engine could be. He conducted a demonstration in which a horse lifted water from a well over a certain period of time and called the corresponding power expended "one horsepower" (Fig. 6–24a). He could then compare his engine (favorably). Assume that water has a mass density of $1.0 \times 10^3 \text{ kg/m}^3$, that the well was 20.0 m deep, and that the horse worked for 8.0 h. How many liters of water did the horse raise from the well?

Setting It Up We sketch the situation in Fig. 6–24b. The water is pulled up vertically through the tension in the rope, with the pulley used to change the direction of the tension T supplied by the horse. We are given the mass density ρ of water, the depth Δy of the well, and the time t the horse worked. The tension in the rope, of magnitude T, is as yet unknown; it will be needed to find the work done and from that the power.

Strategy To determine the work done by the horse, we need T and the (known) distance the water traveled. Assuming the bucket of water does not accelerate, we can use the fact that there is no net force on the bucket to learn that $T = mg$. By dividing the work done by the tension (i.e., by the horse) by the time taken, we can find the power supplied by the horse, but we know this independently (it is 1 hp), so we can solve for the mass raised and hence the volume of water.

◀ **FIGURE 6–24**
(a) As this nineteenth-century representation shows, James Watt's steam engine could do the same work as many horses.
(b) Sketch to help in determining the power supplied by the horse.

[†] A typical large hydroelectric project generates a power of 10^{10} W; household energy consumption in the northern United States typically runs from 250 to 1000 kWh per month.

(continues on next page)

Working It Out   The work done by the tension in lifting a mass $m = \rho V$ of water from the bottom of the well is

$$W = T\,\Delta y = mg\,\Delta y = \rho V g\,\Delta y.$$

Since this work was done over a period of time t, the power is

$$P = \frac{\text{work}}{\text{time}} = \frac{\rho V g\,\Delta y}{t}.$$

We solve for the unknown volume V:

$$V = \frac{Pt}{\rho g\,\Delta y}.$$

All the quantities on the right-hand side are known. The power is, by definition, 1 hp = 746 W, and

$$V = \frac{(746\ \text{W})(8.0\ \text{h} \times 3600\ \text{s/h})}{(1.0 \times 10^3\ \text{kg/m}^3)(9.8\ \text{m/s}^2)(20.0\ \text{m})} = 1.1 \times 10^2\ \text{m}^3.$$

Because there are 10^3 L in 1 m^3, the volume lifted by the horse in liters is 1.1×10^5 L.

What Do You Think?   In the problem statement we used the phrase "in a certain period." So is the horsepower an instantaneous quantity?

---

*6–6 Kinetic Energy at Very High Speeds

We mentioned in Chapter 5 that Newton's laws cease to be applicable in two domains. One is the domain in which speeds approach the speed of light and the other is the domain of quantum physics, which applies largely to atoms and smaller entities. The first domain is discussed in Chapter 39, whose subject is special relativity. In Chapter 39, we will show that the maximum speed attainable by any particle is the speed of light itself. We will also show that the precise expression $\frac{1}{2}mv^2$ for the kinetic energy—the Newtonian form of kinetic energy—of a particle of mass m should be replaced with a more general result,

$$K = mc^2\left(\frac{1}{\sqrt{1 - (v/c)^2}} - 1\right). \tag{6–21}$$

Here c is the speed of light, which has a value of 3×10^8 m/s. This kinetic energy appears in the same work–energy theorem that we used throughout if one also appropriately modifies how work is calculated. In this section, we explore this relativistic form for the kinetic energy.

Equation (6–21) reduces to the usual expression $K = \frac{1}{2}mv^2$ when $(v/c)^2$ is very small (see Problem 77). We call a quantity x "very small" when x can be neglected in comparison with 1, and we denote this by $x \ll 1$. For example, if $v/c = 10^{-2}$, then $(v/c)^2 = 10^{-4}$; this is indeed much less than 10^{-2} if we maintain 1 percent accuracy. Under normal conditions, v/c is much smaller than 10^{-2}. The speed v_E of Earth around the Sun is, by comparison to, say, the speed of an automobile, a large number, yet $v_E/c \cong 10^{-4}$. Molecular speeds in air are also of this order of magnitude.

For particles in cosmic-ray showers and in particle accelerators, v can be very close to c, and the factor $1/\sqrt{1 - (v/c)^2}$ can be very large. In the highest energy accelerators this factor is larger than 10^4, and the relativistic kinetic energy is very much larger than the Newtonian kinetic energy formula would indicate. We note that the formula for K becomes uncontrollably large in the limit $v/c \to 1$. It is indeed an essential tenet of the theory of relativity, which improves on Newton's laws for high-speed particles, that no information—that is, no particles—can be sent at a speed greater than the speed of light. The case of $v = c$ is delicate: With $m = 0$, the expression for K in Eq. (6–21) is ambiguous but not manifestly wrong. The theory of relativity states that massless particles not only can move with the speed of light but must *always* move with the speed of light. An example of such a particle is the elementary particle called the photon, which is the particle that represents light itself in the quantum physics description of nature. Another interesting example is the *neutrino*, which plays an important role in the fundamental structure of matter, in astrophysics, and in cosmology. Experiment shows the neutrino has a mass but that it is a very small fraction of the electron mass. This particle can move very close to the speed of light but not *at* the speed of light. Until the discovery of its mass, the neutrino was thought to move always at exactly the speed of light; once you learn that the mass is not zero, no matter how small, you know that its speed cannot ever reach the speed of light.

Summary

Work W is done by a force that acts on an object when the object moves; a moving object possesses a kinetic energy K given by

$$K = \tfrac{1}{2}mv^2. \tag{6-4}$$

Here K is a scalar quantity formed from the velocity vector, with v^2 (the speed squared) $= v_x^2 + v_y^2 + v_z^2$. The central result of this chapter is the work–energy theorem as it applies to an object under the influence of one or more forces,

$$W_{net} = \Delta K, \tag{6-6}$$

where W_{net} is the net work done by the net force on the object. Here ΔK is the *change* in the kinetic energy of the object as it moves from an initial position $\vec{r}_i$ to a final position $\vec{r}_f$ through a displacement $\Delta \vec{r} = \vec{r}_f - \vec{r}_i$, $\Delta K = K_f - K_i$. The work–energy theorem tells us that when net work—positive or negative—is done on an object, the kinetic energy increases or decreases by the amount of the net work. The net work depends on the net force acting on an object as well as its displacement and, in general, on how the displacement is made. The net work is the work done by the net force; equivalently, it is the sum of the work done by the individual forces that make up the net force. The work done by a force can be expressed in different ways according to the form the force takes. We can enumerate these forms:

For a constant net force in one dimension $\quad W \equiv F\Delta x. \tag{6-3}$

For a constant force in three dimensions $\quad W = F_x\,\Delta x + F_y\,\Delta y + F_z\,\Delta z \tag{6-10}$

$$= \vec{F} \cdot \Delta \vec{r} \tag{6-11}$$

$$= F\,\Delta r \cos\theta, \tag{6-12}$$

where θ is the angle between the vectors $\vec{F}$ and $\Delta \vec{r}$.

For a nonconstant force in one dimension $\quad W = \displaystyle\int_{x_i}^{x_f} F(x)\,dx. \tag{6-14}$

For a nonconstant force in three dimensions $\quad W = \displaystyle\int_{\vec{r}_A}^{\vec{r}_B} \vec{F} \cdot d\vec{r}. \tag{6-16}$

In Eq. (6–16), the displacement is between points A and B. This form for work is the most general one and reduces to the other forms in the appropriate limit. It shows in particular that no work is done on an object that is in uniform circular motion. Both work and kinetic energy are measured in the SI unit of the joule (J).

In some cases, the work done when an object moves between two points depends on the path the object takes; in others, the work done is independent of the path. When the work done is path dependent, we say that the force is nonconservative; friction provides an example. When the work done is path independent, we say that the force is conservative; gravity and the spring force provide examples. Conservative forces are important because all the fundamental forces of nature are conservative.

The work–energy theorem allows us to calculate the speeds of objects when the work done by the forces is known, and it allows us to calculate the net work that must be done if a certain speed is to be achieved. This theorem is often much simpler to use for these purposes than is Newton's second law.

Power is the rate at which work is done:

$$P \equiv \frac{dW}{dt}. \tag{6-18}$$

The SI unit of power is the watt, equivalent to 1 J/s. From the definition of the instantaneous power, we find that in one dimension

$$P = Fv. \tag{6-19}$$

In two or three dimensions, we can generalize this result to

$$P = \vec{F} \cdot \vec{v}. \tag{6-20}$$

Understanding the Concepts

1. We mentioned in the introduction that the work–energy theorem might help us to analyze the motion of a baseball under the influence of both gravity and drag force from the air. How would you do so?

2. You are sitting in an automobile with a ball on your lap. The automobile stops very suddenly and the ball shoots forward out of your lap. To you, its kinetic energy has evidently changed. How would you use the work–energy theorem to analyze the situation?

3. Does it make sense to refer to a force doing negative work when an object moves under its influence? What does negative work mean? In answering, consider what happens when an object is stopped by a force.

4. It certainly seems like work to us when we hold a bag of groceries for a long period of time. Are we expending energy when we hold a bag of groceries for a long period of time? How is the answer to this question consistent with the work–energy theorem?

5. The centripetal forces that cause uniform circular motion do no work because they are perpendicular to the motion. How do such forces fit into the work–energy theorem?

6. A piano can be lifted to the third story of a building by having a crew carry it up the stairs or by using some type of pulley system. Is the same work done in both cases? Assume that friction can be neglected.

7. If the moving crew of Question 6 uses a rope and pulley, it pulls on the rope in the same direction as the force of gravity. Because the crew pulls the rope in the direction opposite to the displacement of the load, is the crew doing negative work on the load?

8. Is the force of a tennis racket acting on a tennis ball a conservative force?

9. A man pushes against the smokestack on a cruise boat. When the ship is stationary, he does no work. When the ship starts to move in the direction in which he is pushing, he appears to be doing work yet he experiences no change in the level of his exertion. Why is this? Keep in mind that the man does not fall down because the force of friction keeps his shoes from sliding backward. What is the work done on the man by the rough deck?

10. You do no net work when you walk at a constant speed. Why do you get tired?

11. A parachutist jumps from a plane and lands safely in a field. Does the net work done on the parachutist depend on the height from which he or she jumps? (The work done by gravity does depend on that height.)

12. No work is done in uniform circular motion. Suppose that you observe circular motion in which the moving object first speeds up and then slows down to its original speed. Is any net work done?

13. The work done by friction on a box that slides across a floor from one spot to another is negative. Can one conceive of a physical situation in which the work done by friction on an object is positive? [*Hint*: In an idealized (no-air-resistance) situation, the net external force on an automobile is friction between tires and road.]

14. No work is done on a bag of groceries while you are holding it stationary. Is work done on the same bag if you are holding it while you move steadily upward in an elevator? What is the difference in the two situations? Are your hands still the origin of the force that does work on the bag?

15. A car with cruise control transports you at a constant speed. The engine does work. How much of that work is done on you?

16. Discuss to what extent the following description of a conservative force is equivalent to the ones given in the text: "In the motion of objects that are subject to conservative forces, there is no energy dissipation involved."

17. One of the entertainments at a carnival is a rotating (vertical) cylinder. The participants step in and place themselves against the interior wall. The cylinder starts to rotate more and more rapidly, and at some point the floor falls away, leaving the customers stuck like flies to a wall. Is any work done on the participants? If so, what force does the work?

18. Tarzan swings from tree to tree on a jungle vine (Fig. 6–25). Is there net work done on him during the motion? If so, what forces do the work?

▲ **FIGURE 6–25** Question 18.

19. A stunt consists of one acrobat standing on the short end of a seesaw whose pivot point is not at its midpoint. A second acrobat leaps down on the long end of the board and flips the first acrobat several meters into the air. How would you determine the work done by the second acrobat in flipping the first one into the air?

20. When a dropped egg hits the ground, it abruptly loses the kinetic energy it had just before it struck the ground. Does this mean that the ground has done work on the egg? If so, what is the sign of this work?

21. Two identical twins work side by side as butchers. They use identical motions and identical hatchets. One brings a hatchet down on some very tender meat and the other on a large bone. Which one does more work per swing?

22. You tow a small child on a sled at a constant speed by pulling the sled with a rope. The rope makes an angle θ with respect to the horizontal. What forces act on the sled and which ones do work?

23. A one-dimensional force acts on an object, changing its velocity from zero to $\vec{v}$. By the work–energy theorem, the work done is $W = \frac{1}{2}mv_f^2 - \frac{1}{2}mv_i^2 = \frac{1}{2}mv^2$. An observer moving with velocity $\vec{v}$ with respect to the original system sees the initial velocity as $-\vec{v}$ and the final velocity as zero. This observer would conclude that $W = -\frac{1}{2}mv^2$. What accounts for the difference?

24. A diver plunges from a 10-m-high diving board into water. How would you determine the average force of resistance of the water that slows down and stops the diver? (Neglect the force of gravity on the diver while she is in the water—we shall see in Chapter 16 how the water's buoyancy takes care of that.)

25. A parachutist jumps off a tower. What measurements would you have to make to determine the work done by the drag force of the air during the entire fall? The drag force is a rather complicated function of the velocity of the jumper.

26. Are the following forces conservative or nonconservative? (a) Air drag on a parachute. (b) The force opposing the fall of a steel ball bearing in a beaker of water. (c) The explosive force causing a bullet to leave a rifle barrel. (d) The force of an ideal trampoline that propels you into the air.

27. How do you know that the drag forces you experience when you swim are not conservative?

28. An object can be said to have a certain kinetic energy. The work–energy theorem relates the change in kinetic energy on the object to the work that is done on it. Does this mean that the object also "has" a certain amount of work?

Problems

6–1 Kinetic Energy and Work

1. (I) An automobile of mass 10^3 kg moves at 1.0 km/h = 0.28 m/s. (a) What is its kinetic energy? (b) At what speeds must a person of mass 80 kg and a bullet of mass 10 g move to have the same kinetic energy as the automobile? (c) What would the speed of the automobile be if the kinetic energy doubled?

2. (I) A construction worker of mass 85 kg rides in an elevator up to the 15th floor, which is 42 m above the ground. The elevator travels with uniform speed. (a) What is the net work done on the worker? (b) What is the work done on the worker by the contact force of the elevator? (c) What is the work done on the worker by gravity?

3. (I) A person lifts a suitcase of mass 10 kg from the floor. Ignore the initial acceleration of the suitcase and suppose that it moves upward at a constant speed between a height $h = 0$ m and $h = 1$ m. (a) What are the forces acting on the suitcase as well as the net force? (b) What is the net work done on the suitcase? (c) Find the work done on the suitcase by the person.

4. (I) A bedroom dresser of mass 38 kg is moved from the first floor of an apartment building to the penthouse on the 44th floor, 130 m higher. (a) What is the work done on the bureau by three men in carrying it up the steps? (b) If the three men take it up on an elevator, how much work is done on the dresser by the normal force of the floor of the elevator?

5. (I) A truck carrying a 66–kg crate accelerates uniformly from rest to 63 km/h in 15 s. Calculate the work done on the crate by the truck.

6. (I) An old piano of mass 180 kg is removed from an apartment building being converted into condominiums. The previous owners found it too much trouble to remove and left it. The workmen decide the easiest thing to do is to drop it out of a double-width window to the ground 25 m below. (a) How much work do the workmen do if they just push it out the window? (b) If the men slowly lower the piano by rope, what is the work done on the piano by the rope's tension? (c) How much work does gravity do in each case?

7. (I) A man pushes a refrigerator of mass 40 kg at uniform speed for a distance of 1.5 m to the kitchen wall. The coefficient of friction between the refrigerator and the floor is $\mu_k = 0.4$. (a) How much work does the man do in moving the refrigerator? (b) What other sources of work done are there? (c) What is the net work done in this process?

8. (I) A person pulls a heavy load of mass 37 kg up the side of a building by using a frictionless pulley. The load travels up a distance of 7.5 m. Take the load to move with constant velocity and ignore any acceleration at the beginning or end of the move. (a) How much work is done on the load by gravity? (b) By the tension of the rope? (c) By the person?

9. (II) Consider the woman who lifts the huge interior lineman in Example 5–4. How much work does she do while pulling down a 2-m length of rope? What is the work done by gravity on the lineman while this is going on?

10. (II) The mass $M = 40$ kg is lifted to a height $h = 4$ m using the system of pulleys shown in Fig. 6–26. The motion is slow and the initial acceleration is negligible. (a) Find the force that must be applied at the free end of the rope. (b) Find the work done on the mass by this force. (c) Calculate the work done on the mass by gravity during the process.

▲ **FIGURE 6–26** Problem 10.

11. (II) A baseball of mass 145 g leaves a pitcher's hand at 96.6 mi/h, but, due to air resistance, it arrives at home plate 60.0 ft away traveling at 95.3 mi/h. Assume that the magnitude of the ball's acceleration is constant and that the ball travels in a straight line (ignore gravity). How much work is done by friction during the flight of the ball?

12. (II) A ball of mass 240 g is dropped from a height of 2 m. (a) What is the work done on the ball by gravity? (b) Suppose that the ball bounces to a height of only 1.5 m. How much work is done by gravity on the ball as it moves from ground level to 1.5 m?

13. (II) A construction worker of mass 75 kg hoists a load of bricks of mass 42 kg by throwing a rope attached to the load over a pulley and letting his weight lift the load. Assuming that there is no friction, what is the work done by gravity during a 2.0-s period?

14. (II) Two masses are connected by a light string over a light, frictionless pulley, as in Fig. 6–27. The table surface is also frictionless. (a) Apply the work–energy theorem for this system to calculate the speed of the masses after the masses have moved a distance Δx starting from rest. Note that the work of the tensions drops out. (b) Use this result to obtain the acceleration of the system.

▲ **FIGURE 6–27** Problem 14.

15. (II) A waterfall of height 40 m has 200 m^3 of water falling every second. How many joules of work are done by gravity every hour? (The mass of 1 m^3 of water is 10^3 kg.)

16. (II) A ball of mass 85 g is dropped from a height of 3.00 m. It bounces back to a height of 2.75 m. Use the work–energy theorem to calculate the change in kinetic energy between the beginning of the contact with the floor and the termination of the contact, assuming that air resistance is totally negligible.

17. (II) Consider the ball of the previous problem. Assume that the energy loss on contact with the floor is proportional to the kinetic energy of the ball as it hits the floor. What will be the height reached by the ball on the second bounce? Can you generalize to the nth bounce?

18. (II) A construction crew is required to pull up a load of mass 106 kg by means of a rope thrown over a pulley. They are to lift the load from rest on the ground to a height of 4 m, and the load should arrive at the end point with a speed of 2.0 m/s. (a) Calculate the work done by the crew if it accelerates the load uniformly over the whole distance. (b) Repeat the calculation for the case in which the acceleration takes place in the first 1 m and the load is pulled with uniform speed the rest of the way.

19. (II) A child has three different sets of cubical blocks (Fig. 6–28). The first set consists of 3 blocks, each 12 cm on a side and of mass 36 g; the second set consists of 6 blocks, each 6 cm on a side and of mass 18 g; the third is a set of 12 blocks, each 3 cm on a side and of mass 9 g. For each set, what is the work the child must do to stack the blocks into a tower 36 cm high? The blocks can be treated as point objects at their *centers* in calculating the work.

▲ **FIGURE 6–28**   Problem 19.

20. (II) Archimedes is supposed to have said: "Give me a fulcrum and I will move the world." As a model (Fig. 6–29), consider a rod of length L. It rests on a sharp rock so that on the longer side the length of the rod is l_1 and on the shorter side it is l_2 (so that $L = l_1 + l_2$). A mass m_1 is placed at the end of the longer side and a mass m_2 at the end of the shorter side. The masses are such that the rod is balanced. Suppose the rod tilts through a *tiny* angle θ so that the mass m_1 is lowered just a little. Use the work–energy theorem to show that the rod will not continue to rotate as an acceleration about the tip of the rock provided that $m_1 l_1 = m_2 l_2$. [*Hint*: For a tiny angle the arc traced by the end point of a rod hinged at one end is $R\theta$, where R is the length of the rod, and the arc is, to a very good approximation, a straight line.]

▲ **FIGURE 6–29**   Problem 20.

21. (II) A mass swings on the end of a rope of length R, rising from a low point to a position $R(1 - \cos\theta)$ above that point (Fig. 6–30). Is work being done on the mass? Ignoring air resistance, what are the forces acting on the mass? Which of those forces does work on the mass, if any? Calculate the work done in any way you choose.

▲ **FIGURE 6–30**   Problem 21.

22. (III) Find the work that must be done by a force lifting against gravity to raise a coiled rope of length L and mass M entirely off a level surface. [*Hint*: Use the method of Problem 19 and divide the rope into more and more segments.]

6–2   Constant Forces in Space

23. (I) What is the scalar product of $\vec{A} = -2\hat{i} + 3\hat{j} - 5\hat{k}$ and $\vec{B} = 5\hat{i} + \hat{j} - 2\hat{k}$?

24. (I) A force $\vec{F} = (-3.1\ \text{N})\hat{i} + (2.7\ \text{N})\hat{j}$ is used to displace an object of mass 17 kg by an amount $\vec{r} = (0.50\ \text{m})\hat{i} + (-0.75\ \text{m})\hat{j}$. What is the work done by the force on the object?

25. (I) An object of mass 0.23 kg is initially at the origin and is acted on by the sole force $\vec{F} = (0.50\ \text{N})\hat{i}$. After a certain amount of time, the object is at a position $\vec{r} = (0.88\ \text{m})\hat{i}$. What is the change in the object's kinetic energy?

26. (I) Show that the vector $\vec{v} = -y\hat{i} + x\hat{j}$ is always perpendicular to the vector $\vec{u} = x\hat{i} + y\hat{j}$.

27. (I) Consider two vectors, $\vec{u} = 3\hat{i} - 4\hat{j} + 7\hat{k}$ and $\vec{v} = -2\hat{i} + 3\hat{j} + z\hat{k}$. What must z be so that $\vec{u}$ and $\vec{v}$ are orthogonal?

28. (I) A person puts a suitcase of mass 11.5 kg into a van, moving the suitcase a total distance of 0.9 m: 0.6 m up and 0.3 m horizontally. How much work is done by the person?

29. (I) A man pulls a sled by a rope, moving his two daughters to the top of a 15° slope. He holds the rope parallel to the slope. If the daughters and the sled have a total mass of 43 kg and the length of the slope is 36 m, how much work does the man do on the sled, assuming that he pulls the sled with uniform velocity? Ignore all friction on the sled.

30. (I) A block of material with mass 1300 kg is used in the construction of a building. During one part of the process of setting the block in place, a complex network of cables acts on it and its motion is transformed from a horizontal motion with speed 15 cm/s to a vertical motion with speed 21 cm/s. What is the net work done on the block during this motion?

31. (II) A skier of mass 72 kg (including skis), starting from rest, slides down a slope at an angle of 18° with the horizontal. The coefficient of kinetic friction is $\mu_k = 0.12$. What is the net work done on the skier in the first 7.0 s of descent?

32. (II) Consider a vector $\vec{A}$ in the xy-plane. Its x- and y-components are A_1 and A_2, respectively. Show that any vector $\vec{B}$ in the same plane that points in a direction perpendicular to $\vec{A}$ must have components $-cA_2$ and cA_1, respectively, where the magnitude of c is the ratio of the lengths ($|c| = B/A$).

33. (II) Sketch the direction of the vector $\vec{e} = (\cos\theta)\hat{i} + (\sin\theta)\hat{j}$. Show that it has unit length and use your sketch to give an expression for the unit vectors $\vec{f}$ that are perpendicular to $\vec{e}$ and that lie in the xy-plane. How many such vectors are there?

34. (II) Consider the vector $\vec{A} = 7\hat{i} + 3\hat{j} - 6\hat{k}$. Find the most general vector in the yz-plane that is perpendicular to $\vec{A}$.

35. (II) Consider the unit vector $\hat{e} = -0.6\hat{i} + 0.8\hat{j}$. What is the magnitude of the projection of vector $\vec{A} = 3\hat{i} - 2\hat{j}$ onto the line along which $\hat{e}$ points?

36. (II) A stone is thrown from a height h_0 above a level field, leaving the hand at a 40° angle. Ignore all effects of air resistance. (a) Compute the work done by gravity as the stone follows its trajectory back to the height h_0. Recall that the motion can be divided into motion in the vertical direction and motion in the horizontal direction. (b) Show, by applying the work–energy theorem, that the speed of the stone when it reaches h_0 again is identical to the speed it had when it left the hand.

37. (II) A force $\vec{F} = (2\hat{i} - 5\hat{j})$ newtons acts on an object that moves from $\vec{r}_1 = (7\hat{i} - 8\hat{j} + 2\hat{k})$ meters to a new position $\vec{r}_2 = (5\hat{i} - 4\hat{j} + 5\hat{k})$ meters. How much work does this force do on the object?

38. (II) A 32-kg crate slides down a plane that makes an angle of 17° with the horizontal, starting from rest at the top. The speed of the crate when it reaches the bottom of the 10-m-long slide is 2.5 m/s. What is the coefficient of friction? How much work is done by the force of friction?

39. (II) A small object is forced to move within a horizontal groove aligned with the x-axis. A constant force $\vec{F} = F_x\hat{i} + F_y\hat{j}$, where each component is constant, acts on the object, making it accelerate within its groove. Find the work done on the object during the period where it moves a horizontal distance L. What is the kinetic energy at the final point assuming that the object started at rest?

6–3 Forces That Vary with Position

40. (I) A spring with spring constant $k = 12$ N/m is attached to a wall at ground level. The end of the relaxed spring is on the floor at a location that we take to be the origin. A mass of 3.0 kg is attached to the end of the spring, and the spring is stretched by 50 cm and released. How much work has the spring done on the mass by the time the mass passes through the origin?

41. (I) A small gizmo is confined to a groove that is aligned with the x-direction. A rod pulls the gizmo in the $+x$-direction. The rod is attached to an apparatus such that the pulling force is 0.3 N when the gizmo is to the left of a point in the groove we label as the origin and 0.7 N when the gizmo is to the right of the origin. What is the work done by the pulling force on the gizmo as it moves from $x = -6$ cm to $x = +7$ cm?

42. (I) A one-dimensional force on a particle is given by $F = \alpha x$, where $\alpha = -3.00$ N/m for $x < 0$ and $\alpha = +7.00$ N/m for $x > 0$ (Fig. 6–31). Calculate the work done by the force on a block when the block is moved from $x_i = -1.50$ m to $x_f = +1.50$ m.

▲ **FIGURE 6–31** Problem 42.

43. (I) A one-dimensional force F depends on the position x of a particle on which it acts as $F = g_1x - g_2x^3$, where g_1 and g_2 are constants. What is the work done in moving the particle from the origin to $x = 2.0$ m?

44. (II) A man pushing a 50-kg crate up a slope that makes an angle of 30° with the horizontal exerts a force parallel to the plane. The coefficient of kinetic friction varies along the slope and is given by $\mu_k = \mu_1 + [(\mu_2 - \mu_1)s/L]$, where s is the distance along the slope starting at the bottom. The largest value of s is L, where $L = 10$ m is the length of the slope; $\mu_1 = 0.2$ and $\mu_2 = 0.3$. What is the work done on the crate by the force as a function of s if the force varies such that the crate is pushed at a constant speed?

45. (II) A spring gun is made by compressing a spring (assumed to be perfect) and latching it. A spring of constant $k = 60$ N/m is used and the latch is located at a distance of 7 cm from equilibrium. The pellets have mass 4 g. What is the muzzle velocity of the gun?

46. (II) An asteroid drops straight toward the Sun; the force on the asteroid due to the Sun has magnitude (constant)/r^2, where r is the distance from the center of the Sun to the asteroid, and is directed to the Sun's center. Given that the surface of the Sun is a distance R_{Sun} from its center and the asteroid starts from rest an infinite distance from the Sun, what is its kinetic energy when it reaches the Sun's surface?

47. (II) A nonstandard spring exerts a force $F = -k_1x - k_2x^3$ to restore itself to equilibrium, where x is the distance from equilibrium. The values of k_1 and k_2 are 5.0 N/m and 15 N/m^3, respectively. Calculate the work done to stretch the spring from 0.10 to 0.20 m.

48. (II) A rocket is scheduled to blast off from Cape Canaveral to study a neighboring solar system. Earth's gravitational force is $F = K/r^2$, where r is the distance from Earth's center and K is a negative constant. What is the minimum work the rocket engine must do so that the rocket leaves the gravitational force of Earth? Assume that the mass of the rocket does not vary in the process. (This assumption is actually very poor, but a more exact treatment must wait until Chapter 8.)

6–4 Conservative and Nonconservative Forces

49. (I) A child swings a streamer toy over her head in a nearly horizontal plane. The toy, of mass 85 g, is at the end of a massless string of length 1.5 m. She starts twirling the toy from rest while she slowly lets out the string to full length and gets the angular speed up to 2 rev/s. How much work has she done?

50. (I) A 74-g ball is tossed straight up in the air, rises to a maximum point, then falls back until it is 0.60 m below the position of the hand that tossed it up. A second ball is simply dropped from the same hand position and also lands 0.60 m below that position. What is the net work done by gravity in the two cases?

51. (I) You are in the process of moving. A 54-kg bed can be brought from ground level to the second floor (4.0 m above ground level) either by pulling it straight up by means of a rope or by dragging it up a frictionless plane inclined at 30° to the horizontal. Calculate the work done in each case by those who move the bed.

52. (I) An object of mass 600 g is suspended from a vertical spring that is attached to the ceiling. Without the mass, the spring is 25 cm long. When the mass is attached to it, the spring is extended to a length of 65 cm. What is the work done by the force of gravity during the extension of the spring?

53. (I) A constant force of 10 N pushes a particle along the x-axis. The position of the particle is represented by $x = 11 \text{ m} - (2 \text{ m/s})t + (0.5 \text{ m/s}^2)t^2$. Find the work done by the force between $t = 0$ s and $t = 1$ s and between $t = 1$ s and $t = 2$ s. Is the force conservative?

54. (II) A force $\vec{F}$ has components $F_x = axy - by^2$, $F_y = -axy + bx^2$, where $a = 2 \text{ N/m}^2$ and $b = 2 \text{ N/m}^2$ (Fig. 6–32). Calculate the work done on an object of mass 4 kg if it is moved in a closed path from (x, y) values of $(0, 1)$ to $(4, 1)$, to $(4, 3)$, to $(0, 3)$, and back to $(0, 1)$ (all coordinates in meters). The path between the points is always the shortest straight one, and all the distances are given in meters.

58. (II) An object of mass m is to be moved from the top of a building of height h to a point on the ground a horizontal distance h from its original location so that the position vector may be chosen to be $h\hat{j}$ at the beginning, and $h\hat{i}$ at the end. Two possible paths are: (a) the object is lowered at constant speed by rope and, after it reaches the ground, it is moved horizontally to the final location; (b) the object is allowed to slide along a straight support that runs from the initial point to the final point. Show that the work done by the force of gravity is the same in both cases.

59. (III) Prove that a force acting in one dimension is conservative if it is a function of position only and not a function of any other information about the motion of an object under its influence. Does this include forces with constant magnitude? In view of your answer, how does the friction force manage to be nonconservative?

60. (III) A small object of mass m is moved up along a track that forms one-quarter of a circle of radius R in the vertical plane (Fig. 6–33). The object moves with a small uniform speed maintained by a tangential force (that varies with the angle) along the track. Calculate the work done by this force in moving the object through a 90° arc from the lowest point to the highest point by (a) direct use of the definition of work and (b) using the work–energy theorem.

▲ **FIGURE 6–32** Problem 54.

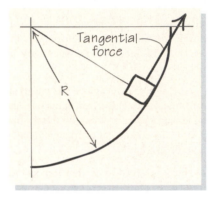

▲ **FIGURE 6–33** Problem 60.

55. (II) A force with an x-component acts on a mass with a strength that varies only with the position x of the mass on the x-axis, according to $|F(x)| = Ax^2$. The sign of the force is negative when x is positive and positive when x is negative, indicating that the force always attracts the mass toward the origin. Compute the work done by this force when $A = 1500 \text{ N/m}^2$ and the mass moves along the x-axis (a) from $x = -5.0$ cm to the origin, (b) from $x = -5.0$ cm to $+5.0$ cm, (c) from $x = +5.0$ cm to $+2.0$ cm, and (d) from $x = -2.0$ cm to -5.0 cm.

56. (II) A child's playground ride consists of four seats, of mass 12 kg each, connected to a vertical axle with spokes of small mass. The seats are placed equidistant in a circle of radius 1.8 m and rotate about the vertical axle. A child of mass 21 kg sits in one of the seats, and his friend pushes the ride to accelerate him from rest to 0.6 rev/s. How much work does the friend do?

57. (II) The net force acting on a particle depends on the position of the particle on the x-axis according to the relation $F = F_0 + Cx$, where $F_0 = 5$ N and $C = -2$ N/m. The particle is initially at rest at the point $x = 0$ m when the force begins to act. (a) Calculate the work done by the force when the particle reaches x values of 1, 2, 3, and 4 m. (b) Determine any positions (other than at $x = 0$ m) where the work done is zero. (c) Is the force conservative?

61. (III) The (one-dimensional) force acting on an object of mass m is given by the expression $F(x) = C|x|$, where $|x|$ is measured in centimeters. (a) How much work is done by the force when the object is moved from $x = -4.0$ cm to $x = +4.0$ cm? Compare your result with the amount of work done if the form of the force law were $F(x) = Cx$. (b) Repeat the calculation if the object moves from $x = 0.0$ cm to $x = 8.0$ cm.

62. (III) Consider a force that acts on an object of mass m which moves in the xy-plane. The force is given by $\vec{F}(x, y) = k_1 x\hat{i} + k_2 y\hat{j}$. Calculate the work done by the object if it moves in a circle of unit radius (given by $x^2 + y^2 = 1 \text{ m}^2$) starting at $x = 0$ m and $y = -1$ m and ending at a point that makes an angle of (a) 90°, (b) 180°, and (c) 360° with the original direction of the position radius vector. [*Hint*: The problem is simplified with polar coordinates r and θ, where $x = r \cos \theta$ and $y = r \sin \theta$.]

6–5 Power

63. (I) Electricity costs about $0.08/kWh. Your monthly electric bill is $26.00. Assuming that your only use of electricity is for light and that you keep your house lit 5 h/day, how many 100-W bulbs do you keep going?

64. (I) How much energy is used by running a fleet of one hundred 80-hp cars around the clock for one month?

65. (I) How long does it take you to climb four flights of steps? Assume that your body is 20 percent efficient and estimate the power you have to generate.

66. (I) Two engines are used to move a mass of 80 kg, starting from rest, a distance of 2.5 m in a straight line along a frictionless flat surface. Engine 1 exerts a constant force of 0.05 N, and engine 2 exerts a constant force of 0.75 N. (a) What is the work done by each engine? (b) What is the average power expended by each engine during the process?

67. (I) A test car of mass 700 kg is moving at a speed of 15 mi/h when it crashes into a wall to test its bumper. If the car comes to rest in 0.3 s, how much average power is expended in the process? (To find the average power, simply imagine Eq. (6–18) for a finite time interval: $P_{av} = \Delta W/\Delta t$, where ΔW is the work done during the time interval Δt.)

68. (I) Assume that a car of mass 1200 kg has an engine with power output of 80 hp. How long would it take to accelerate such a car to a speed of 100 km/h? (Neglect air resistance, which would make this time much larger.)

69. (II) The maximum power of a particular horse is 1 hp. With what speed can this horse pull a sled on level ground if the weight of the sled with its load is 5000 N and the coefficient of kinetic friction is $\mu_k = 0.03$? What is the maximum speed on a 5° upward incline?

70. (II) An accelerator accelerates a proton to 0.99c, where c is the speed of light. If 6.50×10^{10} protons are accelerated every minute, how much power is expended by the accelerator, assuming 5.00 percent efficiency? [For this problem, with v of the order of c, use Eq. (6–21) for the kinetic energy. The *efficiency* is the fraction of the total power that goes into changing the kinetic energy of the protons.]

71. (II) Consider the waterfall in Problem 15. If the waterfall is used to produce electricity in a power station and the efficiency of conversion of kinetic energy of falling water to electrical energy is 60 percent, what is the power production of the station?

72. (II) An escalator moves people from one floor up to another. The height difference between the floors is 4.2 m, and the angle that the escalator makes with the horizontal is 20°. The speed of the escalator is 1.2 m/s, and it is supposed to carry a maximum of 75 passengers, with an average mass of 75 kg. How much power must be generated by the motor that runs the escalator?

73. (II) Trained athletes can exert power for their movements ranging from around 5 hp for 1 s to 0.4 hp or less for periods extending over several hours. (a) A bicyclist is limited by wind resistance, which is roughly of the form $F = Av^2$, where $A = 0.08$ kg/m. Estimate the speed a cyclist can maintain for 1 h. (b) Estimate the time it takes a weightlifter to lift 100 kg a distance of 2 m. (c) Assuming that not too much time is taken up turning the corners, estimate how fast it is possible to climb three flights of steps, a vertical distance of 12 m. You can easily try this one!

*6–6 Kinetic Energy at Very High Speeds

74. (I) A proton is accelerated from rest to a final speed of 0.85c, where c is the speed of light. How much work is done by the accelerator on the proton given that a proton's mass is 1.7×10^{-27} kg?

75. (I) What is the kinetic energy of an electron (mass of 9.1×10^{-31} kg) moving at a speed of 0.9999c? Of 0.9999999999c?

76. (I) (a) How much work does it take to accelerate an electron from a speed of 0.1c to 0.5c? (b) From 0.5c to 0.99c? (c) From 0.99c to 0.999c?

77. (II) In order to show that Eq. (6–21) reduces to the usual form for small v/c, we need an approximation for $1/\sqrt{1-x}$ for small values of x, namely, $1/\sqrt{1-x} \approx 1 + \frac{1}{2}x + \frac{3}{8}x^2$. Check this approximation by calculating both sides of the expression on your calculator. For what value of x is the approximation correct to within 10 percent? To within 1 percent?

General Problems

78. (I) A grocery store pays a monthly power bill of $475. Electricity costs $0.09/kWh. How many joules of energy were used in the month?

79. (I) A ball with mass 100 g is set in motion inside a bowl, with initial speed $v = 2$ m/s. The ball ultimately comes to rest due to friction. How much net work was done by the external forces acting on the ball?

80. (I) A pile driver works by lifting a large mass and dropping it to the ground. A driver used to put pilings in the ground for a tall building has a mass of 6400 kg and is raised to a height of 2.5 m in 0.50 s. (a) How much work is done by the engine each time the weight is lifted? (b) What horsepower engine must be used to run the pile driver?

81. (I) The so-called Domesday Book recorded a general census carried out in England in the year 1086. It catalogued all 6000 waterwheels in the country, each of which had a power output of roughly 2 hp. Waterwheels were the main source of nonanimal energy at the time. By contrast, the power of a Boeing 747 jet airplane at maximum thrust is approximately 1 MW, and at cruise level the power output is roughly 0.3 MW. What fraction of the total nonanimal power of eleventh-century England does a cruising 747 represent?

82. (II) A mass $M = 3$ kg moving without friction in the xy-plane starts at the point labeled by the position vector $\vec{r}_i = 0\hat{i} + 0\hat{i}$ with velocity $\vec{v}_i = (2\hat{i} + \hat{j})$ meters per second. Two forces, $\vec{F}_1 = (2\hat{i} + 7\hat{j})$ newtons and $\vec{F}_2 = (2\hat{i} - 5\hat{j})$ newtons, act on the mass as it moves in a straight line to the point labeled by the position vector $\vec{r}_f = 10\hat{i} + 5\hat{j}$ meters (Fig. 6–34). (a) How much work is done by $\vec{F}_1$ as the mass moves from $\vec{r}_i$ to $\vec{r}_f$? (b) What power is provided by $\vec{F}_1$ at the instant the mass is at $\vec{r}_i$? (c) What is the kinetic energy of the mass when it reaches $\vec{r}_f$?

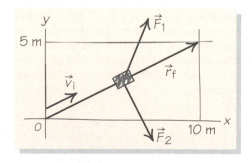

▲ **FIGURE 6–34** Problem 82.

83. (II) A rocket in outer space initially at rest is accelerated uniformly at the rate of 2.0 m/s² for 33 s. (Neglect gravity throughout.) (a) If the mass of the rocket, assumed here to be constant, is 1.0×10^4 kg, how much work is done by the rocket engine? (b) Suppose that the rocket decelerates at such a rate that it comes to rest in 55 s. How much work is done on the rocket during the deceleration?

84. (II) A worker pushes a box of mass $m = 25$ kg in a straight line along a rough floor. The applied force $\vec{F}$ has magnitude 85 N and acts downward at an angle $\theta = 10°$ with respect to the horizontal. The box is initially at rest at the position $x_1 = 0$ m, and it has speed $v_2 = 0.55$ m/s at position $x_2 = 3.50$ m. (a) Use the data to calculate the coefficient of friction. (b) What is the net work done? (c) How much work is done to overcome friction? (d) What is the instantaneous power generated by the worker at $x = x_2$?

85. (II) A block of mass $m = 2.6$ kg is placed on an inclined plane that makes an angle $\theta = 32°$ with the horizontal. It is given an initial speed v_0 up the ramp and slides a distance 1.3 m up the ramp before it comes to a stop. The coefficient of kinetic friction between the block and the ramp is $\mu_k = 0.25$. (a) What are the three forces acting on the block? Give the magnitude and direction of each. (b) What is the work done by each of the three forces during the motion of the block? (c) What was the initial speed v_0?

86. (II) The engine of an automobile requires 45 hp to maintain a constant speed of 80 km/h. (a) What is the resistive force against the automobile? (b) If the resistive force is proportional to the velocity, what must the engine power be to drive at a constant speed of 60 km/h? (c) At 140 km/h?

87. (II) A mass of 4.0 kg is attached to a string tied to a hook in the ceiling. The length of the string is 1.0 m, and the mass is released from rest in an initial position in which the string makes an angle of 30° with the vertical. Calculate the work done by gravity by the time the string is in a vertical position for the first time. [*Hint:* Recall that you can use any path you like to get from the initial to the final point because the work done by gravity is path independent.]

88. (II) A mass of 5 kg is accelerated by applying a force. In each case, calculate the work, in joules, that this force must do on the mass. (a) The mass is brought from rest to a speed of 0.5 m/s. (b) The mass is accelerated from 10 to 10.1 m/s. (c) The speed of the mass is increased from v to $v + 1$ meters per second. (d) If $v \gg 1$ m/s, find an approximation that simplifies the calculation in part (c).

89. (II) A puck of mass $m = 0.2$ kg moves in a circle of radius 0.8 m on a table top and is tied with a massless rope to a tether at the origin. The coefficient of kinetic friction between the puck and the table top is $\mu_k = 0.02$. At $t = 0$ s, the puck is at the point shown in Fig. 6–35 with a velocity in the $+y$-direction of magnitude 10 m/s. (a) How much work is done by the rope on the first revolution? (b) How much work is done by friction on the first revolution? (c) What is the kinetic energy at the end of one revolution?

90. (II) An automobile of mass 1100 kg is brought to a halt by applying the brakes, which lock the wheels. The coefficient of kinetic friction between the wheels and the road is 0.55. The car leaves skid marks 48 m long. (a) What is the force of friction between the car and the road? (b) What is the work done on the car by friction in bringing the car to a halt? Include the sign of the work. (c) What was the speed of the automobile when the brakes were first applied?

91. (II) A mass m is hauled from ground level up an inclined plane that makes an angle θ with the horizontal by means of a rope passing over a frictionless pulley. The mass is pulled along until it reaches a height H. The building of the Egyptian pyramids and many other ancient construction jobs used such ramps. Logs were sometimes used as rollers to reduce friction. (a) Show that if the contact between the mass and the ramp is frictionless, the work done by the tension in the rope (or, equivalently, by the person hauling the rope) is independent of the angle θ. (b) Calculate the work done by the tension in the rope as a function of the ramp angle θ if the coefficient of kinetic friction between the mass and the surface is μ_k.

92. (II) A large laser designed for nuclear fusion can produce 6×10^4 J of energy over a time period of 0.3×10^{-9} s. (a) How much power can such a laser produce during its discharge? (b) How much power is required to reenergize it over a period of 20 min?

93. (II) A pendulum of length L and mass m starts from an initial position in which it makes an angle θ_i with the vertical. Calculate the work done by the force of gravity as the mass moves from θ_i to θ_f and use your result to calculate the speed of the pendulum at the bottom of the swing. Why is it possible to ignore the tension in the string of the pendulum? [*Hint:* The work done by gravity is path independent.]

94. (II) A small dam produces electrical power. The water falls a distance of 18 m to turn a turbine. If the efficiency to produce electrical energy is only 68 percent, at what rate must water flow over the dam to produce 850 kW of electrical energy? (See Problem 70 for the meaning of efficiency.) The power system costs $3.5 million. How many years will it take for the power plant to pay for itself if electricity can be sold for $0.10/kWh? Ignore effects such as inflation and the cost of borrowing money.

95. (II) The force the Sun exerts on a planet in a circular orbit of radius x around the Sun is given by an expression of the form $F(x) = -mK/x^2$, where m is the mass of the planet and K is a constant. How much work must be done by a passing celestial body to move the planet to a radius that is 1 percent larger?

96. (II) Early twentieth-century investigations of the structure of atoms carried out by Ernest Rutherford involved the collisions of alpha particles (α particles) and gold atoms. The most interesting results concern the cases in which an α particle is scattered at a large angle from a gold nucleus. Assume that an α particle approaches a gold nucleus (at rest) head on and that the force between these two objects is repulsive, with magnitude $F = k/r^2$, where $k = 3.65 \times 10^{-26}$ N·m². Suppose that the α particle moves in toward the nucleus from far away under the sole influence of the repulsive force of the nucleus. What is the minimum kinetic energy the α particle must have initially so that it can get within 1.00×10^{-14} m?

97. (II) Consider the motion of an object on the xy-plane. Show that the force whose x- and y-components are respectively $F_x = (3 \text{ N/m})y$ and $F_y = 0$ is not conservative.

▲ **FIGURE 6–35** Problem 89.

◀ The motion of the balls illustrates the conversion of potential energy into kinetic energy. As a ball rises, it slows, losing kinetic energy but gaining potential energy; as it drops, it speeds up, gaining kinetic energy and losing potential energy. If there is no air resistance, once it has been tossed the total energy of a ball—the sum of the kinetic and potential energies—is conserved.

Potential Energy and Conservation of Energy

I n Chapter 6 we looked at the energy an object can possess by virtue of its motion— its kinetic energy. The kinetic energy changes when work is done on the object, as described by the work–energy theorem. Work can be done by both conservative and nonconservative forces, but if the work is done by *conservative forces*, the expression for work takes a form that we can relate to another kind of energy: the *potential energy*. For conservative forces, the work–energy theorem is a simple and powerful principle, the principle of the conservation of energy. We can say this another way: Consider an object in motion on a trajectory under the influence of a conservative force. At any given point the object has a total mechanical energy that consists of the sum of two terms, a potential energy and a kinetic energy. This total energy remains unchanged throughout the motion. The kinetic energy may change, the potential energy may change, but their sum remains constant. The potential energy is a function of position, and each type of conservative force is described by a different potential energy function. Since the kinetic energy depends on speed, the conservation of energy gives us a powerful and simple relation between position and speed.

The balls tossed by a juggler move under the influence of gravity, a conservative force, and clearly illustrate the principle. The initial speed with which they leave the juggler's hand sets the scene. After that, only the force of gravity acts and the conservation of energy provides us with a direct link between height and speed. As the ball rises, it loses speed and its initial kinetic energy decreases, but at the same time it gains in potential energy. At the top of the trajectory the ball has no velocity and all of its initial kinetic energy has been converted to potential energy, which is *given back* as the ball falls. As the ball falls, the potential energy decreases and the kinetic energy increases once more. The potential energy associated with gravity depends on position, in this case height.

A nonconservative force also acts on the tossed balls, air resistance. The work–energy theorem continues to hold, but the work done by the nonconservative force cannot be related to a potential energy. Does that mean that the conservation of energy is only an approximation? Superficially, that is the case. But when we look more closely, as we will in chapters to follow, we will see that it is possible to define other forms of energy associated with nonconservative forces and to extend the principle of the conservation of energy; for example, the tossed ball causes turbulence and heating of the air and ball (small, to be sure) in its passage, and friction causes surfaces to get hot, so we can often associate thermal energy with nonconservative forces. When one looks at a more fundamental level, there is an explanation of nonconservative forces that is both elegant and interesting. On a microscopic scale *there are no nonconservative forces*. When a brake pad gets hot due to the nonconservative force of friction, molecules in the material of the pad and in the surrounding air speed up, so there is microscopic energy of motion. In fact, *all* of the microscopic forces—atomic, molecular, nuclear, and so on—*are conservative*. So although on a macroscopic scale some forces are nonconservative, if we look at nonconservative forces on the microscopic scale we will see that they result from conservative forces. The principle of energy conservation, in which the sum of kinetic and potential energy is unchanged, is exact. Energy conservation is a fundamental law of physics and is also an enormously useful principle for the solution of problems.

7-1 Potential Energy and Conservative Forces

Our starting point is the work–energy theorem, which relates the work done on an object to a change in its kinetic energy, and we will concentrate on the case where the net work done comes from conservative forces. We defined conservative forces (Section 6–4) as those for which the work done in moving an object from one location to another is *independent of the path along which the object is moved*. This simple statement has enormous consequences.

The work done on a given object by a conservative force when the object moves from point A to point B depends only on those points; it does not depend on any points in between: $W = W(A, B)$. We can actually say more than this. Suppose we go from A to B, then from B to C (Fig. 7–1). In this case, the work done by our conservative force does not depend on the intermediate point B, and thus

$$W(A, C) = W(A, B) + W(B, C).$$

But this, in turn, can be true only if W has the form

$$W(P_1, P_2) = U(P_1) - U(P_2), \tag{7–1}$$

where $U(P)$ is some scalar function that depends only on position P. [You can easily check that when W has this form, then $W(A, C) = W(A, B) + W(B, C)$.] We can say that W depends on the end points alone and that it does so as a difference of two values of a function that depends only on position. If we know this function for a given conservative force, then we know how much work that force does as an object moves from one position to another. We have already worked out some cases in Chapter 6. For example, when an object moves from height y_1 to height y_2, the work done by gravity on the object is

$$W(y_1, y_2) = mg(y_1 - y_2).$$

Comparing the work done by gravity with Eq. (7–1), we find

$$U(y) = mgy + U_0, \tag{7–2}$$

▲ **FIGURE 7–1** Three points, A, B, and C, on a mountaineering trip. The work done by gravity depends only on the end point of any segment of the trip, not on the path taken between those points.

POTENTIAL ENERGY OF GRAVITY

where U_0 is a constant. The presence of the constant does not affect the basic relationship shown in Eq. (7–1) because it is the same in both terms and therefore cancels in the subtraction. We call the particular function U for any given conservative force the **potential energy** for that force. Equation (7–2) gives U for gravity, and forces different from gravity (e.g., the spring force) will have other potential energy functions. Although for the moment U is just a way to express the work, we'll show below that U appears on a par with kinetic energy as a piece of a total energy.

The case of gravity is the case of a constant force. Let us now follow the idea for a general one-dimensional conservative force $F(x)$. From Chapter 6, the work done by such a force on an object that moves from point x_0 to point x is

$$W(x_0, x) = \int_{x_0}^{x} F(z)\, dz.$$

Using Eq. (7–1), we now write the work done by our force on an object moving from x_0 to x as $U(x_0) - U(x)$, or in other words

$$U(x) - U(x_0) = -W(x_0, x). \tag{7–3}$$

Note the sign: The potential energy at point x minus its value at some initial point is the *negative* of the work done by the associated force as the object moves to point x from the initial point. We can rewrite this result as

$$U(x) - U(x_0) = -\int_{x_0}^{x} F(z)\, dz. \tag{7–4}$$

This equation defines the potential energy $U(x)$ for us. Once we know the function $F(z)$, we know $U(x)$ aside from an additive number $U(x_0)$ or just U_0, independent of x. Equation (7–4) gives us U as an integral of F. We can also go backward and find F as a derivative of U. Recall that the derivative of an integral with respect to its upper limit is just the integrand evaluated at the upper limit. Applied to Eq. (7–4), we find

$$F(x) = -\frac{dU(x)}{dx}. \tag{7–5}$$

[This equation confirms that the conservative force associated with the potential energy of gravity, Eq. (7–2), is just $-mg$.] It is important to remember that $U(x_0)$ is a constant, no different from the U_0 that appeared in Eq. (7–2). The work done as an object moves between two points is a difference between U evaluated at those points, so U_0 always cancels out. We can choose U_0 to be whatever we like, and we'll see below that this is often a help in solving problems.

CONCEPTUAL EXAMPLE 7–1 An unknown conservative force does negative work on a mass. Does the potential energy of the mass increase, decrease, or remain unchanged?

Answer We identified [Eqs. (7–3) and (7–4)] the work done by our force as $U(x_0) - U(x)$, that is, as the *negative* of the change in the potential energy. That means that when positive work is done, the potential energy decreases, and when negative work is done, the potential energy increases, which gives the answer to the question. There is another way to think about this, namely that the sum of the kinetic and potential energies is conserved, or remains fixed.

The conservation law states that if the kinetic energy increases, then the potential energy must decrease, and if the kinetic energy decreases, then the potential energy must increase to compensate. It is as if the potential energy were a kind of energy reservoir that you can draw on to increase the kinetic energy or refill by decreasing the kinetic energy. We already know from the work–energy theorem that when negative work is done on a mass, its kinetic energy decreases. In the language of the conservation law, this means that potential energy must increase, in accordance with what we already concluded above.

The Conservation of Energy

We are now in a position to establish that energy, consisting of potential plus kinetic energy, is conserved. We suppose that an object moves from point x_0 to point x_1 under the sole influence of a conservative force. From the work–energy theorem [Eq. (6–7)] and Eqs. (7–3) and (7–4), we have

$$K_1 - K_0 = W(x_0, x_1) = U(x_0) - U(x_1),$$

where $K_1 = \frac{1}{2}mv_1^2$ and $K_0 = \frac{1}{2}mv_0^2$. Simple rearrangement gives

$$K_1 + U(x_1) = K_0 + U(x_0). \tag{7–6}$$

This remarkable relation *expresses explicitly the principle of **conservation of energy***. Because our two points x_0 and x_1 can be any two points from the trajectory, Eq. (7–6) shows that $\frac{1}{2}mv^2 + U(x)$ *has the same value at every point in the trajectory of the object*. We call this quantity E, the **total mechanical energy** (or more simply, the **energy**) of the object:

$$E \equiv K + U(x) = \frac{1}{2}mv^2 + U(x). \tag{7–7}$$

TOTAL MECHANICAL ENERGY

We have shown that when only conservative forces act, E is *conserved* during the motion, no matter how it is transformed from potential energy to kinetic energy and back again—$\Delta E = 0$. A football, which moves to a good approximation under the sole influence of gravity, starts with a certain kinetic energy (Fig. 7–2). This kinetic energy is converted to potential energy as the ball rises to its maximum height; the potential energy is converted back to kinetic energy as the ball falls. Throughout the ball's flight, the total energy is always the same. The idea that the total energy is conserved is one of the most powerful ideas in physics as well as one of the most useful.

Applications of the Conservation of Energy

The conservation of E means that any change ΔK in the kinetic energy is compensated by an equal and opposite change ΔU in the potential energy:

$$\Delta K = -\Delta U. \tag{7–8}$$

This equation explains the origin of the term potential energy. Potential energy is energy that *has the potential* to be converted into kinetic energy (and, as we shall see later in the book, into other forms of energy).

Energy conservation is a useful tool for determining the speed of an object when only conservative forces act. If the function $U(x)$ is known and the initial conditions are given, we can determine the speed at any point x on the trajectory of an object. We simply use the fact that the total energy is constant.

We have already remarked that we can add an arbitrary constant U_0 to the potential energy $U(x)$ because this constant will cancel on both sides of Eq. (7–6). In other words, the conservation of energy involves only *changes* in the potential energy, so any

▶ **FIGURE 7–2** Conservation of energy gives us an immediate way to calculate the speed of each ball as a function of height, given its initial speed.

additive constant in the potential energy does not enter into the change. This is explicit in Eq. (7–8), which is the conservation-of-energy equation rewritten to make apparent the fact that only changes in potential energy are important, not the absolute values. The freedom to choose U_0 is equivalent to choosing a particular point where the potential is zero. To see this, consider the case of gravity and suppose that in your coordinate system you have measured height y from sea level. Then if in Eq. (7–2) you choose $U_0 = 0$, the potential energy will be $U(y) = mgy$. For this potential energy, $U(y) = 0$ at $y = 0$—we say the zero of potential energy is at $y = 0$. If instead we wanted the zero of potential energy to be at $y = 3$ m, we could choose $U_0 = -mg \times (3 \text{ m})$. With this choice $U(y) = mgy - mg \times (3 \text{ m})$, and $U(y) = 0$ when $y = 3$ m. The freedom to choose U_0, or equivalently to choose the point where $U = 0$, is a very useful practical tool; often the physical situation suggests a natural choice for the zero of U. This will be seen in a series of examples that follow.

CONCEPTUAL EXAMPLE 7–2   Consider the motion on a flat frictionless table of two masses connected by a spring, as in Fig. 7–3. You observe the motion at sea level, then you take the entire apparatus to the top of a tall building, $y_0 = 100$ m higher, and observe the motion again. A colleague claims that the motion must be different because you have added mgy_0 to the potential energy, so there will be less kinetic energy. Is your colleague right?

Answer   The system has a potential energy that consists of two terms: One involves the interaction between the masses on the table, governed by the spring between them and associated with horizontal motion of the masses; the second is the potential energy of gravity, which we can write as Mgy, where M is the total mass of the system and y is the height above sea level of the table top in the two situations. Changing y to $y + y_0$ just corresponds to adding a constant mgy_0 to the potential energy due to gravity, which, as we have seen, changes nothing. Your colleague is wrong; it does not matter at what altitude the motion takes place. Note that this is consistent with what you would conclude from Newton's second law. Gravity is canceled by the normal force from the table top, and as long as the table top is frictionless, it will play no role whatsoever in the motion of the masses.

What Do You Think?   By changing y to $y + y_0$, from what point are you measuring height? Where is the zero of the potential energy now? *Answers to **What Do You Think?** questions are given in the back of the book.*

Frictionless

▶ **FIGURE 7–3**   Two masses on frictionless table connected by a spring.

Gravity: Equation (7–2) gives the potential energy of a mass m subject to the force of gravity. When only gravity acts, the conserved total energy is

$$E = \tfrac{1}{2}mv^2 + mgy + U_0, \qquad (7\text{–}9)$$

where v is the speed and y is measured upward from some origin.

Suppose we drop a ball from rest from the roof of a building of height H with zero initial velocity and ask for its speed at height y. The value of E at $y = H$, where we know $v^2 = 0$, is $E = mgH + U_0$. A convenient choice for U_0 would be $U_0 = 0$, which corresponds to a potential energy that is zero at $y = 0$ (ground level). With an initial velocity of zero, the initial value of energy is mgH whereas the value at height y is $\tfrac{1}{2}mv^2 + mgy$, where v is the speed at height y. By energy conservation,

$$mgH = \tfrac{1}{2}mv^2 + mgy.$$

We easily solve this for v^2:

$$v^2 = 2g(H - y). \qquad (7\text{–}10)$$

An alternative choice for U_0 might be simpler if instead we wished to calculate how high an object will go when projected upward from the roof with an initial speed v_0. The initial value of E is $mv_0^2/2 + mgH + U_0$. Here a convenient choice might be $U_0 = -mgH$ so that the initial value of energy is $mv_0^2/2$. We have chosen the zero of potential energy at $y = H$ because then $U(y) = mgy + U_0 = mg(y - H)$, and this is zero at $y = H$. Due to energy conservation E has exactly the same value at its maximum height y_{max}, where $v = 0$, and this choice of U_0 gives

$$\tfrac{1}{2}mv_0^2 = \tfrac{1}{2}m(0)^2 + mg(y_{max} - H). \tag{7-11}$$

This is easily solved for y_{max}.

EXAMPLE 7-3 A brick is thrown straight up from 18 m above the ground by a person leaning over the edge of a roof so that the brick will reach the ground on a vertical path. Use the fact that energy is conserved to determine the speed with which the brick is thrown, given that it reaches the ground with a speed of 24 m/s.

Setting It Up We draw the sketch in Fig. 7-4. Included is a y-axis with $y = 0$ at ground level, and the height from which the ball is thrown is at $y = H$. Assuming that the clock starts when the brick is thrown, we know the height H at $t = 0$. We also know the final speed when the brick hits the ground $(y = 0)$ is v_f. We want to determine the initial speed v_0.

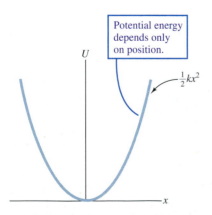

▲ **FIGURE 7-4** Ball thrown initially up falls back down to ground.

Strategy The total energy E at $t = 0$ will be the same as the energy when the brick hits the ground. Equation (7-9) provides us with an expression for the energy. We write out the energy for the two different situations and equate them—this is an equation whose only unknown is v_0, and we can solve for it.

Working It Out The initial energy is $\tfrac{1}{2}mv_0^2 + mgH + U_0$ and the final energy is $\tfrac{1}{2}mv_f^2 + U_0$. Equating these, we have

$$\tfrac{1}{2}mv_0^2 + mgH + U_0 = \tfrac{1}{2}mv_f^2 + U_0.$$

As before, U_0 cancels, and we have an expression that we can solve for the unknown v_0:

$$v_0^2 = v^2 - 2gH$$

or

$$v_0 = \pm\sqrt{(24 \text{ m/s})^2 - 2(9.8 \text{ m/s}^2)(18 \text{ m})} \cong \pm15 \text{ m/s}.$$

Note that energy conservation determines only speed, so either sign for v_0 is possible. Other information is necessary to find the sign; in this case we know the brick was thrown upward, so the plus sign is the right one.

What Do You Think? Suppose the process were reversed and somebody at ground level were to throw the brick upward with speed 24 m/s. Can you say, without doing the calculation, what the speed of the brick is at the initial launching position?

Spring Force: An ideal spring (one with negligible mass and that exerts a force given by Hooke's law) provides another example of a conservative force for which we can calculate a potential energy. You can think of any wind-up toy, in which a turned key effectively compresses a spring, to realize that you can store energy in a stretched or compressed spring. The displacement from the equilibrium position of a mass m attached to the spring is labeled as x. The force on the mass due to the spring is a restoring force, pulling the mass back to equilibrium:

$$F(x) = -kx, \tag{7-12}$$

where k is the spring constant. The "equilibrium point" of the spring is $x = 0$, since there is no force acting on the mass when it is at that point. Equation (7-4) takes the form

$$U(x) = U(x_0) - \int_{x_0}^{x} (-kx)\, dx = U(x_0) + \tfrac{1}{2}kx^2 - \tfrac{1}{2}kx_0^2.$$

It is convenient to choose the potential energy to be zero at the equilibrium point. With this choice, we set $U(0) = 0$, and applying this to the equation above, we see that this means $U(x_0) = \tfrac{1}{2}kx_0^2$, leaving

$$U(x) = \tfrac{1}{2}kx^2. \tag{7-13}$$

POTENTIAL ENERGY OF A SPRING

Potential energy depends only on position.

$-\tfrac{1}{2}kx^2$

▲ **FIGURE 7-5** The potential energy function for a mass attached to a spring. The mass position $x = 0$ is the position where the spring is relaxed, and we have taken the potential energy to be zero there.

This function is illustrated in Fig. 7-5. The total conserved energy is given by the sum of the kinetic and potential energies: $E = (mv^2/2) + (kx^2/2)$.

EXAMPLE 7–4 One end of a massless spring is welded to a flat surface; the other end points upward (Fig. 7–6a). A mass of 1.0 kg is gently set down on top of the spring until the spring is compressed by 17 cm to a new equilibrium position (Fig. 7–6b). What is the spring constant?

Now, the 1.0-kg mass is removed and a 2.0-kg mass is set on top of the spring. The spring is then compressed by hand so that the end of the spring is 42 cm lower than the position of the spring with no mass on top (Fig. 7–6c). The spring is then suddenly released. What is the maximum kinetic energy of the 2.0-kg mass?

(a)   (b)   (c)

▲ **FIGURE 7–6** (a) In our coordinate system, the top of the relaxed spring is at $y = 0$. (b) When a 1.0-kg mass is placed gently on top of the spring, the new equilibrium position is $y = -17$ cm. (c) A 2.0-kg mass is attached to the end of the spring, and the mass is then pushed down to $y = -42$ cm. The mass is then released.

Setting It Up We set $y = 0$ at the equilibrium position in Fig. 7–6a, the top of the unloaded spring. The positive y-direction is up. The motion is all one dimensional, which is why we have not used vector notation. We know the spring position y_1 when we set the first mass m_1 on the spring. We want to find the spring constant k. In the second part, we start with a second mass m_2 at rest at a given height y_2, and we want to find its maximum kinetic energy in the subsequent motion.

Strategy At equilibrium for m_1, the net force is zero. With the knowledge that the only forces acting are gravity (down) and the spring force (up), setting the net force to zero is an equation that can be solved for the spring constant. For the second part, we have a known spring force, hence a known potential energy, containing terms associated with the spring and with gravity. Only these conservative forces act, so we can use the conservation of energy, which relates the values of potential and kinetic energies. The maximum

kinetic energy will occur when the maximum amount of potential energy has been converted to kinetic energy, and this occurs when the overall potential energy has its minimum value. If you look at Eq. (7–5), then you see that the minimum of the potential, which occurs at $dU(x)/dx = 0$ is the place where the force $F(x) = 0$. We'll use this condition to find the minimum point.

Working It Out For the first part, we set the net force to zero. The force due to the spring acts upward, $F_{spring} = -ky_1$ (note y_1 is negative). We add the force of gravity, $-m_1 g$, to obtain the net force $F_{net} = -ky_1 - m_1 g$, and this must be zero. We then solve for the spring constant k,

$$k = -\frac{m_1 g}{y_1} = -\frac{(1.0 \text{ kg})(9.8 \text{ m/s}^2)}{-0.17 \text{ m}} = 58 \text{ kg/s}^2 = 58 \text{ N/m}.$$

For the second question, we want the potential energies associated with each of the forces. We choose both the potential energy of gravity and the potential energy of the spring to be zero at $y = 0$. Thus

$$U_g = mgy, \qquad U_{spring} = \tfrac{1}{2}ky^2,$$

where k is the spring constant. The total energy E of the system in this situation is the sum of the kinetic energy, the spring potential energy, and the potential energy of gravity:

$$E = \tfrac{1}{2}m_2 v^2 + \tfrac{1}{2}ky^2 + m_2 gy.$$

We evaluate E by noting that $v = 0$ m/s at $y = y_2$. Thus

$$E = \tfrac{1}{2}(2.0 \text{ kg})(0 \text{ m/s})^2 + \tfrac{1}{2}(58 \text{ N/m})y_2^2 + (2.0 \text{ kg})(9.8 \text{ m/s}^2)y_2$$
$$= (29 \text{ N/m})(-0.42 \text{ m})^2 + (19.6 \text{ kg} \cdot \text{m/s}^2)(-0.42 \text{ m})$$
$$= -3.1 \text{ J}.$$

Now we look for the point y_3 where the kinetic energy is a maximum, which as described above is the point where the net force is zero, that is, where the spring force and the force of gravity cancel. We have $-ky_3 - m_2 g = 0$, or

$$y_3 = -\frac{m_2 g}{k} = -\frac{(2.0 \text{ kg})(9.8 \text{ m/s}^2)}{58 \text{ N/m}} = -0.34 \text{ m}.$$

At this point, $E = K_{max} + \tfrac{1}{2}ky_3^2 + m_2 gy_3$. By energy conservation, this must equal the initial value of E, -3.1 J, or

$$K_{max} = -\tfrac{1}{2}ky_3^2 - m_2 gy_3 + E = -\tfrac{1}{2}(58 \text{ N/m})(-0.34 \text{ m})^2$$
$$- (2.0 \text{ kg})(9.8 \text{ m/s}^2)(-0.34 \text{ m}) - 3.1 \text{ J}$$
$$= 0.21 \text{ J}.$$

Note that there is conversion not only between kinetic energy and potential energy but also between the two different kinds of potential energy. These conversions all occur within the constraint of a constant total energy.

What Do You Think? Before the mass m_1 was placed on the spring, and with our choice of the zero of potential energy, the total energy was zero. With m_1 added, the spring is compressed. Is the total energy still zero?

The Energy of Systems

The additive nature of energy makes it easy to apply it to entire systems—as long as those systems involve conservative forces. Let's look at an example involving a system composed of two masses connected by a rope.

EXAMPLE 7–5 Consider the Atwood machine introduced in Example 5–5 (Fig. 5–9). The two masses have the values $m_1 = 1.37$ kg and $m_2 = 1.51$ kg. The system is released from rest with m_2 at height $h_2 = 84$ cm from the floor. Use energy conservation to find the speed of m_2 just before it hits the floor.

Setting It Up Figures 7–7a and 7–7b show the initial and final states, respectively. We have included a (vertical) y-axis, with $y = 0$ at the floor. We have also indicated the initial positions h_1 and h_2 of the two masses.

Strategy Energy conservation relates the speed and height of the masses. Assuming no friction at the massless pulley and no air resistance, only gravity acts on this system. Gravity is a conservative force, so we can use the conservation of energy with the potential energy due to gravity. We also assume that the rope is unstretchable, so that the speeds of the two masses will be identical, one moving up as the other moves down.

Working It Out We consider the two masses tied together as a whole, choosing the potential energy of gravity to be zero at $y = 0$; that is, U takes the general form mgy. Before the system is released (Fig. 7–7a), the total energy E_{init} is

$$E_{init} = m_1gh_1 + m_2gh_2.$$

In the final state of the system, just before m_2 hits the floor (Fig. 7–7b), the system contains kinetic as well as potential energy. Each mass has the same speed, v, and therefore the final energy E_{final} is

$$E_{final} = m_1g \times (h_1 + h_2) + \tfrac{1}{2}m_1v_1^2 + m_2g \times (0) + \tfrac{1}{2}m_2v_2^2$$
$$= m_1g \times (h_1 + h_2) + \tfrac{1}{2}(m_1 + m_2)v^2.$$

By conservation of energy, $E_{init} = E_{final}$, and this gives

$$(m_2 - m_1)gh_2 = \tfrac{1}{2}(m_2 + m_1)v^2.$$

Solving for v, we find the speed of m_2 near the floor,

$$v = \sqrt{\frac{2(m_2 - m_1)gh_2}{m_2 + m_1}}$$
$$= \sqrt{\frac{2(1.51\text{ kg} - 1.37\text{ kg})(9.81\text{ m/s}^2)(0.84\text{ m})}{1.51\text{ kg} + 1.37\text{ kg}}}$$
$$= 0.90\text{ m/s}.$$

What Do You Think? Suppose that the pulley in the Atwood machine were attached to the top bar by a spring instead of being fixed to it. Would you expect the motion to be different?

◀ FIGURE 7–7
(a) Initial conditions for Example 7–5.
(b) The situation just before mass m_2 hits the floor.

We have just considered a few examples showing how useful energy considerations can be in analyzing a system. The extension of the energy conservation principle to include many forms of energy provides a powerful tool for the study of complicated systems. All that happens around us can to a very large extent be described in terms of energy flows; the conservation of energy allows us to understand in a very direct way the nature of energy flows, the paths by which the energy moves, and the rate at which it is transferred. Using it, we can account for energy entering or leaving a system by certain routes and at certain rates. For example, if a certain amount of electric lighting is necessary in a new building, how does the heat associated with the light bulbs affect the heating or air conditioning that must be installed? In constructing a refinery, how do the heights at which various pipes must be placed affect the pumping apparatus required? The conservation of energy is a bookkeeping tool that allows engineers to design and analyze the behavior of a system without knowing every last detail of the system's variables. The extension of the energy conservation principle to include many forms of energy is one of the central themes of this book, and we'll come back to it frequently.

7-2 Energy Conservation and Allowed Motion

Let's take a look at the energy conservation equation for the one-dimensional motion of a mass attached to an ideal spring with spring constant k (Fig. 7–8a). To do so, we will use an **energy diagram** of the system (Fig. 7–8b). To create an energy diagram, we mark

(a)

(c)

(b)

(d)

▲ **FIGURE 7–8** A mass on the end of a spring; equilibrium is $x = 0$. (a) First, the spring is compressed by an amount x_0, then released. During one period (or cycle), the mass moves through $x = 0$ to an extension $x = x_0$ and back to $x = -x_0$. (b) Energy diagram for the mass. The potential energy $U = \frac{1}{2}kx^2$ is drawn. The straight line represents the constant total energy E; the kinetic energy K is the difference between E and U. The curve of K is determined purely by the conservation of energy. Points $+x_0$ and $-x_0$ are turning points for the motion, where the kinetic energy and therefore the speed are zero. The mass cannot go outside the range $-x_0 \leq x \leq +x_0$ as the kinetic energy would be negative. (c) As the mass moves, the kinetic and potential energies associated with the mass–spring system change: One increases as the other decreases. (d) Plot of the force of a spring on a mass attached to its end.

the position of the mass (the object whose motion is of interest to us) on the horizontal axis, and then on the vertical axis we draw any of the energies: potential energy, $U = kx^2/2$; kinetic energy, $K = mv^2/2$; or total energy, $E = U + K$. Suppose that the spring is initially compressed by an amount x_0, then released (Fig. 7–8a). Initially, $v = 0$ and $x = -x_0$; at this point, then, $E = kx_0^2/2$. Because E is unchanging, this is the value of E at any value of x. We represent it in the energy diagram by a horizontal line (Fig. 7–8b). We can plot the potential energy as a function of x on the same figure, and the kinetic energy is then automatically represented by a curve such that at each value of x its value is $E - U(x)$. When the spring is initially compressed, the location of the mass is at $-x_0$. The force on the mass, plotted in Fig. 7–8d, is the negative of the slope of the potential energy, Eq. (7–5). The slope of $U(x)$ at $-x_0$ is negative, which means that the force is positive there; the mass thus accelerates to the right under the influence of the spring force.

In the subsequent motion we can, using the energy diagram, follow the relative sizes of the kinetic energy and potential energy. Their sum remains constant at a value of E, but their respective values at any given x differ in a way that one can read off the energy diagram. We can also follow this play of the two types of energy in a different way: Figure 7–8c shows the magnitude of kinetic and potential energies as the sequence of positions of Fig. 7–8a is attained.

With Fig. 7–8c in mind, let us now look again at the mass's motion, starting from its initial position on the end of the compressed spring (negative x). We already stated that under the influence of the right-directed force of the compressed spring, the mass accelerates to the right. As the mass passes the point $x = 0$, the slope of the potential energy curve changes sign, and so therefore does the force (Fig. 7–8d). The mass decelerates, and the kinetic energy, which reached its highest value at the minimum of the potential energy ($x = 0$), falls below the horizontal line representing the constant total energy E. As we go further to the right of $x = 0$, the kinetic energy decreases as the potential energy curve rises. Finally, at $+x_0$, $U(x)$ is equal to E again. The mass cannot go any farther to the right; there, $U(x) > E$ and this would require a negative kinetic energy, an impossibility for the quantity $\frac{1}{2}mv^2$. Thus $+x_0$ and (for exactly the same reason) $-x_0$ are *turning points* for the motion. **Turning points** are those points where the speed drops to zero and the mass changes direction (points A and B in Fig. 7–8b). The mass must therefore remain between $-x_0$ and $+x_0$ throughout the motion. [If the total energy were larger, then the horizontal line of E in Fig. 7–8b would have a greater value (Fig. 7–9), with turning points corresponding to points C and D. A turning point x_0 is characterized by the fact that $E = U(x_0)$, or $K = \frac{1}{2}mv^2 = 0$.]

Let's suppose that the mass moves from $-x_0$ to $+x_0$ and then back to $-x_0$. It then restarts its motion from $-x_0$, and there is nothing to distinguish this second traversal from the first. The motion of the mass attached to the end of a spring is repetitive, or **periodic**. Periodic motion actually occurs for any potential energy function for which

▶ **FIGURE 7–9** When the total energy E of a mass on the end of a spring increases, the turning points move to large values of position. Here the turning points A and B move out to new turning points C and D when the energy increases.

the mass is constrained to move between two boundaries. It is spring-like near *any* minimum of a potential energy function. We will consider this further in Chapter 13, and there we will see many examples.

More about Energy Diagrams

The energy diagram provides a useful visualization. Let us write $U(x) = mgh(x)$. We may think of $h(x)$ as the *height* of the mass at the point x and $U(x)$ as the *gravitational potential energy* there. The motion of a mass in that potential may be visualized as the motion of a cross-country skier (on frictionless snow) on varying terrain whose height is given by $h(x)$. The particular form taken by $h(x)$ depends on the potential in question; for example, for the spring the terrain forms a parabolic bowl with a minimum at $x = 0$ (Fig. 7–5). This form of visualization is useful because we are intuitively familiar with how a skier might move on mountains of various shapes.

Energy diagrams can give us a great deal of insight into qualitative aspects of motion. We'll start with motion in one dimension again, and instead of a skier we treat a small object initially at rest on a frictionless landscape (Fig. 7–10) with a height $h(x)$ that depends on x (a profile). If the object starts on a slope, it will slide downward, converting some of its potential energy into kinetic energy. There are places, however, at which the object starting at rest *will remain* at rest. These points, which must be those for which the force is zero, or equivalently [see Eq. (7–5)] those points for which the potential energy function is flat, are called **equilibrium** points. There are three kinds of equilibrium points (Fig. 7–10). The object starting at rest at the bottom of a valley will not move. If you give this object a tiny bit of kinetic energy, it will move up the side of the valley, but soon it will reach a turning point and slide back, subsequently oscillating about the minimum point. The equilibrium point is said to be **stable** because a displacement from the equilibrium point puts the object in a position where *restoring* forces act to bring the object back to the equilibrium point. Consider next the equilibrium point with the object starting from rest at the top of a hill in our landscape. A little kinetic energy or a starting point a little off the top puts the mass on a slope with forces acting to move the mass further downhill, away from the equilibrium point. In other words, a small departure from the equilibrium point is amplified, and we call such an equilibrium **unstable.** A third possibility is a flat region in $U(x)$, a plateau in our landscape picture. An object at rest at one point on the plateau can be displaced to any other point on the plateau, and there will still be no forces acting on it. We describe this situation as one of **neutral** equilibrium.

Careful examination of the energy diagram can yield a great deal of information about the motion of an object in an arbitrary potential $U(x)$. We illustrate the procedure of extracting this information in the following example.

$U(x)$

Stable, unstable, and neutral equilibrium can be spotted easily on $U(x)$ curve.

Unstable equilibrium

Neutral equilibrium

Potential energy

Stable equilibrium

x

Position

◀ **FIGURE 7–10** Energy diagram for one-dimensional motion, with the potential energy plotted as a function of x. We see three kinds of equilibrium for a mass acted on by the associated force: stable, unstable, and neutral.

EXAMPLE 7–6 Consider the potential energy given by

$$U(x) = \tfrac{1}{4}x^4 - \tfrac{1}{3}x^3 - 3x^2,$$

where the coefficients of the various powers of x have units such that U has units of joules.

(a) Sketch $U(x)$ on an energy diagram.

(b) Find the equibrium points and classify them.

(c) Find the turning points for E values of -2 and 3 J, respectively.

Setting It Up We want to find the equilibrium points corresponding to the given potential and classify them, and we also want turning points (x-values) for two values of E.

Strategy After sketching a figure that corresponds to this potential, we find the equilibrium points—defined as those at which the force vanishes—by letting $dU(x)/dx = 0$ or by seeing where the flat spots, or extrema, of $U(x)$ are found on the plot of $U(x)$ versus x. Maxima are unstable equilibrium points; minima are stable equilibrium points. The turning points are points for which the kinetic energy is zero, that is, $U = E$, and these occur at the values of x where $U = E$, which should be visible on our sketch.

Working It Out (a) Figure 7–11 is a sketch of $U(x)$ with units of x understood. (The process can be carried out with a graphing calculator or in any number of other ways.) In drawing figures such as this there are several things to keep in mind. First, if the potential is a polynomial, as here, the large x behavior, positive or negative, is dominated by the highest power of x. In this case, $U(x) \cong \tfrac{1}{4}x^4$ at large $|x|$, always positive. Second, the degree and form of the polynomial determines the number of zeros. Here $U(x)$ is x^2 times a second-order polynomial. That means there is a double zero at $x = 0$ (which represents an extremum at $x = 0$ but not an axis crossing—the function has no sign change just above and just below $x = 0$) and two more zeros coming from the second-order part. Nevertheless, we see from the graph that the extremum at $x = 0$ is a maximum.

(b) The equilibrium points are defined as those at which the force vanishes, or equivalently the points where $dU(x)/dx = 0$. In this case

$$\frac{dU(x)}{dx} = x^3 - x^2 - 6x = x(x + 2)(x - 3),$$

so that the equilibrium points are $x = -2$, $x = 0$, and $x = 3$. These are the extrema of the graph (the units are understood). The simplest way to classify the equilibrium points is to look at

the sketch of the potential. The points $x = -2$ and $x = 3$ lie at the bottom of valleys and are therefore stable equilibrium points. The point $x = 0$ lies on top of a hill and is therefore a point of unstable equilibrium.

(c) The turning points are those points at which the kinetic energy is zero, that is, $U = E$. Thus we find the turning points by solving the equations $U(x) = -2$ J and $U(x) = 3$ J, respectively, for x. Since U is a fourth-order polynomial, this can only be done numerically.

More insight is obtained by drawing horizontal lines at $E = -2$ J and $E = 3$ J on the energy diagram (Fig. 7–12). We see that for $E = -2$ J there are *four* turning points, corresponding to two different regions in which the kinetic energy can be positive. A mass moving in the left-side valley cannot go over to the right-side valley. It encounters a **potential barrier** and is trapped between the turning points in a **potential well**. The same is true of a mass that starts its motion in the right-side valley. It encounters a potential barrier on the left in the form of the potential's hill around $x = 0$. On the other hand, for $E = 3$ J, there are two turning points on the outer wings of the potential energy curve. The kinetic energy can be positive anywhere between these two turning points, and in particular, it is larger than the potential energy at the intermediate high point at $x = 0$. It can thus cross $x = 0$ with kinetic energy to spare.

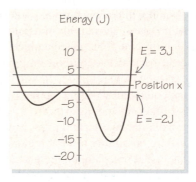

▲ FIGURE 7–12 If the total energy is negative, an object moving under the influence of the potential will be trapped in one of two potential wells on either side of the origin. If the total energy is positive, the object is still trapped, but this time there is only a single trapping region.

Alternative Approach to part (b) Another, more mathematical way of looking at the task in part (b) is to look at the force *in the vicinity* of the equilibrium points. For example, at the point $x = -2 + \Delta$, where Δ is very small, the force is given by calculating $-dU/dx$ at that point and keeping only the term proportional to Δ (i.e., ignoring terms quadratic or higher in Δ). Here this means that $-dU/dx = -10\Delta + 7\Delta^2 - \Delta^3$ and that we ignore Δ^3 and Δ^2 compared to Δ, so that $-dU/dx \cong -10\Delta$. This tells us that for Δ positive—that is, just to the right of the equilibrium point—the force is negative, so that an object will be pulled back to the equilibrium point. For Δ negative, the force is positive, so that again the object will be pushed toward the equilibrium point. Thus we have a stable equilibrium point. Similarly, at $x = 0 + \Delta$, we get $-(\Delta)(2 + \Delta)(-3 + \Delta) \cong +6\Delta$, which, by the same arguments, leads to the conclusion that $x = 0$ is an unstable equilibrium point.

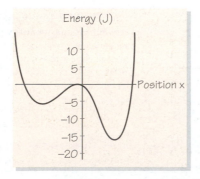

▲ FIGURE 7–11 Plot of the potential energy function of Example 7–6.

What Do You Think? Can you describe the motion of an object that has total energy $E = 0$?

ARE POTENTIAL BARRIERS REALLY IMPENETRABLE?

The answer to this question would appear to be a simple yes, such barriers are indeed impenetrable. If E were less than $U(x)$, which would correspond to being "inside" the barrier, the kinetic energy would be negative, and this is impossible given the form $K = \frac{1}{2}mv^2$. It is therefore very surprising that it is a well-established fact that on a microscopic scale (when we deal with nuclei, atoms, and even large molecules) *barrier penetration*, or *tunneling*, occurs. This is a phenomenon in which a particle can appear on the other side of a potential barrier even though it does not have enough energy to do so classically. On the microscopic scale Newtonian mechanics gives an inadequate description of the dynamics of systems, and in the microscopic domain, one must use another set of physical laws, the laws that form the subject of *quantum mechanics* (Chapter 40). Quantum theory allows for the phenomenon of tunneling and also shows that it is a very improbable process on the macroscopic scale. The physics behind this phenomenon is that matter exhibits wave-like characteristics in quantum mechanics and waves behave in many ways quite differently from particles.

Tunneling through potential barriers is an important aspect of many areas of modern engineering. Of particular interest is the application to scanning tunneling microscopy. This technique, which is based on the fact that electrons can tunnel through the potential barrier that normally keeps them confined to the vicinity of nuclei, allows for a detailed study of surfaces of materials. In another very important set of applications, many modern electronic devices—transistors and other control devices—that appear in computers or other circuits depend on tunneling for their operation.

7–3 Motion in Two or Three Dimensions

As for one dimension, there exists a potential energy that is a function of position for conservative forces acting in two or three dimensions. This potential energy is again defined so that the change in its value in going from one point to another is the negative of the work done by the force between those points. (Remember, for a conservative force the work done will depend only on the starting and ending points of the motion.) We then define the potential energy as before [see Eqs. (7–3) and (7–4)]: It is the negative of the work done, or more precisely

$$U(\vec{r}) - U(\vec{r}_0) = -W(\vec{r}, \vec{r}_0) \tag{7–14}$$

The potential energy is now a function of all three variables x, y, and z or, equivalently, of the position vector $\vec{r}$. The total energy E is given by

$$\begin{aligned} E &\equiv \tfrac{1}{2}m(v_x^2 + v_y^2 + v_z^2) + U(x, y, z) \\ &= \tfrac{1}{2}mv^2 + U(\vec{r}). \end{aligned} \tag{7–15}$$

It is independent of time; that is, it is a constant of the motion—in other words, as before energy is conserved. The value of the constant E is fixed by the initial value of U (at the starting point) and by the initial speed. The potential energy function $U(\vec{r})$ determines the force in a manner analogous to the one-dimensional case of Eq. (7–4), which gives $F(x) = -dU/dx$.

Rather than becoming involved in the calculus of several variables, we can gain some insight into the motion in *two dimensions* by generalizing our picture of the potential energy in one dimension as the profile of a one-dimensional landscape with hills and valleys to a picture of the potential energy in two dimensions as a surface landscape just like the landscape of Earth. Such a landscape can be represented on a topographic map. Figure 7–13 represents such a landscape, which includes peaks, ridges, lake beds, valleys, and saddles. The function $h(x, y) = U(x, y)/mg$ represents the height of the terrain above sea level. The force at the point (x, y) in a given direction is the slope of the terrain in that direction. Near the bottom of a lake, the force in all directions is one that brings a mass back to the bottom—this is what is meant by true minimum. The bottom of a lake bed is therefore a stable equilibrium point. Similarly, at the top of a peak, the terrain slopes downward in every direction, so that although an object could sit at the very top with no force on it, any small movement away from the top brings in a force that moves the object away from the top. Peaks are unstable equilibrium points. Saddles

▲ **FIGURE 7–13** The potential corresponding to a force that depends on the two space coordinates of a plane exhibits many of the features of Earth's surface and is equivalent to a topographic map of Earth's surface.

(mountain passes) present a new situation, as they do not occur in one-dimensional potentials. Saddles occur where the slope is downward on the two sides of a pass and upward along the ridge of the pass. In general, then, the motion can be more complicated than in one-dimensional motion. In one-dimensional motion an object, starting somewhere on a slope with an initial positive kinetic energy, will move up or down, and if the initial kinetic energy is large enough, it will pass over a potential barrier (the total energy is above the top of the barrier). In two dimensions, an object on a slope with some initial kinetic energy can move in a variety of directions, some of which will take the object across ridges and some of which won't. In particular, if the initial velocity of the object is *along a contour line*, that is, a line of constant elevation in our landscape, the potential energy of the object will remain constant. We call a line of constant values of potential energy an **equipotential**. An object will move along an equipotential with constant kinetic energy (constant *speed*).

We can draw an important conclusion from this discussion. We have a unique landscape described by the scalar function $U(x, y)$, and by using our slope argument, we can find the components of the vector force in different directions. For conservative forces these comes from a single function $U(x, y)$ and therefore are related. We can describe motion more economically with a potential energy than with three components of a force.

Potential Energy for Projectile Motion

In Section 7–1, we derived and applied the expression $U(y) = mgy + U_0$ [Eq. (7–11)] for pure vertical motion under the influence of gravity. We can now show that this expression is valid also for arbitrary projectile motion under the influence of gravity.

The force of gravity is conservative in that the work does not depend on the path taken—even when that path contains horizontal components, as we saw when we studied the motion of a mass on a ramp in Section 6–4. Therefore gravity is conservative for motion in more than one dimension. To find the potential energy when an object can move in more than one dimension, we must evaluate the work done when gravity acts. But the work, and hence the potential energy change, depends only on the change in height of the object and not on the object's horizontal motion. For example, $W = mgL \sin \theta$ [Eq. (6–22)] is the work done by gravity in moving a mass a total length L along a ramp that makes an angle θ with the horizontal. But the quantity $L \sin \theta = h$ is the height through which the mass drops (see Fig. 6–20). The work done is thus mgh, which is just the expression that holds when we consider vertical motion only. We conclude that *the potential energy function $U(y) = mgy + U_0$ applies to general projectile motion under the influence of gravity.*

EXAMPLE 7–7 A golfer hits a bad shot and the ball leaves the tee with an initial speed of 28 m/s at an angle of 84° with respect to the horizontal. A bee is cruising innocently at a height of 37 m when it has the bad luck to meet the golf ball. What is the speed of the ball when it hits the bee? Ignore all effects of air resistance.

Setting It Up We sketch the situation in Fig. 7–14 using an *xy*-coordinate system with the origin at the golf tee. The height where the ball and bee meet is $y = h$. We know v_i, the initial speed of the ball, and the initial elevation angle. We want to find the speed v_f of the golf ball when it hits the bee at the known altitude $y = h$.

Strategy Conservation of energy tells us that the initial energy of the struck ball has to equal the energy when the collision occurs, and this will relate speed and height. We let the potential energy be zero at ground-level height ($h = 0$ m), so $U(h) = mgh$.

Working It Out The golf ball has an initial energy $E_i = K_i + U(0) = \frac{1}{2}mv_i^2 + mg \times 0 = mv_i^2/2$, where m is the mass of the ball. The energy of the ball when it meets the bee at height h is $E_f = K_f + U(h) = \frac{1}{2}mv_f^2 + mgh$. Conservation of energy then reads

$$E_i = \frac{1}{2}mv_i^2 = E_f = \frac{1}{2}mv_f^2 + mgh.$$

▲ **FIGURE 7–14** Golf ball hits bee at height of 37 m.

Cancel the factor m and solve for v_f:

$$v_f^2 = v_i^2 - 2gh$$
$$= (28 \text{ m/s})^2 - 2(9.8 \text{ m/s}^2)(37 \text{ m})$$
$$= 59 \text{ m}^2/\text{s}^2,$$

and thus

$$v_f = 7.7 \text{ m/s}.$$

The golf ball is fairly close to the top of its trajectory and has slowed down considerably. Note that we do not need to use the detailed information about the initial angle of the ball or calculate the angle at which the ball is moving when it hits the bee. That is part of the utility of working with the conservation of energy.

What Do You Think? Can we decide with this approach whether the golf ball hit the bee while the ball was still rising or when it had begun to fall back to the ground?

Central Forces

There is a most interesting case that is important in nature: a force with a magnitude that depends only on the radial distance from a fixed point and has a direction aligned along the corresponding radius vector. For example, the force of gravitational attraction on Earth due to the Sun depends on the distance between Earth and the Sun and is aligned along the line between the two objects. The fundamental law of the attraction between particles with electrical charge—the force that allows atoms to exist—is also of this nature. Such forces are called **central forces** and *all central forces are conservative*. Such a force takes the form

$$\vec{F}(\vec{r}) = F(r)\left(\frac{\vec{r}}{r}\right) = F(r)\hat{r}, \tag{7–16}$$

where $\hat{r}$ is a unit vector pointing away from the origin. This force is reminiscent of a one-dimensional force in that it depends on only one space coordinate. The potential energy function of a central force depends only on the radial distance r; the expression for the force in terms of a potential energy function $U(r)$ is

$$\vec{F} = -\frac{dU(r)}{dr}\hat{r}. \tag{7–17}$$

Example 7–8 illustrates how we can apply Eq. (7–17).

EXAMPLE 7–8 The potential energy describing the gravitational interaction (the *gravitational potential energy*) between two point masses such as Earth and the Moon, with masses m_1 and m_2, respectively, is given by the expression

$$U(r) = -\frac{Gm_1 m_2}{r},$$

where G is a constant and r is the distance between the masses. Calculate the force experienced by the Moon due to the presence of Earth. Is the force attractive or repulsive?

Setting It Up We define the unit vector $\hat{r}$ to be in the direction from Earth (taken to be located at the origin) to the Moon (Fig. 7–15).

Strategy This is a straightforward application of Eq. (7–17), which gives the force in terms of the potential.

Working It Out From Eq. (7–17) and the fact that $(d/dr)(1/r) = -1/r^2$, we find

$$\vec{F} = -\frac{Gm_1 m_2}{r^2}\hat{r}.$$

The force is proportional to $1/r^2$ (an inverse-square form), and it is directed opposite to the position vector $\vec{r}$ from Earth to the Moon.

It therefore acts to pull the Moon toward Earth and is an *attractive* force. This force represents the centripetal force that allows the Moon to orbit Earth in a nearly circular orbit (see Chapter 12 for a more complete treatment of the subject).

What Do You Think? What choice has been made for the location(s) where the potential energy vanishes? Why is this a reasonable choice?

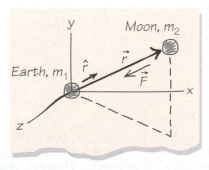

▲ **FIGURE 7–15** The vector $\vec{r}$ pointing radially outward from the origin (the location of mass m_1) is the position vector of mass m_2. The force vector on m_2 acts inward radially.

7–4 Is Energy Conservation a General Principle?

At first glance, it may appear that energy conservation is not all that useful, as many physical systems involve nonconservative forces that dissipate energy such as friction or drag. In the presence of friction, mechanical energy, the sum of the potential and kinetic energies, is not conserved. But as we discussed briefly at the end of Section 7–1, the energy that is lost because of the presence of friction can be identified through experiment as **thermal energy**, that is, the type of energy that raises the temperature of two objects when they are rubbed together. There are still other forms of energy that we will identify in later chapters, such as the energy in sound, in light, in various electrical systems, and in the interaction between molecules. When thermal and indeed all the forms of energy are measured and included as still other components of the total energy of a large system, *the total energy is then conserved*. However, there is a catch. Whereas potential energy can be converted to kinetic energy and vice versa, thermal energy is not a form of potential energy because it is not completely retrievable—it cannot be fully converted to kinetic energy of a mass, for example. A system that contains thermal energy differs from a system for which kinetic and potential energies can be freely converted back and forth. The ways in which mechanical and thermal energy can be converted back and forth are important, and we shall study this in detail in our discussion of thermal phenomena (Chapters 17 to 20).

To illustrate that some forms of energy are not completely retrievable, consider two situations. In one we drop a rubber ball on a stone floor, and in the other we drop a glass. In each situation energy is conserved, but mechanical energy is conserved only in the first case. In the absence of air resistance, the ball will hit the ground and return to the hand—mechanical energy is conserved. What happens to the mechanical energy of the glass? The glass at first merely acquires kinetic energy through loss of potential energy, but when it hits the floor, it shatters. Shards fly all over the place, sliding to a stop under the influence of friction, and your eardrums vibrate because they hear the crash. The glass could bounce up into your hand only if all the initial potential energy it contained could be retrieved and transformed back into kinetic energy. What would it take to retrieve that energy? You would have to run the movie of the crash backward, and you can imagine that getting the energy back into each shard at the right time and the right place so that the glass is reconstructed is effectively impossible. The thermal energy given to the stone floor when the shards slid to a stop would have to be extracted. Even the energy that left as sound would have to be recovered and put back into the glass!

Nevertheless, when we look closely enough, *all the fundamental forces in nature are conservative*, and conservation of energy always holds. ***The conservation of energy is one of the most fundamental principles of physics***. It has received experimental support from all fields of physics, and we shall encounter energy conservation in a variety of applications, including thermodynamics, electromagnetism, relativity, and quantum physics.

Energy Conservation and Nonconservative Forces

Although it is important to realize that all fundamental forces are conservative, we do not often operate at the level of the fundamental forces in our macroscopic world. How then do we deal with the nonfundamental and nonconservative forces, such as friction, that appear in everyday situations?

Suppose both conservative and nonconservative forces act on an object. For example, a skier descending a mountain moves under the influence of both gravity and friction. We can first construct the potential energy as if there were no friction and then ask how much energy is lost due to friction; the loss of energy is attributable to any work done by friction. We can do this because we can divide the work as it appears in the work–energy theorem into a part associated with a potential energy (the work of the conservative forces) and another part that is the work done by nonconservative forces. Thus, if we write W_{nc} as the work done by the net nonconservative force and U as the potential energy of the net conservative force, the work–energy theorem $W_{net} = W_{nc} + W_{conservative} = \Delta K$ becomes

$$W_{nc} = \Delta(K + U) = \Delta E. \qquad (7\text{–}18)$$

Here, we have continued to think of E as the total mechanical energy $K + U$, but E is no longer conserved.

For our skier, if friction is ignored, we have $E = K + mgy = $ constant, where $K = mv^2/2$ is the kinetic energy at a height y above some zero level. When friction is introduced, the initial and final energies E_i and E_f, respectively, differ, with $E_f < E_i$. The deficit is due to the work done by friction. We have

$$W_{\text{friction}} = E_f - E_i = (K_f + mgy_f) - (K_i + mgy_i). \qquad (7-19)$$

The work done by friction is negative, so that the final energy is less than the initial energy.

EXAMPLE 7-9 A ball of mass 10 kg is attached to a 5.0-m-long wire that swings freely from a support. We have a *simple pendulum* (Fig 7–16a). The ball is pulled aside so that the wire makes an angle of 31° to the vertical and it is released from rest. After 10 swings, the maximum angle that the ball reaches is 25° from the vertical. What is the work done by air resistance and any other nonconservative forces acting on the ball during these 10 swings?

Setting It Up The illustration (Fig. 7–16a) shows the starting point for the motion; Fig. 7–16c includes labeling for the angle θ of the pendulum from the vertical and the vertical distance y from the minimum position of the pendulum's swing. We know the ball's mass m, the pendulum length L, and the maximum angle θ_i at time $t = 0$. After 10 swings, the new maximum angle θ_f is given. We want the work done by nonconservative forces during the 10 swings.

Strategy We compute directly the change in the total mechanical energy E (kinetic energy plus potential energy) associated with the conservative forces, in this case gravity, that was lost over the 10 swings. We are given information at moments where E is entirely in the form of potential energy, so that the energy change can be expressed entirely in terms of the difference of two potential energies. This difference must equal the work done by nonconservative forces, in this case air resistance on the ball.

Working It Out The forces acting on the ball are specified in a free-body diagram (Fig. 7–16b). This diagram includes the forces of gravity, the tension of the wire, and the air resistance. Tension acts in a direction perpendicular to the motion, so it does no work and plays no role in any energy considerations. Gravity is conservative and there is a potential energy associated with it. We will first express conservation of energy as if air resistance were not present. If we take the zero of potential energy at the minimum point of the swing and measure the height y from that level, then $U(y) = mgy$. Figure 7–16c shows that

$$y = L(1 - \cos \theta),$$

where L is the wire's length. There is no kinetic energy at the top of the swing, so the total energy is

$$E = U(y_{\text{max}}) = mgy_{\text{max}} = mgL(1 - \cos \theta_{\text{max}}).$$

This expression will apply to the initial and final situations with the maximum angle given by θ_i and θ_f, respectively. The *loss* of energy over the 10 swings is then

$$E_i - E_f = mgL(1 - \cos 31°) - mgL(1 - \cos 25°)$$
$$= mgL(\cos 25° - \cos 31°)$$
$$= (10 \text{ kg})(9.8 \text{ m/s}^2)(5.0 \text{ m})(\cos 25° - \cos 31°)$$
$$= 24 \text{ J}.$$

This energy is lost by virtue of the nonconservative forces; equivalently, it must match the work done by these forces, here air resistance.

What Do You Think? You are told that during the time that the pendulum keeps on swinging it acts as a clock. But a good clock does not slow down. How is it possible that the pendulum described here could make a good clock?

(a)

(b)

(c)

▲ **FIGURE 7–16** (a) Simple pendulum. (b) Force diagram for the mass. (c) Geometry for the determination of the potential energy.

▲ **FIGURE 7–17** When fuel is burned, chemical energy is converted into thermal energy as well as electromagnetic energy—light.

There is a great deal of justifiable concern in the world about producing energy, using energy, and wasting energy. When we talk about energy production, we have in mind electric power generators of various kinds, including coal- or oil-burning plants, nuclear reactors, hydroelectric plants, solar energy plants, and windmills. In each case we are not producing energy but *converting* it. Coal or oil contain chemical energy, a form of potential energy associated with the chemical bonds that tie atoms together into the molecules of the fuel (Fig. 7–17). The burning, which is a chemical reaction, breaks some of these bonds and this releases energy, generally in the form of thermal energy. Some of that energy may be used to convert water to steam, and the pressure of the steam may be used to turn turbines that can generate electrical energy. In nuclear reactors the potential energy is contained in the nuclei that undergo fission—sometimes referred to as *nuclear energy*—which generates rapidly moving particles whose kinetic energy is again available for further conversion. In hydroelectric plants, water in a reservoir is allowed to flow downhill, converting gravitational potential energy to kinetic energy of the water, some of which can turn the wheels of turbines. Similarly, windmills convert the kinetic energy of moving air.

The electrical energy "produced" is not used up when, for example, it lights buildings. But part of the energy is wasted in the sense that even if we put effort into it, the thermal energy given up by a light bulb could only partially be captured and reconverted into energy that could be reused. Whatever use we make of it, energy is always conserved. What we can do is try to ensure that the energy is used as efficiently as possible and the processes we employ have as little deleterious effect on the environment as possible. ■

In this discussion and example, we used the energy loss to find the work done by nonconservative forces. But, of course, we could use this approach in a different way. If we can calculate the work done by friction—we can do this if we know its magnitude and direction and the distance over which it acts—then we can *predict* the energy loss.

Energy Conservation as One of the Underpinnings of Physics

Within this chapter we have seen that energy conservation can be a powerful shortcut to information that can be obtained only laboriously from the equations of motion. It is important to understand that the statement that E in Eq. (7–15) is a constant is entirely equivalent to Newton's laws. It represents a more fundamental formulation of dynamics than Newton's second law, even though the two are equivalent. This is not so easy to see in general, but is quite transparent for motion in one dimension. We write

$$E = \tfrac{1}{2}mv^2 + U(x), \qquad (7\text{–}20)$$

where E is a constant and both x and v are functions of time; they are the position and speed of the object in motion. Because energy is conserved, we have

$$\frac{dE}{dt} = 0. \qquad (7\text{–}21)$$

We apply this to Eq. (7–20) using the chain rule when needed:

$$\frac{dE}{dt} = \frac{1}{2}m\frac{dv^2}{dt} + \frac{dU(x)}{dt} = \frac{1}{2}m\frac{dv^2}{dv}\frac{dv}{dt} + \frac{dU(x)}{dx}\frac{dx}{dt}$$

$$= \frac{1}{2}m(2v)\frac{dv}{dt} + \frac{dU(x)}{dx}v = v\left[m\frac{dv}{dt} + \frac{dU(x)}{dx}\right] = 0.$$

Unless $v = 0$ at all times, the contents of the square brackets must be zero; that is,

$$m\frac{dv}{dt} = -\frac{dU(x)}{dx} = F(x). \qquad (7\text{–}22)$$

This is just our original formulation of Newton's second law.

Physicists and mathematicians studied Newton's equations almost from the time of their discovery. But it is interesting to note that the principle of energy conservation—formulated by Julius Robert Mayer (1842), Hermann von Helmholtz (1847), and James Prescott Joule (1850)—grew instead out of experimental research by engineers on the properties of heat and the conversion of mechanical energy into thermal energy.

Summary

Conservative forces can be specified in terms of a single scalar function called the potential energy. In one-dimensional motion, the potential energy $U(x)$ depends on one coordinate only and is defined by

$$U(x) - U(x_0) = -\int_{x_0}^{x} F(x')\, dx', \tag{7–4}$$

where $U(x_0)\ (= U_0)$ is an arbitrary constant and x_0 is an arbitrary point. In other words, the potential energy $U(x) - U(x_0)$ is the negative of the work done by the force as the object on which it acts moves from x_0 to x. The force is given in terms of U by

$$F(x) = -\frac{dU(x)}{dx}. \tag{7–5}$$

Two important examples of potential energy are associated with gravity and with the spring force. For gravity,

$$U(y) = mgy + U_0, \tag{7–2}$$

where y is the height above ground level. For the spring,

$$U(x) = \tfrac{1}{2}kx^2, \tag{7–13}$$

where x is the distance from the equilibrium point of the spring and the zero of the potential is chosen at the equilibrium point $x = 0$.

For central forces (forces aligned in the radial direction), the potential energy depends only on r, and

$$\vec{F} = -\frac{dU(r)}{dr}\hat{r}. \tag{7–17}$$

For conservative systems, the total mechanical energy—the sum of the kinetic and potential energies—does not change during the motion; thus, the total energy is a constant whose value may be determined at any time during the motion:

$$E = \tfrac{1}{2}mv^2 + U(\vec{r}). \tag{7–15}$$

The fact that E is constant means that any change in the kinetic energy is compensated by an equal but opposite change in the potential energy during the motion. *The total energy is conserved.* The utility of this conservation law is enormous: We can determine the speed when the position of an object is known. Energy diagrams provide us with a systematic way to approach problems that involve energy conservation. With them, we can understand useful concepts such as turning points, potential energy barriers, and stable and unstable equilibria.

Nonconservative (dissipative) forces such as friction or air resistance appear in the macroscopic world. They spoil energy conservation, but since the work–energy theorem always holds, one can use it to find the energy dissipated through friction or drag forces, even if one cannot calculate this energy loss directly. Dissipative forces do not exist on the microscopic level; they are effective forces that come from the action of the microscopic laws. Thus at a deeper level the conservation of energy always holds. Energy is neither created nor destroyed; it can only change from one form to another.

Understanding the Concepts

1. Are drag forces such as air resistance conservative?
2. Discuss why it is possible for the total mechanical energy to be negative, even if the kinetic energy cannot be negative.
3. The potential energy of gravity can be given the form mgh, where h is the height above sea level. That means that if one goes into a deep mine whose bottom lies below sea level, the potential energy becomes negative. What is the meaning of this negative potential energy?
4. If we add a constant term to the potential energy of an object, why doesn't this change the object's motion?

5. Explain why a rubber ball seems never to bounce back to its original height when dropped from rest. What happens to the energy? Explain why the ball can bounce back to a much greater height when it is thrown down.

6. The force $F = -kx$ due to a one-dimensional spring is conservative. Is it still conservative if k is allowed to vary with time? In thinking about this question, it will be helpful to use some of the definitions of conservative forces given in Chapter 6.

7. Is energy conserved when you ingest sugar and then go out and exercise? In what ways, and in which of the various processes for converting food into muscular activity, might the energy conservation principle come into play?

8. For which of the following forces can one write a potential energy function? In each case, justify your answer. (a) The force exerted by an ideal spring. (b) The drag force exerted by the air on an object falling through it (Chapter 5). (c) The normal force exerted by an object sitting on a table top. (d) A force that is the same everywhere in space.

9. Does a real spring exert a truly conservative force? What experiment might you perform to check the answer to this question?

10. Suppose you are given the components of a two-dimensional force in the x- and y-directions as functions of x and y. Can you think of a way of figuring out whether the force is conservative?

11. How, if at all, does experiment demonstrate that the gravitational force is conservative?

12. How would you figure out the energy lost to friction for a skier going down a mountain?

13. A golf ball is dropped with zero initial velocity from a height of 20 m. It bounces on a concrete pad and rises to a maximum height of 19 m. Are any of the forces that the ball experiences during the motion nonconservative?

14. Consider motion with a potential energy of the shape given in Fig. 7–5. Let $U(x)$ describe the height of a smooth slide above the ground. In the absence of friction the motion is easily predicted to repeat with a uniform repetition time. Discuss the motion, starting from the right or the left side, when a small amount of friction is present.

15. When an object slides down a frictionless, curved slope there are normal forces as well as gravitational forces present. How can we assert, without knowing the shape of the slope, that we are dealing with a conservative system of forces?

16. When a force acts in two dimensions, are there two separate laws for the conservation of energy that correspond to motion in each direction?

17. The potential energy is zero at a given point. Is the force necessarily zero or nonzero at that point?

18. An object moves on a rough inclined plane. Can you still use the concept of energy conservation to relate the height and the speed? How?

19. A ball tethered to a vertical rod is struck by a bat every time it passes by a certain point. Every time it goes around the rod, it goes faster, so that on moving back to the starting point it has a different kinetic energy. Why are the forces involved not conservative?

20. The motion of the pendulum of a grandfather clock is slowed by air resistance and bearing friction, yet such clocks can run without stopping for years. What types of mechanisms are typically used to supply the necessary energy?

21. For which, if any, of the following systems does it make sense to say that the energy of the system is conserved? (a) A community of animals living at the bottom of the ocean near a source of hot water and minerals. (b) Earth as a whole. (c) A sealed ecosystem under a huge glass dome, such as has been built in Arizona. (d) A perfectly sealed cylinder of hot gas that can drive a piston.

22. You are watching fireworks. You see a glowing point streak into the air. This is followed by a loud noise and, about a second later, by the appearance of many scattered dots of light that sink in graceful curves before disappearing. Account for all of the energy supplied and spent at each stage of the spectacle.

23. A steel marble is bounced off the top step of a staircase. It rises to 1 m above the step, then falls onto the second step, bounces off, falls onto the third step, and so on, for 39 steps. Assuming that there is no air resistance or other source of energy loss, how high will the marble rise after the last bounce? (Do not do 39 calculations to answer this question!)

Problems

7–1 Energy Conservation

1. (I) An unknown constant force F pushes a 10-kg body from rest on the ground vertically upward. At a height of 2.0 m, the velocity of the object is $\vec{v} = (2.4 \text{ m/s})\hat{j}$. (a) Find the change in the potential energy associated with gravity. (b) What is the net work done, and what is the work done by the unknown force?

2. (I) Consider the expression $U = mgy$ for an object of mass m under the influence of gravity, where y is measured from the ground up. Express U in terms of z, where z is measured in a downward direction from a rooftop 30 m above the ground.

3. (I) A baseball pitcher throws a ball at 95 mi/h off a roof that is 80 m above the ground. How high will the ball be when it is traveling at 120 mi/h?

4. (I) A rock is thrown straight down into a deserted quarry from the edge, which is 45 m above the bottom. The rock has a speed of 42 m/s when it reaches the bottom. What was the rock's initial speed? Ignore air resistance.

5. (I) The spring of a toy gun launches rubber-tipped projectiles with a spring constant of 5 N/m. The spring is compressed by 7 cm with the projectile in place. How much kinetic energy is imparted to the projectile?

6. (II) The potential energy of an archery bow is measured to be $U(x) = bx^2 + cx^3$, where x is the distance the bow string is pulled back from its equilibrium position. When an archer pulls the string a distance x, what force does the archer exert on the string and what force does the string exert on the archer?

7. (II) A package of mass 5.0 kg is subject to a constant force of 8.0 N pointing in the +x-direction. (a) Calculate the potential energy of the package as a function of its position x, defining it such that $U(x)$ at $x = 0.0$ m is zero. (b) Assuming that the package has a velocity of 2.0 m/s at $x = -1.0$ m, calculate the total mechanical energy of the package. (c) What is the speed of the package at $x = 3.0$ m?

8. (II) A spring has a spring constant of 16.5 N/m and obeys Hooke's law. How far must the spring be pulled back if its potential energy is to be 3.77 J? What is the mass of a ball at the end of the spring if the maximum speed of the ball is observed to be 1.71 m/s when the spring is released?

9. (II) The energy of a harmonic oscillator (a mass moving on the end of a spring) is given by $E = \frac{1}{2}mv^2 + \frac{1}{2}kx^2$. Plot contours of constant E on a graph in which x is measured along one axis and v is measured along the perpendicular axis. Choose the parameters $E = 16.0$ J, $m = 2.0$ kg, and $k = 8.0$ J/m². Such a plot is called a *phase plot*; the motion of a system is restricted to the curve corresponding to the energy E.

10. (II) Sketch the potential energy for a 5-kg mass that can move between the ground and a height of 10 m. (a) Assume that zero potential energy is at the ground, (b) at a height of 10 m, and (c) at 4 m.

11. (II) A block of mass 0.528 kg slides with uniform velocity of 3.85 m/s on a horizontal frictionless surface. At some point, it strikes a horizontal spring in equilibrium. If the spring constant is $k = 26.7$ N/m, by how much will the spring be compressed by the time the block comes to rest? What is the amount of compression if the surface is rough under the spring, with coefficient of kinetic friction $\mu_k = 0.411$?

12. (II) An archery bow acts much like a spring displaced from equilibrium when the bow is drawn. Suppose that an archer displaces the string from equilibrium by 47 cm and exerts a force of 65 N. (a) What is the "spring constant"? (b) What is the speed of an arrow of mass 40 g that leaves the bow as the string reaches the equilibrium position? [*Hint*: Use the expression for energy for a mass moving under the influence of a spring.]

13. (II) A cannonball of mass 15 kg is dropped from rest from a height of 6.0 m. It falls onto a large vertically oriented spring that is compressed from its relaxed position when the cannonball lands on it. The spring has spring constant $k = 10^4$ N/m. What is the maximum compression of the spring? How much would the spring be compressed if a man of mass 60 kg jumped onto it from a height of 1.5 m?

14. (II) A mass $m = 0.70$ kg slides along the x-axis of a horizontal frictionless surface with speed $v_x = 2.2$ m/s. It runs into a relaxed spring oriented along the x-axis. This spring had previously been observed to stretch by 3.8 cm when it was oriented vertically with the mass suspended from it. (a) What is the maximum compression of the spring when the mass runs into it? (b) The mass rebounds as a result of having compressed the spring. What is its velocity when it leaves contact with the spring?

15. (II) A spring with spring constant $k = 200$ N/m is used as a launcher for a small block whose mass is 10 g. The block is placed against the compressed spring in a horizontal arrangement on a smooth horizontal surface. The spring, with the block, is compressed 5 cm and then released. (a) Find the speed of the block just as it leaves the spring. (b) The block encounters a rough surface as it leaves the spring. How much work does friction do in bringing the block to an eventual stop? (c) The block slides a distance of 3.5 m before stopping. What is the coefficient of kinetic friction between the block and surface?

16. (II) A person of mass 70 kg jumping on flat ground can raise his center of mass by 1.0 m, whereas the same person jumping on a trampoline, which acts as a type of spring, can raise his center of mass by 2.5 m. What is the potential energy contained in the trampoline at the bottom of the person's 2.5-m jump? Do you need to make an assumption about the mass of the trampoline's elastic surface?

17. (II) Over a large enough height difference, the gravitational acceleration g is not constant but decreases with altitude. If we could approximate this change by $g \rightarrow g_0 - g'y$, where $g' \ll g_0$, where y is the height above sea level, what would be the potential energy of a mass under the influence of gravity? Choose the zero of the potential energy where $y = 0$.

7–2 Energy Conservation and Allowed Motion

18. (II) For a conservative one-dimensional force, show that the sign of the slope of the potential energy function at a position x determines the direction, positive or negative, in which the force acts.

19. (II) When two atoms on a line are far apart, there is no force between them. As they start to move closer, there is an attraction between them, which, at very close distances, turns into a strongly repulsive force. Sketch the potential energy as a function of the distance between the atoms.

20. (II) Figure 7–18 shows the force $F(x)$ that acts on a particle moving along the x-axis. (a) Plot the potential energy of the particle as a function of x. (b) The particle starts its motion at $x = -0.5$ m, with zero initial velocity. How far to the right will the particle travel?

▲ **FIGURE 7–18** Problem 20.

21. (II) Consider the force described in Problem 20. Is there a stable equilibrium point associated with this force? If so, find the turning points that correspond to a given initial total energy.

22. (II) The potential energy for a mass $m = 1.0$ kg moving in one dimension is given by $U(x) = (2.0 \text{ J}) \sin \pi x$. The mass starts at $x = 0$ with an initial velocity $v = +0.71$ m/s. (The plus sign means the motion is in the positive x-direction.) Describe the subsequent motion of the mass. Suppose the initial velocity were $+3.0$ m/s. What would the subsequent motion look like?

23. (II) Consider a force $\vec{F}(x)$ acting along the x-axis that is opposite to the spring force (Hooke's law). In other words, the force has the single vector component $F(x) = +kx$. A mass m is under the influence of this force and of no other forces. Suppose that the mass is placed at rest just to the right of the origin. (a) Which way will the mass move, if at all? (b) Find the speed of the mass as a function of its distance from the origin. (c) Repeat the problem but assume that the mass had been placed just to the left of the origin.

24. (II) Consider the energy diagram in Fig. 7–19. (a) What are the limits of motion for energies E_1 and E_2? Redraw the figure and label it as necessary. (b) Describe the circumstances under which

▲ **FIGURE 7–19** Problem 24.

the particle is always at rest. (c) Find the energies and positions for which motion within turning points is possible. (d) Find the equilibrium positions on your drawing. Are they stable or unstable?

25. (II) Draw a one-dimensional potential energy $U(x)$ diagram with the following characteristics: (a) The particle can never reach negative x. (b) There are three regions in x where the particle can move within turning points. (c) The particle can never reach infinity. (d) The particle has unstable equilibrium positions at 1 and 2 nm.

26. (II) Consider the potential energy $U(x)$ shown in Fig. 7–20. (a) What is the sign of the force at positions 1 through 6? (b) Which positions have the most positive, most negative, and zero force? (c) Find the equilibrium positions and indicate whether they are stable or unstable.

▲ **FIGURE 7–20** Problem 26.

27. (II) The potential energy of an object constrained to move in the x-direction is given by $U(x) = \alpha x^4 + \beta x^2$, where $\alpha = 26 \text{ J/m}^4$ and $\beta = -3.0 \text{ J/m}^2$. Find the equilibrium points and state whether they are stable or unstable.

28. (II) A particle is moving in a potential well described by the potential energy

$$U(x) = -\frac{1.5 \text{ J} \cdot \text{m}^2}{1.2 \text{ m}^2 + (x + 0.80 \text{ m})^2} - \frac{1.5 \text{ J} \cdot \text{m}^2}{1.2 \text{ m}^2 + (x - 0.80 \text{ m})^2},$$

where x is measured in meters. (a) Sketch the shape of the potential energy for $-4 \text{ m} < x < 4 \text{ m}$. (b) The speed of a particle of mass 0.50 kg at $x = -2$ m is 3.2 m/s. Can the particle reach $x = 1.5$ m?

7–3 Motion in Two or Three Dimensions

29. (I) A projectile fired from a gun leaves the barrel at a speed of 500 m/s. The gun is placed 180 m above a level plain. Use energy conservation to calculate the speed of the projectile when it is 16 m above the plain. Neglect all drag effects.

30. (I) What is the gravitational potential energy of the Earth–Moon system? See the appendices for the data you need.

31. (I) A cannonball is fired horizontally with an initial speed of 125 m/s from the top of a cliff that is 68 m above the sea. What is the speed of the cannonball when it hits the water? How is that changed if the cannon is inclined at a 32° angle with the horizontal without any change in initial speed? Ignore all effects of air resistance.

32. (I) A particle has potential energy that depends only on the distance r from some central point. This potential energy has the form $U(r) = U_0 - k/r^2$. What is the corresponding force law?

33. (II) A ball is thrown with initial speed v_0 in a trajectory that makes an initial angle θ with the ground. Air resistance is small. (a) By using the principle of conservation of energy, show that

the speed of the ball when it reaches the height h above the ground is $v_h = \sqrt{v_0^2 - 2gh}$, independent of the angle of the throw. (b) Use the result of part (a) to find the initial speed v_0 that is required if a vertically thrown ball is to reach height H just before turning back. (c) Use the results of part (a) to find the initial speed v_0 required so that a ball thrown at 45° reaches a maximum height H.

34. (II) An object of mass m is subject to two forces: One force acts only in the x-direction and is due to a spring of spring constant k; the other force acts only in the y-direction and is due to a spring with the same spring constant k. (a) What is the potential energy? (b) Use your result to show that the net force is proportional to the distance from the equilibrium point and is directed towards it. (c) Using your result from part (b), show that one possible motion is uniform circular motion in the xy-plane. [*Hint*: Make a simple choice for equilibrium point.]

35. (II) A ball of mass m is thrown with a speed v at an angle θ from the horizontal off the top of a tall building at height h. (a) Show that, in general, it is not possible to tell whether the angle θ is above or below the horizontal when a ball hits the ground just by measurement of the speed of the ball when it hits the ground. (b) By knowing the horizontal distance that the ball travels, show that the ambiguity can be resolved.

36. (II) A skier slides down a hill starting with zero velocity at a height of 43 m above the bottom of the hill. The shape of the terrain is shown in Fig. 7–21. What is the velocity of the skier on top of the second, intermediate hill, whose height is 37 m? What is the skier's velocity at the bottom of the hill? Neglect all frictional effects. Is this neglect reasonable?

▲ **FIGURE 7–21** Problem 36.

37. (II) Suppose that the skier of Problem 36 reaches the bottom of the hill at a speed of 23 m/s. Assuming that the skier, including equipment, has a mass of 75 kg, how much work is done by the resistive forces of friction and drag?

38. (II) A ski jumper starts from rest and follows, with its several ups and downs, the rather bumpy ski jump shown in Fig. 7–22. The

▲ **FIGURE 7–22** Problem 38.

jump track starts at a height $h = 16$ m above the eventual landing point, and the jumper leaves the track while he moves horizontally at a height above the landing point that is exactly one-half h. Assuming that the effects of air resistance are negligible (in real ski jumping they are not), what is the horizontal distance of the edge of the track from the landing point?

39. (II) A particle of mass $m = 30$ g slides inside a bowl whose cross section has circular arcs at each side and a flat horizontal central portion between points a and b of length 20 cm (Fig. 7–23). The curved sides of the bowl are frictionless, and for the flat bottom the coefficient of kinetic friction $\mu_k = 0.21$. The particle is released from rest at the rim, which is 10 cm above the flat part of the bowl. (a) What is the speed of the particle at a? (b) What is the speed of the particle at b? (c) Where does the particle finally come to rest?

▲ **FIGURE 7–23** Problem 39.

40. (II) A toy car of mass M slides down a frictionless track that makes a circular loop of radius R at the bottom (Fig. 7–24). Suppose that the car starts from rest at a height H, with $H > 2R$. (a) What is the car's speed at the bottom of the circle? (b) At the top of the circle? (c) What is the force exerted by the track at the top of the circle? (d) What is the minimum value of H such that the car goes around the loop without falling off under the force of gravity?

▲ **FIGURE 7–24** Problem 40.

41. (II) A block is constrained to move without friction along the x-axis. The block is attached to a spring of equilibrium length L. The other end of the spring is fixed at a point $x = 0$, $y = h$ (Fig. 7–25), where $h < L$. (a) What is the potential energy of the system? (b) What is the net force acting on the block? Sketch both $U(x)$ and the force.

42. (II) The potential energy of two atoms separated by a distance r may be written in the form $U(r) = U_0 \times [(r_0/r)^{12} - 2(r_0/r)^6]$ (Fig. 7–26). Find the separation r at which there is no force between the atoms. What is the magnitude of the potential energy there?

▲ **FIGURE 7–25** Problem 41.

▲ **FIGURE 7–26** Problem 42.

43. (II) The potential energy of a satellite of mass m moving in a circular orbit of radius r about a planet of mass M is given by $U(r) = -GMm/r$, where G is a universal constant. (a) Calculate the force on the satellite. (b) From the equation $\vec{F} = m\vec{a}$, calculate the kinetic energy for a particle in this orbit. (c) What is the total energy?

44. (II) The potential energy of a particle moving in the xy-plane is given by $U(x, y) = a_1 x^2 + a_2 xy + a_3 y^2$, where $a_1 = 3$ J/m^2, $a_2 = -14$ J/m^2, and $a_3 = 1.5$ J/m^2. Calculate the force vector.

45. (III) The ski jumper of Problem 38 has survived the jump and prepares for another jump on a track of a different design: The track takes the jumper through a vertical distance H from its start to its takeoff point, but the jumper can choose the angle at which he leaves the takeoff point. The lip at the edge can be varied so that the jumper leaves the jump at any angle—from taking off horizontally to leaving vertically upward. Show that the angle θ with the horizontal that leads to the maximum-distance jump onto a horizontal plane a distance D below the lip is given by $\sin^2 \theta = 1/[2 + (D/H)]$. Ignore all friction and drag effects. (A real landing area is, in fact, sloped downward to allow the jumper to make a smooth landing.)

7–4 Is Energy Conservation a General Principle?

46. (I) A 10-g Ping-Pong ball is dropped with zero initial velocity from a height of 1.0 m, and it bounces back to a height of 0.90 m. What is the work done by the nonconservative forces in this process?

47. (I) A parachutist jumps off a training tower that is 85 m high. She starts at rest and reaches the ground with a vertical speed of 5.0 m/s. How much work was done by the drag forces acting on her given that her mass is 75 kg?

48. (II) A track consists of a descending ramp, a straight track, and an ascending ramp. The smooth ramps both make an angle of 20° with the horizontal. The coefficient of kinetic friction on the horizontal surface is $\mu_k = 0.18$. An object starts from rest at a vertical height of 1.3 m on the descending ramp. It slides down the ramp, across the horizontal stretch, and up the ascending ramp. It reaches a vertical height of 0.55 m before coming to rest. (a) How long is the horizontal part of the track? (b) The object starts sliding back from the 0.55 m height. How far along the horizontal stretch does it slide?

General Problems

49. (I) Assume that 1 kWh of electric energy costs 12 cents. Estimate the cost of lighting a three-room apartment per day.

50. (I) A rock falls off the edge of a cliff moving initially in the horizontal direction with speed 2 m/s. The cliff is 20 m high. How does the speed of the rock, as it reaches the ground, differ from that of a rock that falls with no horizontal velocity?

51. (I) A furniture mover pushes a crate of mass 60 kg up a rough slope through a vertical distance of 1.0 m at a uniform speed. What is the change in the potential energy of the crate?

52. (I) A diver jumps off a rigid diving platform 5.0 m above the water with an initial upward velocity of 2.2 m/s. Assuming that his takeoff is very nearly in a vertical direction and that there are no drag forces on the diver, with what velocity will the diver hit the water? (Treat the diver as a point particle.)

53. (I) The force on an object of mass m moving along the x-axis is given by $F(x) = -ax + bx^2$, where $a = 3$ N/m and $b = 0.2$ N/m^2. (a) Calculate the potential energy function $U(x)$, letting $U(x) = 3$ J at $x = 0$ cm. (b) Sketch $U(x)$ as a function of x from $x = 0$ m to $x = 4.0$ m in steps of 0.5 m.

54. (II) A massless spring hangs vertically in equilibrium with no mass at its end. When a 2.0-kg mass is connected to the bottom, the new equilibrium position is 5.0 cm lower. The mass is then pulled down and released. It is observed that the speed of the mass is 2.1 m/s when the mass passes the original equilibrium position (before the mass was attached). How far down were the mass and spring pulled together when released?

55. (II) Are the following forces conservative or nonconservative? (a) The force $\vec{F}(x) = (ax + bx^3 + cx^4)\hat{i}$. (b) The force $\vec{F}(x, y) = Ax^2\hat{i} + Bxy\hat{j}$.

56. (II) A ball of mass 50 g falls on a hard surface from a height of 1.0 m. After each bounce it reaches 95 percent of the height from which it falls. How much energy is dissipated at the first bounce, at the second bounce, and at the third bounce? Can you generalize to the nth bounce?

57. (II) A mass moving in the xy-plane is subject to a force in the x- and y-directions whose components are given by $F_x = 2Ax^2y$ and $F_y = Axy^2$. Calculate the work done in going in a counter-clockwise direction around a square that has corners at $(x, y) = (0,0),(1,0),(1,1),(0,1)$. Can you write a potential energy function that gives rise to this force?

58. (II) An anharmonic spring exerts a force, as a function of displacement from equilibrium, given by $F = -(0.9\text{ N/m})x - (1.4\text{ N/m}^3)x^3$ (Fig. 7–27). This force acts on an object of mass 1.25 kg and displaces it from $x = 0$ m to $x = 3$ m. Consider the object being displaced along two different paths: (a) directly from $x = 0$ m to $x = 3$ m and (b) from $x = 0$ m to $x = -2$ m, then to $x = 7$ m, and back to $x = 3$ m. Show that the work done by the force is the same for both paths.

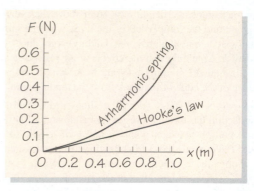

▲ **FIGURE 7–27** Problem 58.

59. (II) A conservative force does 2 J of work in moving a particle from point A to point C via path ABC (Fig. 7–28). The force does -1 J of work to move the particle from D to F, 3 J for E to B, 1 J for E to F, and 1 J for B to C. How much work does the force do as the object moves from C to A, from A to E, from D to C?

▲ **FIGURE 7–28** Problem 59.

60. (II) A particle of mass 50 g leaves, from rest, point a on a loop-the-loop. The heights of the points a, b, c, and d as measured from the table level are 10, 0, 8, and 12 cm, respectively (Fig. 7–29). Ignore friction. (a) What are the speeds of the particle at points b, c, and d? (b) How high up on the other side does the particle rise?

▲ **FIGURE 7–29** Problem 60.

61. (II) A skier skis from rest from a vertical height $h_1 = 18$ m over two successively lower hills of vertical heights $h_2 = 15$ m and $h_3 = 7$ m (Fig. 7–30, see next page). The summit of the third hill fits a circle of radius h_3 centered at height 0 m. Friction with the snow and air resistance are negligible. (a) Find her speeds at x_1, x_2, and x_3. (b) Does the skier leave the surface at x_3? If not, what should h_1 be so that she just leaves the surface at x_3?

▲ **FIGURE 7–30** Problem 61.

62. (II) A pile driver works by lifting a large mass and dropping it to the ground. The mass is 1300 kg, and it is raised to a height of 6 m above the pile for each stroke. The pile driver encounters resistance of a constant force of 2.5×10^6 N on each stroke. Use a combination of energy conservation and the work–energy theorem to determine how many strokes it takes to drive a pile 5 m into the ground. For which forces involved is it possible to use energy conservation, and for which forces is this not possible?

63. (II) The mass of a simple pendulum of length $L = 1$ m is released with the string originally in a horizontal position. (a) Calculate the speed of the mass at its lowest position. (b) What is the speed when the string makes an angle of 45° with the vertical? (c) Determine the tension in the string in both positions if the mass is 0.2 kg.

64. (II) A simple pendulum of length $L = 1.0$ m and mass 0.20 kg is released from the horizontal position. When the mass is at its lowest point, the string hits a nail a distance h above the mass, so that the mass loops around the nail (Fig. 7–31). How large can h be so that the string of the pendulum remains taut even when the mass loops to a point right above the nail? [*Hint*: The string is taut as long as there is tension in the string.]

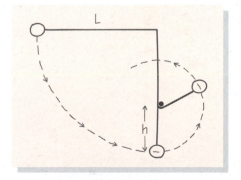

▲ **FIGURE 7–31** Problem 64.

65. (II) An electron is attracted to a proton (the latter is much heavier than the former) with a central force whose magnitude is given by Coulomb's law, $F = C/r^2$, where $C = 2.3 \times 10^{-28}$ kg·m^3/s^2. (a) Is this a conservative force? (b) Write an expression for the potential energy associated with this system. (c) An electron is very far away from a proton and starts from rest. It falls straight toward the proton under the influence of the force. What is the speed of the electron when it is 1.2×10^{-12} m from the proton?

66. (II) A batter hits a baseball and the baseball leaves the bat making a 46° angle with the ground. Air resistance has negligible effect on the trajectory of the baseball, which travels a total horizontal distance of 130 m. (a) What is the speed of the baseball just after it leaves the bat? (b) Use the conservation of the total energy of the baseball to calculate the maximum height to which the baseball rises. (c) What is the speed of the baseball when it has first risen to half its maximum height? (d) When it falls back to half its maximum height?

67. (II) A 3.0-kg block is held against a spring with spring constant $k = 25$ N/cm, compressing the spring 3 cm from its relaxed position. When the block is released, the spring expands and pushes the block upward along a rough surface inclined at a 20° angle (Fig. 7–32). The coefficient of kinetic friction between the block and the surface is $\mu_k = 0.1$. What is the work done on the block (a) by the spring as it extends from its compressed position to its equilibrium position? (b) By friction while the block moves 3 cm as in part (a)? (c) By gravity during the same motion? (d) What is the speed of the block when the spring reaches its equilibrium position? (e) If the block is not attached to the spring, how far up the incline will it slide before it comes to rest? (f) Suppose that the block is attached to the spring so that the spring is extended when the block slides past the equilibrium point. By how much will the spring be extended before the block comes to rest?

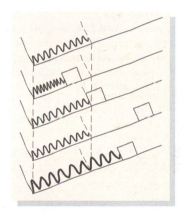

▲ **FIGURE 7–32** Problem 67.

68. (II) A block of wood of mass 1.8 kg is placed on a horizontal table and attached by a massless rope of length 31 cm to a vertical axis that passes through the table. The rope is initially swung around the axis, causing the wood to have a tangential velocity of 3.5 m/s. After one revolution the wood is observed to have a speed of only 2.1 m/s. (a) How much work has friction done on the block of wood during the first revolution? (b) What is the coefficient of kinetic friction between the block of wood and the table? (c) What is the potential energy of the block of wood at the beginning, after one revolution, and when the block of wood comes to rest? (d) How many revolutions does the block of wood make before stopping?

69. (II) Two blocks of mass $m_1 = 5.0$ kg and $m_2 = 2.0$ kg are supported by the system of light frictionless pulleys and massless strings shown in Fig. 7–33 (see next page). Mass m_1 is at rest at a height $h = 0.8$ m above the ground when the system is released. Use the conservation of energy to determine the speed of m_1 when it hits the ground.

▲ **FIGURE 7–33** Problem 69.

70. (II) A representation of the nuclear force between two nucleons (neutron or proton) is given by the *Yukawa potential energy function* $U(r) = -Ae^{-kr}/r$, where $1/k$ has the approximate value of 10^{-15} m and A is a constant. (a) Plot $U(r)$ versus r in steps of 0.2×10^{-15} m up to 2.4×10^{-15} m. Plot U in units of $A \times 10^{15}$. (b) At what distance is the potential energy a minimum? (c) Determine the force $F(r)$. (d) Determine the force at $r = 0.1 \times 10^{-15}$ m and 10×10^{-15} m.

71. (II) Given the relationship between the force and the potential energy, show that Newton's third law is satisfied if the potential energy has the form $U(x_1 - x_2)$ for two particles located at x_1 and x_2, respectively.

72. (II) A 5 kg projectile is fired straight up with an initial speed of 30 m/s and reaches a height that is 90 percent of the height that would have been reached had there been no air resistance. Apply the work–energy theorem in an appropriate form in order to find the work done on the projectile by the force of the air resistance.

73. (II) An object of mass m moves on a horizontal table. It is attached to a central point by a spring so that the radial force acting on the mass is $-kr$, where r is the distance of the mass to the central point. (a) Write an expression for the total energy. If the object is constrained to move in a circle of radius R, what can you say about the velocity? (c) Write an expression for the angular velocity for a given energy for circular motion of radius R.

74. (III) A (uniform) chain with a mass of 7.0 kg and a length of 2.0 m lies on a table with 0.6 m hanging over the edge. How much energy is required to get all of the chain back on the table?

75. (III) A ball at the end of a pendulum of length L is released at rest from an initial position in which the pendulum string is horizontal (Fig. 7–34). The floor is just beneath the low point of the swing. The string is cut after the ball has passed the low point with the string at an angle $\alpha = \theta - 90°$ to the vertical. Find the horizontal distance that the ball travels from the low point before it bounces on the floor.

String is cut at angle θ

▲ **FIGURE 7–34** Problem 75.

76. (III) The shape of the surface of water in a bucket is determined by the condition that the potential energy per unit mass, as determined in the frame in which the water is at rest, is constant at all places on the surface. Assume that the bucket is set into rotation at an angular speed ω (introduced in Chapter 3) about its vertical central axis and that the water rotates with the bucket. Write a formula that gives the shape of the water surface. [*Hint*: Note that the normal force of the liquid must account for the centripetal acceleration.]

◄ The two soccer (football) players collide inelastically with each other, but collide almost elastically with the ball. We will study collisions in this chapter and learn that linear momentum is always conserved in such collisions.

Linear Momentum, Collisions, and the Center of Mass

I n the world around us collisions are commonplace—think of raindrops colliding with the ground or ocean waves with the shore, the collision of a golf club with the ball, or the gentle collision of a mother taking a baby in her arms. Behind these everyday scenes are the incessant motion of air molecules and their collisions. Even the distant interactions between stars in the galaxy can be thought of as collisions—slow ones, for the most part. Behind the complexity of these collisions we can find a great simplification, the conservation of a new quantity called the momentum. The momentum is simply formed as the product of the mass and velocity of a particle, and its importance is associated with the fact that Newton's laws are very simply restated in terms of it. In particular, Newton's third law then shows us that momentum is conserved within an isolated system of particles—a system free from net external forces, such as when two hockey pucks sliding on ice collide—no matter how complex the internal interactions of the particles making up the system. In other words, the total momentum of such a system is constant. The conservation of momentum is enormously useful for understanding the behavior of colliding objects, and there are good practical reasons for wanting to understand collisions—much of our information about the world on the atomic scale and below and about the structure of materials comes from observing collisions in one form or another.

As we study the momentum of a system, we will learn that there is a particular point of the system—the *center of mass*—which moves in an especially simple way. For an isolated system, the center of mass moves without acceleration. When external forces act on the system, the center of mass accelerates according to Newton's second law just as a point object does.

8-1 Momentum and Its Conservation

Newton's second law, $\vec{F} = m\vec{a}$, describes how forces change the motion of objects. In previous chapters this law has been expressed in terms of the mass and the acceleration of an object. Another form of the second law is applicable even if the mass changes, as for an airplane when it consumes fuel. This more general form of the second law is

$$\vec{F}_{\text{net}} = \frac{d(m\vec{v})}{dt}. \tag{8–1}$$

The combination mass times velocity, $m\vec{v}$, is called the **linear momentum**, or just **momentum**, of an object. We denote this quantity by $\vec{p}$:

$$\vec{p} \equiv m\vec{v}. \tag{8–2}$$

LINEAR MOMENTUM

The momentum of an object is a vector whose direction is that of the velocity. Its dimensions are those of a mass times a velocity, namely, $[MLT^{-1}]$; in SI, the units of momentum are kilogram-meters per second.

In terms of momentum, the second law [Eq. (8–1)] has the general form

$$\vec{F}_{\text{net}} = \frac{d\vec{p}}{dt}. \tag{8–3}$$

NEWTON'S SECOND LAW

The kinetic energy of an object can also be expressed in terms of the momentum:

$$K = \frac{1}{2}mv^2 = \frac{p^2}{2m}, \tag{8–4}$$

where p is the magnitude of the momentum. From Eq. (8–3) it can be seen that when a large net force acts on an object, the object's momentum will change rapidly and a small net force will result in a slow momentum change.

CONCEPTUAL EXAMPLE 8–1 The same net force acts on a table tennis ball and a bowling ball. Compare the rates at which their momenta change.

Answer This is a bit like the trick question "Which weighs more, a pound of feathers or a pound of nails?" We did not ask for the rate at which the velocity changes—the acceleration—which will be significantly less for the bowling ball than for the table tennis ball. We asked for the rate of change of momentum, and that is precisely given by the force acting [Eq. (8–3)]. The rate of momentum change is the same for each ball since the net force acting on each is the same.

Conservation of Momentum

When objects exert a force on one another, we say they interact. Let's consider the interaction between objects 1 and 2 in both parts of Fig. 8–1. The two objects may be in contact, as in a collision of two billiard balls (Fig. 8–1a), or they may exert a force on each other at a distance, as in the gravitational attraction between Earth and the Moon, or they may be connected by a spring (Fig. 8–1b). Let $\vec{F}_{12}$ denote the force exerted on object 1 by object 2 and let $\vec{F}_{21}$ denote the force exerted on object 2 by object 1 (the first subscript always labels the object that is acted *upon*). Then, Newton's third law states that

$$\vec{F}_{12} = -\vec{F}_{21}. \qquad (8-5)$$

When it is expressed in terms of momentum, Newton's second law [Eq. (8–3)] tells us that the rate of change of each object's momentum is the force acting on it:

$$\frac{d\vec{p}_1}{dt} = \vec{F}_{12}, \qquad (8-6)$$

$$\frac{d\vec{p}_2}{dt} = \vec{F}_{21} = -\vec{F}_{12}. \qquad (8-7)$$

Addition of these two equations leads to

$$\frac{d\vec{p}_1}{dt} + \frac{d\vec{p}_2}{dt} = \vec{F}_{12} - \vec{F}_{12} = 0;$$

$$\frac{d(\vec{p}_1 + \vec{p}_2)}{dt} = 0. \qquad (8-8)$$

As a consequence,

for zero net external force: $\quad \vec{p}_1 + \vec{p}_2 = \text{a constant.} \qquad (8-9)$

CONSERVATION OF MOMENTUM

We will see below that this result is not confined to two objects. In words, Newton's third law implies that

the sum of the momenta of an isolated system of objects is a constant, no matter what forces act between the objects making up the system.

This is called the **principle of conservation of momentum**. Like the principle of conservation of energy, the conservation of momentum is important both as a general principle and as a powerful tool for solving problems.

▲ **FIGURE 8–1** (a) Two billiard balls at the moment of collision and (b) two masses connected by a spring. In each case, the objects exert forces on one another. Here $\vec{F}_{12}$ is the force exerted on object 1 by object 2, and $\vec{F}_{21}$ is the force exerted on object 2 by object 1. In these collisions, the forces are contact forces although, in general, physical contact is not necessary for two objects to exert forces on one another.

EXAMPLE 8–2 A cue ball moves with a velocity of 1.20 m/s in the +y-direction on a billiard table and strikes an equally massive ball initially at rest (Fig. 8–2a). The cue ball is deflected so that its velocity has a component of 0.80 m/s in the +y-direction and a component of 0.56 m/s in the +x-direction (Fig. 8–2b). What is the velocity of the struck ball immediately after the collision?

Before   After

(a)   **(b)**

▲ **FIGURE 8–2** Two colliding billiard balls. (a) Before the collision. (b) After the collision.

Setting It Up The figures label the initial and final velocities of the cue ball as $\vec{v}_{1i}$ and $\vec{v}_{1f}$, respectively, and the final velocity of the struck ball as $\vec{v}_{2f}$. The given velocities are $\vec{v}_{1i} = (1.20 \text{ m/s})\hat{j}$ and $\vec{v}_{1f} = (0.56 \text{ m/s})\hat{i} + (0.80 \text{ m/s})\hat{j}$.

Strategy We can find the initial and final momenta of the sum of the two balls and use the conservation of momentum, which will relate the velocities in question. The only unknown is $\vec{v}_{2f}$.

Working It Out If m is the mass of each ball, the initial momentum is $\vec{p}_i = m\vec{v}_{1i}$ because the struck ball has an initial velocity of zero. Then the conservation of momentum reads

$$m\vec{v}_{1i} = m\vec{v}_{1f} + m\vec{v}_{2f}.$$

This simplifies to $\vec{v}_{1i} = \vec{v}_{1f} + \vec{v}_{2f}$ or $\vec{v}_{2f} = \vec{v}_{1i} - \vec{v}_{1f}$. Numerically,

$$\vec{v}_{2f} = \vec{v}_{1i} - \vec{v}_{1f} = (1.20 \text{ m/s})\hat{j} - [(0.56 \text{ m/s})\hat{i} + (0.80 \text{ m/s})\hat{j}]$$
$$= (-0.56 \text{ m/s})\hat{i} + (0.40 \text{ m/s})\hat{j}.$$

This corresponds to a final speed of $v_{2f} = \sqrt{(v_{2f,x}^2 + v_{2f,y}^2)} = 0.69 \text{ m/s}$.

What Do You Think? In this example we specified some information about the final state (after the collision) as well as all the information about the initial state. If no information had been supplied about the final state, could the statement that kinetic energy is conserved, if it were true here, have allowed a full solution? *Answers to* ***What Do You Think?*** *questions are given in the back of the book.*

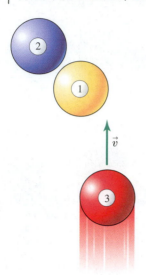

▲ FIGURE 8–3 System of three interacting objects. Any time one ball touches another, there is a force between those two balls.

Conservation of Momentum for a System of Many Objects: The conservation of momentum is not confined to a system of two interacting objects. Suppose that there are three objects in a system on which no external forces act. [For example, you could imagine three billiard balls on a table, the 1 ball and the 2 ball quite close to one another and the 3 ball coming in to interact with both of them (Fig. 8–3).] Then the total force $\vec{F}_1$ on object 1 is given by the sum of the forces on object 1 due to objects 2 and 3:

$$\vec{F}_1 = \vec{F}_{12} + \vec{F}_{13}. \tag{8–10a}$$

Similarly, the total force $\vec{F}_2$ on object 2 is given by the sum of the forces on object 2 due to objects 1 and 3; so

$$\vec{F}_2 = \vec{F}_{21} + \vec{F}_{23}, \tag{8–10b}$$

and similarly,

$$\vec{F}_3 = \vec{F}_{31} + \vec{F}_{32}. \tag{8–10c}$$

(Note that in Fig. 8–3 the 3 ball may not ever directly hit the 2 ball, so that in this case $\vec{F}_{23}$ and $\vec{F}_{32}$ would be zero.) Adding these three equations and using $\vec{F}_{12} = -\vec{F}_{21}$, $\vec{F}_{13} = -\vec{F}_{31}$, and $\vec{F}_{23} = -\vec{F}_{32}$ yield

$$\frac{d\vec{p}_1}{dt} + \frac{d\vec{p}_2}{dt} + \frac{d\vec{p}_3}{dt} = \vec{F}_1 + \vec{F}_2 + \vec{F}_3$$

$$= \vec{F}_{12} + \vec{F}_{13} + \vec{F}_{21} + \vec{F}_{23} + \vec{F}_{31} + \vec{F}_{32} = 0. \tag{8–11}$$

Consequently, the sum of the momenta of the three objects is constant throughout the motion:

$$\vec{P} = \vec{p}_1 + \vec{p}_2 + \vec{p}_3 = \text{a constant}. \tag{8–12}$$

We can easily extend this demonstration to N interacting objects and prove that the sum of the objects' momenta is constant throughout the motion.

8–2 Collisions and Impulse

What happens when objects *collide*? The word "collision" evokes the image of an action with a short, sharp contact, such as the collision between two billiard balls or two automobiles. We can more formally think of a collision between objects as an interaction between them—a set of forces—that is limited in time. (Of course, the word "limited" allows us a lot of leeway.) We will want to think of the colliding objects as otherwise isolated in order to be able to apply the conservation of momentum to the situation. Before we do so, it will pay us to study the idea of briefly acting forces, the kind that occur in collisions, in more detail.

Impulsive Forces

We'll suppose that during a collision the force that alters the motion of the two objects is active for only a short time Δt. We refer to such a force as an **impulsive force**. Over the time Δt the momentum of the object on which the impulsive force acts undergoes a momentum change $\Delta\vec{p}$. We'll call this momentum change the **impulse** $\vec{J}$. (We also say that an object receives or gives an impulse according to whether it is being acted on or is the source of the impulsive force, respectively.) We write

$$\vec{J} \equiv \Delta\vec{p} = \vec{p}_f - \vec{p}_i. \tag{8–13}$$

A little calculus allows us to express the impulse for an impulsive force in terms of that force. The change in momentum is the integral of the rate of change of momentum between the initial and final times that the force acts, and from Newton's second law, the rate of change of momentum is the (impulsive) force $\vec{F}$ that acts on the object:

$$\vec{J} = \int_{t_i}^{t_f} \left(\frac{d\vec{p}}{dt} \right) dt = \int_{t_i}^{t_f} \vec{F} \, dt. \qquad (8\text{–}14)$$

IMPULSE AND FORCE

We have used the fact that the force is zero outside the time interval $\Delta t = t_f - t_i$.

Because the integral of force over time is the area under a curve of force versus time, we can also represent the impulse as the product of the time interval Δt and a quantity that we may call the *average force* $\vec{F}_{av}$ (Fig. 8–4). The average force refers to the average value of the force over the time interval Δt. In this case, the impulse or change in momentum can be written in the form

$$\Delta \vec{p} = \vec{J} = \int_{t_i}^{t_f} \vec{F} \, dt = \vec{F}_{av} \, \Delta t. \qquad (8\text{–}15)$$

Remember that this equation is a *vector* equation. In Example 8–3, however, the vector aspect does not play a crucial role.

▲ **FIGURE 8–4** If an impulsive force acts on an object, its momentum will be changed. That force acts over a time period Δt, varying with time as it acts. The same change in momentum is produced by another force that is constant over Δt and takes the actual force's average value over that time period.

EXAMPLE 8–3 The magnitude of the average force exerted by a bat on a baseball during the time of contact (a period of 2.00×10^{-3} s) is 6660 N. The mass of the baseball is 0.145 kg and its speed is 33.5 m/s just before the bat collides with it. What is the velocity of the ball when it leaves the bat? Assume that the ball leaves the bat along the same line of direction from which it is pitched.

Setting It Up Figure 8–5 illustrates this problem. The x-axis is horizontal and to the right. We know the average force magnitude F_{av} and the time Δt over which it is applied. We are also given the ball's mass m and its initial speed v_0. The motion is one dimensional, in a direction that we label as the x-axis, with the original direction of the ball—the direction of the pitch—in the $+x$-direction. We want to find the final velocity of the ball.

Strategy Because we know both F_{av} and Δt, we can find the ball's impulse—the change in its momentum. Then we can use the value of the impulse to find the ball's final momentum knowing its initial momentum. We can then use its final momentum to find its final speed.

Working It Out The momentum change $\vec{p}_f - \vec{p}_i$ of the ball is given by Eq. (8–15), with $\vec{p}_i = mv_0 \hat{\imath}$. The direction of the force (and hence of the impulse) is in the $-x$-direction, so from Eq. (8–15),

$$\vec{p}_f = \vec{p}_i + \vec{F}_{av} \, \Delta t = (mv_0 - F_{av} \, \Delta t)\hat{\imath}.$$

The final velocity is the final momentum divided by m:

$$\vec{v}_f \equiv v_f \hat{\imath} = \frac{\vec{p}_f}{m} = \frac{mv_0 - F_{av} \, \Delta t}{m} \hat{\imath} = \left(v_0 - \frac{F_{av} \, \Delta t}{m} \right)\hat{\imath}.$$

Thus the final velocity is oriented along the x-axis, with x-component

$$v_f = +33.5 \text{ m/s} - \frac{(6660 \text{ N})(2.00 \times 10^{-3} \text{ s})}{0.145 \text{ kg}} = -58.4 \text{ m/s}.$$

The minus sign indicates that the ball moves in the negative direction, back toward the pitcher.

What Do You Think? Which sort of racket will allow you to return a tennis ball with more velocity: an ordinary strung tennis racket or a solid wooden paddle of the same shape and weight?

▲ **FIGURE 8–5** The direction of the impulse imparted to a ball hit by a bat is to the left, in the $-x$-direction.

HOW ARE THE FOUNDATIONS FOR TALL BUILDINGS MADE?

▲ **FIGURE 8–6** The pile driver is one of the devices that enables us to build large structures.

One way of constructing foundations (Fig. 8–6) is to drive vertical "piles" into the soil down to bedrock and then anchor the building to them. These piles are often made of steel. A pile driver is used to force them into the ground, and momentum and impulse are key to the operation of this device. In Fig. 8–6, there is a weight sitting within a cylinder that encloses the top of the pile. The weight is lifted by exploding fuel and then drops back onto the pile. During the fall, the weight gains momentum. When it reaches the top of the pile, it is brought to rest in an exceedingly short time (determined by the compressibility of the weight and the pile, and the distance the pile moves). Thus the impulse to the dropped cylinder is large, and by Newton's third law there is an equal and opposite impulse to the pile. In turn, the pile delivers an impulse to the rock and soil beneath it, breaking the rock and soil and allowing the pile to move down until friction and normal forces bring it back to a stop. The action is repeated, driving the pile further down with each repetition. ∎

Classification of Collisions

Attempts to understand collisions were carried out by Galileo and his contemporaries. The description of collisions in one dimension were formulated by John Wallis, Christopher Wren (best remembered today as an architect), and Christian Huygens in 1668, and the principle of the conservation of momentum plays a central role in understanding their results.

We can recap our previous discussion on the collision of two objects as follows: The two objects move freely before the collision—no net forces act on either of them—and each has its own constant momentum. During the brief interaction, their individual momenta change because each object experiences an impulsive force due to the other object. After the collision, the two objects are again free but have momenta that differ from those they had before the collision. However, the impulses of the two objects are equal and opposite because the forces each exerts on the other are equal and opposite, so that the change in the momentum of one object is equal and opposite to the change in momentum of the other. In other words, *the sum of their momenta is unchanged*. This feature, the conservation of total momentum of the isolated system, provides a governing constraint. It will hold even if the objects stick together or, at the other extreme, break apart into many pieces.

Suppose that initially object 1 has mass m_1 and velocity $\vec{v}_1$ and object 2 has mass m_2 and velocity $\vec{v}_2$. The total initial momentum is given by

$$\vec{p}_{\text{init}} = m_1\vec{v}_1 + m_2\vec{v}_2. \tag{8–16}$$

Several distinct and interesting possibilities for what the final state can look like present themselves. Figure 8–7 shows these cases and we enumerate them below. In the figure, we have *drawn* the collision in one dimension, although we'll express the conservation of momentum in more general form.

1. The two masses hit each other and stick together, coalescing into one, as in the collision of two blobs of putty or a comet colliding with a planet. Figures 8–7a and b illustrate the before and after for this case. If the mass of the single final object is M and its velocity is $\vec{v}$, momentum conservation reads

$$M\vec{v} = m_1\vec{v}_1 + m_2\vec{v}_2. \tag{8–17}$$

Mass conservation gives us the additional information[†] that $M = m_1 + m_2$.

2. The two masses can remain distinct and unchanged, as in the collision of billiard balls (Fig. 8–7c). We label the final velocities $\vec{v}_1'$ and $\vec{v}_2'$, respectively, and momentum conservation takes the form

$$m_1\vec{v}_1' + m_2\vec{v}_2' = m_1\vec{v}_1 + m_2\vec{v}_2. \tag{8–18}$$

[†]Because special relativity plays an important role in nuclear or subnuclear collisions, the conservation of mass per se is only an approximation, and sometimes a very bad one, in those cases.

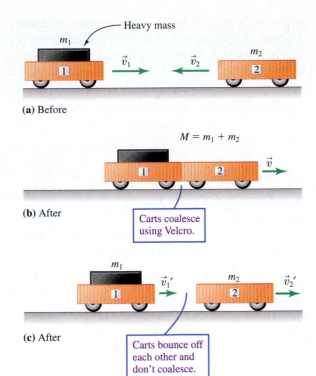

(a) Before

(b) After

Carts coalesce using Velcro.

(c) After

Carts bounce off each other and don't coalesce.

◀ **FIGURE 8–7** A collision between two objects, as shown before the collision in part (a), can have several outcomes. The two objects shatter into several pieces (not shown). (b) The two masses combine as the objects coalesce into one object. (c) Two objects leave the collision, each with the mass of the original objects respectively, drawn for the case of object 1 continuing to move in the original direction.

3. Mass can be transferred from one object to the other such that after the collision one would have two objects with masses m_3 and m_4. As an example, a carbon atom can collide with a molecule of carbon dioxide to make two carbon monoxide molecules. We label the final velocities of masses m_3 and m_4 as $\vec{v}_3$ and $\vec{v}_4$, so that momentum conservation reads

$$m_3\vec{v}_3 + m_4\vec{v}_4 = m_1\vec{v}_1 + m_2\vec{v}_2. \qquad (8\text{–}19)$$

In this case mass conservation would add the information that $m_3 + m_4 = m_1 + m_2$.

4. One or both of the objects can shatter into several pieces (a more complex possibility that we won't deal with in detail).

The vector equations above are indispensable tools for understanding collisions. However, even given all the information about the initial state, they are generally not enough to determine everything about the final state. Only in the case of coalescence [Eq. (8–17)] are the three vector equations sufficient to determine the three components of the single final velocity. In all the other cases, more information is required to understand the details of the motion, and this is usually information about the energy of the objects.

Energy Considerations in Collisions

The degree to which the kinetic energy is conserved in a collision provides us with another piece of information that we can use to help us understand the process. Although for the reasons described in Chapter 7 the *total* energy is always conserved in a collision, some of the energy may be *dissipated* in ways that make it lost to us. For example, when friction is present, some of the energy goes into heating. To take another example, in a collision between two cars, some energy goes into the crumpling of metal, and as a result there is less energy available for motion—the kinetic energy after the collision. If the kinetic energy decrease is known (as in Example 8–4 below), we can use this information to help us calculate the details of the motion after a collision.

The degree to which kinetic energy is conserved provides another way to classify collisions. Kinetic energy is conserved if all the initial kinetic energy of the two colliding objects goes into the kinetic energy of the objects present after the collision, and we call the collision **elastic**. (We are assuming the objects retain their identities as well, as in case 2 above, although that case also includes the possibility that energy is lost.)

When kinetic energy is lost, the collision is **inelastic**—we would also use this term if kinetic energy is gained, something that might happen if the equivalent of an internal spring is released during the collision. The situation in which two objects collide and coalesce [case 1, Eq. (8–17)] is called **perfectly inelastic** because, as we shall see in Section 8–6, it corresponds to the maximum loss of kinetic energy.

EXAMPLE 8–4 A 14,000-kg truck and a 2000-kg car have a head-on collision. Despite attempts to stop, the truck has a speed of 6.6 m/s in the $+x$-direction when they collide and the car has a speed of 8.8 m/s in the $-x$-direction. If 10 percent of the initial total kinetic energy is dissipated through damage to the vehicles, what are the final velocities of the truck and the car after the collision? Assume that all motion takes place in one dimension.

Setting It Up Figure 8–8 shows the collision, with the $+x$-direction to the right. The given masses of the truck and car are M and m, respectively. The known initial velocity component of the truck is V_i and that of the car is v_i. We want to find the final velocity components of the truck V_f and the car v_f if 10 percent of the initial kinetic energy is lost.

▲ **FIGURE 8–8** Head-on collision between large truck and small car.

Strategy We can use the conservation of momentum, which, because the motion is one dimensional, is a single equation for the x-component of velocity. The energy information we have is that the final mechanical energy (all kinetic) is 90 percent of the initial energy. Both the initial momentum and initial energy are easily calculated from the given information. Thus our two conditions should be enough to solve for the two unknowns.

Working It Out Momentum conservation reads

$$MV_f + mv_f = MV_i + mv_i \equiv p_i.$$

The initial total energy is the sum of the kinetic energies of the two vehicles, $K_i = (MV_i^2/2) + (mv_i^2/2)$. The sum of the final kinetic energies is 90 percent of this quantity, so

$$K_f = (MV_f^2/2) + (mv_f^2/2)$$
$$= 0.9[(MV_i^2/2) + (mv_i^2/2)] \equiv 0.9K_i.$$

The two centered equations above are two algebraic equations for the two unknowns V_f and v_f.

Momentum conservation directly yields

$$V_f = \frac{p_i - mv_f}{M}.$$

Substituting this value into the energy relation, we have a quadratic equation for v_f:

$$\frac{1}{2}M\left(\frac{p_i - mv_f}{M}\right)^2 + \frac{1}{2}mv_f^2 - 0.9K_i = 0;$$

$$\frac{1}{2}M\frac{p_i^2}{M^2} - \frac{1}{2}M\left(\frac{2p_imv_f}{M^2}\right) + \frac{1}{2}M\frac{m^2v_f^2}{M^2} + \frac{1}{2}mv_f^2 - 0.9K_i = 0;$$

$$(m^2 + Mm)v_f^2 - 2mv_fp_i + p_i^2 - 2(0.9)MK_i = 0.$$

This equation has solutions

$$v_f = \frac{2mp_i \pm \sqrt{(2mp_i)^2 - 4(m^2 + Mm)[p_i^2 - 2(0.9)MK_i]}}{2(m^2 + Mm)}.$$

When numbers are inserted for the two possible solutions represented by the $\pm$ sign, the solution with a minus sign gives a negative velocity for the car and a positive velocity for the truck. This means that the car continues its motion to the left, going "through" the truck, while the truck similarly goes "through" the car. Since this is not possible, the correct solution has the plus sign. Inserting numbers, we find

$$v_f = 17 \text{ m/s} \quad \text{and} \quad V_f = 2.9 \text{ m/s}.$$

The truck continues in the $+x$-direction with a speed less than its initial speed; the car has completely reversed its direction and is moving even faster than its initial speed. The car has "bounced" from the much more massive truck, much like a tennis ball against a tennis racket. Analysis of this type is used by crash-scene investigators. Adding information such as stopping distance under friction, they can learn, for example, the speeds and directions of the vehicles just before the crash.

What Do You Think? Assuming that they are protected from injury directly associated with the automobile collapsing on them, which would be better for the occupants of the car, an elastic or an inelastic collision?

8–3 Perfectly Inelastic Collisions; Explosions

Perfectly Inelastic Collisions

There is a range of possible collisions from elastic to inelastic to perfectly inelastic, according to how much kinetic energy is lost. The simplest is the case of *perfectly inelastic* collisions in one dimension, which are those in which the objects coalesce as a result of the collision. For example, an asteroid hitting Earth would be a perfectly inelastic collision. Momentum conservation for these collisions are described by Eq. (8–17).

◀ **FIGURE 8–9** A cart with mass m_1 and velocity $\vec{v}_1$ collides with a cart of mass m_2 at rest. The two carts stick together and move on together with velocity $\vec{v}$.

Mass conservation implies that $M = m_1 + m_2$. We can divide Eq. (8–17) by this factor. The velocity of the coalesced object then becomes (in one-dimensional motion)

$$v = \frac{m_1 v_1 + m_2 v_2}{M}. \tag{8-20}$$

Let's analyze this result for some special cases.

If one of the objects (m_2) is at rest ($v_2 = 0$) and the other (m_1) runs into it (Fig. 8–9), then

$$v = \frac{m_1}{M} v_1. \tag{8-21}$$

If $m_1 \gg m_2$, the "composite" object will move with a velocity nearly equal to that of the initially moving object; a car colliding with a bug does not slow down very much. In contrast, when $m_1 \ll m_2$, as when a stationary athlete catches a ball, we get the opposite effect; that is, the athlete will recoil with only a low velocity, just the fraction $m_1/(m_1 + m_2) \cong m_1/m_2$ of the velocity of the ball.

Next, consider the case of a head-on collision in which the two objects have equal and opposite velocities ($v_2 = -v_1$). In this case, Eq. (8–20) becomes

$$v = \frac{m_1 - m_2}{m_1 + m_2} v_1. \tag{8-22}$$

In the special case that $m_1 = m_2$, the two objects have equal and opposite *momenta* because

$$m_1 v_1 + m_2 v_2 = m_1 v_1 + m_1 v_2 = m_1(v_1 + v_2) = 0. \tag{8-23}$$

In that case, the final momentum must be zero and thus $v = 0$, as Eq. (8–22) verifies. The objects collide and come to rest.

Energy Loss in Perfectly Inelastic Collisions: Let's find the change in energy for the collision described above. Before the collision, the total energy E_i is the sum of the kinetic energies of the two objects. The final energy E_f is the kinetic energy of the composite object of mass $M = m_1 + m_2$. The change in energy $\Delta E = E_f - E_i$ is, using Eq. (8–20),

$$\Delta E = \frac{1}{2} M v^2 - \left(\frac{1}{2} m_1 v_1^2 + \frac{1}{2} m_2 v_2^2 \right)$$

$$= \frac{1}{2} \frac{M(m_1 v_1 + m_2 v_2)^2}{M^2} - \left(\frac{1}{2} m_1 v_1^2 + \frac{1}{2} m_2 v_2^2 \right)$$

$$= \frac{1}{2} \frac{m_1^2 v_1^2 + 2 m_1 m_2 v_1 v_2 + m_2^2 v_2^2 - M(m_1 v_1^2 + m_2 v_2^2)}{M}$$

$$= \frac{1}{2} \frac{m_1 m_2 (-v_1^2 - v_2^2 + 2 v_1 v_2)}{M} = -\frac{1}{2} \frac{m_1 m_2}{M} (v_1 - v_2)^2. \tag{8-24}$$

The right-hand side of Eq. (8–24) is always negative, corresponding to an energy loss, or to an inelastic collision. As to why the situation studied here is *perfectly* inelastic, see the Think About This box in Section 8–6.

EXAMPLE 8–5 A dog who jumps into the interior of a stationary ice boat is moving at $v_1 = 32$ km/h when he enters the boat (Fig. 8–10), and his landing on the boat can be regarded as a collision. The dog's mass is 14 kg and that of the boat plus boater is 160 kg. You can assume all the motion is horizontal. (a) Assuming that the ice surface is frictionless, what is the velocity of the boat after the collision? (b) What is the ratio of the energy loss to the initial energy?

▲ **FIGURE 8–10** A dog jumping into the boat changes the boat's momentum.

Setting It Up We specify that the motion is in the x-direction by adding an axis to Fig. 8–10. We know the dog's mass m_1 and that of the boat and person, m_2. We also know the dog's velocity before the collision, or equivalently its x-component v_1, as well as the boat's ($v_2 = 0$). We want the velocity x-component v of the boat after the collision and the ratio of the energy loss ΔE to the initial kinetic energy K_i.

Strategy This is a perfectly inelastic collision, so conservation of momentum applied to the collision that occurs when the dog enters the boat will determine the boat's velocity after the collision. With all speeds known, we are left with a direct calculation of the kinetic energy before and after the collision to find the needed energy ratio.

Working It Out (a) The initial momentum has only the x-component,

$$p_i = m_1 v_1.$$

The final momentum is given by $p_f = Mv$, where $M = m_1 + m_2$. Equating p_f and p_i, we find $Mv = m_1 v_1$, or

$$v = \frac{m_1 v_1}{M} = \frac{(14\ \text{kg})(32\ \text{km/h})}{174\ \text{kg}} = 2.6\ \text{km/h} = 0.72\ \text{m/s}.$$

(b) The initial energy is the kinetic energy of the dog, $K_i = \frac{1}{2} m_1 v_1^2$. The final energy is again all in the form of kinetic energy:

$$K_f = \frac{1}{2} M v^2 = \frac{1}{2} M \left(\frac{m_1 v_1}{M} \right)^2 = \frac{1}{2} \left(\frac{m_1}{M} \right) m_1 v_1^2 = \frac{m_1}{M} K_i.$$

Thus the energy loss is given by

$$\Delta E = K_i - K_f = K_i - \frac{m_1}{M} K_i = K_i \left(1 - \frac{m_1}{M} \right),$$

and the ratio of the energy loss to the initial energy is

$$\frac{\Delta E}{K_i} = 1 - \frac{m_1}{M} = \frac{M - m_1}{M} = \frac{m_2}{M},$$

a number less than 1. The energy has decreased. Numerically,

$$\frac{\Delta E}{K_i} = \frac{160\ \text{kg}}{174\ \text{kg}} = 0.92.$$

What Do You Think? Kinetic energy is lost. Which of the following is correct? (a) It went into gravitational energy. (b) It went into the dog and boater "giving and stretching" as the dog lands. (c) It went into melting the ice.

EXAMPLE 8–6 A 10-g bullet is fired in the $+x$-direction into a stationary block of wood that has a mass of 5.0 kg. The speed of the bullet before entry into the wood block is 500 m/s. What is the speed of the block just after the bullet has become embedded? What distance will the block slide on a surface with a coefficient of friction equal to 0.30?

Setting It Up We have drawn the situation in Fig. 8–11. Positive x is to the right. We label the known mass of the bullet as m_1, its known initial velocity as $\vec{v}_1 = v_1 \hat{i}$, and the known total mass of block and bullet as M. We want to find the distance d the block will slide on a surface for which there is a known coefficient of kinetic (sliding) friction μ_k.

Strategy We can use conservation of momentum to find the velocity of the block and bullet immediately after the collision, then Newton's second law to find the friction-induced negative accelera-

tion of the block. Knowing the acceleration and the initial condition, kinematic relations will give us the distance d the block moves.

▲ **FIGURE 8–11** A bullet fired into a stationary block of wood moves the block.

Working It Out The x-component of the initial momentum is $p_i = m_1 v_1$, while the momentum immediately after the collision is Mv. Momentum conservation gives

$$m_1 v_1 = Mv;$$

$$v = \frac{m_1}{M} v_1 = \frac{10\ g}{5010\ g}(500\ \text{m/s}) = 1.0\ \text{m/s}.$$

We now turn to the problem of finding the acceleration given the value v of the initial speed of the bullet–block composite. The normal force N on the block from the table has magnitude Mg, so the force of friction between block and table is $-\mu_k N = -\mu_k Mg$. (The minus sign indicates that friction points to the left, along the $-x$-direction.) The friction force has constant magnitude and leads

to a constant acceleration a of the block, according to Newton's second law, namely $F_{net} = Ma$ reads $-\mu_k Mg = Ma$, or

$$a = -\mu_k g.$$

The negative sign means the block slows down, traveling a distance d before it stops (Fig. 8–11b). Because the acceleration is uniform, we can use the relation $v_f^2 - v_i^2 = 2ad$ [from Eq. (2–24)]. With $v_f = 0$ and the initial speed $v_i = v$ above, we have

$$d = -\frac{v_i^2}{2a} = \frac{1}{2}\frac{v^2}{\mu_k g} = \frac{1}{2}\frac{(1.0\ \text{m/s})^2}{(0.30)(9.8\ \text{m/s}^2)} = 0.17\ \text{m}.$$

What Do You Think? In what way could we have used the fact that the mass of the bullet is much less than the mass of the block?

Explosions

Imagine that we were to film a perfectly inelastic collision in a frame of reference in which the total momentum is zero. In this reference system, the two objects approach each other and merge, leaving a composite object at rest. If we ran the film in reverse, it would look like a film of an explosion. The "initial" object of mass $M = m_1 + m_2$, at rest, breaks up into two objects, m_1 and m_2, with their total momentum equal to zero,

$$m_1 \vec{v}_1 + m_2 \vec{v}_2 = 0. \tag{8–25}$$

Energy conservation tells us that an explosion is possible if there is an initial potential energy U within the "unexploded" system that can be converted into kinetic energy. For the case we are referring to here, this could be as simple as a compressed spring between two masses that is then released. We'll have

$$U = \tfrac{1}{2}m_1 v_1^2 + \tfrac{1}{2}m_2 v_2^2. \tag{8–26}$$

An explosion can involve many more than two objects in the final state. In the explosion of, say, dynamite, the potential energy is stored in its molecules, in the form we call chemical energy. To take another example, Fig. 8–12 shows the remnants of a stellar explosion far from Earth. In these more complicated cases, we will always have the overriding simplicity that the initial momentum of the system before it explodes is the same as the sum of the momentum of all the fragments after the explosion. Let's next take a look at an explosion that occurs when an unstable atomic nucleus disintegrates—a nuclear decay.

▲ **FIGURE 8–12** A small portion of the Cygnus Loop supernova blast wave passes through clouds of interstellar gas. The collision heats and compresses the gas, which causes the glow. Such images taken by the Hubble Space Telescope reveals the structure of the interstellar medium.

EXAMPLE 8–7 One type of polonium nucleus (symbol ^{210}Po), with mass 3.49×10^{-25} kg, can decay into an α particle (actually a helium nucleus), mass 6.64×10^{-27} kg, and a certain type of lead nucleus (symbol ^{206}Pb), mass 3.42×10^{-25} kg:

$$^{210}\text{Po} \rightarrow \alpha + {}^{206}\text{Pb}.$$

In this process, the final decay products have a kinetic energy of 8.65×10^{-13} J if the polonium nucleus decays at rest, the situation we consider here. What are the speeds of the α particle and the lead nucleus?

Setting It Up We show the decay in Fig. 8–13. With a decay from rest into two bodies, the momenta of the two decay products must go off back to back because that is the only way they can add to zero. We call this direction the x-axis. We label the known masses of the alpha particle, the polonium nucleus, and the lead nucleus as M_α, M_{Po}, and M_{Pb}, respectively, and we denote the known kinetic energy in the final state as Q. We want to find the speeds v_α and v_{Pb} of the alpha particle and lead nucleus, respectively.

Strategy Conservation of (one-dimensional) momentum is a single equation involving both speeds, as is the known final energy value. Therefore we have two equations for the two unknowns.

▲ **FIGURE 8–13** Decay of a nucleus. Before: Original nucleus. After: Two outgoing fragments that result from the decay (explosion).

Working It Out Conservation of momentum and the expression for the kinetic energy read, respectively,

$$M_\alpha v_\alpha = M_{Pb} v_{Pb};$$
$$Q = \tfrac{1}{2}M_\alpha v_\alpha^2 + \tfrac{1}{2}M_{Pb} v_{Pb}^2.$$

These two equations can be solved for the two variables v_α and v_{Pb}.

(continues on next page)

We find that the speeds are

$$v_\alpha = \sqrt{\frac{2Q}{M_\alpha(1 + M_\alpha/M_{Pb})}},$$

$$v_{Pb} = \sqrt{\frac{2Q}{M_{Pb}(1 + M_{Pb}/M_\alpha)}}.$$

Substitution of the known numerical values gives $v_\alpha = 1.60 \times 10^7$ m/s and $v_{Pb} = 3.10 \times 10^5$ m/s. The speed of the α is about 5 percent of the speed of light, and this is where special relativity begins to play a role—we have ignored that here of course.

This decay is a form of radioactivity; the fragments could conceivably be dangerous. A relevant element is the energy carried by the two fragments, and you should be able to calculate the kinetic energy of each one.

What Do You Think? An object at rest explodes into two fragments of unequal mass. Which of the following is true? (a) The lighter fragment could have the same speed as the heavier fragment but most often moves more quickly. (b) The lighter fragment always moves off more quickly. (c) The lighter fragment can move off more slowly, depending on the details of the explosion.

THINK ABOUT THIS . . .

HOW DOES A JET ENGINE WORK?

▲ **FIGURE 8–14** A jet engine propels an airplane through the operation of the conservation of momentum.

Conservation of momentum is one way to understand the operation of a jet engine (Fig. 8–14). Air is brought into the front of the engine by intake fans. The air, containing oxygen molecules (O_2), is then mixed with fuel. Among chemical reactions that occur during the ensuing combustion, there is the production of two water molecules (H_2O) for each O_2. This doubles the volume of that part of the oxygen from the air that combines with the hydrogen. Other combustion reactions leave the number of molecules and hence the volume of the air unchanged, and some part of the air, principally the nitrogen, undergoes no chemical reaction. Because of the water-producing reaction, the net effect is that a bigger volume of gas must leave the engine than entered it. To be able to keep up the continuous action of the engine, the outgoing gas must therefore leave with a velocity greater than the incoming velocity; the combustion provides the necessary kinetic energy to enable this to happen. Furthermore, the mass of the departing gas is larger, since the mass of the fuel used in combustion has been added to it. Thus the outgoing gas has substantially larger momentum in the backward direction than the incoming gas. To conserve the momentum of the entire system, the airplane gains momentum in the forward direction. The force that makes the escaping gas accelerate out the back has its third law partner in a forward force on the airplane. ∎

8–4 Elastic Two-Body Collisions in One Dimension

Let's continue to work in one dimension. As usual, the word "velocity" will mean the velocity *component* in the direction of motion; this can be positive or negative. In an *elastic collision*, there is no mass transfer from one object to another. Further, *all the kinetic energy in the initial state goes into kinetic energy in the final state*. If the final velocities of objects 1 and 2 are denoted by v_1' and v_2', then, in addition to the momentum conservation equation for one dimension,

$$m_1 v_1 + m_2 v_2 = m_1 v_1' + m_2 v_2', \tag{8–27}$$

we have the energy conservation equation

$$\tfrac{1}{2} m_1 v_1^2 + \tfrac{1}{2} m_2 v_2^2 = \tfrac{1}{2} m_1 v_1'^2 + \tfrac{1}{2} m_2 v_2'^2. \tag{8–28}$$

With this information we can find the final velocities of the colliding objects if their initial velocities are known. We rewrite the momentum conservation equation (8–27) as

$$m_1(v_1 - v_1') = -m_2(v_2 - v_2'). \tag{8–29}$$

We use the fact that $v_1^2 - v_1'^2 = (v_1 - v_1')(v_1 + v_1')$ and $v_2^2 - v_2'^2 = (v_2 - v_2')(v_2 + v_2')$ to rewrite the energy conservation equation (8–28) in the form

$$\tfrac{1}{2} m_1(v_1 - v_1')(v_1 + v_1') = -\tfrac{1}{2} m_2(v_2 - v_2')(v_2 + v_2'). \tag{8–30}$$

Dividing both sides of Eq. (8–30) by the two sides of Eq. (8–29) leads to the equation

$$v_1 + v_1' = v_2 + v_2'. \tag{8–31}$$

If we use the letter u to denote the *relative velocity* of the two colliding objects, then

$$u_i = v_1 - v_2 \quad \text{and} \quad u_f = v_1' - v_2'.$$

Using these quantities, Eq. (8–31) can be written in the form

$$u_i = -u_f. \tag{8–32}$$

Equation (8–32) states that *when the collision is elastic, the relative velocity of the colliding objects changes sign but does not change magnitude*. A simple way to remember this result is that the relative velocity behaves like the velocity of a perfectly elastic rubber ball hitting a brick wall.

We may solve Eq. (8–31) for one of the unknown variables, v_2', for example,

$$v_2' = v_1 - v_2 + v_1',$$

and substitute this value into the momentum conservation equation (8–27). We then have

$$m_1 v_1 + m_2 v_2 = m_1 v_1' + m_2(v_1 - v_2 + v_1'),$$

which may be rewritten in the form

$$(m_1 + m_2)v_1' = (m_1 - m_2)v_1 + 2m_2 v_2;$$

$$v_1' = \frac{m_1 - m_2}{m_1 + m_2} v_1 + \frac{2m_2}{m_1 + m_2} v_2. \tag{8–33}$$

A similar calculation leads to the formula

$$v_2' = \frac{2m_1}{m_1 + m_2} v_1 + \frac{m_2 - m_1}{m_1 + m_2} v_2. \tag{8–34}$$

These equations are complicated and it is useful to consider two special cases that simplify them.

1. *Object 2 is initially at rest.* We set $v_2 = 0$, so Eqs. (8–33) and (8–34) become

$$v_1' = \frac{m_1 - m_2}{m_1 + m_2} v_1 \tag{8–35a}$$

and

$$v_2' = \frac{2m_1}{m_1 + m_2} v_1. \tag{8–35b}$$

Let's consider the following situations (in all of which object 2 is initially at rest):

a. The objects have equal masses (Fig. 8–15a). In this case, $v_1' = 0$ and $v_2' = v_1$. The two objects in effect change roles: The moving object comes to rest and the object that was initially at rest moves with the initial velocity of the first object. This effect can be seen vividly in hard billiard shots along a line.

FIGURE 8–15 Two objects collide elastically in one dimension, with the second object initially at rest. (a) If the two masses are equal, the objects simply exchange velocities. (b) If $m_1 \gg m_2$, both objects move off to the right. (c) If $m_2 \gg m_1$, mass m_1 reverses its direction and m_2 moves slowly off to the right.

b. Mass $m_1 \gg$ mass m_2 (Fig. 8–15b). In this case, Eqs. (8–35) yield $v_1' \cong v_1$ and $v_2' \cong 2v_1$. The velocity of the moving object decreases a little, while the object that was at rest picks up almost twice the velocity of the incoming object.

c. Mass $m_2 \gg$ mass m_1 (Fig. 8–15c). In this case, Eqs. (8–35) yield $v_1' \cong -v_1$ and $v_2' \cong (2m_1/m_2)v_1$. The moving object very nearly reverses its velocity, while the object initially at rest recoils with a very small velocity. In the limit that m_2 approaches infinity, the recoil velocity can be neglected and the final velocity of the first object is equal and opposite to its incident velocity. This is just what happens when a tennis ball is bounced off a wall.

2. *The initial total momentum is zero.* The two objects approach each other with velocities such that the initial total momentum is zero, $m_1 v_1 + m_2 v_2 = 0$. Thus

$$v_2 = -\frac{m_1}{m_2} v_1. \tag{8–36}$$

When this value is substituted into Eq. (8–33), we find

$$v_1' = \frac{m_1 - m_2}{m_1 + m_2} v_1 + \left(\frac{2m_2}{m_1 + m_2}\right)\left(-\frac{m_1}{m_2}\right)v_1 = \left(\frac{m_1 - m_2 - 2m_1}{m_1 + m_2}\right)v_1 = -v_1. \tag{8–37}$$

The initial total momentum was zero and so, by momentum conservation, the final total momentum $(m_1 v_1' + m_2 v_2')$ is also zero and

$$v_2' = -\frac{m_1}{m_2} v_1' = \frac{m_1}{m_2} v_1 = -v_2. \tag{8–38}$$

Therefore, in the case where the total momentum is zero, the velocities of the objects are unchanged in magnitude but they change sign. In effect, under these circumstances, each of the objects acts as if it hit an infinitely massive brick wall.

EXAMPLE 8–8 Two spheres with masses of 1.0 and 1.5 kg hang at rest at the ends of strings that are both 1.5 m long. These two strings are attached to the same point on the ceiling. The lighter sphere is pulled aside so that its string makes an angle $\theta_i = 60°$ with the vertical. The lighter sphere is then released and the two spheres collide elastically. When they rebound, what is the largest angle with respect to the vertical that the string holding the lighter sphere makes?

Setting It Up We specify in Fig. 8–16a an initial angle θ_i and in Fig. 8–16b a final angle θ_f. We know the values of the light mass m_1 and the heavy mass m_2. The string length L and the initial angle θ_i are also known. We want the rebound angle, θ_f, of the lighter mass after the collision.

Strategy Conservation of energy, including the initial gravitational potential energy, gives us the speed of the lighter mass before the collision. The collision is elastic, so we have available to us the conservation both of momentum and of energy as they apply to the collision. These will give us the recoil velocity of the lighter mass and we can then apply conservation of energy, including the gravitational potential energy, to find the height to which the lighter mass rises. Geometry then gives us the final angle.

Working It Out We begin by calculating the initial potential energy, which is converted to kinetic energy when sphere 1 swings down to the minimum point (neglecting air resistance). The mass m_1 is raised a distance $L(1 - \cos \theta_i) = L(1 - \cos 60°) = L/2$ above the minimum point. With the potential energy zero when the spheres are hanging vertically, the initial potential energy of the system is $U_i = m_1 gL/2$. Conservation of energy then gives $\frac{1}{2} m_1 v_1^2 = \frac{1}{2} m_1 gL$, or

$$v_1 = \sqrt{gL}.$$

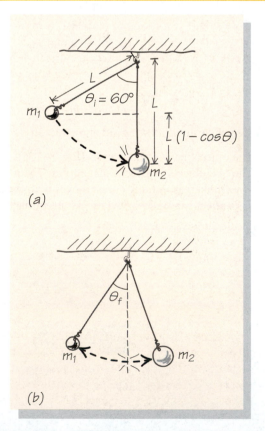

(a)

(b)

▲ **FIGURE 8–16** Two spheres hang from strings of equal length. (a) Sphere m_1 is pulled back to the left at an angle θ_i and released. It collides with sphere m_2, which is at rest. (b) The balls recoil, with sphere m_1 reaching a maximum height characterized by the angle θ_f.

We now treat the collision, which takes place at the bottom of the swing, where $\vec{v}_1$ is horizontal, $\vec{v}_1 = v_1\hat{i}$. (In fact, the collision itself is a one-dimensional one, with all motion aligned with the x-axis, so we'll drop the explicit vector notation and write all our equations for the collision in terms of the x-components of the vectorial quantities momentum and velocity.) The initial momentum is

$$p_i = m_1 v_1 + m_2 v_2 = m_1 v_1.$$

To find the velocities v_1' and v_2', we use conservation of momentum, which states that $p_f = p_i$, or

$$m_1 v_1' + m_2 v_2' = m_1 v_1. \qquad (8\text{–}39)$$

The collision is elastic, so we can also use Eq. (8–32), which is a consequence of energy conservation and which states that the initial relative velocity and the final relative velocity are equal in magnitude but opposite in sign. With

$$v_{\text{rel, initial}} = v_1 - v_2 = v_1,$$

we have a final relative velocity of

$$v_{\text{rel, final}} = v_1' - v_2' = -v_{\text{rel, initial}} = -v_1. \qquad (8\text{–}40)$$

Equations (8–39) and (8–40) are two simultaneous equations that can be solved for v_1' and v_2'. We are interested only in the final velocity of the lighter sphere, and the solution for this quantity is

$$v_1' = \frac{m_1 - m_2}{m_1 + m_2}\sqrt{gL}.$$

With m_2 larger than m_1, this quantity is negative, indicating that the lighter sphere recoils back to the left.

We next find the height to which m_1 recoils. The kinetic energy right after the collision, $m_1 v_1'^2/2$, is converted into gravitational potential energy $U = m_1 gh$ as the sphere rises by h. At the top of the recoil motion, all the kinetic energy is converted to potential energy. As Fig. 8–16b shows, the height risen is $L(1 - \cos\theta_f)$, so we have

$$m_1 gL(1 - \cos\theta_f) = \tfrac{1}{2}m_1 v_1'^2.$$

Numerically,

$$
\begin{aligned}
1 - \cos\theta_f &= \frac{v_1'^2}{2gL} = \frac{(m_1 - m_2)^2}{(m_1 + m_2)^2}gL\frac{1}{2gL} = \frac{(m_1 - m_2)^2}{2(m_1 + m_2)^2} \\
&= \frac{(1.0 \text{ kg} - 1.5 \text{ kg})^2}{2(1.0 \text{ kg} + 1.5 \text{ kg})^2} = 0.020,
\end{aligned}
$$

or $\theta_f = 11°$.

What Do You Think? Describe the same process in the case that the two masses are equal.

8–5 Elastic Collisions in Two and Three Dimensions

In one dimension, the possible motion of colliding objects is limited. When collisions are no longer restricted to lie along a line, as when billiard balls collide on a billiard table, the vector nature of the mathematical equations becomes important. We work here with collisions in which the identities of the two objects are preserved and in which kinetic energy is conserved.

The law of conservation of momentum [Eq. (8–9)] for the collision of two objects of masses m_1 and m_2, with initial velocities $\vec{v}_1$ and $\vec{v}_2$ and final velocities $\vec{v}_1'$ and $\vec{v}_2'$, reads

$$m_1\vec{v}_1 + m_2\vec{v}_2 = m_1\vec{v}_1' + m_2\vec{v}_2'. \qquad (8\text{–}41)$$

The energy conservation law is

$$\tfrac{1}{2}m_1 v_1^2 + \tfrac{1}{2}m_2 v_2^2 = \tfrac{1}{2}m_1 v_1'^2 + \tfrac{1}{2}m_2 v_2'^2. \qquad (8\text{–}42)$$

Let's first consider collisions in two dimensions, where everything happens in a plane, such as the xy-plane shown in Fig. 8–17. The collision of billiard balls on a billiard table represents this case, but with equal masses. Given information about the initial motions, we want to find the magnitudes of the final velocities and the angles they make with the x-axis. Equivalently, we want the x- and y-components of the final velocities of objects 1 and 2. There are four unknowns but only three equations. [Equation (8–41) is a vector equation and actually comprises two equations: one for the x-components and one for the y-components.] Therefore, for a given set of initial velocities, *there is no unique solution for the final velocities* and the final objects can move in a variety of directions with a variety of speeds. Nevertheless, the three equations do impose substantial constraints.

An interesting example of these constraints occurs when the two masses have identical values m and one of the objects is initially at rest. This is very literally the billiards case, in which the projectile is the cue ball and the target is initially stationary. Figure 8–18 shows the geometry—we have chosen object 2 to be initially at rest, $v_2 = 0$. After canceling a common factor of m, we take the square of Eq. (8–41) (the square of a vector equation $\vec{A} = \vec{B}$ implies $\vec{A}\cdot\vec{A} = \vec{B}\cdot\vec{B}$). With $v_2 = 0$, we find

$$v_1^2 = v_1'^2 + 2\vec{v}_1'\cdot\vec{v}_2' + v_2'^2. \qquad (8\text{–}43)$$

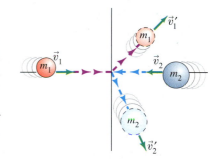

▲ **FIGURE 8–17** An object of mass m_1 and velocity $\vec{v}_1$ collides with another object of mass m_2 and velocity $\vec{v}_2$ in two dimensions.

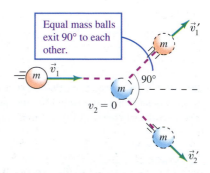

▲ **FIGURE 8–18** An object of mass m_1 and velocity $\vec{v}_1$ collides elastically with an object of the same mass at rest. The angle between the final velocities is 90°.

We can also cancel a common factor of $\frac{1}{2}m$ from Eq. (8–42), leaving, with $v_2 = 0$,

$$v_1^2 = v_1'^2 + v_2'^2. \tag{8–44}$$

Comparing Eqs. (8–43) and (8–44), we conclude that $\vec{v}_1' \cdot \vec{v}_2' = 0$. In other words, the final velocity vectors $\vec{v}_1'$ and $\vec{v}_2'$ are perpendicular to one another. The fact that the angle between the velocities of the outgoing cue ball and the recoiling target ball is a right angle is a fact well known to billiards players (Fig. 8–18). (If spins come in, then this is no longer necessarily true.) It is also worthwhile noting what Eq. (8–44) tells us: The final velocities form the sides of a right triangle whose hypotenuse has magnitude $|v_1|$.

CONCEPTUAL EXAMPLE 8–9 There are two near extremes that describe billiard shots. In the first, the cue ball strikes the target ball nearly head on and the target ball moves rapidly in nearly the original direction of the cue ball. In the second, the cue ball barely grazes the target ball and the cue ball continues with nearly the same speed in its original direction. Qualitatively, what happens to the cue ball in the first case and what happens to the target ball in the second case?

Answer Carried to the limit, these are one-dimensional collisions. In the limit of the first collision, the cue ball stops dead, and in the limit of the second collision the cue ball just misses the target ball, which therefore does not move. The answer to our question must be close to these cases. The billiard balls have equal masses, and the target ball is originally at rest, so that we also have "right angle rules," in which the final velocities obey a right triangle rule [Eq. (8–44)] and the billiard ball motions after the collision make a right angle. Thus in the first case the cue ball moves off slowly at nearly a right angle to the original direction of motion. In the second case the target ball moves slowly off to the side at nearly a right angle to the original direction of motion.

EXAMPLE 8–10 Two billiard balls of equal mass m approach each other along the x-axis; one is moving to the right with a speed of $v_1 = 10 \text{ m/s}$ and the other is moving to the left with a speed of $v_2 = 5 \text{ m/s}$. After the collision, which is elastic, one of the balls moves in the direction of the y-axis (Fig. 8–19). What are the velocities $\vec{v}_1'$ and $\vec{v}_2'$ of the balls after the collision?

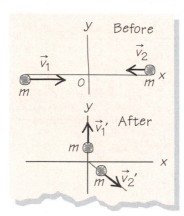

▲ **FIGURE 8–19** Two billiard balls approach, collide, and change directions.

Setting It Up Figure 8–19 shows the situation together with the x- and y-axes. The initial speeds $\vec{v}_1$ and $\vec{v}_2$ are specified in the problem, as is the direction of the final velocity $\vec{v}_1' = v_1' \hat{j}$ of one of the balls—this ball has only one final velocity component. We do *not* know the value of the mass m, which will in fact cancel. We want the final velocities of the two balls, $\vec{v}_1'$ and $\vec{v}_2'$.

Strategy If we use the momentum conservation equation and divide by the common factor of m, we deal with a conservation of velocities. Because the collision is elastic, we can also use the conservation of kinetic energy, from which a common factor of m also cancels. The unknown mass m will cancel from the problem. We will want to count equations to make sure we have enough information to find the unknowns, which are the three final velocity components, one for ball 1 and two for ball 2. Assuming there is enough information, we can algebraically solve for the velocity components.

Working It Out With the cancellation of the mass factors, the momentum and energy conservation equations read

$$\vec{v}_1 + \vec{v}_2 = \vec{v}_1' + \vec{v}_2', \qquad v_1^2 + v_2^2 = v_1'^2 + v_2'^2.$$

The balls move in the xy-plane, so the first equation stands for two component equations and the second equation provides a third. There are three unknowns, the single component of $\vec{v}_1'$ and the two components of $\vec{v}_2'$. The problem is thus solvable. Writing $\vec{v}_2' = v_{2x}'\hat{i} + v_{2y}'\hat{j}$, momentum conservation reads

$$(10 \text{ m/s})\hat{i} + (-5 \text{ m/s})\hat{i} = v_1'\hat{j} + v_{2x}'\hat{i} + v_{2y}'\hat{j}.$$

We now separately equate the coefficients of $\hat{i}$ and of $\hat{j}$; that is, we use the fact that momentum conservation holds in the x-direction and in the y-direction separately. We find

$$v_{2x}' = (10 \text{ m/s}) + (-5 \text{ m/s}) = 5 \text{ m/s} \quad \text{and} \quad v_{2y}' = -v_1'.$$

The energy conservation equation reads similarly

$$(10 \text{ m/s})^2 + (-5 \text{ m/s})^2 = v_1'^2 + (v_{2x}'^2 + v_{2y}'^2),$$
$$100 \text{ m}^2/\text{s}^2 + 25 \text{ m}^2/\text{s}^2 = v_1'^2 + 25 \text{ m}^2/\text{s}^2 + v_1'^2.$$

Thus we have $100 \text{ m}^2/\text{s}^2 = 2v_1'^2$, or $v_1' = \sqrt{50}$ m/s. We also have $v_{2y}' = -\sqrt{50}$ m/s.

Although the collisions of real billiard balls are quite elastic, this is not the case for many other real collisions. A high-speed photograph of a baseball meeting a baseball bat (Fig. 8–20) shows that the ball undergoes significant deformation, which, even though the deformation is not permanent, usually means that there is inelasticity, that is, some energy is lost in the collision.

Problem-Solving Techniques

Collision problems involve an interaction between objects that occurs in a limited space (and a limited time). At some early time, the objects do not exert any forces on one another; at some later time, they again exert no forces on one another. Typically, we need to find some parameters of the final (or initial) motion—the final velocity of one of the objects, for example. The following steps can be useful:

1. We must identify the relevant interaction between the objects involved in a collision. We must therefore be able to assume that the objects are isolated or that during the collision external forces are small compared to the impulsive forces.

2. We must identify the objects involved before the collision and those involved after the collision. The objects that result from a collision may not be the same as the objects before the collision, for example, explosions or collisions in which the colliding objects coalesce.

3. We must identify the quantities that are known and the quantities that are to be found. In particular, a count of the number of known and unknown quantities, including the number of vector components, is helpful.

4. Remember that the conservation of momentum, which is a vector relation, is always applicable if there is no external

force on the colliding system. If the motion is in one, two, or three dimensions, the conservation of momentum gives one, two, or three relations among the momenta (or velocities), respectively.

5. If the collision is known to be elastic, the conservation of energy provides an additional equation that involves the speeds of the objects. If the collision is not elastic, then an equation for the energy is available only if information about the energy loss is available.

6. The number of equations that include the unknowns must match the number of unknowns if the problem is to be solved.

We have thus far concentrated on two-dimensional elastic collisions. What about elastic collisions in three dimensions? If you observe the collision from a frame of reference known as the center-of-mass frame of the two colliding objects (we'll discuss the meaning and use of this inertial frame in Section 8–6), then the incoming objects come in along a line. They would leave along another line in that frame, and since two intersecting lines determine a plane, the collision is two dimensional in this frame. The results established for elastic collisions in two dimensions are thus directly applicable to three dimensions.

8–6 Center of Mass

In a system of many parts, including the two-piece systems we have dealt with to this point in this chapter, there is a particular point with special properties, called the *center of mass*. As we shall see, the importance of this point is that it moves under Newton's second law as if the total mass of the system were concentrated there. In this section we'll learn how to calculate the position of the center of mass and to understand its properties.

▲ **FIGURE 8–20** The softball undergoes a significant deformation during the brief time period that an impulsive force due to the bat acts on it.

In Section 8–1 we saw that in the absence of external forces the sum of the momenta of the N particles that make up a system (an extended object) is unchanging. (There we actually worked this through for $N = 3$, but the procedure we used is easily generalized.) In other words, if $\vec{P}$ is the total momentum of the system, then in the absence of external forces

$$\vec{P} = m_1\vec{v}_1 + m_2\vec{v}_2 + \cdots + m_N\vec{v}_N = \text{a constant.} \quad (8\text{–}45)$$

Here we have labeled the particles, or pieces, that make up the system with subscripts. The momentum of the system is a constant vector—one that does not change in magnitude or direction as a function of time.

We can use this result to locate the center of mass of the extended object. If we also assume that the particle masses do not change as a function of time, Eq. (8–45) can be written as

$$\frac{d}{dt}(m_1\vec{r}_1 + m_2\vec{r}_2 + \cdots + m_N\vec{r}_N) = \text{a constant.} \quad (8\text{–}46)$$

We divide the quantity in parentheses by the total mass, $M = m_1 + m_2 + \cdots + m_N$, to obtain the position vector $\vec{R}$ of the center of mass:

$$\vec{R} \equiv \frac{m_1\vec{r}_1 + m_2\vec{r}_2 + \cdots + m_N\vec{r}_N}{M}. \quad (8\text{–}47)$$

CENTER OF MASS

(a) (b) (c)

▲ **FIGURE 8–21** Three examples of the center of mass (CM). (a) Both balls have equal masses. (b) One ball has twice the mass of the other. (c) Three balls of equal mass.

From its definition, the center-of-mass vector $\vec{R}$ has Cartesian components X, Y, and Z given by

$$X = \frac{m_1 x_1 + m_2 x_2 + \cdots + m_N x_N}{M}, \tag{8–48a}$$

$$Y = \frac{m_1 y_1 + m_2 y_2 + \cdots + m_N y_N}{M}, \tag{8–48b}$$

$$Z = \frac{m_1 z_1 + m_2 z_2 + \cdots + m_N z_N}{M}. \tag{8–48c}$$

Figure 8–21 illustrates some examples of the center of mass for two and three pieces. The following two examples illustrate the calculation of the position of the center of mass. There are some simple ideas that we can extract from the figure and the examples. In particular, we mention here that if the system is made from two pieces, the center of mass will lie along the line between them and a more massive piece will tend to "pull" the center of mass toward it.

EXAMPLE 8–11 Two pointlike masses are placed on a massless rod that is 1.5 m long. The masses are placed as follows: 1.6 kg at the left end and 1.8 kg 1.2 m from the left end. (a) What is the location of the center of mass? (b) By moving the 1.8-kg mass, can you arrange to have the center of mass in the middle of the rod?

Setting It Up Figure 8–22 illustrates this one-dimensional problem. We align the rod with length L along the x-axis, and we place its left end at the origin. The 1.6-kg mass m_1 is placed at $x_1 = 0$ m and the 1.8-kg mass m_2 is placed at $x = 1.2$ m.

Strategy Both parts of this example ask us to calculate in a straightforward way the location of the center of mass in a one-dimensional system. Equation (8–48a) provides the necessary formulation.

▲ **FIGURE 8–22** A rod with two pointlike masses is aligned along the x-axis, with its left end at the origin.

Working It Out (a) The location of the center of mass is given by Eq. (8–48a):

$$X = \frac{m_1 x_1 + m_2 x_2}{m_1 + m_2} = \frac{(1.6 \text{ kg})(0 \text{ m}) + (1.8 \text{ kg})(1.2 \text{ m})}{1.6 \text{ kg} + 1.8 \text{ kg}}$$

$$= 0.64 \text{ m}.$$

(b) The midpoint of the rod is $x_{\text{mid}} = 0.75$ m. Let's place m_2 at a new position x_2 and ask if there is a solution for x_2 along the rod ($x_2 < L$) such that $X = x_{\text{mid}}$. We have

$$X = x_{\text{mid}} = \frac{m_1 x_1 + m_2 x_2}{m_1 + m_2} = \frac{m_1(0 \text{ m}) + (m_2 x_2)}{M} = \frac{m_2 x_2}{M}.$$

We can solve this equation for x_2:

$$x_2 = \frac{M x_{\text{mid}}}{m_2} = \frac{(1.6 \text{ kg} + 1.8 \text{ kg})(0.75 \text{ m})}{1.8 \text{ kg}} = 1.4 \text{ m}.$$

This value is indeed less than L, so if we place the 1.8-kg mass at this point, the center of mass is at the midpoint of the rod.

What Do You Think? The center of mass of the Earth–Moon system is (a) on a line that connects their centers, relatively close to the Moon or even within it; (b) on a line that connects their centers, relatively close to Earth or even within it; (c) along the line that runs through their centers, near Earth's center, on the side away from the Moon; or (d) impossible to tell because each of the two bodies is not a point.

EXAMPLE 8–12 Add a third mass of 2.3 kg to the masses of Example 8–11 at the point shown in Fig. 8–23. Find the center of mass.

▲ **FIGURE 8–23** The location of the center of mass is marked with a circled X.

Setting It Up With the first two masses aligned with the x-axis, we add a y-axis to the previous sketch (See Fig. 8–23) and place the new mass m_3 in it at the coordinates $(x_3, y_3) = (0, 1.1 \text{ m})$.

Strategy Again we have a simple calculation of the center of mass, except that the problem has become a two-dimensional one. Equations (8–48) give both X and Y.

Working It Out With $M = 1.6 \text{ kg} + 1.8 \text{ kg} + 2.3 \text{ kg} = 5.7 \text{ kg}$, we have, from Eqs. (8–48),

$$X = \frac{m_1 x_1 + m_2 x_2 + m_3 x_3}{M}$$

$$= \frac{(1.6 \text{ kg})(0 \text{ m}) + (1.8 \text{ kg})(1.2 \text{ m}) + (2.3 \text{ kg})(0 \text{ m})}{5.7 \text{ kg}}$$

$$= 0.38 \text{ m},$$

$$Y = \frac{m_1 y_1 + m_2 y_2 + m_3 y_3}{M}$$

$$= \frac{(1.6 \text{ kg})(0 \text{ m}) + (1.8 \text{ kg})(0 \text{ m}) + (2.3 \text{ kg})(1.1 \text{ m})}{5.7 \text{ kg}}$$

$$= 0.44 \text{ m}.$$

This point is indicated in Fig. 8–23.

What Do You Think? Could you have alternatively first found the center of mass of the pair 1 and 2, then combined that with object 3 to find the center of mass of the three-body system?

Center of Mass Motion in the Absence of External Forces

The velocity $\vec{V}$ of the center of mass is the time derivative of its position $\vec{R}$ [Eq. (8–47)]:

$$\vec{V} \equiv \frac{d\vec{R}}{dt} = \frac{m_1 \vec{v}_1 + m_2 \vec{v}_2 + \cdots + m_N \vec{v}_N}{M}. \qquad (8\text{–}49)$$

The numerator is the sum of the momenta of the individual pieces, so we may write this result as

$$\vec{V} = \vec{P}/M. \qquad (8\text{–}50)$$

This is of course the same relation between momentum, velocity, and mass obeyed by a point object.

We have already seen that the total momentum of an object is constant in the absence of external forces. Equation (8–50) shows that *in the absence of external forces the center of mass moves with constant velocity.* No matter how complicated the motion of the constituent objects, the motion of its center of mass is constant-velocity motion, corresponding to the motion of a point mass in the absence of net force on it (Fig. 8–24).

▲ **FIGURE 8–24** Note that the center of mass of the complex object (wrench) moves in a straight line (denoted by the white dot) with constant velocity if the net force on the object is zero.

Center of Mass Motion in the Presence of External Forces

What happens when there are external forces present in addition to interparticle forces, the ones that hold a extended object together, rigidly or otherwise? We begin by writing the acceleration of the center of mass. By taking a derivative of the velocity [Eq. (8–50)], we find the acceleration $\vec{A}$ of the center of mass:

$$\vec{A} \equiv \frac{d\vec{V}}{dt} = \frac{d^2\vec{R}}{dt^2}. \tag{8–51}$$

Multiplying by M gives the relation

$$M\frac{d^2\vec{R}}{dt^2} = \frac{d\vec{p}_1}{dt} + \frac{d\vec{p}_2}{dt} + \cdots + \frac{d\vec{p}_N}{dt} = \frac{d\vec{P}}{dt}. \tag{8–52}$$

Let's now consider an external force acting on a system. It will be enough to look at a three-body example rather than the full N-body expression of Eq. (8–52). With the three-object system shown in Fig. 8–25, we have

$$\frac{d\vec{p}_1}{dt} = \vec{F}_{12} + \vec{F}_{13} + \vec{F}_{1,\text{ext}},$$

$$\frac{d\vec{p}_2}{dt} = \vec{F}_{21} + \vec{F}_{23} + \vec{F}_{2,\text{ext}},$$

$$\frac{d\vec{p}_3}{dt} = \vec{F}_{31} + \vec{F}_{32} + \vec{F}_{3,\text{ext}}.$$

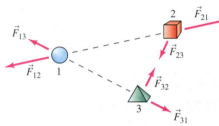

▲ FIGURE 8–25 The internal forces in a three-object system. These forces are labeled by both the object that causes them and the object on which they act. According to Newton's third law, these forces, whether they are attractive or repulsive, are equal and opposite in pairs. The external forces are not shown.

Here $\vec{F}_{i,\text{ext}}$ is the external force on object i and the $\vec{F}_{ij}$ are the internal forces (see Section 8–1). If we add these equations, using Newton's third law—according to which $\vec{F}_{12} + \vec{F}_{21} = 0$, $\vec{F}_{13} + \vec{F}_{31} = 0$, and $\vec{F}_{23} + \vec{F}_{32} = 0$—we find that all the internal forces cancel out,

$$\frac{d\vec{p}_1}{dt} + \frac{d\vec{p}_2}{dt} + \frac{d\vec{p}_3}{dt} = \frac{d\vec{P}}{dt} = \vec{F}_{1,\text{ext}} + \vec{F}_{2,\text{ext}} + \vec{F}_{3,\text{ext}} = \vec{F}_{\text{tot,ext}}. \tag{8–53}$$

Equation (8–52) and the extension of Eq. (8–53) to N objects give the general result

$$\vec{F}_{\text{tot,ext}} = \frac{d\vec{P}}{dt} = M\frac{d^2\vec{R}}{dt^2}. \tag{8–54}$$

For any system with total mass M in which there are both internal and external forces, we have shown that *the center of mass moves like a single point mass of mass M subject to the total external force on the system*. In other words, the system obeys Newton's second law

$$\vec{F}_{\text{tot,ext}} = M\vec{A}, \tag{8–55}$$

EQUATION FOR LINEAR MOTION OF THE CENTER OF MASS

where $\vec{A}$ is the acceleration of the center of mass. The internal forces play no role in the the center-of-mass motion. For example, despite the complicated motion of a twisting, somersaulting diver, the diver's center of mass moves with the same, simple parabolic motion as that of a rock thrown from the diving board with the same initial velocity as the diver's initial center-of-mass velocity.

A bowling ball rolling down an inclined plane consists of some 10^{27} molecules, all interacting through electric forces. Despite the immense complexity of the internal interactions, it is nevertheless possible to treat a bowling ball as a point mass (or as a simple rigid object, as we shall see in Chapters 9 and 10) and to study its motion without being aware of the underlying atomic structure of the ball or the laws that govern the motion of atoms. In deriving Eq. (8–54), we have seen why this is true: Newton's third law makes all the internal forces cancel out in the description of the motion of the center of mass. Thus, when we say we treat the bowling ball as a single object, we mean we describe the motion of the center of mass. This simplification of the overall motion of

◀ **FIGURE 8–26** The center of mass of the hammer (denoted by the small black dot) moves as a point mass under the influence of an external force. In this case, the external force is gravity so the center of mass has a parabolic trajectory.

an object to the motion of its center of mass is a crucial feature of our ability to describe the world around us. In fact, the idealized "point particle" motion that we discussed in the early chapters of this book can now be seen as an exact description of the center-of-mass motion of real objects. And, as we shall see in Chapters 9 and 10, this way of looking at things will allow us to break down the motion of an extended system into motion of the center of mass and, in a highly independent way, motion of the pieces of the system about the center of mass.

Figure 8–26 shows that the motion of a system's center of mass is simple even though there is a rather complex motion of parts of the system. In the next subsection we describe an important practical property of the center of mass: Assuming that our system has no net external force on it, a coordinate system with its origin at the center of mass is an inertial frame within which it is especially simple to treat collisions (interactions) of the constituents of the system.

Conservation of Momentum in Different Inertial Frames

The center of mass of a system with no external forces acting on it moves with constant total momentum, a result that generalizes Newton's first law to systems. As we saw in Chapter 4, Newton's first law is closely associated with the equivalence of inertial frames. Our generalization of the first law is accordingly a generalization of the idea of inertial frames.

We can start with a frame of reference whose origin is placed at the center of mass and moves with it. In this frame the center of mass is at rest—this is called the **center-of-mass frame**—and the total momentum is zero [Eq. (8–50)]. We can then go to another inertial frame by adding any constant velocity to the entire system (i.e., to each part of the system). In this way one can move back and forth between these inertial frames, observing the behavior of the system in the different frames. We can use this fact to great advantage in problem solving for collisions because calculations involving colliding objects may be simpler in one inertial frame than in another. In particular, the center-of-mass frame presents simplifications.

An example best illustrates how to use this technique.

EXAMPLE 8–13 Two objects of equal mass collide in what we call the laboratory frame. In this frame, one of them is moving with momentum $\vec{p} = p\hat{i}$ and the other is at rest. Assuming that you have the same two distinct objects after the collision, describe this collision in the center-of-mass frame and use your result to find the general form of the final momenta in the laboratory frame.

Setting It Up We sketch what is described in Fig. 8–27a, including xy-coordinates. We are asked here first to describe the two-body-to-two-body collision in the center-of-mass frame, then to take this general description and transform it to the laboratory frame.

Strategy We have objects moving in one inertial frame and want to see how they look in another. As we described above, moving between such frames involves a shift by a constant velocity; in

other words, we can add a constant overall velocity $\vec{v}$ to every individual particle in the problem. Although in this problem all the masses are the same, that will not always be true. In general, if the particles have initial momenta $\vec{p}_1 = m_1\vec{v}_1$ and $p_2 = m_2\vec{v}_2$, respectively, then, as viewed from the new frame, their momenta will be $m_1(\vec{v}_1 + \vec{v})$ and $m_2(\vec{v}_2 + \vec{v})$, respectively. The center-of-mass (or, as we see below, the center-of-momentum) frame is obtained if we can find $\vec{v}$ such that the sum of the new momenta is zero. This condition reads $m_1(\vec{v}_1 + \vec{v}) + m_2(\vec{v}_2 + \vec{v}) = 0$, and it is satisfied with the choice

$$\vec{v} = -\frac{m_1\vec{v}_1 + m_2\vec{v}_2}{m_1 + m_2}.$$

(continues on next page)

We can see that this is indeed the center-of-mass frame by comparison with Eq. (8–49), which shows that we are viewing the collision from the point of view of an observer moving along with the center of mass of the colliding objects. But for our purposes the most important feature is that this is the frame in which the particles come in along a straight line with equal and opposite momenta, and must leave with equal and opposite momenta along another straight line. This fact tells us as much as we can know about the final state. Particle 1 will have a final center-of-mass frame momentum $m_1\vec{v}'_{c2}$ and the momentum of particle 2 will be $m_2\vec{v}'_{c2} = -m_1\vec{v}'_{c1}$ in the center-of-mass frame.

Finally we take these most general center-of-mass frame final momenta back to the laboratory frame. Thus the momentum of particle 1 in the laboratory frame will be $m_1(\vec{v}'_{c1} - \vec{v})$, and so forth.

Let us now return to the special conditions of this problem, that $m_1 = m_2 = m$ and that particle 2 is initially at rest. If we call $\vec{p}$ the initial laboratory momentum of particle 1, then by substitution into the equations above we see immediately that the center-of-mass momentum of particle 1 is $\vec{p}/2$ and that of particle 2 is $-\vec{p}/2$. For the equal-mass case (and *only* for this case), we can go from laboratory frame to center-of-mass frame by adding to the individual momenta a constant momentum, $-\vec{p}/2$, and we go back from the center-of-mass frame to the laboratory frame by adding to the individual momenta a constant momentum, $+\vec{p}/2$.

Working It Out   We suppose we have made the transformation described above to the center-of-mass frame. As above the final momenta in this frame have the form $\vec{p}'$ and $-\vec{p}'$ (Fig. 8–27c). Without information about the conservation of kinetic energy, we cannot say much more about the magnitude p' of these momenta, and any angle is possible. The only requirement is that in the center-of-mass frame the final objects go off back to back.

To go back to the laboratory frame, we add the momentum $+\vec{p}/2$ to every part of the system. Thus in the laboratory frame the final momenta must be of the form $\vec{p}' + \vec{p}/2$ and $\vec{p}' - \vec{p}/2$, respectively (Fig. 8–27d).

What Do You Think?   We stated that any angle is possible for the final objects in the center-of-mass frame. Is that true also for the laboratory frame? If not, what determines the restriction?

▲ **FIGURE 8–27**   (a) View before collision from the frame in which ball 2 is at rest. (b) View before collision from the frame in which the total momentum is zero. (c) View after collision from the frame in which the total momentum is zero. (d) View after collision from the frame in which ball 2 was originally at rest.

Now that we have learned about the center-of-mass frame, we can understand why the term perfectly inelastic is appropriate in the case that two colliding objects "stick" and form a single final object. In the center-of-mass frame, this object has zero total momentum; hence it is at rest. The collision is correctly termed *perfectly inelastic* because in the center-of-mass frame there is *no* final kinetic energy. *All the energy of motion* in the initial state of the system goes into the "sticking together" of the objects. If the final state consists of more than one object, however slowly separating, there is some kinetic energy in the center-of-mass frame and the inelasticity will be less than "perfect."

Center of Mass of a Continuous Mass Distribution

We know that if we look closely enough matter is composed of discrete masses in the form of atoms. But to a detector that is not sufficiently acute to see atoms, such as our eyes, a solid object looks and behaves like a continuum of matter; that is, a *continuous mass distribution*. We may define the center of mass of a continuous distribution of

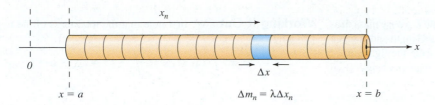

▲ **FIGURE 8–28**   A continuous mass is divided into N tiny mass segments of equal length Δx (where N is large). This technique allows us to find the center of mass. The nth segment, x_n, has mass Δm_n.

mass by a simple generalization of the center-of-mass vector equation, Eq. (8–48). Consider a thin rod of mass M lying along the x-axis (Fig. 8–28). We may take one end to lie at $x = a$ and the other at $x = b$. Now let's divide the interval between a and b into a large number N of tiny segments. The segments are of equal length, Δx, but they are not necessarily of equal mass, because the rod need not have a uniform mass distribution. The position of the nth segment is labeled x_n, and we call the mass of this tiny segment Δm_n. We may write this mass in terms of the local mass density:

$$\Delta m_n = \left(\frac{\Delta m_n}{\Delta x} \right) \Delta x. \tag{8–56}$$

The masses of all the segments add up to the total mass of the rod, so that we have

$$\Delta m_1 + \Delta m_2 + \cdots + \Delta m_n = \left(\frac{\Delta m_1}{\Delta x} + \frac{\Delta m_2}{\Delta x} + \cdots + \frac{\Delta m_n}{\Delta x} \right) \Delta x = M. \tag{8–57}$$

If we were to use infinitesimal intervals, this equation would read

$$M = \int_a^b \left(\frac{dm}{dx} \right) dx. \tag{8–58}$$

According to Eq. (8–48a), we have

$$X = \frac{\Delta m_1 x_1 + \Delta m_2 x_2 + \cdots + \Delta m_n x_n}{M}$$

$$= \frac{1}{M} \left[\left(\frac{\Delta m_1}{\Delta x} \right) x_1 + \left(\frac{\Delta m_2}{\Delta x} \right) x_2 + \cdots + \left(\frac{\Delta m_n}{\Delta x} \right) x_n \right] \Delta x.$$

As $N \to \infty$, this equation becomes an integral:

$$X = \frac{1}{M} \int_a^b \left(\frac{dm}{dx} \right) x \, dx.$$

It is usual to write the mass density dm/dx of a one-dimensional object as λ. If λ is a constant, it can come out of the integral; if it is not a constant, $\lambda = \lambda(x)$, then it must remain under the integral. (You could easily imagine that this happens because the thickness of our "one dimensional stick" varies along its length.) In this notation, the position of the center of mass is

$$X = \frac{1}{M} \int_a^b x \lambda(x) \, dx. \tag{8–59}$$

Equation (8–59) allows us to calculate the center of mass of a one-dimensional object, as in Example 8–14.

EXAMPLE 8–14 Consider a rod of length L whose mass density (mass per unit length dm/dx) varies along its length and is given by $\lambda = C(1 + ax^2)$, where x is the distance from the light end of the rod and C is a constant with dimensions of mass per length (Fig. 8–29). Calculate the center of mass of the rod.

▲ **FIGURE 8–29** A rod has varying density along its length.

Setting It Up In Fig. 8–29 we align the rod with the x-axis as shown.

Strategy This is a straightforward application of the center-of-mass formula for an object of varying density. Equation (8–59) provides us with the tool to find the center-of-mass position. To use it, we must also calculate the total mass M of the rod.

Working It Out We first note that the total mass of the rod is given by

$$M = \int_0^L \lambda(x)\, dx = \int_0^L C(1 + ax^2)\, dx$$

$$= C\left(x + \frac{ax^3}{3}\right)\Bigg|_0^L = C\left(L + \frac{aL^3}{3}\right).$$

Now, from Eq. (8–59),

$$X = \frac{1}{M}\int_0^L xC(1 + ax^2)\, dx = \frac{1}{M}C\left(\frac{x^2}{2} + \frac{ax^4}{4}\right)\Bigg|_0^L$$

$$= \frac{1}{M}C\left(\frac{L^2}{2} + \frac{aL^4}{4}\right) = \frac{C[(L^2/2) + (aL^4/4)]}{C[L + (aL^3/3)]}$$

$$= \left(\frac{L}{2}\right)\frac{1 + (aL^2/2)}{1 + (aL^2/3)}.$$

If $a = 0$, then the rod is uniform, and the center of mass is at $L/2$, as expected. In the case of nonuniformity, the center of mass is closer to the more massive end. Note that the parameter C of the mass density cancels.

What Do You Think? In all the examples we have thus far looked at, including this one, the center of mass lies within the object. Can you think of an object for which the center of mass lies outside the object?

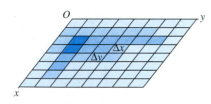

▲ **FIGURE 8–30** We have placed a two-dimensional object in the xy-plane and divided it up into squares, each of area $\Delta x\, \Delta y$. The mass of a square at a given point is the two-dimensional density σ times the area. The shading of the squares in this figure represents the value of σ, with dark blue representing high density and light blue representing low density; we see the density varying from point to point.

Continuous Objects in Two and Three Dimensions: Suppose that instead of a one-dimensional object we deal with a two-dimensional one such as a sheet of metal with an area A. We divide the object into tiny elements of area $\Delta A = \Delta x\, \Delta y$ and mass Δm (Fig. 8–30). Suppose that the two-dimensional mass density, or mass per unit area, for the element at the point (x, y) is $\sigma(x, y)$. This quantity is analogous to the one-dimensional mass density λ; σ has dimensions of mass per unit area. If we take the area element small enough so that σ is a constant within that area, then the mass of that area element is

$$\Delta m = \sigma\, \Delta x\, \Delta y.$$

Remember, the value of σ can vary from place to place, as the varying colors within Fig. 8–30 indicates. Thus σ depends on x and y. We can find the total mass by summing the masses over all the elements, and, in the calculus limit, this summation takes the form[†]

$$M = \lim \sum \Delta m = \lim \sum \sigma\, \Delta x\, \Delta y = \int_{\text{Surface}} \sigma(x, y)\, dx\, dy. \qquad (8\text{–}60)$$

Similarly, we find the coordinates of the center of mass by summing over the contributions of the elements:

$$MX = \lim \sum x\, \Delta m = \int_{\text{Surface}} x\sigma(x, y)\, dx\, dy, \qquad (8\text{–}61a)$$

$$MY = \lim \sum y\, \Delta m = \int_{\text{Surface}} y\sigma(x, y)\, dx\, dy. \qquad (8\text{–}61b)$$

For a three-dimensional object, we must know its three-dimensional mass density $\rho(x, y, z)$, measured in mass per unit volume. The dependence on x, y, z indicates that the mass density can vary from point to point. The total mass is

$$M = \int_{\text{volume}} \rho(x, y, z)\, dx\, dy\, dz, \qquad (8\text{–}62)$$

[†]The subscript "Surface" on the single integral sign denotes a two-dimensional integration over the area A; a subscript "volume" similarly indicates an integration over the full three-dimensional space.

while the center of mass is at the position

$$M\vec{R} = \int_{\text{volume}} \vec{r}\rho(x, y, z)\, dx\, dy\, dz. \qquad (8\text{–}63)$$

While the integrals above look difficult, they are normally done as a sequence of single integrations and therefore present nothing fundamentally new. The following example illustrates this.

CONCEPTUAL EXAMPLE 8–15 A square sheet of metal of area 1 m² sits as shown (Fig. 8–31a). Its mass density (mass per unit area) is constant. Find the location of the center of mass.

Answer There is a two-dimensional integral to be performed here. We approach it by taking the square and breaking it up into a series of vertical strips (Fig. 8–31b). We find the center of mass of each strip, then add the effect of each strip as if it were entirely concentrated at its own center of mass (see the subsection below for more on this point). Each strip is like the rod of Example 8–14, but with a constant (linear) mass density, so the center of mass of each strip is at its center, that is, at $y = \frac{1}{2}$ m (Fig. 8–31c). (We could have calculated this as a single integration.) When we look at this arrangement, the center of mass of each vertical strip forms a horizontal strip of uniform linear mass density, and the center of mass of that strip (again, by a single integration) is at its center, that is, at $x = \frac{1}{2}$ m. The center of mass is therefore at $(x, y) = \left(\frac{1}{2}\,\text{m}, \frac{1}{2}\,\text{m}\right)$. We have broken our double integration down into a series of two single integrations.

What Do You Think? Could you have found this answer even more quickly?

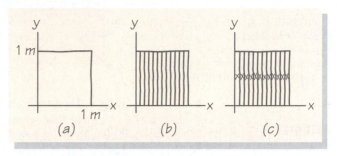

▲ **FIGURE 8–31** To find the center of mass of a sheet of metal, as in (a), we can divide the metal sheets into vertical strips, as in (b), find the center of mass of each strip (a one-dimensional problem), then replace each strip by all of its mass located at the strip's center of mass, and treat this as a horizontal one-dimensional line of masses, as in (c).

A Few Tips for Finding the Center of Mass

We have described the calculation of the center-of-mass of systems composed of discrete point masses and of continuous systems. In either case there are a few simple ideas to keep in mind that will help you to do the calculations—we have already used these in our calculations above, although we did not single them out.

Symmetry Helps: The center of mass of two equal masses lies midway on the line between them, the center of mass of a square flat uniform plate lies at the center of the square, and the center of mass of a sphere lies at its center if the distribution of its mass is spherically symmetric (no direction from the center different from any other direction).

How Subsystems Enter: You may have a system that you can think of as composed of subsystems, themselves distributions of mass. The center of mass of the whole system can be found by first finding the center of mass of the subsystems and then combining these into a final center of mass as if each subsystem were a particle (a point mass) with the total subsystem mass located at the subsystem's center of mass. (We used this in Conceptual Example 8–15 when we treated each strip as if it were a particle located at its own center of mass.) This result follows because of the simple linear nature of the calculation of the center-of-mass position and can be easily understood with a three-particle example in one dimension: Three particles with masses m_1, m_2, and m_3 are placed along the x-axis at positions x_1, x_2, and x_3, respectively. The center-of-mass position is

$$X = \frac{m_1 x_1 + m_2 x_3 + m_3 x_3}{m_1 + m_2 + m_3}.$$

Now we consider the "two-body" system consisting of particle 1 alone and particles 2 and 3 taken as a subsystem located at X_{23}, the center of mass of the 2–3 subsystem. We have $X_{23} = (m_2 x_2 + m_3 x_3)/(m_2 + m_3)$, and the expression of the full center of mass in this view is

$$X = \frac{m_1 x_1 + (m_2 + m_3)X_{23}}{m_1 + (m_2 + m_3)}.$$

It is no trouble whatever to verify that our two expressions are identical, and this result easily generalizes.

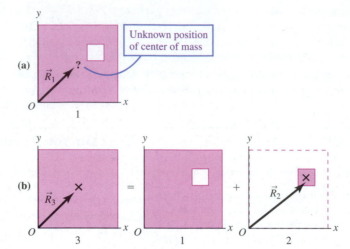

▶ **FIGURE 8–32** To find the center of mass of an object (1) with a hole in it, think of the object without the hole (3) as consisting of the original object plus another object that would fill the hole (2).

How to Deal with Holes: Sometimes it is easier to think of an object in terms of what isn't there than what is. We can use the paragraph above, How Subsystems Enter, to deal with this; again we'll proceed by example. Consider the flat plate with a hole cut in it, as in Fig. 8–32a. We'll label this system 1. We can think of this distribution entering as one piece of a full plate, without the hole, labeled as 3 in Fig. 8–32b; we find the center of mass of plate 3 by combining plate 1 with the material from the hole, labeled as 2. We know the center of mass of plate 3 and plate 2—by symmetry these points are at their centers and are marked $\vec{R}_2$ and $\vec{R}_3$ in Fig. 8–32b. We have

$$\vec{R}_3 = \frac{m_1\vec{R}_1 + m_2\vec{R}_2}{m_1 + m_2},$$

and we can solve this to find the unknown center of mass of the system with the hole,

$$\vec{R}_1 = \frac{(m_1 + m_2)\vec{R}_3 - m_2\vec{R}_2}{m_1}.$$

This technique is quite useful.

*8–7 Rocket Motion

A rocket is a system that undergoes a kind of continuous explosion (see Section 8–3). To understand rocket motion, imagine an astronaut who finds himself or herself at rest with respect to the vehicle but separated from it. No amount of bodily contortions will change the fact that the astronaut's center of mass remains at rest in the absence of external forces. However, the astronaut could, for example, throw a hammer in a direction away from the space vehicle. Then he or she will move toward the space vehicle and safety (Fig. 8–33). Momentum conservation lies behind this motion. The total momentum of the astronaut–hammer system is initially zero, and this remains true after the hammer is thrown. A rocket moves through the same principle. The ejected rocket fuel acts as a series of thrown hammers. The presence or absence of an atmosphere has nothing to do with the basic principle of operation.

A rocket is propelled by the ejection of hot gases from burning chemical fuel. By the "rocket" we mean the forward section containing the payload and any remaining unburned fuel. Two important parameters describe the rocket (Fig. 8–34). First there is the

▲ **FIGURE 8–33** An astronaut can get a small push back toward the space shuttle by throwing a hammer in the opposite direction.

▶ **FIGURE 8–34** A rocket expels gases that have a speed u_{ex} with respect to the rocket. If we know the rate at which the gases are expelled, conservation of momentum allows us to find the velocity $\vec{v}$ of the rocket.

exhaust speed of the hot gases, u_{ex}, which by definition is positive. Keep in mind that the exhaust speed involves a burning reaction within the rocket and that it has a given value *with respect to the rocket*. Second, there is the rate of change of mass dm/dt of the rocket, that is, the rate at which the rocket expels the exhaust gases. Don't confuse these parameters; you could have two rockets that each burn fuel by the same reaction and for which u_{ex} is identical, but one could burn it at a very different rate than the other and therefore expel exhaust at a very different rate. In our discussion below, we'll suppose that all the motion is in one direction. If this direction is along a particular axis, any "velocity" is the component of the velocity along that axis. That component could be positive or negative.

Suppose now that the initial mass of the rocket with all its fuel is m_0; at some later time t, its velocity with respect to some ground-based (inertial) observer is v while its mass has been reduced to m. In the next small time interval Δt, the observer sees that a small bit of exhaust gas, mass Δm, has been ejected backward relative to the rocket; the observer sees this bit of gas moving with velocity $v - u_{ex}$, while he sees the rocket itself now traveling at a velocity $v + \Delta v$ and with a decreased mass $m - \Delta m$.

The conservation of momentum requires that the momentum change of the rocket be matched by the momentum carried away by the gas. In the absence of external forces, the total momentum at time t must equal the total momentum at time $t + \Delta t$. We have $P(t) = mv$, while $P(t + \Delta t) = (\Delta m)(v - u_{ex}) + (m - \Delta m)(v + \Delta v)$. The last two terms are the momentum of the exhaust gas and the rocket, respectively. Thus $P(t) = P(t + \Delta t)$ reads

$$mv = (\Delta m)(v - u_{ex}) + (m - \Delta m)(v + \Delta v)$$
$$= (\Delta m)v - (\Delta m)u_{ex} + mv + m\,\Delta v - (\Delta m)v - (\Delta m)(\Delta v),$$

or, neglecting the doubly small term $(\Delta m)(\Delta v)$,

$$m\,\Delta v - (\Delta m)u_{ex} = 0. \tag{8–64}$$

Divide Eq. (8–64) by the time interval Δt and take the limit $\Delta t \to 0$. Rearranging, we find $u_{ex}(dm/dt) = m(dv/dt)$. Now we are looking for an equation for the rocket. But the quantity dm/dt refers to the (positive) rate at which the ejected fuel mass increases. We can get an equation for the rocket itself by noting that the rate at which the ejected fuel mass changes is the negative of the rate at which the rocket mass changes. Thus if we change the sign of the dm/dt term we can interpret dm/dt as the rate of mass change of the rocket itself:

$$-u_{ex}\frac{dm}{dt} = m\frac{dv}{dt}. \tag{8–65}$$

This equation looks just like Newton's second law $F = ma$. The left side is the force term, the **thrust** of the rocket, a positive quantity. It contains both of our rocket parameters.

Equation (8–65) may be written in a form that leads directly to a solution for the speed of the rocket as a function of its remaining mass by dividing both sides by mu_{ex}:

$$-\frac{1}{m}\frac{dm}{dt} = \frac{1}{u_{ex}}\frac{dv}{dt}. \tag{8–66}$$

We now use

$$\frac{d(\ln m)}{dt} = \left(\frac{1}{m}\right)\left(\frac{dm}{dt}\right). \tag{8–67}$$

[This formula follows from the calculus results in Appendix IV–7. We use $d(\ln x)/dx = 1/x$. We must replace x by m and use the chain rule of calculus.] Equations (8–66) and (8–67) imply

$$\frac{d}{dt}\left(\ln m + \frac{v}{u_{ex}}\right) = 0.$$

The quantity in parentheses must then be a constant,

$$\ln m = -\frac{v}{u_{ex}} + \text{a constant.} \tag{8–68}$$

Equation (8–68) describes the rocket's speed as a function of its mass. Assuming the rocket starts from rest at $t = 0$, we have at that time $m = m_0$ and $v = 0$. That means that the constant in Eq. (8–68) is given by $\ln m_0$. Equation (8–68) is then rearranged to give

$$v = u_{ex}[\ln m_0 - \ln m] = u_{ex} \ln\left(\frac{m_0}{m}\right), \tag{8–69}$$

ROCKET SPEED

where in the last step we have used the fact that $(\ln a) - (\ln b) = \ln(a/b)$. This is a fundamental equation in rocket propulsion.

What does Eq. (8–69) tell us? First of all, we note that the speed v is proportional to u_{ex}. This dependence is to be expected because u_{ex} is the only quantity with the dimensions of speed that appears in the problem. Next, we observe that the coefficient of u_{ex} must be a dimensionless quantity, and therefore it must be a function of m/m_0, which is the fraction of the rocket mass left over after time t. Because the logarithm is a very slowly varying function of its argument, the most productive way to get a sizable speed is by making u_{ex} large. Chemical rockets can generate gas exhaust speeds up to a maximum of about 4000 m/s, and thus the typical speeds of rockets are also in the range of 10^3 m/s. Much higher exhaust speeds can be achieved with nuclear reactions, but rockets that use nuclear energy for propulsion in this way present enormous environmental problems. A different way to increase the speed of the rocket would be to carry more fuel, that is, increase m_0/m. But large values of m_0/m help only moderately; for example, $\ln 10 = 2.3$, so changing the mass ratio by a factor of 10 improves the speed by a factor of 2.3.

Rocket Motion in the Presence of Gravity

The previous discussion involves a rocket free of all external forces. When gravity acts on the rocket, Eq. (8–69) is modified. In Problem 92 we outline the steps that lead to the conclusion that the velocity (assumed one-dimensional as before, here vertical) includes a new term, $-gt$, in addition to the term proportional to $u_{ex} \ln(m_0/m)$ due to the fuel exhaust. (The new term corresponds to free fall under the influence of gravity, as we already know from Chapter 2. Thus Eq. (8–69) is modified to

$$v = u_{ex} \ln(m_0/m) - gt. \tag{8–70}$$

EXAMPLE 8–16 A Saturn V rocket (the vehicle that sent humans to the Moon) of mass 2.5×10^6 kg takes off from Earth in a vertical direction. It burns fuel at a uniform rate of 1.6×10^4 kg/s for a duration of 2 min. If the exhaust speed of the gas is given by 3.0 km/s, what is the speed of the rocket immediately after the combustion ceases?

Setting It Up We know the rocket's initial mass m_0 as well as the rate dm/dt at which it loses mass through the burn; we also know the burn time t and the exhaust speed u_{ex}.

Strategy This is a direct application of the equation that describes rocket speed, Eq. (8–70). All the parameters of the right side of that equation are known, except that we need to calculate the final mass, a calculation that involves a known starting mass and a given rate of mass loss for a given time.

Working It Out The mass m in Eq. (8–70) is the final mass of the rocket after the gas has been burned,

$$m = m_0 - \frac{dm}{dt}t = 2.5 \times 10^6 \text{ kg} - (1.6 \times 10^4 \text{ kg/s})(120 \text{ s})$$

$$= 0.58 \times 10^6 \text{ kg.}$$

We also have

$$\ln\left(\frac{m_0}{m}\right) = \ln\left(\frac{2.5 \times 10^6 \text{ kg}}{0.58 \times 10^6 \text{ kg}}\right) = \ln\left(\frac{2.5}{0.58}\right) = 1.5.$$

Thus, from Eq. (8–70),

$$v_f = (3.0 \times 10^3 \text{ m/s})(1.5) - (9.8 \text{ m/s}^2)(120 \text{ s})$$

$$= 3.2 \times 10^3 \text{ m/s.}$$

Our calculations are correct only near the surface of Earth. Earth's gravitational force on the rocket decreases as the rocket moves away from the surface, so in effect g decreases as the rocket moves away.

What Do You Think? If you were interested in sending a rocket such as this far far away, would it help to launch it from the Moon rather than from Earth?

*8–8 Momentum Transfer at High Energies

High-energy physics is the study of the properties of the elementary particles that make up matter at the most fundamental level. These properties are studied with the aid of high-energy accelerators. There, projectile particles such as protons or electrons are accelerated to high speeds and allowed to collide with target particles of the same type. The results of such collisions are then analyzed by means of sophisticated detectors in which the energies and momenta of the particles that emerge after the collisions are precisely measured. The patterns revealed in these measurements are used to probe the underlying structure of the colliding particles.

THINK ABOUT THIS . . .

HOW DO WE KNOW THAT ATOMS HAVE A NUCLEAR STRUCTURE?

The fact that the atom has a massive central structure was discovered in experiments carried out by Ernest Rutherford in the first decade of the twentieth century. These experiments involved the scattering of projectiles known as α particles (later these were discovered to be themselves the nuclei of helium atoms) from atoms. Imagine a BB gun fired at two targets, one of cotton candy and the other of cotton candy but with a marble implanted at its center. A BB shot (the projectile) will hardly be deflected by the first target. Most of the time, the same will be true for the second target. Occasionally, however, the projectile will penetrate the second target close to its center and, upon colliding with the marble, will be deflected through a large angle. An analysis of the collisions in the two cases would allow us to deduce that the second target had a hard, compact center (the marble). The pattern of many projectile deflections would even allow us to say whether the marble is spherical or cubical. Rutherford's experiments were similar, with the nucleus playing the role of the marble in the cotton candy. His projectile was occasionally deflected through large angles, and analysis of the pattern revealed a dense small center, the atomic nucleus.

There is a benefit in increasing the speed—or momentum—of the projectile in scattering experiments, a benefit that the rules of classical physics could not have predicted. For collisions of atomic or subatomic particles, the rules of *quantum mechanics* apply. This is a field of study developed in the first part of the twentieth century, with some surprising consequences. One consequence of quantum mechanics is that high momenta (or high energy) are *required* to study the regions close to the center of a target. More precisely, the **Heisenberg uncertainty relation**, discovered by physicist Werner Heisenberg, describes a relation between the magnitude of the momentum change, $\Delta\vec{p}$, of a target particle in a collision and the size of the region around the center of the target that can be studied. Let us denote by r_{coll} the smallest radius of a spherical region around the center of a target that can be studied in a collision. Then the uncertainty relation is

$$|\Delta\vec{p}|r_{\text{coll}} \cong \hbar, \qquad (8\text{–}71)$$

where $\hbar \cong 10^{-34}$ J $\cdot$ s is Planck's constant divided by 2π. According to Eq. (8–71), the larger the value of $\Delta\vec{p}$, the smaller r_{coll} can be.

Experiments with elementary particles—like the BB gun fired at cotton candy—have been carried out and have revealed much about the structure of these particles. We have already mentioned the structure of the atom itself in the Think About This box. More recently, experiments involving protons colliding at momenta that could probe down to distances of 10^{-16} m revealed that the proton, which is itself a component of the nucleus, contains "pellets" known as *quarks*. When a next generation of accelerators is built, we will discern even more detail. Judging by the enormous difference between the physics of atoms and the physics of nuclei, we can expect some surprises as we probe more and more deeply.

Summary

The momentum (or linear momentum) of an object of mass m moving with velocity $\vec{v}$ is defined by

$$\vec{p} \equiv m\vec{v}. \tag{8–2}$$

The rate of change of momentum with time is given by Newton's second law:

$$\vec{F} = \frac{d\vec{p}}{dt}, \tag{8–3}$$

where $\vec{F}$ is the net force acting on the object. The momentum of a collection of objects with momenta $\vec{p}_i$ (where $i = 1, 2, \ldots, N$) is their vector sum, and the net external force on the system determines the rate of change of the total momentum. In particular, if the only forces present are the internal forces exerted by the objects on each other, then the total momentum does not change with time; that is, momentum is conserved. In the case of two objects, this relation is

$$\text{for zero net external force: } \vec{p}_1 + \vec{p}_2 = \text{a constant.} \tag{8–9}$$

One consequence of Eq. (8–3) is that the impulse $\vec{J}$, the change in momentum of an object influenced by a force that acts over a limited time (an impulsive force), is an integral of the force over time:

$$\Delta\vec{p} = \vec{J} = \int \vec{F}\, dt = \vec{F}_{av}\, \Delta t. \tag{8–15}$$

The impulse does not involve the detailed time dependence of the force.

Collisions involve the (generally brief) interactions between pieces of an isolated system. Since the system is isolated, the total momentum is conserved, and the vector equations that express this conservation law is one set of constraints on the kinematics of the particles involved in the collision. We can also distinguish collisions by whether energy is or is not dissipated in the process. Accordingly, collisions may be elastic, inelastic, or perfectly inelastic. In elastic collisions—the collisions of billiard balls is a close approximation—the sum of the kinetic energies is conserved. In inelastic collisions the sum of kinetic energies in the initial state, before the collision, differs from the sum of kinetic energies in the final state. The change is generally a decrease due to the presence of dissipative forces such as friction, or by deformations, but an increase in kinetic energy is also possible, as in explosions. The maximum amount of kinetic energy is lost in perfectly inelastic collisions, where the objects that collide merge into a single object.

The center of mass of a collection of objects with masses m_i and position vectors $\vec{r}_i$ $(i = 1, 2, \ldots, N)$ is the mean position of a system's mass, which is defined by

$$\vec{R} \equiv \frac{m_1\vec{r}_1 + m_2\vec{r}_2 + \cdots + m_N\vec{r}_N}{M}. \tag{8–47}$$

In the absence of external forces, the center of mass moves with uniform velocity $\vec{V}$, given by

$$\vec{V} \equiv \frac{d\vec{R}}{dt} = \frac{m_1\vec{v}_1 + m_2\vec{v}_2 + \cdots + m_N\vec{v}_N}{M} = \frac{\vec{P}}{M}. \tag{8–49, 8–50}$$

The total mass of the system is $M = m_1 + m_2 + \cdots + m_N$. The center of mass for a continuous object may be calculated in terms of the mass density. For a one-dimensional object with mass density $\lambda(x)$, it is given by

$$X = \frac{1}{M}\int_a^b x\lambda(x)\, dx, \tag{8–59}$$

where the mass is related to $\lambda \equiv dm/dx$ by

$$M = \int_a^b \lambda(x)\, dx. \tag{8–58}$$

For a two-dimensional object with mass density $\sigma(x, y)$, the corresponding expressions are

$$X = \frac{1}{M}\int_{\text{Surface}} x\sigma(x, y)\, dx\, dy \tag{8–61a}$$

and

$$Y = \frac{1}{M}\int_{\text{Surface}} y\sigma(x, y)\, dx\, dy, \tag{8–61b}$$

where

$$M = \int_{\text{Surface}} \sigma(x, y)\, dx\, dy. \qquad (8\text{--}60)$$

Similar expressions hold for a three-dimensional object with mass density $\rho(x, y, z)$.

When a net external force $\vec{F}_{\text{net}}$ acts on a system, the center of mass continues to move in a simple way, as described by Newton's second law: The position of the center of mass $\vec{R}$ obeys the equation

$$\vec{F}_{\text{tot, ext}} = \frac{d\vec{P}}{dt} = M\frac{d^2\vec{R}}{dt^2} = M\vec{A}. \qquad (8\text{--}54,\ 8\text{--}55)$$

Here $\vec{A}$ is the acceleration of the center of mass.

The concept of momentum conservation is useful for the description of rocket motion. For a rocket that is not influenced by external forces, whose mass is initially m_0 and whose mass at time t is m, the speed is

$$v = u_{\text{ex}} \ln\left(\frac{m_0}{m}\right). \qquad (8\text{--}69)$$

Here, u_{ex} is the exhaust speed of the gas relative to the rocket.

Understanding the Concepts

1. A comet enters the solar system, is deflected, and then leaves the solar system in a direction different from the direction by which it entered. Is momentum conserved in this situation?

2. A common toy consists of a series of five balls that touch and form a line; each ball is suspended as a pendulum. When ball 1 is pulled away and then released, it strikes the line and ball 5 rises in a motion like the reverse of ball 1. How can you explain this motion?

3. Two pucks collide on a table top and eventually come to rest. Does this mean that their center of mass comes to rest? If so, does this violate the principle that the center of mass of an isolated system moves with constant velocity?

4. Describe a physical object for which the center of mass is not actually inside the object.

5. A diver leaps off the diving board, performing a difficult series of maneuvers. Can he cause his center of mass to perform a midair loop?

6. You have two masses, labeled 1 and 2, moving in space, and their momenta are equal and opposite, $\vec{p}_1 + \vec{p}_2 = 0$. Pick out the statements that are necessarily true.
   (a) The two masses are going to collide.
   (b) The masses could be moving in two planes that are parallel to one another.
   (c) The masses are moving parallel to one another but not necessarily along the same line.
   (d) The masses are moving along the same line.

7. The center of mass of a championship-level high jumper passes below the bar even though the jumper passes above the bar. How is this possible?

8. A particle collides with another particle at rest. If there are two particles observed to come out of the collision, do the momenta of the two final state particles and the initial incoming particle have to lie in a plane? Why or why not? What if the final system consists of three particles?

9. A piece of machinery is modeled as a thin circle of iron, radius R, lying in the xy-plane and centered at the origin, with a second point mass on the x-axis at the point $x = R$. Is the center of mass within the circle? How would you find the position of the center of mass of this system with the least calculation?

10. A vase falls to the floor and shatters. Is momentum conserved in the collision? What objects need to be taken into account in describing the conservation of momentum?

11. As a tennis racket hits a tennis ball, the racket continues to move forward. Is this consistent with the conservation of momentum?

12. If a tennis player wants more power, should he or she choose a racket with more or less tension on the strings? If the player wants more control? [*Hint:* Think about how long the ball is on the strings in each case.]

13. The mass on the end of a pendulum swings in part due to the effects of gravity on the mass. Does momentum have to be conserved in the interaction of this mass with Earth? If so, does this mean that Earth moves back and forth along with the mass?

14. A very small child and a large adult want to use a see-saw, a board with a pivot at its midpoint. By comparison to the position of the child, should the adult sit (a) closer to the midpoint, (b) farther from the midpoint, or (c) at the same distance from the midpoint? Justify your answer.

15. We have said that an impulsive force acts for only a short time. What decides whether a time is "short" or not?

16. Cricket players catch balls hit as hard as baseballs but they do not use padded mitts. How do they avoid injuring their hands?

17. An amusing experiment involves the use of a golf ball (or any other small, light ball) and a large ball. If you keep the large ball stationary on the ground and drop the small ball onto it, the small ball will bounce back to almost the height from which you dropped it. If, however, you take both balls and, holding the smaller ball just above the larger one, drop them together, you will find that the small ball might hit the ceiling with some force! Try this and explain what is going on.

18. If you have the misfortune to be in an automobile collision, you are better off in a more massive car (all other things being equal). Why?

19. In films with naval battles, especially involving pirate ships, the guns that fire always roll back. Why is that?

20. If you have the misfortune to be in an automobile collision, you are better off in a car that tends to crumple on impact rather than a car that holds together stiffly. Why?

21. A closed railroad car is at rest on a flat stretch of track. A cannon located inside the car at the front end points to the rear. What is the motion of the railroad car when the cannon is fired and the shell is absorbed by the rear wall?

22. A large, closed crate contains many pigeons that sit on the floor of the crate. A sudden noise makes them all fly up and hit the top of the crate at the same time. Will they be able to lift the crate off the ground?

23. Drag forces act when a parachute opens. The parachute and its load will slow down to a terminal velocity. Why is the drag force velocity dependent? Think of the parachute as colliding with a lot of tiny air molecules.

24. Suppose two identical billiard balls moving with equal and opposite velocities along a line have a head-on collision. How will this look to an observer moving with a uniform velocity equal to that of one of the billiard balls?

25. A billiard ball strikes the cushion of a billiard table at an angle of 45°. Assuming that there is no energy loss (and no effects due to the spin of the ball), what can you say about the angle at which the billiard ball bounces off? What would happen if there were some energy loss?

Problems

8–1 Momentum and Its Conservation

1. (I) Calculate the magnitudes of the momenta of (a) a 40-g arrow traveling at a speed of 110 km/h; (b) a 145-g baseball traveling at a speed of 35 m/s; (c) a 72-kg sprinter running at 22 mi/h; (d) a 95-kg tackler running 100 m in 12.5 s.

2. (I) A 1.65-kg mass falls vertically downward from a roof 27.5 m high. What is the momentum of the object after 1.6 s given that the initial velocity is zero?

3. (I) Calculate the magnitudes of the momenta of (a) a man of mass 70 kg, running 6 m/s; (b) a freight car of 100,000 kg, moving 60 m/s; (c) a car of 1100 kg, moving 25 mi/h; (d) a proton moving at 2×10^5 m/s; (e) a feather of 10 g in an airless container that has fallen 10 cm due to gravity.

4. (II) An object of mass m is constrained to move in a circle of radius R by a central force F. (a) What is the magnitude of the momentum of the object? (b) Suppose that the force has magnitude $F = Kv$, where K is a constant and v is the speed of the object. Calculate R in terms of K and the momentum of the object. (This kind of force acts on a charged object in a uniform magnetic field.)

5. (II) A 7-kg rifle is used to fire a 10-g bullet that travels with a speed of 700 m/s. (a) What is the speed of recoil of the rifle? (b) How much energy does it transmit to the shoulder of the person using the rifle as it stops?

6. (II) Two objects, of masses m and M, respectively, move in circular orbits that have the same center in such a way that they are always at opposite sides of a diameter (Fig. 8–35). If the force that gives rise to this motion is a force of attraction between the two objects acting along a line joining them, using momentum conservation and Newton's second law, (a) show that they move with the same angular speed (see Chapter 3) and (b) calculate the ratio of the radii of the two circular orbits.

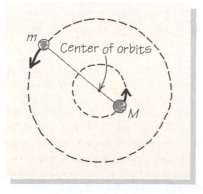

▲ **FIGURE 8–35** Problem 6.

7. (II) Two objects are moving in the xy-plane. The first, with mass 2.4 kg, has a velocity $\vec{v}_1 = (-2.0 \text{ m/s})\hat{i} + (-3.5 \text{ m/s})\hat{j}$; the second object, with mass 1.6 kg, has velocity $\vec{v}_2 = (1.8 \text{ m/s})\hat{i} + (-1.5 \text{ m/s})\hat{j}$. (a) What is the total momentum of the system? (b) If the system observed at a later time shows that the 2.4-kg object has $\vec{v}_1' = (2.5 \text{ m/s})\hat{i}$, what is the velocity of the 1.6-kg object? (c) Consider again the initial situation. Now suppose that there has been a mass transfer so that the first object now has a mass of 2.1 kg. The total mass is conserved. What is $\vec{v}_1'$ if the velocity of the second object is $(-2.5 \text{ m/s})\hat{j} + (1.3 \text{ m/s})\hat{k}$? (d) Calculate the sum of the kinetic energies in the initial configuration and in the configurations of parts (b) and (c). Compare and discuss them.

8. (III) (a) A billiard ball collides head on with a second billiard ball that is touching a third billiard ball placed directly behind it. Describe the subsequent motion of all three billiard balls. Assume that all collisions are elastic. (b) A moving billiard ball hits the two balls at rest precisely between the two touching balls at rest. The velocity of the moving ball is perpendicular to the line between the centers of the two balls at rest. Again, all collisions are elastic. Describe the subsequent motion now.

8–2 Collisions and Impulse

9. (I) A baseball of mass 0.15 kg moving horizontally with a momentum of 4.0 kg · m/s is struck head on by a baseball bat with an impulse of 10 N · s. What is the speed of the baseball after it is struck?

10. (I) If a 32-g arrow moving at 160 km/h penetrates a block of wood suspended by a rope, what impulse is delivered to the block? What velocity will the wood block acquire if its mass is 3.5 kg? Ignore the mass of the arrow compared to that of the wood block.

11. (I) A 145-g baseball traveling at a speed of 36 m/s hits a bat and moves back along its incoming trajectory with a speed of 45 m/s. What is the impulse delivered to the ball by the bat? If the duration of the bat–ball collision is 7.0×10^{-4} s, what is the average force exerted by the bat on the ball during this period?

12. (I) A 440-g ball is dropped from a height of 3.2 m onto a hard floor and bounces back to exactly that height. (a) What is the impulse received by the ball? (b) What is the average force on the ball during the 0.008 s during which, an independent measurement shows, it was in contact with the floor?

13. (I) A fire rescue unit uses a tightly woven net to catch an 80-kg person who jumps out of a burning building from a height of 11 m. What is the impulse transmitted to the net? If the net sinks 70 cm as it slows down the jumper, what is the average force exerted on the jumper by the net?

14. (II) (a) Estimate the possible impulse received by a nail when it is hit by a hammer. (b) Estimate the possible impulse received by a football when it is kicked in a kickoff. The ball has mass of about 250 g and travels a horizontal distance of some 50 m. Note the difference between parts (a) and (b); in the latter, you have additional information on the momentum the ball receives.

15. (II) A high jumper of mass 55 kg clears 6 ft 7 in. What is the impulse transmitted to the jumper by the ground? (It has been found that the center of mass of the high jumper remains *below* the bar. In this problem, assume that the center of mass starts at 3 ft 10 in and reaches the height 6 ft 7 in.)

16. (II) A communications satellite of mass 850 kg can be ejected from the cargo bay of the space shuttle by means of springs. A particular satellite is ejected at 0.45 m/s. (a) What impulse does the spring provide? (b) If the spring operates over a time period of $\Delta t = 0.85$ s, what average force does the spring provide?

17. (II) A ball of mass 260 g is dropped from a height of 2.0 m. It hits the ground and rebounds to a height of 1.4 m. Assuming that the ball is in contact with the ground for 0.004 s, what is the average force exerted on the ball during the contact?

18. (II) A golf ball of mass 0.05 kg placed on a tee is struck by a golf club. The speed of the golf ball as it leaves the tee is 100 m/s. (a) What is the impulse? (b) If the time of contact between the club and the ball is 0.02 s, what is the average force? (c) If the force decreases to zero linearly with time during the 0.02 s, what is the value of the force at the beginning of the contact?

19. (II) A catcher catches a 45-m/s pitch and his glove recoils 0.25 m. The mass of the ball is 0.14 kg. Assume that the deceleration of the ball during the catching time, Δt, is constant and equal to some a_{av}. (a) Find the magnitude of the average force F_{av} that the catcher exerts. (b) How much work does he do? (c) Find Δt.

20. (II) Superman rushes to save Lois Lane, who has fallen (has been pushed?) from a window 65 m above a crowded street. Superman swoops down in the nick of time, arriving when Lois is 1.0 m above the street and stopping her just at ground level. Lois has a mass of 52.5 kg. Ignore air resistance throughout. (a) What is the impulse that Lois receives as Superman catches her? (b) If the force that Superman supplies in stopping Lois is constant, how long does it take for Lois to come to rest? (c) What is the average force Superman applies to Lois? Compare this to the force of gravity on her. Draw your own conclusions about these last-minute rescues.

21. (II) A ball of mass 150 g is dropped from a height of 60 cm onto the first step at the top of a staircase. Each step is 20 cm above the next. The ball bounces perfectly elastically, but it has a small horizontal velocity, so it hits the second step on the next bounce, and then the third, and so on. Assume that the size of each step is such that the ball always bounces onto the next step down. What is the impulse transmitted to the nth step?

22. (II) A golfer playing on a level course hits a ball that leaves the tee at a 30° upward angle and hits the ground 175 m away. What was the impulse given to the ball by the swinging club? If you need the information, you can estimate the mass of the golf ball.

23. (II) Two apparently identical balls of the same mass 120 g are dropped from rest 1.5 m above a floor. One of them is a "super-ball," and when this ball hits the floor, it bounces up such that its speed immediately after leaving the floor is equal to its speed just before it hits. The other ball hits the floor with a thud and just stops there. What is the impulse received by the floor in each case?

24. (II) A brother and sister live near a lake. In the middle of winter, when the lake is solidly and smoothly frozen, they play catch by sliding a rock back and forth between them across the ice. Ignore all friction. The mass of the two children is 40 kg each, and the rock has a mass of 2.0 kg. Each time the rock is given a push by the child who has just caught it, it leaves the hand with a speed of 1.0 m/s (relative to the hand). (a) The children are initially at rest, and the sister first slides the rock toward her brother. What is the initial motion of the sister after she releases the rock? (b) The brother receives the rock. What is his initial velocity after he has received it? (c) The brother then slides the rock to his sister. What is his velocity now? (d) The sister catches the rock. What is her velocity now? (e) Will either child ever have a speed such that the rock will never reach the other player? In other words, can the game go on forever?

8–3 Perfectly Inelastic Collisions; Explosions

25. (I) A 250-g cart moves on an air track (a one-dimensional system in which motion is friction free) at 1.2 m/s. It collides with and sticks to another cart of mass 500 g, which was moving in the opposite direction at 0.80 m/s before the collision. What is the velocity of the composite cart after the collision?

26. (I) An object of mass m_0 with speed v_0 hits another object of mass m at rest. The two masses stick together in the process. Find the fractional change in kinetic energy.

27. (I) A snowball of mass 400 g is thrown with a speed of 10 m/s and hits the loosely placed head of a snowman. If the snowball sticks to the head, which is of mass 5 kg, with what initial speed will the now enlarged head recoil?

28. (I) A fireworks rocket is shot straight up in the air. Its "payload," mass 3.80 kg, explodes just as it reaches its peak height and separates into two fragments before eventually exploding again into spiraling colors. The first fragment, of mass 1.1 kg, heads straight down with an initial speed of 15 m/s. What is the velocity of the second fragment immediately after the explosion?

29. (I) Two objects each move with speed v in opposite directions along a line. They meet and have a perfectly inelastic collision. After the collision, the composite object moves along the same line with a speed of $v/2$. What is the ratio of the masses of the two objects?

30. (I) Two automobiles have a perfectly inelastic collision. The automobiles, which are identical models except for the color and which contain only drivers of identical masses, meet at an intersection. Each was moving with speed 12 mph, one coming from the south, the other from the east. What is the velocity of the final composite object?

31. (II) An asteroid of diameter 1.5 km and density of 7.1×10^3 kg/m^3 traveling at 25 km/s hits the Moon in a completely inelastic collision. Estimate the maximum fractional change in velocity of the Moon in its orbit as a result of this impact.

32. (II) An ^{241}Am nucleus at rest emits an alpha particle (a ^{4}He nucleus). The energy released during the process is 6 MeV = 9.6×10^{-13} J. (a) What are the speeds of the alpha particle and the remaining ^{237}Np nucleus? (b) What is the kinetic energy of the ^{237}Np nucleus? (The masses of the nuclei may be taken to be $241m_0$, $4.0m_0$, and $237m_0$, respectively, where $m_0 = 1.66 \times 10^{-27}$ kg.)

33. (II) In a *ballistic pendulum*, a bullet of mass m and speed v embeds in a block of mass M suspended by a string (Fig. 8–36). The block and bullet, of total mass $m + M$, then move as a pendulum and the maximum height h that it reaches can easily be

measured. This apparatus can be used to measure the speed of the bullet. (a) What fraction of the bullet's kinetic energy is lost in the collision? (b) Give a formula for measuring v in terms of m, M, g, and h.

▲ **FIGURE 8–36** Problem 33.

34. (II) A bullet, mass 18 g, strikes a ballistic pendulum (see Problem 33) whose target block has a mass of 1.8 kg. The block is observed to rise to a height of 4.5 cm. What was the bullet's speed?

35. (III) Two persons, one of mass m_1 and the other of mass m_2, are sitting initially at rest on a bobsled of mass M on a frozen lake. Answer the following questions for a frame fixed to the lake. (a) What is the final velocity of the bobsled if the person of mass m_1 jumps off the rear of the bobsled with horizontal velocity v followed a few seconds later by the second person, who also jumps off in the same way with the same horizontal velocity with respect to the sled? (b) What is the velocity of the bobsled if the person of mass m_2 jumps off first? (c) What is the velocity if both jump off at the same time?

8–4 Elastic and Inelastic Collisions

36. (I) An object with velocity $(1.4 \text{ m/s})\,\hat{i}$ and mass 0.30 kg collides with an object whose velocity is $(-2.5 \text{ m/s})\,\hat{i}$ and whose mass is 0.15 kg. The motion takes place in one dimension. (a) What are the final velocities of the objects if the collision is elastic? (b) What is the total initial kinetic energy in the collision?

37. (I) On an air track, a 0.4-kg mass m_1 moves at 3.0 m/s in the positive direction. It approaches a stationary mass m_2 of 0.8 kg. They collide and, after the collision, the velocity of mass m_2 is 1.6 m/s in the positive direction. (a) What is the velocity of mass m_1 after the collision? (b) Is this collision elastic or inelastic? If the latter, what percentage of the maximum possible kinetic energy loss occurs?

38. (I) A bullet of mass 70 g is moving horizontally at a speed of 450 m/s when it strikes a 2.6-kg block at rest. (a) The block is made of wood; the bullet penetrates and stops. What is the initial speed of the bullet–block combination? (b) The block is made of a very hard steel, as is the bullet, and the collision is perfectly elastic. What is the resulting velocity of the block? [*Hint*: Use the approximation that the mass of the bullet is much less than the mass of the block.]

39. (II) A block of mass 126 g is moving along the $+x$-axis with a speed of 0.875 m/s. Just ahead of it is a 9.66-kg mass moving in the same direction with the same speed. At some point, the large mass hits a wall and bounces off the wall perfectly elastically (Fig. 8–37). What is the return speed of the small mass after its elastic collision with the large mass?

▲ **FIGURE 8–37** Problem 39.

40. (II) A machine gun in automatic mode fires 20-g bullets with $v_{\text{bullet}} = 300$ m/s at 60 bullets/s. (a) If the bullets enter a thick wooden wall, what is the average force exerted against the wall? (b) If the bullets hit a steel wall and rebound elastically, what is the average force on the wall?

41. (II) Two spheres of masses 400 and 600 g, respectively, are suspended from the ceiling by massless strings 1.00 m long. The lighter sphere is pulled aside through an angle of 80° and let go. It swings and collides elastically with the second sphere at the bottom of the swing. How high will the lighter of the spheres swing (in terms of an angle)?

42. (II) A pendulum of mass $m = 0.37$ kg and length $l = 0.95$ m is released from a horizontal position. At the bottom of its swing, it collides elastically with a mass $M = 0.56$ kg. (a) What is the velocity of mass M right after the collision? (b) To what height above its low point does mass m rebound?

43. (II) A superball has collisions that are nearly perfectly elastic. A superball of mass M is dropped from rest from a height h (where $h \gg$ the size of the superball) together with a smaller marble of mass m; the marble is initially just a little above the top of the superball and remains right over it throughout the fall. The superball hits the floor first and immediately rebounds elastically, colliding with the marble. (a) What is the speed of the superball and the marble just before the superball hits the floor? (b) Just after the superball rebounds from the floor but before it hits the marble? (c) What is the velocity of the marble after the superball hits it in the head-on collision? (d) How high does the marble go after its collision with the superball, assuming the marble has stayed in line with the superball so that all the motion is vertical? (e) What is the answer to part (d) in the limit $M \gg m$?

44. (II) Consider a perfectly elastic collision in one dimension between a ball of mass M and another of mass m, where $M \gg m$. The light ball is initially at rest, and the heavy ball has a given initial velocity v_i and final velocity v_f. What is the fractional velocity change $(v_i - v_f)/v_i$ of the large mass? Use $M \gg m$ to find an approximate expression.

8–5 Elastic Collisions in Space

45. (I) A ball of mass 1.2 kg moves along the positive x-axis with a speed of 2.4 m/s. Another ball (mass 0.80 kg) moves along the negative x-axis with a speed of 3.6 m/s. After colliding with each other, the lighter ball moves at a speed of 1.8 m/s along a line that makes an angle of 60° with the positive x-axis. What are the speed and direction of the heavier ball? Is the collision elastic?

46. (I) Two objects of equal mass approach each other with equal but opposite velocities along the x-axis. After a collision, one particle has velocity $v_1\hat{i} + v_2\hat{j}$. What is the velocity of the other particle?

47. (I) A billiard ball with velocity $\vec{v} = (2.50 \text{ m/s})\hat{i}$ strikes a stationary billiard ball of the same mass. After the collision, the first billiard ball has velocity $\vec{v}_1 = (0.50 \text{ m/s})\hat{i} + (-1.00 \text{ m/s})\hat{j}$. What is the velocity of the second ball? Is the collision elastic?

48. (I) Suppose there are two billiard balls of equal mass; one is at rest and one is moving with a speed of 3.5 m/s. They collide elastically. After the collision, one of the two balls is measured to be moving with speed 2.3 m/s. What is the speed of the other ball?

49. (I) A billiard ball moving at 3.0 m/s collides with another billiard ball at rest. The balls move off at right angles to one another. If the first ball continues with a speed of 1.5 m/s, what is the speed of the ball that was initially at rest?

50. (II) Two objects with masses 2.0 and 3.0 kg move toward each other, both with speeds $v_0 = 5.0$ m/s. They collide head on and stick together. (a) Calculate their final velocity. (b) Calculate the amount of kinetic energy lost during the process. (c) Suppose the two masses approach each other at 90° before the collision (e.g., along the x- and y-axes). What will be the kinetic energy loss in this case?

51. (II) In a target shooting game, wooden blocks are thrown into the air and shot in flight. A block of 0.80 kg has a speed of 10 m/s at the top of its trajectory when it is hit by a bullet from below at an angle of 60° from the horizontal (Fig. 8–38). The mass of the bullet is 5.0 g and its speed is 550 m/s when it hits the block. The bullet is embedded in the block. What is the velocity of the block immediately after impact?

▲ **FIGURE 8–38** Problem 51.

52. (II) An air puck with mass 0.15 kg and velocity $(-1.7 \text{ m/s})\hat{i} - (2.0 \text{ m/s})\hat{j}$ on a frictionless table collides with a second air puck of mass 0.22 kg and velocity $(3.6 \text{ m/s})\hat{i}$ (Fig. 8–39). As a result of the collision, the first air puck comes to rest. What is the kinetic energy of the second air puck after the collision?

▲ **FIGURE 8–39** Problem 52.

53. (II) In a court hearing, a police expert reconstructs an accident in the following way (Fig. 8–40): A sports car of mass $m_1 = 1000$ kg collided with a parked pickup truck of mass $m_2 = 1500$ kg. From the skid marks, it is estimated that the speed of the pickup immediately after the collision was 21.6 m/s at an angle of 33.7° with the direction of the road. The sports car left a mark at an angle of 60° with the road but did not stop, leaving the scene badly damaged. Can the expert determine from these data if the sports car was speeding and its driver fully responsible?

▲ **FIGURE 8–40** Problem 53.

54. (III) A proton of mass m and kinetic energy K scatters elastically from an alpha particle at rest (with mass four times that of the proton). The proton is deflected at 30° from its original direction. (a) At what angle does the alpha particle recoil? (b) Draw a momentum diagram of the collision. (c) What are the final energies of the two particles?

8–6 Center of Mass

55. (I) A massless, rigid bar that is 2.24 m long connects two tiny spheres (of negligible radii compared to the length of the bar) of masses 895 and 478 g, respectively. Locate the system's center of mass.

56. (I) Point masses of 0.15, 0.40, and 0.25 kg are located in the xy-plane at the origin, at $\vec{r}_2 = (35 \text{ cm})\hat{j}$, and at $\vec{r}_3 = (15 \text{ cm})\hat{i} + (58 \text{ cm})\hat{j}$, respectively. Where is their center of mass?

57. (I) Redo Example 8–12 by placing the origin at the location of mass m_3, which was added to the original system.

58. (I) What is the velocity of the center of mass of the system in Problem 8, both before and after the collision (both parts), if the speed of the moving billiard ball is v?

59. (I) Two children of masses 25 and 30 kg, respectively, stand 2.0 m apart on skates on a smooth ice rink. The lighter of the children holds a 3.0-kg ball and throws it to the heavier child. After the throw the lighter child recoils at 2.0 m/s. With what speed will the center of mass of the two children and the ball move?

60. (II) A uniform iron bar that is 0.75 m long with mass 7.5 kg is placed along the y-axis. From the midpoint of that bar a 0.25-m-long bar of negligible mass extends perpendicular to it in the x-direction; a 1.5-kg point mass is placed at its end. Where is the center of mass of this system? [*Hint*: First calculate the center of mass of the iron bar, $\vec{R}_1$, then treat that bar as a point mass of mass 7.5 kg at $\vec{R}_1$.]

61. (II) A mallet forms a symmetric T-shape. The top of the T is a uniform iron block of mass 4.0 kg. The wooden handle is uniform, 1.2 m long, and has a mass of 1.8 kg. Where is the mallet's center of mass? (See the hint in Problem 60.)

62. (II) Where is the center of mass of a uniform, L-shaped iron rod of sides 1.1 and 0.25 m, respectively? (Assume that the rod is so narrow that the dimensions of the outer bend are the same as those of the inner bend of the L.)

63. (II) By using arguments based on symmetry, show that the center of mass of a uniform, thin, spherical shell made of any material of uniform density is at the center of the shell. Using this result, show that the center of mass of any sphere whose density varies arbitrarily only with distance from the center of the sphere but not with angle (in other words, a spherically symmetric density) is at the center of the sphere.

64. (II) Calculate the center of mass of the T-shaped object shown in Fig. 8–41. Assume that the mass density is uniform.

▲ **FIGURE 8–41**   Problem 64.

65. (II) Find the center of mass of a flat, semicircular object of radius R. Assume that the density is uniform.

66. (II) A cylinder 30 cm in diameter and 1.00 m long has a hole drilled in it. The hole is 10 cm in diameter, 30 cm long, and it is symmetrically aligned along the axis of the cylinder. Where is the center of mass of the drilled cylinder?

67. (II) A set of four point masses are arrayed on the xy-plane. A mass of 1 kg is placed at the origin, two similar masses are at the points $(x, y) = (1\,\text{m}, 0)$ and $(0, 1\,\text{m})$, respectively, and a fourth mass of 2 kg is at the point $(1\,\text{m}, 1\,\text{m})$. Where is the center of mass? Suppose that the fourth mass were 1 kg rather than 2 kg. Find, without a detailed calculation, the location of the center of mass in this case.

68. (III) Show that the center of mass of a planar object can be found by dividing the object into a convenient number of sections of known mass M_i, finding the center of mass of each section, and then treating each section as a collection of point masses M_i located at the center of mass of that section.

69. (III) Calculate the center of mass of a stick extending from $x = 0$ to $x = L$; assume that the density of the mass distribution is given by $\lambda(x) = K/(x^2 + a^2)$, where K is a constant with dimensions $[ML]$.

70. (III) Find the center of mass of a flat triangular object of uniform density with sides of lengths a, b, and c.

8–7   Rocket Motion

71. (I) The thrust of a rocket is 10×10^6 N. If the speed of its engine's exhaust gas is 3.0 km/s, how fast is the rocket's mass changing?

72. (I) An orbiter must fire its thruster rockets to increase its speed. If the mass of the orbiter is about 20,000 kg and the exhaust gas velocity is 2600 m/s, how much mass must be discarded to increase the speed from 5800 to 5900 m/s?

73. (I) After ejecting a communication satellite, the space shuttle must make a correction to account for the change in momentum.

One of the thrusters with $u_{ex} = 10^3$ m/s is used to increase the orbital velocity by 10 m/s. What percent of the mass of the space shuttle must be discarded?

74. (I) A rocket that uses a fuel with an exhaust speed of 2700 m/s is shot straight up and reaches a final speed $v = 3400$ m/s when the fuel is exhausted, which happens 120 s after the launch. What fraction of the initial mass of the rocket with all its fuel is the mass of the portion of the rocket that remains? Do not ignore gravity, but assume the force of gravity to be the same throughout the burn period.

75. (II) The *burnout velocity* is the final velocity of a rocket when all the fuel is burned away. If m_0 is the total initial mass of the rocket and m_{fuel} is the mass of the fuel, show that the burnout velocity is given by

$$v = u_{ex} \ln\left(\frac{m_0}{m_0 - m_{\text{fuel}}}\right) - gt,$$

where g is assumed to be constant. If the ratio m_{fuel}/m_0 is 70 percent, $t = 90$ s, and u_{ex} is 2800 m/s, what is the burnout velocity?

76. (II) The rocket considered in Problem 75 carries human passengers. Consequently, its initial upward acceleration may not exceed $2.5g$. What is the largest initial burnout rate (the mass expelled over unit time) that satisfies this condition? (The mass of the rocket is 60,000 kg.)

77. (II) A wooden block of mass 0.40 kg is placed on a fence 1.0 m high. A bullet of mass 10 g is fired horizontally into the block. If the speed of the bullet is 450 m/s, how far will the block, with the bullet in it, land from the fence? Ignore any friction between the fence and the block.

78. (II) A wooden block of mass 0.40 kg is placed on a fence 1.0 m high. A bullet of mass 10 g is fired horizontally into the block with an initial speed of 450 m/s. It passes right through the block, and in doing so gives the block enough kick so that the block lands 55 cm from the fence. Assuming that the land behind the fence is completely flat, how far from the fence will the bullet land? Ignore any friction between the fence and the block and any air drag on the block or bullet.

79. (II) To test certain aspects of rocket propulsion, a student stands initially motionless on frictionless ice. He has 100 watermelon seeds that he can spit horizontally in the $-x$-direction, thereby propelling himself in the $+x$-direction. The speed he can give one seed with respect to himself is u, but if he tries to spit two seeds simultaneously, he can give them only speed $u/2$. More generally, if he tries to spit n seeds simultaneously, he can give them speed u/n. The student has mass M and each seed has mass m. (a) To maximize his speed after spitting all 100 seeds, is the student better off spitting out all 100 seeds at once, spitting out the seeds one at a time, or something in between? Assume that the ratio m/M and even $100\,m/M$ can be taken to be small. (b) Let $M = 50$ kg, $m = 1$ g, and $u = 3$ m/s. What is the maximum final speed of the student?

80. (III) Because the final velocity of rockets is limited by the exhaust velocity and by the mass ratio of total mass to the rest of the rocket minus the fuel, engineers have developed multistage rockets in which the fuel tank of each stage is discarded before the next stage is ignited. Consider a two-stage rocket for which m_0 is the total mass of the rocket at liftoff, m_1 is the mass of the fuel in the first stage, m'_1 is the mass of the first stage that is discarded, and m_2 is the mass of the second-stage fuel. Find an equation that gives the final velocity of the rocket; assume that gravity is constant.

General Problems

81. (I) A 48-kg figure skater (skater 1) moving at 6 m/s is picked up by her 82-kg partner (skater 2), who is skating in the same direction at 9 m/s. They then coast off together. Ignore all friction. (a) What is the velocity of their center of mass before he picks her up? (b) After he picks her up? (c) What were the velocities of skaters 1 and 2 in the center-of-mass frame before he picked her up? (d) During a brief period of 0.05 s, they crash into the barrier and collapse together. What average force is exerted on the barrier?

82. (II) A railroad car of mass 16 Mg (1 Mg $= 10^3$ kg) is released from rest in a railway switchyard and rolls to the bottom of a slope 1.8 m below its original height. At the low point, it collides with and sticks to another car of 8 Mg. The two cars roll off together up another slope to a position a height h above the low point, where they come to a stop. Ignoring the effects of friction, find h.

83. (II) A cannon of mass 800 kg shoots a 5-kg projectile at an angle of 35° from the horizontal. The projectile has an initial speed of 800 m/s. The cannon is on wheels. (a) Draw a diagram showing the momentum vectors. (b) With what velocity does the cannon recoil along the horizontal ground? (c) What happens to the recoiling cannon's component of momentum perpendicular to the ground?

84. (II) A small bag of sand is suspended by a thread of length $L = 1.2$ m. A bullet of mass $m = 8.0$ g moving with speed $v_0 = 600$ m/s in the horizontal direction hits the bag and leaves with a speed $v = 250$ m/s (Fig. 8–42). The bag swings to a maximum angle of $\theta = 40°$. What is the mass of the sandbag?

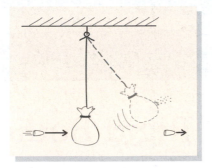

▲ **FIGURE 8–42** Problem 84.

85. (II) Two masses slide without friction down the sides of a hemispherical bowl. They each start with zero velocity at the lip, which is a height h above the bottom. They move in opposite directions and collide elastically at the bottom of the bowl. Suppose that the masses are m and M, respectively, with $m < M$. (a) By how much will the lighter mass overshoot the lip after the collision? (b) How high will the masses move if they coalesce on contact?

86. (II) A boxcar of length 9 m and height 3.0 m is at rest on frictionless rails. Inside the boxcar, whose empty mass is 4200 kg, a tank containing 1800 kg of water is located at one end. The tank is 1.5 m long and 2.0 m high. At some point, the walls of the tank start to leak and the water fills the floor of the boxcar uniformly. Assuming that all the water stays in the boxcar, describe how the breakage of the tank manifests itself to an outside observer.

87. (II) A steer of mass 500 kg runs at 30 kph past a cowboy of mass 75 kg, who jumps on in an attempt to bring down the steer. Assuming that the cowboy is not dragging his feet, what is the speed of the steer after the cowboy jumps on?

88. (II) In the Olympic Games of the year 2020, a new sport is introduced—cyclo-shooting. A bicyclist moving in the +x-direction fires a gun in the y-direction in an attempt to hit a target lined up with the position $x = 0$ m. The mass of the cyclist plus the bicycle is 65 kg, and the cyclist is originally moving with speed 35 km/h. The gun shoots a projectile, with mass 15 g, at a speed of 1600 km/h. The target is a distance 30 m from the bicycle track. (a) At what x-value should the cyclist fire in order to hit the target? Ignore air resistance. (b) Immediately after she fires, the cyclist moves in a new direction due to the recoil. What is the new value of her velocity, both magnitude and direction?

89. (II) A sphere of styrofoam has radius R. A cavity of radius $R/2$ centered a distance $R/2$ directly above the center of the sphere is hollowed out and filled with a solid material of density five times the density of styrofoam (Fig. 8–43). Where is the center of mass of the new sphere?

▲ **FIGURE 8–43** Problem 89.

90. (III) Two identical spherical billiard balls each with a radius of 3 cm move toward each other with velocities $(1.5 \text{ m/s})\hat{i}$ and $(-1.1 \text{ m/s})\hat{i}$. The center of one of them moves along the +x-axis ($y = 0$ m), and the center of the other one moves in the −x-direction, 1.0 cm below the x-axis ($y = -1.0$ cm). Assuming that the collision is perfectly elastic, what are the final velocities of the two billiard balls? Pretend that the objects slide without friction. Note that the impulse is directed along the line that joins the two centers at the moment of collision.

91. (III) Two students sit on opposite ends of a 4-m-long sled that is initially at rest on frictionless ice. Each student has a mass of 65 kg; the sled's mass is 30 kg. The student at one end slides a 3-kg object on the sled across to the other at a uniform speed of 6 m/s relative to the sled (the object moves friction free on the sled). (a) What is the sled's speed relative to the ice before the second student catches the object? (b) After the second student catches the object? (c) Over what distance does the sled move while the object slides across? (d) Over what distance does the center of mass move while the object slides across?

92. (III) When a rocket moves vertically in the presence of gravity, the motion is different than when gravity is not present. If you go back to the derivation of Eq. (8–65), you will see that the starting point of that derivation, the condition that $P(t + \Delta t) - P(t) = 0$, where P is the total momentum of the rocket plus its exhaust gases, would be replaced by $P(t + \Delta t) - P(t) = -F_g \Delta t$. Here F_g is the magnitude of the force of gravity and the minus sign in front of F_g signifies that the "up" direction is positive in our one-dimensional derivation. Follow the steps that led to Eq. (8–65) to find the equation of motion when gravity is included [the analog of Eq. (8–65)] and show that the solution of the equation of motion is Eq. (8–70).

The wind acting on the sails is trying to rotate the boat in one direction; the weight of the sailors acts to rotate it in the opposite direction. The description and dynamics of rotations are an important element in understanding the behavior of extended systems such as this sailboat.

Rotations of Rigid Bodies

To this point our discussion of dynamics has treated objects from crates to planets as if they were all pointlike (particles). Our study of an extended object, or of a system of particles, has reached the point where we can show that under the influence of a net external force the center of mass moves as if the entire system were concentrated at that single point. But for an extended object that rotates there is more to its motion than just the motion of its center of mass. Consider the wheels of a skateboard, a tossed blackboard eraser, the flow of water down the drain of a bathtub, or the motion of Earth. All these systems rotate, or spin, about an axis, as do many other objects from galaxies to molecules, and we must be able to analyze this sort of motion if we are to understand the dynamics of our world.

In this chapter, we study the rotational motion of **rigid bodies**, which are objects of fixed form that do not distort or deform as they move. Blackboard erasers, basketballs, Earth, and compact disks are, at least to a good approximation, all rigid bodies, and all can have a rotational aspect to their motion. (The water flowing down a drain is not a rigid object, but a rotation is an important part of its description—the motion of these more complicated nonrigid systems will be explored more fully in Chapter 10.) Here we start with a simple description (kinematics) of the rotational motion of rigid objects and then continue with the causes of their rotations (dynamics). Even though rotations appear to encompass many new concepts, these motions are in fact described by the application of Newton's laws to collections of particles. No new physical laws are involved.

9–1 Simple Rotations of a Rigid Body

Motions of a Rigid Body

Suppose we have a rigid body that has an axis within it about which the body can rotate and that is permanently fixed in space. What kind of motions are possible? The characteristic feature of a rigid body is that any point P of the object remains at a fixed distance from any other point and hence at a fixed distance from the axis. To see the effect of this fact, consider a clock hand with an axis passing through the end of the hand at the clock center and perpendicular to the face. If this axis is fixed, the possible motion corresponds to each point of the hand moving in circles about the fixed axis (Fig. 9–1). The particular constant distance of our point is the radial distance R to the center. In simple language, the clock hand rotates about this axis.

In Fig. 9–1, we have established a set of fixed axes with their origin at the center. At some time t_1, point P has coordinates (x_1, y_1); at a later time, point P has moved to where its coordinates are now (x_2, y_2). *Both x and y change with time, and this is complicated.* Things are simpler if we work with polar coordinates. We can describe point P on the hand with an angle θ measured from the x-axis and a radial distance R from the origin. In this description, *the only variable that changes with time is the angle θ.* Our point P has undergone an *angular displacement*.

Perhaps you are wondering if the fact that the clock hand moves in a plane has a special role. Consider then a more complicated rigid object, with the fixed axis again passing through it, as in Fig. 9–2. Any point in the object—here, a pot on a potter's wheel—that lies along the axis is fixed in space. A point P that is off the axis will always be at a fixed perpendicular distance from the axis. The motion of the pot consists of rotations about the axis. And as the discussion of the clock hand shows, to describe rotations about a fixed axis, we need only specify a single angle. Other examples of this limited motion are provided by Earth's rotation about its axis or by a turning merry-go-round.

To repeat, what distinguishes a rigid body is that when it moves through a certain angle in the rotational motion about an axis, all points in the rigid body turn about the axis by exactly the same angle. Even if the motion is not uniform, it is still just circular motion. We need only one variable—the angle of circular rotation θ.

Angular Velocity and Angular Acceleration

To describe the motion mathematically, let's orient the z-axis along the rotation axis (Fig. 9–3a). Consider the position vector $\vec{r}$ from the origin to point P. Take the projection of $\vec{r}$ on the xy-plane to go to point P' and let $\theta(t)$ be the angle this projection makes with the x-axis at some time t (Fig. 9–3b). This angle, which in Section 3–5 was denoted ϕ, is measured in *radians* (2π rad $= 360°$). Figure 9–3b also shows that at a later time $t + \Delta t$ the projection point P' has moved from angle $\theta(t)$ to angle $\theta(t + \Delta t)$.

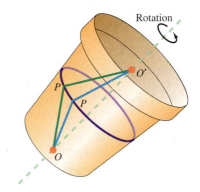

▲ **FIGURE 9–1** A flat object in the *xy*-plane rotates about a fixed axis perpendicular to the plane. Here, the object is the hand of a clock, which rotates through the angle $\theta_2 - \theta_1$. The fixed point is at the center of the clock face.

▲ **FIGURE 9–2** A three-dimensional object—a clay pot on a potter's wheel—rotates about the axis defined by the line between the two fixed points O and O'. The point P remains the same distance from these points or from any point on the axis as the pot rotates about the axis.

◀ **FIGURE 9–3** (a) A pot on a potter's wheel rotates about the z-axis. (b) The pot as viewed looking down the z-axis. The angle of rotation θ, which varies with time, is defined to be the angular displacement from the x-axis of some point—any point—in the pot. It is positive when the pot rotates counterclockwise.

Arrow indicates sense of rotation.

(a)

(b) View from above

The average angular velocity, ω_{av}, over a finite time interval Δt is obtained from the change in the angle over that time interval:

$$\omega_{av} = \frac{\theta(t + \Delta t) - \theta(t)}{\Delta t}. \tag{9-1}$$

Every point in our rigid body moves with the same average angular velocity. It will equally be true that every point in the rigid body will move with the same *instantaneous angular velocity* (or just the *angular velocity*) $\omega(t)$, a quantity that we define by taking the limit $\Delta t \to 0$ of Eq. (9–1) [see also Eq. (3–46), where we discuss the magnitude, or angular speed]:

$$\vec{\omega} = \frac{d\theta}{dt}\hat{\omega}. \tag{9-2}$$

This expression is a vectorial one, and we did not use the term "velocity" casually: The angular velocity is a vector. Its direction, which is contained in the unit vector denoted $\hat{\omega}$ in Eq. (9–2), is defined conventionally but unambiguously as follows. We start by saying that it is aligned with the axis of the rotation. This leaves an ambiguity; if the axis of the rotation is vertical, is the angular velocity up or down? We break this ambiguity by the following rule: If *the fingers of your right hand curl along with the rotation* [*that is, in the direction of the angle change that appears in the numerator of Eq. (9–2)*], *your thumb will give the direction of the angular velocity.* This is a so-called *right-hand rule* (Fig. 9–4). As Fig. 9–4 shows, the movement of the rotating object is in the *xy*-plane with a counterclockwise motion as seen from above. With the right-hand rule, the extended thumb of the right hand points in the $+z$-direction (along the axis of rotation), and this is the direction of the angular velocity $\vec{\omega}$.

If we look back at Fig. 9–1, we see that for a rotation by $\Delta\theta$, the arc distance s traced out by a point a distance r from the axis of rotation is given by $s = r\,\Delta\theta$; similarly the linear speed of that point will be $v = r\omega$.

The units that describe angular velocity are *radians per second* (rad/s). Because a radian (the ratio of an arc length to the radius of the arc) is a dimensionless measure, the dimensions of ω are $[T^{-1}]$.

Suppose that the angular velocity is a *constant;* that is, $\omega(t) = \omega_0$ in some fixed direction that defines an axis. Then the angle θ that describes rotation around the axis changes linearly with time:

$$\theta = \theta_0 + \omega_0 t, \tag{9-3}$$

where θ_0 is the value in radians of the angle θ at time $t = 0$. A full rotation is made over a time T, which is the *period* of the motion. The angle θ goes from θ_0 to $\theta_0 + 2\pi$ in time T, so $\omega_0 T = 2\pi$, or

$$T = \frac{2\pi}{\omega_0}. \tag{9-4}$$

▶ **FIGURE 9–4** Right-hand rule. The thumb of the right hand points along the axis of rotation (the *z*-axis) when the fingers curl in the direction of rotation (from the *x*-axis to the *y*-axis). This will also specify the direction of the angular velocity vector.

Thumb points in *vector* direction of rotation and angular velocity $\vec{\omega}$.

Fingers curl along rotation.

The *frequency f* measures the number of times per second that point P returns to its original position in uniform rotation:

$$f = \frac{1}{T} = \frac{\omega_0}{2\pi}. \tag{9–5}$$

FREQUENCY AND PERIOD

For nonuniform motion, we also speak of an instantaneous frequency, $f(t) = \omega(t)/2\pi$. For more on these relations you might want to refer back to Section 3–5.

Angular Acceleration

If the angular velocity is not constant, we can define an average angular acceleration:

$$\vec{\alpha}_{av} = \frac{\vec{\omega}(t + \Delta t) - \vec{\omega}(t)}{\Delta t}. \tag{9–6}$$

The instantaneous angular acceleration, or more simply the **angular acceleration**, is defined as the instantaneous rate of change of the angular velocity:

$$\vec{\alpha} = \frac{d\vec{\omega}}{dt} = \frac{d^2\theta}{dt^2}\hat{\omega}, \tag{9–7}$$

The last step applies when the angular velocity changes in magnitude but not direction. Note that this equation has the same structure as the equation for linear motion,

$$\vec{a} = \frac{d\vec{v}}{dt} = \frac{d^2\vec{r}}{dt^2}.$$

The direction of the angular acceleration vector is specified by the rate of change of the angular velocity vector, exactly as the linear acceleration vector is the rate of change of the linear velocity vector. As for the angular velocity, the angular acceleration is the same for all the points in the rigid body. The angular acceleration has units of radians per second squared, and its dimension is $[T^{-2}]$, again because radians are dimensionless.

Constant Angular Acceleration: Provided that the angular acceleration and the angular velocity are parallel (this is the case for most situations we'll deal with here), a constant angular acceleration means that the angular velocity has a constant time derivative and an unchanging direction. (If we understand that the equations below should contain a unit vector in the relevant direction, we can now dispense with the over-arrow notation. All the vector information will be contained in the *sign* of the respective quantities.) With constant angular acceleration the angular velocity changes linearly with time; that is,

$$\omega = \omega_0 + \alpha t, \tag{9–8}$$

where ω_0 is the angular velocity at $t = 0$. The angle is then given by

$$\theta(t) = \theta_0 + \omega_0 t + \tfrac{1}{2}\alpha t^2. \tag{9–9}$$

This solution corresponds to the familiar linear motion equation $x(t) = x_0 + v_0 t + \tfrac{1}{2}at^2$, valid for constant linear acceleration a. Just as the position of a car moving on a line and undergoing constant linear acceleration changes quadratically with time, so the angle of the wheels of that car change quadratically with time if the wheel undergoes constant angular acceleration.

We may eliminate the time from Eqs. (9–8) and (9–9) to find that

$$\omega^2 = \omega_0^2 + 2\alpha(\theta - \theta_0), \tag{9–10}$$

which is the analog of the linear motion expression $v^2 = v_0^2 + 2a(x - x_0)$.

EXAMPLE 9–1 A pulley is rotating at a frequency of 32 rev/min. A motor speeds up the wheel so that 30 s later the frequency of the motion is 82 rev/min. (a) What is the average angular acceleration in radians per second squared during that period? (b) How far will a point 0.30 m from the center of the pulley have traveled during the acceleration period assuming that the acceleration is constant?

Setting It Up The rotating pulley is shown in Fig. 9–5, with the direction of rotation shown. We label the given initial and final rotational frequencies as f_i and f_f and the time interval Δt. We want the initial and final angular speeds ω_i and ω_f, the average angular acceleration α_{av}, and the distance D traveled by a point a given radial distance r from the center of the pulley under the assumption that the angular acceleration is constant.

Rotation

0.3 m

P

Pulley

Engine

▲ **FIGURE 9–5**

Strategy We first will need to change the frequencies to angular speeds. This is accomplished by recalling that each revolution consists of an angular displacement of 2π radians. Thus we multiply frequency by 2π to find angular speed. We can then use this in the definition of average angular acceleration [Eq. (9–6)] to find the average angular acceleration. As for the distance traveled by a point, we can find the angular displacement under the assumption of constant angular acceleration using Eq. (9–9), then use the relation between angular change, radius, and arc length to find the distance traveled.

Working It Out (a) A rotation frequency of $f_i = 32$ rev/min corresponds to an angular speed of $\omega_i = (32 \text{ rev/min})(2\pi \text{ rad/rev}) = 64\pi \text{ rad/min}$, or $64\pi \text{ rad}/60 \text{ s} = 3.4 \text{ rad/s}$. Similarly, after 30 s the angular velocity ω_f is $(82 \text{ rev/min})(2\pi \text{ rad/rev})(1 \text{ min}/60 \text{ s}) = 8.6 \text{ rad/s}$. Thus the average angular acceleration is

$$\alpha_{av} = \frac{\omega_f - \omega_i}{\Delta t} = \frac{(8.6 - 3.4) \text{ rad/s}}{30 \text{ s}} = 0.17 \text{ rad/s}^2.$$

(b) We calculate the angular displacement $\Delta\theta$ during the period by using Eq. (9–9), which applies when there is uniform acceleration. In our case $\omega_i = 3.4 \text{ rad/s}$ and $\alpha_{av} = 0.17 \text{ rad/s}^2$, so Eq. (9–9) gives

$$\Delta\theta = \theta - \theta_i = \omega_i t + \tfrac{1}{2}\alpha_{av}t^2$$
$$= (3.4 \text{ rad/s})(30 \text{ s}) + \tfrac{1}{2}(0.17 \text{ rad/s}^2)(30 \text{ s})^2$$
$$= 180 \text{ rad}.$$

The distance traveled is therefore

$$D = r\,\Delta\theta = (0.30 \text{ m})(180 \text{ rad}) = 54 \text{ m}.$$

The unit "radian" is deleted from the final value because radians are dimensionless.

What Do You Think? How does the answer to part (a) depend on the details of the speed-up? *Answers to **What Do You Think?** questions are given in the back of the book.*

Acceleration of a Point in a Rotating Rigid Body

Consider the potter's wheel of Fig. 9–6, and, more specifically, a point within it located at a distance r from the axis of rotation. The point moves in a circle; its velocity has a magnitude of $r\omega$ and a direction that is tangential to the circle of radius r. We learned in Section 3–5 that a point moving in a circle of radius r with speed v undergoes an acceleration in the radial direction toward the center of the circle—we refer to the point's radial acceleration. The magnitude of that acceleration is v^2/r. In terms of the angular speed $\omega = v/r$, this acceleration has magnitude

$$a_{radial} = \frac{v^2}{r} = \omega^2 r. \tag{9–11}$$

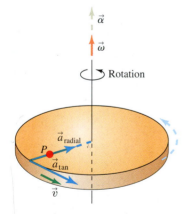

$\vec{\alpha}$

$\vec{\omega}$

Rotation

$\vec{a}_{radial}$

P

$\vec{a}_{tan}$

$\vec{v}$

▲ **FIGURE 9–6** A rotating potter's wheel undergoes an angular acceleration. Any point P must have an acceleration with both a centripetal component (this component keeps the point on its circular path) and a tangential component.

If the potter's wheel had a changing angular speed, or in other words an angular acceleration α, then our given point would also have a linear acceleration in the tangential direction, a tangential acceleration, of magnitude

$$a_{tan} = \alpha r. \tag{9–12}$$

Figure 9–6 illustrates these acceleration components for a point in a rotating potter's wheel. If a point in a rigid body accelerates, there must be forces involved. Both the radial acceleration and any tangential acceleration occur at least in part through the interatomic forces that fix the relative positions of the atoms of the rigid body. We'll discuss this in more detail later.

CONCEPTUAL EXAMPLE 9–2 A lump of clay rotates on a potter's wheel as the potter speeds it up. Consider two points, P_1 close to the rotation axis and P_2 far from the rotation axis. Which of these points has the larger radial acceleration a_{radial} and which the larger tangential acceleration a_{tan}?

Answer As we have tried to emphasize, the angular variables are common for all points in a rigid object; all points move through the same angle in a given time, all points move with the same angular velocity, and all points have the same angular acceleration. Therefore, to make a comparison of the accelerations of different points

within a rotating object, it is best to write their accelerations as much as possible in terms of angular speed and acceleration. For a_{radial}, Eq. (9–11) provides us with the necessary expression: The angular speed is the same for the two points, so $a_{radial} \propto r$. For the tangential acceleration, we can use Eq. (9–12): The angular acceleration is the same for the two points, so $a_{tan} \propto r$. Point P_2, which sits at a larger radial distance, has both the larger radial and tangential acceleration components.

What Do You Think? Which of the two points has the larger angular acceleration?

9–2 Rotational Kinetic Energy

Except for points on the rotation axis, every point of a rotating object such as that shown in Fig. 9–7a moves, so the object has **rotational kinetic energy** K. To analyze it, we take the hamburger package in Fig. 9–7a, which rotates with angular velocity ω about an axis that we label as the z-axis. In Fig. 9–7b we divide the package into elements labeled by the indices i, with each element having mass Δm_i. The total kinetic energy of the object is given by

$$ K = \sum_i K_i = \frac{1}{2} \sum_i \Delta m_i v_i^2. $$

Because each mass element is rotating about the z-axis with angular velocity ω, the speed of an element is given by $v_i = R_i \omega$. The length R_i is the *perpendicular distance* between the mass element and the rotation axis, as shown in Fig. 9–7b. Substitution of this value of v_i into the expression for the kinetic energy gives

$$ K = \frac{1}{2} \left(\sum_i \Delta m_i R_i^2 \right) \omega^2. \qquad (9\text{–}13) $$

The factor ω has come out of the sum because it is the same for all the elements of the object. The kinetic energy of our rotating object is thus proportional to ω^2. We may write

$$ K = \tfrac{1}{2} I \omega^2, \qquad (9\text{–}14) $$

KINETIC ENERGY OF ROTATION

where

$$ I \equiv \sum_i \Delta m_i R_i^2 \qquad (9\text{–}15) $$

ROTATIONAL INERTIA

(a)

(b)

◀ **FIGURE 9–7** (a) The hamburger package has kinetic energy associated with its rotation on the turntable. (b) We break the package into elements in order to analyze the kinetic energy in terms of rotational variables.

The quantity I is a property of the rigid object called the **rotational inertia** or the **moment of inertia**. It has dimensions of $[ML^2]$ and is measured in units of kilogram-meters squared in SI. From its definition, we see that I sums the contributions of each element of the object; for a single element or single particle, then,

$$I = mR^2, \qquad (9\text{--}16)$$

where R is the perpendicular distance from the single element to the axis of rotation. According to Eq. (9–15), the rotational inertia can be defined with respect to any axis we choose (we refer to the reference axis), not just the axis of rotation. Moreover, I *depends on the axis from which R_i is measured.*

Rotational inertia plays the same part in rotational motion as mass does in translational (linear) motion. It differs from the mass in that it depends on the choice of a particular axis. The kinetic energy of a particle in translational motion is one-half its mass times the velocity squared; the kinetic energy of an object rotating about a fixed axis is one-half its rotational inertia times its angular velocity squared; that is, $K_{\text{translation}} = mv^2/2$ and $K_{\text{rotation}} = I\omega^2/2$. Just as the mass measures the resistance of an object to changes in velocity, we shall see that the rotational inertia measures the resistance of an object to changes in its angular velocity.

THINK ABOUT THIS . . .

CAN YOU RUN A CAR USING ROTATIONAL KINETIC ENERGY?

▲ **FIGURE 9–8** This 14-foot flywheel was used in a steam pumping engine to pump water in the Southend Waterworks Company in Great Britain. It was installed in 1931, but is no longer operational.

A *flywheel* is a rotating disk with a large rotational inertia and the ability to reach a large angular velocity; a flywheel thereby stores energy, and various means can be used to recover this energy. Flywheels (see Fig. 9–8) have a long history; for instance, from the start of the industrial revolution flywheels were used to accumulate energy from turning waterwheels, then that energy was distributed through systems of belts to machine tools. In their early configuration, San Francisco's cable cars used three flywheels 15 ft in diameter with a mass of 25 tons to keep the cables moving steadily as cable cars gripped and ungripped the cables. By the 1950s Zurich was running a bus system in which energy to run the buses came from flywheels mounted within them—the buses' flywheels were about $1\frac{1}{2}$ m in diameter, had a mass of 1500 kg, and turned at 3000 rpm. In the Zurich buses, the flywheel powers an electric generator and the electricity runs electric motors used to drive the wheels. Periodically the flywheel is sped up by connection to the local electrical grid at the stops. In a search for nonpolluting automobiles, the possibility that a rapidly spinning flywheel in an automobile could power that car is a very real one. Of course, the electricity that would have to be used to bring the flywheel up to speed would have to be generated somewhere, but the pollution of that plant could at least be localized.

We have defined the rotational inertia by treating an object as composed of discrete elements. Most real objects are more conveniently thought of as continuous, and a treatment of this type requires us to take a limit in which the elements have infinitesimal volume and infinitesimal mass. In this case, the sum becomes an integral over the volume of the object. We can write this continuum expression by labeling the mass element by the differential notation dm, and thus

$$I = \int R^2 \, dm. \qquad (9\text{--}17)$$

To make this formal expression useful, we employ the *mass density* function ρ. The mass density $\rho(\vec{r})$ at the location $\vec{r}$ of the ith element is defined as

$$\rho(\vec{r}) \equiv \lim_{\Delta V_i \to 0} \frac{\Delta m_i}{\Delta V_i} = \frac{dm}{dV}.$$

Note that as the volume ΔV_i of the ith element approaches zero, the mass of that element also approaches zero, leaving a finite ratio ρ. Having defined the density, we can write the mass element as $dm = \rho \, dV$. The total mass is found from the mass density by an integration over the volume V (the single integral sign with the subscript "Volume" denotes a three-dimensional integration over the volume):

$$M = \sum_i \Delta m_i \to \int dm = \int_{\text{Volume}} \rho \, dV. \qquad (9\text{--}18)$$

$$\Delta m = \lambda \, \Delta x$$

◀ **FIGURE 9–9** A one-dimensional
object is aligned with the x-axis. A
segment of width Δx is a displacement x
from the origin.

In this language the rotational inertia of a continuous object is

$$I = \int R^2 \, dm = \int_{\text{Volume}} \rho R^2 \, dV. \qquad (9\text{–}19)$$

The integrals referenced here may look difficult because they refer to such a general situation. But in many particular cases of interest in this book—for example, the solid cylinder below—we can use symmetry to reduce the integrals to simple one-dimensional integrals. At worst, the integrations can be formulated as a series of simple one-dimensional integrals (see Example 9–7).

An equation like Eq. (9–19) applies to one-dimensional (linear) objects or two-dimensional (flat) objects. In these cases, we replace the mass per unit volume ρ by a mass per unit length $\lambda \equiv dm/dx$ or a mass per unit area $\sigma \equiv dm/dA$, respectively. Thus

for a one-dimensional object: $dm = \lambda \, dx$;

for a two-dimensional object: $dm = \sigma \, dA$;

for a three-dimensional object: $dm = \rho \, dV$.

The densities λ, σ, and ρ may vary from point to point of the object.

To evaluate the rotational inertia about an axis perpendicular to a one-dimensional object, we can align our object with the x-axis, and

$$I = \int \lambda x^2 \, dx. \qquad (9\text{–}20)$$

The distance x is measured from the reference axis, which passes through the origin, as shown in Fig. 9–9, and the integral is carried out over the length of the object. When the object is two dimensional, we have

$$I = \int_{\text{Surface}} \sigma r^2 \, dA. \qquad (9\text{–}21)$$

We have taken the reference axis perpendicular to the object and integrated over its entire area (Fig. 9–10). The quantity r^2 is the square of the distance of the integration element to the axis about which I is measured.

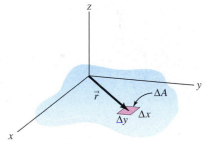

▲ **FIGURE 9–10** A two-dimensional object is aligned in the xy-plane. A segment of area ΔA is a position $\vec{r}$ from the origin.

9-3 Evaluation of Rotational Inertia

A description of the rotational motion of an object requires us to evaluate its rotational inertia. For continuous objects, this evaluation involves integrations. We will see how this works in a series of examples. Table 9–1 contains a summary of the results as well as rotational inertias of other objects for which the calculations are not carried out in this section.

TABLE 9–1 • Rotational Inertias of Simple Solids

$\frac{1}{12}ML^2$	MR^2	$\frac{1}{2}MR^2$	$\frac{2}{5}MR^2$
Solid rod about perpendicular axis through center	Cylindrical shell about central axis	Solid cylinder about central axis	Solid sphere about any diameter
$\frac{1}{3}ML^2$	$\frac{1}{2}M(R_1^2 + R_2^2)$	$\frac{1}{4}MR^2 + \frac{1}{12}ML^2$	$\frac{2}{3}MR^2$
Solid rod about perpendicular axis through end	Hollow cylinder about central axis	Solid cylinder about diameter through center	Thin spherical shell about any diameter

▲ **FIGURE 9–11** A dumbbell consists of two point masses, m_1 and m_2, connected by a rigid massless rod of length L. The object is aligned along the x-axis, with m_1 at the origin. The axis of rotation is perpendicular to the rod, passing through the center of mass.

Dumbbell: Almost the simplest rotating object that we can contemplate is a dumbbell consisting of two masses compact enough to be treated as points, m_1 and m_2, connected by a massless rigid rod of length L (Fig. 9–11). We'll find here the rotational inertia about an axis that goes through the center of mass and is perpendicular to the rod.

If we put mass m_1 at the origin and mass m_2 at $x = L$, Eq. (8–48a) tells us that the location X of the center of mass is

$$X = \frac{m_1 x_1 + m_2 x_2}{m_1 + m_2} = \frac{(m_1)(0) + (m_2)(L)}{m_1 + m_2} = \frac{m_2 L}{M},$$

where M is the total mass. Our reference axis is therefore perpendicular to the rod and passes through the point $x = m_2 L/M$. As measured from the center of mass, the coordinates of m_1 and m_2 are $-m_2 L/M$ and $L - (m_2 L/M) = m_1 L/M$. As in Fig. 9–11, the rotational inertia about the axis passing through the center of mass and perpendicular to the axis of the dumbbell is therefore

$$I = m_1 \left(-\frac{m_2 L}{M} \right)^2 + m_2 \left(\frac{m_1 L}{M} \right)^2 = \frac{(m_1 m_2^2 + m_2 m_1^2) L^2}{M^2}$$

$$= \frac{m_1 m_2 L^2 (m_2 + m_1)}{(m_1 + m_2)^2} = \frac{m_1 m_2}{m_1 + m_2} L^2. \tag{9–22}$$

If the masses are equal, then $m_1 m_2/(m_1 + m_2) = m/2$; if $m_2 \gg m_1$, then it is a good approximation to write $m_1 m_2/(m_1 + m_2) = m_1$, the mass of the light point mass. These results are easy to understand physically. In the case of equal masses m, the center of mass is the midpoint of the rod, a distance $L/2$ from each mass, and the rotational inertia about this axis is then $I = 2m(L/2)^2 = mL^2/2$. In the case of one very light mass, $m_1 \ll m_2$, to a good approximation the center of mass lies very near to the mass m_2, and m_2 no longer contributes to the rotational inertia about this point. The lighter mass m_1 is a distance L from the axis, and $I = m_1 L^2$.

Note finally that if the reference axis is the line connecting the two point masses, then each mass lies directly on the axis. Their respective distances to the axis is zero, and the rotational inertia about this axis is zero. This axis also passes through the center of mass!

EXAMPLE 9–3 A dumbbell consists of point masses 2.0 and 1.0 kg attached by a rigid massless rod of length 0.6 m. Calculate the rotational inertia of the dumbbell (a) about an axis perpendicular to the rod of the dumbbell and going through the center of mass and (b) about an axis perpendicular to the rod of the dumbbell and going through the 2.0-kg mass.

Setting It Up Figure 9–12 serves as a suitable sketch. The masses are treated as points. The rod lies along the x-axis, and the rotation axes referred to in the question are also drawn. With the coordinate system of the sketch, the larger mass m_1 is at the origin, $x_1 = 0$ m, and the smaller mass m_2 is at $x_2 = L$, L being the known length of the rod.

Strategy If we are to find a rotational inertia about an axis through the center of mass, we must first find the center-of-mass location; we can do this with Eq. (8–48a). Once that is done, both parts of the problem are simple substitution into the expression for the rotational inertia, Eq. (9–16).

Working It Out (a) The center of mass is located at [Eq. (8–48a)]

$$X = \frac{m_1 x_1 + m_2 x_2}{m_1 + m_2} = \frac{(2.0\,\text{kg})(0\,\text{m}) + (1.0\,\text{kg})(0.6\,\text{m})}{2.0\,\text{kg} + 1.0\,\text{kg}} = 0.2\,\text{m}.$$

Now we find the rotational inertia about the axis through this point perpendicular to the dumbbell rod:

▲ **FIGURE 9–12** Dumbbell with two masses.

$$I_{cm} = m_1(x_1 - X)^2 + m_2(x_2 - X)^2$$
$$= (2.0\,\text{kg})(0\,\text{m} - 0.2\,\text{m})^2 + (1.0\,\text{kg})(0.6\,\text{m} - 0.2\,\text{m})^2$$
$$= 0.2\,\text{kg} \cdot \text{m}^2.$$

(b) For the axis through m_1, m_1 itself does not contribute to the rotational inertia because it is on the axis. Mass m_2 is a distance L from the axis. Thus

$$I = m_2 L^2 = (1.0\,\text{kg})(0.6\,\text{m})^2 = 0.4\,\text{kg} \cdot \text{m}^2.$$

CONCEPTUAL EXAMPLE 9–4 In the discussion of the dumbbell to this point, we assumed that the masses were point masses. Would the calculation of the position of the center of mass or of the rotational inertia change if the masses were not point masses but were instead uniform spheres of some radius?

Answer The mass distribution changes when the masses are no longer pointlike. This has no effect on the calculation of the center of mass. The center of mass is linear in the positions of the elements involved, and as we established in Chapter 8, we could first find the center of mass of each sphere, which would be at the center of the sphere, then use those as a two-pointlike-mass input to the overall center of mass. Thus the result for the center of mass position is unchanged. The rotational inertia would change, however, because that quantity involves distances squared. This is most easily understood with a simplified example: Place an object of mass M at the right end of a massless rod of length L and find the rotational inertia about an axis perpendicular to the rod and through the rod's left end. If the object is pointlike, $I = ML^2$. But if instead the object were composed of a pointlike mass $M/2$ a distance $L - \Delta L$ from the left end and a pointlike mass $M/2$ a distance $L + \Delta L$ from the left end, with $\Delta L \ll L$, we would find

$$I = (M/2)(L - \Delta L)^2 + (M/2)(L + \Delta L)^2 = M[L^2 + (\Delta L)^2].$$

This differs from the single-pointlike-mass case. The fact that it differs by an amount that is second order in the "size" of the object tells us that it is a good approximation to treat a real dumbbell as a dumbbell with pointlike masses at the end.

Thin Cylinder: The thin-walled cylinder (a cylindrical shell) shown in Fig. 9–13 has radius R, height h, and total mass M distributed uniformly around the ring. Let's find the rotational inertia about the central axis of the cylinder. The word "thin" means that we can consider all of the mass to be the same distance R from the axis. Thus, the rotational inertia is simply

$$I = MR^2. \tag{9–23}$$

Solid Cylinder: We can very often simplify the calculation of the rotational inertia of an object by breaking the object up into appropriate pieces. The rotational inertia of a uniform solid cylinder about its central axis can be calculated by breaking the solid cylinder into a series of concentric thin cylinders and using our result for the thin cylinder. We start with the mass density. The solid cylinder has exterior radius R, height h, and mass M, so its mass density ρ is

$$\rho = \frac{M}{V} = \frac{M}{\pi R^2 h}. \tag{9–24}$$

Now we divide the solid cylinder into thin cylindrical shells. The jth shell is a distance r_j from the axis and has (small) thickness Δr and mass Δm_j (Fig. 9–14). The cylindrical shell labeled j in Fig. 9–14 contributes an amount $\Delta m_j r_j^2$ to the rotational inertia I of the solid cylinder. We sum these contributions to find I:

$$I = \sum_j (\Delta m_j) r_j^2. \tag{9–25}$$

Just as we remarked in the discussion leading to Eq. (9–19), we can carry out this sum once we find the mass of each shell. The mass of a given volume is ρ times that volume; for the thin shell labeled j, the volume is approximately the shell thickness times the shell circumference times the height, $\Delta V_j = (\Delta r)(2\pi r_j)h$. Thus $\Delta m_j = \rho(\Delta r)(2\pi r_j)h$, and

$$I = \sum_j (\Delta m_j) r_j^2 = \sum_j \rho(\Delta r)(2\pi r_j) h r_j^2 = 2\pi h \rho \sum_j (\Delta r) r_j^3.$$

In the limit that the cylindrical shells are very thin, the sum, which includes shells from $r = 0$ to $r = R$, becomes an integral:

$$I = 2\pi h \rho \int_0^R r^3 \, dr = 2\pi h \rho \frac{R^4}{4}.$$

When we substitute Eq. (9–24) for ρ, we find

$$I = 2\pi h \frac{M}{\pi R^2 h} \frac{R^4}{4} = \frac{1}{2} MR^2.$$

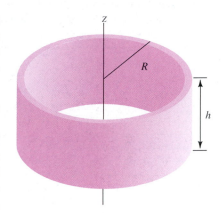

▲ **FIGURE 9–13** A thin cylinder of mass M has height h and radius R. We want to find its rotational inertia about the z-axis, which is the symmetry axis of the cylinder.

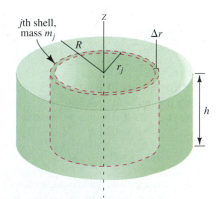

▲ **FIGURE 9–14** To find the rotational inertia of a thick cylinder of mass M, height h, and radius R about its symmetry axis, we break it into a series of thin concentric cylinders. We know the rotational inertia of the thin cylinders from Eq. (9–23) and can sum these inertias to find the rotational inertia of the solid cylinder.

Note that we have reduced the volume integration (a three-dimensional integration) to a one-dimensional one. How did we do this? All the mass at a given distance—meaning at all angles—contributed equally. In other words, symmetry has allowed us to perform the angle integration "automatically." In addition, there is no dependence on height, so the integration over height was also done automatically. This sort of simplification is very often possible when the object involved has some symmetry.

Note that the value of I is less than that of a thin cylinder of the same radius and mass. The weighted sum for the solid cylinder includes mass pieces that give a smaller contribution because they are closer to the axis. Note too that the result is independent of the height and applies for a thin disk as well as a tall cylinder.

CONCEPTUAL EXAMPLE 9–5 Will the rotational inertia of a thick cylinder about its central axis be (a) less than, (b) greater than, or (c) the same as the rotational inertia of a thin cylinder of the same outer radius, length, and total mass?

Answer (a) There is relatively more of the mass of the thin cylinder at the outer radius, and this "large radius" material is weighted more heavily in the calculation of the rotational inertia. Indeed, we can use a dimensional analysis to write the rotational inertia about their axes of a set of cylinders of the same outer radius, height, and mass in the form $I = CMR^2$. Here C is a dimensionless parameter that runs from 1 (all mass concentrated at outer radius, as in the thin cylinder) to 0 (all mass concentrated along the central axis). By the reasoning here, we can conclude that

$$C_{\text{thin cylinder}} > C_{\text{thick cylinder}} > C_{\text{solid cylinder}}.$$

▲ **FIGURE 9–15** Calculating the rotational inertia about an axis perpendicular to the rod passing through the center of mass. This is at the midpoint if the rod's density is constant.

Rod: Next we consider the rotational inertia of a uniform rod of length L about an axis that is perpendicular to it and passes through its center of mass (Fig. 9–15). As this is a thin rod, it can be approximated as a one-dimensional mass distribution, and Eq. (9–20) can be used. The center of mass is the midpoint of the rod, and we will place the origin of our coordinate system at this point. The mass density is the total mass divided by the rod's length, $\lambda = M/L$. Consider the mass dm contained in a slice of the rod of length dx at a distance x from the axis. We have $dm = \lambda\, dx$, so the rotational inertia about the axis is given by

$$I = \int_{-L/2}^{L/2} x^2 \lambda\, dx = \lambda \int_{-L/2}^{L/2} x^2\, dx = \lambda \left. \frac{x^3}{3} \right|_{-L/2}^{L/2} = \frac{ML^2}{12}.$$

If λ is not a constant but is symmetric about the midpoint, so that the midpoint is still the center of mass, then λ is replaced by $\lambda(x)$, and the rotational inertia about the center of mass is the integral

$$I = \int_{-L/2}^{L/2} x^2 \lambda(x)\, dx.$$

EXAMPLE 9–6 Find the rotational inertia of a uniform thin rod of length L about an axis perpendicular to the rod and passing through one end of it.

Setting It Up We sketch the situation in Fig. 9–16. The x-axis is along the rod with $x = 0$ at the end that is the site of the rotation axis. The rod, of length L and mass M, has a constant mass density (mass per unit length) $\lambda = M/L$.

Strategy This is a direct application of the expression for the rotational inertia of a continuous object in one dimension, Eq. (9–21). The fact that the rotation axis passes through one end of the rod makes it useful for the origin of our x-axis to pass through that end as well.

Working It Out With the rod running along the x-axis from $x = 0$ to $x = L$,

$$I = \int_0^L x^2 \lambda\, dx = \lambda \int_0^L x^2\, dx = \left. \frac{\lambda}{3} x^3 \right|_0^L = \frac{\lambda L^3}{3} = \frac{1}{3}ML^2. \quad (9\text{–}26)$$

The mass density is a constant, which explains why it came out of the integration.

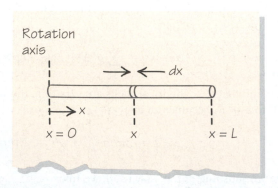

▲ **FIGURE 9–16** The axis of rotation passes through the end at $x = 0$ and is perpendicular to the rod.

◀ **FIGURE 9–17** To find the rotational inertia of any three-dimensional object about a given axis of rotation, first find the rotational inertia about the parallel axis through the center of mass. The parallel-axis theorem gives a general relation between these two rotational inertias. Here, the object in question is a tennis racket.

Parallel-Axis Theorem

We have worked out the rotational inertia for several simple systems and for several specific axes. Yet how useful are such exercises, as after all there is a different answer for every axis? There is, however, a natural set of axes for every object, and these are axes that go through the center of mass of the system. It may be sufficient to calculate the rotational inertia of objects about these axes because the rotational inertia about a set of other axes can be obtained from the **parallel-axis theorem**. This theorem, whose proof we outline below, states that there is a simple relation between the rotational inertia of an object about an axis through the object's center of mass and the rotational inertia about any other parallel axis (Fig. 9–17). More precisely, the parallel-axis theorem states that

$$I_{pa} = I_{cm} + Md^2, \qquad (9\text{–}27)$$

PARALLEL-AXIS THEOREM

where I_{cm} and I_{pa} are the rotational inertias between the axis through the center of mass and the second axis, respectively, M is the total mass of the object, and d is the perpendicular distance between the axis through the center of mass and the parallel axis.

As a simple example, let's take a dumbbell with equal masses m and consider two parallel axes each perpendicular to the connecting rod: One passes through the center of mass and one through one of the masses (Fig. 9–18). For the axis through the center of mass, we have $I_{cm} = m(L/2)^2 + m(L/2)^2 = mL^2/2$, while direct calculation for the axis through the end gives $I_{end} = mL^2$. What does the parallel-axis theorem give for I_{end}? The distance d in Eq. (9–27) is $L/2$, while $M = 2m$, so the parallel-axis theorem gives $I_{end} = I_{cm} + M(L/2)^2 = mL^2/2 + (2m)(L/2)^2 = mL^2$, the same result given by the direct calculation.

The theorem shows that the rotational inertia of any object is smallest if the axis of rotation goes through the object's center of mass ($d = 0$). This means that the kinetic energy of rotational motion of an object for fixed angular velocity ω is smallest if the rotation is about an axis containing the center of mass.

You can get an idea of how a proof of the parallel-axis theorem works by looking at a two-dimensional object (Fig. 9–19). The reference axes in question will be perpendicular to the page. We have set the origin of the coordinate system at the position O of the center of mass, and the second axis passes through the point P; we have chosen the x-axis along the line OP. A particular element is picked out with mass dm; the rotational inertia about the axis through P is then $I_P = \int r'^2 \, dm$. But $r'^2 = d^2 + r^2 - 2r'd \cos \theta$, so that

$$I_P = \int (r^2 + d^2 - 2rd \cos \theta) \, dm = \int r^2 \, dm + d^2 \int dm - 2d \int r \cos \theta \, dm.$$

The first term on the right is just the rotational inertia about the axis through the center of mass, I_{cm}. The second term is Md^2. The factor $r \cos \theta$ in the integral of the third term is x, and the integral $\int x \, dm$ is by definition the x-component of the center of mass. But the center of mass is at the origin, so this term is zero. We are left with Eq. (9–27), the parallel-axis theorem.

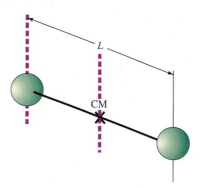

▲ **FIGURE 9–18** Two parallel axes, one through the center of mass and the other through one end, can be drawn through this dumbbell of equal masses. By using the rotational inertia for the center of mass axis and then the parallel-axis theorem, we can easily find the rotational inertia about the axis through one end.

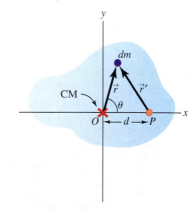

▲ **FIGURE 9–19** For establishing the parallel-axis theorem, we set two axes perpendicular to the page, one through the center of mass at O and the other through the point P.

EXAMPLE 9–7 Consider a uniform, thin, rectangular sheet of metal of mass M with width a and length b. Calculate the rotational inertia about an axis that is perpendicular to the sheet and passes through one of the corners.

Setting It Up Choose the corner through which the axis passes as the origin and line up the width along the x-axis and the length along the y-axis, as shown in Fig. 9–20.

Strategy This problem is another direct application of the calculation of a rotational inertia for a continuous object. The object is two dimensional, so Eq. (9–21) is relevant. That equation is expressed in terms of the two-dimensional mass density (mass per unit area) σ, which in this case is uniform.

Working It Out To apply Eq. (9–21), we note that $0 \le x \le a$ and $0 \le y \le b$, that r^2 in the integral is $x^2 + y^2$, and that the two-dimensional uniform mass density σ is

$$\sigma = \frac{M}{\text{area}} = \frac{M}{ab}.$$

Also, the area element dA in Eq. (9–21) can be written as $dx\,dy$—we have broken the area down into little rectangles each of area width times height. To sum over the whole area (integrate), we take consecutive steps in which we first fix x and sum the rectangle heights dy over all y-values and then sum the rectangle widths dx over all x-values. Thus

$$I_{\text{corner}} = \int_0^a dx \int_0^b dy\, \sigma(x^2 + y^2)$$

$$= \frac{M}{ab} \int_0^a dx \int_0^b dy(x^2 + y^2)$$

$$= \frac{M}{ab}\left(\int_0^a x^2\, dx \int_0^b dy + \int_0^a dx \int_0^b y^2\, dy \right)$$

$$= \frac{M}{ab}\left[\left(\frac{a^3}{3}\right)b + \left(\frac{b^3}{3}\right)a \right]$$

$$= \frac{1}{3}M(a^2 + b^2).$$

Note that this is a case where we did not directly reduce a multidimensional integral to a one-dimensional integral, as we did, for example, in the calculation of the rotational inertia of the solid cylinder. However, note that the integration procedure above is in fact a sequence of one-dimensional integrals.

What Do You Think? Does it matter if the sheet is thick rather than thin?

▲ **FIGURE 9–20** We calculate the rotational inertia about the z-axis, which passes through one corner.

EXAMPLE 9–8 Use the result of Example 9–7 to calculate the rotational inertia of the same sheet of metal about an axis that is perpendicular to the plane of the sheet and passes through the sheet's center of mass.

Setting It Up Figure 9–21 sketches the situation. We include the same coordinate system as in Fig. 9–20.

Strategy Comparison of Figs. 9–21 and 9–20 suggests that this is a problem for which the parallel-axis theorem is applicable. To use it, we must find the distance d between the two axes.

Working It Out The center of mass is located at the midpoint of the rectangle, that is, at $x = a/2$, $y = b/2$. The distance to the corner is given by

$$d^2 = \left(\frac{a}{2}\right)^2 + \left(\frac{b}{2}\right)^2.$$

Thus, by the parallel-axis theorem, $I_{\text{corner}} = I_{\text{cm}} + Md^2$:

$$I_{\text{cm}} = I_{\text{corner}} - Md^2 = \frac{1}{3}M(a^2 + b^2) - M\left[\left(\frac{a}{2}\right)^2 + \left(\frac{b}{2}\right)^2\right]$$

$$= M(a^2 + b^2)\left(\frac{1}{3} - \frac{1}{4}\right) = \frac{1}{12}M(a^2 + b^2).$$

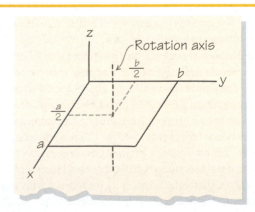

▲ **FIGURE 9–21** As for Example 9–7, except that we calculate the rotational inertia about an axis passing through the center of the plate.

What Do You Think? Why is the answer to this example much smaller than that of the previous example?

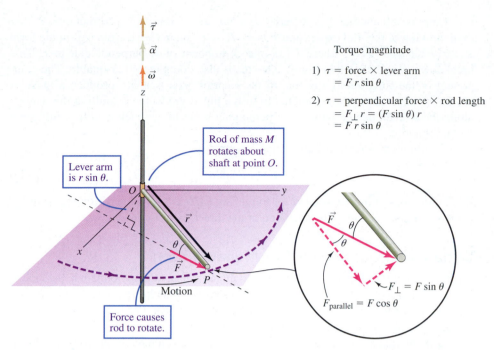

Torque magnitude

1) $\tau = \text{force} \times \text{lever arm}$
   $= F \, r \sin \theta$

2) $\tau = \text{perpendicular force} \times \text{rod length}$
   $= F_\perp \, r = (F \sin \theta) \, r$
   $= F \, r \sin \theta$

Rod of mass M rotates about shaft at point O.

Lever arm is $r \sin \theta$.

Motion

Force causes rod to rotate.

$F_\perp = F \sin \theta$

$F_{parallel} = F \cos \theta$

▲ **FIGURE 9–22** A rod is attached to a shaft and can rotate in the plane perpendicular to the shaft. The shaft acts as a rotation axis. A force lying in the plane in which the rod can move is applied to the rod, causing a rotation. The effectiveness of the force in causing a rotation depends on the strength of the force, on where along the rod it is applied, and on its direction, specified by the angle θ. The lever arm is $r \sin \theta$.

9–4 Torque

We have described rotations; now, let's turn to the question of what causes them. The analogies we have established between rotational motion and the motion of point particles (we refer to this as *linear motion*) will be useful here. Newton's second law describes the dynamics of linear motion by the equation $\vec{F} = m\vec{a}$. We'll see here that for rotational motion there is a dynamical equation for angular acceleration with a quantity analogous to force called **torque**, symbol $\vec{\tau}$. Just as a force provides a push or a pull, torque provides a twist. Thus it takes a torque to accelerate the rotation of a wheel or to turn a screw.

It is useful to keep a definite example in mind as we go through our discussion, the rotation of a rod attached at a point O to a shaft that can rotate (Fig. 9–22). The rod, which has length r and mass M, makes a right angle with the shaft. The motion is entirely in the plane perpendicular to the shaft. Suppose now that a force $\vec{F}$ acts in that plane on the end of the rod, at point P in Fig. 9–22. This force causes the system to undergo rotational motion, and in what follows we describe the effectiveness of the force in bringing about changes in rotational motion by making a suitable definition of the torque.

How effective a force is in setting an extended system (such as that of Fig. 9–22) into rotational motion about a point O depends on two features. First, a larger force will be more effective in exerting torque than would an otherwise similar force of smaller magnitude, so we expect the torque to be proportional to the force. Second, a force is most effective at producing a torque if it is applied at a point P in Fig. 9–22 that is *far* from the rotation axis (i.e., far from point O in Fig. 9–22) and is at *right angles* to the line between the rotation axis and the point of application of the force (line OP in Fig. 9–22). In other words, the torque τ is also reasonably proportional to $r_\perp \equiv r \sin \theta$ in Fig. 9–22, where $\vec{r}$ is the vector from point O to point P and θ is the angle between $\vec{r}$ and $\vec{F}$. The factor $r \sin \theta$ is known as the **lever arm** or **moment arm**. You can see from Fig. 9–22 that the lever arm is found by extending the line along which the force acts and measuring the closest distance of this line to the rotation axis. Plumbers who use long wrenches to free stubborn pipes are quite familiar with the effectiveness of a large lever arm (Fig. 9–23). Pulling the proportionality to both F and $r_\perp$ together, we would write $\tau = r_\perp F$.

▲ **FIGURE 9–23** A plumber taking advantage of a long lever arm. A long wrench is more effective than a short one for freeing a stubborn pipe joint.

There is still another way to verify that our idea of torque is a correct one. Rather than thinking of the trial expression $\tau = r_\perp F = (r \sin \theta)F$, we arrange it in the form $\tau = r(F \sin \theta) = rF_\perp$, where $F_\perp$ is the component of $\vec{F}$ perpendicular to $\vec{r}$. This breakdown is also shown in Fig. 9–22. This is also completely reasonable: The component of the force *along* the line to the rotation axis will not produce a twist. A plumber's apprentice who attempted to turn a pipe by pushing or pulling the wrench along the wrench handle rather than perpendicular to the wrench handle would soon be out of a job!

To summarize, our trial expression for torque is

$$\tau = rF \sin \theta. \tag{9–28}$$

The dimensions of this expression for torque are $[ML^2T^{-2}]$, and in SI it is measured in newton-meters. We discuss the question of the direction of torque below.

Dynamical Equation for Rotational Motion

To this point, our expression for torque is no more than an attempt to include the factors we know intuitively are effective at producing a twist. To make this concept a more useful one, we want to connect it *quantitatively* to a dynamical equation for rotations. We do this by applying Newton's second law.

Both internal and external forces can act on a given element of an object. In linear motion, internal forces do not cause any acceleration because they come in pairs that by Newton's third law cancel. In the same way, these pairs of canceling internal forces cannot cause a twist because they are aligned with each other, so we shall consider only external forces. And as Figs. 9–24a and b show, a pair of equal and opposite external forces can produce a twist if they are not aligned.

Suppose, then, that a net external force $\vec{F}_i$ acts on an element of a rod labeled by i (Fig. 9–25); this is similar to the situation described in Fig. 9–22. We suppose that the force is perpendicular to the rod; this is good enough because any component along the rod has zero lever arm and is ineffective at producing a twist. The direction of the force is therefore purely *tangential*. The ith element is a distance r_i from the rod's attachment point to the fixed shaft and has mass Δm_i. The torque on element i about the origin therefore has magnitude $\tau_i = r_i F_i$. At the same time, we know from the second law that F_i produces an instantaneous tangential acceleration a_i of element i whose magnitude is $a_i = F_i / \Delta m_i$. Because the acceleration a_i is tangential, we can use Eq. (9–12), $a_i = \alpha r_i$. In other words, because the force is perpendicular to the rod, the instantaneous acceleration of the mass element is associated with an *angular* acceleration of the mass element. Because of internal forces within the (rigid) rod, the entire rod undergoes the same angular acceleration, so there is no need for a subscript i on the quantity α. Combining, we have Newton's second law:

$$F_i = \Delta m_i \, r_i \alpha;$$

hence, the torque on the element i has magnitude

$$\tau_i = r_i F_i = \Delta m_i \, r_i^2 \alpha.$$

The *net* torque τ comes from summing the torques on each element and therefore has magnitude

$$\tau = \sum_i \Delta m_i \, r_i^2 \alpha = \left(\sum_i \Delta m_i \, r_i^2 \right) \alpha.$$

The quantity in parentheses is the rotational inertia I about the shaft [Eq. (9–15)]. We have found the dynamical relation we were looking for:

$$\tau = I\alpha. \tag{9–29}$$

EQUATION FOR ROTATIONAL MOTION

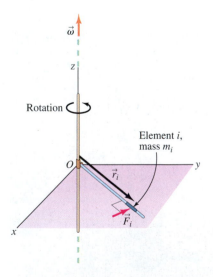

▲ **FIGURE 9–24** (a) A pair of equal and opposite forces that do not act along the same line exert a torque on an object. (b) There is no torque only if the same pair of forces act along the same line.

▲ **FIGURE 9–25** A force of magnitude F_i acts on an element i of a rod attached to a rotating shaft. The force is perpendicular to the rod.

When there is an angular acceleration about an axis perpendicular to the plane of rotation, it is dynamically determined by the torque about the same point according to Eq. (9–29). This equation is analogous to Newton's second law, which determines the linear acceleration of an object in terms of the object's inertial mass and the force acting on the object. The major difference is that all the quantities in Eq. (9–29)—torque, angular acceleration, and rotational inertia—refer to a given axis.

It is worth emphasizing that we have not introduced any new laws of physics here. All the laws of rotational motion follow from Newton's original laws.

Newton's second law is a vector relation and we have already stated that the angular acceleration is a vector. Where is the vector aspect of the torque? Figure 9–26 shows that we can use a right-hand rule to define correctly the direction of the torque: If the fingers of the right hand are aligned along the line perpendicular to the rotation axis and extending to the point where the force is applied (in other words, along the vector $\vec{r}$) and then curled in the direction of the force, the thumb points in the direction of the torque vector $\vec{\tau}$. The procedure is illustrated in Fig. 9–27, and we can see that the torque is always in the same direction as the angular acceleration. The vector equation of motion takes the form

$$\vec{\tau}_{\text{net}} = I\vec{\alpha}, \qquad (9\text{--}30)$$

where, as usual, the word "net" means the vector sum of all the individual torques acting.

In this chapter, the vector nature of torques will not play a very important role, because we are restricting ourselves to rotations about a single axis. It is important, however, to become used to thinking of the dynamics of rotational motion in terms of vectors. In the more general treatment given to rotations in Chapter 10, the vector nature of torque is critical.

▲ **FIGURE 9–26** The direction of the torque can be determined by a right-hand rule applied to the vectors $\vec{r}$ and $\vec{F}$. The fingers of the right hand are aligned along $\vec{r}$ and then curled toward $\vec{F}$, with the angle between $\vec{r}$ and $\vec{F}$ always taken to be less than 180°. The torque is then along the thumb.

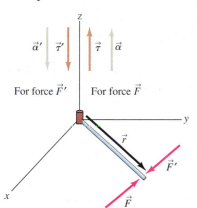

◀ **FIGURE 9–27** The direction of the angular acceleration of the rod, $\vec{\alpha}$, and of the torque on the rod, $\vec{\tau}$, depends on the direction of the force. When the force is in one direction ($\vec{F}$), the torque $\vec{\tau}$ and angular acceleration point along $+z$; when the force is in the other direction ($\vec{F}'$), the torque $\vec{\tau}'$ and angular acceleration point along $-z$.

CONCEPTUAL EXAMPLE 9–9

Look at Fig. 9–28. What is the direction of the torque on the large ship about its center of mass? Supposing that the pushing force applied by the tugboat is at its maximum, is the magnitude of this torque also a maximum?

Answer The torque will make the large ship turn in a clockwise fashion as viewed from above. Using a right-hand rule for the sense of this rotation, the right thumb will point down to Earth's center, and this is the direction of the torque. The force is applied at the very end of the ship, and this is as far from the ship's center of mass as one can get. In addition, the force is apparently applied at right angles to the line from the center of mass. Thus the torque has maximum magnitude.

▲ **FIGURE 9–28** Tugboat rotating ship.

EXAMPLE 9–10 A massless rod of length 0.83 m connects two small spheres of mass 0.25 kg each. The rod is constrained to rotate about an axis perpendicular to the rod and passing through its midpoint. The initial angular velocity has magnitude 2.1 rad/s, and the rod rotates counterclockwise. A tangential force of magnitude 9.6 N acting in the counterclockwise directon is applied to one of the spheres (Fig. 9–29). If the force is applied for 2.0 s, what is the final angular speed of the rod?

Setting It Up As Fig. 9–29 shows, we place the z-axis along the rotation axis, with $+z$ in the direction of the initial angular velocity. We know the rod length L, the masses m, and the initial angular speed ω_0. We also know the magnitude F of the tangential force and the time t over which it acts. We want the final angular speed ω.

Strategy We are given enough information to find the torque. With it, we can use the dynamical equation relating torque and angular acceleration, and because the torque is constant, so is the angular acceleration—we can solve for its value from the dynamical equation. The kinematics of constant angular acceleration will then lead us to a final angular velocity linearly changing with time.

Working It Out The angular acceleration about the axis points in the same direction as the torque about that axis. The rotation axis is at the midpoint, a distance $L/2$ from the point where the force is applied, so the torque has magnitude

$$\tau = (L/2)F.$$

Equation (9–29) then gives an angular acceleration in terms of the torque, constant in magnitude for a constant-magnitude torque, $\alpha = \tau/I$. We have just found τ, but we also need the rotational inertia of the two-sphere system about the axis,

$$I = m\left(\frac{L}{2}\right)^2 + m\left(\frac{L}{2}\right)^2 = \frac{mL^2}{2}.$$

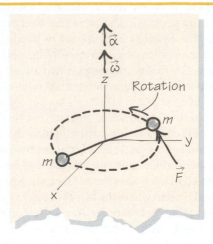

▲ **FIGURE 9–29** Force produces torque which causes angular acceleration.

Thus we have

$$\alpha = \frac{\tau}{I} = \frac{LF/2}{mL^2/2} = \frac{F}{mL}.$$

With a constant angular acceleration, the angular speed has a linear dependence on the time,

$$\omega = \omega_0 + \alpha t = \omega_0 + \frac{F}{mL}t.$$

At time $t = 2.0$ s, this gives

$$\omega = 2.1 \text{ rad/s} + \frac{9.6 \text{ N}}{(0.25 \text{ kg})(0.83 \text{ m})}(2.0 \text{ s}) = 95 \text{ rad/s}.$$

What Do You Think? True or false: If the applied force had half the magnitude but were applied for twice the time, the resulting final angular speed would be the same.

Free-Body Diagrams Revisited

The discussion of torque suggests that to extend the usefulness of free body-diagrams to situations with rotations of extended objects we need to take into account the location at which forces act. A description of appropriately extended free-body diagrams and advice on how to approach problems that involve rigid-body motion are given in the Problem-Solving Techniques box.

Problem-Solving Techniques

1. Identify and isolate the rigid body on which external forces act.

2. Identify the external forces that act on the object as well as where they act.

3. Prepare an extended free-body diagram: All forces should be included, as well as *information on where they act* on the rigid body. Where the force acts is of paramount importance in understanding torques. Note that gravity acts as if it were applied to the center of

mass of the object [see Eq. (9–36) and the following text].

4. Identify a single convenient axis, which will normally be the axis about which the rigid body rotates.

5. Find the torques about this axis that result from the acting forces. The direction of positive torque should be clearly understood using the right-hand rule as necessary.

6. Express the dynamical equations of motion that correspond to the *net* torque. This step generally requires knowledge (or calculation) of the rotational inertia about the axis of rotation. The solutions of the dynamical equations describe the rotational motion of the rigid body.

7. The net force governs the linear motion of the center of mass.

Let us look at an example. A bucket of water of mass *m* is connected to a rope of negligible mass (Fig. 9–30a). The rope is wrapped around a pulley of mass *M*, radius *R*, and rotational inertia *I* about its axis. The pulley is free to rotate without friction about its axis, which is horizontal and fixed to the wall. The bucket is released from rest and drops, making the pulley turn. We want to use extended force diagrams to find the equations that would allow you to find the angular velocity of the pulley after the bucket has fallen for a time *t*.

From Fig. 9–30a we see that there are two objects to consider, the bucket and the pulley. We will therefore identify the forces on and draw extended free-body diagrams for both of these objects. Figure 9–30a also reminds us that there is a connection, namely an unstretchable rope, between the pulley and the bucket, which results in the angular acceleration of the pulley being kinematically connected to the linear acceleration of the bucket.

We next identify all the forces on *each* object and draw corresponding extended force diagrams. Once that is done, we can express Newton's second law and any torque equations. Finally, we will use the fact that the rotational motion of the pulley is connected to the linear motion of the bucket because of the unstretchable rope connecting them. Once we have written all the corresponding expressions, we can count equations and unknowns to ensure that we can carry through the solution.

The forces acting on the pulley are a reaction force that keeps the pulley in place, F_N, at its axis, the force of gravity, magnitude *Mg*, and the tension of the rope, magnitude *T*. These appear in the extended free-body diagram for the pulley, Fig. 9–30b. The forces on the bucket are the rope tension and the force of gravity on the bucket, magnitude *mg*, and the extended free-body diagram for the bucket is Fig. 9–30c, although the "extended" free diagram for the bucket is really just an ordinary free-body diagram.

With the free-body diagrams drawn, we can write the dynamical equations. All the forces are vertical, so there is a single (*y*-component) Newton's law equation for each object—this axis is identified in Fig. 9–30a. There is a torque equation only for the pulley, and if we calculate angular quantities about the pulley axis, we can see that only the rope tension produces a torque. This is also only a single equation as the angular velocity in Fig. 9–30b will point along the axis of rotation—in our drawing the right-hand rule shows that it will be out of the page. Explicitly we recognize that the center of mass of the pulley does not accelerate, as it is fixed to the wall, so the net force acting on it must be zero:

$$\text{second law for pulley: } F_N - T - Mg = 0. \tag{9–31}$$

The torque acting on the pulley has magnitude *TR*; hence

$$TR = I\alpha, \tag{9–32}$$

with α the pulley's angular acceleration magnitude and *I* the given rotational inertia of the pulley about its axis. As the torque is directed out of the page, α will also be directed out of the page, corresponding to an increasing angular speed. Finally, for the bucket, there is an (as-yet-unknown) acceleration *a*, which we assume is downward by writing the acceleration term of the second law as $-ma$:

$$\text{second law for bucket: } T - mg = -ma. \tag{9–33}$$

(The minus sign on the right-hand side will take care of itself; if the equations were to give us a value for *a* that is negative, that would mean that the actual acceleration is upward.)

We now have three dynamical equations but four unknowns: F_N, *T*, *a*, and α. We have not yet brought in a last element, a *kinematic* relation between *a* and α. This relation comes from the fact that if the bucket moves at a speed *v*, then any point on the rope moves at the same speed, and thus the outer rim of the flywheel moves at that speed. This means that the pulley rotates about its axis at an angular speed ω such that

$$v = \omega R.$$

If we take a time derivative of this equation, we find

$$a = \alpha R. \tag{9–34}$$

▲ **FIGURE 9–30** (a) Bucket suspended from a pulley. The pulley turns as the bucket drops. (b) Extended free-body diagram for the pulley. (c) Free-body diagram for the bucket.

Equations (9–31) to (9–34) are enough information to solve for the unknown a, which will be a constant. We can then use the fact that the speed v increases linearly with time, and once we have v, we have $\omega = v/R$.

EXAMPLE 9–11 Referring to the discussion of the falling bucket above, suppose that the bucket and pulley masses are $m = 12$ kg and $M = 88$ kg, respectively, and that the pulley is a uniform cylinder of radius $R = 0.50$ m. Find the angular velocity of the pulley after the bucket has fallen for 5 s.

Setting It Up See Fig. 9–30 for a suitable diagram.

Strategy This is a matter of solving Eqs. (9–31) to (9–34) for one of the quantities (e.g., the angular acceleration of the pulley), then using kinematics to obtain the angular speed as a function of time. (Actually only three of the equations are needed.) We'll also need to find the rotational inertia of the pulley, contained in Eq. (9–32). As for virtually all problems, we first solve algebraically, plugging in numbers only at the end.

Working It Out First the rotational inertia of the pulley: The cylinder is a uniform solid, so $I = \frac{1}{2}MR^2$ (see Table 9–1). Next we solve our equations for the desired unknown, here the angular acceleration α, a process that can be done in a number of ways. The simplest is to solve Eq. (9–33) for T, $T = m(g - a)$. Now we substitute $a = \alpha R$ in this result and insert it into the torque equation (9–32):

$$TR = m(g - \alpha R)R = I\alpha.$$

This equation can be immediately solved for α, namely,

$$\alpha = \frac{mgR}{mR^2 + I} = \frac{mgR}{mR^2 + (M/2)R^2} = \frac{mg}{[m + (M/2)]R}$$

This angular acceleration is constant, implying that the angular velocity changes linearly with time:

$$\omega = \omega_0 + \alpha t = \omega_0 + \frac{mgt}{[m + (M/2)]R}.$$

By the initial conditions of the problem, $\omega_0 = 0$ rad/s, and after 5 s,

$$\omega = \frac{(12 \text{ kg})(9.8 \text{ m/s}^2)(5 \text{ s})}{[(12 \text{ kg}) + (88 \text{ kg}/2)](0.50 \text{ m})} = 21 \text{ rad/s}.$$

What Do You Think? The fact that both the angular velocity vector of the pulley and its angular acceleration vector are directed out of the page means that the angular velocity is (a) constant in time, (b) changing quadratically with time, (c) changing in magnitude while unchanging in direction, or (d) none of the above.

CONCEPTUAL EXAMPLE 9–12 Consider once again the arrangement of the falling bucket that turns the pulley shown in Fig. 9-30. Will the angular frequency increase more quickly or less quickly if more of the mass M of the pulley is concentrated near the outer radius R?

Answer If more of the mass M of the pulley is concentrated near the outer radius R, the rotational inertia of the pulley about its axis will be larger. The increased rotational inertia for torque is like increased mass for force: In each case the motion in response is more "sluggish." For a given tension in the rope the angular acceleration will be smaller. We can therefore conclude that the angular speed will increase more slowly if the mass of the pulley is concentrated at the edge. In equations, the rotational inertia of the pulley will take the general form CMR^2, where $C = 0$ corresponds to all the mass concentrated at the pulley axis, $C = \frac{1}{2}$ is a uniform solid cylinder, and $C = 1$ corresponds to all the mass concentrated at the outer radius R. You can trace through the algebra of the previous example and see that ω increases more slowly for $C = 1$.

Gravity and Extended Objects

Gravity acts on every point of an object at once. Here, we argue that the torque on an extended object due to gravity can be computed as if the entire mass of the object were concentrated at its center of mass. It is enough to work with an easily generalized example, a rod of length L and mass M fixed to a pivot at its upper end (Fig. 9–31a). We divide the rod into a series of pieces labeled by the subscript i. The pieces, which have

▶ **FIGURE 9–31** The force of gravity rotates a rod about the point O. (a) The differential element of force $\vec{F}_i$ acts on a differential mass Δm_i located a distance s_i down the rod. The circled cross indicates that the torque τ is directed into the plane of the page. (b) The net effect of the sum of the torques acting on each piece is a net torque that is the same as would be present if the entire mass of the rod were concentrated at the rod's center of mass.

(a)

(b)

mass Δm_i, are small enough so that we can consider each of them to be a given distance s_i down the rod. We will compute the net torque due to gravity about the pivot point assuming the rod makes an angle θ to the vertical. The net torque τ is the vector sum of the torques on each piece, τ_i. All the torques are directed into the page, so this sum is the simple algebraic sum of the magnitudes τ_i. In turn, the moment arm for the force of gravity on the piece labeled by i is $s_i \sin \theta$, so the torque τ_i is

$$\tau_i = (s_i \sin \theta)\, \Delta m_i\, g.$$

The net torque then has magnitude

$$\tau = \sum_i \tau_i = \sum_i (s_i \sin \theta)(\Delta m_i\, g) = \left[\sum_i (s_i)(\Delta m_i) \right](\sin \theta) g. \qquad (9\text{–}35)$$

By definition (see Chapter 8), the quantity in brackets is MR, where R is the distance of the center of mass from the reference point. Thus the net torque due to gravity is (Fig. 9–31b)

$$\tau = R(Mg) \sin \theta. \qquad (9\text{–}36)$$

Compare this result with Eq. (9–28), which is our general form for the torque. *The torque due to gravity on an extended object of total mass M may be represented by the torque due to gravity acting on a particle of mass M located at the object's center of mass.* This important and very useful result is a general one that holds for any object, not just a rod, as you could prove by breaking up an arbitrary object into pieces as we did for the rod.

We could add as a corollary that gravity cannot exert torque on an object about the object's center of mass. This can be helpful in problem solving.

The same reasoning can be applied to energy considerations. *When an extended object of mass M moves in any way, we can find the change in its gravitational potential energy by finding the change in height, Δh, of the center of mass. The change in gravitational potential energy is then $Mg(\Delta h)$.*

9–5 Angular Momentum and Its Conservation

We have uncovered a series of analogues between rotation and linear motion—the analogous descriptive variables appear in analogous positions in expressions for energy and the dynamical equation for rotational motion is analogous to Newton's second law. The analogue between linear and rotational motion can be of further use to us. In our original formulation of Newton's second law, the quantity $m\, d\vec{v}/dt$ appears. In Chapter 8 we found it useful to define the linear momentum $\vec{p} \equiv m\vec{v}$ and to formulate a more general form of Newton's second law:

$$\vec{F} = \frac{d\vec{p}}{dt}.$$

Similarly, it is useful to define the **angular momentum** $\vec{L}$ of a symmetrical object that rotates about its symmetry axis (or an axis parallel to the symmetry axis) with angular velocity $\vec{\omega}$ by

$$\vec{L} \equiv I\vec{\omega}. \qquad (9\text{–}37)$$

ANGULAR MOMENTUM

Here, I is the rotational inertia of the object with respect to the rotation axis. We see here that angular momentum is a vector aligned with $\vec{\omega}$. For uniform rotational motion about an axis, the angular momentum does not change in either magnitude or direction.

To gain some idea of the possible range of angular momenta in nature, you might try calculating the angular momentum of Earth's motion about its axis of rotation given that its mass is 6×10^{24} kg and its radius is 6.4×10^6 m. With a uniform mass density,

$I = 2/5MR^2 = 10^{38}$ kg·m². The value of ω comes from the known period T, 1 day, namely $\omega = 2\pi/T$. After conversion of T to seconds, you find $\omega = 7.3 \times 10^{-5}$ rad/s. Combining, the magnitude of Earth's angular momentum is $L = I\omega = 7 \times 10^{33}$ kg·m²/s. (In fact, the actual value of the angular momentum is about 20 percent less than this, and this is because the distribution of mass is not uniform but is more concentrated at the center, reducing I from the value we found by 20 percent.) At another extreme there is a smallest nonzero value of angular momentum that occurs in nature, about 10^{-34} kg·m²/s. This is a significant fact associated with quantum mechanics that we shall discuss in the next chapter and elsewhere.

In terms of angular momentum, the dynamical equation for rotational motion takes the more general form

$$\vec{\tau} = \frac{d\vec{L}}{dt}. \tag{9-38}$$

EQUATION FOR ROTATIONAL MOTION

Because angular momentum is so fundamental to rotational motion, it is useful to write the energy in terms of it. We have for a simple rotation

$$K = \frac{1}{2}I\omega^2 = \frac{1}{2}\frac{(I\omega)^2}{I} = \frac{L^2}{2I}. \tag{9-39}$$

Angular momentum shares with momentum (linear momentum is simply momentum) the important property that it is independent of time for a system that is "left alone," that is, a system on which there is no (net) torque due to external forces. (Note that the external torque may be zero even when the external force is not zero, depending on where the external force is applied and on its direction. Conversely, it is possible to have a net torque even though the net force is zero.) When the net torque on an object is zero, its angular momentum is independent of time—the angular momentum is *conserved*. For rigid bodies, the rotational inertia is constant, and so the conservation of angular momentum tells us that the angular velocity is constant in time. The conservation of angular momentum in more complicated systems is a powerful tool. The following example shows how collisions can be analyzed with the help of this conservation law. In Chapter 10 we'll show in much more detail how we can use the conservation of angular momentum.

EXAMPLE 9–13 An engineer works with a turntable that accepts a stack of disks. The engineer starts with a model in which a freely rotating turntable, friction free, of rotational inertia I_1 about the axis passing through its center and perpendicular to the center is rotating with an angular speed ω_i when a disk of rotational inertia I_2 about the same axis is dropped from rest onto the turntable, with its center on the turntable's center. Friction between disk and turntable brings them to a common final angular speed ω_f. What is ω_f?

Strategy We recognize that this is a colliding system upon which no external torques act so that angular momentum is conserved. The initial angular momentum is associated with the turntable—the disk initially has none. As for the final angular momentum, we recognize that when the turntable and the disk rotate together about a common axis, their net rotational inertia is the sum of their individual rotational inertias about the same axis, $I_1 + I_2$. We can then set the initial angular momentum to the final angular momentum and solve for ω_f.

Working It Out The initial angular momentum is $I_1\omega_i$ and the final angular momentum is $(I_1 + I_2)\omega_f$. We set them equal and easily solve for ω_f:

$$\omega_f = \frac{I_1\omega_i}{I_1 + I_2}.$$

Note that this is less than ω_i: The disk speeds up but the turntable slows down.

What Do You Think? Is energy conserved in the process described here?

Parallels between Rotational and Linear Motion

Throughout this chapter, we have emphasized the similarities and the differences between rotational motion and linear motion. The major differences are that the vectors that describe the parameters of rotational motion are all measured with respect to an axis, and a right-hand rule applies to specify their direction. The similarities, or parallels, between these two types of motion are many. In fact, in Chapter 10 we'll accumulate a few more than we already have, and there we'll summarize the analogues in the form of a table. Although the description of rotational motion may appear to be complicated, keep in mind that no fundamentally new physical laws are involved. Everything we have discussed to this point is derived from application of Newton's laws to extended systems of point masses.

THINK ABOUT THIS...

CAN ANGULAR MOMENTUM BE TRANSFERRED IN COLLISIONS?

In Chapter 8 we described how linear momentum is transferred in collisions. Angular momentum can similarly be transferred. Billiard players certainly know how to put spin on a billiard ball when the cue stick collides with it. If you have an air table available, you can easily see a demonstration of the effect with pucks that have "sticky" sides. Another good example is in a context completely familiar to you. In the normal operation of an automobile, angular momentum contained in the turning engine is transferred to the wheels when a clutch, or its equivalent in an automatic transmission (Fig. 9–32), is engaged—the engagement is a kind of soft inelastic collision. (Indeed, the very word "transmission" describes the process of transfer of the engine's behavior to the wheels.) The transmission shafts and the rear wheels, which start at rest before the clutch begins its engagement, have rotational inertia, and as the clutch is engaged, some of the engine's angular momentum is transferred to the shafts and wheels. The angular momentum of the engine is accordingly reduced. The total energy of the engine/transmission-wheel system is also reduced. This is analogous to the completely inelastic collision of a moving object with a stationary one, in which momentum is conserved but kinetic energy is reduced.

▲ FIGURE 9–32 An automobile transmission is a device to transmit the rotational motion of the engine to a rotational motion of the wheels.

9–6 Rolling

Kinematics of Rolling

Our discussion so far has dealt with the kinematics of pure rotation. *Rolling* is a type of motion that is a combination of rotational and linear motion. A rolling wheel undergoes pure rotation about its axis, but to an observer on the ground the axis is moving (Fig. 9–33); as a consequence, the observer sees a point on the wheel's rim undergo motion that is a kind of connected linear and rotational motion. Our first task is to describe rolling mathematically and to discover in just what way the connection between the linear and rotational motion is realized.

Consider a wheel of radius R that rolls without slipping or skidding in a straight line on a horizontal surface (Fig. 9–34a). The center of the wheel moves at uniform speed v. In rolling without slipping, the connection between the speed v and the magnitude of the angular velocity, ω, of the wheel's rotation is

$$v = R\omega. \qquad (9\text{–}40)$$

ROLLING MOTION CONNECTION

◀ FIGURE 9–33 Rolling cylinder. A small light at the center and one at the edge of the cylinder show that the center of mass moves linearly, whereas a point along the edge has rotational motion and traces out a *cycloid*.

▶ **FIGURE 9–34** (a) A wheel of radius R rolls without slipping in the x-direction. Its plane lies in the xy-plane and its center is at the position $\vec{r}$. (b) If the center of the wheel is at $x = 0$ at $t = 0$ and moves at speed v along the $+x$-direction, then the angular speed about the center has magnitude $\omega = v/r$.

A simple way to see this result is to imagine a chalk mark made on the rim of the wheel that rolls on a surface. The wheel makes chalk marks on its path as it rolls, and the distance between chalk marks made in one revolution is $2\pi R$ (the wheel's circumference). If the time taken to move this distance is T, then the center of mass will have traveled a distance vT (Fig. 9–34b). The distances are equal, so $vT = 2\pi R$, or $v = 2\pi R/T$. But $2\pi/T$ is the angular speed ω, and we end up with Eq. (9–40). You can easily see that if there is slipping then Eq. (9–40) no longer holds, by thinking of the extreme in which there is no friction whatever between the wheel and the surface: The wheel will never "grip" and there will be no rotation at all.

CONCEPTUAL EXAMPLE 9–14 The motion of a rolling wheel may be viewed as a rotation about the contact point with the road (Fig. 9–35) in that the point of contact of the wheel is *instantaneously* at rest. If the linear speed of the wheel hub is v (the speed of the hub is the speed of the vehicle), find the angular speed of the wheel about the contact point.

Answer The angular velocity about that point is determined by the fact that the wheel hub, at a distance R from the point of contact, has speed v (Fig. 9–35). Thus the instantaneous angular speed of the hub about the contact point is $\omega = v/R$. The wheel is rigid, so *every* point on the wheel has the same angular speed about the contact point.

What Do You Think? Could you use your answer to find the linear speed of the top of the wheel?

(a)

(b)

▶ **FIGURE 9–35** We work in the frame of reference in which the point of contact between the rolling bicycle wheel and the road is at rest. In this frame, the center of the wheel moves with velocity $\vec{v}$ and the top of the wheel with velocity $2\vec{v}$. Both points rotate about the contact point.

Energy in Rolling

With respect to the point of contact, the motion of the wheel is a *pure* rotation about the point of contact, and the total kinetic energy is

$$K = \tfrac{1}{2} I_{\text{contact}} \omega^2, \tag{9–41}$$

where I_{contact} is the rotational inertia of the wheel about an axis through the point of contact. By the parallel-axis theorem, the rotational inertia I_{contact} may be written in terms of the rotational inertia about the central axis through the center of mass:

$$I_{\text{contact}} = I_{\text{cm}} + MR^2, \tag{9–42}$$

where M is the mass of the rolling object. Thus the total kinetic energy of the rolling object is

$$K = \tfrac{1}{2}(I_{cm} + MR^2)\omega^2 = \tfrac{1}{2}I_{cm}\omega^2 + \tfrac{1}{2}Mv^2, \qquad (9\text{-}43)$$

where we have used $v = R\omega$. Equation (9–43) expresses an important result: *The kinetic energy of an object that rolls without slipping is the sum of the kinetic energy of rotation about its center of mass* $(I_{cm}\omega^2/2)$ *and the kinetic energy of the linear motion of the object as if all the mass were at the center of mass.*

EXAMPLE 9–15 Calculate the total kinetic energy of a uniform solid ball of mass 2 kg and radius 10 cm that rolls without slipping on a flat surface at a speed of 0.8 m/s.

Setting It Up We have a solid (and constant-density) sphere of known mass M and radius R; it rolls with a uniform speed v (its center of mass moves with that speed). We want the total kinetic energy, that is, the linear plus rotational kinetic energy.

Strategy The kinetic energy for a rolling object is given by Eq. (9–43). First we find—it will suffice to look in Table 9–1—the rotational inertia of a uniform solid sphere about an axis through its center, then use the fact that $v = \omega R$ for a rolling object. The problem can then be solved by inserting the values into Eq. (9–43).

Working It Out From Table 9–1, the rotational inertia of a solid ball about an axis through its center is $\tfrac{2}{5}MR^2$. From Eq. (9–43), the total kinetic energy is then

$$K = \tfrac{1}{2}\left(\tfrac{2}{5}MR^2\right)\omega^2 + \tfrac{1}{2}Mv^2 = \tfrac{1}{5}Mv^2 + \tfrac{1}{2}Mv^2 = \tfrac{7}{10}Mv^2$$
$$= 0.7(2\text{ kg})(0.8\text{ m/s})^2 = 0.9\text{ J}.$$

(In the first step we have replaced ω by v/R.) The kinetic energy associated with the rotation is 40 percent of the energy associated with the linear motion.

What Do You Think? Would you expect the percentage of energy in rotation to be larger for larger I assuming the mass and radius of the rolling object stays the same?

Dynamics of Rolling

Consider an object of mass M and radius R that can roll without slipping down a plane that makes an angle θ with the horizontal. The object is symmetric about the central axis but is otherwise unrestricted; for example, it could be a solid cylinder, a hollow cylinder, or a sphere. In Fig. 9–36 we choose the shape to be a solid cylinder, but our analysis does not depend on this choice. Let us use the dynamical equations of motion to find the angular velocity of the object after its center of mass has traveled a distance ℓ down the incline, starting from rest.

The forces are included in Fig. 9–36, which thereby serves as a free-body diagram. The force of gravity, $M\vec{g}$, may be separated into a component normal to the plane, $Mg \cos \theta$, and a component parallel to the plane, $Mg \sin \theta$. Both components act on the center of mass of the rolling object, as we learned in Section 9–4. There is also the contact force $\vec{F}_N$, normal to the plane and acting at the point of contact, which just cancels the component of gravity normal to the plane. Finally, there is the force of friction, $\vec{f}$, which acts at the point of contact and points back up along the plane. It is *static* friction that acts here; there is no relative motion of the rolling object and the ramp at the point of contact. The existence of the static friction is crucial for there to be rolling. Without it the surface would not "grip," and, round or not, the object would slide rather than roll. However, remember that up to some maximum the magnitude of static friction takes a value determined by the dynamics.

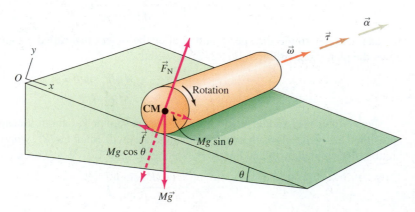

▲ **FIGURE 9–36** A cylinder of radius R rolls down an inclined plane. Static friction causes the cylinder to roll without slipping. The rotational motion can be analyzed about any axis, but a convenient one passes through the geometrical center of the cylinder and another is formed by the line of contact between the cylinder and the plane. The forces acting are friction $\vec{f}$, the normal force $\vec{F}_N$, and gravity $M\vec{g}$. We have decomposed the force of gravity into components along and perpendicular to the inclined plane.

To be able to apply the equations of motion to the rolling object, we want to find the net torque acting on the object. We will use the axis through the center of the object as a reference axis, as shown in Fig. 9–36. Friction is the *only* force of all the forces acting on the object that exerts a torque about the center of mass. Both the normal force and gravity have no lever arm through this point, gravity because it effectively acts at that point and the normal force because if you extend the line along which it acts (perpendicular to the surface) that line passes through the center of mass.

The net force along the plane determines the linear acceleration, magnitude a in this direction. According to Newton's second law,

$$Ma = Mg \sin \theta - f. \tag{9–44}$$

Because it is static friction that is involved, its magnitude is not uniquely determined until we include the dynamical equation of rotation. The torque due to friction has magnitude

$$\tau = fR. \tag{9–45}$$

By the right-hand rule, the torque's direction is along the axis of rotation $\vec{\omega}$ in Fig. 9–36. The dynamical equation that determines the rate of change of angular momentum about the axis of the cylinder comes from Eq. (9–29), with the torque given by Eq. (9–45):

$$I\alpha = \tau = fR. \tag{9–46}$$

This equation implies that the friction force is given by $f = I\alpha/R$. Substituting into Newton's second law, Eq. (9–44), we find

$$Ma = Mg \sin \theta - \frac{I\alpha}{R}. \tag{9–47}$$

Finally, we use Eq. (9–40), $v = R\omega$, to supply a relation between a and α. The time derivative of that equation immediately gives us $a = R\alpha$. Substitution into Eq. (9–47) then gives us a single equation for the angular acceleration:

$$M\alpha R = Mg \sin \theta - \frac{I\alpha}{R},$$

with the algebraic solution

$$\alpha = \frac{MgR \sin \theta}{MR^2 + I}. \tag{9–48}$$

Only geometric factors appear on the right, so the angular acceleration is a constant. With a constant angular acceleration, the linear dependence of angular velocity on time is

$$\omega = \omega_0 + \frac{MgR \sin \theta}{MR^2 + I} t. \tag{9–49}$$

Here, the angular velocity at $t = 0$ is ω_0. If the object starts from rest, then $\omega_0 = 0$. Similarly, the linear acceleration is constant, implying a center-of-mass velocity that increases linearly with time according to

$$v = \omega R = v_0 + \frac{MgR^2 \sin \theta}{MR^2 + I} t. \tag{9–50}$$

We can also determine the speed gained after the object has rolled a total distance ℓ down the plane. According to Eq. (2–25d), we have

$$v^2 = v_0^2 + 2a\ell = v_0^2 + 2\frac{MgR^2 \sin \theta}{MR^2 + I} \ell. \tag{9–51}$$

For an object of radius R and mass M that is symmetric about a central axis, dimensional analysis tells us that the rotational inertia will equal CMR^2, where C is a pure number determined by the geometry of the object. Thus Eq. (9–49) becomes

$$\omega = \omega_0 + \frac{(g/R) \sin \theta}{1 + C} t,$$

or equivalently

$$v = v_0 + \frac{g \sin \theta}{1 + C} t. \tag{9-52}$$

Similarly,

$$v^2 = v_0^2 + 2a\ell = v_0^2 + 2\frac{g \sin \theta}{1 + C}\ell. \tag{9-53}$$

It is instructive to compare this last result with the speed gained by a point object that moves down the same ramp without friction: $v^2 = v_0^2 + 2g\ell \sin \theta$. The rolling object moves less rapidly because *some of the potential energy goes into rotational motion*. If there is some slipping, then this same reasoning about energy tells us that we end up in an intermediate situation, with speed greater than in the case of no slipping but less than in the case of pure sliding (no friction at all). Note also the striking result that *the speed of the center of mass is independent of both M and R*. We are familiar with the speed independent of mass—that is the typical behavior under gravity. The lack of dependence on R, which you may find surprising, is something you easily test.

Let's look at a few special cases. The smallest possible value of C is zero, corresponding to an object with all its mass concentrated at its central axis; this is just the case of a sliding object since there is no rotational energy in this case. The case $C = \frac{1}{2}$ is a solid cylinder (see Table 9–1). The case $C = 1$ corresponds to a thin hollow cylinder with all its mass concentrated at the outside. Still larger values of C are possible if we consider an object like a yo-yo, where the rolling takes place on the inner cylinder. Thus the factor $1 + C$ increases from a minimum value of 1, and the rate at which ω, or v, increases with time for symmetric rolling objects is always less than that of a sliding object.

EXAMPLE 9–16   A ramp of length 1.5 m is set at an angle of 5.0° to the horizontal. Two objects are initially at rest at the top of the ramp and simultaneously start rolling without slipping. Object 1 is a solid cylinder of mass 0.65 kg and radius 4.7 cm. Object 2 is a hollow, thin-walled tube of the same radius and of mass 0.13 kg. How much time does it take for each object to arrive at the bottom of the ramp?

(a)

Setting It Up   We sketch the ramp and the rolling objects in Fig. 9–37a. The coordinate system is positioned at the top of the ramp with the x-axis down the ramp. The ramp length is d and its angle is θ. The solid roller has mass M_1 and the hollow one mass M_2; both have the same radius R. All five of these variables are known.

Strategy   Because there is no slipping, we can use the direct connection between linear motion and angular motion of the rolling objects. We can also use directly the results of the discussion above, but with different values of I for the two different objects. We therefore know the values of the linear acceleration in each case. This acceleration is constant, so kinematics can give us the travel time.

Working It Out   The time required to travel a given straight-line distance d under constant acceleration a when the initial velocity is zero is expressed by the constant-acceleration equation $d = at^2/2$, or $t = \sqrt{2d/a}$. Next we need to find the value of the acceleration. Equation (9–48) with $a = \alpha R$ gives

$$a = \frac{MgR^2 \sin \theta}{MR^2 + I},$$

where M is the object's mass. For the case of the solid cylinder, $I = MR^2/2$. The center of the solid cylinder then has linear acceleration

$$a_1 = \frac{g \sin \theta}{1 + \frac{1}{2}} = \frac{2}{3} g \sin \theta.$$

(b)

◀ **FIGURE 9–37**   The rate at which a rolling object accelerates depends on its rotational inertia. In a race between a solid cylinder and a hollow cylinder, both starting from rest and rolling down an inclined plane, the solid cylinder comes in first.

The hollow cylinder has a rotational inertia about its axis of $I = MR^2$; thus, its linear acceleration has magnitude

$$a_2 = \frac{g \sin \theta}{1 + 1} = \frac{1}{2} g \sin \theta.$$

(continues on next page)

In each case, the linear acceleration is independent of the mass and the radius, so this information is not needed. The only relevant difference between the solid cylinder and the hollow tube is that they have different rotational inertias because their masses are distributed differently. Numerically,

$$a_1 = \tfrac{2}{3}g \sin \theta = (0.67)(9.8 \text{ m/s}^2)(\sin 5.0°) = 0.57 \text{ m/s}^2;$$
$$a_2 = \tfrac{1}{2}g \sin \theta = (0.5)(9.8 \text{ m/s}^2)(\sin 5.0°) = 0.43 \text{ m/s}^2.$$

The descent time of the solid and hollow cylinders respectively are

$$t_1 = \sqrt{\frac{2d}{a_1}} = \sqrt{\frac{2(1.5 \text{ m})}{0.57 \text{ m/s}^2}} = 2.3 \text{ s}$$

and

$$t_2 = \sqrt{\frac{2(1.5 \text{ m})}{0.43 \text{ m/s}^2}} = 2.6 \text{ s}.$$

It is easy to try this experiment and confirm that a solid cylinder reaches the bottom of a ramp faster than a hollow one, as Fig. 9–37b shows.

What Do You Think? How would a thick cylinder of the same radius as the thin cylinder do if it became the third entry in this race? What if the thick cylinder has radius $2R$?

Summary

In this chapter we have explored the fact that a rigid body can undergo rotations as well as linear motion. Rotational motion can be described using quantities that are analogous to the quantities used to describe linear motion. The linear motion of an object can be described by specifying the motion of the center of mass. A rotation about a fixed axis can be described with a single angle θ. The angular velocity $\vec{\omega}$ describes the rate of change of this angle, while the angular acceleration $\vec{\alpha}$ describes the rate of change of the angular velocity:

$$\vec{\omega} = \frac{d\theta}{dt}\hat{\omega}, \tag{9–2}$$

$$\vec{\alpha} = \frac{d\vec{\omega}}{dt}. \tag{9–7}$$

The directions of these vectors are determined by a right-hand rule. The period T and frequency f of rotational motion are related by

$$f = \frac{1}{T} = \frac{\omega}{2\pi}. \tag{9–5}$$

The rotational kinetic energy of a rigid body rotating about an axis is

$$K = \tfrac{1}{2}I\omega^2, \tag{9–14}$$

where I is the rotational inertia of the rotating object with respect to that axis. For a discrete system of point masses Δm_i that is a perpendicular distance R_i from an axis,

$$I \equiv \sum_i \Delta m_i R_i^2. \tag{9–15}$$

In particular, for a single mass m that is a perpendicular distance from the axis of rotation, the rotational inertia is

$$I = mR^2. \tag{9–16}$$

The definition in Eq. (9–15) can be extended to continuous objects: for an object with mass density ρ,

$$I = \int_{\text{Volume}} \rho R^2 \, dV, \tag{9–19}$$

where R^2 is the perpendicular distance of an internal point of the object to the axis.

Rotational inertia plays a role in rotational motion analogous to the role played by mass in linear motion. A useful tool for its evaluation is the parallel-axis theorem. This theorem states that the rotational inertia of an object about a given axis is

$$I = I_{\text{cm}} + Md^2. \tag{9–27}$$

In this equation, I_{cm} is the rotational inertia about an axis that goes through the center of mass and is parallel to the given axis, M is the mass of the object, and d is the perpendicular distance between the axes.

Torque τ is the cause of changes in rotational motion and is analogous to force, the cause of changes in linear motion. For a rigid body rotating about a fixed axis with rotational inertia I about that axis, the equation of motion is analogous to Newton's second law:

$$\vec{\tau}_{\text{net}} = I\vec{\alpha}, \tag{9–30}$$

where $\vec{\alpha}$ is the angular acceleration. With constant torque, the angular velocity grows linearly with time.

The net torque is expressed in terms of the net force by the expression

$$\tau = rF \sin \theta, \tag{9-28}$$

where r is the distance from the point of application of the force to the axis of rotation and θ is the angle between the net force and r. The combination $r \sin \theta$ is the lever arm, or perpendicular distance from the axis to the line along which the force acts—the line of action. The direction of the torque is specified by a right-hand rule. The torque due to gravity acts as if it were applied to the center of mass.

The technique of free-body diagrams for solving linear motion can be extended to rotational motion by specifying the point of application of the force.

The equation of motion takes an alternative form

$$\vec{\tau} = \frac{d\vec{L}}{dt}, \tag{9-38}$$

where $\vec{L}$ is the angular momentum—a quantity analogous to the linear momentum in linear motion. For rotations of a symmetric, rigid body about a symmetry axis, the angular momentum is

$$\vec{L} = I\vec{\omega}. \tag{9-37}$$

When the net torque is zero, the angular momentum is constant; thus, the angular momentum is conserved for isolated systems.

When an object rolls, linear motion is combined with rotational motion. Objects with radius R that roll without slipping have an angular velocity about their axes, ω, that is related to the speed v of the center of mass of the object:

$$v = R\omega. \tag{9-40}$$

The kinetic energy of a rolling object is the sum of its rotational kinetic energy about its axis of rotation, $I_{cm}\omega^2/2$, and the kinetic energy of its linear motion:

$$K = \tfrac{1}{2}(I_{cm} + Mr^2)\omega^2 = \tfrac{1}{2}I_{cm}\omega^2 + \tfrac{1}{2}Mv^2. \tag{9-43}$$

Understanding the Concepts

1. A record turntable rotates in the clockwise sense when seen from above. In what direction is the angular velocity of the turntable? If you had used a left-hand rule rather than a right-hand rule, which direction would you have chosen for the direction of the angular velocity vector? Is the choice between a right-hand or a left-hand rule purely conventional?

2. In the discussion of rolling without slipping, we saw that the kinetic energy of the rolling object is greater than the kinetic energy associated with the linear motion alone. Are there cases of rolling, with or without slipping, for which this is not true?

3. In a juggling act a juggler spins a plate rapidly in its plane and then supports it on a stick, walking around with it, and hardly paying attention to the exact angle that the stick makes with the vertical. How does the trick work?

4. Suppose that you have a set of spherical objects with the same total mass but different radial distributions of mass. Which objects have the larger rotational inertias about an axis through the center: the spheres with more mass at the center or the spheres with more mass toward the outer surface?

5. A paddle wheel that propels a Mississippi River boat dips down into the water. Discuss the direction of motion of the portion of the paddle wheel under the water according to an observer on the shore and then according to an observer on the boat.

6. If you were to hold an arm out stiffly at a right angle while grasping a rock, the effect of the force of gravity on the rock would translate to a torque on you about, say, your center of mass. Does this mean you would rotate?

7. A ball starts from rest down the inside of a parabolic bowl and rolls without slipping. At the bottom, the surface is a frictionless surface. The ball then moves up the other side of the bowl to a certain height. Is this height higher than, lower than, or the same as the height from which the ball started?

8. Solid and hollow cylinders both roll from rest down an inclined plane. Explain the difference in speeds at the bottom.

9. You have two cylindrical cans, one filled with congealed wax and the other with water. The cans are identical, and the total masses of the two cans with their contents are the same. If you roll the two down an inclined plane, which will reach the end first?

10. Devise a method to determine the rotational inertia of a sphere whose density can vary with the distance from its center.

11. Two skaters holding on to opposite ends of a rope circle a common point between them. In order to double their angular velocity about that point, should they lengthen or shorten the distance between them? By how much?

12. You have a flat outline of the continental United States cut out in $\frac{1}{4}$-in plywood. Devise a method to find the center of mass of this outline.

13. When canoeing in rapidly flowing water, it is possible for the opposite ends of the canoe to become pressed against rocks such that the canoe is aligned across the flow of the stream. Why is this bad?

14. When a landing airplane first touches the runway, a puff of smoke comes from the wheels; the smoke stops after a moment and the airplane can roll to a halt. Describe these events in terms of the relevant forces acting.

15. What limits the amount of energy a flywheel could store in its rotational motion?

16. A solid and a hollow cylinder are rolled from rest down an inclined plane. At the bottom, the center of the solid cylinder is moving more rapidly, even if the two cylinders have the same radius and mass. Is this a violation of the conservation of energy?

17. Why did the wheels used to steer large sailing ships have large radii?

18. The accuracy of guns improved greatly when their barrels were rifled, making the bullets emerge with a rapid rotation about their axes. Can you explain why this might improve accuracy?

19. When you switch to a longer wrench, you have not suddenly become stronger. So why is it easier to loosen a pipe with a long wrench than with a short wrench?

20. We stated that the rotational inertia of a dumbbell about an axis that runs through the rigid rod connecting the two masses is zero. What approximations (or idealizations), if any, are made in coming to this conclusion?

21. Attach one end of a given spring to a fixed pivot point on a frictionless table. Attach a mass to the other end and start the mass in circular motion about the pivot point (Fig. 9–38). What happens when the revolution time is decreased by some external means and why?

22. When a wheel of radius R rolls without slipping, the relationship between the speed of the center of the wheel, v, and the angular speed, ω, is $v = R\omega$. How would this relationship change if there were some slipping?

23. A driver presses on the accelerator, increasing the angular speed of the driven wheels by a factor of 2. Assuming that there is no slipping, by what factor does the linear speed of the automobile increase?

▲ **FIGURE 9–38** Question 21.

24. Suppose that you swing a rock at the end of a string in a circle in a vertical plane. What happens to the rock if the string suddenly breaks? Does it suddenly lose its angular momentum?

25. You have two wheels with rotational inertias in the ratio 2 : 1. Initially, the wheels are at rest. If the same torque is applied to both of them for the same length of time, what will be the ratio of their angular momenta at the end of that time interval?

26. You have two wheels with rotational inertias in the ratio 2 : 1. Initially, the wheels are at rest. If the same torque is applied to both of them for the same length of time, what will be the ratio of their rotational kinetic energies at the end of that time interval?

27. A wheel rotates about a central axis. When the angular speed increases beyond a certain value, the wheel breaks. Why?

Problems

9–1 Simple Rotations of a Rigid Body

1. (I) When a phonograph turntable is switched off, it comes to rest from its original $33\frac{1}{3}$ rev/min in 5.15 s. What is the magnitude of the angular deceleration assuming that it is constant while the turntable comes to rest?

2. (I) How many revolutions does the turntable of Problem 1 make between the moment it is switched off and the moment it stops?

3. (I) A skater does a pirouette at the rate of 2.0 rev/s and then stops within $\frac{3}{4}$ rev. Assume that the angular deceleration is constant and calculate its magnitude.

4. (I) A turntable slows down from 15 rev/s to rest in 45 s. What is the average angular acceleration of the turntable?

5. (I) What is the angular velocity of Earth's rotation about its axis? What is the angular velocity of Earth in its orbital motion around the Sun?

6. (I) A carousel has a 7-m radius and requires 8 s for a single revolution at full speed. A carousel pig sits at a distance of 3 m from the axis and a carousel horse sits at a distance of 6 m. (a) What is the period T for a single revolution of the pig? (b) Of the horse? (c) What is the angular frequency of the motion of the pig? (d) Of the horse? (e) What is the velocity of the pig? (f) Of the horse? (g) What is the centripetal acceleration of the pig in its motion around the axis? (h) Of the horse?

7. (II) A centrifuge whose maximum rotation rate is 10,000 rev/min can be brought to rest in 4.00 s. (a) What is the average angular acceleration of the centrifuge? (b) What is the distance that a point on the rim travels during the deceleration time assuming that the radius of the centrifuge is 8 cm and that the acceleration is uniform?

8. (II) A CD turntable is rotating at 313 rev/min in a clockwise direction, viewed from above. (a) What is its angular velocity, both direction and magnitude, in radians per second? (b) The turntable is switched off and comes to rest in 0.55 s. What is the average value of the angular acceleration, direction, and magnitude during that period?

9. (II) A carousel initially at rest has an angular acceleration of 0.4 rad/s² and accelerates for 5 s. It then rotates at a constant angular velocity for 30 s before slowing down at the same rate with which it accelerated. (a) What is the average acceleration during the first 20 s? (b) How many total revolutions does it make? (c) How far does a child sitting on a horse 3 m from the center travel?

10. (II) A thread is wrapped around a cylindrical spool of radius 1.5 cm whose central axis is fixed on a support (Fig. 9–39). The

▲ **FIGURE 9–39** Problem 10.

thread is pulled off at a constant rate, causing the spool to spin at a constant rate; it takes 2 s to pull off 3 m of thread. What is the angular velocity of the spool while the thread is pulled off?

11. (II) A more careful measurement of the unwrapping of the thread in Problem 10 shows that the spool accelerates from rest at a steady rate in the 2 s it takes to pull off 3 m of thread. (a) Give a formula for the position of the hand that pulls the thread as a function of time. What is the value of the constant (linear) acceleration? (b) What is the value of the angular acceleration of the spool? (c) Give a formula for the magnitude of the angular velocity as a function of time.

12. (II) A vacuum pump is connected to its electric motor by a belt drive (Fig. 9–40). The motor rotates at a rate of 1200 rev/min and the diameter of the motor shaft is 1.5 cm. How large should the pulley be if it is designed for a speed of 33 rev/s?

▲ **FIGURE 9–40** Problem 12.

13. (II) The path of the tip of a needle on a phonograph record may be described by the formula $r = r_0 - (\theta\rho/2\pi)$, where r_0 is the outer starting radius, θ is the angle (in radians) that a fixed radial line in the record makes with the needle arm, and ρ is the spacing between grooves. Assuming that the record turns at a rate of $33\frac{1}{3}$ rev/min and that the radial distance traveled by the needle is 9 cm in 20 min, what is ρ?

14. (II) The angular acceleration of a wheel starting from rest has magnitude $C_1 t + C_2 t^3$, where $C_1 = 48$ rad/s³ and $C_2 = -9.5$ rad/s⁵. (a) What is its angular velocity at 3.0 s? (b) How many revolutions has the wheel made after 2.0 s? (c) When will the wheel be at rest again?

15. (II) By marking a point on its edge, the angle θ that measures the rotation of a turntable is found to increase with time as t^3 during a certain time interval. How did the angular acceleration change with time, if at all, during that interval?

16. (II) The angular velocity of a wheel that can rotate in the xy-plane around a fixed axle oriented along the z-axis varies with time as $\vec{\omega} = (w_1 t - w_2 t^2)\hat{k}$, where w_1 and w_2 are constants. Assuming that the wheel starts at $t = 0$ with a mark on its edge sitting on the positive x-axis, where is the mark the next time the wheel comes to rest?

17. (III) A rigid solid undergoes rotational motion about an axis. Its angular velocity has magnitude $\omega = \alpha t$, where α is constant. As the angular speed ω increases with time, the period T decreases. (a) Show that the rate of change of the period is described by the equation $dT/dt = -2\pi/\alpha t^2$. (The period is infinite at $t = 0$ because the rotation has not yet started, and a measurement of the time for 1 rev at that rate would be infinite!) (b) Show that the change in the period between $t = t_1$ and $t = t_2$ is given by

$$T(t_2) - T(t_1) = \frac{2\pi}{\alpha}\left(\frac{1}{t_2} - \frac{1}{t_1}\right).$$

9–2 Rotational Kinetic Energy

18. (I) A metal ball of mass 350 g at the end of a 1.75-m long wire rotates with an angular speed of 85 rev/min (85 × 2π rad/min). What is the rotational kinetic energy of the ball?

19. (I) Two identical balls are spinning on a flat surface. Ball 1 has three times the angular speed of ball 2. What is the ratio of their kinetic energies?

20. (I) Measurements of the amount of energy used by an electric motor to speed up a wheel from rest show that to bring the wheel from rest to an angular speed of 3.7 rad/s the motor expends 7600 J. In a second use of the motor and wheel, the motor expends 9200 J to bring the wheel up from rest to an unmeasured angular speed. The motor and wheel have practically negligible amounts of friction or other type of damping. What is the unmeasured angular speed?

21. (II) A string is wrapped around a cylindrical spool of radius 1 cm. The axis of the spool is fixed. A length of string of 0.8 m is pulled off in 1.5 s at a constant tension of 20 N. What is the rotational inertia of the spool?

22. (II) What is the rotational kinetic energy of a dumbbell consisting of two equal (compact) masses of 1.5 kg each connected by a massless rod of length 0.62 m when the dumbbell rotates about an axis through the center of and perpendicular to the rod at 36 rev/min? What is the rotational kinetic energy if the dumbbell rotates with the same angular velocity about a parallel axis through one of the masses?

23. (II) A ball of mass 0.75 kg is attached by a 1.5-m-long rope to the top of a rod. The ball swings in a circle at the rate of 25 rad/s with the rope making an angle of 30° with the vertical. What is the rotational kinetic energy of the ball? What is it when the angle is 60°?

9–3 Evaluation of Rotational Inertia

24. (I) A pipe made of aluminum with a density of 2.7 g/cm³ is a right cylinder 16 cm long whose outer diameter is 5.0 cm and whose inner diameter is 4.0 cm. What is the rotational inertia about the central axis of the pipe? Note that the rotational inertia of the thick cylinder can be expressed as the rotational inertia for a solid cylinder of radius R_2 minus the rotational inertia of the solid cylinder of radius R_1.

25. (I) What is the rotational inertia of a uniform 4.0-kg iron rod 0.25 m long about (a) an axis through its center point and perpendicular to the rod and (b) an axis through an end point and perpendicular to the rod?

26. (II) A dumbbell consists of two point masses 0.10 and 0.15 kg connected by a (nearly massless) rod of length 35 cm. Find the rotational inertia of this object about an axis that is parallel to the rod and a distance 10 cm away from it.

27. (II) A neutron star has a constant density of 6×10^{17} kg/m³ and a mass five times that of our Sun. Compare its rotational inertia with that of Earth (assume constant density). In both cases the reference axis is an axis through the center of the sphere; Table 9–1 gives the rotational inertia for such an axis.

28. (II) Mass m_1 sits at the point $(x, y, z) = (0$ m, 0 m, 0 m$)$, and mass m_2 at point $(0$ m, 1 m, 0 m$)$. (a) Where is the center of mass? (b) What is the rotational inertia about an axis through the center of mass and parallel to the z-axis? (c) Parallel to the y-axis? (d) Parallel to the x-axis? (e) What is the rotational inertia about an axis through the origin and along the z-axis? (f) Verify the parallel-axis theorem for this system using the results of parts (b) and (e).

29. (II) Find the rotational inertia about the symmetry axis of a thick cylinder of mass M. Take the inner radius to be R_1 and outer radius to be R_2.

30. (II) Calculate the rotational inertia of a section of a right circular cylinder of radius R that subtends an angle θ_0 at the origin (Fig. 9–41) when the reference axis is at the origin and perpendicular to the section.

▲ **FIGURE 9–41** Problem 30.

31. (II) Calculate the rotational inertia about the central axis of the solid cone of mass M illustrated in Fig. 9–42, of opening half-angle α and height H.

▲ **FIGURE 9–42** Problem 31.

32. (II) A thin stick of length $L = 1.6$ m is denser at one end than at the other: Its mass density is $\lambda = (0.40 \text{ kg/m}) - (0.070 \text{ kg/m}^2)x$, where x measures the distance from the heavier end of the stick. The stick rotates with period $T = 1.1$ s about an axis perpendicular to the stick through the heavy end. Determine the rotational kinetic energy of the stick.

33. (II) Calculate the rotational inertia of a sphere of radius R and mass M about an axis through the center of the sphere; assume that the density is not uniform but is given by ρ_1 for $0 \le r \le R_1$ and by ρ_2 for $R_1 \le r \le R$ (Fig. 9–43).

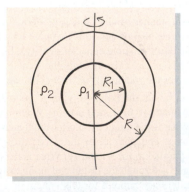

▲ **FIGURE 9–43** Problem 33.

34. (III) Use Eq. (9–19) to show that the rotational inertia about an axis through the center of a uniform, constant-density sphere of mass M and radius R is $\frac{2}{5}MR^2$.

35. (III) In Problem 89 of Chapter 8, we described a styrofoam sphere of radius R. A cavity of radius $R/2$ centered a distance $R/2$ directly above the center of the sphere was hollowed out and filled with a solid material of density five times the density of styrofoam. In that problem, the location of the center of mass of the composite sphere was determined. What is the rotational inertia of the composite sphere about a horizontal axis through the center of mass? Express your result in terms of the total mass M of the composite sphere and its radius R. (*Hint*: You may view the mass as consisting of a large sphere of radius R and density ρ and a small sphere, off-center, of radius $R/2$ and density 4ρ. Use the parallel-axis theorem.)

36. (III) A square of mass m is made of thin wire with sides of length a. Calculate the rotational inertia about the axes shown in Fig. 9–44.

▲ **FIGURE 9–44** Problem 36.

9–4 Torque

37. (I) An instruction manual calls for a bolt to be tightened to 20 ft · lb. All you have is a 10-in wrench. How much force do you need to apply to the end of the wrench to tighten the bolt as required?

38. (I) A plumber of mass 74 kg just loosens a rusted-in bolt with the help of a 45-cm-long wrench. He places the wrench in a horizontal position and applies torque by hanging from the end of the wrench. What is the torque applied?

39. (I) The assistant of the plumber in Problem 38 meets a similar situation but aligns the wrench at an angle of 50° from the vertical (Fig. 9–45). What must his mass be in order for him to use the same technique to loosen the same bolt?

▲ **FIGURE 9–45** Problem 39.

40. (I) A flywheel of rotational inertia $I = 53$ kg · m² rotates with angular speed 4.0 rad/s. A tangential force of 6.5 N is applied at a distance of 0.36 m from the center in such a way that the angular speed decreases. How long will it take for the wheel to stop?

41. (II) A uniform rod of length L lies along the x-axis. A force F_{1y} is applied to one end of it and a force $-F_{2y}$ is applied to the other end of it. How large is the torque on the rod about its center of mass?

42. (II) A uniform rod 1.1 m long with mass 0.7 kg is pivoted at one end, as shown in Fig. 9–46, and released from a horizontal position. Find the torque about the pivot exerted by the force of gravity as a function of the angle that the rod makes with the horizontal direction.

▲ **FIGURE 9–46** Problem 42.

43. (II) A seesaw pivots as shown in Fig. 9–47. (a) What is the net torque about the pivot point? (b) Give an example for which the application of three different forces and their points of application will balance the seesaw. Two of the forces must point down and the other one up.

▲ **FIGURE 9–47** Problem 43.

44. (II) A two-dimensional object placed in the xy-plane has several forces acting on it. Find the torques about points A and B in Fig. 9–48.

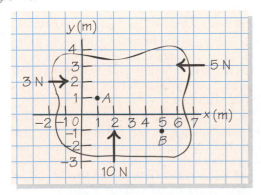

▲ **FIGURE 9–48** Problem 44.

45. (II) An aluminum casting is finished with a band sander. The sandpaper is stretched by two cylindrical rollers, one of them

driven by a motor (Fig. 9–49). The rollers have diameter of 16 cm. How large is the torque that has to be applied to the driven cylinder if the coefficient of kinetic friction between aluminum and sandpaper is 1.2 and the force applied to push the casting against the sandpaper is 4 N?

▲ **FIGURE 9–49** Problem 45.

46. (II) A massless rope is wrapped around a hollow cylinder of radius 12 cm whose central axis is fixed in a horizontal position. A mass of 4.0 kg hangs from the rope and, starting from rest, moves 180 cm in 2.0 s. What is the mass of the cylinder?

47. (II) A wheel of radius 24.6 cm whose axis is fixed starts from rest and reaches an angular velocity of 4.15 rad/s in 2.68 s due to a force of 13.4 N acting tangentially on the rim. (a) What is the rotational inertia of the wheel? (b) What is the change in the angular momentum during the 2.68 s? (c) How many revolutions does the wheel make? (d) How much rotational kinetic energy does the wheel have after 2.68 s?

48. (III) A motorcycle has a mass of 500 kg, the wheels have a diameter of 60 cm, and the centers of the wheels are separated by 1.5 m. Assuming that the weight is distributed uniformly over the wheels, that the wheels roll without sliding, and that the coefficient of static friction between the wheels and the road is $\mu = 0.5$, calculate the torque about the center of the front wheel exerted by the forces between the road and the wheels when there is a maximum braking.

9–5 Angular Momentum and Its Conservation

49. (I) A student sits on a piano stool and holds the axle of a bicycle wheel that rotates with angular velocity of magnitude 4π rad/s, pointing upward. The wheel's axis of rotation goes through the axis of the stool, which is at rest. The rotational inertia of the wheel about its axis is 1.2 kg·m², and the rotational inertia of the student and stool about the stool's axis is 8 kg·m². The student suddenly flips the shaft of the wheel so that its angular velocity points down. How fast and in what direction will the student and stool rotate? Ignore friction.

50. (I) Two identical tops spin with angular velocities 35π rad/s up and 25π rad/s down, respectively, about vertical axes on a table. The tops bump into one another and separate. After the collision, one of the tops has an angular velocity of 30π rad/s in its original direction. What is the angular velocity of the other top?

51. (I) A spherically symmetric celestial object rotates at 3.24592 rev/s. Through some mechanism that does not involve an application of external torque, a change of total mass, or a change of shape, its radius decreases rapidly and uniformly. As a consequence, the rate of revolution changes to 3.24608 rev/s. What is the fractional change in the radius?

52. (II) A child of mass 25 kg stands at the edge of a rotating platform of mass 150 kg and radius 4.0 m. The platform with the child on it rotates with an angular speed of 6.2 rad/s. The child jumps off in a radial direction. (a) What happens to the angular speed of the platform? (b) What happens to the platform if, a little later, the child, starting at rest, jumps back onto the platform? (Treat the platform as a uniform disk.)

53. (II) Suppose Earth's radius increased *uniformly* by 0.001 percent without any change in mass. What would be the change in the length of the day?

54. (II) An old-fashioned record player drops a second record onto a first one that is spinning at a fixed rate on a coasting turntable. The turntable itself is a uniform disk, as are the records, and the turntable is 8 times the mass of a single record. By what percentage does the angular speed of the turntable change when the second record drops?

55. (II) Compact disks and long-playing records are made from the same material. The former have a diameter of about 12 cm; the latter, about 32 cm. When in use, records spin at $33\frac{1}{3}$ rev/min and compact disks spin at, say, 400 rev/min. What is the ratio of the angular momentum of a compact disk in use to that of a record? Assume that a compact disk has half the thickness of a record.

9–6 Rolling

56. (I) Bicycle racers sometimes use solid wheels in order to cut down the drag force between the air and the spokes of an ordinary wheel. This can be an important effect because drag forces rise quite rapidly as the speed of an object through the air increases. If the radius of a wheel is 35 cm and the speed of the bicycle relative to the ground is 22 mi/h (9.8 m/s), what is the speed relative to the ground of the end of the spoke closest to the rim for (a) a spoke leading to the contact point with the ground, that is, a spoke pointing vertically down? (b) A spoke pointing vertically up? (c) A spoke that is horizontal and points forward? (d) A spoke that is horizontal and points backward?

57. (II) A cylindrical shell starting from rest rolls down an inclined plane that makes an angle of 20° with the horizontal. How far will the shell travel in 4 s? How far would a solid cylinder travel in the same time?

58. (II) A homogeneous cylinder of mass 1.20 kg and diameter 25 cm rolls down an inclined plane that makes an angle of 25° with the horizontal. What is the speed of the axis of the cylinder by the time the cylinder has rolled 1.00 m down the plane?

59. (II) The homogeneous cylinder of the preceding problem rolls down an inclined plane that makes an angle of 12.5° with the horizontal. How much further does it move down the ramp before it reaches the same speed that it did in rolling 1.00 m in the preceding problem?

60. (II) The following objects all roll without slipping and have uniform density, mass M, and radius R and the speed of the center of mass in each case is v. Find the ratio of the rotational kinetic energy to the total kinetic energy for (a) a solid cylinder, (b) a hollow cylinder, and (c) a solid sphere.

61. (II) Figure 9–50 shows a disk attached to an axle placed on an incline made of two parallel bars. The radius of the disk is 0.12 cm and its mass is 0.8 kg; the radius of the axle is 0.020 m and its mass is 0.10 kg, not including the part inside the disk. Calculate the acceleration of the system if the incline makes an angle of 5° with the horizontal and the axle rolls without slipping.

▲ **FIGURE 9–50** Problem 61.

62. (II) A hollow cylinder moves down an inclined plane of length ℓ and angle θ. The cylinder has uniform density, mass M, and radius R. It is initially at rest at the top of the plane. Calculate and compare the times taken to reach the bottom if the cylinder rolls without slipping as opposed to the case in which the cylinder slips all the way down the plane without rolling. In the latter case, the coefficient of friction must be zero.

General Problems

63. (I) A basketball player shoots a desperate last shot, spinning the ball at an angular speed of 15 rad/s to give it "action." The ball is shot with an initial velocity of 4.1 m/s at an elevation angle of 45° and leaves the player's hands 1.7 m off the floor. Unfortunately, the shot misses the backboard, the rim, and the net. How many revolutions has the ball made when it hits the floor? Ignore air resistance.

64. (I) *Estimate* the angular momentum of a spinning ice skater.

65. (II) Earth's radius is 6.4×10^6 m. Assume that its density is not uniform; that is, the inner core has a density of 8.0×10^3 kg/m³ and the outer mantle has a density of 3.0×10^3 kg/m³. Given that the rotational inertia of Earth is 8.3×10^{37} kg·m², calculate the radius R at which the density changes. [*Hint*: Go back to Problem 33.]

66. (II) A solid cylinder of mass 17 kg and radius 33 cm rotates at 300 rev/min about its central axis. What is the rotational kinetic energy of this motion? Suppose that a 0.8-kg mass is attached at one point on the rim of the cylinder. If the additional mass can be treated as a point mass and the rotational speed is unchanged, what is the percentage change in the rotational kinetic energy?

67. (II) A solid uniform cylinder of mass M and radius R is projected up an incline of angle θ. It rolls without slipping from an initial speed v_0 of the center of mass. What distance s does the center of the cylinder travel before it starts to fall back?

68. (II) The lid of a box is balanced vertically on its hinges. The slightest displacement from the vertical leads to its falling shut. If the lid is of uniform density and it is 30 cm wide, what will be its angular velocity when it does fall shut, that is, when it is in a horizontal position?

69. (II) A spool within a piece of machinery consists of a solid cylinder of radius R_1 capped at each end by solid disks of a larger radius R_2. The spool can turn freely about its axis. A mass m_1 can be hung from a light rope wrapped about the inner cylinder and a second mass m_2 can be hung similarly, but in the opposite sense, from one of the disks (Fig. 9–51, see next page). What is the value of m_2 such that the spool will not turn? How would your answer change if the central cylinder were hollow, with internal struts for support?

▲ **FIGURE 9–51** Problem 69.

70. (II) When a bicycle rider accelerates, he must accelerate his own and his bicycle's linear motion as well as the angular motion of the wheels. Suppose that the cyclist has mass 55 kg; the bicycle (not counting wheels), 8.0 kg; and both wheels together, 1.8 kg. Assume that the wheels, each of radius 30 cm, have all their mass concentrated in the (thin) rim. (a) At 25 km/h, what fraction of the kinetic energy of the rider plus his bicycle is in linear motion and what fraction is in rotational motion? (b) Suppose that the cyclist loses 3.0 kg on a diet. What percentage of the original force is required to accelerate the system uniformly from 0 to 25 km/h in 10 s? (c) Suppose that instead of going on a diet the cyclist replaces his wheels with ones of total mass 1.2 kg. Now what is the percentage of the original force required to accelerate the system uniformly from 0 to 25 km/h in 10 s?

71. (II) A hollow cylinder of radius 15 cm and mass 3.0 kg rolls without slipping at a constant speed of 1.6 m/s. (a) What is its angular momentum about its symmetry axis? (b) What is its rotational kinetic energy? (c) What is its total kinetic energy?

72. (II) A pulsar (the remnant of a star after a supernova explosion) has a mass of 3.7×10^{30} kg and a radius of 17 km. It rotates with a period of 0.15 s. (a) Assuming a spherical shape and constant density, what is the kinetic rotational energy of the pulsar? (b) If the period changes by 1 part in 10^8 in 1 yr, what is the rate of energy loss of the pulsar?

73. (II) A yo-yo has mass M and external radius R. The central stem has negligible mass and a radius r. The string is pulled horizontally on the lower side with a constant force F, while the yo-yo rests on a rough horizontal surface (Fig. 9–52). What is the maximum value of F for which the yo-yo will roll without slipping assuming that the coefficient of static friction between the yo-yo and the surface is μ?

▲ **FIGURE 9–52** Problem 73.

74. (II) A cylinder of known mass M, radius R, and rotational inertia I is placed on an inclined plane with angle θ (Fig. 9–53). A string is wound around the cylinder and pulled up with a tension T parallel to the inclined plane. The coefficient of static friction is large enough to prevent slipping. (a) Find the tension T_0 needed to keep the cylinder in equilibrium. (b) Find the acceleration of the cylinder if the tension is known and is different from T_0.

▲ **FIGURE 9–53** Problem 74.

75. (II) Suppose that the yo-yo described in Problem 73 is released from rest while the upper end of the cord is held steady. What is the tension in the cord during the yo-yo's downward motion?

76. (II) A thin rod of mass M and length ℓ is lying on a frictionless table. It is given an impulse of 4.5 N · s by a force of magnitude F on one end at an angle of 35° to the rod. How far will the center of mass of the rod travel when the rod has completed 12 rev?

► The complicated and violent swirl of stars and gas—much of it not glowing or hidden—that make up this galaxy move in such a way that the angular momentum of the entire galaxy is conserved.

More on Angular Momentum and Torque

While the rigid body rotations we studied in Chapter 9 are certainly observable in the world, many other real physical systems, such as the collections of stars that make up galaxies, a whirlpool, or a figure skater executing a complex maneuver, are not rigid, and their motions are equally interesting. Just as the concept of linear momentum and its conservation provides us with important simplifications for systems in linear motion, angular momentum and its conservation are extremely useful concepts for systems of particles that do not form rigid bodies. In this chapter we deal in more detail with the vectorial nature of rotational motion, angular momentum, and torque. This study will allow us to extend and expand the correspondence between the dynamics of linear motion and rotational motion.

10–1 Generalization of Angular Momentum

In Section 9–5 we attributed angular momentum to rigid systems in pure rotation. Here we shall see that we can define this quantity for any system, and that, as for the rigid object, angular momentum is conserved if the system is isolated. Our technique for doing this is to define angular momentum for the simplest possible system—the point particle moving freely in space—and then to extend the definition, essentially by superposition, to a larger system composed of pointlike constituents. As in Chapter 9, the angular momentum of even a point mass moving in space must be defined with respect to a fixed origin.

(b)

◀ **FIGURE 10–1** (a) A point mass m moves freely in a plane with velocity $\vec{v}$. (b) To find its angular momentum with respect to a point O we match its motion at point A, when it is closest to O, to a particle of the same mass moving in a circle about O.

What may be surprising is the idea that a free particle—one with no forces on it—has something to do with rotations. Such an object moves along a straight line with constant linear momentum (or just momentum) $\vec{p}$; as for example a hockey puck sliding along the ice. (Strictly speaking, the puck has no *net* force on it, but because, as we shall see, it also has no net torque, it behaves as the truly free object.) Consider our free particle, together with a point that we label as the origin O. As long as this point is not on the line formed by the trajectory, it and the trajectory form a plane, which we label the *xy*-plane (Fig. 10–1a). Our aim here is to define the particle's angular momentum with respect to the point. We will start by insisting that the angular momentum $\vec{L}$ that we assign to the particle is constant, because there are no forces, and thus no torques, acting on the particle. As the particle moves, its point of closest approach to O is a distance d from the origin. At this point, which we label A on Fig. 10–1, the line of motion is perpendicular to the line from the origin; further, the particle's velocity at point A is *instantaneously* identical to that of a second particle of the same mass moving in a circular path in the *xy*-plane about point O (Fig. 10–1b). We already know how to find the angular momentum of the second particle: Its angular momentum is in the $+z$-direction and has magnitude

$$L = I\omega = (md^2)(v/d) = mvd = pd.$$

Now, since the motion of the free particle is indistinguishable at the point of closest approach to the motion of the second particle moving in a circular path, we assign the free particle the instantaneous value of the angular momentum of the second particle. And since the angular momentum of the free particle does not change, this is the angular momentum that we assign to the freely moving particle for all times:

$$\vec{L} = pd\,\hat{k}. \tag{10–1}$$

As Fig. 10–2 illustrates, the freely moving particle may at some time be at point P, a distance r from the origin. Here, the momentum makes an angle θ with respect to the line from O. Geometry shows that $d = r\sin\theta$ so that

$$\vec{L} = pd\,\hat{k} = (pr\sin\theta)\hat{k}. \tag{10–2}$$

Equation (10–2) is the basis for a more general definition of the angular momentum of a point particle with respect to a point O. If our particle has linear momentum of magnitude p and the vector from O to our particle's location has magnitude r (i.e., r is the

▲ **FIGURE 10–2** At a later time mass m has moved to point P. Its angular momentum, however, is the same as it was in Fig. 10–1.

▶ **FIGURE 10–3** A particle of the same mass and velocity as that of the free particle in Fig. 10–2 at the same point P. Its angular momentum at that point is the same as that of the freely moving particle.

Particle at this *instant* has same angular momentum as particle in Fig. 10–2.

Trajectory

particle's distance from O), with the vectors $\vec{r}$ and $\vec{p}$ making an angle θ, then we assign the particle an angular momentum as in Eq. (10–2). (Note that we now determine angular momentum with respect to a point rather than an axis.) This is our definition whether the particle is free or not. The difference between a free particle and one on which a torque acts is that for the free particle, Eq. (10–2) is the angular momentum for *all times*; whereas for the particle on which a torque acts, Eq. (10–2) is only the *instantaneous* value of $\vec{L}$, at the moment picked out in Fig. 10–2. It does not matter if the particle is moving in a circle, in a straight line, or on any other trajectory (Fig. 10–3).

The torque, suitably defined, will determine how the angular momentum changes. Generally, we'll want to define torque with respect to the same point O that serves for the definition of angular momentum. Note that a force may act, changing the linear momentum, but if that force exerts no torque on the particle about point O, Eq. (10–2) remains the constant value of $\vec{L}$.

It is worthwhile to emphasize what our definition entails in terms of vectors, and to understand that a right-hand rule is involved in the determination of the direction of angular momentum. We have a vector $\vec{r}$ from the origin (the reference point) to our particle, which has momentum $\vec{p}$. The direction of the angular momentum is perpendicular to both $\vec{r}$ and $\vec{p}$ (two vectors always form a plane, and you can always find the direction perpendicular to the plane), pointing along the direction of the thumb of the right hand if the fingers curl from $\vec{r}$ to $\vec{p}$. You should check that this is indeed how Eq. (10–2) works. Below we'll find a more convenient way to say this.

We arrived at our result by thinking of a freely moving particle as momentarily equivalent to an object in circular motion. Therefore it is not surprising that it is possible—although we'll forgo the exercise—to start from Eq. (10–2), apply it to the point-like components that make up a rigid object, all rotating about a fixed axis with angular speed ω, and recover the results of Chapter 9, namely

$$\vec{L} = I\vec{\omega}. \tag{10–3}$$

Here I is the rotational inertia about the rotation axis, and both $\vec{L}$ and $\vec{\omega}$ are aligned with the rotation axis according to the right-hand rule. In other words, our new definition of the angular momentum about a point—Eq. (10–2)—reduces to the earlier form, Eq. (9–37), for the angular momentum about an axis in the case of a rotating rigid body.

CONCEPTUAL EXAMPLE 10–1 A powerful rocket is fired straight up. What is its angular momentum with respect to Earth's center? Answer the same question about a little ball that a baby tosses initially parallel to Earth's surface. Ignore Earth's rotation.

Answer To find the angular momentum of an object with respect to some reference point, we can extrapolate the line of its momentum and find the closest approach of that line to the reference point. That is the moment arm d for the angular momentum. In this case, "straight

up" means a line that, if extrapolated back toward Earth's center, passes directly through the center. Thus the moment arm is zero, and so is the angular momentum of the rocket with respect to Earth's center. For the case of the ball, which we might suppose has mass m and an initial speed v, the motion is perpendicular to the line from Earth's center, a distance R_E from the center, where Earth's radius is R_E. The moment arm is R_E, and the angular momentum of the ball with respect to Earth's center is mvR_E.

EXAMPLE 10–2 In an engineering design, a light but stiff rod of length R is attached at an angle θ to a shaft along the z-axis; it is used to rotate a mass M about the shaft (Fig. 10–4a). The mass moves with speed v. Describe the angular momentum of the mass with respect to the attachment point of the rod.

Setting It Up The origin of the coordinate system, placed at the point of attachment of the rod to the shaft, is the reference point for the angular momentum. In Fig. 10–4b we have drawn a side view with the velocity of the mass into the page; the page is in the yz-plane. This second drawing will be useful for understanding the geometry of the situation.

Strategy In identifying the vectors $\vec{R}$ from the origin O to the mass M and $\vec{p}$—the momentum vector, which points along the direction of the velocity—we can work out both the magnitude and the direction of the angular momentum. $\vec{L}$ must be perpendicular to each, as determined by a right-hand rule, and the magnitude L is the product of the magnitudes of $\vec{R}$ and $\vec{p}$ times the sine of the angle between them, as in Eq. (10–2).

Working It Out From our discussion of the direction of the angular momentum, perpendicular to both $\vec{R}$ and $\vec{p}$, we can draw the angular momentum vector (Fig. 10–4b). Note in particular that $\vec{L}$ makes the angle θ with the horizontal. $\vec{R}$ and $\vec{p}$ are perpendicular to each other, hence the angular momentum has magnitude $L = Rp = RMv$. In addition, Fig. 10–4b makes it evident that $\vec{L}$ has both z-component and a radial component toward the shaft, given by

$$L_z = L \sin \theta = L\frac{d}{R} = (RMv)\frac{d}{R} = dMv;$$

$$L_{\text{radial}} = -L \cos \theta = -L\frac{\sqrt{R^2 - d^2}}{R} = -\sqrt{R^2 - d^2}Mv.$$

The minus sign in L_{radial} indicates it points to the rotation axis.

We can anticipate from our discussion in Chapter 9 that because the angular momentum is not constant (the radial component rotates so as to point to the shaft), there must be a torque about point O. This classic engineering problem tells us that the attachment to the shaft must be appropriately constructed to supply the torque. We'll describe such a torque in Conceptual Example 10–4 in Section 10–2.

What Do You Think? How would the result change if the attachment point were in the plane formed by the rotating mass? *Answers to What Do You Think? questions are given in the back of the book.*

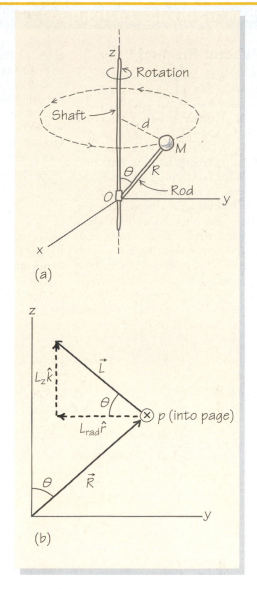

(a)

(b)

▲ **FIGURE 10–4** (a) A rod attached at the origin with a mass at its end rotates about the z-axis. In (b) we draw a side view of (a).

Angular Momentum as a Vector Product

Vector products ("The Vector Product", p. 284) are useful for many of the quantities and relations involved in rotation. One of the most important examples is the angular momentum itself about some origin O of a point mass with momentum $\vec{p}$, which is given by

$$\vec{L} \equiv \vec{r} \times \vec{p}, \tag{10–4}$$

ANGULAR MOMENTUM OF A POINT MASS

where $\vec{r}$ is the position vector of the point mass with respect to O (Fig. 10–5). We can reduce this result to that previously given in Eq. (10–2) and described there. The magnitude of $\vec{L}$ is $rp \sin \theta$, where θ is the angle between $\vec{r}$ and $\vec{p}$, and $r \sin \theta$ is the perpendicular distance d of Eq. (10–2). The vector product described in the box has the property that the angular momentum vector $\vec{L}$ defined in Eq. (10–4) is perpendicular to *both* $\vec{r}$ and $\vec{p}$, with the direction given by the right-hand rule (see Figs. 10–6b, c). Note that $\vec{L}$ is zero when $\vec{r}$ and $\vec{p}$ are parallel; that is, when the straight-line extension of the vector $\vec{p}$ passes through the reference point O. Example 10–3 illustrates the use of Eq. (10–4).

Angular momentum $\vec{L}$ is perpendicular to plane of $\vec{r}$ and $\vec{p}$.

▲ **FIGURE 10–5** Vector relation for the angular momentum $\vec{L}$ of a point particle about a point O.

Problem-Solving Techniques

The Vector Product

In addition to the scalar product discussed in Chapter 6, we can define a product of two vectors—$\vec{A}$ and $\vec{B}$—that is itself a vector. The **vector product** (or **cross product**) of $\vec{A}$ and $\vec{B}$ is defined to be perpendicular to both $\vec{A}$ and $\vec{B}$; it is denoted $\vec{A} \times \vec{B}$. Any two nonparallel vectors form a plane that we can define by choosing the x-axis along $\vec{A}$ and then defining the y-axis so that $\vec{B}$ has only x and y components. The direction of the vector product of $\vec{A}$ and $\vec{B}$ is then perpendicular to this plane (Fig. 10–6a). A right-hand rule is used to define the direction of the vector product because there is an ambiguity as to whether the direction is into or out of the plane. The direction of $\vec{A} \times \vec{B}$ is specified as follows

(Figs. 10–6b, c): Point the fingers of your right hand in the direction of $\vec{A}$ and curl toward $\vec{B}$ according to the rule that the angle through which the fingers are curled to reach $\vec{B}$ must always be less than 180° (it will be hard to get the fingers to follow the opposite of this rule!). The direction of your thumb then indicates the direction of the vector product $\vec{A} \times \vec{B}$. The magnitude of the vector product is given by

$$|\vec{A} \times \vec{B}| = AB \sin \theta, \qquad \text{(B1–1)}$$

where θ is the angle (less than 180°, so that the sine is positive) between $\vec{A}$ and $\vec{B}$.

An immediate consequence of the right-hand rule is that

$$\vec{A} \times \vec{B} = -\vec{B} \times \vec{A}. \qquad \text{(B1–2)}$$

Thus the vector product is not commutative. It also follows from Eq. (B1–1) that *the vector product of two vectors that are parallel (or antiparallel) to each other is zero:* When $\theta = 0°$ or 180°, $\sin \theta = 0$. The vector product attains its maximum magnitude when $\vec{A}$ and $\vec{B}$ are perpendicular to each other because then $\sin \theta = 1$. Contrast this to the scalar product, which is zero when the two vectors are perpendicular and which has a maximum magnitude when they are parallel.

The unit vectors $\hat{i}$, $\hat{j}$, and $\hat{k}$ (Fig. 10–6d) along the x-, y-, and z-axes, respectively, obey the relations

$$\hat{i} \times \hat{j} = -\hat{j} \times \hat{i} = \vec{k}, \qquad \text{(B1–3a)}$$
$$\hat{j} \times \hat{k} = -\vec{k} \times \hat{j} = \hat{i}, \qquad \text{(B1–3b)}$$
$$\vec{k} \times \hat{i} = -\hat{i} \times \vec{k} = \hat{j}. \qquad \text{(B1–3c)}$$

Also $\hat{i} \times \hat{i} = \hat{j} \times \hat{j} = \hat{k} \times \hat{k} = 0$. We can expand the vectors $\vec{A}$ and $\vec{B}$ into their components and find

$$\vec{A} \times \vec{B} = (A_x \hat{i} + A_y \hat{j} + A_z \hat{k}) \times (B_x \hat{i} + B_y \hat{j} + B_z \hat{k})$$
$$= A_x B_y (\hat{i} \times \hat{j}) + A_x B_z (\hat{i} \times \hat{k}) + A_y B_x (\hat{j} \times \hat{i}) + A_y B_z (\hat{j} \times \hat{k})$$
$$+ A_z B_x (\hat{k} \times \hat{i}) + A_z B_y (\hat{k} \times \hat{j}).$$

With the results of Eqs. (B1–3), we find that

$$\vec{A} \times \vec{B} = (A_y B_z - A_z B_y)\hat{i} + (A_z B_x - A_x B_z)\hat{j} + (A_x B_y - A_y B_x)\hat{k}. \qquad \text{(B1–4)}$$

(a) (b) (c) (d)

▲ **FIGURE 10–6** (a) Two vectors $\vec{A}$ and $\vec{B}$ and their vector product $\vec{A} \times \vec{B}$. (b), (c) A right-hand rule specifies the direction of the vector product. (d) The process is broken down to vector products between Cartesian unit vectors.

EXAMPLE 10–3 A tether ball is tied by a rope of length R to a central pole. As the ball whirls around the pole in a (nearly) horizontal plane, the rope gradually winds around (or unwinds from) the pole, shortening (or lengthening) the amount of rope between the ball and the pole (Fig. 10–7). The motion is spiral rather than circular: At the time t, when the ball is at a distance nearly equal to R and rotating around the pole with angular speed nearly equal to ω, the distance is more precisely $r = R - A\omega t$, where A is a constant with dimensions of length and $A\omega t$ is small compared to R. Find the ball's angular momentum with respect to the pole at time t.

Setting It Up The plane of the rope is the xy-plane, and z is along the pole. We denote by θ the angle made by the rope as it goes around the pole (Fig. 10–7). We know the ball's motion in detail and are asked to compute its angular momentum.

Strategy Because the tether ball does not follow a circular path, we must use the more general form $\vec{L} = \vec{r} \times \vec{p}$ for the angular momentum. We first write out the x- and y-coordinates of the ball as a function of time and use these to calculate the velocities and thus the momentum. Finally, we can use the calculation of the linear momentum vector to find the angular momentum.

Working It Out   We have

$$x = r \cos \theta = (R - A\omega t) \cos(\omega t);$$
$$y = r \sin \theta = (R - A\omega t) \sin(\omega t).$$

The momentum components of the ball are therefore

$$p_x = Mv_x = M\frac{dx}{dt}$$
$$= M[(R - A\omega t)(-\omega \sin(\omega t)) + (-A\omega) \cos(\omega t)];$$
$$p_y = Mv_y = M\frac{dy}{dt}$$
$$= M[(R - A\omega t)(\omega \cos(\omega t)) + (-A\omega) \sin(\omega t)].$$

▲ **FIGURE 10–7**   As the rope winds around the pole, it shortens.

We can now find $\vec{L}$ using Eq. (10–4). From Eq. (B1–4) in "The Vector Product" box on page 284, we have

$$\vec{L} = (xp_y - yp_x)\hat{k}$$
$$= M\left\{\begin{array}{l}[(R - A\omega t) \cos(\omega t)][(R - A\omega t)(\omega \cos(\omega t)) + (-A\omega) \sin(\omega t)] \\ -[(R - A\omega t) \sin(\omega t)][(R - A\omega t)(-\omega \sin(\omega t)) + (-A\omega) \cos(\omega t)]\end{array}\right\}\hat{k}$$
$$= M(R - A\omega t)^2 \omega[\cos^2(\omega t) + \sin^2(\omega t)]\hat{k}$$
$$= M(R - A\omega t)^2 \omega\hat{k}.$$

$\vec{L}$ always lies in the z-direction because the motion is, by assumption, always in the xy-plane. But the magnitude of the angular momentum changes with time—albeit slowly ($R \gg A\omega t$). As a check, note that if $A = 0$ (pole diameter zero), the angular momentum has the familiar magnitude corresponding to uniform circular motion of a point mass, namely $MR^2\omega = MvR$.

What Do You Think?   The rope holding the ball suddenly breaks cleanly. By how much is the angular momentum of the ball changed from the moment just before the rope breaks to the moment just after?

10–2 Generalization of Torque

Torque, introduced in Chapter 9, has a magnitude that is given by a lever arm times the magnitude of the applied force, and a direction specified by a right-hand rule. The lever arm is the distance to the reference axis times the sine of the angle between the force and the line to the axis. This has an immediate expression in terms of a vector product: The torque $\vec{\tau}$ with respect to any reference point O due to a force $\vec{F}$ applied at some point P of an object is

$$\vec{\tau} = \vec{r} \times \vec{F}, \qquad (10\text{–}5)$$

THE TORQUE

where $\vec{r}$ is the position vector from O to the point of application (P) of the force (Fig. 10–8). The object in question can be an extended system or it can even be a point mass located at P. The reference point O about which the torque is defined can be located inside or outside the object. Aligning the fingers of the right hand along $\vec{r}$ and curling them toward $\vec{F}$, the torque is in the direction of the thumb. In Fig. 10–8 this is upward.

▶ **FIGURE 10–8** Vector relation for the torque about a point O for a force applied to the corner of a book, i.e., applied at a specific point on an extended object.

The test as to whether this is correct is to see whether the dynamical relation between torque and angular momentum, Eq. (9–39), continues to be satisfied:

$$\vec{\tau} = \frac{d\vec{L}}{dt}. \qquad (10\text{–}6)$$

DYNAMICAL EQUATION FOR ROTATION

To check that Eq. (10–6), which is the rotational analog of Newton's second law, is indeed satisfied, we take Eq. (10–4), which is our generalized expression for the angular momentum of a point mass, and compute its rate of change. The rate of change of angular momentum can be found by using the chain rule of calculus (Appendix IV–7):

$$\frac{d\vec{L}}{dt} = \frac{d(\vec{r} \times \vec{p})}{dt} = \left(\frac{d\vec{r}}{dt} \times \vec{p}\right) + \left(\vec{r} \times \frac{d\vec{p}}{dt}\right). \qquad (10\text{–}7)$$

The first part of Eq. (10–7) drops out because it is of the form $\vec{v} \times (m\vec{v})$, and the vector product of two parallel vectors is zero. For the second part of Eq. (10–7) we use Newton's second law in the form $d\vec{p}/dt = \vec{F}$ and obtain the equation

$$\frac{d\vec{L}}{dt} = \vec{r} \times \vec{F} = \vec{\tau},$$

which is just the desired Eq. (10–6).

We argued in Section 10–1 that it was possible to show that our expression for the angular momentum of a point object, $\vec{L} = \vec{r} \times \vec{p}$, reduces to the form $I\vec{\omega}$ for symmetric rigid bodies. Similarly, $\vec{\tau} = d\vec{L}/dt$ reduces to the form $I\vec{\alpha}$.

We can also use Eq. (B1-4) to express torque in terms of the components of $\vec{r}$ and $\vec{F}$. To do this, write $\vec{r} = x\hat{i} + y\hat{j} + z\hat{k}$. Then Eq. (B1–4) immediately leads to

$$\vec{\tau} = \vec{r} \times \vec{F} = (yF_z - zF_y)\hat{i} + (zF_x - xF_z)\hat{j} + (xF_y - yF_x)\hat{k}. \qquad (10\text{–}8)$$

CONCEPTUAL EXAMPLE 10–4 In Example 10–2 we studied a mass on the end of a light stiff rod attached at an angle to a vertical axis and rotating about it. The angular momentum calculated with respect to the attachment point O of the rod had a constant vertical component but a changing radial component (Fig. 10–9a). To change the radial component of angular momentum there must be a torque about point O. Where does that torque come from, what is the direction of the torque, and how does all this affect the engineering design of the attachment of the rod to the shaft?

Answer We start with an extended free-body diagram (Fig. 10–9b) and find the net force on the mass. Only gravity and a contact force $\vec{N}$

from the rod act on the mass; the vertical component of the contact force merely cancels gravity, allowing the mass to stay in a horizontal plane. The net force on the mass is therefore the horizontal component of the contact force, which is radial and must point inward to make the mass move in its circular path; we'll label it as $\vec{N}_{\text{radial}}$. As we see below, this force will supply the needed torque.

We now verify that there is a torque due to this net force about point O. In Fig. 10–9c, which picks out the radial component of the contact force, the cross product $\vec{r} \times \vec{N}_{\text{radial}}$ would be out of the page. This follows from the fact that both $\vec{r}$ and $\vec{N}_{\text{radial}}$ are in the page plane and the cross product is perpendicular to the plane made by the two of them, together with use of the right-hand rule. You can verify by

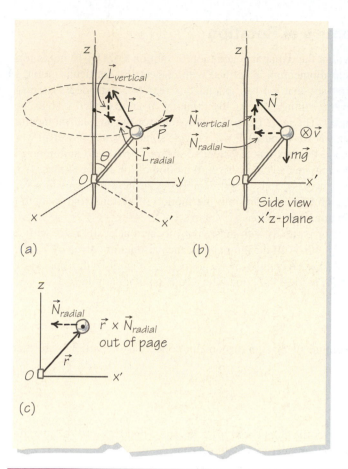

(a)

(b) $\vec{N}_{vertical}$, $\vec{N}_{radial}$, $\vec{N}$, $\otimes \vec{v}$, $m\vec{g}$

Side view
x'z-plane

(c) $\vec{N}_{radial}$, $\vec{r} \times \vec{N}_{radial}$ out of page, $\vec{r}$

repeating the exercise of Example 10–2 that this is indeed the direction of the change in the radial component of mass's angular momentum on this part of the trajectory.

How does this affect the engineering design for the attachment point? The point is fixed. But the normal force from the rod acting on the mass means that the mass exerts a normal force on the rod, and this in turn means that the shaft must supply a horizontal (and vertical) force on the rod to cancel properly the force on the rod due to the mass. In particular, the vertical force from the shaft must be fixed, while the horizontal force from the shaft will have to change with time, and the attachment must be constructed strongly enough to accommodate this.

What Do You Think? How does your answer about torque change if the attachment point is in the plane of the motion ($\theta = 90°$)—that is, the rod is perpendicular to the shaft?

◀ **FIGURE 10–9** In part (a) we are reminded of the results of Example 10–2. (b) A free-body diagram for the mass. The direction of the contact force from the rod is only approximate here, but it must have an upward vertical component to cancel gravity and it must have an inward component to enable the mass to move in a circle. (c) The torque supplied by the radial component of the normal force.

EXAMPLE 10–5 A stone of mass $m = 1$ kg is dropped from an outstretched arm of length $\ell = 0.8$ m (Fig. 10–10). Find the net torque on the stone, taking the reference point O to be the person's shoulder, at the moment the stone has dropped exactly $d = 1$ m. Ignore air resistance.

▲ **FIGURE 10–10** The origin of the coordinate system is placed at the shoulder. $\vec{F}_g$ is the force of gravity on the falling stone.

Setting It Up The coordinate system chosen (Fig. 10–10) sets the rock's motion parallel to the z-axis. The origin of this coordinate system is point O, so we want the torque on the stone about the origin.

Strategy The net torque is determined by the net force, and in this case the only force acting is gravity, $\vec{F} = -mg\,\hat{k}$. Given this information, we can find the net torque by the "lever arm" method described in Section 9–4 or by the equivalent, but more formal, vector product expression of Eq. (10–5). We'll do both.

Working It Out 1 We extend the line of the force in Fig. 10–10 and observe that the distance of closest approach to O is the arm length ℓ; this is the lever arm. The magnitude of the torque is the product of lever arm and force magnitude, i.e., ℓmg, with direction given by the right-hand rule to be in the $-x$-direction, that is, into the page. *The answer is independent of the distance the rock has fallen.* Numerically, the magnitude is $\tau = (0.8$ m$)(1$ kg$)(9.8$ m/s$^2) = 8$ N · m.

Working It Out 2 In the more formal approach, we use the definition of the vector product in terms of components. We have, from Eqs. (10–5) and (B1–3b),

$$\vec{\tau} = \vec{r} \times \vec{F} = \ell\hat{j} \times (-mg\,\hat{k}) = -\ell mg(\hat{j} \times \hat{k}) = -\ell mg\,\hat{i},$$

the same result as in the first method.

What Do You Think? The torque is constant in time and independent of the distance the stone has fallen. This is because (a) there is no time dependence in the force; (b) the moment arm is independent of time; (c) the moment arm is independent of the distance fallen; (d) all of the above.

▲ **FIGURE 10–11** A system of many particles labeled by the index i. The position of particle i with respect to an origin O is $\vec{r}_i$, and the displacement of particle i from the center of mass is $\vec{\rho}_i$.

10–3 The Dynamics of Rotation

In Section 10–2 we saw that the dynamical equation of motion Eq. (10–6) holds for a point mass whose angular momentum is defined with respect to a particular point. In this section we shall first see that we can extend this result to extended systems—whirlpools, spinning divers, rolling objects—by thinking of an extended system as a collection of point particles. The more general expressions for the angular momentum and torque set out in Sections 10–1 and 10–2 will provide some new and powerful ways to think about rotations and the motions of extended objects, both rigid and nonrigid. We will see how we can separate the simple, overall motion of the center of mass from rotations about the center of mass, and we will learn to treat these rotations in a very general way. We will thereby gain insight into the motion of systems and be able to solve problems involving complex motions.

Consider a collection of particles with masses m_i, located at positions $\vec{r}_i$ with respect to a point O, as in Fig. 10–11. If the momenta of the particles are given by $\vec{p}_i$, then the particle with the label i has an angular momentum $\vec{L}_i = \vec{r}_i \times \vec{p}_i$ with respect to O. If the force acting on the ith particle is $\vec{F}_i$, then the torque on that particle is

$$\vec{\tau}_i = \vec{r}_i \times \vec{F}_i = \frac{d\vec{L}_i}{dt}. \qquad (10\text{–}9)$$

The total angular momentum of and torque on the collection of particles are then, respectively,

$$\vec{L} = \sum_i \vec{L}_i \quad \text{and} \quad \vec{\tau} = \sum_i \vec{\tau}_i. \qquad (10\text{–}10)$$

By summing Eq. (10–9) over the component particles and using Eqs. (10–10), we find the rate of change of the total angular momentum to be

$$\vec{\tau} = \frac{d\vec{L}}{dt}. \qquad (10\text{–}11)$$

This is the same as Eq. (10–6) and shows that we can expand the rotational version of Newton's second law to any extended system.

THINK ABOUT THIS . . .

WHAT IS THE ROLE OF INTERNAL FORCES IN THE TORQUE ON A SYSTEM?

▲ **FIGURE 10–12** Two mass points, labeled 1 and 2, within an extended object, and the internal forces they exert on one another.

In Chapter 8 we were able to establish that internal forces play no role in how the linear momentum of an extended system changes under the influence of forces. Similarly, only external torques—torques associated with external forces—determine how a system changes its angular momentum. It will be enough to look at just two particles (or two small mass regions), with labels 1 and 2; our result can then be generalized to include more pieces of the larger systems. For the torque on particles 1 and 2, respectively, we have

$$\vec{\tau}_1 = \vec{r}_1 \times (\vec{F}_1^{\text{ext}} + \vec{F}_{12})$$

and

$$\vec{\tau}_2 = \vec{r}_2 \times (\vec{F}_2^{\text{ext}} + \vec{F}_{21}),$$

where $\vec{F}_{ij}$ is the (internal) force exerted on particle i due to particle j (Fig. 10–12) and $\vec{F}_i^{\text{ext}}$ is the external force on particle i.

The total torque is the sum of the two terms above. The internal forces contribute to the total torque through the terms

$$(\vec{r}_1 \times \vec{F}_{12}) + (\vec{r}_2 \times \vec{F}_{21})$$
$$= (\vec{r}_1 \times \vec{F}_{12}) - (\vec{r}_2 \times \vec{F}_{12})$$
$$= (\vec{r}_1 - \vec{r}_2) \times \vec{F}_{12},$$

where we have used Newton's third law, $\vec{F}_{12} = -\vec{F}_{21}$. Now $\vec{r}_1 - \vec{r}_2$ is the vector leading from particle 2 to particle 1, and we would generally expect the interparticle forces to lie along this line (even if there is no general rule, we require this). Thus the vector product $(\vec{r}_1 - \vec{r}_2) \times \vec{F}_{12}$ is zero—the two vectors entering are parallel to each other—and the contribution to the total torque from the internal forces is zero. In other words, *only the external forces appear in the total torque $\vec{\tau}$* that is to be used in Eq. (10–11). Just as for linear motion, we can find the rotational motion of extended bodies without knowing the details of how their individual parts interact with each other.

The Role of the Reference Point

Throughout we have emphasized that rotational quantities—angular momentum, torque, rotational inertias—are defined with respect to a reference point. While some reference points—for example, the center of mass itself—are more "natural" choices than others, there is no physical principle that demands one choice over another. To put this another way, physical motion does not depend in any way on the choice of this point. What is true is that rotational problems can often be solved more easily by the choice of a good reference point; the right choice may, for example, simplify the geometry of a situation. Here we find some results that allow us to vary the reference point for rotational quantities. After we present the results, we illustrate their usefulness with an example.

Let us first recall the fact [Eq. (9–37)] that for a rigid body with rotational inertia I about an axis through the center of mass, the dynamical equation for rotation takes the form

$$\vec{\tau} = \frac{d\vec{L}}{dt} = I\frac{d\vec{\omega}}{dt} = I\vec{\alpha}. \tag{10–12}$$

The quantities in this equation all are assigned relative to the specified axis, or as we would now say, with respect to the location of the center of mass itself.

We now extend the "center of mass" results with a series of three relations that allow us instead to consider rotations about a second point A. These results apply to nonrigid as well as rigid systems. Let $\vec{R}$ be a vector from the center of mass to point A. Then, as we prove in a separate optional section, the following three results hold:

1. The angular momentum about A, which we call $\vec{L}_A$, is the sum of two terms: (a) the angular momentum $\vec{R} \times \vec{P}$ about A of a point mass that is carrying the whole mass (M) of the object, as though it were located at the center of mass and moving with the velocity $\vec{V}$ of the center of mass; and (b) the angular momentum $\vec{L}_{cm}$ of the object about the center of mass:

$$\vec{L}_A = (\vec{R} \times \vec{P}) + \vec{L}_{cm}. \tag{10–13}$$

Here, $\vec{P} = M\vec{V}$.

2. The total torque about A is the sum of two terms: (a) the torque about A due to the total external force applied to the center of mass; and (b) the torque about the center of mass:

$$\vec{\tau}_A = (\vec{R} \times \vec{F}_{tot}) + \vec{\tau}_{cm}. \tag{10–14}$$

3. The rate of change in each term of the total angular momentum in Eq. (10–13) is equal to the corresponding term of the total torque; that is,

$$\vec{R} \times \vec{F}_{tot} = \frac{d}{dt}(\vec{R} \times \vec{P}) \quad \text{and} \quad \vec{\tau}_{cm} = \frac{d}{dt}\vec{L}_{cm}. \tag{10–15}$$

*How to Get Equations (10–13) Through (10–15) for Angular Momentum and Torque

We break our extended object into a set of discrete pieces labeled with the subscript i. The center of mass position $\vec{R}$ with respect to an origin O was defined in Eq. (8–47) by

$$\vec{R} = \frac{1}{M}\sum_{i=1}^{N} m_i\vec{r}_i,$$

where $M = \sum_i^N m_i$ is the total mass, N is the number of pieces, and $\vec{r}_i$ is the position of the ith piece. We introduce the vector $\vec{\rho}_i$ (Fig. 10–11), which is the position vector of particle i as measured from the center of mass:

$$\vec{r}_i = \vec{R} + \vec{\rho}_i.$$

The angular momentum with respect to point O may now be written in the form

$$\vec{L} = \sum_i \vec{r}_i \times \vec{p}_i = \sum_i (\vec{R} + \vec{\rho}_i) \times \vec{p}_i = (\vec{R} \times \sum_i \vec{p}_i) + (\sum_i \vec{\rho}_i \times \vec{p}_i)$$

$$= (\vec{R} \times \vec{P}) + (\sum_i \vec{\rho}_i \times \vec{p}_i),$$

where $\vec{P} = \sum_i \vec{p}_i$ is the total momentum of the system. The term $\sum_i \vec{\rho}_i \times \vec{p}_i$ is the angular momentum $\vec{L}_{cm}$ about the center of mass. We have thus proved Eq. (10–13).

There is a similar decomposition of the total torque:

$$\vec{\tau} = \sum_i \vec{r}_i \times \vec{F}_i = \sum_i (\vec{R} + \vec{\rho}_i) \times \vec{F}_i = (\vec{R} \times \sum_i \vec{F}_i) + (\sum_i \vec{\rho}_i \times \vec{F}_i)$$

$$= (\vec{R} \times \vec{F}_{tot}) + (\sum_i \vec{\rho}_i \times \vec{F}_i).$$

The second term in the final equality is the torque about the center of mass, and so we have demonstrated Eq. (10–14).

To derive the third result, we take the rate of change of the $\vec{R} \times \vec{P}$ term in Eq. (10–13). We find that it is the same as the first term of Eq. (10–14):

$$\frac{d(\vec{R} \times \vec{P})}{dt} = \left(\frac{d\vec{R}}{dt} \times \vec{P}\right) + \left(\vec{R} \times \frac{d\vec{P}}{dt}\right) = \left(\frac{1}{M}\vec{P} \times \vec{P}\right) + (\vec{R} \times \vec{F}_{tot})$$

$$= \vec{R} \times \vec{F}_{tot}.$$

This in turn implies that the rate of change of the total angular momentum about the center of mass is equal to the torque about the center of mass due to all the forces:

$$\frac{d\vec{L}_{cm}}{dt} = \vec{\tau}_{cm}.$$

Why It Is Useful to Be Able to Choose the Reference Point

The results above are of practical importance because they provide alternative ways to approach problems involving both linear and rotational motion. We can choose convenient points of reference for torque, angular momentum, and rotational inertia. This is illustrated with a cylinder rolling down an incline, a problem we first examined in Chapter 9 (Fig. 10–13a). There we used a point on the symmetry axis of the cylinder as a reference. Let's now choose our reference point to be point O, the point of contact between the cylinder and the plane. Why choose this point? The answer is that it is easiest to compute net torque about a point through which the largest number of individual forces act. As the extended free-body diagram (Fig. 10–13b) shows, such forces have a zero lever arm for this point and so they do not contribute to the torque about the point. In this case, only gravity produces a torque about the contact point; moreover, only the component of gravity that is parallel to the plane—of magnitude $mg \sin \theta$—contributes.

EXAMPLE 10–6 A cylinder, mass M and radius R_0, whose mass density is symmetric about its axis, rolls down a ramp making an angle θ to the horizontal. (You can think for definiteness of any wheel; this is certainly a problem of practical interest.) Assuming its angular speed ω about its axis at $t = 0$ is ω_0, find the angular speed at a later time t.

Setting It Up Figure 10–13 illustrates the problem. While a rotational inertia for the cylinder will enter the problem, we don't have the precise mass distribution, so it will have to remain as an unknown quantity.

Strategy In keeping with the express aim of this example, we compute all angular quantities with respect to the instantaneous

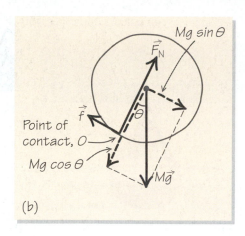

◀ **FIGURE 10–13** (a) A cylinder rolls without slipping down an inclined plane. (b) Force diagram for the rolling cylinder.

contact point O between the cylinder and the ramp (Fig. 10–13b). As described prior to the example, with this reference point only gravity produces a torque, and this is easily calculable. The angular momentum in this situation is a little more complex, and we turn to the technique described above and expressed by Eq. (10–13). This expression utilizes the momentum $\vec{P}$ of the center of mass, as well as the angular momentum about the symmetry axis of the cylinder, which has magnitude $L = I_{cm}\omega$. Finally, once we know the torque and hence the rate of change of the two terms in the angular momentum about the axis A, we will have the angular acceleration and can use kinematic relations to find angular speed as a function of time.

Working It Out Using the right-hand rule, we see that the torque due to gravity about point O points into the page and has magnitude

$$\tau = (Mg \sin\theta)R. \qquad (10\text{–}16)$$

Equation (10–13) gives us the angular momentum about O. This expression contains two terms. First, there is the angular momentum of the total mass as though that mass were placed at the object's center of mass, $\vec{R} \times \vec{P} = \vec{R} \times (M\vec{V})$. $\vec{R}$ and $\vec{V}$ are perpendicular to each other, and the magnitude of $\vec{R}$ is R_0, so this term has magnitude

$$MVR_0 = M(\omega R_0)R_0 = MR_0^2\omega. \qquad (10\text{–}17)$$

Second, there is the angular momentum about the symmetry axis passing through the center of mass, magnitude $L = I_{cm}\omega$. This is in the same direction as the $\vec{R} \times \vec{P}$ term above, so the magnitude of the total angular momentum about the point of contact is the sum of Eq. (10–17) and $I_{cm}\omega$:

$$L = (MR_0^2 + I_{cm})\omega. \qquad (10\text{–}18)$$

Equations (10–16) and (10–18) are the torque and angular momentum, respectively, about the same reference point. The dynamical equation $dL/dt = \tau$ is therefore applicable:

$$(MR_0^2 + I_{cm})\frac{d\omega}{dt} = MgR_0 \sin\theta;$$

$$\alpha = \frac{d\omega}{dt} = \frac{MgR_0 \sin\theta}{MR_0^2 + I_{cm}}.$$

This equation states that the magnitude of angular acceleration α is constant, so the angular speed increases linearly with time:

$$\omega(t) = \omega_0 + \alpha t = \omega_0 + \frac{MgR_0 \sin\theta}{MR_0^2 + I_{cm}}t. \qquad (10\text{–}19)$$

This is the desired answer.

Comparison with the Chapter 9 treatment of this problem is useful. For a symmetric rolling object, $I_{cm} = CMR_0^2$, where C is a numerical constant (Section 9–3). Then Eq. (10–19) tells us that the linear speed $v = \omega R_0$ of the cylinder center is

$$v(t) = v_0 + \frac{g \sin\theta}{1 + C}t. \qquad (10\text{–}20)$$

This result is the same one found in Eq. (9–52). In Chapter 9 we used the center of mass as a reference point, while here we use the contact point of the rolling cylinder and the plane. Compare this example with the discussion of Eq. 9–52, and consider which, if any, is the preferable approach. Having a choice of reference point is always valuable.

Angular Impulse

The angular impulse for the motion of extended systems is useful in the same way as the linear impulse for the motion of point masses. The linear impulse, $\Delta\vec{p} \equiv \vec{J} = \vec{F}\,\Delta t$, describes the change in momentum when a force $\vec{F}$ acts for a short duration Δt. Similarly, the dynamical equation for rotational motion tells us how the change in the angular momentum of a system, $\Delta\vec{L} \equiv \vec{J}_\tau$ (the **angular impulse**), is described by a torque $\vec{\tau}$ acting for a duration Δt:

$$\Delta\vec{L} \equiv \vec{J}_\tau = \vec{\tau}\,\Delta t. \qquad (10\text{–}21)$$

Here, both the torque and angular momentum are measured with respect to the same reference point.

▶ **FIGURE 10–14** The angular impulse delivered to a punching bag. In parts (a) and (b) we draw two views of the situation.

We can illustrate how forces lead to angular impulses with the punching bag of Fig. 10–14. Let the reference point for torque and angular momentum be O. The force, $\vec{F}$, is applied briefly at $\vec{r}$. The torque is then $\vec{\tau} = \vec{r} \times \vec{F}$; multiplying by Δt, we see that the angular impulse is given by

$$\vec{J}_\tau = \vec{\tau}\,\Delta t = \vec{r} \times (\vec{F}\,\Delta t) = \vec{r} \times \vec{J}. \tag{10–22}$$

EXAMPLE 10–7 A law-court expert is called upon for a case in which a stationary automobile is struck from the side. To model this situation, he treats the automobile as a uniform rod of mass M and length ℓ at rest on a frictionless surface. An impulse of magnitude $F\,\Delta t$ is applied at right angles at a distance $\ell/6$ from one end of the rod (Fig. 10–15). Describe the subsequent motion of the rod.

FIGURE 10–15 The automobile is modeled by a uniform rod. (a) Before: An impulse is given to the stationary rod a distance $\ell/6$ from the top. (b) After: The rod moves with a combination of a linear center-of-mass motion and a rotation about the center of mass.

Setting It Up In Fig. 10–15 the rotation axis goes through the center of mass. Let us label the rod's center of mass momentum and velocity as $\vec{P}$ and $\vec{V}$, respectively, and the angular momentum and angular speed about its center of mass as $\vec{L}$ and $\vec{\omega}$, respectively. We want to find these quantities after the collision (subscript f).

Strategy The motion may be decomposed into motion of the center of mass and rotation about the center of mass. The initial momentum of the center of mass is zero, and therefore its final momentum is given by the linear impulse. We can translate information of the momentum change to details of the speed change, through the mass of the automobile. The angular impulse is equal to the change in angular momentum in time Δt. The rotational inertia of the rod about the rotation axis will then allow us to relate the change in angular momentum to the change in angular speed.

Working It Out With $P_i = 0$,

$$\Delta\vec{P} = \vec{P}_f - \vec{P}_i = \vec{P}_f = \vec{F}\,\Delta t,$$

and the final center of mass speed is

$$\vec{V}_f = \frac{\vec{P}_f}{M} = \frac{\vec{F}\,\Delta t}{M}.$$

Similarly, the angular impulse is equal to the change in angular momentum in time Δt. The initial angular momentum is zero, and the distance from the point of application of the force to the center of mass is $(\ell/2)-(\ell/6)$. Thus

$$\Delta\vec{L} = \vec{L}_f - \vec{L}_i = I\vec{\omega}_f = (F\,\Delta t)\left(\frac{\ell}{2} - \frac{\ell}{6}\right)\hat{k} = \frac{(F\,\Delta t)\ell}{3}\hat{k},$$

pointing down into the street. $I = M\ell^2/12$ (see Table 9–1), so the angular speed can be found:

$$\frac{M\ell^2\omega_f}{12} = \frac{(F\,\Delta t)\ell}{3} \quad\text{or}\quad \omega_f = \frac{4}{M\ell}(F\,\Delta t) = \frac{4}{\ell}V_f$$

The final linear and angular motions are shown in Fig. 10–15.

What Do You Think? Suppose the car was struck at its center rather than off center. Would the center of mass move off with (a) more (b) less or (c) the same velocity as in the example?

The fact that angular impulse, torque, and angular momentum are vectors has some consequences that are not very intuitive, as we can see in the following conceptual example. Keep in mind as you look at this example that the laws that govern rotational motion follow directly from the laws for linear motion.

CONCEPTUAL EXAMPLE 10–8 A bicycle wheel is rotating about an extended axle (Fig. 10–16), with its initial angular momentum $\vec{L}_i$ aligned horizontally in the xy-plane, and with one end of the axle attached to a pivot about which it can move freely. The other end of the axle is struck by a downward hammer blow, receiving an impulse $\vec{J} = -J\hat{k}$ (Fig. 10–16a). What is the immediate response of the wheel to the impulse, assuming that gravity can be neglected in the short run?

Answer The change in the wheel's angular momentum is the angular impulse. We can find this quantity using the fixed end (pivot point) of the axle as a reference point. If the vector from the fixed end to the point where the impulse is applied is $\vec{\ell}$, then the angular impulse is

$$\Delta\vec{L} = \vec{J}_\tau = \vec{\ell} \times \vec{J}.$$

By a right-hand rule, the direction of this vector lies in the xy-plane as shown (Fig. 10–16b). Even though the blow comes from above, *the free end of the wheel tends to rotate back toward the y-axis!* The wheel moves in a direction perpendicular to the direction of the blow. The result we have found here is a surprising consequence of the vector nature of angular momentum and torque. This counterintuitive motion is echoed by a similar phenomenon in the motion of rotating objects (gyroscopes) under the effect of gravity (Section 10–8).

(b)

(c)

▶ **FIGURE 10–16**
(a) Rotating wheel precesses about string. (b) A rapidly rotating bicycle wheel is spinning with its angular velocity, $\vec{\omega}$, and its initial angular momentum, $\vec{L}_i$, aligned as shown. One end of the axis is fixed on a pivot point, and the other end is given an impulse $\vec{J} = -J\hat{k}$. (c) When a corresponding angular momentum change, $\Delta\vec{L}$, is added to the initial angular momentum, the result is that the free end of the wheel tends to rotate toward the y-axis.

(a)

10–4 Conservation of Angular Momentum

The general dynamical equation for rotational motion is $\vec{\tau}_{net} = d\vec{L}/dt$, where the torque always refers to *external* forces. If there is no net external torque on a system—rigid or otherwise—the angular momentum of the system is *conserved*,

$$\frac{d\vec{L}}{dt} = 0, \tag{10–23}$$

meaning that the angular momentum is constant during the motion of the system. This fact has a variety of powerful consequences that we shall explore in this section.

Angular Momentum and Central Forces

Central forces—forces directed along a line from a given source—provide examples of forces with no torque about a special point. In this case, the point is the source of the force itself. For example, the Sun is responsible for a central force on the planets, and this gravitational force is directed toward the Sun itself. This force exerts no torque on the planets about the Sun, and the angular momentum of the planets as they orbit the Sun is therefore constant. Because both gravitation and electrostatic forces (such as the force holding electrons to nuclei in atoms) are central, central forces are of particular importance in the physical world. Example 10–9 illustrates how to find angular momentum for an object under the influence of a central force.

THINK ABOUT THIS...
HOW DOES A SUBMARINE NAVIGATE 'SILENTLY'?

▲ **FIGURE 10–17** The heart of a navigating device based on conservation of momentum for a rotating sphere.

Nuclear submarines can navigate with great accuracy without communication with anyone or anything above the surface, an obvious necessity given what submarines do. Similarly, even before the advent of the Global Positioning System or other satellite-based communication, transatlantic flights could fly across great distances without communication with the ground. The technology used for silent navigation makes use of the fact that the rotation axis of an isolated rotating object (one with no external torques acting on it) will not change. One version of such a device consists of a small metal sphere levitated by a magnetic field so that it has no material contact (Fig. 10–17). The sphere is given a high initial angular velocity with the aid of special magnetic fields. Once the rotation is established, the sphere's rotation axis remains aligned in a constant spatial direction. As the sphere does not touch anything, there are no mechanical bearings, and so there are no vestigial torques that can cause the sphere's axis to rotate. The sphere is engraved with black and white marks that can be read with the aid of photocells. When the submarine changes direction, the position of these marks relative to a casing attached to the submarine will change. The photocells and an associated computer keeping track of the marks will observe the sphere's rotation axis move relative to the submarine and can measure the exact degree of the motion. The measurement is a way to keep track of any changes in the submarine's direction of motion throughout the voyage. Such devices can provide positioning within meters over great distances. ■

EXAMPLE 10–9 A comet, mass 10^{15} kg, moves in a highly eccentric orbit about the Sun, as shown in Fig. 10–18. At its closest approach to the Sun, 10^6 km, the comet is measured to be moving at a speed of 6×10^6 m/s. Assuming that the comet can be treated as a point mass, what is the angular momentum of the comet with respect to the Sun at the moment of closest approach?

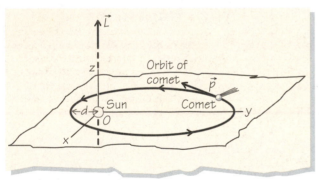

▲ **FIGURE 10–18** The direction of the angular momentum is found from a right-hand rule to be up out of the plane of the orbit, in the +z-direction.

Setting It Up We know the comet's mass M, the distance d of its closest approach to the Sun, and its speed v at that point.

Strategy The angular momentum is given by the linear momentum $p = Mv$ times the distance of closest approach, i.e., magnitude $L = Mvd$, so the problem can be solved through simple substitution. We can also use a right-hand rule to find the direction of the angular momentum.

Working It Out The angular momentum has magnitude

$$L = Mvd = (10^{15} \text{ kg})(6 \times 10^6 \text{ m/s})(10^6 \text{ km})[(10^3 \text{ m})/(1 \text{ km})]$$

$$= 6 \times 10^{30} \text{ kg} \cdot \text{m}^2/\text{s}.$$

The direction of the angular momentum is given by a right-hand rule. In Fig. 10–18, this direction is up out of the plane of the orbit, in what is labeled the +z-direction. (Below we'll see why angular momentum is conserved for central forces, so that the angular momentum at the moment the comet is closest to the Sun is its angular momentum throughout its orbital motion.)

What Do You Think? Suppose the comet of the example has an orbit that is much less eccentric (i.e., more nearly circular) with speed at closest approach of 12×10^6 m/s and a distance of closest approach 0.5×10^6 m. Does the comet have (a) more angular momentum, (b) the same angular momentum, or (c) less angular momentum than in the example?

The torque on a particle is zero if the force and the displacement vector $\vec{r}$ from the point of reference (the origin) to the point of application of the force are *parallel* (or antiparallel), because then the vector product between $\vec{r}$ and $\vec{F}$ is zero. This applies for central forces, so *for central forces the angular momentum about the origin is conserved*, as in Example 10–9.

Consider an object moving under the influence of a central force, and take the origin of the coordinate system to be at the source of this force. For example, let us think about the comet of Example 10–9 moving under the influence of the Sun. Let the initial position and velocity of the comet be $\vec{r}_0$ and $\vec{v}_0$, respectively. The initial direction of the angular momentum is given by $\vec{r}_0 \times \vec{v}_0$, that is, perpendicular to the plane formed by $\vec{r}_0$ and $\vec{v}_0$. Because the angular momentum is constant and its direction does not change, *the comet's motion is always confined to the initial plane formed by $\vec{r}_0$ and $\vec{v}_0$*. To obtain the magnitude of the angular momentum, consider the path shown in Fig. 10–19a.

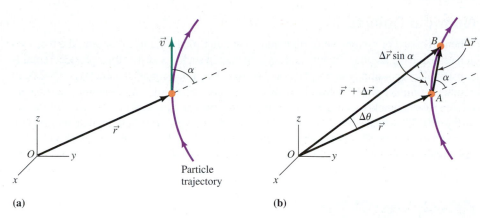

◀ **FIGURE 10–19** The trajectory of a particle with constant angular momentum. (a) The velocity at any given time is tangent to the trajectory and makes an angle α with the position vector. (b) In a small time interval, the particle moves from point A to point B and is displaced by $\Delta\vec{r}$. The angle $\Delta\theta$ is approximately $\Delta\theta = (|\Delta\vec{r}|\sin\alpha)/r$.

At any given moment t, the comet is at position $\vec{r}$ and is moving with a velocity $\vec{v}$ tangent to the trajectory. The magnitude of the angular momentum is given by

$$L = |\vec{r} \times \vec{p}| = |\vec{r} \times m\vec{v}| = rmv \sin\alpha, \tag{10–24}$$

where α is the the angle between $\vec{r}$ and $\vec{v}$. If the object's position changes from $\vec{r}$ to $\vec{r} + \Delta\vec{r}$ in a small time interval Δt as in Fig. 10–19b, and if we replace $\vec{v}$ by $\Delta\vec{r}/\Delta t$, we find

$$L = mr\frac{|\Delta\vec{r}|\sin\alpha}{\Delta t} = mr\frac{r\,\Delta\theta}{\Delta t}. \tag{10–25}$$

We have used the geometric relation $|\Delta\vec{r}|\sin\alpha = r\,\Delta\theta$. In the limit $\Delta t \to 0$, Eq. (10–25) becomes

$$L = mr^2\frac{d\theta}{dt} = mr^2\omega, \tag{10–26}$$

where ω is the instantaneous angular velocity about the origin. Thus $r^2\omega$ is constant throughout the motion when, as here, the angular momentum is constant. For uniform circular motion both r and ω are constant, and there is nothing new in Eq. (10–26)— this is simply the circular motion described in Section 3–5. If r is not constant, then Eq. (10–26) can be interpreted as follows: Consider the time interval Δt and the geometry of Fig. 10–19b. For infinitesimal displacements Δr, the area of the triangle OAB is given by

$$\Delta A = \frac{1}{2}(r\,\Delta\theta)r = \frac{1}{2}r^2\,\Delta\theta, \tag{10–27}$$

from which it follows that $dA = \frac{1}{2}r^2\frac{d\theta}{dt}dt = \frac{1}{2}r^2\omega\,dt$. In other words,

$$\frac{dA}{dt} = \frac{1}{2}r^2\omega = \frac{L_0}{2m}, \tag{10–28}$$

where we have used Eq. (10–26) and L_0 is the constant value of the angular momentum. Whatever the trajectory of a particle moving under the influence of a central force (which depends on the detailed form of the central force), it follows from angular momentum conservation that *the rate at which the radius vector sweeps out an area is constant*. In the context of the gravitational force (Chapter 12), this is known as **Kepler's second law**. With that force a comet moves in an elliptical orbit with the Sun located at one of the focal points; this orbit sweeps out equal areas in equal lengths of time (Fig. 10–20). The comet must move rapidly when it is near the Sun compared with its speed at large distances from the Sun.

Another example of a particle that moves with no torque on it is a freely moving particle such as a puck sliding on ice. The reference point is any point whatever, and this motion too satisfies the condition of Eq. (10–28), i.e., its position vector, as measured from any point, sweeps out each area at a uniform rate. This is demonstrated in Fig. 10–21.

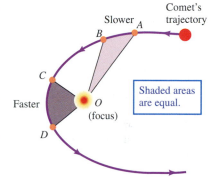

▲ **FIGURE 10–20** If a particle moves with constant angular momentum, then its trajectory sweeps out equal areas in equal times. If the motion of a comet from A to B takes time Δt and the motion from C to D takes the same amount of time, then the particle must move more quickly in traveling from C to D.

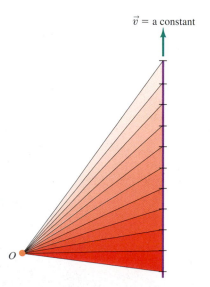

▲ **FIGURE 10–21** The triangles that a uniformly moving particle sweeps out with respect to any point O have equal areas.

Nonrigid Objects

When an object or system is not rigid and its angular momentum is constant, the rotational inertia and the angular velocity can each change in such a way that their product remains constant. This possibility commonly occurs with divers and figure skaters (Fig. 10–22). In another example, water going down a drain holds its angular momentum constant by swirling more quickly as it approaches the drain. We can also mention the case of the cat in Fig. 10–22, which even when dropped back down with no initial angular momentum can land on its feet. This it does by rotating and extending different parts of its body while the angular momentum remains zero (as it must—the cat has no means to do anything about its angular momentum once it has been dropped).

(a)   (b)

▲ **FIGURE 10–22**   (a) A cat can maneuver in midair while conserving angular momentum. (b) Although a figure skater spins more slowly when her arms are extended than when her arms are close to her body, her angular momentum remains constant.

EXAMPLE 10–10 In a common classroom experiment, a student sits on a spinning stool with weights in each hand. An idealized version of this experiment is as follows: a solid cylinder of diameter 0.5 m and mass 50 kg is oriented vertically and spins freely about its axis with a period of 3 s. Two massless rods are attached horizontally to the cylinder, with their ends 1 m from the surface, and there is a mass 2 kg at the end of each rod (Fig. 10–23a). The rods are drawn into the cylinder by an internal mechanism until the 2-kg masses are at the surface of the cylinder (Fig. 10–23b). What are the initial and final angular velocities?

Setting It Up We know the diameter d and mass M of the spinning cylinder, as well as the length D of each of the attached rods; the masses m at the end of each cylinder are also given, as is the initial period of rotation T. We want to find the initial and final angular speeds, ω_0 and ω_f, respectively.

Strategy The angular momentum $L = I\omega$ about the symmetry axis of the cylinder is constant because there are no external torques. (The force pulling in the "arms" is *internal* and *radial*, each property by itself sufficient to ensure that there is no torque about the central axis.) We can calculate the initial and final rotational inertia I_0 and I_f by simple geometry, and we can find ω_0 from its kinematic connection with period. The condition that L is conserved then determines ω_f. Physically, we can anticipate as a check that as the masses are pulled in, I decreases; therefore, ω must increase.

Working It Out The conservation of angular momentum reads

$$I_0\omega_0 = I_f\omega_f,$$

or

$$\omega_f = \omega_0 \frac{I_0}{I_f}.$$

I_0 is the rotational inertia of a solid cylinder ($MR^2/2$ from Table 9–1) of mass $M = 50$ kg and radius $R = d/2 = 0.25$ m, plus the rotational inertia of the two masses of 2 kg a distance $D + R = 1.25$ m from the axis:

$$I_0 = \tfrac{1}{2}MR^2 + 2m(D + R)^2.$$

I_f is found the same way, except the two masses are a distance R from the axis:

$$I_f = \tfrac{1}{2}MR^2 + 2mR^2 = [(M/2) + 2m]R^2.$$

The initial angular velocity is found from the period, $\omega_0 = 2\pi/T = 2\pi/(3 \text{ s}) \cong 2$ rad/s. Thus

$$\begin{aligned}
\omega_f &= \omega_0 \frac{(MR^2/2) + 2m(D + R)^2}{[(M/2) + 2m]R^2} \\
&= (2 \text{ rad/s}) \frac{(0.5)(50 \text{ kg})(0.25 \text{ m})^2 + 2(2 \text{ kg})(1.25 \text{ m})^2}{(0.5)(50 \text{ kg})(0.25 \text{ m})^2 + 2(2 \text{ kg})(0.25 \text{ m})^2} \\
&= 9 \text{ rad/s}.
\end{aligned}$$

The angular velocity has increased by a factor of 4.5. The movement of the end weights into the axis is remarkably effective at decreasing the rotational inertia! You can perform this demonstration in class (Figs. 10–23c and 10–23d).

(a) (b) (c) (d)

▲ **FIGURE 10–23** (a) An object rotates with masses at the end of a rod. (b) When the masses are brought in closer to the rotation axis, the conservation of angular momentum requires the rotation to speed up. (c), (d) A demonstration of this.

An Astrophysical Example: A supernova occurs when a massive star stops burning because it has used up its fuel and undergoes a collapse due to gravitational forces, gaining enough energy in the process to blow off most of its material in a stupendous explosion (Fig. 10–24). What is left behind is a *neutron star*: an extremely dense sphere of matter with a mass of several solar masses and a diameter of perhaps 10 km (compared with 10^6 km for the original star). If the original star is rotating, as most stars do, at some rate on the order of several days, and if the outer layers are blown off nearly radially, then the neutron star will have a good part of the original angular momentum, and because it has so much mass packed into a tiny sphere, its angular speed will be very large. Many neutron stars so produced are known to have periods on the order of milliseconds.

▲ **FIGURE 10–24** At the center of the outwardly moving remnants of the Crab Nebula supernova, a neutron star spins rapidly.

CONCEPTUAL EXAMPLE 10–11 Suppose that a spinning bicycle wheel with its angular velocity pointing up (the wheel spins counterclockwise as seen from above) is handed to a student sitting on a stool that can freely rotate about the vertical axis (Fig. 10–25a) but is otherwise attached to the ground. The student turns the wheel upside down, so that its angular velocity points downward. Which of the following is true? (a) The student will be forced to the right and off the stool by an angular impulse; (b) the stool (with the student) will also pick up a downward angular velocity; (c) the stool (with the student) will pick up an upward angular velocity; (d) the wheel is isolated from the student and nothing happens to the student.

Answer The answer is (c). An upward angular velocity corresponds to an upwardly directed angular momentum. Because the system of student, stool, and wheel is isolated as far as vertical rotations are concerned, the z-component of angular momentum is conserved. Before the wheel is turned, the total angular momentum is oriented up. After the wheel is turned, the wheel has an angular momentum oriented down, so the student and stool must have enough angular momentum oriented up to keep the total angular momentum unchanged. That means the student-stool spins counterclockwise— the original direction of rotation of the wheel (Fig. 10–25b). This new spin direction is generated to conserve the total angular momentum (Fig. 10–25c).

If you trace the motion through, you can see that there was a moment when the wheel was oriented with its angular momentum in the horizontal direction. At that moment the only way the entire system could have no net angular momentum in the horizontal plane is for the student and stool to rotate over! The attachment point of the stool will in fact exert a compensating torque to keep things upright—in other words, the system is not truly isolated from its surroundings for this situation.

What Do You Think? Suppose the student now flips the wheel back to its starting orientation. What will the result be?

(a) (b) (c)

▲ **FIGURE 10–25** (a) A student on a stool that can rotate holds a freely rotating wheel. (b) When the wheel is inverted, the stool begins to turn in order to hold the total angular momentum constant. (c) To the now negative angular momentum of the wheel must be added a positive angular momentum from the student. Only in this way can the total angular momentum be conserved.

10–5 Work and Energy in Angular Motion

Rotating systems of particles, even those that are nonrigid, have energy, and work has to be done on them to change their energy. For example, the motor must do work to start an airplane's propeller, and gravity does work on a ball that rolls down a hill. Both of these examples are symmetric, rigid objects, the subject of Chapter 9. There we saw that if the object has rotational inertia I and angular speed ω about the axis of rotation, its energy is

$$K = \frac{1}{2}I\omega^2. \tag{10–29}$$

The angular velocity is a vector, which can be expressed as

$$\omega = \omega_x \hat{i} + \omega_y \hat{j} + \omega_z \hat{k} \tag{10–30}$$

and

$$\omega^2 = \vec{\omega} \cdot \vec{\omega} = \omega_x^2 + \omega_y^2 + \omega_z^2. \tag{10–31}$$

If a torque is present, the rotational kinetic energy of the body will change because the angular speed changes. We calculate the rate of change of energy, that is, the *instantaneous power* that must be supplied by whatever changes the energy,

$$\frac{dK}{dt} = \frac{1}{2}I\frac{d}{dt}\omega^2 = I\frac{d\vec{\omega}}{dt}\cdot\vec{\omega} = I\vec{\alpha}\cdot\vec{\omega} = \vec{\tau}\cdot\vec{\omega},\tag{10–32}$$

where we have used Eq. (10–12), $I\vec{\alpha} = \vec{\tau}$. Equation (10–32) is the analog of the linear motion equation for power that we saw in Chapter 6: $\dfrac{dK}{dt} = \vec{F}\cdot\vec{v}$.

The Work–Energy Theorem for Rotations

It is fairly straightforward to derive a work–energy theorem for rotational motion *about a fixed axis*. To define the work, we first follow the analog with linear motion once more. Just as the infinitesimal work in linear motion is defined as $dW = \vec{F}\cdot d\vec{x}$, the infinitesimal work done in rotating a rigid body through an infinitesimal angle $d\theta$ about the axis is defined to be

$$dW \equiv \vec{\tau}\cdot d\vec{\theta}.\tag{10–33}$$

INFINITESIMAL WORK DONE BY TORQUE

The infinitesimal angle is a vector whose direction is defined by a right-hand rule: If the fingers curl in the direction of the infinitesimal angle, the thumb direction gives the vector direction. When the torque points along the axis of rotation, this reduces to the form

$$dW = \tau\, d\theta.\tag{10–34}$$

Exactly as in Chapter 6, the test as to whether this is a reasonable definition for the work is that it leads to an appropriate work–energy theorem. To verify this, we use $\tau = I\, d\omega/dt$ and $d\theta = \omega\, dt$, and hence

$$W = \int_{\theta_0}^{\theta} \tau\, d\theta = \int_0^t I\frac{d\omega}{dt}\omega\, dt = \int_0^t I\frac{1}{2}\frac{d\omega^2}{dt}\, dt$$

$$= \frac{1}{2}I\int_{\omega_0^2}^{\omega^2} d\omega^2 = \frac{1}{2}I(\omega^2 - \omega_0^2) = K - K_0.\tag{10–35}$$

The work done is indeed the change in kinetic energy. We have found the usual work–energy theorem for rotational motion about a fixed axis when the torque is directed along the axis.

The Energy of an Extended Object in Motion

We next consider the energy of an extended object that may or may not be rigid. First, we break up the object into a number of individual pieces with masses m_i located at $\vec{r}_i$ relative to some origin. We introduce the center of mass position $\vec{R}$ and the coordinates $\vec{\rho}_i$ relative to the center of mass according to $\vec{r}_i = \vec{\rho}_i + \vec{R}$, as in Fig. 10–11. When we take a derivative with respect to time, we have

$$\vec{v}_i = \frac{d\vec{r}_i}{dt} = \frac{d\vec{R}}{dt} + \frac{d\vec{\rho}_i}{dt} = \vec{V} + \vec{u}_i,\tag{10–36}$$

where $\vec{u}_i \equiv d\vec{\rho}_i/dt$ are the velocities of the masses m_i with respect to the center of mass and $\vec{V}$ is the velocity of the center of mass. We can then show that the kinetic energy of the system takes the form

$$K = \sum_i \frac{1}{2}m_i\vec{v}_i^2 = \frac{1}{2}MV^2 + \frac{1}{2}\sum_i m_i\vec{u}_i^2.\tag{10–37}$$

(We do not prove this result here.) The total kinetic energy splits into two parts: the kinetic energy of the total mass of the object moving with the velocity of the center of mass, and the kinetic energy of the motion relative to the center of mass. By now this should be a familiar type of split.

The kinetic energy of Eq. (10–37) takes on a familiar form for a *rigid* body. In that case, the vectors $\vec{\rho}_i$ have a fixed magnitude, and only the angular variable changes. For rotations with angular speed ω about an axis through the center of mass, the speed of any point that is a radial distance r from the axis is given by $v = \omega r$. Thus $u_i = \omega \rho_i$, and $u_i^2 = \omega^2 \rho_i^2$. We can substitute this expression into Eq. (10–37) to find that

$$K = \frac{1}{2}MV^2 + \frac{1}{2}I_{cm}\omega^2. \qquad (10\text{–}38)$$

Equation (10–38) shows that *the total kinetic energy of a rigid body consists of the kinetic energy of the total mass moving with the velocity of the center of mass together with the rotational kinetic energy of the object rotating about an axis passing through the center of mass*. We had already derived this result [see Eq. (9–43)] for the special case of rolling objects. Now we see that it is more general. This result is useful in calculations involving a combination of rotational and linear motion, as Example 10–12 shows.

EXAMPLE 10–12 A spool of thread of total mass M, rotational inertia I about its axis, and radius R falls and unwinds under the force of gravity. By using energy considerations, find the speed of the spool's center of mass after it has unwound a length h of thread.

Setting It Up Figure 10–26 is a sketch of the situation.

▲ **FIGURE 10–26** A spool of thread unwinds as it falls.

Strategy This example asks explicitly for the use of the conservation of energy. The falling object is subject to a conservative force, gravity, so that it has both a potential energy and a kinetic energy, the sum of which is conserved. As the spool falls it loses potential energy and gains kinetic energy. This situation is different from that of a falling object in that the kinetic energy consists of both rotational and linear terms. Because the thread unwinds, the motion is effectively that of a rolling object, and the center of mass speed v and the angular speed ω of the spool about the symmetry axis are related by $v = R\omega$.

Working It Out We begin with the total kinetic energy:

$$K = \frac{1}{2}Mv^2 + \frac{1}{2}I_{cm}\omega^2.$$

With the constraint $v = R\omega$,

$$K = \frac{1}{2}MR^2\omega^2 + \frac{1}{2}I_{cm}\omega^2 = \frac{1}{2}(MR^2 + I_{cm})\omega^2.$$

For the potential energy of gravity, we set the zero of the potential energy at the initial height of the spool. Thus if the spool falls a distance h, the potential energy is $-Mgh$ and the spool will have gained a kinetic energy Mgh. With the spool starting from rest, the kinetic energy after the spool has fallen a distance h is

$$K = Mgh.$$

Thus

$$\tfrac{1}{2}(MR^2 + I_{cm})\omega^2 = Mgh.$$

We can solve this expression for ω and hence for v:

$$v = R\omega = R\sqrt{\frac{2Mgh}{MR^2 + I_{cm}}}.$$

Alternative Approach A simple application of the parallel-axis theorem shows that the quantity in parentheses in the expression for the kinetic energy is the rotational inertia about the point at which the thread separates from the spool. So we can find the kinetic energy by treating the motion as pure rotation about that point.

What Do You Think? Imagine two spools with the same mass and radius but different mass distributions: Spool A is solid, while spool B is hollow. Which of these two spools would have a larger speed after dropping as in the example a distance h?

10–6 Collecting Parallels Between Rotational and Linear Motion

Through both Chapters 9 and 10 we have emphasized two things: First, there are thus far no new laws of physics to describe rotational motion. Second, there are a series of analogies between the expressions for linear and rotational motion. Of course, these two things are not separate. The analogies appear because aggregate systems are made up of individual pieces that separately obey the laws of linear motion discovered by Newton. At this point it is useful to gather all the analogies into Table 10–1. Study of this table is well worthwhile.

TABLE 10–1 • Analogies Between Linear and Rotational Motion

Linear Motion	Rotational Motion
Infinitesimal linear displacement: $d\vec{r}$	Infinitesimal angular displacement: $d\vec{\theta}$
Velocity: $\vec{v} = \dfrac{d\vec{r}}{dt}$	Angular velocity: $\vec{\omega} = \dfrac{d\theta}{dt}\hat{\omega}$
Acceleration: $\vec{a} = \dfrac{d\vec{v}}{dt}$	Angular acceleration: $\vec{\alpha} = \dfrac{d\vec{\omega}}{dt}$
Momentum: $\vec{p} = m\vec{v}$	Angular momentum: $\vec{L} = I\vec{\omega} = \vec{r} \times \vec{p}$
Force: $\vec{F} = \dfrac{d\vec{p}}{dt}$	Torque: $\vec{\tau} = \dfrac{d\vec{L}}{dt} = \vec{r} \times \vec{F}$
Impulse: $\Delta\vec{p} = \vec{F}\,\Delta t$	Angular impulse: $\Delta\vec{L} = \vec{\tau}\,\Delta t$
Kinetic energy: $\frac{1}{2}\,mv^2$	Kinetic energy: $\frac{1}{2}\,I\omega^2$
Work: $\int \vec{F} \cdot d\vec{r}$	Work: $\int \vec{\tau} \cdot d\vec{\theta}$
Power: $\vec{v} \cdot \vec{F}$	Power: $\vec{\omega} \cdot \vec{\tau}$

*10–7 Quantization of Angular Momentum

In systems with atomic dimensions or smaller, the effects of quantum physics become important. One quantum effect is that *angular momentum can have only certain discrete values*—we say that it is *quantized*. Consider an electron that is orbiting the nucleus of an atom, under the influence of the central force between the electron and the nucleus. This force is of the same form as the gravitational force, and if the atom were a classical system (i.e., described with the Newtonian physics we have studied to this point) the electron would behave like a planet orbiting the Sun—in particular, it would have an angular momentum $\vec{L}$ that points along an axis perpendicular to the plane in which it orbits. In Newtonian physics, angular momentum can have any magnitude. Niels Bohr proposed in 1913 that this is in fact incorrect and that the angular momentum can have only a component in the direction perpendicular to the plane of motion (here the z-direction) given by

$$L_z = n\hbar. \qquad (10\text{–}39)$$

In this quantum mechanical expression, n has *only* the integer values $0, \pm 1, \pm 2, \ldots$, and $\hbar$ is Planck's constant, h, divided by 2π, with value $\hbar \simeq 10^{-34}$ J·s.

An angular momentum of 10^{-34} J·s is not one that is within our common experience; a "typical" macroscopic angular momentum is closer to 1 J·s than 10^{-34} J·s. Perhaps you can get an idea of just how small this value of angular momentum is by imagining a uniformly rotating solid wheel of radius 1 cm and mass 20 g—a rotational inertia of $\frac{1}{2}MR^2 = 10^{-6}$ kg·m^2. An angular momentum of magnitude $L = 10^{-34}$ J·s would mean an angular speed of $\omega = L/I = 10^{-28}$ s^{-1}, which would correspond to a full revolution of the wheel in a time $T = 2\pi/\omega \cong 2 \times 10^{22}$ yr, or about a factor of 10^{12} larger than the lifetime of the universe! To take a second example, and one that is relevant to the real world, imagine a mass m in circular motion of radius of 10^{-10} m about a center and moving with a speed of 1% of the speed of light ($v \cong 3 \times 10^6$ m/s). If the angular momentum of the mass about the center had magnitude $L = 10^{-34}$ J·s, then the mass would be $m = L/(vR) \cong 3 \times 10^{-31}$ kg. This is nearly the electron mass, and the example corresponds to the atom! In the atomic and molecular worlds, angular momenta of order $\hbar$ really do appear, and Bohr's quantization rule has a very noticeable effect.

We have enough evidence to believe that angular momenta is really quantized, and Eq. (10–39) must apply to macroscopic as well as microscopic systems. So why don't we notice that angular momentum is restricted to certain values in everyday systems? The answer is that the incredibly small step between allowed values is too small to detect in such systems. The typical macroscopic value for angular momentum of 1 J · s is 10^{34} units of $\hbar$. The quantization is not observable in our typical macroscopic system, because its observation would require a determination of the radius or the angular velocity to an accuracy of one part in 10^{34}, an accuracy far beyond the capacities of our most refined instruments. In molecular or atomic systems, however, angular momenta take on values of one or several units of $\hbar$. The difference between, say, 5 and 6 units of $\hbar$ represents 10 to 20 percent of the whole. The fact that the angular momentum is quantized—it has only the discrete values dictated by quantum mechanics—is easily detectable in experiments that measure the energies of the molecule, such as those described next.

Quantization of Energy

In Eq. (9–43) we showed that the kinetic energy of a system with an angular momentum could be expressed in terms of the angular momentum. As we shall see in our more detailed work later, this can be extended to include the total energy. But if the angular momentum is quantized, then we are forced to conclude that *the allowed values of energy of a system such as an electron orbiting an atomic nucleus are quantized as well*.

When the system has one of its possible values of energy, corresponding to different values of n, it is said to be in an **allowed state**. The possible values of energy can then be labeled by the particular value of n—as E_n—and we then refer to a particular **energy level**. These energy levels are detectable through a central fact of our physical world: Atoms emit electromagnetic radiation (light) when an electron makes a transition from one energy level to another one of lower energy. The energy of the radiation is just the difference between the final and initial energy levels—energy is conserved in the transition process. But if the final and initial levels are quantized, then so is the energy of the radiation. We add one more ingredient to this recipe: The energy of electromagnetic radiation is associated with the frequency of that radiation, and our eyes perceive different frequencies as different colors.

Putting all this together, we conclude that as a result of the quantization of angular momentum, light is emitted by atoms in only certain, characteristic frequencies. The frequency of any such radiation is easily measurable by a variety of techniques. It was in fact the observation that atoms and molecules emit light with discrete frequencies that provided one of the keys to the development of quantum mechanics.

The consequences of the discreteness of atomic (and molecular) energies are momentous. For example, there is a minimum energy—an *energy gap*—required to move an atomic system such as a hydrogen atom up the "quantum ladder" from a lowest energy level to higher energy levels. It is the relatively large size of the energy gap that explains the stability of atoms. It is difficult to excite a hydrogen atom. The fact that even large molecular systems have energy gaps is crucial to the stability of biological systems. If there were no energy gaps, even the slightest perturbation would change molecular systems in important ways.

*10–8 Precession

If you have ever played with a spinning top, you will recall that its motion is quite complex. The top spins about its symmetry axis, but if the axis is not oriented to start in the vertical direction, the top of the symmetry axis sweeps out a circle about a vertical through the point where the tip touches the ground, a phenomenon called **precession** (Fig. 10–27). Still another type of motion is a complex up-and-down bobbing motion of the symmetry axis while it rotates about the vertical, a movement called *nutation* (Fig. 10–28). The movement depends on how the motion starts and also on the mass distribution of the top; some tops will actually flip over. While the

(a)

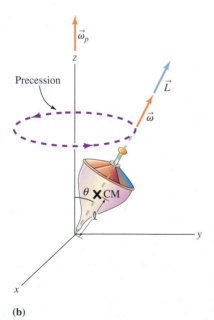

(b)

▲ **FIGURE 10–27** (a) Simple precession. (b) The precession of a top is governed by Eq. (10–41), where ℓ is the distance between the pivot point and the top's center of mass.

general motion of tops and their close cousins—gyroscopes—is beyond the scope of this book, we can show you how precession arises. It is helpful to precede this discussion with an example that reminds us of how a torque on an already spinning object can have some rather unintuitive consequences.

◀ **FIGURE 10–28** The interplay of gravity, angular momentum, and torque lead to perturbations on the precession of a top, resulting in the up-and-down bobbing motion called nutation.

EXAMPLE 10–13 In making a science-fiction movie about an asteroid that collides with Earth, a studio constructs a model Earth attached firmly to a support fixed at the South Pole and spinning rapidly about the (vertical) south-to-north axis with angular speed ω pointing out of the North Pole. A model asteroid approaches horizontally and strikes a grazing blow with the model Earth just at the North Pole. Describe the new motion of the model Earth.

Setting It Up We show the collision in Fig. 10–29a, with a coordinate system in which the north direction is to $+z$ and the asteroid approaches from the $+y$-direction. The model Earth spins with initial angular velocity $\vec{\omega}$ and angular momentum $\vec{L}_i$. We want to describe the change in both of these quantities as a result of the collision.

Strategy Because the model Earth is fixed at its pivot at the South Pole, we calculate the torque on the model Earth about this point. The collision delivers a linear impulse $\vec{J}$ to the model Earth. We can then use Eq. (10–22) to find the angular impulse $\vec{J}_\tau$ about the support point. This angular impulse describes the change in angular momentum, $\Delta\vec{L}$, of model Earth. The new angular momentum will be the old angular momentum plus this change.

Working It Out Given the direction of the asteroid's motion, the linear impulse delivered to model Earth is $\vec{J} = -J\hat{j}$. Then from Eq. (10–22) the angular impulse about the support point is $\vec{J}_\tau = \vec{r} \times \vec{J}$, where $\vec{r}$ is the vector from the fixed point to the point where the impulse is delivered. Because $\vec{r}$ is in the $+z$-direction, the vector product $\vec{r} \times \vec{J}$ is in the $+x$-direction (Fig. 10–29b). But this describes the change in angular momentum, $\Delta\vec{L}$, produced by the angular impulse. The initial angular momentum $\vec{L}_i$ is in the $+z$-direction, and when $\Delta\vec{L} = (+\Delta L)\hat{i}$ is added to this, the new angular momentum $\vec{L}_f$ corresponds to Earth having tilted toward the $+x$-axis (Fig. 10–29b). Note that this is *not* the direction from which the model asteroid comes.

What Do You Think? Will the magnitude of the model Earth's angular momentum change?

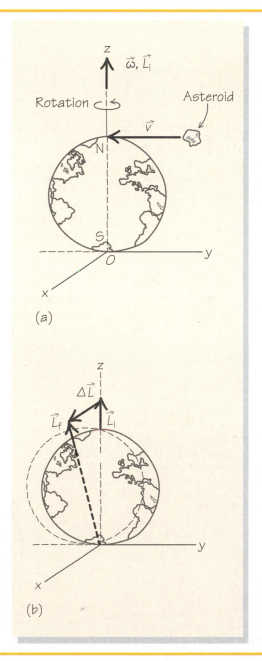

▶ **FIGURE 10–29** (a) An asteriod is imagined to hit Earth near the North Pole. (b) The collision of the model asteroid along the $-y$-direction causes the spinning model Earth to tilt toward the $+x$-direction.

Torque on a Spinning Top

In Example 10–13, there is a torque, and hence a change in angular momentum, which is perpendicular to an existing angular momentum. As a result the existing angular momentum changes its direction. This is reminiscent of something that we have seen in linear motion, where there are two rather different ways that linear momentum can change in response to forces. If the net force is parallel to the linear momentum, the momentum will change in magnitude but not direction, and this means a change in speed. But another possibility, which is realized when the acting net force is perpendicular to the momentum, is that the momentum can change *direction* rather than magnitude, as in circular motion. The same is true for rotational motion. If the applied torque is parallel to the angular momentum, then we have changes in the magnitude of the angular momentum, and this represents the majority of the cases we have studied. But if the torque is perpendicular to the existing angular momentum, then we can make that angular momentum change direction without changing its magnitude. With the net torque supplied by gravity, this is what makes a top precess.

To see how this works, consider a top (here in the form of a wheel) of rotational inertia I rotating with angular velocity ω about a horizontal shaft of length ℓ, which has one end pivoted at point A and the other end free. The system is shown at $t = 0$ in Fig. 10–30a. If the sense of rotation of the wheel is as shown, then the angular momentum about the pivot initially has magnitude $L = I\omega$ and points in the x-direction. The force of gravity, magnitude Mg and pointing in the $-z$-direction, acts on the center of mass of the wheel. The torque about point A due to this force has magnitude

$$\tau = Mg\ell \qquad (10\text{–}40)$$

and points in the $+y$-direction by the right-hand rule. As

$$\Delta \vec{L} = \vec{\tau}\, \Delta t,$$

the angular momentum after a short time interval Δt is given by

$$\vec{L} = I\omega \hat{i} + Mg\ell(\Delta t)\hat{j}. \qquad (10\text{–}41)$$

Thus the angular momentum vector has rotated slightly in a counterclockwise direction, as seen from above. As Fig. 10–30b shows, the new angular momentum makes an angle

$$\Delta \theta = \frac{Mg\ell\Delta t}{I\omega} \qquad (10\text{–}42)$$

with the original direction. This is the angle of the shaft, and so we have shown that the direction of the shaft changes with time in a steady way. When the shaft of the spinning wheel steadily rotates with the pivot point fixed, we say that the shaft *precesses* (Fig. 10–27a). The angular velocity of the precession of the angular velocity vector determined from Eq. (10–42) is

$$\omega_p = \frac{\Delta \theta}{\Delta t} = \frac{Mg\ell}{I\omega} \qquad (10\text{–}43)$$

▶ **FIGURE 10–30** (a) A wheel rotates with angular velocity $\vec{\omega}$ about a massless horizontal shaft that can pivot about point A. The force of gravity, $\vec{F}_g = m\vec{g}$, acts at the center of the wheel. (b) As seen from above, the shaft has rotated an angle $\Delta \theta$ after a time Δt. The angular momentum vector has therefore changed direction.

(a)

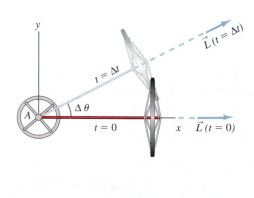

(b)

and is known as the *angular frequency of precession*. The precession frequency is, in fact, a vector that points in the $+z$-direction due to the right-hand rule. The precession frequency turns out to be independent of the angle the initial value of $\vec{L}$ makes with the vertical (see Problem 45). This formula applies for any top—not just a rotating wheel—if ℓ is the distance of the top's center of mass from the support point.

Our description of the precession of the wheel's shaft is only approximately correct. A more elaborate treatment involving energy considerations would show how to explain the up-and-down bobbing motion of the shaft called nutation (Fig. 10–28). The effect is small if the angular velocity of the wheel is large or, more precisely, if $\omega_p \ll \omega$. Indeed, the nutation becomes more important when the top slows down, and the wobble exaggerates as the top finishes its motion. Correctly applied, the laws of rotational motion described in this chapter are sufficient to account for all these features.

Summary

A particle moving with momentum $\vec{p}$ has an angular momentum about a point given by

$$\vec{L} = \vec{r} \times \vec{p}, \tag{10–4}$$

where $\vec{r}$ is the position vector from the point to the particle. This result involves the vector product of the vectors $\vec{r}$ and $\vec{p}$. If θ is the angle between these two vectors, the vector product has magnitude $rp \sin \theta$, and its direction—perpendicular to both $\vec{r}$ and $\vec{p}$—is determined by a right-hand rule. The angular momentum of a system of particles, rigid or otherwise, about some point is a sum over terms like Eq. (10–4).

Extended systems obey a dynamical equation that follows from Newton's second law. It states that the rate of change of angular momentum about some reference point O is given by

$$\vec{\tau} = \frac{d\vec{L}}{dt}, \tag{10–11}$$

where $\vec{\tau}$ is the net torque on the system. This net torque is a sum of terms of the type

$$\vec{\tau} = \vec{r} \times \vec{F}, \tag{10–5}$$

where $\vec{r}$ is the vector from the reference point to the point where one of the pieces, $\vec{F}$, of the net force is applied.

There is a simple way to relate the motion of a system whose rotational aspects are described about the center of mass O to a description about some other point A.

(1) The angular momentum consists of two parts. The angular momentum about A is the sum of the angular momentum about A of a point mass carrying the whole mass of the object, as if it were located at the center of mass and moved with the velocity of the center of mass, plus the angular momentum of the object about the center of mass,

$$\vec{L} = (\vec{R} \times \vec{P}) + \vec{L}_{cm}. \tag{10–13}$$

(2) Similarly, the total torque about A is the sum of the torque about A due to the total external force applied to the center of mass and the torque about the center of mass,

$$\vec{\tau}_A = (\vec{R} \times \vec{F}_{tot}) + \vec{\tau}_{cm}, \tag{10–14}$$

and (3) the rate of change of each term of the total angular momentum equation is equal to the corresponding term of the total torque.

In the absence of net torque, the angular momentum of a system is constant throughout the motion. This fact is an important tool for the study of the motion of nonrigid extended systems. The torque will naturally be zero in the absence of external forces, but it also is zero when only central forces are present. The conservation of angular momentum as applied to the motion of a single particle moving under the influence of a central force implies in particular that the areas swept out by position vectors from the source of the force to the moving object in equal times are equal.

The kinetic energy of rotational motion is given by $K = I\omega^2/2$, and the change in K equals the work done—just as in linear motion. For rotational motion in which the torque lies along the angular velocity, this relation (the work–energy theorem) reads

$$W = \int_{\theta_0}^{\theta} \tau \, d\theta = K - K_0, \tag{10–35}$$

where K_0 is the initial rotational energy at the angle θ_0. In many problems both rotational and linear motion are present. The total energy of a rigid body may be written as the sum of the linear kinetic energy, calculated as if all the mass of the object were concentrated at the center of mass, and the rotational energy of the object about the center of mass:

$$K = \frac{1}{2}MV^2 + \frac{1}{2}I_{cm}\omega^2. \qquad (10\text{–}38)$$

Quantum physics tells us that angular momentum is quantized in the form

$$L_z = n\hbar = n\frac{h}{2\pi}, \qquad (10\text{–}39)$$

where h is Planck's constant and $n = 0, \pm1, \pm2, \ldots$. The consequences of this quantization are important at the atomic scale and include the result that not all energy values are permitted for atoms and molecules.

Many subtle effects are associated with the effect of torque on angular momentum, and we illustrated one of them: the precession of angular momentum for a spinning top. If the top is spinning while inclined from the vertical, then gravity produces a torque on the top that causes its axis to precess about the vertical with a precession frequency

$$\omega_p = \frac{Mg\ell}{I\omega}. \qquad (10\text{–}43)$$

Understanding the Concepts

1. Angular momentum is sometimes given in units of joule-seconds $(J \cdot s)$. Is this a correct SI unit for angular momentum?
2. A comet is heading at high speed straight into the center of the Sun. Why is the angular momentum of the comet with respect to the Sun zero?
3. Why is a cyclist more stable on a rapidly moving bicycle than on one that is almost stationary?
4. When a quarterback throws a football, he tries to put quite a bit of spin on it. Since some of the energy put into the ball then goes into rotational motion, this means that there is less energy for translational motion. Why is it nevertheless done?
5. In taking a fast corner on a bicycle, it is safer if you crouch as low as possible. Why is that?
6. Films from Skylab show the astronauts reorienting themselves by spinning their arms as they float weightlessly. Is this consistent with the conservation of angular momentum?
7. Relief from the heat is often provided by large fans that are enclosed in rectangular structures. If you try to move one of these structures while the fan is on, you will encounter some difficulties. Explain what these might be.
8. A long, flexible, heavy bar can be very useful to a tightrope walker. Why?
9. Why is it easier on your back for you to lift heavy objects by bending your knees, keeping your back straight, and straightening your knees, rather than by picking up the object as you bend over from your waist with your legs straight?
10. An astronaut floating in a space station holds the axle of a rotating wheel. When the astronaut is vertical and the axle points away from her, she suddenly rotates the axle to point up. What happens to the astronaut?
11. A diver prepares for a complex set of midair maneuvers. In springing off an elastic diving board, the diver wishes to acquire angular momentum relative to her center of mass. What should she strive for in the takeoff?
12. In Example 10–3, we studied a tether ball and found that as the rope tying it to the pole winds, the angular momentum decreases. Where does the corresponding torque come from?
13. When a cue hits a billiard ball off center, the ball will move as well as spin. What is the sense of rotation, and how does this follow from considerations of the angular impulse?
14. It is possible to tell whether an egg is hard-boiled without cracking it open by setting it in rotation on a table. Discuss what you would find, and why. Compare the behavior of the hard-boiled egg to that of a raw one.
15. If you were to tie a rock to a rope and swing it in a horizontal circle above your head, you could rather easily do so without spinning around yourself. Is this a violation of the conservation of angular momentum?
16. The center of mass of an object accelerates as the result of an impulse (a brief force). If the object is extended, the impulse may also be an angular impulse. Is it possible to have an angular impulse without a linear impulse?
17. A comet falls straight toward the Sun, under the influence of gravitation, a central force, having started at rest with respect to the Sun. What is the angular momentum of the comet with respect to an axis through the Sun?
18. A diver executes a series of midair maneuvers. To do so, is it necessary for the diver to give herself some angular momentum about her center of mass?
19. We saw that the torque–angular momentum relation is a consequence of Newton's laws. Is it also true in Newtonian mechanics that the conservation of angular momentum follows from Newton's laws?
20. You are given a stool of known rotational inertia that can rotate with minimal friction. You are also given a stopwatch, a very light meter stick, and two known masses that can slide along the meter stick. How would you use this apparatus to measure the rotational inertia about the rotational axis of a person sitting on the stool?
21. The propeller of a single-engine airplane rotates clockwise, as seen from the cockpit. The plane makes a slow turn to the right. What else happens?
22. If only one set of handbrakes on your bicycle works when you descend a steep downhill slope, which set would you prefer?

23. In a conceptual example, we discussed a student, initially at rest, on a stool free to rotate about a vertical axis. The student holds a spinning wheel with angular momentum oriented up. What would have happened if instead of reversing the wheel completely, the student gave the wheel a turn such that its spin axis became horizontal?

24. A woman stands on the edge of a freely rotating platform. She walks toward the center along a radius. Will the speed of rotation of the platform change? If so, in what way, and what is the source of the torque?

25. A horizontal platform is rotating at a certain speed. A boy jumps onto the platform from an overhanging tree branch. He lands with both feet straddling the center and remains standing. Will the platform speed up, slow down, or neither?

26. To prevent rolling, a boat can be stabilized by attaching a large flywheel to the sides (Fig. 10–31). The attachment point is above the waterline. Both ω and the rotational inertia, I, about the axis are large. (a) A wave hits the boat on the side. The wave would tend to rock the boat or rotate it about its longitudinal axis without the flywheel. With the flywheel installed, how does the boat react to this wave? (b) A wave comes straight at the bow (the front). It would lift the bow of the boat without the flywheel. With the flywheel, what happens?

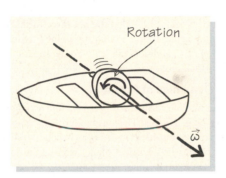

▲ **FIGURE 10–31** Question 26.

27. In an industrial machine a cylinder is spinning without friction about its (fixed) axis, angular speed ω_i (Fig. 10–32). A small movement of its axis is allowed so that it comes into contact with a second identical cylinder, also free to spin about its fixed axis but initially at rest. As a result of friction between the surfaces, both end up spinning with equal and opposite angular velocity, magnitude ω_f. The net final angular momentum about, say, the axis of the first cylinder is therefore zero. What happened to the angular momentum?

▲ **FIGURE 10–32** Question 27.

28. When jugglers perform juggling acts, the props they use are frequently put into a spinning motion when they are tossed into the air. Why is this a good idea?

29. A baseball is thrown horizontally, hits a rough floor, and bounces. The ball is spinning in the direction of its motion (like a rolling bicycle wheel). Which of the following is true? When the ball leaves the floor, it does so at (a) the same angle at which it struck the floor; (b) a larger angle with respect to the floor; (c) a smaller angle with respect to the floor.

30. Think about the demonstration in which someone spinning on a stool pulls in his or her arms and speeds up because of angular momentum conservation. In the process, does the energy of rotation decrease, remain constant, or increase?

Problems

10–1 Generalization of Angular Momentum

1. (I) An airplane of mass 2000 kg located 100 km north of New York City is flying 200 km/h in an easterly direction. (a) What is its angular momentum with respect to New York City? (b) What if it is flying in a northeasterly direction?

2. (I) What is the angular momentum about the origin of a particle of mass 270 g at position $\vec{r} = (0.1\hat{i} - 0.5\hat{j} + 0.2\hat{k})$ m, moving with a velocity of $\vec{v} = (12\hat{i} - 7\hat{j} - 3\hat{k})$ m/s?

3. (I) You are standing on the corner of Main Street and Elm, watching the cars on Main pass at a steady 10 m/s. You stand 5 m from the line of traffic. You watch a red convertible, whose mass is about 1000 kg, from the moment it is one block away, a distance of 200 m. Treat the car as pointlike, and assume that Main Street is straight. (a) What are the car's angular momentum, magnitude, and direction, with respect to you when it is one block away? (b) What is its angular momentum with respect to you when it passes your position on the corner?

4. (I) Assuming that each object is pointlike, how large is the angular momentum (a) of Earth about the Sun; (b) of the Moon about Earth? (c) Compare these results to the angular momentum of Earth (no longer pointlike) about its own axis; assume constant density.

5. (I) A bicycle travels east. The mass of the wheel, 1.8 kg, is uniformly distributed along the rim, with the mass of the hub and spokes negligible. If the radius of the wheel is 18 cm and the wheel rotates at the rate of 4.2 rev/s, what is the direction and the magnitude of the angular momentum of the wheel about its axis? (Neglect the width of the rim.)

6. (II) Calculate the angular momentum about the origin of a particle of mass m moving along the trajectory $y = ax + b$ with uniform speed v.

7. (II) A rock of mass 60 g is thrown with initial horizontal speed $v_x = 25$ m/s off a building from a height of 30 m. Calculate the angular momentum of the rock about the line along the edge of the roof as a function of time.

8. (II) A unicycle has a wheel of mass 1.5 kg, rotational inertia 0.28 kg·m² about the axle, and radius 0.38 m. What is its angular momentum with respect to a point on the road if the wheel rolls without slipping with an angular velocity of 2.5 rad/s?

9. (II) The position vector of an object of mass m subject to two constant forces that act at right angles is given by $\vec{r} = \left(\frac{1}{2}at^2\right)\hat{i} + (vt)\hat{j} + \left(\frac{1}{2}bt^2 - wt\right)\hat{k}$. Calculate the angular momentum of this object about the origin.

10. (II) An object of mass m moves in a path given by $\vec{r} = (x_0 + \rho\cos[\omega t])\hat{i} + (y_0 + \rho\sin[\omega t])\hat{j}$. What is the angular momentum of the object about the origin?

11. (II) Consider two objects whose position vectors are given by $\vec{r}_1$ and $\vec{r}_2$ and whose momenta are given by $\vec{p}_1 = m_1\vec{v}_1$ and $\vec{p}_2 = m_2\vec{v}_2$, respectively (Fig. 10–33). Show that in the special case that the center of mass of the two bodies is at rest at the origin (that is, $\vec{P} = \vec{p}_1 + \vec{p}_2 = 0$, and the position of the center of mass is $\vec{R} = 0$), the sum of the angular momenta of the two objects about the center of mass equals the angular momentum of a single object of mass $\mu = m_1m_2/(m_1 + m_2)$, rotating in circular motion about the origin at a distance $r = r_2 - r_1$. The quantity μ is called the *reduced mass*. [Hint: Introduce $\vec{r} \equiv \vec{r}_2 - \vec{r}_1$, and express $\vec{p}_1$ in terms of $d\vec{r}/dt$.]

▲ **FIGURE 10–33** Problem 11.

12. (II) Earth is not a point object but is a sphere with a rotational inertia of 9.8×10^{37} kg·m² about its axis. Assume that Earth's axis of rotation is parallel to the axis of the orbital motion of Earth around the Sun and that the Sun is so massive that it can be considered to be fixed. Calculate (a) the rotational inertia of Earth about the axis of its orbital motion around the Sun; (b) the total angular momentum of Earth about that same axis. (c) Calculate the fractional difference between your result for part (b) and the angular momentum you would find for part (b) if Earth were pointlike; explain why the difference is small.

13. (II) A square, 20 cm on the side, is made of very light sticks. Four identical masses of $m = 0.1$ kg form the corners of the square. The square rotates with an angular velocity of 8 rad/s about an axis perpendicular to its plane through the center of the square. (a) Calculate the rotational inertia of the system about the rotation axis and use it to find the angular momentum about this axis. (b) Use the general definition of angular momentum to calculate the angular momentum of each mass with respect to the center of the square, and add these up. Compare the results of (a) and (b).

14. (II) Consider the square studied in Problem 13. Calculate the angular momentum of each particle about a point on the axis of rotation 14 cm below the plane of the square. Compare the total angular momentum calculated in this way with the results of Problem 13.

15. (II) Three identical masses m are attached to the corners of an equilateral triangle of sides d. Calculate the angular momentum (a) if the triangle rotates at an angular velocity ω about the center of mass around an axis perpendicular to the plane of the triangle; (b) if the triangle rotates with angular velocity ω about one of its sides; (c) if the triangle rotates about an axis going through one of its vertices to the midpoint of the opposite side.

10–2 Generalization of Torque

16. (I) A construction worker of mass 72 kg stands at the end of a 3.4-m-long (massless) horizontal mast attached to a building. What is the magnitude of the torque exerted on the hinge that holds the mast fixed in position?

17. (I) What is the torque about the origin on a particle positioned at $\vec{r} = (3\hat{i} - \hat{j} - 5\hat{k})$ m, exerted by a force of $\vec{F} = (2\hat{i} + 4\hat{j} + 3\hat{k})$ N?

18. (I) A flagpole 2.2 m long is attached to a building. It makes an angle of 20° with the horizontal. A mass of 18 kg is suspended from the end. What is the torque acting on the point of attachment to the building due to the suspended mass?

19. (I) A 1.0-m-long massless stick lies on a table. A force of 200 N is applied to one end of the stick for 0.1 s at an angle of 45° to the stick, and in a direction pointing away from the center of the stick. What is the torque about the other end of the stick during the brief period that the force is acting?

20. (II) Consider two forces $\vec{F}$ and $-\vec{F}$ that act at different points on an extended object. Show that the net force of this combination is zero and that the torque about any point P is independent of the location of P, and depends only on the separation of the two points at which the forces act.

10–3 The Dynamics of Rotation

21. (I) A point mass M is attached to a turntable at a distance R from the center (Fig. 10–34). The turntable rotates with constant angular speed ω about its axis. If the axis is horizontal so that the turntable rotates in a vertical plane, what is the torque that the force of gravity on the mass exerts about the axis as a function of time? Assume that the mass is at the topmost position at $t = 0$.

▲ **FIGURE 10–34** Problem 21.

22. (II) The position of a ball of mass m thrown from a building is given by $\vec{r} = (v\cos\theta)t\hat{i} + [(v\sin\theta)t - gt^2/2]\hat{j}$, measured from the point from which the ball was thrown. What is the torque about the origin that the force of gravity exerts on the ball?

23. (II) A ball of mass m slides at speed v on a frictionless horizontal surface and bounces elastically from a wall. The initial path of the ball makes an angle θ with the wall (Fig. 10–35, see next page). Find the initial and final angular momenta of the ball about the point A. What causes the change in angular momentum?

▲ **FIGURE 10–35** Problem 23.

24. (II) A pulley system is used to lift a heavy mass. How much force must be applied to lift the object in Fig. 10–36 at a steady speed? Neglect friction at the axle.

▲ **FIGURE 10–36** Problem 24.

25. (II) What is the vector product of $\vec{A} = 2\hat{i} - 4\hat{j} + 5\hat{k}$ and $\vec{B} = \hat{i} + 3\hat{j} - 2\hat{k}$?

26. (II) Show that the magnitude of the vector product of two vectors is the area of the parallelogram for which the two vectors form adjacent sides.

10–4 Conservation of Angular Momentum

27. (I) A playground merry-go-round of diameter 3.4 m and rotational inertia 120 kg·m^2 is pushed with no one on it by three children to an angular speed of 2.5 rad/s. Two of the children, of mass 25 kg each, jump on the edge of the merry-go-round, coming radially in. What is the new angular speed?

28. (I) A uniform disk rotating without friction about its (vertical) central axis, with total mass 38 kg and radius 1.7 m, acts as a turntable. Its angular speed is $\omega = 0.075$ rad/s. A person of mass 71 kg jumps straight down onto the rotating turntable. The person lands 0.9 m from the axis. What is the new angular speed of the turntable?

29. (I) A firetruck, mass 6000 kg, passes a parked car on a straight street at $t = 0$ with speed 15 m/s. A physics graduate student finds that, 10 s later, the angular momentum of the firetruck with respect to the parked car is twice its value at $t = 0$. What is the speed of the truck at $t = 10$ s?

30. (I) A skater twirls at 0.7 rev/s with her arms extended and holds a 3-kg mass in each hand; each mass is 0.8 m from the axis of rotation. She pulls the masses in along the radial direction until they are 0.4 m from the axis of rotation. Assuming that the rotational inertia of the arms is negligible and that the rotational

inertia of the skater without the masses is 2.3 kg·m^2, what is the speed of rotation after the masses have been pulled in?

31. (II) A bug of mass $m = 2.0$ g walks around a horizontal turntable, which may be viewed as a uniform cylinder of mass $M = 0.24$ kg. If both the turntable and the bug are initially at rest, how much does the turntable rotate relative to the ground while the bug makes one full circle relative to the turntable?

32. (II) A small mass of 17 g slides down a frictionless slope starting from rest at 1.1 m above the ground level. When the slope reaches the bottom, it levels off, and the mass strikes the bottom of a vertical uniform bar of mass 0.2 kg and length 20 cm, pivoted at its midpoint, and sticks to it (Fig. 10–37). With what angular speed will the bar start its rotation?

▲ **FIGURE 10–37** Problem 32.

33. (II) A gob of clay, mass 100 g, falls from rest a distance 75 cm before striking and sticking to the edge of a wheel free to rotate about a horizontal axis through its center (Fig. 10–38). The wheel can be approximated as a solid disk of mass 10 kg and radius 50 cm. What is the angular speed of the wheel with the gob of clay attached?

▲ **FIGURE 10–38** Problem 33.

10–5 Work and Energy in Angular Motion

34. (I) A flywheel has radius 1.2 m and mass 680 kg, almost all of which is concentrated on the rim. It is spinning at $\omega = 4.5$ rad/s about its axis when a torque is applied along the axis, producing an angular acceleration of 0.3 rad/s^2. How much time does it take for ω to increase to 6 rad/s, and how much work is done by the torque during this time?

35. (I) An airplane engine develops 240 hp while turning the propeller at 3200 rev/min. What is the torque exerted on the propeller axis by the engine?

36. (II) Repeat the problem of the cylinder rolling down the inclined plane treated in Section 9–6, but this time use energy techniques to learn the speed of the cylinder as a function of distance traveled.

37. (II) The work done to slow down a spool used in a manufacturing process during a certain period is 1200 J. The spool is fixed on an axis and has rotational inertia of 0.033 kg · m² about that axis. The spool is rotating at an angular speed of 490 rad/s before the specified period. What is the spool's angular speed at the end of the period?

38. (II) A lawn roller consists of a cylinder—essentially solid—of mass 150 kg and radius 60 cm, pulled by a light handle oriented 30° to the horizontal and attached at the axis of the roller. Assume that the roller starts from rest and that a force of magnitude 55 N acts through the handle. Use the work–energy theorem to find the speed of the center of mass after the roller has moved 2 m.

39. (II) A cylinder of mass 0.2 kg rolls without slipping down an inclined plane of 15°. What is its rotational kinetic energy after it rolls 80 cm?

40. (II) A particular flywheel used for the storage of energy is a solid steel cylinder of density 8 g/cm³. The cylinder has a radius of 1.2 m, is 45 cm thick, and spins about its axis at a frequency of 260 rev/min. (a) What is the rotational inertia of the flywheel? (b) What is the work the flywheel can do in being brought to a halt if there is no energy loss to frictional forces?

*10–6 Quantization of Angular Momentum

41. (II) Consider an object of mass m that moves in a circular orbit caused by a central force given by $F = -kr$. Suppose that the Bohr quantization condition is applied to this motion. What are the allowed quantized radii, velocities, and kinetic energy values? [Hint: The acceleration for circular motion is v^2/r, and $E = (mv^2/2) + U(r)$.]

42. (II) Repeat the calculation of Problem 41 for an object that moves in a circular orbit caused by a central constant attractive force, obtained from the potential energy $U(r) = Cr$.

43. (II) The energy of the hydrogen atom, when quantized, is given by $E_n = -(13.6 \text{ eV})/n^2$, where $n = 1, 2, 3 \ldots$ Particles of energy 2.0 eV repeatedly pass through a gas of hydrogen atoms in the $n = 1$ state but never excite them out of this lowest-energy state. Explain. What energies would excite the hydrogen atoms?

44. (II) A proton has mass 1.67×10^{-27} kg and "radius" 1.3×10^{-15} m. (We put the radius in quotation marks because the proton is not a classical object with a radius, like a baseball.) It also can be thought of as having angular momentum with respect to an internal axis of 0.5×10^{-34} kg · m²/s. Take as a model that the proton is a uniform sphere. (a) What, according to the model, is the angular frequency of the proton's rotational motion? (b) What is the speed with which the outermost portion of the proton, on its equator, moves? Compare this to the speed of light, 3×10^8 m/s. (c) What is the energy associated with the rotational motion of the proton?

*10–7 Precession

45. (II) By repeating the derivation of the precession frequency of a top, but with the top making an angle ϕ with respect to the vertical, show that the precession frequency is *independent* of the angle ϕ.

46. (II) A student sits on a piano stool that is not rotating. She holds a vertical shaft on which a bicycle wheel of rotational inertia I is mounted and rotates with an angular speed ω oriented upward. She wants to tilt the wheel away from herself in a radial direction. What is the direction of the torque that she must exert? Suppose she succeeds in reversing the direction of the shaft by 180°. What will the speed and direction of rotation of the student and the piano stool be, assuming that the rotational inertia of the student and stool together is I^*?

47. (II) A wheel with massless spokes has mass 1 kg and radius 10 cm and is mounted on one end of a massless axle (Fig. 10–39). The axle rests on a pivot at a point 16 cm from the mounting point and 10 cm from the wheel. At the other end a mass of 0.8 kg is attached. The wheel spins at an angular frequency of 10 rad/s. What is the rate of precession?

▲ **FIGURE 10–39** Problem 47.

General Problems

48. (I) A 110-g ball is thrown out of a second-story window 6.0 m above the ground. The initial velocity of the ball is horizontal, with magnitude 4.5 m/s. What is the angular momentum of the ball, as a function of time, about the point on the ground directly below the window?

49. (II) A solid cylinder of mass 0.85 kg and radius 4.2 cm initially at rest rolls down a plane inclined at 28° with the horizontal and 1.5 m long. Use energy conservation to calculate the angular velocity of the cylinder at the bottom of the ramp; assume that all the kinetic energy of the cylinder is in rolling motion (that is, there is no sliding).

50. (II) A constant-density cylinder of mass 0.5 kg and radius 4 cm can rotate freely about an axis through its center. It has thread wound around an attached axle of radius 0.5 cm that also runs through its center (Fig. 10–40). The thread is attached to a mass of 1 kg, which slides down an inclined plane with an acceleration of 0.1 m/s². What is the coefficient of kinetic friction between the block and plane?

▲ **FIGURE 10–40** Problem 50.

51. (II) A child of mass 32 kg stands at the center of a platform of radius 2 m and rotational inertia 450 kg · m². The circular platform rotates about a frictionless shaft with angular speed of 0.8 rad/s. The child walks in a radial direction until he reaches the rim. What will the angular velocity of the platform be when that happens? What is the change in energy of the platform plus child? Identify the source of the work responsible for the change in rotational kinetic energy.

52. (II) A wooden board 4 m long and of mass 20 kg lies on a frictionless surface. A mass of 30 kg slides along the frictionless surface, strikes the board at right angles near one of its ends, and stays attached to it (as in a perfectly inelastic collision). Describe the subsequent motion of the board and mass together.

53. (II) A door hangs on frictionless hinges. A ball of mass 35 g moving with a velocity of 45 m/s strikes the door at the edge opposite to the hinged side and bounces back with a velocity of 35 m/s. Assuming that the door has a mass of 3.0 kg and is 85 cm wide, what will be the angular velocity acquired by the door?

54. (II) A bullet of mass 15.0 g and velocity 350 m/s passes through a wheel at rest (Fig. 10–41). The wheel is a solid disk of mass 3.0 kg and radius 18 cm. The bullet passes through the wheel at a perpendicular distance of 14 cm from the center, and the bullet's final velocity is 270 m/s. What are the wheel's angular velocity, angular momentum, and kinetic energy? Is energy of motion conserved?

▲ **FIGURE 10–41** Problem 54.

55. (II) Figure 10–42 shows the Atwood's machine treated in the problems of Chapter 5, with a rope of negligible mass 2 m long. Earlier the pulley was treated as massless. Now suppose that the pulley can be approximated by a solid disk of radius 0.1 m and mass 2 kg. The system is released from rest with the 4 kg mass 1.5 m from the floor and the 1 kg mass on the floor. (a) What is the speed of either block just before the 4 kg mass hits the floor? (b) How long does it take the 4 kg mass to reach the floor?

▲ **FIGURE 10–42** Problem 55.

56. (II) Two weights of mass 1.2 kg and 0.85 kg, respectively, are connected by a massless string that passes over a pulley. The pulley is a hollow cylinder of radius 18 cm and mass 0.45 kg. What are (a) the acceleration of the system and (b) the time that it takes for the larger mass to descend a distance of 1.6 m if the weights start from rest?

57. (II) A ball of mass M, radius R, and uniform density falls from rest from the top of a hemispherical bowl (Fig. 10–43). The left side of the bowl is frictionless, but the right side has a large coefficient of friction with the ball, and for all practical purposes the ball immediately rolls without slipping. How far up the right side of the bowl does the ball reach? Explain your answer in light of our assumptions about the very short distance in which sliding changes to rolling.

▲ **FIGURE 10–43** Problem 57.

58. (II) A solid door of mass $M = 15$ kg and width $\ell = 120$ cm is hit at a right angle by a mud ball of mass $m = 0.3$ kg, which, as Fig. 10–44 shows, hits the door at the edge with speed $v = 12$ m/s and sticks. (a) What is the rotational inertia of the door about the hinges? (b) What is the angular velocity of the door after having been struck? (c) What fraction of the initial energy does the moving door–mud ball system retain?

▲ **FIGURE 10–44** Problem 58.

59. (II) A 75-kg bank robber is escaping on a 450-kg motorcycle, runs out of gas, and thereafter coasts (friction free) at 80 km/h. As he passes under an overpass, a 75-kg policeman drops vertically onto the back of the seat and hangs on. Ignore the impulse due to the sudden change of the force of friction on the road on the translational and rotational motion. (a) Find the final velocity of the motorcycle. (b) What fraction of the initial kinetic energy (motorcycle plus robber) is lost? Ignore the rotational energy in the wheels. (c) Redo part (b) but include the effect of the rotational energy of the wheels, which each have a rotational inertia about their axes of 3 kg · m² and a radius $r = 0.5$ m.

60. (II) Electric power is used to speed up a centrifuge whose rotational inertia is $1.2 \text{ kg} \cdot \text{m}^2$; 1.6 kW of power were used to make the centrifuge accelerate at a steady rate from rest to 17,000 rev/min. If the electricity use was 100 percent efficient, how much time is required to speed up the centrifuge?

61. (II) A point mass $m = 0.2$ kg is attached to a string, which passes through a hole in a table and rotates in a circle of radius $r = 0.8$ m with an angular velocity of 40 rad/s. What mass M must be attached to the end of the string under the table to maintain this motion? Suppose that mass M is slowly increased by an amount that makes it descend a distance 0.1 m. What is the amount of the increase of M? What will the new angular velocity of the point mass be? [*Hint*: Use angular momentum conservation.]

62. (II) Show that $(\vec{r} \times \vec{p}) \cdot (\vec{r} \times \vec{p}) = r^2 p^2 - (\vec{r} \cdot \vec{p})^2$. [*Hint*: It is convenient, without any loss of generality, to assume that both $\vec{r}$ and $\vec{p}$ lie in the xy-plane.] Use this result to express the kinetic energy of a particle in terms of the momentum in the radial direction and of the square of the angular momentum.

63. (II) A cylindrical shaft of radius 5 cm is connected by a band to a solid cylindrical flywheel of mass 300 kg and of radius 0.35 m (Fig. 10–45). A motor brings the shaft up to a rotational rate of 1400 rev/min. Calculate the amount of work done by the motor, neglecting the rotational inertia of the shaft.

5 cm radius

300 kg
35 cm

1400 rev/min

▲ **FIGURE 10–45**   Problem 63.

64. (II) A thin rod of mass M, length ℓ, and constant density is standing on end on a rough table that forms the xy-plane. The rod begins to fall, with its top moving in the $+x$-direction, but as it falls, its point of contact does not move. As the rod hits the table, what are its (a) angular velocity, (b) angular momentum, and (c) kinetic energy?

65. (II) An object of mass M moves in a circular planar orbit about a center of gravitational attraction. The force of attraction has magnitude $F = K/r^2$, where r is the radius of the circle, and it is

directed toward the center. Calculate (a) the velocity, (b) the radius, (c) the period, T, and (d) the acceleration of the object, all in terms of the angular momentum L, M, and K.

66. (II) An object of mass m moves in a plane with its path described by the radius vector $\vec{r} = \hat{i} A \cos \omega_1 t + \hat{j} B \sin \omega_2 t$. Calculate the angular momentum about the origin. In what direction will the angular momentum point? Under what circumstances will the angular momentum be constant?

67. (II) A hurricane is a vast swirl of Earth's atmosphere. Using your knowledge of the size of such storms, the depth of the atmosphere, the speed of the winds, the density of air, and so forth, estimate the kinetic energy contained as well as the angular momentum. Compare your estimate of the angular momentum with that of Earth itself (see Section 9–5).

68. (III) A putty ball of mass $m = M/5$ is thrown with velocity $\vec{v} = v\hat{i}$ and hits the top of the thin rod in Problem 64 as the rod stands vertically. If the putty ball makes a completely inelastic collision, and if again the point of contact between the rod and the table does not move, what are the angular velocity, angular momentum, and kinetic energy of the system as it hits the table?

69. (III) A particular top can be approximated as a solid cylinder of mass 100 g and radius 2 cm. A string of negligible mass and length 1 m is wound around the top, which is started by pulling horizontally on the string with a constant force of magnitude 0.6 N. The top starts from rest at point O, and the string is pulled off. Ignore all friction between the top and the table on which it moves. (a) What is the final velocity of the center of mass of the top? (b) the final angular velocity of the top about its center of mass?

70. (III) A uniform solid cylinder of radius R and mass M rests against a vertical curb of height h, where $h < R$ (Fig. 10–46). The cylinder is mounted through its axis on a frictionless horizontal axle. You exert a horizontal force of magnitude F on the axle, pushing the cylinder against the curb. What is the minimum value of F that will cause the cylinder to roll up over the curb?

▲ **FIGURE 10–46**   Problem 70.

◀ The tightrope artist Jay Cochrane, seen here walking towards the tower of Casino Niagara with the American Falls in the background, uses the large bar to control the position of his center of mass and minimize any bad effects of the torque acting on the system formed by him and the bar. Cochrane completed the 70 m walk more than 40 stories high without a net in less than ten minutes.

Statics

In the previous two chapters, we have seen how extended objects or systems can have a rich array of motions, both linear and rotational. In this chapter, we are going to look at rigid objects—or at least systems that are approximately rigid—that are *motionless*. This is the area of physics known as **statics**. Although an object may be motionless, this does not mean that there are no forces or torques acting upon it. In fact it is likely that the object is motionless *because* forces and torques act upon it. Consider the tightrope walker in the opening photograph and the rock in Fig. 11–1. What forces and torques do you think are acting upon them? Statics has its most notable use in the field of structural engineering, in the design of bridges, buildings, and other structures. Statics also plays an important part in the analysis of the role of muscle, tendon, and bone in living systems. We can also use this opportunity to begin our study of the properties of solids, as rigid objects are formed from them. What we shall see is that solids themselves are not truly rigid, and the knowledge of how solids deform in response to forces acting on them is one step in understanding in what ways extended systems are truly rigid. In fact, well-engineered structures are never truly rigid, and part of the genius of good design lies in the degree to which one can avoid the overbuilding that true rigidity would require.

▶ 11–1 Static Conditions for Rigid Bodies

As long as a building does not move, we can say that the net force and the net torque on it is zero. But a building is not an indivisible thing. Its components exert torques and forces on one another, and even one end of a heavy beam can be said to exert forces on

▲ **FIGURE 11–1** Among the forces and torques acting on this rock are those due to contact forces and to gravity. This static marvel has remained balanced for a very long time, suggesting that the equilibrium is stable.

313

(a)

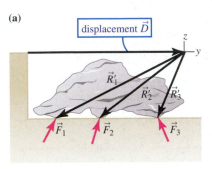

(b)

▲ **FIGURE 11–2** (a) A series of forces act at several places on this object, each at a different position measured with respect to an origin. To keep the figure clear, we have left off gravity. (b) The same forces act but the origin that measures their positions of action has shifted.

a neighboring part of the beam. The complete analysis of a structure, with all its internal forces, can be quite difficult. The problem becomes much simpler if we can think of the building as *rigid*. Under this approximation, it is only the external forces that determine whether the structure will stand or tip. This may be an approximation, but it is often a reasonable one in practice—in other words, many of the structures that we are interested in come close to being rigid.

We know that external forces acting on a rigid object have two effects: First, no matter where they are actually applied on the object, their vectorial sum produces a linear acceleration of the center of mass. Second, depending on where they are applied, they may produce torques that act to rotate the entire object. In the previous two chapters, we were interested in the resulting motion. In this chapter, we will look at the conditions for *no center-of-mass acceleration and no angular acceleration about any point*; although it is irrelevant from the point of view of the dynamical equations, we will also normally have *no center-of-mass velocity and no angular velocity about any point*. The object in question is then said to be in equilibrium. This is a term that we first met in Chapter 7, where we looked at the question of stable, unstable, and neutral equilibrium in terms of the potential energy. In statics, we normally are in situations where we want to avoid an unstable equilibrium, and certainly an engineer who is designing a bridge will carry through an analysis which will reveal the nature of the equilibrium conditions and ensure that the structure is in a stable equilibrium. In this chapter we won't look systematically into this question, although we will have occasion to remark on it from time to time.

Suppose that we have a set of external forces $\vec{F}_i$, each member of the set labeled by the index i, that act at various points described by the position vectors $\vec{R}_i$ on a rigid body of mass M (Fig 11–2a). Then the full dynamical equations for the object are

$$\vec{F}_{\text{net}} = \sum_i \vec{F}_i = M\vec{A} \qquad (11\text{–}1)$$

and

$$\vec{\tau}_{\text{net}} = \sum_i (\vec{R}_i \times \vec{F}_i) = \frac{d\vec{L}}{dt} = \vec{\alpha}. \qquad (11\text{–}2)$$

The points at which the forces act have position vectors $\vec{R}_i$ measured with respect to any convenient origin. Here, $\vec{A}$ is the acceleration of the center of mass due to the net force $\vec{F}_{\text{net}}$; $\vec{L}$ is the angular momentum of the object whose change is due to the net torque, $\vec{\tau}_{\text{net}}$, about the point chosen as the origin. We can recall, apropos of the torque equation, that a torque has a magnitude given by the product of the force and the *lever arm*, also known as the *moment arm*. Recall from Section 9–4 that to find the lever arm about a certain point O, extend the line of the force, and the closest, or perpendicular, distance between the extension line and point O is the lever arm. The direction of the torque is given by a right-hand rule. These review remarks are summarized in Figs. 9–22 and 9–26.

We can now write down the conditions for statics. When the linear acceleration $\vec{A} = 0$ in Eq. (11–1) the net force is zero:

$$\sum_i \vec{F}_i = 0. \qquad (11\text{–}3)$$

FORCE EQUILIBRIUM CONDITION

and when the angular acceleration $\vec{\alpha} = 0$ in Eq. (11–2), the net torque is zero:

$$\sum_i (\vec{R}_i \times \vec{F}_i) = 0. \qquad (11\text{–}4)$$

TORQUE EQUILIBRIUM CONDITION

The key to statics then is that the vector sums of forces and torques acting on an object are zero. We shall sometimes refer to these two equations as the equilibrium conditions.

CONCEPTUAL EXAMPLE 11–1 Rock climbing is a demanding activity, and it is important for a climber to be able to rest from time to time. Consider the climber in Fig. 11–3 resting in a "chimney." What are the forces that keep the climber in equilibrium?

Answer Gravity is certainly one force acting. The other forces acting on the climber at rest are either contact forces that act perpendicular to the rock surfaces or friction that acts along those surfaces. As the chimney walls are vertical, the normal forces will not oppose gravity. However, that does not mean they are not present and that they are not important to this problem. In particular, the maximum magnitude of static friction is proportional to the normal force, and friction can act vertically in this case and in a direction that cancels gravity. By pushing with his feet and back into the walls, the normal forces at the walls are increased, and that has the positive effect of increasing the friction forces on the climber. These forces will cancel gravity and allow the climber to be in equilibrium. The technique of pushing the walls within cracks such as this is instinctive and effective.

▲ **FIGURE 11–3** A rock climber is resting in a chimney. Note that the forces acting on him are in equilibrium and allow him to be at rest.

The Condition of No Torque Is Independent of the Choice of Reference Point

When we say that there is no angular acceleration [Eq. (11–4)], to which reference point are we referring? We can show that if an object has no angular acceleration about any one point, then, as long as the object is not in linear acceleration, it will not have an angular acceleration about any other point. Thus Eq. (11–4) applies to *any* choice of origin when Eq. (11–3) holds.

In order to prove this important result, we reconsider the system of Fig. 11–2a but with an origin displaced by $\vec{D}$ from the old origin (Fig. 11–2b). We'll find the net torque about this new origin due to the forces, labeled by an index i, shown in the figure; we'll suppose these are the only forces acting. We also assume that the vector sum of these forces is zero, so that there is no linear acceleration. If the force labeled i is applied at point $\vec{R}_i$ with respect to the original origin, then it is applied at the point $\vec{R}_i'$ in the new system:

$$\vec{R}_i' = \vec{R}_i - \vec{D}.$$

The condition for no rotational acceleration is

$$0 = \sum_i (\vec{R}_i \times \vec{F}_i) = \sum_i [(\vec{R}_i' + \vec{D}) \times \vec{F}_i] = \sum_i (\vec{R}_i' \times \vec{F}_i) + \sum_i (\vec{D} \times \vec{F}_i)$$

$$= \sum_i (\vec{R}_i' \times \vec{F}_i) + (\vec{D} \times \sum_i \vec{F}_i).$$

But there is no net force—the sum over the forces in the second term of the right-hand side is zero. Thus

$$\sum_i (\vec{R}_i \times \vec{F}_i) = \sum_i (\vec{R}_i' \times \vec{F}_i) = 0,$$

and this equation shows that if there is no torque about one origin, then there is no torque about any other origin.

Because a static object has no net torque about any point, we can place the point about which we calculate torques wherever the calculation will be easiest. A good choice leads to considerable simplification, as the calculation of the torques about some points may be trivial—the torque about the point where a force is applied, or anywhere along the line of that force, is zero because the lever arm is zero for that force.

CONCEPTUAL EXAMPLE 11–2 Consider a hard ball being pushed against a curb by a horizontal force $\vec{F}$ that is applied at precisely curb height (Fig. 11–4a). The ball itself has a radius equal to the curb height. Will this force cause the ball to climb over the curve?

(a)

Answer A force can make the ball climb the curb if it causes the ball to rotate about the point P, and this can happen only if there is a net torque about P. In the present case, the contact force of the curb point, $\vec{F}_C$, and the external force, $\vec{F}$, are both directed so that extensions of them pass through point P. These forces therefore produce no torque about P, as the extended free-body diagram in Fig. 11–4b shows. Thus the external force cannot make the ball climb the curb, and the ball is static.

▶ **FIGURE 11–4** (a) The ball has a radius equal to the height of the curb, and a horizontal force $\vec{F}$ is applied at exactly curb height. (b) Extended force diagram (we have ignored the vertical forces). The forces shown can have no torque about point P.

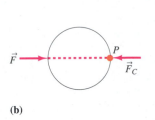

(b)

▶ 11–2 Gravity and Rigid Bodies

Gravity acts in all the static situations we consider in this chapter, and it is an important element in the analysis of large structures, so it will pay us to study it more carefully. The mass of a large structure is distributed, and gravity therefore acts all across the structure. In Section 9–4 we showed that the result of an analysis of this situation is quite simple: *Gravity acts as though it were applied to a concentrated (point) mass equal to the total mass M at the center of mass of the extended object*. Thus, gravity produces a torque as though the gravitational force acted on the center of mass, which is sometimes called the *center of gravity* in this context.

EXAMPLE 11–3 Consider a rectangular book with a uniform mass density and a length L. The book is lying on a table with one side parallel to the table edge, hanging off that edge by an amount ℓ (Fig. 11–5a). How large can ℓ be before the book rotates off the edge and falls?

Setting It Up Figure 11–5b is a free-body diagram for the problem and includes a coordinate system with its origin at the table edge.

Strategy In the extended free-body diagram of Fig. 11–5b, gravity, $m\vec{g}$, acts on the book's center of mass, which is at a point midway along its length. If the book falls off the table, it will do so by rotating about the table edge. The force equilibrium condition, Eq. (11–3), can be satisfied through the cancellation of gravity by a normal force, $\vec{F}_N = -m\vec{g}$. For the torque equilibrium condition, Eq. (11–4), we want to calculate the torque about an axis passing through the origin along the edge of the table. The book will not fall off the table if we can maintain the torque equilibrium condition. Note that just at the point of rotating off the table, the normal force acts *at* the table edge (Fig. 11–5c).

Working It Out As long as the center of mass of the book lies above the table surface, we can satisfy the torque equilibrium condition, Eq. (11–4), if $\vec{F}_N$ acts at the same point as the force of gravity, as shown in Fig. 11–5b. If, however, the book's center of mass is beyond the edge of the table, then the second equilibrium condition can no longer be satisfied. The normal force will be acting at the edge of the table; at this point, the lever arm for the torque of the normal force is zero. The force of gravity will have a torque for rotations about the edge. Thus the equilibrium condition can no longer be satisfied, and the book falls (Fig. 11–5c) by rotating off the edge of the table. We conclude that the equilibrium conditions *can* be satisfied if the center of mass of the book lies above the table surface, and the largest possible value of ℓ is $L/2$.

What Do You Think? If you observe a book that is resting more than halfway over the table, what can you conclude? *Answers to* **What Do You Think?** *questions are given in the back of the book.*

▶ **FIGURE 11–5** (a) A book rests on a table top with a length ℓ hanging off. (b) The center of mass of the book is over the table surface. (c) The center of mass of the book is beyond the table edge.

THINK ABOUT THIS . . .

IS CEMENT AN INDISPENSABLE ELEMENT OF CONSTRUCTION?

An interesting generalization of Example 11–3 involves the static conditions for a stack of several books of mass m on a table, shown in Fig. 11–6. If you stack enough books, it is possible to make a static pile with the topmost book lying as far from the table as you like (see Problem 13). Even with a stack of only four books, the topmost book can be made to lie completely off the table edge. While it is static, the extreme case is in unstable equilibrium, in the sense that a fly walking too far to the edge could cause the pile to fall over. These results can be applied to other constructions; for example, if building blocks of stone are used judiciously it is possible to build stable structures without the need for cement (usually used in the form of mortar between bricks and stones). A crucial ingredient for such structures is the keystone (Fig. 11–7), which, in effect, allows the joining of the topmost edges of two stacks of books. With the keystone, neither pile need be stable by itself; with it the structure is truly stable—the equilibrium is a stable one—in the sense that the structure will continue to stand even when the external forces change a little, as in an earthquake. Mortar often represents the weak point of a structure, and many ancient building structures have lasted precisely because good use was made of the possibility of stable equilibrium without mortar. Among these are numerous Egyptian, Greek, and Roman monuments.

▲ **FIGURE 11–6** In an extension of Example 11–3, it is possible to show that, with as few as four books, the topmost book in a stable pile can lie entirely over the edge of the table.

◀ **FIGURE 11–7** The arch is a stable architectural element that has been used since ancient times, here at the Roman ruins in Djemila, Algeria.

CONCEPTUAL EXAMPLE 11–4 You have a cutout map of the United States made of thick cardboard. Use the fact that gravity effectively acts on the center of mass of the map to devise an experimental method to determine its center of mass. [*Hint*: First show that because gravity effectively acts on the center of mass, a suspended object will be in stable equilibrium when the center of mass lies directly below the suspension point. Then use this fact to devise your method of finding the center of mass.]

Answer We know that a simple pendulum has its lowest potential energy when the pendulum bob lies directly below the support. This is therefore a stable equilibrium point. An analysis in terms of torque about the point of suspension shows that only gravity exerts a torque, and that torque has the effect of rotating the bob back to the equilibrium point. Knowing that gravity acts on a suspended object as if all the mass were concentrated at the center of mass, the same analysis shows that only gravity exerts a torque about the suspension point and that the suspended object has a stable equilibrium with its center of mass directly below the suspension point.

We can now find the center of mass of our map with the aid of a plumb bob. The map is suspended twice by hanging it from two holes punched in it in different locations (Fig. 11–8). A plumb bob is dropped from each suspension point and a vertical line is drawn on the object (Figs. 11–8a, b). The center of mass lies at the intersection of the two lines. Hanging the object from a third point, as in Fig. 11–8c, is a good check on the technique.

What Do You Think? You have a little floor stand with a vertical shaft coming to a point, and you want to locate the center of mass of the map by a different method, this time with the map in a horizontal orientation. How would you proceed?

(a)

(b)

(c)

▲ **FIGURE 11–8** Here's a technique to find the center of mass of this cutout map of the United States. (a) Suspend the map from any point and drop a plumb line from that point. The center of mass lies somewhere along the line. (b) The same is true for a second plumb line dropped from a second point, and the intersection of the two lines is the location of the center of mass. (c) The plumb line dropped from a third point does indeed pass through the center of mass.

11–3 Applications of Statics

How many beams of a given size are required to support a roof of a given weight? Can the beams be reduced in size and still safely support the roof? Where should the beams be located? The equations for statics are often used to determine the forces ("loads") on such components, which in turn are used to find the location and size of components needed for construction on various scales, or to determine the limits of safe construction. In Examples 11–5 and 11–6, we look at considerations like these for some everyday situations.

EXAMPLE 11–5 In order to handle a hot pizza of mass 1.0 kg, a cook slides a light (i.e., of negligible mass) but stiff spatula of length 2.00 m from the right hand to the center of the pizza, which he then carries horizontally with two hands (Fig. 11–9a). His right hand is at one end of the pole (point A); his left hand is at point B, a distance 0.50 m farther down the pole; and the pizza is at the other end. What forces must the worker's hands exert on the pole so that it maintains its horizontal position?

Setting It Up The origin of our coordinate system is placed at point A (at the right hand). The figure labels the distance ℓ between the hands, the total length L of the pole, and the pizza mass m, as well as the unknown forces due to the hands.

Strategy The external forces acting on the pole are gravity and contact forces from each hand, and our strategy will be to use the static conditions to find the contact forces. In the free-body diagram for our extended object (Fig. 11–9b), we assumed that the hands will exert upward forces. But, as is usual with free-body diagrams, it is not necessary that the forces drawn correspond precisely in direction and magnitude to the forces to be determined—these values are determined through the static conditions for these forces. Having identified the forces and drawn the free-body diagram, our goal is simply to solve the static condition equations for those forces. Point A is a simple point about which to evaluate the torque because one of the forces runs through it.

Working It Out A net force on the pole of zero, Eq. (11–3), reduces to one equation for the vertical components of the force. A net torque on the pole of zero, Eq. (11–4), is also one equation because the torques about point A are all into or out of the page. With two equations we can find the two unknown forces exerted by the hands. Dropping the vector indications, our static conditions are

$$\text{Force: } F_A + F_B - mg = 0$$
$$\text{Torque: } \ell F_B - Lmg = 0.$$

Note that F_B and mg must exert torques in opposite directions.
We solve these equations for the unknown forces:

$$F_B = mg\left(\frac{L}{\ell}\right);$$

$$F_A = mg - mg\frac{L}{\ell} = mg\left(\frac{\ell - L}{\ell}\right).$$

Because $\ell < L$, F_A is negative: Hand A must in fact exert a downward force. In effect, point B, which lies between the two ends of the pole, acts as a pivot point about which the pole may rotate. The force at A must point in the *same* direction as the force of gravity on the mass to allow the torque to be zero. Note the coefficients in the forces that must be exerted by the hands. For example, F_B is much larger than mg if L is much larger than ℓ. It is difficult to hold a mass at the end of a long horizontal pole!

Inserting numbers, we find

$$F_A = (1.0\text{ kg})(9.8\text{ m/s}^2)\frac{0.50\text{ m} - 2.00\text{ m}}{0.50\text{ m}} = -29\text{ N};$$

$$F_B = (1.0\text{ kg})(9.8\text{ m/s}^2)\frac{2.00\text{ m}}{0.50\text{ m}} = 39\text{ N}.$$

What Do You Think? Assume that the distance ℓ is fixed. If the pole were shorter, would the forces exerted by the pizza cook have to be (a) smaller, (b) larger, or (c) unchanged?

(a)

(b)

◀ **FIGURE 11–9**
(a) A cook holds a pizza with a 2-m long spatula. (b) Extended force diagram. The object on which the forces act is the massless rod with the pizza at the end. The problem solution shows that the actual direction of $\vec{F}_A$ points down.

Problem-Solving Techniques

Suggestions for Statics Problems

1. After clear identification of the object in question, prepare an extended free-body diagram for it. This will include all the forces and where they act. Follow the first three steps described in the Problem-Solving Techniques box in Chapter 9.

2. Choose an origin and axes. In particular, the torques due to the forces may be simpler to compute about some origins than about others; an origin through which more than one force passes is good. Further, the more forces are aligned with coordinate axes, the easier the analysis.

3. Be clear about the direction you have chosen for positive torque.

4. Write down the conditions for statics that follow from steps 1 through 3—Eqs. (11–3) and (11–4).

Problem-Solving Techniques *(continued)*

5. Count the resulting equations as well as the unknown forces to be determined to make sure that the number of equations corresponds to the number of un-

knowns and the problem does indeed have a unique solution.

6. In solving the equations for statics, carry through an algebraic rather than a

numerical solution as far as possible. In this way, you can make checks in various limits or for special cases and see whether the results are reasonable.

EXAMPLE 11–6   A crane whose cabin and engine are effectively fixed to Earth is used to lift a 5300-kg mass (Fig. 11–10a). The 10.0-m-long arm of the crane is supported at its base, at point *B*, by a strong but friction-free pivot, and at its top, at point *A*, by a supporting cable. The arm and the supporting cable make angles of 45° and 32°, respectively, with the horizontal. The mass is lifted by a line from a point on the arm 0.52 m from point *A*. Ignoring the mass of the arm (reasonable compared to the 5-ton mass being lifted) and assuming the suspended mass is not accelerating, compute the tension in the supporting cable.

Setting It Up   We place the origin of the coordinate system origin at the bottom of the crane arm, point *B* (Fig. 11–10b). (In the discussion of strategy we'll see why this is a good choice.) We know the mass *m* that is to be lifted, the arm length *L*, the length ℓ from the end of the arm to where the cable lifting the heavy mass is attached, the

angle θ_T between the horizontal and the supporting cable, and θ_{arm} that the crane arm makes with the horizontal. These quantities are all labeled in Fig. 11–10b. We want the tension magnitude, *T*, in the supporting cable.

Strategy   Figure 11–10b is an extended free-body diagram for the arm, which is the system in equilibrium. The forces acting on the arm are (1) the tension, $\vec{T}$, acting at *A* along the cable; (2) the weight of the suspended mass; and (3) an unknown contact force, $\vec{F}_N$, at the bottom pivot point. The contact force points in a direction to be determined, although it lies within the plane of the page. We must choose a reference point for the torque equilibrium equation. If this point is at the application point of any of the forces, then there will be no torque due to that force. Since the contact force has unknown direction, it will be easiest to choose the reference point to be point *B*, where the crane arm is attached (Fig. 11–10b). We set the origin there. We'll also want to count the equilibrium equations, Eqs. (11–3) and (11–4), to ensure that there are enough to allow us to solve for the magnitude of $\vec{T}$.

Working It Out   From the free-body diagram (Fig 11–10b), we see that the force equation has two vector components:

$$x\text{-component: } F_{Nx} - T \cos \theta_T = 0, \qquad (11\text{–}5)$$

$$y\text{-component: } F_{Ny} - T \sin \theta_T - mg = 0. \qquad (11\text{–}6)$$

The forces all produce torques perpendicular to the page, so the torque equation has a single vector component perpendicular to the page:

$$-(L - \ell)mg \sin(\theta_{arm} + 90°) + LT \sin(-\theta_{arm} + 180° + \theta_T) = 0. \quad (11\text{–}7)$$

There are three equations: two for the *x*- and *y*-components of the net force and one for the torque. There are three unknowns: the two components of $\vec{F}_N$ and the tension magnitude *T*. The number of equations match the number of unknowns, and the problem has a solution.

Without $\vec{F}_N$ in the torque equation, Eq. (11–7) alone is sufficient to solve for *T*:

$$T = \left(\frac{L - \ell}{L}\right)\frac{\sin(\theta_{arm} + 90°)}{\sin(-\theta_{arm} + 180° + \theta_T)}mg$$

$$= \left(\frac{10.0 \text{ m} - 0.52 \text{ m}}{10.0 \text{ m}}\right)\frac{\sin(45° + 90°)}{\sin(-45° + 180° + 32°)}(5300 \text{ kg})(9.8 \text{ m/s}^2)$$

$$= 1.5 \times 10^5 \text{ N}$$

A useful check on this result is that if $\ell = L$, the mass is hung directly from point *B* and, from physical considerations, we expect no tension to be required; this is indeed the case. It is also true that if $\theta_T = \theta_{arm}$—so that the cable is also attached at the pivot point at *B*—then there is no way that *T* can provide a torque of the opposite sign from the torque coming from the weight, and the arm cannot be stabilized. Indeed, we cannot solve for *T* when the angles are equal.

What Do You Think?   We refer to a crane effectively fixed to Earth at the start of this example. What might happen if it is not fixed to Earth?

(a)

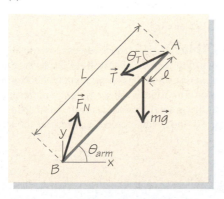

(b)

▲ **FIGURE 11–10**   (a) Example 11–6. (b) Extended force diagram. The object on which the forces act is the arm of the crane. The relevant angle between the gravitational force and the position vector from point *B* along the crane's arm is given by $\theta_{arm} + 90°$. The relevant angle between the tension and the position vector from point *B* along the crane's arm is given by $-\theta_{arm} + 180° + \theta_T$. The sines of these angles enter into the expression for the torque about point *B* due to the forces of gravity and tension, respectively.

An analysis like that of Example 11–6 could be useful if you were to design a crane and had to choose the size of the supporting cable. Notice that for this type of crane the tension in the cable must be much greater than the weight itself. For safety it would also be necessary to solve for the contact force to ensure that the pivot itself is sufficiently strong.

As an exercise, choose some other point—say, the other end of the arm—to be the origin and verify that the answer is independent of the choice of origin (see Problem 24).

EXAMPLE 11–7 A ladder of length 3.0 m is leaning against a wall at an angle of 58° as in Fig. 11–11a. Its eight rungs are spaced 0.33 m apart. The ladder's mass is insignificant compared to the 85-kg mass of a window washer who is climbing the ladder. The co-efficient of static friction between the rubber feet of the ladder and the floor is 0.51, while you can ignore friction between the top of the ladder and the wall. Is the ladder safe from slipping if the window washer climbs to the seventh rung?

Setting It Up The extended free-body diagram, Fig. 11–11b, sets a coordinate system with origin at the base of the ladder. We know the ladder length L, the angle θ the ladder makes with the ground, the coefficient of static friction μ_s between the ladder and the floor, the mass m of the window washer, and the spacing d be-tween rungs. Some of these are labeled in Fig. 11–11b. We want to know if static friction will be large enough to maintain static equilib-rium when the mass m is centered at a given spot.

Strategy The extended object on which the forces act is the lad-der along with the washer. We start by drawing the extended free-body diagram. The forces acting on this system are the normal forces $\vec{F}_{\text{floor}}$ and $\vec{F}_{\text{wall}}$, the friction force $\vec{f}$ between floor and ladder, and, given that the ladder is much lighter than the window washer, gravi-ty acting on the washer. A crucial piece of information in solving this problem is that the static friction force takes on its *maximum value* just before slipping. Finally, to apply the torque equation, we need a reference point, and the chosen origin at the ladder base is a useful one because two forces (friction and the normal force of the floor) act here, and neither will exert torque about that point. (The answer will not in the end depend on the choice of reference point for torque, but we want to make the calculation as simple as possible.) The forces act in the plane of the page, and the torque is perpendicu-lar to the page, so we'll have three (equilibrium) equations for the two normal forces and friction. We compare the calculated force of static friction with its maximum value,

$$f_{\text{max}} = \mu_s F_{\text{floor}}. \tag{11–8}$$

If the calculated magnitude of friction exceeds this maximum value, then the ladder will slip.

Working It Out Placing the washer at the nth rung of the ladder (from the bottom), and noting that friction acts to the left, the statics equilibrium condition for the forces are, in the x- and y-directions, respectively,

$$F_{\text{wall}} - f = 0, \tag{11–9}$$

$$F_{\text{floor}} - mg = 0. \tag{11–10}$$

For the torque, Fig. 11–11b shows that the lever arm for the force of gravity about the origin is $nd \cos \theta$, whereas the lever arm for the normal force of the wall is $L \sin \theta$. These tend to rotate the system in opposite directions. The torque equilibrium condition is thus

$$mgnd \cos \theta - F_{\text{wall}} L \sin \theta = 0. \tag{11–11}$$

Equations (11–9) and (11–11) can be solved for the contact force at the wall and the friction, and we find for friction

$$f = \frac{mgnd \cot \theta}{L}. \tag{11–12}$$

Equation (11–10) gives the contact force at the floor directly. The condition that the friction force at the floor not exceed the maximum value of static friction is

$$\frac{mgnd \cot \theta}{L} \leq \mu_s F_{\text{floor}} = \mu_s mg,$$

$$nd \leq \mu_s L \tan \theta.$$

The higher the washer climbs, the closer the ladder comes to slip-ping. The right-hand side of the inequality is $\mu_s L \tan \theta = (0.51)(3.0 \text{ m})(\tan 58°) = 2.4 \text{ m}$, so with $d = 0.33 \text{ m}$, the condition can be written as

$$n \leq (2.4 \text{ m})/(0.33 \text{ m}) = 7.3.$$

At the 8th rung the ladder slips.

What Do You Think? The ladder comes closer to slipping as the window washer climbs higher because (a) the lever arm of the washer's weight with respect to the contact point with the ground de-creases; (b) the lever arm of the washer's weight with respect to the contact point with the ground increases; (c) the normal force F_{floor} becomes too large; (d) the normal force F_{wall} becomes too large.

▶ **FIGURE 11–11** (a) The window washer is standing vertically, such that his center of mass is over the rung on which he stands. (b) Extended force diagram for the ladder. The lever arm for the force of gravity about the origin is $nd \cos \theta$, and the lever arm for the normal force of the wall is $L \sin \theta$.

(a)

(b)

EXAMPLE 11–8 The biceps muscle is responsible for bending your arm. It acts through a kind of lever system (Fig. 11–12a), with an upward force where the biceps attaches to the bone and a downward force at the attachment point of the triceps. The elbow–hand distance a, and x, the distance from the biceps attachment point to the elbow, are important parameters. If a book of mass M is held in your hand with your forearm horizontal and your upper arm vertical, what upward force does the biceps have to exert on the forearm bones (radius and ulna) to remain in equilibrium? Ignore the mass of the forearm bones, and assume the hand-forearm forms a single system.

Setting It Up The extended force diagram of Fig. 11–12b, discussed further below, includes specification of the parameters a and x. We label the magnitude of the desired biceps force F_B.

Strategy The three vertical forces on the forearm are the force $\vec{F}_H$ from the upper arm acting at the elbow, the force $\vec{F}_B$ of the biceps on the forearm, and the weight of the mass, each included in the extended free-body diagram in Fig. 11–12b. For the torque, we choose the reference point to be the elbow, although the hand would do just as well. For this situation there is one condition for equilibrium from the force equi-

librium equation (all forces are oriented vertically), and one from the torque equilibrium equation. These two conditions should be enough to solve for the two unknown force magnitudes, including F_B.

Working It Out We have for the torque equilibrium condition

$$0 = (Mg)a - F_B x.$$

We see that this condition alone determines F_B, and we have no need to write the force equilibrium condition. (That is because the force F_H exerts no torque when the reference point is the elbow.) We immediately solve:

$$F_B = Mg\frac{a}{x}.$$

The force exerted by the contracting muscle must be a large factor greater than the mass's weight. Some typical values for an arm are $a = 30\text{ cm}$ and $x = 4\text{ cm}$, for which the factor is $(30\text{ cm})/(4\text{ cm}) = 8$.

What Do You Think? Consider two people of the same mass, height, and muscle strength. One of them can do pull-ups much more easily than the other. What is a possible explanation?

(a)

(b)

▲ **FIGURE 11–12** (a) The bicep muscle bends the arm at the elbow. (b) Extended force diagram for forearm/book.

Statics is an important element in understanding the mechanical aspects of living things such as ourselves, and the structure treated in Example 11–8 is representative of the study of this sort of question. Living animals are remarkably adaptable, many creatures being capable of a wide variety of activities. When we apply equilibrium conditions to flexible objects such as animals, we approximate them as a collection of rigid parts (bones) that have variable orientations (Fig. 11–13).

(a)

(b)

◀ **FIGURE 11–13** Biological systems can change the relative orientations of their different parts to allow them to be stable. (a) At least four different external forces act on the professor. (b) These forces add to a net force of zero.

▲ **FIGURE 11–14** Construction cranes are a familiar feature of the urban landscape.

A common crane used at construction sites (Fig. 11–14) consists of a flimsy-looking tower constructed from three or four planes arranged so that the tower has a triangular or square cross section. Each plane consists of a lattice of triangular forms (this is called a ge-odetic construction). At the top of this tower is a horizontal boom that can lift the large masses involved in the construction. The tower can rotate and the boom can move the mass radially inward or outward, all in order to position the object at the correct location; the object is finally set down. (While these towers are tall, they do not have to be exces-sively so; they are gradually moved up as a skyscraper under construction gets taller.) The strength of such a tower is insufficient to withstand the torque on it due to a large mass at the end of the boom, so there is a balance

weight at the far side of the crane that moves in an opposite direction from the lifting equipment and the object being moved. This exerts a torque on the tower equal and oppo-site to the torque from the weight of the large mass. The motion of the counterweight is controlled by sensors at the top of the tower, which ensure that the boom produces no net torque at the top of the tower, and this in turn ensures that the tower does not tip over.

Another technique is applied to the portable cranes that are carried by trucks. These cranes have sensors that measure the downward force of the crane and truck body on each wheel. If this force reaches a lower limit, indicating that a wheel is lifting off the ground, movement of the crane boom is halted, with operator con-trols being overridden. Thus the danger that the crane might tip over is avoided. ■

Underdetermined Systems

We sometimes face problems in static systems where the forces cannot be uniquely de-termined by the conditions of force and torque equilibrium. This occurs when there are fewer equations for the forces to be determined than there are forces, and such systems are said to be *underdetermined*. When a system is underdetermined, we often have a good deal of freedom to change parameters in significant ways.

As an illustration, consider a uniform square table of mass M with four light legs on a horizontal surface (Fig. 11–15a). The top has sides of length L and, if we take the axes shown with the origin at leg 1, legs 1, 2, 3, and 4 make contact with the ground at the re-spective points $(x, y) = (0, 0)$, $(L, 0)$, $(0, L)$, and (L, L). The center of mass of the table is at its geometric center, the point $(x, y) = (L/2, L/2)$.

The forces acting on the table are the four contact forces $\vec{F}_{N1}, \ldots \vec{F}_{N4}$ acting up-ward on the legs, and the force of gravity, $M\vec{g}$, acting on the center of mass. Figure 11–15b is the free-body diagram. The forces all act in the vertical, or z-, direc-tion. We calculate the torque about the origin. The torque vectors due to the forces are perpendicular to these forces and therefore lie in the xy-plane. Our static conditions thus consist of one force equation (for the z-component) and two torque equations (for the x- and y-components). These equations are, respectively,

(a)

(b)

▲ **FIGURE 11–15** (a) A uniform four-legged table. (b) Extended force diagram for the table.

$$F_{N_1} + F_{N_2} + F_{N_3} + F_{N_4} - Mg = 0,$$

$$LF_{N_3} - \frac{L}{\sqrt{2}}Mg \sin 45° + \sqrt{2}LF_{N_4} \sin 45° = 0,$$

$$-LF_{N_2} + \frac{L}{\sqrt{2}}Mg \sin 45° - \sqrt{2}LF_{N_4} \sin 45° = 0. \qquad (11\text{–}13)$$

With $\sin 45° = 1/\sqrt{2}$, the last two (torque) equations simplify to

$$F_{N_3} - \frac{1}{2}Mg + F_{N_4} = 0, \qquad (11\text{–}14)$$

$$F_{N_2} - \frac{1}{2}Mg + F_{N_4} = 0. \qquad (11\text{–}15)$$

We have only three equations to determine the four unknown forces $\vec{F}_{N_i}$. We could have some additional information to help us. For example, if the floor were rigorously flat and if every leg had exactly the same length, so that the table really was symmetric, then we could say that the contact forces must be the same. But lacking this information, we cannot solve uniquely for the individual contact forces. The flip side of the coin is that there is a good deal of freedom in our ability to change parameters and still have a stable table. One leg—any leg—could be a millimeter shorter than the others and the table would still be stable against falling over even if it wobbled at that point. Indeed, we could even remove any one leg completely and still have a table in equilibrium. Then Eqs. (11–13) through (11–15) give three equations for the three remaining unknown contact forces. But a four-legged table is in stable equilibrium—it remains in equilibrium if there are small changes in parameters like the length of the legs—while a three-legged table is in unstable equilibrium for some motions, meaning that a small change could make a big difference in the situation. If a teacup were placed just off the center of the table toward the corner with the missing leg, the table would tip over in that direction.

In engineering language, we might say that an underdetermined system is a "robust" system. But we could also say that it is an overbuilt system in that it could be made lighter and simpler in various ways. Which is more desirable is a matter of budget and tolerance for instability. You certainly wouldn't put a square table with three legs in your dining room!

11–4 Solids and How They Respond to Forces

To this point we have assumed that all the objects we have dealt with are rigid. Of course that is only an approximation, and not just because there is some give between different elements that make up a structure. Even the solid materials that make up our "rigid" objects deform in various ways when a force is applied. While solids may appear at first glance to resist any kind of deformation, close examination reveals that they exhibit responses to external influences. When compressed, their volumes change slightly; when put under tension or heated, they stretch.

The explanation for this behavior lies in the underlying atomic structure of the solid, an extremely important ingredient in our understanding of the physical world. Solids are aggregates of atoms or molecules for which the interatomic forces lead to an organized, three-dimensional grouping of atoms, called a *lattice structure*. Figure 11–16, a schematic two-dimensional representation of such a structure, depicts the interatomic forces as springs. Because the lattice structure is one of stable equilibrium, the use of spring forces is at least approximately correct. A **crystal** is an object whose atoms have been arranged in a lattice structure. Different crystals are determined by the nature of the interatomic forces between atoms or molecules and by the shapes and orientations of these constituents, that is, how they fit together; atoms can be arranged to form crystal structures in many different ways. Figure 11–17 shows the three-dimensional structure of one particularly simple type of crystal structure, the *simple cubic lattice*.

Even one kind of simple atom can form different crystals—both graphite and diamonds, two solids with very different properties, are formed of carbon atoms alone. If geometrically complicated molecules rather than simple atoms make up a crystal, or if several different kinds of atoms are present, the number of possible crystal structures and, perhaps more important, how the crystal structure influences the bulk appearance and behavior, increases rapidly. The way different atoms assemble into crystals depends also on the conditions under which they are formed. The formation processes for diamond and graphite are quite different. In a certain sense, the formation of a crystal by atoms is like a three-dimensional jigsaw puzzle with identical pieces, *which has more than one solution*!

As important as crystalline structure is to the macroscopic behavior of a solid, so are its lattice imperfections, called **defects** or dislocations. Defects are closely associated with the way a crystal grows. Figure 11–18 shows a simple two-dimensional crystal that is easy to grow and observe: a raft of bubbles that float on the surface of a liquid. Figure 11–18a is a perfect crystalline form, whereas a defect has been introduced in Fig. 11–18b. Can you recognize it? A more general type of defect is associated with grains and grain boundaries. A **grain** is a region in a solid where the crystal structure is perfect. Chance plays a role in the organization of atoms when crystal growth starts, and as an object cools and changes over from a liquid to a solid, a lattice may start to form in one region with one

▲ **FIGURE 11–16** Schematic diagram of atoms connected by bonds in a lattice structure.

▲ **FIGURE 11–17** Sketch of a simple cubic lattice.

▶ **FIGURE 11–18** (a) A perfect crystalline raft of bubbles. Bubbles will arrange themselves in crystalline patterns (in this case, two-dimensional), just like the atoms in real solids. (b) A defect in a raft of bubbles.

(a)　　　　　　　　　　　　　**(b)**

orientation and in another in another region. Once a lattice starts growing, it serves as a template for the lining up of other atoms, but if there are two growth centers with different orientations, these will not combine to a single crystal. Rather, there will be two grains. The boundaries where these grains meet involve special kinds of defects. Figure 11–19a shows such a boundary. A real solid is made up of many grains (Fig. 11–19b). Another class of defects occurs when single lattice sites are empty or are occupied by an impurity (an atom chemically different from those that make up the bulk of the crystal), or when impurities occupy spaces between lattice sites. Defects are related to a crystal's color, luminescence, transport properties, and mechanical properties. The study of the nature of defects is a major field of research and has taught us much about the behavior of solids.

▶ **FIGURE 11–19** (a) A close look at the grain boundaries in a crystal, as illustrated by a raft of bubbles. (b) An overview of numerous grains in a crystal.

(a)　　　　　　　　　　　　　**(b)**

Our simple picture of lattice structure illustrates why a solid is not truly rigid. Figure 11–20 suggests how a cubic structure might react to compression (Fig. 11–20a), to stretching (Fig. 11–20b), and to a *shear* (Fig. 11–20c). Just how hard it is to

(a)　　　　　　　**(b)**　　　　　　　**(c)**

▲ **FIGURE 11–20** Schematic diagram of lattice distortions: (a) compressed in the horizontal direction; (b) stretched in the horizontal direction; (c) subjected to shear forces, which act in opposite directions at the top and bottom.

compress, stretch, or shear a solid depends on its interatomic forces, sometimes referred to in shorthand as the interatomic *bonds*. In the final analysis, it is these forces, along with the grain and defect structure, that differentiate steel from a pie crust. (For more on this, see the "Think About This" feature on the strength of solids on p. 327.)

In principle, all properties of solids can be described by describing the atomic structure of solids, but in practice it is simpler to describe the properties of solids using observed macroscopic behavior. This is the point of view we take in the remainder of this section, where we develop empirical relations about the deformations of solids. Remember that even if these rules are sometimes called "laws," they are more properly rules of thumb, valid only under restricted conditions.

Stresses and Strains

Stress is a measure of the force that acts on a solid object, whereas **strain** is a measure of deformation, the object's response to that stress. Our aim here is to connect these quantities. To do so, we shall assume that the applied force is small in some sense, which implies that the response is correspondingly small. (Another type of response, more properly described as catastrophic, is associated with large forces. When a heavy hammer strikes a piece of ice, the ice shatters. Or a girder may buckle if it is loaded too much. This aspect the behavior of materials is as important as the small response to small forces, and we'll discuss it later.)

We define the stress on a solid object, S, as the external force per unit area that is exerted on it:

$$S \equiv \frac{F}{A},\qquad (11\text{--}16)$$

STRESS DEFINED

where F is the magnitude of the force that acts perpendicularly to an area A of the object. (Actually we are oversimplifying here. We really should include the possibility that the force is not perpendicular to the surface in question—see our discussion of shear below. However, what we have done is adequate for the kind of discussion carried out in this chapter.) As an example of this, we might have a force that acts on a solid rod along its length and tends to stretch it (Fig. 11–21)—this particular force is a *tension*. When the rod is subjected to a force along its length and tends to shorten it, you have a compression. These are the same sorts of forces but with opposite sign, and the usual convention is that tension corresponds to a positive stress, and compression to a negative stress. Stresses of this type are called *longitudinal stresses*. You will quite commonly have this sort of situation to deal with in the beams that make up a structure. Stress has the same dimensions as pressure and has SI units of newtons per square meter. As this unit is rather small for ordinary engineering applications, a more practical unit for stress is the meganewton per square meter (MN/m^2).

The *response* to the stress associated with either a tension or a compression is called the *compressional strain, e*. It is defined as *the fractional change in the length of the solid object along the direction of the compressional force*:

$$e \equiv \frac{\Delta L}{L}.\qquad (11\text{--}17)$$

COMPRESSIONAL STRAIN DEFINED

Note that e is dimensionless. It is negative if the stress is due to compression, and positive if it is due to tension.

The result of a tension on the rod in Fig. 11–21 is more than just an increase in length; when it stretches, the bar also *shrinks* in its lateral dimensions—the width h and depth w (or the radius if the rod has a circular cross section). If the force is a compression, the bar will instead bulge in the lateral directions. The fractional shrinking or bulging is proportional to the compressional strain by an amount characteristic of the

▲ **FIGURE 11–21** Equal but opposite forces applied evenly to opposite ends of a piece of material cause the material to stretch.

material, and it is described by the positive constant called *Poisson's ratio*, σ, after Siméon Poisson, who worked in the first part of the 19th century:

$$\frac{\Delta h}{h} = \frac{\Delta w}{w} = -\sigma e. \qquad (11\text{--}18)$$

A typical value of σ for solid material is 0.3, and this constant must in fact always be less than 0.5. You can show (see Problem 60) that if it were to exceed 0.5, the volume of the solid would increase when a uniform pressure acts *in* on it, and this would be quite at odds with our experience.

If an external force F of constant magnitude is applied everywhere perpendicular to the entire surface of a solid object, as for example for rocks below Earth's surface, the volume V of the object changes. The volume decreases if the force is inward and increases if the force is outward. The *volume stress, p,* is defined as stress was defined in Eq. (11–16): $p \equiv F/A$, where A is now the *total* surface area of the solid. For solids the volume stress can equally well refer to an outward force on the body, as when an object is hollow and contains a gas of high pressure. The accompanying response is a *volume strain, e_V,* defined by the fractional change in volume due to a volume stress: $e_V \equiv \Delta V/V$. If Poisson's ratio were small, the volume strain would be just a special case of the compressional strain; a bar that stretches while its cross-sectional area A remains constant has a volume $V = AL$, and when such a bar is stressed longitudinally, the change in volume is $\Delta V = A \, \Delta L$. When these quantities are substituted in the definition for e_V, we obtain Eq. (11–17) for e.

There is a third kind of stress and its corresponding strain that we want to mention: shear stress and shear strain. Suppose that a uniform force acts *along* a face of a solid object, as in Fig. 11–22a. A tangential force like this is called a *shear force*. If the bottom surface of the solid is fixed (by some kind of contact force—glue, say), the result is a distortion of the object (Fig. 11–22b). The quantity F/A is the *shear stress* on this solid, and the *shear strain* is the fractional amount by which the upper surface moves, $\Delta L/L$. The volume of an object is essentially unchanged under shear stress, in contrast to what happens under a longitudinal stress.

▲ **FIGURE 11–22** (a) An object fixed along its bottom surface has a force applied. (b) Shear stress from uniform force distorts the object.

The Relation Between Stress and Strain

Now we turn to the question of "laws" that relate a stress and its corresponding strain. As early as the seventeenth century, Robert Hooke realized that to a good approximation stress and compressional strain are proportional to each other. A clear understanding of this proportionality was first provided by Thomas Young in 1800, who wrote down the relation

$$\frac{S}{e} = Y. \qquad (11\text{--}19)$$

LINEAR RELATION BETWEEN STRESS AND STRAIN

This equation applies to the relation between (small) longitudinal stresses (compression or tension) and the corresponding strain. The parameter Y is called **Young's modulus,** or the *elastic modulus*. Its value, which is a measure of the "stiffness" of a material, varies with the material. Y is analogous to the *spring constant, k,* in the spring force $F = -kx$. In fact, Young's modulus is a constant only for *small* strains because the atomic bonds behave just like simple springs, and we know that if a spring is stretched too much it will no longer exhibit a force that varies linearly with the amount of stretch. We all know about materials for which the linear relation between stress and strain fails: bread dough or licorice, materials that certainly don't show a decreased strain when the stress is decreased, i.e., they don't go back to their original shapes when the force "lets up." Because e is dimensionless, Young's modulus has the same dimensions as S and therefore has units of force per area—meganewtons per square meter, for instance. Table 11–1 lists values of Young's modulus for various solid materials. For small stresses and strains, the value of Young's modulus is generally independent of whether the material is under tension (positive strain) or under compression (negative strain). However, for some materials this is not a very good rule, as we shall discuss in the "Think About This" feature below.

TABLE 11–1 • Young's Moduli for Various Solids

Material	Y (MN/m^2)
Rubber	7
Wood	14,000
Concrete	17,000–30,000
Bone	9,000–21,000
Glass	70,000
Aluminum	73,000
Steel	210,000
Diamond	1,200,000

EXAMPLE 11–9 A vertical steel rod of length 2.000 m and diameter 2.0 cm is fixed at the top and has a 9500-kg mass hanging from its lower end. Given that Young's modulus for this particular type of steel is 250,000 MN/m^2, calculate the elongation of the rod. What is the strain?

Setting It Up We have sketched the situation in Fig. 11–23.

Strategy The vertical force due to the hanging mass acts across the cross section of the rod; the force is a tension, so we can find the

▲ **FIGURE 11–23** A large mass causes a slight elongation to a steel rod supporting its weight.

stress using Eq. (11–16). Given the stress, Eqs. (11–17) and (11–19) express the amount of stretch ΔL of the rod of length L and Young's modulus Y in terms of the stress.

Working It Out The magnitude of the tension on the bar is mg, so that according to Eq. (11–16) the stress is $S = mg/A$, where A is the cross-sectional area of the rod. We then use Eqs. (11–17) and (11–19) to find

$$\Delta L = eL = \frac{S}{Y}L = \frac{mg}{AY}L.$$

With $A = \pi r^2$, where $r = \frac{1}{2}(2.0 \text{ cm}) = 1.0 \text{ cm}$ is the radius of the cross section,

$$\Delta L = \frac{mg}{\pi r^2 Y}L = \frac{(9.5 \times 10^3 \text{ kg})(9.80 \text{ m/s}^2)}{3.14(1.0 \times 10^{-2} \text{ m})^2(2.5 \times 10^{11} \text{ N/m}^2)}(2.000 \text{ m})$$

$$= 2.4 \times 10^{-3} \text{ m}.$$

The 2.4-mm elongation gives a strain of

$$\frac{\Delta L}{L} = \frac{2.4 \times 10^{-3} \text{ m}}{2.000 \text{ m}} = 1.2 \times 10^{-3}.$$

What Do You Think? We assumed the cross section was constant over the stretch. Was that a good approximation in this case?

The relation between volume stress and volume strain takes exactly the same form as the relation between ordinary stress and strain, for small strains. The proportionality constant $B = $ (volume stress)/(volume strain), which is analogous to Young's modulus, is called the **bulk modulus**. We have

$$B \equiv -\frac{F/A}{\Delta V/V} = -\frac{p}{e_V}. \qquad (11–20)$$

The minus sign means that an *inward* pressure on a solid, which is positive, implies a *decrease* in volume, $\Delta V < 0$. There is a relation between Young's modulus and the bulk modulus that must involve Poisson's ratio. Only two of these three constants are independent of one another (see Problem 59), and since many solid materials have nearly the same Poisson ratio, the bulk modulus of different materials tends to track the Young's moduli, listed in Table 11–1. Finally, there is a *shear modulus G*, which is given by the ratio of shear stress to shear strain. Liquids have a shear modulus of zero, meaning that they have an arbitrarily large shear strain for even small shear stresses. Liquids do not resist shear at all.

THINK ABOUT THIS . . .
WHAT MAKES A SOLID STRONG?

An important property of solids is their strength or, more specifically, their *tensile strength*, which is also known as *fracture stress*. This quantity is the stress required to break a material into two pieces—by pulling it along an axis, for example. The tensile strength of different materials varies a great deal, and this is a criterion in the selection of materials for a variety of tasks, such as in the manufacture of household goods, or in civil engineering. Steel may break when the tension is 3000 MN/m^2, whereas a piece of concrete may be pulled apart by a stress of only 5 to 10 MN/m^2. The tensile strengths for a variety of materials are listed in Table 11–2. The values of tensile strength are a factor of 100 to 1000 times smaller than the values of Y (Table 11–1) for the same materials. This huge difference reflects the importance of even tiny cracks and imperfections in the lattice structure of a solid.

(continues on next page)

TABLE 11–2 • Tensile Strengths for Various Solids

Material	Tensile Strength (MN/m^2)
Steel piano wire	3000
Steel	400–1500
Cast iron	70–250
Aluminum (pure)	70
Aluminum alloys	140–550
Copper	140
Titanium alloys	700–1400
Spruce, along grain	100
Spruce, across grain	3
Glass	700–170
Brick	5
Cotton	350
Spider silk	240
Human tendon	100
Rope	80

The tensile strength is a kind of *critical stress*. We may define a corresponding *critical strain* as $(\Delta L/L)_c$, such that

$$\text{tensile strength} \equiv Y\left(\frac{\Delta L}{L}\right)_c.$$

TENSILE STRENGTH DEFINED

The factor of 100 to 1000 difference between Young's modulus and the tensile strength for a given solid means that a fractional extension of a rod by 1 percent, or even by 0.1 percent, will break the rod. This is because the pervasive existence of cracks weakens the material. When a stretched rod breaks, the ultimate reason is that the interatomic forces between atoms are overcome. Without cracks, the stress on a section of rod between two layers of atoms is carried by many bonds, but when cracks are present, the stress is concentrated on fewer bonds. The force F in the expression F/A for the stress is the same, but the real area A over which the force acts is very much smaller than the apparent area. This means that the stress experienced by the material is much larger than the stress obtained by using the apparent area. For extremely thin—micrometer-size—fibers, which are nearly perfect crystals, measurements give a tensile strength of around $0.2Y$. This value is

what we would expect upon evaluating $(\Delta L/L)_c$ from our understanding of interatomic forces. The discrepancy between the values of the tensile strength for bulk materials and tiny perfect crystals illustrates the importance of imperfections in the structure of solids.

In some materials the failure under compression is not the same as the failure due to tension. This is often because of imperfections, such as tiny cracks in the material. Compression tends to reduce the cracks, whereas tension magnifies them. (To take an extreme case, a stack of bricks may be viewed as one brick with cracks going right through it. The stack behaves as a solid for compression but has no resistance at all to tension.) Materials such as cast iron or concrete tend to behave this way, and for them the tensile strength is in effect larger for compression than for tension. The response of concrete to stress means that it is relatively weak under tension, and this is remedied by the addition of rebar (iron rods that form reinforcing bars) to make *reinforced concrete*. The rods have a high tensile strength under tension but can catastophically buckle under compression. Reinforced concrete works because its two components have compensating properties against failure and is therefore strong under both tension and compression.

Summary

An extended object is static when the net external force and the net external torque about any origin are both zero:

$$\sum_i \vec{F}_i = 0 \tag{11–3}$$

and

$$\sum_i (\vec{R}_i \times \vec{F}_i) = 0. \tag{11–4}$$

In order to apply these conditions, it is necessary to know both what the external forces acting on an object are and *where* they act. Gravity acts as though it were applied to the center of mass of a rigid body. We can say that the weight of an object acts on its center of mass.

The static conditions are most often used to learn under what conditions a structure can be held static or to learn what forces various components of a structure must endure. Some structures are underdetermined, which means that it is not possible to determine uniquely the forces acting through the consideration of the static conditions alone.

Solids consist of atoms arranged in regular crystalline structures. They are held in this structure by the interatomic forces, which can be considered springlike. As a consequence, solid bodies may stretch or be compressed when subjected to external forces, and are not truly rigid. For longitudinal (pulling or pushing) forces, the force per unit area, or stress, S, is related to the fractional change in length, or strain, e, by Y, Young's modulus:

$$\frac{S}{e} = Y. \tag{11–19}$$

A similar relation applies to volume stress, p, and volume strain, e_V, which are related by the bulk modulus, B:

$$B = -\frac{p}{e_V}, \tag{11–20}$$

and finally there is still another relation for shear stress and strain. Beyond the small deformation that occurs for small stresses, solids break and buckle. The real tensile strength of solids, which is a measure of the amount of strain at which they fracture, is smaller than one would expect with a description based purely on interatomic forces and a perfect crystal structure. That is because of the many imperfections that occur in the crystalline structure of solids. The nonrigidity of solids is evidently reflected in the fact that real structures are never perfectly rigid.

Understanding the Concepts

1. It is a common (and quite useful) piece of advice for those who are learning mountaineering techniques to "stand away" from the mountain on steep slopes rather than to follow one's instincts and "hug" the mountainside. Explain why this is so, using your knowledge of torques and forces on a climber standing on the steep slope of a mountain.

2. A baby pulls straight down with all his might on the flush handles of the closed drawer of a bureau (Fig. 11–24). Can he cause the bureau to tip over?

Force

▲ **FIGURE 11–24** Question 2.

3. Why does a rope from which an object is suspended line up with the vertical?

4. Is it possible for an object not to be in equilibrium even though the net force on the object is zero? If you answered yes, give at least one example.

5. You are sitting quietly in a porch rocker that is suspended by chains from the ceiling. Are you in a stable or unstable equilibrium?

6. Consider Example 11–7. Does it help (in the sense of allowing the washer safely to climb higher) if the ladder is massive? You may want to think about this by imagining the ladder is *very* massive.

7. Bridges and buildings are not really rigid. Does this mean that nothing we have said in this chapter is relevant to them? How can an object be "approximately" rigid?

8. Look at the rock structure in Fig. 11–1. If your task were to topple the rock with a single minimum force, where would you apply it and why?

9. A massive bar is maintained horizontal by three point supports placed at three specified positions along the length of the bar. True or false: You can determine the contact forces at each of the three supports.

10. Is the height of the four-legged table discussed in Section 11–3 irrelevant to the calculation of the equilibrium conditions for the table?

11. A motorcycle with its weight equally distributed over the wheels is resting on level ground with its front wheel bumped up against a curb, perpendicular to the curb. The motorcycle is driven by its rear wheel. Describe the equations that determine the motion of the motorcycle. Can the cycle climb the curb?

12. A pendulum suspended from the roof of an accelerating rail car makes a nonzero angle with the vertical. The pendulum is not swinging. Is this a case of stable equilibrium?

13. For some objects (U-shaped objects, for example), the center of mass is outside the object itself. For such objects, can we still think of gravity as though it acts on the center of mass?

14. The doughnut-shaped space station in Stanley Kubrick's film *2001: A Space Odyssey* rotates about its axis of symmetry (see Fig. 5–27a). Is the rotating station in stable equilibrium, unstable equilibrium, or neither?

15. Suppose you firmly hold the shoelaces of your shoe, one at an angle of 30° with the horizontal, the other at an angle of 60° with the horizontal. You then push straight down with your foot. One of the shoelaces breaks. Which one will it be?

16. Does the method of finding the center of mass of a flat object, as discussed in this chapter, work even if the density of the object is not constant?

17. You want to hold a beam with its end against a frictionless wall. Is it possible to do so by running a rope from the far end of the beam to any attachment point on the wall?

18. When a cart has to be pushed uphill, it is better to exert one's force by pushing on the top of the wheel rather than at the axle of the wheel. Why is that?

19. Can a ladder placed against a rough vertical wall remain standing when the floor is so smooth that there is no friction between it and the feet of the ladder?

20. Identical twins are placed at opposite ends of a seesaw pivoted about its midpoint. No forces other than those due to the twins, the pivot, and gravity act on the seesaw. What determines the inclination to the horizontal made by the seesaw when it is balanced? Is the equilibrium stable, unstable, or neutral?

21. Suppose that the seesaw of Question 20, still pivoted about its midpoint, has a sharp downward bend at the midpoint. If there is an equilibrium, will it be stable, unstable, or neutral?

22. Why is cement unsuitable as a construction material for a boiler?

23. In Chapter 5 of the Book of Exodus, the Israelites in captivity complain to Pharaoh that they are being asked to make bricks without straw. Why was it a good idea to put straw in the clay that was allowed to dry in the hot sun?

24. Why does the amount of stretch of a bar on which a tension acts depend on the overall length of the bar? [*Hint:* Think of breaking the bar into two pieces of equal length and of how much each piece would stretch under the same force.]

25. If it is possible to make diamonds from graphite by using high pressure, how would it be possible to make graphite from diamonds?

26. On a macroscopic level, solids are distinguishable from liquids by their resistance to shear. Is this distinction likely to be a sharp one?

27. What kind of crystal structure might have little resistance to shear forces in some directions and much resistance in other directions? Would such a crystal have direction-dependent resistance to stretching?

Problems

11–2 Gravity and Rigid Bodies

1. (I) A uniform board of mass 80 kg and length 3.6 m is placed on top of a pivot 1.2 m from one end. What mass must be put at that end to allow the board to balance?

2. (I) A 20-kg board 2.5 m long is supported in a horizontal position at the two ends. A 70-kg worker stands 1.2 m from one end. What forces are exerted by the board on the two support points?

3. (I) Two workmen each carry one end of a 2.2-m-long ladder of mass 24 kg. The ladder is tapered so that its center of mass is 0.9 m from the wider end. What are the forces exerted by the ladder on the two workmen?

4. (I) A rail of length 3.0 m and mass 8.0 kg runs horizontally between two scales; a bowling ball of mass 5.5 kg is allowed to roll at a steady speed of 0.15 m/s from the left scale to the right scale (Fig. 11–25). During the time the ball is moving, how do the readings on the two scales change?

▲ **FIGURE 11–25** Problem 4.

5. (I) A projectile has broken into two parts. At time t_0, the two parts are located at the following points: m_1 at $(x, y, z) = (3, 0, 0)$ and m_2 at $(0, 0, 3)$, where all distances are measured in meters. The mass m_1 is twice the mass m_2. What is the location at time t_0 of the point that follows a parabolic trajectory (assuming that there is such a point)?

6. (II) Two people of unequal strength must carry a uniform beam of length L while holding it horizontal. The weaker of the two holds the beam at one end. (a) How far from the other end must the stronger person hold the beam in order to support three-quarters of the weight? (b) Is there a way in which the stronger person can carry the beam at one end and still support more than half the weight of the beam?

7. (II) A playground seesaw is balanced at its midpoint. Two children, weighing 25 kg and 40 kg, respectively, want to balance on the seesaw. If the children are separated by a distance of 2.8 m, how far from the pivot point will the lighter child sit?

8. (II) Consider a seesaw whose total mass is 8 kg and total length is 3.50 m. Suppose the seesaw is placed off center on the pivot point so that the pivot point is 24 cm from the center of the seesaw (Fig. 11–26). How far from the center will the children of Problem 7 have to sit if the lighter child sits on the longer part of the seesaw? (Their separation is still 2.80 m.)

▲ **FIGURE 11–26** Problem 8.

9. (II) In order to hang a load of mass $M_1 = 30$ kg from the horizontal, flat roof of a building, a plank of length $L = 2.4$ m is placed on the roof (Fig. 11–27). One end is held in place with a chunk of concrete of mass $M_2 = 15$ kg, and the other supports the load M_1 with a light rope. How far can the end of the plank reach without tipping over? Neglect the mass of the plank.

▲ **FIGURE 11–27** Problem 9.

10. (II) A uniform beam of mass 80 kg and length 3.0 m rests on two pivots, one at the left edge and one 2.4 m from the left edge. How far to the right of the right pivot can a mass of 150 kg be placed without the beam tipping?

11. (II) A rectangular piece of plywood (60 cm × 120 cm) lies in the horizontal xy-plane. The surface mass density of the plywood is 3 kg/m². Calculate the torque about one of the corners due to the force of gravity.

12. (II) A uniform book of mass 1.0 kg is placed such that 60% of the book is hanging over the edge of a table, with a paperweight of mass m is centered on top of the 40% of the book that is on the table. What is the minimum mass of the paperweight such that the book doesn't fall off the table?

13. (II) Two books are stacked at the edge of the table, with their lengths perpendicular to the table edge. If the width of each book is L, how far out from the table edge can the top book's extreme edge be placed without the books falling down? How does this generalize to three books?

11–3 Applications of Statics

14. (I) A door 90 cm by 195 cm of mass 14 kg hangs on two hinges: One is attached to the bottom of one side of the door, and the other to the top of the same side. What are the horizontal forces exerted on the door by the hinges?

15. (I) A football player is at the top of a pushup. The angle that the (rigid) torso makes with the floor is 25°. His arms are perpendicular to his torso and his center of mass is located at a point 3/8 of the distance from the shoulders to the feet. Assuming that the mass of his head can be neglected, what is the force, in terms of the player's weight, along his arms?

16. (I) A student wants to place a flower pot on a board that juts out from a window so it will get more sunlight (Fig. 11–28). The flower pot has a mass of 3.5 kg and needs to be 50 cm from the windowsill. The student can only place a nail into the sill 4 cm from the edge. Neglect the mass of the board and find out how much force the nail must exert to hold the board in place. What is the nature of the force "exerted" by the nail?

▲ **FIGURE 11–28** Problem 16.

17. (I) A uniform rod, mass 12 kg and length 1.5 m, rests on two points, one at its left end and one at the center point. What are the contact forces on the rod at these points? Comment on the stability of the situation.

18. (II) An 8.5-m extension ladder of mass 26 kg is propped up against a wall, touching at a point 8.0 m above the level ground. A man of mass 75 kg climbs 7 m up the ladder to repair a window. The ladder rests against a frictionless wall, but the ground has friction. Determine all the forces on the ladder.

19. (II) Consider a ladder of mass 10 kg and length 4 m, leaning against a vertical wall at an angle of 30° with the vertical. The coefficient of friction between the ladder and the floor is $\mu_s = 0.40$, and there is no friction between the wall and the ladder. A man of mass 80 kg climbs up the ladder. (a) How high can he climb before the ladder begins to slip? (b) Work out your calculation by taking the torques about the three points A, B, and O in Fig. 11–29, and show that the resulting equations are independent of the choice of point.

▲ **FIGURE 11–29** Problem 19.

20. (II) Consider the ladder in Example 11–7 (see Fig. 11–11a). The 3.0-m-long ladder is placed against a frictionless wall, making the same 58° angle to a different horizontal surface; this time, much to his dismay, the same 85-kg window washer finds himself starting to slip when he steps to the second rung. What is the coefficient of static friction between the ladder and floor?

21. (II) Using a uniform strut, a rigid brace hinged at the floor, a person holds a 30-kg engine in equilibrium while it is being repaired. The strut has a mass of 12.5 kg. A smooth rope passes over a pulley at the end of the strut (Fig. 11–30). (a) What is the force exerted on the rope by the person? (Specify the direction of this force by calculating the angle θ that the rope makes with the horizontal.) (b) What forces are exerted by the strut?

▲ **FIGURE 11–30** Problem 21.

22. (II) A piece of plywood leans against a wooden wall with which it has a coefficient of static friction of $\mu_s = 0.28$. (a) If the coefficient of static friction between the board and the floor is 0.35, what is the minimum angle that the board can make with the floor yet still not slip? (b) What happens if the coefficient of friction between the board and the floor is zero?

23. (II) A lawn mower of mass m is at rest on a rough slope, coefficient of static friction μ_s between wheels and ground, as shown in Fig. 11–31. Find the largest slope angle θ for which the mower will not (a) slip down the slope and (b) tip over. You should be able to show that what determines whether the mower tips or slides is the same for any angle.

▲ **FIGURE 11–31** Problem 23.

24. (II) Re-solve the problem posed in Example 11–6, this time using as a reference point for the torque equation the high end of the crane arm. You should find the same result as in the example solution.

25. (II) A stepladder consisting of two ladders of mass M and length L is held together by a crossbar attached to the midpoints of the two ladders (Fig. 11–32). What force is exerted on the crossbar by each ladder if the length of the crossbar is $L/2$? Assume that any friction between the ladders and the floor is negligible.

▲ **FIGURE 11–32** Problem 25.

26. (II) A flagpole of mass 6 kg and length 2.4 m is hinged at a wall and supported in a horizontal position by a cable attached to the free end (Fig. 11–33). The cable makes an angle of 25° with the horizontal. What is the tension in the cable? What is the vector force exerted on the hinge at the wall?

▲ **FIGURE 11–33** Problem 26.

27. (II) Two pulleys are mounted on the same axis. A rope is attached to the large pulley of diameter 20 cm. A car engine of mass 300 kg is hung from the small pulley of diameter 8.0 cm (Fig. 11–34). With what force must a person pull the rope to hold up the engine?

▲ **FIGURE 11–34** Problem 27.

28. (II) A frictionless pivot joins two uniform boards of the same mass m and length L, which are then placed to form a symmetrical "tent" on a rough horizontal ground surface (Fig. 11–35). The coefficient of static friction between the board ends and the ground is μ. What is the largest angle θ that each board makes with the vertical to the ground such that the arrangement will not slip?

▲ **FIGURE 11–35** Problem 28.

29. (II) A desk of height 0.82 m, length 1.54 m, and mass 43 kg is pushed across a horizontal floor at a steady speed with a horizontal force $\vec{F}$ applied at the top (Fig. 11–36). The coefficient of kinetic friction between the legs and the floor is $\mu_k = 0.45$. What is the friction force at each leg, and what is $\vec{F}$? Assume that the two right legs each support the same forces, as do the left legs.

▲ **FIGURE 11–36** Problem 29.

30. (II) Repeat Problem 29, but this time assume that the desk is being pushed down a slope of 3.5°. The force applied is parallel to the sloping floor. The center of mass of the desk is in the middle of its long dimension and 0.38 m down from its top surface.

31. (II) Figure 11–37 (see next page) is a side view of a seat used for babies. It enables a baby to sit at the edge of a table by means of four points of contact; in the side view shown, two of these points are visible—point A at the top of the table, and point B beneath the table. The other two points are aligned directly behind. The center of mass of the baby plus the seat can be approximated by a mass m at point C. (a) Taking into account only forces in the xy-plane (the plane of the figure), calculate the forces *on the table* at points A and B. (Remember that there are four contact points, not two. Assume that the symmetric legs share the force equally.) (b) What happens for $\ell_2 \to 0$? for $\ell_1 \to 0$? (c) Work out numerical values for the forces for $m = 10$ kg, $\ell_1 = 20$ cm, and $\ell_2 = 30$ cm.

FIGURE 11–37 Problem 31.

32. (II) Consider the baby seat of Problem 31. The table has a total mass $M = 18$ kg, all concentrated in the (uniform) top, and the center of the table is 70 cm from the vertical extension of point P. For the geometry of the baby seat and the tabletop shown in Fig. 11–38, what mass m of the baby plus the seat will cause the table to tip over by a rotation at point P? Can the possible movements of a baby seriously destabilize the situation?

FIGURE 11–38 Problem 32.

33. (II) A trap door 1.8 m square with mass 20 kg is hinged at one edge and is attached to a rope at the opposite edge (Fig. 11–39). The trap door makes an angle of 55° with the horizontal, and the rope is perpendicular to the door. (a) What is the tension in the rope? (b) What is the force vector acting on the door at the hinge?

FIGURE 11–39 Problem 33.

34. (II) A beam of mass 12 kg is hinged at the top, and its bottom end is pulled to the side by a horizontal rope. The beam makes an angle of 15° with the vertical. What is the tension in the rope, and what is the force exerted on the beam at the suspension point?

35. (II) A variant of the crane in Example 11–6 is shown in Fig. 11–40. The pivot at point B is frictionless, and the beam, of length 3.00 m, has mass 100 kg. The rope makes an angle of 30° with the horizontal and can withstand a tension of 10,000 N before breaking. If this arrangement is used to lift masses from the point shown on the beam, what is the maximum mass that can be so raised?

FIGURE 11–40 Problem 35.

36. (II) A uniform door is attached to the wall by two hinges. Draw a force diagram for the door and describe the various forces that can act. Describe the conditions for static equilibrium, and find the forces that are solvable. Is this an underdetermined system? If so, which forces are underdetermined?

37. (II) Consider the four-legged table discussed in Section 11–4, with leg 1 slightly shorter than the others. Suppose that a tea cup were placed on the diagonal between legs 1 and 4, at location

$$(x, y) = \left(\frac{L}{2} + x, \frac{L}{2} + x\right).$$

(a) Find a satisfactory solution to the magnitude of the contact forces on the legs if $x > 0$. Neglect the masses of the legs. (b) Repeat part (a) for $x < 0$.

38. (II) A solid block of mass M, height h, and base width w stands on a rough floor. By applying a large enough force F perpendicular to one side of the block, it is possible to topple it: the block will rotate on one edge so that the edge remains in contact with the floor without sliding. Through what critical angle must the box be rotated before it falls over? What is the work that must be done to tip it over?

39. (II) One way to keep a hollow cylinder of length L, mass M, and radius R from rolling down a rough (coefficient of static friction μ) inclined plane of angle θ is to exert a torque on it about its center that would tend to rotate the cylinder in the direction opposite from its rotation sense as it rolls. How large should this torque be in terms of the parameters given here?

40. (II) A large, spherical satellite is held in place in the bay of a space station with six ropes: four are attached and equally spaced around the equator and the other two are attached at the poles. (a) What equations describe the situation in which the satellite is held motionless with respect to the station? (b) Is this an underdetermined system? (c) Is the system underdetermined when all the ropes are under identical tension?

41. (III) A seaside tower is supported by a cable (Fig. 11–41, see next page). A horizontal wind that increases with height often blows from the left, exerting a horizontal force to the right on the tower. The horizontal force on a unit length of the tower increases with the height h according to force/unit length $= \alpha h$, where h is the height from ground level, and, if h is measured in meters, $\alpha = 50$ N/m². The total height of the tower is 20 m. What is the tension in the cable?

▲ FIGURE 11–41 Problem 41.

42. (III) The mythical Greek king Sisyphus, pushing a large round rock up a mountain, wishes to take a rest and supports the rock with a horizontal rope attached to the top of the rock (Fig. 11–42). The coefficient of static friction between the rock and the slope is $\mu_s = 0.6$. (a) What is the largest value of θ for which this method of support is possible? (b) If the mass of the rock is 1088 kg, what tension must the rope be able to support on this maximum slope? (c) How does the tension vary as a function of θ for angles less than the maximum angle? [*Hint*: It is easiest to take moments about the point of contact.]

▲ FIGURE 11–42 Problem 42.

11–4 Solids and How They Respond to Forces

43. (I) The crystal structure of copper is face-centered cubic; i.e., the structure is built of cubes of edge length a, with copper atoms at the corners as well as the face centers of the cube (Fig. 11–43). The lattice length $a = 0.361$ nm. (a) Calculate the diameter of a copper atom. (Assume that atoms are described by spheres that are centered on the corners and centers, and that they just touch). (b) Show that the structure can also be described as a stacking of triangular layers of atoms, perpendicular to the maximum diagonal of the cube.

▲ FIGURE 11–43 Problem 43.

44. (I) The building block of a simple cubic lattice is a cube with atoms at the corners. If the atoms are represented by spheres of diameter d that are centered on the corners and just touch, what is the diameter of the largest impurity that will not displace any of the existing atoms that can be put at the center of the cube? [Figure 11–44 shows what a diagonal slice across the cube looks like. *Hint*: The longer side of the rectangle is $\sqrt{2}$ times the shorter side.]

▲ FIGURE 11–44 Problem 44.

45. (II) The volume filling of a crystal structure can be defined as the ratio of the volume of the atoms as represented by touching spheres to that of the crystal unit. Calculate this ratio for simple cubic and face-centered cubic lattices. [Refer to Problems 43 and 44.]

46. (I) A concrete pier of area 0.85 m² and height 3.6 m is built to hold up a bridge (Fig. 11–45). The load on the pier is 100 tons (0.91×10^5 kg). $Y = 17{,}000$ MN/m² for the concrete. How far will the concrete compress?

▲ FIGURE 11–45 Problem 46.

47. (I) A mine elevator is supported by a steel cable 2.0 cm in diameter. The mass of the elevator and its contents is 800 kg. By how much is the cable stretched when the elevator is 250 m below Earth's surface?

48. (I) A steel bolt must withstand forces up to 1000 N. If a safety factor of 3 is to be taken into account, what must the minimum diameter of the bolt be? (With a safety factor of 3, the bolt should be able to support 3×1000 N before it snaps.)

49. (I) Calculate the critical strain for steel piano wire. You will need to consult the tables. How far would a wire 1 m long stretch before breaking?

50. (I) Assume that the piano wire of Problem 49 is 0.85 mm in diameter. Calculate the weight it could hold before fracturing.

51. (II) A crate of mass 30 kg slides across the ground. The coefficient of kinetic friction between the crate and the ground is 0.3. The physical contact area between the crate and the ground (which is less than the area of the whole crate) is 0.35 m². Calculate the shear stress on the crate.

52. (II) A steel wire of diameter 5.0 mm and length 2.0 m stretches 0.30 mm when a load of 60.0 kg is hung from it. What is its Young's modulus? How much mass can the wire hold before it may fracture? Use the tensile strength of steel from the tables.

53. (II) The 3-km-long cables on a large suspension bridge are stretched from their equilibrium length by 3 m. Estimate the change in the equilibrium separation between any two adjacent atoms along the cable.

54. (II) A steel bar of length 3.5 m is placed in a structure where it is subject to extreme stress under tension. The area of the bar is 25 cm^2, and it has room to stretch by 3 mm. It cannot stretch any more than this without butting up against a much stronger part of the structure. Does the bar break before it stretches the 3 mm? Use the Young's modulus and tensile strength of the tables.

55. (II) A uniform beam of length 2.0 m and mass 10 kg is freely pivoted at one end about a point fixed to the wall (Fig. 11–46). It is held in a horizontal position by a steel cable of diameter 2.0 mm. The cable makes an angle of 30° with the horizontal. A load of 30 kg is suspended from the end of the beam. What will be the angle of the beam relative to the horizontal? [Hint: The distortion is small, so you may calculate the tension in the cable without taking into consideration the slight change in angle.]

▲ FIGURE 11–46   Problem 55.

56. (II) A 0.5-m-long piece of metal is compressed longitudinally by 0.1 mm. Its Poisson's ratio is 0.32. Calculate its volume strain.

57. (II) Poisson's ratio for steel is 0.3. What is the new diameter of the steel rod used in Example 11–9 after the rod has been stretched?

58. (III) According to our discussion of Poisson's ratio, a bar of length L and cross-sectional area A under compression at its ends expands at the sides. Find the amount of pressure that must be exerted on the sides to keep them fixed when there is a compression F on the two ends. Find the total amount of compression in the longitudinal direction under these conditions, and show that a bar with constrained sides is stiffer under compression than the same bar with unconstrained sides. (Keep the terms up to order σ^2 only.)

59. (III) Let the strain on a rod along the direction of applied stress be e_1. Because the stretching produces a slight reduction in the diameter of the rod, we may speak of an induced transverse strain, given by $e_{tr} = -\sigma e_1$, where σ is Poisson's ratio (Fig. 11–47). Recall that Y is defined in terms of longitudinal stress and strain. A cube of the rod's material immersed in a liquid, so that the pressure on all sides is p, suffers a volume change given by $\Delta V/V = 3p(1 - 2\sigma)Y$. (a) Derive this result. (b) Show that when Pois-

son's ratio is not zero, the relation between Young's modulus and the bulk modulus under uniform pressure is given by

$$B = \frac{1}{3}\frac{Y}{1 - 2\sigma}.$$

▲ FIGURE 11–47   Problem 59.

60. (II) By using the results of Problem 59, show that Poisson's ratio, σ, must be less than 0.5 because, if it exceeds 0.5, the total volume of the solid would increase when a uniform pressure is applied to it.

61. (III) Compressibility is defined by $-\Delta V/pV$, the reciprocal of the bulk modulus. Using the equation for $\Delta V/V$ in Problem 59, calculate the compressibility of silver, given that $\sigma = 0.38$ and $Y = 7.9 \times 10^4 \text{ MN/m}^2$.

62. (III) Show by analogy with a simple spring, for which the force is $F = -kx$, that a volumetric potential energy associated with strain—that is, potential energy per unit volume of a wire or rod of length L and cross-sectional area A—is given by $u = \frac{1}{2}$ (stress) (strain). [Hint: Recall the work–potential energy connection.]

General Problems

63. (II) A 30-cm-wide shelf is supported at the wall by a cable at each end of the shelf placed at 45° to the wall (Fig. 11–48). A 20-kg sack of potatoes is placed and centered on the shelf. What is the tension of the cable if we ignore the mass of the shelf?

▲ FIGURE 11–48   Problem 63.

64. (II) The cross section of an A-frame house is shown in Fig. 11–49 (see next page). The total height of the apex is 5.0 m, and the 1.5-m-long crossbeam is two-thirds of the way up the roof line. The crossbeam must support two roof beams, each of mass

3000 kg. (a) Does the crossbeam push the roof beams out or does it pull them in? (b) What is the force exerted by the roof beams on the crossbeam?

▲ **FIGURE 11–49** Problem 64.

65. (II) A uniform beam of length L and mass M is freely pivoted at one end about an attachment point in a wall. The other end is supported by a horizontal cable also attached to the wall, so that the beam makes an angle θ_0 with the horizontal (Fig. 11–50). (a) What is the tension in the cable? (b) The cable snaps. What is the angular acceleration of the beam about its pivot point immediately afterward? (c) What is the angular velocity of the beam as it falls through the horizontal position?

▲ **FIGURE 11–50** Problem 65.

66. (II) A uniform board of length 2.4 m and weight 47 N has one end on the ground. With the aid of a horizontal force applied at the upper end by means of an attached horizontal rope, the board is held at an angle θ with respect to the vertical (Fig. 11–51).

▲ **FIGURE 11–51** Problem 66.

The coefficient of static friction between the end of the board and the ground is $\mu_s = 0.32$. What is the range of angles the board can make with the vertical and still be in static equilibrium? How does the tension in the rope vary with the angle within the angle's possible range?

67. (II) A sign is to be constructed from a piece of plywood in the shape of a 30° right triangle of mass 15 kg. It is to be attached to a wall, as shown in Fig. 11–52. The lower attachment point is a frictionless pivot, and the upper point is a rope that can be reeled in or out to make the bottom of the sign and the rope itself horizontal. What is the tension in the rope?

▲ **FIGURE 11–52** Problem 67.

68. (II) A car is lifted vertically by a jack placed at the car's rear end 40 cm off the central axis, so that the weight of the car is supported by the jack and the two front wheels (Fig. 11–53). The distance between the front wheels is 1.60 m, the distance from the axis connecting the two wheels to the center of mass of the car is 80 cm, and the distance from the rear of the car to the center of mass is 2.10 m. What fraction of the car's weight is carried by each of the wheels, and what fraction is carried by the jack? (Note that the weight on the wheels will *not* be symmetrically distributed.)

▲ **FIGURE 11–53** Problem 68.

69. (II) A roller of radius 30 cm and mass 80 kg is pulled by a force F applied horizontally to the axle. How large must F be in order for the roller to climb a step 15 cm high?

70. (II) Consider the roller of Problem 69. Assume that the force F can be applied in any direction. What direction minimizes the magnitude of F necessary to make the roller climb the step?

71. (II) A chest of drawers 58 cm wide, 1.6 m long, and 1 m high is pulled by a horizontal rope attached at a height of 58 cm to the midpoint of the long side. The force is such that the chest moves with uniform velocity. (a) Express the force on the legs on the side with the rope, and the force on the legs on the opposite side, in terms of the coefficient of kinetic friction μ_k. (b) For what value of μ_k will the chest topple over?

72. (II) A wooden box of uniform density and dimensions $h = 1.4$ m, base 0.3 m $\times$ 0.3 m, and mass 50 kg stands on a rough surface with coefficient of static friction $\mu = 0.70$. When a large enough force is applied perpendicularly to one side, the box will either slide or topple over, depending on the height above the ground that the force is applied to. What is the highest point at which you can apply the force to slide the box rather than toppling it?

73. (II) A spherical nut is placed between the handles of a nutcracker (Fig. 11–54). As the handles are closed, the nut slides away from the hinge until the angle α decreases to the point at which $\tan(\alpha/2) = \mu$, where μ is the coefficient of friction between the nut and the inner surface of the nutcracker. Show that no matter what the force on the handles is, the nut will not slide any farther.

▲ **FIGURE 11–54** Problem 73.

74. (II) A box 0.6 m long and 1.5 m tall is placed on a flatbed truck. The truck accelerates at a rate of 1.2 m/s^2. Will the box topple over?

75. (II) A cylinder of mass M and radius R rests on an inclined plane (Fig. 11–55). It is held in place by a horizontal string that is attached to the edge of the cylinder. If the angle that the plane makes with the horizontal is θ, and the coefficient of static friction between the cylinder and the plane is μ_s, what is the smallest value of μ_s that will maintain this position as an equilibrium position?

▲ **FIGURE 11–55** Problem 75.

76. (II) A centrifugal governor consists of light rods pivoted at points A, B and C, as shown in Fig. 11–56. They are loaded with masses M, m, and m respectively, and the whole apparatus rotates about the vertical axis. Depending on the angular velocity, the mass M slides up or down, and its position can be used to control the flow of steam in a steam engine. Derive a relationship between the angular velocity and the angle the arms make with the vertical.

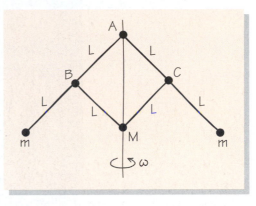

▲ **FIGURE 11–56** Problem 76.

77. (II) A rubber tether ball is attached by a string to the wall as shown in Fig. 11–57, with the string tangential to the ball. The mass of the ball is 150 g, the diameter is 12 cm, and the string is 90 cm long. The string is attached at a position 11 cm from the wall. What is the smallest possible value of the coefficient of static friction between the ball and the wall such that the ball will not fall?

▲ **FIGURE 11–57** Problem 77.

78. (III) Three steel ball bearings of mass m and radius 1.33 cm are placed in a closed vertical tube of diameter 4.00 cm that is closed on the bottom. Find the forces on the middle ball bearing.

79. (III) Two steel ball bearings of mass m and radius 2 cm are placed in a vertical tube of diameter 6 cm that is closed on the bottom. Find all the forces on the two balls.

▶ Why doesn't this satellite fall toward Earth under the influence of gravity? Well, it actually does, but because the satellite has a high velocity in the direction perpendicular to Earth's radius, it "falls around the Earth" and has a stable orbit.

Gravitation

While the forces that we see and experience every day are ultimately derived from the more basic forces that govern the interactions of matter, in most cases the basic force itself is well hidden. Examples include friction, tension, and contact forces, which are all ultimately explained by the forces that hold the components of matter together—the electromagnetic force—although it would be very difficult to explain all the characteristics of these "effective" forces from first principles. There is, however, one important exception, and that is gravity. Earth's gravity, as we experience it every day, is a simple manifestation of the force of universal gravitation; as far as we know today, gravitation is one of the truly fundamental forces. Gravitation governs not only the motion of a falling apple but also the majestic orbital motions of the Moon around Earth, the planets around the Sun, and the stars in their voyage around the galaxy. In 1687 Isaac Newton published the realization that the same force governs both the falling apple and the orbiting Moon in his book *Principia*; this was one of the great intellectual leaps in human history. Fortunately, the law of gravitation is a force law for which the equations of motion can be solved; we can use it to give a full description of the motion of astronomical bodies. Gravitation is conservative, and we can use potential energy to help analyze many situations involving the gravitational force.

From the start, the study of the motion of celestial objects was motivated by philosophical and esthetic considerations ("the harmony of the spheres"), but there are also good, practical reasons to want to understand gravity and gravitation. The measurement and understanding of local variations of the strength of gravity has taught us a lot about the Earth's interior structure and is an important prospecting tool for oil and minerals.

Satellites, whose orbits are determined by the gravitational force, are an indispensable part of the world's communications system, and we rely on the satellites of the Global Positioning System (GPS) for navigation. Indeed, the GPS relies for its functioning not just on Newtonian gravitation but on Einstein's general theory of relativity, a subject that had once been thought to have no practical consequences. Finally, much of our exploration of the solar system relies on understanding complicated orbital motions. From the earliest times, people have looked into the sky and considered the Sun, Moon, and stars. This interest in observing our celestial surroundings has led to a deep understanding of gravitation and its consequences.

12–1 Early Observations of Planetary Motion

The earliest astronomical observations of the night sky led ancient peoples to divide the points of light they saw into two classes: the so-called fixed stars, which move each day in nearly perfect circles around Earth, and the planets (in Greek, the "wanderers"), which move in what appear superficially to be complicated, erratic patterns in the night sky. The first interpretations of these observations placed Earth at the center of the universe, in a **geocentric** frame. Earth was pictured as being surrounded by a rotating spherical shell in which the stars were fixed. The more complicated motion of the planets, Sun, and Moon were explained by placing the planets upon moving, transparent shells within the outer, fixed-star shell. These shells had to move in complicated ways if the observations were to be consistent with observation. In the second century A.D., Claudius Ptolemy made the most detailed formulation of these notions, and to explain the various paths taken by planets, he had to depart from the simple spherical shell description and construct paths of planetary motion that form circles that themselves lie on circles around Earth (Fig. 12–1). His theory was based on a culturally imposed belief that circular motion is in some sense "perfect" and that superposed circular motions embodied that perfection. As his *epicycles* (the circles on circles used to describe the movements of the planets) gave a reasonable description of the motion of the planets, the idea proved serviceable for over 1400 years.

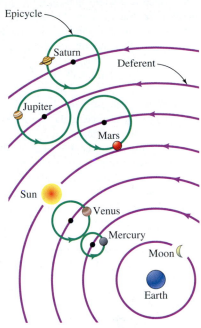

▲ **FIGURE 12–1** Planetary motion based on the Ptolemaic system, which was the generally accepted framework for celestial phenomena from the second century to the 1600s. According to Ptolemy, Earth is the center of the world system. Planets move in small circular paths called epicycles, and the centers of the epicycles move around Earth on large circles called deferents.

The Copernican Picture

In 1543 Nicolaus Copernicus introduced a revolutionary and controversial view of the motion of the planets in a **heliocentric** frame, with the Sun rather than Earth at the center of the solar system. Copernicus continued to insist on describing all motions with circles, and because the true motions of the planets about the Sun are not circles, epicycles were still needed in the Copernican description to accommodate the observations of planetary motion (Fig. 12–2). The epicycles meant that the Copernican description was about as unwieldly as the Ptolemaic description, so in the light of continued philosophical objections, there was no compelling reason to accept the Copernican view; the Copernican hypothesis was not accepted for almost a century. During this transitional period the construction of more refined instruments (protractor-like instruments called quadrants, as there were still no telescopes) allowed Tycho Brahe to improve the knowledge of planetary **orbits**, or paths, to an accuracy of less than half a minute of arc by the end of the sixteenth century.

◀ **FIGURE 12–2** The Sun and Moon move west to east relative to fixed stars from night to night, but some planets do a retrograde "loop" from east to west every once in a while. The yellow dashed line traces a planetary position relative to the stars over a period of several months. The simplest Ptolemaic universe does not easily explain the retrograde motion of Mars shown here; "epicycles" are required (Fig. 12–1). It is simpler to explain this motion with the Copernican description.

CONCEPTUAL EXAMPLE 12–1 As Earth rotates, Mars moves across the sky from east to west, as do the Sun, the Moon, the other planets, and the stars. Most of the time Mars moves across the sky a little more rapidly than the background stars, but every once in a while, and for a period of a few months, Mars exhibits *retrograde motion*, during which it moves more slowly than the stars, in effect moving backward with respect to them. How can you explain this motion, given our understanding that Earth and Mars each orbit about the Sun? Do the Sun and the Moon ever exhibit retrograde motion?

Answer The effect of Earth's daily rotation about its axis is responsible for our seeing the overall east-to-west motion, and this motion applies to Sun, Moon, planets, and stars alike. In thinking about the answer to the question, we'll subtract out this aspect of the motion. Earth also orbits about the Sun, and there is an additional moving platform effect. The stars are so distant that they remain fixed. The planets move with respect to the stars because they themselves are orbiting the Sun and we are close enough to them to clearly observe that motion. Most of the time we see them moving "forward" in their orbits against the backdrop of the fixed stars,

meaning each day they will be a little farther along their paths than they were the previous day. But if Earth in its orbit overtakes them, as happens with Mars, then they will appear to move "backward" against the stars. For a useful analogy, think about yourself in a car on a highway. The distant landscape hardly seems to move, while you see other cars on the highway move against the landscape because of the relative motion between you and them. Cars moving in your direction move forward against the landscape. The analogy to retrograde motion of a planet occurs when you pass another car. From your point of view the car you overtake will move backward with respect to the landscape. A similar thing happens with Mars seen from Earth. Earth has a solar orbit inside that of Mars, and Earth moves faster in its orbit than does Mars. As we on Earth overtake Mars on an inside track, we see it move backward against the fixed background. The Moon and Sun have no such motion with respect to us, because Earth orbits the Sun, and the Moon orbits Earth. While this retrograde motion is simple in the Sun-centered picture, it is explicable, though in a more complex way, in the Ptolemaic system with epicycles. Indeed, a glance at Fig. 12–2 reveals possible retrograde motion for Mars, and the occurence of retrograde motion is not enough to kill the Ptolemaic picture.

(a)

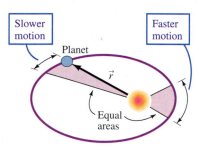

(b)

▲ **FIGURE 12–3** (a) Kepler's first law specifies that the planets move in elliptical orbits with the Sun at one focus. The perihelion is the distance of closest approach and the aphelion is the farthest distance between a planet and the Sun. (b) Kepler's second law specifies that the radius vector from the Sun to a planet sweeps out equal areas in equal times. The two shaded areas shown are equal if the same time is taken to sweep out each of them. This means that the planet moves more quickly over a part of the orbit closer to the Sun than over a more distant part.

Kepler's Laws

After Brahe's death in 1601, his assistant Johannes Kepler inherited the data that Brahe had accumulated. Kepler spent some 20 years analyzing these data, looking for mathematical regularities. In a crucial breakthrough he concluded that the idea of circular orbits should be discarded and replaced with elliptical orbits. **Kepler's laws** summarize Kepler's most important conclusions:

1. Planets move in planar elliptical paths with the Sun at one focus of the ellipse (Fig. 12–3a).

2. During equal time intervals the radius vector from the Sun to a planet sweeps out equal areas (Fig. 12–3b).

3. If T is the time that it takes for a planet to make one full revolution around the Sun, and if R is half the major axis of the ellipse (R reduces to the radius of the planet's orbit if that orbit is circular), then

$$\frac{T^2}{R^3} = C, \tag{12–1}$$

where C is a constant *whose value is the same for all planets.*

Kepler's laws were so simple that once they were known it was no longer possible for scientists to cling to pre-Copernican ideas.

We found the origin of Kepler's second law in Chapter 10: Eq. (10–28). The law follows from the conservation of angular momentum. We saw in Chapter 10 that the angular momentum of a body under the influence of a central force is conserved. Here angular momentum conservation, and hence the "equal area in equal times" rule, is a consequence of the fact that *the gravitational force between the Sun and each planet is central*; that is, the force acts along the line between the Sun and the planet. In fact, Kepler's second law can be taken as evidence that the gravitational force is central. Conservation of angular momentum also implies that the trajectories of planets must lie in a plane—the plane that is perpendicular to the direction of the fixed angular momentum vector.

▶ 12–2 Newton's Inverse-Square Law

Newton recognized the importance of Kepler's laws. He realized that since the planets do not move in straight lines, they must be subject to a net force. He concluded from Kepler's second law that the net force on a planet must point from the planet to the Sun—it must be

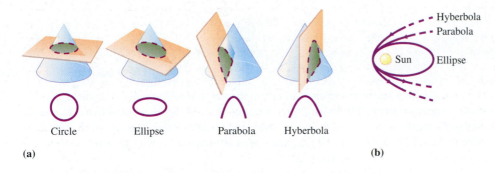

Circle   Ellipse   Parabola   Hyberbola

(a)

Hyberbola
Parabola
Sun   Ellipse

(b)

◀ **FIGURE 12–4** (a) Slicing a cone with a plane produces conic sections. The conic sections are the familiar geometric curves shown here. (b) The orbits of astronomical objects such as planets can follow all the curves drawn in (a).

central. He also found that the elliptical paths described by Kepler can be explained by an *inverse-square force law*: If the source of the force is at the origin O and the mass on which the force acts is at position $\vec{r}$, then the force has magnitude k/r^2, where k is a constant.

Newton showed a more general result as well: The possible paths of a mass under the influence of a force of this form had to be **conic sections**. Conic sections are the shapes obtained if a cone is sliced by a plane. As can be seen in Fig. 12–4a, if a cone is sliced by a plane parallel to the cone's base, the conic section is a **circle**. If the plane is tilted a little, the conic section is an **ellipse**. If the slice is cut parallel to the slope of the cone, the conic section turns out to be a **parabola**. If the slice is steeper than the slope of the cone, the conic section is a **hyperbola**. The planets from which Kepler discovered his first law move in closed elliptical orbits, but there are known astronomical bodies that move along parabolic and hyperbolic trajectories as well (Fig. 12–4b).

By postulating a central inverse-square force law, Newton was able to explain not only Kepler's first two laws but Kepler's third law as well. Here we'll see how Kepler's third law follows for a circular orbit of radius R. From above, the force has magnitude k/R^2, and we know from Chapter 3 that the acceleration for an object moving with speed v in a circle of radius R is v^2/R. (We are assuming the Sun is fixed in space, an assumption justified as we'll see later by its large mass.) We can now apply Newton's law of motion $F = ma$ for the case of a planet of mass m, so that F is replaced by k/R^2 and ma is replaced by mv^2/R:

$$\frac{k}{R^2} = \frac{mv^2}{R}. \tag{12–2}$$

In terms of the *period*, T, of the planet's orbital rotation, we have $v = \text{circumference}/T$, so that $T = 2\pi R/v$. Squaring, we have

$$T^2 = \frac{4\pi^2 R^2}{v^2}.$$

If we now substitute $v^2 = kR/mR^2 = k/mR$ from Eq. (12–2), we find

$$T^2 = \frac{4\pi^2 R^2}{k/mR} = \left(\frac{4\pi^2 m}{k}\right)R^3. \tag{12–3}$$

This is just the statement of Kepler's third law as it applies to a circle, with the additional condition that the constant $C = (4\pi^2 m/k)$ in Eq. (12–1) should be the same for all planets, which it will be if k is proportional to the mass of the planet.

The Law of Universal Gravitation

The constant k just described involves the force between two objects; by Newton's third law, there is a symmetry between these objects as they appear in the force law. If k is proportional to one of the masses, it must also be proportional to the other one. Then k must have the form GmM, where m is the mass of one object, M is the mass of the other, and G is a proportionality constant known as the **gravitational constant**. In 1686 Newton put all this together into what is now known as **Newton's law of universal gravitation**: The force of gravitation acting on a point mass of mass m due to another point mass of mass M is an attractive force with an inverse-square form,

$$\vec{F} = -\left(\frac{GmM}{r^2}\right)\hat{r}. \qquad (12\text{–}4)$$

LAW OF UNIVERSAL GRAVITATION

Here $\hat{r}$ is the unit vector in the direction from the mass M to the mass m, and r is the distance separating them. The minus sign indicates that the force is attractive. Of course, Newton's third law states that if Eq. (12–4) is the force exerted on m by M, then its negative is the force exerted on M by m. Mass M is attracted to m with the same magnitude of force by which m is attracted to M.

We can now return briefly to Kepler's third law, specializing to circular orbits. According to Newton, the constant C in Kepler's third law is given by

$$C = \frac{4\pi^2 m}{k} = \frac{4\pi^2 m}{GmM} = \frac{4\pi^2}{GM}. \qquad (12\text{–}5)$$

Kepler's third law, Eq. (12–3), thus takes the form

$$T^2 = \frac{4\pi^2 R^3}{GM}. \qquad (12\text{–}6)$$

This is a remarkable equation. It provides a quantitative realization of Kepler's third law[†], and it shows that its analog with a different value of M applies equally for the motion of satellites and moon around a planet as well. In fact, once we know G, we can use it to obtain the mass of a planet using the period of one of its moons and the moon's distance from the center of the planet.

The Gravitational Constant: The constant G characterizes the strength of the gravitational force and must be determined from experimental data. Its dimensions are $[L^3 M^{-1} T^{-2}]$, and its units in SI are N · m^2/kg^2. The constant G in Eq. (12–4) cannot be measured independently unless the masses of the two objects involved are known; for this reason, G cannot be measured from astronomical objects such as Earth, the Moon, or the Sun. The value of G was first determined by Henry Cavendish in 1798 in an experiment whose basic idea is shown in Fig. 12–5. Two masses m at the end of a rod of

▶ **FIGURE 12–5** (a) In experiments modeled after those of Cavendish, the attraction of the smaller masses to the larger ones twists the fiber from which the rod is suspended. A light shining on a mirror on the rod indicates the small rotation by reflection on a distant screen. Switching the positions of the large masses reverses the rotation. (b) The apparatus used in a Cavendish-type experiment.

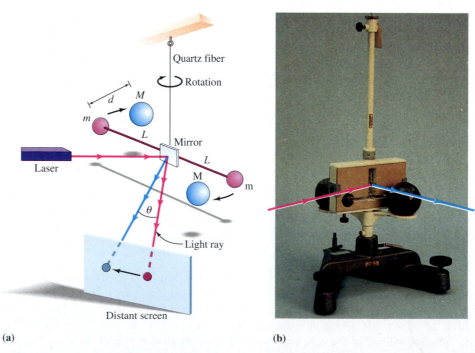

Quartz fiber

Rotation

Laser

Mirror

Light ray

Distant screen

(a)

(b)

[†]Although we have expressed it for a circular orbit, Newton's mathematical techniques were powerful enough to show it holds for elliptical orbits with R being the semimajor axis of the ellipse. Most of the planetary orbits are close enough to circular for us to approximate them that way.

negligible mass and length $2L$ are suspended by a wire or a quartz fiber. Large masses M are placed near the masses m, with their centers of mass separated by a distance d. The magnitude of the torque about the point of suspension due to the force of gravitation on the masses is counterbalanced by a second torque due to the resistance of the fiber to twisting and proportional to the angle of deflection θ. When this resistance is known, the torque due to the gravitational force, and hence G, can be determined. The most recent measurements lead to the value

$$G = 6.673 \times 10^{-11} \text{ N} \cdot \text{m}^2/\text{kg}^2, \qquad (12\text{–}7)$$

with an uncertainty of about 0.06 percent. The magnitude of the gravitational force between two 10-kg masses separated by 10 cm is only about 7×10^{-7} N. We only sense the force of gravity when large masses such as the Earth are involved. The Cavendish experiment is difficult because the gravitational force is weak, making G one of the least accurately measured of all the fundamental physical constants.

EXAMPLE 12–2 Calculate the mass of the Sun, assuming that Earth's orbit around the Sun is circular, with radius $r = 1.5 \times 10^8$ km.

Strategy Kepler's third law in its extended form, Eq. (12–6), relates the given distance of Earth from the Sun, the period of Earth's orbit, and the mass of the Sun. We know for ourselves that the period of the orbit is one year. We can thus directly solve Eq. (12–6) for M_S.

Working It Out Equation (12–6) gives

$$M_S = \frac{4\pi^2 r^3}{GT^2}.$$

Given the value of G, the radius of the orbit, r, and the period, T, which is $365 \text{ days} = (365 \text{ d})(24 \text{ h/d})(60 \text{ min/h})(60 \text{ s/min}) = 3.15 \times 10^7$ s, we find that

$$M_S = \frac{4\pi^2(1.5 \times 10^{11} \text{ m})^3}{(6.67 \times 10^{-11} \text{ N} \cdot \text{m}^2/\text{kg}^2)(3.15 \times 10^7 \text{ s})^2}$$
$$= 2.0 \times 10^{30} \text{ kg}.$$

This is a factor of 3×10^5 larger than the mass of Earth.

What Do You Think? Mars is farther from the Sun than is Earth. Assuming that its orbit is also circular, is Mars's year (a) longer than Earth's, (b) shorter than Earth's, or (c) could it be either, depending on additional information? *Answers to **What Do You Think?** questions are given in the back of the book.*

EXAMPLE 12–3 Calculate how high above Earth's equator a satellite must be to stay above the same point at all times.

Setting It Up Visualization of this problem is particularly helpful. Figure 12–6 shows a satellite in circular orbit above the equator with Earth *turning beneath it* in its daily rotation. With this figure we can understand immediately that a satellite will be stationary above a particular point if its period is the same as Earth's—24 hours.

Strategy We'll start with the assumption, implicit in the problem statement, that the orbit is circular, with radius r to be determined. As described above, we want a period of 24 hours. Given the period, we can find the distance r using the explicit form of Kepler's third law given in Eq. (12–6).

Working It Out We solve Eq. (12–6) for the radius r as a function of period,

$$r^3 = \frac{GM_{\text{Earth}}T^2}{4\pi^2}.$$

Numerically the orbital radius is

$$r = \left[\frac{(6.67 \times 10^{-11} \text{ m}^3/\text{kg} \cdot \text{s}^2)(5.976 \times 10^{24} \text{ kg})(24.0 \times 3600 \text{ s})^2}{4\pi^2} \right]^{1/3}$$
$$= 4.22 \times 10^7 \text{ m} = 26,200 \text{ miles}.$$

This is 22,300 miles above the surface of Earth, or seven times Earth's radius.

What Do You Think? Is it possible to place a satellite that remains permanently above Washington, D.C.?

▲ **FIGURE 12–6** A sketch of a satellite in an orbit whose plane is the plane of the equator. Earth rotates as the satellite orbits.

▲ FIGURE 12–7 The potential energy of the gravitational force is negative and has a $1/r$ dependence. We define it to be zero at $r = \infty$.

The Potential Energy Associated with Newton's Gravitational Force

The gravitational force is central and depends only on the distance of the influenced object from the force center. It is therefore *conservative* and can be derived from a potential energy function (Section 7–1). We know the value of having such a function—the conservation of energy principle is very useful. We can show here that the potential energy of a system of two point masses interacting with each other through the gravitational force, Eq. (12–4), is

$$U(r) = -\frac{GmM}{r}. \qquad (12\text{–}8)$$

GRAVITATIONAL POTENTIAL ENERGY

A graph of this function is shown in Fig. 12–7, where we have used numbers appropriate to the Earth/Moon system, with the point corresponding to the actual lunar orbital radius also shown. As always, we are free to choose the point where the potential energy is zero, and we have chosen the potential energy to be zero at infinity, the point where the force itself is zero.

To derive Eq. (12–8), we recall the definition of potential energy from Chapter 7, Eq. (7–4):

$$U(r) - U(\infty) = -\int_{\infty}^{r} \vec{F}(\vec{r}') \cdot d\vec{r}',$$

where we have chosen the point x_0 in Eq. (7–4) at ∞. The force points from the location of mass m to the origin (the location of M), and as any integration path gives the same result (the force is conservative), we can choose the path to go directly along a radial direction, so that $\hat{r}' \cdot d\vec{r}' = dr'$. We thus obtain

$$U(r) - U(\infty) = -\int_{\infty}^{r} \frac{-GmM}{r'^2}\, dr' = -\frac{GmM}{r'}\bigg|_{\infty}^{r} = -\frac{GmM}{r}.$$

By choosing $U(\infty) = 0$, Eq. (12–8) follows.

The Superposition Principle

Newton's gravitational force, like all other forces, is vectorial and obeys the **principle of superposition**: When several objects of varying masses exert gravitational forces on a point mass m, the total force is the (vectorial) sum of the individual forces. This is equivalent to the fact that potential energies are *additive*: If we have two masses M_1 and M_2, then the potential energy U of m in the presence of the two masses is the *sum* of the potential energies U_1 and U_2. If the source of the gravitational force or potential energy for a point mass is an extended continuous object, we can apply the principle of superposition through integration.

EXAMPLE 12–4 A satellite is to be sent to the position between the Moon and Earth where there is no net gravitational force on an object due to those two bodies. Locate that point.

Setting It Up Figure 12–8 (next page) displays the position of the satellite between Moon and Earth. Using the fact that the Moon is much less massive than Earth, we have drawn the satellite as closer to the Moon than to Earth; a quantitative calculation is needed to confirm this. We denote the known Earth–Moon distance as d and the Earth–satellite distance as x; the Moon–satellite distance is then $(d - x)$. We want the value of x for which the net force on the satellite is zero.

Strategy The forces on the satellite due to the Moon and Earth are respectively $\vec{F}_M$ (to the right) and $\vec{F}_E$ (to the left). Each of these can be expressed as functions of the unknown x. We set the magnitudes of the two forces equal and solve for x.

Working It Out The equality of the force magnitudes reads

$$F_M = F_E$$
$$\frac{GmM_E}{x^2} = \frac{GmM_M}{(d - x)^2}.$$

The satellite mass m cancels, and, on rearranging, we have

$$(d - x)^2 = x^2 \alpha,$$

where the ratio $\alpha = M_M/M_E = 0.0123$ (see Appendix III–1.1). This quadratic equation for x has the following solution with x between 0 and d:

$$x_0 = \frac{1 - \sqrt{\alpha}}{1 - \alpha}\, d = \frac{1 - \sqrt{0.0123}}{1 - 0.0123}\, d = 0.900\, d.$$

Zero net force occurs nine-tenths of the way to the Moon.

What Do You Think? How would you approach the problem of finding out whether the zero point determined in this example is a stable point of equilibrium? (A stable point is one for which the forces are such that a little movement away from the point would push you back to that point.)

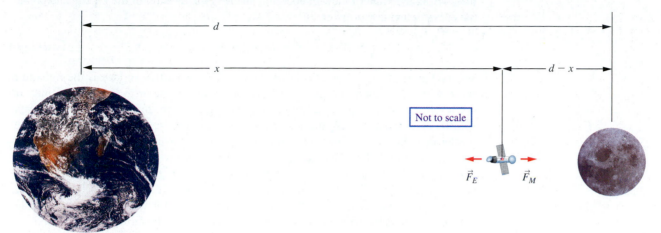

▲ **FIGURE 12–8** The satellite is placed at a position on the line between Earth and the Moon and is affected by gravitational forces from both bodies.

THINK ABOUT THIS . . .

HOW DO WE KNOW EARTH'S MASS?

The force of local gravity is just Newton's law of universal gravitation at a (nearly) fixed distance $r = R_E$. (You may be wondering how this can be, since Earth is not a point object to us. In fact, as we will discuss in detail in Section 12–4, the gravitational force due to a spherically symmetric object—Earth is a good approximation of one—acts as if the mass of the object were concentrated at its center.) Given this fact we can determine Earth's mass, M_E, from the size of the acceleration due to gravity, g, and Earth's radius, R_E. The force exerted on an object of mass m on Earth's surface by Earth's gravitational attraction is given by

$$F = \frac{GmM_E}{R_E^2} = mg.$$

We thus obtain

$$M_E = \frac{gR_E^2}{G} = \frac{(9.80 \text{ m/s}^2)(6.37 \times 10^6 \text{ m})^2}{6.673 \times 10^{-11} \text{ N} \cdot \text{m}^2/\text{kg}^2}$$
$$= 5.96 \times 10^{24} \text{ kg},$$

where we have used the measured values of g, R_E, and G. This mass determination shows that Earth has an average density of 5.5 g/cm³.

12–3 Planets and Satellites

The gravitational force [Eq. (12–4)] determines the trajectories (or orbits) followed by astronomical bodies and other properties of their motion. The orbit of an object subject to the gravitational force due to a single source can be derived analytically from Newton's second law. This is certainly not true for *every* force law! We have already remarked that Newton showed that these orbits were conic sections.

It was in thinking about orbits that Newton was able to cement his revolutionary idea that an apple and the Moon obey the same force law. What is surprising about Newton's idea is that the motion of an apple and the Moon look, superficially, so different. But Newton was able to reconcile this apparent difference with the simple thought experiment shown in Fig. 12–9, taken directly from Newton's work. If you shoot a cannonball horizontally from a mountaintop, it will fall to the ground on a parabolic path, the motion we would expect from an apple. But if you increase the initial speed of the cannonball, it will go farther, eventually going far enough that the surface of the spherical Earth "falls away" beneath it. At a sufficiently high speed, the path of the falling cannonball will never catch up with Earth's surface falling away, and the motion will be a circular orbit. The Moon in its orbit, Newton reasoned, is "falling" the same way as the cannonball. This is precisely how the satellite can continue to orbit the Earth in the chapter opener photo.

In the remainder of this section, we'll describe various aspects of orbital motion.

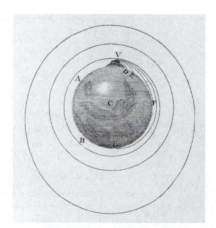

▲ **FIGURE 12–9** Newton's cannon shoots a projectile horizontally from a mountaintop. If the projectile has just the right speed, it will attain a circular orbit at the mountaintop's height. This figure is from Newton's work *A Treatise of the System of the World*, published posthumously.

Escape Speed

It is simplest to approach the orbit of a single object—we refer to such objects as satellites—from the point of view of energy. The energy is the sum of the kinetic and potential energies and is thus given by

$$E = K + U = \frac{1}{2}mv^2 - \frac{GmM}{r}. \quad (12\text{–}9)$$

The **escape speed**, v_{esc}, of a projectile launched from Earth's surface is the minimum speed with which the projectile must leave the surface in order to leave the vicinity of Earth forever—that is, travel an infinite distance from Earth. In other words, the escape speed is the speed at $r = R_E$ that gives zero speed at infinite r. (If the initial speed at R_E is greater than the escape speed, then the projectile moves away with nonzero speed even when it is arbitrarily far.) An object given the escape speed will have zero kinetic energy at $r = \infty$, and the potential energy is defined to be zero at that point, so the total energy will be $E = 0$ at infinity. Because the total energy is conserved, we can set $E = 0$ everywhere; then the condition that $E = 0$ at the surface is the condition that determines v_{esc}. Let the mass of the projectile be m and Earth's mass be M_E. At Earth's surface, $r = R_E$, and

$$E = \frac{1}{2}mv_{\text{esc}}^2 - \frac{GM_Em}{R_E} = 0.$$

If we solve this equation for v_{esc}, the factor m cancels, and we find that

$$v_{\text{esc}} = \sqrt{\frac{2GM_E}{R_E}}. \quad (12\text{–}10)$$

With the known values of G, M_E, and R_E, the escape speed from Earth's surface is $v_{\text{esc}} = 1.12 \times 10^4 \text{ m/s} = 11.2 \text{ km/s}$.

CONCEPTUAL EXAMPLE 12–5 The escape speed is independent of the direction in which the object leaves Earth's surface. Why is this?

Answer The simple response to this question is that Eq. (12–10) contains only radial quantities, not angular ones. The potential energy associated with a conservative central force such as gravity depends *only* on radial distance from the center. The kinetic energy depends only on speed. Thus although the object may leave the surface in different directions, its speed depends on its initial speed and its radial distance. Physically we can understand this result as fol-

lows: When one leaves the surface, the angular aspect appears important—certainly you can tell if your takeoff is vertical or at a 20° angle to the vertical or at any other angle. But when one is very far away, that aspect becomes irrelevant. You look back and, whatever angle you took off at, the distant Earth looks like a point object from which you have gone out in a radial direction.

What Do You Think? Did it require the *Apollo* astronauts a higher escape speed to leave the Earth originally or to leave the Moon on the return trip?

Types of Orbits

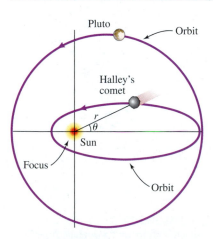

▲ **FIGURE 12–10** For negative total energies, the value of r is limited and the orbit is elliptical (or circular). In this figure we see two elliptical orbits, one not very elongated (that of Pluto) and one quite elongated (that of Halley's comet).

We saw earlier that the orbits, or trajectories, of objects (satellites) under the influence of the gravitational force can take one of four forms: circular, elliptic, hyperbolic, or parabolic (Fig. 12–4). A related way of classifying orbits is through the sign of the total energy. Consider Eq. (12–9) once more:

$$E = K + U = \frac{1}{2}mv^2 - \frac{GmM}{r}.$$

Why are the form of the orbit and the sign of the energy closely related? If the total energy E is positive, then the object's kinetic energy is greater than its potential energy at any radius r, and as r becomes infinite, $|v| > 0$. This corresponds to an orbit that never closes, and it is, in fact, a hyperbolic trajectory. When the energy of the object is exactly zero, we have the special case of a parabolic orbit, also an open trajectory. The object starts at infinity with a speed of zero, swings past the source of the gravitational force (for example, the Sun) in a parabolic orbit, and slows down as it moves away from the source, ending up with zero velocity when r becomes infinitely large. This is exactly how we determined the escape speed. For negative energies the potential energy is always greater than the kinetic energy of the object. Then r cannot become too large, because if it did, then v^2 would have to be negative, which is impossible. In that case the orbit is limited and is elliptical (Fig. 12–10) or circular. As applied to the Sun, the planets all have slightly elongated elliptical orbits; Halley's comet (Fig. 12–11), like many other comets, follows a very elongated elliptical orbit. The comet's potential energy is greatest at the farthest point from the Sun, and its kinetic energy least at this point. Conversely, at the nearest point to the Sun the comet will have its greatest velocity and kinetic energy and its least (most negative) potential energy. Circular orbits are the simplest of all, as Example 12-6 illustrates. Only the direction of the motion changes while the potential energy and kinetic energy remain the same throughout. In summary, when $E < 0$, the satellite has a bound orbit, and when $E \geq 0$, the object follows an unbound orbit.

▲ **FIGURE 12–11** Halley's comet, shown here in a recent passage near Earth, has an elongated elliptical orbit with a period of 76 yr.

EXAMPLE 12–6 Imagine that a small moon of Jupiter, of mass m, moves in a circular orbit of radius r under the influence of the gravitational force due to Jupiter, mass $M \gg m$ (Fig. 12–12). Calculate the total energy of the moon as a function of r.

Strategy Equation (12–9) gives the total energy, but it is a function of both v and r. To eliminate the speed, we recall that the acceleration (which is centripetal) has magnitude v^2/r for a circular orbit and that the force (also centripetal) has magnitude GmM/r^2. We can then find v as a function of r by equating the force on m with m multiplied by the acceleration. Once we know the speed as a function of radius, we can calculate the energy as a function of radius.

Working It Out The second law for the moon is

$$\frac{GmM}{r^2} = \frac{mv^2}{r}.$$

(All directions are centripetal.) We solve for the speed squared,

$$v^2 = \frac{GM}{r}.$$

We can then use this in the expression for the energy, Eq. (12–9),

$$E = \frac{1}{2}mv^2 - \frac{GmM}{r} = \frac{1}{2}m\frac{GM}{r} - \frac{GmM}{r} = -\frac{1}{2}\frac{GmM}{r}.$$

The total energy of the orbiting moon is just one half the potential energy for a circular orbit. It is negative, as is appropriate for a closed orbit. Nothing we did depends on the fact that this is a Jovian moon, and it is generally true that for a circular orbit the total energy is one half the potential energy.

What Do You Think? Which parameter best characterizes the difference between bound and unbound orbits? (a) speed, (b) kinetic energy, (c) potential energy, (d) total energy.

▲ **FIGURE 12–12** Here Europa is a distance r from Jupiter's center. The innermost moon, Io, can be seen by its bright, brown-yellow surface against Jupiter's background. This photo was taken by Voyager 1 in 1979 as it passed Jupiter on its way out of our solar system.

EXAMPLE 12–7 A satellite moves in a circular orbit around Earth, taking 90.0 min to complete 1 revolution. We are given the following information: Earth's radius is $R_E = 6.37 \times 10^6$ m; the distance from the Moon to Earth is 3.84×10^8 m; the Moon's orbit is circular; the period of the Moon's rotation about Earth is 27.32 d; and Earth's gravitational force acts as if all of Earth's mass were concentrated at its center. With this information, calculate the height of the satellite above Earth.

Setting It Up We label the known periods of Moon and satellite as T_M and T_S, respectively. The Moon's orbit radius is R_{ME}. All these are given in the problem statement. Let the satellite orbit radius be R_S; it determines the height of the satellite above Earth through $h = R_S - R_E$.

Strategy Once again, Kepler's third law comes into play, but this time applied to Earth as the force center. Both the Moon and a satellite in circular orbit around Earth behave like planets in circular orbit around the Sun. Here the simpler form of Kepler's third law is useful [Eq. (12–1)], that the cube of the radius of the orbit is proportional to the square of the orbital period. In other words, the ratio $(\text{radius})^3/(\text{period})^2$ is the same for both the Moon and the satellite.

Working It Out The ratio equality reads

$$\frac{R_S^3}{T_S^2} = \frac{R_{ME}^3}{T_M^2}.$$

This equation can be solved for R_S,

$$R_S = R_{ME}\left(\frac{T_S}{T_M}\right)^{2/3}.$$

The period of the satellite must be in the same units as that of the Moon. We thus have $T_S = 90.0$ min $= 1.5$ h $= (1.5$ h$)/(1$ d$/24$ h$) = 0.0625$ d. This gives us the numerical result

$$R_S = (3.84 \times 10^8 \text{ m})\left(\frac{0.0625}{27.32}\right)^{2/3} = 6.67 \times 10^6 \text{ m}.$$

The height is then

$$h = R_S - R_E = (6.67 - 6.37) \times 10^6 \text{ m}$$
$$= 3 \times 10^5 \text{ m} = 300 \text{ km}.$$

Many satellites use this orbit, which is one that lies just above Earth's atmosphere. Small air drag makes the orbit long-lasting, and it takes less energy to put a satellite in a lower than in a higher orbit.

What Do You Think? When the radius increases, so does the period. Does the speed also increase?

Properties of Noncircular Orbits: The hyperbolic, parabolic, and elliptical orbits corresponding to different signs of total energy are closely related; as stated in Section 12–2, they are all *conic sections* (see Fig. 12–4). Let's focus our attention on the solar system, although we should keep in mind that our results will be applicable to, say, satellites orbiting Earth as well. The Sun, which is very heavy in comparison to the planets, plays a special role in the orbits. For closed orbits in particular, the Sun sits at one *focus* of the ellipse, a fact noted by Kepler in his first law and illustrated in Fig. 12–3a. The *semimajor axis* in Fig. 12–3a is denoted as a, and the *semiminor axis* as b. The equation describing an ellipse is

$$\frac{x^2}{a^2} + \frac{y^2}{b^2} = 1. \tag{12–11}$$

The point where the planet makes its closest approach to the Sun ($r = r_{min}$) is called the *perihelion*; the point where the farthest distance between a planet and the Sun is attained ($r = r_{max}$) is the *aphelion*. (For Earth satellites the corresponding words are perigee and apogee.) From Fig. 12–3a, we have

$$\text{semimajor axis} = a = \tfrac{1}{2}(r_{min} + r_{max}). \tag{12–12}$$

The *eccentricity e* of the orbit is a dimensionless measure of the elongation of the orbit and is proportional to the difference between r_{max} and r_{min}:

$$e = \frac{r_{max} - r_{min}}{2a}. \tag{12–13}$$

Note that when $r_{max} = r_{min}$, the orbit reduces to a circular one of radius a and the eccentricity is zero. The most extreme possible orbit corresponds to $r_{min} = 0$, when $r_{max} = 2a$, and

$$e_{extreme} = \frac{2a}{2a} = 1.$$

The eccentricity therefore varies between zero and one. The role of these quantities in the shape of the possible elliptical orbits is detailed in Fig. 12–13, which also includes the possible non-elliptical orbits.

▶ **FIGURE 12–13** The set of possible orbits for objects whose distance of closest approach to the Sun are the same. Two ellipses are included, and the more elongated of these has the larger value of eccentricity.

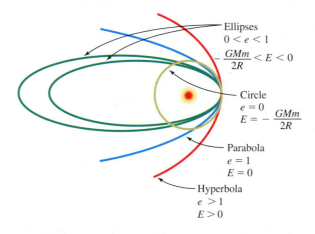

Ellipses
$0 < e < 1$
$-\dfrac{GMm}{2R} < E < 0$

Circle
$e = 0$
$E = -\dfrac{GMm}{2R}$

Parabola
$e = 1$
$E = 0$

Hyperbola
$e > 1$
$E > 0$

EXAMPLE 12–8 Consider the motion of a comet in an elliptical orbit around a star. The eccentricity of the orbit is given by 0.80, and the distance between the perihelion and the aphelion is 1.0×10^{10} km. (a) Find the distances of nearest and farthest approaches of the comet. (b) If the speed of the comet is 81 km/s at perihelion, what is its speed at aphelion?

Setting It Up We show the motion of the comet about the star in Fig. 12–14 (next page). We are given the eccentricity, which we label by e, as well as the sum of the aphelion and perihelion distances, which we label by $2a$. For part (a) we are asked to find the distances r_{max} and r_{min} labeled in the figure.

Strategy Part (a) is simply a matter of solving equations that relate the unknowns, r_{max} and r_{min}, to the knowns, $2a$ and e. Equations (12–12) and (12–13) apply; we solve them for r_{max} and r_{min} in terms of a and e. For part (b) we note that at both perihelion and aphelion the orbit is at right angles to the vector $\vec{r}$ from the star to the comet. The angular momentum is therefore just the product of r with mv, where v is the speed. Because the angular momentum is conserved in motion under the influence of gravitation, we must have

$$mv_{perihelion}r_{min} = mv_{aphelion}r_{max}. \tag{12–14}$$

Since after part (a) everything in this equation except $v_{aphelion}$ is known, we can solve for it.

Working It Out (a) If we add Eq. (12–12) to a times Eq. (12–13), r_{min} cancels, and

$$r_{max} = a + ae = a(1 + e). \qquad (12\text{–}15)$$

In turn,

$$r_{min} = 2\left(a - \frac{r_{max}}{2}\right) = 2\left[a - \frac{a(1 + e)}{2}\right] = a(1 - e). \quad (12\text{–}16)$$

Numerically, with $e = 0.80$ and $2a = 1.0 \times 10^{10}$ km,

$$r_{max} = \tfrac{1}{2}(1.0 \times 10^{10} \text{ km})(1.80) = 9.0 \times 10^{9} \text{ km};$$
$$r_{min} = \tfrac{1}{2}(1.0 \times 10^{10} \text{ km})(0.20) = 1.0 \times 10^{9} \text{ km}.$$

(b) We solve Eq. (12–14) to find the speed at aphelion:

$$v_{aphelion} = v_{perihelion}\frac{r_{min}}{r_{max}} = (81 \text{ km/s})\left(\frac{1.0 \times 10^{9} \text{ km}}{9.0 \times 10^{9} \text{ km}}\right)$$

$$= 9.0 \text{ km/s}.$$

We used the conservation of angular momentum to relate the speed at perihelion to the speed at aphelion in this example. The use of energy conservation would also allow us to find the star's mass (see Problem 38).

What Do You Think? A comet comes in toward the Sun, comes very close to the Sun's surface, and moves out again. Is this enough information to deduce that the comet is in an elliptical orbit?

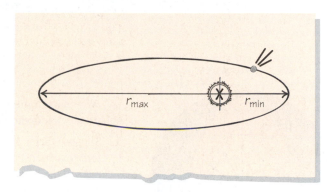

▲ **FIGURE 12–14** A comet's orbit, with the minimum and maximum distances from the star (perihelion and aphelion, respectively).

12–4 Gravitation and Extended Objects

We have viewed the planets and the Sun on the scale of the entire solar system and have treated them as point masses. But they are not really pointlike. This problem is especially evident if we think about how the gravitational force from Earth acts on an apple. To an apple, Earth most certainly is an extended object, yet we treated the motion of the apple as if all the mass of Earth were concentrated at Earth's center. We can study how extended objects behave by using the principle of superposition (Sec. 12–3). Generally in this section we consider spherical objects, the (approximate) form of many large astronomical objects.

The Gravitational Force Due to a Spherically Symmetric Object

We claimed in Section 12–2 that when an extended object is spherically symmetric, the force it exerts on a point mass outside the extended object is the same as the force that would be exerted if the entire mass of the extended object were concentrated at its center. It is this assertion that allowed us to relate the acceleration due to gravity at Earth's surface, g, to Earth's mass and radius, and thereby weigh Earth itself (see the feature Think About This ... How do we know Earth's mass?). In fact, for any spherically symmetric object[†] of mass M and radius R, the acceleration due to gravity at its surface is

$$g = \frac{GM}{R^2}. \qquad (12\text{–}17)$$

Thus we would expect the acceleration of gravity to be different at the surfaces of the different planetary objects, such as the Moon.

Many mass distributions are spherically symmetric, or nearly so. For such systems the mass density depends only on the distance from the center of the distribution. The mass density at the center could vary with radial distance, but it must not depend on angle. Earth conforms well to these requirements (Fig. 12–15). Earth's core is denser than the outer layers—it most likely consists primarily of iron. (The small deviations from spherical symmetry come from the fact that, because of its rotation, Earth bulges a little in the equatorial region; there is an additional distortion that makes the planet look a little pear-shaped. In addition, there are local regions of slightly greater or lesser density, due for example to the presence of ore bodies.)

Inner core

▲ **FIGURE 12–15** A view of Earth's internal structure. The inner core is substantially denser than the rest.

[†]This means its density varies only with distance from the center.

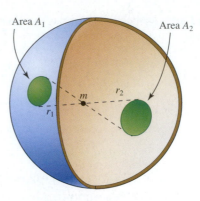

▶ **FIGURE 12–16** The gravitational force on an object of mass m inside a spherical shell. Regions from opposite sides of the shell exert equal and opposite forces on the object, and the result is that there is no net force on the mass m.

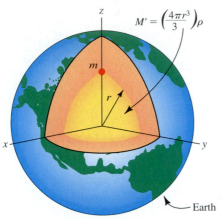

▲ **FIGURE 12–17** The gravitational force on an object with mass m located inside Earth depends on the total amount of mass M' inside the sphere whose density is ρ and whose radius r is the distance of mass m from the center.

We can summarize the principal conclusions about the gravitational force due to spherically symmetric systems as follows.

1. Suppose that a point mass is *outside* a spherically symmetric object. The gravitational force experienced by the point mass is identical to the force that would arise if the whole mass of the spherical object were concentrated at its center. In other words, the force exerted on the point mass is the same as the force that would be exerted on it if the uniform spherical object (total mass M) were a point mass M located at the center of the sphere.

2. Suppose that a point mass is somewhere *inside* a thin spherical shell of constant density (Fig. 12–16). Then there is *no* gravitational force on the mass. This conclusion holds for a point mass inside an arbitrarily thick shell as well, as long as the mass density of the shell depends *only* on the distance from the geometric center of the shell. We'll describe the reasoning behind this surprising result later.

These two important facts show that the gravitational force on a point mass m within Earth at a distance r from the center, for example, would be due to a mass M' concentrated at Earth's center; where M' is the mass contained within Earth only up to the radius r (Fig. 12–17).

EXAMPLE 12–9 Suppose that a tunnel is drilled through our planet along a diameter. Assume that Earth's mass density is uniform and is given by ρ. Describe the force on a point mass m dropped into the hole as a function of the distance of the mass from the center.

Setting It Up As in Fig. 12–18a, the tunnel passes through Earth's center. In addition to the quantities that are directly given, we can assume we know either Earth's mass M_E or its radius R_E (given one, we can find the other using ρ).

Strategy The gravitational force on the point mass m is due only to the mass $M(r)$ of the material contained within a radius r, where r is the distance from the point mass m to Earth's center. The force is attractive, toward the center. We must calculate $M(r)$ in order to find the force, and that mass is the density ρ times the volume of the sphere of radius r.

Working It Out The magnitude of the force at a radius $r \leq R_E$ is

$$F(r) = -\frac{GmM(r)}{r^2}.$$

Since the volume of a sphere of radius r is $(4/3)\pi r^3$, we have

$$M(r) = \left(\frac{4\pi r^3}{3}\right)\rho.$$

Thus, as is sketched in Fig. 12–18b,

$$F(r) = -\frac{Gm}{r^2}\left(\frac{4\pi\rho r^3}{3}\right) = -\left(\frac{4\pi Gm\rho}{3}\right)r \quad \text{for} \quad r \leq R_E.$$

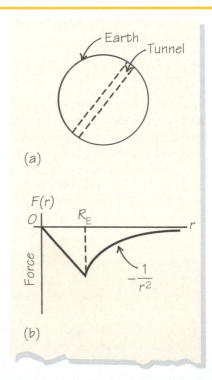

(a)

(b)

▲ **FIGURE 12–18** (a) A tunnel passes through Earth's center. (b) A graph of the variation of the force with r shows that the force of gravity increases linearly from Earth's center out to the surface (assuming constant density), and dies away with the inverse square of distance from Earth's center at distances from the center greater than Earth's radius.

As we shall describe in Chapter 13, the fact that $F \propto r$ means that inside Earth the point mass acts as if it were moving under the influence of a spring, or as a pendulum. This motion is oscillatory, and the point mass moves from one end of the tunnel to the other and back.

What Do You Think? Suppose Earth's density were not uniform but rather more mass was concentrated at the center, which is in fact the case. The force on the object as it moved through the tunnel would (a) be unchanged; (b) vary linearly with r but with a different coefficient; (c) fall off more rapidly than r as $r \rightarrow 0$; (d) fall off less rapidly than r as $r \rightarrow 0$, and perhaps even grow.

The two summarizing conclusions about the force of gravity for spherically symmetric objects will not be derived mathematically until Chapter 23, where we give a simple but indirect proof. Direct proof would follow from integrating the contributions of the force from different pieces of our extended object, but it is an unnecessary diversion at this point. However, we can gain a physical picture of why a spherically symmetric shell that surrounds a point mass exerts no force on the point mass.[†]

Figure 12–16 shows the point mass placed off center within a thin spherical shell. Consider a double cone making an opening angle θ at the point mass and, in particular, the gravitational effects of the two circular sections that the cone cuts on the shell. The point mass is a distance r_1 from each point in the area on the left, A_1, and a different distance r_2 from each point in the area on the right, A_2. If the density of the shell is ρ and its thickness is τ, then the force F_1 that attracts the point mass to the left-hand area is

$$F_1 = \frac{Gm}{r_1^2} \rho \tau A_1,$$

and the force F_2 that attracts the point mass to the right-hand side is

$$F_2 = \frac{Gm}{r_2^2} \rho \tau A_2.$$

Now we note that areas A_1 and A_2 are proportional to r_1^2 and r_2^2, respectively, so that

$$\frac{A_1}{r_1^2} = \frac{A_2}{r_2^2}.$$

This means that the two forces are independent of the distance of the point mass to the circular sections and the forces cancel because they pull in opposite directions. In effect, as the point mass moves closer to one side, the inverse-distance-squared factor in the force increases, but the amount of mass seen decreases by the same factor; thus, the pull of every sector of the shell is the same and cancels.

This curious effect is very special to the inverse-square law—a law that also holds for electric charges. Indeed, it was Benjamin Franklin who noticed that there is no force on an electric charge surrounded by a shell of the opposite charge, and it was on the basis of this observation that Joseph Priestley in the eighteenth century first suggested that electric forces obey the inverse-square law. We will consider electrical charge further in Chapter 21.

Dark Matter

Astronomers have inferred the presence of what is known as *dark matter* by using the properties we described above. Dark matter gets its name from the fact that it emits no radiation and hence cannot be observed with telescopes. The discovery of dark matter has had important implications for our understanding of the overall structure of the universe. To understand how dark matter is revealed, consider a mass m outside of a spherically symmetric mass distribution of total mass M and moving in circular motion at a distance r from the center of the distribution. For that motion, the acceleration is centripetal, with magnitude v^2/r, and because of our result that the net gravitational force from the distribution is the same as if it were all concentrated at the center, Newton's second law reads

$$\frac{mv^2}{r} = \frac{GmM}{r^2}$$

[†] Strictly speaking, our arguments hold only for small values of θ. But generalization is possible.

or

$$v \propto \frac{1}{\sqrt{r}}.$$

This result is now tested for clouds of visible ionized hydrogen that circulate around galaxies, outside the visible ("shining") distribution of matter—glowing stars and visible dust—in those galaxies. If the only matter making up the galaxies were the visible matter, we would expect the measured speed of the hydrogen clouds to drop off with distance from the galaxy as $1/r^{1/2}$. But instead the speeds of the hydrogen clouds orbiting about the galaxy remain constant with distance from the center of the galaxy. This only makes sense if there is matter that goes beyond the visible boundaries of the galaxy and which influences the orbit of the hydrogen clouds. More precisely, suppose that the *total* (not just the visible) galactic mass enclosed in a sphere of radius r is $M(r)$, and assume that the mass is symmetrically distributed. Then the orbital motion of a cloud at a distance r from the center of the sphere is governed by that mass,

$$\frac{mv^2}{r} = \frac{GmM(r)}{r^2},$$

so that

$$v \propto \sqrt{\frac{M(r)}{r}}.$$

If the speed v is a constant for large r, as the observation of the hydrogen clouds indicates, we can conclude that $M(r)$, grows linearly with r. But any material at the distance of the glowing hydrogen clouds themselves, which are outside the radius of the visible galaxy, is not luminous—it does not shine. Although the details are somewhat different, studies of clusters of galaxies also lead to the conclusion that dark matter must be present to explain the motions of galaxies within a cluster. Altogether the various pieces of orbital evidence tell us that the mass of shining matter is roughly a factor of 10 smaller than the mass of dark matter. What exactly is this dark matter? No one knows at this point, and this question holds center stage in much current research.

How g Varies with Altitude

Equation (12–17), together with the result that a symmetric, spherical Earth behaves like a centered point mass, implies that g varies with altitude. Suppose that we measure altitude, h, from sea level, and that Earth's radius at sea level is R_E. From Eq. (12–17) we find

$$g(h) = \frac{GM}{(R_E + h)^2} = \frac{GM}{R_E^2[1 + h/R_E]^2}. \tag{12–18}$$

Now, if the ratio $h/R_E \ll 1$ (and even for the top of Mount Everest, $h/R_E \cong 1.5 \times 10^{-3}$), then we can use the approximation $1/(1 + x)^2 \cong 1 - 2x$ (for $x \ll 1$; see Appendix IV–10), so $g(h) \cong (GM/R_E^2)[1 - 2h/R_E)]$, or

$$\frac{g(h)}{g(0)} \cong \frac{(GM/R_E^2)[1 - (2h/R_E)]}{(GM/R_E^2)} = 1 - \frac{2h}{R_E}. \tag{12–19}$$

To get a feeling for this effect, we can see from Eq. (12–19) that, at the top of the Sears Tower in Chicago, with $h = 443$ m, g is 99.99 percent of its value at sea level; even at the top of Mount Everest, with $h = 8848$ m, g is 99.74 percent of its value at sea level (Fig. 12–19).

Other effects modify Eq. (12–17). Earth is not a perfectly uniform sphere. Not only is it not spherical but it contains lumps of higher or lower density that affect g. Measurements of g are routinely made to many significant figures, more than enough to reveal all the effects we have discussed as well as one associated with Earth's rotation (Conceptual Example 12–10).

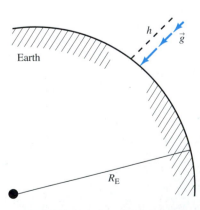

▲ **FIGURE 12–19** As we move away from the surface, the value of g decreases.

CONCEPTUAL EXAMPLE 12–10 Earth's rotation has an effect on g. Which of the following statements are true, if any? (a) The effect is largest at the equator, where g increases over its zero-rotation value; (b) The effect is largest at the poles, where g decreases; (c) The effect is largest at the equator, where g decreases; (d) The effect is largest at the poles, where g increases.

Answer (c) The effect is due to the fact that a rotating frame is an accelerating frame, with fictitious forces present (see Chapter 5).

In other words, there is a "centrifugal force" that makes it appear to the observer in the rotating frame that he or she is being pushed outward. If the frame rotates with angular velocity ω, then the acceleration that produces the effect is proportional to $\omega^2 r$. Thus the effect is absent at the poles, where the r factor is 0 and is largest at the equator, for which $r = R_E$. Since the "force" is outward, it tends to cancel the acceleration of gravity—g is decreased. Quantitatively, it is not hard to calculate, using the $\omega^2 r$ value, that at the equator g is 99.57 percent of what it would be if Earth did not rotate.

(a) (b)

◀ **FIGURE 12–20** Alma beach in the Fundy National Park, New Brunswick, Canada, at (a) high and (b) low tide. The Bay of Fundy has some of the largest tidal differences in the world.

Tidal Forces

Why do the oceans on Earth experience two tides each day? To answer this, we need to consider how gravity acts on an extended object. If the object it acts on is large, the strength of the gravitational force will be different on different parts of the object, and when different parts of the object experience different forces, they will accelerate differently. This fact can cause the object to distort in shape or even fall apart if there are no compensating internal forces that hold the object together. We say that such distortions are due to **tidal forces**, a term we apply to the *difference* of gravitational forces across an extended object. Newton, who first correctly developed the idea of tidal forces, showed that just as the Moon attracts the water nearest it more strongly than Earth as a whole, making a bulge toward the Moon, so the Moon also attracts Earth as a whole more strongly than the water on the *far* side of Earth, leaving behind a second bulge on the far side. The result is two high tides per day (Fig. 12–20). We can make this argument a little more quantitative using the same reasoning that gave us Eq. (12–19).

While we are going to apply tidal forces to understand ocean tides on Earth, there are other astronomical situations where tidal forces play an important role. A revealing if fanciful example occurs if we apply Eq. (12–18), the variation of g with height, to an object of height $h = 2$ m moving on a grazing orbit under the influence of a *neutron star*, an object of very high surface gravity. We take realistic numbers for the star, $M = 2 \times 10^{30}$ kg (one solar mass) and $R = 10$ km $= 10^4$ m. Then the difference between g at the top of the object farthest from the star and g at the bottom, 2 m nearer to the star, is

$$\frac{GM}{R^2[1 + (h/R)]^2} - \frac{GM}{R^2} \cong -\frac{GMh}{R^3} \cong 27 \times 10^7 \text{ m/s}^2 = 2.7 \times 10^7 \, g_{\text{Earth}}.$$

No object that we know of could remain intact under this sort of differential acceleration. The tidal forces are just too strong.

Let us turn now to ocean tides on Earth. These are correlated, though imperfectly, with the position of the Moon, and we'll concentrate on the Moon's role. (The Sun is important too, although we'll ignore it at first.) We begin by finding the tidal forces on Earth due to gravitational attraction to the Moon (Fig. 12–21). Let the distance from the Moon, mass M_M, to Earth's center O be D and Earth's radius be denoted by R_E. Then the acceleration experienced by a piece of matter at Earth's center is in the direction of the Moon, with magnitude

$$a_O = \frac{GM_M}{D^2}.$$

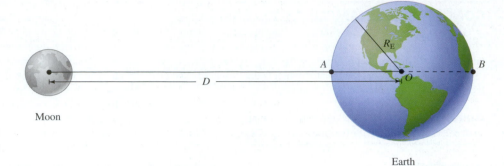

FIGURE 12–21 A sketch helpful in calculating tidal forces on Earth due to the Moon.

The acceleration experienced by a mass at point A on Earth's surface (Fig. 12–21) is in the same direction but has magnitude

$$a_A = \frac{GM_M}{(D - R_E)^2} = \frac{GM_M}{D^2\left(1 - \frac{R_E}{D}\right)^2}.$$

We can now use the fact that $R_E \ll D$, and approximate this just as we did for Eq. (12–19):

$$a_A = \frac{GM_M}{D^2}\frac{1}{\left(1 - \frac{R_E}{D}\right)^2} \cong \frac{GM_M}{D^2}\left(1 + \frac{2R}{D}\right) = \frac{GM_M}{D^2} + 2\frac{GM_M R_E}{D^3}.$$

The first term on the right is just a_O, and the second term shows that a_A is a little larger in magnitude than the acceleration of a point at Earth's center—the Moon is a little closer and has a little larger effect. We can interpret the second term as the acceleration *relative to Earth's center* of the piece of matter closest to the Moon; it is in a direction toward the Moon. In other words, the piece of matter at A experiences a differentially larger force that tends to move it toward the Moon, away from Earth's center. Similarly, the acceleration experienced by a mass bit at point B has magnitude

$$a_B = \frac{GM_M}{(D + R_E)^2} \cong \frac{GM_M}{D^2} - 2\frac{GM_M R_E}{D^3}.$$

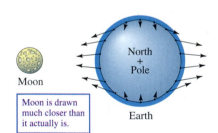

North + Pole

Moon

Earth

Moon is drawn much closer than it actually is.

(a)

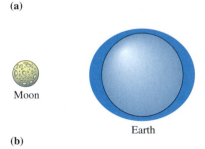

Moon

Earth

(b)

FIGURE 12–22 (a) The arrows indicate how the forces from the Moon at Earth's surface differ from the average value of the force on Earth. The arrows along the plane perpendicular to the line connecting Earth and the Moon should be much smaller than are drawn here; the tidal forces there are about 1/60 of the tidal forces along the line from Earth to the Moon. (b) The resulting water distribution has two bulges.

A mass bit at point B, the point farthest from the Moon, will have a smaller acceleration than that of Earth toward the moon, that is, it will accelerate *away* from Earth's center and from the Moon with an acceleration relative to Earth's center of magnitude $2(GMR/D^3)$. This acceleration is interpreted by someone measuring things with respect to Earth's center as due to a force tending to move the mass bit away from the Moon. Of course, in applying this to oceans on Earth, we should make the same calculation for every spot on Earth. We could make a map of the differential forces—the tidal forces—and we would find that tidal forces occur everywhere on Earth's surface (Fig. 12–22a), and they all tend to reinforce a piling up of water at points A and B (Fig. 12–22b).

What is the numerical size of the corrections? The accelerations due to the tidal forces are of order $2R/D$ of the Earth's acceleration due to the Moon, and this is approximately a factor 1/30. The Sun also contributes to tidal forces. The Sun is about 2.7×10^7 times more massive than the Moon, but its distance from Earth is also a factor 3.9×10^2 larger. Since the correction to the acceleration is proportional to M/D^3, the tidal effects due to the Sun are $(2.7 \times 10^7)/(3.9 \times 10^2)^3 \cong 0.45$ times smaller than those due to the Moon.[†] This is still significant, and it shows up most clearly when

---

[†]The actual gravitational force from the Sun is almost 200 times that from the Moon—it is the tidal forces due to the Sun that are less than those due to the Moon.

the Sun, Moon, and Earth are aligned, when the tides are about 50% higher than average (spring tides), and when the line between Earth and the Sun is perpendicular to the line between Earth and the Moon, when the tides are about 50% lower than average (neap tides).

If water could flow arbitrarily easily, the high tides would always be directly under the Moon. But water is viscous, and a kind of friction acts within it to oppose its flow. For this reason the high tides lag the Moon's position. The friction has the additional effect of slowing down the rate of Earth's rotation. Since angular momentum is conserved (there are no external torques), the Moon must accordingly move away from Earth. This increase in the Moon's orbital distance from Earth is about 0.5 m per year. Lunar ranging—the measurement of the distance between the Moon and Earth by means of reflection of laser light from reflectors placed on the Moon combined with extremely high precision time measurements—confirms detailed calculations of the effect.

Ultimately Earth will slow its rotation until one side always faces the Moon, much like the Moon presents only one of its faces to Earth. At that point there will be no more tides and no further slowing. One can show that this will occur when the Moon has receded to a distance 1.44 times its present distance to Earth, a very long time at a rate of 0.5 m per year. One could then use Kepler's third law to estimate the length of the lunar month. Because Earth will always be presenting the same face to the Moon, this will also be the length of the day!

THINK ABOUT THIS . . .

WHAT DETERMINES HOW BIG MOUNTAINS CAN GET?

Mount Everest and the other very tall mountains on Earth are about 10 km high. Is this an accident? Why shouldn't the tallest mountains be 100 km, or 1 km high? In an entertaining demonstration of how to pull in knowledge from many areas of physics, the late Victor Weisskopf, who was a very talented teacher, argued that 10 km is no accident. Although we can't explain every step of the argument at this point, it is nonetheless well worth thinking about. Tectonic forces—the movements of great crustal plates at Earth's surface—cause uplift of parts of the crust. Think then about a block of matter of height h. If this block gets too high, then its mass will be so large that it will cause plastic deformation of the base: The base will melt and run out, and the block will settle back down. In effect, the melting of a layer of thickness d is equivalent to moving a layer of thickness d from the top of the mountain to the bottom, and we know the energy involved is proportional to gh. In order to convert this to a maximum height, we must realize that if this number is big enough, it will cause the rock to melt. We have to know how much energy is involved in raising the rock temperature to the melting point, about 2000°C, then melting it, and this can be obtained from a knowledge of its atomic structure or looked up in tables. When these numbers are brought in, the maximum height h on Earth is 14 km, a number very close to the measured values.

If Mars is made of the same basic material as on Earth, then this argument allows us to estimate the maximum height of Mars mountains: $g_{Mars} h_{max\ Mars}$ must be the same as $g_{Earth} h_{max\ Earth}$, and we need only find g_{Mars}. The result of this exercise (see Problem 56) is a maximum height about 30 km, and this also corresponds very well with the observed highest mountain, about 25 km.

12–5 A Closer Look at Gravitation

When we have applied the law of gravitation, it has always been in the context of the attraction exerted on one object by another with a much larger mass object. We have considered planets or comets moving around the Sun, tennis balls falling on Earth's surface, and so on. We assumed that the massive object was at rest and that the orbits being studied (for example, the ellipses) were those of the light object. We know that just as Earth exerts a force on the Moon, the Moon exerts a force on Earth. The Moon moves in an orbit

▲ FIGURE 12–23 In a double star system with stars of equal mass, the center of mass is midway between the stars. Each star orbits around this point.

▲ FIGURE 12–24 An orbit that does not close on itself. The orbit is said to precess.

around Earth, but does Earth move around the Moon? The answer to this question is yes. When two objects move under the influence of the gravitational forces they mutually exert on each other, they move about their common center of mass. This is a property we first observed in the context of momentum. If we regard the two objects as an isolated system, then there is no external force on that system, and its center of mass remains unaccelerated. When one of the two objects is much heavier than the other, the center of mass is closer to the heavier object and may even be within that object. In the Earth–Sun system, the Sun's motion is a very small orbit about a point very close to the center of (and well within) the Sun. Cases in which the center of mass is well separated from the two objects in question occur in double star systems with stars of comparable masses. These stars each move in an orbit around a point between them (Fig. 12–23).

Effects of Other Objects

The superposition principle tells us that the orbit of any planet is affected not only by the Sun but also by the presence of all other planets—although to a much smaller extent. The largest effects are due to the most massive planets, Jupiter and Saturn. We can write the potential energy of a planet of mass m in the form

$$U = -Gm\left(\frac{M_S}{r_S} + \frac{M_1}{r_1} + \frac{M_2}{r_2} + \cdots\right), \tag{12–20}$$

where the terms in parentheses represent the contribution of the Sun and those of the other planets. Thus the net force on a planet is no longer a pure $1/r^2$ force directed exactly at the Sun; there are small corrections to it. As a consequence, the orbits are no longer exact ellipses that close on themselves, but instead, the orbits *precess* (Fig. 12–24), which means that the perihelion gradually moves, and the orbit never actually repeats. Almost all the major figures in nineteenth-century mathematics worked on the problem of computing orbits subject to perturbations from additional masses; in fact, accurate orbits were computed even before the advent of large-scale computing machines. Two interesting historical events are worth mentioning in this connection. First, the calculation of the orbit of Uranus, with the inclusion of all the perturbations, did not fit the observed orbit. In 1845 both John Adams (an undergraduate at Cambridge University) and Urbain Le Verrier in France calculated the potential effects of a hypothetical new planet and published their results. Adams's work was ignored, whereas Le Verrier was more successful in mounting a search for the new planet, which culminated in the discovery of the planet Neptune in 1846. Second, the calculation of the precession of the perihelion of the planet Mercury was also carried out to great accuracy by Le Verrier. The result of the comparison of observation and theory left a discrepancy in the rate of precession of the perihelion that amounted to only 43″ of arc per century (out of an observed total of some 5600″ of arc per century). Both theory and experiment were so good that there was no doubt of the existence of this discrepancy, but explanation of the discrepancy had to await Einstein's theory of gravitation.

Equality of Inertial and Gravitational Masses

The parameter m in $\vec{F} = m\vec{a}$ describes a property of an object that is properly called the *inertial mass* [see Eq. (5–1)]. It is a constant that characterizes the object, and it appears as a coefficient of the acceleration in response to *any* force. The parameter m that appears in the expression for the gravitational force that is exerted on that object is the *gravitational mass*, and there is no a priori reason why the inertial and gravitational masses should be equal.

The equality of inertial and gravitational masses to one part in 10^{11} has been demonstrated by Robert Dicke and Vladimir Borisovitch Braginsky. Newton had already measured the equality of these quantities to an accuracy of one part in 10^3, and Loránd von Eötvös carried out measurements in the period from 1890 to 1922 to an accuracy of one part in 10^9. The equality of inertial and gravitational masses made a great impression on Einstein and led toward his formulation of the equivalence principle, a cornerstone of the general theory of relativity (see Section 12–6).

Is Newton's Law of Gravitation Right over All Distances?

We know to a good degree of accuracy that interactions between celestial objects obey an inverse-square law, but these phenomena occur on scales ranging from hundreds of kilometers to much greater distances. The Cavendish experiment and its modern versions verify the inverse-square law at distances ranging from less than centimeters to meters. But there remain large gaps in the range of distances for which the law has been well tested, and these are all the targets of current experiment. Such experiments are often motivated by theoretical thinking, and there are some current ideas that suggest the law may fail in some as-yet untested distance range. Even without the theoretical motivation, Newton's predictions are subject to the same experimental verification as are those of any other scientist, and the passage of 300 years does not mean that we should stop asking questions about the validity of scientific ideas.

▶ *12–6 Einstein's Theory of Gravitation

As accurate as it is for astronomical scales, Newton's theory of gravitation was superseded in 1915 by a still more accurate description: Albert Einstein's theory of gravitation. The Einstein theory, also known as the **general theory of relativity**, reduces to Newton's theory for objects that move with a speed $v \ll c$, where c is the speed of light, and for gravitational potential energies small compared to mc^2. These conditions are satisfied except in extreme circumstances, such as in regions very close to masses on the order of stellar masses and larger, depending on the size of the star, or for measurements at a precision only possible in the 20th century, so Newton's theory has always been adequate in virtually all its applications.

Einstein's theory arose out of his attempts in 1915 to combine Newton's theory with the special theory of relativity (see Chapter 39). While working on this problem, Einstein had what he described as "the happiest moment of my life" when he realized the importance of the fact that a freely falling person does not feel (and has no way to measure) his or her own weight. This idea was generalized in 1907 to form the **equivalence principle**. According to this principle, no experiment can distinguish between the following two situations: (1) a physical system at rest that is subject to a uniform gravitational force; and (2) a physical system that is uniformly accelerating in the absence of gravity. A simple example can help illustrate what is implied. Suppose an observer stands in an elevator and experiences a force on his feet. He can interpret this effect as being due to an upward acceleration of the elevator. Because the observer, of mass M, experiences an acceleration, a, he must be subject to an upward force of magnitude $F = Ma$. In fact, this force is the contact force that the elevator floor exerts on his feet. A second interpretation is that the elevator is at rest, but there is a uniform gravitational force acting to pull him downward. He again feels the upward normal force, F_N, of the floor and, because the elevator and thus the observer are at rest, that force must just cancel the gravitational force. Thus $F_N = Mg$, where g is the acceleration due to gravity. Notice that we have used the same mass M in both descriptions. Strictly speaking, the mass M in $F = Ma$ is the *inertial* mass, whereas it is the *gravitational* mass in the force-identity equation $F_N = Mg$. The gravitational and inertial masses were observed by Newton to be identical, but he was unable to draw deep conclusions from this observation. Einstein's equivalence principle states that if g has the same numerical value as a, there is no way for the observer to distinguish the two cases; thus, the inertial mass and the gravitational mass are *required* to be equal.

Predictions of the Equivalence Principle

Light Falls under the Influence of Gravity: Light falls just like matter does. To see this, let's consider the elevator again. A beam of light shines across the elevator in a horizontal direction from one side just as the elevator accelerates upward (Fig. 12–25). Because it takes light some time to travel across the elevator, it will hit the opposite wall closer to the floor than a horizontal line parallel to the floor would indicate. More important, for constant acceleration, the amount by which the elevator moves upward is proportional to the square of the time. That means that a *series* of measurements made by an occupant of

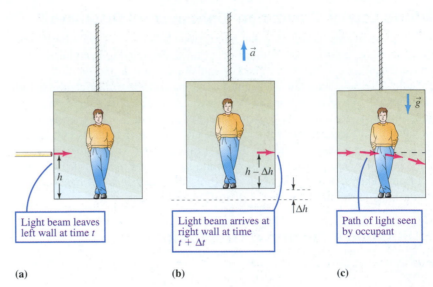

► **FIGURE 12–25** (a) A beam of light shines across an elevator at a time t when the elevator starts to rise. (b) At time $t + \Delta t$ the light hits the opposite side of the elevator at a point lower than it would have hit had the elevator not been moving. (c) The observer in the elevator cannot know whether the elevator is undergoing an upward acceleration $\vec{a}$ or whether a local gravitational acceleration of the same magnitude is present. We conclude from thought experiments like this that light must be bent when gravitational forces are present.

| Light beam leaves left wall at time t | Light beam arrives at right wall at time $t + \Delta t$ | Path of light seen by occupant |

(a) (b) (c)

the elevator that located the light beam relative to the floor would show that the light beam follows a parabolic path, just as a falling object would under the effect of gravity. If we are to interpret this observation from the point of view that the elevator and its contents feel a gravitational force, then the observed deflection of the light must be due to the gravitational force. Light, in this sense, does not behave any differently than matter, except that it moves faster. The first observation of the deflection of light due to gravitation was made during a solar eclipse in 1919. At such a time, pairs of stars whose light passes very close to the Sun become visible. The angular spread between the stars when their light comes around the two sides of the Sun can be compared with their observed angular spread when they are seen away from the Sun (Fig. 12–26). The confirmation of the predicted effect brought Einstein's theory to the public's attention.

► **FIGURE 12–26** During a solar eclipse, it is possible to observe the light of two stars on either side of the Sun. Because the Sun's gravity bends the light, the stars appear to be farther apart than they actually are.

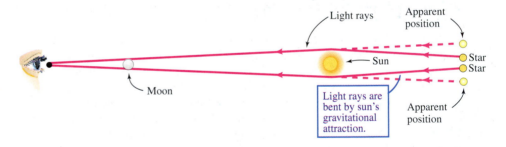

Gravitational Lenses: Astronomers today routinely observe what is known as *gravitational lensing*. In an example, consider the observation of two identical images of *quasars* (extraordinarily bright sources of light) very close to one another as seen from Earth (Fig. 12–27). Because there is good evidence that quasars are billions of light-years away, any two quasars are literally quite disconnected from one another; thus, it is highly improbable that any two could be as nearly identical as the members of the observed pair appear to be. The correct explanation is that there is only one quasar, but its light passes close to an extremely massive object, which bends light going

► **FIGURE 12–27** A double image results when light from a single quasar is bent as it passes on either side of an object with strong gravity, such as a galaxy or a large black hole. The effect is greatly exaggerated here.

around it just as a lens would; this gives rise to a double image; the effect is said to be due to a *gravitational lens*. An even more spectacular demonstration of this effect is seen in a ring image of the source (Fig. 12–28). (Can you see how the ring is produced?) To produce such a large effect, the light must be bent by a galaxy whose mass equals that of about 300 billion suns. Observation of lensing is so routine that it is now used as a tool to learn about the otherwise invisible intermediate objects that cause the effect.

Black Holes: The fact that light falls when it passes near masses implies that it is possible to imagine a large enough mass, localized in a small enough region, for which the speed of light is smaller than the escape speed from the surface of the mass. Such a mass would not be directly visible, because light or matter could not escape from it. It would manifest itself only through the gravitational force it exerts. Such a mass forms a *black hole*. Astonomers have accumulated impressive evidence for both small and large black holes. The "small" ones have masses of one or several solar masses. For example, there are pairs of stars only one of which can be seen directly. The second star is invisible, but its presence and properties can be deduced from the motion of the visible star and other features of the system. Some such pairs emit X-rays copiously, and the characteristics of the X-rays indicate that the invisible second star is in fact a black hole and the X-rays come from matter falling into it. The strong gravitational force necessary to produce such dramatic effects can come only from a black hole. Even more spectacularly, many galaxies, including our own, contain at their centers black holes with a mass of millions of Suns.

Precession of Planetary Orbits: The equivalence principle forms the foundation of the general theory of relativity, which is the full theory of gravitation. The mathematical application of the theory leads to subtle corrections to Newton's gravitational force law. The corrections to the Newtonian form predict, among other things, that the perihelion of Mercury, even in the absence of other planets, should precess by $43''$ of arc per century, an amount that is in agreement with observation when the effects of other planets on the motion of Mercury are taken into account. Other predictions of the general theory of relativity are being confirmed by recent experiments.[†]

▲ **FIGURE 12–28** This photograph, taken by the Hubble Space Telescope, of a massive, compact galactic cluster illustrates gravitational lensing. The lensing effects are evident in the arclike pattern. They are caused when light from an object far beyond the cluster passes near the cluster, which magnifies, distorts, and brightens that light on its way to our eyes. The image tells us a great deal both about the matter far beyond the cluster and about the cluster itself.

Summary

Astronomical observations led Kepler to three laws for planetary motion:

1. Planets move in planar elliptical paths with the Sun at one focus of the ellipse.
2. During equal time intervals, the radius vector from the Sun to a planet sweeps out equal areas.
3. If T is the time that it takes for a planet to make one full revolution around the Sun, and if R is half the major axis of the ellipse (R reduces to the radius of the orbit of the planet if that orbit is circular), then

$$\frac{T^2}{R^3} = C, \tag{12–1}$$

where C is a constant whose value is the same for all planets.

Newton showed that these laws are a consequence of a law of universal gravitation, which states that any two point masses m and M, separated by a distance r, will attract each other with a force that is along the radius vector connecting the masses. The force on mass m due to mass M has the inverse-square form

$$\vec{F} = -\left(\frac{GmM}{r^2}\right)\hat{r}, \tag{12–4}$$

where $\hat{r}$ is the unit vector pointing from mass M to mass m. The constant G has been measured, and its value is $G = 6.673 \times 10^{-11} \text{ N} \cdot \text{m}^2/\text{kg}^2$. The gravitational force is conservative and may be derived from a potential energy:

$$U(r) = -\frac{GmM}{r}. \tag{12–8}$$

[†]There is a wealth of literature on this fascinating subject, easily uncovered with a search on the Internet.

Masses that exert gravitational forces are not always pointlike. The most important case of an extended object is a spherically symmetric mass distribution, such as Earth or the Sun. In this case the gravitational force is the same as if all the mass of the extended object were concentrated at the center of the spherical distribution. Another consequence of the formula given in Eq. (12–8) is that the gravitational force exerted on an object anywhere inside an arbitrarily thick spherical shell, with a mass distribution that depends only on the distance from the center, is zero.

The gravitational force between two objects acts on both objects; for example, not only does Earth revolve around the Sun but the Sun also revolves around Earth. The tides are explained when this is taken into account in the Earth–Moon system. More precisely, when a gravitational force acts between two objects, both objects revolve around the center of mass of the two-object system.

The Newtonian theory of gravity is a limiting case of a more accurate and fundamental theory of gravity: Einstein's general theory of relativity. That theory is based on the equivalence principle, which states that no experiment can distinguish between a uniform acceleration and the effects of a uniform gravitational force. One consequence is the equality between the inertial mass (m_i in $F = m_i a$) and the gravitational mass (the masses that appear in the law of universal gravitation). Another consequence is the fact that light falls just like ordinary matter when it is subject to gravity.

Understanding the Concepts

1. What are some forces, other than those mentioned in this chapter, that are not fundamental? From what fundamental forces are they derived?

2. Any projectile fired with enough initial speed will eventually escape Earth, regardless of the direction in which it is fired. How do you reconcile this statement with the fact that the height above Earth to which a cannonball will rise depends on the angle at which it is fired?

3. To a good approximation, the Sun and the Moon both move with respect to Earth in the plane of Earth's equator. Does this mean there would be little or no tide at the North Pole if there were surface water at the North Pole?

4. The European satellite launch area is in French Guiana, because less energy is required to launch rockets into orbit from there than from a point in Europe. Explain why this is so.

5. How can Earth's rotational motion be used to minimize the fuel needed to boost a satellite into a given orbit around the planet? How can Earth's orbital motion be used to minimize the fuel needed to boost a satellite into a given orbit around the Sun?

6. The space shuttle orbits at an altitude of about 400 km. At that distance the acceleration of gravity due to Earth is very close to its surface value. Why then do the astronauts float around inside the space shuttle?

7. The satellites of Jupiter follow Kepler's third law: The square of their periods, divided by the radius of their orbits cubed, is a constant. Is this the same constant as for the planets moving around the Sun?

8. In our discussion of the tidal forces near a neutron star [Section 12–4], we had our unfortunate subject passing a neutron star in a close orbit. Why did we do this rather than having him or her simply stand on the surface? [*Hint*: What would the value of *g* itself be on the surface?]

9. Describe the path of a celestial object whose angular momentum with respect to the Sun is zero.

10. What gravitational force would a 1-kg piece of tungsten feel at the center of Earth?

11. If the gravitational force were a central force proportional to $1/r^3$ rather than $1/r^2$, the planetary orbits would no longer be closed (unless they are circular). Would the planets still sweep out equal areas in equal times?

12. We often hear that Earth satellites burn up when they leave their orbit and return to Earth. Why don't satellites burn up as they go up into orbit?

13. The same side of the Moon always faces Earth. What does this tell us about the rotational motion of the Moon? As we saw in our discussion of tides, there is a reason for this.

14. The acceleration of Earth due to the Moon is about 175 times smaller than that due to the Sun. How can it be that Earth's acceleration due to the Sun is so much larger than its acceleration due to the Moon yet the tidal forces due to the Moon are larger than those due to the Sun?

15. When astronauts float in the bay of their spacecraft, they are in outer space, orbiting the Earth. Is gravity acting on them?

16. Scientists and engineers have proposed that a base on the Moon would be useful for interplanetary launches. This would be advantageous compared to a launch from Earth's surface or from an orbit close to Earth because (a) there is no air on the Moon; (b) it would cost less net energy; (c) one wouldn't need to take as much fuel to the Moon's orbit distance in one launch.

17. Very careful measurements of the orbits of satellites can help teach us about Earth's internal structure. How can we use such measurements to study regions with a mass density that is higher or lower than that of Earth's average mass density?

18. If Earth were a perfect sphere, would you weigh more or less at the equator than at the poles?

19. Earth is not a perfect sphere. What types of observations might we use to learn this fact?

20. When a satellite is in circular orbit around Earth, there is a direct relationship between its angular momentum and the radius of the orbit. Suppose that such a satellite collides elastically with a meteor that was heading directly toward the center of Earth. Because the impulse is in a radial direction, the angular impulse is zero. What happens? Draw some diagrams, using the conservation laws that you know. [*Hint*: If you get stuck, review Kepler's first law.]

21. Consider the apparatus in Fig. 12–29 (see next page). Assume that the friction is normal sliding friction and that the rope and pulley are ideal. The motion can depend on (a) M_1 and M_2 separately; (b) the ratio M_1/M_2; (c) neither; the motion is independent of either mass; (d) the product $M_1 \times M_2$.

▲ **FIGURE 12–29** Question 21.

Rough surface

22. In the text we spoke of friction between Earth and its shell of water as slowing down the rotation frequency of Earth. Would this still occur if the Moon were not present?

23. The "day" on Jupiter is 9h 50 min at the equator, and 9h 55 min at its poles. What does this tell you about the rigidity of the surface of Jupiter?

24. Is it possible for friction from the tides to slow down Earth's rotation and still conserve Earth's angular momentum (with its oceans)? The origin of the tides involves the gravitational force

due to the Moon on Earth's oceans, so Earth and its oceans do not form an isolated system.

25. A satellite is in circular orbit around Earth. How much work is done on the satellite by the gravitational force of Earth during one orbit?

26. Assume that Earth is perfectly spherical and that its density depends only on the distance from its center. A large asteroid makes a close pass to Earth. Could such an asteroid (which is nevertheless small compared to Earth itself) change the rate of Earth's rotation without actually colliding?

27. You are in a spaceship very far away (say, 100 Earth radii) from Earth. Could you move to a region where the gravitational force due to Earth is less by ejecting some material from the spaceship in the proper direction?

28. In Section 12–3 we described Newton's thought experiment demonstrating how a cannonball shot horizontally off a mountaintop could attain a circular orbit if its initial speed increases to some large-enough value. Describe the motion of the cannonball if its initial speed exceeds this value.

29. Edgar Rice Burroughs's character Tarzan discovers that Earth is hollow. Tarzan finds an entry and discovers a whole new civilization, with modernistic buildings, people walking around or driving modernistic vehicles, and so on, in Earth's interior. What is wrong with this picture?

30. Spring tides (especially high tides) occur both when the Sun and Moon are aligned on the same side of Earth and when they are aligned on opposite sides of Earth. How do you explain these facts?

Problems

12–1 Early Observations of Planetary Motion

1. (I) Use the data listed in the appendices to calculate C in Eq. (12–1).

2. (I) In planetary tables we find that Jupiter's satellites Io and Europa each follow nearly circular orbits: Io's orbit has a mean radius of 422,000 km and Europa's orbit has a mean radius of 671,400 km. The period of Io is 152,854 s. What is Europa's period?

3. (II) Consider an object of mass m, moving in a circular orbit, subject to a central attractive force whose magnitude is given by $F(r) = h/r^3$. (a) What are the dimensions of h? (b) Show that the angular momentum for the motion is uniquely determined by h and m. (c) What is the resulting relation between period and radius analogous to Kepler's third law for this force?

4. (III) Angular momentum is conserved for a radial, or central, force. Show that the orbits due to a radial force lie in a plane.

12–2 Newton's Inverse-Square Law

5. (I) A man of mass 95 kg is dancing with his wife, who has a mass of 68 kg. Assume that each person's mass is concentrated at their respective centers of mass, which are separated by 48 cm. (a) What is the gravitational attraction between them? (b) Which person has the greater gravitational attraction toward the other?

6. (I) Calculate the gravitational attraction between a proton and an electron in a hydrogen atom if the radius of the atom is 0.6×10^{-10} m. The masses can be found in Appendix II.

7. (I) A Cavendish experiment involves the force between two spheres of 1 kg each whose centers are separated by 40 cm. Using the known value of G, find the gravitational force between these spheres. Compare this force to the weight of a fly.

8. (I) What is the acceleration due to gravity on the surface of (a) the Moon $(R = 1.74 \times 10^3$ km, $m = 7.35 \times 10^{22}$ kg); (b) Mars $(R = 3.40 \times 10^3$ km, $m = 6.42 \times 10^{23}$ kg); (c) Jupiter $(R = 7.14 \times 10^4$ km, $m = 1.90 \times 10^{27}$ kg); (d) the Sun $(R = 6.96 \times 10^5$ km, $m = 1.99 \times 10^{30}$ kg)?

9. (I) A satellite orbits Earth in 90 minutes. What is the radius of its motion around the center of Earth?

10. (I) What is the period of a satellite circling the Moon at a height of 90 km above the Moon's surface?

11. (I) What is the period of a satellite circling Earth at a height of 300 km above Earth's surface?

12. (I) What is the surface gravity (the value of g) on a spherical asteroid of diameter 30 km and density 5400 kg/m^3?

13. (I) Two identical satellites move in circular orbits around Earth. One has twice the kinetic energy of the other. The radius of the faster one's orbit is three Earth radii. What is the radius of the slower one's orbit?

14. (II) A weight lifter can lift 138 kg on Earth. What mass could the same weight lifter lift on (a) the Moon, (b) the Sun (use the data in Problem 8)?

15. (II) The height achieved in a jump is determined by the initial vertical velocity that the jumper is able to achieve. Assuming that this is a fixed number, how high can an athlete jump on Mars if she can clear 1.85 m on Earth?

16. (II) There is a point on the line joining two astronomical bodies where there is no gravitational force on a rocket. Find this point for (a) the Earth–Sun system, (b) a binary system of stars, one of which is one solar mass and the other two solar masses.

12–3 Planets and Satellites

17. (I) Pluto has the most eccentric orbit of all the planets, with $e = 0.25$. Its semimajor axis is 39.5 AU. What is $(r_{max} - r_{min})/(r_{max} + r_{min})$ for Pluto's orbit?

18. (I) The semimajor axis of Earth's orbit is 149.6×10^6 km, while its eccentricity is 0.017. What is the maximum distance between Earth and the Sun as Earth traces out its orbit?

19. (I) Determine the escape speed of an object from the Sun's surface.

20. (I) What are the escape speeds on the surface of (a) the Moon, (b) Mars, and (c) Jupiter? (Use the data in Problem 8.)

21. (I) Consider the asteroid of Problem 12. What is the escape speed from the surface of that satellite?

22. (I) The radius of a neutron star is 750 times smaller than Earth's radius, and its mass is 1.8×10^5 times larger than Earth's mass. What is the escape velocity from the surface of a neutron star? (Ignore the fact that, at high speeds, one should not really use $mv^2/2$ for the kinetic energy.)

23. (II) If the asteroid of Problem 12 rotates with an angular speed ω about an axis, material on the equator will have a tendency to be thrown off. How slowly must it rotate so that material that is not attached permanently just barely stays on the surface?

24. (II) A rocket is sent vertically upward from Earth's surface with an initial speed of 6.8 km/s. How far above Earth's surface will it go before falling back? Ignore atmospheric friction.

25. (II) Astronomers discover a meteorite at a distance of 80,000 km from the center of Earth. The meteorite is moving directly toward Earth with a velocity of 2000 m/s. What will be the velocity of the meteorite when it hits Earth's surface? Ignore all drag effects.

26. (II) Determine the minimum energy needed to allow an unmanned rocket of mass 3800 kg to leave the Moon's surface and arrive at a point very far away. Ignore the effect of Earth.

27. (II) Calculate the distance from Earth's center to a satellite in circular orbit with a period that is (a) one-third the Moon's period; (b) three times the Moon's period; (c) one-thousandth of the Moon's period.

28. (II) Using the information in Example 12–3, determine a satellite's time of revolution about Earth if it is in a circular orbit 370 km above Earth. How would your answer change if the mass of the satellite were to double?

29. (II) Determine the velocity with respect to Earth of a satellite of mass 500 kg in a circular orbit 200 km above Earth's surface. What is its kinetic energy? What is its angular momentum?

30. (II) A geosynchronous communications satellite orbits Earth, always positioned above the same point on the equator. (a) What is the period and angular velocity of the satellite? (b) What is the radius of the orbit? (c) Show that Kepler's third law applies to the orbits of the satellite and the Moon.

31. (II) The mass of Mars is 6.42×10^{23} kg, and its radius is 3393 km. What is the period of a satellite in a circular orbit 95 km above the surface of Mars?

32. (II) The Moon goes around Earth once in 27.3 d. What is the distance between the Moon and Earth?

33. (II) Use the mass of the Sun, given in Problem 8, to estimate the distance from Earth to the Sun. Assume that the orbit is circular.

34. (II) Consider a satellite on a circular polar orbit, one whose plane is the same as a plane made by a great circle passing through the North and South Pole. What is the altitude of the orbit with a 24-hour period? When the satellite passes repeatedly over the equator, over how many different equatorial points does it pass? Similarly, when the satellite passes repeatedly over a given latitude, over how many different points does it pass? How frequently does it pass directly over a given point?

35. (II) What is the minimum speed (relative to Earth) required for a rocket to send it out of the solar system? Note that you need to make use of Earth's speed to arrive at your result.

36. (II) A satellite is fired off horizontally with an initial speed $v_0 = 10.5$ km/s from the North Pole. Ignore air resistance. (a) What is the maximum distance from Earth's center attained by the satellite? Use both the conservation of energy and the conservation of angular momentum about the center of Earth. (b) What is the maximum distance attained if the satellite is fired vertically with the same speed? (c) if the satellite is fired at an angle of 41° with the same speed? (d) Sketch the motion of the satellite in the three cases.

37. (II) A small package is fired off Earth's surface with a speed v at a 45° angle. It reaches a maximum height h above the surface at $h = 6370$ km, a value equal to Earth's radius itself (Fig. 12–30). What is its speed when it reaches this height? Ignore any effects that might come from Earth's rotation.

▲ **FIGURE 12–30** Problem 37.

38. (II) In Example 12–8 we used the conservation of angular momentum to find the speed of a comet at perihelion and aphelion around a star. Use the fact that the energy at these two points must be equal to find the mass of the star.

39. (II) An object of mass 3×10^{15} kg approaches the solar system (Fig. 12–31). When it is very far away—where the gravitational potential energy can be neglected in comparison with its kinetic energy—the object moves with a velocity of 12 km/s in a straight line. By straight-line extrapolation, the closest this line would come to the Sun is 3×10^8 km. The point of the object's nearest approach to the Sun is characterized by the fact that the radius vector from the object to the Sun is perpendicular to

▲ **FIGURE 12–31** Problem 39.

the tangent to the path at that point. (a) Sketch the orbit of the object. (b) Use conservation of energy and of angular momentum to calculate the velocity of the object at the point of nearest approach. (c) Calculate the distance of nearest approach.

40. (II) A satellite is in a circular orbit of radius R_1 around Earth. Small rockets aboard the satellite change its direction so that it has an elliptical orbit. The change causes the satellite to lose half its orbital angular momentum, but the total energy remains constant. In terms of R_1, what are the perigee and apogee distances of the new orbit with respect to the center of Earth?

41. (II) A satellite of mass 300 kg is in circular orbit 2000 km above Earth's surface ($M_E = 6 \times 10^{24}$ kg, $r_E = 6370$ km). (a) What is the orbital speed of the satellite? (b) What is the angular momentum of the satellite? (c) A rocket engine is fired, reducing the speed of the satellite to half its initial value, but leaving the direction of motion unchanged. What is the new angular momentum? (d) Does the satellite crash as a result of the maneuvers in part (c)? Explain your answer.

42. (III) The distance of closest approach of Halley's comet to the Sun is 8.9×10^{10} m. Its period is 76 yr. What is the nature of its orbit? Calculate the following: (a) semimajor axes; (b) eccentricity; (c) aphelion distance (farthest distance from Sun).

43. (III) A satellite of mass 2000 kg is in circular orbit about Earth at a distance of 300 km above the surface. (a) What is the speed of the satellite in its orbit? (b) What is the angular momentum of the satellite about the center of Earth? (c) What is the total energy of the satellite in its orbit? (d) Controllers back on Earth wish to move the satellite to a new orbit 500 km above the surface. They propose to do this by briefly firing a rocket engine on the satellite for several seconds in the direction of the center of Earth; that is, the force on the satellite is directly away from Earth's center. What is the torque on the satellite about the center of its orbit? (e) Can the new orbit be circular?

12–4 Gravitation and Extended Objects

44. (I) What is the approximate difference between the value of g at sea level and the value on top of a 14,000-ft-high mountain? Assume that Earth has a constant density, and Earth's radius is the radius at sea level.

45. (I) The right-hand-side of Eq. (12–19) is an approximation. The exact form, given in Eq. (12–19), is $g(h)/g(0) = R_E^2/(R_E + h)^2$. Verify the accuracy of the approximation by calculating the ratio $g(h)/g(0)$ for $h = 10{,}000$ m (the altitude of a cruising passenger jet) according to the approximate and the exact forms.

46. (I) How much does the acceleration of gravity due to Earth decrease from sea level to (a) the 37,000-foot elevation flown by a jet airplane, and (b) the position of zero net gravitational attraction on the line between Earth and the Moon (see Example 12–4)?

47. (II) An object of mass m falls freely toward a large sphere of radius R and mass M from a great distance, starting from rest. The object has no angular momentum with respect to the large sphere. It arrives at the large sphere at a spot where there is a small hole and passes within. The large sphere turns out to be hollow. How long does it take for the object to make the trip from one side of the large sphere to the other?

48. (II) A deep hole in Earth reaches a depth of one half of Earth's radius (Fig. 12–32). How much work is done when a 1-kg mass is slowly lifted from the bottom of the hole to Earth's surface?

▲ **FIGURE 12–32** Problem 48.

49. (II) Jupiter is about 5.2 times as far from the Sun as is Earth, and its diameter is 11.2 times that of Earth. From what you know about the Sun's effect on tides on Earth, what can you say about tides on Jupiter due to the Sun?

50. (II) How big is the acceleration of Earth due to the Moon compared to that due to the Sun? Use the astronomical tables and estimate the average acceleration of Earth due to Jupiter, compared to that due to the Moon.

51. (II) What is the speed of the mass in the tunnel through Earth in Example 12–9 as it passes through the center of Earth?

52. (III) Rather than a tunnel through Earth's center, as in Example 12–9, consider a tunnel drilled along a chord of Earth, meaning that it passes a perpendicular distance d away from the center of Earth (Fig. 12–33). Find the potential energy of a mass placed in such a tunnel as a function of (a) its distance r from the center of Earth and (b) its distance x from the midpoint of the tunnel.

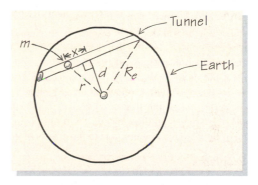

▲ **FIGURE 12–33** Problem 52.

12–5 A Closer Look at Gravitation

53. (I) The mean Earth–Sun separation is 1.50×10^8 km, Earth's radius is 6.37×10^3 km, the mass of the Sun is 1.99×10^{30} kg, and Earth's mass is 5.98×10^{24} kg. Given these data, locate the center of mass of the Earth–Sun system.

54. (I) At a certain moment, the Sun, Jupiter, and Saturn are lined up, with Jupiter between the Sun and Saturn. Suppose the orbits of Jupiter and Saturn were circular, with mean distances from the Sun of 7.78×10^8 km and 1.42×10^9 km, respectively. The mass of Saturn is 0.029 percent of that of the Sun. What is the ratio of the gravitational force on Jupiter due to Saturn to the gravitational force on Jupiter due to the Sun, at the specified moment?

55. (II) Some delicate gravitational measurements must be accurate to one part in 10^{11}. Suppose a 20-ton truck drives by the laboratory at a distance of 20 m. By what angle will this deflect a pendulum from the vertical (defined as the equilibrium position in the absence of the truck)? Should one worry about trucks near the laboratory?

56. (II) The maximum height of mountains on Earth can be estimated by noting that a mountain will sink if the resultant decrease in potential energy can be absorbed by the melting of the rock underneath the mountain. The energy associated with melting is a number that depends on the properties of rock. The result is that for any planet in which the mountains are made of the same material as Earth, the maximum height is given by $gh = C$, a constant. (a) Assuming that the rocks on Mars are more or less the same as those on Earth, how high would you expect the highest mountains on Mars to be? (You may find it interesting that Mount Everest on Earth is about 10 km high, while Mount Olympus on Mars is about 25 km high.) (b) The highest mountains on the Moon, where g is about 1/6 of g_{Earth}, are about 2 km in height. What does this tell you about the Moon?

57. (II) A binary star system consists of two stars, each of mass M, orbiting around their common center of mass with radii R from the center of mass. Determine the period of revolution.

*12–6 Einstein's Theory of Gravitation

58. (I) Fighter pilots are able to withstand accelerations up to $7g$ for a short period. A jet dives toward Earth and then pulls up in a parabolic orbit. Draw a force diagram showing the various forces on the pilot at the bottom of the orbit. If the dive was at night, could the pilot tell the difference between an increased value of gravitational force and the effect of the contact forces on him at the bottom of the dive?

59. (II) A very sharply defined laser beam, directed horizontally, enters a hotel room at height h. At what height does the light beam hit the opposite wall, which is 8 m from the first wall? Compare the difference in heights to the size of an atom. Is this a feasible experiment with which to test Einstein's theory of gravitation?

60. (II) An elevator of width w in free space is accelerated upward with acceleration g. A ray of light, traveling with speed c, enters through a pinhole on one side of the elevator, at right angles to the side at the moment the elevator starts to accelerate. It will strike the opposite wall at a somewhat lower height. What is the angle of deflection of the light? According to the equivalence principle, a passenger in the elevator could not distinguish this bending of light from a bending due to the effects of gravity.

General Problems

61. (I) A spaceship of the future is cylindrical in shape, with a radius of 60 m. In order to simulate terrestrial gravity on the inside surface of the cylinder, the spaceship is made to rotate about its axis. What is the angular velocity of the spaceship about its axis?

62. (I) (a) What is the acceleration g due to gravity on a planet with the same density as Earth but with 1.6 times the radius? (b) The orbital period and radius of Jupiter's moon Ganymede are 7.16 d and 660,000 mi, respectively. What is the period of the moon Io, whose orbital radius is 262,000 mi? (c) Planets A and B are both in circular orbits around a star. Planet A has two-thirds the orbital speed of planet B. What is the radius of A in terms of the radius of B?

63. (II) The Little Prince (a character in a book by Antoine de Saint-Exupery) lives on the spherically symmetric asteroid B-612 (Fig. 12–34). The density of asteroids, including B-612, is 5.2×10^3 kg/m³. Assume that the asteroid does not rotate. The Little Prince noticed that he felt lighter whenever he walked quickly around his asteroid. In fact, he found that he became weightless and started to orbit the asteroid like a satellite whenever he speeded up to 2 m/s. (a) Estimate the radius of the aster-

▲ **FIGURE 12–34** Problem 63.

oid from these data. (b) What is the escape speed for the asteroid? (c) Suppose that B-612 does rotate about an axis such that the length of the day there is 12 h. Can the Little Prince take advantage of this rotation when he wants to orbit his asteroid?

64. (II) A neutron star has a mass of 5.4×10^{30} kg and a radius of 12 km. (a) Calculate the acceleration due to gravity at the surface of the neutron star. (b) What is the difference between the gravitational forces acting on the top and the bottom of a tiny dumbbell held vertically on the surface (that is, with one end on the surface of the neutron star and the other 1 mm above the surface)? The dumbbell consists of two 1 g point masses connected by a massless connector of length 1 mm.

65. (II) Suppose that, instead of a $1/r^2$ dependence, an attractive central force varied with distance as $1/r^n$. (a) Would such forces support a circular orbit? (b) Find the resulting relation between period and radius analogous to Kepler's third law for this force.

66. (II) Consider a cluster of galaxies that fills a sphere of radius R and average mass density ρ. (There are so many galaxies that you can assume uniform density.) There is a galaxy of mass M at the edge of this sphere (Fig. 12–35). (a) Write an expression for the energy of the galaxy. (b) In the big-bang model of the origin of the universe, the velocity of the galaxy is directed radially outward from the center of the sphere; the galaxy's speed is $v = HR$, where $H = (15 \text{ km/s})/(10^6 \text{ ly})$ is the Hubble parameter. For what critical density ρ_c of the large cluster will the galaxy be able to escape to infinity with a final velocity of zero?

▲ **FIGURE 12–35** Problem 66.

67. (II) A satellite in low circular Earth orbit is subject to a very small constant friction force, f, due to the thin atmosphere. As it spirals in, it slowly decreases its radius. Find the decrease in radius per revolution under the assumption that the orbit is approximately circular with radius r. Find the changes in potential energy, total energy, and kinetic energy per orbit. [*Hint*: If you find the kinetic energy increasing, you are on the right track!]

68. (II) The electric force between an electron and a (much more massive) proton is attractive and of magnitude e^2/r^2, where $e^2 = 2.3 \times 10^{-28}$ N·m^2. The electron circles the proton in a circular orbit of total energy $E = -1.6 \times 10^{18}$ J. (We have taken zero potential energy to be at infinite separation.) (a) What is the radius of the orbit? (b) What is the period of the orbit? (c) If another electron were in another circular orbit around the proton, with an orbit radius three times as large as the first, what would its total energy and period be?

69. (II) An astronaut of mass 115 kg (including equipment) finds himself drifting away from his orbiting space ship at 0.05 m/s. He throws a 3-kg wrench in the direction of his drift and comes to rest relative to the ship 1 m from its surface. The ship is a sphere of radius 12 m and mass 10^5 kg. (a) At what speed does he throw the wrench? (b) How many hours must he wait for the gravitational attraction of the ship to pull him to its surface, assuming that the force of gravity is approximately constant in the region of interest?

70. (II) Suppose that an object of mass m is placed at the point at which the gravitational attraction of the Moon is just canceled by that of Earth; further, suppose that the object is displaced by a small distance x along a line perpendicular to the line connecting the centers of Earth and the Moon (Fig. 12–36). What are the magnitude and direction of the net force on the object as a function of x? Calculate your answer by using the approximation $(r^2 + x^2)^n = r^{2n}[1 + (nx^2/r^2) + \cdots]$, valid for $x^2/r^2 \ll 1$.

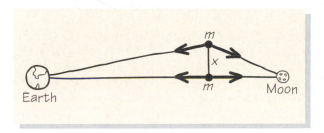

▲ **FIGURE 12–36** Problem 70.

71. Jupiter's moon Io has a radius of 1815 km, a mass 4.7×10^{-5} that of Jupiter, and an orbital radius about Jupiter that is 5.95 times Jupiter's own radius (Fig. 12–37). Given that Jupiter's mass is 318 times that of Earth and that its radius is 26 times larger than Earth's, calculate the acceleration due to gravity on Io at (a) the point nearest Jupiter and (b) the point farthest from Jupiter.

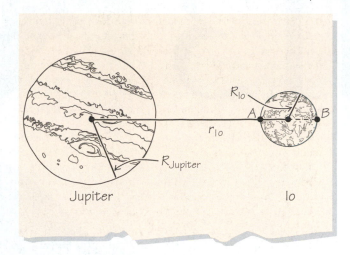

▲ **FIGURE 12–37** Problem 71.

72. (III) Astrologers claim that a person's life is influenced by the position of the planets at the moment of that person's birth. To check whether this influence could be due to gravity, compare the following two quantities: the change in the gravitational force on a baby in a hospital due to the change in the position of Jupiter from one day to the next, and the change in the gravitational force due to the presence or absence of a 4-ton truck parked near the hospital at a distance of 75 m. Jupiter has a mass of 1.90×10^{27} kg; its mean distance from the Sun is 0.78×10^9 km, and its period is 11.9 yr. Assume a circular orbit for Jupiter, and a circular orbit of radius 1.5×10^8 km for Earth. Choose the region of closest approach of the two planets for convenience.

73. (III) Three stars, each the mass of the Sun, form an equilaterial triangle. Each moves in a circular orbit about the center of mass of the system because of the gravitational force exerted by the other two stars. (a) Is such an arrangement possible? (b) If so, what is the period of the motion, assuming that the side of the triangle is an Earth–Sun distance? (c) Is the system stable?

74. (III) When the first nuclear weapons were detonated, concern was expressed in some quarters that a huge nuclear chain reaction would be set up, blowing Earth to pieces. Show that the energy that would be required to disassemble Earth completely into pieces totally separate from each other is $\frac{3}{5}GM_E^2/R_E$. [*Hint*: Imagine that layers of Earth are peeled off one by one, like layers of an onion.]

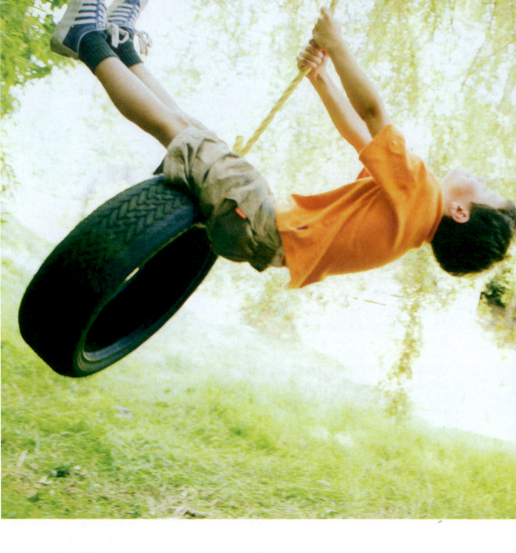

▶ A backyard swing provides an example of oscillatory motion. Such motion occurs everywhere in the physical world, from vibrations in molecules to oscillations in the shape of the Sun.

Oscillatory Motion

Rhythmic motion—also known as periodic motion—is a common occurrence in the physical world. The very concept of time arises from the observation that certain motions, such as the human heartbeat and the cycling of the seasons, repeat themselves in a reliable and regular way. An important class of periodic motions involves what are known as restoring forces, forces that act to bring an object back to an equilibrium point. As we have already seen in Chapter 7, such restoring forces have potential energy functions with minima at the equilibrium point. Objects in this sort of motion oscillate, and oscillatory motion is the central subject of this chapter. The most basic type of oscillatory motion is omnipresent in nature: **simple harmonic motion**. This motion occurs when the strength of the restoring force is directly proportional to the object's displacement from the equilibrium point. Everyday examples are the motion of a mass on the end of a spring and the motion of a pendulum. The position of an object in simple harmonic motion varies with time as a sine or a cosine. While the spring force is an example that we will use repeatedly, simple harmonic motion is of universal importance because virtually any small oscillatory motion about a stable equilibrium point is simple harmonic motion.

We'll also see the effects of dissipative forces in this chapter—which not surprisingly cause the motion to progressively die out—and the effects of an oscillatory driving force. The presence of the driving force illustrates the remarkable feature known as resonance, in which the motion can become catastrophically large if the frequency of the driving force is just right.

13–1 The Kinematics of Simple Harmonic Motion

Simple harmonic motion, which describes the small repeating motion followed by a mass on the end of a spring or a pendulum, is a simple form of **oscillatory** motion. The word "harmonic," signifying agreement and accord, reveals that humankind have always seen beauty in this motion. In the back-and-forth of simple harmonic motion, the position $x(t)$ of an object is of the form $\sin(\omega t)$ or $\cos(\omega t)$, where the coefficient ω is the **angular frequency**. Both sines and cosines repeat themselves periodically as time t passes. The trigonometric functions are functions of a dimensionless argument, an angle measured in radians (or, sometimes, degrees). Thus the coefficient of the time must have the dimensions $[T^{-1}]$. We'll see later that the angular frequency ω is a fundamental property of the motion, determined by the inertia of the moving objects and the restoring force acting on them.

How do we figure out whether the motion of a mass on the end of a spring is described by a sine or by a cosine? Let's look at a graph of $\sin\theta$ versus θ next to a graph of $\cos\theta$ versus θ (Fig. 13–1). Both functions repeat every time the angle θ changes by 2π rad. When $\theta = 0$, the sine function is zero, whereas the cosine function is $+1$, but this is only a matter of placing the axis. Indeed, the functions are *identical* if the origin of the θ axis is shifted. We can specify such a shift of θ by an angle we call the **phase**, δ. By what angle δ would θ have to be shifted so that the $\sin\theta$ curve in Fig. 13–1a is coincident with the $\cos\theta$ curve of Fig. 13-1b? If δ is chosen properly, the function $\sin(\omega t + \delta)$ can represent $\sin(\omega t)$, $\cos(\omega t)$, or anything in between. The phase simply makes explicit the "starting" point for harmonic motion. Both sine and cosine have the same shape, but displaced, and the phase sets the amount of displacement.

Another quantity that characterizes oscillatory motion is how far the moving object gets from the equilibrium position before it turns around. In the case of a point mass in simple harmonic motion in, say, the x-direction, the motion is symmetric from one side to the other, and the maximum distance of displacement to the right of the equilibrium point equals the maximum distance of displacement to the left. We call this distance the **amplitude**, A. It is by definition positive. The sine function is dimensionless and varies between -1 and $+1$. But $x(t)$ has dimensions of length. To express $x(t)$, we therefore have to multiply the harmonic sine (or cosine) function by a constant with dimensions of length, and this constant is the amplitude A described above. The resulting expression for the position of an object in simple harmonic motion is

$$x(t) = A\sin(\omega t + \delta), \qquad (13\text{–}1a)$$

SIMPLE HARMONIC MOTION

and we can immediately confirm that A describes the magnitude of the maximum excursion away from the point of zero displacement (Fig. 13–2). An alternative form of this expression turns out to be very useful. We can use the basic trigonometry rule $\sin(x + y) = \sin x \cos y + \cos x \sin y$ to rewrite $\sin(\omega t + \delta)$ as

$$x(t) = [A\cos\delta]\sin(\omega t) + [A\sin\delta]\cos(\omega t).$$

sin θ

(a)

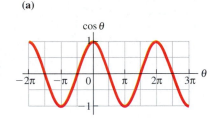

cos θ

(b)

▲ **FIGURE 13–1** Plots of (a) $\sin\theta$ and (b) $\cos\theta$, both as a function of θ.

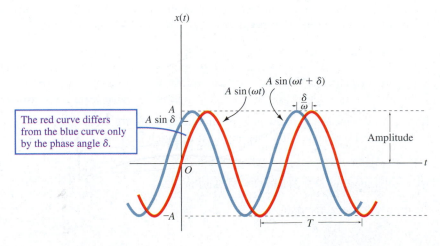

The red curve differs from the blue curve only by the phase angle δ.

◄ **FIGURE 13–2** In simple harmonic motion the phase, δ, corresponds to a sliding of the curve of displacement versus time to earlier or later times. The amplitude and period of the motion are also shown.

The original quantities A and δ are two constants that characterize the motion, and we can think of the two quantities in square brackets as combinations of these two constants that equally well characterize the motion. In other words, simple harmonic motion can alternatively be expressed as

$$x(t) = a_1 \sin(\omega t) + a_2 \cos(\omega t). \tag{13-1b}$$

Comparing the intermediate step with Eq. (13–1a), we find relations that can be used to connect the constants a_1 and a_2 to the constants A and δ,

$$a_1 = A \cos \delta, \, a_2 = A \sin \delta.$$

Inverting, we can express A and δ in terms of a_1 and a_2:

$$A^2 = a_1^2 + a_2^2, \tan \delta = a_2/a_1.$$

Which of these two forms, Eq. (13–1a or b), is more convenient depends on the circumstances, and we'll sometimes use one and sometimes the other.

Properties of Simple Harmonic Motion

Three independent parameters appear in simple harmonic motion and describe the motion: the amplitude A, the phase δ, and the angular frequency ω. The amplitude and the phase are determined by specifying the position $x(t)$ at $t = 0$ and the maximum magnitude of $x(t)$. It follows in this case from Eq. (13–1a) that $x(0) = A \sin \delta$, while $|x_{max}| = A$. These two equations give A and δ in terms of $x(0)$ and $|x_{max}|$. Or one may know the position $x(t)$ and velocity $v(t) = dx/dt$ at an initial time $t = 0$. In this case we say that A and δ are determined by the **initial conditions** for the motion. It follows from Eq. (13–1a) that $x(0) = A \sin \delta$. We can also use Eq. (13–1a) to find the velocity $v(t) = dx/dt = A\omega \cos(\omega t + \delta)$ [see Eq. (13–7)], so that $v(0) = A\omega \cos \delta$. The two expressions $x(0)$ and $v(0)$ are enough to specify both amplitude and phase provided that ω is known. A similar analysis can be done for Eq. (13–1b), in which the two constants a_1 and a_2 are determined by initial conditions. The fact that ω needs to be known here suggests that we should turn to that constant next.

The angular frequency ω is a measure of the repetition time for the motion, i.e., the time for one full cycle of the motion. We call this time the **period** T. The sine function repeats itself either when the angle increases by 2π rad (see Fig. 13–1) or, because δ is a constant, when ωt increases by 2π. Thus the period satisfies $\omega T = 2\pi$. We can solve for the period:

$$T = \frac{2\pi}{\omega}. \tag{13-2}$$

PERIOD OF SIMPLE HARMONIC MOTION

Thus the value of the angular frequency ω determines the period. In Chapter 3, where we described uniform circular motion, we defined the **frequency**, f, as the number of full oscillations per unit time, or equivalently the inverse of the period. A period of 5 s means a frequency of one complete repeat of the motion every five seconds, while a period of 0.5 s means a repeat frequency of two per second, and so forth:

$$f = \frac{1}{T}. \tag{13-3}$$

FREQUENCY OF SIMPLE HARMONIC MOTION

If the period is measured in seconds, the frequency is measured in s^{-1}. In SI, the unit s^{-1} is the **hertz** (Hz), named after the physicist Heinrich Hertz:

$$1 \text{ Hz} = 1 \text{ s}^{-1}. \tag{13-4}$$

By comparing Eqs. (13–2) and (13–3), we find that

$$f = \frac{\omega}{2\pi}. \tag{13-5}$$

CONCEPTUAL EXAMPLE 13–1 Your classmate states that if the acceleration of a mass acted on by a spring is proportional to the displacement from the equilibrium point of the mass, then the farther the mass gets from the equilibrium, the larger the acceleration, and the mass will soon be accelerating so much that it will be in the next county in a few minutes. Is he right? How would you correct him?

Answer It is indeed true that the acceleration is proportional to the displacement, but as a look at Eq. (13–8) verifies, *there is a*

crucial minus sign in the relation. This sign keeps the motion within bounds. If the displacement is to the right of the equilibrium point, the acceleration is to the left, tending to send the mass back to the left; if the displacement is to the left, the acceleration is to the right, tending to send the mass back to the right. Without the minus sign, your classmate is correct. In that case, the position is an exponential function of time rather than oscillatory.

We will see in Section 13–2 that the angular frequency ω can be identified with the angular speed, a quantity we have already defined and used in Sections 3–5 and 9–1 in connection with circular motion. Inversion of Eq. (13–2) or (13–5) gives

$$\omega = \frac{2\pi}{T} = 2\pi f. \tag{13–6}$$

When the position is specified as a function of time, the velocity and the acceleration are determined by taking successive derivatives. As a consequence of Eq. (13–1a), we have (see Appendix IV–7)

$$v(t) = \frac{dx}{dt} = \frac{d}{dt}[A\sin(\omega t + \delta)] = \omega A\cos(\omega t + \delta). \tag{13–7}$$

One further derivative gives the acceleration as a function of time:

$$a(t) = \frac{dv}{dt} = -\omega^2 A\sin(\omega t + \delta) = -\omega^2 x(t). \tag{13–8}$$

The acceleration is proportional to the displacement. Since we will argue that virtually all stable equilibrium situations, from the back and forth of a rocking chair to the oscillation of a spider on his web in the breeze, are associated with simple harmonic motion; thus, the proportionality of the acceleration and the displacement is a universal property of motion near equilibrium.

Relations Among Position, Velocity, and Acceleration in Simple Harmonic Motion

In Fig. 13–3 we plot the position, velocity, and acceleration of an object in simple harmonic motion over two full periods, starting with $x(t) = A\sin(\omega t)$. (For convenience, the phase has been taken to be zero. The relations discussed here are not affected by the phase.)

The photo in Fig. 13–4 represents the up-and-down motion of a ball on a spring, presented so you can follow the ball's vertical position as a function of time. This motion matches the motion described in Fig. 13–3. In Fig. 13–3a the object is at the origin at $t = 0$. As we see in Fig. 13–3b, the velocity at $t = 0$ is maximum in magnitude and is positive, while Fig. 13–3c shows that at this time the acceleration is zero, so that the velocity is not changing. After one-quarter of the period ($\omega t = \pi/2$), the object has moved to the right-hand extreme of its motion and is ready to turn around. The velocity is zero at this turnaround point, but the acceleration has actually reached a maximum in magnitude and is negative, indicating that the velocity will be turning to the left and will become

(a)

(b)

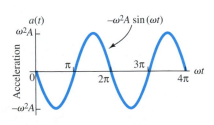

(c)

▲ **FIGURE 13–3** Starting with (a) the graph of position $x(t) = A\sin(\omega t)$, a single derivative gives (b) the velocity $v(t) = \omega A\cos(\omega t)$. One further time derivative gives (c) the acceleration $a(t) = -\omega^2 A\sin(\omega t)$. We have in each case plotted two cycles, or periods, for zero phase; the curves repeat after every period. Note also, as the upper scale indicates, the correspondence between the number of periods and ωt as a multiple of 2π.

◀ **FIGURE 13–4** A photograph of the simple harmonic motion of the mass on the end of a spring.

negative. (Think of a ball thrown in the air; at the maximum height—the turnaround point—the velocity is zero even if the acceleration is nonzero and directed toward Earth.) After one-half the period ($\omega t = \pi$), the object once again passes through the origin, this time moving to the left. The acceleration is again zero. The three-quarter mark ($\omega t = 3\pi/2$) is at another turnaround, characterized by a maximum negative value of x—the object is at its left-hand extreme—and zero velocity. The acceleration is maximum and positive, meaning that the velocity is becoming positive, and the object will subsequently move back to the right. Finally, after one full period ($\omega t = 2\pi$), the object has come back to its starting point, moving to the right through the origin with its largest positive velocity and zero acceleration. The situation at $t = 2\pi/\omega$ is identical to what it was at $t = 0$.

EXAMPLE 13–2 A cork floating on a pond moves in simple harmonic motion, bobbing up and down over a range of 4 cm. The period of the motion is $T = 1.0$ s, and a clock is started at $t = 0$ s when the cork is at its minimum height. What are the height and velocity of the cork at $t = 10.5$ s?

Setting It Up We draw a graph of the motion in Fig. 13–5, which is along a z-axis whose origin is the midpoint of the motion. The maximum value of z is $z_{max} = 2$ cm, and the minimum value is $z_{min} = -2$ cm, which is the location at $t = 0$.

Strategy We must find an expression for position and velocity as a function of time given the information in the problem, and then evaluate these at $t = 10.5$ s. For position, we'll use the general form of Eq. (13–1b), which requires two constants and knowledge of ω. We are given T, and that will determine ω directly. With Eq. (13–1b) we can find the velocity by taking the derivative of the position. To evaluate constants of our expressions, we can use the facts that at $t = 0$, $z = z_{min}$ and $v = 0$.

Working It Out We know the period, T, and from Eq. (13–6), $\omega = 2\pi/T$. The motion (position) takes the general form

$$z(t) = a_1 \sin(\omega t) + a_2 \cos(\omega t).$$

With a single derivative, we also get the velocity:

$$v(t) = a_1\omega \cos(\omega t) - a_2\omega \sin(\omega t).$$

To find the constants a_1 and a_2, we use the initial conditions. As stated above, these read, $z = z_{min}$ and $v = 0$ at $t = 0$. The second equation above is simple to apply: v can only be 0 at $t = 0$ if the constant $a_1 = 0$. Applying this, the condition that $z = z_{min}$ at $t = 0$ then gives immediately $a_2 = z_{min}$. Finally, $\omega = 2\pi/T = 2\pi/(1\text{ s}) = 2\pi$ rad/s. In summary,

$$z(t) = a_2 \cos(\omega t)$$

and

$$v(t) = -a_2\omega \sin(\omega t)$$

with $a_2 = -2$ cm and $\omega = 2\pi$ rad/s.

The second part of Fig. 13–5 shows the velocity of the cork.

It is straightforward to plug $t = 10.5$ s into these expressions. We can also employ some simple reasoning to make a shortcut to the numerical answer. Since both z and v repeat themselves every period, the values of z and v at 10.5 s are the same as at 0.5 s (0.5 period). Moreover, after half a period, the cork moves from the bottom of the motion to the top, i.e., z will move from z_{min} to z_{max} and the velocity will once again be zero as the motion of the cork turns around. Thus

$$\text{for } t = 10.5\text{ s}, z = z_{max} = +2\text{ cm} \quad \text{and} \quad v = 0\text{ m/s}.$$

What Do You Think? For what value(s) of z does the acceleration of the cork have maximum magnitude? For what value(s) of z does the acceleration have minimum magnitude? *Answers to **What Do You Think?** questions are given in the back of the book.*

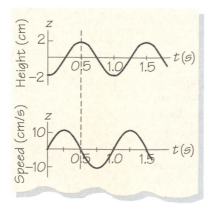

▲ **FIGURE 13–5** The height and speed of a bobbing cork in a pond.

▲ **FIGURE 13–6** The relation between uniform circular motion and simple harmonic motion is evident in the piston-linkage connection on the train wheel and the resulting motion.

13–2 A Connection to Circular Motion

In Chapter 3 we discussed another kind of periodic motion: uniform circular motion. The photograph in Fig. 13–6 of the wheels and driving piston of a steam engine suggests that circular motion has a simple connection to harmonic motion, and we next demonstrate this connection in more detail. Figure 13–7 shows uniform circular motion for a point moving in the xy-plane a constant distance R from the origin. The motion is described by an angle θ, measured from the x-axis, that varies linearly with time:

$$\theta = \omega t + \delta. \tag{13–9}$$

The phase, δ, is just the value of θ at time $t = 0$.

If we were to look at a side view of the uniform circular motion of a pin stuck on a rotating turntable, we would see the pin oscillate in simple harmonic motion. Figure 13–7

▲ **FIGURE 13–7** Uniform circular motion in the *xy*-plane, and its projection onto the *y*-axis. The projection represents simple harmonic motion, easily visible in a plot of *y* versus $\theta = \omega t + \delta$, as on the right.

indicates the *projection* of the circular motion on the *y*-axis, but you could easily project onto both *x* and *y*. Simple trigonometry gives us these projections:

$$x = R \cos \theta = R \cos(\omega t + \delta); \tag{13–10}$$

$$y = R \sin \theta = R \sin(\omega t + \delta). \tag{13–11}$$

Thus uniform circular motion corresponds to simple harmonic motion in both the *x*- and *y*-directions. A cosine rather than sine appears in *x*, but as we discussed above, this is just the standard form with a different phase. We can use the trigonometric identity $\sin[\theta + (\pi/2)] = \sin\theta \cos(\pi/2) + \cos\theta \sin(\pi/2) = \cos\theta$ to replace the cosine in Eq. (13–10) with a sine function, and we thereby obtain

$$x = R \sin\left(\omega t + \delta + \frac{\pi}{2}\right). \tag{13–12}$$

Both the *x*- and *y*-motions are now in the standard form of Eq. (13–1a). The two motions have a phase that differs by exactly $\pi/2$ (90°), and the sign of this phase difference specifies the direction—clockwise or counterclockwise—of the corresponding uniform circular motion (see Problem 22).

13–3 Springs and Simple Harmonic Motion

Having described simple harmonic motion—the kinematics—and armed with our knowledge of Newton's second law, we now can turn to the cause of the motion. Springs give rise to simple harmonic motion. Let's restrict ourselves to one-dimensional motion and dispense with vector notation. The spring force on a mass displaced by *x* from the equilibrium position of the spring is a restoring force linearly dependent on *x*, the form known as *Hooke's law*:

$$F = -kx. \tag{13–13}$$

This form is valid provided the spring is not overly stretched or compressed, in which case it loses its "springiness" and distorts—this is why we have spoken about "small" motions about the equilibrium point. Here *k* is the *spring constant*. It is the minus sign in Eq. (13–13) that indicates that the force is a *restoring force*. A displacement in the +*x*-direction gives rise to a force that acts in the −*x*-direction and vice versa. Figure 13–8a shows a series of possible starting points for the motion. Let us choose the third one, where the mass is released at $t = 0$ from an extended position. The resulting motion is shown in Figure 13–8b over a complete period of the motion. Newton's second law provides us with the connection between the force and the acceleration; namely, $F = -kx = ma$. Thus the acceleration of a mass on the end of a spring is proportional to its displacement, with a minus sign:

$$a = -\frac{k}{m}x. \tag{13–14}$$

An acceleration proportional to the position, with a minus sign, is just the kinematic characteristic that we found in Section 13–1 for simple harmonic motion. Comparison of

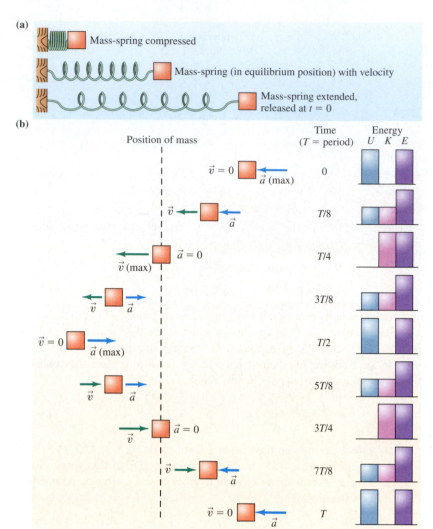

(a)

Mass-spring compressed

Mass-spring (in equilibrium position) with velocity

Mass-spring extended, released at $t = 0$

(b)

► **FIGURE 13–8** (a) Some possible starting points for the motion of a mass on the end of a spring. (b) The simple harmonic motion of the mass when it is released from the stretched position. The speed is lowest (and the acceleration is highest) when the displacement from equilibrium is a maximum, and the speed is highest (and the acceleration is lowest) when the displacement is a minimum. We can also see the play between kinetic and potential energy; one is large where the other is small.

Eqs. (13–8) and (13–14) yields the important result that the angular frequency is determined by the mass and the spring constant:

$$\omega^2 = \frac{k}{m};$$
(13–15)

$$\omega = \sqrt{\frac{k}{m}}.$$
(13–16)

ANGULAR FREQUENCY FOR MASS ON A SPRING

In turn, Eqs. (13–2) and (13–3) give the period and the frequency of the oscillations:

$$T = 2\pi\sqrt{\frac{m}{k}} \quad \text{and} \quad f = \frac{1}{2\pi}\sqrt{\frac{k}{m}}.$$
(13–17)

Remarkably, *the period of the motion is independent of the amplitude*. The same is then true for the frequency.

The spring is the prototype of dynamical systems moving back and forth about a stable equilibrium—virtually all such systems exhibit simple harmonic motion. All these systems reduce to a mass on the end of a spring, in that the *form* of the force is the same as that of the spring, a restoring force linear in some variable.

CONCEPTUAL EXAMPLE 13–3 The spring constant k of a mass-spring system is doubled. By what factor does m have to change so that (a) the acceleration at $x = 0$ is unchanged; (b) the acceleration at $x = A$ [A is the original amplitude] is unchanged; (c) the velocity at $x = A$ is unchanged; (d) the period of the motion is unchanged?

Answer (a) The acceleration is proportional to x; at $x = 0$, the acceleration remains zero, regardless of the values of k or m.

(b) From Eq. (13–14), the original acceleration at $x = A$ is $a = -(k/m)A$. If k is doubled, doubling m will leave a unchanged.

(c) The velocity at the extremes of the motion—i.e., $x = A$—is zero, and this is independent of the values of k or m.

(d) The period is inversely proportional to the angular frequency, which is in turn a function of k/m. So double m to leave the period unchanged.

EXAMPLE 13–4 A mass $m = 0.50$ kg moves along the x-direction under the influence of a spring with spring constant $k = 2.0$ N/m. The origin of the x-axis is at the equilibrium point of the mass. At $t = 0$ s, the mass is at the origin and moving with a speed of 0.50 m/s in the $+x$-direction. (a) At what time t_1 does the mass first arrive at its maximum extension? (b) What is this maximum extension?

Setting It Up We note specifically that we are given initial conditions, in this case the position and velocity at $t = 0$ s.

Strategy We can give a description of the position at all times, then substitute specific times. The motion as a function of time is given by either of the two Eqs. (13–1)—we'll use Eq. (13–1a) here. With k and m known, we can find the angular frequency, ω. And the initial conditions will be sufficient to find the two remaining parameters A and δ in Eq. (13–1a). In part (a) we want the time to go from the origin to the maximum extension, and this is just a quarter period—for that we need only ω. For part (b) the parameter A *is* the maximum extension.

Working It Out From Eq. (13–16), the angular frequency, ω, is

$$\omega = \sqrt{\frac{k}{m}} = \sqrt{\frac{2.0 \text{ N/m}}{0.50 \text{ kg}}} = \sqrt{4.0 \text{ s}^{-2}} = 2.0 \text{ rad/s}.$$

(a) The time to go from the equilibrium position to the maximum extension is $T/4$:

$$t_1 = \frac{1}{4}T = \frac{1}{4}\frac{2\pi}{\omega} = \frac{1}{4}\frac{2\pi \text{ rad}}{2.0 \text{ rad/s}} = \frac{1}{4}3.1 \text{ s} = 0.78 \text{ s}.$$

(b) We use the information about x and v at $t = 0$ s to find the amplitude. Writing $x(t) = A \sin(\omega t + \delta)$, we have $x(t = 0) = A \sin \delta = 0$. This implies that $\delta = 0$. We use this, in turn, for the value of v at $t = 0$, $v(t = 0) = A\omega \cos(0) = A\omega$. (The argument of the cosine is zero because both t and δ are zero.) Thus

$$A = \frac{v(t = 0)}{\omega} = \frac{0.50 \text{ m/s}}{2.0 \text{ rad/s}} = 0.25 \text{ m},$$

which is the maximum excursion of the mass from the origin.

What Do You Think? If the speed at $t = 0$ were doubled, then the time to reach the maximum extension would be (a) doubled (b) the same (c) halved.

Additional Constant Forces

Suppose we start with a spring force and we add a constant force to it that acts along the same line. How different is the motion of an object under the influence of both these forces from the motion with the spring force alone? The answer is, remarkably little. The only thing that changes is the equilibrium point. As we have seen, the original (one dimensional) spring force always takes the form $F_{\text{spring}} = -k(x - x_0)$ (the sign takes into account the vector nature in one dimension), here aligned with the x-axis. The quantity $x - x_0$ is the displacement of the mass from its equilibrium point at $x = x_0$. The period of this spring, or indeed any spring, is independent of the equilibrium point. Now imagine adding (also acting along the x-axis) a constant force F_c. We can *always* write F_c in the form

$$F_c = kx_1,$$

where k is the same spring constant as for the original spring and $x_1 \equiv F_c/k$. That means the net force takes the form

$$F_{\text{net}} = F_{\text{spring}} + F_c = -k(x - x_0) + kx_1 = -k(x - [x_0 + x_1]).$$

This is again a spring force, with the same spring constant as the original spring force. Thus the motion will have the same frequency, but a shifted equilibrium point, $x_0 + x_1$ instead of x_0.

This behavior is exhibited by a mass hanging vertically from a spring. The supplementary constant force is that of gravity. The frequency of the simple harmonic motion will be the same whether the spring is hanging vertically or not. For the hanging case, and assuming the spring itself is much less massive than the mass attached to its end, the equilibrium position will be lowered by an amount Δy proportional to the additional weight of the mass, as in Fig. 13–9. More precisely, we have

$$mg = k\,\Delta y, \quad \text{or} \quad \Delta y = mg/k.$$

The harmonic motion is measured from the new equilibrium position.

(a) (b) (c)

No gravity | With gravity | Free-body diagram

▶ **FIGURE 13–9** A mass on the end of a spring is suspended vertically. (a) If its equilibrium length would place it at height y_0 in the absence of gravity, then (b) it will be stretched an additional amount, Δy, to a new equilibrium position, y_1, under the influence of gravity. (c) Free-body diagram for the mass.

13–4 Energy and Simple Harmonic Motion

We examined energy considerations for the spring force in Chapter 7, where we found that the work done by a spring force in moving a mass from one position to another is independent of the path taken by the mass. That means that the spring force is conservative and has a potential energy function $U(x)$ associated with it. The total energy E (the sum of kinetic energy, K, and potential energy) is *conserved* throughout any motion.

In Section 7–1 we computed the potential energy $U(x)$ of an object attached to a spring and found

$$U(x) = \frac{1}{2}kx^2. \tag{13–18}$$

POTENTIAL ENERGY FOR MASS ON A SPRING

In Eq. (13–18) zero potential energy has been chosen at the equilibrium position of the spring, $x = 0$. The kinetic energy is simply

$$K = \frac{1}{2}mv^2. \tag{13–19}$$

Because both x and v are known for simple harmonic motion from Eqs. (13–1) and (13–7), the variation in time of U and K can be plotted. If we write the argument $\omega t + \delta$ as θ, we have

$$U = \frac{1}{2}kA^2 \sin^2 \theta, \tag{13–20}$$

and using $\omega^2 = k/m$ [Eq. (13–15)],

$$K = \frac{1}{2}mA^2 \omega^2 \cos^2 \theta = \frac{1}{2}kA^2 \cos^2 \theta. \tag{13–21}$$

Figure 13–10 is a plot of the potential and kinetic energy functions as θ varies between 0 and 2π, which corresponds to a complete cycle. Both $\sin^2 \theta$ and $\cos^2 \theta$ vary between 0 and 1; when $\sin^2 \theta$ is a minimum, $\cos^2 \theta$ is a maximum and vice versa. Thus U and K each vary between 0 and $kA^2/2$. Suppose that an object attached to a spring starts at the origin and moves to the right, motion you can follow on the graphs of Fig. 13–10. At the origin the potential energy is zero and K is a maximum. As the mass moves to the right, it slows until it has reached its turnaround point at one-quarter cycle, where the velocity and hence K are zero. Because x is at its maximum here, U is also a maximum. The mass now moves to the left, gaining speed until the speed is a maximum as it passes

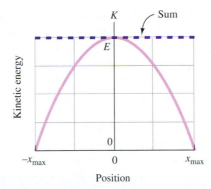

FIGURE 13–10 The potential energy and the kinetic energy of a mass in simple harmonic motion plotted over one cycle (a) as a function of θ, with the origin at the equilibrium point, and (b) as a function of displacement x. When one is a maximum, the other is a minimum, and their sum, the total energy, is conserved.

through the origin once more. Here, after one-half cycle, K is a maximum and U is a minimum. Finally, at the left-hand turnaround point, K is a minimum and U is a maximum. *The energy flows back and forth between U and K.*

The Total Energy

The total energy, $E = U + K$, must be constant. We have (again, $\theta = \omega t + \delta$)

$$E = \frac{1}{2}kA^2 \sin^2 \theta + \frac{1}{2}kA^2 \cos^2 \theta$$

$$= \frac{1}{2}kA^2[\sin^2 \theta + \cos^2 \theta]. \qquad (13\text{–}22)$$

Because the sum of $\sin^2 \theta$ and $\cos^2 \theta$ is unity for any θ, E is indeed constant in time:

$$E = \frac{1}{2}kA^2. \qquad (13\text{–}23)$$

TOTAL ENERGY OF MASS ON A SPRING

The dependence of energy on the square of the amplitude is typical of simple harmonic motion.

EXAMPLE 13–5 A mass m attached to a spring of spring constant k is stretched a length X from its equilibrium position and released with no initial motion. (a) What is the maximum speed attained by the mass in the subsequent motion? (b) At what time is this speed first attained?

Strategy For part (a) the conservation of energy is a useful tool. Initially all the energy is potential, and the maximum speed occurs later, when all the potential energy is converted to kinetic energy. Once we know the maximum kinetic energy, we also know the maximum speed. For part (b) we are asked about time, and we need more

information than energy alone can supply. However, we can use our knowledge that in spring motion the potential energy is zero when the mass passes through the origin, and that time is one-quarter period later than the time it is at a maximum extension, which in this case is the starting point of the motion.

Working It Out (a) Just before the mass is released from rest at a position $x = X$, all of its energy is potential energy; that is, the total energy is

$$E = \frac{1}{2}kX^2.$$

(continues on next page)

This agrees with Eq. (13–23) because the maximum displacement of the motion is, by definition, the amplitude of the motion. E is the value of the energy at all times. When the maximum speed is attained, all the energy is in the form of kinetic energy:

$$\frac{1}{2}mv_{max}^2 = E = \frac{1}{2}kX^2.$$

We solve for v_{max}:

$$v_{max} = \sqrt{\frac{k}{m}}X = \omega X.$$

(b) The maximum speed is attained when $x = 0$ (zero potential energy). The mass is released at the maximum value of x, so the first time the mass passes through the origin is one-quarter period later:

$$t = \frac{T}{4} = \frac{1}{4}\frac{2\pi}{\omega} = \frac{1}{4}2\pi\sqrt{\frac{m}{k}} = \frac{\pi}{2}\sqrt{\frac{m}{k}}.$$

What Do You Think? We asked for the first time the maximum speed is attained, implying that this maximum speed is attained a second time at least. How many times is the maximum speed attained?

It's Not Just About Springs

The motion described in this chapter is of universal importance because *almost all systems that are in stable equilibrium exhibit simple harmonic motion when they depart slightly from their equilibrium position.* Everyday experience bears this out in a qualitative way. For example, a marble nudged a little from its stable equilibrium at the bottom of a bowl rolls back and forth, a child's swing will move back and forth through the stable equilibrium position when it is disturbed, and an automobile rocks up and down on its worn shock absorbers. It is obvious that this motion is oscillatory, and as we'll argue below, it is also simple harmonic motion as long as the amplitude of the oscillations is small enough. Table 13–1 gives a sampling of the range of periods of mechanical systems that move in simple harmonic motion.

TABLE 13–1 • Periods of Mechanical Systems in Simple Harmonic Motion	
Mechanical System	**Period (s)**
Sloshing of water in a tidal basin or large lake	10^2 to 10^4
Large structures (bridges, buildings)	> 1
Strings or air columns of musical instruments	5×10^{-2} to 10^{-4}
Piezoelectric crystals, ultrasound generators	10^{-5} to 5×10^{-1}
Vibrations in molecules	10^{-14}

The discussion of energy in this chapter tells us why simple harmonic motion occurs in these situations. For a spring, and indeed for every case of stable equilibrium, a mass is confined to a *potential energy well* (Fig. 13–11). A potential energy well has a *minimum* on a graph of potential energy versus a position variable. For a spring the position variable x is the stretch of the spring, and the minimum potential energy occurs at zero stretch, the position of stable equilibrium. In this case the potential energy function is parabolic in x; it is proportional to x^2. The reason that almost any oscillation about a stable equilibrium point is simple harmonic motion is that in most cases any minimum in a potential energy–versus–position curve is a parabola close enough to the minimum point, at least if the amplitude of the motion is not too large.

The Taylor expansion (Appendix IV–8) is a very general mathematical result that allows us to see why the minimum of a potential energy well forms a parabola and explains why simple harmonic motion is universal near equilibrium. Suppose we apply the Taylor expansion to a potential energy function near a minimum. Let's label the position of the minimum as the origin, $x = 0$. Then the Taylor expansion says that

$$U(x) = \{U(0)\} + x\{U'(x)|_{x=0}\} + (x^2/2)\{U''(x)|_{x=0}\} + \cdots$$

where we have labeled differentiation with respect to x with a prime. The quantities in curly brackets in this expression are constants, and the variable x no longer appears in them. The constant $U(0)$ plays no physical role, and as we know, we can always replace it by 0. (This is implicit in the expression $U = \frac{1}{2}kx^2$ that applies for the spring itself.) The first derivative of U at $x = 0$ is zero because that is a minimum point. Thus, if we keep the first nonzero term in the Taylor expansion—and this is a good approximation if x remains small, so our result refers to small oscillations—we find

$$U(x) \cong (x^2/2)U''(0). \tag{13–24}$$

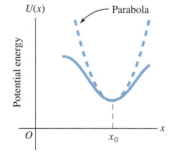

▲ FIGURE 13–11 A potential well, in which potential energy has a minimum at $x = x_0$. This point is a point of stable equilibrium. The dashed line is a parabola that matches the minimum of the well.

[Here we have used the notation $\{U''(x)|_{x=0}\} = U''(0)\}$. This is indeed in the form of a spring force, with $U''(0)$ playing the role of the spring constant. Thus the force takes the general form near the equilibrium point:

$$F(x) = -\frac{dU}{dx} \cong -U''(0)x. \qquad (13\text{--}25)$$

The force is proportional to the displacement and in a direction opposite to the displacement. It is the familiar linear restoring force of the spring.

We can conclude that almost all stable equilibrium behavior is simple harmonic motion close to the equilibrium point. (The "almost" is present as it is conceivable that a force might have a potential for which the term $U''(0)$ is zero. This requires, however, a restoring force of a very special form, and for these cases you would have to go to the term of order (x^3) in U to find the leading term. Figure 13–11 shows how a minimum on a potential energy curve can be approximated by a parabola, the dashed curve in Fig. 13–11.

EXAMPLE 13–6

A mass m on a frictionless table is attached to two pegs by springs with spring constants k_1 and k_2, respectively. The mass can move along the straight line between the pegs. The separation between the pegs has been arranged so that each spring is in its relaxed position, neither stretched or compressed, when the mass is placed at an equilibrium position. What is the motion of the mass when it is displaced from this position? In particular, assuming the motion is periodic, what is the period?

Setting It Up In Fig. 13–12 we show in part (a) the mass at equilibrium, at the point $x = 0$ where there is no force on the mass from either spring; in part (b) the mass is displaced to position $x \neq 0$ as indicated. We measure x positive to the right. The physical situation implies a point of equilibrium at $x = 0$. When the mass moves away from $x = 0$, the forces tend to send it back to that point, so it is stable in this position. We want to show that when the mass is displaced from $x = 0$, the net force is a linear restoring force, and then find the period of the harmonic motion.

Strategy The motion is one-dimensional, along the line between the pegs. We find the net force acting on the mass, which is a force composed of the forces from the two springs. From the general dis-

cussion of stable equilibrium, we expect that the net force will be proportional to the displacement x, and the coefficient will give us the net, or effective, spring constant. From this we can deduce the period of the motion. To calculate the net force, we simply add the two forces, taking into account their signs.

Working It Out We let positive values of force be to the right, which takes care of the vector aspect of this problem. From Fig. 13–11 we see that for the displacement shown, the force from the left-hand spring is $F_1 = -k_1x$, while the force from the right-hand spring is similarly $F_2 = -k_2x$. [You can check that the signs are correct: With x positive (to the right), the left-hand spring is stretched and its force is to the left, while the right-hand spring is compressed and its force is also to the left.] Adding, the net force on the mass is

$$F_{\text{net}} = F_1 + F_2 = -(k_1 + k_2)x.$$

Thus the two springs together act as a single spring with effective spring constant

$$k_{\text{eff}} = k_1 + k_2.$$

The motion is simple harmonic motion, with period

$$T = 2\pi\sqrt{\frac{k_{\text{eff}}}{m}} = 2\pi\sqrt{\frac{k_1 + k_2}{m}}.$$

Alternative Strategy A different strategy utilizes the potential energy in the two springs. For the displacement of the figure, the potential energy in springs 1 and 2 are

$$U_1 = \frac{1}{2}k_1x^2 \quad \text{and} \quad U_2 = \frac{1}{2}k_2x^2,$$

respectively—we have chosen the zero of potential energy at $x = 0$ for each spring. The total potential energy is then

$$U = U_1 + U_2 = \frac{1}{2}(k_1 + k_2)x^2.$$

This is a harmonic oscillator potential energy for a spring with spring constant $k_1 + k_2$, so we get a period corresponding to spring constant $(k_1 + k_2)$, the same result we obtained using forces.

▲ FIGURE 13–12 In (a) the mass is at its equilibrium position (no net force acts on it). In (b) the mass is no longer in the equilibrium position, and it feels a force from both springs.

What Do You Think? Suppose that the initial separation between the pegs were larger than that of the problem. Would there still be harmonic motion for movement on the line between the pegs?

▲ **FIGURE 13–13** The simple pendulum, illuminated by a strobe light at equal time intervals. Half of a complete cycle is imaged. The pendulum bob moves faster near the bottom of its swing and more slowly near the ends. Here the motion is periodic but not harmonic.

(a)

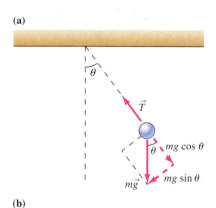

(b)

▲ **FIGURE 13–14** (a) Geometry of the simple pendulum. (b) Force diagram for the simple pendulum. The tension, $\vec{T}$, is along the radial direction, but $m\vec{g}$ has components along both the radial direction and tangent to the arc of the circle traced out by the motion of the mass.

13–5 The Simple Pendulum

A particularly important example of simple harmonic motion is the ordinary pendulum, most commonly observed in some clocks. For centuries, such clocks were the most accurate way to measure time, and the pendulum clock, now more a curiosity or a work of art than a scientific instrument, was crucial in setting standards for time. When we idealize the form of the pendulum to a point mass suspended from a massless string of length ℓ, as in Fig. 13–13, we have the **simple pendulum**. The mass moves along the arc of a circle traced out by the end of the taut string (Fig. 13–14a).

Let's first look at how Newton's second law applies. Suppose that the string makes an angle θ with the vertical. The force diagram for the mass, Fig. 13–14b, includes the force of gravity, $m\vec{g}$, and the tension of the string, $\vec{T}$. The tension is perpendicular to the path of the motion. Its only role in the motion is to constrain that motion to lie along the arc of a circle of radius ℓ. The position s of the mass along the arc of the circle is given by

$$s = \ell\theta, \tag{13–26}$$

where s is measured from $\theta = 0$. The angle θ varies with time, and we wish to determine just how it varies.

To obtain the velocity along the arc of the circle, we differentiate s with respect to time. Because ℓ is a constant, we find

$$v = \frac{ds}{dt} = \ell\frac{d\theta}{dt}. \tag{13–27}$$

The tangential acceleration, which is the component of the total acceleration along the arc of the circle, is in turn associated with changes in the magnitude of this velocity:

$$a = \frac{dv}{dt} = \ell\frac{d^2\theta}{dt^2}. \tag{13–28}$$

This component of the total acceleration (which should *not* be confused with the centripetal component of the acceleration) is due to the tangential force component

$$F_t = -mg\sin\theta. \tag{13–29}$$

The sign of this force component is important. It is negative when θ is itself positive—when the mass is on the right side of the vertical—and in the positive direction when the mass is on the left side of the vertical. This means that *the force of gravity always acts to bring the mass back to the vertical*. This is enough to ensure that the motion will be oscillatory but not enough to guarantee simple harmonic motion: In simple harmonic motion the force must be *linear* in the dynamical variable itself—in this case, the angle.

Using Eq. (13–28) for the acceleration and Eq. (13–29) for the force, we see that Newton's second law takes the form

$$m\ell\frac{d^2\theta}{dt^2} = -mg\sin\theta.$$

Canceling the mass from this equation, we get

$$\ell\frac{d^2\theta}{dt^2} = -g\sin\theta. \tag{13–30}$$

Equation (13–30) would satisfy our requirement for simple harmonic motion if instead of $\sin\theta$, the angle θ itself appeared on the right-hand side. But $\sin\theta$ is in fact very close in value to θ when θ is small. To see this, refer to Fig. 13–15a for $\sin\theta$ and θ, which gives

$$\sin\theta = \frac{x}{\ell} \quad \text{and} \quad \theta = \frac{s}{\ell}, \tag{13–31}$$

where s is the arc length corresponding to angle θ. We can see from Fig. 13–15b that x comes closer and closer to the value of s as θ becomes smaller. Thus, for small θ, $\sin\theta \cong \theta$. This can also be seen from the Taylor expansion for the sine function:

$$\sin\theta = \theta - \frac{\theta^3}{3!} + \frac{\theta^5}{5!} - \cdots. \qquad (13\text{–}32)$$

When θ is small, the terms of order θ^3, θ^5, and so forth can be ignored, justifying our approximation. For example, when $\theta = 0.2$ rad (about 11°), the difference between $\sin\theta$ and θ is about 1 percent, but when $\theta = 0.1$ rad (about 6°), the difference is only about 0.1 percent.

The small-θ approximation demonstrates that a pendulum will have true simple harmonic motion only for small excursions. With the small-excursion approximation $\sin\theta \cong \theta$ taken as an equality, Eq. (13–30) becomes for small θ

$$\ell\frac{d^2\theta}{dt^2} = -g\theta. \qquad (13\text{–}33)$$

As long as the approximation is good, this equation is *precisely* the equation for simple harmonic motion, as we can see if we compare it to Newton's second law for the spring, $md^2x/dt^2 = -kx$. We need change only some variable names and constants to go from the equation of simple harmonic motion for a mass-spring system to that for a pendulum. Using horizontal arrows to indicate name substitutions,

$$x \to \theta, \quad k \to g, \quad \text{and} \quad m \to \ell \qquad (13\text{–}34)$$

summarizes the changes needed. The solution to the motion of the simple pendulum for small angles is then taken directly from the solution for the motion of the spring:

$$\theta = \theta_0 \sin(\omega t + \delta) \quad \text{with} \quad \omega = \sqrt{\frac{g}{\ell}}. \qquad (13\text{–}35)$$

The quantities θ_0 and δ are determined from initial conditions. θ_0 is the amplitude of the angular motion, the maximum angle attained. Again, it is the dynamics of the motion that determines the angular frequency ω—you cannot adjust it by a change of initial conditions. The period and frequency of the pendulum's motion come from Eqs. (13–2) and (13–3):

$$T = 2\pi\sqrt{\frac{\ell}{g}} \quad \text{and} \quad f = \frac{1}{2\pi}\sqrt{\frac{g}{\ell}}. \qquad (13\text{–}36)$$

Just as for a mass on an ideal spring, *the period of the small-amplitude pendulum is independent of the amplitude.*

(a)

(b)

▲ **FIGURE 13–15** Comparing the horizontal displacement x of the mass of a simple pendulum with the arc length s traced out by the motion. These two lengths differ by a larger percentage (a) when θ is large than (b) when θ is small.

THINK ABOUT THIS . . .

WHY WERE PENDULUM CLOCKS SO USEFUL?

Why did the pendulum play such a central role in the development of time standards? The meaning of standards is that they are easily reproducible—two persons can have two different clocks and still agree on what constitutes a given time period. The pendulum clock fills this bill because its period is independent of amplitude as long as we stick to small amplitudes, so that the first term of Eq. (13–32) dominates, and $\sin\theta \cong \theta$. The fact that $\sin\theta$ is a series in odd powers rather than all powers of θ means that the first correction to $\sin\theta \cong \theta$ is two powers of θ down rather than just one, and $\sin\theta \cong \theta$ is therefore an especially good approximation. The result of all this is that one can construct pendulum clocks that keep the same time without worrying that their amplitudes have to

match. One only needs to make sure that the length of different pendula are the same, and this can easily be arranged by screw mechanisms below the mass to allow small adjustments to the length. Once two pendulum clocks have pendula of the same length, one does not have to worry about whether they run at the same amplitude. Similarly, a given clock will keep consistent time even if perturbations cause its amplitude to change (although the driving mechanism is meant to hold the amplitude constant). The time is *not* consistent, however, if there are effects, in particular changing temperatures, that change the length of the pendulum, and indeed clockmakers struggled to construct pendulums whose length would not change under varying conditions.

EXAMPLE 13–7 A simple pendulum 2.00 m long is suspended vertically in a region where $g = 9.81 \text{ m/s}^2$. The point mass at the end is displaced from the vertical and given a small push, so its maximum speed is 0.11 m/s. What is the maximum horizontal displacement of the mass from the vertical line it makes when at rest? Assume that all the motion takes place at small angles.

Setting It Up Figure 13–15b can help us with the geometry of this situation. In the figure we have labeled the pendulum length ℓ, the angle θ of the swing, and the horizontal displacement x of the mass, with $x = 0$ at the bottom of the swing. We can also use it to solve x in terms of θ. We have drawn a sequence of the motion in Fig. 13–16. We want the maximum horizontal distance $x = x_{max}$ reached by the swinging mass, assuming always that the swing covers only small angles. We are given the maximum speed v_{max}.

Strategy θ varies harmonically, with a known angular frequency ω. By calculating v, we find a formula in which the amplitude appears. We have information on the maximum velocity v_{max}, which occurs at the bottom of the swing ($\theta = 0$), and this will allow us to find the amplitude of the swing. Knowing the amplitude, we can find the maximum value of θ and, using geometry, the maximum value of x.

Working It Out We know $\theta = \theta_0 \sin(\omega t + \delta)$, where ω is the angular frequency. The arc length s traced by the mass at the end of the string is given by $s = \ell\theta$, where ℓ is the string's length. Thus s also varies harmonically: $s = A \sin(\omega t + \delta)$, where $A = \ell\theta_0$. The velocity is the rate of change of s (understood to be in the tangential direction), so we write $v = ds/dt$. Thus we have

$$v = \frac{d}{dt}[A \sin(\omega t + \delta)] = A\frac{d}{dt}[\sin(\omega t + \delta)]$$

$$= A\omega \cos(\omega t + \delta).$$

From this expression, we see that v varies harmonically, with amplitude $A\omega$. The maximum magnitude of v, v_{max}, is thus $A\omega = 0.11 \text{ m/s}$. We can find ω from Eq. (13–35):

$$\omega = \sqrt{\frac{g}{\ell}} = \sqrt{\frac{9.81 \text{ m/s}^2}{2.00 \text{ m}}} = 2.21 \text{ rad/s}.$$

Thus from $v_{max} = A\omega$,

$$A = \frac{v_{max}}{\omega} = \frac{0.11 \text{ m/s}}{2.21 \text{ rad/s}} = 0.050 \text{ m}.$$

As we can see from Fig. (13–15b), this is to a good approximation also the horizontal displacement of the mass as long as the angle is small. Our result for maximum displacement, 5.00 cm, is indeed small compared to the length, so our small-angle approximations are good.

What Do You Think? If the mass had been pushed a little harder, the maximum angle would have been larger, and at some point you could no longer trust the small-angle approximation. How would you go about deciding whether the small-angle approximations still hold at the larger angle?

▲ **FIGURE 13–16** Sequence of motion for a simple pendulum.

The Energy of the Simple Pendulum

Let's now consider the simple pendulum from the point of view of energy. The kinetic energy, K, which is a function of θ, is found by expressing the speed as a function of θ. We can use Eq. (13–27):

$$K(\theta) = \frac{1}{2}mv^2 = \frac{1}{2}m\ell^2\left(\frac{d\theta}{dt}\right)^2. \qquad (13\text{–}37)$$

The only force with a component along the motion of the mass is gravity. This force is conservative, and to find the associated potential energy function, U, we can express the height h gained by the mass in terms of θ. From Fig. 13–15a, we see that $h = \ell - \ell \cos \theta$, so that

$$U = mgh = mg\ell(1 - \cos \theta). \tag{13–38}$$

Here we have taken zero potential energy to be at the bottom of the swing, at $\theta = 0$. In a small-angle approximation, $\cos \theta \cong 1 - (\theta^2/2)$, and

$$U(\theta) \cong mg\ell\left(1 - 1 + \frac{1}{2}\theta^2\right) = \frac{1}{2}mg\ell\theta^2. \tag{13–39}$$

Comparing Eqs. (13–37) and (13–39) for the kinetic and potential energies with their counterparts for the spring, Eqs. (13–19) and (13–18), we see that they have the same form, with just the name changes of Eq. (13–34).

CONCEPTUAL EXAMPLE 13–8 Two teams of students are each given a stopwatch and a simple pendulum whose length, around 1 m, has been precisely measured. They are asked to measure the acceleration of gravity, g. One team pulls their pendulum back a small amount and measures the time for 1 oscillation. The second team releases the pendulum from the same point as the first and lets it swing for a minute, carefully counting the number of oscillations as the amplitude drops to a low value. Which team will come up with the more accurate value of g?

Answer We can assume that the initial displacement in each case is small enough so that the period is correctly given by Eq. (13–36). In that case the precision of each team is limited by the ability of the member with the stopwatch to start and stop the watch at the beginning or end of a full swing. Reflex times for stopwatches are on the order of a tenth of a second, and we'll take this as a measure of the uncertainty Δt in the measurement of a time period. The percentage error made by the first team in its measurement of the period T is $\Delta t/T$, and we can estimate from experience [or from Eq. (13–36)] that a pendulum with length of around 1 m has a period of around 2 s. Thus the percentage error in the time measurement is roughly 5%. That translates into a 10% error in the measurement of g, as g depends on T^2 (see Section 1–3). If the second team keeps track of the number of oscillations, which is just an integer number N and not likely subject to an error, they will have a fractional error in their time measurement of $\Delta t/(NT)$. Over a minute, say, the pendulum will make 30 swings, so the error is (5%)/30, or only about 0.2%. The error in g is diminished by the same factor of 1/30. It is often more accurate to make a cumulative measurement.

13-6 More About Pendulums

The Physical Pendulum

A pendulum is not limited to a massless string with a pointlike mass at the end of it. When a suspended, swinging object has some other form, we call it a *physical pendulum*. Any object can be suspended from any point on it to act as a physical pendulum. For example, when you are walking, the leg that is not in contact with the ground swings and acts as a pendulum. Such examples illustrate the point that oscillatory motion is a general characteristic of motion about a stable equilibrium. We can approach such problems by recasting them as a spring problem, through analysis of either the potential energy or the dynamical equation, and then use the results we have already found for the spring.

From previous chapters, we have built up the necessary tools to handle the physical pendulum: We study the torque on it, τ, and the corresponding angular acceleration. Figure 13–17 illustrates a physical pendulum allowed to pivot through some horizontal axis, called y, which defines the vertical plane of the swinging. We take the y-axis to be into the page and through the oscillation point. We need to know the rotational inertia, I, about the pivot axis, the total mass M of the object, and the distance r from the center of mass to the pivot axis. The stable equilibrium point for this object is $\theta = 0$, when the center of mass hangs directly below the axis. When $\theta \neq 0$, only the force of gravity, which acts on the center of mass, exerts a (restoring) torque. From Eq. (9–29) the equation of motion governing the behavior of the object is

$$\tau = I\alpha, \tag{13–40}$$

where α is the angular acceleration about the pivot point. The magnitude of α is $d^2\theta/dt^2$.

▲ **FIGURE 13–17** A swinging sign is an example of a physical pendulum. The center of mass of the oscillating sign executes simple harmonic motion about the stable equilibrium position, $\theta = 0$.

The torque is given by $\vec{\tau} = \vec{r} \times \vec{F}$, where $\vec{r}$ is the vector from the pivot to the center of mass and $\vec{F}$ is the force of gravity acting on the center of mass. According to the right-hand rule, this torque is along the $+y$-axis (when θ is positive, as shown); the angular acceleration $\vec{\alpha}$ must therefore also be along the $+y$-axis, which corresponds in Fig. 13–17 to an angular acceleration that brings the object back to equilibrium. (Both $\vec{\tau}$ and $\vec{\alpha}$ would change signs if the object were drawn with θ on the other side of the vertical axis.) The magnitude of the torque is

$$\tau = rF \sin \theta, = rMg \sin \theta, \qquad (13\text{–}41)$$

so the equation of motion, Eq. (13–41), becomes

$$Mgr \sin \theta = -I \frac{d^2\theta}{dt^2}. \qquad (13\text{–}42)$$

Equation (13–42) has exactly the same form as the equation governing the motion of the simple pendulum, which is Eq. (13–30). If the motion is restricted to small angles, then $\sin \theta$ can be replaced with θ itself and, as for the simple pendulum, simple harmonic motion follows: $\theta = \theta_0 \sin(\omega t + \delta)$. To find ω, it is necessary only to replace ℓ and g for the simple pendulum with I and Mgr, respectively, for the physical pendulum. Thus, from Eqs. (13–35) and (13–36),

$$\omega = \sqrt{\frac{Mgr}{I}}, \qquad (13\text{–}43a)$$

$$T = 2\pi \sqrt{\frac{I}{Mgr}}. \qquad (13\text{–}43b)$$

Because I is always M multiplied by some length squared, M will cancel from Eqs. (13–43). *The period of a physical pendulum is independent of its total mass—only how the mass is distributed matters.* We can verify that the period of the physical pendulum as expressed in Eq. (13–43b) reduces to the simple pendulum when the swinging object is a point mass m on a massless string of length ℓ. In that limit, $I = m\ell^2$, $r = \ell$, $M = m$, and T reduces to the appropriate value,

$$T = 2\pi \sqrt{\frac{m\ell^2}{mg\ell}} = 2\pi \sqrt{\frac{\ell}{g}}.$$

EXAMPLE 13–9 A thin, uniform rod of mass M and length L swings from its end as a physical pendulum. What is the period of the oscillatory motion for small angles? Find the length ℓ of the simple pendulum that has the same period as the swinging rod.

Setting It Up Figure 13–18 illustrates the situation.

Strategy Equation (13–43b) gives the period of a physical pendulum swinging in small angles and thus is directly applicable here. Its use requires us to know the rotational inertia of a rod about an axis through its end. This quantity was calculated in Eq. (9–26) and is listed in Table 9–1. For the second part, we compare the period for our pendulum with that of a simple pendulum of length ℓ and solve for ℓ.

Working It Out From Chapter 9 the rotational inertia of our rod is

$$I = \frac{1}{3}ML^2.$$

As we can see in Fig. 13–18, the distance of the center of mass from the end is $r = L/2$, and Eq. (13–43b) then gives the period of the motion as

$$T_{\text{rod pendulum}} = 2\pi \sqrt{\frac{I}{Mgr}} = 2\pi \sqrt{\frac{ML^2/3}{MgL/2}} = 2\pi \sqrt{\frac{2}{3}\left(\frac{L}{g}\right)}.$$

The length ℓ of a simple pendulum of the same period is determined by

$$T_{\text{simple pendulum}} = 2\pi \sqrt{\frac{\ell}{g}} = 2\pi \sqrt{\frac{2}{3}\left(\frac{L}{g}\right)};$$

$$\ell = \frac{2}{3}L.$$

What Do You Think? If a toy monkey is attached to the bottom of the rod, does the period (a) increase; (b) decrease; (c) remain unchanged; or (d) change in a way that cannot be determined from the information given?

▲ **FIGURE 13–18** A thin, uniform rod swings as a physical pendulum.

13–7 Damped Harmonic Motion

Almost all physical systems, including masses, springs, and other oscillating systems, are affected by friction or drag (resistive) forces. These forces tend to remove energy from a moving system and thereby slow it down, or damp its motion. The universality and importance of these forces is evident in the world around us: Without "pumping," a child on a swing slows down to a stop over a few minutes, and a marble in a bowl will not oscillate indefinitely. When energy is lost in this way, the motion is said to be *damped*. From Eq. (13–23) we can see that a decrease in energy implies a decrease in amplitude. If a clock pendulum is to maintain a given amplitude, energy must be supplied from, say, a wound spring or hung weights.

What are the *quantitative* effects on a mass-spring of these nonconservative, or dissipative, forces? We can answer in the case of a drag force $\vec{F}_d$ that is proportional to velocity (Fig. 13–19):

$$\vec{F}_d = -b\vec{v} = -b\frac{d\vec{x}}{dt}, \qquad (13\text{--}44)$$

where b is the *damping coefficient* (or *damping parameter*). The minus sign indicates that this force is always opposite to the direction of motion. For this case the equations of motion are solvable in analytic form, and we can use these solutions as a guide to the behavior of other damped systems. For a one-dimensional system, the equation of motion is

$$-kx - b\frac{dx}{dt} = m\frac{d^2x}{dt^2}. \qquad (13\text{--}45)$$

By finding the function $x(t)$ that satisfies this equation, we will have found the position of a mass on a spring with damping. To solve differential equations such as Eq. (13–45), we attempt trial solutions that are educated guesses and see if they work. As a guess to the solution of Eq. (13–45), we keep a sinusoidal component but also incorporate a decreasing term due to damping. Our trial solution is

$$x = Ae^{-\alpha t}\sin(\omega' t + \delta), \qquad (13\text{--}46)$$

where A, δ, and α are constants to be determined. We use a frequency ω' rather than ω_0, the frequency in the absence of drag, because we want to allow for the possibility that the frequency is changed by the damping. (In this context and in the context of the next section, ω_0 is called the **natural frequency**. As we have emphasized throughout, any

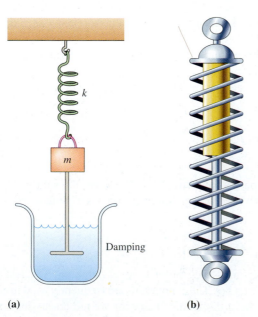

Damping

(a)

(b)

◀ **FIGURE 13–19** (a) A damped oscillator. (b) An automobile shock absorber is a damped harmonic oscillator.

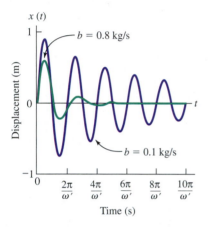

(b)

▲ **FIGURE 13–20** Damping of simple harmonic motion by a drag force $\vec{F} = -b\vec{v}$. Plotted is the function $x(t) = Ae^{-bt/2m} \sin(\omega' t)$ versus t, with $\omega' = \sqrt{k/m - b^2/4m^2}$. The values of the parameters are those chosen in the text.

physical system will have a natural frequency for motion near a stable equilibrium point.) We check whether the trial solution is a good one by seeing if the equation of motion, in this case Eq. (13–45), is satisfied when the trial solution is plugged into it. This procedure is left to Problem 77. The result is that the trial solution is satisfactory provided that

$$\alpha = \frac{b}{2m} \tag{13–47}$$

and that

$$\omega' = \sqrt{\frac{k}{m} - \frac{b^2}{4m^2}} = \sqrt{\omega_0^2 - \frac{b^2}{4m^2}}. \tag{13–48}$$

Thus *the damping factor α and the modified angular frequency ω' are determined by the equations of motion.* (A and δ remain undetermined.) It is easy to check that Eq. (13–46) reduces to standard simple harmonic motion when the damping coefficient $b = 0$.

If we plot our solution [Eq. (13–46)], we see that the falling exponential function forms a kind of *envelope*, marked in Fig. 13–20a by dashed lines. The damping thereby modulates what would otherwise be simple harmonic motion, such that the amplitude of the motion decreases as time goes on. The argument of the exponential is directly proportional to b; that is, to the size of the drag force. Figure 13–20b shows the motion described by Eq. (13–46) for two values of b, one larger than the other. Here we have taken $k = 1$ N/m, $m = 1$ kg, and $A = 1$ m, and in Fig. 13–20b the two values of b are $b = 0.1$ kg/s, and $b = 0.8$ kg/s. In each case the decrease of the amplitude with time is clearly visible, but the motion for $b = 0.1$ kg/s is weakly damped compared to that for $b = 0.8$ kg/s. For both these systems the oscillations continue to some extent and the motion is said to be **underdamped.** The difference between ω' and the angular frequency $\omega_0 = \sqrt{k/m}$ of the spring without drag is harder to see.

As b increases, the angular frequency ω' decreases (and hence the period of the motion increases) until, from Eq. (13–48), $\omega' = 0$ when $b^2 = 4mk$. We refer to this value of b as the *critical* value b_c,

$$b_c = \sqrt{4mk} = \sqrt{4m^2\omega_0^2}. \tag{13–49}$$

When $b = b_c$ the system is said to be **critically damped**; there are no oscillations at all, as is the case when $b > b_c$ and the system is **overdamped.**

Damping is quite frequently introduced to systems that would otherwise oscillate in undesirable ways. Automobile shock absorbers are present for just this purpose (Fig. 13–19).

The quantity $b/2m$, which appears multiplying t in the exponential factor in Eq. (13–46), has dimensions of inverse time because the argument of an exponential must be dimensionless. We therefore define the **lifetime** (or the *mean life*), τ, of the damped oscillator by

$$\tau \equiv \frac{m}{b}. \tag{13–50}$$

In terms of τ, the exponential envelope has the form $e^{-t/2\tau}$. The larger the value of τ, the slower the exponential falloff. Still another nomenclature employs the dimensionless **Q factor**, defined by

$$Q \equiv \omega_0\tau. \tag{13–51}$$

Because Q is proportional to τ, it too measures the amount of damping. The less the damping, the larger are the values of τ and Q. These notations are frequently used in connections with electric circuits that exhibit the kind of behavior we are discussing here.

EXAMPLE 13–10 A gong struck with a hammer can be modeled as a damped harmonic system, with its frequency of oscillation measured by the musical tone emitted and the "loudness" of its sound measured by the amplitude squared (see Chapter 14). A certain gong is struck, and after 9.0 s the loudness has dropped to 0.85 times the original loudness. How much more time will have elapsed before the loudness is 0.25 times the original loudness?

Setting It Up We are given information on the rate of falloff of the loudness, but not a direct value for the lifetime τ characteristic of the system. If, however, we can find τ, then we would know all we need about the loudness at any time t.

Strategy In the damped systems we are studying, the amplitude decrease is governed by the exponential factor $e^{-t/2\tau}$, hence the loudness varies with time as $e^{-t/\tau}$. The information on the loudness at 9.0 s will allow us to find τ; once we have that, the loudness as a function of all time is determined.

Working It Out Denoting the loudness as $L(t)$, we have $L(t) = L(0)\, e^{-t/\tau}$. At 9.0 s, $L = L(0) \times (0.85)$, and therefore

$$e^{-(9.0\,\text{s})/\tau} = 0.85.$$

We invert to find τ. The inverse of the exponential is the natural logarithm, ln (see Appendix IV–6). If we take the natural log of both sides of this equation, we have

$$\ln\!\left[e^{-(9.0\,\text{s})/\tau}\right] = -\frac{9.0\,\text{s}}{\tau} = \ln(0.85) = -0.16$$

and hence $\tau = (-9.0\,\text{s})/(-0.16) = 55$ s.

Knowing τ, we want to find the time for which $L(t)/L(0) = e^{-t/\tau} = 0.25$. We insert $\tau = 55$ s and take the natural log of both sides:

$$\ln\!\left[e^{-t/(55\,\text{s})}\right] = -\frac{t}{55\,\text{s}} = \ln(0.25) = -1.4,$$

or $t = (1.4) \times (55\,\text{s}) = 76$ s.

What Do You Think? Can we tell anything about the period of oscillation from the data given?

13–8 Driven Harmonic Motion

Systems such as a mass on the end of a spring may be subject to external forces. We saw the effect of a constant force (gravity) on an oscillator in Section 13–5. Another common situation involves a repeating external force (we refer to a driving force), as occurs, for example, in any pendulum clock or in a child being pushed on a swing. The simplest example of such a force is a *harmonic driving force*, in which the driving force varies sinusoidally with time (Fig. 13–21); we can treat this case in a straightforward manner. The resulting motion of the mass is called **driven harmonic motion**. (Damped harmonic motion is undriven.)

The most important consequences of driven harmonic motion involve **resonances**. In resonance phenomena, the amplitude of simple harmonic motion grows enormously when the frequency of the driving force matches the natural frequency of the oscillating system. You may have read about the possibility of a bridge being driven to large oscillations and collapsing if soldiers march over it in step with the right rhythm. Less spectacular examples are familiar: the trampoline jumper who, by timing her jumps, can make them much more effective; the swinging child who gets a regular push or who "pumps" in a regular fashion; or the coffee that sloshes out over the edge of the cup when you walk at just the wrong pace. Resonance phenomena are even more important in microscopic situations, as in the action of a microwave oven, where microwave radiation drives the electrons of water molecules with a natural frequency of the molecular system. Resonances also occur in acoustical phenomena, and many musical instruments are driven in harmonic resonance in order to produce their notes. Finally, electric circuits often behave like driven harmonic systems even though they are not mechanical.

It is easy to demonstrate the resonance phenomenon for yourself on a swing or with the simple aid of a loose rubber band tied to a mass such as a kitchen utensil. By extending the suspended band and releasing it, you can get a good idea of the natural frequency of the system and observe the amplitude steadily reduced by the damping. If you then tie the band to one hand, you can drive the system by moving your hand up and down in an approximation of harmonic motion—first with a frequency less than, then greater than, and finally equal to the natural frequency. At resonance, when the hand frequency equals the natural frequency of the system, the oscillation amplitude is dramatically large.

▲ **FIGURE 13–21** A car passing over a rough road is an example of driven, damped harmonic motion. The washboard road provides the force, the shock absorber provides the damping, and the spring provides the harmonic motion.

Equations of Motion for Driven Harmonic Motion

To see how resonance arises, consider a mass subject to a spring force, a drag force proportional to the speed, and an external harmonic driving force F whose time variation is determined by the angular frequency ω, where $F(t) = F_0 \sin(\omega t)$. F_0 is a type of amplitude for the driving force. It is useful to recall the natural frequency of the un-encumbered mass on the spring, $\omega_0 = \sqrt{k/m}$. All motion is in the x-direction, so the x-component of Newton's second law, $\vec{F} = m\vec{a}$, is

$$-kx - b\frac{dx}{dt} + F_0 \sin(\omega t) = m\frac{d^2x}{dt^2}. \tag{13–52}$$

Equation (13–52) appears to be more complicated than the equation of motion with a drag force alone, but if we consider the equation further, we can see that simplification is possible. By definition, the driving force has gone on and will go on forever—we mean, of course, for a long time—and we would expect that eventually the mass would have to move as the driving force dictates—that *the motion of the mass is simple harmonic motion with the frequency of the driving force.* Any transient effects due to friction will have long since died out exponentially. Imagine, for example, a spring with a mass on its end suspended from a harmonically moving hand. In time the mass will move with the frequency of your hand motion, even if the motion of the mass and your hand are not in phase. Therefore we expect that after long times the solution to the equation of motion is

$$x = A\sin(\omega t + \delta). \tag{13–53}$$

This solution can be verified by substitution into Eq. (13–52), and the amplitude A and phase δ are *determined* in this substitution. This is a complicated exercise in algebra, which we forgo. In particular, the physically interesting amplitude of oscillation is

$$A = \frac{F_0}{\sqrt{m^2(\omega^2 - \omega_0^2)^2 + b^2\omega^2}}. \tag{13–54}$$

AMPLITUDE OF HARMONICALLY DRIVEN SYSTEM

This amplitude indeed displays the remarkable property of *resonance*: *It is peaked when the driving frequency, ω, nears the natural frequency, ω_0.* In fact, if there were no damping ($b = 0$), the amplitude would become infinite when $\omega = \omega_0$. This is not a re-alistic physical situation, because it corresponds to the spring being stretched to infinite length. A real spring will snap rather than accept an infinite stretch; in other words, some form of damping will ultimately occur. But it does illustrate that, at resonance, the response of a harmonic system to a driving force can be catastrophically large.

THINK ABOUT THIS . . .

HOW CAN WE REDUCE THE WIND-DRIVEN SWAY OF SKYSCRAPERS?

The engineering of modern skyscrapers uses driven harmonic motion to *reduce* a natural harmonic movement. This is a kind of "anti-resonance" phenomenon. Think about the child on a swing getting a regular push. You know that not only should the driving force have the same frequency as the natural frequency but that it should be applied with a phase that gives a kick when the mass is ready to take the kick as a positive help to its motion. This means that the phase of the driving force should match that of the motion. But if we want to inhibit the motion of an object, we can instead give a regular kick when the kick tends to hinder the natural motion. For example, if you push the child backward just when he or she is coming through at the bottom of the swing, the motion will be inhibited. An application of this idea occurs in skyscrapers, which without a "tuned mass damper" can sway alarmingly in the wind. The damper is a large mass placed within the building that can be moved back and forth with a phase that tends to damp the movement of the building. The first such damper in the United States was placed within the 280-m-tall Citicorp building in New York. A 400-metric ton mass of concrete is slid back and forth within the building, ensuring a relative stability. Today many tall buildings use dynamical damping of this type.

Properties of Resonance

If we plot amplitude as a function of ω for several values of b (Fig. 13–22), we can clearly see the effect of resonance near the natural frequency ω_0. We choose $F_0 = 0.01$ N, $\omega_0 = 1$ rad/s, and $m = 1$ kg. The damping coefficient b is given the values 0.01, 0.05,

◀ **FIGURE 13–22** The amplitude of a harmonic system under the influence of a harmonic driving force with frequency ω. The natural frequency of the harmonic system is $\omega_0 = 1$ rad/s. There is a drag force with damping coefficient b. Resonance occurs at the peak (near $\omega = \omega_0$) and is strongest when b is smallest.

and 0.2 kg/s. From the plot, we can see that the position of the peak amplitude shifts slightly from $\omega_0 = 1$ rad/s when b increases from a small value. From Eq. (13–54) the peak amplitude generally occurs at

$$\omega_{max}^2 = \omega_0^2 - \frac{1}{2}\left(\frac{b^2}{m^2}\right). \tag{13–55}$$

That the peak is less sharp for larger b is sometimes expressed as the fact that the *width* of the peak is a measure of b. More precisely, it is possible to compute the *total width at half-maximum* $\Delta\omega$ of the peak, which we find by evaluating the amplitude where it is one half its peak value and measuring the spread of frequencies to which this corresponds, $\Delta\omega$ (Fig. 13–23). For small b, this width can be shown to be

$$\Delta\omega \cong \frac{2b}{m}. \tag{13–56}$$

Thus the sharpness $\Delta\omega$ of the resonance peak is a direct measure of the damping coefficient, b, divided by m: The smaller the damping, the sharper the resonance peak.

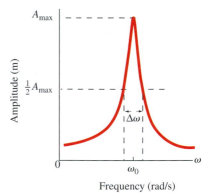

▲ **FIGURE 13–23** The width at half-maximum of a resonant peak near ω_0 is $\Delta\omega$, and is a measure of the damping coefficient of the system.

*Resonance and Uncertainty

We have seen that the amplitude for a driven, damped harmonic oscillator has a resonance peak. We have also seen that the undriven damped harmonic oscillator has a characteristic decay time. These phenomena are closely related, and this relation has important consequences for our ability to construct systems with resonances, such as radio tuners or filters for screening electronic noise.

Equation (13–46) can be used to find the rate at which the energy is dissipated in an undriven damped oscillator. The energy is proportional to the amplitude *squared* and therefore falls as $e^{-t/\tau}$, where $\tau = m/b$ is the lifetime of the motion. The width in frequency of the driven harmonic oscillator is given in Eq. (13–56), and we see that it is inversely proportional to τ. From Eqs. (13–50) and (13–56) we have

$$\tau\,\Delta\omega \text{ is on the order of 1.} \tag{13–57}$$

This equation is known as the **uncertainty principle**: It expresses the impossibility of measurements that are arbitrarily precise in both time and frequency. Strictly speaking, we have derived it only for a special damping force. But it actually represents a very general property. It states that if the damping time of a harmonic oscillator is large, then the resonance width is small, and vice versa. The weaker the damping of a harmonic oscillator, the more sharply it responds to, or selects, a harmonic driving force of the right frequency. What is the significance of this result? One of the important uses of the resonance phenomenon, both in mechanical systems and in electric circuits, is that it allows us to select, or filter, certain frequencies by letting these frequencies act as a driving force on our selector. For example, the tuner on a radio selects a given station by letting in only frequencies near the central frequency of the station. Equation (13–57) sets strong limits on our ability to design filters that can respond to only a narrow range of driving frequencies. The transient effects of such filters die away more slowly with time as the frequencies they select are more and more limited. The results of the uncertainty principle reflect a fundamental property of systems and cannot be avoided with clever design. We'll see later in our discussion of quantum-mechanical phenomena that the uncertainty principle plays an important role in physical systems for which quantum mechanics is important.

Summary

Simple harmonic motion is characterized by a particular type of periodic behavior; namely, sinusoidal time dependence of the position of a moving object. In one dimension this dependence can be written as

$$x(t) = A \sin(\omega t + \delta). \tag{13–1a}$$

The amplitude, A, measures the maximum displacement from equilibrium. The phase, δ, is the angle by which the motion is shifted from $x = 0$ at $t = 0$; the angular frequency, ω, measures how rapidly the motion repeats. Alternatively, the motion can be written as

$$x(t) = a_1 \sin(\omega t) + a_2 \cos(\omega t). \tag{13–1b}$$

The period, T, and frequency, f, of the simple harmonic motion are related to ω by

$$T = \frac{2\pi}{\omega} \tag{13–2}$$

and

$$f = \frac{1}{T}. \tag{13–3}$$

In simple harmonic motion the velocity and the acceleration are also sinusoidal. In particular, the acceleration is proportional to the displacement but opposite in sign:

$$a = -\omega^2 x. \tag{13–8}$$

Uniform circular motion is closely related to simple harmonic motion. It is just the result of simultaneous simple harmonic motion of equal amplitudes and frequencies in the x- and y-directions, 90° out of phase.

Because acceleration is proportional to displacement in simple harmonic motion, the force that leads to this motion is a restoring force proportional to the displacement. The prototype of this is the force exerted on a mass by a spring, Hooke's law:

$$F = -kx, \tag{13–13}$$

where k is the spring constant and x is the displacement of the mass from the equilibrium position of the spring. The resulting angular frequency of the motion of a mass m subject to this force is

$$\omega = \sqrt{\frac{k}{m}}. \tag{13–16}$$

The period of the motion is independent of the amplitude.

The potential energy of a mass subject to a spring force is

$$U(x) = \frac{1}{2}kx^2. \tag{13–18}$$

It oscillates with time as $\sin^2(\omega t + \delta)$, whereas the kinetic energy oscillates with time as $\cos^2(\omega t + \delta)$, and the total energy, E, which is their sum, remains constant with time:

$$E = \frac{1}{2}kA^2. \tag{13–23}$$

The simple pendulum, consisting of a mass on the end of a light string of length ℓ, also undergoes simple harmonic motion when it is not allowed to swing too far. In this case, it is the angle with respect to the vertical that varies sinusoidally, and the period of the motion is independent of the mass as well as the amplitude:

$$T = 2\pi\sqrt{\frac{\ell}{g}}.$$ (13–36)

Almost all small departures from stable equilibrium situations exhibit simple harmonic motion. Two examples are a bead rolling back and forth across the bottom of a bowl and a physical pendulum. The period of a physical pendulum depends on the geometric factor contained in its rotational inertia about the axis through which it swings, but does not depend on the object's mass.

The presence of additional velocity-dependent drag, or resistive forces, causes the amplitude of a particle that moves under the influence of springlike forces to decrease, or damp. When a system that by itself would move in simple harmonic motion is driven by a force with sinusoidal time dependence, the system moves with the frequency of the driving force. The amplitude of the resulting motion of the system shows resonant behavior when the frequency of the driving force equals the natural frequency of the system. The width of the resonance peak is inversely related to the exponential rate of falloff of the undriven system due to damping—a result known as the uncertainty principle.

Understanding the Concepts

1. Not all periodic motion is harmonic. Sketch the pattern of motion of a ball that falls on a flat surface and bounces back, in the idealized case of a perfectly elastic bounce.
2. Describe an experiment to determine the spring constant of a spring.
3. If the amplitude of simple harmonic motion is doubled, what happens to the maximum kinetic energy?
4. Will a pendulum clock lose time or gain time when it is taken from sea level to the top of a mountain? Ignore all damping effects.
5. In the previous question we suggested that all damping effects be ignored. Suppose that you could not ignore damping effects due to the air, which becomes less dense as you gain in altitude. Why might you find it difficult to predict whether a pendulum clock gains or loses time when it is taken from sea level to the top of a mountain?
6. In Section 13–3 we showed that when a constant force is added to a spring force and the two forces are aligned in the same direction, the motion remains periodic and with the same period; only the equilibrium point shifts. Is the same true when the constant force is aligned with a different direction than that of the spring force?
7. A damped harmonic oscillator driven by a harmonic external force maintains a steady oscillatory motion. Is energy lost to friction in the motion? If so, what keeps the oscillator moving?
8. The length of a simple pendulum is doubled, and the mass at the end is halved. What happens to the period?
9. Grandfather clocks with long pendulums have a screw device at the end to adjust the period and keep the clock in time. Sometimes it is necessary to make a winter–summer adjustment to account for changes in pendulum length associated with temperature changes. If the clock is running slow, how should the length of the pendulum be changed?
10. What is the phenomenon that allows you to increase the amplitude of your motion when you swing on a swing?
11. Suppose you stand on a swing instead of sitting on it. Will your frequency of oscillation increase or decrease?
12. It can be shown that if the frequency, the energy, or the amplitude of a harmonic oscillator changes very slowly then, although the energy E is no longer constant, the ratio E/ω, where ω is the angular frequency, does not change. Suppose you have a pendulum whose maximum angle of deflection is θ_0. If the bob at the end of the pendulum consists of fine sand in a spherical container, and the sand leaks very slowly out the bottom, what will happen to the motion?

13. Make use of the assertion in Question 12 to discuss what happens to the period and amplitude of a pendulum in which the length is very slowly decreased by a factor of 1.5.
14. We say the period of a physical pendulum doesn't depend on the pendulum's mass. Does it depend on the distribution of the mass? In what way?
15. Discuss if and why each of the following is an example of simple harmonic motion, damped harmonic motion, and/or driven harmonic motion: (a) leaves blowing on a tree limb; (b) children seesawing; (c) a child swinging on a playground swing; (d) a car bouncing up and down after hitting a large pothole in the road; and (e) water sloshing back and forth in a tub.
16. Consider the pendulum of Example 13–9. Suppose you could adjust the place along the rod from which the rod could be hung, and that you hung it from a spot very near to the center of mass of the rod than in the example. Would the period be shorter or longer?
17. A mass is suspended from two ropes of equal length, attached to different points on a ceiling. Would small oscillations about the point of stable equilibrium represent simple harmonic motion?
18. A simple pendulum acts as a *conical pendulum* when the mass moves in a horizontal circle (Fig. 13–24). What force keeps it moving in a circle? Does the total energy consist of potential and kinetic energies that each vary in this case?

▲ **FIGURE 13–24** Question 18.

19. Consider a pendulum with *large*-angle oscillations. Will the motion still be periodic? Will the period be longer or shorter than for a small-angle oscillation?

20. Suppose that there were a harmonic driving force on a mass attached to a horizontal spring and that very little damping occurs. What could supply the energy that would allow the amplitude to become enormously large near resonance?

21. What are you doing when you adjust your bounce up and down on a diving board to get a big boost for your dive?

22. Suppose you studied a pendulum in a place where a steady horizontal wind blew in the direction parallel to the swing. How would the period of the pendulum differ from its value in a protected place?

23. A large sled is sliding down a snowy hill. Resting on the sled are a mass and a spring; one end of the spring is attached firmly to the sled, and the other end is attached to the mass. There is little friction between the mass and the sled, and the mass is set into oscillation. Will the sled accelerate smoothly down the hill?

Problems

13–1 The Kinematics of Simple Harmonic Motion

1. (I) What phase, δ, is necessary if $\sin(\theta + \delta) = \cos(\theta)$? This can be shown graphically, with Fig. 13–1, or by using a trigonometric identity for the sine of the sum of two angles.

2. (I) A simple harmonic motion along the x-direction has the following properties: The maximum amplitude $= 0.04$ m, the time between the maximum and minimum x-values $= 0.30$ s, and $x = 0.01$ m for $t = 0.10$ s. Find the period, the angular frequency, and the general equation of motion.

3. (I) A spring has the speed $v = 0.4 \sin(\omega t + \pi)$ m/s, where $\omega = 2.00$ rad/s. Plot x, v, and a as functions of time for three periods of motion.

4. (I) The angular frequency of a mass on the end of a spring in simple harmonic motion is 3.827 rad/s. What is the period of the motion?

5. (I) The amplitude of the motion of a mass attached to a spring is $A = 2.84$ m, while the maximum speed of the mass is 4.36 m/s. What is the period of the motion?

6. (I) The maximum speed of a mass attached to a spring is $v_{\max} = 0.371$ m/s, while the maximum acceleration is 1.05 m/s². What is the maximum displacement of the mass?

7. (I) What is the position of the mass in Problem 6 at time $t = 1$ s, given that the mass is precisely at the origin $(x = 0 \text{ m})$ at $t = 0$ s?

8. (II) The expression for the position of a simple harmonic oscillator is given by $x(t) = B \cos(\omega t) + C \sin(\omega t)$. (a) Show that this can be written in the form given in Eq. (13–1a), and express the constants B and C in terms of the constants A and δ. (b) Express B and C in terms of the position x_0 and the speed v_0 at time $t = t_0$. (c) What is the maximum speed in terms of B and C?

9. (II) A harmonic oscillator operates at a frequency of 813.52 Hz. What is the amplitude for which the maximum acceleration is 183.25 m/s²?

10. (II) A small object is placed on a horizontal platform that vibrates vertically with an amplitude of 3 cm. The frequency of the vibration is slowly increased. At what frequency will the object start bouncing on the platform?

11. (II) The motion of a mass can be described by the function $x(t) = A \sin(\omega t + \delta)$, where $\omega = 2.0$ rad/s and $\delta = 0.40$ rad. Express the motion as a cosine function.

12. (II) A professor pacing back and forth is observed by the class to move back and forth along the x-axis in a rough approximation to simple harmonic motion. Relative to the center of the classroom, the motion is between the two extremes $+3$ m and -3 m. The

professor was at $x = -0.3$ m at $t = 0$ s, and the motion is observed to repeat itself six times in 90 s. What are the amplitude, phase, period, frequency, and angular frequency of the motion?

13. (II) A spring has spring constant 0.50 N/m and a 0.20 kg mass on its end, which has a maximum speed of 2.0 m/s. (a) What are the angular frequency and period of the system? (b) What is the amplitude of the motion?

14. (II) A sailing ship rolls sideways in simple harmonic motion, with the period given by $T = 5.0$ s (Fig. 13–25). The tip of a 25-m mast travels a maximum of 2.5 m from the vertical position. What is the speed of the tip of the mast at the instant it is in a vertical position?

▲ FIGURE 13–25 Problem 14.

15. (II) A particle undergoing simple harmonic motion travels a total distance of 6.98 cm during one cycle of 1.71 s. (a) What is the average speed of the particle? (b) What are its maximum speed and acceleration?

16. (II) A mass $m = 0.35$ kg is attached to an ideal spring of spring constant $k = 12$ N/m. All motion takes place in the (horizontal) x-direction, and the equilibrium position of the mass is defined to be $x = 0$ m. The mass is then displaced to $x = 0.040$ m and released from rest. (a) Write an expression for the function $x(t)$ that describes the subsequent motion. (b) At what position x does the maximum positive acceleration of the mass occur? (c) What is the magnitude of the maximum acceleration? (d) What is the maximum speed of the mass?

17. (II) An object oscillates with an angular frequency of 3.0 rad/s. Its initial displacement from equilibrium is $+3.0$ cm, and the initial velocity is 5.0 cm/s in the direction of the equilibrium point. (a) Find the displacement as a function of time. (b) How soon will the displacement be $+3.0$ cm again? (c) At what times will the object move with a speed of 5.0 cm/s? (d) Sketch x and v as functions of t.

13–2   A Connection to Circular Motion

18. (I) Find the phase angle δ for uniform circular motion when $x = -R$ and $y = 0$ at $t = 0$.

19. (I) A particle moves in the xy-plane so that its x and y coordinates are described by Eq. (13–10) and Eq. (13–11); namely,

$$x = R\cos(\omega t + \delta), \quad y = R\sin(\omega t + \delta).$$

Show that the distance of the particle from the origin is a constant, and find that constant.

20. (I) When a certain uniform circular motion in the xy-plane is projected onto the x- and y-axes, this projection gives $x(t) = R\sin[\omega t + \delta - (\pi/2)]$ and $y(t) = R\sin(\omega t + \delta)$. (a) Show that the uniform circular motion is clockwise. (b) What would happen if instead of an extra phase $-\pi/2$ in $x(t)$, there was an extra phase $+\pi/2$ in $y(t)$?

21. (I) Write down a formula that describes the motion of Earth around the Sun as seen by a distant observer in the plane of Earth's orbit. Assume a circular orbit.

22. (II) A small object is placed at the outer edge of a turntable of diameter 8 in. The turntable rotates clockwise at 45 rev/min. (a) What is the projection of the object's motion on the x-axis, assuming that at $t = 0$ the projection of the motion places the object at $x = 0$? (The point $x = 0$ corresponds to the center of the record.) (b) Give the amplitude, angular frequency, and largest speed of the projection of the motion. (c) What is the acceleration of the projection of the motion on the x-axis? (The answer to this last part is another way to see why the acceleration in uniform circular motion is centripetal.)

23. (II) Show that if the motion along the x- and y-axes of a mass moving in the xy-plane is $x(t) = R_1\cos(\omega t + \delta)$ and $y(t) = R_2\sin(\omega t + \delta)$, then this motion traces out an ellipse of axis lengths R_1 and R_2. (An ellipse is the curve described by $[x^2/a^2] + [y^2/b^2] = 1$, where a and b are the axis lengths.)

24. (II) An object on a frictionless table is attached to a peg by a spring. How should you start it off, if you want it to move in a circle? What should you do to make it move in a straight line? (Pretend that the object can go right through the peg anchoring one end of the spring.)

13–3   Springs and Simple Harmonic Motion

25. (I) A mass is attached to a vertical spring. When it is pulled down 6.0 cm and released, it starts upward with an acceleration of 40 cm/s². What is the period of the motion? With what speed does it pass the equilibrium point?

26. (I) A 4.0 kg ball is suspended motionless under the influence of gravity from a spring with a force constant of $k = 300$ N/m. (a) How much is the spring stretched? (b) The same spring is sent to the Moon and suspended in the same way. By how much is the spring stretched in this case?

27. (I) A student has a spring with $k = 200$ N/m and wants to build a horizontal mass-spring system with period 1.0 s. What mass should the student use at the end of the spring?

28. (I) The spring on a scale is compressed by 2.45 cm when a 60-kg man stands on it. What is the spring constant of the spring?

29. (I) The period of a 45-g mass attached to the end of a spring is 3.1 s. Find the spring constant of the spring.

30. (I) A mass moves along the x-axis under the influence of a spring whose equilibrium position is at the origin. The mass moves between limits of -3.5 cm and $+3.5$ cm, and the period of its motion is 1.3 s. (a) What is the period if the mass is doubled?

Halved? (b) What is the period if the amplitude is increased to 4.0 cm? Decreased to 1.0 cm?

31. (II) The plate at the base of a floor sander moves back and forth in simple harmonic motion at a frequency of 20 oscillations/s, and the amplitude of the motion is 0.80 cm. If the mass of the oscillating plate is 1200 g, what is the maximum value of the driving force?

32. (II) A spring is placed in a vertical position by suspending it from a hook at its top. A similar hook on the bottom of the spring is measured to be 45 cm above a tabletop. A mass of 180 g, of negligible size, is then suspended from the bottom hook, which is now found to be 20 cm above the tabletop. The mass is then pulled down a distance of 15 cm and released (Fig. 13–26). Find (a) the spring constant, k; (b) the angular frequency, ω; (c) the position of the bottom hook after 5 s.

▲ FIGURE 13–26   Problem 32.

33. (II) A spring is suspended from the ceiling. A mass of 25 g is attached to it, and the spring stretches by 12 cm. Ignore damping. What is the period of the oscillation of a 75-g mass attached to the spring?

34. (II) Two identical springs of spring constant k are attached end to end to make one longer spring. Show that this new spring has spring constant $k/2$. The springs are said to be attached *in series*. The case of n springs attached in series gives a spring n times as long, with a spring constant $k_n = k/n$.

35. (II) A small object of mass $m = 0.060$ kg is held in place by two springs (Fig. 13–27). The one acting on the left has spring constant $k_1 = 100$ N/m; the one acting on the right has spring constant $k_2 = 200$ N/m. The object is moved away from its equilibrium position by 1.0 cm to the right and released at time $t = 0$. What is the displacement of the object as a function of time?

▲ FIGURE 13–27   Problem 35.

13–4   Energy and Simple Harmonic Motion

36. (I) What is the energy of a 1.2-kg mass that moves with amplitude 14 cm on a flat, frictionless table and is attached to a spring of spring constant 375 N/m?

37. (I) A trampoline acts like a spring of spring constant 1200 N/m. A mass is placed on the trampoline. The trampoline surface is depressed by a total of 30 cm. What is the period of oscillation of the system if this mass is pushed beyond the equilibrium displacement?

38. (I) A mass of 1.2 kg, attached to a spring, is in simple harmonic motion along the x-axis, and its period is $T = 2.5$ s. If the total energy of the spring and mass is 2.7 J, what is the amplitude of the oscillation?

39. (I) A spring with a 1 kg fish at its end and with spring constant $k = 2$ N/m is compressed 3 cm from equilibrium and then released. Use the conservation of energy to find the maximum speed of the fish.

40. (I) A puck of mass 350 g moves horizontally with speed $v = 0.88$ m/s on a frictionless surface toward the end of a relaxed spring for which $k = 140$ N/m, and compresses the spring. By how much is the spring compressed? How would your answer change if the mass of the puck were doubled?

41. (I) A mass at the end of a string moves in a circle in a vertical plane. The only forces acting on it are the tension of the string, which is central, and the force of gravity. If the length of the string is 1.2 m and the angular velocity at the top of the circle is $\omega = 2.2$ rad/s, what is the angular velocity at the bottom of the circle?

42. (II) Consider a mass m, moving along the x-axis, with potential energy given by $U(x) = \frac{1}{2}m\omega^2 x^2$. Show that the motion of this mass is simple harmonic motion with angular frequency ω by using $dE/dt = 0$, where E is the (constant) total energy of the object.

43. (II) A point mass m on a turntable that rotates with angular frequency ω is located a distance d from the center of the turntable. Show that the energy of the point mass is the sum of the energies of the harmonic motions in the x- and y-directions.

44. (II) A mass m of 0.50 kg is attached to the end of a spring and released from rest at $t = 0$ s from an extended position $x_{max} = 13$ cm. All other forces acting on the spring cancel, so that the motion of the mass is due entirely to the effect of the spring. After its release, the mass's speed drops back to zero for the first time after 0.45 s. Find the maximum speed of the mass.

45. (II) A mass m attached to the end of a spring is released from rest at $t = 0$ s from an extended position x_{max}. The mass $m = 0.2$ kg, and $k = 1$ N/m. After 0.5 s, the speed of the mass is measured to be 1.5 m/s. Calculate x_{max}, the maximum speed of the motion, and the total energy.

46. (II) An object of mass m has potential energy given by $U(x) = U_0(x^2 - a^2)^2$. (a) What are the stable equilibrium positions of the object? (b) What is the angular frequency of its motion about a stable equilibrium point, when the object is displaced by an amount that is small compared with a? [Hint: Let $x = x_{eq} + z$, and keep only up to z^2 terms in $U(z)$.]

47. (II) Imagine a tunnel that has been drilled through Earth: a smooth, straight tunnel with a frictionless interior (Fig. 13–28). The deepest point of the tunnel is at depth d, and the coordinate x measures the distance along the tunnel from its deepest point to an arbitrary point P a distance ℓ from Earth's center. The known parameters are the mass and radius of Earth, M_E and R_E, respectively, as well as G. Earth is assumed to have uniform density. (a) What is the total mass of that portion of Earth that lies within the distance ℓ in terms of ℓ, M_E, R_E, and G? (b) What is the gravitational force, in direction and magnitude, acting on a ball of mass m at point P? (c) What is the *total* force, in direction and

magnitude, acting on the ball *as a function of x* and of the constants of the problem? Why is there no net force acting perpendicular to the tunnel? (d) What is the period of the motion if the ball is released at rest at an entrance to the tunnel? Ignore air resistance. (e) What is the period of a satellite in circular orbit around Earth at a radius equal to Earth's radius?

▲ **FIGURE 13–28** Problem 47.

48. (III) A bead of mass $m = 40$ g slides without friction at the bottom of a bowl whose bottom traces out the arc of a circle of radius $R = 15$ cm. If the bead moves in a plane that passes through the bottom of the bowl, its position can be described by an angle θ measuring its angular displacement from the bottom. (a) Express the potential energy as a function of θ for very small angles θ. [Hint: The potential energy is a linear function of height. Reexpress the energy in terms of θ, and use an approximation suitable for small θ.]

   At $t = 0$ s the mass is released from rest from a very small angle $\theta_0 = 0.1$ rad. Parts (b)–(e) refer to this initial condition. (b) Give an expression $\theta(t)$ describing the subsequent motion. In particular, what is the frequency of small oscillations at the bottom of the bowl? (c) What is the total energy of the system? (d) the velocity of the bead at $t = 0.1$ s? (e) the acceleration of the bead at $t = 0.2$ s?

49. (III) A small metal block of mass m is placed on a smooth horizontal table and constrained to move along a frictionless, rectilinear groove. The block is attached to one end of a spring (of spring constant k) whose other end is fastened to a pin P (Fig. 13–29). Let length ℓ be the equilibrium length of the spring and the perpendicular distance to the groove. The spring is now pulled a distance x_0 from the equilibrium position and released. Show that, if the displacement along the groove $x \ll \ell$, the restoring force on the block is proportional to x^3, so the motion is not simple harmonic (although it will still be periodic). [Hint: For $x \ll \ell$, $\sqrt{x^2 + \ell^2} \simeq \ell + (x^2/2\ell)$.]

▲ **FIGURE 13–29** Problem 49.

13–5 The Simple Pendulum

50. (I) A simple pendulum 1.20 m long is suspended in a location where g is 9.82 m/s². What is the period of the pendulum?

51. (I) A simple pendulum has a frequency of 0.342 Hz. The length of the pendulum string is 2.12 m. What is the local value of g?

52. (I) A thin wire 3.88 m long is attached to the ceiling of a lecture room. A lead ball is attached at the bottom. The wire is displaced by an angle of 0.055 rad and released. Express the angular displacement of the lead ball as a function of time, given that $g = 9.80$ m/s².

53. (II) You need to measure the height of a room. You have a watch but no meter stick. You also have your human brain, not to mention an opposable thumb. A long pendulum with a point mass at the end extends from the ceiling to the floor and has a period $T = 3.0000$ s. (a) What is the height of the room? (b) You take the same pendulum to the top floor of a skyscraper and measure a period $T = 3.0002$ s. What is the height of the skyscraper?

54. (II) The difference in temperature between summer and winter causes the length of the pendulum in a clock to change by one part in 30,000. What time-difference error will this make in one week?

55. (II) A small lead ball of mass 2 kg is suspended at the end of a light string 1 m in length. A small peg, 0.5 m below the suspension point, catches the string in its swing (Fig. 13–30). The ball is set swinging through small angles. (a) What is the period of this pendulum? (b) The ball is started swinging on the side that does not catch the peg, at an initial height 0.05 m above the low point. How high does it rise on the side where the peg restricts the pendulum length to 0.5 m?

▲ **FIGURE 13–30** Problem 55.

56. (II) A mass m moves as a simple pendulum, the massless string holding it making an angle θ with the vertical $\theta(t) = \theta_0 \sin(\omega t)$. It receives a tangential momentum impulse of magnitude J, to the left ($-\theta$-direction) when the pendulum is at the angle $\theta = \theta_0$ (i.e., at $t = T/4$). Find the new amplitude and phase of the motion.

57. (II) A simple pendulum of length L with a bob of mass m is undergoing small oscillations of maximum angle α. Express the total energy of the pendulum in terms of these parameters. Suppose the maximum angle were not small. Find in that case a corresponding expression for the total energy.

13–6 More About Pendulums

58. (I) A small door to allow a dog to pass in and out of the house has a mass of 0.45 kg and is 28 cm wide and 40 cm tall. The door is hinged along the top of the 28-cm width. What is the frequency of oscillation of the door?

59. (I) A student wants to build a pendulum out of a circle of plywood as shown in Fig. 13–31. The circle has a radius of $R = 10$ cm and the plywood has a mass of 200 g. What is the period of the motion?

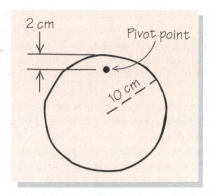

▲ **FIGURE 13–31** Problem 59.

60. (II) A uniformly dense book of dimensions 6 cm × 15 cm × 21 cm has a mass of 1.8 kg. You pick it up close to one corner with a pair of calipers and let it swing. What is the angular frequency? You can do this experiment with your own textbook.

61. (II) A thin, uniform rod of mass M and length L oscillates with small amplitude in a vertical plane about a pin that pierces the rod a distance y from the upper end. Calculate the period of the motion as a function of y. [*Hint*: You need to use the parallel-axis theorem to calculate the rotational inertia to be used in Eq. (13–43b).]

62. (II) A straight wire 80 cm long is bent in the middle into an L-shape and balanced with the two ends down on a knife edge. With what frequency will it oscillate about its equilibrium position?

63. (II) A thin uniform disk of mass M and radius R hangs from a nail driven straight into the disk at a distance ℓ from the center. (a) What is the rotational inertia of the disk about the nail? (b) What is the equation of motion for small oscillations of this pendulum about the point where the nail is driven in? [*Hint*: Use Newton's law for the torque about the point in question.] (c) What is the period, T, of oscillations about the point of suspension? (d) What is T in the limit where ℓ goes to zero?

64. (II) A uniform circular disk of radius R and mass M is attached at its center to one end of a massless, rigid rod of length L. The other end of the rod is attached to the ceiling and pivots freely about that point. The system thus makes a physical pendulum. What is the period of small oscillations of the pendulum?

65. (II) A *torsion pendulum* consists of a dumbbell suspended from its center by a wire that resists being twisted (Fig. 13–32). The dumbbell has length ℓ and has masses m on each end. It remains in the horizontal plane as it twists back and forth about the equilibrium position, $\theta = 0$ rad. For small oscillations about this equilibrium position, the torque exerted by the wire has the form

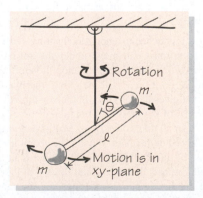

▲ **FIGURE 13–32** Problem 65.

$\tau = -\gamma\theta$, where γ is the *torsion constant*. (a) For $m = 80$ g, $\ell = 30$ cm, and $\gamma = 2 \times 10^5$ g $\cdot$ cm^2/s^2, what is the period of the oscillation? (b) If the system is started from rest at $\theta = 0.1$ rad, what is the total energy of the system? (c) What is the maximum speed of either mass, given the initial conditions of part (b)?

66. (II) A slender, uniform rod of mass m and length ℓ can be pivoted on a frictionless horizontal support at any point along the rod's length. The rod then moves as a physical pendulum for small oscillations about the vertical equilibrium position. Suppose that the pivot point is at a point $z\ell$ from the end of the rod, where z is a fraction between 0 and 1. Using an equation for the angular acceleration of the rod as a function of the angle θ, which measures the departure from the vertical, find the period of small-angle oscillations. Check the answer by specifying $z = 0.5$, where the period should become large.

13–7 Damped Harmonic Motion

67. (I) Consider the gong struck in Example 13–10. What is the loudness of the sound (as a fraction of the original loudness) after 2 min? How much time does it take for the sound to have a loudness 1/10,000 of the original loudness? Your ear is a very good detector of loudness, and you would have no trouble hearing the gong for this entire time.

68. (I) A mass on a spring with a natural angular frequency $\omega_0 = 3.6$ rad/s is placed in an environment in which there is a damping force proportional to the speed of the mass. If the amplitude is reduced to 0.35 times its initial value in 12.9 s, what is the angular frequency of the damped motion?

69. (I) The damping coefficient of a damped harmonic oscillator can be adjusted. Two measurements are made: First, when the damping coefficient is zero, the angular frequency of motion is 3880 rad/s. Second, a static measurement shows that the effective spring constant of the system is 184 N/m. To what value should the damping coefficient be set in order to have critical damping?

70. (II) Suppose we have a mass m attached to a spring of mass constant k. The mass slides on a rough surface with coefficient of sliding friction μ. The mass starts from rest with the spring extended in the positive direction to a distance A. Describe the subsequent motion. How large must μ be so that we get an "overdamped" situation? You can assume that static friction plays no role here.

71. (II) Using your knowledge of the approximate mass of an automobile as well as the frequency with which you feel it oscillate when you go over a bump, estimate the value of the effective spring constant of the suspension. Judging how long it takes for the oscillations to die out, estimate the value of the damping constant b.

72. (II) A mass m is attached to a horizontal spring, with spring constant k, and the mass can move under the spring's influence along the x-axis on a frictionless tabletop. It is initially at rest at the equilibrium position $x = 0$. At $t = 0$, the mass is struck a brief blow that gives it a speed v_0 in the $-x$-direction. (a) What is the position x of the mass as a function of time? (b) Suppose that there is a small drag force $f_D = -bv$, where b is a constant and v is the speed of the mass. What is the position x of the mass as a function of time? (c) Suppose that rather than a drag force proportional to speed, there is a small force f of kinetic friction, constant in magnitude but always opposing the motion, acting on the object. How, qualitatively, does the motion differ from that described in part (b)?

73. (II) A harmonic oscillator with natural period $T = 2.0$ s is placed in an environment where its motion is damped, with a damping force proportional to its speed. The amplitude of the oscillation drops to 70 percent of its original value in 5.0 s. What is the period of the oscillator in the new environment?

74. (II) Consider a damped harmonic oscillator. The damping, proportional to the speed, is sufficiently weak so that it is a good approximation to view the amplitude as constant over the duration of a cycle. What is the energy of the oscillator at time t if its original energy at time $t = 0$ is E_0? [*Hint*: For one cycle, the usual energy formula involving the square of the amplitude can be used.]

75. (II) A spring with $k = 12.0$ N/m and an attached bob oscillates in a viscous medium (Fig. 13–33). A given maximum, of $+6.0$ cm from the equilibrium position, is observed at $t = 1.5$ s, and the next maximum, of $+5.6$ cm, occurs at $t = 2.5$ s. What will the position of the bob be at 3.0 s and at 4.8 s? What was its position at $t = 0$ s?

▲ **FIGURE 13–33** Problem 75.

76. (II) For what mass m of the bob in Problem 75 will the spring system have its critical value? What are the lifetime and Q factor in this case?

77. (II) Show by direct substitution that Eq. (13–46) is a solution to the equations of motion of the damped harmonic oscillator, Eq. (13–45), provided that Eqs. (13–47) and (13–48) hold. [*Hint*: You will need to show that the sine and cosine terms *separately* vanish.]

13–8 Driven Harmonic Motion

78. (I) A long, flexible, and very light strip of metal, such as a saw blade, is clamped to the edge of a table, extending horizontally over the edge like a miniature diving board. When a gob of putty is placed on the end, the strip sags and comes to a new equilibrium position with its tip a distance 0.6 cm below its original position. If the end of the strip is then lightly tapped, its tip will oscillate in simple harmonic motion. With what frequency should the tip be tapped to make the strip oscillate with maximum amplitude?

79. (I) A mass of 0.5 kg is suspended from a spring, which stretches by 8 cm. The support from which the spring is suspended is set into sinusoidal motion. At what frequency would you expect resonant behavior?

80. (II) Show that resonance occurs for driven harmonic motion when $\omega = \sqrt{\omega_0^2 - (b^2/2m^2)}$. The resonant frequency occurs when the amplitude has a maximum as a function of frequency.

81. (II) A particular spring has a spring constant of 86 N/m and a mass of 0.548 kg at its end. When the spring is driven in a viscous medium, the resonant motion occurs at an angular frequency of 12.2 rad/s. What is (a) the damping parameter due to the viscous medium? (b) The lifetime of the system? (c) The sharpness of the resonance peak?

82. (II) Consider the driven, damped harmonic oscillator discussed in Section 13–8. (a) Calculate the power dissipated by the damping force. (b) Calculate the average power loss, using the fact that the average of $\sin^2(\omega t + \delta)$ over a cycle is one-half. [*Hint*: Recall that $P = Fv$.]

General Problems

83. (II) When you bounce up and down on a diving board of negligible mass, you find yourself and the board bouncing with the maximum amplitude when your bounce frequency is once every 1.2 s. How much will the board deflect vertically when you stand on the end without bouncing? Assume that the damping that must be present is small.

84. (II) Consider simple harmonic motion of a mass on the end of a spring, $x = (0.35 \text{ m}) \sin(\omega t + \delta)$. At $t = 0$ s the position is -0.080 m, and the velocity is 2.1 m/s in the $-x$-direction. The total energy of the motion is 6 J. What is the value of the (a) phase, δ; (b) frequency, f; (c) acceleration at $t = 0$ s; (d) spring constant, k; (e) mass, m?

85. (II) A mass m on the end of a spring oscillates with angular frequency ω. The mass is removed, the spring is cut in two, and the mass is reattached. What is the new angular frequency?

86. (II) When a 100-g mass is placed gently onto a foam-rubber mattress, the mass sinks 4.0 cm into it. Suppose a 200-g mass is dropped from a height of 30.0 cm onto the same slab of foam rubber. How far will this mass sink into the mattress?

87. (II) A pendulum whose period is exactly 1 s is taken to the Moon, whose mass is 7.35×10^{22} kg and whose radius is 1738 km. What is the period there?

88. (II) A spring of equilibrium length 30 cm has one end anchored, and a mass is attached to the other end. The mass is set in uniform circular motion in a plane. The angular frequency of the rotational motion is two-thirds the natural angular frequency of the spring, ω_0. (a) By how much is the spring extended by the motion? (b) Derive the general result for $\omega = \alpha\omega_0$.

89. (II) In the spectacular sport of bungee jumping, a light elastic cord (a bungee cord) is tied tightly around the ankles of someone who jumps from a bridge of height H (the other end of the cord is attached to the bridge). The length of the cord is calculated so that the jumper, of mass m, will not quite reach the surface of the water below the bridge before he or she springs back up. Suppose that the cord behaves like a spring of spring constant $10 \, mg/H$, where g is the acceleration due to gravity. (a) How long must the cord be so that a jumper just touches the water before being pulled back up? Neglect the height of the jumper and any effects due to friction. (b) Friction damps the up-and-down motion of the jumper that results after the initial jump. How far above the water would the jumper be when the oscillations have ceased? Express your answer as a fraction of H.

90. (II) A block of mass $m = 0.80$ kg is dropped onto a horizontal platform supported by a spring with spring constant $k = 3.5$ N/m. The mass of the platform and spring are negligible compared with m. The speed of the block when it hits the platform is 1.7 m/s. (a) Find the maximum compression of the spring. (b) How long does it take to reach the lowest point after the block hits the platform?

91. (II) The following systems, illustrated in Fig. 13–34, exhibit simple harmonic motion. What is the period of each motion? (a) A toy tightrope walker, with geometry as shown, whose body has mass much less than that of the barbell weights, each of mass 50 g. The toy sways from side to side in harmonic motion. (b) A mass m attached to two parallel springs, each of spring constant k. Ignore gravity. (c) A spring with spring constant k and a mass m attached

▲ **FIGURE 13–34** Problem 91.

to each end. (d) A mass m hanging vertically from a spring, of spring constant k, under the influence of gravity.

92. (II) A simple pendulum of mass $m = 0.28$ kg and length $L = 0.65$ m is attached to a cart of mass $M = 1.0$ kg (Fig. 13–35). The mass of the pendulum support is negligible. The cart can roll freely on a horizontal surface. At $t = 0$, the pendulum bob is released from rest when the string makes an angle of $10°$ with the vertical. Assume that the resulting motion of the cart relative to the ground is simple harmonic motion. Determine the amplitude of the motion of the cart. Why would you expect the cart to move in simple harmonic motion?

▲ **FIGURE 13–35** Problem 92.

93. (II) A bar 5.0 m long with a mass of 12 kg has a sharp bend (totaling $14°$) at its midpoint (Fig. 13–36). The bar rests on a pivot placed under the bend. Twins of mass 32 kg each are seated at opposite ends of the bar. What is the period of small oscillations of this modified seesaw?

▲ **FIGURE 13–36** Problem 93.

94. (II) When a rope of length L is stretched by an amount x, the rope acts like an imperfect spring that exerts the force $F = -f_1(x/L) - f_2([x/L]^2)$, where $f_1 > 0$, $f_2 > 0$, and $f_1 \gg f_2$. The motion is restricted to the x-direction. (a) Describe qualitatively what happens when a large bucket hanging vertically at the end of the rope is slowly filled with water. (b) Find the potential energy function for the rope and discuss what happens when the bucket, at rest in equilibrium, is set in motion with an initial vertical velocity v_0. What happens when the initial velocity is increased?

95. (II) The potential energy of a diatomic molecule whose two atoms have the same mass, m, and are separated by a distance r is given by the formula $U(r) = (A/r^2) - (e^2/r)$, where A and e^2 are positive constants. (a) Find the equilibrium separation r_0 of the two atoms. (b) Show that if the atoms are slightly displaced, so that their separation is $r_0(1 + x)$, then they will undergo simple harmonic motion about the equilibrium position. Calculate the angular frequency of the harmonic motion. Use

$$(1 + x)^{-2} = 1 - 2x + 3x^2 - \ldots \quad \text{and}$$
$$(1 + x)^{-1} = 1 - x + x^2 - \ldots \quad \text{for} \quad x \ll 1.$$

(Remember the reduced mass.)

96. (II) A small mass m is placed midway on the line between two large masses M. The large masses are separated by a distance L. (a) What gravitational force does the small mass experience? (Assume that the system is isolated from other masses.) (b) The small mass is displaced a distance y in a direction perpendicular to the line between the masses. What is the direction of the net force on that mass? (c) Evaluate the magnitude of the net force for $y \ll L$. Give an expression that incorporates the inequality. (d) What is the motion of m when it is displaced from the midway point?

97. (II) A thin, circular hoop of mass m and radius R hangs on a small peg (Fig. 13–37). What is the frequency of small oscillations with the center of mass moving back and forth along the x-direction?

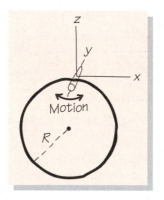

▲ **FIGURE 13–37** Problem 97.

98. (II) A machine gun fires bullets in a direction perpendicular to the plane of a target 180 m away from the gun. The bullets leave the muzzle at a speed of 370 m/s. The machine gun is set up on a platform that is vibrating with a frequency of 4.0 Hz and an amplitude of 8.50 cm in a horizontal direction parallel to the plane of the target. Describe the distribution of bullet impact points on the target. The bullets are fired with a frequency much greater than 4 Hz.

99. (III) A solid, uniform cylinder of mass m and radius R is fitted with a frictionless axle along the cylinder's long axis. A horizontal spring (spring constant k) is attached to this axle (Fig. 13–38). Under the influence of the spring, the cylinder rolls back and forth without slipping on a horizontal surface. What is the frequency of this motion?

▲ **FIGURE 13–38** Problem 99.

◄ The Canadian goose is generating a series of waves that spread in circles from where the goose sits. These waves represent a kind of organized disturbance on the water surface that are induced by a primary disturbance at the center of the spreading circles. It is only the disturbance that spreads; there is no flow of water away from the goose.

Waves

Waves are familiar to all of us. We can see ocean waves by means of light waves. Music may originate as waves on piano strings, be transported as waves of electrical current in a stereo system, form waves on the cone of a loudspeaker, and reach our ears as sound waves in air, which in turn form waves within the mechanical structure of our ears and are carried as waves of electrical impulses to our brains. Waves may appear as traveling waves, which move in some direction, as ocean surf moves toward a beach. They may appear as standing waves, like the vibrations of a guitar string. They may appear in one-dimensional systems, such as a Slinky spring; two-dimensional surfaces, such as the surface of a pond; or three-dimensional media, such as the air around us.

Consider a struck bell. To a very good approximation it is a rigid body, yet its ringing shows that it is vibrating just as a guitar string does. Even what look like the hardest of objects distort, if only by small amounts, in response to internal and external forces—indeed, a bell made of the hardest steel may ring the best. The resulting waves are generated within continuous media. In Chapters 14 and 15 we study waves on a guitar string, on the sea, or in the air, all of which involve continuous media. Eventually we shall encounter (in Chapter 34) electromagnetic waves such as light waves, in which no medium at all is required!

▶ 14–1 Types of Waves

In Chapter 13 we emphasized that when a point mass is displaced from its stable equilibrium position, the displacement gives rise to a restoring force, which leads to oscillatory motion. Moreover, the motion is generally harmonic as long as the displacement is small. The same physical principle applies to extended systems, but this time disturbances can be transmitted from one part of the system to another through internal restoring forces. A simple example of this is a system composed of a series of identical pendulums hanging in line side by side and connected to their neighbors on each side by identical springs (Fig. 14–1a). If the pendulums are at rest, we have a stable equilibrium situation. Let us now displace one of the pendulums from its vertical position and let it swing in a direction perpendicular to the line of pendulums (Fig. 14–1b). What we observe is that the stretched spring tends to displace the neighboring pendulums from their equilibrium position, and they too start to swing, with their motion lagging the originating pendulum. These pendulums in turn affect their neighbors—both the originating

(a) Pendulum at rest. Each bob mass is connected with its nearest neighbor by a spring.

(b) A central bob mass is pulled in the y direction, pulling a group of neighboring masses along.

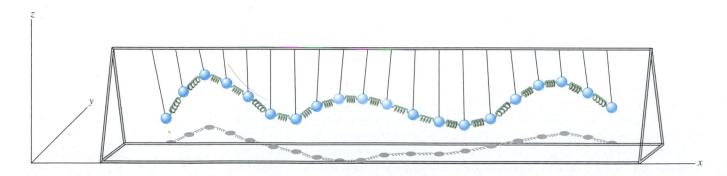

(c) As the central mass swings back the neighboring masses induce pulses of motion going in both directions away from the initiating disturbance.

▲ **FIGURE 14–1** Each pendulum in this line of pendulums swings in a plane parallel to the *yz*-plane. The line itself is along the *x*-axis. The pendulums are coupled by springs, and an initial swing of any given pendulum will be transmitted from one pendulum to the next, propagating a disturbance along the line.

pendulum and their other neighbors. What one eventually observes is a wave of displacements that travels from the originating pendulum in both directions.

This little toy models motions that can occur along a taut rope. Imagine that you are holding one end of the rope. If you rapidly displace your end of the rope to the side a little, the displacement affects the adjacent part of the rope, which in turn disturbs and displaces its neighboring part of the rope. Each section of the rope that is displaced acts on the adjacent sections to pull them in the direction of its own displacement, and by Newton's third law the adjacent sections act back on the displaced section to pull it back into alignment with the rest of the rope—the forces on the displaced section are *restoring forces*. The result is that the sideways displacement effectively travels down the rope, section by section, and gives rise to a traveling **transverse wave**.

How, beyond the generalities we have discussed so far, is the series of pendulums connected by springs a model for the rope? Below we'll show quantitatively how tension within the rope acts like the springs connecting the pendulums to provide the necessary restoring forces. At a deeper level, springlike atomic forces between neighboring atoms or molecules in the rope provide restoring forces analogous to those of the springs between the pendulums.

The motions that generate a moving disturbance can also occur in a *longitudinal* direction (along the line of the system). A series of point masses connected by springs and lying in a line (Fig. 14–2) presents a simple illustration of how such a **longitudinal wave** is transmitted. Again one needs to look no further than Newton's laws to understand the wave. A displacement of one of the masses to the right causes the spring on the left to stretch and the one on the right to compress. These act to displace the neighboring masses, and again a *traveling wave* of compressions and extensions (rarefactions) results (Fig. 14–3). This system is a model for longitudinal waves in general, including sound waves in air and longitudinal waves that travel through solids (also referred to as sound).

▲ **FIGURE 14–2** The line of masses connected by springs can propagate a longitudinal disturbance. A displacement of the first mass along the *x*-direction, which is the direction of the line, will propagate from mass to mass through the springs.

Stretched   Stretched

Compressed   Compressed

(a)

◀ **FIGURE 14–3** (a) The point masses of the system illustrated in Fig. 14–2 are displaced longitudinally (along the equilibrium line). Some springs are compressed, some are stretched. (b) Longitudinal waves on a Slinky spring.

← → Motion   Longitudinal waves

(b)

Keep in mind another feature of the propagating disturbances that we have described here. In transverse waves, like those that can be generated on a rope, all the actual motion is to the side (transverse). In longitudinal waves, like sound, no material is displaced a distance farther than the back-and-forth displacements of the system elements. Yet the disturbance, or wave, travels along the medium for what may be very long distances indeed. The sound wave that an earthquake generates within the Earth may involve actual atomic motions that are limited to motions on the order of several atomic spacings, yet the disturbances that the sound represents may propagate all the way to Earth's opposite side. And it is not just a "disturbance" that is traveling; we'll see below that a traveling wave carries energy and momentum.

Standing waves represent another manifestation of propagating disturbances, but in an apparently very different form from traveling waves. A harp string under tension is fastened at both ends and can be set into transverse motion with a pluck. An organized, or regular, pattern of displacements from the equilibrium position of the string results, and the pattern is *periodic*. Just as for the simple harmonic motion we have already enountered, the motions of a standing wave are not just periodic but are *harmonic* when the displacements are not too large (Figure 14–4). The vibrating harp string is an example of a transverse standing wave, but longitudinal standing waves, such as sound in a closed organ pipe, are also possible. Standing waves have a distinct property—*only certain oscillation frequencies are allowed*. This is why a plucked harp string or a given organ pipe has a certain pitch.

▲ **FIGURE 14–4** The vibrating strings of a harp form an example of transverse standing waves, which can be established on strings fixed at both ends.

(a)

Transverse waves

(b)

FIGURE 14–5 (a) The point masses of the system illustrated in Fig. 14–2 are displaced transverse to the equilibrium line (dashed). The stretching of a spring acts to bring the system back to the equilibrium line. (b) Transverse waves on a Slinky spring.

So far we have been discussing waves in a one-dimensional system, such as the displacement of a spring (Fig. 14–5) from its straight-line equilibrium shape, or the compression and rarefaction in the closeness of the spirals in a horizontally laid out Slinky (Fig. 14–3), and it is such waves that we will be mostly dealing with in this chapter and in Chapter 15. However, two- and three-dimensional systems also support waves. For example, standing waves can be generated on a drumhead—a two dimensional surface—in the same way that they can be generated on a string. These standing waves are transverse, consisting of an organized harmonic motion transverse to the drumhead (Fig. 14–6). Traveling waves in two-dimensional systems form **wave fronts**, which are lines that are formed by a series of propagating crests (or any other identifiable part of the wave)—we'll see later that these fronts are perpendicular to the direction of motion of the wave. This is a familiar phenomenon that we see in the waves at the beach or in the spreading ripples from a goose floating on a lake (see Fig. 14–7, or the photo on page 397). The ripples on a still lake move radially away from a central point, and the wave fronts are expanding circles. Just as for one-dimensional traveling waves in ropes or wires, water waves travel, but the actual movement of the water is entirely local (limited), as is easily seen by putting a cork in a pond of water (Fig. 14–8). (This figure illustrates that the local water motion, which is what the cork traces out for us, is a combination of transverse and longitudinal.) For a three-dimensional system such as air or Earth's interior, the wave fronts formed by traveling waves move out as spheres from a central point.

We begin with a description of the dynamical equation that leads to waves. Newton's second law relates the acceleration of a small portion of the medium to the forces that arise from its displacement from equilibrium. The propagation of a disturbance through the system is governed by the dynamical equation, which in this context

time = t time = $t + T$

FIGURE 14–6 Standing waves on a drumhead. The *nodes*, points where the drumhead is not in motion, form lines rather than individual points as they did for the string. The lines shown are not necessarily nodes. Can you determine the nodes for each pair of vibrations?

FIGURE 14–7 The spreading waves resulting from a disturbance of a water surface form circles centered on the disturbance.

is called a **wave equation**. We will see that the name is in fact justified: The solutions to the dynamical equations properly describe waves. The wave equation in its various forms can be used to describe the mechanical waves of this chapter and the next; it will reappear when we discuss electromagnetic waves in Chapter 34. In particular, the wave equation will allow us to understand how to find the speed with which waves travel and the discrete vibration frequencies of standing waves.

14–2 The Wave Equation

Let us apply Newton's second law to a string under tension, neglecting the forces due to gravity in this analysis. Figure 14–9 shows a rope displaced transversely from its equilibrium position along the horizontal x-axis. We write the sideways displacement as the function $z(x, t)$. It changes with time, and it changes from place to place along the x-axis. It is important to note here that there are two space coordinates involved—z, the transverse displacement, and x, the distance along the rope—and that to understand wave motion, you need to keep them distinct in your thinking. Because z varies along the rope, we cannot treat the entire rope, but only a tiny segment of it. We denote the length of the segment by Δl. We assume that the rope is uniform with a mass per unit length μ. Thus the mass of the element is $\mu \, \Delta l$. Since the transverse position of the rope segment is described by z, its (transverse) acceleration is

$$a = \frac{d^2 z(x, t)}{dt^2}. \tag{14–1}$$

x labels the point on the rope where the transverse displacement occurs, and in this expression we differentiate with respect to t while holding x fixed. The conventional way of denoting this is to replace d^2/dt^2 by the mathematically correct $\partial^2/\partial t^2$. The *partial derivative* sign ∂ indicates only that the function whose derivative is taken depends on more than one independent variable, and that for the partial derivative with respect to, say, t, all the remaining variables (here x) are regarded as constants.

Let us now consider the net force $\vec{F}$ acting on the element of the rope. We see in Fig. 14–9 that all the forces that act on our element come from the tension of the rope. The force that acts on the left side of the element at x has two components. If the angle that the rope makes with the horizontal direction at x is θ_1, then the horizontal component of this force is $-T \cos \theta_1$ and the vertical component of the force is $-T \sin \theta_1$, where T is the magnitude of the tension. The minus signs are present because we define the positive directions as that of the positive x-axis, and the vertical pointing upward in our figure. The force that acts on the right side of the rope, at $x + \Delta x$, has a horizontal component that acts to the right, and therefore is $+T \cos \theta_2$, where θ_2 is the angle the

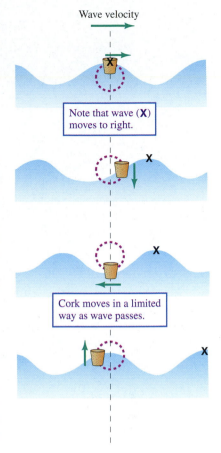

Wave velocity

Note that wave (**X**) moves to right.

Cork moves in a limited way as wave passes.

▲ **FIGURE 14–8** A cork floats on the ocean surface, where the water is not too shallow, as a wave passes. The X marks the crest of the passing wave, the dashed circle marks the path followed by the cork, and the arrow marks the direction of motion of the cork at the particular point in the wave's passage.

Wave pulse moving down rope

(a)

z

Expanded view of wave pulse above

$\vec{T}_2$

θ_2

$\Delta \ell$

Δz

θ_1

$\vec{T}_1$

Δx

x $x + \Delta x$

x

(b)

◀ **FIGURE 14–9** A string has been transversely displaced. We have isolated a segment of length $\Delta \ell$, as well as the tension forces acting on it. Due to the curvature, the direction through which the tension acts is different on each side. The magnitudes of $\vec{T}(x)$ and $\vec{T}(x + \Delta x)$ are the same.

rope makes at $x + \Delta x$. The vertical component acts in the upward direction and it is $+T \sin \theta_2$. Before we write down the net force on the element, it is helpful to relate the angle θ to the displacement at point x and to make some approximations. The first observation is that if we make the rope segment small enough, the angles θ_1 and θ_2 won't differ by much, and unless we are dealing directly with their difference, we can just call them θ. Next we note that $\tan \theta$ is the slope of the piece of rope at x.

$$\tan \theta = \frac{dz(x, t)}{dx}. \tag{14–2}$$

As in Eq. (14–1), the derivative is with respect to just one variable, in this case x (t is fixed here), so that we should really replace dz/dx by $\partial z/\partial x$. Next we note that the displacement is *small* and the angle θ itself is small (the rope is *nearly* flat), and we may replace $\tan \theta$ by $\sin \theta$ and $\cos \theta$ by unity. With these approximations in hand, let's look at the net force. The horizontal component of the net force is

$$F_x = T \cos \theta_2 - T \cos \theta_1 = T(\cos \theta_2 - \cos \theta_1) = T[\cos \theta(x + \Delta x) - \cos \theta(x)].$$

We have here recognized that θ_2 is θ at $x + \Delta x$ while θ_1 is θ at x. If we divide the right hand side by Δx and multiply it by the same quantity, and use the fact that

$$\frac{d \cos \theta}{dx} = \frac{\cos \theta(x + \Delta x) - \cos \theta(x)}{\Delta x},$$

we see that

$$F_x = T \Delta x \frac{d \cos \theta}{dx} = T \Delta x \left(-\sin \theta \frac{d\theta}{dx} \right). \tag{14–3}$$

This is *doubly* small, since θ (hence $\sin \theta$) as well as its derivative are small. The horizontal component of the net force is thus negligible compared to the vertical component of the force (see below). This confirms our assertion that the motion is really up and down, and we can drop this term. Using the exact same techniques, the vertical component of the net force is upward,

$$F_{\text{vert}} = T \sin \theta_2 - T \sin \theta_1 = T[\sin \theta(x + \Delta x) - \sin \theta(x)] = T \Delta x \frac{d \sin \theta}{dx}. \tag{14–4}$$

Using Eq. (14–2) and the small-θ approximation, we have

$$\sin \theta \cong \tan \theta = \frac{\partial z(x, t)}{\partial x}. \tag{14–5}$$

Thus

$$F_{\text{vert}} = T \Delta x \frac{d}{dx} \frac{\partial z(x, t)}{\partial x} = T \Delta x \frac{\partial^2 z(x, t)}{\partial x^2}. \tag{14–6}$$

(We have again written the *partial* derivative: We are differentiating with respect to x, while t remains fixed.) We are now close to our final result. We write Newton's second law for the vertical motion and use Eqs. (14–6) for the force, Eq. (14–1) for the acceleration, and the fact that the mass of the piece of rope is

$$\mu \, \Delta l \cong \mu \, \Delta x,$$

because for small angles $\Delta l = \Delta x / \cos \theta \cong \Delta x$. Both the force and the mass of the segment are proportional to Δx, which we cancel, leaving

$$T \frac{\partial^2 z(x, t)}{\partial x^2} = \mu \frac{\partial^2 z(x, t)}{\partial t^2}, \quad \text{or} \quad \frac{\partial^2 z(x, t)}{\partial x^2} = \frac{\mu}{T} \frac{\partial^2 z(x, t)}{\partial t^2}. \tag{14–7}$$

We can go one step further. Look at the dimensions of the ratio μ/T:

$$\left[\frac{\mu}{\text{Tension}} \right] = \frac{[M/L]}{[ML/T^2]} = \left[\frac{T^2}{L^2} \right] = \frac{1}{[\text{velocity}^2]}. \tag{14–8}$$

This dimensional argument suggests that the combination μ/T will play a role as a speed, and accordingly we define

$$v \equiv \sqrt{\frac{T}{\mu}}. \tag{14–9}$$

Equation (14–7) now takes its final form, the **wave equation**.

$$\frac{\partial^2 z(x, t)}{\partial x^2} = \frac{1}{v^2} \frac{\partial^2 z(x, t)}{\partial t^2} \tag{14–10}$$

THE WAVE EQUATION

This equation may be viewed as a generic equation for the motion of all small displacements away from the equilibrium position in a one-dimensional continuous medium. That includes not only traveling waves but standing waves as well. It, or something very much like it, governs the wavelike behavior of many different systems that we described above.

What is the meaning of the parameter v that appears in the wave equation? We shall see in Section 14-4 that it is *the speed with which disturbances, or waves, propagate along the rope*. Here we will just note that this quantity depends on *internal* properties of the medium, in this case the tension and mass density of a rope.

14–3 Periodic Waves

If the hand in Fig. 14–9 shakes the rope just once, then the resulting disturbance will propagate down the rope. If instead the hand regularly—periodically—repeats the motion, then a continuous, periodic wave will be generated (Fig. 14–10). Such periodic disturbances are the ones that will interest us the most in this chapter. To establish that these periodic disturbances are indeed what occur, we will want to show that they satisfy the wave equation. In other words, we want to investigate periodic solutions of the wave equation—that is, solutions that repeat every time interval τ, the *period* of the motion[†]. Mathematically this is expressed by the condition that

$$z(x, t) = z(x, t + \tau). \tag{14–11}$$

If you were to look at this wave through a narrow vertical slit located at some point x, you would see the rope rising and falling with a definite period. Figure 14–10 reveals a displacement along the rope with regularly spaced peaks and troughs, with consecutive peaks separated by a distance which we label λ and which we refer to as the wavelength. This observation is expressed mathematically as

$$z(x, t) = z(x + \lambda, t). \tag{14–12}$$

Our waves are periodic in both time and space.

We know a very simple set of periodic functions the trigonometric functions, and we have seen how they enter into motion in Chapter 13. In that chapter we saw that a function periodic in time could have, for example, the form $\cos \omega t$. The quantity ω, which has dimensions of inverse time (so that the argument of the cosine is dimensionless), is

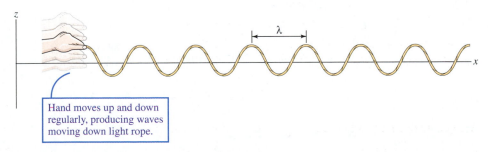

◄ **FIGURE 14–10** A periodic wave on a rope has a disturbance that repeats at regular intervals. The distance over which the pattern repeats is the wavelength λ.

Hand moves up and down regularly, producing waves moving down light rope.

[†]We use τ rather than T, as in Chapter 13, to avoid confusion with the tension T, which plays a role in many of our illustrations of wave motion.

the angular frequency familiar from Chapter 13. The **period** τ comes from the requirement that $\cos \omega t = \cos \omega(t + \tau) = \cos(\omega t + \omega \tau)$, or

$$\omega\tau = 2\pi. \tag{14–13}$$

The period is thus given in terms of ω by

$$\tau = 2\pi/\omega. \tag{14–14}$$

Note that in terms of the **frequency** f, which is the inverse of the period ($f = 1/\tau$), we have

$$\omega = 2\pi f. \tag{14–15}$$

We had already argued above how you can interpret the period or frequency for the traveling wave on a rope described above: The period is the time it takes for a complete back-and-forth motion at any given point on the rope, and the frequency measures how often this motion repeats per unit time.

All the remarks above hold true for $\sin \omega t$ as well, and therefore for any linear combination of $\sin \omega t$ and $\cos \omega t$.

Functions that repeat in space can be treated in exactly the same way. Consider, for example, $\cos kx$. The quantity k, which has dimensions of inverse length (so that the argument of the cosine is dimensionless), is analogous to ω. We call it the **wave number**. Our function is of course repetitive: $\cos kx = \cos k(x + \lambda) = \cos(kx + k\lambda)$ provided that

$$k\lambda = 2\pi. \tag{14–16}$$

This establishes that the repetition length λ, which we call the **wavelength**, is expressed in terms of k by

$$\lambda = 2\pi/k. \tag{14–17}$$

Now we combine the periodicity in time and in space to find a wave that is periodic in both. Take, for example,

$$z(x, t) = A \cos kx \cos \omega t. \tag{14–18}$$

In this form, or in other combinations of trigonometric functions of the arguments kx and ωt, the value of $z(x, t)$ runs between $+A$ and $-A$. The quantity A is therefore called the **amplitude**. The most general such combination is

$$z(x, t) = A \cos kx \cos \omega t + B \cos kx \sin \omega t + C \sin kx \cos \omega t + D \sin kx \sin \omega t. \tag{14–19}$$

This is a sum of different waves of the type shown in Eq. (14–18) and sketched in Fig. 14–10. Each one is characterized by a different amplitude A, B, C, or D. (There is no simple way of characterizing an overall amplitude.) It is not hard to check that Eq. (14–19) will satisfy the wave equation, *although only for special values of k and ω*. If, for example, we take just the last term and substitute it into the wave equation in the form

$$\frac{\partial^2 z}{\partial x^2} - \frac{1}{v^2}\frac{\partial^2 z}{\partial t^2} = 0,$$

we find

$$\frac{\partial^2(\sin kx \sin \omega t)}{\partial x^2} - \frac{1}{v^2}\frac{\partial^2(\sin kx \sin \omega t)}{\partial t^2}$$

$$= -k^2 \sin kx \sin \omega t - \left[\left(-\frac{\omega^2}{v^2}\right)\sin kx \sin \omega t\right]$$

$$= \left(-k^2 + \frac{\omega^2}{v^2}\right)\sin kx \sin \omega t = 0.$$

This can only be satisfied if k and ω are related:

$$k^2 = \frac{\omega^2}{v^2}. \tag{14–20}$$

Using Eqs. (14–15) and (14–16), we see that this is equivalent to

$$\lambda f = v. \qquad (14\text{–}21)$$

The same conclusion is reached with each of the terms in the more general solution form, Eq. (14–19). Thus Eq. (14–21) is a general relation between the wavelength, the frequency, and v, which as we shall see in Section 14–4 is *the speed with which waves travel*.

EXAMPLE 14–1 A disturbance along a long string aligned initially along the x-axis has a wave form that along a small part of it takes the form $z(x, t) = a(x - vt)^3$, where $z(x, t)$ is the transverse displacement of the string from the x-axis. Here a is a small constant. (Figure 14–11 illustrates what is meant here: The solid part of the curve is the one we are concentrating on, and that part is cubic in the variable $x - vt$.) (a) Does this part of the disturbance satisfy the wave equation? (b) If so, describe the position of the spot on the string whose displacement is zero.

▲ **FIGURE 14–11** A disturbance which over a limited range has the form $z(x, t) = a(x - vt)^3$ plotted vs x at $t = 0$.

Setting It Up We regard Fig. 14–11 as a snapshot at $t = 0$.

Strategy For part (a) we need only take derivatives and check that Eq. (14–10) is satisfied. For (b) we look for the position x as a function of t for which $z = 0$; this requires setting $z = 0$ and solving the resulting expression to find x as a function of t.

Working It Out (a) We have

$$\frac{\partial^2}{\partial x^2}[a(x - vt)^3] = \frac{\partial}{\partial x}[3a(x - vt)^2] = 6a(x - vt)$$

$$\frac{\partial^2}{\partial t^2}[a(x - vt)^3] = \frac{\partial}{\partial t}[3a(-v)(x - vt)^2] = 6a(-v)^2(x - vt).$$

Taking the first term minus $(1/v^2) \times$ the second term, we find zero, so the wave equation is satisfied.

(b) The transverse displacement is zero at one spot only, when $x - vt = 0$. This point is at $x = 0$ when $t = 0$, and *it moves at a speed v, $x = vt$*. This verifies our claim that the wave equation allows solutions in which the disturbance propagates with speed v, as we will see in more detail in later sections.

What Do You Think? Does this displacement function really satisfy all the conditions that we insisted on to establish the wave equation? *Answers to **What Do You Think?** questions are given in the back of the book.*

14–4 Traveling Waves

Transverse Waves

We have seen that the wave equation has solutions that are periodic in both space and time for a transverse displacement on a rope—by transverse we mean that the disturbance is perpendicular to the rope's alignment. If we call the rope's alignment direction the x-axis and the direction perpendicular to the rope the z-axis, the fully periodic solution $z(x, t)$ is described by combinations of the products of trigonometic functions in x and in t, as in Eq. (14–19). At this point we want to show that some solutions of this type represent trigonometric shapes (sines or cosines) that move along the rope with a fixed speed, and we refer to these waves as **traveling waves**.

We first consider a solution formed by two terms from Eq. (14–19), with equal and opposite coefficients:

$$z(x, t) = A \sin kx \cos \omega t - A \cos kx \sin \omega t.$$

If we use the trigonometric identity $\sin(a - b) = \sin a \cos b - \cos a \sin b$, we see that the above solution has the form

$$z(x, t) = A \sin(kx - \omega t). \qquad (14\text{–}22)$$

We can now show that this represents a traveling wave. Consider first what is happening at time $t = 0$. We have $z(x, 0) = A \sin kx$, which is drawn in Fig. 14–12a. This function has a *node* (the technical name for a zero of the function) at $x = 0$. Where is this node a short time Δt later? The node is located for any time at the zero of the argument in the solution in Eq. (14–22). So if our node is displaced by Δx, we can find that value by requiring that

$$k\,\Delta x - \omega\,\Delta t = 0.$$

(a) $t = 0$

(b) $t = \Delta t$

▲ **FIGURE 14–12** (a) The harmonic traveling wave of Eq. (14–22) is plotted at the fixed time $t = 0$. (b) As time passes, this pattern moves to the right.

This gives for the new location $\Delta x = (\omega/k)\,\Delta t$, a little to the right. This same reasoning works for *any* point on the sine-curve, hence the entire distortion of the rope looks like the drawing of Fig. 14–12b, which is just the $t = 0$ curve (Fig. 14–12a) displaced an amount Δx to the right. Equivalently, the speed with which the node (or any other point) has moved is

$$\frac{\Delta x}{\Delta t} = \frac{\omega}{k} = \frac{2\pi f}{2\pi/\lambda} = \lambda f = v. \qquad (14\text{–}23)$$

Thus the constant v that appears in the wave equation is the *speed* of the traveling wave, just as stated at the end of Section 14–2. This wave travels in the positive x direction, since x increases as t increases.

We can easily construct waves that travel in the negative x direction. A solution made up of a different combination of terms in Eq. (14–19) will do the trick, namely

$$A \cos kx \sin \omega t + A \sin kx \cos \omega t = A \sin(kx + \omega t). \qquad (14\text{–}24)$$

(We have again used a trigonometric identity.) Here the argument of the sine function will remain the same as t increases only if Δx *decreases*. The wave moves to the left.

We can write our two solutions, Eqs. (14–22) and (14–24), in a way that makes the role of the velocity evident:

$$A \sin(kx - \omega t) = A \sin\left[k\left(x - \frac{\omega}{k}t\right)\right] = A \sin[k(x - vt)] \qquad (14\text{–}25a)$$

$$A \sin(kx + \omega t) = A \sin\left[k\left(x + \frac{\omega}{k}t\right)\right] = A \sin[k(x + vt)]. \qquad (14\text{–}25b)$$

In each case, the shape of the disturbance is a sine function, and the angle that is the argument of the sine is proportional to $x - vt$ and $x + vt$, respectively. As we have already described, these correspond to right- and left-moving harmonic waves of speed v. We can also see by following the argument above that the particular shape of the wave (its **waveform**), here a sine function, is completely independent of the discussion that shows that the waveform moves to the right or the left. The advancing of the shape, which is what characterizes traveling waves, depends only on the fact that the argument of the waveform is $x - vt$ or $x + vt$. We'll confirm this in more detail at the end of this section. Finally, we emphasize again that the value of v is determined by internal properties of the wave medium—here the tension and mass density.

EXAMPLE 14–2   A person standing at the narrow entrance to a harbor sees regular sinusoidal water waves moving into the harbor. He counts 42 wave crests in 1.0 min, and he estimates the distance between the crests to be 3.0 m. Write an expression for the height of the wave where the person is standing. What are the wavelength, wave number, frequency, angular frequency, and speed of these waves?

Setting It Up   Figure 14–13 labels an appropriate axis as well as the wavelength λ. Note that the maximum height (amplitude) of the waves is not given, so we simply treat that as an unknown parameter h_0, where h_0 labels the maximum wave height above a steady background water level.

Strategy   For the first part of the question we can use any of the trigonometric expressions with argument $kx - \omega t$ (right-moving waves, as in the sketch). Since we are asked only for the wave height where the person is standing, we substitute $x = 0$. We can find the wave number k directly in terms of the wavelength λ. The frequency involves counting waves per second, which is given information, and the speed is always the product of the frequency and the wavelength.

Working It Out   Let $h(x, t)$ be the height of the waves above the average water level. Then using, say, Eq. (14–22), $h(x, t) =$

▲ FIGURE 14–13

$h_0 \sin(kx - \omega t)$. This function at the observer's position is

$$h(x, t) = -h_0 \sin(\omega t).$$

The wavelength is the distance between crests, $\lambda = 3.0$ m. The wave number, k, is given in terms of λ by Eq. (14–16):

$$k = \frac{2\pi}{\lambda} = \frac{2\pi}{3.0}\,\text{m}^{-1} = 2.1\,\text{m}^{-1}.$$

The frequency of the waves is $f = 42$ per min, or

$$f = (42 \text{ min}^{-1})\frac{1 \text{ min}}{60 \text{ s}} = 0.70 \text{ s}^{-1}.$$

The angular frequency, ω, and the frequency, f, are related by Eq. (14–15):

$$\omega = 2\pi f = (2\pi \text{rad})(0.70 \text{ s}^{-1}) = 4.4 \text{ rad/s}.$$

Finally, the speed v is given by Eq. (14–21),

$$v = \lambda f = (3.0 \text{ m})(0.70 \text{ s}^{-1}) = 2.1 \text{ m/s}.$$

What Do You Think? Suppose the observer sees waves in the same system but with a wavelength of 2.0 m. Can he state with confidence whether the frequency of the waves will be less or greater than that of the 3.0-m waves she has observed?

CONCEPTUAL EXAMPLE 14–3 We have several wires available with different mass densities. Tests indicate that for the particular construction of these wires the maximum tension that the wires can support before breaking increases as the square of the linear mass density. Which wire can produce a traveling wave of the greatest speed?

Answer We already know that the speed of traveling waves on the wire is given by Eq. (14–9), $v = (T/\mu)^{1/2}$. Thus for a given wire of linear mass density μ, you get the greatest speed when the tension is highest. We therefore set the tension in each wire at its largest value, namely $T = \kappa\mu^2$, where κ is a constant with the dimensions of $(\text{N m}^2/\text{kg}^2)$—if the tension were higher, the wire would break. The maximum wave speed for each wire is then

$$v = (T/\mu)^{1/2} = (\kappa\mu^2/\mu)^{1/2} = (\kappa\mu)^{1/2}.$$

We conclude that the greatest wave speed can be achieved with the wire of largest linear mass density.

EXAMPLE 14–4 A very long wire is held under tension by suspending weights from an end that passes over a pulley. The speed of transverse traveling waves on the wire is 51 m/s. The wire is replaced by another with three times the mass density, and twice as much weight is suspended from the end of the new wire. What is the speed of the traveling waves now?

Setting It Up With Fig. 14–14 it is clear that the tension T in the wire equals the suspended weight mg. We are given neither the initial tension nor mass density, but since the question only involves the velocities, we only need the ratio of these quantities. We are given the initial speed v_1 and want the final wave speed v_2.

Strategy We can write down the expression for v_1 in terms of T and μ, Eq. (14–9). The expression for v_2 involves the same expression with T multiplied by 2 and μ multiplied by 3.

Working It Out Equation (14–9) gives $v_1 = (T_1/\mu_1)^{1/2}$. The second speed v_2 is

$$v_2 = \sqrt{\frac{2T_1}{3\mu_1}} = \sqrt{\frac{2}{3}}\sqrt{\frac{T_1}{\mu_1}} = \sqrt{\frac{2}{3}}v_1.$$

Numerically, $v_2 = (0.67)^{1/2} (51 \text{ m/s}) = 42 \text{ m/s}$. Increasing the tension tends to increase the speed, whereas increasing the mass den-

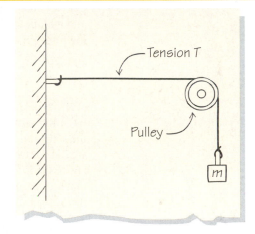

▲ **FIGURE 14–14** A wire is held under tension by the gravitational force mg.

sity tends to decrease the speed. In this case, the mass density is increased by a larger factor than is the tension, so the speed decreases.

What Do You Think? How would you adjust the tension and/or mass density so that the wave speed drops to zero? What is the physics behind your answer?

Longitudinal Waves

A Slinky spring provides us with a familiar example of a longitudinal traveling wave. Figure 14–3 showed regions of rarefaction and compression along the coil that move along the system. The actual motion within the spring is along the same line as the spring itself, which is also the line along which the wave propagates. A traveling wave like this is generated by a hand moving longitudinally back and forth at the end of the spring. Another example of a longitudinal wave is a wave of compression and rarefaction traveling in a metal rod—we'll refer to such waves as **sound** waves. We can treat this case quantitatively using the stress and strain properties of the rod. (You may want to review Section 11–4.) We first derive a wave equation for this kind of wave in the same way as we did for transverse waves on a string. We'll find an analogous equation, with v determined this time by a mass factor analogous to the mass density of the string and a force factor analogous to the tension in the string.

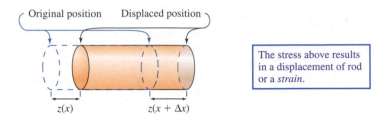

FIGURE 14–15 The (a) stresses on and (b) resulting elongation of a section of a rod lying between the positions x and $x + \Delta x$.

Start with a rod lying along the x-axis of cross-sectional area A and concentrate on a segment of the rod between x and $x + \Delta x$ (Fig. 14–15a). Just as the net force on a string segment originates with tension from one side and from the other, the net stress S (force per unit area) on our segment comes from the right and the left. It will vary with position and with time. Since generally the force on the right is $F(x + \Delta x) = AS(x + \Delta x)$ and the force on the left is $F(x) = AS(x)$, the net force on the segment is

$$\Delta F = F(x + \Delta x) - F(x) = A[S(x + \Delta x) - S(x)] = A\left[\frac{\partial S(x)}{\partial x}\right]\Delta x.$$

(We have immediately gone to a partial derivative notation, anticipating independent variation with time.) Let's denote the longitudinal displacement of the material within the rod from its equilibrium position by $z(x, t)$. (Note that we use the same symbol z as we did for transverse waves to represent the actual displacement of material.) The force above leads to an acceleration and hence displacement of the material. The mass Δm of material in our segment is the mass density ρ multiplied by the volume, $\Delta m = \rho A\, \Delta x$; the acceleration of the material is given by the second time derivative of the displacement $z(x, t)$, $a = \partial^2 z(x, t)/\partial t^2$. Newton's second law, $(\Delta m)a = \Delta F$, reads

$$\rho A\, \Delta x\frac{\partial^2 z}{\partial t^2} = A\left(\frac{\partial S}{\partial x}\right)\Delta x$$

or

$$\rho\frac{\partial^2 z}{\partial t^2} = \frac{\partial S}{\partial x}. \tag{14–26}$$

We now relate the stress S (a force factor) to the strain e (a fractional displacement of the material). What is the strain here? The net change in the length Δx of our little segment of the rod is the difference between the displacements measurable at the two ends of the section, $z(x + \Delta x) - z(x)$ (Fig. 14–15b). Thus the strain is given by the ratio of the difference of displacements to the original length, Δx:

$$e = \frac{z(x + \Delta x, t) - z(x, t)}{\Delta x} = \frac{\partial z}{\partial x}.$$

Equation (11–19) states that $S = eY$, where Y is Young's modulus, a characteristic constant of the material:

$$S = Y\frac{\partial z}{\partial x}.$$

When this result is substituted into Eq. (14–26), we find that

$$\rho\left(\frac{\partial^2 z}{\partial t^2}\right) = Y\left(\frac{\partial^2 z}{\partial x^2}\right). \tag{14–27}$$

This result has the form of the wave equation, Eq. (14–10), and the speed can be identified as

$$v = \sqrt{\frac{Y}{\rho}}. \qquad (14-28)$$

If you compare this to the speed for transverse waves on a string, Eq. (14–9), you will see that the form, as well as the kind of factors that appear, are entirely analogous.

We can get some feeling for the speed of sound waves in solids by calculating v for a steel rod. For steel, $\rho = 7.9 \times 10^3 \text{ kg/m}^3$, while Young's modulus is $Y = 200{,}000 \text{ MN/m}^2$. Therefore

$$v = \sqrt{\frac{Y}{\rho}} = \sqrt{\frac{2.0 \times 10^{11} \text{ N/m}^2}{7.9 \times 10^3 \text{ kg/m}^3}} = \sqrt{0.25 \times 10^8 \text{ m}^2/\text{s}^2} = 5.0 \times 10^3 \text{ m/s}.$$

THINK ABOUT THIS . . .

WHY ARE LONGITUDINAL WAVES IN SOLIDS CALLED SOUND?

We think of sound in terms of our hearing, and this type of sound does involve waves, propagating in air from the source of the disturbance to our ears. The method by which they propagate is not identical to the way "sound" propagates in a solid. But the dynamics of a sound wave in air has certain similarities to the longitudinal waves in solids described above, and that is why we use the same name for both. Both are longitudinal waves of compression and rarefaction. In the case of solids, the dynamics that keeps the disturbance moving has to do with the atoms of the solid, densely packed, interacting through what for all practical purposes are little springs—actually electrical forces at the atomic level. In air the disturbance is propagated through collisions of fairly sparse moving molecules (see Chapter 19). The number we found above for the numerical value of the speed of sound in steel is, typically for sound in dense media, much larger than the speed of sound in air. It is in excellent agreement with experimental data. We can get an idea of the variety of values of sound waves from Table 14–1, which

gives the speed of longitudinal waves—sound—in various media.

TABLE 14–1 • The Speed of Sound Waves in Various Media[†]

Medium	Speed of Sound (m/s)
Hydrogen	1284
Air	330
Liquid mercury (20°C)	1450
Methyl alcohol	1189
Water	1402
Polyethylene	920
Lead	1210
Silver	2700
Aluminum	5000
Beryllium	12,870

[†]*Longitudinal waves. Temperature = 0°C, pressure = 1 atm unless otherwise stated.*

How the Wave Speed Depends on the Medium

Once we have established that waves traveling through a certain medium obey a wave equation, we can try to determine the wave speed, which, as we've said before, depends on properties of the medium. For the string $[v = (T/\mu)^{1/2}]$, the tension T acts as a restoring force that tends to bring the string back to its equilibrium position. The mass density μ describes the reaction of the string to this restoring force. For sound waves in solids $[v = (Y/\rho)^{1/2}]$, the Young's modulus is the "spring constant" of a stress, and the mass density ρ describes how "nimbly" the solid responds to this force. Generally, this suggests that the speed of mechanical waves is a function of an internal force factor (an elastic restoring force) divided by a mass factor characteristic of the system.

A dimensional analysis helps us to extend this analysis. The tension is a force and has dimensions $[MLT^{-2}]$. The mass density (mass per unit length) has dimensions $[ML^{-1}]$. Therefore

$$\left[\frac{\text{tension}}{\text{mass density}}\right] = \left[\frac{MLT^{-2}}{ML^{-1}}\right] = [L^2 T^{-2}].$$

Thus if v is a function of the ratio of the restoring force factor to the mass factor, *the particular function must be a square root*:

$$\text{wave speed} = \sqrt{\frac{\text{restoring force factor}}{\text{mass factor}}}. \tag{14–29}$$

WAVE SPEED

If we restrict ourselves to bulk (macroscopic) properties of the string, Eq. (14–9) is the *only* dimensionally correct relation for wave speed. Indeed, the *only* bulk properties of the string that we know are the tension, the total mass, the length, and the thickness; the first three of these are figured into the wave speed. (For the true mass density, or mass per unit *volume*, the thickness of the string would also enter.) As we discussed in Chapter 1, dimensional analysis does not permit a calculation of constant factors such as 2 or π, but it will give the system's dependence on the bulk properties, and it will give a correct estimate of orders of magnitude.

The Speed of Sound in Air

We can use Eq. (14–29) to estimate the speed of sound waves in a gas (such as air). We must find a restoring force factor and a mass factor for the gas. We would not expect the total amount of gas to enter, because the speed is not dependent on the total amount of gas. That leaves us with only one choice for the mass factor, namely the mass density ρ_0. The restoring force factor is the pressure p_0 within the gas. We will look at this quantity in detail in Chapter 16, but here we can describe pressure as follows. Within the gas, molecules are constantly in motion. If a tiny membrane is placed into the gas, it is bombarded on each side by many molecules. When the molecules collide elastically with the membrane, they transfer momentum to it. A transfer of momentum is equivalent to a force, so we can characterize the gas by a force per unit area, the **pressure**, that acts on any surface in contact with the gas. Pressure has dimensions of force per area, or $[MLT^{-2}][L^{-2}] = [ML^{-1}T^{-2}]$, whereas density has dimensions of mass per volume, or $[ML^{-3}]$. The ratio p_0/ρ_0 has dimensions $[L^2T^{-2}]$, a speed squared. We thus estimate for the speed of sound in a gas

$$v_{\text{sound}} \cong \sqrt{\frac{p_0}{\rho_0}}. \tag{14–30a}$$

This result was first found by Newton and is appropriately called Newton's formula.

A more detailed calculation yields a better value for the speed of sound,

$$v_{\text{sound}} = \sqrt{\frac{\gamma p_0}{\rho_0}}. \tag{14–30b}$$

Here γ is a constant with the approximate value 1.4 for air—we'll see in Chapters 17–20 that γ has something to do with the thermal properties of the gas. (Thermal properties enter because local compression and rarefaction of a gas result in local heating and cooling.) The numerical correction to our estimate due to the factor γ is of the order of 20%. For normal atmospheric conditions at sea level, $p_0 \cong 1.0 \times 10^5 \text{ N/m}^2$, $\rho_0 \cong 1.3 \text{ kg/m}^3$, and

$$v_{\text{sound}} \cong 330 \text{ m/s} = 740 \text{ mi/h}, \tag{14–31}$$

or about 1000 ft/s.

Shear Waves in Solids

Just as small longitudinal displacements within a solid lead to longitudinal waves, small transverse displacements within a solid lead to transverse waves. This is because solids resist shear forces, and the wave speed of transverse waves—also called *shear waves*— is described by using the shear modulus, G, of Section 11–4. (This correctly implies that bulk gases or fluids do not support transverse waves because they do not resist shear forces—there is no restoring force for shear displacements.) Application of Eq. (14–29) leads to

$$v_{\text{shear}} = \sqrt{\frac{G}{\rho}}. \qquad (14\text{–}32)$$

In most cases, the shear modulus is lower than Young's modulus, and the speed of shear waves is less than that of longitudinal waves.

THINK ABOUT THIS . . .

HOW DO WE KNOW WHAT'S INSIDE EARTH?

Both sound and shear waves can travel through Earth. When they are initiated by earthquakes, we refer to them as *seismic waves*. This phenomenon is the basis for an important set of tools used in geological exploration and research. The varying arrival times of the seismic waves from an earthquake at a seismograph—a device whose vibrations detect seismic waves—gives information about the location of the earthquake. More important, the study of these waves has allowed us to build up a picture of the inside of the Earth. One of the ingredients in the analysis is the fact that at a boundary between two media, waves are reflected or transmitted, a subject we'll be discussing in Chapter 15. In the case of Earth, these boundaries are places where the composition or state of the internal material changes. Seismic waves from an earthquake can follow many paths, reflecting from various internal layers and boundaries between different materials (Fig. 14–16), and information on Earth's internal structure comes from the arrival time and strength of waves generated by distant earthquakes. We have learned that Earth has an overall structure that can be divided into four main parts: crust, mantle, outer core, and inner core. We know for another example that Earth's outer core is liquid because shear waves do not pass through it. The search for petroleum also relies heavily on evidence from wave propagation within Earth. In this case the waves are generated by explosions judiciously set off in the vicinity of the area to be explored. This tool can reveal in detail the structure of the interior deep below the surface.

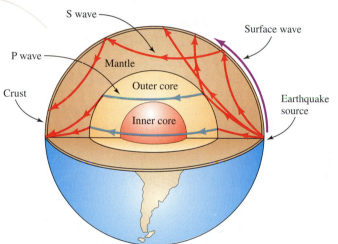

◀ **FIGURE 14–16**

Several types of waves are possible in solids. Seismic waves, produced in events such as earthquakes, include all types. The P waves marked here stand for primary, geologists' notation for the sound waves, while the S waves stand for secondary, geologists' notation for shear waves. These waves travel at different speeds; moreover, as we have remarked, fluids do not support shear waves. Hence P and S waves convey different information. ∎

The Propagation of Arbitrary Waveforms

We have thus far concentrated on traveling *harmonic* waves, of generic form

$$z(x, t) = A \sin(kx - \omega t) = A \sin k\left(x - \frac{\omega}{k}t\right) = A \sin k(x - vt).$$

We will now show that a waveform of *any shape* can be propagated provided it has argument $x - vt$:

$$z(x, t) = f(x - vt). \qquad (14\text{–}33)$$

To prove this, we need to show that this function is a solution of the wave equation, which is a simple mathematical exercise (see Problem 15). Here $f(x - vt)$ is any function we like; it describes the shape of whatever waveform is propagating.

We'll illustrate the procedure with a simple waveform, one that we call a *pulse*, as in Fig. 14–17. (This is a term normally reserved for a nonrepetitive waveform of limited

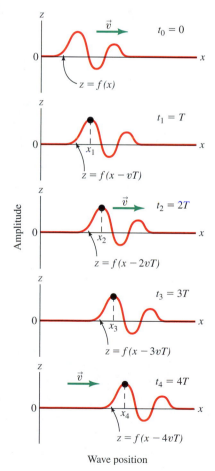

▲ **FIGURE 14–17** The curve that is formed by the pulse represents some function f of the variable $x - vt$. As time goes on (here $t_2 > t_1$), the curve moves to the right with speed v.

duration.) First we single out a particular point on the function, such as a peak, and mark this point with a black dot. At time t_1 this point is at an x-value of x_1. If t_1 increases to t_2, the argument of f will have the same value—the point will still be at the black dot—only if x increases to compensate for the decreased value of the term $-vt$. In order for the point to stay at the same argument, we must have

$$x_2 - vt_2 = x_1 - vt_1,$$

which can be solved for x_2:

$$x_2 = vt_2 + x_1 - vt_1 = x_1 + v(t_2 - t_1). \tag{14–34}$$

Eq. (14–34) describes a point moving to the right with speed v. Since the argument applies for any point of the pulse, all points of the pulse move in the same way—the waveform $f(x - vt)$ moves as a whole. Similarly, $f(x + vt)$ represents a waveform moving to the left. In summary, $f(x - vt)$—or $f(x + vt)$—*represents a waveform that moves to the right—or left—with speed* v.

CONCEPTUAL EXAMPLE 14–5 Suppose you laid out a very long rope with a small bulge in it (as in Fig. 14–17), poured water on it, and then let it freeze. You have a frozen waveform, which you then arrange to be carried at a uniform velocity past an observer. Mathematically this is identical to a traveling pulse. But is it really a traveling wave? Is a "wave" of cheering, popular at football games, a traveling wave?

Answer No, to both questions. In the first case, matter is moving in the direction of propagation. With a real wave, a bit of colored string attached to some part of the rope would stay at some level,

then rise and fall, but always stay at the same lateral position relative to the observer (e.g. right in front of him). With the fake frozen wave, the string would be moving along with the fixed velocity v. We can add that in the fake wave, there are no forces acting to accelerate portions of the rope, only static forces that hold it rigid. As to the second question, the stadium "wave" does not have the defect of the moving frozen rope: Matter does not travel along the stadium wave. Nevertheless, it is not like the waves treated here in the sense that no power is carried by the wave. No forces, other than moral persuasion, act between neighboring elements.

14–5 Energy and Power in Waves

In waves, masses are in motion, so energy must be present. This is obvious to anyone who has stood within the water waves approaching a beach (Fig. 14–18). In a traveling wave, energy is delivered from one end of the system to the other even though no mass is transferred in the motion. It is an important observation that in waves *energy is transmitted without the transport of mass*. Moreover, the energy can be delivered at a steady rate, and this leads us to look at the power transmitted by traveling waves.

The Power Delivered by Waves

Consider right-moving transverse traveling waves on a uniform rope that has a mass per unit length μ and tension T. As you recall from Chapter 7, the power P delivered at a point x along the rope is given by $P = \vec{F} \cdot \vec{v}_{\text{transverse}}$, where $\vec{v}_{\text{transverse}}$ is the (transverse) velocity of the rope at that point and $\vec{F}$ is the force acting on the portion of the rope to the right of x. (Don't confuse this velocity with the velocity $\vec{v}$ with which the wave propagates!) Only the transverse component of the force enters into the scalar product. To find the transverse component of the force, refer to Fig. 14–19, which shows that the transverse component of force at x acts downward (*negative* sign) with magnitude

$$T \sin \theta \cong T \tan \theta = T \frac{\partial z}{\partial x}.$$

We have used the small distortion approximation here—the angle θ is small—and we have remembered to use the partial-derivative notation again, as a reminder that z in fact depends on two variables, x and t. The transverse velocity of the rope at x is given by

$$v_{\text{transverse}} = \frac{\partial z}{\partial t},$$

where the sign of the time derivative determines whether the sign of the velocity is positive (moving up) or negative (moving down). Thus the power delivered to the rope at point x is

$$P = \vec{F} \cdot \vec{v}_{\text{transverse}} = -T \frac{\partial z}{\partial x} \frac{\partial z}{\partial t}. \tag{14–35}$$

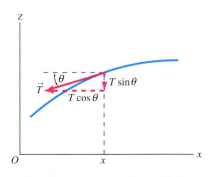

▲ **FIGURE 14–18** Waves carry energy, a fact well known to anyone who has spent time at the seashore, or, especially, who has tried to surf them.

▲ **FIGURE 14–19** A segment of a distorted string is subject to a tension force from either side whose direction depends on the distortion. We are interested in this case in the vertical component.

It is simple to calculate the derivatives involved here for a harmonic traveling wave, $z = A \sin(kx - \omega t)$. We find

$$P = Tk\omega A^2 \cos^2(kx - \omega t) = \mu v \omega^2 A^2 \cos^2(kx - \omega t). \qquad (14\text{--}36)$$

To find the last form, we have used Eq. (14–9), which gives $T = \mu v^2 = \mu(\omega/k)v$, where v is the wave speed.

The power in Eq. (14–36) oscillates with time, and it is useful to think instead about the time average of the power delivered. We denote this average with angular brackets. To find its value, we need the average value taken by the cosine squared over a long time period, namely 1/2. Thus

$$\langle P \rangle = \tfrac{1}{2}\mu\omega^2 A^2 v. \qquad (14\text{--}37)$$

Energy Transport

There is no mechanism for energy loss in a rope such as we have described. (Of course, if we have to take into account the fact that the rope vibrates within the air and that there is air resistance, there is energy loss. We ignore this possibility among others.) Thus the power supplied by whatever generates the waves in the first place is transmitted from one piece of the rope to another. This power is the rate P of energy transport through the rope. We can treat this issue by thinking of an energy density u—an energy per unit length—in the vibrating rope. This energy density must also take the form of a traveling wave, which travels down the rope. How is the energy density u related to the power P? We note that the quantity $v\,\Delta t$ is the length of wave that moves past a given point in a brief time interval Δt. All the energy in that length, namely the energy density times the length, or $uv\,\Delta t$, will be delivered; the rate of energy delivery P is this quantity divided by Δt. In other words, $uv = P$, or

$$u(x, t) = P/v = \mu\omega^2 A^2 \cos^2(kx - \omega t). \qquad (14\text{--}38)$$

We have shown that the power delivered and the energy density on the rope are each quantities that travel with the medium's wave speed. These quantities are always positive. Moreover, *they are proportional to the squares of both the amplitude and the frequency of the wave*. Although we have derived our results for traveling waves on a rope, the quadratic dependence of the rate of energy transport, of power, or of energy density on amplitude and frequency is a general feature of *all* waves.

EXAMPLE 14–6 A long rope of mass density 150 g/m is tied at one end. A person holds the other end with a horizontal force of 95 N so that the equilibrium position of the rope is horizontal. The person then shakes the held end of the rope up and down such that sinusoidal traveling waves are generated with a frequency of $1.0\ \text{s}^{-1}$ and an amplitude of 5.0 cm. What average power must be delivered by the hand that is shaking the rope?

Setting It Up Figure 14–20 shows the wave produced when the rope is shaken up and down. We are given all the parameters that enter into the calculation of average power, namely the tension $T = 95\ \text{N}$, the mass density $\mu = 0.15\ \text{kg/m}$, the frequency $f = 1.0\ \text{Hz}$, and the amplitude $A = 0.050\ \text{m}$.

Strategy The average power delivered to the rope is the average power carried by the rope in its wave motion. Thus Eq. (14–37) can be used to calculate the power. In this expression only the angular frequency and the wave speed are not given directly. The frequency gives the angular frequency, while we find the wave speed from the given tension and the given mass density using Eq. (14–9), and only the numerical details need to be worked out.

Working It Out To apply Eq. (14–37), we need $\omega = 2\pi f = 6.3\ \text{rad/s}$, and $v = (T/\mu)^{1/2} = [(95\ \text{N})/(0.15\ \text{kg/m})]^{1/2} = 25\ \text{m/s}$. Then Eq. (14–37) gives

$$P = \tfrac{1}{2}(0.15\ \text{kg/m})(6.3\ \text{s}^{-1})^2(0.050\ \text{m})^2(25\ \text{m/s}) = 0.19\ \text{W}.$$

This is a tiny fraction of 1 hp, easily delivered by the hand holding the rope.

What Do You Think? In stating this problem, we made the unspoken assumption that gravity does not act on the rope. Since the rope has mass, it is subject to gravity. Would you expect inclusion of the force of gravity to lead to a significant adjustment to the solution?

▲ **FIGURE 14–20** A person shaking a rope produces traveling waves.

Our results for the energy density, together with the conservation of energy, determine how the amplitude (wave height) of a periodic wave originating at a point, in this case the shouter's mouth, varies with distance from that point. For a one-dimensional system, the amplitude and the average energy density remain unchanged as the wave propagates—you might say that the energy has nowhere else to go. For waves in two and three dimensions the energy spreads out as it moves away from the source. We can visualize this if we think of the two-dimensional case of a toe periodically dipped into a pond. The moving toe is the source of the energy contained in the ripples that spread outward in circles from the toe. If a given ripple has radius r, the energy in the ripple is spread evenly over the circle, which has length $2\pi r$. Thus the total energy density falls off as $1/r$, while the amplitude falls off as $r^{-1/2}$ because the energy density is proportional to the square of the wave's amplitude. Similarly, a shout in open air is the source of waves that spread outward from the source as a sphere. Sound energy is spread over the surface of area $4\pi r^2$ of a sphere of radius r. Thus the energy density at a point decreases as $1/r^2$, and the amplitude of the sound waves from the sound decreases as its square root—that is, as $1/r$.

14–6 Standing Waves

When you pluck a guitar string, it is the string as a whole that vibrates, with the vibration largest in the middle of the string and smallest—indeed, zero—at the stopping fret and the bridge. As we remarked in Section 14–1, this doesn't seem much like the traveling waves we have been discussing. In fact, as we shall now see, it is a very closely related phenomenon.

Let us return to the very general harmonic solution to the wave equation, Eq. (14–19), which we rewrite here for convenience,

$$z(x, t) = A \cos kx \cos \omega t + B \cos kx \sin \omega t + C \sin kx \cos \omega t + D \sin kx \sin \omega t. \tag{14–19}$$

Can we find solutions that describe the vibrating guitar string by choosing the parameters A, B, C, and D correctly? If the string runs from $x = 0$ to $x = L$, such solutions would give the string's ends fixed at $x = 0$ and $x = L$, that is, $z(0, t) = z(L, t) = 0$ *for all time*. The requirement that $z(0, t) = 0$ means that $A \cos \omega t + B \sin \omega t = 0$ for all times, and this is only possible if $A = B = 0$, and leaves us with

$$z(x, t) = (C \cos \omega t + D \sin \omega t) \sin kx.$$

The term $C \cos \omega t + D \sin \omega t$ can always be written in the form $z_0 \cos(\omega t + \phi)$. To do this, simply write, as is always possible, $C = z_0 \cos \phi$ and $D = -z_0 \sin \phi$. Then

$$C \cos \omega t + D \sin \omega t = z_0 \cos \phi \cos \omega t - z_0 \sin \phi \sin \omega t = z_0 \cos(\omega t + \phi).$$

Our wave now has the form

$$z(x, t) = z_0 \cos(\omega t + \phi) \sin kx. \tag{14–39}$$

We must still enforce the requirement that the end of the string at $x = L$ does not vibrate, i.e. $z(L, t) = 0$ for all time, or by comparison with Eq. (14–39),

$$\sin kL = 0. \tag{14–40}$$

This is a condition on the k-values that we'll investigate immediately below. We call the form Eq. (14–39) a **standing wave**, since at any point between $x = 0$ and $x = L$, the guitar string moves harmonically up and down (transverse to the string alignment) with time, with an amplitude $z_0 \sin kx$ that varies from point to point along the string. Nothing "propagates" in this solution to the wave equation.

What does Eq. (14–40)—the equation that fixes the end of the string—imply? This equation can be satisfied if the argument of the sine function is any positive integer multiple of π (negative integer multiples only reverse overall sign of the wave), or if k takes on the discrete values

$$k_n = \frac{n\pi}{L} \quad \text{for } n = 1, 2, 3, \ldots. \tag{14–41}$$

(The case $n = 0$ gives just a flat or undistorted string—we say the string is *unexcited.*) This is a condition on the wavelength of the wave, since $k = 2\pi/\lambda$, and Eq. (14–41) then implies that the allowed wavelengths also carry an index n:

$$\lambda_n = \frac{2\pi}{k_n} = \frac{2}{n}L \quad \text{for } n = 1, 2, 3, \ldots. \tag{14–42}$$

When the string is set into motion with values of k that correspond to a certain n-value, we say the string is *excited* into the wave labeled by n. Equation (14–42) can be rewritten as $L = \frac{1}{2}n\lambda_n$, providing a convenient way to recall the possible values of the wavelength. For a standing wave *the total length of a string (or any one-dimensional medium with fixed ends that supports waves) can be divided into an integral number of half-wavelengths*. The allowed waves are called **modes**—possible vibration patterns—of the standing waves, and they are labeled by the value of n. $n = 1$ labels the *fundamental mode* (see Fig. 14–21), also called the *first harmonic*. $n = 2, 3, \ldots$ are called the second, third, $\ldots$ harmonics. Note that for odd values of n, the number of wavelengths that can fit on the string is a whole number plus one half, while for even values of n, a whole number of wavelengths fits into the string. For $n > 1$ there are *nodes*, locations along the x-axis where the string rests motionless with no transverse displacement. Although these nodes are not caused by a physical clamping of the string as at the end points, they otherwise behave as if the string were fixed at these points. Not counting the end points, there are $n - 1$ of these nodes—for example, one node in the $n = 2$ case of Fig. 14–21. We can also see the presence of *antinodes*, locations where the displacement is a maximum. How many nodes and antinodes are there in the standing wave with $n = 4$ in Fig. 14–21?

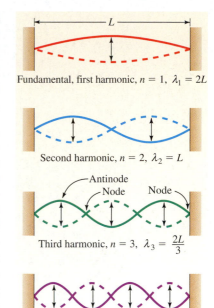

Fundamental, first harmonic, $n = 1$, $\lambda_1 = 2L$

Second harmonic, $n = 2$, $\lambda_2 = L$

Antinode — Node   Node

Third harmonic, $n = 3$, $\lambda_3 = \frac{2L}{3}$

Fourth harmonic, $n = 4$, $\lambda_4 = \frac{L}{2}$

▲ **FIGURE 14–21** Some modes of transverse standing waves on a string with fixed ends. Higher modes—those characterized by larger values of n—have more nodes and antinodes. The arrows indicate the extremes of the motion as the string moves back and forth.

CONCEPTUAL EXAMPLE 14–7 Consider two strings, both fixed at each end, one twice as long as the other. Do they have any allowed wavelengths for standing waves in common?

Answer If the length of the shorter string is L, then for the shorter string the possible wavelengths are $\lambda_1 = 2L/n_1$, and for the longer string, length $2L$, we have $\lambda_2 = 4L/n_2$. The question then be-

comes: For what values of n_1 and n_2 can we have $\lambda_1 = \lambda_2$? The quickest way to see this is to eliminate L by writing a ratio

$$\frac{\lambda_2}{\lambda_1} = \frac{4L}{n_2} \times \frac{n_1}{2L} = \frac{2n_1}{n_2}.$$

This ratio must equal 1. We thus see that for every *even* value of n_2 there is an n_1, half its value, that leads to the equality of the wavelengths. The n values are integers, so this is the only solution.

EXAMPLE 14–8 Standing waves can be excited on a guitar string of length 1.00 m whose ends are fixed at $x = 0$ and $x = L$. Find the two smallest values of the wave number and the two longest wavelengths allowed for standing waves. Sketch the space dependence of these waves.

Strategy The possible wave numbers are given by Eq. (14–41). The allowed values of k grow with n, so the smallest wave numbers will correspond to $n = 1$ and $n = 2$, and we will use these values in our calculation. The wavelength is inversely proportional to the wave number, so the two smallest wave numbers are also the two longest wavelengths.

Working It Out The smallest values of k and the largest values of λ are the two cases $n = 1$ and $n = 2$ of Eqs. (14–41) and (14–42), respectively. These are

$$k_1 = \pi/L = 3.14 \text{ m}^{-1} \quad \text{and} \quad k_2 = 2\pi/L = 6.28 \text{ m}^{-1}$$

and

$$\lambda_1 = 2L = 2.00 \text{ m} \quad \text{and} \quad \lambda_2 = 2L/2 = 1.00 \text{ m}.$$

The space dependence of the waves is given by the function $\sin(k_n x)$ times an arbitrary amplitude z_0. These are sketched in Fig. 14–22.

$n = 1$

$n = 2$

▲ **FIGURE 14–22** Standing waves on a guitar string.

What Do You Think? Suppose the guitar string is additionally clamped at $x = 20$ cm, for example by a finger against the fret. How do you think the results will change to the left of the clamping point? To the right of the clamping point? Will there be certain wavelengths such that the pattern looks like a single string?

The Frequencies of Standing Waves

The waves we have described above cannot have arbitrary wavelengths, and we'll now see that they cannot have arbitrary frequencies. The relation between frequency and wavelength follows from the wave equation, Eq. (14–7). If we substitute Eq. (14–39) into the the wave equation rewritten in the form $\dfrac{\partial^2 z(x,t)}{\partial x^2} - \dfrac{1}{v^2}\dfrac{\partial^2 z(x,t)}{\partial t^2} = 0$, we find for the left side

$$\frac{\partial^2}{\partial x^2}\sin kx \cos(\omega t + \phi) - \frac{\mu}{T}\frac{\partial^2}{\partial t^2}\sin kx \cos(\omega t + \phi)$$

$$= -k^2 \sin kx \cos(\omega t + \phi) - \frac{\mu}{T}[-\omega^2 \sin kx \cos(\omega t + \phi)]$$

$$= \left(-k^2 + \frac{\mu\omega^2}{T}\right)\sin kx \cos(\omega t + \phi).$$

This can equal 0 (the wave equation is satisfied) only if $k^2 = \mu\omega^2/T$ or, in terms of the wavelength and the wave speed $v = (T/\mu)^{1/2}$ (which here has no apparent connection to an actual speed),

$$\omega = \frac{2\pi}{\lambda}\sqrt{\frac{T}{\mu}} = \frac{2\pi v}{\lambda}. \tag{14–43a}$$

Equivalently this gives a condition on the period $\tau = 2\pi/\omega$ of the oscillations,

$$\tau = \lambda\sqrt{\frac{\mu}{T}} = \frac{\lambda}{v}. \tag{14–43b}$$

What do Eqs. (14–43a, b) tell us about the motion of the string?

1. For a particular string, the standing waves with smaller wavelengths also have smaller oscillation periods.

2. For a given wavelength and string length, a less dense string vibrates more rapidly than a denser one.

3. For a given wavelength and string length, a string under more tension vibrates more rapidly than a string under less tension.

Harmonics

For the string with fixed ends, $\lambda_n = (2/n)L$, and the frequency, angular frequency, and period can have the discrete values

$$f_n = \frac{n}{2L}\sqrt{\frac{T}{\mu}}, \tag{14–44a}$$

$$\omega_n = \frac{n\pi}{L}\sqrt{\frac{T}{\mu}}, \tag{14–44b}$$

$$\tau_n = \frac{2L}{n}\sqrt{\frac{\mu}{T}}. \tag{14–44c}$$

The frequency f_1, the lowest frequency, is known as the **fundamental frequency**, and $\lambda_1(= 2L)$ is the **fundamental wavelength**. The frequency $f_n = nf_1$ is called the **nth harmonic** of the string, and the full set is called the harmonic series; Fig. 14–21 depicts some harmonics. (We will return to these topics in the next section.)

EXAMPLE 14–9 The time taken to complete 100 full vibrations (100 periods) in the fundamental mode of a piano wire is 0.5 s. The wire length is 2 m, and the total mass of the wire is 25 g. What is the tension on the wire? What value should the tension have in order to halve this time?

Setting It Up We have the length L and total mass M of the wire as well as the period τ_1 of the fundamental mode.

Strategy The period, which in this case is 1/100th of 0.5 s, of any mode is directly related to the tension and mass density [Eq. (14–44c)]. Although it is not given directly, the mass density is $\mu = M/L$.

Working It Out Equation (14–44c) gives

$$\tau_1 = 2L\sqrt{\frac{\mu}{T}}.$$

This may be solved for the tension, T:

$$T = \frac{4\mu L^2}{\tau_1^2} = \frac{4L^2}{\tau_1^2}\frac{M}{L} = \frac{4ML}{\tau_1^2}.$$

With $L = 2.0$ m, $\tau_1 = (0.50 \text{ s})/100 = 5.0 \times 10^{-3}$ s, and $M = 0.025$ kg

$$T = \frac{4(0.025 \text{ kg})(2.0 \text{ m})}{2.5 \times 10^{-5} \text{ s}^2} = 8.0 \times 10^3 \text{ N},$$

equivalent to the average weight of ten men.

▲ **FIGURE 14–23** This piano tuner adjusts the natural frequencies of the strings on this piano by adjusting their tensions.

Since the period is proportional to $T^{-1/2}$, we can halve the period if the tension is increased by a factor of 4, to 3.2×10^4 N.

What Do You Think? Suppose the frequency of a particular piano note is to be shifted upward by 1%. Will the piano tuner in Fig. 14–23 (a) increase the tension by 1%? (b) decrease it by 1%? (c) decrease it by 2% or (d) increase it by 2%? Do all the strings in a piano differ only in the tensions?

THINK ABOUT THIS . . .
WHAT DO STANDING WAVES LOOK LIKE IN TWO AND THREE DIMENSIONS?

Traveling waves in two and three dimensions, such as ocean waves or sound waves in air, all behave according to the appropriate wave equation. Just as for waves on a string, if the medium supporting them has boundary conditions that restrict the displacement, standing waves will occur. Drumheads are an example of standing waves on a two-dimensional surface. The boundary conditions for a drumhead require that the vibrations vanish on the circular rim of the drum. Figure 14–6 shows a few of the modes possible on a circular drumhead. The modes on an elliptical drumhead differ from those on a circular drumhead, as the boundary conditions are different. Your microwave oven, which sets up standing waves within its interior volume, provides an example of three-dimensional standing waves. The boundary conditions determining the allowed values of k and the dynamics determining the allowed values of ω are more complicated when the system is a surface or a volume rather than a one-dimensional system like the string, but the common feature of all these systems is that the allowed wavelengths and frequencies are discrete and can be labeled by integers. ◼

Energy and Standing Waves

There is no energy flow in a standing wave such as the vibrations of a guitar string, as there is no continuing source of energy input into the string. Nevertheless, there are internal forces leading to (changing) transverse velocity, and the motion contains energy; the motion simply does not translate into an energy flow. To understand why, we look to the formula for power. As we saw in Eq. (14–35), the power is given by $P = -T(\partial z/\partial x)(\partial z/\partial t)$. For a traveling wave of the form $z(x, t) = A \sin(kx - \omega t)$, we find $P = A^2 Tk\omega \cos^2(kx - \omega t)$. This is always positive and is itself a traveling wave. But for a standing wave of the form $z(x, t) = A \sin kx \cos \omega t$, we get

$P = A^2 Tk\omega[\cos kx \sin kx] \times [\cos \omega t \sin \omega t] = (ATk\omega/4) \sin 2kx \sin 2\omega t$. This is another standing wave. While it does not vanish, it oscillates in time, and since the time average of $\sin 2\omega t$ vanishes, the average power in the standing wave is zero.

14–7 More About Sound

In Section 14–4 we described sound waves in a metal bar and, more briefly, in air. Sound waves in air are generated by the mechanical motion of some object in contact with the air—the vibration of vocal cords, an oscillating violin string, or the collapsing hot-air column formed by lightning. A traveling sound wave transports energy through the air, which can then be picked up by another mechanical system such as an eardrum. In this section we look more closely at these waves.

The Nature of Sound Waves

Let's look at sound in air on a microscopic level. The molecules in air are widely spaced so the intermolecular forces that act as springs in solids or liquids play a very limited role here. Except for brief, random collisions with each other, air molecules move freely. How, then, can a wave propagate? Consider a pipe oriented along the x-axis as in Fig. 14–24. At one end of the pipe is a speaker. When the speaker diaphragm moves to the right, the air molecules within a thin slice of the pipe to the right of the diaphragm are subject to a force that acts along the $+x$-direction. These air molecules acquire a net momentum in the $+x$-direction in addition to their random motion. Some of the molecules in the original slice will then move to an adjacent region, increasing the number of molecules there and creating a region of *compression*, where the molecular density is higher. When the diaphragm moves back to the left, fewer molecules remain in this region, which corresponds to a region of *rarefaction* where the molecular density is lower. We thus have adjacent regions of compression and rarefaction, as in Fig. 14–24. In the region of compression, there are more molecules and therefore more molecular collisions. The air molecules that bounce out of the region of more collisions will fill in the rarefied region. The molecules to which the net momentum had been transferred will move in the $+x$-direction, so the region of compression moves to the right in the form of a pulse. If the diaphragm undergoes periodic motion, regions of compression and rarefaction are propagated in the $+x$-direction as a periodic wave in air. Figure 14–24 shows the molecular pattern of a sound wave in air. The displacement of the molecules is aligned with the propagation direction of the wave, so sound waves in air are *longitudinal* waves.

▶ **FIGURE 14–24** When a piston—here a speaker diaphragm—at one end of an air-filled pipe moves along the $+x$-direction, there is a force on the air molecules to the right of the piston, initially supplied by collisions of molecules with the moving piston, and subsequently by collisions between molecules. If the piston moves back and forth regularly, the result is alternate regions of higher and lower density propagating along the pipe, which are longitudinal waves.

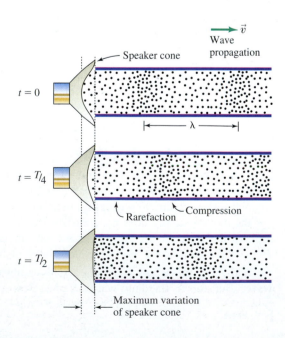

There is no net movement of molecules in the direction of the wave. It is rather the alteration of the density from the average (or equilibrium) density that propagates. If ρ is the density of air (or any other gas), then sound waves may be described in terms of $\Delta\rho = \rho - \rho_0$, the deviation from the equilibrium density ρ_0. A harmonic sound wave traveling in the $+x$-direction would be described by

$$\Delta\rho = A_\rho \sin(kx - \omega t), \qquad (14\text{-}45)$$

where A_ρ is the maximum deviation of the density from ρ_0. Generally, $A_\rho \ll \rho_0$, so we may speak of a nearly constant mass density, even in the presence of sound waves. The speed of the sound waves obeys the relation $v = \omega/k$, and it is given by Eq. (14-30b), $v = \sqrt{\gamma p_0/\rho_0}$, where ρ_0 is the equilibrium density of the gas, p_0 is the pressure, and γ is a numerical constant equal to 1.4 in air.

Standing Sound Waves

Standing sound waves can be generated if the necessary boundary conditions exist. For example, a pipe that is much narrower than the wavelength of a wave will quite easily generate standing waves that we can view as one dimensional. If the pipe is closed at one end, then there can be no displacement of gas molecules across that end. The closed end of the pipe is then like the fixed end of a string. If both ends are closed, we have the equivalent of a string fixed at both ends. As Fig. 14-25 indicates in a schematic way, the wavelength for these standing waves takes on only the discrete values allowed by Eqs. (14-41) and (14-42). The frequencies are determined in terms of the wavelengths by Eq. (14-21),

$$v_{\text{sound}} = \lambda f. \qquad (14\text{-}46)$$

As for a vibrating string, the discrete integer n labels the modes [Eqs. (14-44)]. When $n = 1$, we have the fundamental frequency (discussed in Section 14-6); when $n > 1$, we speak of the nth harmonic.

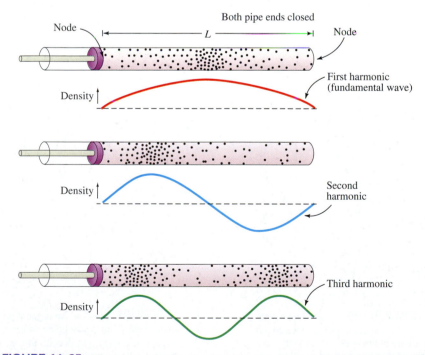

▲ FIGURE 14-25 There is a node of molecular displacement at each end of a closed pipe. Thus the longest possible wavelength in this situation is the one that has one half of the standing sound wave within the pipe. The longest wavelength represents the fundamental frequency, corresponding to the first harmonic. The next two harmonics are also shown here. The curves represent the amplitudes of the density wave which, you should keep in mind, is a wave in the departure of the density from its equilibrium value described by the horizontal dashed line.

EXAMPLE 14–10 A pipe closed at both ends has length $L = 4.0$ m. Find the fundamental frequency for standing sound waves, as well as the frequencies of the next two modes.

Setting It Up Figure 14–25 shows the fundamental and the next two harmonics.

Strategy As Eq. (14–42) states, the fundamental wavelength, λ_1, for a system with two closed ends is twice the length of the system: $\lambda_1 = 2L$. Higher modes have shorter wavelengths, enumerated in this case by $n = 2, 3$: $\lambda_2 = \lambda_1/2$, $\lambda_3 = \lambda_1/3$. The relation between frequency and wavelength is given by Eq. (14–46). Thus the frequencies in question are determined by

$$f_n = \frac{v_{\text{sound}}}{\lambda_n} = \frac{v_{\text{sound}}n}{2L}.$$

Working It Out Taking v_{sound} from Eq. (14–30b), we find

$$f_1 = \frac{330 \text{ m/s}}{2(4.0 \text{ m})} = 41 \text{ Hz}, \quad f_2 = 2f_1 = 83 \text{ Hz}, \quad f_3 = 3f_1 = 124 \text{ Hz}.$$

Figure 14–25 shows how the standing wave of fundamental frequency fits within the pipe—it has a node at each end, but none in between.

What Do You Think? The standing waves in this example run from end to end of the pipe. Do we have to assume that the pipe is infinitely narrow to justify this? If not, are there standing waves that go across the pipe?

Other boundary conditions are possible for standing sound waves. Consider the organ pipe in Fig. 14–26. When one end of a pipe is open, the pressure at that end takes on the value of the pressure of the exterior. There is no variation of the pressure there, which corresponds to a pressure node. This can be shown to mean that the displacement has an amplitude maximum (or *antinode*) rather than a node at the open end. Figure 14–27 shows a similar pipe, but this time open at both ends. When both ends are open, there is an antinode at each end (Fig. 14–27), in which the first few harmonics are drawn.

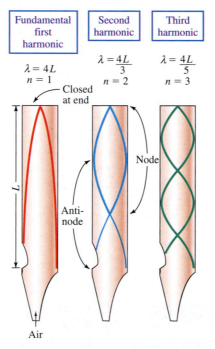

▲ **FIGURE 14–26** A standing sound wave in an organ pipe open at one end has a maximum at the open end. In this situation, one-quarter of the standing wave with the longest possible wavelength—the fundamental—fits within the pipe. The next harmonic is also shown. The curves represent the amplitudes of the density wave.

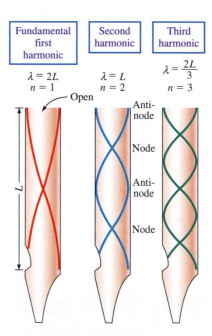

▲ **FIGURE 14–27** An organ pipe open at each end has standing waves with antinodes at each end, hence a different harmonic series than for the organ pipe closed at one end.

EXAMPLE 14–11 An organ pipe with the dimensions of the pipe in Example 14–10 is closed at one end and open at the other. Find the fundamental frequency and next harmonic.

Setting It Up We are asked to find frequencies just as in the previous example. What is different about this situation? Here the boundary conditions have been changed, and we must take this into account.

Strategy We must have a node at one end and an antinode at the other. The longest possible standing wave that fits this description is easily drawn (Fig. 14–26). The figure shows that the wavelength is four times the length of the pipe, $\lambda_1 = 4L$. This is double the wavelength of the fundamental mode when both ends are closed. Equivalently, the corresponding frequency is half the fundamental frequency for both ends closed.

Working It Out The fundamental frequency is

$$f_1 = \frac{v_{\text{sound}}}{\lambda_1} = \frac{v_{\text{sound}}}{4L} = \frac{330 \text{ m/s}}{4(4.0 \text{ m})} = 21 \text{ Hz}.$$

The next harmonic is not the harmonic with a frequency twice the fundamental frequency; such a wave would have a node at the open end. Rather, it is the harmonic with three times the fundamental frequency, also illustrated in Fig. 14–26. For this wave, three-quarters of one wavelength fits into the pipe. Mathematically, notice that at L the magnitude of the solution $\sin kx$ must have its maximum value, so that $\sin kx = \pm 1$. Thus $kx = \pi/2, \; 3\pi/2, 5\pi/2 \ldots = (2n + 1)\pi/2$, with $n = 0, 1, 2, \ldots$

What Do You Think? If you take an empty bottle and blow across the open end, you get a deep sound. Why? Suppose you were to blow across a cardboard tube with both ends open. What would you hear?

The Excitation of Standing Sound Waves: We can excite standing waves in an organ pipe if we vibrate the air within at the standing wave frequency. One way to do this is to direct externally generated sound waves of that frequency to the pipe. This process must also go in the other direction—the standing waves within the pipe must generate traveling sound waves of the same frequency if the members of an audience are to perceive the standing waves as sound. The couplings between these different types of waves normally involve rather complicated mechanisms. The many different ways that standing waves in musical instruments are excited (bow, single and double reed, mouthpiece, and so forth) indicates the rich variety of physical phenomena involved.

In each case, the excitation of waves involves a transfer of energy. This transfer of energy, from the energy of an external source—such as the air forced across a clarinet mouthpiece—to the energy of the standing wave within the tube of a clarinet, takes place in ways closely related to the *resonance* phenomenon discussed in Chapter 13. If a tuning fork is placed near the end of an organ pipe and if the frequency of the tuning fork is the same as the fundamental frequency of the pipe, strong standing waves will rapidly be established within the pipe. Resonance of this type is responsible for the fact that when a trumpet plays a note whose frequency matches the natural frequency of a delicate goblet, standing waves can be induced in the goblet strong enough to make it break (Fig. 14–28).

Hearing Sounds

The combination of ear and brain acts as a very sensitive detector of sound with many interesting properties. The most basic quantities it detects are pitch and loudness. *Pitch* is a measure of frequency, and the human ear can hear frequencies ranging from about 20 Hz to 20,000 Hz. *Loudness* is a measure of the power carried in a wave.

Because sound waves in air form two-dimensional fronts (Fig. 14–29), it is appropriate to define the average power per unit area rather than the total power in the wave.

▲ **FIGURE 14–28** The glass breaks because its natural frequencies are resonant with a frequency (or with several) of the sound wave, which allows energy to be transferred easily from the sound to the glass.

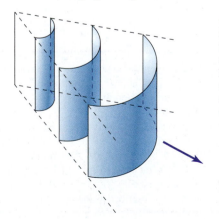

◀ **FIGURE 14–29** Waves generated in a three-dimensional medium form two-dimensional fronts. A front is defined as traced out by, say, maxima of a harmonic wave. In one-dimensional waves, the front would be a point moving along the medium and in two-dimensional waves the front is a line.

► FIGURE 14–30 Typical noise levels at the source. We have here plotted intensity, measured in decibels, as a function of frequency. The frequency axis is divided into three parts based on the central, audible, portion of the spectrum—each is plotted on a logarithmic scale.

This average power per unit area is called the **intensity**, I. The intensity measures energy per unit time per unit area and has SI units of watts per square meter (W/m^2). Intensity, like power, is proportional to the square of both the amplitude and the frequency. Intensity itself is not a particularly convenient measure physiologically; a measure better suited to the human ear/brain combination is the *decibel scale*, in which the measure of intensity is a dimensionless quantity, β, with units of decibels (dB). It is defined by

$$\beta = 10 \log_{10}\left(\frac{I}{I_0}\right), \tag{14–47}$$

where

$$I_0 = 10^{-12} \text{ W/m}^2.$$

A logarithmic scale such as this easily covers a large range of intensities (Fig. 14–30). The smallest detectable intensity is I_0 itself, for which $\beta = 0$, and the largest intensity still perceived as sound is about $I = 1 \text{ W/m}^2$, or $\beta = 120$ dB. Higher intensities are not perceived as sound, but rather as something painful. The range of intensities detectable as sound by the ear–brain combination is thus an impressively large factor of 10^{12}, as Figure 14–30 shows. Curiously enough, the range of sound intensities detectable by the human ear is about the same as the range of light intensities detectable by the human eye.

CONCEPTUAL EXAMPLE 14–12 A football fan sits in the upper deck of a large football stadium and has a portable sound device to cheer for his team. The device emits its sound equally in all directions, and it has the property that the sound is loud enough that someone sitting 300 m away on the other side of the stadium can hear it above the crowd noise of 80 dB. Why do you not want to be sitting 1 m in front of the fan?

Answer The intensity I is proportional to $1/D^2$ where D is the distance from the device (see "Think About This.... Why does a

distant shout sound fainter than a nearby one?" in Section 14–5), so that whatever the intensity at 300 m, the intensity at 1 m will be $(300)^2$ i.e. 90,000 times larger than the intensity of the ambient noise in a football stadium.

We can work this out quantitatively in terms of β. At 300 m the intensity is such that β is larger than 80. If $\beta = 80$, then $I/I_0 = 10^{12}I = 10^{8/10} = 10^8$, and $I = 10^{-4} \text{ W/m}^2$. At 1 m, the intensity is then $I = (300)^2 \times (10^{-4} \text{ W/m}^2) = 9 \text{ W/m}^2$, with a corresponding decibel level of $\beta = 10 \log_{10}(9 \times 10^{12}) = 120 + 10 \log_{10}(9)$. This number exceeds what we described as the bearable limit of sound.

14–8 The Doppler Effect

Traveling waves move with a finite velocity. It is therefore possible for an observer who measures the waves (the receiver) to move relative to them or for the generator of traveling waves (the source) to move relative to the waves it produces. The movement of the source or of the receiver affects the measured frequencies. An extreme case illustrates the effect: A receiver moves along with the wave crests and measures a wavelength, just as if he or she were stationary with respect to the medium, by extending a yardstick from one crest to another. But this receiver measures no frequency at all, for the frequency measures the rate at which crests pass the observer.

It is not hard to see the result of motion of receiver or source in the case of sound waves. The perception of a sudden change in the pitch of an automobile horn (a source) as a vehicle passes is a common experience. In 1842 Christian Doppler gave an explanation for this phenomenon, as well as for a corresponding phenomenon for light. The **Doppler effect** is the shift in frequency and wavelength of waves that results from a source moving with respect to the medium, receiver moving with respect to the medium, or even a moving medium. We'll treat the various possibilities in order.

Case A: A Moving Source: Consider a point source of traveling waves, such as a siren wailing at some frequency f_0 (period $\tau_0 = 1/f_0$). These waves spread away from the source at a wave speed v that is characteristic of the medium alone and is unaffected by the source's motion. If the source is stationary with respect to the medium, these waves are symmetric on all sides of the source, and the wavelength is $\lambda = v/f_0$ [from Eq. (14–46)]. But if the source moves with velocity $\vec{v}_s$ with respect to the medium, then the wave fronts will be squeezed together in the direction of $\vec{v}_s$ and spread apart in the opposite direction (Figs. 14–31a, b). The larger circles in Fig. 14–31a and b correspond to emission at some earlier time, when the source was farther back in the direction $-\vec{v}_s$. Thus the wavelengths are shorter along the direction $+\vec{v}_s$ and longer in the opposite direction.

Consider, for simplicity, the source to be moving directly toward one observer and directly away from another observer (Fig. 14–31c). For the case that the source moves toward the observer, during one period τ_0 it moves a distance $v_s \tau_0 = v_s/f_0$. The wavelength is then decreased by this amount:

$$\lambda' = \lambda - \frac{v_s}{f_0} = \frac{v - v_s}{f_0}. \tag{14–48}$$

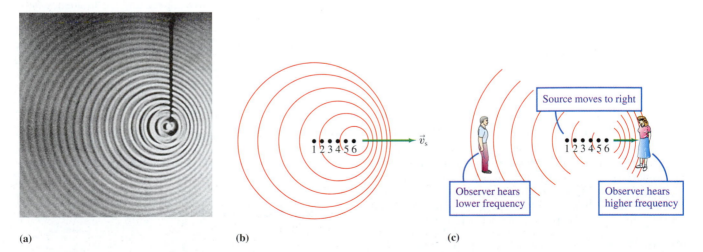

(a) (b) (c)

▲ FIGURE 14–31 (a) The Doppler effect is seen in the waves of a ripple tank. Here the "dipper," the device that generates the circular waves by rhythmically touching the surface, is moving slowly to the right, so that successive centers of the spreading ripples are farther to the right. (b) The dots along the direction of the velocity $\vec{v}_s$ of the source represent the source position when the wave centered on them was emitted. Dot 1 corresponds to the outermost wave, dot 2 to the second from the outermost wave, and so forth. (c) Two observers at rest on either side of the moving source detect the waves, and observe different frequencies.

Because the wave speed remains unaffected throughout, the frequency with which the observer receives the waves becomes

$$f' = \frac{v}{\lambda'} = f_0\left(\frac{v}{v - v_s}\right) = \frac{f_0}{1 - (v_s/v)}. \tag{14-49}$$

When the source moves toward the observer, the wavelength *decreases* and the frequency *increases*—the pitch will be higher in the case of sound. For the case that the source moves away from the observer, we simply reverse the sign of v_s:

$$\lambda' = \lambda + \frac{v_s}{f_0} = \frac{v + v_s}{f_0}; \tag{14-50}$$

$$f' = \frac{f_0}{1 + (v_s/v)}. \tag{14-51}$$

When the source moves away from the observer, the wavelength *increases* and the frequency *decreases*—for sound, the pitch drops. A manifestation of this phenomenon is familiar to anyone who has heard a siren drop to a lower pitch as the vehicle with the siren approaches, then passes.

Case B: A Moving Observer: Figure 14–32 depicts an observer moving with velocity $\vec{v}_r$ with respect to a wave-carrying medium, within which a stationary source is exciting waves with frequency f_0. We have illustrated the case of an observer moving directly *toward* the source. Such an observer will measure exactly the same wavelength λ as if he or she were at rest, but the observer sees the wave crests pass by more quickly. The observer measures a *modified wave speed*

$$v' = v + v_r. \tag{14-52}$$

The wavelength and wave speed provide the modified wave frequency that the moving observer measures:

$$f' = \frac{v'}{\lambda} = \frac{v + v_r}{\lambda} = f_0 + \frac{v_r}{\lambda} = f_0\left(1 + \frac{v_r}{v}\right). \tag{14-53}$$

When an observer moves *toward* the source of traveling waves, the wavelength is unchanged, the wave speed *increases*, and the frequency *increases*. We reverse the sign of v_r and repeat the derivation to learn that when an observer moves *away from* a wave source, the wavelength is again unchanged and the wave speed and the frequency both *decrease*:

$$v' = v - v_r; \tag{14-54}$$

$$f' = \frac{v - v_r}{\lambda} = f_0 - \frac{v_r}{\lambda} = f_0\left(1 - \frac{v_r}{v}\right). \tag{14-55}$$

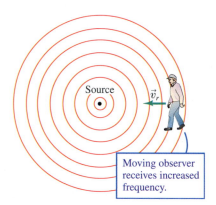

Moving observer receives increased frequency.

▲ **FIGURE 14–32** A receiver (an observer) moving with velocity $\vec{v}_r$ receives waves from a stationary source. In the situation shown, the observer hears an increased frequency.

EXAMPLE 14–13 While moving along a straight road, you first approach, and then pass, a stationary car just at the side of the road. The car's horn is being continuously sounded. Having perfect pitch, you recognize that the frequency you hear approaching the car is 853 Hz, while you hear a note of 741 Hz as you recede from the car. Use the value 330 m/s for the speed of sound in air to find the frequency the occupant of the car hears and your own speed.

Setting It Up We call your speed v_r, f_0 the frequency of the horn at rest, f_+ the frequency as you approach the horn, and f_- the frequency as you recede from it.

Strategy The numbers can be found with a proper application of the Doppler effect. As you move toward the car, f_+ is increased over f_0 according to Eq. (14–53), $f_+ = f_0\left(1 + \dfrac{v_r}{v}\right)$, where v is the speed

of sound in air. Similarly, as you move away, you hear a decreased frequency $f_- = f_0\left(1 - \dfrac{v_r}{v}\right)$. These are two equations for the two unknowns f_0 and v_r.

Working It Out We solve the two equations by taking their sum and difference:

$$f_+ + f_- = 2f_0$$
$$f_+ - f_- = 2f_0(v_r/v) = (f_+ + f_-)(v_r/v).$$

The first equation immediately gives us f_0 as the average of the two frequencies,

$$f_0 = \tfrac{1}{2}(f_+ + f_-) = \tfrac{1}{2}(853 \text{ Hz} + 741 \text{ Hz}) = 797 \text{ Hz}.$$

This is the frequency the car's occupant hears. The second equation gives us v_r:

$$v_r = v\frac{f_+ - f_-}{f_+ + f_-} = (330 \text{ m/s})\frac{853 \text{ Hz} - 741 \text{ Hz}}{853 \text{ Hz} + 741 \text{ Hz}} = 23.2 \text{ m/s}.$$

What Do You Think? Suppose you are driving a car toward the base of a massive cliff, and you honk your horn, which has a frequency f_0. A few seconds later you hear an echo—the sound reflected by the cliff. Do you expect to hear a higher or a lower frequency? How would you calculate it, knowing your speed? You may assume that a sound wave does not change its frequency when it is reflected.

Case C: A Moving Source and a Moving Observer: If all motion is in the same line, which we label the x-axis, then we can simply combine the results of cases A and B. Any velocity, including that of the wave itself, is positive if it is to the right and negative if it is to the left. With this reminder about the sign of the velocity, we drop the absolute-value signs. The effect of the moving source is to change the wavelength but not the wave speed, and the effect of the moving observer is to change the wave speed but not the wavelength. This can be expressed as

$$\lambda' = \frac{v - v_s}{f_0};$$
$$v' = v - v_r. \tag{14-56}$$

The modified frequency f' is then

$$f' = \frac{v'}{\lambda'} = \left(\frac{v - v_r}{v - v_s}\right)f_0. \tag{14-57}$$

Source and observer approach one another when v_s and v have the same sign and v_r has the opposite sign. In this case, f' is *increased* over f_0. When the source and observer are moving away from each other, v_r and v have the same sign and v_s has the opposite sign. The perceived frequency is *decreased*. Equations (14–56) and (14–57) include cases A and B as special cases. Note that Eq. (14–57) is *not* symmetric between the source and the observer. If we know the relative speed, we can tell what is moving—the source or the observer—by measuring the at-rest and in-motion frequencies.

When the speeds of the observer and the source are *small* compared to v, as is often the case when the Doppler effect applies to sound waves, then Eq. (14–57) is *approximately*

$$f' \cong \left(1 + \frac{v_s - v_r}{v}\right)f_0. \tag{14-58}$$

In this approximation, the Doppler shift (the change in frequency) depends only on the relative velocity between source and observer, and it is no longer possible to say who is moving—the source or the observer.

Case D: A Moving Medium: We can consider a fourth possibility: that the medium is itself in motion with respect to some fixed reference frame. Waves generated in a stream provide an example. The source is fixed with respect to the bank, and the stream is moving with speed v_m (Fig. 14–33). Let's take a look at how an observer on the bank downstream

▲ **FIGURE 14–33** Traveling waves are generated at a fixed point, here marked with a cross, in a moving medium. The spreading waves move downstream; we can describe their positions by thinking of their centers as having moved downstream.

$\vec{v}_m$
Stream flow

from the source sees the waves. There are two effects: First, the observer sees the crests moving by at an increased speed $v' = v + v_m$. Second, the wavelength is also increased because a downstream crest moves an additional distance before a second crest is emitted. From Fig. 14–33, it is easy to deduce that the wavelength increases from λ to $\lambda' = \lambda[1 + (v_m/v)]$. The frequency measured by the observer is $f' = v'/\lambda' = (v + v_m)/\lambda[1 + (v_m/v)] = (v/\lambda)[(v + v_m)/(v + v_m)] = v/\lambda = f$.

Similarly, an observer on the bank upstream from the source will see both the wave speed and the wavelength decreased, and in the same ratio, $[1 + (v_m/v)]$, so the frequency $f' = v'/\lambda'$ is unchanged. In other words, because both the speed and the wavelength are changed in the same way by the movement of the medium, *the frequency is unchanged by the movement of the medium*. The fact that the frequency is unchanged is the reason why it is possible to give outdoor concerts even under windy conditions. The intensity of the sound may be decreased if the wind is blowing in the wrong direction, but at least the pitch remains true.

Uses of the Doppler Shift

You are well aware of the use of the Doppler shift in police work—many speeders are caught this way. A radar beam (electromagnetic wave) is bounced from a moving car. The reflected wave is coming from a moving source, and the device's receiver translates the shifted frequency to the car's speed. Doppler radar can also be used in meteorological studies, where frequency shifts in radar waves reflected from moving objects such as raindrops convey a great deal of information about the dynamics of storms. Probably the most important use of the Doppler shift is in astronomy, where the measurement of the light emitted by distant stars exhibits a Doppler shift according to the motion of the star. The detection of the shift is greatly facilitated by the fact that certain characteristic discrete frequencies are emitted in starlight. By observing that the speed and direction with which stars and galaxies move has a correlation with their distance, Edwin Hubble was able to show that the universe is expanding.

▶ *14–9 Shock Waves

When a source of traveling waves moves with a speed exceeding the wave speed, we encounter a new situation in which *shock waves* can form. This can be seen from Eqs. (14–49) or (14–57), which show that when $v_s = v$, the observed frequency becomes infinite. When $v_s = v$, the waves emitted directly in front of the source pile up on top of one another, so the wavelength is zero. Thus the time separation between the waves is zero, which corresponds to an infinite frequency. When v_s exceeds v, the source outruns the motion of the waves themselves; thus, instead of the squeezed-together wave fronts that occur for $v_s < v$ (see Fig. 14–31), the wave fronts are actually left behind.

The geometry of this situation can be analyzed as follows (Fig. 14–34). Suppose that the wave source is at point x_0 at $t = 0$ and emits a crest. In a time t_2, the source has traveled to x_2, a distance $v_s t_2$ from x_0. In this same time, the crest has moved out to a circular wave front a distance vt_2 from x_0. The perpendicular to the line drawn from x_2 tangent to this circle makes an angle θ_0 with respect to the line perpendicular to the motion of the source, and

$$\cos \theta_0 = \frac{vt_2}{v_s t_2} = \frac{v}{v_s}. \tag{14–59}$$

All the wave fronts emitted as the source moves lie along this tangent, which results in a piling up of wave fronts that reinforce each other and create a large-amplitude wave front. For instance, consider the crest emitted at t_1 when the source is at x_1, a distance $v_s t_1$ from x_0. The geometry of the figure gives

$$\cos \theta_1 = \frac{v(t_2 - t_1)}{v_s(t_2 - t_1)} = \frac{v}{v_s}.$$

Thus $\theta_1 = \theta_0$ and the tangent is formed by all the waves emitted during the voyage of the source. This line—a linear wave front called a **shock wave**—moves out with a

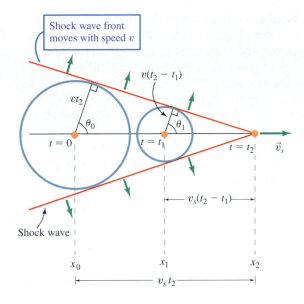

▲ FIGURE 14–34 When the wave speed v is less than the source speed v_s, a linear wave front forms along the red lines. This front is called a shock wave. The source rhythmically emits waves that form outgoing circles in two dimensions or spheres in three dimensions; their crests are marked in blue in the figure. At $t = 0$ the source is at the center of the larger circle, at a later time t_1 at the center of the smaller circle, and at a still later time t_2 about to emit another wave crest. Successive circles form the continuous line of the shock wave.

speed v, maintaining the angle given by Eq. (14–59) (Fig. 14–35). When the wave is propagating in a three-dimensional medium, such as air, a cone is formed instead. We should stress that *there is only one such wave front.* A cone whose apex is at any location on the trajectory of the source represents just an earlier or later location of the cone that appears in Fig. 14–34. The wave front will pass an observer located on the ground just once.

When $v < v_s$, Eq. (14–59) determines without ambiguity the angle of the shock wave. The ratio v_s/v is called the *Mach number*; the larger the Mach number, the closer θ_0 comes to 90°—that is, the more acute is the angle of the line or cone. The sonic boom that occurs whenever an airplane moves faster than the speed of sound is a manifestation of the arrival of a shock wave of sound at an observer's ear. A second type of shock wave is associated with the movement of an electrically charged particle through some medium such as a gas or Plexiglas. The speed of light in this medium can be less than the speed of light in a vacuum, and a charged particle, whose speed cannot exceed that of light in a vacuum, can nevertheless move through the medium faster than the speed of light in that medium. Light waves then propagate along a shock-wave front whose angle is given by Eq. (14–59). These waves are called *Cherenkov radiation*, after Pavel Cherenkov (Fig. 14–36). When the speed of light in a given medium is known, measurement of the angle of the shock wave gives the speed of the charged particle. This is a useful measurement tool in experiments on particle physics.

▲ FIGURE 14–35 A bullet passing through the hot gases above a candle flame produces a shock wave because its speed is faster than the speed of sound in air. The ratio of these speeds determines the angle of the shock wave.

◀ FIGURE 14–36 The blue glow is Cherenkov radiation. It is produced when a charged particle moves through a medium at a speed greater than the speed of light in the medium. In this case the medium is the water of a nuclear reactor and the particles are the product of the radioactivity of the reactor elements.

▲ FIGURE 14–37 The wakes of these motorboats on the Connecticut River are a consequence of the fact that the speed of a water wave depends on its wavelength.

The *wake*, or track (Fig. 14–37) that extends out behind an ocean liner, a row boat, or a paddling duck is not the same as the shock wave that is responsible for a sonic boom. Upon observation, we see that the wake forms the same angle for *all* these examples, even though the ocean liner moves much more swiftly than the duck. This wake forms a V with the boat at the point and an angle of 39° between the arms of the V (the linear wave fronts). A wave in deep water (the depth is much greater than the wavelength) behaves differently from all the other waves we have studied in this chapter, because *its speed depends on its wavelength*. The wake of a ship, and other differences between that wake and the shock waves discussed here, depend on the variation of speed with wavelength, a phenomenon called *dispersion*. We'll talk again about dispersion when we deal with light waves in materials.

Summary

Extended bodies will often respond elastically to external and internal forces, and as a result they can support the organized motions we call waves. Waves can be classified as traveling waves, in which limited internal motion propagates progressively with some definite speed and in some definite direction through the medium, or as standing waves, in which the motion of the entire medium has the same sinusoidal time dependence. Both traveling and standing waves are further classified into either longitudinal waves or transverse waves. In transverse waves the motion within the medium is perpendicular to the direction in which the waves are organized. In longitudinal waves the motion is parallel to the wave direction.

The dynamics that allow waves to propagate are described by the wave equation, which for the particular case of transverse waves on a string of mass density μ under tension T takes the form

$$\frac{\partial^2 z(x,t)}{\partial x^2} = \frac{\mu}{T}\frac{\partial^2 z(x,t)}{\partial t^2}. \tag{14–7}$$

Here the string is aligned with the x-axis, and z describes the transverse displacement of a point on the string. More generally, wave motion can be described by the wave equation, generically of the form

$$\frac{\partial^2 z(x,t)}{\partial x^2} = \frac{1}{v^2}\frac{\partial^2 z(x,t)}{\partial t^2}. \tag{14–10}$$

The quantity v is the speed with which disturbances propagate.

For *harmonic* traveling waves, the motion of any point in the medium is described by

$$z(x,t) = A\sin(kx - \omega t). \tag{14–22}$$

The parameters of this wave are the amplitude A, the angular frequency ω, and the wave number k. The wavelength, λ, the length over which a wave repeats itself, is related to the wave number by Eq. (14–16), $\lambda = 2\pi/k$. The parameters of frequency f and period τ are also used to describe wave motion. These parameters are all related by $\lambda f = \lambda/\tau = \omega/k$, from Eq. (14–21). The speed of the traveling wave is

$$v = \lambda f = \frac{\lambda}{\tau} = \frac{\omega}{k}.$$

This speed, here described in terms of its kinematic relation to the parameters of $z(x,t)$, is fixed by the wave equation. For example, comparison of Eqs. (14–7) and (14–10) shows that for a string of mass density μ under tension T, the speed is

$$v = \sqrt{\frac{T}{\mu}}. \tag{14–9}$$

Waves carry energy, and traveling waves carry energy and momentum along the direction of their motion, even if the medium has no net motion in this direction. The power delivered by the wave is proportional to the square of the amplitude A and of the angular frequency ω. Its average value is

$$\langle P \rangle = \tfrac{1}{2}\mu\omega^2 A^2 v. \tag{14–37}$$

Traveling sound waves in solids, liquids, or gases are longitudinal waves. The speed of sound in air is given by

$$v_{sound} = \sqrt{\frac{\gamma p_0}{\rho_0}}, \qquad (14\text{–}30b)$$

where γ is a constant close to 1, and p_0 and ρ_0 are the pressure and the density of the medium, respectively. For normal atmospheric conditions at sea level, the speed of sound in air is roughly 330 m/s.

For standing waves, v is once again fixed by dynamical variables such as tension and mass density. The possible values of k and λ are determined by the boundary conditions of the medium and are generally discrete. The frequency of such waves is then determined from $f\lambda = v$, so that the angular frequency is

$$\omega = 2\pi f = 2\pi \frac{v}{\lambda}.$$

We see that the wave speed v affects the properties of standing waves even though they are not propagating.

The wave speed for mechanical waves takes the general form

$$v = \sqrt{\frac{\text{restoring force factor}}{\text{mass factor}}}. \qquad (14\text{–}29)$$

The Doppler effect describes what happens to the frequency and wavelength of traveling waves emitted by a source moving with velocity $\vec{v}_s$ and detected by a receiver moving with velocity $\vec{v}_r$. If v is the speed of a wave within the medium, then the frequency perceived by the receiver is

$$f' = \left(\frac{v - v_r}{v - v_s}\right) f_0. \qquad (14\text{–}57)$$

Here v_r and v_s are positive (negative) when the corresponding motion is along (against) the motion of the wave.

Understanding the Concepts

1. Give some examples of transverse waves and longitudinal waves and discuss the nature of the mechanical motion in each wave.

2. Meteorologists use something called "Doppler radar" to study the motion of clouds, tornadoes, and so on. What do you think is going on?

3. When a gong is sounded, it produces sound waves in air. If the same gong were struck in a vacuum, would it continue to vibrate for less time, for the same amount of time, or for more time than it does when struck in the same way in air? Explain your reasoning.

4. In old cowboy movies, you have probably seen someone place his ear on railroad rails to tell if a train is coming. Why is this done, and how does it work?

5. You have two ropes—one of them half the length and twice the mass density of the other—both under the same tension. The ropes are firmly attached to walls at each end. Compared to those for the shorter rope, the allowed wavelengths for the fundamental and first harmonic modes ($n = 1$ and 2, respectively) for the longer rope are (a) both doubled; (b) $n = 1$ doubled, $n = 2$ the same; (c) the same.

6. Would you expect the speed of sound (a) in hydrogen and (b) in argon to be greater than or less than the speed of sound in oxygen if all gases are under the same pressure?

7. When a lecturer in a classroom demonstration takes a lungful of helium and speaks, his or her voice is comically higher in pitch. Why is that?

8. There is a biological reason why a church does not need an organ pipe longer than 4.0 m. What is that reason? There are much longer pipes. What is their effect?

9. Captain Kirk is somewhere between galaxies. When a gong sounds in a neighboring spaceship, Kirk reacts to the sound. What is wrong with this scenario?

10. Suppose that a string supports a standing wave with a node at a particular location. How would things change if the point of the node were clamped physically?

11. In the expression for $h(x, t)$ in Example 14–2, the wave height at the observer's position varies sinusoidally, with a minimum (a trough) at $t = 0$. How would you write a sinusoidal function that would allow you to set the wave height at an arbitrary value (between $+h_0$ and $-h_0$) at $t = 0$?

12. What prevents sound waves from being transverse waves in a dilute gas such as air? In solids, transverse "sound" waves are generally possible. What can we learn about the nature of the intermolecular forces in such a solid?

13. Are the nodes actually motionless when a longitudinal standing wave is established on a Slinky?

14. Suppose the supports that hold the two ends of a piano wire are just simply pulled apart, so that the length of the wire increases by a small amount. How will this affect the fundamental period?

15. The speed of deep-water waves of wavelength λ is given by the formula $v = \sqrt{g\lambda/2\pi}$. When there is a storm somewhere on the ocean, it manifests itself by large breakers (breaking waves) at the shore. Initially you observe that the breakers arrive every 20 s at the shore. Somewhat later you observe that they are coming every 10 s. How do you interpret this?

16. A child concocts a telephone with two cans that are open at one end and connected by a string attached through a small hole to the closed ends. When the string is made taut by pulling on the cans at opposite ends, two children talking into the open ends can carry out a soft conversation over rather large distances. How does this happen?

17. The two supports that hold the end of a piano wire are pulled apart. If you wanted the fundamental frequency (pitch) of the wire to remain unchanged, what should you do about the tension, if anything?

18. A common toy consists of a series of five balls that touch and form a line; each ball is suspended as a pendulum. When ball 1 is pulled away and then released, it strikes the line and ball 5 rises in a motion like the reverse of ball 1. If there is a small gap between the balls at equilibrium, then a series of collisions occurs in motion, each with an exchange of velocities, until the last ball swings up. The phenomenon does not change even if the balls touch when at rest. If we describe the phenomenon as a collision of the first ball with an object four times its mass (this is how we describe the four touching balls), we do not get the observed motion of the balls. Why is the last description inadequate?

19. Suppose a rope on which standing waves are generated has its already substantial tension increased by a factor of 4. Do you expect that the velocity will be *exactly* doubled? If not, in which direction will it depart from doubling?

Problems

14–4 Traveling Waves

1. (I) A traveling wave is described by a displacement function $y = A \sin(ax - bt)$, where $a = 0.30 \text{ m}^{-1}$ and $b = 0.020 \text{ s}^{-1}$. Plot the displacement versus t at both $x = 1.0$ m and $x = 12$ m.

2. (I) A wave is described by $y = A \sin(kx + \omega t)$, where $k = 3.0 \text{ m}^{-1}$, $\omega = 2.0 \text{ s}^{-1}$, and $A = 0.40$ m. Determine (a) the wavelength, (b) the frequency, (c) the maximum amplitude, (d) the period, and (e) the wave number.

3. (I) Consider a long length of the wire used in Example 14–4. The wire is taut under a tension of 55 N. What is the mass density of the wire?

4. (I) The French train known as the TGV travels at 300 km/hr. What is the frequency with which the cars of this train pass a given point at that speed, given that the cars are 30 m in length?

5. (I) A harmonic wave with wavelength 0.27 m moves with speed 13 m/s. Find the wave number k and the frequency f of this wave.

6. (I) A rope of density μ_1 is attached to another rope, of density μ_2 (Fig. 14–38). A traveling wave of speed v_1 is sent down the first rope. What will the speed in the second rope be?

▲ **FIGURE 14–38** Problem 6.

7. (I) The values of Young's modulus, Y, for aluminum and copper are $Y_{Al} = 7.0 \times 10^{10} \text{ N/m}^2$ and $Y_{Cu} = 11 \times 10^{10} \text{ N/m}^2$. Their mass densities are $2.70 \times 10^3 \text{ kg/m}^3$ and $8.96 \times 10^3 \text{ kg/m}^3$, respectively. Calculate the speed of elastic wave propagation in these two media.

8. (I) A mountain climber of mass 68 kg hangs from a nylon rope 28 m below the top of a cliff. The total mass of the rope is 1.3 kg; assume that this mass is small enough so that the variation in equilibrium tension along the rope can be ignored. How long does it take a wave pulse to travel up the rope from the man to the top of the cliff?

9. (I) The speed of sound in a material of density $2.7 \times 10^3 \text{ kg/m}^3$ is measured to be 5.1×10^3 m/s. What is Young's modulus for that material?

10. (I) The shear modulus of copper is $G = 4.2 \times 10^{10} \text{ N/m}^2$. Given that the density of copper is $8.93 \times 10^3 \text{ kg/m}^3$, what is the speed of a shear wave in copper?

11. (I) The density of pure gold is 19.3 g/cm^3, and the speed of sound in a small sample is measured to be 2.04×10^5 cm/s. What is Young's modulus for gold in dynes/cm^2?

12. (I) Given that the ratio of densities of steel and lead is 7.86 to 11.3, and that the ratio of Y values is 200,000 MN/m^2 to 16,000 MN/m^2, what is the ratio of the speeds of sound in these two materials?

13. (II) Use Eq. (14–22) for the traveling wave $z(x, t)$ and the wave equation to show that $Tk^2 = \mu\omega^2$.

14. (II) A perfect triangular pulse travels along a string with a velocity of 4 m/s. The shape of the pulse at $t = 0$ is shown in Fig. 14–39. Plot the displacement and velocity of the segment at $x = 3$ m as a function of time. Is such a pulse physically possible?

▲ **FIGURE 14–39** Problems 14, 16.

15. (II) Show by substitution that the general form of a traveling wave, $f(x - vt)$, satisfies the wave equation. [*Hint:* Let $s = x - vt$ and use the chain rule for differentiation.]

16. (II) Between $x = 0$ cm and $x = 10$ cm, a very long wire (so long that the boundary effects do not matter) is distorted from its straight-line equilibrium form into a triangular shape like that of Fig. 14–39, with the peak 4.0 cm above the original position. The tension and density of the wire are such that waves travel along the wire at a speed of 400 m/s. (a) What is the shape of the wire at the initial time $t = 0$? (b) What is the shape of the wire at an arbitrary time t, assuming that the wave travels in the $-x$-direction?

17. (II) A scuba diver has to send a message to save an oil platform in the Gulf of Mexico. He realizes that sound travels faster in steel than in air or water. With only one second left for the message to be received on the platform, he finds a steel mooring cable that is 5.0 km long, which runs directly to the platform. Will the message reach the platform in time? (The density of steel is $7.9 \times 10^3 \text{ kg/m}^3$ and $Y = 210,000 \text{ MN/m}^2$ for steel.)

18. (II) Young's modulus for Earth can be approximated by $5.5 \times 10^5 \text{ MN/m}^2$, and the corresponding shear modulus by $2.0 \times 10^5 \text{ MN/m}^2$. A first signal of an earthquake arrives at a detection station at $t = 0$ min, and a second signal arrives 2.0 min later. Assume that the successive signals are those of longitudinal waves and shear waves, respectively, that follow the same straight-line path through the planet. Assume further that the average density of Earth is 5.5 g/cm^3. How far from the station did the earthquake take place?

19. (II) Earthquake waves within Earth are of two types: S (transverse) waves, which can be supported within Earth's mantle but not in the liquid part of the core, and P (longitudinal) waves, which propagate in both the mantle and the core (see Fig. 14–16). The speeds of S waves and P waves in the mantle are approximately 4.0 km/s and 8.0 km/s, respectively. The speeds of these waves can be calculated from the Young's and shear moduli. Take the density of Earth's mantle to be 5.0 g/cm^3. Calculate the Young's and shear moduli. If the difference between the time of arrival of the two waves is 27 min, how far away was the earthquake? Ignore Earth's thin crust, and assume that the waves travel in a straight line.

20. (II) A steel rod of length 0.80 m is rigidly supported at one end. Calculate the frequencies of the lowest three longitudinal modes, and the lowest three transverse (shear) modes, given that $Y = 2.1 \times 10^{11}$ N/m^2 and $G = 8.1 \times 10^{10}$ N/m^2.

21. (II) Consider a closed loop chain of mass density μ kg/m, which is spinning at a high angular velocity ω till it moves like a hoop. Find the tension in the chain in terms of μ, ω, and the radius R.

22. (II) Assume that experiment has shown that the velocity of waves in shallow water is independent of the wavelength of these waves. Use dimensional analysis to show that the speed of these waves is proportional to the square root of the depth D. Express the speed in terms of the relevant constants. [*Hint:* The relevant force here is that due to gravity.]

23. (II) Assume that the expression for the velocity of shallow-water waves in terms of g and depth D that you derived in Problem 22 is exact. A speed boat that travels close to the speed of such waves tends to build up a large wave just at its bow and thus will have trouble going any faster. At what speed in kilometers per hour will this happen in a harbor in which the water is 5.0 m deep?

24. (II) The considerations of Problem 22 apply to all water waves for which the wavelength is much larger than the depth. For tidal waves (or *tsunami:* waves caused by underwater earthquakes), the wavelengths are enormous, many hundreds of kilometers. What is the speed of a tidal wave in water that is 5.0 km deep?

25. (III) When n springs—each of spring constant k and equilibrium length ℓ—are attached in a line, the resulting system is a spring with spring constant $k_n = k/n$ and length $L = n\ell$. If there is a mass m at the connection point between each spring (the springs themselves are assumed to be massless), then the resulting system can support longitudinal, or compressional, waves (see Fig. 14–3b). By imagining an arbitrarily long series, you can see that the speed of these waves is independent of the total length of the spring, and hence of n. Use the available constants k, ℓ, and m to estimate the speed of longitudinal waves. This system is an excellent model for compressional waves on a real spring. [*Hint:* Dimensional analysis can be useful here.]

26. (III) A set of tiny beads is connected in a line by springs, as in Problem 25. The equilibrium length of the springs is 3.0 cm. The beads have mass 5.0 g each, and the spring constant of the springs is 20 N/m. Longitudinal waves can be generated on the string of beads by moving the end of the string along the direction of the string. Estimate the speed of such longitudinal pulses. [*Hint:* Dimensional analysis can be useful here.]

14–5 Energy and Power in Waves

27. (I) A wave of wavelength 0.70 m is sent moving down a rope of density 12 g/cm, under a tension of 800 N. The amplitude of the wave is 5.0 cm. What is the average power transported by the rope? Suppose that the wavelength is doubled; what is the average power transported by the rope now?

28. (I) Calculate the kinetic energy and potential energy contained in one wavelength of a harmonic traveling wave on a string whose mass density is μ and whose tension is T.

29. (I) The power delivered per unit area by a sound wave is given by an expression similar to that of Eq. (14–36). In this case, μ is replaced by the density, ρ, of the gas in which the sound propagates: $P/\text{area} = \text{intensity} = \frac{1}{2}\rho\omega^2 A^2 v$, where A is the amplitude of the displacement of air molecules from their average positions. Given an 8600 Hz sound wave with an intensity of 5.0×10^{-7} W/m^2, what is A? The density of air is 1.30 kg/m^3, and the speed of sound in air is 330 m/s.

30. (II) A sound wave has an intensity of 1.7×10^{-6} W/m^2. Given that the speed of sound in air is 330 m/s, what is the energy density of the sound wave?

31. (II) Obtain an expression analogous to Eq. (14–36) for the rate of energy transport in a left-moving wave. In particular, show that the power remains positive.

32. (II) Harmonic water waves of a given amplitude are generated on a water surface by rhythmic dipping of a pointer at one spot. The waves thus generated spread in circular wave fronts. If R is the distance from the generation point, show that conservation of energy implies that the amplitude A must vary with R, and that this variation is of the form $A \propto 1/\sqrt{R}$.

33. (III) We derived the power or rate of energy transport for a sinusoidal wave [Eq. (14–36)]. Obtain an expression for the power delivered by an arbitrary right-moving wave of the form $f(x - vt)$.

14–6 Standing Waves

34. (I) The third harmonic of a certain piano wire ($\mu = 0.010$ kg/m, $T = 500$ N) has a frequency of 600 Hz. What is the wavelength of the fundamental?

35. (I) A guitar string has end points 54 cm apart. What are the three largest wavelengths of standing waves this string can support? Sketch the standing waves for these wavelengths.

36. (I) A string that is fastened at the two end points, $x = \pm 60$ cm, oscillates in its fundamental mode (no nodes except at the walls). The period of the oscillation is 1 s, the displacement at $x = 0$ cm at $t = 0.0$ s is 2.0 cm, and at $t = 0.1$ s, 3.0 cm. Express $z(x, t)$ for the displacement.

37. (I) A standing wave is formed on a string for which three and one half wavelengths fit into the total length of the string, which is 2.7 m. The wave has a period of 0.10 s. If the string has a mass density of 220 g/m, what is the tension in the string?

38. (I) The tension on a wire is 2840 N. The wire is 2.6 m long and the fundamental frequency is 310 Hz. What is the mass of the wire?

39. (I) Two wires of the same length and under the same tension have fundamental frequencies that are in the ratio 3:1. What is the ratio of their masses?

40. (II) A string with fixed ends is in a standing wave that vibrates in five segments (Fig. 14–40, see next page). The frequency of this mode is 120 Hz. (a) Find the fundamental frequency of the string. (b) The tension in the string is reduced by a factor of 9. What is the new fundamental frequency?

▲ FIGURE 14–40 Problem 40.

41. (II) A steel string under tension vibrates in its lowest mode at a frequency of 1200 Hz. A string of the same material and length but three times the thickness vibrates in its lowest mode at a frequency of 400 Hz. What is the ratio of the tensions of the two strings?

42. (II) Figure 14–41 shows a pulse at $t = 0$ traveling to the right with a speed of 0.50 m/s. (a) Sketch, properly scaled, the shape of the rope at $t = 4$ s. (b) Plot the displacement of the point at $x = 5.0$ m as a function of time.

▲ FIGURE 14–41 Problem 42.

43. (II) A rope has a mass density of 40 g/cm and is under tension of 200 N. The rope is fixed at one end and is connected to a rod at the other end that sets up standing waves by moving slightly. The rope is 1.0 m long. Determine the three lowest possible frequencies and the corresponding angular frequencies and periods.

44. (II) A string with length L and linear density μ is under tension T. One end is fixed; the other end remains horizontal while it moves periodically up and down. Standing waves with an antinode (a point of maximum displacement) can be induced in this way, as in Fig. 14–42. Find the possible wavelengths and frequencies.

▲ FIGURE 14–42 Problem 44.

45. (II) The mass density of copper is 8.92×10^3 kg/m³. What is the tension on a copper wire 1.00 mm in diameter and 60.0 cm long, fastened at each end, when the frequency of vibration in the $n = 3$ mode is 870 Hz?

14–7 More About Sound

46. (I) Sound waves in air travel at 330 m/s, whereas light travels at 3.0×10^8 m/s. You can determine how far away a lightning strike is by counting the number of seconds between the time you see the lightning flash and the time you hear the thunder-

clap. How far are you from the lightning bolt for each 1 s difference between the flash and the thunderclap?

47. (I) Using the information in Table 14–1, compute the densities of air and hydrogen.

48. (I) A telephone receiver is at your left ear, with the sound of the conversation at a level of 58 dB. At the same time, your little sister's screams, at a level of 93 dB, enter your right ear. What is the ratio of the sound intensities that enter your two ears?

49. (I) A series of noises have intensities 8.3×10^{-4} W/m², 3.5×10^{-6} W/m², and 7.2×10^{-9} W/m², respectively. What is the sound level of each in decibels?

50. (I) A sound suddenly appears to the ear to be twice as intense as usual. How has the amplitude of the pressure wave (the magnitude of the maximum pressure in the wave) changed?

51. (II) A sound source emits sound with equal intensity in all directions. What is the power output of the source if the intensity of the sound measured at 5.0 m from the source is 83 dB?

52. (II) Two sounds have levels of 35 dB and 53 dB, respectively. What is the ratio of the power of the two noises? What is the ratio of the amplitudes of the pressure waves they make?

53. (II) In an ideal fluid, the speed of sound is given by $v = \sqrt{B/\rho}$, where B is the *bulk modulus*, a measure of how easy it is to compress the fluid. Sound waves in a fluid are only longitudinal because an ideal fluid cannot bounce back from the application of transverse forces, sometimes called shearing forces. Using the information given in Table 14–1, calculate the bulk modulus of water.

54. A source of sound waves is operating in midair, far from any reflecting surfaces. If, 1.0 m from the source, the sound intensity level is 120 dB, what is the intensity 1000 m from the source? What is the total power output of the source?

55. (II) Warm air has a higher density than cold air, and the density at Earth's surface can be taken to be inversely proportional to the temperature, measured on the Kelvin scale K (see Chapter 17). In this scale water freezes at close to 273K. Given that the speed of sound at 273K is 330 m/s, what is its speed at room temperature, roughly 295K?

14–8 The Doppler Effect

56. (I) A musician with perfect pitch—the capacity to identify pitch by ear—has run a red light and is being pursued by a police car whose siren emits at the single frequency $f = 400$ Hz. The musician's car is moving east at 60 mi/h (27 m/s), and the police car follows at 80 mi/h. The speed of sound is 330 m/s, and the air is still. What does the musician say the frequency of the siren is?

57. (I) A horn emits sound at a frequency of 160 Hz. What frequency is heard by an observer moving away from the source at 26.0 m/s?

58. (I) A child is blowing a whistle while riding her bicycle toward you on the sidewalk. You have set up an accurate audio receiver that measures the frequency of the whistle sound to be 793 Hz. You estimate the bicycle to be moving at 7 mph (3.1 m/s). What would you measure the frequency of the whistle to be at rest?

59. (I) A whistle emits sound with a frequency of 333 Hz. The whistle is mounted on a train that travels at 140 km/h in the $-x$-direction toward an observer at rest. (a) What frequency is heard by the observer? (b) Suppose that the whistle is at rest and the observer moves in the $+x$-direction at 140 km/h. What frequency is heard by the observer?

60. (II) A truck is traveling at 60 mi/h (27 m/s) down the interstate highway where you are changing a flat tire. As the truck

approaches, the truck driver sounds an air horn, which has a frequency of 185 Hz. (a) What frequency do you hear? (b) Suppose that you were driving your car at 50 mi/h while the truck driver blows his air horn after he passes you. What frequency would you hear?

61. (II) The frequency of light emitted by a star moving away from us (the observer) with speed v will be *red-shifted* (its frequency will be lowered) by an amount properly described by the treatment in Section 14–7, provided $v/c \ll 1$, where c is the speed of light ($c = 3.0 \times 10^8$ m/s). How fast is a galaxy receding from us if the observed frequency is 0.95 times the natural frequency of the light?

62. (II) You are moving southward on a divided highway and, heading northward on the other side, a fire engine passes with its (one-note) siren screaming. You are close enough to each other as you pass to assume that you and the fire engine are moving on the same line. Your speedometer reads 45 mi/h, and being a trained musician, you recognize that the frequency you hear before you pass the fire engine is 910 Hz and 680 Hz after you pass. Use the value 330 m/s for the speed of sound in air to find the frequency emitted by the siren and the speed of the fire engine.

63. (II) An ambulance siren is the source of sound emitted with a frequency of 1600 Hz. The ambulance is driving toward the base of a large cliff at 95 km/h (Fig. 14–43). What is the frequency of the reflected sound wave heard at the ambulance?

▲ **FIGURE 14–43** Problem 63.

64. (III) Show that for $v \ll c$, where c is the speed of a wave in a certain medium, the frequency shift for an observer at rest and a source moving away at speed v is the same as that for a source at rest and the observer moving at speed v in the opposite direction.

65. (III) A tuning fork with frequency 440 Hz is dropped from a tower of height 100 m by an enterprising student. She can hear the fork sounding as it falls. Given that the speed of sound in air is 330 m/s and that $g = 10$ m/s^2, what frequency does the student hear as a function of time? Do not forget that it takes time for sound to travel back to the student.

*14–9 Shock Waves

66. (I) An airplane moving at three times the speed of sound (Mach 3) flies horizontally and at a very low altitude. It emits sound continuously with a frequency that the pilot hears as 300 Hz. (a) What angle does the plane's shock wave cone make with the vertical? (b) What is the frequency of the sound heard by a stationary observer after the plane has passed? (Assume that the plane is moving directly away from the observer.)

67. (I) The angle made by the shock wave associated with the sonic boom of a low-flying airplane is $\theta_0 = 70°$. What is the speed of the airplane?

General Problems

68. (I) Two children stretch a rope between two trees. In spite of their efforts, the tension turns out to be only 20 N. The mass density of the rope is 48 g/cm. If they strike the rope hard with a stick, how fast does the resulting wave travel to the other tree?

69. (I) The human voice system, which includes the vocal cords, the mouth, and the lips, is a device that can change its shape and size and so is capable of emitting sound with a good deal of complication. There is nevertheless a fundamental frequency. Assume the vocal cavity is approximately 14 cm long and determine the fundamental frequency of the human voice. Assume a model where the vocal tract is open at both ends (larynx and mouth) and use 350 m/s as an approximation to the speed of sound at the human body temperature.

70. (II) A string is stretched between a pulley and a wave generator consisting of a plate vibrating up and down with small amplitude and frequency 120 Hz (Fig. 14–44). A standing wave with two intermediate nodes is generated when the string has a mass of 215 g attached to it. How large a load is needed to generate standing waves with one and with four nodes?

▲ **FIGURE 14–44** Problem 70.

71. (II) A string is fastened to a wall at $x = 0$ cm and is free to move on another wall at $x = 60$ cm, with the constraint that it is always perpendicular to the wall there (Fig. 14–45). What are the permissible values of the wavelength?

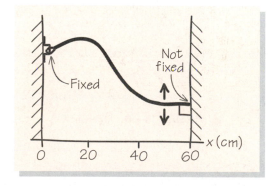

▲ **FIGURE 14–45** Problem 71.

72. (II) Transverse traveling waves are generated on a uniform wire of constant tension and a given mass. By what factor will the required power be increased or decreased if (a) the given mass of wire is doubled in length and the angular frequency remains constant? (b) the amplitude is doubled and the angular frequency is halved? (c) both the wavelength and the amplitude are doubled? (d) the given mass of wire is doubled in length and the wavelength is halved?

73. (II) The sound intensity at a rock concert reaches the pain threshold of 120 dB. Normal conversation has an intensity of 55 dB. What is the ratio of the power delivered to the ear at a rock concert compared to that delivered by a typical conversation?

74. (II) The left-hand end of a long, taut string is moved harmonically up and down with amplitude 0.14 m. This motion produces a wave of wavelength $\lambda = 2.5$ m that travels in the $+x$-direction at speed $v = 245$ m/s. (a) What is the frequency, f, of the traveling wave? (b) What is the maximum transverse velocity of a point along the string? (c) What is the maximum transverse acceleration of a point along the string?

75. (II) The density of neon is approximately 0.90 kg/m³ at standard atmospheric pressure and temperature. The value of γ for neon is 5/3. A loudspeaker emits a note with a frequency of 400 Hz in air. What will the frequency and wavelength of that note be if the loudspeaker is put into a chamber filled with neon?

76. (II) A wire is attached to two walls 1.0 m apart. The wire oscillates in such a way that a point on the wire 25 cm from one end is held fixed at its equilibrium position. What are the possible wavelengths for the first three allowed modes?

77. (II) A tuning fork with frequency $f = 512$ Hz is held by someone who swings it vigorously in a circle in the horizontal plane. The radius of the circle is 1.0 m, and the frequency of revolution is 3.0 rev/s. (a) What are the maximum and minimum frequencies that a second person would hear? (b) Which part of the rotation corresponds to the highest frequency the second person hears, and which part corresponds to the lowest?

78. (II) A 1.20-m-long steel bar is clamped at points 0.20 m from one end and 0.60 m from the other. If the bar is tapped at the end farther from the clamp, what frequency should one hear? What if the 0.60-m clamp is removed and the bar is tapped at the free end?

79. (II) A tuning fork with a frequency of 440 Hz is held just above the top of a uniform tube containing water (Fig. 14–46). The tube can excite the column of air above the water, whose level can be changed by a spigot at the bottom of the tube. As the water is drained out, the sound intensity of the fork is enhanced when the air column has a length of 0.60 m and again when the air column has a length of 1.0 m. Use these data to find the speed of sound in air.

80. (III) A square sheet of metal with sides of length L is clamped on all four sides. It is convenient to put these sides at $x = 0$, $x = L$, and $y = 0$, $y = L$ (Fig. 14–47). The vertical displacement at any point on the sheet is given by $z(x, y, t)$. and it satisfies the wave equation

$$\frac{\partial^2 z}{\partial x^2} + \frac{\partial^2 z}{\partial y^2} = \frac{1}{v^2}\frac{\partial^2 z}{\partial t^2}.$$

Show that a standing wave of the form $z(x, y, t) = A \sin(kx) \sin(qy) \cos(\omega t)$ satisfies the wave equation, and find the values of k and q that satisfy the boundary conditions and the constraints of the wave equation.

▲ **FIGURE 14–47** Problem 80.

81. (III) The relation between angular frequency and wave number for a stiff wire takes the form $\omega^2 = (T/\mu)k^2 + \alpha k^4$, where α is a measure of the stiffness. Suppose that such a stiff wire is clamped at $x = 0$ and $x = L$. Express the frequency for the two lowest modes of the wire. For which frequency is the stiffness a more important effect?

▲ **FIGURE 14–46** Problem 79.

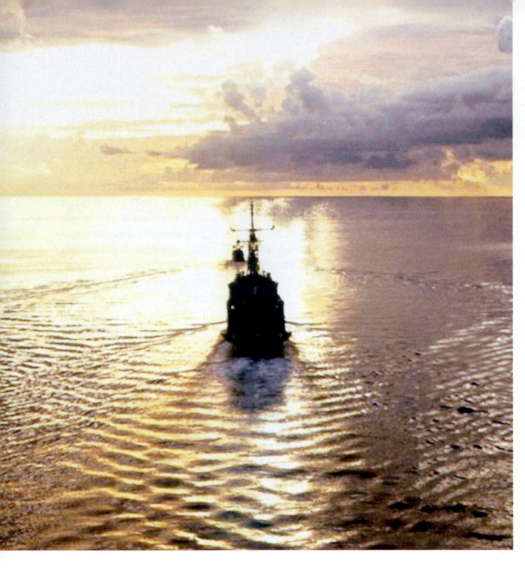

◄ The water waves behind the ship display an important characteristic: They add or subtract algebraically, creating interference patterns.

Superposition and Interference of Waves

The harmonic waves discussed in Chapter 14 are idealizations. Waves in nature or in technology are rarely as simple as harmonic waves. These include water waves, sound in air, waves in material, waves within or on musical instruments, the waves generated by a doctor's ultrasound machine, or the vast gamut of electromagnetic waves that include light. The surface of the ocean is a chaotic welter of waves moving in various directions. A guitar string is in fact set into a very complicated motion when it is plucked. A single shake of the end of a rope sends a single pulse rather than a sine function traveling down the rope.

Nevertheless, we have the tools to understand these more complex waves in terms of simple harmonic waves: Harmonic waves can be combined, or superposed, to produce the full variety of waves observed in nature. Superposed waves, which sometimes reinforce and sometimes (partially or totally) cancel each other, are said to interfere. This interference may produce regular patterns of reinforcement and cancellation; such patterns are seen most easily in light waves as the colors in oil spots or in the pattern of ocean waves as they enter a harbor.

Our goal in this chapter is to understand the different ways in which waves can combine. We will see how the standing waves we encountered in Chapter 14 are formed by the superposition of traveling waves, how the pulsing phenomenon known as beats arises, how there can be well-organized interference patterns in space, and how waves of any shape can be formed by the superposition of harmonic waves.

15–1 The Superposition Principle

The waves that appear in nature are complex. Nevertheless, our work on harmonic waves is relevant because most elastic media that undergo small displacements in response to forces exhibit **linearity**. Linearity means that if two waves can travel through a medium, then a third wave whose form is the algebraic sum of these waves is possible. Thus if a stretched rope supports two different waves in the form $z_1(x, t)$ and $z_2(x, t)$, then it can also support the wave that is described by $z(x, t) = z_1(x, t) + z_2(x, t)$. In other words, *the sum of two solutions of the wave equation for the system is also a solution.* When linearity applies, the sum of any number of waves can add to form another wave, and a wave multiplied by any constant is also an allowed wave. Of course, when we say the "sum" or "multiple" of waves, we are talking about the displacement involved, the quantity we have often called z that describes the wave. This could be a transverse displacement on a rope, a longitudinal displacement in a solid, or the pressure change or density change in a gas. This fact is known as the **superposition principle**.

Let us see how a sum of two solutions of the wave equation can also be a solution to that equation. We saw in Chapter 14 that for transverse waves on a string the wave equation reads [Eq. (14–10)]

$$\frac{\partial^2 z}{\partial x^2} = \frac{1}{v^2} \frac{\partial^2 z}{\partial t^2}, \tag{15–1}$$

where x is the distance along the string and $z(x, t)$ is the transverse displacement of the string at point x and time t. Equation (15–1) is *linear* because the function z appears linearly; that is, as z rather than as z^2 or as z to some other power. If $z_1(x, t)$ and $z_2(x, t)$ are two different solutions satisfying some linear equation such as the wave equation, then the sum $z_1(x, t) + z_2(x, t)$ also satisfies that equation. Superposition works only for waves of small amplitude because when a medium is stretched or distorted too far, the restoring force is no longer purely springlike, and the dynamical equation becomes nonlinear.

Interference

The superposition principle allows us to add sinusoidal waves of different amplitudes, wavelengths, and frequencies, each of them moving in different directions. These sums are *algebraic* sums, complete with the possibilities of algebraic cancellations (Fig. 15–1). This cancellation is known as **interference**, and the result of the interference is a rich palette of patterns called **interference phenomena**.

A simple and striking example of interference is the case of two transverse harmonic waves that each travel to the right on a rope. Let the waves have the same amplitude and be 180° (π rad) out of phase. The two waves have the respective forms

$$z_1(x, t) = z_0 \sin(kx - \omega t) \tag{15–2}$$

and

$$z_2(x, t) = z_0 \sin(kx - \omega t + \pi) = -z_0 \sin(kx - \omega t). \tag{15–3}$$

The superposition of these two transverse waves adds to zero! This is an extreme case of **destructive interference**, which is interference that occurs at any point where two waves have the opposite sign. Although interference may be *fully* destructive (waves adding to zero at a given point), keep in mind that less than total cancellation is possible. **Constructive interference** occurs when the waves add in such a way that they reinforce. If, for example, our two waves z_1 and z_2 had exactly the same phase, we would find a summed solution with the same x- and t-dependence of either wave but with amplitude $2z_0$. Here too, less than maximal levels of constructive interference are possible.

Coherence

What conditions are necessary for waves from two different sources (for example, the sound waves coming from two loudspeakers) to exhibit an interference pattern? If the two sources put out a series of waves with different frequencies and with phases that change randomly, then there is no *consistent* interference pattern. Any pattern would change so

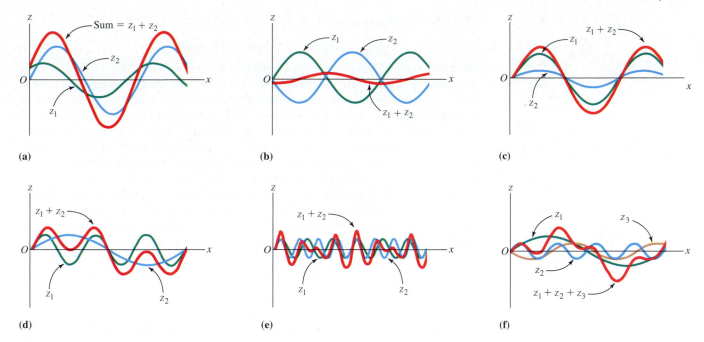

▲ **FIGURE 15–1** The superposition of (a) two waves of the same frequency, with a slight phase difference between them; (b) two waves of the same amplitude and frequency, with a phase difference of nearly 180° between them; (c) two waves of the same frequency and phase but different amplitudes; (d) two waves of the same amplitude but frequencies that differ by a factor of 3; (e) two waves of the same amplitude but slightly different frequencies, resulting in beats; (f) three waves of different amplitudes and frequencies.

rapidly that we could not recognize it. *An interference pattern occurs only when the waves have a definite, stable relation between their frequencies and phases.* The waves are then said to be **coherent**. When, in contrast, we have waves without a definite, stable relation between their frequencies and phases, the waves are **incoherent**. Sources such as piano strings or lightbulbs or the human voice that are not point sources are generally incoherent with other sources such as other piano strings or other voices. If the form of vibration of the piano string is complex, as it generally is, even waves coming from different parts of the string will not generally be coherent with each other. You won't observe interference by listening to a rock band! This is discussed further at the end of Section 15–4.

15–2 Standing Waves Through Interference

Standing waves can be viewed as the result of the superposition of traveling waves moving in opposite directions. Imagine a very long string with two waves that move in opposite directions but are otherwise identical, each a solution of the wave equation:

$$\text{for the left-moving wave: } z_l(x, t) = A \sin(kx + \omega t); \quad (15\text{–}4a)$$

$$\text{for the right-moving wave: } z_r(x, t) = A \sin(kx - \omega t). \quad (15\text{–}4b)$$

We have taken the wavelengths, frequencies, phases, and amplitudes to be equal. (You may want to review the meaning and relations between these quantities at this point; see Section 14–3.) According to the principle of superposition, the resultant displacement is

$$z_1(x, t) = z_r(x, t) + z_l(x, t) = A[\sin(kx - \omega t) + \sin(kx + \omega t)]. \quad (15\text{–}5)$$

If we use the trigonometric identity 5.3 of Appendix IV–4, with $a = kx - \omega t$ and $b = kx + \omega t$, then we obtain

$$\sin(kx - \omega t) + \sin(kx + \omega t)$$

$$= 2 \sin\left[\frac{(kx - \omega t) + (kx + \omega t)}{2}\right] \cos\left[\frac{(kx - \omega t) - (kx + \omega t)}{2}\right]$$

$$= 2 \sin(kx) \cos(\omega t).$$

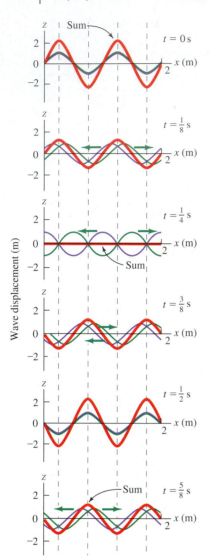

In the last line, we have used the fact that $\cos(-a) = \cos(a)$. Thus

$$z(x, t) = 2A \sin(kx) \cos(\omega t). \qquad (15\text{–}6)$$

This expression is identical to our expression for standing waves, Eq. (14–39), except for an (unimportant) phase difference and a factor of 2 that can be taken as part of the amplitude. It is now easier to understand why the wave speed v appears in the equation for a standing wave, which after all is the wave equation itself, Eq. (15–1): It is the speed of the two traveling waves that are superposed to form the standing wave. In Fig. 15–2 we plot a time sequence of two traveling waves and their superposition. The superposition produces a standing wave whose period and wavelength are the same as those of the traveling waves that compose it.

Note that a standing wave would not be formed if the amplitudes were not of the same magnitude. (The phases have no special requirement to meet, although as they change, the location of nodes would move.) We can see the effect of different amplitudes by, for example, doubling the amplitude in the right-moving wave [Eq. (15–4b)]. Then the sum would give

$$A[\sin(kx - \omega t) + \sin(kx - \omega t) + \sin(kx + \omega t)]$$
$$= 2A \sin kx \cos \omega t + A \sin(kx - \omega t).$$

The last term is a leftover wave that travels to the right. This means that creating pure standing waves is not a trivial matter. Of course, standing waves are created on the strings of musical instruments all the time—this is because, as we'll see below, in the reflection of waves from a place where the string is fixed, such as a fret, the conditions for a standing wave are automatically satisfied.

◀ **FIGURE 15–2** Superposition of two traveling waves produces a standing wave, plotted in red. Shown are the waves $\sin(kx - \omega t)$, moving rightward, and $\sin(kx + \omega t)$, moving leftward. The amplitude of each wave is 1 m, the wavelength is 1 m, and the period is 1 s. Thus, $k = 2\pi$ m^{-1} and $\omega = 2\pi$ rad/s. Arrows indicate the direction of motion of the two traveling waves. Time progresses in $\frac{1}{8}$-s intervals starting at $t = 0$ s and ending at $t = \frac{5}{8}$ s. Note that the peaks of the standing wave remain at exactly the same x-values.

EXAMPLE 15–1 Two traveling waves move in opposite directions in a one-dimensional medium and form a standing wave. The angular frequency of that standing wave is 2.9 rad/s, and the velocity of each of the traveling waves is 1.8 m/s. For comparison purposes you also have a string attached tightly to two walls whose separation can be adjusted. In particular, they are adjusted so that the largest wavelength of the standing waves (lowest mode) that can be supported on your string is the same as the wavelength of the superposition of the two traveling waves. What is that separation?

Setting It Up The medium with the traveling waves is not specified, but we can imagine it too is a string. This system and the comparison string are both shown in Fig. 15–3. We are given the angular frequency ω as well as the speed v of the traveling waves that form the standing wave in the medium. The wavelength λ, which is labeled in the figure, is unknown; we can find the wall separation L in terms of it.

Strategy We first must find the wavelength of the standing wave set up within the medium, given that we know the angular frequency and the wave speed. For this we can use the relation $v = f\lambda = (\omega/2\pi)\lambda$. We then want to adjust the separation L of the walls so that the lowest, or fundamental, mode (longest wavelength) of the string connected to the walls has wavelength λ. For that we need only recall that the longest wavelength on a string connected at its two ends is twice the string length, Eq. (14–42) with $n = 1$. Thus we set $L = \lambda/2$.

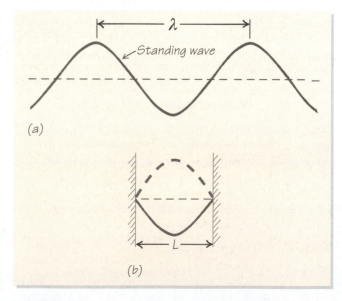

▲ **FIGURE 15–3** Part (a) shows the standing wave on the one-dimensional system, while (b) shows the string attached to walls.

Working It Out   We find

$$\lambda = 2\pi v/\omega = (2\pi)(1.8 \text{ m/s})/(2.9 \text{ rad/s}) = 3.9 \text{ m}.$$

In turn,

$$L = \lambda/2 = 2.0 \text{ m}.$$

What Do You Think?   Is the frequency of the wave on the string necessarily that of the standing wave in the medium? *Answers to **What Do You Think?** questions are given in the back of the book.*

Interference of Incident and Reflected Waves

In order to understand the production of standing waves, it is helpful to consider an *incident*, or incoming, harmonic traveling wave that encounters a wall. To carry this wave, we can stretch a long rope aligned with the *x*-axis attached to a wall at the point $x = 0$ and extending to the left. Since the string is tied to the wall at that point, the displacement $z(x, t)$ is always zero at $x = 0$. A harmonic incident wave is initiated somewhere on the far left and arrives at $x = 0$. When the wave arrives at the wall it is *reflected*. The energy initially carried to the right bounces back and travels back down the string to the left as a reflected wave. The reflected wave has the same frequency *f* as the incoming wave, as you can see by thinking about the reflection process: A peak of the incident wave will produce a peak of the reflected wave, and it will take the same period for the reflected wave to produce its next peak as it did for the next peak of the incident wave to arrive at the wall. Also, the rigid wall absorbs no energy, so the reflected wave must have the same amplitude as the incident wave. Finally, the reflected wave moves with the same speed as the original incident wave because the wave speed *v* is a property of the string. These are just the conditions that give us Eq. (15–5), and they lead to Eq. (15–6): The reflected wave interferes with the rightward-moving incident wave to produce a standing wave with the same frequency and wavelength. This is why the idea that a standing wave on a string is a superposition of traveling waves is more than merely a mathematical construct.

One feature of the reflection that creates a standing wave deserves special notice. The standing wave formed on the string must have a node at the fixed end $x = 0$. This means that the incident wave and its reflection must *cancel* at $x = 0$ (and hence form a node at that point). This requires that *the sinusoidal wave that reflects from a fixed end has a displacement with a sign opposite to that of the incoming wave and is therefore inverted*. Whether reflected waves are inverted or not is an interesting subject and one that we'll take up again in Section 15–5.

15–3 Beats

Beats occur when two waves of nearly the same frequency are superposed; a long-wavelength (low-frequency) interference pattern is produced. Musicians use beats to tune their instruments, and the phenomenon is of scientific importance because it provides a sensitive technique for measuring frequency (or period) differences. A simple visual analogue to beats can be constructed with the help of two combs with slightly different tooth spacings; the slightly different spacings are analogous to slightly different periods. In Fig. 15–4b, two combs are superposed. Where the teeth lie directly on

(a)

(b)

◀ **FIGURE 15–4**   An analogy indicating beats. (a) Two combs have slightly different spacings. (b) When superposed, the combs produce an interference pattern known as a moiré pattern.

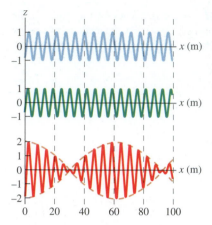

▲ FIGURE 15–5 Two traveling waves of identical amplitude, $z_0 = 1$ m, and nearly the same frequency and wavelength, at $t = 1$ s. The blue traveling wave has $\omega = 1$ rad/s and $k_1 = 1$ m^{-1}. The green traveling wave has $\omega_2 = 0.9$ rad/s and $k_1 = 1.1$ m^{-1}. Beats are visible in their superposition (the red wave), formed by the envelope shown as a dashed line. The vertical lines trace the regions of constructive and destructive interference.

top of one another, the spaces are visible. Some distance away, the teeth will lie out of phase and no spaces are visible, farther still and the spaces will once again be visible. The superposed combs exhibit a pattern that repeats over a much greater distance than the separation of the teeth. The large spacing of the pattern observed when the two combs are superposed is equivalent to the large period of the interference pattern observed when waves with two slightly different periods are superposed.

Consider two traveling waves of slightly different frequencies, as in the blue and green waves in Fig. 15–5. The two waves have equal amplitudes of 1 m. The first wave has $\omega_1 = 1.00$ rad/s and $k_1 = 1.00$ m^{-1}, and the second has $\omega_2 = 0.90$ rad/s and $k_2 = (1/0.90)$ m$^{-1} = 1.11$ m^{-1}. To the eye, the two waves are barely distinguishable from each other. Yet their superposition has a striking visual signature: *A very different wave—one with a long wavelength—is formed* (the red wave in Fig. 15–5 and the close-up view of Fig. 15–6).

Here is how this pattern is formed: Both waves start out together at $x = 0$ m, in phase, and hence interfere constructively. The amplitude of the superposed wave is therefore large at this point. Because the wavelengths of the two waves differ slightly, they begin to develop a phase difference as we move along x, and at some point along x (at about 30 m), they are almost perfectly out of phase. Whereas they started out with maximum on maximum or minimum on minimum, they now are such that the maximum of one is at the same point as the minimum of the other. At this point they interfere destructively, and the amplitude of the superposed wave is small. The cycle repeats itself as the waves slowly come together to interfere constructively again, at about 63 m, as for the combs of Fig. 15–4.

Calculation of the Beat Frequency

To analyze this situation more quantitatively, start with the two waves

$$z_1(x, t) = z_0 \sin(k_1 x - \omega_1 t) \quad \text{and} \quad z_2(x, t) = z_0 \sin(k_2 x - \omega_2 t). \quad (15\text{–}7)$$

The angular frequencies and wave numbers differ a little. A convenient way to account for this fact is to deal with the sum and the difference of the frequencies. The frequency difference will be a small quantity. If $\omega_1 > \omega_2$, and $k_1 > k_2$, we define

$$\delta\omega \equiv (\omega_1 - \omega_2) \quad \text{and} \quad \Omega \equiv \frac{1}{2}(\omega_1 + \omega_2), \quad (15\text{–}8)$$

and similarly,

$$\delta k \equiv (k_1 - k_2) \quad \text{and} \quad K \equiv \frac{1}{2}(k_1 + k_2). \quad (15\text{–}9)$$

It is easy to show that the ratios $\delta\omega/\delta k$ and Ω/K are given by the wave speed $v = \omega_1/k_1 = \omega_2/k_2$ in the medium. Both $\delta\omega$ and δk are positive and small.

We can now reexpress the original frequencies and wave numbers in terms of their sums and differences:

$$\omega_1 = \Omega + \frac{1}{2}\delta\omega \quad \text{and} \quad \omega_2 = \Omega - \frac{1}{2}\delta\omega, \quad (15\text{–}10)$$

▶ FIGURE 15–6 An enlarged look at the superposition of the red wave of Fig. 15–5. The dashed envelope is a traveling wave of wavelength $\lambda_{\text{beat}} = 126$ m.

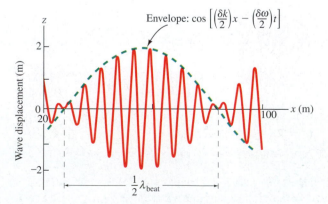

$$k_1 = K + \frac{1}{2}\delta k \quad \text{and} \quad k_2 = K - \frac{1}{2}\delta k. \tag{15-11}$$

When Eqs. (15–10) and (15–11) are substituted into the waves of Eq. (15–7), we find that

$$z_1(x, t) = z_0 \sin\left[\left(K + \frac{\delta k}{2}\right)x - \left(\Omega + \frac{\delta\omega}{2}\right)t\right]$$

$$= z_0 \sin\left\{(Kx - \Omega t) + \frac{1}{2}[(\delta k)x - (\delta\omega)t]\right\}, \tag{15-12}$$

$$z_2(x, t) = z_0 \sin\left[\left(K - \frac{\delta k}{2}\right)x - \left(\Omega - \frac{\delta\omega}{2}\right)t\right]$$

$$= z_0 \sin\left\{(Kx - \Omega t) - \frac{1}{2}[(\delta k)x - (\delta\omega)t]\right\}. \tag{15-13}$$

To sum these two terms and find the resultant superposition, we expand the trigonometric functions with identity 5.3 of Appendix IV–4, and we use the fact that $\cos(-a) = \cos(a)$. The result is

$$z(x, t) = z_1(x, t) + z_2(x, t) = 2z_0 \sin(Kx - \Omega t) \cos\left[\left(\frac{\delta k}{2}\right)x - \left(\frac{\delta\omega}{2}\right)t\right]. \tag{15-14}$$

Equation (15–14) states that the sum of the original two waves can be reformulated as a *product* of two traveling waves. The wavelength and frequency of the first of these traveling waves, which is the sine term in Eq. (15–14), are given by the *average* of the corresponding quantities for the original two waves. The envelope formed by the second traveling wave is the dashed curve in Fig. 15–5. This envelope is a wave traveling with the same speed as the sine wave but with a much longer wavelength. The envelope wave is called a *beat* of the original two waves. The coefficient of x in the cosine factor in Eq. (15–14), $\delta k/2$, gives the wave number of the beats, whereas the coefficient of the time, $\delta\omega/2$, gives the angular frequency. In turn, we can use Eqs. (15–8) and (15–9) to find the wavelength and frequency of the beats:

$$\frac{1}{\lambda_{\text{beat}}} = \frac{k_{\text{beat}}}{2\pi} = \frac{k_1 - k_2}{4\pi} = \frac{1}{2}\left(\frac{1}{\lambda_1} - \frac{1}{\lambda_2}\right); \tag{15-15}$$

$$f_{\text{beat}} = \omega_{\text{beat}}/2\pi = (\omega_1 - \omega_2)/4\pi = \tfrac{1}{2}(f_1 - f_2). \tag{15-16}$$

BEAT WAVELENGTH AND FREQUENCY

The beat has a wave number and frequency that are one half the *differences* of the original wave numbers and frequencies. Because this difference is small by assumption, the wave number of the envelope is small and the wavelength is *large*. Similarly, the frequency of the beat is *small*. In deriving this result, we tacitly assumed $f_1 > f_2$, so that f_{beat} was positive. Note, however, that the beat factor in Eq. (15–14) is a cosine factor, and the cosine is unchanged if its argument changes sign. This means that if $f_2 > f_1$, f_{beat} would be given by $(f_2 - f_1)/2$ rather than $(f_1 - f_2)/2$: The beat frequency is always positive.

Figure 15–6 is an enlarged version of the superposed waves plotted in Fig. 15–5. Recall that the original two waves have $k_1 = 1.0 \text{ m}^{-1}$ and $k_2 = 1.1 \text{ m}^{-1}$. From Eq. (15–15), the beat wavelength is then

$$\lambda_{\text{beat}} = \frac{4\pi}{|(k_1 - k_2)|} = \frac{4\pi}{|(1.00 \text{ m}^{-1} - 1.10 \text{ m}^{-1})|} = 126 \text{ m}.$$

Only a little over half a wavelength is plotted in Fig. 15–6, and we can indeed see that $\lambda_{\text{beat}}/2 \cong 63 \text{ m}$. The frequency of the beat is, in this case,

$$f_{\text{beat}} = (\omega_1 - \omega_2)/4\pi = (1.00 \text{ rad/s} - 0.90 \text{ rad/s})/4\pi = 0.0080 \text{ Hz}.$$

Beats in Sound: When we hear beats of sound, the combination of ear and brain has a definite physiological response. In addition to the note of average frequency, we hear a second note with the pitch of the beat frequency if the beat frequency exceeds 25 Hz or so. If the beat frequency is less than this, however, we hear regular pulses in amplitude. Because the ear interprets loudness as proportional to the wave intensity, which is in turn proportional to the amplitude squared, the ear hears a pulse every time the beat passes through either a maximum or minimum. The result of this is that *the frequency of the perceived pulse is twice the beat frequency.*

CONCEPTUAL EXAMPLE 15–2 Two tuning forks labeled as identical actually differ in frequency by approximately 0.5%. By determining the beat frequency, can you determine precisely the difference in frequencies? Can you determine which tuning fork has the higher frequency? Suppose you had some electrical tape. Knowing that winding the tape around the upper end of the prongs of a tuning fork reduces the frequency, could you then determine which tuning fork has the higher frequency?

Answer Since the beat frequency is one half of the frequency difference, an accurate measurement of the beat frequency determines the frequency difference accurately. However, the beat frequency only determines the absolute value of the frequency difference, so that this measurement alone cannot determine which tuning fork has the higher frequency.

The tape can be used to determine which tuning fork has the higher frequency. Suppose the two tuning forks are distinguished by color, say blue and green. Suppose we put the electrical tape on the blue one. If the blue one has the lower frequency, then the tape lowers it further, and the beat frequency increases. If the blue one has the higher frequency, then the tape reduces the difference and that will be noticed in a measurement of the beat frequency.

What Do You Think? Why would putting electrical tape on the end of the tuning fork prongs reduce the frequency?

EXAMPLE 15–3 Two sinusoidal waves traveling along a string with equal amplitudes and with periods of 19 and 20 seconds, respectively, are observed to have a maximum displacement at a particular point and time (say $x = 0$ and $t = 0$). After how many seconds will the resultant displacement be zero at $x = 0$? When will this occur for the second time?

Setting It Up We sketch the two waves as a function of time at the fixed point $x = 0$ (Fig. 15–7). They coincide at $t = 0$, where they are each a maximum. The sketch extends to beyond $t = 5$ s, the time when the wave with 20-s period crosses the axis, while the wave with 19-s period will have crossed it at $t = 4.75$ s.

▲ **FIGURE 15–7** Two traveling waves have the same amplitude, but slightly different frequencies.

Strategy We deal at a given space point ($x = 0$) with the superposition of two traveling waves with slightly different periods and equal amplitudes. The first step is to write expressions for the two waves at $x = 0$, taking into account the fact that they are both at their maximum at $t = 0$. We then add the two waves and use trigonometric identities to express the sum as a product of single waves, one involving a beat frequency and one involving an "average" frequency, just as in the text above. Because the oscillation in the "average" frequency is faster, we expect that it is the average frequency term that will vanish first. At that point it is a matter of numerically identifying the time when the harmonic function involving the average frequency is zero. We then need to convert periods to angular frequencies ω and then work out the details.

Working It Out A harmonic wave at a particular point is a function only of t. If it takes its maximum value at $t = 0$, it must be proportional to $\cos \omega t$. We have two waves, and since their amplitudes are equal, we describe them by $A \cos \omega_1 t$ and $A \cos \omega_2 t$, respectively.

Now we work out the sum, breaking out sums and differences of the frequencies:

$$A(\cos \omega_1 t + \cos \omega_2 t) = A\left[\cos\left(\frac{\omega_1 + \omega_2}{2}t + \frac{\omega_1 - \omega_2}{2}t\right) \right.$$
$$\left. + \cos\left(\frac{\omega_1 + \omega_2}{2}t - \frac{\omega_1 - \omega_2}{2}t\right) \right]$$
$$= 2A \cos\left(\frac{\omega_1 + \omega_2}{2}t\right) \cos\left(\frac{\omega_1 - \omega_2}{2}t\right).$$

The first cosine term contains the average frequency, the second the beat. Our expression will be zero when either $\frac{\omega_1 + \omega_2}{2}t = \frac{\pi}{2}$ or $\frac{\omega_1 - \omega_2}{2}t = \frac{\pi}{2}$, whichever happens earlier. These correspond respectively to $t_{1,\text{av}} = \frac{\pi}{\omega_1 + \omega_2}$ or $t_{1,\text{beat}} = \frac{\pi}{\omega_1 - \omega_2}$. As must be the case, $t_{1,\text{av}}$ is smaller. The sum of the two waves will again be zero at *either* $t_{1,\text{beat}}$ or at the second zero of the average frequency factor $t_{2,\text{av}}$, whichever occurs first. We have $\frac{\omega_1 + \omega_2}{2}t_{2,\text{av}} = \frac{3\pi}{2}$ or $t_{2,\text{av}} = \frac{3\pi}{\omega_1 + \omega_2}$. We will find which occurs first numerically, but since the frequencies of the two waves are almost the same, we might expect the beat period to be very large, and $t_{2,\text{av}}$ to be first.

We start with the relation $\omega = 2\pi f = 2\pi/T$ to calculate $\omega_1 = 2\pi/(19 \text{ s}) = 0.33 \text{ rad/s}$ and $\omega_2 = 2\pi/(20 \text{ s}) = 0.31 \text{ rad/s}$. We substitute these into our expressions for $t_{1,\text{av}}, t_{2,\text{av}}$, and $t_{1,\text{beat}}$. We find that these are, respectively, 4.9 s, 15 s, and 1.6×10^2, and the beat period is much larger than the times associated with the "average" wave.

What Do You Think? Would these results be changed if at $x = 0$ each wave had their maxima at $t = 5$ s rather than 0 s? Would they change if each wave had a $t = 0$ maximum at a different space point than $x = 0$?

15–4 Spatial Interference Phenomena

Sound waves, light waves, and water waves all exhibit *interference patterns* in space.[†] *Any* waves that propagate in two or three dimensions can show spatial interference phenomena provided that the interfering waves are coherent. For example, if a laser beam is separated into two beams by passing it through two holes (to ensure two laser light sources of the same wave characteristics), it is possible to produce a regular series of intense spots where the two beams interfere constructively. A shallow tank of water—what is often called a ripple tank—is another useful tool for demonstrating interference patterns, in this case of water waves (Fig. 15–8). Spatial interference phenomena are behind many precision distance measurements, as we'll see when we study the interference of light waves (Chapters 37 and 38).

To see how spatial interference phenomena arise, imagine a single source of harmonic traveling waves with wavelength λ and frequency f, such as a harmonically driven loudspeaker, and two pipes that carry the sound waves away from the loudspeaker (Fig. 15–9). The pipes fix the path that the waves can follow, and the two paths do not necessarily have the same length even if they end up at the same place. At the source, i.e., the beginning of the respective paths, the waves are in phase. For example, if there is a wave crest at the start of path 1, then the same crest is at the start of path 2. These two crests each travel down their paths at the same speed. If the paths are the same length, $L_1 = L_2$, then they will both arrive at the observation point P at the same time. They will still be in phase at P, and they will interfere constructively. Under these circumstances, the adjacent troughs will arrive together, as will the next crests, and so forth. The interference will always be constructive.

If the path lengths are different, however, then these same crests will not arrive at the same time. Whether the interference is constructive or destructive depends on the *difference* in path length, $\Delta L = L_2 - L_1$. Suppose path 2 is longer, so that ΔL is positive. If ΔL is one full wavelength, $\Delta L = \lambda$, then the second crest will arrive at P along path 1 just when the first crest arrives at P along path 2. The two waves will be in phase and the interference will be constructive. If $\Delta L = \lambda/2$, then when the crest arrives at P from path 2, it meets a trough from path 1. In this case the interference is destructive, and no disturbance is recorded at P. The situation can be summed up as follows:

for constructive interference: $\Delta L = n\lambda, n = 0, \pm 1, \pm 2, \ldots$ (15–17)

for destructive interference: $\Delta L = (n + \frac{1}{2})\lambda, n = 0, \pm 1, \pm 2, \ldots$ (15–18)

CONDITIONS FOR INTERFERENCE

The effect just described also occurs with two sources at different distances from an observation point. Imagine two separate sources emitting harmonic waves of the same wavelength in a given medium. We could for example make the same rhythmic disturbance at two different points on a water surface. At each point, a series of spreading ripples is created that interfere with one another. In Fig. 15–10, we have two separated sources, S_1 and S_2, each producing waves with the same wavelength λ and frequency f. We assume that these waves are *in phase*; that is, a crest always emerges from S_1 at the same moment that a crest emerges from S_2. These waves are coherent. An observation point P is located a distance L_i from source S_i; the path difference is $\Delta L = L_2 - L_1$. The conditions for constructive interference are simply that the crests, or the troughs, of the waves from the two sources arrive at P together. For this two-source situation, the same conditions respectively apply for constructive and destructive interference as before; namely, Eqs. (15–17) and (15–18). Since different places in space are different distances from the sources, there will be constructive interference at some points in space and destructive interference at others.

▲ **FIGURE 15–8** Two sources that produce waves in a ripple tank form an interference pattern, visible as regions of constructive and destructive interference radiating out from the sources.

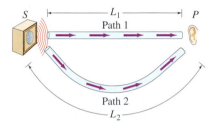

▲ **FIGURE 15–9** Unless they propagate on a string or on other one-dimensional media, waves from a source S may follow different paths to get to a given observation point P. Here two pipes of different lengths determine two different paths between S and P.

[†]Not to be confused with beats, in which we are primarily interested in time dependence at a particular point.

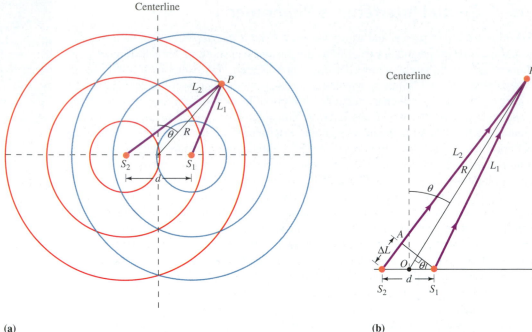

(a)　　　　　　　　　　　　　　　　　　**(b)**

▲ **FIGURE 15–10** (a) The water waves generated by two coherent sources in phase spread away in circles from these sources. Whether the interference is constructive or destructive at observation point P depends on the difference in path lengths. The circles are (arbitrarily) drawn on the wave crests. (b) The geometry necessary to determine the path difference $\Delta L = L_2 - L_1$ and hence for the pattern of interference maxima and minima. The observation point P is far away compared to the separation of the two sources, d.

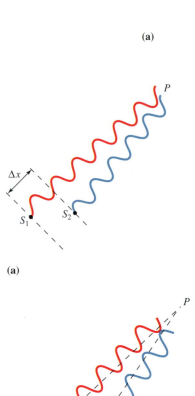

(a)

(b)

▲ **FIGURE 15–11** The conditions for maxima and minima in the interference of coherent waves of the same wavelength emitted by two sources. (a) Maxima correspond to the difference in path length equal to an integer multiple of λ, while (b) minima occur when the path difference is an integral multiple of λ plus $\lambda/2$.

Locations of Maxima and Minima

The observation point P can be anywhere that the waves propagate. Some locations will have constructive interference, some destructive, and some points will have an intermediate level of interference, depending only on the distances from the sources. The locations where the interference is constructive are said to exhibit interference *maxima*, and the locations where the interference is destructive exhibit interference *minima*. Finding the spatial pattern of the maxima and minima is an exercise in geometry—of the type sketched out in Fig. 15–11.

We can get a sense of patterns of this type with a visual analogue to wave interference. Figure 15–12a shows a set of finely spaced gray concentric circles around a point representing a wave source. Think of the gray circles as simplified crests and the clear areas between the circles as simplified troughs. When two such wave sources are present, as they are in Fig. 15–12b (too closely spaced to see easily in that figure), we can think of constructive interference as an overlap of the gray areas (crests) with themselves, giving very dark areas, or of the clear areas (troughs) with themselves, leaving a clear area; we can think of destructive interference as a uniform gray where the gray areas and the clear areas overlap. In fact, bands of maxima and minima are clearly visible in the figure, even if the centers of the waves are not.

To find the patterns of maxima or of minima defined by Eqs. (15–17) and (15–18), the necessary geometric information is the wavelength, λ, and the separation of the two sources, d. The geometry is shown in Fig. 15–10. We have put an origin, O, midway between the sources and measured the distance R from point P to the origin, as well as the angle θ between the center line and the line from the origin to P. The center line is the line perpendicular to and bisecting the line of the sources. Note that at the angle $\theta = 0$, where $L_1 = L_2$, we have $\Delta L = 0$; *there is a maximum everywhere along the center line.* The regions of the other maxima and minima are simple to see when $R \gg d$; that is, when P is very far from the pair of sources. This case is illustrated in Fig. 15–10b, which shows, to a good approximation, that all three lines—the line from S_1 to P (S_1P), the line from the origin to P, and the line from S_2

(a)   (b)

◀ **FIGURE 15–12** (a) Concentric circles representing waves propagating out from a source. (b) Overlay of the circles that spread from two adjacent sources. Regions of constructive and destructive interference, marked as max (maxima) and min (minima), are clearly visible. We have measured $\lambda = 5$ mm and $d = 16$ mm on the original of this reproduction, so the ratio $\lambda/d = 0.32$. The first minimum should occur at $\sin \theta_1 = \lambda/2d = 0.16$, or $\theta_1 = 9°$, and this is the value of the angle found in the figure.

to P (S_2P)—make the same angle θ with the center line. Then ΔL can be measured by drawing a line from S_1 perpendicular to S_2P. This line meets S_2P at point A in the figure. S_2S_1 is perpendicular to the center line, and AS_1 is perpendicular to S_2P, so the angle formed by lines AS_1 and S_2S_1 is θ. Because $AP = S_1P$, we have $\Delta L = L_2 - L_1 = S_2A$, and $\sin \theta = \Delta L/d$, or

$$\Delta L = d \sin \theta. \tag{15–19}$$

As a result of this approximation, for $R \gg d$ we can restate Eqs. (15–17) and (15–18) for the maxima and minima in terms of the angle θ by using Eq. (15–19):

$$\text{for maxima: } \sin \theta = n\frac{\lambda}{d}, n = 0, \pm 1, \pm 2, \ldots; \tag{15–20}$$

$$\text{for minima: } \sin \theta = \left(n + \frac{1}{2}\right)\frac{\lambda}{d}, n = 0, \pm 1, \pm 2, \ldots. \tag{15–21}$$

These equations determine a set of angles for which there are interference maxima and minima. Positive θ values are to the right of the center line and negative values are to the left. The array of angles is symmetric about the center line, so we can concentrate on one side. On the right side $n = 0$ is a maximum at $\theta = 0$, which is along the center line itself. As we increase θ, the first minimum occurs at $\sin \theta = \lambda/2d$. The second maximum is at $\sin \theta = \lambda/d$, the second minimum at $3\lambda/2d$, and so forth. In Fig. 15–12b these angles are marked on the pattern made by overlaying two sets of concentric circles, one set slightly displaced from the other. In the situation shown here $\lambda = 5$ mm and $d = 16$ mm. The first three minima should then be at $\sin \theta_1 = 0.16$, $\sin \theta_2 = 0.47$, and $\sin \theta_3 = 0.78$, or $\theta_1 = 9°$, $\theta_2 = 28°$, and $\theta_3 = 51°$, in agreement with Fig. 15–12b.

Any unchanging phase difference between two harmonic waves—coherent waves—will lead to an interference pattern. Above, the two sources have the same frequency and no phase difference between them. But suppose we change things a little. Source S_1 could consist of a sharp object dipping into water with some fixed frequency, and S_2 consists of a similar object dipping into the water with the same frequency but a time interval τ later. This changes the relative phase of the two waves (by $\omega\tau$), but the relative phase is still fixed, so that we still have coherence. Let us now change source S_2 as follows. Its frequency is still the same as that of S_1, but the sharp object dips into the water in different positions around what was previously the fixed position of S_2. This means that for each wave, the interval ΔL is slightly different. If ΔL were much smaller than the wavelengths involved, then there would still be an interference pattern, but a blurred one. If ΔL is of the order of the wavelength or more, then the interference pattern is gone: There are still points of total constructive or destructive interference, but these vary from wave to wave, and no steady pattern of the type seen in Fig. 15–8 will appear. This is an example of incoherence—no interference pattern appears when incoherent waves superpose.

Example 15–4 describes the interference pattern that occurs when two waves of the same wavelength have phases differing by π.

EXAMPLE 15–4 Two loudspeakers, S_1 and S_2, are separated by 0.50 m. These speakers form sound waves of the same amplitude at a frequency of 4400 Hz. The amplifier emits the two waves 180° out of phase. A set of chairs is arranged in a semicircle 30.0 m from the midpoint of the two speakers. Find the amplitude of the wave at the chair on the perpendicular bisector (center line) of the line between the speakers. At what distance to the right of this central chair is there a first maximum in sound intensity?

Setting It Up The situation is sketched in Figure 15–13. We have marked on the figure the known speaker separation d and the radius R of the semicircle of chairs surrounding them. The speakers emit with known frequency f. The unknown distance D where a maximum occurs is also marked.

▲ **FIGURE 15–13** The geometry is the same as that illustrated in Fig. 15–10b.

Strategy The first step is a geometrical one, the calculation of the difference in the lengths of the paths from the two speakers to a point on the semicircle. This can be done exactly, but in the approximation that $(d/R)^2 \ll 1$ we can treat the chairs not far from the center line

as if they were on a straight line, perpendicular to the center line, and we can approximate the sine of the small angle θ by θ alone. We then need to translate that difference in path length into a statement about destructive/constructive interference. Because the two speakers emit the waves 180° out of phase, the rules in Eqs. (15–20) and (15–21) are interchanged: *Minima become maxima and vice versa.* Thus, along the center line, for which the path difference ΔL is zero, the interference is now destructive. We want the smallest angle θ for which ΔL implies constructive interference, and that translates geometrically into D.

Working It Out With the phase change taken into account, the rules are

$$\text{for minima:} \quad \sin \theta = n\frac{\lambda}{d}, n = 0, \pm 1, \pm 2, \ldots;$$

$$\text{for maxima:} \quad \sin \theta = \left(n + \tfrac{1}{2}\right)\frac{\lambda}{d}, n = 0, \pm 1, \pm 2, \ldots.$$

The first maximum ($n = 0$) is at an angle given by

$$\sin \theta = \frac{\lambda}{2d}.$$

If the semicircle of chairs is a distance R from the speakers, then $\theta = D/R$. For small θ, $\sin \theta \cong \theta$, and the value of D for which the first maximum occurs is

$$\sin \theta = D/R = \lambda/2d, \quad \text{or} \quad D = (\lambda/2d)R = (v/2fd)R.$$

We have used the relation $\lambda f = v$, where v is the wave speed. The speed of sound is roughly 330 m/s, so

$$D = \frac{(330 \text{ m/s})}{2(4400 \text{ Hz})(0.50 \text{ m})}(30.0 \text{ m}) = 2.3 \text{ m}.$$

We see that $D \ll R$, so our small-angle approximation is justified.

What Do You Think? Suppose the listener in the chair at which the first maximum is heard rolls radially away from the midpoint of the speakers at a rate of 3 m/s. What properties of the sound will have changed when he reaches a distance of 60 m?

CONCEPTUAL EXAMPLE 15–5 The spatial interference pattern of maxima and minima that we have studied forms lines corresponding to particular angles from a center line. These angles depend on the wavelength λ of the interfering waves and the separation d of the sources. When these angles are small, we say the pattern is *narrow*; when the angles are large, we say the pattern is *broad*. If you had a particular pattern and you wanted to make it broader, and you could control both λ and d, what would you do?

Answer The angular distance between adjacent angles is given by $\sin \theta_{n+1} - \sin \theta_n = (\lambda/d)$. Thus increasing λ and/or decreasing the source separation d make this distance increase. We say that the pattern *spreads* (or is broadened). The pattern *shrinks*, or *narrows*, as λ decreases and/or d increases. For example, as the separation of our two loudspeakers decreases, the pattern of interference maxima and minima spreads in angle.

What Do You Think? By how much can the pattern spread?

15–5 Pulses

When an incident ocean wave meets a wall, the wave rebounds and forms a reflected wave that moves back toward the sea. If the wall is low, then part of the incident wave can pass over the top of it and continue as a transmitted wave. This demonstrates that waves can both reflect from, and be transmitted across, boundaries. A simple lecture demonstration of reflection is obtained by attaching a long rope to a wall and holding the other end fairly taut flipping the wrist. The result is a bump traveling to the wall and

HOW DO NOISE-CANCELING HEADPHONES WORK?

The noise-canceling headphones worn by workers in helicopters or in other very noisy environments (Fig. 15–14) make use of the interference phenomena. They contain both speakers and a microphone located close to the entrance of the ears. Sound received at the microphone is amplified and supplied to the speaker with a phase difference, such that the sound from the speaker cancels the original signal as completely as possible. Maximum cancellation is set to occur within the ear, nominally at the surface of the eardrum. Communication from the radio or intercom of the aircraft is not subject to cancellation and is therefore audible to the crew members.

▶ **FIGURE 15–14** An electronic noise cancellation headset such as this one can reduce noise in the low-frequency range by some 10–15 dB over conventional headsets.

▲ **FIGURE 15–15** A pulse is generated on a light spring on the right side and moves to the left until it encounters a wall, where the spring is attached. The displacement at this point is zero, and the pulse is then reflected from the wall towards the right with an inverted pulse as explained below.

then traveling back to the other end (Fig. 15–15). If the held rope is connected to another rope of different mass density, with the second rope connected to the wall, then both the reflection and transmission are visible at the two-rope boundary.

The bump generated in the ropes is a **pulse**, and the discussion of reflection and transmission is somewhat easier to visualize in terms of pulses. What are pulses? Pure harmonic waves (cosines or sines), or any truly periodic waves, are idealizations in that they are infinitely long. Pulses are disturbances that have a beginning and an end, whether short or not. For us to be able to deal with pulses we need only be sure that they obey the wave equation. This is not much of a limitation, basically only requiring for us that the amplitudes involved are not too great. Aside from that, we saw in Chapter 14 that *any* function of the form $F(x \pm vt)$ satisfies the wave equation. The wave equation is *linear*, i.e., the sum of two solutions is also a solution. So the sum of two pulses is another acceptable pulse.

We are interested in pulses because they are the way waves occur in the physical world. Communication of any kind—the passage of information—is by means of pulses. The sound of a drum, a flash of light, and a ripple caused by a pebble dropped into a pond are all pulses. In Section 15–6 we will see that pulses are formed from the superposition of harmonic waves, but in this section we will use the superposition of pulses to provide us with insight into wave interference.

Collisions Between Pulses

Consider a pulse moving to the right along a string (Fig. 15–16). In the figure the velocity of the transverse motion is represented by vertical arrows. There is upward transverse motion along the leading edge of the pulse. The transverse motion vanishes at the peak, and there is downward transverse motion along the trailing edge until the string again comes to rest after the pulse has passed.

Consider next the superposition of two pulses of identical shape on the string: Pulse 1 is positive and moves to the right; pulse 2 is negative and moves to the left. Figure 15–17 illustrates a sequence in which the pulses approach each other, superpose, then continue on. At the moment when their respective peaks would be at exactly the same position, $t = 0$ s, the superposition of the two pulses gives a flat string. This situation is in fact no more peculiar than that of a ball that has been thrown vertically and is momentarily at rest at the peak of its flight. Even if the ball is momentarily at rest, its acceleration remains constant. Similarly, even though the string is flat momentarily,

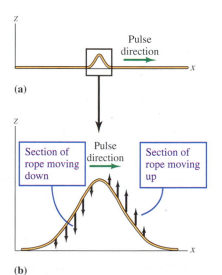

▲ **FIGURE 15–16** A pulse moving to the right (horizontal arrow) on a string. The vertical arrows indicate the velocities of elements of the string.

it may still be moving. Not only do the displacements of the pulses superpose but so do their time derivatives, which measure their transverse velocities. The velocities of the string are indicated in sequence in Fig. 15–17. When the peaks superpose, the leading edge of pulse 1 is superposed on the trailing edge of pulse 2. Both edges are moving upward, so there is a region just right of the center of the superposed peaks moving upward rapidly. Similarly, the region just left of the center of the superposed peaks, consisting of the trailing edge of pulse 1 and the leading edge of pulse 2, is moving downward rapidly. Even though the string is flat for a brief moment, it is not at rest!

Reflection

In our discussion of two pulses of opposite sign that meet and interfere, the central point between the two pulses—marked on Fig. 15–17 by a dot—never moves at all. It is clear that the colliding pulses must be of opposite sign and equal magnitude in order to leave a central point undisplaced. Recall now that there is also a single point that remains undisplaced when an incident wave or pulse reflects from the fixed end of a string (or rope). The stationary point there is the point fixed to the wall. Because only pulses of opposite sign can lead to this fixed point, we then argue that the pulse that reflects from a fixed wall is inverted. To phrase all this in another way, we think of the reflection of a pulse as the meeting of two pulses, with one pulse coming from the other side of the wall. The string and the inverted pulse or wave on the other side of the wall are purely imaginary, just an aid to picturing the form of the reflected wave. Figure 15–18 shows a sequence of a pulse reflecting from the fixed end of a string. The pulse moving away from the boundary is the *reflected* pulse.

Suppose that instead of being fixed at the wall, the end of a rope under tension is *free to move in the vertical direction*. This could be arranged by fixing the rope to a support that allows it to slide freely, as in Fig. 15–19a. The analogous sequence of reflection is shown in Fig. 15–19b. As before, the reflected pulse matches the incident pulse

▲ **FIGURE 15–17** The superposition of two pulses, one moving to the right and one to the left. The center point between the two waves, at the dot, remains motionless throughout.

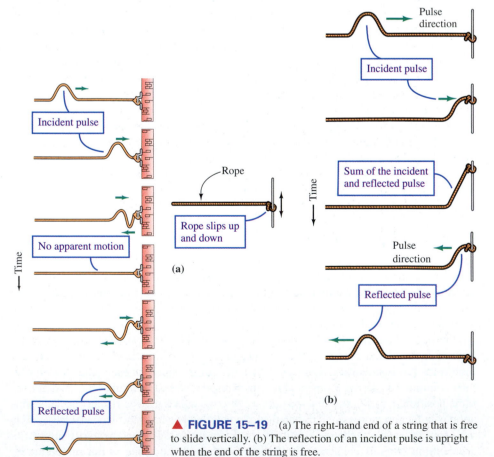

▶ **FIGURE 15–18** An incident pulse reflecting from the fixed end of a string is inverted. If the pulse is symmetric, then at some point the string is flat, because the reflected pulse cancels the incoming pulse by destructive interference.

▲ **FIGURE 15–19** (a) The right-hand end of a string that is free to slide vertically. (b) The reflection of an incident pulse is upright when the end of the string is free.

in shape. This time, however, the reflected pulse must be right-side up rather than inverted because the end of the string has been thrown upward by the incident pulse. In this case, we can also visualize an (imaginary) traveling wave on the far side of the slide support moving to the left. We see from the sketch that when the two pulses meet and combine at the slide support, the pulse height at the support is twice as high as the original amplitude. (What has happened here is that the free end gets a vertical impulse when the pulse arrives. Because there is no mass to be accelerated farther down the string—there is no string to the right of the spot—the end can move an additional amount.) Just as the fixed end of the string was an argument that the reflected wave was inverted, the large size of the displacement at the end of the string in this case is an argument that when the end of the string is not fixed, the reflected wave is right-side up. This is easily confirmed by experiment.

The conclusions we have drawn about whether reflected pulses are inverted or not can also be used to establish the form of reflected harmonic waves, as we see in the following example.

EXAMPLE 15–6 A harmonic sound wave travels in the $+x$-direction. We can describe this wave by using the quantity z that measures the longitudinal displacement of a layer of air from its equilibrium position. z will be a function of the longitudinal position x and will take the form $z_i = A_i \cos(kx - \omega t)$. (We have managed to set the clock so that there is no additional phase here.) The wave encounters a brick wall at $x = 0$, and because no air can cross the brick wall, the boundary condition $z = 0$ at $x = 0$ holds. What is the form of the reflected wave, and what is the form of the superposition of the incident and reflected waves?

Strategy First of all, we need to construct a generic form of the reflected wave. This will have a form similar to the incident wave, with the change from $(kx - \omega t)$ to $(kx + \omega t)$, with a different amplitude A_r and with an unknown phase. We next impose the condition that $z = 0$ at $x = 0$. The expectation is that this condition, which must hold *at all times*, will fix the reflected amplitude and phase. Once we have established the form of the reflected wave, we can study how it adds to the incident wave.

Working It Out The reflected wave has form $z_r = A_r \cos(kx + \omega + \phi_r)$. The boundary condition stipulates that the superposition of the two waves (the net wave) must be zero at $x = 0$: $z_i + z_r = 0$ at $x = 0$ or, for all t,

$$A_i \cos(k \times 0 - \omega t) + A_r \cos(k \times 0 + \omega t + \phi_r)$$
$$= A_i \cos(-\omega t) + A_r \cos(\omega t + \phi_r) = 0.$$

We can expand the second cosine by using the identities in Appendix IV–4—here, $\cos(\omega t + \phi) = \cos \omega t \cos \phi - \sin \omega t \sin \phi$.

Then we separate the coefficients of $\sin \omega t$ and $\cos \omega t$ that result:

$$(A_i + A_r \cos \phi_r) \cos \omega t + (-A_r \sin \phi_r) \sin \omega t = 0.$$

If this equation is to hold for all t, the coefficients of both $\sin \omega t$ and $\cos \omega t$ must be zero. For example, if the expression is to be zero at $t = 0$, where $\sin \omega t$ is zero automatically, the coefficient of $\cos \omega t$ must be zero; if the expression is to be zero at $t = \pi/2\omega$, where $\cos \omega t$ is zero automatically, the coefficient of $\sin \omega t$ must be zero. Thus

$$A_i = -A_r \cos \phi_r \quad \text{and} \quad 0 = A_r \sin \phi_r.$$

The second equation has as its most interesting solution that $\phi_r = 0$. Then the first equation will give $A_r = -A_i$.

Pulling things together, the reflected wave takes the form

$$-A_i \cos(kx + \omega t).$$

This is like the incident wave, except propagating to the left and inverted. The sum of the incident and reflected wave is therefore

$$z_{\text{total}} = z_i + z_r = A_i[\cos(kx - \omega t) - \cos(kx + \omega t)]$$
$$= 2A_i \sin kx \sin \omega t.$$

The important results to note are that (a) $z = 0$ at $x = 0$, as required, and (b) the conditions that determine the reflected amplitude and phase *do not depend on the value of* ω.

What Do You Think? How is this wave different, if at all, from the standing waves we saw in Chapter 14?

Transmission

Reflections of waves and pulses are familiar to us through sound echoes and the bouncing of water waves from boundaries. Less familiar, but equally important, is the passage of waves and pulses through boundaries that separate different media. This is the process of *transmission*. The transmission of waves occurs when sound waves penetrate from one medium to another—for example, from water to air, and when light waves pass between air and glass (Chapter 35).

In this chapter we will discuss transmission in terms of harmonic waves or pulses moving along strings. Up to this point, we have supposed that one end of a string was either tightly fixed or completely free. Imagine instead that our string is attached to a new string that is heavier or lighter than the original one (Fig. 15–20). If the new string is sufficiently heavy, then it acts like the wall; if the new string is sufficiently light, then

▲ **FIGURE 15–20** A string is attached to a heavier string (top) and to a lighter string (bottom).

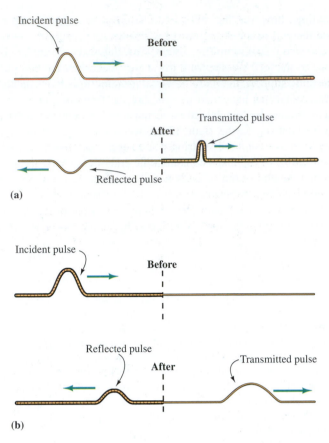

▲ **FIGURE 15–21** (a) A pulse is incident on a heavy string from a lighter one. The reflected pulse is reduced in amplitude and inverted. The transmitted pulse has an amplitude less than that of the incident pulse and moves more slowly than either the incident or reflected pulse. (b) A pulse is incident on a light string from a heavier one. The reflected pulse is reduced in amplitude and right-side up. The transmitted pulse has an amplitude less than that of the incident pulse and moves more rapidly than either the incident or reflected pulses.

it serves as a device that allows the end of the original string to move freely. The situations in Fig. 15–21 are between these extremes, and there we observe both a wave reflected back along the original string and a transmitted wave carried by the new string. We can most easily understand this by considering a pulse traveling along the first string toward the boundary.

The Amplitudes of Reflected and Transmitted Pulses: Figure 15–21a shows what happens when a pulse is incident on a heavier string and Fig. 15–21b shows what happens when a pulse is incident on a lighter string. Consider first the *reflected* pulse. In view of our discussion of the string with fixed or free ends, it is not surprising that when the new string is heavier than the original string, the reflected pulse is inverted, whereas when the new string is lighter, the reflected pulse is right-side up. The amplitude of the reflected pulse decreases to zero in the limit where the new string has a mass density equivalent to that of the original string because, in that case, there is no boundary. This behavior is both smooth and reasonable. What can we say about the transmitted pulse other than that it has the same sign as that of the incident pulse? To say more, we use energy conservation.

Consequences of the Conservation of Energy: If we have an incident pulse that is partly reflected and partly transmitted, then the sum of the energies in these two pulses equals the energy of the incident pulse. If there is no transmitted pulse, as is the case when the new string is either infinitely heavy or infinitely light, then the energy in the reflected pulse matches the energy in the incoming pulse. The transmitted and reflected waves must share the energy of the incident wave and each therefore has an amplitude reduced from that of the incident wave. We'll illustrate the quantitative consequences in Example 15–7.

The Change in Velocity with a Change in Medium: Consider again two different strings attached to each other. The tension throughout the two strings is uniform, as there is <u>no</u> horizontal acceleration. The speed, however, is not uniform. With $v = \sqrt{T/\mu}$, where μ is the mass density of the string, we see that when the new string is heavier (per unit length), the velocity of a harmonic wave is smaller; when it is lighter, the velocity is larger.

We'll state without proof that if the wave speed in the medium containing the incident wave is greater than the wave speed in the medium containing the transmitted wave, then the reflected wave is inverted but is otherwise similar in shape to the incident wave. The transmitted wave is always right-side up, but its shape will be stretched if the wave speed is higher in the second medium and compressed if the wave speed is lower in the second medium (Fig. 15–21). This information is further summarized in Table 15–1.

TABLE 15–1 • Properties of Reflected Waves and Transmitted Waves[†]		
A Wave in Medium 1 Is Incident on a Boundary That Separates Medium 1 from Medium 2		
	Reflected Wave	**Transmitted Wave**
Medium 1 is less dense than medium 2 (wave speed in 1 is higher than in 2)	Inverted	Right-side up, Compressed
Medium 1 is denser than medium 2 (wave speed in 1 is lower than in 2)	Right-side up	Right-side up, Stretched

[†]*See Fig. 15–21.*

EXAMPLE 15–7 A string with mass density μ_1 is attached at $x = 0$ to a second string, with mass density μ_2. The strings are under a common tension T. An incident harmonic traveling wave of amplitude A_i arrives on the first string. A reflected harmonic wave of amplitude A_r results on the first string, and a transmitted harmonic wave of amplitude A_t results on the second string (Fig. 15–22). The strings behave elastically so there is no mechanism for energy loss. Use the fact that energy is conserved to relate the three amplitudes.

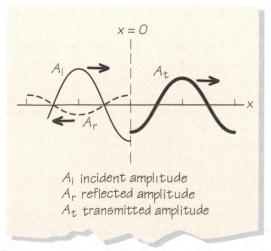

A_i incident amplitude
A_r reflected amplitude
A_t transmitted amplitude

▲ **FIGURE 15–22**

Setting It Up Figure 15–22 labels the amplitudes of the incident, reflected, and transmitted wave. The reflected wave is drawn as having changed sign, which is appropriate if $\mu_2 > \mu_1$, but nothing in what follows depends on that.

Strategy We can solve this using energy conservation alone. However, we can equally well deal with the energy carried per unit time by the three waves—that is, the power—and as we know how to

write expressions for the average power in a traveling wave, Eq. (14–37), this is an effective way to proceed. We will therefore write down expressions for the incident, reflected, and transmitted power. Generically the average power is given by $P = \mu\omega^2 A^2 v/2$, where μ is the mass density, ω the angular frequency, A the amplitude, and v the velocity of the harmonic wave. The statement of energy conservation is equivalent to the statement that the incident power is equal to the sum of the reflected and the transmitted power. We also need to use the condition that the frequencies are the same for all three waves—they are the same because they simply represent a counting of the rate at which crests reach and leave the boundary. (If the frequency changed, wave crests would be lost or would pile up somewhere.) When we apply the energy conservation condition, we find a relation for the amplitudes.

Working It Out The statement that the average incident power is equal to the sum of the average reflected and transmitted power is

$$\frac{\mu_1\omega_i^2 A_i^2 v_1}{2} = \frac{\mu_1\omega_r^2 A_r^2 v_1}{2} + \frac{\mu_2\omega_t^2 A_t^2 v_2}{2}.$$

Here we have noted that the speed of both the incident and reflected waves is v_1, whereas the speed of the transmitted wave is v_2. In particular, $v_1 = \sqrt{T/\mu_1}$ and $v_2 = \sqrt{T/\mu_2}$. The (identical) frequencies and the factors of $\sqrt{T}$ and of $\frac{1}{2}$ cancel, and we are left with the relation between the three amplitudes:

$$\sqrt{\mu_1}A_i^2 = \sqrt{\mu_1}A_r^2 + \sqrt{\mu_2}A_t^2.$$

Note that when there is no transmitted amplitude—either because there is a wall at $x = 0$ or because the end of the first string is free—then the relation is $A_i^2 = A_r^2$. Because this relation depends on the *squares* of the amplitudes, it tells us nothing about whether the reflected wave is right-side up (positive A_r) or inverted (negative A_r).

What Do You Think? What other conditions do you think would be available for the determination of the amplitudes?

▲ **FIGURE 15–23** The colors in this ultrasound image of a human fetus are the result of computer processing of the waves that reflect back to the source/receiver.

Our eyes primarily detect light reflected from the objects around us, and it is with the information in these reflected light waves that we interpret the world around us. The light arrives at Earth's surface having traveled through space from the Sun and having been transmitted into the atmosphere. Many technologies use the phenomena of reflection and transmission. For example, geophysicists use sound waves to explore Earth's interior (see the Chapter 14 "Think About This . . . How do we know what's inside Earth?"). We can empha-

size here the important role of reflection and transmission across rock boundaries. Doctors use ultrasound (sound of very high frequencies) to make interior images of the human body (Fig. 15–23), and the same techniques are used for investigating imperfections in materials. Radar and sonar devices emit electromagnetic waves or sound waves, then detect the arrival of their reflections, allowing us to learn about the reflecting object. Finally, Doppler shifts of reflected waves tell us about the speed of the reflecting object.

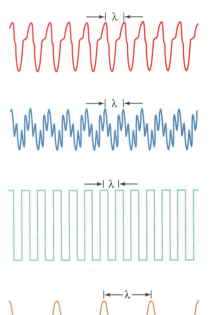

▲ **FIGURE 15–24** Some nonharmonic periodic waves.

*15–6 Fourier Decomposition of Waves

We have studied two types of waves: single pulses, which are not periodic phenomena, and harmonic (or sinusoidal) waves, which are. In another important category we might put periodic waves that are not harmonic (Fig. 15–24). The superposition principle, together with a remarkable mathematical result called *Fourier's theorem*, shows that the waves in these three categories are very closely related. The substance of the theorem is (1) *any periodic (not necessarily sinusoidal) wave can be approximated by a superposition of purely harmonic (or sinusoidal) waves with different frequencies in the form of a sum,* and (2) *any pulse can be approximated by a superposition of purely harmonic waves in the form of an integral.*

Fourier's theorem does more than just state a possibility; it also lays out the rules for finding what the harmonic components are that make up a wave and the relative weights of these components—what we call the *decomposition* of the wave. Even better, it shows how only a few harmonic components can approximate any periodic wave. Suppose that the wave we wish to fit with harmonic forms has period T or, equivalently, frequency $f = 1/T$, and angular frequency $\omega = 2\pi f$ (where f and ω are the *fundamental* frequency and *fundamental* angular frequency of the wave). For such waves, the period in time is related by properties of the medium (in particular, the wave speed) to a period in space—the wavelength. Thus, for a periodic wave, we can refer either to the periodic time dependence at a particular point or to a snapshot in time of a wave in space. Let's look at the periodic time dependence at a particular point, $f(t)$, of the wave we want to fit. The fit is known as a *Fourier expansion* and takes the general form

$$f(t) = \sum_{n=1}^{\infty} A_n \sin(n\omega t + \phi_n). \tag{15–22}$$

The only frequencies that enter into the sum are integer multiples of the fundamental frequency, $f_n = nf$. These frequencies are called harmonics of the fundamental frequency, with second harmonic referring to $n = 2$, third harmonic referring to $n = 3$, and so on.

If, instead of concentrating on the time dependence, we take a snapshot of a periodic wave $g(x)$ in space, then that wave, however complex in form, has a fundamental wavelength λ or a fundamental wave number $k = 2\pi/\lambda$. The Fourier expansion of our wave takes the form

$$g(x) = \sum_{n=1}^{\infty} A'_n \sin(nkx + \theta_n). \tag{15–23}$$

Here, the higher harmonics have wave numbers that are integer multiples of the fundamental wave number. The Fourier expansion tells exactly how to calculate the constants in Eqs. (15–22) and (15–23) with a set of rules. While we won't detail the expansion rules here, we do emphasize that these rules are not at all ambiguous.

If the infinite sum in Eq. (15–22) or (15–23) is truncated (cut off), the resulting finite sum forms an approximation to the original waveform. Such approximations are quite useful. We may want to learn about the important frequencies in a particular waveform so that we can suppress or enhance certain modes. For example, in electronic sound equipment, we may want to boost or suppress certain components of a note without changing its pitch, thus changing the quality of the sound.

Fourier Decomposition of a Triangular Wave

To demonstrate Fourier decomposition, let's try to fit the triangular wave illustrated in Fig. 15–25 with harmonic functions. (Such waves might be used in electronics applications where they are generated precisely by the superposition of harmonic waves.) The triangular wave has a period T and an angular frequency $\omega = 2\pi/T$. The rules for the constants that appear in the expansion require the phases ϕ_n to equal $\pi/2$. This means that only cosines appear in the expansion. This is reasonable because the triangular waveform is symmetric about $t = 0$, and the cosine also has this symmetry. The cosine is thus a "natural" expansion function for the triangular wave. Cosines that reach negative maxima at $t = \pm T/2$ would make it easy to build the sharp negative trough of the triangular wave at these points. This feature is shared by $\cos(\omega t)$, $\cos(3\omega t)$, $\cos(5\omega t)$, and so forth, but *not* by $\cos(2\omega t)$, $\cos(4\omega t)$, and so on. The calculation of A_n bears this out: $A_2, A_4, \ldots$ all turn out to be zero. When the expansion rules are applied, the exact Fourier expansion of our wave is

$$
z(t) = \frac{8}{\pi^2} z_0 \sum_{n=\text{odd}} \frac{1}{n^2} \cos(n\omega t)
$$

$$
= 0.81 z_0 \left[\cos(\omega t) + \frac{1}{9}\cos(3\omega t) + \frac{1}{25}\cos(5\omega t) + \cdots \right]. \quad (15\text{–}24)
$$

Figure 15–26 illustrates this Fourier expansion. Figure 15–26a shows the individual terms $z_1 = 0.81 z_0 \cos(\omega t)$ and $z_2 = (0.81/9) z_0 \cos(3\omega t)$. Figure 15–26b shows how first one term and then two terms of Eq. (15–24) match the triangular wave.

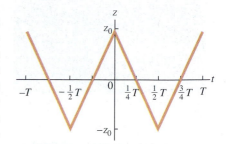

▲ **FIGURE 15–25** A triangular wave of period T. The time dependence of the transverse displacement z of a string is plotted at some fixed point in space.

(a)

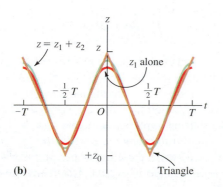

(b)

◀ **FIGURE 15–26** Fourier analysis of the triangular waveform of Fig. 15–25. (a) The first two contributing harmonics, $z_1 = 0.81\, z_0 \cos(\omega t)$ and $z_1 = (0.81/9)\, z_0 \cos(3\omega t)$. (b) The triangular waveform, the expression z_1 alone, and an approximation containing the first two terms of Eq. (15–24). The approximation with z_1 alone is already a good one, and the two-term expression is an even better one.

The $\cos(\omega t)$ term alone does reasonably well in approximating the triangular wave. By the time the second term is included, the approximation is at worst good to 10%.

The Fourier theorem is of great importance in engineering and physics because of its relevance to the solution of many different problems. In addition to the applications in vibrations and wave motion that are simply sketched out here, heat-flow problems and the solutions of a great class of problems in electricity and magnetism across the entire gamut of electrical engineering require the Fourier expansion.

Fourier Analysis and the Ear

Our ears work very well as Fourier analysis devices. When the note A is produced by an electronic "tuning fork," a pure harmonic wave of period $T = 1/(440\text{ Hz}) = 2.27 \times 10^{-3}$ s emerges. The same note played on a piano, a violin, a trombone, an oboe, or a saxophone contains a harmonic wave of the same fundamental frequency. But each instrument emits its own complicated waveform (Fig. 15–27). In the language of Fourier analysis, this means that a note played by any of those instruments contains vast numbers of differing harmonic components. The ear is capable of picking out the fundamental frequency; we can all agree when the same note is played by a piano and an electric guitar. It is the higher components by which the ear differentiates between different instruments or between different voices. Even different violins have very different higher harmonic contents, a very subtle but important matter in the construction of violins. Although the physical differences between Stradivarius violins and others of lesser value are so slight that attempts to duplicate "Strads" fail, the sounds of Strads are generally more appealing than those of ordinary violins. A Fourier analysis shows that better violins have a far greater content of higher harmonics. Our ears can even distinguish between individual violins made by the same master!

*15–7 Pulses and the Uncertainty Principle

When we treat pulses with Fourier's theorem, we discover a fundamental relation between the width of the pulse and the frequencies of the periodic waves of which it is composed. This relation is known as the **uncertainty principle**.

The Fourier Decomposition of Pulses

In Section 15–6 we used Fourier's theorem to talk about general periodic waves. At first sight, pulses would seem to have little to do with such periodic waves. But the last drawing in Fig. 15–24 suggests a way to think of pulses as periodic. Here we see a periodic wave that is a series of short, repeated pulses, with a very long period, T, between the pulses or, equivalently, separated by a very long wavelength, λ. This means that the fundamental angular frequency, ω (or the fundamental wave number, k), is very small, because $\omega = 2\pi/T$ (or $k = 2\pi/\lambda$).

A true pulse in time is like the periodic sequence shown in the limit that $T \rightarrow \infty$, or $\omega \rightarrow 0$. Similarly, a true pulse in space is like the periodic sequence shown in the limit that $\lambda \rightarrow \infty$, or $k \rightarrow 0$. In the Fourier expansion, the sum is taken over integer multiples of angular frequencies (wave numbers). Therefore, in the $\omega \rightarrow 0$ limit, *all* frequencies enter into the sum and *the sum in the Fourier expansion is converted into an integral.* More concretely, the statement of the theorem is

$$P(t) = \int_0^\infty d\omega A(\omega) \cos \omega t + \int_0^\infty d\omega B(\omega) \sin \omega t$$

for any pulse function $P(t)$. The functions A and B describe the relative amounts of different frequencies. While the integral runs formally over all values of ω, it is really $A(\omega)$ and $B(\omega)$ that determine the range of frequencies entering the integral.

Just what can we say about this range? Although the exact answer to this question depends on the shape of the pulse, we can state an important principle. *If a pulse is limited in time (or space), then it is also limited in the range of angular frequencies (or wave numbers) that enter into it.* (As we'll see below, these limits are inversely related; the narrower the range in time, the larger the range in angular frequencies and vice

(a) Trumpet

(b) Synthesized trumpet

(c) Guitar

(d) Synthesized guitar

▲ **FIGURE 15–27** The characteristic sounds of different instruments are associated with frequencies above that of the sound of the "note" being played. These higher frequencies are known as overtones, and their presence makes the waveform of the sound different than the purely harmonic sine or cosine form. A synthesizer electronically generates the characteristic pattern of overtones to trick the ear. Shown in Figs. (a) to (d) is the note of F, played by a trumpet, a synthesized trumpet, a guitar, and a synthesized guitar, respectively.

versa.) In other words, the angular frequencies that contribute are in a narrow range, and we can estimate that range by the following argument. The different frequencies that contribute to a pulse interfere constructively in the region where the pulse is strong and interfere destructively in the region where the pulse drops to zero. Suppose that a pulse lasts a time Δt; we say it has a **width** in time of Δt. Suppose also that the contributing frequencies go from a minimum angular frequency ω_{min} to a maximum ω_{max}, with $\Delta\omega = \omega_{max} - \omega_{min}$. We want these waves to be out of phase at the beginning of the pulse, to be in phase at the center of the pulse, and then to be out of phase once more at the end of the pulse. If in time span Δt the number of periods of the waves of frequency ω_{max} differs by just one from the number of periods of the waves of frequency ω_{min}, then waves that are out of phase at the beginning of the pulse will be in phase at the center and out of phase at the end. Now the number of periods of a wave of angular frequency ω contained in a time span Δt is

$$\frac{\Delta t}{T} = \Delta t \frac{\omega}{2\pi}.$$

Our condition is then $\dfrac{\Delta t}{2\pi}(\omega_{max} - \omega_{min}) \cong 1$, or $(\Delta t)(\Delta\omega) \cong 2\pi$. In fact, this approximate equality holds for some pulses, and a more detailed mathematical study shows that for others it is an inequality that is usually stated as

$$(\Delta t)(\Delta\omega) \geq 2\pi. \tag{15–25}$$

This equation is a **reciprocal relation**; if Δt is large, $\Delta\omega$ is small, and vice versa. A very sharp pulse in time contains a very broad range of harmonic components. Similarly, a pulse that is very broad in time will have only a limited range of angular frequencies.

A more commonly used reciprocal relation follows if we use the frequency $f = \omega/2\pi$ in place of the angular frequency, ω:

$$(\Delta t)(\Delta f) \geq 1. \tag{15–26}$$

RECIPROCAL RELATION FOR TIME AND FREQUENCY

It is also worthwhile to see how the reciprocal relation works when we study a pulse in space rather than time. Here, the breakdown of the pulse is in terms of a continuous range of wave numbers (or wavelengths) rather than angular frequencies. Suppose that the pulse has width Δx in space. Then the same method of estimating the range of wavelengths, $\Delta\lambda$ of the contributing harmonic waves, can be used. The condition would then be

$$\Delta x\left(\frac{1}{\lambda_{min}} - \frac{1}{\lambda_{max}}\right) \cong 1$$

or, in terms of the wave number, $k = 2\pi/\lambda$, $(\Delta x)(\Delta k) \cong 2\pi$. Again, the approximate equality is actually an inequality,

$$(\Delta x)(\Delta k) \geq 2\pi. \tag{15–27}$$

RECIPROCAL RELATION FOR SPACE AND WAVE NUMBER

Here, $\Delta k = k_{max} - k_{min}$ is the range of wave numbers that must enter into the construction of a pulse whose width in space is Δx. Again, the smaller Δx is (i.e., the narrower the pulse), the larger the range of wave numbers.

Uncertainty Relations

In the context of the microscopic world of atoms, the reciprocal relations we have discussed are called **uncertainty relations**; the fact that such relations come into play is the *uncertainty principle*. In atomic physics quantum mechanics is important, and the key idea in quantum mechanics is that all matter has wavelike characteristics. Although the words "uncertainty principle" are often used only in connection with quantum mechanics, the principle really applies to any wavelike phenomenon. The trade-off in the context of quantum mechanics involves the position and speed of microscopic particles. If you measure the position of a particle at any given time with high precision, then you can know the speed—or the momentum—only with poor precision, and vice versa. The momentum is closely related to a wave number in quantum mechanics, which we'll discuss further in Chapter 40.

THINK ABOUT THIS . . .

DOES THE RECIPROCAL RELATION HAVE PRACTICAL CONSEQUENCES?

There is a very real place where the reciprocal relation is relevant: communication by signal construction. Pure harmonic waves cannot convey information, because they have neither a beginning nor an end. Signals and information transfer are necessarily in the form of pulses. For example, all movement of information within and between computers is done with sequences of pulses in binary form. Rapid communication of information requires that the pulses carrying the information be very narrow. If not and, in particular, if pulses overlap to make wider pulses, information is lost in the confusion between the pulses. According to the reciprocal relation, this means that a very broad spectrum of frequencies must go into the construction of these narrow pulses: The narrower the pulse, the broader the range of frequencies must be. But this costs money—in a very literal sense. The energy density and power in harmonic waves of frequency ω are proportional to ω^2, as discussed in Section 14–5. The reciprocal relation forces a trade-off: A sharper pulse, which is more effective in transferring information, is more expensive.

Summary

The algebraic sum, or superposition, of two solutions of the wave equation is also a solution. The collection of physical phenomena that follow from this fact fall under the general heading of wave interference. When two waves that are superposed have opposite algebraic signs at a particular point or a particular moment of time, they interfere destructively. When they have the same sign, they interfere constructively.

Standing waves can be considered to be the result of the superposition of traveling waves moving in opposite directions. Standing waves on a string can be decomposed into the superposition of traveling waves with their own reflections from the ends.

When two waves with nearly equal frequencies, f_1 and f_2, are superposed, the result is that the original wave is modulated by a wave of much smaller frequency and larger wavelength called a beat. The beat frequency is

$$f_{\text{beat}} = \tfrac{1}{2}(f_1 - f_2). \tag{15–16}$$

Two waves of the same wavelength are coherent if there is a constant phase difference between them. If the distances that the waves travel differ by ΔL, an interference pattern that depends only on ΔL and on the phase difference is traced out. If the phase difference at the source is zero, the interference pattern is

$$\text{for constructive interference: } \Delta L = n\lambda, \; n = 0, \pm 1, \pm 2, \ldots; \tag{15–17}$$

$$\text{for destructive interference: } \Delta L = \left(n + \tfrac{1}{2}\right)\lambda, \; n = 0, \pm 1, \pm 2, \ldots. \tag{15–18}$$

Suppose that two coherent wave sources of wavelength λ and zero phase difference are separated by a distance d that is much smaller than the distance to the observation point. Then there is an interference pattern that is a function of the angle θ made with the perpendicular bisector of the line separating the wave sources:

$$\text{for maxima: } \sin\theta = n\frac{\lambda}{d}, \; n = 0, \pm 1, \pm 2, \ldots; \tag{15–20}$$

$$\text{for minima: } \sin\theta = \left(n + \frac{1}{2}\right)\frac{\lambda}{d}, \; n = 0, \pm 1, \pm 2, \ldots. \tag{15–21}$$

The pattern described by these equations broadens (the angle θ corresponding to a particular minimum or maximum increases) as the wavelengths involved increase and/or the separation between the sources decreases. This is a very general feature of interference patterns.

Waves reflect from boundaries and are transmitted past boundaries that separate one wave-carrying medium from another. A reflected wave is either inverted or right-side up, compared to the original wave, according to whether the medium from which the wave reflects is denser or less dense than the original medium.

Fourier's theorem shows that any wave, both periodic and nonperiodic (pulses), can be approximated as closely as we like by a superposition of purely harmonic waves with different frequencies. This result allows for the construction of arbitrary waveforms from harmonic waves and for the analysis of physical systems in terms of harmonic waves.

A pulse can also be treated by Fourier analysis. If a pulse has a time span of Δt, then the range of frequencies, Δf, that comprise that pulse is limited:

$$(\Delta t)(\Delta f) \geq 1. \tag{15–26}$$

There is a similar relation between the spatial extent of the pulse, Δx, and the range in wave numbers, Δk, that enter into the pulse's decomposition:

$$(\Delta x)(\Delta k) \geq 2\pi. \tag{15–27}$$

These relations hold whenever wave phenomena are present and explain why sharp information-carrying signals are expensive to generate.

Understanding the Concepts

1. A standing wave on a string can be constructed from two traveling waves that move in opposite directions. Can a traveling wave be constructed from the sum of two standing waves?

2. When transverse positive and negative pulses that have the same symmetric shape and size but travel in opposite directions meet, is it necessary that there be a moment when the string or wire on which they move is flat? If so, how do the pulses "know" to continue moving on the string?

3. Two people on each end of a long rope send off a wave pulse. If both wave pulses are on the same side of the rope, describe what happens when the pulses meet. What about when the pulses are on opposite sides? What happens when one pulse is oriented at a 90° angle to the second pulse?

4. If a higher harmonic than the fundamental one is excited on the string attached between the two walls, will it be the frequency or the velocity that changes?

5. Listening to the beat frequency between an unknown and a standard (known) tuning fork tells you only about the difference between the two frequencies. The unknown frequency could be larger or smaller than the known frequency. With a piece of chewing gum at your disposal, can you devise a method to determine the unknown frequency?

6. A wave with truly sharp edges, such as the triangular wave discussed in Section 15–7, can never exist on a string in nature. What might limit how sharp the changes can be?

7. Consider a sinusoidal traveling wave that moves down a sequence of successively lighter strings tied together end to end. The wave starts at the end containing the heaviest string. Qualitatively describe what happens to the frequency and wavelength of the wave by the time it reaches the lightest end of the system.

8. We saw what interference pattern two speakers close to one another make near the line perpendicular to and bisecting the line that connects the speakers. Will there be an interference pattern even far from this center line, where the small angle approximation that we used to derive the pattern is not valid?

9. A pebble dropped into still water produces a pulse. The speed of water waves of different wavelengths differ from one another. Why does a dropped pebble produce a series of spreading concentric waves rather than a single wave front?

10. How do you reconcile the principle of conservation of energy with the observation that, in one region, the wave medium is not in motion when waves interfere destructively?

11. In Section 15–1 we referred to two waves that move to the right and cancel; they give no wave at all. Yet each wave alone has an energy density. What happens to the energy? [Hint: What are the implications of the superposition principle for the energy density?]

12. Musicians often use beats to tune to some standard, such as an oboe. Can a trumpet player tell from the beat frequency whether one particular note is sharp (at a higher frequency) or flat (at a lower frequency) compared to the same note played on the oboe?

Problems

15–1   The Superposition Principle

1. (I) Show that the superposition of the two traveling waves $\sin(kx - \omega t)$ and $\cos(kx - \omega t)$ can be written in the form $A \sin(kx - \omega t + \phi)$, and find A and ϕ.

2. (I) Two traveling waves have the forms $z_1 = A \sin(kx - \omega t)$ and $z_2 = A \sin(kx - \omega t + \pi/2)$, respectively. Sketch the superposition of these two waves.

3. (I) Work out the result of the superposition of the two waves of Problem 2 in the form $z = z_1 + z_2 = A \sin(kx - \omega t + \delta)$.

4. (II) Consider the superposition of two harmonic waves, $\psi(x, t) = A \sin(kx - \omega t) + 3A \cos(kx - \omega t + \theta)$. (a) For what values of θ will this superposition result in the maximum constructive interference? (b) the maximum destructive interference? (c) If this superposition is written in the form $B \sin(kx - \omega t + \alpha)$, express B and α in terms of A and θ.

5. (II) Two sinusoidal waves with the same frequency travel down a long rope (Fig. 15–28, see next page). The waves have amplitudes 3 cm and 2 cm, respectively, but the second wave trails the first wave by a phase of $\pi/2$. Determine the amplitude of the resultant motion.

▲ **FIGURE 15–28** Problem 5.

6. (II) Four sinusoidal waves have the same frequency, but the first two waves have amplitudes twice that of the other two waves. The phases of the four waves are 0, $\pi/2$, π, and $3\pi/2$, respectively. Plot the resulting superposed wave, and describe its motion.

7. (II) Two sine waves, $A \sin(\omega t + kx)$ and $A \sin(\omega t + kx + \delta)$, combine to form a sine wave with the same amplitude A. For what values of δ of the second wave is this possible?

15–2 Standing Waves Through Interference

8. (I) Two waves are described by $\psi_1 = (2.5 \text{ cm}) \sin(kx - \omega t)$ and $\psi_2 = (2.5 \text{ cm}) \sin(kx + \omega t)$, where $k = 3\pi/2 \text{ cm}^{-1}$ and $\omega = 2\pi \text{ s}^{-1}$. (a) Show that the addition of the two waves produces a standing wave. (b) Determine the amplitude and frequency of the resulting standing wave.

9. (I) Consider traveling waves moving in opposite directions: one has the form $z_r(x, t) = z_0 \sin(kx - \omega t)$ and the other has the form $z_l(x, t) = z_0 \cos(kx + \omega t)$. Do these waves add to a standing wave? If so, what is the expression for the standing wave, and what are the locations of the nodes? Note that $\cos \theta$ can be written as $\sin[\theta + (\pi/2)]$.

10. (II) Show by a specific construction that two standing waves can add up to give a traveling wave. What standing wave would you have to add to $A(\cos kx)(\cos \omega t)$ to get a wave traveling to the left?

11. (II) Two traveling waves that move in opposite directions along the x-axis interfere to produce a standing wave of the form $6 \sin kx \cos \omega t$, where $k = 3\pi \text{ m}^{-1}$ and $\omega = 22\pi \text{ s}^{-1}$. (a) What are the frequencies and speeds of the traveling waves? (b) What are the forms of the traveling waves?

12. (II) One harmonic of an organ pipe has a single node at 3/4 of the length of the pipe and an antinode at the open end (see Fig. 14–26). Write an expression to describe this mode as a standing wave. Find two traveling waves that produce the same result. The length of the pipe is 0.80 m and the speed of sound is 330 m/s.

13. (II) Show that the sum of a wave that travels to the right and a wave that travels to the left with the same amplitude, wave number, and frequency is a standing wave, independent of phase. [*Hint*: Consider $\psi(x, t) = A \sin(kx - \omega t) + A \sin(kx + \omega t + \theta)$; define $u \equiv kx + \theta/2$ and $v \equiv \omega t + \theta/2$, and use the appropriate identities.]

14. (II) The most general forms of two traveling waves that move in opposite directions along the x-axis with the same amplitude, wavelength, and frequency are $z_r(x, t) = z_0 \sin(kx - \omega t + \phi_1)$ and $z_l(x, t) = z_0 \sin(kx + \omega t + \phi_2)$. What is the form of the standing wave that results from their superposition?

15. (II) A harmonic wave traveling to the right along the x-axis has its maximum amplitude at the time $t = 0$ s at the point $x = 0$ m. Superimposed on this is a wave traveling to the left with equal amplitude, wavelength, and frequency, such that the net standing wave has nodes at $x = 3$ m and at $x = 6$ m. What is the largest possible value of the wavelength?

15–3 Beats

16. (I) You have a tuning fork of unknown frequency and a tuning fork of frequency 552.0 Hz. Combined, the two produce a pulse frequency ($f_{\text{pulse}} = 2f_{\text{beat}}$) of 1.8 Hz. What are the possible values of the frequency emitted by the first tuning fork?

17. (I) You have a tuning fork of frequency 512 Hz and look for another tuning fork of smaller frequency so that there is a perceived beat frequency of 6 Hz between them. What frequency should the other tuning fork have?

18. (I) Two violinists play their (identical) A-strings, nominally 440 Hz. A beat is heard at a pulsing of 2 Hz. What is the magnitude of the fractional difference in wave speed, $(v_1 - v_2)/v_1$, for the two strings? [*Hint*: The length L of a violin string is the same from violin to violin.]

19. (I) Two strings of the same material on a pair of guitars are in tune with a frequency of 414 Hz. The string of one of them is tightened slightly, and you hear a beat frequency of 8 Hz. What is the frequency of the tightened string?

20. (I) Two speakers are wired so that they emit sound coherently. Because of a manufacturing error, the same electrical impulse that leads one to emit sound of wavelength 27.10 cm leads the other one to emit sound of wavelength 26.71 cm. What is the pulse frequency ($f_{\text{pulse}} = 2f_{\text{beat}}$)?

21. (II) Two identical piano wires have a frequency of 512 Hz when under the same tension. As a consequence of slippage, the tension of one of the wires changes slightly, so that its frequency decreases slightly. If 4 beats/s are heard when the wires both vibrate, what is the fractional change in the tension of the lower-frequency string?

15–4 Spatial Interference Phenomena

22. (I) Two speakers that emit waves with the same wavelength λ are placed at $x = L$ and $x = -L$, respectively. Assuming that the speakers emit waves with equal amplitudes A and that the phase of the wave on emission is zero at $t = 0$ in both cases, what is the form of the traveling waves that travel toward a microphone located between them at $x = 0$?

23. (I) Find the waveform picked up by the microphone in Problem 22; that is, the superposition of the two waves arriving at the microphone.

24. (I) Two speakers separated by 80 cm are fed with the same pure sinusoidal wave of 750 Hz. A microphone is placed 4.0 m away along the center line between the speakers (Fig. 15–29). It picks up an intensity maximum. It is then moved slowly out to the side. (a) How far out to the side, in meters, does the microphone have to be moved to pick up a first minimum? (b) How much farther does the microphone now have to be moved to pick up a second maximum? (c) How do the answers to parts (a) and (b) change if the speakers are now separated by just 10 cm? The approximation $\sin \theta \cong \theta$ is not valid here, although the angles of the lines from the speakers to the microphone are nearly equal.

▲ **FIGURE 15–29** Problem 24.

25. (I) Two coherent sources are 3.0 cm apart and make harmonic ripples of the same frequency (see Fig. 15–8). Consider the ripples along a straight line L_1 parallel to the line L_2 that connects the sources and 40.0 cm away from line L_2. If the distance between the central maximum at L_1 and the next maximum on L_1 is 8.0 cm, what is the wavelength of the ripples?

26. (I) Sound of frequency 600 Hz is emitted in phase from two speakers 1.4 m apart (Fig. 15–30). A microphone can move along a line parallel to the line that joins the two speakers and is 10.0 m from the base of the speakers; it measures sound intensity, which can then be plotted versus distance. The interference produced is constructive in the center. Where is the first minimum in the sound intensity curve along the track of the microphone?

▲ FIGURE 15–30 Problem 26.

27. (II) Two coherent wave sources 5 cm apart make ripples in a water tank (see Fig. 15–8). The wavelength of the waves is 3 cm. (a) Determine the angle between the perpendicular bisector of the line segment connecting the two wave sources and the direction of the first noncentral maximum. (b) What is the direction of the first maximum if the two sources operate with opposite phases? (c) How will the answers to (a) and (b) change if the wavelength is changed to 6 cm?

28. (II) Harmonic light waves are emitted by a source equidistant between two narrow vertical slits 0.18 mm apart in a screen (Fig. 15–31). The light waves detected at a screen parallel to the screen containing the slits but 35 cm away have the usual maximum at the point opposite the midpoint of the two slits, and the first minimum occurs at a point 0.6 mm from the maximum. What is the wavelength of the light?

▲ FIGURE 15–31 Problem 28.

29. (II) Coherent waves emerge from two slits separated by a distance $2a$. A screen is placed at a distance L away from the slits, and the first interference minimum is found at a distance d from the center point. Use the small-angle approximation to show that the relationship between the wavelength (λ) and the other lengths is given by $\lambda = 4ad/L$.

30. (II) Two tuning forks vibrating in phase with frequency 440 Hz are placed 85 cm apart. A detector is moved in a circle of radius 540 cm around the center of the tuning forks (Fig. 15–32). Plot the approximate variation of the square of the net amplitude as a function of angle. Neglect the small variation in the magnitude of the amplitude that results from the sources not being at the same distance from the detector. Take the velocity of sound to be 330 m/s.

▲ FIGURE 15–32 Problem 30.

31. (II) An oscillator, located at $(x, y) = (-2 \text{ cm}, 0 \text{ cm})$, that dips in and out of water in a tank produces a water wave of the form $\psi_1 = (3 \text{ cm}) \sin(\pi r_1 - \pi v t)$, where r_1 is the distance from the oscillator; a second oscillator, located at $(x, y) = (+2 \text{ cm}, 0 \text{ cm})$, produces a water wave of the form $\psi_2 = (3 \text{ cm}) \sin(\pi r_2 - \pi v t)$, where r_2 is the distance from the second oscillator. (a) Where on the line between the sources (the x-axis) are the interference maxima located? (b) What is the form of the wave along the line $x = 0 \text{ cm}$? (Take all times in seconds.)

32. (II) Two tuning forks that vibrate coherently (in phase) at 512 Hz are placed 0.35 m apart on a horizontal surface. A short circular wall of radius 8 m, centered on the midpoint between the sources, is placed around them. Describe the interference pattern of maxima and minima at the wall.

33. (II) Suppose that two in-phase sources of surface waves in water generate water waves that spread out in circles with equal amplitudes. (a) Show that the intensity of the disturbance or the energy generated in the water is, at a maximum, four times the intensity for each wave taken separately. (b) Similarly, show that there is no intensity at a minimum. (c) Show that the intensity at any given point, if there were no wave interference, is twice the intensity from each wave. (d) Do the different answers to parts (a), (b), and (c) mean that the conservation of energy principle must be abandoned for waves? Explain your answer.

15–5 Pulses

34. (I) A student pulls a rope taut and shakes it once, sending a wave pulse down the rope. The pulse is described by a *Gaussian function*,

$$z(x, t) = z_0 \exp\left[-\frac{(x - (1 \text{ m/s})t)^2}{(0.1 \text{ m})^2}\right].$$

Where is the center of the pulse at time t?

35. (I) At $t = 0$, a sawtoothed pulse has the shape $z = 0$ for $x < -a$ and for $x > 0$; $z = k(x + a)$ for $-a < x < 0$. Sketch this pulse. Suppose that the pulse moves at speed v in the $+x$-direction. What is the algebraic expression for the moving pulse?

36. (II) Consider a train of pulses of the type described in Problem 35. Here, at $t = 0$, the displacement is $z = 0$ for $x < -a + nb$ and for $x > nb$; $z = k(x + a - nb)$ for $-a + nb < x < nb$; n is any integer—positive, negative, or zero—and $b > a$. (a) Sketch this wave train of pulses. (b) Suppose that the entire train of pulses moves at speed v in the $+x$-direction. What is the algebraic expression for the train? (c) What are the period and wavelength of this nonharmonic wave?

37. (II) Consider a transverse wave that moves to the right on a string, in the form of a *Gaussian function*, namely

$$z(x, t) = z_0 e^{[-(x-vt)^2/\alpha^2]}.$$

By taking a time derivative, compute the transverse velocity of this pulse, and sketch it as a function of $(x - vt)/\alpha$. Do the same for the acceleration. Note that any time derivative is automatically a function of $x - vt$ and is itself therefore a quantity that forms a traveling wave.

38. (II) By using the results of Problem 37, *estimate* the energy in a Gaussian pulse that moves to the right. What would the total energy in a harmonic wave be? One of the reasons that real waves have beginnings and ends and are therefore, strictly speaking, pulses, is that pulses involve only a finite amount of energy.

39. (II) Two transverse pulses travel along a string at speed v in opposite directions. The pulse that moves to the right approximates a square with sides of length ℓ projecting above the equilibrium level of the string. The pulse that moves to the left approximates a square with sides of length $\ell/2$ projecting below the equilibrium level. Assume that the superposition principle holds, and describe the approximate shape of the string as the pulses collide.

40. (I) A pulse shaped like an upright two-dimensional pyramid, or an isosceles triangle, travels to the left along a string under tension (Fig. 15–33). The left-hand end of the string is free to move up and down. Sketch a sequence of drawings that show how the pulse reflects from the free end.

▲ **FIGURE 15–33** Problem 40.

41. (I) A string with linear mass density μ_1 is attached to another string, with linear mass density μ_2. A wave with amplitude A that travels along the μ_1 string will at the attachment point give rise to a reflected wave with amplitude B and a transmitted wave with amplitude C, where

$$B = \frac{\sqrt{\mu_2} - \sqrt{\mu_1}}{\sqrt{\mu_2} + \sqrt{\mu_1}} A, \quad C = \frac{2\sqrt{\mu_1}}{\sqrt{\mu_2} + \sqrt{\mu_1}} A.$$

For $\mu_1 = 5.0 \times 10^{-2}$ kg/m and $\mu_2 = 3.8 \times 10^{-2}$ kg/m, what are B and C in terms of the amplitude of the incoming wave?

42. (I) Use the formulas in Problem 41 to show that the average power of the reflected wave plus the average power of the transmitted wave add to the average power of the incident wave.

43. (I) A string with mass density μ_1 is attached to a string with mass density μ_2. The amplitude of a reflected pulse is 85 percent of the amplitude of the incident pulse and is inverted. What is the ratio of μ_1 to μ_2? (See Problem 41.)

44. (II) When a symmetric pulse is incident on the fixed end of a rope, the reflected and incident pulse interfere, and there is a moment when the rope is completely flat. Is this a violation of the conservation of energy? (The energy of waves is discussed in Chapter 14.) Where is the energy at the moment the rope is flat?

45. (III) A string with mass density μ_1 is attached at the point $x = 0$ to a string with mass density μ_2. A tension T is maintained in both strings. A transverse wave of the form $A \cos(kx - \omega t)$, moving to the right, is generated far to the left in the first string. Use the fact that the string is unbroken at $x = 0$ and that no sharp kinks can develop there to obtain the forms of the reflected and transmitted waves. [*Hint*: The stated boundary conditions are sufficient to determine all the parameters of the reflected and transmitted waves. These conditions are (a) that the sum of the incident and reflected waves at $x = 0$ on string 1 equals the transmitted wave at $x = 0$ (unbroken string), and (b) that the derivative with respect to x of the sum of the incident and reflected waves at $x = 0$ on string 1 equals the derivative with respect to x of the transmitted wave at $x = 0$ (no kinks).]

*15–6 Fourier Decomposition of Waves

46. (III) Figure 15–34 shows a square wave $f(t)$ with period $T = 2\pi$ (or angular frequency $\omega = 1$) and amplitude $= 1$ unit. The wave's Fourier expansion is

$$f(t) = \frac{4}{\pi} \left[\sin(\omega t) + \frac{1}{3} \sin(3\omega t) + \frac{1}{5} \sin(5\omega t) + \cdots \right].$$

Plot an approximation to the square wave that corresponds to the first, the first two, and the first three terms in this series. (It helps in this to have access to computer software or to a calculator that can plot graphs.)

▲ **FIGURE 15–34** Problem 46.

*15–7 Pulses and the Uncertainty Principle

47. (I) A synthesizer, which is an electronic device for generating and combining harmonic waves, generates an electromagnetic pulse of width 10^{-6} s. Use Eq. (15–26) to find the range of frequencies that the synthesizer must utilize to generate the pulse.

48. (II) Waves are usually not infinite but come in wave pulses caused by the turning on and turning off of the source. Suppose a laser with wavelength 550 nm is switched on for one second. What is the average frequency and the width of the range of frequencies in this pulse?

General Problems

49. (I) Two harmonic waves are given by $y_1 = A \cos(kx - \omega t)$ and $y_2 = A \sin[kx - \omega t + (\pi/3)]$, where $k = 5\pi$ m^{-1}, $\omega = 800\pi$ s^{-1}, and $A = 4.0$ cm. (a) What is the frequency of each wave? What is the wavelength of each wave? What is the speed of each wave? (b) What is the amplitude of the wave that is the superposition, $y_1 + y_2$, of these two waves?

50. (I) Two rectangular pulses 4 cm long and of amplitude 3 cm but opposite sign are traveling down a string from opposite directions. At $t = 0$ s, the pulses approach each other with their centers 16 cm apart. The wave speed on the string is 1 cm/s. (a) Describe and sketch the shape of the string at $t = 0, 2, 5, 8, 10$, and 15 s. (b) What has happened at $t = 8$ s? What is the energy at this time?

51. (I) Two violinists play their (identical) A-strings, nominally 440 Hz. A beat is heard at a pulsing of 2 Hz. What is the fractional difference in tensions $(T_1 - T_2)/T_1$ of the two strings? [*Hint*: The length of a violin string is the same from violin to violin; when a violinist tunes his or her instrument, the string tension is changed.]

52. (II) The police are busy on the day that you hear two police cars sounding their one-note sirens as you wait at a light (Fig. 15–35). The police cars are moving at 67.1 mi/h (30.0 m/s) and their drivers hear their own sirens emitting at a frequency of 650 Hz. One police car passes you while the other is still approaching. Do you hear a beat? If so what is it?

▲ **FIGURE 15–35** Problem 52.

53. (II) Consider two sources of sound that both emit waves with a frequency of 1800 Hz. These sources are placed on a turntable of radius 50 cm that rotates with an angular speed of 120 rev/min. Calculate the pulse frequency ($f_{\text{pulse}} = 2f_{\text{beat}}$) that results from the opposite Doppler shifts of the two sources, heard at a large distance from the turntable as a function of time. Ignore the fact that the sources are not quite equidistant from the detector of the sound. The speed of sound in air is 330 m/s.

54. (II) A guitar string has a fundamental frequency of 240 Hz. Another apparently identical guitar string is stretched beside the first one. (a) During successive stretching, each time tightening the second guitar string, the beat frequencies are 10, 5, and 1 Hz, consecutively. During the tightening, the beat frequency never passes through zero. What are the fundamental frequencies of the second guitar string for each case? (b) By what percentage should the tension be increased for the final adjustment to make the guitar strings have identical frequencies?

55. (II) A loudspeaker (source) that emits sound waves uniformly in all directions is 10 m from a microphone (receiver). Both the source and the receiver are 20 m from a smooth, rigid wall that reflects the sound waves from the source (Fig. 15–36). As the frequency of the waves from the source is smoothly increased from zero, there is a

Source

Receiver

▲ **FIGURE 15–36** Problem 55.

first interference maximum detected by the receiver and subsequently a series of further maxima. List the frequencies of the three maxima that occur above 80 Hz. Assume that sound reflects from the walls as would a ball in an elastic collision.

56. (II) Consider two wave sources on a water surface. They generate identical waves of $\lambda = 1$ cm. What possible values of the separation of the sources, d, can you take to ensure that there is no disturbance to the side of the sources (that is, on the extension of the line joining the sources)? When this happens, is there a disturbance on that same line but *between* the sources? If so, describe it; if not, explain why not. Neglect the change in amplitude as a function of distance from the source; that is, treat this as a one-dimensional problem.

57. (II) The speed of the propagation of light in different media depends on the properties of the media. The formula is given by $v = c/n$, where $c = 3.0 \times 10^8$ m/s and n is the *index of refraction* ($n > 1$ for all media). Consider light of frequency 0.50×10^{15} Hz entering from empty space, with $n = 1.00$, into glass, with $n = 1.52$. What is the wavelength of the light in empty space? What is the wavelength of the light in glass?

58. (II) Waves are generated in phase in a water tank. The waves move along the x- and y-axes, respectively, with equal amplitude, so that the total wave is given by $\psi(x, y, t) = \sin(kx - \omega t) + \sin(ky - \omega t)$. Show that the pattern of interference maxima is given by the line $x - y = 2n\pi/k$ (where $n = 0, \pm 1, \pm 2 \dots$).

59. (II) Which of the following equations are linear? Explain your reasoning:

(a) $\dfrac{\partial^2 \psi}{\partial t^2} = v^2 \dfrac{\partial^2 \psi}{\partial x^2}$

(b) $\dfrac{\partial \psi}{\partial t} = -K \dfrac{\partial^2 \psi}{\partial x^2}$

(c) $\dfrac{\partial^2 \psi}{\partial t^2} = a\psi^2 + b \dfrac{\partial \psi}{\partial x}$

(d) $\dfrac{\partial \psi}{\partial t} = 2\psi \dfrac{\partial \psi}{\partial x} + 3\psi$

(e) $\dfrac{\partial^2 \psi}{\partial t^2} = -k \ln \psi$

60. (II) Two sources emit sound coherently and in phase at a frequency of 1250 Hz with the same intensity. They are 55.0 cm apart on a horizontal surface. A circular wall of radius 220 cm (centered on the midpoint between the sources) is placed around them. Describe systematically the interference pattern of maxima and minima at the wall.

61. (III) Two waves of the form $z_1(x, t) = z_0 \sin(k_1 x - w_1 t)$ and $z_2(x, t) = z_0 \sin(k_2 x - w_2 t)$, respectively, propagate at speed v in a nondispersive medium. The frequency difference $\Delta f = f_1 - f_2$ is small. What function describes the envelope of the resulting superposition of the two waves? Express this envelope as a wave, and give its frequency and wavelength.

62. (III) Sinusoidal waves are generated on an infinitely long rope. One wave, $y_1(x, t)$, moves to the left and has amplitude y_0, wave number k_1, and angular frequency ω_1; the other, $y_2(x, t)$, moves to the right with amplitude y_0, wave number k_2, and angular frequency ω_2. Each wave has the same phase. (a) Express $y_i(x, t)$ for each of the waves. (b) Assuming that the rope is a nondispersive medium, what is ω_2 in terms of k_1, k_2, and ω_1? In parts (c) and (d), assume that $k_2 = k_1 + \delta k$ and $\omega_2 = \omega_1 + \delta\omega$. Take the speed of traveling waves on the rope to be v. (c) What is $\delta\omega$ in terms of δk and v? (d) Superpose these waves, assuming that δk and $\delta\omega$ are small. Use the notation $k_{\text{av}} = (k_1 + k_2)/2$ and $\omega_{\text{av}} = (\omega_1 + \omega_2)/2$. Show that the result of the superposition is two waves, one moving at the low speed $v_l = v\,\delta k/2k_{\text{av}}$ and wavelength $2\pi/k_{\text{av}}$, and one moving at the high speed $v_r = 2vk_{\text{av}}/\delta k$ and the long wavelength $4\pi/\delta k$.

▶ The remote controlled submarine robot, *Jason Junior*, is peeking through a window of the *Titanic* in 1986 during the investigation of the sunken ship. *Jason Junior* was attached to the three-man submersible *Alvin* (which was just outside the *Titanic* at a depth of 4000 m) by a 200 foot cable and eventually entered *Titanic* and explored extensively.

Properties of Fluids

W aves on the sea, smoke rings, and a whirlpool of water flowing down a drain are demonstrations of the wide variety of phenomena visible in fluid flow. Although fluids can support waves, as we have seen, a little observation suggests that the response of fluids to forces goes well beyond wave motion. The physical laws involved are those we have already discussed, but applied in a new, more complex setting. Our job here will be to develop descriptive and dynamical tools for handling both static fluids and fluids in motion. Conservation laws for momentum, energy, and mass will be very helpful in this task. With these tools and a series of approximations that turn out to be surprisingly useful, we can arrive at a description of the behavior of fluids that allows us to understand many interesting properties.

16–1 States of Matter

Let us start with a review of the states of matter. The particular form that matter takes, whether solid, gaseous, or liquid, is determined ultimately by the forces between its molecules, as well as conditions such as temperature (Fig. 16–1). The typical force between a pair of molecules as a function of their separation will look something like Fig. 16–2. Generally these forces are strongly repulsive at very small separations. The force becomes attractive at a separation of about 10^{-10} m and then tapers off virtually to zero at a distance of a few times 10^{-10} m. How does this information allow us to understand the three states of matter?

◀ **FIGURE 16–1** The three states of matter, which for water are ice, liquid water, and water vapor, are very different on the macroscopic scale. These differences are reflections of differences on the microscopic scale. In solids, the molecules are arranged in ordered arrays called crystals. In liquids, the spacing of the molecules is not very different than in solids, but the order of the lattice is lost. The molecules are close enough together in each case that intermolecular forces are important. In gases the molecules are so widely spaced that their interactions come into play only very briefly in collisions.

Gases: If molecules are, on average, sufficiently far apart, then the forces between them are small. In effect, their only interaction occurs when the molecules occasionally collide with each other. The molecules behave as a **gas**. We will study gases further in Chapters 17 through 20 and only note here that gases are easily compressed and that the shape of a body of gas is easily deformed.

Solids: If molecules are sufficiently close together that the forces due to neighboring molecules are significant, matter behaves in a way that depends on the detailed properties of the intermolecular forces. In some cases the forces act to organize the molecules into a **lattice**, which is a regularly repeating three-dimensional structure in which the molecules or atoms can move only by small displacements (Fig. 16–3). This is a typical **solid**: a substance whose structure resists forces that would deform its shape. We described the lattice structure as well as some properties of solids in Chapter 11.

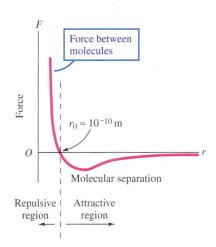

▲ **FIGURE 16–2** Qualitative behavior of the force between a pair of molecules as a function of their separation, r. Near $r = r_0$, the point at which the force is zero, the force is springlike (restoring). For $r < r_0$, the force is strongly repulsive; for $r > r_0$, it is attractive.

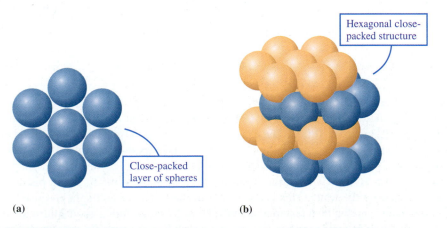

▲ **FIGURE 16–3** A lattice structure characteristic of solids. Here we show a hexagonal close-packed structure in which (a) one layer is fit together and (b) the layer above is nested into the depressions in the first layer. The forces between the molecules are springlike for small displacements from the equilibrium positions.

Liquids: In liquids the molecules are closely spaced and the intermolecular forces between the molecules at their typical spacing are strong, but for a variety of reasons they do not lead to an organized lattice structure. For such aggregates of molecules, there is no resistance to deformation. Unlike the molecules of gases, the molecules of a liquid are close enough to each other so that the repulsive intermolecular forces are responsible for an aggregate that resists compression. As a result, liquids have a definite volume but an easily changeable shape. A major difference between gases and liquids is in their compressibility.

The distinction among the three forms of matter is not absolute. If the forces are large enough, even rock will flow, and gases also share many properties with liquids. A collection of molecules can form a solid, a liquid, or a gas, depending on external circumstances. Still other states are possible in the right circumstances. For example, in one type of *liquid crystal* rod-shaped molecules line up in such a way that seen from end on they are ordered like a crystalline solid, whereas seen from the side there is no such ordering. This is a state of matter that in some ways looks like a solid and in others looks like a liquid. We say that matter appears in different **phases** or states; the particular phase depends on external conditions. Water is solid at low temperatures, but at atmospheric pressure it can become a liquid at 0° C and a gas at 100° C. At very high temperatures, such as those found within the Sun, the molecules themselves may dissociate into negatively charged electrons and positively charged ions. That kind of matter has very different properties because electrical forces between the constituents play an important role. Matter that is electrically dissociated into positive ions and negatively charged electrons is called a *plasma*. This state of matter is one that we don't directly deal with very often, and we won't discuss it further.

We shall define a **fluid** to be a substance that is not resistant to the action of shear forces (see Section 11–4). Essentially a fluid undergoes deformation without elastic forces to bring it back to its original shape. This description involves how a material responds to external forces and applies to both liquids and gases. Therefore, except for phenomena in which compressibility is important, our description of the behavior of fluids *applies to gases as well as to liquids.*

▶ 16–2 Density and Pressure

We start our study of fluids with **hydrostatics**—the study of fluids at rest. Density and pressure are the two important variables for such fluids, and in this section we'll look at them in some detail.

Density

A quantity of matter has a mass M and a volume V. The average **mass density** ρ of the matter is

$$\rho = \frac{M}{V}. \tag{16–1}$$

Table 16–1 gives densities of some typical substances. Densities vary with temperature; the density of water at sea level is $0.9998 \times 10^3 \, \text{kg/m}^3$ at 0°C, but it decreases to $0.9584 \times 10^3 \, \text{kg/m}^3$ at 100°C. The densities of liquids can also change under a change of the pressure on them ("compression" if the effect is to try to decrease the volume), although the effect is often small. The fractional density change of water in a cylinder is only about 1.5×10^{-8} when a force per unit area of $3000 \, \text{N/m}^2$ acts on a piston that acts to compress the water. This is equivalent to the force per unit area produced by a 75-kg person standing on a piston of area $0.25 \, \text{m}^2$. It is therefore a good approximation to treat water as **incompressible**. Finally, we can define the **specific gravity** of a substance as the ratio of the mass density of that substance to the mass density of water; thus, for example, the specific gravity of glycerin is 1.26.

TABLE 16–1 • Mass Densities

Material	Phase	Density (kg/m³)	Material	Phase	Density (kg/m³)
Hydrogen (0°C, 1 atm)	Gas	0.090	Liquid mercury	Liquid	13.6×10^3
Air at sea level	Gas	1.29	Sun's core	Plasma	1.6×10^5
Chlorine (0°C, 1 atm)	Gas	3.21	Cork	Solid	0.25×10^3
Ethyl alcohol	Liquid	0.79×10^3	Applewood	Solid	0.745×10^3
A good olive oil	Liquid	0.92×10^3	Ice	Solid	0.917×10^3
Water	Liquid	1.00×10^3	Quartz	Solid	2.65×10^3
Glycerin	Liquid	1.26×10^3	Steel alloys	Solid	7.6×10^3–8.9×10^3
Earth's core	Liquid	9.5×10^3	Iridium	Solid	22.4×10^3

EXAMPLE 16–1 A neutron star is a star that is much smaller than our Sun and has the density of an atomic nucleus. A typical neutron star has a radius of 10 km, but a mass of 2×10^{30} kg (the mass of the Sun). How much would a 1 cm³ volume of such a star weigh under the influence of Earth's gravity?

Strategy We use the given radius R to find the volume V, which together with the given mass M allows us to find the density. For the weight of 1-cm³ of a neutron star on Earth's surface, we use the density we have calculated to find the mass of this volume and hence the weight.

Working It Out The density is

$$\rho = \frac{M}{V} = \frac{M}{4\pi R^3/3} = \frac{2 \times 10^{30} \text{ kg}}{4\pi(10^4 \text{ m})^3/3} = 0.5 \times 10^{18} \text{ kg/m}^3$$
$$= 0.5 \times 10^{12} \text{ kg/cm}^3.$$

Thus a volume $V_{cube} = 1 \text{ cm}^3$ (10^{-6} m^3) of the material of a neutron star has a mass $m = \rho V_{cube} = 0.5 \times 10^{12}$ kg, and it weighs $mg = (0.5 \times 10^{12} \text{ kg})g = (0.5 \times 10^{12} \text{ kg})(9.8 \text{ m/s}^2) = 0.5 \times 10^{13}$ N. For comparison, 1 cm³ of water on Earth's surface has a mass of 1 g and weighs 0.01 N. A neutron star is incredibly dense!

What Do You Think? If the radius of a solid sphere doubles, and its mass also doubles, then is the density (a) less than (b) unchanged from (c) greater than the original density? *Answers to What Do You Think?* questions are given in the back of the book.

Pressure

Pressure is a measure of the force acting on the surface of an object within a fluid, even a fluid at rest. Swimmers know from experience that liquids do exert such forces. If you dive into deep water, your eardrum may hurt or even rupture due to the pressure of the water on it. Why then doesn't a sheet of glass placed under water shatter? It is because there is as much pressure on one side of it as on the other, and indeed it is a characteristic of the pressure in a static fluid that it acts upon every surface, perpendicular to the surface.

As we see more precisely below, the pressure is defined as the force per unit area. We can in principle determine the force exerted by a fluid by using the spring-balance system shown in Fig. 16–4. The plate on the spring balance is in contact with the fluid, while the spring attached to the other side of the plate is inside an enclosure where there is a vacuum, so that there is no force on the plate from the inside due to any pressure. Only the spring force resists the pressure from the fluid, and the force from the fluid is therefore measured by observing the compression of the spring. Our device would allow us to define the average pressure p on any surface of area A as the *force per unit area* acting perpendicular to that surface:

$$p \equiv \frac{F}{A}. \tag{16–2}$$

The device described above is not a very realistic one. The **manometer** shown in Fig. 16–5 is a realistic way to do the measurement. As we shall see later in this chapter, the difference in the mercury levels in the two sides of the U-shape is sensitive to the pressure in the fluid compared to the pressure in the atmosphere. With such an instrument one could see how pressure varies with position, and since it does vary, a more precise local definition of the pressure is in order. We consider a small area ΔA located at point

▲ **FIGURE 16–4** A device that would allow the measurement of pressure in a liquid. The fit between the piston and cylinder walls allows the piston to move but is tight enough so that a near vacuum can be maintained on the side with the spring. The compression of the spring provides a measurement of the force, from which the pressure is determined.

Open end

▶ **FIGURE 16–5** A more realistic device to measure pressure. The end of the tube is covered by a flexible membrane that is pushed inward if the pressure in the fluid increases. This pushes the column of mercury higher on the right side of the U-shaped tube than on the left side.

Flexible membrane on end

Mercury

$\vec{r}$, and take the limit $\Delta A \rightarrow 0$. If we denote the force perpendicular to this area as ΔF, we find

$$p = \lim_{\Delta A \to 0} \frac{\Delta F}{\Delta A} = \frac{dF}{dA}. \tag{16–3}$$

PRESSURE

This is the pressure at point $\vec{r}$. Pressure is measured in SI units as *pascals* (Pa), named after Blaise Pascal, who studied pressure in the mid-seventeenth century:

$$1 \, \text{Pa} = 1 \, \text{N/m}^2. \tag{16–4}$$

Other common units for pressure are pounds per square inch (lb/in^2): $1 \, \text{lb/in}^2 = 6.9 \times 10^3$ Pa; the atmosphere (atm), which is the atmospheric pressure at sea level: $1 \, \text{atm} = 1.01 \times 10^5$ Pa $= 14.7 \, \text{lb/in}^2$; the bar: $1 \, \text{bar} = 10^5$ Pa; and the torr: $1 \, \text{torr} = 133.32$ Pa.

If you took your manometer and moved the membrane about, you would discover two important facts about pressure in static fluids: First, pressure varies only with depth, increasing as one goes deeper. As we shall explain in the next section, this is a consequence of the presence of gravity. Second, you would learn that *when a fluid is at rest, the pressure at any given point must be the same in all directions.* The microscopic origin of pressure is the bombardment of the surface on which the pressure acts by the individual molecules of the substance (Chapter 19). The pressure is the same in all directions because there is no preferred direction for the random motion of molecules. For fluids at rest, the pressure is then simply a positive scalar quantity. Note also that the pressure at a static interface between two fluids, say a layer of oil on water, is the same on both sides of the interface; if it were not, there would be a net force at the interface and a resulting motion of the interface.

16–3 Pressure in a Fluid at Rest

The gravitational force causes the pressure of fluids at rest to be larger at ocean depths than at Earth's surface. In order to establish exactly how pressure varies with depth for an incompressible fluid, consider a liquid of fixed density ρ at rest in a container. Imagine isolating a very thin cylinder of that liquid (Fig. 16–6). (The top, base, and walls of the cylinder are purely imaginary; their sole purpose is to help us visualize the forces acting on an element of the fluid.) The cylinder's top, of area A, is horizontal and lies a distance y below the surface of the liquid. The bottom of the cylinder is at a depth $y + \Delta y$. Note that we measure the depth y downward from the surface, so it increases as we go down. We denote the pressure at the top of our cylinder by p and the pressure at the bottom by $p + \Delta p$. We want to find an expression for Δp.

The downward force on the cylinder, which is applied at the top surface by pressure from the surrounding fluid, is pA; the upward force, which is applied at the bottom surface by the pressure there, is $(p + \Delta p)A$. Additionally, there is a downward force on the cylinder due to gravity; namely, the weight of the liquid inside the cylinder. That

Pressure above and below imaginary cylinder is different

Area A

p

$p + \Delta p$

y

O

y

$y + \Delta y$

▲ **FIGURE 16–6** An imaginary thin cylinder of fluid is isolated to show the forces that act to keep it in equilibrium.

force has a magnitude of $(\Delta m)g$, and the mass of the cylinder Δm is the density times the volume of the cylinder. Thus, the downward force of gravity is $\rho(A\Delta y)g$. All together, we find

$$F_{up} = (p + \Delta p)A,$$
$$F_{down} = pA + (\Delta m)g = pA + \rho(A\Delta y)g.$$

Because the cylinder does not accelerate, the net force must cancel:

$$F_{net} = F_{up} - F_{down} = (p + \Delta p)A - pA - \rho(A\Delta y)g = 0.$$

We cancel the pA terms, divide by A, and rearrange terms to find that

$$\frac{\Delta p}{\Delta y} = \rho g.$$

In the limit $\Delta y \rightarrow 0$, this result becomes

$$\frac{dp}{dy} = \rho g. \qquad (16\text{–}5)$$

According to Eq. (16–5), the pressure change is positive for positive Δy. Pressure therefore increases with increasing depth (Fig. 16–7). In physical terms this phenomenon is a measure of the increased amount, and hence increased weight, of the liquid above any given point as the depth of the point increases. There is a greater weight of water above a diver at an ocean depth of 50 m than above a diver at a depth of 10 m.

Equation (16–5) looks just like the formula for the constant acceleration g of an object acted on by the constant force of gravity, $dv/dt = g$. When we studied projectile motion in Chapters 2 and 3, we saw that this equation has a solution $v = v_0 + gt$. If the fluid is incompressible (as water is, but gases are generally not), then the density does not depend on the depth y, and we can directly compare to write the solution to Eq. (16–5), which gives the pressure as a function of depth:

$$p = p_0 + \rho g y. \qquad (16\text{–}6)$$

Alternatively, we can integrate Eq. (16–5) to find this result. The constant p_0 is the value of the pressure at the surface, where $y = 0$; for a body of fluid such as the ocean, this is just atmospheric pressure. The pressure increase with depth y is equal to the weight per unit area of the overlying fluid. The total mass of the fluid above the cylinder is $\rho V = \rho A y$, so its weight is $\rho A y g$, and thus the weight per unit area is $\rho y g$.

The variation of atmospheric pressure with depth (from the top of the atmosphere) is not at all like that of water, Eq. (16–6). The difference is that gases are compressible, and this changes the form of the equation obeyed by the pressure, and with it its variation with depth. The atmospheric pressure will be treated in Chapter 19.

Diver blows to equalize pressure

▲ **FIGURE 16–7** Divers experience the increased pressure of increased water depth. To equalize the pressures of the surrounding water and the sinus cavities, a diver blows air into his or her sinuses.

CONCEPTUAL EXAMPLE 16–2 Look at Fig. 16–8: the level of the liquid in all the connected vessels is the same. Why?

Answer The derivation that led us to Eq. (16–6) is completely independent of the horizontal location of the imaginary cylinder in the water tank of Fig. 16–6. Equation (16–6) itself shows that the pressure in a fluid depends *only* on the depth—not the horizontal position. In our series of vessels the pressures at the fluid surfaces are all equal to atmospheric pressure (the surface is in contact with the atmosphere and is therefore at the same pressure). According to Eq. (16–6), a given pressure corresponds to a given height, and so the level of the top of the fluid will be the same for each vessel.

◀ **FIGURE 16–8** The surface pressure is equal in each individual vessel.

We can see from Eq. (16–6) that if p_0 is changed as a result of some external effect, p is changed by exactly the same amount. In other words, *a change in pressure applied to any point in a fluid at rest results in the same pressure change at every point in the fluid.* This statement is commonly known as **Pascal's principle**.

EXAMPLE 16–3 Many automobile repair shops have a hydraulic jack (Fig. 16–9), which is a device used to lift cars. This device has two connecting vertical cylindrical columns filled with fluid, a wider one that sits beneath the car to be lifted and a narrower one that is pushed down on by an operator. In this problem the wider column has a radius of 20.0 cm and lifts a car of mass 1000 kg. The narrower column has a radius of 2.5 cm, and the operator pushes down on a piston in it by applying air pressure from a pump. (a) What pump pressure is needed to lift the car? (b) If the car is lifted up a distance of 2.00 m, what downward distance must the piston in the narrower cylinder move?

▲ **FIGURE 16–9** A schematic view of the hydraulic lift used in garages.

Setting It Up Figure 16–9 labels two of the variables with which we are going to deal, both given, the areas A_1 and A_2 of the wider and narrower columns, respectively. We denote as d_1 the given vertical distance that the car, mass m, moves. We are going to find the pump pressure p that will allow the car to be raised and the distance d_2 that the piston in the narrow cylinder must move.

Strategy (a) We can lift the car with an upward force of magnitude mg; this has to be supplied by the fluid pressure p, and since this pressure acts over an area A_1, the upward force is pA_1. That will determine what p has to be. (This pressure is in addition to the pressure

that is already present and allows the car to remain stationary.) By Pascal's principle this additional pressure p can be applied at the narrower cylinder—at the pump—and it will be transmitted throughout the fluid. (b) There are various ways to approach this part. In one method we note that the car has acquired a potential energy mgd_1 in being raised. According to the work–energy theorem, this energy must come from the work done at the narrower tube. If the force acting at the narrower tube is $F_2 = pA_2$, the work done is $d_2 F_2$. We match this to the energy gain of the car to give us an equation for d_2. In another method, we use the fact that the fluid is incompressible, so the same volume of fluid $d_1 A_1$ that has been added to the wide cylinder must come from the smaller cylinder. Since the volume in the smaller cylinder is $d_2 A_2$, equating $d_2 A_2$ to $d_1 A_1$ gives us an equation for d_2. We'll use the second method, which is very simple.

Working It Out We have $pA_1 = mg$, or $p = mg/A_1$. In terms of the given radius $R_1 = 0.20$ m of the wider tube,

$$p = \frac{mg}{\pi R_1^2} = \frac{(1000 \text{ kg})(9.8 \text{ m/s}^2)}{\pi (0.20 \text{ m})^2} = 7.8 \times 10^4 \text{ Pa}.$$

This is the necessary pump pressure, a number 80% of the ambient atmospheric pressure. (The ambient pressure alone cannot lift the car as it acts downward on the topmost surfaces of the car and upward on the bottom surfaces, and so forth, and the net force from the ambient pressure is zero. The pump pressure here is really an additional pressure, and it acts in one particular direction because of the orientation of the plate that supports the car.)
(b) We set $d_2 A_2 = d_1 A_1$ and solve for d_2:

$$d_2 = \frac{A_1}{A_2} d_1 = \frac{\pi R_1^2}{\pi R_2^2} d_1 = \frac{(20.0 \text{ cm})^2}{(2.5 \text{ cm})^2} (2.0 \text{ m})$$
$$= 8^2 \times (2.0 \text{ m}) = 1.3 \times 10^2 \text{ m}.$$

In the narrower tube there will be some sort of ratcheting mechanism which allows the vertical motion to be carried out in many strokes. Many air pumps have such mechanisms.

What Do You Think? Why do bicycle and car tires have such thin valves?

EXAMPLE 16–4 A column of mercury ($\rho_{Hg} = 13.6 \times 10^3$ kg/m^3) in a vertical tube that is open at the bottom rests in an open pool of mercury. The column is sealed off at the top after all the air has been evacuated from the tube, creating a vacuum above the mercury. What is the height of the column of mercury?

Setting It Up In Fig. 16–10 we see the height H of the mercury column that is asked for in this problem. We are explicitly given only the density of mercury, and more information will be required to find the height of the mercury column. The clue is in the remark that the pool of mercury is open at the bottom. That means that the pressure of the atmosphere acts on it, and so that is the pressure in the mercury at the level of the pool.

Strategy The fact that the tube above the mercury column is evacuated means that there is no pressure above the column. (In reality, a little mercury will have evaporated, but the pressure that it exerts is negligible.) However, the surface of the open pool, and therefore in the interior of the column at the same height, is that of the atmosphere—call this pressure p_0. If the mercury column has a cross section A, then there is an upward force at the bottom, $p_0 A$, that counteracts the weight of the mercury in the column, mg. We can express the mass m

of mercury in the column as the density of mercury times the column volume V, and $V = HA$. In this way, H is determined.

▲ **FIGURE 16–10** A column of mercury balanced by atmospheric pressure.

Working It Out The expression for the balancing of the forces is $p_0 A = mg$. But the volume of the column of mercury is $V = HA$, hence the mass is $m = \rho_{Hg}V = \rho_{Hg}HA$, and we have

$$p_0 A = \rho_{Hg}HAg.$$

The area cancels from both sides. We solve the remaining equation for H:

$$H = \frac{p_0}{\rho_{Hg}g}.$$

Numerically,

$$H = \frac{1.01 \times 10^5 \text{ Pa}}{(13.6 \times 10^3 \text{ kg/m}^3)(9.8 \text{ m/s}^2)} = 0.76 \text{ m}.$$

This example describes the working principle of a **barometer**, a device used to measure atmospheric pressure. One atmosphere of pressure will support a column of mercury that is 760 mm high.

What Do You Think? What would the height of the mercury column be if the vacuum at the top of the tube were instead a fraction of 1 atm?

EXAMPLE 16-5 A U-shaped tube of uniform cross section is open at both ends. First, amyl bromide is poured into the tube. (Amyl bromide is a fluid that does not mix with water.) Then water is carefully added to one side of the U, so there is water at the top of one column of fluid and amyl bromide is at the top of the other. The two columns of fluid meet at a point that we label $h = 0$, as in Fig. 16-11. What is the ratio of the heights of the fluids above $h = 0$ in each column? The density of water is $\rho_W = 1.00 \times 10^3$ kg/m^3 and that of amyl bromide is $\rho_{ab} = 1.26 \times 10^3$ kg/m^3.

▲ **FIGURE 16-11**

Strategy The atmospheric pressure p_0 is the same at the open surface of both columns (we neglect the weight of any amount of air that adds to the pressure over the lower column; air pressure changes very slowly with height). The pressure in the two fluids at $h = 0$, where they meet, is equal—otherwise there would be movement. We can write the expression for the pressure in each fluid at this point and set them equal, and from this we can determine the height ratios.

Working It Out We measure heights from the point where the fluids meet and denote h_{ab} as the height of the amyl bromide column and h_w as the height of the water column. The pressures p_{ab} and p_w at $h = 0$ in the respective columns is the air pressure at the top plus the pressure due to the weight of the fluid above,

$$\text{for the amyl bromide, } p_{ab} = p_0 + \rho_{ab}h_{ab}g;$$

$$\text{for the water, } p_w = p_0 + \rho_w h_w g.$$

Setting these pressures equal to one another, we have

$$\rho_{ab}h_{ab}g = \rho_w h_w g.$$

We rearrange this equation to find

$$\frac{h_{ab}}{h_W} = \frac{\rho_W}{\rho_{ab}} = \frac{1.00 \times 10^3 \text{ kg/m}^3}{1.26 \times 10^3 \text{ kg/m}^3} = 0.794.$$

The water column is higher than the amyl bromide column. This makes sense physically because amyl bromide is denser than water; the shorter column of amyl bromide weighs the same as the taller column of water. Note that this result is independent of the amount of either fluid.

What Do You Think? What would happen if we used heptyl ether, another fluid that does not mix with water and has a mass density of 0.81×10^3 kg/m^3?

16-4 Buoyancy and Archimedes' Principle

A cork floats in water while a brick sinks. Why? Experience tells us that this has something to do with the density of the material (and not the mass; a small brick sinks just like a large one). Let's consider the situation in Fig. 16-12a, a block of material of volume $V = Ah$ and density ρ (the mass is thus $m = \rho V$) partly immersed in water, which has a density of ρ_W. What forces do you think are acting on the block in this equilibrium position? In particular, focus on the lower surface of the partly submerged block, as in Figure 16-12b. The upward force is due to the pressure of the liquid, as given by Eq. (16-6): $p = p_0 + \rho_W gy$, where y is the vertical submerged distance of the block. The upward force on the block is therefore

$$F_{up} = pA = p_0 A + \rho_W gyA.$$

(a)

(b)

(c)

(d)

▲ **FIGURE 16–12** (a) A block of material of density ρ is partly immersed in water, density ρ_w. The volume of the block is $V = Ah$, and the volume of water displaced by the block is $V_d = Ay$. (b) Force diagram of the forces acting on the lower surface of the block, which is partly immersed, as in part (a). The net force is $F_{net} = 0$, and the block floats. (c) When $\rho = \rho_w$, the block is totally submerged, and the block floats because $F_{net} = 0$. (d) When $\rho > \rho_w$, the block sinks because $F_{net} > 0$.

The downward force has two components: that due to the atmospheric pressure p_0 above the block, and the weight of the block. The first downward component is $F_{atm} = p_0 A$, and the second is the force of gravity on the block, $F_g = mg = (\rho Ah)g$. Thus the total downward force is

$$F_{down} = p_0 A + \rho ghA.$$

As the block is in equilibrium—floating but not moving—the net force will be zero. Therefore,

$$F_{net} = 0 = F_{down} - F_{up} = p_0 A + \rho ghA - p_0 A - \rho_W gyA = \rho ghA - \rho_W gyA.$$
$$(16–7)$$

Dividing by gA, we obtain

for a floating object: $\rho h = \rho_W y$;

$$\frac{\rho}{\rho_w} = \frac{y}{h}. \qquad (16–8)$$

If $\rho < \rho_W$, then $y/h < 1$ and only a fraction of the block is submerged. In the limiting case that $\rho = \rho_W$, the block is totally submerged and $y = h$ (Fig. 16–12c). In this case, the block would float just under the surface because the upward and downward forces cancel. If $\rho > \rho_W$, then the block must sink (Fig. 16–12d) because the two forces can no longer cancel completely. In this case, once the block is fully submerged ($y \geq h$), the net (downward) force is

$$F_{net} = \rho ghA - \rho_W ghA > 0. \qquad (16–9)$$

We can interpret the difference between the weight of the block ρghA and F_{net} as an *upward* **buoyant force**:

$$F_{buoy} = F_g - F_{net}. \qquad (16–10)$$

When the block is partly submerged and floating, we have $F_{net} = 0$, and Eq. (16–10) gives

for partial submergence: $F_{buoy} = \rho ghA - 0 = \rho ghA = \rho_W gyA,$

where we have used Eq. (16–8) in the last step. When the block is totally submerged (at any depth where $y \geq h$), F_{net} is given by Eq. (16–9), and we find

for total submergence: $F_{buoy} = \rho ghA - (\rho ghA - \rho_W ghA) = \rho_W ghA.$

If we think of $V_{submerged}$ as the volume of the object beneath the water ($V_{submerged} = yA$ or hA, according to whether the object is partly or totally submerged), then we can combine our results into the single expression (Fig. 16–13)

$$F_{buoy} = \rho_W g V_{submerged}. \qquad (16–11)$$

THE BUOYANT FORCE

The buoyant force opposes the downwardly directed force of gravity, ρghA, on the block. For a floating body, the buoyant force and the force of gravity cancel exactly. When $\rho > \rho_W$, the buoyant force opposes the force of gravity even though it does not cancel the force of gravity completely; a sinking object descends with less acceleration than it would under the influence of gravity alone. When a metal sphere is immersed in

Case I: Floating: not completely submerged

$\vec{F}_{net} = 0$ $\vec{F}_{buoy}$

$V' < V$
$F_b = \rho_w g V' = F_g = \rho g V$
$\dfrac{V'}{V} = \dfrac{\rho}{\rho_w}$

$\vec{F}_g$

Case II: Barely floating, completely submerged

$\vec{F}_{net} = 0$ $\vec{F}_{buoy}$

$V' = V$
$\rho = \rho_w$

$\vec{F}_g$

Case III: Sinking, even if completely submerged

$\vec{F}_{net}$ down

$\vec{F}_{buoy}$

$V' = V$
$\rho > \rho_w$

$\vec{F}_g$

V = Total volume of object
Displaced volume V'
Fluid

▲ **FIGURE 16–13** By comparing the buoyant force F_{buoy} to the force of gravity F_g, we determine whether a body sinks, just floats, or pops up.

a bucket of water and suspended from a spring balance, the balance will indicate a lower weight than it would for the same sphere in air. The difference in these weights is the buoyant force.

CONCEPTUAL EXAMPLE 16–6 Think about the hot-air balloon in Fig. 16–14a and the steel ball floating in mercury in Fig. 16–14b. Explain both phenomena.

Answer Above we saw that if a material is less dense than the medium in which it is is submerged, the (upward) buoyant force on it exceeds the force of gravity acting downward. The object will float or, if completely submerged, it will rise. This describes the hot-air balloon: Hot air is less dense than cooler air, and the balloon is completely submerged in the cooler air, so the net force on the balloon is

upward. More precisely, we can use Eq. (16–7) with $h = y$, ρ_w replaced by the density of cool air and ρ equal to the density of hot air (very nearly the density of the entire balloon). The net upward force is thus $F_{net} = Vg(\rho_{cool} - \rho_{hot})$.

The steel ball is in a situation like that of Fig. 16–13a, floating on the surface of a medium denser than it is. Perhaps that seems odd because we think of steel as being quite dense. That is true, but mercury is a liquid of extraordinarily high density. The steel ball floats easily because its density, 7.8 g/cm^3, is less than that of mercury, 13.6 g/cm^3.

◀ **FIGURE 16–14** (a) A hot air ballon is floating in air. (b) The heavy steel ball floats in mercury.

(a)

(b)

Archimedes' Principle

Regardless of whether the block is partly or totally submerged, Eq. (16–11) shows that the buoyant force is equal to the weight of the liquid (here, water) displaced by the block:

$$W_{\text{liq}} = m_{\text{liq}}g = \rho_{\text{liq}}V_{\text{liq}}g = \rho_{\text{liq}}yAg,$$

where $y = h$ when the block is totally submerged. This fact was discovered in the third century B.C. by Archimedes, a great scientist and mathematician of ancient Greece, and is known as **Archimedes' principle**:

> **The buoyant force on an immersed object equals the weight of displaced fluid.**

We may check the validity of this principle by considering the behavior of a block of iron, with a mass M and a density $\rho_{\text{Fe}} = 7.86 \times 10^3 \text{ kg/m}^3$, in water. The upwardly directed buoyant force is the weight of water displaced, $F_{\text{buoy}} = mg = \rho_W Vg$, where m is the mass of displaced water and V is its volume. The downward force of gravity on the iron block is $F_g = Mg = \rho_{\text{Fe}}Vg$. Because $\rho_{\text{Fe}} > \rho_W$, the net force is downward and the block sinks. The net downward force—effectively the weight of iron in water—is

$$F_{\text{net}} = F_g - F_{\text{buoy}} = \rho_{\text{Fe}}Vg - \rho_W Vg = \left(1 - \frac{\rho_W}{\rho_{\text{Fe}}}\right)\rho_{\text{Fe}}Vg$$

$$= \left(1 - \frac{\rho_W}{\rho_{\text{Fe}}}\right)F_g. \tag{16–12}$$

We can check the magnitude of the net force, and verify Archimedes' principle, by weighing an object in water (Fig. 16–15). Archimedes is said to have discovered this technique when he was asked to ascertain that a crown was made entirely of gold without marring the crown; he did this by weighing the crown in air and in water, thereby measuring its density [Eq. (16–12)]. It is said that he found the density of the crown to be less than that of gold, so that the gold in the crown must have been diluted with another metal.

The crown's owner would have not been very happy if Archimedes had been required to melt the crown down and recast it as a rectangular block to make his test! Although our calculations for buoyancy and Archimedes' principle were carried out for a uniform rectangular block, the results do not depend on the shape or the unifor-

(a)

(b)

(c)

▲ **FIGURE 16–15** By Archimedes' principle, the magnitude of the buoyant force on the sphere, F_{buoy}, equals the weight of the water the sphere displaces. The scale measures the difference of the magnitudes of the force of gravity on the sphere and the buoyant force on it. (a) No water is displaced; the scale measures the weight of the sphere. (b) Half the volume of the sphere is displaced. (c) The entire volume of the sphere is displaced. The buoyant force in this case is about 10 N.

mity. An object of any shape can be thought of as being composed of small rectangular blocks. Similarly, an object of varying density can be treated as a collection of many elements, each of constant density. The end result is that all that matters is the weight of the liquid that is displaced by the object.

EXAMPLE 16–7 A metal sphere weighs 29.4 N in air and 18.5 N in water. What is its average density?

Setting It Up In contrast to the situation of Fig. 16–14, the sphere sinks. When its weight is measured while it is beneath the surface of the water, there is a nonzero reading (Fig. 16–15c). We label the weight of the metal sphere in air $F_g = mg$ and in water W_w.

Strategy Equation (16–12) applies directly; it gives the weight of the sphere in water. Since W_w is proportional to F_g, it is simplest to eliminate the weight in air by taking the ratio W_w to mg. The resulting relation can be solved for the unknown average density ρ_{sphere} of the sphere.

Working It Out We have

$$\frac{W_w}{F_g} = \frac{\left(1 - \dfrac{\rho_w}{\rho_{\text{sphere}}}\right)mg}{mg} = \left(1 - \frac{\rho_w}{\rho_{\text{sphere}}}\right).$$

We solve for ρ_{sphere}:

$$\rho_{\text{sphere}} = \rho_w \frac{1}{1 - \dfrac{W_w}{mg}} = \rho_w \frac{1}{1 - \dfrac{18.5}{29.4}}$$

$$= \rho_w \times 2.70 = 2.70 \text{ g/cm}^3.$$

At this point you could also solve for the unknown mass of the sphere: Knowing the sphere density, the expression for W_w would give us the mass.

What Do You Think? If the metal sphere had been weighed in salt water rather than fresh water and given the same weight reading, would the density of the metal sphere have been (a) greater, (b) smaller, (c) the same, (d) can't tell. (Salt water is denser than fresh water.)

EXAMPLE 16–8 A rectangular tub made of a thin shell of cement has length $\ell = 1$ m, width $w = 80$ cm, depth $d = 60$ cm, and mass $M = 200$ kg. The empty tub floats in a lake. How many people of mass $m = 80$ kg each can stand in the tub before it sinks?

Setting It Up The tub is shown in Fig. 16–16. The depth of the bottom of the tub is y, and we denote the unknown number of passengers by N.

▲ **FIGURE 16–16** A tub sinks to a water depth y. When $y > d$, the depth of the tub, the tub sinks.

Strategy With the tub submerged to a distance y, we can find the volume $\ell w y$ of displaced water and then the weight of the displaced water. This will allow us to find the buoyant force. The buoyant force will cancel the weight of the tub with its passengers up to the point that the immersed depth is equal to the tub's height, $y = d$. This is the point at which the boat is barely floating. Beyond this the tub will fill with water, increasing its average density, and it will sink to the bottom. So we set the buoyant force at $y = d$ to the total weight, and this

relation will determine that total weight and in turn the maximum number of passengers N_{max}.

Working It Out With the tub submerged a vertical distance y, the mass of the displaced water is $\rho(\ell w y)$, where ρ is the water density, $1 \text{ g/cm}^3 = 10^3 \text{ kg/m}^3$. Then the weight of the displaced water is the mass times g, and this is the buoyant force,

$$F_{\text{buoy}} = \rho(\ell w y)g.$$

This force cancels the total weight of the tub and N passengers, $(M + Nm)g$. The equality of the buoyant force and the total weight,

$$\rho(\ell w y)g = (M + Nm)g,$$

is what determines y. We see from this relation that y increases as N does (the tub moves lower in the water), until the critical value $y = d$ is reached, at which point any additional mass sinks the tub. We thus set $y = d$ to find N_{max}. With $\rho(\ell w d)g = (M + N_{\text{max}}m)g$, we solve for N_{max}:

$$N_{\text{max}} = \frac{\rho \ell w d - M}{m}.$$

Numerically,

$$N_{\text{max}} = \frac{(10^3 \text{ kg/m}^3)(1 \text{ m})(0.80 \text{ m})(0.60 \text{ m}) - 200 \text{ kg}}{80 \text{ kg}} = 3.5.$$

With three people the tub does not sink; with four people she goes down. Cargo ships were built out of concrete during both World Wars because of the high demand for steel. The older ships also served a useful purpose when they were deliberately sunk during the Normandy invasion of World War II to serve as a breakwater that would improve conditions for landing ships.

What Do You Think? If this tub had been weighed in water, what would the scale read?

16–5 Fluids in Motion

The marvelous fluid flow patterns observed in nature go far beyond the particle and rigid body motions we have studied to this point. **Hydrodynamics** describes fluids in motion, including the flow of streams, the eddies of oars in a pond, mudslides, smoke rings, ocean waves and tides, arterial blood flow, Jupiter's (and Earth's) atmosphere, the rotational flow of galaxies, and nuclear fission. Numerical calculations performed by supercomputers are necessary to solve the most complex types of behavior, but a remarkable variety of phenomena can be described with a very simplified set of equations. This simplified set involves a number of assumptions that we describe here.

We assume, as we have already discussed, that the fluid under consideration is incompressible. This does not mean that we exclude phenomena that involve gases, such as airflow; but it does mean that we will discuss these phenomena only under circumstances in which compressibility does not play a role.

We also assume that we are dealing with a fluid at constant temperature. (We will learn more about temperature and heat flow in Chapters 17 through 20.) Phenomena such as the **convection** of fluids—in which a liquid in the bottom of a vessel is heated, rises, cools, and falls in a circulating pattern—will not be considered.

We assume **steady flow**, in which the velocity $\vec{v}$ and the pressure p of a fluid at a given point do not depend on time. The velocity and pressure may vary from point to point, but we assume that any changes are smooth ones.

The description of how velocity can change from point to point deserves a more detailed discussion and leads us to some further simplifying assumptions. Consider a fluid in motion. We could represent the direction of the velocity at some point by an arrow at that point, with the length of the arrow proportional to the magnitude of the velocity. For steady (time-independent) flow, there is a *local velocity $\vec{v}(\vec{r})$*. If we start at some point, then a tiny element of the fluid is at that point, and it will move in the direction of the velocity vector. An instant later the element is at an adjacent point, which has a slightly different velocity vector. This vector dictates the element's direction and speed of motion in the next instant and determines the next step of the motion. In this way, the motion of a tiny element of fluid can be traced. (That we can follow such an element is an idealization that approximates reality if the element is small enough.) The line of motion traced out by our element is called a **streamline**. Flow described by streamlines is called **laminar flow**; in laminar flow each element of fluid travels along a smooth, well-defined path. Under certain circumstances fluid flow ceases to be laminar and becomes turbulent. **Turbulent flow**, as in the flow of a white-water river, involves rapidly fluctuating velocities and pressures. Turbulence is a characteristic of many flow patterns that we observe in nature, but there is no simple theory that describes it. We'll look further at turbulence, an exciting area of research in engineering and physics, in Section 16–9, but we'll assume for now that we are dealing with laminar flow.

We will also assume here that we deal with **irrotational flow**. When streamlines close in on themselves, such as those shown in Fig. 16–17, the flow is called **rotational**, and in irrotational flow this does not occur. We are consequently unable to discuss here vortices, or whirlpools (Fig. 16–18), such as those that arise in water running out of a bathtub, smoke rings, or even the much simpler motion of coffee stirred by a spoon in a cup. A mathematical expression of the statement that the flow is irrotational is that the **circulation**, defined as the integral of the velocity along any *closed* path in the fluid, is zero.

To simplify matters further, we make one more assumption: We neglect the analogue of friction in the fluid. That there is such an analogue is obvious from simple observations. The flow of honey is different than the flow of water. The friction that enters here is an internal friction which is called **viscosity**. In neglecting viscosity, we are neglecting internal energy losses, and the conservation of energy can then be used profitably.

To summarize, in the next two sections we make the following assumptions about fluid flow:

1. The fluid is incompressible.
2. The temperature does not vary.

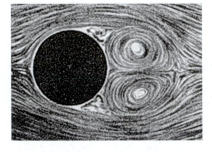

▲ **FIGURE 16–17** Streamlines for a cylinder pulled steadily through a fluid, showing circulation in the fluid in the form of the two vortices on the right.

▲ **FIGURE 16–18** The water moves faster near the center of the vortex than it does far from the center. The water is in rotational flow.

3. The flow is steady, so that the velocity and pressure do not depend on time.

4. The flow is laminar rather than turbulent.

5. The flow is irrotational, so there is no circulation.

6. There is no viscosity in the fluid.

In Section 16–9 we will look at some of the important differences between this idealized flow and the flow in real fluids. In particular, we'll come back to the subjects of viscosity and turbulence.

16–6 The Equation of Continuity

For a fluid in motion, we can focus on a group of adjacent streamlines, which we say form a *bundle*, and follow them (Fig. 16–19). (We must work with a bundle because a single streamline is just a one-dimensional line—it contains no mass, hence no energy, no momentum, and so forth.) By their definition, streamlines cannot cross. If they did, the crossing point would be a point where the fluid has two different velocities simultaneously, and that is impossible. And because streamlines never cross, the bundle, which forms a thin tube of nonuniform cross section known as a **streamtube**, has the property that no fluid passes from the inside to the outside of it. We can then choose a streamtube thin enough so that the velocity across a section drawn at right angles to the tube is the very nearly the same across the whole cross section. The volume of fluid that crosses the section of area A_1 (Fig. 16–19) in a given time t is the volume of a cylinder whose base is A_1 and whose length is $\ell_1 = v_1 t$, where v_1 is the speed across that section; the volume of fluid is thus $v_1 t A_1$. If the density of the fluid at this section is ρ_1, then the mass of fluid crossing the section in the given time t is

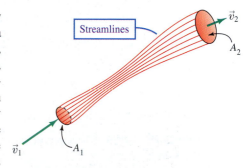

▲ **FIGURE 16–19** Flow contained in a bundle of streamlines (streamtube). The velocity perpendicular to the surface of area A_1 is $\vec{v}_1$, and that perpendicular to the surface of area A_2 is $\vec{v}_2$.

$$m_1 = \rho_1 v_1 t A_1. \qquad (16\text{–}13)$$

Similarly, the mass of fluid crossing a section of area A_2 at right angles to the bundle of streamlines with speed v_2 and density ρ_2 in the same time t is

$$m_2 = \rho_2 v_2 t A_2.$$

But because fluid neither enters nor leaves the tube by the side walls, the *conservation of mass* implies that $m_1 = m_2$, or, if we cancel the time t,

$$\rho_1 v_1 A_1 = \rho_2 v_2 A_2. \qquad (16\text{–}14)$$

EQUATION OF CONTINUITY

This equation, the **equation of continuity**, is true even for fluids whose densities can vary with position. Note that $m/t = \rho v A$ is the mass flow per unit time.

When we restrict ourselves to incompressible flow, we have $\rho_1 = \rho_2$, and Eq. (16–14) simplifies to the very useful equation

$$v_1 A_1 = v_2 A_2. \qquad (16\text{–}15)$$

CONSERVATION OF RATE OF FLOW, OR FLUX

Since $\rho v A$ is the mass flow per unit time, for constant density the combination vA is the *volume flow per unit time*—usually we call this just the rate of fluid flow. Equation (16–15) is a conservation law for the rate of flow. It states that the product of speed and area of any bundle of streamlines is the same anywhere along the bundle. The speed of fluid flow thus varies inversely with the streamtube cross section A. The rate of flow—the product (velocity perpendicular to an area) × (area)—is also known as the **flux**, Φ,

$$\Phi \equiv vA. \qquad (16\text{–}16)$$

Equation (16–15) describes the **conservation of flux**. Perhaps you have observed the conservation of flux while entering or leaving a building on a windy day: The wind speed is greatest in the smallest doorway.

EXAMPLE 16–9 Water flows at a rate of 0.25 L/s through a garden hose of inside diameter 2.0 cm. What is the speed of the water in the hose? The nozzle of the hose is a circular opening of diameter 1.0 cm. What is the speed of the water when it emerges?

Setting It Up Figure 16–20 indicates the area A_{hose} of the hose and the area A_{noz} of the nozzle. We denote the known flow rate as Φ. We want the water speeds v_{hose} in the hose and v_{noz} in the nozzle—the latter is also the speed just outside the nozzle.

▲ **FIGURE 16–20** The speed of the water must be larger in the narrow opening of the nozzle.

Strategy The given flux Φ is the product vA. The fact that flux is conserved allows us to say that product vA is the same at the nozzle as in the hose. We can then solve for v for either the hose or nozzle in terms of Φ and the appropriate area.

Working It Out We have immediately $v_{hose} = \Phi/A_{hose}$ and $v_{noz} = \Phi/A_{noz}$, where Φ is the same in the two sections. We can note that since the nozzle is half the diameter of the hose, the area of the nozzle is one quarter the area of the hose, so that the speed in the nozzle is four times higher than in the hose. This will save a little numerical work. The quantitative results are

$$v_{noz} = \frac{(0.25 \text{ L/s}) \times (10^3 \text{ cm}^3/\text{L})}{\pi(0.50 \text{ cm})^2} = 3.2 \times 10^2 \text{ cm/s} = 3.2 \text{ m/s},$$

and $v_{hose} = \frac{1}{4}v_{noz} = 0.87 \text{ m/s}.$

What Do You Think? If the nozzle were of a type that adjusts the diameter of the hose when twisted, should the nozzle diameter be reduced or increased to increase the distance reached by the water stream?

16–7 Bernoulli's Equation

In principle, the motion of a fluid can be understood by dividing the fluid into small elements and applying Newton's second law to each of them. Alternatively, providing that only conservative forces are present, we can use the conservation of energy. If there is no viscosity in a fluid, and hence no energy dissipation, and if the fluid is incompressible, it is straightforward to formulate the conservation of energy for fluids. This is a simple and important tool for the study of fluid motion. In the application here, we will use the conservation of energy in the form of the work–energy theorem and will end with an equation that relates speed, pressure, and height for an ideal fluid.

We again consider a streamtube—a bundle of streamlines of fluid—this time under the influence of gravity. Concentrate on the fluid forming a plug between two points along the tube: point 1, which has height h_1 and cross-sectional area A_1 and point 2, which is at height h_2 and has area A_2 (Fig. 16–21). The velocity perpendicular to the cross sections will have magnitudes v_1 and v_2, respectively, and we take our tube to be thin enough so that the velocity does not vary across the slices. The fluid to the left of the plug exerts pressure p_1 on the plug to the right, while the fluid to the right of the plug exerts pressure p_2 on the plug to the left. There will be, say, flow to the right if the force to the right, p_1A_1, is larger than the force to the left, p_2A_2.

Let's examine what happens during a time interval dt. The plug of fluid that starts out in the region between A_1 and A_2 at $t = 0$ will advance—in Fig. 16–21 we have supposed the movement is to the right. The distance by which it advances at point 1 is $v_1\,dt$, and the distance by which it advances at point 2 is $v_2\,dt$. We calculate the net work done in advancing the fluid in the tube by looking at the forces acting at either end. The work done by the force of the fluid to the left of point 1 is the force there times the displacement. The force is p_1A_1, so this work is

$$W_1 = (p_1A_1)(v_1dt). \tag{16–17}$$

▲ **FIGURE 16–21** A streamtube of fluid at different heights. In time Δt, the fluid moves to the right under the influence of a difference in pressures between p_1 and p_2. In effect, the shaded slice in region 1 is transferred to the shaded slice in region 2.

The work done by the force of the fluid to the right of point 2 can be calculated similarly, but there is a minus sign because the force (and thus the pressure) acts to the left while the displacement is to the right, so that $\vec{F} \cdot \vec{s} = -Fs$. Thus

$$W_2 = -(p_2A_2)(v_2\,dt).$$

The net work done on the streamtube plug between regions 1 and 2 is then

$$W_{net} = W_1 + W_2 = (p_1 - p_2)vA\,dt, \qquad (16\text{–}18)$$

where in the last line we have made use of Eq. (16–15), which expresses the constancy of the flux vA at any point along the tube: $vA = v_1A_1 = v_2A_2$.

We have now determined the amount of work done in moving the fluid in the plug during the time dt. To apply the work–energy relation, we next find the change in energy. If we examine Fig. 16–21, we can see that in the time interval dt, the effect of the motion is to transfer a small slice of fluid from region 1 to region 2. This transfer has two effects on the energy: The potential energy changes as the slice is lifted from h_1 to h_2, and the kinetic energy changes as the speed changes from v_1 to v_2. The change in potential energy is

$$\Delta U = mg\,\Delta h = (\rho vA\,dt)g(h_2 - h_1). \qquad (16\text{–}19)$$

Again we have used the constancy of vA. The change in kinetic energy is

$$\Delta K = \frac{1}{2}(\rho vA\,dt)(v_2^2 - v_1^2). \qquad (16\text{–}20)$$

Because $W_{net} = \Delta K + \Delta U$, we find, from Eqs. (16–18) through (16–20),

$$(p_1 - p_2)vA\,dt = \frac{1}{2}(\rho vA\,dt)(v_2^2 - v_1^2) + (\rho vA\,dt)g(h_2 - h_1).$$

After the factor $vA\,dt$ is divided out, we are left with

$$p_1 - p_2 = \frac{1}{2}\rho(v_2^2 - v_1^2) + \rho g(h_2 - h_1).$$

This equation can be rearranged into a conservation law, with one side dependent on the quantities defined in region 1 and the other side dependent on the quantities defined in region 2:

$$p_1 + \tfrac{1}{2}\rho v_1^2 + \rho g h_1 = p_2 + \tfrac{1}{2}\rho v_2^2 + \rho g h_2. \qquad (16\text{–}21)$$

In other words,

$$\text{along a streamline: } p + \tfrac{1}{2}\rho v^2 + \rho g h = \text{a constant.} \qquad (16\text{–}22)$$

BERNOULLI'S EQUATION

Our derivation deals with the motion of one particular streamline. But it can be shown in general that *for irrotational flow, the constant in Eq. (16–22) is the same for all streamlines* within a fluid. Equation (16–22) is **Bernoulli's equation**, derived in 1738 by Daniel Bernoulli, one of a remarkable family of scientists and mathematicians. In Section 16–8 we discuss applications of this formula.

Problem-Solving Techniques

Bernoulli's equation expresses conservation of energy for a fluid; as such, its use is similar to the use of all conservation laws. Keep in mind the following points:

1. When we say the right-hand side of Eq. (16–22) is a constant, we mean that it takes on the *same* value for all points within the fluid. If we know the value of $p + \frac{1}{2}\rho v^2 + \rho g h$ at one point, then that is the value it takes at all points.

2. In its typical application, we can find the value of the constant in Bernoulli's equation because we know the value of all the variables at some point. Then we can use the value of the constant to find the unknown value of a variable at another point.

3. When one of the three terms in the expression $p + \frac{1}{2}\rho v^2 + \rho g h$ is the same at all points, then we can ignore it and

concentrate on using Bernoulli's equation to learn about the other variables. For example, in Section 16–8 we first apply Bernoulli's equation to the case of a fluid at rest. In that case, v is the same everywhere ($v = 0$), and we can use Bernoulli's equation as a conservation law for the quantity $p + \rho g h$.

(a)

(b)

▲ **FIGURE 16–22** By blowing over the top of the paper, the pressure above the paper is reduced according to Bernoulli's principle. Below the paper the pressure is unchanged, and there is a net force that makes the paper rise.

16–8 Applications of Bernoulli's Equation

Bernoulli's equation is useful for describing a variety of phenomena. First, we observe that if a fluid is at rest, so that $v = 0$ in Eq. (16–22), then

$$p + \rho g h = \text{a constant.} \tag{16–23}$$

This is just the hydrostatic relation between pressure and height that we obtained in Eq. (16–6).

The Bernoulli Effect

For flow at constant height, the conservation of energy [Eq. (16–22)] implies that

$$p + \tfrac{1}{2}\rho v^2 = \text{a constant.} \tag{16–24}$$

Thus the pressure must be lower in a region in which a fluid is moving faster. You can observe this phenomenon—known as the **Bernoulli effect**—in a simple experiment. If you blow air across the top of a single sheet of paper (Fig. 16–22), the pressure above the paper is reduced. However, the pressure is unchanged below the paper; thus, the paper rises. The Bernoulli effect also enters into the operation of a chimney. When the wind blows over the top of a chimney, the pressure is decreased at the top by the Bernoulli effect, and the air in the tube rises. This is an important element in a chimney's "draw."

THINK ABOUT THIS . . .

HOW IS FLUID SPEED IN A CLOSED PIPE MEASURED?

How does an engineer running a refinery know the speed of the fluids that pass through the pipes? Equation (16–24) is applied in the *Venturi flowmeter*, a device designed for just this purpose (Fig. 16–23a). Figure 16–23b is a simplified version of the flowmeter. The large pipe, with cross-sectional area A_1, is constricted smoothly to a smaller area A_2. A U-shaped tube containing a different fluid connects the two segments of pipe. The speeds in the larger and smaller sections of the pipe are v_1 and v_2, respectively; similarly, the pressures in the two sections are p_1 and p_2, respectively. Then Eq. (16–24) tells us that

$$p_1 + \frac{1}{2}\rho v_1^2 = p_2 + \frac{1}{2}\rho v_2^2;$$

or

$$p_1 - p_2 = \frac{1}{2}\rho(v_2^2 - v_1^2).$$

The continuity equation, Eq. (16–15), tells us that $v_2 = v_1 A_1/A_2$, so

$$p_1 - p_2 = \frac{1}{2}\rho v_1^2\left[\left(\frac{A_1}{A_2}\right)^2 - 1\right]. \tag{16–25}$$

The difference in pressures is determined from the difference in heights, h, of the two columns of fluid in the U-shaped tube (see Section 16–3). Equation (16–25) is then used to find speed v_1 in terms of this pressure difference.

Fluid must move faster in smaller area to conserve flux.

Area A_1, pressure p_1

Area A_2, pressure p_2

$\vec{v}_1$

$\vec{v}_2$

Column 1

h

Column 2

(a)

(b)

▲ **FIGURE 16–23** (a) The Venturi flowmeter. (b) The speed v_1 of a fluid in a pipe is measured by the difference in heights, h, of column 1 and column 2 of a different fluid. The number of measuring tubes that tap the pipe through which the fluid flows can be varied according to the measurement desired.

EXAMPLE 16–10 A giraffe needs a strong heart because of its long neck. Suppose that the difference in height between the aortic valve (the place where arterial blood comes out of the heart) and the head of a giraffe is 2.50 m, and that the artery leading from near the aortic valve to the head has constant cross section all the way to the head. What is the minimum pressure at the aortic valve? Blood is an incompressible fluid with density $\rho = 1.0$ g/cm^3.

Setting It Up See Fig. 16–24 for a sketch. We will use Bernoulli's equation with constant speed in order to relate pressure at the heart to pressure at the head and the height change, and for that we will need to know the minimum value of pressure at the giraffe's head. For the survival of the giraffe this value must be more than $p = 0$ and we would have to ask a veterinarian what the true minimum at the head is. For our purposes we'll just assume that $p = 0$ at the head is the smallest physical value allowed.

Strategy Because blood is incompressible and the arterial area is constant, the equation of continuity (flux conservation) implies that the speed of the blood will be the same at the head as at the heart. Bernoulli's equation in the form of Eq. (16–23) (the form appropriate to constant speed) is a relation between the heart height and pressure and the head height and pressure. Only the heart pressure is unknown in this expression, and we can solve for it.

Working It Out Let's use subscripts 1 and 2 for the heart and head, respectively. Then, with $p_2 = 0$, Eq. (16–23) reads $p_1 + \rho g h_1 = \rho g h_2$, which can be solved for p_1:

$$p_1 = \rho g (h_2 - h_1).$$

▲ **FIGURE 16–24** A giraffe's long neck requires a strong heart for arterial blood flow.

Numerically, $p_1 = (1.0 \text{ g/cm}^3)(10^2 \text{ cm/m})^3(10^{-3} \text{ kg/g})(9.80 \text{ m/s}^2)$ $(2.50 \text{ m}) = 2.5 \times 10^4$ Pa. Various factors, in particular the loss of energy due to friction between the blood and the arterial walls and the desire to have a larger value than 0 for the head pressure, make this a low estimate. Comparing this to the *peak* output pressure of the human heart, 1.6×10^4 Pa, we can see that the giraffe's long neck requires a strong heart.

What Do You Think? What would be the effect on the giraffe if the artery diameter narrowed down as it approached the brain?

Fluid Motion with Constant Speed

In the Bernoulli effect, height is constant but the speed of the flowing fluid is not. But we can easily imagine reversing this and confronting constant speed but changing height. To fix the speed, we need only confine our moving fluid to a pipe of unchanging cross section. If the fluid is incompressible, then the equation of continuity, Eq. (16–15), shows that the speed cannot change. However, the pipe could perfectly well move from one height to another, and in this situation the pressure will change. For this case Bernoulli's equation is $p + \rho g h = $ a constant, identical to the static case expressed by Eq. (16–23) (see Example 16-10).

Flow from a Tank

Another interesting phenomenon of practical importance in fluid flow is the speed of flow of a liquid from an opening in a tank (Fig. 16–25a). If we make a small hole in the side of an open tank of water at a depth h below the surface, we can apply Bernoulli's equation, Eq. (16–22), to the fluid at different locations in the tank. We denote the height above the hole by the variable y, so that the height at the water surface is $y = h$ and at the opening $y = 0$. Equation (16–22) states that the value of $p + \frac{1}{2}\rho v^2 + \rho g y$ is the same at point 1 (on the upper surface, where the pressure $p = p_0$) and point 2 (just outside the hole, where $p = p_0$ because the sides of the stream of fluid flowing from the tank are in equilibrium with the atmosphere). The speed v_2 at point 2 is to be determined. We equate the values of $p + \frac{1}{2}\rho v^2 + \rho g y$ at the two points and obtain

$$p_0 + \frac{1}{2}\rho v_1^2 + \rho g h = p_0 + \frac{1}{2}\rho v_2^2 + \rho g \times 0.$$

We can neglect the small velocity v_1 of the liquid at the top of the tank. We solve for the speed at point 2 (the hole) to find that

$$v_2 = \sqrt{2gh}, \tag{16–26}$$

(a)

(b)

▶ **FIGURE 16–25** The escape of fluid through a hole in a tank. (a) The curved lines are streamlines for real fluid flow. (b) The pressure at the bottom of the tank is greater than the pressure at the top, so a stream of water leaving the tank has greater speed if it leaves from a hole that is lower on the tank.

as is consistent with Fig. 16–25b. This is just the speed of a single particle falling through a height h under the influence of gravity, as in Eq. (2–27). We find the same result for the falling liquid as for a falling object because they both are consequences of the conservation of energy. To find the rate of fluid flow from the hole,[†] we use the expression for the flux, Eq. (16–15), namely $\Phi = vA$.

EXAMPLE 16–11 A tank of cross-sectional area 0.07 m² is filled with water. A tightly fitting piston, with a total mass 10 kg, rests on top of the water. A circular hole of diameter 1.5 cm is opened at a depth of 60 cm below the water level of the tank. What is the initial rate of flow of water out of the hole?

Setting It Up See Fig. 16–26 for a sketch. The sketch labels the tank area A, the hole area a, the piston mass m, and the hole depth h; all three quantities are given in the question. We'll also need to assume that the atmospheric pressure p_0 does not change from the top of the tank to the place where the hole is cut.

▲ **FIGURE 16–26** The speed of the water flowing out of the small hole depends on the height h of the water above it.

Strategy This is a typical situation where Bernoulli's equation, which relates pressure, speed, and height, is useful: We know some of these quantities at one point and want to learn about them at another point. Here, the pressure, speed, and height are known at the

top, as are the pressure and height at the hole. We can then use Bernoulli's equation to learn about the speed at the hole. Then given the water speed v, we can find the flow rate va.

Working It Out At surface level, the total pressure $P = p_0 + p_1$, where p_0 is the atmospheric pressure and p_1 is the pressure due to the piston resting on the water, $p_1 = mg/A$. The water at the surface, which we set at height h, is essentially stationary. At the hole, which here is at height 0, the pressure just outside the hole is p_0, as in the discussion above. An application of Bernoulli's equation at the top of the water and at the opening leads to the equality

$$p_0 + p_1 + \rho g h = p_0 + \tfrac{1}{2}\rho v^2.$$

We cancel p_0 from both sides of the equation and solve for v^2:

$$v^2 = \frac{2(p_1 + \rho g h)}{\rho} = \frac{2\left(\dfrac{mg}{A} + \rho g h\right)}{\rho} = 2g\left(\frac{m}{\rho A} + h\right).$$

In turn, the flow rate is $\Phi = va = v\pi(d/2)^2$, where d is the hole diameter. If $\rho = 10^3$ kg/m³, we find

$$v^2 = 2(9.8 \text{ m/s}^2)\left[\frac{10 \text{ kg}}{(10^3 \text{ kg/m}^3)(0.07 \text{ m}^2)} + 0.6 \text{ m}\right] = 15 \text{ m}^2/\text{s}^2,$$

or $v \cong 4$ m/s, and $\Phi = (4 \text{ m/s})\pi(7.5 \times 10^{-3} \text{ m})^2 = 7 \times 10^{-4}$ m³/s = 0.7 L/s.

What Do You Think? Let's assume that we want the water coming out of the hole to land as far away as possible. How would you determine where to put the hole? Discuss, qualitatively, how far the stream would land according to whether the hole is near the top or bottom of the tank.

*Lift

▲ **FIGURE 16–27** Schematic version of streamlines over an airplane wing section. There is a streamline that ends on the front of the wing at the stagnation point and a streamline that starts at the rear edge of the wing.

The upward force that keeps a heavy airplane in the sky is an example of **lift**. The Bernoulli effect—the fact that when the speed of a fluid increases, pressure in the fluid decreases—helps us to understand the origin of lift. However, there are some subtleties, and circulation, viscosity, and turbulence all play an important role in lift. Thus it is necessary to relax some of the simplifying assumptions listed at the end of Section 16–5 to provide a correct description of this phenomenon. Consider the cross section of a wing shown in Fig. 16–27. We visualize the situation from the reference frame of the wing, so that we think of the air moving toward the wing, which is at rest, rather than the wing moving through stationary air. (This is, in fact, just what happens in a wind tunnel in which problems such as this one are studied.) Air approaches the wing with some speed v along a set of streamlines. Some of the streamlines go below the wing, some go above. The wing is designed so that the air speed above the wing is greater than that below the wing, which can happen only if there is circulation in the clockwise direction[‡]. Circulation occurs only around the wing.

[†]A more careful treatment replaces the rate of flow with the expression kvA, where the coefficient k is less than 1. Its correct value for a simple small hole is 0.62. In the examples and problems, we use $k = 1$ for simplicity.
[‡]No air actually moves forward under the wing; rather, the speed beneath the wing is less than the speed above the wing.

The flow elsewhere is irrotational, which means that the constant appearing in the Bernoulli equation, Eq. (16–22), is the same for all the streamlines that flow near the wing, below or above. From Eq. (16–24) we see that the pressure on top of the wing, p_t, must therefore be less than the pressure below, p_b, and there is a net upward force, or lift, on the wing. The net upward pressure, $p_{net} = p_b - p_t$, is determined from Eq. (16–24):

$$p_b + \tfrac{1}{2}\rho v_b^2 = p_t + \tfrac{1}{2}\rho v_t^2,$$

where here ρ is the air density. We thus find that

$$p_{net} = p_b - p_t = \frac{1}{2}\rho(v_t^2 - v_b^2) = \rho\left(\frac{v_t + v_b}{2}\right)(v_t - v_b) \cong \rho v(v_t - v_b), \quad (16\text{--}27)$$

where v is the average of the top and bottom speeds. Because the difference between the top speed and the bottom speed is small, both are nearly equal to the average speed, which we call the *airstream speed v*.

The expression in Equation (16–27) can be used to calculate the lift. If the wings have a width w and a tip-to-tip span S, then their total area is wS (we are ignoring the fuselage), and the lift is given by

$$L = wSp_{net} = wS(p_b - p_t) \cong wS\rho v(v_t - v_b).$$

Experiments show that for speeds much slower than the speed of sound, $v_t - v_b$ is proportional to the airstream speed v, with a proportionality constant K (in general, $K \ll 1$), which depends on the wing design and the wing angle relative to the airplane direction. Thus we can finally write the lift as

$$L \cong wS\rho Kv^2. \qquad (16\text{--}28)$$

A typical value of the constant K is 0.12, so for a small plane with a total wing area of 30 m^2 and an air speed of 200 km/h $= 56$ m/s and with air density of about 1.3 kg/m^3, the lift is given by

$$L = (0.12)(30 \text{ m}^2)(1.3 \text{ kg/m}^3)(56 \text{ m/s})^2 = 1.6 \times 10^4 \text{ N}.$$

Thus the mass that can be lifted is $(1.6 \times 10^4 \text{ N})/(9.8 \text{ m/s}^2) \cong 1600$ kg. The fact that air is compressible plays little role if the relative speed of the wing and the air is much smaller than the speed of sound in air, $v_s \cong 330$ m/s. (The design and operation of the wings of supersonic aircraft presents a different set of problems.)

To give the flow around the wing a nonzero circulation, it is necessary to have some turbulence. The turbulence is induced by the viscosity of the air and further enhanced by the placement of small protuberances (ailerons) on the wings.

THINK ABOUT THIS...

WHY DO RACING CARS HAVE WINGS?

In contrast to airplanes, racing cars are designed so that the air beneath the car moves faster than the air above it, and a downward pressure is exerted on the car. This downward pressure is produced or augmented by both front and rear wings, sometimes called *spoilers*. These act like upside-down airplane wings, producing a reverse lift that pushes the car vertically down onto the road. The increased contact force with the road is very useful to a racing car, because static friction between the wheels and the track is proportional to the contact force. Increased friction allows more torque to be supplied to the wheels without them spinning, and in this way larger accelerations and speeds are possible. Moreover, the increased friction allows better steering control.

---

*16–9 Real Fluids

Our discussion has mostly been limited to idealized kinds of flow of idealized fluids. Some of our assumptions apply only approximately, at best, to real fluids. Nonviscous flow, for example, has been called the flow of "dry water." Here we give some qualitative notions of how real fluids behave under ordinary circumstances.

(a)

(b)

▲ **FIGURE 16–28** Velocity profiles for laminar flow in a pipe: (a) idealized fluid flow that does not take into account viscosity; (b) realistic fluid flow (Poiseuille flow).

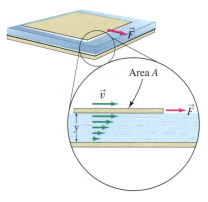

Area A

▲ **FIGURE 16–29** Two glass plates have a film of fluid between them; one slides against the other. The top plate moves with speed v. Fluid near the top plate moves with the same speed as that plate, while fluid near the bottom plate is at rest.

Viscosity

Figure 16–28a is a picture of the velocity profile of pipe flow for fluids idealized as we have described. For real fluids the liquid near the surface of the pipe is at rest, and the velocity profile is actually parabolic (Figure 16–28b). Such flow is referred to as **Poiseuille flow**, after Jean Poiseuille, who studied blood circulation. The reason for this effect is the internal friction of the fluid, or *viscosity*. Viscosity comes from intermolecular collisions. The molecules of a fluid have a random motion, to be described in Chapter 19. If two adjacent layers of fluid flow with slightly different speeds, the random sidewise intrusion of some slower molecules into the faster stream will tend to slow down the faster stream, whereas intrusion of faster molecules into the slower stream will tend to speed up the slower stream.

Viscous flow can be analyzed by taking two glass plates with a film of fluid between them and sliding one plate over the other. The easier it is to slide the plates, the less viscous is the fluid between them. To quantify this effect, take the smaller plate to have an area A; it is separated from the other plate by a distance y (Fig. 16–29). If the upper plate is to move with a uniform velocity $\vec{v}$, a force must be applied—just as if a brick is made to slide along a rough surface at constant speed. This force, F, is proportional to A and inversely proportional to y. In addition, the force increases linearly with speed (like the damping force on a mass at the end of a spring or like air drag). A formula that reflects these factors will have a proportionality constant η that will differ in different materials,

$$F = \frac{\eta v A}{y}. \tag{16–29}$$

The coefficient η is the **coefficient of viscosity**. It is measured in newton seconds per square meter or equivalently kilograms per meter second. Typical values are $10^{-3}\ \text{kg}/(\text{m} \cdot \text{s})$ for water and $830 \times 10^{-3}\ \text{kg}/(\text{m} \cdot \text{s})$ for glycerin, both at 20°C. We mention the temperature because, as anyone who has poured syrup over pancakes knows, viscosity is strongly temperature-dependent.

Turbulence

A fluid flow marked by complex and chaotic movement is referred to as turbulent. At the onset of turbulence, the whole notion of neatly separated streamlines ceases to make sense (Fig. 16–30). A dimensionless parameter known as the **Reynolds number**, Re,

▶ **FIGURE 16–30** Turbulence, an important feature of fluid flow, is seen in the steam and smoke from the eruption of Mount Augustine in Alaska.

determines when turbulent flow occurs, and this parameter contains the fluid speed. It is defined as

$$Re = vL\left(\frac{\rho}{\eta}\right), \qquad (16\text{–}30)$$

where ρ is the density of the fluid, v is its speed, L is a length associated with the flow, such as the pipe's diameter for flow in a pipe, and η is the viscosity coefficent. When the Reynolds number increases past about 2000 to 3000, the flow becomes turbulent. For water flowing through a pipe 1 cm in diameter, $Re \cong (10^4 \text{ s/m})v$ so the flow becomes turbulent when v reaches a mere 0.3 m/s. Turbulence is very common, but fortunately a small amount of turbulence does not change the applicability of Bernoulli's equation for short distances of flow (just as a small amount of friction does not disturb energy conservation over short times), so Eq. (16–22), although not exact, is still a very good and useful approximation.

THINK ABOUT THIS . . .

OF WHAT USE ARE THE DIMPLES ON A GOLF BALL?

When an object such as a golf ball moves through the air, there is drag on it. We can identify two different sources of drag. There is drag associated with turbulence, in which energy is lost in churning the air. There is also a type of drag known as pressure drag associated with laminar (smooth) flow, in which to satisfy continuity equations the air flow leaves a region *behind* the moving object where there is a lower pressure than the pressure before it. Pressure drag can be more important than turbulent drag at high speeds. The dimples on a ball increase the turbulence around the ball, and while this means more turbulent drag, it has a second important effect: it decreases the size of the region with reduced pressure behind the ball and thereby decreases the pressure drag. Experiment shows that this is a winning strategy: a dimpled ball will travel farther. Moreover, the increased turbulence due to the dimples can also increase a lift associated with the spin of the golf ball. The dimples may look cute, but they also have a solid positive effect on the flight of the ball.

Summary

Liquids are distinguished from gases in that the molecules that form liquids are closer together and impose strong forces on each other. Liquids are distinguished from solids in that the molecules in liquids do not form a rigid lattice. Both liquids and gases deform in response to external forces and therefore flow as fluids. The density of a material of mass M and volume V is unchanging in incompressible fluids:

$$\rho = \frac{M}{V}. \qquad (16\text{–}1)$$

The fluids we study in this chapter are for the most part incompressible.

Pressure, measured in SI in pascals, expresses the way that fluids transmit forces. The pressure in a fluid is the (pushing) force per unit area on any area A:

$$p \equiv \frac{F}{A}; \qquad (16\text{–}2)$$

and is perpendicular to A. When a fluid is at rest, the pressure at any given point is the same in all directions. For an incompressible fluid in Earth's gravity, the pressure increases with the depth y according to

$$p = p_0 + \rho gy, \qquad (16\text{–}6)$$

where p_0 is the pressure at $y = 0$. A change in pressure applied to a fluid at rest is transmitted without change to every part of the fluid, a statement known as Pascal's principle. This principle is the basis of hydraulic lifts and brakes.

For an object immersed in a fluid, there is an upward buoyant force equal to the weight of the displaced fluid,

$$F_{\text{buoy}} = \rho g V_{\text{submerged}}. \qquad (16\text{–}11)$$

This is Archimedes' principle.

Nonviscous (friction-free) fluids in smooth (nonturbulent) motion follow paths traced out by streamlines. The conservation of mass for a fluid moving in this way is stated as the continuity equation,

$$\rho_1 v_1 A_1 = \rho_2 v_2 A_2. \tag{16-14}$$

Here ρ and v are the density and speed at some point in a tube of streamlines; A is the corresponding cross section of the tube. If the fluid is incompressible, the (constant) density cancels, and the resulting expression is the conservation of flow rate, or flux:

$$v_1 A_1 = v_2 A_2. \tag{16-15}$$

The conservation of energy for a moving fluid under the influence of gravity is Bernoulli's equation:

$$\text{along a streamline: } p + \tfrac{1}{2}\rho v^2 + \rho g h = \text{a constant.} \tag{16-22}$$

When the fluid is at rest, this reduces to Eq. (16–6); when the fluid flow is horizontal, this reduces to

$$p + \tfrac{1}{2}\rho v^2 = \text{a constant.} \tag{16-24}$$

This equation is the basis of the Venturi flowmeter, which measures the speed of fluids in pipes, as well as the basis for understanding the flow of fluids from holes and the lift on wings.

It is possible to go quite far in understanding fluids with a number of simplifications. The relaxation of the simplifying assumptions brings in rotational flow, viscosity, and turbulence. The behavior of real fluids remains a field of active study.

Understanding the Concepts

1. How does a hot-air balloon get off the ground?
2. We weigh a brick twice with a bathroom scale. The first time, the scale and the brick are inside a tub of water, and the second time, they are outside the tub. Water can enter the scale. Does the scale indicate a difference in weight in the two cases?
3. One accurate method of measuring the fat content of your body is to measure your weight twice: once while you are immersed in a tank of water, and once when you are out of the water. Explain how such a method might work.
4. The most exciting white water for canoeing often occurs when the river narrows. Why is this?
5. If an empty balloon is weighed on a scale, a certain value is obtained. If a child now blows up the balloon and waits for several minutes so that the temperature of the air in the balloon is the same as the room temperature, what value will the scale give for the balloon's weight?
6. Suppose that a certain fluid has no viscosity. Will the acceleration of objects falling under gravity within this fluid be independent of the mass of the objects?
7. You have two objects in a bathtub—one floating high up, the other just barely floating. Does the buoyant force cancel gravity in each of these cases?
8. Water-storage tanks for communities normally are placed high off the ground rather than underground. This is obviously a safety hazard, so why is it done?
9. Mercury is poisonous. Why, then, is it typically used in barometers to measure atmospheric pressure?
10. The "antilift" effect for a racing car, the opposite of the lift effect on airplanes, is said to be so strong that a car could race on an upside-down track. Is this a plausible statement? Make some rough estimates of the surface area of a racing car, its mass, and its velocity, and assume that the difference between the air speeds under the car and over the car is about 20 m/s.
11. In Example 16–10 we suggested that the minimum pressure at the giraffe's head must be $p = 0$. But why couldn't the minimum value of pressure be negative? Comment on the signs of the three terms in Bernoulli's equation, Eq. 16–21.
12. One way to get the water out of a stopped-up sink is to use a long piece of rubber tubing. You put one end in the water and suck on the other end until you have filled the tubing. You then close off your end, bring that end to a point below the bottom of the sink, and reopen that end (Fig. 16–31). The water will flow out through the tubing until the sink is empty. You have made use of the *siphon mechanism* (see Problem 73). Explain how it works.

▲ **FIGURE 16–31** Questions 12, 13, and Problem 73.

13. The siphon shown in Fig. 16–31 and described in Question 12 is limited in the height h_0 that it can overcome. What determines this limit?
14. Suppose mercury were less dense than it is. Would the height of a barometer column be (a) taller or (b) shorter?
15. Do sailboats make use of any of the aspects of Bernoulli's laws, such as those we described for airplanes? In particular, do sails need to be curved in order to work?
16. A common classroom demonstration employs a table tennis ball in a funnel. An air hose is connected to the tube of the funnel and air is blown through. If the funnel is held upside down, the ball does not fall out. Why?

17. Which weighs more, a ton of Styrofoam or a ton of lead? Which of the two has the larger volume? How might you estimate the density of each?

18. People find it very easy to float in the Dead Sea. Why?

19. The point of a cone-shaped bottle (pointed end up) is removed to provide an opening. The pressure of the liquid at the very bottom of the bottle is the sum of the atmospheric pressure and the weight of the liquid above. If we consider a part of the bottom that is not directly below the opening, we might argue that there is less pressure because, above that point, there is less liquid and no atmosphere. This argument violates Pascal's principle. What has the argument left out?

20. In Example 16–11 we described water flowing from a hole cut in a tank. What happens to the flow with the passage of time? In particular, is the rate of flow constant? Is the distance from the tank to the landing point of the stream constant?

21. Suppose that you put a mixture of oil and water into a centrifuge that rotates at high speeds (the oil is less dense than the water and does not mix with it). The two liquids separate. Why, and which component will be farthest from the center of the centrifuge?

22. An old proposal for transcontinental travel along a fixed latitude is the following: Take a balloon high into the stratosphere. Wait until Earth has rotated the desired distance beneath you, then descend. Comment on the feasibility of this mode of transport.

23. A snorkel is a breathing tube meant to be used while you swim under water; it runs from your mouth to the water surface. Why is the length of a snorkel limited?

24. A pressure of 8 kPa across the eardrum is enough to cause serious pain, and 17 kPa will rupture the eardrum. What fraction of an atmosphere is this? Given your answer, why is it that merely stepping outside is not dangerous?

25. In the course of a lazy summer morning, ice cubes floating in a pitcher of water melt. What happens to the water level in the pitcher? (Ignore evaporation.)

26. Suppose that you put a kilogram weight and a jar half-filled with water on a kitchen scale. The scale reads 2.5 kg. Now you place the kilogram weight *inside* the jar, and no water spills out. You might argue that because the kilogram weight weighs less in water than in air, the scale would read less than 2.5 kg, but the reading remains 2.5 kg. What is wrong with your argument?

Problems

16–2 Density and Pressure

1. (I) Jupiter has a radius $R = 7.14 \times 10^4$ km, and the acceleration due to gravity at the surface is $g_J = 22.9$ m/s^2. Use these data to calculate Jupiter's average density.

2. (I) The numbers given in Problem 1 have the following values for some of the other planets: Venus: $R = 6.05 \times 10^6$ m, $g = 8.86$ m/s^2; Mars: $R = 3.39 \times 10^6$ m, $g = 3.73$ m/s^2; Uranus: $R = 2.54 \times 10^7$ m, $g = 9.12$ m/s^2; Neptune: $R = 2.48 \times 10^7$ m, $g = 12.0$ m/s^2. Calculate the densities of these bodies.

3. (I) The density of a nucleus is about 2×10^{17} kg/m^3. The amount of water in a large lake is 10^{13} m^3. If this amount of water were compressed to nuclear density, how many liters of water would there be?

4. (I) A platinum sphere has a diameter of 0.96 cm. What is the diameter of an aluminum sphere of the same mass, given that the densities of platinum and aluminum are 21.4×10^3 kg/m^3 and 2.70×10^3 kg/m^3, respectively?

5. (I) Atoms of an imaginary two-dimensional solid are arranged in a square lattice. The distance of neighboring atoms is a. Assume that the solid is acted upon by a shear force and deformed by a *small* angle f without any change in the nearest neighbor distance (Fig. 16–32). Calculate the change in the distance between atoms situated at opposite corners of a square.

6. (I) The density of mercury is 13.6 times that of water. Compare the height of a column of water to that of a column of mercury, assuming that the pressure exerted by the weights of the respective liquids at the bottom of each column are the same.

7. (I) A circus clown stands on a pair of stilts that each have a square cross section of 4.0 cm per side. If the mass of the clown plus the stilts is 68 kg, what pressure is exerted on the floor?

8. (II) A wedge of mass 15 N and opening angle $\theta = 55°$ is floating on the surface of water in the symmetrical position shown in Fig. 16–33. Calculate the force of the water that acts on each of the two surfaces.

▲ **FIGURE 16–33** Problem 8.

9. (II) A hollow stainless steel sphere of radius 20 cm is evacuated so that there is a vacuum inside. (a) What is the sum of the magnitudes of the forces that act to compress the sphere? (b) There is a circular hole of diameter 4 cm on the side of the sphere for access to the inside. Calculate the force needed to pull a flat plate off the hole when the sphere is evacuated. Do you think that you could remove such a plate by pulling on it?

10. (II) During a hurricane, the atmospheric pressure changes dramatically. Explain why it is recommended that house windows be kept slightly open during a hurricane. What is the net force on a wall that is 300 ft^2 in area when the pressure on one side is 14.7 lb/in^2 and the pressure on the other is 14.0 lb/in^2?

▲ **FIGURE 16–32** Problem 5.

11. (II) A hollow metal pyramid with a square base that is 15 cm on each side has a mass of 1.8 kg. What is the pressure exerted by the pyramid on the table on which it stands? Suppose that the temperature in the room rises so that the metal expands. Will the pressure increase or decrease as a result of the expansion?

16–3 Pressure in a Fluid at Rest

12. (I) Using the densities in Table 16–1, what is the height of a fluid column in a barometer that has water as its fluid, in an air pressure of 1 atm? One that has alcohol? (Neglect the vapor pressure of the fluid above the column.)

13. (I) What is the pressure at depths of 1 m, 10 m, 100 m, and 10 km under the surface of the ocean? Take $\rho = 1.03 \times 10^3$ kg/m^3 for the density of sea water and $p_0 = 1.01 \times 10^5$ Pa for the atmospheric pressure at the ocean surface. (Ignore the fact that, to this accuracy, density varies with depth.)

14. (I) Recreational scuba divers rarely go deeper than 20 m. What is the water pressure at this depth? By what factor does your result differ from sea-level water pressure?

15. (I) What is the pressure 1 cm below the surface of a column of mercury exposed to air? 3 cm below the surface?

16. (I) An oceanic research vehicle operates at a depth of 1.20 km under water. Assuming that the pressure inside the vehicle is 1.0 atm, what is the force on a small window of dimensions 10 cm × 10 cm?

17. (II) Assume that a spaceship lands on Venus with the cabin pressure maintained at 1 atm. What would be the height of a column of mercury in a barometer in the spaceship on Venus? Repeat for a spaceship landing on Neptune. (Use the data given in Problem 2.)

18. (II) Tests in pressure test chambers show that mental processes are affected when the gauge pressure exceeds 35 lb/in^2. *Gauge pressure* is the pressure in excess of atmospheric pressure. This is a well-known effect that limits the depth at which scuba divers should breathe pure air. In sea water, whose density is 1.03 g/cm^3, to what depth should a diver be limited?

19. (II) A hydraulic jack is used to lift a 1200-kg car on a piston of diameter 30 cm (Fig. 16–34). How large a force is needed to push down the smaller piston if its diameter is 2 cm? By how much is the car lifted with a single push, assuming that the small piston moves 0.50 m?

▲ **FIGURE 16–34** Problem 19.

20. (III) The vertical wall of a small dam is 32 m long and it holds back a body of water that is 5 m deep (Fig. 16–35). Calculate the force acting on the wall due to the pressure of the water. Determine the effective point of action of this force. [*Hint*: Recall the method used to calculate the center of mass of a nonuniform object in Chapter 8.]

▲ **FIGURE 16–35** Problem 20.

16–4 Buoyancy and Archimedes' Principle

21. (I) A bathysphere (a spherical, watertight research facility) 2.6 m in diameter has a mass of 9400 kg. It is released from a submarine at a depth of 20 m below the surface. Will the bathysphere sink or rise?

22. (I) Use the densities in Table 16–1 to show that ice floats in water. What fraction of the volume of a floating ice cube will be above the surface?

23. (I) A cubical box, contents unknown, floats in water; 16 percent of the volume of the box is above water. What is the average density of the box and its contents?

24. (I) Over 10,000 icebergs are calved from Greenland each year. These icebergs are a serious danger to Atlantic shipping because most of the iceberg remains under water, invisible. If 10% of the iceberg sticks out of the water, what is an iceberg's density?

25. (II) A supertanker can carry 2.2×10^5 m^3 of oil with a density of 0.88 g/cm^3. When fully loaded its mass is 2.3×10^8 kg, and the dimensions of its hull are approximately 400 m long, 60 m wide, and 38 m high. Given that the density of seawater is 1.03 g/cm^3, how deeply is the hull submerged in the water?

26. (II) A hemispherically shaped bowl is floating on the surface of water; the mass of the bowl is 0.6 kg (Fig. 16–36). Water is poured into the bowl; when 4.3×10^3 cm^3 of water has been added, the bowl just sinks. What is the outer radius of the bowl?

▲ **FIGURE 16–36** Problem 26.

27. (II) Find the volume of the smallest block of ice which when floating on salt water will carry a person of mass 80 kg. Assume that the density of salt water is 1.03×10^3 kg/m^3. You will have to look up the density of ice.

28. (II) 75% of the volume of a sphere floating in a pool of mercury is above the surface. Given that the density of mercury is 13.6×10^3 kg/m^3, find the average density of the sphere.

29. (II) A group of Scouts intend to make a raft to float down the Ohio River. The mass of four of them plus their equipment is 400 kg. They find trees with an average diameter of 20 cm and a specific gravity of 0.80. Determine the minimum area of the log raft that would keep them dry.

30. (II) A temporary bridge is supported on floating empty drums that are 1.0 m in diameter and 1.8 m long. The design criterion is

that the drums may only be submerged to 2/3 of their diameter under a maximum load of 5000-kg trucks, 7.5 m long, that follow each other bumper to bumper in a single line. How far apart should the drums be placed? Neglect the weight of the bridge and of the drums.

31. (II) A valuable preserved biological specimen is weighed by suspending it from a spring scale. It weighs 0.45 N when it is suspended in air and 0.081 N when it is suspended in a bottle of alcohol. What is its density?

32. (II) Consider a spherical balloon filled with helium, with a density of 0.18 kg/m³. The density of air is 1.3 kg/m³. What must the radius of the balloon be in order to lift a load of 130 kg (including the mass of the balloon)?

33. (II) Olive oil floats on water. Take ρ_1 to be the density of the oil and ρ_2 to be the density of the water. Consider an oil–water interface across which a bouillon cube of density ρ_3 floats; use the geometry shown in Fig. 16–37. We know that for the oil to float on the water, ρ_1 must be less than ρ_2. What is the condition on ρ_3 so that the cube floats at the oil–water interface? How much of the cube will be in the water?

▲ **FIGURE 16–37** Problem 33.

34. (II) A beaker of mass 350 g contains 2.8 kg of water and rests on a bathroom scale. A 5.2-kg block of aluminum, of specific gravity 2.7, is suspended from a spring scale and completely submerged in the water. Find the readings on both scales.

35. (II) Balloons of mass 3.5 g can be filled with helium to make spheres of radius 21 cm. How many of these balloons must a child of mass 32 kg hold in order to float off Earth's surface? Assume that helium has a density of 0.18 kg/m³.

36. (II) An air-filled lead balloon of radius $R = 0.1$ m is totally submerged in a tank of water (Fig. 16–38). What is the thickness t of the lead skin of the balloon if the balloon neither rises nor sinks? The density of lead is 11.3×10^3 kg/m³.

▲ **FIGURE 16–38** Problem 36.

37. (II) (In this problem we ask you to draw the figure that is described here.) When a body floats it is because the downward force of gravity just cancels the upward force of buoyancy. This alone is not enough; one also wants to know if the floating is stable, that is, whether the floating object might tip over. In thinking about whether, say, a tilting boat will right itself, it is important to note that gravity acts as a single downward force passing through the center of mass of the body (denote this point by M) while the buoyancy acts as a single upward force *passing through the center of mass of the displaced fluid* (a point denoted by B). In equilibrium B and M lie on the vertical, with B usually below M. To determine stability, we look at the body tilted. Denote the new center of mass of the displaced fluid by C. Draw a vertical line up through C. If this line intersects MB below M then the torques due to the forces of gravity and the force of buoyancy act in the same direction, and the body will topple. (a) Draw a diagram to show that a wooden log will not float vertically in water. (b) Draw a diagram to show why having several people stand up in a canoe is likely to tip it over.

38. (III) A sphere of radius R of material with an average density of 0.82 g/cm³ is immersed in a body of water. What is the height of that portion of the sphere that projects above the surface?

16–6 The Equation of Continuity

39. (I) Water flows at a speed of 1.3 m/s through a hose of diameter 1.5 cm and emerges from a nozzle of radius 0.50 cm. With what speed does the water emerge? How much water emerges from the nozzle per second?

40. (I) Water flows down a channel 1.0 m deep and 0.5 m wide at a rate of 2 metric tonnes/s. At some point the channel widens to 0.8 m. How fast does the water flow in the wider channel?

41. (I) A steady stream of automobiles moves along the eastbound side of a highway in a region where there is neither exit nor entrance. Where two eastbound lanes are available, the automobiles are spaced so that there are 66 cars per mile in each lane, and the average speed is 45 mi/h. In a region to the west, where three eastbound lanes are available, the automobiles are spaced to give 35 cars per mile in each lane. What is the average speed in the region where three eastbound lanes are available?

42. (II) A lawn sprinkler is connected to a garden hose of inside diameter 2.5 cm that has a water-flow rate of 0.5 L/s. The sprinkler has four nozzle openings, each with a diameter of 7 mm. (a) Calculate the speed of the water that emerges from the nozzles. (b) How could the nozzles be arranged to cause the sprinkler to rotate? What principle could you use to calculate the torque on the sprinkler? What are the limitations, if any, on the speed of rotation of the sprinkler?

43. (II) Students would like to spray water 18 m across their neighbor's yard onto a patio. They have a hose of inside diameter 1.5 cm that can spray a distance of only 1.5 m. What size nozzle do they need?

44. (II) When water drains out of a hole in the bottom of a tank, the stream radius contracts as the speed of the water in the stream increases. Assume that the stream starts with zero water velocity at the top of the tank, at a height H above the hole, and ignore the various instabilities that will eventually cause the stream to break up. Find the distance below the hole at which the radius of the stream is 80 percent of the hole radius.

16–8 Applications of Bernoulli's Equation

45. (I) A bilge pump is used to pump water out of a leaking ship (Fig. 16–39). The pump's hose has a diameter of 3.0 cm, and the pump moves water through the hose up and out of a porthole 5.0 m above the waterline at a speed of 4.0 m/s. Calculate the power of the pump.

▲ **FIGURE 16–39** Problem 45.

46. (I) What pressure must a pump generate to get a jet of water to leave a nozzle at a speed of 8 m/s? Can this jet reach a height of 8 m above the nozzle? If so, what angle should the nozzle make with the horizontal so that the jet reaches 8 m at its maximum height?

47. (I) Wind gusts at 60 mi/h past a ship's cabin in which the air is at rest and the pressure is 1 atm. What are the pressure outside the cabin and the net pressure on the walls past which the wind blows?

48. (I) A stream of water sprays like a fountain vertically from a small hole in a pipe, reaching a height of 2.4 m. Assuming that the water in the pipe is static, what is the gauge water pressure in the pipe? (See Problem 8.)

49. (I) A folded piece of notebook paper can exhibit lift in the air flow from a fan of a window air conditioner. A plastic straw is taped perpendicular to the plane of the paper through the paper's middle, and a thread runs through the straw allowing the paper to move up and down the thread. The folded paper has an area of 300 cm^2. The air speed is about 4 m/s and the proportionality constant K in the approximate lift formula $KA\rho v^2$ is taken to be $K = 0.14$. Here, A is the surface area of the wings, ρ is the air density, and v is the air speed. How much lift is possible? Given the mass of the paper and straw to be 8 g, will the paper rise?

50. (II) A fountain shoots a vertical jet of water at a rate of 0.33 L/s to a height of 1.6 m. (a) What is the initial speed of the jet, and what is the radius of the hole out of which the jet passes? (b) What pressure must the pump of the fountain supply? (Assume that it sits just below the emerging jet.) (c) At height 1.2 m, what is the speed of the jet, and what is the radius of the column of water? Ignore effects of turbulence; in particular assume that the jet does not break up.

51. (II) Heavy rains have flooded your favorite professor's basement to a depth of 15 cm. This basement is 7.5 m × 12 m in area. You have a hose of diameter 1.2 cm and can run the hose from the water in the basement to a level 3 m below the level of the basement floor, into a low portion of the yard. When the hose is filled with water, it acts as a siphon (see Question 9) and empties the basement. (a) What is the speed of the stream of water leaving the hose? (b) Approximately how long does it take to empty the basement?

52. (II) A liquid of density 1.05×10^3 kg/m^3 flows in a horizontal pipe. The cross-sectional area in one part of the pipe is 42 cm^2. When the liquid enters another part of the pipe with cross-sectional area 56 cm^2, the pressure as measured by a gauge is 0.4×10^4 Pa higher than it was in the first part. Calculate the velocities of the liquid in the two parts of the pipe.

53. (II) A large horizontal pipe of diameter 10 cm contains water flowing with a speed of 0.3 m/s. The pipe branches into four pipelets, each of diameter 2 cm. The four pipelets each run horizontally after an initial height change of 3.5 m (Fig. 16–40). (a) What are the speeds of the fluid in the horizontal portion of each pipelet? (b) If the pressure in the large pipe is $p = 2.5$ atm, what is the pressure in the pipelets?

▲ **FIGURE 16–40** Problem 53.

54. (II) A uniform glass tube with internal cross-sectional area A is formed into a U-shape and partly filled with a volume V of an incompressible fluid of density ρ. The two ends of the tube are open to the atmosphere. At equilibrium, the level in each vertical segment of the tube is $h = 0$. By applying additional pressure to one end of the tube, the levels of the liquid are displaced, so one side is at height h_0 and the other is at height $-h_0$. The additional pressure is suddenly released, and the liquid level starts to oscillate. (a) Use Bernoulli's equation to show that the force on the fluid in the tube is proportional to the displacement of the fluids from their equilibrium levels, so the force is harmonic. (b) Find the "spring constant" of the harmonic motion described in part (a) and use it, together with the mass in motion (ρV), to calculate the oscillation frequency.

55. (II) Water flows through a large horizontal pipe of diameter D at speed v_0. The pressure in the pipe is p_0. The pipe branches into two horizontal pipelets of diameter d_1 and d_2, respectively, both at the same height as the large pipe. Write the equations that would allow you to find the pressures p_1 and p_2 and the speeds v_1 and v_2 in the two pipes. Show that one of these four quantities must be known to be able to solve for the others.

56. (II) A vertical cylinder of height 1.0 m and radius 10 cm is filled with water and placed at the edge of a table. A 1 cm diameter hole is drilled in the side wall near the base. Water starts flowing out, forming a stream that passes over the edge of the table. Neglecting any turbulence, calculate the velocity of the water as it leaves the cylinder. What will be the diameter of the stream of water at a point 1.0 m below the opening?

57. (III) The surface of a bucket of water centered on a rotating turntable will form a parabola with its minimum on the rotation axis. Prove that this is the case by the consideration of a small mass element at the surface, located at the point (y, r), where y is the height from the eventual minimum on the axis and r is the radial distance from the axis. This is in equilibrium, and subject to (a) the atmospheric pressure normal to the surface of the liquid at the location of the mass, (b) the force of gravity, and (c) the forces associated with the rotation. Show that the slope dy/dr at the location of the mass is proportional to r, so that $y = y_0 + (\text{constant})r^2$. What is the value of the constant?

58. (III) An open water tank stands on a plane surface. The water surface in the tank is a height h above the plane. A small hole is opened up at a depth y below the surface of the water. (a) Show that the jet of water will hit the plane surface a distance D from the tank, where $D = \sqrt{4y(h - y)}$. (b) Show that the hole should be placed at a depth $y = h/2$ for the jet to cover a maximum horizontal distance.

*16–9 Real Fluids

59. (II) The retarding force on a sphere of radius R that moves with speed v through a fluid of viscosity η is proportional to η and depends on R and v, in the form $R^a v^b$. Use dimensional analysis to determine the powers a and b.

60. (III) The result of the solution of Problem 59, the retarding force on a sphere of radius R that moves with speed v through a fluid of viscosity η, is $F = 6\pi\eta Rv$, known as *Stokes's law*. Such a sphere, with mass density ρ_m, falls through glycerin, whose density is ρ. The forces on the sphere are the retarding Stokes's law force, the buoyant force, and the force of gravity. (a) Derive a formula for the value of the velocity v_t for which the net force is zero. (b) How does v_t depend on the radius of the sphere? (c) If the viscosity of glycerin is $8.3 \text{ kg}/(\text{m} \cdot \text{s})$ and its density is $\rho = 1.26 \times 10^3 \text{ kg/m}^3$, what is the terminal velocity v_t of a sphere of iron, which has $\rho = 8.5 \times 10^3 \text{ kg/m}^3$ and a diameter of 5 cm? Of an iron sphere with a radius of 1 mm?

General Problems

61. (II) A rectangular container is divided into two parts by a movable vertical partition. This partition is sealed so that no water can seep from one side to the other. Initially, the wall is held in the middle of the 30-cm length of the container. One side is filled with 2 kg of water to a depth of 10 cm; the other side is attached to a spring in equilibrium (unstretched, uncompressed). The spring constant is 180 N/m. The partition is now released and the spring compresses. What is the position of the partition at maximum compression? Assume that the process takes place without turbulence so that no energy is dissipated.

62. (II) A hollow sphere of radius 0.08 m and total mass 0.35 kg is placed in water. One very small lead pellet of mass 0.08 kg is attached to the inside surface of the sphere. When the sphere is floating in equilibrium, the lead pellet is oriented on the bottom. If the sphere is rotated slightly, so that the pellet is moved off to the side, and then released, the sphere will rock, and the pellet will move like a pendulum. Calculate the period of small oscillations for this motion. Assume that the center of the sphere does not move.

63. (II) A depth of about 100 ft (≈ 30 m) is as deep as a scuba diver should go without special precautions because of an effect that causes disorientation. What is the water pressure at a depth of 30 m? Does a scuba diver take in the same mass of air in each breath at 30 m that he or she would at a depth of 5 m? Explain.

64. (II) A cylindrical bucket of fluid of density ρ sits at the center of a turntable that rotates with some fixed angular velocity, ω. Show that the pressure variation in the fluid at a horizontal distance r from the axis is given by $p = p_{\text{axis}} + \frac{1}{2}\rho\omega^2 r^2$. [*Hint:* If you place yourself within the rotating fluid and at rest relative to it, the problem is one of hydrostatics. The only force in the horizontal direction is a fictitious force away from the axis of the rotation, due to the fact that your frame of reference is not inertial. This force is the centrifugal force.]

65. (II) Late in the seventeenth century, King Louis XIV of France had a series of fountains constructed at Marly, some 20 km from his chateau at Versailles. These fountains were fed from the Seine River, whose altitude is some 150 m below that of Marly, through a remarkable series of pumps. The tallest of these fountains was the Grand Jet, which is calculated to have risen a height of 37 m from its base. (a) At what speed does this jet leave the orifice at the fountain's base? (b) The rate of flow in the Grand Jet is measured to be 0.051 m^3/s. Compare this to an estimate of the rate of flow from a garden hose. (c) Given the rate of flow in part (b), calculate the area of the orifice from which the Grand Jet comes. (d) What is the pressure of the water just

behind the orifice from which the Grand Jet comes? Beyond the crude calculations outlined here, the height to which a jet of water will rise can be affected by the addition of a short length of conical pipe above the level of the orifice at the base.

66. (II) A cylindrical tank is filled with water to a height h_1 above its base. Unfortunately, someone has cut a hole in the bottom through which fluid escapes in a vertical stream into your room. (a) What is the speed of the escaping stream at the hole? (b) You insert a flexible hose into the hole. If the other end of the tube is held at the same height as the base of the tank and is directed horizontally, what is the speed of the water as it leaves the hose? (c) You hold the tube, whose end is still at the height of the base of the tank, so that the stream is directed straight up. How high does the escaping stream shoot?

67. (II) An inventor proposes to you the following perpetual motion machine (a machine that can do work "forever" without your having to add energy to it) (Fig. 16–41). A cylindrical tower of height h filled with water rests on the ground. Blocks of wood are inserted at the bottom of the tower. They float to the top, where they are removed. The potential energy that they acquired in floating to the top can be converted into work. They can, for example, be dropped down for re-use, with the kinetic energy they acquire in falling used to turn the wheels of a turbine. When they reach the bottom, the heavier blocks are again inserted, and the process is repeated. How can you prove to the mad inventor that at best this machine can do no net work? [*Hint:* Calculate the work required to insert a block, which requires in effect that a mass of water whose volume is that of the block be raised to the top. Compare this to the energy the block acquires in being accelerated to the top under the influences of gravity and the buoyant force. Neglect any drag caused by the water.]

Do not buy this machine! Perpetual motion machines are impossible to construct (see Chapters 18 and 20). A close analysis of the details of such machines always reveals this fact.

Water sealed
exit port

Sealed
entry
port

Insertion
mechanism

▲ **FIGURE 16–41** Problem 67.

68. (II) Consider the machine discussed in Problem 67. Why does the machine still fail if you replace the entry port by an exchange chamber (Fig. 16–42, see next page) with two sliding doors, so that the insertion of the blocks does not require that work be done?

69. (II) A cubical block of wood 0.20 m on each side floats in water, with 8.0 cm of the block above the surface of the water (Fig. 16–43). (a) What is the density of the wood? (b) Suppose that the block of wood is pushed down below the equilibrium level by a distance Δ. What is the magnitude of the restoring force that pushes the wood back toward the equilibrium level? (c) In general, if dissipation of energy is ignored, the block will overshoot the equilibrium level and rise to a height Δ above that level, and is then subjected to a downward force. How large is that force?

▲ **FIGURE 16–42** Problem 68.

(d) Recall from Chapter 13 that a force that is proportional to the displacement Δ from equilibrium (of the form $F = -k\Delta$) and acts on an object of mass m leads to oscillations with the frequency $f = (k/4\pi^2 m)^{1/2}$. Calculate that frequency for the block that bobs in the water.

▲ **FIGURE 16–43** Problem 69.

70. (II) The ticket-window opening of a movie theater, where money goes into the cashier's hands and tickets emerge, is a rectangle 12 cm × 15 cm. A breeze of speed 3 m/s comes out of this opening. Sometimes, as a result, paper money goes flying. (a) What is the difference in air pressure between the inside and the outside of the theater? (b) Suppose that the opening were the size of a door, 0.9 m × 2.1 m. What would the speed of the breeze be in this case, assuming the air pressure difference were the same as in part (a)?

71. (II) A water-storage tank is an upright cylinder of height 8 m and radius 6 m. The storage tank is full of water but is vented to the atmosphere. The bottom of the tank is placed 30 m off the ground. A pipe of diameter 12 cm runs vertically down from the tank and goes 1 m underground before turning horizontal. (a) What is the water pressure in the bottom of the tank? (b) in the horizontal pipe underground? (c) The water flow in the pipe is 80 L/s. How fast does the water level in the tank drop? (d) A saboteur drills a hole of diameter 6 mm near the bottom of the tank. How fast does the water shoot out? (e) What volume flow of water is lost out the hole?

72. (II) A *Pitot tube* is a device for measuring flow velocities (Fig. 16–44). It is as useful for measuring the flow velocity of blood in an artery as for measuring the air speed of a jet airplane. The fluid passes the opening at point B, creating a pressure p_B inside. The fluid enters the opening of the tube and comes to a halt, creating a pressure p_A. Apply Bernoulli's equation as needed to show that the flow velocity is given by $v = \sqrt{2(p_A - p_B)/\rho} = \sqrt{2gh\rho_\ell/\rho}$, where ρ is the density of the fluid being measured, ρ_ℓ is the density of the fluid shown in the device, and h is the difference in height of the liquids.

▲ **FIGURE 16–44** Problem 72.

73. (II) With a siphon, it is possible to transfer fluids from one container to another. Figure 16–31 shows a siphon that must be filled with fluid to start the transfer. If the speed with which the fluid leaves the higher container is zero, and if the fluid has a fixed density ρ, find an expression for the speed of the fluid as it enters the lower container, in terms of the heights h_0 and h_1.

74. (III) A tank of surface area A is filled with water to a height x above the position of a hole of surface area σ, and the water flows out. Assume throughout that the fluid speed is zero at the top surface. (a) How much water flows out of the hole in the first time interval Δt after the hole is opened? (b) How large is the drop in the water level, Δx, in that time interval? (c) Use the results of parts (a) and (b) in the limit of small Δt to express a relation of the form $dx/dt = f(x)$. This is a differential equation for the height as a function of time. Take care to get the right signs, noting that x decreases with time, so that Δx is a negative quantity. (d) Show that this equation is solved by the formula

$$x(t) = \left[\sqrt{x_0} - \frac{1}{2}\left(\frac{\sigma}{A}\right)t\sqrt{2g} \right]^2.$$

(e) How long will it take for the water to drop to the level of the hole?

75. (III) The ice-cream cone shown in Fig. 16–45 is filled with melted ice cream of density 1.2 g/cm^3. The cone has a diameter of 6 cm at the larger end and is 10 cm long. Find the pressure at the bottom of the cone. If a small hole of diameter 1 mm is opened at the bottom, the ice cream starts to run out. Ignoring the viscosity of the melted ice cream, find the amount of time it takes the ice cream to run out. Assume that the fluid speed is zero at the top.

▲ **FIGURE 16–45** Problem 75.

◀ That the properties of materials depend on temperature is no surprise. Whether we are referring to the steel being worked in this photograph or to the properties of the air over the polar ice cap, we would fail to pick out essential features if we did not recognize the role of temperature.

Temperature and Ideal Gases

Many properties of the physical world depend on temperature. For example, the compound water behaves quite differently at temperatures below its freezing point than it does at temperatures above its boiling point. Temperature has something to do with energy, as we can see from the fact that hot steam can lift a piston, and the hotter the steam, the better it is able to perform the work. How much energy does hot steam contain, and how does that energy get into the steam? How can we quantify "hotness" or "coldness?" To answer these questions, we will consider both the macroscopic properties and microscopic behavior of gases, liquids, and solids, focusing primarily on gases. These properties lead us to a good definition of temperature. We will see how temperature relates to energy, how that energy is like the mechanical energy we have earlier studied, and how efficiently that energy can be used.

The historical development of thermal physics was based on the properties of bulk matter—such as density, pressure, volume, and temperature—before atomic theory was well established. The study of thermal phenomena received its strongest impetus from engineers who studied the steam engines that powered the Industrial Revolution of the 19th century. Their treatment of thermal phenomena on the macroscopic scale is known as **thermodynamics**. Nineteenth-century physicists also learned how these macroscopic phenomena could be understood in terms of the behavior of the atomic constituents of matter. This work, known as **statistical mechanics**, brings in an entirely new idea—that it is the statistical behavior of these constituents that is reflected in thermal behavior in the macroscopic world. All these topics will be discussed in Chapters 17 through 20 under the general heading of thermal physics. Thermal phenomena are pervasive in the physical world, and our understanding of them has had a profound effect on the way we live.

17–1 Temperature and Thermal Equilibrium

We call a physical system a **thermal system** whenever we are interested in its temperature-dependent properties. By virtue of the fact that it is hot, steam can lift a piston and do work on its surroundings, so that a piston and a cylinder full of hot steam (Fig. 17–1) form a good example of a system that is worth keeping in mind as we proceed, even if we will be more interested in the work that thermal systems do in later chapters. We all have an intuitive feeling for temperature—for example, we can usually tell when one system is hotter than another or, in other words, that its temperature is higher. Our aim here is to develop a more precise notion of temperature.

Central to the idea of temperature is the notion of thermal equilibrium. Suppose that we pour cold water into a bucket of hot water. Experience shows that after a short time the water becomes lukewarm throughout. Until that point, we cannot say that the water has a particular temperature. When the temperature and temperature-dependent quantities are no longer changing, we can say that **thermal equilibrium** has been reached (Fig. 17–2). At this point the temperature becomes a variable that characterizes the whole bucket of water; that is, the whole thermal system. Similarly, the gas in a container eventually comes to thermal equilibrium if its container is in thermal equilibrium with its surroundings, and two containers of gas at different temperatures will come to thermal equilibrium if they are allowed to mix. One mechanism for this process is collisions between the constituent molecules of the gas, but as we shall see there are other mechanisms and mixing is not necessary. We shall see in some detail that the temperature is a measure of how fast the molecules of the gas are moving, and in collisions between fast-moving and slow-moving molecules, the fast-moving ones will generally slow down and the slow-moving ones speed up, thus bringing the ensemble to a common temperature.

If two thermal systems are in thermal equilibrium with one another, then *they have the same temperature*. We can add that if two systems are each in thermal equilibrium with a third, then they are in thermal equilibrium with one another. The observation that if system *A* is in thermal equilibrium with system *B*, and if system *B* is in thermal equilibrium with system *C*, then *A* is in thermal equilibrium with *C* is sometimes called the **zeroth law of thermodynamics.**

If we put an ice cube into a bathtub full of hot water, the temperature of the tubful of water changes very little; if we add a teaspoon of hot water to a swimming pool, the temperature change is virtually undetectable. These observations allow us to define a **thermal reservoir** (or **heat bath**) as a thermal system so large that it maintains a constant temperature when it interacts with other thermal systems. A thermal reservoir is of importance because it allows us to bring a given (smaller) system to a predetermined temperature, the temperature of the reservoir, simply by bringing the system into **thermal contact** with the reservoir.

Thermal contact is established between thermal systems in three important ways. We can illustrate them with two metal plates in an enclosure, one red hot and the other cold. Thermal contact by *conduction* occurs when the two plates actually touch. "Something" passes between them, which with foreknowledge we can describe as the agitation of the molecules of the hot plate passing through collisions to the agitation of the molecules of the cold plate. When you sit on a cold metal bleacher seat at a football game, there is thermal contact by conduction between you and the seat. Thermal contact by *convection* occurs when there is a fluid (a liquid or a gas) between the plates. The fluid acts as a third thermal system whose temperature increases in the region near the hot plate through its own thermal contact with the plate. The fluid then circulates in currents to the colder plate, increasing its temperature. The reason for this circulation has to do with thermal properties of the conducting fluid; for example, as we saw in Chapter 16, hot gas is less dense than cold gas, so that it tends to rise. If you live in a

▲ **FIGURE 17–1** A bottle of hot steam with a piston at one end provides us with a system that can do work by virtue of the thermal properties of the steam. In particular, the hotter the steam, the farther it can move the piston.

▲ **FIGURE 17–2** The vendor's stand is not in thermal equilibrium with its environment. The steam is evidence of an interaction with the surroundings that would eventually cool the stand down to where it would be serving frozen hot dogs.

house with radiators, then you are warmed by the air that starts near the radiator and then circulates in the room. Similarly, if your house has hot air heat, the fact that hot air rises explains why heating ducts are placed on or near the floor. Finally, in thermal contact by *radiation*, a hot plate emits electromagnetic waves—the "radiation" in question—while a cold plate absorbs them. (We'll say a good deal more about electromagnetic radiation in the context of thermal physics in Section 17–5.) This form of thermal contact operates even with the plates separated by some distance and even across a vacuum. You experience this form of thermal contact if you tan on the beach or if you stand in the vicinity of a fire.

In contrast to the descriptions above, if two fluids at different temperatures are separated by a double-walled, silvered glass flask with a vacuum between the two glass surfaces, the fluids will take a very long time to reach the same temperature. The vacuum prevents thermal contact by conduction or convection, and the silvered surfaces of the glass walls reflect radiation back to the heat source and prevent thermal contact by radiation. The liquids are **thermally isolated** from one another. Don't confuse thermal contact with mechanical contact. If the pistons in an automobile engine were not in mechanical contact with the drive train, the car could not go very far, but the manufacturer goes to some length to keep the hot cylinders and their contents out of close thermal contact with most of the remainder of the vehicle.

THINK ABOUT THIS . . .

IS IT EASY TO MAKE A POT OF TEA IN A SPACE STATION?

That depends on the method you use. Aside from the fact that you would have to make sure the pot was closed to keep the water from drifting out, the technique you use for boiling water would have to change from what you do at home. On Earth the heating of a pot of water proceeds as follows: The water at the bottom becomes heated first, by conduction, and then a process of convection begins. The liquid moves in a circular pattern. The hot water rises, because on heating it expands, and since the density of the hot water is lower, there is a buoyant force. The hot water then pushes the cold water from above it to the bottom and the process continues until the water is uniformly heated. On a space station, the fact that the density of hot water is less than that of cold water makes no difference, because without an "up" or "down" as on Earth, there is no buoyancy. So some external stirring mechanism is needed to help you boil the water. Or you could wait for the entire pot to heat by conduction, a slower process. We might guess that microwaving a container of water would be a good way to proceed, since with a good microwave oven the heating of the water occurs uniformly throughout the volume.

Thermometers

A **thermometer** is a device that measures the temperature of a thermal system quantitatively, generally by coming to thermal equilibrium with that system. A thermometer must satisfy several criteria. It should have a minimal effect on the system being measured. (A small room thermometer brought inside from the cold will have this property.) The thermometer must make good thermal contact with the system being measured. Like a meter stick for measuring length, the thermometer should have a reproducible and easily read scale.

There are several properties of matter that are suitable for thermometry. Ordinary fever thermometers rely on the fact that fluids, such as mercury, expand and contract when their temperatures change. We can therefore make a thermometer by marking numbers on a column of mercury that measure the mercury's volume. Other properties, such as the pressures and volumes of gases, the volumes of liquids or solids, reflective properties of liquid crystals, electrical and magnetic properties of solids, and the color of light emitted by hot objects, all vary with temperature and can be used as thermometers (Fig. 17–3). Not all of the properties of matter that depend on temperature, however, are suitable for making thermometers: The brittleness of many materials depends on temperature, but we would not make a thermometer that works by seeing how easy it is to smash a material with a hammer! Table 17–1 lists some useful thermometers.

▶ **FIGURE 17–3** (a) This thermometer contains a (red) fluid that expands at higher temperatures and contracts at lower temperatures. (b) This digital thermometer uses the temperature dependence of electronic properties of materials. It is calibrated in the Celsius scale, in which the temperature of ice melting in pure water is 0 degrees.

(a)

(b)

TABLE 17–1 • Some Thermometers

Thermometer	Physical Property Measured	Comments
Ideal gas	Pressure and volume of dilute gas	See Section 17–2
Mercury bulb	Expansion or contraction of fluid	Good where fluid does not change phase
Bimetallic strip	Difference in expansion of two metals	
Resistance	Electrical resistance	See Chapter 26
Thermocouple	Electrical voltage across different metals	Most widely used thermometer in industry
Paramagnetic	Magnetic properties of matter	Useful at ultracold temperatures; see Section 31–5
Optical pyrometer	Color of emitted light	Useful at high temperatures; see Section 17–5

CONCEPTUAL EXAMPLE 17–1 A friend proposes that you check the temperature of boiling water that you have prepared in order to make some hot tea by dumping 10 large ice cubes of a calibrated size in it and seeing how long it takes for the ice cubes to melt. What is wrong with this thermometer?

Answer This is indeed a thermometer, and assuming that the initial temperature of the ice is known, a measurement of the melting time would indeed give you a way to measure the initial temperature of the water. However, this thermometer would change the temperature of the very system it is supposed to measure. By the time you complete your measurement, your water would be too cold to make your tea!

17–2 Ideal Gases and Absolute Temperature

Consider a gas in a container of fixed volume with a gauge that can read the pressure of the gas (Fig. 17–4). If we dip the container into thermal reservoirs at various temperatures, we find that as the gas becomes hotter, its pressure increases. The pressure correspondingly decreases when the gas becomes colder.

We can alternatively arrange to allow the volume to vary while keeping the pressure fixed, as in Fig. 17–1, with masses of various sizes sitting on the top of the piston. We find that the hotter gas has a larger volume for a fixed pressure, while the colder gas has a smaller volume for that same pressure. In fact, if we take all this into account, we find the following relation between the pressure p, the volume V, and the amount of gas n:

▲ **FIGURE 17–4** A container of gas of fixed volume with means to measure the pressure of the gas. This system forms a constant volume thermometer whose pressure is a measure of the temperature of the system with which the gas is in thermal contact.

The product of pressure and volume divided by the amount of gas in the bottle increases or decreases according to whether the thermal reservoir in contact with the bottle is hotter or colder, respectively. We accordingly define temperature T as proportional to the product of pressure and volume, divided by the number of moles, that is, $T \propto pV/n$.

The *constant volume gas thermometer* that we are describing here comes extremely close to being a universal thermometer because experiment shows that the proportionality constant is the same for all gases as long as their *densities are sufficiently low.* Dilute (or low-density) gases, which we call **ideal gases**, will enable us to give an unambiguous definition of temperature, a definition that can be employed by anyone who follows the simple set of instructions that we'll give below. In other words, we will have a thermometer that meets the requirements we have laid out. Just how dilute the gas needs to be is simply a question of the accuracy demanded of the thermometer. The lower the density of the gas, the more 'ideal' the behavior of the gas. Gas thermometers accurate enough for laboratory use may contain gas hundreds of times less dense than air at sea level.

Figure 17–5 plots the pressure of a fixed amount of a dilute gas at fixed volume against our temperature scale. The dashed part of the curve, near where T and p drop to zero, is an extrapolation and is drawn as a straight line.

We have not yet set our temperature scale. To do so, work with a constant-volume ideal gas thermometer and define the temperature $T = 0$ as the point where the pressure, p, would become *zero* (if that were possible), so

$$\text{for constant volume: } p = (\text{a constant}) \times T. \tag{17–1}$$

The point $T = 0$ is sometimes called **absolute zero**. Although experiments have come quite close to absolute zero, this temperature cannot in fact be reached and it corresponds to an extrapolation. We need to specify the constant in this relation, and we do so by defining T at some other point. This second point, called a *defined point*, was chosen in 1968 by international agreement to be the **triple point** of water. The triple point of water is the state at which water vapor, liquid water, and frozen water coexist. This can only happen at a single pressure p_{tp} ($= 4.58$ mm of mercury) and temperature; that temperature is defined as 273.16K. The letter K stands for kelvin, and the temperature scale so defined is the **Kelvin scale**, after Sir William Thomson, Lord Kelvin, who did important work on thermal phenomena and on electricity and magnetism during the mid- to late nineteenth century.) To summarize, temperature in the Kelvin scale, T, is defined by a constant-volume ideal-gas thermometer as

$$T = \left(\frac{p}{p_{tp}}\right)273.16\text{K}, \tag{17–2}$$

▲ **FIGURE 17–5** As p varies in an ideal gas of constant volume, T is defined by drawing a straight line.

▲ FIGURE 17–6 The Kelvin temperature scale is defined by a constant-volume ideal-gas thermometer. The two fixed points are $T = 0K$, located where p extrapolates to zero, and $T = 273.16K$, located at the pressure of the triple point of water, where $p = p_{tp}$.

where p is the pressure of the ideal gas in the thermometer. Figure 17–6 illustrates Eq. (17–2). The assignment of 273.16K as the triple point makes the Kelvin scale compatible with the Celsius scale, to be described below.

Table 17–2 gives us a feeling for the Kelvin scale by listing some temperatures that occur in nature. We call your attention to the numbers that refer to the origin of our universe in the so-called Big Bang, which occurred some 13 billion years ago. No one was around at that time to take notes, but remarkably these values can be estimated with some confidence from laboratory data.

TABLE 17–2 • Temperatures Occurring in Nature

Physical System	Temperature (K)
Lowest temperature reached in the lab	$\cong 10^{-9}$
Temperature of background radiation that fills the universe	3
Liquid helium at 1 atm	4.2
Coldest recorded outdoor temperature on Earth	185
Average surface temperature on Mars	218
Freezing point of water	273.15
Typical sea-level temperature on Earth	300
Body temperature	310
Hottest recorded outdoor temperature	331
Boiling point of water	373
Melting point of gold	1335
Surface of the Sun	6000
Interior of the Sun	10^7
Helium fusion reaction	10^8
Interior of hottest stars	10^9
Universe at 3 min after the Big Bang	5×10^8
Universe at 1 s after the Big Bang	10^{10}
Universe at 10^{-6} s after the Big Bang	10^{14}
Universe at 10^{-12} s after the Big Bang	10^{16}

EXAMPLE 17–2 Estimate the pressure in a hypothetical closed bottle of air that, starting with sea-level temperature and pressure, has been heated to the temperature of the surface of the Sun. Express your answer in atmospheres.

Setting It Up "Typical" sea-level temperature and pressure are 300K and 1 atm, respectively. As Table 17–2 shows, the temperature at the Sun's surface is 6000K.

Strategy We assume that the gas in the bottle is ideal over the temperature range involved. According to Eq. (17–1), the ratio of the two temperatures in question—300K and 6000K—is the same as the ratio of the two pressures.

Working It Out If we refer to the temperature and pressure on Earth and the Sun with the subscripts E and S, respectively, we have

$$\frac{p_E}{T_E} = \frac{p_S}{T_S};$$

$$p_S = p_E\left(\frac{T_S}{T_E}\right) = 1\ \text{atm}\left(\frac{6000\text{K}}{300\text{K}}\right) = 20\ \text{atm}.$$

Ideal-gas thermometers are not very practical at high temperatures, as it is not easy to construct a suitable container for temperatures much above 2000K, even at atmospheric pressure.

What Do You Think? According to Fig. 17–5, there also seems to be a difficulty in producing ideal-gas thermometers at low temperatures and pressures. Why is this so? *Answers to **What Do You Think?** questions are given in the back of the book.*

Other Temperature Scales

By changing both the slope and the intercept in Fig. 17–6, we can construct other temperature scales. Equivalently, if two points on the linear curve are assigned to special temperatures, a new scale emerges. The temperature in the **Celsius scale**—once known as the centigrade scale—is denoted by t_C and measured in units of °C. It assigns the value $t_C = 0°C$ to the freezing point of water at 1 atm of pressure (the *ice point*) and $t_C = 100°C$ to boiling water at 1 atm. From Fig. 17–7, we see that

$$t_C = T - 273.15, \tag{17-3}$$

where T is the temperature in the Kelvin scale.[†] Note that the slopes of the Kelvin and the Celsius scales are exactly the same.

Other scales differ from the Kelvin scale in slope as well as in zero level. The **Fahrenheit scale**, in which the temperature t_F is measured in °F, is an example. Different scales are useful for different purposes, and the Fahrenheit scale was invented with humans in mind. The ice point of water is 32°F, the temperature of the human body is about 100°F, and the steam point of water is 212°F. The relation between the Fahrenheit scale and the Kelvin scale is

$$t_F = \frac{9}{5}T - 459.67, \tag{17-4}$$

with T again measured in kelvins. We can easily find the relation between any two scales (Figs. 17–7 and 17–8).

17–3 Thermal Expansion

Observation shows that a solid expands when its temperature increases. A simple demonstration of this fact is shown in Fig. 17–9. Empirically, this expansion is described by the **coefficient of thermal expansion**, α, defined by

$$\alpha \equiv \frac{1}{L}\frac{dL}{dT}. \tag{17-5}$$

The quantity $\alpha\Delta T$ is the fractional change $\Delta L/L$ in the length L of a solid due to a change ΔT in its temperature. In principle, the coefficient α is a function of temperature T, but tends to vary quite slowly with T. The analogous **coefficient of volume expansion**, β, is simply related to α. It gives the fractional change in the volume V due to a temperature change ΔT: $\beta\Delta T = \Delta V/V$, or

$$\beta \equiv \frac{1}{V}\frac{dV}{dT}. \tag{17-6}$$

We can show that $\beta = 3\alpha$ by considering a cube of volume $V = L^3$. We have

$$\beta = \frac{1}{V}\frac{dV}{dT} = \frac{1}{L^3}\left(3L^2\frac{dL}{dT}\right) = \frac{3}{L}\frac{dL}{dT} = 3\alpha. \tag{17-7}$$

This result applies as long as the material is *isotropic*; that is, its properties are the same in any direction within the material. Unless a uniform solid is a single crystal, with consequent preferred directions, we would normally expect it to be isotropic. Thermal expansion also occurs in liquids, which are more correctly described as isotropic than are solids. The volume expansion coefficient β is usually given when liquids are involved. Both β and α are measured in units of K^{-1}. Table 17–3 lists some thermal expansion coefficients.

TABLE 17–3 • Coefficients of Thermal and Volume Expansion (at 20°C)

Material	$\alpha(K^{-1})$	Material	$\beta(K^{-1})$
Aluminum	23.0×10^{-6}	Ethyl alcohol	11.2×10^{-4}
Copper	16.7×10^{-6}	Gasoline	9.5×10^{-4}
Iron	19×10^{-6}	Mercury	1.82×10^{-4}
Pyrex glass	3.2×10^{-6}	Water	2.07×10^{-4}
Fused quartz	0.42×10^{-6}		
Steel	10.5×10^{-6}		

[†]The difference of 0.01 between 273.16 in Eq. (17–2) and 273.15 here is the difference between the temperature of water at its triple point and water at its ice point.

▲ **FIGURE 17–7** Comparison of the Kelvin and Celsius temperature scales.

▲ **FIGURE 17–8** Comparison of the Celsius and Fahrenheit temperature scales.

▲ **FIGURE 17–9** The expansion joint in this bridge allows the material of the bridge to undergo expansion in hot weather and contraction in cold weather.

CONCEPTUAL EXAMPLE 17–3 The unheated ball in Fig. 17–10 passes through the ring; heated it will not, because its linear dimensions—in this case its radius—have increased. What will happen if the ring is heated instead of the ball?

Answer In thermal expansion, *all* the linear dimensions increase, including the dimensions of the holes. For the ring, both inner and outer radii are linear dimensions. The holes—increase in size just as if they were made of the material in question, and the ball will go through.

(a)

(b)

(c)

▲ **FIGURE 17–10** Thermal expansion: The linear dimensions of a material expand proportionally with the temperature increase. In this instance, the linear dimension of the metal ball is its radius. (a) It just fits through a ring at room temperature, (b) is heated, and (c) becomes too big to fit through the ring.

CONCEPTUAL EXAMPLE 17–4 We learn by experience that we can loosen a metal lid that is stuck on a glass jar by pouring hot water over the lid. Why does this work?

Answer As the lid's temperature rises, it expands. Movement occurs where the glass and metal are stuck and the lid releases (Fig. 17–11). In fact, you could dip the entire system (lid and glass container) in hot water, and the different expansions of the metal and the glass will lead to the same result. Note that the same difference in the thermal expansion is used in making a good seal: lids are placed on jars when the contents are hot.

▶ **FIGURE 17–11** By pouring hot water over this lid, you can take advantage of thermal expansion to free it from the jar.

EXAMPLE 17–5 A steel bridge is 600 m long. How much allowance must be made for linear expansion between the temperatures of $+40°C$ and $-40°C$?

Setting It Up We know the values of L, ΔT, and the coefficient of expansion α for steel; we are interested in the length change ΔL.

Strategy This is a straight application of Eq. (17–5), which gives ΔL in terms of L and ΔT.

Working It Out We take α from Table 17–3: $10.5 \times 10^{-6}\ \text{K}^{-1}$ for steel (α varies little with temperature in the range $+40°C$ to

$-40°C$). Eq. (17–5) rearranged in the form $\Delta L = \alpha L \Delta T$ then gives

$$\Delta L = (10.5 \times 10^{-6}\ \text{K}^{-1})(600\ \text{m})(80\text{K}) = 5 \times 10^{-1}\ \text{m} = 50\ \text{cm}.$$

Expansion joints that allow for this much thermal expansion are distributed at regular intervals along bridges (Fig. 17–9). If there are N joints, then each joint need accommodate a length for $\Delta L/N$.

What Do You Think? Would it be more pleasant to drive over the steel bridge in this example at $+40°C$ or $-40°C$?

EXAMPLE 17–6 You want to fill an aluminum container with gasoline in the dead of winter, at a temperature of $-20°C$. The filled container is to be stored into the middle of summer, when the temperature may reach $T = +40°C$. How much unfilled space must you leave in the container to allow for expansion when summer comes around?

Setting It Up We know from Table 17–3 that $\beta_{gas} = 9.5 \times 10^{-4}\text{ K}^{-1}$ is the coefficient of volume expansion for gasoline. The coefficient of volume expansion for aluminum can be found from α using Eq. (17–7); from the table $\beta_{alu} = 3\alpha_{alu} = 6.9 \times 10^{-5}\text{ K}^{-1}$. We want to find the difference in the fractional volume changes of the gasoline and the container.

Strategy Both the gasoline and the aluminum of the container will expand when T increases. We calculate and compare the fraction volume change $\Delta V/V = \beta \Delta T$ for both. With $\beta_{gas} > \beta_{alu}$ the gasoline expands more than the container, and the difference is the fraction that would be lost if the container were full; it is thus the fraction

that should be left unfilled. If β_{gas} had turned out to be smaller than β_{alu}, the fractional volume increase of the gasoline would be smaller than that of the container, in which case there would be no need for expansion space.

Working It Out The temperature change ΔT is positive, $\Delta T = +60°C = +60\text{K}$. Then for the gasoline

$$\Delta V/V = \beta_{gas}\Delta T = (9.5 \times 10^{-4}\text{ K}^{-1})(60\text{K}) = 0.057.$$

For the container the volume change is

$$\Delta V/V = \beta_{alu}\Delta T = (6.9 \times 10^{-5}\text{ K}^{-1})(60\text{K}) = 0.004.$$

The container volume change is negligible compared to the nearly 6% volume change of the gasoline. You should fill the container only to the 94% level in winter if you want to avoid a dangerous situation.

What Do You Think? Can you safely fill the container to the brim in summer?

17–4 The Equation of State of Gases

A thermal system such as a gas in a bottle is described by its pressure p, temperature T, volume V, and the amount (or mass) of the gas. We have already discussed the first three variables. A useful unit for the amount of material in a thermal system is the **mole** (abbreviated mol), which represents either an amount of gas with a certain mass (according to the chemical species of the material) or a count of the number of molecules. (You should be familiar with this unit from your chemistry studies.) One mole of a given gas is defined to be the amount of gas with a mass in grams equal to the *atomic* (or *molecular*) *weight* of the gas. For example, 1 mol of helium gas has a mass of 4 g; the atomic weight of helium is 4 units of atomic mass. One mole of gas *always* contains *Avogadro's number* of molecules—$N_A = 6.022 \times 10^{23}$ molecules. The total number of molecules in a container, N, can be written in terms of the number of moles, n, as

$$N = nN_A. \tag{17–8}$$

Variables such as p, T, V, and n describe thermal systems, and we refer to them as **thermodynamic variables**.

CONCEPTUAL EXAMPLE 17–7 In an experiment, a tiny known volume V of oil is placed on calm water and subsequently spreads over an area A, forming a film of oil one molecule thick (a monolayer). The oil has a known density ρ and molecular weight M. How can you use this information to (a) estimate the size of a molecule of the oil, and (b) estimate Avogadro's number N_A. What additional assumptions are necessary for your estimates? (This experiment was apparently first performed by Benjamin Franklin.)

Answer (a) If we make the additional assumption that the oil molecules in the monolayer are as close as they are in the original

drop, then we estimate the thickness t of the monolayer as follows: The original drop forms a "pancake" of volume At equal to the original volume V. Thus $t = V/A$. If we now assume further that the molecules themselves are compact—that we can, for example, treat them as little cubes of size l—then we can identify $l = t = V/A$. (b) For an estimate of N_A, we note that the number N of our molecules is given by $N = V/l^3$. But we have another way to get N: If $m = \rho V$ is the (known) mass of the drop, then $N : N_A = m : M$. (Remember, N_A is the number of molecules in a mass M of the material.) In other words, $N_A = N(M/m)$.

Ideal gases—think of dilute gases here—obey *Boyle's law*, discovered in 1662 by Robert Boyle. Boyle's law states that for a dilute gas held at constant temperature, the product of pressure p and volume V divided by the number of moles n is a constant:

$$\text{for constant } T: \frac{pV}{n} = \text{a constant.} \tag{17–9}$$

BOYLE'S LAW

We previously defined temperature as a linear function of pressure if volume is constant. We can combine this result [Eq. (17–1)] with Boyle's law into the **ideal gas law**:

$$pV = nRT, \qquad (17\text{–}10)$$

where R is a proportionality constant. The ideal gas law is an example of an **equation of state**, a relation between the thermodynamic variables of a thermal system. Such a relation holds for any thermodynamic system. The ideal gas law is the equation of state for an ideal gas. Measurements of ideal gases give

$$R \cong 8.314 \frac{\text{N} \cdot \text{m}}{\text{mol} \cdot \text{K}} = 8.314 \text{ J/mol} \cdot \text{K}. \qquad (17\text{–}11)$$

R is known as the **universal gas constant**. If we count the number of molecules N of a gas rather than the number of moles n, we can use Eq. (17–8) in order to rewrite Eq. (17–10) as

$$pV = NkT, \qquad (17\text{–}12)$$

where the constant k, **Boltzmann's constant**, is given by

$$k = \frac{R}{N_A} \cong 1.381 \times 10^{-23} \text{ J/K}. \qquad (17\text{–}13)$$

Boltzmann's constant, named for Ludwig Boltzmann, one of the important figures in nineteenth-century thermal physics, provides the bridge between temperature and energy. To see how, note that the quantities pV/n and pV/N, from Eqs. (17–10) and (17–12) respectively, have the dimensions of [force/length^2][length^3] = [force][length], which is the same as work or, equivalently, energy. *The dimensions of RT and kT are those of energy.* This is already reflected in the units of R and k [Eqs. (17–11) and (17–13)]. The significance of this observation is that temperature, as defined by the ideal gas law, must be closely related to the energy of a thermal system. This *thermal* (or *internal*) energy is the subject of Chapter 18.

CONCEPTUAL EXAMPLE 17–8 Before you drive through West Texas on Interstate 10 toward El Paso in the summer, you are advised not to pressurize your automobile tires to the maximum allowed pressure. Why?

Answer First, you should know that the air temperature is likely to be high and that people drive very fast on the wide-open roads of Texas. The tires will get quite hot. However, tires are constructed with steel belts so that they do not expand in volume past a certain amount. If the volume stays constant, and the temperature increases considerably, then the pressure in the tires will increase substantially, and they may become unsafe against a blowout. This is good advice for driving anywhere in summer.

EXAMPLE 17–9 Calculate the volume occupied by 1.0 mol of an ideal gas at 20°C and at atmospheric pressure. Use this volume to calculate the mass density of air, whose molecular weight M is 29 g/mol.

Strategy The ideal gas law gives us the volume for n moles of gas, $V = nRT/p$. We then use the M of air ($= 29$ g/mol) to find the density of air, $\rho = nM/V$. Here we'll want to set $n = 1.0$.

Working It Out The temperature T is 20°C $= (20 + 273)$K $= 293$K, and T must be in K for the ideal gas law. The volume is

$$V = \frac{nRT}{p} = \frac{(1.0 \text{ mol})(8.3 \text{ J/mol} \cdot \text{K})(293\text{K})}{1.0 \times 10^5 \text{ Pa}} = 2.4 \times 10^{-2} \text{ m}^3.$$

As for the mass density of air,

$$\rho = \frac{29 \text{ g}}{2.4 \times 10^{-2} \text{ m}^3} = \frac{(29 \text{ g})(10^{-3} \text{ kg/g})}{2.4 \times 10^{-2} \text{ m}^3} = 1.2 \text{ kg/m}^3.$$

What Do You Think? Where does the molecular weight 29 g/mol used in this example come from?

Changing the Thermodynamic Variables of a Gas

Thermal systems change as the thermodynamic variables change. We say that a *thermal transformation* or, alternatively, a *thermal process*, occurs. For example, when the cylinder of gas in Section 17–2 was put in contact with a hotter thermal reservoir, the gas' temperature changed; so does the gas' volume in the set-up in Fig. 17-1. In this transformation the temperature, the volume, and the pressure all changed subject to the constraints of the equation of state. There are many different ways to change a thermal system. For example, in an **isothermal** transformation the temperature remains fixed while the pressure and volume both change. In an **isobaric** transformation the pressure is held fixed while both temperature and volume change. Such transformations are most easily shown on diagrams in which one variable is plotted against another. For example, in a *p–V* diagram, pressure is plotted against volume. When such a diagram shows various *phases* of a material—whether it is solid, liquid, or gas—it is referred to as a *phase diagram*. Figure 17–12 is a *p–T* phase diagram for water. We'll be looking at phase diagrams throughout, concentrating on phase changes—transitions across phase boundaries—in Chapter 18. Careful: the line that separates phases on a phase diagram is not a transformation!

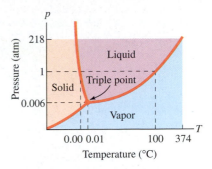

▲ **FIGURE 17–12** Phase diagrams are useful for the different phases of a material, such as those for water, shown here: liquid, solid (ice), and vapor (steam). The intersection of these phases is the triple point of water.

CONCEPTUAL EXAMPLE 17–10 Referring to Fig. 17–12, what changes can be made to water to allow you to go from solid ice to liquid water without any change in temperature? How might this be useful for understanding ice skating?

Answer Consider the almost vertical line near 0.00°C on the phase diagram in Figure 17–12. In the region of 1 atm, increasing the pressure—moving up along the constant temperature line, say at 0°C—takes you from the solid phase (ice) to the liquid one. It is the fact that the slope of the line describing the solid-liquid phase separation is negative on the *p–T* diagram that allows this to happen. As an ice skater moves over the ice, the weight of his or her body increases the pressure under the blade on the ice. If the temperature is not too low, this increased pressure will be enough to melt the ice, forming a film of water between the ice skate and ice, in effect lubricating the motion of the blade over the ice. A narrow blade makes the pressure higher and the effect more pronounced. A glance at the diagram shows that this effect is more difficult to realize at lower temperature.

EXAMPLE 17–11 Exactly 1 mol of an ideal gas is taken through the sequence of changes shown on the *p–V* diagram in Fig. 17–13. The change $A \rightarrow B$ is an isothermal transformation. In $B \rightarrow C$ the transformation is isobaric, that is, the gas is compressed in volume in such a way that the pressure remains fixed. (a) If $p_A = 5.00$ atm and $V_A = 8.00$ L, what is the temperature T_A? (b) If $V_B = 40.0$ L, what is p_B? (c) If $T_C = T_A/8$, what is V_C?

Setting It Up Figure 17–13 provides a very useful picture of what is happening. Diagrams such as this one are very important in problems involving thermal transformations.

Strategy We can apply the equation of state for the ideal gas, $pV = nRT$, with $n = 1$ mol. This gives us T_A immediately.

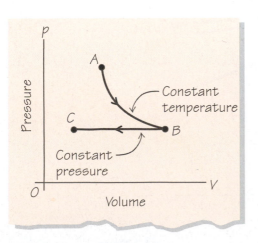

▲ **FIGURE 17–13**

Then, since $T_B = T_A$, we can use the ideal gas law to find p_B. Finally, since $p_C = p_B$ and T_C is given, we can again use the ideal gas law to find V_C.

Working It Out (a) We know p_A and V_A; the temperature T_A is then given by

$$T_A = \frac{p_A V_A}{nR} = \frac{(5.00 \text{ atm})(8.00 \text{ L})}{(1 \text{ mol})(8.314 \text{ N} \cdot \text{m/mol} \cdot \text{K})}.$$

We make some unit conversions in order to cancel units:

$$T_A = \frac{(5.00 \text{ atm})(1.01 \times 10^5 \text{ Pa/atm})(8.00 \text{ L})(10^{-3} \text{ m}^3/\text{L})}{(1 \text{ mol})(8.314 \text{ N} \cdot \text{m/mol} \cdot \text{K})} = 486 \text{K}.$$

(b) $T_B = T_A$ from the conditions stated in the problem. We know V_B and T_B, and hence we can compute p_B:

$$p_B = \frac{nRT_B}{V_B} = \frac{nRT_A}{V_B} = \left(\frac{nR}{V_B}\right)\left(\frac{p_A V_A}{nR}\right) = p_A \left(\frac{V_A}{V_B}\right)$$

$$= (5.00 \text{ atm})\left(\frac{8.00 \text{ L}}{40.0 \text{ L}}\right) = 1.00 \text{ atm}.$$

(c) Since we know $p_C = p_B$ and we know T_C, we can calculate V_C. It is given by

$$V_C = \frac{nRT_C}{p_C} = \frac{nR}{p_C}\left(\frac{T_A}{8}\right) = \frac{nR}{8p_C}\left(\frac{p_A V_A}{nR}\right) = \frac{p_A V_A}{8p_C} = \frac{p_A V_A}{8p_B}$$

$$= \frac{(5.00 \text{ atm})(8.00 \text{ L})}{8(1.00 \text{ atm})} = 5.00 \text{ L}.$$

What Do You Think? Suppose you had followed a straight line from A to C on the diagram of Fig. 17–13. What would have been constant: p, V, or T?

HOW DO WE GET THE FIZZ INTO DISPENSED DRINKS?

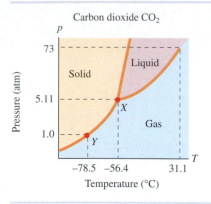

Carbon dioxide CO_2

A fizzy drink from a carbonated beverage dispenser is made by mixing a "syrup" of sugar and flavorings with water and carbon dioxide as it is dispensed. It is the carbon dioxide that gives the drink its fizz. But how is the carbon dioxide handled? At a pressure of 1 atm, carbon dioxide exists only in the solid phase (at low temperatures) and in the gas phase (at normal temperatures). This is visible in the p–T phase diagram for car-

◀ **FIGURE 17–14** Phase diagram for carbon dioxide. At a pressure of 1 atm, only the solid and gas phases exist.

bon dioxide (Fig. 17–14). Both forms are inconvenient to use; the gas is bulky, and the solid is difficult to incorporate in mixing systems. Carbon dioxide is therefore transported and delivered at pressures greater than 40 atmospheres, pressures at which the room-temperature phase is liquid. This liquid can then be conveniently stored, transported efficiently through pipes, controlled by automatic valves, and so forth. When it is needed for incorporation into soft drinks near room temperature, it is released to one atmosphere of pressure, at which point it changes to the gas phase, becoming the bubbles in your drink. ■

How Close Do Real Gases Come to Being Ideal?

When is a real gas "sufficiently dilute" to obey the ideal gas equation of state? Of course, this is a question without a definite answer, because there is no sharp line dividing ideal gases from nonideal ones, just a matter of tolerance to approximation. But we can approach this question by asking how well Boyle's law is satisfied as we vary the density of a gas. Consider a series of identical closed tubes. A pressure gauge can measure the pressure of any gas in the tubes. Put different amounts (a different number of moles, n) of the same gas in each tube and immerse them in the same thermal reservoir. If Boyle's law is satisfied exactly, the (measured) ratio pV/n will be the same in each tube, independent of the pressure in the tube. One can also imagine a set of thermal reservoirs with different temperatures. In that case the more stringent test for ideal gas behavior is that the ratio pV/nT is the same in each tube.

Experiment shows that for monatomic gases,[†] such as helium, the ratio pV/nT is independent of the pressure over a wider range of pressures and therefore fits the ideal gas limit better than do polyatomic gases such as O_2, or carbon dioxide, or ammonia. However, *all* gases give precisely the same value of pV/n when this quantity is extrapolated to *zero density* (Fig. 17–15). All gases are indeed ideal in this limit. A more appropriate definition of temperature than that presented in Eq. (17–2) is obtained by *extrapolating the pressure ratio* in Eq. (17–2) to zero density ρ:

▶ **FIGURE 17–15** As the pressure (and hence the density) of any gas drops to zero, the ratio pV/nT goes to the same constant, which is the universal gas constant. This behavior characterizes the ideal gas.

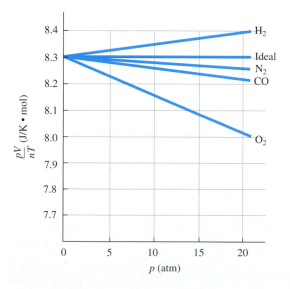

[†]The molecule of a monatomic gas consists of one atom; the molecule of a polyatomic gas consists of several.

$$T = (273.16\text{K}) \lim_{\rho \to 0} \frac{p}{p_{tp}}. \qquad (17\text{–}14)$$

KELVIN TEMPERATURE SCALE DEFINED

Equation (17–14) provides us with the desired universal definition of temperature because it holds in the domain where *every* gas is ideal.

The van der Waals Equation of State

In 1873 Johannes D. van der Waals proposed an equation of state that describes the properties of many real gases better than the ideal gas law. His formula, known as the *van der Waals equation*, takes the form

$$\left[p + a\left(\frac{n}{V}\right)^2 \right] \left(\frac{V}{n} - b \right) = RT. \qquad (17\text{–}15)$$

The constants a and b are both positive and specific to a given gas. The van der Waals equation includes a correction for intermolecular forces through the constant a and a correction for finite molecular size through the constant b. In contrast, the ideal gas law can be derived at the molecular level through assumptions that the gas molecules are noninteracting point particles. (See Chapter 19 for more on this.) We can verify that the van der Waals equation reduces, as it must, to the ideal gas law when the gas is very dilute: In the limit as $n \to 0$, the term $a(n/V)^2$ is small compared to p, and the term V/n is large compared to b. Thus, in this limit, Eq. (17–15) reduces to $pV/n = RT$.

The van der Waals equation of state is a considerable empirical improvement over the ideal gas law for polyatomic gases that are about as dilute as the air we breathe. It describes all gases reasonably well, even when a gas is nearly a liquid; that is, at very low temperatures and at densities approaching those of real liquids.

*17–5 Blackbody Radiation

Most of us have enjoyed an evening before a roaring blaze that slowly dies down during the course of the evening. As the fire dies, its color changes from brilliant white to bright yellow, to orange, and finally to a dull red. The change is more marked if you look into a cavelike hole within the embers, a spot where the temperature is steadier because it is protected from the vagaries of drafts. This phenomenon is the basis of a thermometer called the *optical pyrometer* (Table 17–1). This thermometer works because the color or, rather, the range of colors emitted by a hot object is characteristic of the temperature of that object. This thermometer works best for relatively hot objects with temperatures in the range of several hundred kelvins.

The light emitted by hot objects is **electromagnetic radiation**. Radiation exists in the form of waves, and all frequencies are possible in principle. One small range of frequencies corresponds to visible light, and different frequencies within this range are perceived as different colors, but even if we cannot literally see the radiation at other frequencies, we often use it (radio, infrared, microwave, infrared, X-rays, etc). All matter, at any temperature, absorbs and emits electromagnetic radiation, and across the full range of frequencies. The basic physical emission mechanism is that atoms go into *excited states* when the temperature of the matter they comprise is raised. They then radiate energy when they return to their normal states. For example, the atoms heated by an electric current in a lightbulb filament radiate visible light, from red to orange to yellow to green to blue to violet, as the frequency of the radiation increases from 0.5×10^{15} Hz to approximately 1.0×10^{15} Hz. The filament also radiates a great deal of energy in the form of longer wavelength radiation, and that is why the bulb is hot as well as brilliant.

In some ways, radiation in an enclosure acts like a dilute gas in a bottle; in particular, the radiation has a temperature just as a gas does. However, there is an important difference: Radiation achieves thermal equilibrium through the fact that it is absorbed and emitted by the atoms of the walls of the enclosure rather than through collisions among its components (as do the molecules that make up a dilute gas). The radiation in the enclosure at thermal equilibrium at a given temperature is a mix of *different*

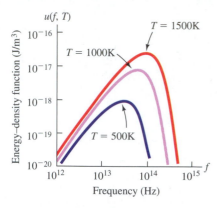

u(f, T)

▲ **FIGURE 17–16** The energy density of electromagnetic radiation in thermal equilibrium at the indicated temperatures. The maximum energy density occurs for increasing frequency as the temperature increases. The maximum energy density increases very rapidly with T for a given f; note that this is a log–log plot.

frequencies, and what is important here is that in the mix the proportion of high frequencies to low frequencies rises with temperature. The average frequency of the radiation associated with a white-hot object is greater than the average frequency associated with the object when it is glowing red. We can give a function that describes the radiation at *different* frequencies associated with a given temperature. Such a function is denoted by $u(f, T)$, where $u(f, T)$ *df is the radiation energy per unit volume, or energy density, with a frequency between f and f + df*. We say that the radiation is at a given temperature when it is in thermal equilibrium, meaning that the radiation interacts with an object or system at that temperature over a long period and that the object and the associated radiation is isolated from its surroundings. Radiation of this type is known as **blackbody radiation**, and the object that emits and absorbs such radiation is known as a blackbody. Figure 17–16 shows this function for different values of temperature. *The measurement of $u(f, T)$ provides a measurement of T.*

We sometimes refer to this radiation as being "emitted" when we look into the enclosure; that is, we can see the radiation that escapes from the "bottle" of radiation. This picture most exactly describes your looking deeply into a furnace through a small window or a campfire through a little opening in the pile of wood. If the window is large, then so much radiation escapes to the outside that the system may cool down, which is why we refer to a small window. But equivalently, if we have a way to maintain a high temperature, then we just say that the blackbody radiation is emitted from an object at that temperature. An example is provided by the filament of a lightbulb, which would quickly cool if we didn't maintain an electric current through it that holds it at a white-hot temperature.

The form of the energy–density function $u(f, T)$ was a subject of intense study in the latter part of the 19th century. The work of Gustav Robert Kirchhoff, Wilhelm Wien, Lord Rayleigh, and James Jeans set the stage for the breakthrough of Max Planck, who determined in 1900 that

$$u(f, T) = \frac{8\pi h}{c^3} \frac{f^3}{e^{hf/kT} - 1}. \tag{17–16}$$

This formula, known as the **Planck formula**, perfectly fits the data on the energy density of radiation for a wide range of frequencies and temperatures. In addition to Boltzmann's constant, k [introduced in Eq. (17–13)], and the Kelvin temperature, T, the formula contains the speed of light, $c = 3 \times 10^8$ m/s, which is the speed of electromagnetic waves of *every* frequency. A new fundamental constant also appears in this formula. The new constant h, called *Planck's constant*, was found by fitting the formula to the observed energy density, and its value is $h = 6.625 \times 10^{-34}$ J·s. This is an awfully small number, and many of the effects associated with it are visible only on an atomic scale, but its presence in the formula for blackbody radiation shows that it can perfectly well have a macroscopic effect. Note that the quantity hf has dimensions of energy as does the quantity kT. (Indeed, these quantities must have the same dimensions because their ratio is the argument of the exponential function.)

The discovery of the Planck formula was the first step on the path to the basic theory of matter and radiation, called *quantum theory*. This theory is based on a set of ideas that revolutionized the physical sciences in the twentieth century. (The other great revolution in 20th-century physics—Einstein's theory of relativity—also has its origin in the investigation of electromagnetic radiation.)

An interesting consequence of the Planck formula is that the total power radiated through a small hole in an enclosure containing radiation in thermal equilibrium at temperature T is proportional to the fourth power of the temperature. The **Stefan–Boltzmann formula**, which was discovered experimentally by Josef Stefan in 1879 and inferred from thermodynamics by Boltzmann in 1884 before the discovery of the Planck formula, states that the radiated power (energy per unit time) per unit area emitted through the hole is given by $P(T)/A$, usually written as

$$E(T) = \sigma T^4, \tag{17–17}$$

where $\sigma = 5.67 \times 10^{-8}$ W/(m^2·K^4). The T dependence in Eq. (17–17) can be obtained by integrating Eq. (17–16) over all frequencies f (see Problem 76).

EXAMPLE 17–12 A surface that is "white hot" emits about 10 times more power than a "red hot" surface of the same area. What does this tell us, quantitatively, about the relative temperatures?

Strategy Power and temperature are related through the Stefan–Boltzmann formula, Eq. (17–17). By setting the ratio of the powers to 10, we have an equation for the ratio of the temperatures.

Working It Out From Eq. (17–17),

$$\frac{P(\text{white hot})}{P(\text{red hot})} = 10 = \frac{\sigma T_{\text{white}}^4}{\sigma T_{\text{red}}^4} = \left(\frac{T_{\text{white}}}{T_{\text{red}}}\right)^4$$

This implies that the temperature of the "white hot" surface exceeds that of the "red hot" surface by a factor of

$$\frac{T_{\text{white}}}{T_{\text{red}}} = 10^{1/4} = 1.78.$$

Let's examine the Planck formula for the energy–density function, Eq. (17–16). If the frequencies are small, $hf \ll kT$, then the exponential function in $u(f, T)$ has a small argument, and we can use the expansion

$$\text{for small } x: e^x \cong 1 + x + \cdots$$

In this case, the denominator of $u(f, T)$ is

$$e^{hf/kT} - 1 \cong 1 + \frac{hf}{kT} - 1 = \frac{hf}{kT},$$

and the function $u(f, T)$ takes the approximate form

$$\text{for } hf \ll kT: u(f, T) \cong \frac{8\pi h}{c^3} f^3 \frac{kT}{hf} = \frac{8\pi f^2}{c^3} kT. \qquad (17\text{–}18)$$

This expression is independent of Planck's constant and is thus independent of quantum physics. It was, in fact, first found by Lord Rayleigh in 1900 by using a derivation based on classical physics. The energy density in Eq. (17–18) rises with increasing frequency as f^2 and linearly with temperature.

If, however, the frequencies are large, $hf \gg kT$, then the exponential argument in $u(f, T)$ is large and is the exponential function itself. The denominator then becomes

$$e^{hf/kT} - 1 \cong e^{hf/kT}.$$

(Here we have used the fact that for $hf/kT \gg 1$, $\exp(hf/kT) \gg 1$.) Then the energy density is approximately

$$\text{for } hf \gg kT: u(f, T) \cong \frac{8\pi h}{c^3} f^3 e^{-hf/kT}. \qquad (17\text{–}19)$$

Because the exponential function falls with f much faster than the polynomial factor f^3 grows, $u(f, T)$ decreases as f grows in this range of f. Note that h plays a crucial role in this part of the energy–density function, indicating that quantum physics is important.

As we have seen, the energy–density function increases with f for $hf \ll kT$ and decreases with f for $hf \gg kT$. Such a function has an intermediate maximum, which we can estimate as occurring at about $hf \cong kT$. The underlying physics of the high-frequency side of the curve of the energy–density function in Fig. 17–16 is quantum physics because it involves Planck's constant, but the low-frequency side can be explained by classical physics. The maximum moves to higher values of f as T increases. We can therefore say that the importance of quantum physics decreases as the temperature grows, and this is generally true.

EXAMPLE 17–13 The surface of the Sun is observed to emit light most strongly at a frequency of 2×10^{14} Hz. Estimate the temperature of the Sun's surface (Fig. 17–17).

Setting It Up We can safely assume the radiation is blackbody; thus we want to find the blackbody temperature that gives a maximum in the energy–density function $u(f_0, T)$ at a given frequency f_0.

Strategy As we argued above, the frequency f_0 for which u is a maximum is given by $hf_0/kT \cong 1$. We can solve this for T. (Alternatively, we could go beyond this estimate and calculate the maximum of the energy–density function, Eq. (17–16), by taking

◀ FIGURE 17–17
The temperature of the Sun's surface is not constant but varies somewhat in active areas, such as regions with Sun flares.

(continues on next page)

the derivative of u with respect to f and setting the derivative to zero. The simpler estimate will be enough here.)

Working It Out We have

$$T \cong \frac{hf_0}{k} = \frac{(6.63 \times 10^{-34} \text{ J} \cdot \text{s})(2 \times 10^{14} \text{ Hz})}{1.38 \times 10^{-23} \text{ J/K}} = 10^4 \text{K}.$$

The more precise alternative solution described above gives a solar surface temperature of 6000K (see Problem 80).

What Do You Think? The interior of the Sun is at a much higher temperature than the surface. What happens to the radiation, characteristic of the higher internal temperatures, that must be present in the interior?

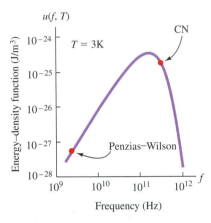

▲ **FIGURE 17–18** The energy density of electromagnetic radiation in thermal equilibrium at $T = 3$K. The Penzias–Wilson frequency measurement and the frequency associated with interstellar cyanogen, labeled CN, are marked by red dots.

▲ **FIGURE 17–19** Satellite-based measurements of the cosmic background radiation show remarkable agreement with the Planck formula. The data shown for $T = 2.735$K fit the smooth blackbody curve very well indeed.

The Discovery and Measurement of the Background Radiation of the Universe

In 1964 Arno Penzias and Robert Wilson used a large radio antenna to study the radio waves emitted within our galaxy. They found a puzzling anomaly, which they at first thought might be a problem with their antenna: A background electromagnetic radiation that comes from no particular source is present. Their antenna was sensitive only to the particular frequency $f = 4.08 \times 10^9$ Hz, but later measurements over other frequencies showed that this background radiation was consistent with the Planck formula.

Penzias and Wilson were not aware of work by Ralph Alpher, George Gamow, and Robert Herman in the late 1940s, or of work by James Peebles that was contemporary with their own, in which a cosmological, or large-scale, model of the universe predicted the existence of this radiation. As described by this model, the early universe expanded from an initial catastrophic event—the Big Bang. As the universe expanded, it cooled, maintaining thermal equilibrium between electromagnetic radiation and matter up to a point, some 100,000 years after the Big Bang, when the temperature was about 3000K. At that point, the model shows that electromagnetic radiation and matter decoupled; they were no longer in thermal equilibrium. However, and quite remarkably, as the universe continued to expand, the Planck formula still described the radiation, but with a temperature that dropped as the universe expanded. *A measurement of the present-day temperature of the radiation provides us with a measurement of the rate of expansion of the universe; hence, it provides us with a measurement of the time that has elapsed since the radiation was no longer in thermal equilibrium with matter.* Presently the temperature of the radiation, as characterized by the Planck formula, is about 3K, and this radiation is the oldest directly observable relic of the explosion in which the universe is formed.

A measurement of a single frequency is not a measurement of an entire spectrum! Penzias and Wilson's frequency measurement occurs at one value of frequency on the classical, or low-frequency, side of the Planck formula, as Fig. 17–18, the energy–density function for $T = 3$K, indicates. Since 1964 many other Earth-based measurements have been made by radio antennas sensitive to a great variety of frequencies. Because Earth's atmosphere absorbs radiation of high frequency, these measurements are all made on the classical side of the Planck formula. Fortunately, one other technique can make a measurement on the quantum, or high-frequency, side of the peak of energy density, and measure the amount of light emitted by interstellar cyanogen, the CN molecule. This light is emitted with a strength that can be explained only if the cyanogen is in the presence of background radiation of a certain temperature. The frequency measurement of CN, first made in 1941, is marked on Fig. 17–18. Today we have a series of satellite measurements of the background radiation. Measurements can be made with extraordinary precision across a broad range of the spectrum, including across the maximum of the energy–density curve. We show the results in Fig. 17–19; a temperature of about 2.7K fits the curve perfectly. (The actual temperature can be determined to six significant figures!) Still other measurements reveal that the temperatures in different directions differ ever so slightly from one another. Some regions are somewhat hotter, while others are cooler. These temperature inhomogeneities may well be relics of the instabilities that led to the formation of the great clusters of galaxies present today. Indeed, the precise nature of the inhomogeneities provides further evidence for the Big Bang as well as other cosmological (universe-wide) features. The Big Bang model of the universe, which predicts the presence of this background blackbody radiation, has a remarkable amount of experimental support.

Summary

The temperature of a system in thermal equilibrium is defined on the Kelvin scale by the behavior of the pressure, p, of a dilute gas in a constant-volume gas thermometer:

$$T = (273.16\text{K}) \lim_{\rho \to 0} \frac{p}{p_{tp}}, \qquad (17\text{--}14)$$

where p_{tp} is the pressure of the gas in the thermometer at the triple point of water (where the gaseous, liquid, and solid phases coexist); thus 273.16K is chosen to be the temperature at the triple point. The other fixed point of the Kelvin scale is built into the thermometer: $T = 0\text{K}$, absolute zero, is the point where the pressure extrapolates to zero. Two other temperature scales in common use are the Celsius scale and the Fahrenheit scale. The Celsius temperature is given in terms of the Kelvin temperature by

$$t_C = T - 273.15. \qquad (17\text{--}3)$$

Any sufficiently dilute gas behaves as an ideal gas, meaning that its thermodynamic variables are related by the ideal gas law, an equation of state:

$$pV = nRT = NkT, \qquad (17\text{--}10, \; 17\text{--}12)$$

where the universal gas constant is

$$R \cong 8.314 \, \text{J/mol} \cdot \text{K} \qquad (17\text{--}11)$$

and Boltzmann's constant is

$$k = \frac{R}{N_A} \cong 1.381 \times 10^{-23} \, \text{J/K}, \qquad (17\text{--}13)$$

with N_A Avogadro's number. n and N measure the quantity of gas, in moles or in numbers of molecules, respectively. The quantities kT and RT have dimensions of energy.

Real gases are ideal only in the limit that they are very dilute. The small departures from ideal behavior are accurately reflected for many real gases by the empirical van der Waals equation of state,

$$\left[p + a\left(\frac{n}{V}\right)^2 \right]\left(\frac{V}{n} - b\right) = RT. \qquad (17\text{--}15)$$

Materials expand when heated, and they are characterized by a coefficient of thermal expansion, α, defined by

$$\alpha \equiv \frac{1}{L}\frac{dL}{dT}. \qquad (17\text{--}5)$$

Electromagnetic radiation displays the characteristics of a thermal system. The total power emitted per unit area by radiation in equilibrium with another thermal system at temperature T, which is known as blackbody radiation, is given by the Stefan–Boltzmann formula,

$$P(T)/A = E(T) = \sigma T^4, \qquad (17\text{--}17)$$

where $\sigma = 5.67 \times 10^{-8} \, \text{W/(m}^2 \cdot \text{K}^4)$. Blackbody radiation contains a range of frequencies f that have an energy per unit volume (or energy density) per unit frequency, given by

$$u(f, T) = \frac{8\pi h}{c^3} \frac{f^3}{e^{hf/KT} - 1}, \qquad (17\text{--}16)$$

a result known as the Planck formula. The constant h is Planck's constant.

Electromagnetic radiation with this frequency range is observed to occur throughout the universe at a characteristic temperature of 3K. This is strong evidence for the Big Bang theory of the origin of the universe.

Understanding the Concepts

1. Estimate to within an order of magnitude the number of molecules in a breath of air.

2. Why is mercury used in most thermometers rather than a more common liquid such as water?

3. If a scuba diver takes the same volume of air for each breath and breathes at a constant rate, why is the tank of air used up much faster at greater water depths?

4. A student proposes to make a steel sphere in which helium gas can be inserted at various pressures. The student will insert the sphere into various temperature baths (boiling water, ice water, liquid nitrogen, dry ice-methanol) and measure the pressure at each temperature. Then he or she will change the amount of helium in the sphere and do it again. If the student plots pressure vs. temperature for these separate amounts of helium, where, if anywhere, should they intersect?

5. How can a triple point of water exist? If ice, water, and vapor all existed together, would the ice cool the vapor, making it water? Or would the vapor and water melt the ice?

6. Why is it advisable to measure the pressure in automobile tires when the car has not been driven for some time?

7. Devise an experiment to prove the experimental fact that hot objects radiate more energy per unit area than cold objects do.

8. It is well known that chameleons can change their skin color. In this chapter we spoke about the different colors associated with light emitted from objects of different temperature. Is this how a chameleon changes its color—by changing its temperature?

9. Why is it important to check the pressure in automobile tires before the winter season?

10. All gases obey the ideal gas law when they are made sufficiently dilute. Is this true when the temperature is lowered sufficiently?

11. In recent years an attempt has been made to convert Americans to the use of the metric system, so that, for example, two sets of tools would not be necessary for a mechanic who works on both American and European cars. Would it be of equal practical importance to convert Americans from the use of Fahrenheit to Celsius temperature scales?

12. When we observe a fire, we are observing radiation that has escaped the region of the hot fire. Is this a mechanism for cooling the fire? In what way could a fire be enclosed so that it remains hot?

13. In the experiment shown in Fig. 17–10, it often happens that the ball gets stuck in the ring while performing the experiment. Why

do you think this happens? Could you dislodge the ball by heating up either the ball or the ring?

14. Suppose that we have no idea that gases consist of molecules. Could we still use the low-pressure limit of a gas to define temperature?

15. Substances in thermal contact reach thermal equilibrium after some time. Can you think of any experimental test that will indicate just when that time will have arrived?

16. Water expands when it freezes. You also have heard that 90% of icebergs are hidden beneath the surface. Why is the latter statement true? Can you estimate by how much water expands on freezing?

17. If the molecules of gases move around a lot, why do we never see a container of gas shake?

18. The pressure of a gas produces an outward force on the walls of a container. Why is there no force at an angle to the walls?

19. The Chamber of Commerce of a town plagued by temperature extremes devises a way to avoid bad publicity: It defines a new temperature scale, the G scale, such that (a) −40°F is defined to be +20°G; (b) +120°F is defined to be −40°G; and (c) the scale is linear. Is there anything wrong with this scale?

20. If a doughnut-shaped piece of a solid that expands under heat is heated, does the hole expand or shrink?

21. Water is one of the relatively rare substances that becomes less dense rather than more dense when it freezes. Describe what might happen to the ocean if ice were denser than liquid water.

Problems

17–1 Temperature and Thermal Equilibrium

1. (II) Which of these pairs of systems are in thermal equilibrium with one another? (a) A roast reaching the rare stage and the oven in which it sits. (b) The point of a meat thermometer in the roast of part (a) and the roast. (c) A sunbather getting a tan and the air around him. (d) A bather in a hot tub and the water around her. (e) An ice cube in a glass of water at 0°C. (f) Molten iron ore in a blast furnace and the walls of the furnace.

2. (II) You must measure the different temperatures of a series of bottles that contain 1 L of an unknown fluid. Which one of the following techniques would make a suitable thermometer, and why or why not? (a) The time it takes 1L of ice cubes dumped into the fluid containers to melt. (b) The melting time for ice cubes sealed in full 1-L plastic bags that are dumped into the bottles. (c) The melting time for ice sealed in full 1-mL plastic bags that are dumped into the bottles. (d) All of the above techniques, but the fluids in the bottles are gently stirred during the measurement.

3. (II) Thermodynamic variables are either extensive or intensive. A variable is *extensive* if, when two identical thermodynamic systems are combined into one, the variable of the combined system is double its original value in each system; otherwise, it is *intensive*. Which of the following variables are extensive, and which are intensive: (a) volume, (b) temperature, (c) pressure, (d) number of moles?

4. (II) A physicist needs to measure room temperature. Not having anything better at hand, she decides to make use of the thermal expansion of water to construct a thermometer (Fig. 17–20). She calibrates her apparatus to a dependable thermometer at 14°C and 38°C, and divides the range between these temperatures into 24 equal intervals. What will be the true temperature when her

▲ **FIGURE 17–20** Problem 4.

equipment reads 28°C? She knows that the volume of water can be described in this temperature range by the formula

$$V = V_0 \times (1 - 2.525 \times 10^{-4} + 4.98 \times 10^{-6}T + 4.94 \times 10^{-6}T^2),$$

where T is the temperature in °C.

17–2 Ideal Gases and Absolute Temperature

5. (I) At what temperature values are the following scales the same: (a) the Fahrenheit and the Celsius, (b) the Celsius and the Kelvin, (c) the Fahrenheit and the Kelvin?

6. (I) The title of Ray Bradbury's science-fiction novel *Fahrenheit 451* refers to the temperature at which paper ignites. What is that temperature on the Celsius scale?

7. (I) An oral fever thermometer ranges from 35.5°C to 42.5°C. What is the corresponding range on the Fahrenheit scale?

8. (I) An ideal gas in a container is heated from 17°C to 120°C. By what factor does the pressure increase?

9. (I) A very dilute gas confined to a closed 1-L container is put on the burner of a stove and the temperature is raised from 271K to 349K. By how much does the pressure, which is originally at 1Pa, change?

10. (I) If an ideal gas thermometer could be constructed to measure the temperature of a hot star (see Table 17–2), by what factor would the pressure in the thermometer differ from p_{tp}?

11. (II) (a) Determine the temperature of the surface of Venus (730K) in the Celsius and the Fahrenheit scales. (b) Determine the temperature of the boiling point of liquid nitrogen (77K) in the Celsius and the Fahrenheit scales. (c) A comfortable room temperature is 75°F. Determine this temperature in the Kelvin and Celsius scales. (d) Heat pumps become less efficient than furnaces for heating homes when the temperature drops below 36°F. New laws may require that all temperatures be posted in the Celsius and the Kelvin scales as well as the Fahrenheit scale. What is this temperature in these other systems?

12. (II) In 1701 Isaac Newton proposed a linear temperature scale in which ice water has a temperature of 0°, whereas the human body in good health is assigned a temperature of 12°. In this scale, what is the temperature of someone with hypothermia (chilling)? What is the temperature of boiling water?

13. (II) A temperature scale—the Reaumur scale—developed in the early eighteenth century sets the ice point at 0°R and the boiling point of water at 80°R. Give the conversion formula from the Reaumur scale to the Kelvin scale and to the Fahrenheit scale.

14. (II) For the new temperature scale described in Question 19, what are the general conversion rules from °G to °F and vice versa? What is the conversion rule from °C to °G?

15. (II) The Rankine scale for temperature, named for William Rankine, is used in engineering applications. It has units of °R and is defined by $t_R = 9T/5$, where T is the temperature in the Kelvin scale. Find the value of t_R for the temperatures 100°C, 4.2K, 6000K, 32°F, −30°F, and −25°C.

16. (II) A bottle that has a freely sliding vertical piston and contains an ideal gas is used as a thermometer for water: The bottle is submerged in a water bath (whose temperature is to be measured) until thermal equilibrium is reached (Fig. 17–21). When the bottle is in contact with ice water, the volume inside the bottle is 0.35 L. The bottle is removed from the ice water and put in a second bath. This time, the enclosed volume is 0.38 L. What is the temperature of the second bath, in °C?

▲ FIGURE 17–21 Problem 16.

17. (II) One Sunday morning a family takes an automobile trip to Grandma's. At the start of the trip, the temperature is 288K (15°C), and the gauge pressure in the tires is 32 lb/in² (psi). (The gauge pressure is the excess over 14.5 psi, the exterior air pressure.) After an hour's ride over an interstate highway, the gauge pressure in the tires is 38 psi. What is the temperature of the air in the tires,

assuming that air behaves as an ideal gas? Neglect any changes in volume of the tires.

17–3 Thermal Expansion

18. (I) By how much will a steel rod of length 1.5 m and diameter 1.75 cm elongate if it is heated by 100°C?

19. (I) How much farther than a steel rod will a 1 m long aluminum rod expand when each undergoes a temperature increase of 40°C?

20. (I) The coefficient of thermal expansion for Invar is $0.70 \times 10^{-6}\,\mathrm{K}^{-1}$. (Invar is a steel alloy.) Given that an Invar pendulum clock keeps perfect time at a room temperature of 20°C, how much time will the clock gain or lose per day when it is in a room at 30°C? (For the time-keeping properties of a pendulum, see Chapter 13.)

21. (I) The area of the circle formed by a circular loop of metal is found to increase by 1.6% when the temperature changes by 10°C. What is the coefficient of thermal expansion of the metal?

22. (II) When the temperature of mercury changes by ΔT, the fractional volume change, $\Delta V/V$, is given by $\Delta V/V = \beta \Delta T$, where $\beta = 1.8 \times 10^{-4}\,\mathrm{K}^{-1}$ is the coefficient of thermal expansion. Consider a thermometer containing a bulb attached to a thin, cylindrical capillary tube. If the bulb and capillary tube contain 0.2 cm³ of mercury, how large must the diameter of the capillary tube be so that a 1K change in temperature corresponds to a 2 mm change in the mercury level in the capillary tube? (Neglect the change in volume of the glass bulb when the temperature is raised by 1K.)

23. (II) A copper bowl of volume 1500 cm³ is filled to the brim with water. Assume that the coefficient of volume expansion is $5.10 \times 10^{-5}\,\mathrm{K}^{-1}$ for copper and $2.07 \times 10^{-4}\,\mathrm{K}^{-1}$ for water. How much water will spill out when the temperature of the system is raised from 20°C to 50°C?

24. (II) Show that the area A of a plate whose coefficient of thermal expansion is α undergoes a fractional change $dA/A \cong 2\alpha\,dT$ when the temperature changes by a small amount dT.

25. (II) Look at the experiment done in Fig. 17–10. (a) How hot do we have to heat the brass ball of 2.028 cm diameter at room temperature to increase the diameter by 0.004 cm, so it will not pass through the ring? (b) What is the coefficient of volume expansion β for brass? (c) By what fraction does the volume expand for this temperature change?

26. (II) A metal bar is 1.75 m long with a coefficient of thermal expansion of $1.34 \times 10^{-5}\,\mathrm{K}^{-1}$. It is rigidly held between two fixed beams. When the temperature rises, the metal bar takes on the shape of the arc of a circle (Fig. 17–22). What is the radius of curvature of the circle when the temperature rises by 40°C? [Hint: Use the small-angle approximation $\sin\theta \cong \theta - \theta^3/6$.]

▲ FIGURE 17–22 Problem 26.

27. (I) In the design of a mercury thermometer, it is desirable for each °C change to correspond to a 1 cm expansion of the metal. What is the cross section of the mercury column if the mercury at the lowest reading on the thermometer is contained in a bulb whose volume is 0.7 cm³? Ignore any possible expansion of the glass tube.

28. (II) In space, where heat transfer by air is missing, large temperature differences can develop. A space telescope (Fig. 17–23) consists of a main mirror and an instrument unit, held together by titanium rods. The optical axis of the mirror (perpendicular to the mirror, through its center) hits the middle of the receiving window of the instrument unit. Suppose that the temperature of two neighboring rods increases by 20°C due to direct sunlight, with the temperature of the other two remaining unchanged. How far will the optical axis of the mirror move on the instrument unit, given that the rods are 5 m long and that $\alpha = 8.5 \times 10^{-6} \text{ K}^{-1}$?

▲ FIGURE 17–23   Problem 28.

29. (II) As we know from Chapter 13, the time expressed by a grandfather clock depends on the length of the pendulum. (a) Which is the best material for the pendulum of a clock to be used outdoors: aluminum, copper, or steel? (See Table 17–3.) (b) If the temperature of the clock ranges from −20°C to +30°C over the course of a year, and if the clock ticks with frequency $f = 1.00000$ Hz at 5°C, what is the yearly frequency range for each of the three materials?

30. (III) A strip of brass 1.5 mm thick and 12 cm long is glued back-to-back to a strip of Invar with the same dimensions. When the temperature is increased by ΔT, the composite strip tends to curl into an arc of a circle (Fig. 17–24). Find the radius of curvature of this circle for $\Delta T = 8$°C, given that the coefficient of thermal expansion is $19 \times 10^{-6} \text{ K}^{-1}$ for brass and $0.7 \times 10^{-6} \text{ K}^{-1}$ for Invar.

▲ FIGURE 17–24   Problem 30.

31. (II) The coefficient of volume expansion, β, for ethyl alcohol is $1.12 \times 10^{-3} \text{ K}^{-1}$. Suppose that a cylindrical copper container, with a coefficient of thermal expansion of $1.67 \times 10^{-5} \text{ K}^{-1}$, is filled to the very brim with ethyl alcohol at 5°C. What percentage of the alcohol will spill if the temperature is raised to 25°C?

17–4   The Equation of State of Gases

32. (I) A container holds gas at a pressure of 1.0 atm and a temperature of 300K. Half the gas leaks out while the temperature is raised to 340K. What is the pressure in the container?

33. (I) A dilute gas in a container fitted with a piston occupies a volume of 2500 cm³. If the pressure is increased by 50% and the temperature in kelvin is decreased by 15%, what volume will the gas occupy?

34. (I) A 1000 L container is filled with an ideal gas at 1.2 atm and 310K. How many moles are in the container? How many gas molecules?

35. (I) What is the volume occupied by three moles of an ideal gas at 30K and 10^{-7} atm in the upper atmosphere?

36. (I) What is the mass density of steam at a temperature of 100°C and a pressure of 1 atm, given that the molecular weight (the number of grams of 1 mol of the material) of water is 18 g/mol?

37. (I) What is the mass density of air (average molecular weight 29 g/mol), at a pressure of 0.31 atm and a temperature of −43°C? (These conditions correspond roughly to those at the top of Mount Everest.)

38. (I) A helium balloon is filled with 150,000 cm³ of helium gas at *STP* (standard temperature and pressure: 0°C and 1 atm). How many moles and molecules of helium (molecular weight 4.0 g/mol) are contained in the balloon?

39. (I) A container of ideal gas whose volume can be changed by the movement of a plunger is placed in a large bath that maintains the temperature of the gas at a fixed value of 295K. The plunger is pulled out so that the volume is increased by a factor of 1.78. By what factor is the pressure changed in the vessel, if at all?

40. (I) The atomic weight A of an elemental substance is closely related to the mass of the atoms of that substance. Atoms consist of electrons, of negligible mass, and a total of A protons and neutrons, which each have nearly the same mass M. The total mass of an atom is therefore approximately AM, and N_A atoms have a total mass AMN_A. By the definition of Avogadro's number, N_A, this mass is A g. Given the value of N_A, what is M in grams?

41. (I) A container at 0°C holds a mixture of gases: 80% nitrogen (N_2 molecular weight 28 g/mol) and 20% oxygen (O_2 molecular weight 32 g/mol) by weight. Assuming that for a dilute gas each constituent behaves as if it alone occupied the volume, what is the partial pressure due to each constituent of the mixture, given that the total mass of gas in the container is 8.0 g and that the volume is 4.0 L? What is the total pressure in the container?

42. (I) Consider a container of an ideal gas at a fixed pressure. The temperature is 290K. What is the fractional change in volume if the temperature changes to 296K? Can you derive a general formula that holds for all ideal gases?

43. (II) Given that the molecular weight of water (H_2O) is 18 g/mol and that the volume occupied by 1.0 g of water is 10^{-6} m³, use Avogadro's number to find the distance between neighboring water molecules. Assume for simplicity that the molecules are stacked like cubes.

44. (II) The lowest pressure achievable with a rotary vane pump is on the order of 10^{-2} N/m^2. What is the number density of molecules of the gas in a chamber that holds gas at this pressure and at a temperature of 20°C?

45. (II) The pressure of an ideal gas in a closed container is 0.60 atm at 35°C. The number of molecules is 5.0×10^{22}. (a) What are the pressure in pascals and the temperature in kelvins? (b) What is the volume of the container? (c) If the container is heated to 120°C, what is the pressure in atmospheres?

46. (II) Freon has van der Waals values of $a = 1.08 \text{ J} \cdot \text{m}^3/\text{mol}$ and $b = 10^{-4} \text{ m}^3/\text{mol}$. (a) Calculate the pressure inside a 0.40 L container of 1 mol of freon at 15°C. (b) What is the percentage difference between this pressure and the pressure you would have found had Freon been ideal? (c) What would your answers be if there were 100 mol inside the container?

47. (II) The Dieterici equation of state is

$$p\left(\frac{V}{N} - b\right) = kT \exp\left(-\frac{aN}{kTV}\right).$$

(a) Find the equation of state to which this reduces if $a = b = 0$. (b) By expanding the exponential function, find the equation of state to first order in a. (c) Sketch the isothermal curve on a p–V diagram, together with the corresponding curve for the ideal gas. In making your sketch, you can assume a and b are small.

48. (II) The molecular weight of H_2 is 2.0 g/mol. Consider H_2 gas in a container at STP (see Problem 38). (a) What volume is occupied by 30 g? (b) What mass and volume are occupied by 0.50 mol and 10 kmol?

49. (II) Use the ideal gas law to calculate the volume occupied by 1 mol of ideal gas at 1 atm pressure and 0°C. Given that the average molecular weight of air is 28.9 g/mol, calculate the mass density of air, in kg/m³, at the above conditions.

50. (II) Use the result of Problem 49 and Avogadro's number for the number of molecules in 1 mol, 6.02×10^{23}, to estimate the number of molecules in your lecture hall.

51. (II) What is the mass density of helium gas (atomic weight 4.00 g/mol) at 1 atm pressure and a temperature of 8K?

52. (II) A child takes 20 puffs of air to blow up a spherical balloon. The diameter of the balloon is 23 cm. The temperature is 23°C, and the exterior atmospheric pressure is 1 atm. (a) How many molecules are in each of the child's puffs? (b) If the surrounding air and the balloon are heated to 28°C, what size will the balloon be?

53. (II) In an experiment, a vacuum of 10^{-10} atm is achieved in a bottle. If the bottle is at room temperature (30°C), what is the number of molecules in the bottle per cubic centimeter?

54. (II) At altitudes above 14,000 ft, pilots must breathe air with enriched oxygen because the density of the atmosphere is decreased. The pressure of air is only about 0.26 atm at a typical height for flying of 32,000 ft. If the airplane is not pressurized, what must the fraction of oxygen in the air be in order for a pilot to breathe the same amount of oxygen as at sea level?

55. (II) An ideal gas is contained in a tank at 120 atm of pressure and 263K. (a) If half the mass of gas is drawn off and the temperature then rises by 50K, what is the new pressure? (b) Suppose instead that the temperature first rises by 50K and then half the mass of the gas is drawn off. What is the new pressure?

56. (II) For safety reasons, compressed gases for laboratory use are often stored outdoors and transferred to where they are needed via a system of tubes and valves (Fig. 17–25). Suppose that a 150 L tank contains hydrogen at 25 atm at −5°C at the beginning of an experiment. If 190 L are used at 25°C and 1 atm pressure, what is the pressure of the remaining hydrogen? The temperature outside does not change during the measurement.

57. (II) The cylinder of a manual bicycle tire pump is 35 cm long; its inside diameter is 2.5 cm. When the piston is pulled up, the valve to the tire is closed and air at atmospheric pressure enters the cylinder through the gap between the piston and the wall of the

▲ FIGURE 17–25 Problem 56.

cylinder, which acts like another valve (Fig. 17–26). When the piston is pushed down, the air in the cylinder is compressed. The valve to the tire opens when the pressure in the tire is reached, and air is transferred to the tire during the rest of the stroke. The last 5% of the volume of air remains in the cylinder and the hose. Suppose that you are pumping up a tire that already contains air at 2.5 atm gauge pressure. How much air is transferred to the tire at this pressure? What is the volume of this air at atmospheric pressure? Sketch a p–V diagram of the process. Assume that the temperature is constant.

▲ FIGURE 17–26 Problem 57.

58. (II) Figure 17–27 represents the variation of the volume and pressure of gas contained in a cylinder with a movable piston. The curved sections of the curve are isotherms. Calculate the temperature, pressure, and volume at the points 1, 2, 3, and 4 given that $p_1 = 1.00$ atm, $V_1 = 1.00$ L, $T_1 = 0.0°C$, $V_2 = 3.5V_1$, and $p_4 = 3p_1$.

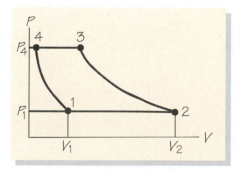

▲ FIGURE 17–27 Problem 58.

59. (II) The specific volume of a gas is its volume per unit mass; that is, the inverse of the density. Compute the specific volume and density of oxygen gas, O_2, at $T = 15°C$ and 1.0 atm of pressure, assuming that oxygen gas is ideal.

60. (II) An ideal gas is confined in a steel cylinder at 25°C and a pressure of 7.0 atm. (a) If the cylinder is surrounded by boiling water and allowed to come to equilibrium, what will the pressure of the gas be? (b) If the gas is allowed to escape until the pressure again reaches 7.0 atm, what fraction of the original gas, by weight, will escape? (c) If the temperature of the remaining gas now drops to 20°C, what is its pressure?

61. (II) A cylinder is closed at one end by a movable piston. The cylinder contains 300 cm³ of air at 20°C and 1.0×10^5 Pa pressure. The cylinder undergoes the following changes: (a) the gas is heated in such a way that its volume doubles but its pressure remains constant; (b) the volume is kept constant and the temperature is changed until the pressure increases by 30 percent; (c) the gas cools, and the piston position is adjusted to keep the pressure constant until the initial volume is reached. Calculate the volume, pressure, and temperature after each step.

62. (II) The law of Joseph Gay-Lussac and Jacques Charles (1802) states that any ideal gas undergoes the same fractional increase in the product of pressure, p, and volume, V, when the gas is brought from one temperature to another. What is the fractional increase in pV for an ideal gas brought from the temperature of ice melting at 1 atm of pressure to the temperature of water boiling at the same pressure?

63. (II) Calculate the volume of 24 g of the gaseous form of ethyl ether, $C_4H_{10}O$, at a temperature of 120K and pressure of 0.080 atm. Assume that the gas is ideal.

64. (II) A tank of volume 0.80 m³ contains 125 mol of helium gas at 25°C. Assuming that the helium behaves like an ideal gas, what is the pressure in the tank?

65. (II) It is a scorching summer day with an air temperature of 308K. What is the value, in joules, of RT for 1.0 mol of air? Repeat this for a cold winter's night when the air temperature is 258K. Compare these results to the kinetic energy that a ball with the mass of 1.0 mol of air (28.9 g) acquires when it falls 1.0 m under the influence of gravity.

66. (II) At the beginning of a compression stroke, the cylinder in a combustion engine contains 1.0 L of air at atmospheric pressure and a temperature of 20°C (Fig. 17–28). At the end of the stroke, the air has been compressed to a volume of 60 cm³, and the total pressure is 35 atm. What is the temperature of the air when it is so compressed?

▲ **FIGURE 17–28** Problem 66.

67. (II) The steam point of water is the point where water vapor (steam) and boiling water coexist at a pressure of 1.00 atm.

When a constant-volume gas thermometer calibrated to read a temperature of 273.16K at the triple point of water is used to measure the temperature of the steam point, the result depends slightly on the pressure of the gas in the thermometer because the gas is not precisely ideal (Fig. 17–29). In particular, if oxygen is used in the thermometer, the temperature of the steam point is determined to be 373.35K when the gas pressure is 0.40 atm, and 373.25K when the pressure is reduced to 0.20 atm. What is the temperature of the steam point?

▲ **FIGURE 17–29** Problem 67.

68. (II) Carbon dioxide obeys the van der Waals equation of state with $a = 5.96 \times 10^6$ atm·cm⁶/mol² and $b = 98.6$ cm³/mol. We know that 0.30 moles of CO_2 occupies 2.5×10^3 cm³ at 1.7 atm of pressure. (a) Find the temperature of the gas. (b) Redo your calculation, but this time assume that CO_2 is an ideal gas. What is the percentage difference between the two answers?

69. (II) There is a certain temperature T_B (the *Boyle temperature*) for which a van der Waals gas behaves as if it were ideal. (a) Show that an expression for this temperature is $T_B = (a/bR)$ $[1 - (bn/V)]$. (b) The constants a and b of the van der Waals equation of state are $a = 0.140$ m⁶·Pa/mol² and $b = 4.00 \times 10^{-5}$ m³/mol for argon gas. What is the value of the Boyle temperature of argon gas when the gas is so dilute that the term bn/V in the expression for the Boyle temperature can be ignored?

70. (III) An isotherm of 1 mol of a gas that obeys the van der Waals equation of state and is plotted on a p–V curve falls monotonically for large T. As T is decreased and new isotherms are plotted, the curve begins to show a kink, and at a temperature known as the *critical temperature*, T_c, there is a flat spot in the curve, where the slope is zero (Fig. 17–30). At this point, $p = p_c$ and $V = V_c$. Find T_c, p_c, and V_c in terms of the constants a and b.

▲ **FIGURE 17–30** Problem 70.

*17–5 Blackbody Radiation

71. (I) According to the Stefan–Boltzmann formula, the energy of radiation emitted by a blackbody per unit area per unit time is proportional to T^4, where T is the temperature in the Kelvin scale. Compare the energy emitted by a tungsten filament at 3200°C and the same filament at room temperature (25°C).

72. (I) The intensity of radiation from a small source decreases with the square of the distance from the source. Consider the energy of radiation that reaches your face after being emitted by solid aluminum at room temperature (25°C). At what distance will this energy of radiation be the same as the energy emitted by the same amount of molten aluminum (800°C) that reaches your face when you are 20 m from the aluminum? (Your answer will surprise you. Can you explain why? A look at Fig. 17–16 should help.)

73. (I) The dominant frequency of radiation emitted by an object is related to its temperature by $hf \cong kT$. Find the dominant radiation frequency emitted by (a) an object in interstellar space at 3K; (b) a body of water at 280K; (c) an electric stove heating unit (800K); (d) melting tantalum (3000K).

74. (I) The human eye is most sensitive to yellow light with a wavelength of approximately 550 nm. What is the temperature of an incandescent bulb filament that radiates most of its energy as a blackbody at this wavelength?

75. (II) The temperature of your skin is about 35°C. Calculate the wavelength at which the Planck radiation curve has a maximum for this temperature, and therefore the wavelength at which your body radiates the most energy. Use the more acccurate estimate indicated in Example 17–13.

76. (II) Show that the integral over all frequencies of the Planck formula given by

$$U(T) = \int_0^\infty \left(\frac{8\pi hf^3}{c^3} \right) \frac{1}{e^{hf/kt} - 1} df$$

gives a result that is of the form (a constant) $\times T^4$. [Hint: Change variables from f to hf/kT.] The energy emitted per unit area per unit time, $P(T)$, is proportional to $U(T)$, and thus $P(T)$ is also proportional to T^4, as in the Stefan–Boltzmann formula, Eq. (17–17).

77. (II) The surface of the Sun is at a temperature of 6000K. At what rate is energy radiated from the whole surface of the Sun, given that the radius of the Sun is $R = 6.95 \times 10^8$ m?

78. (II) Assume that the radiation emitted from the Sun moves radially outward from the Sun and that no radiation is absorbed between the Sun and Earth. How much energy in the form of radiation will fall per second on an area of 1 m² on Earth, if that area is perpendicular to the straight-line path of the radiation? The distance from the Sun to Earth is 1.5×10^{11} m.

79. (II) The average surface temperature of Earth is 290K. How much energy per second is radiated by Earth's surface, assuming that Earth simulates a blackbody. Compare this result with the amount of radiative energy that reaches Earth from the Sun. Use the result of Problem 78.

80. (III) The energy density per unit frequency in the frequency range from f to $f + df$ in blackbody radiation is $u(f, T) \, df$, where $u(f, T)$ is given by Eq. (17–16). An alternative way to express the blackbody radiation is to give $u'(\lambda, T) \, d\lambda$, the energy density per unit wavelength in the wavelength range from λ to $d\lambda$. (a) Use the fundamental wave relation $c = \lambda f$, where c is the speed of light, to show that $u(f, T) \, df = u[(c/\lambda), T]c \, d\lambda/\lambda^2$, so $u' = u[(c\lambda), T]c/\lambda^2$. (b) Assuming that the temperature is fixed, use the results of part (a) to find an equation for the λ_{max} for which $u'(\lambda, T)$ has a maximum. (c) The result of part (b) is a transcendental equation. Solve it for $\lambda_{max}T$. For what range of temperatures does λ_{max} fall within the visible spectrum, $\lambda \cong 450$ nm to 650 nm?

81. (III) Find the constant coefficient of T^4 in the Stefan–Boltzmann formula, Eq. (17–17), given that the relation between $P(T)$ and the total energy density $U(T)$ calculated in Problem 76 is $P(T) = cU(T)/4$, where c is the speed of light. [Hint: An appropriate entry in a table of integrals is

$$\int_0^\infty [x^3/(e^x - 1)] \, dx = \pi^4/15.]$$

General Problems

82. (II) Some astronomers estimate that toward the center of the galaxy, the average interstellar gas is molecular hydrogen, at a density of roughly 1 molecule per cubic centimeter and a temperature of between 10 and 20K. Estimate the pressure of this gas. Compare this with atmospheric pressure (1 atm = 760 torr) and very good laboratory vacuums (10^{-10} torr) that are achieved only with some difficulty.

83. (II) Use Archimedes' principle (Section 16–4) to estimate how much helium is needed for a balloon to lift a payload of 230 kg (including the mass of the balloon) in air at 1.0 atm pressure and 10°C. You will need the molecular weight of helium (4.00 g/mol) and that of air (29 g/mol).

84. (II) A constant-volume gas thermometer is placed in contact with ice water, and the pressure in the thermometer is observed to be 0.76×10^5 N/m². If the same thermometer is then placed in a pot of boiling water, the pressure rises to a new value of 0.90×10^5 N/m². The thermometer is calibrated in °C. What temperature will this thermometer read if it has a pressure reading of 0.84×10^5 N/m²?

85. (II) Suppose some measurable property Z of a material you have is sensitive to temperature and that you wanted to make a thermometer with a Celsius scale based on it. You therefore write $T_C = a + bZ$. How do you establish the values of a and b? Would your scale necessarily coincide with the Celsius scale that you would get from an ideal gas?

86. (II) The human eye can detect colors over the electromagnetic wavelength range of about 300 to 700 nm. (a) Use the results of Section 17–5 to calculate what temperatures correspond to the maximum energy densities ($hf \cong kT$) for this range of frequencies? (b) The answer to Problem 80 is $\lambda_{max}T = 2.9 \times 10^{-3}$ m·K. Calculate the temperatures using this relation. (c) What is the percentage difference, if any, for the ends of the range in the two calculations?

87. (II) A cylindrical vessel with a tight but movable piston is placed in a vertical position so that the piston, whose area is 70 cm², is subject to atmospheric pressure. When the gas in the vessel is heated from a temperature of 20°C to 80°C, a 0.5 kg mass must be placed on top of the piston to hold the piston at the position it occupied at the lower temperature (Fig. 17–31, see next page). What is the volume of 0.2 mol of the gas?

▲ **FIGURE 17–31** Problem 87.

88. (II) An air bubble of volume 8 cm³ rises from the bottom of a lake 15 m deep. The temperature at the bottom is 8°C. The bubble rises to the surface, which is at 20°C and at atmospheric pressure. Assuming that the bubble is in thermal equilibrium with the surroundings at all times, calculate the volume of the bubble at the surface.

89. (II) In Chapter 14 we stated that the speed of sound in air is given by $v_{sound} = \sqrt{\gamma p / \rho_0}$, where the constant $\gamma \cong 1.4$ for air near STP (see Problem 38), p is the pressure, and ρ_0 is the density of the air. Reexpress the speed of sound in terms of the temperature. What is the difference in the speed of sound at the coldest and hottest times of day in a desert?

90. (II) A bottle containing air is closed with a watertight yet smoothly moving piston. The bottle with its air has a total mass of 0.30 kg. At the surface of a body of water whose temperature is a uniform 285K throughout, the volume of air contained in the bottle is 1.5 L (Fig. 17–32). Recall that the pressure of water

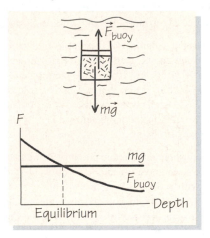

▲ **FIGURE 17–32** Problem 90.

increases with depth below the surface, D, as $p = p_0 + \rho g D$, where p_0 is the surface pressure and $\rho = 1.0$ kg/L. The bottle is submerged. (a) What is the volume of the air in the bottle as a function of depth? (b) Calculate the buoyant force on the bottle as a function of depth. (c) At what depth do the buoyant force and the force of gravity cancel? (d) Is the depth at which the force on the bottle is zero, calculated in part (c), a stable or an unstable equilibrium?

91. (II) Suppose that, instead of the Kelvin scale, an alternative temperature scale—call it the Kelvin′ scale, with the units K′—were defined in which the zero level remains unaffected but the temperature at the triple point of water is taken to be 500K′. Calculate the value of Boltzmann's constant in units of J/K′.

92. (II) Consider the radiation emitted from a furnace through a hole of area 1 cm². Assume that all the radiation is absorbed by a beaker that contains 42 g of water. Given that it takes 4.2 J of energy to raise the temperature of 1 g of water by 1°C, how long will it take for the water to be heated by 8°C if the furnace temperature is 1600K? Assume that there are no energy losses due to the beaker.

93. (III) The van der Waals equation of state, Eq. (17–15), can be put into a form that is an expansion in powers of the density. Dividing by RT and then by $[(V/n) - b]$ in Eq. (17–15), we can write

$$ p + \frac{\alpha(n/V)^2}{RT} = \left(\frac{V}{n} - b\right)^{-1} = \frac{n}{V}\left(1 - b\frac{n}{V}\right)^{-1}. $$

(a) Show that, for $bn/V < 1$, this equation has the approximate form

$$ \frac{p}{RT} = \frac{n}{V} + b\left(\frac{n}{V}\right)^2 - \frac{\alpha}{RT}\left(\frac{n}{V}\right)^2. $$

(b) The result in part (a) is of the general form

$$ \frac{p}{RT} = \frac{n}{V} + B_2(T)\left(\frac{n}{V}\right)^2, $$

where, in this case, the function $B_2(T)$ is $B_2(T) = b - (a/RT)$. This is an example of a *virial expansion* for the equation of state, and the function $B_2(T)$ is known as a *virial coefficient*. Show that $B_2(T)$ starts out negative as T increases from some small value, passes through zero, and then becomes positive. The value of T at which $B_2(T)$ is zero is the Boyle temperature, where a real gas satisfying the van der Waals equation behaves as though it were ideal (see Problem 69).

◀ Part of a large plant near Geyserville, in northern California, that uses thermal energy from within Earth to do work and generate electric power. While such a plant does not require a separate fuel source that must be replenished, any electric energy that it has produced has depleted at least that much thermal energy from the underground thermal source.

Heat Flow and the First Law of Thermodynamics

Steam contains thermal energy, and as a result steam can push a piston in a cylinder or turn a turbine and can thereby do work. The observation that we can transfer thermal energy to a mechanical system and do work is important. In this chapter we will see how thermal energy is related to other forms of energy with which we are familiar and how it can be used to do mechanical work. We will see that as the temperature of a thermal system increases, so does the amount of thermal energy in the system.

Changes in the energy of thermal systems (thermal energy) and its relation to changes in the energy of the systems' surroundings is summarized in two important laws of physics—the first and second laws of thermodynamics—which we examine here and in Chapter 20, respectively. The first law of thermodynamics describes thermal energy as just another form of energy—like kinetic energy or potential energy. The principle of the conservation of energy that we developed in Chapter 7 is now expanded to include the thermal energy of a system. This energy can be accounted for in mechanical form: It is the kinetic and potential energy of the system's microscopic atomic constituents (Chapter 19). In other words, the first law of thermodynamics is nothing more than the conservation of energy principle applied to thermal systems. The second law of thermodynamics states limitations on how much of a system's thermal energy can be converted to other forms of energy.

It is difficult to overestimate the importance of the conservation of energy as a scientific principle. The discovery that thermal energy is just another form of energy, and the consequent expansion of the principle of conservation of energy, constitutes one of the truly important scientific advances of the 19th century.

18–1 Changes in Thermal Systems

As we saw in Chapter 17, thermal systems in equilibrium, such as a container of hot steam, are described by only a few thermodynamic variables. For a gas these variables are the temperature T, the pressure p, the volume V, and the number of moles n (or the number of molecules, N). Thermodynamic variables describe the *state* of the thermal system, and an equation of state relates these variables. For example, the equation of state for an ideal gas [Eqs. (17–7) and (17–9)] is $pV = nRT = NkT$, where R is the universal gas constant and k is Boltzmann's constant. But other equations of state are equally possible and can all be handled the same way. If we assume that n is fixed, the remaining three variables are connected by an equation of state, and we can very usefully describe the state as a point on a two-dimensional plot, such as a p–V diagram, a p–T diagram, or a V–T diagram. Changes, or transformations, in the state of a thermal system can be represented as curves on such diagrams—at least if thermal equilibrium is established at every stage of the transformation. We have already described and used such curves in Section 17–4 (see Example 17–11).

To maintain thermal equilibrium at each step of a transformation, changes must be made slowly. Just how slowly depends on the internal dynamics that lead to equilibrium. For example, it is not difficult to compress a gas in such a way that it maintains thermal equilibrium at each stage. To see why, consider a piston compressing a gas. When molecules bounce from the inward-moving piston, they are speeded up (thereby gaining energy) and distribute their additional energy to other molecules through collisions. The time required for the effect of the push to be evenly spread through the gas is characterized by the size of the container divided by the *speed of sound* in the gas. That is because the speed of sound is a measure of the spread of an organized disturbance through the gas through successive collisions. The speed of sound is large compared to typical mechanical motion, such as the movement of the piston; equivalently, the intermolecular collision time is short compared to the time for mechanical motion.

Reversible and Irreversible Processes

A change in which thermal equilibrium is maintained throughout, such as the sufficiently slow compression of a piston in an insulated cylinder, or the cooling of a bottle of soda in a refrigerator, is called a **reversible process**. Such changes can always be reversed by changing the external conditions in such a way that the thermodynamic variables return to their original values following the same path, only reversed, on the graph of thermal variables. As an example, the volume of a cylinder of gas held at a fixed temperature is increased from V_1 to V_2 in a controlled way when a piston slowly moves out of the cylinder with an external force balancing the pressure, as in Fig. 18–1a. If the piston is slowly pushed back in, as in Fig. 18–1b, the external conditions are reversed, and the thermodynamic variables of the gas recover their initial values. These changes are reversible, and Fig. 18–1c shows the transformation as a curve on a p–V diagram. Because the thermal system remains in equilibrium throughout a reversible process, reversible processes always connect equilibrium states.

Conversely, a gas expanding to fill a vacuum through an open stopcock (Fig. 18–2a) provides an example of an **irreversible process**, a process in which the thermal system's changes cannot be retraced. The expansion from V_1 to V_2 occurs rapidly and without control. Nothing you do with the stopcock will cause the released gas to reverse itself and reconcentrate within the original volume. We can mark the original and final points on a p–V diagram (Fig. 18–2b), but we cannot mark the points in between, because the system is not in thermodynamic equilibrium during the expansion. The particular irreversible process described here, in which the gas expands freely, is known as **free expansion**. Other examples of irreversible thermal transformations include the melting of ice cubes in lemonade and the heating of a brake shoe as it rubs against the brake drum. Unless we state otherwise, in this chapter *we consider only reversible changes of state.*

(a)

(b)

◀ **FIGURE 18–1** The volume of a gas in a cylinder is reversibly changed when a piston is (a) slowly pulled or (b) slowly pushed. The temperature is held fixed during the transformations. (c) Constant-temperature reversible transformation of a gas between V_1 and V_2 on a p–V diagram. The curve is an *isotherm.*

(c)

(a)

(b)

▶ **FIGURE 18–2** (a) A gas initially confined to the left-hand side of a container expands to fill the container's total volume when a stopcock is opened. (b) Irreversible expansion of a gas from V_1 to V_2. The path is not well defined, and the dots serve only to indicate that there has been some irreversible change.

CONCEPTUAL EXAMPLE 18–1 A gas in contact with a thermal reservoir at temperature T is expanded reversibly from V_1 to V_2. The trace of points made by this change on a p–V diagram is an isotherm. Describe the isothermal curve if the gas is ideal.

Solution Because we have a gas in contact with a thermal reservoir, the temperature T must remain constant. Therefore, the ideal gas law, $pV = nRT$, reveals that the product of p times V must be a constant, because nRT is constant. We can see that as the volume increases from V_1 to V_2, the pressure must decrease from p_1 to p_2 in order to keep the product pV constant as shown in Figure 18–1c. The curve that corresponds to $pV = $ a constant is a hyperbola. Figure 18–3 shows a family of hyperbolas, each curve corresponding to a different constant temperature.

What Do You Think? What would such a curve look like on a p–T diagram? *Answers to **What Do You Think?** questions are given in the back of the book.*

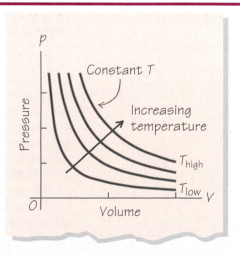

▲ **FIGURE 18–3** An isotherm for an ideal gas is a hyperbola, $p \propto TV^{-1}$. The isotherms corresponding to different temperatures are indicated.

18–2 Heat Flow

We suggested in Section 17–3 that the temperature of a system is closely related to its **internal energy**, or **thermal energy**. We will learn here that the internal energy of a thermal system is a function of the thermodynamic variables; in particular, the higher the temperature of a system, the higher the internal energy. Changes in the internal energy of a thermal system can manifest themselves in only a few ways: *The system can transfer its internal (thermal) energy by changing the temperature (or phase) of another system, or it can use its internal energy to do mechanical work on its surroundings, or it can do both.*

When the temperature of a thermal system in contact with a neighboring system changes, we say that there has been a **heat flow** into or out of the system.[†] We will use the term *heat flow* rather than *heat* in order to discourage the idea that a thermal system has a certain amount of "heat." Heat flow is more like work than energy: Work is done *on* or *by* a mechanical system, and we do not speak of a system containing a certain amount of work. It would be equally misleading to think of a system as containing a certain amount of "heat." Heat flow into, and/or mechanical work done on, a system may each contribute to raising the temperature of a thermal system. We employ the symbol Q for heat flow or, often, ΔQ.

As far as units, we start with the historical unit of heat flow, the calorie (cal). (We are going to have to learn some further physics before we can sensibly relate this to SI units with which we are familiar.) Provisionally, 1 cal is defined to be the amount of heat flow required to raise the temperature of 1 g of water at atmospheric pressure from 14.5°C to 15.5°C. Another unit for heat flow used in industrial applications is the *British thermal unit* (Btu), which is defined as the heat flow required to raise the temperature of a 1-lb weight of water by 1°F, averaged over the temperature range 32°F to 212°F. The British thermal unit is related to the calorie by 1 Btu = 252.02 cal.

A major aim of this chapter is to establish the fact that heat flow is a transfer of energy and that thermal systems have an internal energy related to the system's thermal properties. In the remainder of this section we will set the stage for this by learning about some properties of heat flow that are independent of its identification with energy transfer.

Heat Capacity

When there is a heat flow to or from a thermal system, the temperature of the system changes. The connection between heat flow and temperature change is the **heat capacity**, C. A *small* heat flow, dQ, *into* a thermal system will change the temperature by a small amount dT, with a proportionality constant that is the heat capacity C:

$$dQ = C\,dT. \tag{18–1a}$$

This can also be written for a *finite* heat flow ΔQ and a *finite* temperature change ΔT if the heat capacity does not depend on the temperature—a situation that is a good approximation for many physical situations:

$$\Delta Q = C\Delta T. \tag{18–1b}$$

In these equations the heat flow, whether large or small, can be positive or negative. The quantity C is always positive; thus, the sign of the temperature change indicates the sign of the heat flow. Equations (18–1a, b) state that there is a positive heat flow into a system when its temperature increases and there is a negative heat flow into the system when its temperature decreases. The value of the heat capacity depends on the particular material and is proportional to the amount (mass) of material. It may also depend on the temperature, in which case a finite heat flow is related to a finite temperature change through an integral.

We can eliminate the dependence of heat capacity on the amount of material by defining the **specific heat**, c, as the heat capacity of 1 g of the material—the heat capacity per unit mass of the material—and the **molar heat capacity**, c', as the heat capacity of 1 mol of the material. If the mass in grams of a thermal system consisting of n moles is m, then

$$C = mc = nc', \tag{18–2}$$

[†]We will see later that we can also have a heat flow without a temperature change when the system does work or undergoes a phase change.

and if A is the molecular weight (the mass of 1 mol, in grams), then

$$c' = Ac. \tag{18–3}$$

The units of heat capacity are cal/K; those of specific heat are cal/g·K; and of molar specific heat, cal/mol·K.

CONCEPTUAL EXAMPLE 18–2 You have a substance with a large specific heat—water is an example. For a given positive heat flow out of a given amount of the substance, the temperature will (a) rise a lot; (b) rise a little; (c) fall slightly; (d) fall a lot.

Answer If we remove energy (heat flow) from a substance with a large specific heat capacity, the temperature of the substance will fall only slightly. The substance has a "large capacity" for holding its temperature. (The word "slightly" should take into account the amount of the substance.) Water has a relatively large specific heat, which is why large bodies of water have strong moderating effects on the climate in their vicinity.

Path Dependence of Heat Flow

The heat capacity is defined in Eq. (18–1) in terms of a temperature change. But the conditions for a temperature change are not unique—for example, the volume or the pressure, or neither, can be held constant during the temperature change. The heat capacity generally depends on how the temperature change is made and on the values of the thermodynamic variables p, V, and T. This dependence is indicated by writing $C = C(p, T)$ or $C = C(V, T)$. We cannot use Eqs. (18–1a, b) to find a net heat flow unless we know more about how either pressure or volume changes as temperature changes. We say that the heat flow *depends on the path* along which temperature changes.

Let us examine two cases of special interest: In the first case the volume is kept constant, and in the second the pressure is held constant. For these cases, we label the heat capacity with the subscript V or p, and Eq. (18–1a) becomes

$$\text{for constant volume:} \quad dQ = C_V \, dT; \tag{18–4}$$

$$\text{for constant pressure:} \quad dQ = C_p \, dT. \tag{18–5}$$

Generally, C_V differs from C_p, as we will see later when we find C_V and C_p for an ideal gas. If these quantities are independent of temperature, then the generalizations of the finite change Eq. (18–1b) applies. Indeed, the heat capacities do turn out to be independent of temperature for monatomic ideal gases, which describe at least roughly all gases that are not too hot, and that will simplify things. Note that the subscript must be used on c and c' as well. (The volume change with temperature for liquids is insignificant compared with that of gases; for solids, pressure is not a useful variable. Therefore, the subscript is not used for these cases.)

One can show (to see why, see Section 18–7) that the heat capacity for constant pressure is larger than the heat capacity for constant volume:

$$C_p \geq C_V. \tag{18–6}$$

This inequality holds even for nonideal gases.

EXAMPLE 18–3 An ideal gas undergoes transformations that take it from point A to point B on a p–V diagram by the two different paths, α and β, shown in Fig. 18–4. Points A and B have the same temperature, $T = T_0$ (they lie on an isotherm). Assuming that C_V and C_p are constants, what is the heat flow into the gas (a) for path α (through point 1 at temperature T_1) and (b) for path β (through point 2 at temperature T_2)?

Strategy Figure 18–4 labels the various thermodynamic variables. We can use Eqs. (18–4) and (18–5) to find the heat flow along the various legs of the transformation. But because both C_V and C_p are constants, we can use these equations in "Δ" form, that is, we can use them for finite heat flows and finite temperature changes.

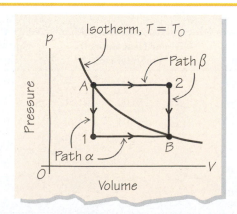

▲ **FIGURE 18–4** An ideal gas undergoes a transformation between points A and B by two different paths.

(continues on next page)

Working It Out (a) Along path α, there are two contributions: from point A to point 1 (constant volume), and from point 1 to point B (constant pressure). Thus

$$\Delta Q_\alpha = C_V \Delta T_{A \to 1} + C_p \Delta T_{1 \to B}$$
$$= C_V(T_1 - T_0) + C_p(T_0 - T_1) = (T_0 - T_1)(C_p - C_V).$$

(b) For path β, there are also two contributions: from point A to point 2 (constant pressure), and from point 2 to point B (constant volume).

$$\Delta Q_\beta = C_p \Delta T_{A \to 2} + C_V \Delta T_{2 \to B}$$
$$= C_p(T_2 - T_0) + C_V(T_0 - T_2) = (T_2 - T_0)(C_p - C_V).$$

The factor $(T_0 - T_1)$ is not equal to the factor $(T_2 - T_0)$. Thus the two answers are different, and the heat flow depends on the path that is followed.

What Do You Think? There is no temperature difference between points A and B. Why isn't the heat flow zero?

Calorimetry

The measurement of heat capacities is called **calorimetry**. Calorimetry, or more properly *adiabatic calorimetry*, is based on the fact that *when a system is mechanically and thermally isolated, heat flow is conserved within the system.* (A system is said to be *mechanically isolated* if it cannot do mechanical work on its surroundings.) A heat flow into one part of such a system must be matched with a corresponding heat flow out of another part. In such circumstances, heat flow is present only as long as there is a temperature difference between parts of the system; it ceases when equilibrium is established throughout the system.

Our definition of the calorie shows that the specific heat for water at 15°C and at 1 atm of pressure is

$$c_{H_2O} = 1 \text{ cal}/(g \cdot °C).$$

The measurement of the specific heats of other substances is based on that of water, as in Example 18–4.

EXAMPLE 18–4 A block of iron (mass 100 g) is heated in an oven to 500K and then plunged into a closed, thermally insulated container of 500 g of water at 292K. The block and the water come to an equilibrium temperature of 297K. What is the specific heat of iron, c_{Fe}? Assume that the specific heats of water and iron do not vary significantly over the temperature ranges in question.

Strategy There is a positive heat flow ΔQ_{water} into the water—it is positive because the iron is initially hotter than the water and the water temperature increases. Similarly the iron cools, so the heat flow into it, ΔQ_{Fe}, is negative. The entire system of iron and water together is isolated, so that

$$\Delta Q_{system} = 0 = \Delta Q_{water} + \Delta Q_{Fe}.$$

We next relate the heat flows to the specific heats. There are two things to keep in mind: First, the heat capacities C are independent of temperature here, so we can use Eq. (18–1b) for finite temperature shifts. These temperature changes are all specified in the problem statement. Second, we are interested in the specific heats c, so we can use the relation $C = mc$, where m is the amount of material, also specified.

These two facts allow us to write $\Delta Q = mc\Delta T$ for both water and iron in the equation above, and that is enough to solve for c_{Fe}.

Working It Out We have

$$0 = \Delta Q_{water} + \Delta Q_{Fe}$$
$$= m_{water} c_{water} (\Delta T)_{water} + m_{Fe} c_{Fe} (\Delta T)_{Fe}$$
$$= (500 \text{ g})(1 \text{ cal}/g \cdot K)(297K - 292K)$$
$$+ (100 \text{ g}) c_{Fe}(297K - 500K)$$
$$= 2500 \text{ cal} + (-2.03 \times 10^4 \text{ g} \cdot K) c_{Fe}.$$

We solve for c_{Fe} to find that

$$c_{Fe} = \frac{2500 \text{ cal}}{2.03 \times 10^4 \text{ g} \cdot K} = 0.12 \text{ cal}/g \cdot K.$$

What Do You Think? What if we had used an aluminum block instead of iron in this experiment? Would the final system temperature be (a) higher, (b) lower, (c) no change? (Look at Table 18–1 for the specific heat of aluminum.)

TABLE 18–1 • Molar Heat Capacities and Specific Heats of Some Materials at Constant Volume[†]

Material	Molar Heat Capacity (cal/mol · K)	Specific Heat (cal/g · K)	Material	Molar Heat Capacity (cal/mol · K)	Specific Heat (cal/g · K)
Helium gas	3.00	0.75	Silicon	4.78	0.17
Argon gas	30.00	0.75	Aluminum	5.82	0.22
N₂ gas	4.94	0.176	Copper	5.85	0.092
O₂ gas	4.97	0.155	Iron	6.00	0.107
SO₂ gas	7.52	0.118	Lead	6.32	0.031
Diamond	1.46	0.12	Water	18.0	1.00
Beryllium	3.85	0.43	Mercury	6.7	0.033

[†] *Gases at standard temperature and pressure.*

Example 18–4 shows how the specific heat of iron (or of many materials) can be measured. Calorimetry is used most often in the opposite sense: Given the specific heats, temperature changes are calculated in situations where there is a heat flow from one system to another. Table 18–1 gives the molar heat capacities and specific heats *at constant volume* for a variety of materials at room temperature. In this table, the gases are at standard pressure. In Chapter 19 we will see that the values in Table 18–1 can be explained in terms of the microscopic constituents and structure of the material.

THINK ABOUT THIS . . .

WHAT ARE THE ORIGINS OF THE TERMS CALORIE, CALORIMETRY, CALORIC?

The prefix "calor" is derived from the Latin word *calor*, meaning heat in its colloquial sense. The French scientist Lavoisier coined the term "caloric" in his caloric theory of 1787; this theory proposed that a massless substance (a fluid) flows from one material to another when an object is heated. The fluid was called the *caloric*. The unit of heat flow was later called the *calorie*. Caloric was thought never to be created or destroyed, and all substances had caloric that could be absorbed or released. The term calorimetry refers to the science of measurements associated with heat flow; the term is still used today even though we know the idea of the caloric is wrong. A calorimeter is a device used to measure heat flow.

Count Rumford (Benjamin Thompson) played an important role in discrediting the caloric theory. While he was supervising the boring of cannon barrels in Bavaria in 1798, Rumford observed and quantified the continual heating of the metal (Fig. 18–5). He reasoned that if the heating were due to the transfer of a fluid—as proposed by the caloric theory—then the fluid eventually would be used up and no more heating should occur. He found that heat was caused by the process of friction and that there was a continual heat flow as long as the boring continued. He even used the hot chips of metal that were bored out to boil water!

▲ FIGURE 18–5 Count Rumford observed that the temperature of cannon barrels rose steadily as they were being hollowed out. He concluded that the caloric theory could not be correct.

Phase Changes and Heat Flow

When there is a **change of phase** in a system, there is no temperature change even though there is a heat flow. For example, a solid state may change to a liquid state by melting, or a gaseous state may change to a liquid state by condensation (Fig. 18–6). It takes a certain heat flow to convert 1 g of ice at its melting temperature of 0°C to liquid water at 0°C (Fig. 18–7). The heat flow necessary to bring about the transformation is called the **latent heat of fusion**, L_f. Measurements show that $L_f = 79.6$ cal/g for ice.

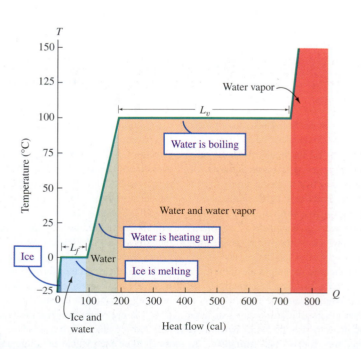

▲ FIGURE 18–7 Phase changes for 1 g of water. Note that the temperature remains constant as heat flow is added during melting and vaporization.

(a)

(b)

▲ FIGURE 18–6 (a) Mist forms when water droplets condense from water vapor. (c) Gallium melts at the temperature of the human body. When it does so, it absorbs heat flow from its surrounding environment.

During the melting process, there is no change of temperature and, to a good approximation, there may be little change of pressure or volume. The heat flow goes into changing the molecular arrangement of the substance (see Fig. 16–1). Similarly it takes a certain heat flow to convert 1 g of water at its vaporization temperature, 100°C (at 1 atm pressure), to 1 g of steam. The heat flow necessary for this is called the **latent heat of vaporization**, L_v, whose value for water is 540 cal/g. Most of this heat flow goes into overcoming the intermolecular forces that hold the molecules close together and make a liquid much denser than a gas. Some latent heats are given in Table 18–2.

TABLE 18–2 • Some Latent Heats		
Material	**Melting Temperature (K)**	**Latent Heat of Fusion (kJ/mol)**
H_2	14	0.12
O_2	54	0.44
Water (H_2O)	273	6.0
Ethyl alcohol (CH_3CH_2OH)	159	4.8
Mercury	234	2.3
Lead	600	4.8
Copper	1357	13

Material	**Boiling Temperature[†] (K)**	**Latent Heat of Vaporization (kJ/mol)**
H_2	20	0.92
O_2	90	6.8
Water (H_2O)	373	41
Ethyl alcohol (CH_3CH_2OH)	351	39
Mercury	630	59
Lead	2023	178
Copper	2839	300

[†] *Boiling takes place at atmospheric pressure.*

THINK ABOUT THIS . . .

WHY ARE LATENT HEATS OF VAPORIZATION SIGNIFICANTLY LARGER THAN LATENT HEATS OF FUSION?

As we described in Section 16–1, the arrangement of atoms and molecules of a substance differs between the solid and liquid phase. Broadly speaking, solids are crystalline arrangements; the atoms or molecules that make up solids form a lattice structure. The forces between the nearest neighbors are springlike. With the input of a little energy, which we can anticipate is associated with a heat flow, the atoms vibrate about their equilibrium positions on the lattice, and as the energy input increases, the amplitude of these oscillations increases. At larger separations the forces become weaker, and the melting point is reached when the oscillations become of large enough amplitude to take the atoms beyond the point where the forces are springlike. At this point, which is where the substance has gone to the liquid phase, the atoms are no longer tied to a particular neighbor and can move in random directions. In the liquid phase there are still, however, bonds between neighboring atoms, and the density of liquids remains essentially the same as the density of the corresponding solid. So in this simplified picture the energy needed to melt a solid is determined only by the requirement that the *directional order* be destroyed—that the molecules leave their neat lattice arrangement. In vaporization the molecules form a gas, and (as long as we are not near the critical point) the distance between them increases enormously. To achieve this, it is necessary to *break the bonds* between neighboring atoms or molecules, and this takes much more energy than a rearrangement without actual separation.

CONCEPTUAL EXAMPLE 18–5 One of the games that children used to play was to take a toy tin soldier, put it in a spoon over a flame till the soldier melted, and then drop it into a pan of water. The shape of the solid mass was then used to cast shadows or to "read," like tea leaves. The temperature of the water naturally rose. Would it have risen more, less, or the same if the tin soldier had been heated to the point of almost melting and then dropped in the water?

Answer The just-melted tin and the not-quite-melted tin are at very nearly the same temperature. Nevertheless, they have quite dif-ferent effects on the water. The water temperature rises because there is heat flow to it from the hot tin. But when a liquid solidifies (freezes), as does the melted tin, there is heat flow in the form of a la-tent heat of fusion. The not-quite-melted soldier does not have this additional thermal energy. There is thus more heat flow in the case of the melted tin soldier than in the case of the solid tin soldier, and the temperature of the water will rise higher when the melted soldier is dropped in rather than the not-quite-melted one.

EXAMPLE 18–6 One hundred grams of ice at 0°C is dropped into 200 g of liquid water at 49°C. The system is thermally isolated (see Chapter 17). After a period of time, the ice is entirely melted, leaving 300 g of water at 6°C. Assume that the specific heat of liquid water has the constant value $c = 1 \text{ cal/g} \cdot °C$. Calculate the latent heat of fusion of water.

Setting It Up As in Fig. 18–8, we label the mass of ice as m_1, its initial temperature as T_i, the initial mass of liquid water as m_2, and its initial temperature as T_i'. The final temperature of the system is T_f. All these quantities are given.

▲ **FIGURE 18–8** Ice dropped into water changes the temperature of the system.

Strategy With the system thermally isolated, the temperature cannot change as a result of thermal contact with the surroundings. Any heat flow must occur within the system between the water and the melting ice, and the net heat flow to the outside must be zero. We can summarize this condition as

$$\Delta Q_1 + \Delta Q_2 + \Delta Q_3 = 0,$$

where ΔQ_1 is the heat flow needed to melt the ice, ΔQ_2 is the heat flow necessary to raise the temperature of an amount m_1 of liquid water (from the melted ice) to the T_f, and ΔQ_3 is the heat flow nec-essary to cool an amount m_2 of liquid water (the initially present liq-uid) to T_f. The specific heat of water is assumed to be a constant, so we can use the finite heat-flow version of Eq. (18–1). By definition the latent heat of fusion, L_f, is the heat flow necessary to melt 1 g, so if m_1 is measured in grams, we have

$$\Delta Q_1 = m_1 L_f, \Delta Q_2 = cm_1(T_f - T_i), \Delta Q_3 = cm_2(T_f - T_i').$$

The only unknown is L_f, so to solve for it, we use the condition that $\Delta Q = 0$ for the system.

Working It Out Let's calculate ΔQ_2 and ΔQ_3:

$$\Delta Q_2 = (1 \text{ cal/g} \cdot °C)(100 \text{ g})(6°C - 0°C) = 600 \text{ cal}$$
$$\Delta Q_3 = (1 \text{ cal/g} \cdot °C)(200 \text{ g})(6°C - 49°C) = -8600 \text{ cal}.$$

Now we have

$$m_1 L_f + \Delta Q_2 + \Delta Q_3 = 0$$

or

$$L_f = \frac{-\Delta Q_2 - \Delta Q_3}{m_1} = \frac{-600 \text{ cal} + 8600 \text{ cal}}{100 \text{ g}} = 80 \text{ cal/g}.$$

You can find $\Delta Q_1 = L_f \times (100 \text{ g}) = 8.00 \cdot 10^3 \text{ cal}$. This is much larger than the value associated with raising the temperature of the melt-ed ice; we have $L_f/c = 80$. Latent heats are typically quite significant.

What Do You Think? Describe what would be different if the latent heat of fusion of water were (a) much larger, (b) much smaller than it is. Think about lakes in Minnesota or about iced drinks for your answer.

18–3 Heat Flow in Materials

When one end of a solid beam of length L is maintained at a temperature T_1 and the other end is kept at a lower temperature T_2, there will be a steady heat flow from the higher-temperature end to the lower-temperature end. Experiment shows that the rate of heat flow is proportional to the cross section A of the beam and to the tempera-ture difference $\Delta T = (T_1 - T_2)$ between the two ends; it is inversely proportional to the length of the beam L. This rate is thus described by an equation of the type

$$\frac{\Delta Q}{\Delta t} \propto A \frac{\Delta T}{L}. \tag{18–7}$$

Heat flow is only linearly dependent on the temperature difference ΔT if ΔT is small. This poses no difficulty if we think about the rate of heat flow for a thin slice of length

TABLE 18–3 • Thermal Conductivities for Various Materials

Material	κ (kW/m·K)
Aluminum	0.21
Copper	0.39
Lead	0.035
Silver	0.42
Water (at 0°C)	5.65×10^{-4}
Water (at 20°C)	5.99×10^{-4}
Air	2.6×10^{-5}
Asbestos	0.80×10^{-4}
Brick	6.3×10^{-4}
Glass	10.5×10^{-4}

Δx rather than for the entire beam at once. In that case ΔT is small and the linear relation between ΔT and the rate of heat flow is valid. In this "thin-slice limit" we replace L by Δx and rewrite Eq. (18–7) in derivative form as

$$\frac{dQ}{dt} = -\kappa A \frac{dT}{dx}. \qquad (18\text{–}8)$$

The coefficient κ is the **thermal conductivity**, a constant that varies from material to material and may have a temperature dependence. The SI units of κ are W/(m·K), or J/(m·s·°C). (Note that a difference of 1°C is identical to a difference of 1K. The appearance of watts in the units of κ anticipates that heat flow is an energy transfer.) We present some typical values of thermal conductivity in Table 18–3. When κ is large, the material is said to be a good thermal conductor, and when κ is small, the material is a poor thermal conductor. The minus sign in Eq. (18–8) indicates that if the *temperature gradient dT/dx is negative* (i.e., the temperature is lower for larger values of x), there is positive heat flow toward larger x. In other words, heat flow goes from a hot region to a cold region. The differential equation Eq. (18–8) applies to liquids and gases as well as to solids, and Table 18–3 lists values of κ for materials of all three states.

EXAMPLE 18–7 An igloo, a hemispherical enclosure built of ice ($\kappa = 1.67$ J/m·s·°C), has an inner radius of 2.50 m. The thickness of the ice is 0.30 m. At what rate must thermal energy be generated to maintain the air inside the igloo at 5°C when the outside temperature is −40°C? Ignore all thermal energy losses through the ground or by air currents.

Setting It Up Given the configuration and dimensions of an igloo, as in Fig. 18–9, we want the rate of heat flow out of it for a given inside-outside temperature difference. This much heat flow has to be supplied from a heater to keep the inside temperature steady.

0.3 m — Ice

2.5 m

▲ **FIGURE 18–9**

Strategy The right-hand side of Eq. (18–8) tells us the rate at which heat flow must be supplied to the interior of the igloo. The temperature gradient is given by $(T_{in} - T_{out})/(\text{ice thickness})$. The surface area A is that of half a sphere. Because the ice thickness is small compared to either the inner or outer radius, we could use either radius or anything in between; in fact, we'll use a radius halfway between the inner and outer radius.

Working It Out The appropriate radius is $R_{med} = 2.50$ m + 0.15 m = 2.65 m. Then

$$\frac{dQ}{dt} = -\kappa A \frac{dT}{dx} = -\kappa A \frac{\Delta T}{\Delta x} = -\kappa (2\pi R_{med}^2) \frac{T_{in} - T_{out}}{\text{ice thickness}}$$

$$= -2\pi (1.67 \text{ J/m·s·°C})(2.65 \text{ m})^2 \frac{(5°C) - (-40°C)}{0.30 \text{ m}}$$

$$= -1.1 \times 10^4 \text{ J/s} = -11 \text{ kW}.$$

This igloo requires an 11 kW heater just to keep things at a chilly 5°C. Note the use of units involving joules rather than calories. For comparison, a room-size kerosene heater puts out some 5 kW.

What Do You Think? Purely from a thermal point of view, why is it best to make igloos hemispherical?

Thermal Resistance

The **thermal resistance**, or **R value**, is used as an alternative to the thermal conductivity rating; it is inversely proportional to thermal conductivity. For a slab of a given material of thickness L, R is defined by

$$R \equiv \frac{L}{\kappa}. \qquad (18\text{–}9)$$

Whereas κ depends only on the type of material, *R depends also on the thickness of the material*. Therefore, R characterizes the effectiveness of a thermal barrier of given thickness. A large R value corresponds to a small thermal conductivity and/or a large thickness; a piece of material with such characteristics is a good *thermal insulator*, whereas a piece of material with a low R value is a poor insulator. The R value has SI units of $m^2 \cdot K/W$. When it is used in connection with insulation in construction in America, however, it is often quoted in units of $ft^2 \cdot h \cdot °F/Btu$, where $1\ ft^2 \cdot h \cdot °F/Btu = 0.18\ m^2 \cdot K/W$. The R value of a 6-in piece of fiberglass insulation is $22\ ft^2 \cdot h \cdot °F/Btu$, compared to $0.32\ ft^2 \cdot h \cdot °F/Btu$ for a 3/8-in piece of plasterboard.

In terms of R, the relation between thermal energy flow and temperature difference from Eqs. (18–8) and (18–9) is

$$\frac{\Delta Q}{\Delta t} = \frac{1}{R} A\ \Delta T, \qquad (18\text{–}10)$$

where A is the area across which thermal energy is transported. Because R refers to a given material of a given thickness, it is interesting to study the effective R values of different pieces of material combined in different ways—after all, that is how a house is constructed.

Thermal Resistance in Series: Let's first consider combining two conducting materials of the same area A *in series*, as shown in Fig. 18–10. Suppose that the left-hand piece has an R value of R_1 and the right-hand piece R_2 and that the temperature varies from T_c (cold) on the left to T_h (hot) on the right. We want to find the effective R value of the combined piece, R_{eff}. When thermal energy flows at a steady rate through the two materials, as shown in Fig. 18–10, then energy conservation implies that the rate of transport of thermal energy across one piece must equal the rate of transport across the second piece, and hence also the rate across the combined piece. Application of this idea (see Problem 25) implies that

$$\text{for two solids in series: } R_{eff} = R_1 + R_2. \qquad (18\text{–}11)$$

This result generalizes for many slabs of material that are combined in series.

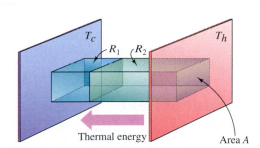

Thermal energy   Area A

◀ **FIGURE 18–10** Thermal resistance, or R values, for two materials placed "in series"—one behind the other. Thermal energy passes through the materials as shown.

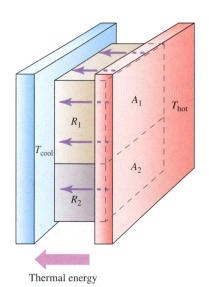

Thermal energy

▲ **FIGURE 18–11** Thermal resistance, or R values, for two materials placed "in parallel." Thermal energy passes through the materials as shown.

Thermal Resistance in Parallel: Now suppose that instead of being placed in series, two slabs are placed *in parallel*, as shown in Fig. 18–11. In this case the temperature difference on the right-hand and left-hand sides of both slabs is the same, and the total rate of transport of thermal energy is the sum of the transport rates through the two slabs. In Problem 27, this fact gives the result

$$\text{for two solids in parallel: } \frac{1}{R_{eff}} = \frac{1}{A_1 + A_2}\left(\frac{A_1}{R_1} + \frac{A_2}{R_2}\right). \qquad (18\text{–}12)$$

This result generalizes for more than two slabs placed in parallel.

EXAMPLE 18–8 A hole in a wall has been temporarily patched on the inside with a 1 sq-ft piece of plywood for which $R = 1\ ft^2 \cdot h \cdot °F/Btu$. The rest of the wall is insulated with fiberglass insulation for which $R = 22\ ft^2 \cdot h \cdot °F/Btu$. What is the R value of the repaired wall, whose total area is $100\ ft^2$?

Setting It Up The sketch of this situation, Fig. 18–12, shows that we have two materials placed in parallel. We know A_1 and R_1, the area and R value of the plywood, and A_2 and R_2, the area and R value of the fiberglass, respectively. We want R_{eff}, the R-value of the wall formed as in the figure.

(continues on next page)

Strategy This is an application of the formula for materials in parallel; Eq. (18–12) applies. We require only a numerical evaluation.

▲ FIGURE 18–12 Hole in wall is patched with plywood.

Labels in figure:
Plywood patch (1 ft²)
Fully insulated wall
10 ft
10 ft

Working It Out We have

$$\frac{1}{R_{eff}} = \frac{1}{A_1 + A_2}\left(\frac{A_1}{R_1} + \frac{A_2}{R_2}\right)$$

$$= \frac{1}{100\ \text{ft}^2}\left(\frac{1\ \text{ft}^2}{1\ \text{ft}^2 \cdot \text{h} \cdot {}^\circ\text{F/Btu}} + \frac{100\ \text{ft}^2 - 1\ \text{ft}^2}{22\ \text{ft}^2 \cdot \text{h} \cdot {}^\circ\text{F/Btu}}\right)$$

$$= \frac{1}{18\ \text{ft}^2 \cdot \text{h} \cdot {}^\circ\text{F/Btu}}.$$

Thus

$$R_{eff} = 18\ \text{ft}^2 \cdot \text{h} \cdot {}^\circ\text{F/Btu}.$$

Note that because the R values already take the thickness of the materials into account, the plywood is not necessarily the same thickness as the fiberglass insulation. In fact, plywood is only a fraction of an inch thick, whereas the fiberglass is 6 in thick. The rest of the thickness may be taken up by air, but because this air will be continuous with the outside environment, we have ignored its insulating properties.

What Do You Think? Having 1% of the area in plywood changed the R value by almost 20%. If instead of 1% the plywood area were 10%, the R value would have changed by about a factor of 3. Why such a big effect?

In Example 18–8 a hole in a wall that is patched with little insulation and takes up only 1% of the area of the wall reduces the R value by some 18%. This effect is typical of thermal insulators combined in parallel. In particular, if one of the elements of a set of insulators combined in parallel has a very small R value, Eq. (18–12) shows that the effective R value is very small. Small holes can be very costly sources of thermal energy loss. Indeed, the best window has a thermal resistance of about one-tenth of normal wall and ceiling insulation. Thus once wall and ceiling energy losses become small, the only way to conserve energy further is to reduce window area. This is already practiced in many commercial buildings in cold climates.

18–4 The Mechanical Equivalent of Heat

Placing a thermal system in thermal contact with a second thermal system at a different temperature is one way to produce a temperature change in the first system, via heat flow in this case. For example, we can heat water by placing it in a pot on a hot stove. In this case a heat flow is responsible for the temperature change. In the first half of the nineteenth century the major discovery was made that the temperature of a thermal system can also be raised by performing mechanical work on that system. For example, when an object is rubbed so that work is done on it by friction, its temperature is raised, just as in Count Rumford's late 18th century observation (Fig. 18–5). Such observations and the realization of their significance took place through a slow buildup of data rather than through a single breakthrough experiment. The history is thus rather tangled, with many participants. Rumford argued that Lavoisier's caloric theory could not be correct. Rumford's result was not widely accepted at the time, but in the mid-19th century the careful experiments of James Joule established with precision that a given amount of mechanical work done on a thermal system produces a temperature rise that corresponds to a given heat flow, as defined by Eq. (18–1).

Joule's Experiments

In an idealized version of Joule's experiments, the temperature of a thermally isolated quantity of water is measurably raised by a set of paddles that churn the water (Fig. 18–13). We know that a drag force is present to resist the motion of the turning paddles,

Labels in figure 18–13:
m
$W_g = mgh$
h
Water
Rotation
Falling mass causes paddles to rotate, thereby raising water temperature.

▲ FIGURE 18–13 Schematic diagram of the apparatus used in Joule's classic experiments. Churning paddles raise the temperature of a thermally isolated container of water.

and work must be performed to keep them rotating. The amount of work done can be measured precisely by attaching the paddles to a falling mass that keeps the paddles moving, and then measuring the distance the mass falls.

When the paddles have done a certain amount of work on the water, the temperature of the system (water plus paddles) will have risen by a definite amount. And once the temperature has been raised, *there is no way to tell* whether the temperature increase occurred because work has been done on the system or because of heat flow into it. A series of experiments of this type establishes that there is a specific temperature rise for a specific amount of work; therefore, there is a precise equivalent between work done on a system, W, and the heat flow into it, ΔQ. Because *no* experiment can be performed that can tell whether a given temperature change was produced by mechanical work or by heat flow, we can assert that *heat flow, like work, is an energy transfer*. This result is known as the **mechanical equivalent of heat**. It means that the unit of heat flow, the calorie, must be convertible to the unit of energy—that is, there is a direct relation between the SI unit of work, the joule, and the unit of heat flow, the calorie, and Joule's experiment allows one to find the numerical relation:

$$1 \text{ cal} = 4.185 \text{ J}. \qquad (18\text{–}13)$$

THE MECHANICAL EQUIVALENT OF HEAT

This relation may seem to be a simple unit conversion; however, it is far more. Until Joule's experiment, the calorie was unrelated to any other units familiar to us. Once his work was completed, one could say that the calorie has the dimensions of energy and that the SI unit of heat flow is the joule. Equation (18–13) thus states the crucial result of a basic experiment. It should not be confused with, say, the statement that 1 Btu is 252.00 cal, which is merely the relation between different units.

Today the *numerical definition* of the calorie is taken to be Eq. (18–13); it is no longer based on a measurement of the temperature change of water, as it was in Section 18–2. Such a definition is possible only because of the physical equivalence between heat flow and work. The calorie continues to be a widely used unit.

CONCEPTUAL EXAMPLE 18–9 Two *very* thin-walled (but unbreakable and undentable!) containers are dropped from a tall building. One of them holds water, the other an equal mass of mercury. Which of the liquids inside will be warmer after the container hits the ground?

Answer When the containers land, *kinetic energy* is converted into thermal energy in each case. The mass of each container is the same, so each has the same kinetic energy just before landing. But since the specific heat of water is higher than that of mercury (by a factor of 30; see Table 18–1), the water will undergo a smaller temperature change as a result. Note that the temperature does not rise gradually as the container falls; the conversion to thermal energy takes place rapidly when the container hits the ground and the fluid is strongly shaken.

What Do You Think? Would the heating be the same if the containers were flung from the height of about 1 m, with a velocity comparable to the one they reach in falling from the building?

EXAMPLE 18–10 The paddles of the apparatus shown in Fig. 18–13 are driven by a mass m of 0.50 kg falling at a constant speed. The mass of the thermally isolated water in the apparatus is 250 g. How much is the temperature of the water raised if the driving mass falls a distance $h = 2.0$ m? Ignore the mass of the paddles.

Strategy According to the work–energy theorem, the work W done on the water is the change in energy of the falling mass m. Since the mass falls at constant speed, that change is the change in the potential energy of m, namely mgh. But this work is equivalent to a heat flow, $\Delta Q = W$, and if we note that ΔQ is related to the temperature change ΔT by the specific heat (which we assume constant) according to Eq. (18–1), we can solve for ΔT.

Working It Out The specific heat of water has the constant value $c_{H_2O} = 1.00$ cal/g $\cdot$ °C, so Eq. (18–1) reads $m_{H_2O}c_{H_2O}\Delta T = mgh$, or

$$\Delta T = \frac{mgh}{m_{H_2O}c_{H_2O}} = \frac{(0.50 \text{ kg})(9.8 \text{ m/s}^2)(2.0 \text{ m})}{(250 \text{ g})(1.00 \text{ cal/g} \cdot °C)}(4.185 \text{ cal/J})$$
$$= 0.16°C.$$

What Do You Think? It looks as though you could reverse the entire process: You turn the crank in Fig. 18–13 in reverse and thereby raise the weight. Will the thermal process also reverse, leaving you with cooler water?

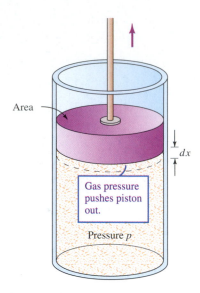

▲ FIGURE 18–14 The volume of a cylinder increases when a piston of area A moves out a distance dx under the influence of the pressure, p, of the gas contained in the cylinder.

(a)

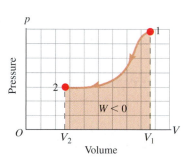

(b)

▲ FIGURE 18–15 The work done by a gas on its surroundings (like that of the piston–cylinder system of Fig. 18–14) in moving reversibly from point 1 to point 2 is the area under the curve of p versus V. (a) If the volume V_2 is greater than V_1 (the gas expands), the work the gas does on its surroundings is positive. (b) If the volume V_1 is greater than V_2, the gas is compressed, and the work it does on its surroundings is negative.

18–5 Work Done by Thermal Systems

In the last section, we saw that work can be done on a thermal system. By a simple sign reversal, we can think of this as work done by the thermal system on its surroundings—negative work, in the case of the churning paddles of Joule's experiments. Thermal systems, particularly gases, which can expand greatly, can also do *positive* work on their surroundings.

Consider a piston of area A pushed out by the pressure p of the gas contained in a cylinder (Fig. 18–14). By the definition of pressure, the magnitude of the force F on the piston is pA. Moreover, if the piston moves out, this force acts in the direction of the displacement, which we take to have infinitesimal magnitude dx. The infinitesimal work done by the gas in moving the piston is

$$dW = \vec{F} \cdot d\vec{x} = pA \, dx. \tag{18–14}$$

Because $A \, dx = dV$ (the infinitesimal change in the volume of the gas), this expression can be rewritten as

$$dW = p \, dV. \tag{18–15}$$

Equation (18–14) is specific to the linear motion of a piston with the geometry of Fig. 18–14, but Eq. (18–15) is a more general form that holds when a gas changes its volume by dV in pushing against its surroundings in any direction. Note the sign carefully: dW, the work the gas does *on* its surroundings, will be positive or negative according to whether dV is positive or negative.

As for the heat flow, we must specify the way in which the finite volume change is made in the integration of Eq. (18–15). Thus, *the net work done by the gas depends on the path in a p–V, p–T, or V–T diagram*. The integral expressing the work done in going from some state 1 to another state 2 is

$$W = \int_{V_1}^{V_2} p \, dV. \tag{18–16}$$

This integral is defined only when the path $1 \rightarrow 2$ is specified, meaning that how p is changing as we follow the path has to be specified.

If the path of the transformation $1 \rightarrow 2$ is drawn on a p–V diagram, as in Fig. 18–15a, then the work done by the gas has a simple interpretation: *It is the area under the curve of p versus V*, which is shaded in the figure. The sign of the work done by the gas is determined by whether V_2 is larger than V_1, as in Fig. 18–15a, or smaller than V_1, as in Fig. 18–15b. In Fig. 18–15b, the net work done *by* the gas is negative; in other words, there is positive work done *on* the gas to compress it. All this follows the normal rules of calculus, where $\int_{x_1}^{x_2} F(x) \, dx$ is the area under a curve of $F(x)$ versus x.

If we think of the work done by a thermal system as the area under a curve of p versus V, the fact that the work done depends on the path of the change becomes evident. The path describes the curve between two states, 1 and 2, and the area under different curves may certainly differ.

Cyclic Transformations

The operation of the pistons in an automobile engine is repetitive.[†] This is an example of a **cyclic transformation**: a change in which a thermal system follows a path that takes it back to its original state. The importance of such processes rests on the fact that all engines work in cycles. The net work done by the thermal system in a cyclic transformation can be calculated by starting anywhere along the path shown on a p–V diagram (Fig. 18–16a). (Note that we are assuming that all the changes treated here are reversible ones; the very fact that we are tracing the path indicates that.) Let's divide the cycle of Fig. 18–16a into two parts, as in Fig. 18–16b and c. The first part is the change $B \rightarrow A$,

[†]If you are not familiar with the details of such an engine, the only important aspect here is the repetitive nature of an engine's action. But you can also peek ahead to Chapter 20 if you are curious.

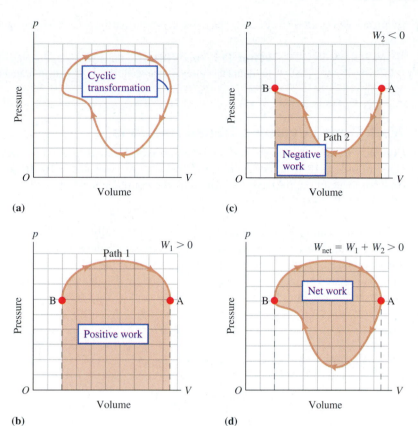

(a)

(b)

(c)

(d)

Net work

◀ **FIGURE 18–16** In a closed cycle, the work done by a gas on its surroundings is the area enclosed by the curve of p versus V. To show this, the full cycle is broken into two paths.

along path 1 (Fig. 18–16b), and the second part is $A \rightarrow B$, along path 2 (Fig. 18–16c). The net work W_{cycle} adds algebraically (Fig. 18–16d) and is thus given by

$$W_{\text{cycle}} = \underset{\text{path 1}}{W_{B \rightarrow A}} + \underset{\text{path 2}}{W_{A \rightarrow B}}.$$

As we saw in Chapter 6 during our discussion of work, $\displaystyle\int_{x_1}^{x_2} f(x)\, dx = -\int_{x_2}^{x_1} f(x)\, dx$,

when the two integrals are taken over the same path. Thus the integral over path 2, which goes from larger to smaller values of x, is the negative of the area under that path. The net work is then the area under path 1 minus the area under path 2:

$$W_{\text{cycle}} = \text{area enclosed by cyclic path.} \qquad (18\text{–}17)$$

Because the cycle is traced in a clockwise direction, the work done by the gas along path 1 is larger than the work done on the gas along path 2. The net work done on the surroundings is positive; if this cycle were traced in a counterclockwise direction, the net work done would be negative.

EXAMPLE 18–11 A thermal system is taken around the cycle shown in Fig. 18–17. Along the paths $A \rightarrow B$ and $C \rightarrow D$, the volume is held fixed at V_1 and V_2, respectively. From $B \rightarrow C$ and from $D \rightarrow A$, the pressure is held fixed at p_2 and p_1, respectively. What is the work done by the thermal system in tracing out one cycle?

Strategy We can solve the total work around the cycle by calculating the work over each step using Eq. (18–16) and summing the work done in each step, or we can do what is suggested in Eq. (18–17) and evaluate the area enclosed by the p–V diagram. We shall use the second method, which is simpler because it summarizes the steps of the first method. Nevertheless, it is useful to make one comment on the first method. Two steps of the cycle, the ones from $A \rightarrow B$ and the ones from $C \rightarrow D$, are constant volume segments. For these legs the differential element $dV = 0$, and the integral is zero, as there is no area beneath a vertical line on the p–V diagram.

▲ **FIGURE 18–17** A thermal system undergoes a cyclic transformation.

(continues on next page)

Working It Out The area enclosed by the rectangle in Fig. 18–17 is the width $(V_2 - V_1)$ times the height $(p_2 - p_1)$:

$$W_{cycle} = (p_2 - p_1)(V_2 - V_1).$$

Because the cycle is traced out in a clockwise manner, W_{cycle} is positive.

What Do You Think? Suppose p_2 is doubled. The work is (a) doubled; (b) more than doubled; (c) less than doubled.

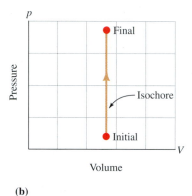

▲ **FIGURE 18–18** (a) An isobar, which represents a constant-pressure transformation. (b) An isochore, which represents a constant-volume transformation.

Work Done in Various Transformations

Constant Pressure: An isobaric (constant-pressure) transformation is drawn on a p–V diagram in Fig. 18–18a. For such transformations the work done by the gas is just the constant value of the pressure multiplied by the volume change,

$$W = p\Delta V. \qquad (18\text{–}18)$$

How does an isobaric transformation occur? Generally, if the volume of a gas is to increase without a change in pressure, the temperature must increase. Let's calculate this temperature change for an ideal gas. Suppose that the initial temperature, pressure, and volume are T, p, and V and that V is increased by dV while p is held constant. It follows from the ideal gas law, $T = pV/nR$, that

$$\text{for constant } p: dT = \frac{p\,dV}{nR}. \qquad (18\text{–}19)$$

Substituting T/V for p/nR, Eq. (18–19) becomes

$$\text{for constant } p: dT = \frac{T}{V}dV \quad \text{or} \quad \frac{dT}{T} = \frac{dV}{V}. \qquad (18\text{–}20)$$

Conversely, if the heat flow is such that the fractional increase in temperature is equal to the fractional increase in volume, as in Eq. (18–20), then for an ideal gas the pressure must be constant, and the heat flow is $C_p\,dT$.

Constant Volume: When the volume is kept fixed, the curve of the transformation is said to be an **isochore** (Fig. 18–18b). Thermal systems do no work—positive or negative—in an isochoric transformation because there is zero area under the curve on a p–V diagram. The pressure and temperature both change, however, so such a transformation must be accompanied by a heat flow, just as for the isobaric and the isothermal transformations. In this case the heat flow is $dQ = C_V\,dT$.

Adiabatic: Finally, there is a particular curve that will be followed if the gas does work while it is thermally isolated from its environment. Imagine a cylinder of gas enclosed in a thermal insulator, such as Styrofoam, with only the mechanical movements of a piston connecting it to its surroundings. Because of the thermal isolation, there is no heat flow dQ into or out of the system. When $dQ = 0$, the temperature can change only if work is done on the system. Reversible transformations of a thermal system in which there is no heat flow to the system are called **adiabatic** transformations. In Section 18–8 we will explore the precise curve followed by adiabatic transformations for ideal gases.

Internal Energy

In an adiabatic transformation a thermal system has no thermal contact with its surroundings but does work on them. The work–energy theorem, as formulated in Chapters 6 and 7, can be applied to the system, which we can accordingly view as having a thermal (or internal) energy that can change if work is done on or by the system. We call this thermal energy[†] U. In particular, when a system does work on its surroundings in an adiabatic transformation, *it changes its internal (thermal) energy by the amount of work it does*:

$$\text{for an infinitesimal adiabatic transformation: } dU = -dW; \qquad (18\text{–}21)$$

$$\text{for a finite adiabatic transformation: } U_B - U_A = -W_{A \rightarrow B}. \qquad (18\text{–}22)$$

[†]Don't confuse U with a potential energy!

In other words, any changes in a thermal system's internal energy are matched by mechanical work done if a transformation is adiabatic. If we want to study how the internal energy of a thermal system changes when we go from one thermodynamic state to another, we should look at adiabatic transformations because then we can avoid the complications of heat flow. *The adiabatic process allows us to learn just how the internal energy depends on the thermodynamic variables.*

The thermal (internal) energy is a function of p and V, p and T, or T and V. The thermal energy is associated with the energy of the molecules of a thermal system; it is simply a function of the thermodynamic variables of the system, not of how those variables might have changed. Changes in internal energy, like changes in the potential energy of a conservative mechanical system, are *not* path dependent. This contrasts sharply with work or heat flow, which *are* path dependent.

In Eq. (18–21) both dU and dW are written as differentials. The symbol dU has its usual meaning as a differential, in that a finite change in internal energy is

$$\int_A^B dU = U_B - U_A, \tag{18–23}$$

which is independent of the path between A and B. There is no equivalent expression for work, because the integral of dW *does* depend on the path taken. In a strict mathematical sense, dW is not a true differential, but there is no need for a more complicated notation, provided we keep in mind that the integral of dW is path dependent.

The minus signs on the right-hand sides of Eqs. (18–21) and (18–22) occur because W always denotes the work done *by* the system *on* its surroundings. When the system does positive work in an adiabatic transformation, its internal energy U decreases; when the work done is negative, the internal energy increases.

Let's now reconsider the Joule experiments that determined the mechanical equivalent of heat (Section 18–4). The water whose temperature is raised by the churning paddles is thermally isolated. The transformation is irreversible; it is not possible to "unchurn" the water and use it to raise the fallen weight. Nevertheless, the (positive) net work done by the paddles does indeed determine the (positive) change in the internal energy of the water. *How* a change is made is irrelevant for the internal energy, because the internal energy is purely a function of state. Thus Eq. (18–22) holds for both reversible and irreversible transformations of thermally isolated systems.

18–6 The First Law of Thermodynamics

The **first law of thermodynamics** is a statement of the conservation of energy for thermal systems. It is essential in understanding any thermal system—particularly those in which fuel is burned and work is done, such as engines or power plants. We have seen that heat flow and work are equivalent ways to change a system's internal, or thermal, energy. The first law simply recognizes this by stating that the change in the internal energy of a thermal system in going from one state to another is the negative of the work that the system does on its surroundings, plus the heat flow into the system, during that transformation. In other words, if we label our initial and final states by A and B, respectively,

$$\Delta U = U_B - U_A = -W_{A \to B} + Q_{A \to B}. \tag{18–24}$$

FIRST LAW OF THERMODYNAMICS

This equation is the primary result of this chapter. The minus sign in front of W indicates that W is the work done *by* the thermal system, and the plus sign in front of Q indicates that Q is the heat flow *into* the system. In Fig. 18–19 we give another view of the meaning of these terms.

We should mention that there is no universal agreement as to the sign convention for work W. Engineers and physicists tend to use the convention stated here; engineers and physicists are primarily interested in what the system does to its surroundings—for example, how much work an engine is capable of doing. But chemists prefer to say that W is positive when it is done *on* the system, a convention that makes the sign in Eq. (18–24) + instead of −. That is because chemists are more interested in the thermal system itself than its effects on its surroundings.

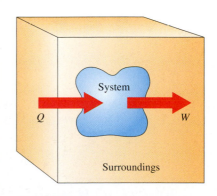

▲ **FIGURE 18–19** In the first law of thermodynamics, W represents the work done by the system, while Q represents the heat flow into the system. Either term can be negative or positive, depending on the transformations involved.

Because U is a function of state only, ΔU is the same whether the transformation $A \rightarrow B$ is reversible or irreversible. Similarly, the work done by the system can be measured by observing its effect on its surroundings—for example, how high the system lifts a mass—independently of whether the transformation $A \rightarrow B$ is reversible or irreversible. The heat flow $Q_{A \rightarrow B}$ can be determined from the first law of thermodynamics if the transformation is irreversible or determined independently by making a calorimetric measurement.

The first law of thermodynamics is often written as

$$\Delta U = -\Delta W + \Delta Q \tag{18–25a}$$

or in a differential form as

$$dU = -dW + dQ. \tag{18–25b}$$

While both dW and dQ depend on the path, dU is independent of path. Remarkably, the path dependence of the two separate terms cancels in the difference $-dW + dQ$.

The First Law of Thermodynamics in Closed Cycles

Engines operate and perform work in closed cycles. Thus it is of practical importance to be able to apply the first law of thermodynamics to a closed cycle. There is no change in internal energy over a cycle, because the system comes back to its original state. The work done by the thermal system over the cycle must then match the heat flow into the system, by the first law:

$$Q_{\text{cycle}} = W_{\text{cycle}}. \tag{18–26}$$

Here Q_{cycle} and W_{cycle} are the net heat flow into, and net work done, by the system over the entire cycle.

CONCEPTUAL EXAMPLE 18–12   Suppose that a thermal system follows the cycle $A \rightarrow B \rightarrow C \rightarrow D \rightarrow A$, as drawn in Fig. 18–17 and treated in Example 18–11. What is the total heat flow into the system after one cycle is completed?

Answer   To repeat the argument above, the internal energy change over one cycle is zero because the initial point is the same as the final point. The fact that the system passes through a complicated succession of intermediate states is irrelevant. Then we apply

Eq. (18–26), $Q_{\text{cycle}} = W_{\text{cycle}}$, together with the results of Example 18–11, where the work done for this cycle is calculated:

$$Q_{\text{cycle}} = (p_2 - p_1)(V_2 - V_1).$$

What Do You Think?   Suppose that instead of the rectangular path of Fig. 18–17, an arbitrary closed curve were followed. What can you say about the sign of the heat flow into the system?

How the Energy Changes in Constant-Volume and Constant-Pressure Transformations

No work is done in reversible changes under constant volume ($dW = pdV$, and $dV = 0$), so Eq. (18–25) becomes

$$\text{for constant volume: } dU = dQ_V = C_V\, dT. \tag{18–27}$$

We have emphasized the fact that the volume is constant by the subscript on dQ. No subscript is necessary for dU, because the change in U is independent of the path taken. Similarly, we can express the first law of thermodynamics for reversible changes under constant pressure as

$$\text{for constant pressure: } dU = -dW_p + dQ_p = -pdV + C_p\, dT. \tag{18–28}$$

Equations (18–27) and (18–28) can both be integrated over paths with constant V and constant p, respectively, to find the net internal energy change under a finite volume or pressure change. Keep in mind that in such an integration the heat capacities may depend on temperature, so they cannot be removed from under the integral sign.

18–7 Internal Energy of Ideal Gases

Joule performed an experiment in 1845 to study the properties of U for ideal gases. He demonstrated that *for ideal gases, the internal energy is a function of temperature alone*: $U = U(T)$. The equation of state adds to this by saying that U is equivalently a function only of the product pV.

(a)

(b)

◀ **FIGURE 18–20** Schematic diagram of Joule's experiment. When Joule opened the stopcock to allow free expansion, he observed no temperature change for a dilute (ideal) gas.

Joule's experiment is as follows: Fill one part of a two-part container with a dilute sample of gas (if it is sufficiently dilute, we can be sure it is ideal). The other part of the container is evacuated, and the two parts are connected by a tube with a stopcock (Fig. 18–20a). The container as a whole is sealed and does no work on the outside world. The container is submerged in a thermally isolated container of water or some other fluid whose temperature can be monitored. This fluid, initially at temperature T_1, acts as a calorimeter: If its temperature changes because of some change in the state of the gas in the container, then there has been a heat flow into or out of the gas.

The stopcock is now opened, and the gas undergoes free expansion into the previously empty part of the container. The crucial feature of free expansion is that the gas does no work, because it does not cause mechanical movement in its surroundings. The new temperature of the fluid, T_2, is then measured. When the gas is sufficiently dilute, the experiment shows that $T_2 = T_1$ (Fig. 18–20b). *The temperature of an ideal gas undergoing free expansion remains constant.* This result means that there has been no heat flow into the gas. There is no work done and no heat flow in free expansion. Thus, according to the first law of thermodynamics, the internal energy, U, remains unchanged. But because the volume of the gas does change, U must be independent of volume. Therefore, for an ideal gas, U is a function only of temperature:

$$U_{\text{ideal}} = U(T). \qquad (18\text{–}29)$$

ENERGY OF AN IDEAL GAS

This result follows directly from experiment, but it can also be derived from the second law of thermodynamics, which we will explore in Chapter 20.

For real gases—which are not perfectly ideal—there is a deviation from the result of Eq. (18–29), and temperature changes during free expansion. In other words, U is a function of density as well as temperature. We typically ignore this effect.

Let's now couple the fact that the internal energy of an ideal gas depends only on the temperature with Eqs. (18–27) and (18–28). Because U depends *only* on T, the integral of Eq. (18–27) gives

$$U(T_B) - U(T_A) = \int_{T_A}^{T_B} C_V \, dT. \qquad (18\text{–}30)$$

This equation must hold for *all* values of the volume because the left-hand side of the equation is independent of volume. Hence Joule's experiment shows that C_V must itself be independent of volume. Equation (18–30) expresses only a *change* in energy, so it is equivalent to the form

$$U(T) = \int_0^T C_V \, dT + \text{a constant}. \qquad (18\text{–}31)$$

The constant is arbitrary because only *changes* in U matter, not the actual values of $U(T)$ and $U(0)$. We therefore take the constant to be zero. Finally, we can use the experimental fact that over a wide temperature range, C_V is independent of temperature. We can remove C_V from behind the integral sign and find

$$\text{for } C_V \text{ independent of temperature: } U(T) = C_V T. \qquad (18\text{–}32)$$

The internal energy of an ideal gas is simple indeed, as long as C_V is constant with temperature.

EXAMPLE 18–13 An ideal gas in contact with a thermal reservoir at temperature $T = T_0$ does work by expanding against a piston. That work is 10 J. How much heat flow goes into the gas during its expansion?

Setting It Up The path, which for an ideal gas making an isothermal transformation is a hyperbola, is already drawn in Fig. 18–1c.

Strategy The first law of thermodynamics gives us the heat flow, Q, in terms of the work W done by the system and the change in internal energy, ΔU: $Q = \Delta U + W$. Because the gas is ideal, U is a function only of temperature, and therefore in an isothermal transformation such as this one, U is unchanged. This allows us to solve for the heat flow Q.

Working It Out With $\Delta U = 0$, we have

$$Q = W = 10 \text{ J}.$$

What Do You Think? Where does the heat flow come from?

A Relation Between C_p and C_V for Ideal Gases

We can express C_p in terms of C_V. Equations (18–27) and (18–28) (respectively $dU = C_V\, dT$ and $dU = -p\, dV + C_p\, dT$) are two differential expressions for the energy change, dU. Because Eq. (18–28) specifies constant pressure, the term $p\, dV$ that occurs on the right-hand side can also be written for an ideal gas as $p\, dV = nR\, dT$. Then both Eqs. (18–27) and (18–28) involve only terms proportional to dT on the right-hand side. Because the energy shift dU is independent of whether the infinitesimal transformation is made at constant V or at constant p, the two expressions for dU must be equal:

$$C_V\, dT = -nR\, dT + C_p\, dT.$$

We cancel the common factor dT from this equation to find that

$$C_p = C_V + nR, \tag{18–33}$$

RELATION BETWEEN C_p AND C_V

which is a result that applies to ideal gases. As C_V is a function of T only, C_p is also; and to the extent that C_V is in fact constant with T, so is C_p. Equation (18–33) is supported by experiment for ideal gases and is a good approximation for real gases. To go further than this relation—actually to compute quantities such as C_V—it is necessary to consider the microscopic dynamics of gases. This calculation will be made in Chapter 19.

18–8 More Applications for Ideal Gases

In order to understand the behavior of engines or, more generally, of gaseous thermal systems, we want to express the changes that occur in terms of the thermodynamic variables. In principle, this is possible as long as the transformations are reversible. Better, if we can approximate real gases by ideal gases, then all the transformations we have discussed can be expressed mathematically, as can the corresponding work done, heat flow, energy changes, and related quantities. Some of these mathematical expressions have already been given. For example, the shape of the p–V curve that represents an isothermal transformation at $T = T_0$ is a hyperbola, $p = nRT_0/V$. The relation $C_p = C_V + nR$ [Eq. (18–33)] is a consequence of the fact that the gas is ideal.

Of the four reversible transformations we discussed earlier—constant-volume (isochoric), constant-pressure (isobaric), constant-temperature (isothermal), and adiabatic transformations—the first two are simple to treat. Let's now look at the other two in more detail.

▲ **FIGURE 18–21** An ideal gas expands from V_1 to V_2 while its temperature is held fixed, and in doing so does work on its surroundings.

Isothermal Transformations of an Ideal Gas

When the temperature is constant, the work done by an ideal gas is simple to compute. We see from Fig. 18–21 that

$$W = \int_{V_1}^{V_2} p\, dV = \int_{V_1}^{V_2} \frac{nRT_0}{V}\, dV = nRT_0 \int_{V_1}^{V_2} \frac{dV}{V}. \tag{18–34}$$

The integral of $1/x$ is the natural logarithm, $\ln x$ (see Appendix IV-6). Because $\ln x_1 - \ln x_2 = \ln(x_1/x_2)$, we have, for the work done by an ideal gas in an isothermal transformation:

$$W = nRT_0 \ln\left(\frac{V_2}{V_1}\right). \tag{18–35}$$

As expected, W is positive when the gas expands $(V_2 > V_1)$ and negative when the gas contracts $(V_2 < V_1)$.

EXAMPLE 18–14 Suppose that 0.12 mol of an ideal gas, in contact with a thermal reservoir that holds the temperature fixed at $T_0 = 9.8°C$, has an initial volume of 1.3 L and does 14 J of work. What is the final volume, and what is the final pressure?

Strategy We can use Eq. (18–35) to find the final volume V_f because every other quantity in the equation is known. (We'll need to convert T_0 to the Kelvin scale.) Then we can use the ideal gas law, $pV = nRT_0$, to find the final pressure of the gas.

Working It Out To start, we have $T_0 = 9.8 + 273.15 \cong 283K$. Then from Eq. (18–35)

$$\ln\left(\frac{V_f}{V_i}\right) = \frac{W}{nRT_0} = \frac{14\ \text{J}}{(0.12\ \text{mol})(8.314\ \text{J/mol}\cdot\text{K})(283K)} = 0.050,$$

or

$$\frac{V_f}{V_i} = \exp(0.050) = 1.05;$$

$$V_f = 1.05\ V_i = 1.05 \times (1.3\ \text{L}) = 1.4\ \text{L}.$$

The final pressure is given by the equation of state:

$$p_f = \frac{nRT_0}{V_f} = \frac{(0.12\ \text{mol})(8.314\ \text{J/mol}\cdot\text{K})(283K)}{(1.4\ \text{L})(10^{-3}\ \text{m}^3/\text{L})}$$

$$= 2.0 \times 10^5\ \text{Pa} = 2.0\ \text{atm}.$$

The pressure has decreased from an initial value of $p_i = p_f(V_f/V_i) = 2.2$ atm.

What Do You Think? Did (positive) thermal energy come out of the reservoir, or did it go into the reservoir, or was there no thermal energy flow at all?

The internal energy of an ideal gas is unchanged in an isothermal transformation. Thus by the first law of thermodynamics, Eq. (18–35) also expresses the heat flow into the gas during an isothermal transformation.

Adiabatic Transformations of an Ideal Gas

In an adiabatic expansion, the first law of thermodynamics becomes $dU = -dW$. Because dW is positive, dU is negative. U is proportional to T, so the temperature will fall in an adiabatic expansion and rise in an adiabatic compression. When a gas expands in a thermally isolated container, it cools; when it is compressed, it heats up (as you can feel when you compress a bicycle pump).

The curves that describe this transformation are derived in the next subsection. They are given by

$$pV^\gamma = p_0 V_0^\gamma = \text{a constant}, \tag{18–36}$$

where (p_0, V_0) are the initial values of pressure and volume, respectively, of the ideal gas that undergoes an adiabatic transformation to new values (p, V). The constant γ is given by

$$\gamma \equiv \frac{C_p}{C_V} = \frac{C_V + nR}{C_V} = \frac{c_V' + R}{c_V'}, \tag{18–37}$$

where c' is the molar heat capacity, from Eq. (18–2). We have used Eq. (18–33), which relates C_p to C_V. Because $C_p > C_V$, γ is greater than unity (a result that is true even for nonideal gases). In terms of V and T, the transformation follows the curves

$$TV^{\gamma-1} = \text{a constant}. \tag{18–38}$$

Compare Eq. (18–36) with the curve that describes an isothermal transformation, $pV = \text{a constant}$, or $p \propto V^{-1}$. Figure 18–22 contains an isotherm and an adiabatic curve for a monatomic ideal gas, for which $\gamma = 1.67$. The isothermal curve is a hyperbola, $p \propto V^{-1}$. The adiabatic curve $p \propto V^{-\gamma}$ falls more steeply than the hyperbola as V increases, because $\gamma > 1$.

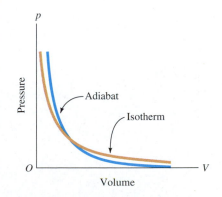

▲ **FIGURE 18–22** For ideal gases, adiabatic curves ($p \propto V^{-\gamma}$), called adiabats, are steeper than isotherms ($p \propto V^{-1}$) on a p–V diagram.

*How to Derive the Adiabatic Transformation of an Ideal Gas on a *p–V* Diagram

To derive the curves that are described by Eq. (18–36), we start with the first law of thermodynamics in its differential form [Eq. (18–25)], with the heat flow, dQ, set to zero: $dU = -dW$. The infinitesimal work done is still given by $dW = p\,dV$. We also know that dU is proportional to the temperature change, dT. In particular, dU can be calculated by differentiating Eq. (18–31) with respect to the temperature, T. Because C_V is constant for an ideal gas, we have $dU/dT = C_V$, or

$$dU = C_V\,dT.$$

Thus the adiabatic transformation satisfies

$$C_V\,dT = -p\,dV. \tag{18–39}$$

All three thermodynamic variables appear here, but because of the ideal gas law, $nRT = pV$, only two are independent. To find the curve of p versus V, we eliminate T in favor of p and V. The ideal gas law implies that

$$nR\,dT = p\,dV + V\,dp.$$

(See Appendix IV–7.) Equation (18–39) can thus be rewritten as

$$C_V \frac{p\,dV + V\,dp}{nR} = -p\,dV.$$

Solving for $V\,dp$, we find

$$V\,dp = -\frac{C_V + nR}{C_V} p\,dV. \tag{18–40}$$

The numerator on the right-hand side of Eq. (18–40) is the heat capacity at constant pressure, C_p, as Eq. (18–33) shows. Thus the right-hand side may be written as $-(C_p/C_V)p\,dV = -\gamma p\,dV$, with γ defined in Eq. (18–37).

With the definition of γ, we rewrite Eq. (18–40) in the form

$$\frac{dp}{p} = -\gamma \frac{dV}{V}.$$

Each side of this equation may be integrated separately, with the lower limits on the integrals determined by the initial state (p_0, V_0, T_0). Thus

$$\int_{p_0}^{p} \frac{dp}{p} = -\gamma \int_{V_0}^{V} \frac{dV}{V}.$$

Each integral gives a logarithm, and

$$\ln\left(\frac{p}{p_0}\right) = -\gamma \ln\left(\frac{V}{V_0}\right).$$

Because $A \ln x = \ln(x^A)$, it follows that

$$\ln\left(\frac{p}{p_0}\right) + \ln\left[\left(\frac{V}{V_0}\right)^{\gamma}\right] = 0.$$

We use $\ln x_1 + \ln x_2 = \ln(x_1 x_2)$ to find

$$\ln\left[\left(\frac{p}{p_0}\right)\left(\frac{V}{V_0}\right)^{\gamma}\right] = 0.$$

We now take the exponential of both sides of the equation to find

$$pV^{\gamma} = p_0 V_0^{\gamma} = \text{a constant}.$$

This is Eq. (18–36). By using the ideal gas law, we can show that the corresponding equation in terms of T and V is Eq. (18–38).

Variation of Atmospheric Temperature with Height

If you have taken a long-distance commercial flight, you know that the air high up is quite a bit colder than the air at ground level. Adiabatic transformations explain the variation of atmospheric temperature with height rather well. Air is transported from high altitude to low altitude and vice versa by large atmospheric wind currents. Air is a sufficiently good thermal insulator, so the transport of air between different altitudes is, to a good approximation, adiabatic. When air rises, the pressure decreases, so that the air undergoes adiabatic expansion; its temperature therefore falls. When air is carried to lower altitudes, it undergoes adiabatic compression, and its temperature rises.

If we assume that air is an ideal gas, we can quantify these concepts. Suppose that pressure and temperature are related by an adiabatic transformation. The adiabatic curve for p versus T is derived from, for example, Eq. (18–36) with $V \propto T/p$. For the p-versus-T adiabatic curve, this gives the equation $p^{1-\gamma}T^\gamma = $ a constant. If we raise both sides to the power $1/\gamma$ and solve for T, we find that

$$T = (\text{a constant})p^{(\gamma-1)/\gamma}. \qquad (18\text{--}41)$$

If we know how pressure varies with height, then Eq. (18–41) can tell us how temperature varies with height. But how does pressure vary with height? In Chapter 16 we found that pressure decreases linearly with increasing height for an *incompressible* fluid, such as water. Air, which is assumed to obey the ideal gas law in this case, is certainly not incompressible, but we can use the methods of Section 16–3 to find an equation for the way pressure varies with height. There we considered an imaginary horizontal slab cut out of the fluid, with its base at height h, where the pressure is p. The top of the slab, where the pressure is $p + dp$, is at height $h + dh$. Note that we expect dp to be negative because pressure should decrease with height. According to Eq. (16–5),

$$\frac{dp}{dh} = -\rho g,$$

where ρ is the density of the fluid. From this equation, we see that as h decreases, p increases due to the increasing weight of the overlying fluid (in this case, air). For water, ρ is constant. For air, ρ is not constant but is given by

$$\rho = \frac{m}{V} = \frac{nM}{V} = M\left(\frac{p}{RT}\right),$$

where M is the mass of 1 mol of air (its molecular weight). The equation for pressure as a function of height becomes

$$\frac{dp}{dh} = -\left(\frac{Mg}{RT}\right)p. \qquad (18\text{--}42)$$

Rather than solving this equation directly (see Problem 61), we can convert it into an equation for temperature as a function of height (see Problem 77). This is done by using the adiabatic curve, Eq. (18–41). Temperature is thus found to change linearly with height as

$$T = T_0 - \left(\frac{mgh}{R}\right)\left(\frac{\gamma - 1}{\gamma}\right). \qquad (18\text{--}43)$$

Because $\gamma > 1$, temperature does indeed drop with increasing height—as anyone who has been to the mountains can appreciate.

How does Eq. (18–43) compare with observation? The molecular weight of air is approximately 0.029 kg/mol, $\gamma_{\text{air}} \cong 1.4$, and $g \cong 9.8$ m/s^2. Then if we take the derivative of T [Eq. (18–43)] with respect to h, we find that

$$\frac{dT}{dh} = -\frac{(0.029 \text{ kg/mol})(9.8 \text{ m/s}^2)}{8.3 \text{ J/}^\circ\text{C} \cdot \text{mol}}\left(\frac{1.4 - 1}{1.4}\right) = 0.010^\circ\text{C/m}.$$

We have estimated that temperature drops about 10°C/km. The experimental result is approximately 6.5°C/km. The main sources of error in our calculation are the neglect of the effects of water condensation and, closer to the surface, the effects of local topography, such as mountain ranges. In the vicinity of flat deserts and high in the atmosphere, the value calculated is a fairly good approximation.

Summary

Reversible changes of state are changes that are made sufficiently slowly that the temperature, T, the pressure, p, and the volume, V—the thermodynamic variables that describe a thermal system such as a gas—are well defined at all times. The three variables are related by the equation of state of the gas. Reversible transformations, in contrast with irreversible transformations, can be traced out on a plot of any one of the three variables versus either of the other two. In this chapter we dealt with isothermal, isobaric, isochoric, and adiabatic changes.

When the temperature of a system changes, then the internal energy of the system will have changed. This may happen because the system has done work on its surroundings—for example, it pushed a piston and raised a weight—and/or because there has been a heat flow into the system. Infinitesimal heat flow into a system, dQ, is related to the infinitesimal temperature change of the system, dT, by the heat capacity, C, which depends not only on the system's composition but on how the temperature change is made:

$$\text{for constant volume: } dQ = C_V \, dT; \qquad (18\text{--}4)$$

$$\text{for constant pressure: } dQ = C_p \, dT. \qquad (18\text{--}5)$$

It is generally true that

$$C_p \geq C_V. \qquad (18\text{--}6)$$

Because dQ depends on how a transformation is made, the net heat flow in a finite transformation is path dependent. Heat flow is measured in calories. Calorimetry, the measurement of heat capacities, is based on the fact that heat flow is conserved in a mechanically and thermally isolated system.

The specific heat of a material, c, is the heat capacity of 1 g of the material; its molar heat capacity, c', is the heat capacity of 1 mol of the material.

Solids conduct thermal energy in a characteristic way. The rate of transport of thermal energy depends on the temperature drop per unit length and the area of the material through which the thermal energy flows. The thermal conductivity, κ, is defined by

$$\frac{dQ}{dt} = -\kappa A \frac{dT}{dx}, \qquad (18\text{--}8)$$

where the left-hand side represents the flow of thermal energy per unit time, A is the area, and dT/dx is the temperature drop per unit length. The reciprocal of κ is related to the thermal resistance, or R value, of the insulating material, with

$$R \equiv \frac{L}{\kappa}. \qquad (18\text{--}9)$$

where L is the thickness of the material. The R values of two insulating slabs of the same area are additive when the slabs are connected in series, whereas the quantities area/R are additive when the slabs are connected in parallel.

Heat flow, like work, is an energy transfer. Any transformation of a thermal system caused by a heat flow can be made equally well by doing work on the system. This means that heat has an exact mechanical equivalent and that the calorie can be expressed as an energy unit:

$$1 \text{ cal} = 4.185 \text{ J}. \qquad (18\text{--}13)$$

A gas undergoing an infinitesimal transformation does an infinitesimal amount of work on its surroundings, given by

$$dW = p \, dV. \qquad (18\text{--}15)$$

As for heat flow, the net work done in a finite transformation is path dependent. If the path is adiabatic ($dQ = 0$), the work done by the system is the negative of the internal energy change,

$$\text{for an infinitesimal adiabatic transformation: } dU = -dW; \qquad (18\text{--}21)$$

$$\text{for a finite adiabatic transformation: } U_B - U_A = -W_{A \to B}. \qquad (18\text{--}22)$$

An adiabatic transformation is a reversible transformation in which there is no heat flow between a system and its surroundings.

The internal energy (thermal energy), U, is a function of state. Any changes U undergoes in thermal transformations are path independent. The first law of thermodynamics specifies that the change in the internal energy of a thermal system in going from state A to state B is the negative of the work it does on its surroundings, plus the heat flow into the system:

$$\Delta U = U_B - U_A = -W_{A \to B} + Q_{A \to B}. \tag{18–24}$$

The work in this equation, W, is the work done *by* the thermal system, and Q is the heat flow *into* the system. The differential form of the first law of thermodynamics is written as

$$dU = -dW + dQ. \tag{18–25}$$

The internal energy of ideal gases is a function of temperature alone:

$$U_{\text{ideal}} = U(T). \tag{18–29}$$

To the extent that C_V is independent of temperature, the dependence is linear,

$$\text{for } C_V \text{ independent of temperature: } U(T) = C_V T. \tag{18–32}$$

It is also true that for an ideal gas

$$C_p = C_V + nR. \tag{18–33}$$

The curve of an adiabatic transformation can be calculated for an ideal gas by using the ideal gas law. It is given on a p–V diagram by

$$pV^\gamma = \text{a constant}, \tag{18–36}$$

where

$$\gamma \equiv \frac{C_p}{C_V} = \frac{C_V + nR}{C_V} = \frac{c'_V + R}{c'_V}. \tag{18–37}$$

This equation can be used to predict the variation of temperature with height in the atmosphere.

Understanding the Concepts

1. When a bicycle tire is pumped up, the end of the pump near the valve feels hotter. What is happening?
2. A gas expands adiabatically, doing 500 J of work. By how much does the internal energy of the gas change?
3. We stated that the system of Example 18–6 was thermally isolated. Shouldn't we have added that it is mechanically isolated as well?
4. If a bucket of water is carried uphill, raising the water's potential energy, will its temperature change?
5. Which of the following are reversible processes: (a) the slow inflation of a balloon with a bicycle pump; (b) the heating up of a drill bit used to bore a hole in a log; (c) the slow stretching of a wire by an external force, carried out at a constant temperature?
6. Thermopane windows are windows made of two glass panes with air between them. Explain why they are better thermal insulators than a simple window pane of twice the thickness.
7. You want to make a reversible transformation between 2 points A and B on a p–V diagram. Assume that $V_B > V_A$. On what path is the least amount of work done by the gas?
8. To survive a cold night in the woods, you are advised to make a thick bed of dry leaves to sleep on and a thick cover of dry leaves to sleep under. Why does this work?
9. Why are cross-country skiers told to wear several layers of clothing rather than a single, thick coat?
10. Why is free expansion not an example of an adiabatic process?
11. Would it make sense for you to shake a container of soup in order to heat it? Would the fact that the soup is in an insulating container change your answer?
12. Should the cooling fluid for an engine have a large or a small heat capacity?
13. What use can you think of for fluids with very large specific heats? With very small specific heats?
14. You glue together two strips of metals that have different coefficients of thermal expansion, back-to-back. What will happen to such a composite strip when it is heated or cooled? Can you propose some technological use for such strips?
15. When a mass m_1 of water at temperature T_1 is mixed with a mass m_2 of water at temperature T_2, with $T_1 > T_2$, the temperature change of the hotter mass, ΔT_1, is related to the temperature change of the colder mass, ΔT_2, by $\Delta T_1 / \Delta T_2 = -m_2/m_1$. In an experiment in which water and mercury are mixed, it was found that 100 g of mercury acts like 3.3 g of water. Explain both the formula and the observation about mercury.
16. A well-insulated container is divided into two separate compartments by an adiabatic wall. One compartment contains gas at a pressure p_1 and temperature T_1; the other contains gas at a pressure p_2 and temperature T_2. What happens when the partition is removed?
17. Why is the latent heat of vaporization for water so much larger than the latent heat of fusion?

18. When a freeze threatens, how can spraying fruit trees with water protect the fruit from frost damage?

19. A cyclic transformation shows up as a closed curve in a p–V plot. Suppose a cyclic transformation differs from the first one only in that the value of V is double at every step. How are the values of the net work related for the two transformations?

20. Why does perspiring during exercise help you to cool off?

21. For an ideal gas, the heat capacities are independent of temperature, and it is easy to relate temperature change to heat flow. Would it be impossible to relate temperature change to heat flow if the heat capacities varied with temperature?

22. Can all thermodynamic transformations be plotted on a graph of one thermodynamic variable versus another?

23. A change in a thermal system is reversible if (a) there is no friction and (b) the change is carried out slowly. Explain these constraints.

24. In the evening, after a hot summer day, why is it much cooler in the suburbs than downtown?

25. In Example 18–10 you have an apparatus that is capable of measuring ΔT to 0.02°C. If you want a better than 20% accuracy measurement, you need at least a 0.10°C temperature increase. Would you be better off with mercury instead of water in the experiment? (See Table 18–1.)

26. The center of a continent has more temperature extremes than do coastal areas at the same latitude. What effect contributes significantly to this?

27. People who drive cars in cold climates observe that the car's engine sometimes overheats if they keep the engine idling. This does not happen while they are driving, even though the engine puts out much more power then. Why?

Problems

18–1 Changes in Thermal Systems

1. (II) Are the following thermal transformations reversible or irreversible? Show the changes on a diagram when one is requested. (a) The sudden release of a gas from one portion of a container, of volume V_1, into a second portion, previously under vacuum. The total final volume, the volume of the whole container, is V_2 (p–V diagram). (b) As in part (a), but rather than a sudden release, the release is made a little at a time until there is no further flow of gas (p–V diagram). (c) The compression of a gas in thermal contact with a heat reservoir by loading weights, a little at a time, onto a piston connected to the gas (V–T diagram). (d) The slow heating of a gas by the release, a little at a time, of a warmer gas into the original sample of gas (p–T diagram). (e) The slow heating of a gas by placing it in thermal contact with a series of ever hotter thermal reservoirs (p–T diagram). (f) The slow rusting of a nail. (g) The release of electrical energy as lightning. (h) The release of chemical energy in the operation of a rechargeable battery to lift a weight.

18–2 Heat Flow

2. (I) (a) Using Table 18–1, calculate the heat capacity of a 5.7-kg aluminum rod. (b) Assuming that heat capacity is constant with temperature, how much heat flow is required to raise the temperature of the rod from 14°C to 58°C?

3. (I) A water heater heats 0.8 L of water/min from 20°C to 50°C. If the efficiency of the heater is 80%, how much power is consumed by the heater?

4. (I) The latent heat of fusion of oxygen (O_2) is 0.44 kJ/mol. Given that the atomic weight of molecular oxygen is 32 g/mol, find the latent heat of fusion in units of (a) kJ/kg and (b) cal/g.

5. (I) Consider an ideal gas initially in a state (p, V) and finally in a state (p, $2V$). If the transformation occurs at a constant pressure, what is the heat flow into the gas in terms of p and V?

6. (II) A 16 kg lead brick is at room temperature, 25°C. The melting point of lead is 327°C. Using the specific heat for lead, given in Table 18–1, calculate the heat flow to the brick needed to bring the brick to the melting temperature.

7. (II) A 300 cm^3 glass is filled with 100 g of ice at 0°C and 200 g of water at 25°C. (a) Characterize the content of the glass after equilibrium has been reached. Neglect heat transfer to and from the environment. (b) Repeat your calculations for 50 g of ice and 250 g of water.

8. (II) In Example 18–4 an experiment in calorimetry was described in which the assumption is made that the heat capacity of water and iron are constant over the temperature range of the experiment. Describe a series of experiments that would establish whether or not these heat capacities are indeed constant.

9. (II) The quantity 0.60 kg of thermally isolated water at an initial temperature of 291K serves as a calorimeter. Into it can be placed 1.20 kg of metal that has been heated to 373K. Find the specific heats and molar heat capacities of the metals if the equilibrium temperature is (a) 316K; the metal is aluminum. (b) 305K; the metal is iron. (c) 297K; the metal is lead.

10. (II) A calorimeter consists of 400 g of water at 25°C. A 500 g piece of copper at 100°C is thrown into the water, and the equilibrium temperature is found to be 28.3°C. What are the specific heat of copper in cal/g · K and the molar heat capacity, given that 1 mol of copper has a mass of 63.5 g?

11. (II) The quantity 200 g of thermally isolated water at an initial temperature of 293K serves as a calorimeter. Into it are placed, *successively*, 200 g of iron, then 200 g of silver, each of which has been heated to 400K. Using the known specific heats of iron and silver, find the final equilibrium temperature. (Use $c_{Ag} = 0.0557$ cal/g · K and $c_{Fe} = 0.112$ cal/g · K.)

12. (II) An axe head consisting of 1.8 kg of iron is left outdoors one cold winter's night and is brought indoors when the outside temperature is a brisk 240K. The room into which it is brought is initially at a nice, comfortable 293K and 1.0 atm of pressure. The volume of the room, which is well insulated, is 38 m^3. Assuming that the axe head comes to thermal equilibrium with the air in the room, by how much is the temperature of the room lowered? (Ignore the thermal interaction with furniture, walls, and so forth. Use $c_{air} = 0.172$ cal/g · K, $c_{Fe} = 0.112$ cal/g · K, and 28.8 g/mol for the molecular weight of air.)

13. (II) A thermally isolated system consists of 1 mol of helium gas at 100K and 2 mol of a solid at 200K; the gas and the solid are separated by an insulating wall of negligible mass. If the wall is removed, what is the equilibrium temperature of the system? Assume that at the given temperatures, the heat capacity of the solid is given by $3R/2$ per mole.

18–3 Heat Flow in Materials

14. (I) A single window pane of thickness 6.0 mm has an area of 1.0 m^2. The inside and outside temperatures are 73°F and 13°F,

respectively. Calculate the rate of thermal energy loss through the window. (See Table 18–3.)

15. (I) Compare the rate of thermal energy transfer across a piece of asbestos that is 1.0 m² in area and 1.0 cm thick with that of a piece of glass of the same dimensions for a temperature difference of 60°C.

16. (I) The glass from a basement window is replaced by a sheet of asbestos that is 0.80 cm thick. The window is 60 cm wide and 35 cm high, and there is no air leakage around the edges. Given that the heat loss through the surface is 0.006 J/s, what is the temperature difference between the outside and the inside of the basement?

17. (I) A brick wall 5.0 m by 5.0 m in size and 0.10 m thick separates the inside of a house from the outside, where the temperature is 285K (12°C). A heater supplies energy inside at a rate of 0.35 kW. What is the inside temperature? Ignore all surfaces except the brick wall in question.

18. (I) Your refrigerator can be thought of as a box with six sides of total area 2.5 m². The effective R value of the walls is 1.5 m²·K/W. The temperature inside is 5°C, while the temperature outside is 25°C. Calculate the rate of heat loss.

19. (I) A house of dimensions 36 ft × 21 ft × 9 ft is built of bricks that are 9.0 in thick. The inside and outside temperatures are 76°F and 25°F, respectively. Calculate the rate of thermal energy loss through the brick wall. (See Table 18–3.)

20. (II) The R value of brick is about 1.1 ft²·h·°F/Btu across the brick's width. A dog who expends 57 W of energy lives in a brick doghouse of total effective area 1.6 m². How much temperature variation between the inside and outside of the doghouse can the dog withstand if only its body heat warms the house? Assume that the door locks out drafts and that the heat conduction through the door is identical to that through the walls.

21. (II) Pine wood 1 in thick has an R value of 1.3, compared with 3.0 for fiberglass insulation of the same thickness. (The units of R are ft²·h·°F/Btu.) Calculate the rate of thermal energy loss of a 6-in thickness of these same materials across a 4.0 ft × 8.0 ft section of wall for a temperature difference across the wall of 35°F. If 10 percent of the wall section is made of wood studs and the space between studs is filled with fiberglass, calculate the effective R value of the wall section (Fig. 18–23).

▲ **FIGURE 18–23** Problem 21.

22. (II) Consider a wall of thickness L and area A. Part of the area, A_1, consists of a material with thermal conductivity κ_1 and the remaining area, $A - A_1$, consists of a material with thermal conductivity κ_2 (Fig. 18–24). Calculate how much thermal energy flows through each part of the wall, and calculate an effective thermal conductivity for the wall as a whole.

▲ **FIGURE 18–24** Problem 22.

23. (II) A wall of area A consists of two slabs: one of thickness L_1 and thermal conductivity κ_1, the other of thickness L_2 and thermal conductivity κ_2, as shown in Fig. 18–25. The temperatures are T_1 and T_2, respectively, on the left-facing side and the right-facing side of the wall. (a) Calculate the effective thermal conductivity of the wall in terms of the parameters given. (b) Calculate the temperature at the interface of the two slabs.

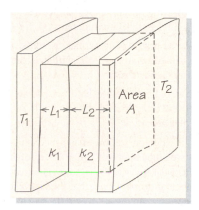

▲ **FIGURE 18–25** Problem 23.

24. (II) Two pieces of material are arranged in series as in Fig. 18–10. If the temperature to the left is T_1 and the temperature to the right is T_2, find the temperature between the pieces, T_{int}, as a function of T_1, T_2, and the R values of the two pieces.

25. (II) Equation (18–11) states that when two slabs of conducting material are combined in series, as pictured in Fig. 18–10, their R values are additive. Show that this is so by using the principle that the rate of flow of thermal energy is the same for each slab and for the two slabs combined:

$$\frac{dQ}{dt} = \frac{1}{R_1}A(T_1 - T_{int}) = \frac{1}{R_2}A(T_{int} - T_2)$$

$$= \frac{1}{R_{eff}}A(T_1 - T_2).$$

26. (II) A trapper lives in a cave. At the cave's entrance, the trapper has constructed an exterior wall from lumber 6.0 cm thick. The wood has a thermal conductivity, κ, of 0.35 W/m·K. Assume that all thermal energy loss occurs through the wall, which forms a rectangle of dimensions 2.5 m × 2.0 m. This wall, together with a stove, is capable of maintaining the cave's interior at a temperature of 8°C when the outside temperature is −25°C. What is the power output of the stove?

27. (II) Equation (18–12) gives the effective R value for two slabs of conducting material combined in parallel, as pictured in Fig. 18–11. Show that this equation is valid by using the principle that the temperature difference on opposite sides of both slabs is the same, and that the rate of flow of thermal energy across the combined slabs is the sum of the rate of flow of thermal energy across each slab.

28. (III) Repeat Example 18–7 but do not assume that the igloo's thickness is small compared to its radii. You may assume that the surfaces of constant temperature form concentric hemispheres. [Hint: Break up the hemisphere of ice into thin hemispherical shells and find the R value of all these hemispheres combined in series. Note that the area of the hemispherical shells is not constant, so Eq. (18–8) must be modified appropriately.]

29. (III) A long, cylindrical hot-water pipe of radius R_1 is at temperature T_1. It is wrapped with insulation of thickness α and of thermal conductivity κ; the outer surface of the insulation is at temperature T_2. Derive an expression for the rate of heat loss per unit length of insulated pipe. [Hint: The temperature within the insulation varies with radial position r. Break the insulation into thin cylindrical shells and find the areas of these shells as a function of their radii.]

18–4 The Mechanical Equivalent of Heat

30. (I) How many kilojoules of energy are required to raise the temperature of a 3.0 kg bar of aluminum by 32°C?

31. (I) One hundred grams of ice at 0°C is added to 500 grams of thermally insulated water at 24°C. What is the temperature of the water after the ice melts?

32. (I) A 50 g lead bullet traveling at 200 m/s stops in a target. How much would the bullet's temperature rise if 80% of the energy were to go into heating it?

33. (I) A primitive demonstration of the mechanical equivalent of heat consists of repeatedly dropping a bag of lead shot from a given height and measuring the temperature change in the bag after a number of drops. Given that the heat capacity of lead is 128 J/kg·K, how many times would the bag have to be dropped from a height of 2.0 m to change its temperature by 5°C? Neglect all heat flows from the lead.

34. (II) Suppose that a kilogram of water drops, starting from rest, from a height of 13 m. When the water hits the ground suppose that half of the kinetic energy it attains in falling is converted into thermal energy within the water. By how much will the temperature of the water rise? Neglect the effect of air resistance.

35. (II) Julius Robert Mayer, who first enunciated the principle of the conservation of energy, estimated the mechanical equivalent of heat by considering a quantity of air confined in a cylinder by a movable piston. If the volume is fixed, 0.172 cal must be supplied to raise the temperature of 1 g of air by 1.0°C. If the piston is not fixed but exerts a constant pressure of 1 atm $(= 1.01 \times 10^5 \text{ N/m}^2)$, 0.241 cal are needed to raise the temperature of 1 g of air by 1°C. If the volume change experienced by 1 g of air when the temperature changes by 1°C at constant pressure is $2.83 \times 10^{-6} \text{ m}^3/\text{g} \cdot \text{K}$, what is the relation between the calorie and the joule?

36. (II) The amount of solar energy delivered to a horizontal surface in Washington, D.C., averaged over a full year, is 160 W/m². Assuming a 10% efficiency for absorption and conversion to usable heating, how many liters of water can be heated from 0°C to 20°C in a 24 h period by a 3 m² solar panel?

37. (II) Angel Falls, in Venezuela, has the highest free waterfall in the world, at 2,648 ft. The water at the top of the falls is at a temperature of 10°C. (a) What is the temperature of the water in the pool at the base of the falls? As a first approximation, ignore the effects of air resistance, so that there is no terminal velocity. Also, ignore the effects of evaporation. [Hint: The kinetic energy that the water gains in falling, thereby losing potential energy, goes into turbulent motion at the base of the falls.] The exact terminal velocity is a complicated function of the precise configuration of the waterfall, such as whether it forms a tight tube or whether it is a broad, dispersed spray. (b) Repeat part (a) for Niagara Falls, which is approximately 50 m high. (c) Return to Angel Falls. Discuss the effect of air resistance and the fact that there is a terminal velocity. Would you expect the water to be heated as much as in part (a)? (d) It takes a positive heat flow into a liquid, the latent heat of vaporization, to cause it to evaporate. What is the sign of the temperature change in the waterfall, when you take into account the effect of evaporation of the falling droplets?

38. (II) Work is done on a thermally isolated container of 1.5 kg of water under 1 atm of pressure, initially at a temperature of 295K. Assume that the specific heat of water does not vary with temperature, and that churning paddles do work on the water at a rate of 40 W. How long must the paddles continue to turn to bring the water to its boiling point, 373K?

18–5 Work Done by Thermal Systems

39. (I) The volume of a gas in a cylinder slowly increases by a factor of 3 while its pressure decreases from 9×10^5 Pa to 1×10^5 Pa, as shown in Fig. 18–26. How much work is done on the piston by the gas during the expansion?

▲ FIGURE 18–26 Problem 39.

40. (I) Express the work done by an ideal gas at constant temperature T_0 as a function of the initial pressure, p_i, and final pressure, p_f.

41. (I) If a small amount of water is evaporated at atmospheric pressure until a volume of 3.0 m³ is occupied, how much work is done? Ignore the volume taken up by any remaining liquid water.

42. (I) Gas in a cylinder is allowed to expand from a volume of 0.85×10^{-3} m³ to a volume of 0.95×10^{-3} m³, while the pressure changes linearly with the volume from an initial pressure of 4.8 atm to a final pressure of 4.2 atm. What is the work done by the gas?

43. (II) Half a mole of an ideal gas is compressed isothermally at 293K so that the pressure is increased from 1.0 atm to 3.6 atm. What are the initial and final volumes of the gas, and how much work is done by the gas during the compression?

44. (II) Two liters of nitrogen gas at an initial temperature of 320K expand at a constant pressure of 5 atm against a piston, thereby doing work to raise a total mass of 100 kg. The gas is allowed to expand to a volume of 5.0 L. (a) How high can the mass be lifted? (b) What is the final temperature of the gas?

45. (II) A gas is compressed from V_1 to V_2 by reversibly following the path shown in Fig. 18–27, $p = p_0 - \beta V$, where β is a constant. (a) What is the dimension of β? (b) Determine the work done by the gas during the transformation by integrating the infinitesimal work along the path, $dW = p\,dV$. (c) Check your answer by evaluating the area under the p–V curve in the figure.

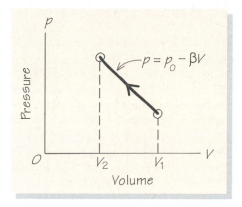

▲ **FIGURE 18–27** Problem 45.

46. (II) Figure 18–28 shows a reversible path for a gas being taken from state A to state B. Both states have the same pressure. (a) What is the work done in taking this path? (b) Draw a second reversible path that connects the same points by which no work is done. If the gas is ideal, to what minimum temperature is the gas taken on this path? (Real gases cannot be taken to this temperature.) (c) Draw a third path in which the work done is greater than the work done in following the original path.

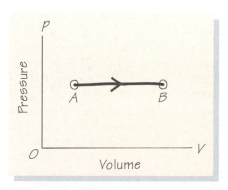

▲ **FIGURE 18–28** Problem 46.

47. (II) Two moles of an ideal gas are carried around the thermodynamic cycle shown in Fig. 18–29. The cycle consists of (1) an isothermal expansion $A \to B$ at a temperature of 700K, with the pressure at A given by $p_A = 8.0$ atm; (2) an isobaric compression $B \to C$ at 3.0 atm; and (3) an isochoric pressure increase $C \to A$. What work is done by the gas per cycle?

48. (III) An external force compresses 0.10 mol of an ideal gas in thermal isolation. The curve of this transformation on a p–V diagram is $pV^{1.4} =$ a constant. The gas initially has a volume $V_i = 1.6$ L and a temperature of 275K. When the compression is

▲ **FIGURE 18–29** Problem 47.

finished, the temperature has increased to 380K (Fig. 18–30). (a) What are the final volume and final pressure? (b) How much work was done on the gas to compress it?

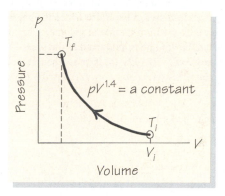

▲ **FIGURE 18–30** Problem 48.

18–6 The First Law of Thermodynamics

49. (I) A certain engine follows a closed thermodynamic cycle in which the heat flow into the system is measured to be 633 cal. Assuming that all transformations of the cycle can be approximated as reversible ones, what is the net work done, in joules, by the engine in one cycle?

50. (I) An ideal gas undergoes a reversible transformation from an initial state $(p_i = 1.0$ atm, $V_i = 350$ cm$^3)$ to a final state $(p_f = 2.0$ atm, $V_f = 175$ cm$^3)$. The transformation is carried out in two steps: The gas is compressed at constant pressure, and when the final volume is achieved, the pressure is increased while the volume is held fixed. Use the first law of thermodynamics to calculate the heat flow into the gas during this transformation. Assume (as is shown in Section 18–7) that, for an ideal gas, the internal energy U is a function of T only.

51. (II) Consider the p–V diagram shown in Fig. 18–31. If the system is taken from point A to point B via path $A \to C \to B$, 40,000 cal of heat flows into the system, and the system does 20,000 cal of work. (a) How much heat flows into the system along path $A \to D \to B$ if the work done is 7000 cal? (b) When the system returns from point B to point A along the curved path, the work done is 15,000 cal. Does the system absorb or liberate heat, and how much?

52. (II) The latent heat of vaporization for water at 1 atm and 100°C is 540 cal/g. How much of that heat is converted into the mechanical work needed to change the volume of 1 g of water into 1 g of steam? [*Hint:* Calculate the volume of 1 g of steam.]

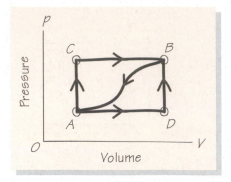

▲ **FIGURE 18–31** Problem 51.

▲ **FIGURE 18–33** Problem 55.

18–7 Internal Energy of Ideal Gases

53. (I) A thermal system consisting of 1 mol of an ideal gas forms a cycle that contains a leg AB in which pressure decreases linearly as the volume increases, a leg BC in which the gas is compressed at constant pressure until the initial volume is attained, and a leg CA in which the volume is held fixed as the pressure increases to its initial value (Fig. 18–32). (a) Express the internal energy at A, B, and C in terms of p_A, p_B, V_A, and V_B. (b) How much work does the gas do over the cycle?

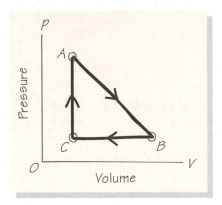

▲ **FIGURE 18–32** Problem 53.

54. (II) An ideal monatomic gas, with $c_V' = 3R/2$, undergoes a change of state. On a plot of pressure versus volume, it follows straight lines from an initial point (p_0, V_0) to $(p_0, 2.5V_0)$ to $(1.5p_0, 2.5V_0)$. (a) Draw the p–V plot. (b) By how much does the internal energy change? (c) How much work is done by the gas on its surroundings? (d) How much heat is added to the gas?

55. (II) Five grams of N_2 gas (molecular weight 28 g/mol) are allowed to expand from an initial state $(p_i, V_i) = (1.2 \times 10^5 \text{ N/m}^2, 3.5 \times 10^{-3} \text{ m}^3)$ to a final state $(p_f, V_f) = (1.8 \times 10^5 \text{ N/m}^2, 5.0 \times 10^{-3} \text{ m}^3)$ along a path that forms a straight line on a p–V plot (Fig. 18–33). (a) What are the initial and final temperatures? (b) What is the change in the internal energy of the gas? (c) What is the work done by the gas? (d) What is the heat flow into or out of the gas during the process?

56. (II) One mole of an ideal gas in contact with a thermal reservoir at 350K is under 10 atm of pressure. The pressure is slowly increased to 25 atm. (a) What are the initial and final volumes of the gas? (b) How much work is done *on* the gas? (c) What is the change in internal energy of the gas? (d) What is the heat flow into the gas?

57. (II) According to a calorimeter in contact with 0.20 mol of an ideal gas, the gas does 8.0 J of work while 5.0 J of heat is added to it. If the molar specific heat of the gas, c_V', is 20.8 J/mol · K, independent of temperature, by how much is the temperature of the gas changed in the process described? Is it warmed or cooled?

58. (II) One mole of an ideal gas, $\gamma = 5/3$, is at STP (0°C, 1 atm), and 350 J of heat is added at constant pressure. (a) What are the initial and final internal energies of the gas? (b) What is the work done by the gas? (c) Repeat parts (a) and (b) for the same amount of heat added at constant volume.

18–8 More Applications for Ideal Gases

59. (I) How much work does it take to compress 30 mol of an ideal gas at a fixed temperature of 15°C to half its volume?

60. (I) One mole of ideal gas undergoes an adiabatic transformation in which its temperature changes by 25K. How much work is required to bring about this transformation, given that the constant-volume molar heat capacity of the gas is $c_V' = 30.00$ cal/mol · K?

61. (I) Equation (18–42) is an equation for pressure as a function of height for the atmosphere, which is treated as an ideal gas. Show that, if T is independent of height, this equation has the solution $p(h) = p_0 e^{-Mg(h-h_0)/RT}$, where p_0 is the pressure at some reference height $h = h_0$.

62. (I) A monatomic ideal gas with $\gamma = 1.67$ undergoes an adiabatic compression starting from an initial temperature of 248K and an initial volume of 3.01 L. The final volume is 2.22 L. What is the final temperature?

63. (II) The temperature of the (dry) air at the bottom of a mountain valley is 26°C. The air moves as a brisk wind to the ridge 600 m above the valley floor. What is the temperature of the air on the ridge?

64. (II) The wildfires of the California coast are often fed by desert winds, which heat up to about 100°F in the desert at an altitude of 500 m, then rush down to sea level. What is the temperature of the air in these winds at sea level?

65. (II) Consider two thermal states of an ideal gas, A and B, together with the possible reversible paths that connect them. Is there an adiabatic path that connects these states when (a) $V_A = V_B$? (b) $p_A = p_B$? (c) $T_A = T_B$?

66. (II) An ideal gas is contained in a cylinder with a movable piston of cross sectional area 300 cm². The external pressure is held constant at 1.01×10^5 Pa. An external force lifts the piston from a height of 30 cm to a height of 60 cm. What is the work done by the applied force on the gas if the process takes place isothermally?

67. (II) An ideal gas expands adiabatically from an initial state p_0, V_0 to a final state p_f, V_f. Such an expansion is characterized by the p–V curve $pV^\gamma =$ constant, where $\gamma > 1$ is a constant. Show that the work done by the gas is $W = (\gamma - 1)^{-1}(p_0V_0 - p_fV_f)$.

68. (II) A diatomic ideal gas such as air, for which $\gamma = 1.4$, expands adiabatically to 40 times its original volume. (a) By what factor does the temperature change? (b) By what factor does the pressure change?

69. (II) One liter of gas characterized by $\gamma = 1.4$ is allowed to expand adiabatically to twice its volume. How does this pressure change compare with the pressure change that would occur for an isothermal expansion?

70. (II) How much work is done by 1.00 L of helium when it expands adiabatically from sea-level pressure (1.00 atm) to the point where the pressure is 0.10 atm? Helium is monatomic, with $c'_V = 3R/2$.

71. (II) A particular adiabatic transformation of an ideal gas occurs as the pressure increases by a factor of 5 while the volume decreases by a factor of 3. What is the ratio of C_V to C_p? How much work is done on the gas during the compression, in terms of the initial pressure and volume? By how much does the internal energy of the gas change?

72. (II) Use the result of Problem 61 to find the pressure at the height of Mt. Everest, which is about 10 km. Find also the temperature at this height. How much volume does 1.0 mol of air occupy? Calculate the density of air at this altitude and compare it with the density of air at sea level. The molecular weight of air is 29.2 g/mol.

73. (II) A gasoline engine of volume 25 cm^3 intakes air at 25°C and 1.0 atm. The engine compresses the air adiabatically to 30% of the original volume. Find the final pressure and temperature and how much work is performed on the gas. You can assume that the air is ideal.

74. (II) The compression ratio (the ratio of the maximum volume to the minimum volume) of the cylinder of a diesel engine is 15:1. The working gas is air (for which $\gamma = 1.4$), which enters the cylinder at room temperature, approximately 300K. The compression is so rapid that there is no heat flow through the cylinder walls, and the compression can be said to be adiabatic. To what temperature is the air heated?

75. (II) In Problem 17–57 the operation of a bicycle pump was studied on the assumption that the air in the pump is in thermal equilibrium with the environment. Repeat the analysis for the case when the pump is pushed down so quickly that no heat exchange occurs and the process is adiabatic. Assume that the temperature of the environment is 22°C. (The real process actually lies between the very slow isothermal and the very rapid adiabatic case).

76. (III) Consider the *Carnot cycle* shown in Fig. 18–34. One mole of an ideal gas with volume V_A at temperature T_1 undergoes an isothermal expansion to a volume V_B. This is followed by an adiabatic expansion to a volume V_C and then an isothermal compression at temperature T_2 to volume V_D; the cycle is closed by an adiabatic compression ending with volume V_A at temperature T_1. You will need to know the form of the transformation curve followed on each leg, and the heat flow to the gas for each leg. For the adiabatic legs, you know the heat flows, but for the isothermal legs, you may want to find the work done directly and use your knowledge of the energy change of an ideal gas when the temperature is constant. You also need to know the starting and ending points of the curves. What is the net heat, expressed in terms of the given variables, supplied to the gas during the cycle?

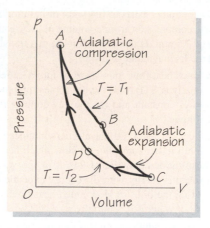

▲ **FIGURE 18–34** Problem 76.

77. (III) In Eq. (18–43) we state that by following an adiabatic curve, the temperature of Earth's atmosphere drops linearly with height h from the surface, and give the coefficient of h. Here this result is derived. The starting points are Eq. (18–41), which expresses the relation between T and p in an adiabatic transformation of an ideal gas, as well as Eq. (18–42), which expresses the pressure of the atmosphere as a function of height. (a) Take the derivative with respect to height of Eq. (18–41), using the chain rule of differentiation (see Appendix IV–7) for the right-hand side, to show that

$$\frac{dT}{dh} = a \text{ constant} \left(\frac{\gamma - 1}{\gamma}\right) p^{[(\gamma-1)/\gamma]-1} \frac{dp}{dh}.$$

(b) Show that this result can be rewritten as

$$\frac{dT}{dh} = \left(\frac{\gamma - 1}{\gamma}\right) T p^{-1} \frac{dp}{dh}.$$

(c) Substitute Eq. (18–42) for dp/dh into the result derived in part (b) to find that

$$\frac{dT}{dh} = -\left(\frac{Mg}{R}\right)\left(\frac{\gamma - 1}{\gamma}\right).$$

(d) Show that Eq. (18–43) is the solution of the differential equation derived in part (c).

78. (III) An experiment shows that the amount of work done is 88.9 J when a certain ideal gas is compressed adiabatically from an initial pressure of 1.00 atm and an initial volume of 1.00 L to a final volume of 0.50 L. In a second experiment, the same amount of gas is again compressed from 1.00 L to 0.50 L, but this time it is in thermal contact with a thermal reservoir at $T = 20$°C, and 189 J of work is required for the compression. (In the second experiment, the initial pressure is not necessarily 1 atm.) (a) How many moles of gas are there? (b) What is γ for the gas? [*Hint*: Calculate the work done in the adiabatic compression for several values of γ between 1.2 and 2.] (c) What is C_V for the gas?

General Problems

79. (I) Five liters of oxygen at one-half atmospheric pressure and 273K is heated at constant pressure until its volume has tripled. It is then compressed isothermally back to 5 L. (a) Draw the process on a p–V diagram. (b) What is the final pressure of the gas? (c) How much work is done by the gas during the entire process?

80. (II) A system that consists of 0.40 mol of helium gas undergoes an isobaric compression from a volume of 1.20 L to 0.70 L at a pressure of 2.5 atm. Is there heat flow in this transformation? If so, what are its magnitude and sign?

81. (II) How much heat flow must there be into a system of 10 L of oxygen initially at STP if (a) the gas triples in volume while the pressure stays constant and (b) the gas doubles in pressure at constant volume?

82. (II) A camper carries a 5.0 gal plastic container of water up a 40-m-high hill to her campsite. (a) How much heat flow does it take to raise the temperature of 2.0 qt of the water by 65°C? (b) Compare this energy with the energy expended to carry the entire 5 gal up the hill. (c) How much heat flow will it take to boil the 2 qt of water if it is initially at 23°C?

83. (II) Consider the adiabatic and isothermal processes that each pass through the point (p_0, V_0) on a p–V diagram (Fig. 18–35). Which of the processes has the greater slope at this point? Compute the ratio of the slopes at this point.

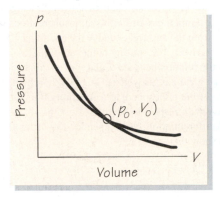

▲ **FIGURE 18–35**   Problem 83.

84. (II) A 1-mol sample of an ideal gas for which $c'_V = 3R/2$ follows the closed thermodynamic cycle shown in Fig. 18–36. There are three legs in the cycle: an isothermal expansion, $A \rightarrow B$; an isobaric compression, $B \rightarrow C$; and a constant-volume increase in pressure, $C \rightarrow A$. The temperature during the isothermal leg is $T = 340K$, $p_A = 1.8$ atm, and $p_B = p_C = 1.2$ atm. What are (a) V_A? (b) V_B? (c) The work done by the system during leg $A \rightarrow B$? (d) Q for leg $A \rightarrow B$? (e) T_C? (f) the net work done by the system over the complete cycle?

▲ **FIGURE 18–36**   Problem 84.

85. (II) The energy in the form of solar radiation that falls on Earth, after reflection is taken into account, is 956 J/m² s. The effective area that faces the Sun is that of a disk whose radius is the same as that of Earth. In equilibrium Earth radiates at a rate of σT^4 where σ is given in Section 17–5. T is the temperature of Earth. Under equilibrium conditions, what is T?

86. (II) Consider a van der Waals gas (Section 17–4). Find the work done by such a gas in an isothermal expansion from V_0 to V_f. Is this more or less than the work done by an ideal gas in an isothermal transformation from V_0 to V_f at the same temperature? (The signs of the constants a and b for the van der Waals gas are discussed in Section 17–4.)

87. (II) A 100-g piece of copper is heated from 0°C to 100°C at atmospheric pressure. What is the change in its internal energy?

88. (II) A pan contains 300 g of water at 23°C. (a) What heat flow does it require to raise the temperature of the water to 100°C and vaporize it at 1 atm? (b) How much work is done by the water as it changes to the gaseous phase? (c) What is the total change in internal energy of the water?

89. (II) A nuclear power plant requires 1000 MW (megawatts) of cooling. If environmental concerns limit the temperature rise of the water to 12°C, how much water flow is required to cool the plant?

90. (II) Heat flow is transferred to 0.30 mol of an ideal gas at a temperature of 500K and pressure of 2.5 atm, resulting in an isothermal expansion of the volume by a factor of 5. How many calories are involved in the heat flow?

91. (II) One mole of N_2 gas initially at a temperature of 300K and pressure of 1 atm is compressed isothermally until its volume is reduced by a factor of 10. (a) What is the final pressure of the gas? (b) What is the work done by the gas? (c) How much heat flow is transferred?

92. (II) An electric immersion heater is placed in a coffee cup to heat the water to boiling. The cup initially contains 0.18 L of water at 72°F. The heater is rated for 400 W. Neglect heat loss from the water and the cup. How long does it take to heat the water up to boiling? How long after this does it take to boil the water away completely?

93. (II) The internal energy of a van der Waals gas (Section 17–4) is

$$U = \frac{3}{2}nRT + \frac{an^2}{V} + \text{constant}.$$

(a) Find C_p as a function of V and T. (b) Find C_V as a function of V and T. (c) What is the value of the parameter $\gamma \equiv C_p/C_V$ for this gas? Check that your result reduces properly in the low-density limit.

94. (II) A thermally isolated container contains 1.0 L of water at 20°C. A 100-g block of aluminum at 80°C is placed into the container. (a) What is the final temperature of the water? Next 500 g of ice at $-20°C$ is placed in the container. (b) What is the final temperature of the water, iron, and ice (if any)? $c_{\text{ice}} = 0.467$ cal/g · K.

95. (II) A glass contains 150 cm³ of water at 70°F. Four ice cubes of 25 g each of temperature 10°F are dropped into the water. Neglecting heat-flow loss, what is the final temperature of the water? The specific heat of ice in this temperature range is 2.04 J/g · K.

96. (II) The temperature of espresso coffee (mostly water) can be increased by blowing 100°C steam into it. How much steam (in grams) is needed to heat up a 20-cm³ cup of espresso from 50°C to 80°C? What is the volume of this quantity of steam, assuming that the steam is an ideal gas?

97. (II) One mole of an ideal gas is carried around the thermodynamic cycle shown in Fig. 18–37. The cycle consists of an isothermal expansion at a temperature of 400°C with an initial pressure of 5 atm, leading to a doubling of the initial volume. This is followed by a pressure drop of a factor of 2 at constant volume, and then by an isobaric compression until the initial volume is restored. The cycle

is completed by a constant-volume pressure increase to the initial value of 5 atm. Calculate the values of p, T, and V not given, and the work done by the gas during one cycle.

▲ **FIGURE 18–37** Problem 97.

98. (II) In experiments carried out between 1759 and 1762, Joseph Black started with a glass cup of mass 32 g, containing 467 g of water at 88°C. Black took a piece of ice of mass 404 g at 0°C and put it in the cup. The final equilibrium temperature was measured to be 12°C. How many calories are needed to melt 1 g of ice; that is, convert it to water at 0°C? That quantity is the latent heat of fusion. What would the equilibrium temperature be if no energy were required to convert ice to water at 0°C? Ignore the heat capacity of the glass.

99. (II) An ideal gas for which $c_V' = 5R/2$ is carried around a cycle $a \rightarrow b \rightarrow c \rightarrow a$. There are 2.4 mol of the gas in the cycle. The expansion $a \rightarrow b$ is a straight line on a p–V diagram with $T_b = T_a = 540$K, $b \rightarrow c$ is a constant-pressure segment with $p = 1.8$ atm, and $c \rightarrow a$ is a constant-volume segment at 15 L. (a) What is p_a? (b) What is V_b? (c) What is the work done in segment $a \rightarrow b$? (d) What is T_c? (e) What is the change in internal energy in segment $c \rightarrow a$? (f) What is the net work done during the entire cycle?

100. (II) The trapper of Problem 18–26 tires of looking at the blank entrance wall and decides to cut a window in it. The window will have a sheet of mica 1.0 mm thick with a thermal conductivity of 0.85 W/m·K. How large a window can he construct if he keeps the same stove and plans to keep the temperature inside the cave at 5°C?

101. (II) Cold air blows through a crack in the wall of the previous problem, so the trapper covers the crack with a piece of wood 5 mm thick. The thermal conductivity of the wood is 25×10^{-2} W/m·K. If the area of the crack is 6 cm^2 and the temperature difference between outside and inside is 30°C, what is the heat loss through the covered crack?

102. (III) A cyclic process with 0.40 mol of ideal gas is represented by a circle on the p–V diagram (if the appropriate scale is chosen) (Fig. 18–38). Calculate (a) the amount of heat transferred to the gas between the minimum and maximum volumes, as given in the figure; and (b) the largest and smallest internal energies during the cycle, given that $c_V' = 5R/2$.

▲ **FIGURE 18–38** Problem 102.

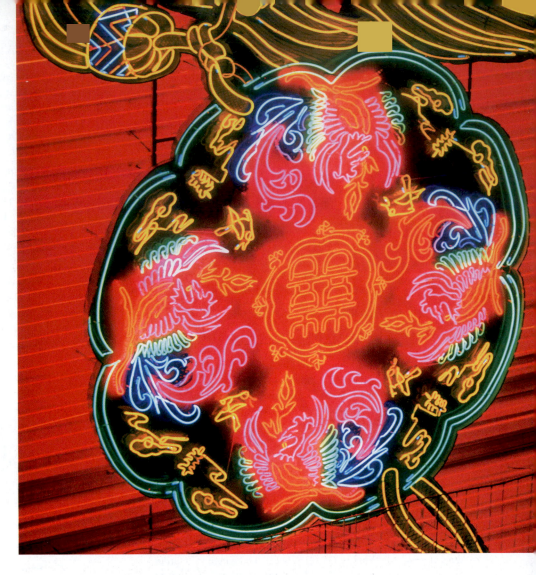

▶ The properties of the molecules that make up the gases within these lights determine the colors produced when electrical effects act. Indeed, a more complete understanding of the thermodynamic properties of these and other systems depends critically on an understanding of these systems at the molecular level.

The Molecular Basis of Thermal Physics

If we could watch a film of the molecular motion of a dilute gas, we would see the molecules rushing around at different velocities, spaced widely compared to their sizes, and bouncing elastically off each other and off the walls of the container in a gigantic three-dimensional game of billiards. Huge numbers of molecules make up even the smallest measurable quantity of a gas. The behavior of such a vast number of molecules can be expressed only in terms of averages. The **kinetic theory** of ideal gases, which is a statistical treatment of the large ensemble of molecules that make up a gas, is the focus of this chapter. We will use kinetic theory to understand the behavior of ideal and nonideal gases. Along the way, we will learn about the velocities of molecules and their statistical distributions, and about how such quantities as energy and momentum can move through a gas—transport phenomena. The kinetic theory of gases is remarkable because it explains so much of the thermodynamics we have already studied with the simplest of assumptions.

Statistical techniques for the study of large collections of particles have been the subject of intensive work in physics, chemistry, astronomy, and other fields of physical science. Early work was done by James Clerk Maxwell in the 1850s and by Ludwig Boltzmann and Josiah Willard Gibbs in the latter half of the 19th century. Research in this area falls under the general heading of **statistical physics**, a vast field that also includes many aspects of the physics of liquids and solids. Statistical techniques allow us to explain a great many of the physical properties observed in bulk matter. Statistical physics has been used to explain new and often unexpected phenomena: phase changes, superconductivity, complex crystalline structure, and chaotic behavior, to name just a few.

19–1 A Microscopic View of Gases

A volume of air the size of a birthday balloon contains some 10^{23} molecules. This is an easy number to calculate, given that the air around you is at least approximated by an ideal gas, so that the number of molecules of a volume V of a gas at standard temperature and pressure is $N = pV/kT$, where k is Boltzmann's constant. Equivalently, the number of moles is pV/RT, where R is the universal gas constant. As we know from Chapter 17, the ratio R/k gives us Avogadro's number N_A, some 6.02×10^{23} molecules, the number of molecules in 1 mol of any gas. The properties of individual molecules throw additional light on this number. Atoms and molecules consist of electrons bound to nuclei by electrical forces. Nuclei consist of protons and neutrons. These have almost equal masses, about 1.67×10^{-27} kg $= 1.67 \times 10^{-24}$ g. The atomic weight A of a type of molecule is the total number of protons and neutrons, so that the mass of a single molecule, in grams, is $A \times (1.67 \times 10^{-24}$ g$)$. Now one mole of a substance of atomic weight A has, by definition, a mass of A grams, so to find the number of molecules N_A in one mole, we divide the mass A g of a mole by the mass of one molecule: $N_A = (A \text{ g})/(A \times 1.67 \times 10^{-24} \text{ g}) = 6.02 \times 10^{23}$ molecules.

Avogadro's number is incomprehensibly large. But just how crowded are the molecules? It follows from the ideal gas law that, at standard temperature and pressure, the volume of 1 mol of gas is 22.4 L $= 22.4 \times 10^{-3}$ m^3. This means that, on average, each molecule occupies a volume of $(22.4 \times 10^{-3}$ m$^3)/(6 \times 10^{23}) \cong 4 \times 10^{-26}$ m^3. Using a radius r of 10^{-10} m for a typical molecule, we calculate the molecular volume to be $(4/3)\pi r^3 \cong 4 \times 10^{-30}$ m^3. A molecule therefore takes up only $(4 \times 10^{-30}$ m$^3)/(4 \times 10^{-26}$ m$^3) = 10^{-4}$ of the volume available to it! The cube root of this number, namely about $1/20$, gives the ratio of the typical molecular radius to the mean spacing between molecules. If we were to scale molecules to the size of, say, sheep, with a radius of 0.5 m, the molecules would be spaced some 10 m, or 30 ft, apart.

As a gas becomes more dilute, molecular collisions occur less frequently. A dilute (or ideal) gas is a collection of independent molecules that move about with differing velocities and relatively rare collisions. However rare, collisions play an important role. They are the means by which gases come to thermal equilibrium. When a hot gas (consisting of rapidly moving molecules) is mixed with a cold gas (composed of more slowly moving molecules), elastic collisions between fast molecules and slow molecules slow down the fast molecules and speed up the slow ones (see Chapter 8). The entire system comes to equilibrium at an intermediate temperature.

Our treatment of gases will be based on a statistical picture, in which averages of quantities such as speed appear rather than their values for individual molecules. Why do averages tell us anything at all about a gas? After all, if we toss 10 coins ("a set of tosses") many times, the average number of heads in each set of tosses is 5, but we have very little confidence that we will actually get heads 5 times in just 10 tosses. (Try it!) It is only because the number of molecules we deal with is so large that we can be confident that average quantities really do represent the behavior of gases. Suppose that the number of coin tosses (N) is very large. The average number of heads is $0.5N$. But more important, in any given trial (a set of N tosses), the probability that we will get heads more than the average number of times depends on N. For example, the probability that we will get heads more than 51% of the time is about 38 in 100 for $N = 100$. This probability drops to about 2 in 100 for $N = 10^4$, and is on the order of 1 in 10^{88} for a million tosses! It is generally true that for systems with a large number of independent components (for example, the number of tosses in a trial), large deviations from the average value of a variable are unlikely. The molecules in a gas can be viewed as moving independently, and when a gas that contains 10^{24} or so molecules in thermal equilibrium is sampled, the chances that the measured value of any randomly varying quantity will differ from its average value are tiny indeed.

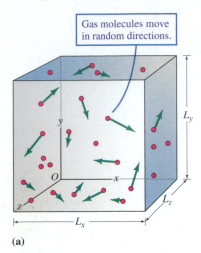

Gas molecules move in random directions.

(a)

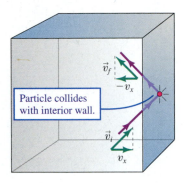

$\vec{v}_f$

$-v_x$

Particle collides with interior wall.

$\vec{v}_i$

v_x

(b)

▲ **FIGURE 19–1** (a) The dimensions and orientation of a box that contains a gas. (b) A typical molecule, whose x-component of velocity is v_x, has an elastic collision with the right-hand wall. The x-component of the velocity changes its sign in the collision, but the y- and z-components are unchanged.

19–2 Pressure and Molecular Motion

Consider a container of dilute gas that consists of N independently moving molecules each of mass m. In kinetic theory these molecules move with a variety of speeds in a variety of directions, and pressure is due to molecular collisions with the walls of the container. Each time a molecule hits and rebounds from the wall, there is a momentum transfer to the wall, and since a rate of momentum transfer is a force, there is accordingly a force on the wall. To calculate the pressure we need to calculate the *average* rate of momentum transfer.

Some Average Values in a Gas

Suppose that the dilute gas in the container above is in thermal equilibrium at temperature T. The container has volume V and is oriented in a Cartesian coordinate system, as shown in Fig. 19–1a. The molecules in the box are moving in random directions and *the average velocity of the molecules is zero* because as many move in one direction as move in any other. In other words, the average x-, y-, or z-components of the velocity are zero. We will use angular brackets, $\langle \ \rangle$, to indicate the average value, so

$$\langle v_x \rangle = \langle v_y \rangle = \langle v_z \rangle \quad \text{so that} \quad \langle \vec{v} \rangle = 0. \tag{19–1}$$

While the average velocity may be zero, the same is not true for the average *speed*. The speed is by definition positive, and its average value is likewise positive. We will concentrate on the average value of the velocity squared $\langle v^2 \rangle$, or more precisely on $\sqrt{\langle v^2 \rangle}$, a quantity we call the **root-mean-square (rms) speed**. Note that $\langle v^2 \rangle$ is *not* the same as the square of the average value of the velocity: $\langle v^2 \rangle \neq \langle \vec{v} \rangle^2$; in fact, the latter quantity is zero because $\langle \vec{v} \rangle$ is zero.

We can relate $\langle v^2 \rangle$ to the internal energy of a dilute gas by kinetic theory. Forces between molecules are important only when the molecules are close together, and this happens relatively rarely in dilute gases. Thus the internal energy, U, consists mainly of the kinetic energies of the molecules. In other words, U is given by

$$U = N\langle K \rangle = N\left(\frac{1}{2}m\langle v^2 \rangle\right), \tag{19–2}$$

where $\langle K \rangle$ is the average kinetic energy per molecule and N is the total number of molecules. This equation provides us with a link to the temperature of the gas, because, as we learned in Chapter 18, the internal energy of an ideal gas is a function of temperature only.

The average of a sum of terms is the sum of the average of those terms, so

$$\langle v^2 \rangle = \langle v_x^2 + v_y^2 + v_z^2 \rangle = \langle v_x^2 \rangle + \langle v_y^2 \rangle + \langle v_z^2 \rangle. \tag{19–3}$$

The gas as a whole is not moving, so there must be an equivalence among the three directions. The average value of the x-component of the velocity squared must be the same as that of the y-component and that of the z-component. Thus

$$\langle v_x^2 \rangle = \langle v_y^2 \rangle = \langle v_z^2 \rangle. \tag{19–4}$$

Using this relation in Eqs. (19–3) and (19–2), it is possible to relate the average of the components of the velocity squared to the thermal energy of the gas:

$$\langle v^2 \rangle = 3\langle v_x^2 \rangle;$$
$$\langle v_x^2 \rangle = \frac{1}{3}\langle v^2 \rangle = \frac{2}{3}\frac{U}{mN}. \tag{19–5}$$

Next we will relate $\langle v_x^2 \rangle$ to the pressure of the gas.

The Origin of Pressure

Pressure arises from the multiple collisions the molecules of a gas have with the walls that contain the gas. For a quantitative calculation, we compute the momentum transfer to a wall due to a single collision, and then find the number of molecules that strike the wall per unit time. We use this to find the average momentum transfer to the wall per unit time, which is the force on the wall. Pressure is then the force per unit area. Figure 19–1b shows a molecule colliding elastically with the right-hand wall of the box of Fig. 19–1a. The wall is oriented parallel to the yz-plane, so only the x-component of the velocity changes. If the velocity before the collision is

$$\vec{v}_i = v_x \hat{i} + v_y \hat{j} + v_z \hat{k},$$

then the velocity after the collision is

$$\vec{v}_f = -v_x \hat{i} + v_y \hat{j} + v_z \hat{k}.$$

The momentum change of the molecule, $\Delta \vec{P}_{\text{mol}}$, is

$$\Delta \vec{P}_{\text{mol}} = m\vec{v}_f - m\vec{v}_i = -2mv_x \hat{i};$$

the momentum transfer to the wall, $\Delta \vec{P}$, is the negative of this:

$$\Delta \vec{P} = 2mv_x \hat{i}. \tag{19–6}$$

This sign is reasonable: The wall is pushed to the right by the collision.

We have found the momentum transfer from a single molecule. As we continue, we think of this molecule as an "average" molecule; it has the average x-component of velocity, which we continue to write as v_x for now. The next question is: How many collisions with the wall occur per unit time? Consider Fig. 19–2a: The number of molecules with an x-component of velocity of magnitude v_x that strike an area A in a time interval dt is the number of molecules contained in an imaginary cylindrical volume whose base, which is against the wall, is the area A and whose length is $v_x \, dt$. Molecules that travel toward this area but are farther away than the cylinder height will not reach the wall in the infinitesimal time interval dt. Note also that by making the time interval short enough, we can safely assume that the molecules do not collide with each other before they reach the wall. If the number of molecules per unit volume, the number density, is N/V, then the total number of molecules in our cylindrical volume is $(N/V)(v_x \, dt)(A)$. Thus the number of collisions with the wall in time dt is

$$\text{\# of collisions} = N_{\text{coll}} = \frac{1}{2}\frac{N}{V}(v_x \, dt)A. \tag{19–7}$$

The factor $1/2$ is present because only half of the molecules are moving to the right; the other half are moving to the left. Only the ones that move to the right contribute to the pressure in this case. The number of collisions must be multiplied by the individual collision momentum transfer ΔP_x, the x-component of Eq. (19–6), to find the total momentum dP_x transferred in the time interval dt:

$$dP_x = (2mv_x)\left(\frac{N}{2V}\right)(v_x \, dt)A = mv_x^2 \frac{N}{V} A \, dt.$$

In turn, the total momentum transfer per unit time—the force exerted on area A—is

$$\frac{dP_x}{dt} = F_x = mv_x^2 \frac{N}{V} A. \tag{19–8}$$

The pressure on the wall, p, is the force per unit area:

$$p = \frac{F_x}{A} = mv_x^2 \frac{N}{V}. \tag{19–9}$$

Finally, recall that we have employed the average x-component of velocity squared for each molecule. We make this explicit by employing the notation $\langle v_x^2 \rangle$ rather than v_x^2. Then using Eq. (19–5), we find

$$p = m\langle v_x^2 \rangle \frac{N}{V} = m\left(\frac{2}{3}\frac{U}{mN}\right)\frac{N}{V} = \frac{2}{3}\frac{U}{V}; \tag{19–10}$$

$$pV = \frac{2}{3}U. \tag{19–11}$$

PRESSURE IN TERMS OF ENERGY

This derivation is an important one. We have used the microscopic properties of a gas to find a relation between macroscopic thermodynamic variables.

(a)

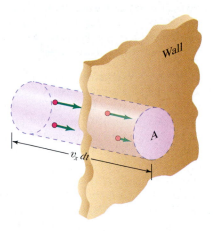

(b)

▲ **FIGURE 19–2** (a) All the molecules that will collide with an area A of a wall in a time interval dt are contained in a cylinder of area A and length $v_x \, dt$. (b) Only the molecules in the cylinder whose x-component of the velocity is directed towards the wall will contribute to the pressure. The figure only shows the x-component of the velocity.

EXAMPLE 19–1 Exactly 1 mol of helium gas has an internal energy of 3600 J. It is contained within a cube of sides 0.50 m. (a) Compare helium under these conditions with air at standard temperature and pressure (STP). If air is ideal at STP, can the same be said for helium? (b) Approximately how many times per second do the walls suffer collisions from the molecules of the helium gas?

Setting It Up For part (a) we want a "comparison"; we find the physically measurable quantities for air at STP and the same quantities for the helium. Specification of STP means specification of temperature and pressure. For part (b) we must pass from macroscopic properties to the microscopic ones.

Strategy We start with air at STP (0°C; 1 atm). Given p and T, we can find the number density N/V from the ideal gas relation $pV = NkT$. Next we turn to helium. Given the internal energy, we can use Eq. (19–11) to compute the pressure; given the volume and the number of moles, we can find the number density. If this is smaller than the number density for air at STP, the helium will be an ideal gas and we can then compute the temperature using the ideal gas law. To find the number of collisions, we know from Eq. (19–7) that in time Δt the number of collisions with one wall of area A is $(N/2V)(v_x \Delta t)A$. We can find an appropriate value for v_x from the rms speed, which is calculable from the total energy. We must take into account the fact that there are six sides to the box.

Working It Out (a) We start with the number density of air. Rearranging the ideal gas law,

$$\text{air:}\ \frac{N}{V} = \frac{p}{kT} = \frac{1.01 \times 10^5\ \text{Pa}}{(1.38 \times 10^{-23}\ \text{J/K})(273\text{K})} = 2.68 \times 10^{25}\ \text{m}^{-3}.$$

We can find the number density of helium from the fact that there are 6.02×10^{23} molecules (1 mol) in the given volume of $(0.5\ \text{m})^3$:

$$\text{helium:}\ \frac{N}{V} = \frac{6.02 \times 10^{23}}{(0.50\ \text{m})^3} = 4.8 \times 10^{24}\ \text{m}^{-3}.$$

This is about one sixth that of air at STP, so we can treat the helium as ideal.

We use Eq. (19–11) for the helium pressure: $p = \dfrac{2}{3}\dfrac{U}{V} =$

$\dfrac{2}{3}\dfrac{3600\ \text{J}}{(0.50\ \text{m})^3} = 1.9 \times 10^4\ \text{Pa}$, which is about 0.2 atm. The temperature can be computed from the ideal gas law:

$$T = \frac{pV}{nR} = \frac{(1.9 \times 10^4\ \text{N/m}^2)(0.50\ \text{m})^3}{(1\ \text{mol})(8.31\ \text{J/mol})} = 2.9 \times 10^2\text{K}.$$

This is in the range of room temperature. Our container of helium has a temperature close to that of the air at STP, is about six times less dense, and has a pressure one fifth that of air.

(b) The collision rate (# of collisions per second) with a wall of area A perpendicular to the x-axis is $N_{\text{coll}} = (N/2V)v_x A$; for v_x we take $\langle v_x^2 \rangle^{1/2}$, and we can use (symmetry) $\langle v_x^2 \rangle = \langle v^2 \rangle/3$. Finally, we can find $\langle v^2 \rangle$ from $U = N\left(\frac{1}{2}m\langle v^2 \rangle\right)$. Pulling this all together, we have

$$\frac{N_{\text{coll}}}{\Delta t} = 6\frac{N}{2V}A\sqrt{\langle v^2 \rangle/3} = 6\frac{N}{2V}A\sqrt{\frac{2U}{3mN}}.$$

The factor 6 takes into account the number of walls. The quantity mN, the mass of one molecule times the number of molecules, is the total mass of a gas, M. There is 1 mol of helium, so the mass is $4.0\ \text{g} = 4.0 \times 10^{-3}\ \text{kg}$. Thus

$$\frac{N_{\text{coll}}}{\Delta t} = \frac{6(6.02 \times 10^{23})}{2(0.50\ \text{m})^3}(0.50\ \text{m})^2\sqrt{\frac{2(3600\ \text{J})}{3(4.0 \times 10^{-3}\ \text{kg})}}$$

$$= 2.8 \times 10^{27}\ \text{collisions/s}.$$

What Do You Think? If the internal energy is doubled, how will the wall collision rate change? *Answers to **What Do You Think?** questions are given in the back of the book.*

19–3 The Meaning of Temperature

The microscopic picture of a gas provides us with a simple and enlightening interpretation of temperature. The fact that the energy is the sum of the kinetic energy of all the molecules, $U = \frac{1}{2}Nm\langle v^2 \rangle$ [Eq. (19–2)], on the one hand, and that the energy is related to the pressure, $U = (3/2)pV$ [Eq. (19–11)], on the other, provides a link to temperature. From the ideal gas law $pV = nRT = NkT$, where here n is the number of moles of the gas, R is the universal gas constant, and k is Boltzmann's constant. Thus

$$U = \frac{3}{2}nRT = \frac{3}{2}NkT. \tag{19–12}$$

Equations (19–2) and (19–12) together give a microscopic interpretation of temperature:

$$kT = \frac{2}{3}\frac{U}{N} = \frac{2}{3}\langle K \rangle. \tag{19–13}$$

TEMPERATURE AS ENERGY

The temperature of an ideal gas is a measure of the average kinetic energy of the constituents. The number of molecules has canceled from this expression, so T is (correctly) independent of the amount of gas.

Figure 19–3 gives another view of the close ties between energy and molecular motion, showing at the microscopic level how the change in energy of a system is associated with the work done by the system.

Macro   Micro   Macro   Micro

$\vec{F}$ $\vec{F}$ Movement

Insulation $\vec{F}' < \vec{F}$ Gas molecules

I.   II.

Initial   Initial $Q = 0$ Gas   Molecules

$pV = nRT$ $\frac{1}{2}mv^2_{\text{rms}} = \frac{3}{2}kT$ Reduce p transfer energy to

$W > 0$ piston to move it;

$\Delta U < 0$ v_{rms} drops and

$\Delta T < 0$ $\Delta T < 0$

◀ **FIGURE 19–3** Macroscopic and microscopic views of the work done by an insulated gas. Molecules colliding with the recoiling piston lose energy. Thus the temperature drops.

CONCEPTUAL EXAMPLE 19–2 You have two identical bottles, one containing neon gas and the other containing the same number of atoms of argon, each in thermal equilibrium with the same thermal reservoir. Both gases can be regarded as ideal. Atoms of argon have twice the mass of atoms of neon. How does the average kinetic energy of the neon atoms differ from that of the argon atoms? How do the measurable thermodynamic variables T, p, and V compare?

Answer Both bottles of gas are at the same temperature (this is what equilibrium with the same thermal reservoir means). We can conclude from this that since the kinetic energy per atom is precisely a measure of temperature [Eq. (19–13)], it is the same for the two gases. The volumes of the two gases are the same, since the bottles containing them are identical. It follows from the ideal gas law, rewritten as $p = NkT/V$, that with N, T, and V the same for the two gases, the pressure will be the same.

EXAMPLE 19–3 1.00 mol of helium gas is contained in a large volume at 31.5K. What is the internal energy of the gas? What is the average value of the x-component of molecular velocity squared, $\langle v_x^2 \rangle$?

Strategy Because the gas is in a large volume, it is dilute and behaves as an ideal gas. We may therefore use Eq. (19–12) in the form $U = (3/2)nRT$ to find the internal energy. Every quantity on the right-hand side of the equation is known. As for $\langle v_x^2 \rangle$, that is dependent on the temperature but independent of the number of molecules: We can use $m\langle v^2 \rangle/2 = 3kT/2$, then $\langle v_x^2 \rangle = \langle v^2 \rangle/3$. Finally, the mass m of a helium atom is obtained by noting that 1 mole = N_A atoms have a mass of 4.00 g, that is, $m = (4.00 \times 10^{-3} \text{ kg})/N_A$.

Working It Out For the internal energy we have

$$U = (3/2)nRT = (3/2)(8.31 \text{ J/K})(31.5\text{K}) = 393 \text{ J}.$$

For the square of the x-component of the velocity we have

$$\langle v_x^2 \rangle = (1/3)(3kT/m) = (kT)[N_A/(4.00 \times 10^3 \text{ kg})].$$

We can calculate this or alternatively recognize that $N_A k = R = 8.31 \text{ J/mol} \cdot \text{K}$ to get

$$\begin{aligned} \langle v_x^2 \rangle &= RT/(4.00 \times 10^{-3} \text{ kg}) \\ &= (8.31 \text{ J/mol} \cdot \text{K})(31.5\text{K})/(4.00 \times 10^{-3} \text{ kg}) \\ &= 6.5 \times 10^4 \text{ m}^2/\text{s}^2. \end{aligned}$$

The square root of this quantity is about 250 m/s. Even at temperatures as low as 31.5K, the helium atoms are moving quite rapidly.

What Do You Think? A few argon atoms are added to your gas. After a little while, will they, on average, move (a) faster than, (b) slower than, or (c) at the same speed as the helium atoms?

THINK ABOUT THIS . . .

WHAT DOES ABSOLUTE ZERO MEAN AT THE MOLECULAR LEVEL?

Absolute zero, the zero of the Kelvin temperature scale, is the point at which the pressure drops to zero. In the microscopic view of an ideal gas, the temperature is zero when the average kinetic energy of the ideal gas is zero. Pressure vanishes because the molecules no longer move around and bounce against the walls.

Interpretation of the van der Waals Gas

Knowing how the microscopic properties of an ideal gas relate to its macroscopic behavior enables us to understand the origin of the van der Waals equation of state, which we encountered in Section 17–4, Eq. (17–15). This equation describes gases away from the dilute gas limit more precisely than does the ideal gas equation of state. As you may recall, the van der Waals equation takes into account the fact that the gas molecules are not point particles and also provides a correction for intermolecular forces—neither of these effects are included in the ideal gas law. Here we can see in more detail how these corrections are made.

The equation is

$$\left[p + a\left(\frac{n}{V}\right)^2\right]\left(\frac{V}{n} - b\right) = RT. \tag{17–15}$$

It reduces to the ideal gas law when the constants a and b are zero. Table 19–1 gives us a sample of some values for these constants, which have a simple meaning in the kinetic theory. Gas molecules are not truly like pointlike billiard balls. Billiard balls exert a force on one another only during a very brief time period when they touch, whereas real molecules have forces that act between them over distances larger than their radii. These forces are repulsive at short range, so that the molecules bounce off one another, but are slightly attractive at longer distances. The attractive component of the intermolecular force acts to make a gas more compact. This translates into a reduced pressure: The term $a(n/V)^2$ in Eq. (17–15) represents the effect of the long-range attraction between molecules. If we solve Eq. (17–15) for p, we see it is reduced by $a(n/V)^2$. As for the term proportional to b, molecules, like billiard balls, take up some space. This term is present because of the strong repulsion between molecules at a characteristic radius. By appearing as a term subtracted from V, the constant b measures the volume unavailable for the motion of molecules because it is space already occupied by other molecules—it is the volume taken up by one mole of molecules. Example 19–4 illustrates how we can use our interpretation of b to estimate the size of molecules.

TABLE 19–1 • Some a and b Coefficients that Appear in the van der Waals Equation of State

	a (Pa · m^3)	b (m^3/mol)
Helium	3.46×10^{-3}	23.71×10^{-6}
Neon	2.12×10^{-2}	17.10×10^{-6}
Hydrogen	2.45×10^{-2}	26.61×10^{-6}
CO_2	3.96×10^{-1}	42.69×10^{-6}
Water vapor	5.47×10^{-1}	30.52×10^{-6}

EXAMPLE 19–4 Measurements show that nitrogen gas obeys the van der Waals equation of state with the constant $b = 3.94 \times 10^{-5}$ m^3/mol. What is the size of a nitrogen molecule?

Strategy We argued above that the parameter b represents the volume occupied by one mole of molecules. We can therefore find the volume of a single molecule by dividing b by Avogadro's number N_A. Setting the result equal to $4\pi R^3/3$ allows us to calculate the radius R of the molecule. (The problem statement gives no indication of the shape of the molecule, so we will assume it is spherical.)

Working It Out The volume of a single molecule is

$$V = \frac{b}{N_A} = \frac{3.94 \times 10^{-5} \text{ m}^3/\text{mol}}{6.02 \times 10^{23} \text{ molecules/mol}}$$
$$= 0.654 \times 10^{-28} \text{ m}^3/\text{molecule}.$$

We now set this equal to $4\pi R^3/3$, and solve for R:

$$R = \left(\frac{3V}{4\pi}\right)^{1/3} = \left(\frac{3 \times 65.4 \times 10^{-30} \text{ m}^3}{4 \times 3.14}\right)^{1/3} = 2.50 \times 10^{-10} \text{ m}.$$

This radius is in fair agreement with more direct atomic measurements.

What Do You Think? Table 19–1 shows that all of the b values are within a factor of two of each other. What does this tell you about molecules? Why are the a values all positive?

19–4 Probability Distributions

Just as every student in a class does not receive the average grade on a given test, the molecules in a gas do not each have the average velocity or the average position. Instead, each molecule has its own velocity and position. Two functions describe the probability that a molecule has a particular velocity or a particular position. The **velocity distribution function** describes how many molecules have one velocity, how many have another, and so forth. The **position distribution function** similarly describes how the molecules are distributed in space. We need to know something about these distributions to further develop our understanding of the thermal properties of gases.

We all have a certain familiarity with the notions of probability and averages. Perhaps you have seen films or even played games of chance such as roulette. In this section, we will sharpen these notions so that we can use them in the analysis of gases and other thermal systems.

A Case Study: The Distribution of a Set of Test Grades

Consider a quiz taken by a class of 100 students. The grades can run from 0 to a maximum score of 50. The number of students who receive each grade is shown in Fig. 19–4, where the number of students with grade g, N_g, is plotted as a function of the grade g. The total number of students here is $N = 100$, and if we add up all the N_g values, we get all of the students taking the quiz:

$$\sum_g N_g = N. \tag{19–14}$$

Given such a distribution of grades, we can find the average grade. We take the number of people with a given grade, multiply it by the grade, add these products for each grade, and divide by the total number of students. We denote the average grade by $\langle g \rangle$. According to the above technique, we find

$$\langle g \rangle = \frac{1}{N}\left(\sum_g g N_g \right). \tag{19–15}$$

In the case illustrated, $\langle g \rangle$ is 28. We could also ask for the average of the *square* of the grade $\langle g^2 \rangle$ given by

$$\langle g^2 \rangle = \frac{1}{N}\left(\sum_g g^2 N_g \right). \tag{19–16}$$

In the case illustrated, $\langle g^2 \rangle = 908.6 = (30.1)^2$. Note that $\langle g^2 \rangle \neq \langle g \rangle^2$, analogous to the remark we made about molecular velocities in Section 19–2. The quantity $\langle g^2 \rangle$ contains different information than does $\langle g \rangle$.

What is the probability that any one student, chosen at random, has a grade of 38 points? This probability is the number of students with a grade of 38 divided by the total number of students—in this case, 6/100. More generally, the probability of finding a student who received a grade g is

$$P_g = \frac{1}{N} N_g; \tag{19–17}$$

we refer to P_g as the **probability distribution** for the grades. It can run from a minimum of 0 to a maximum of 1, with 1 being the proverbial "sure thing." As a consequence of Eq. (19–14), the probability distribution is such that

$$\sum_g P_g = 1. \tag{19–18}$$

That the sum of probabilities is 1 is known as a **normalization condition**. It expresses the fact that it is a "sure thing" that *something* will happen, in this case that the student will get some grade. As we see from Eqs. (19–15) and (19–16), averages are obtained from weighted sums over P_g:

$$\langle g \rangle = \sum_g g P_g, \tag{19–19}$$

$$\langle g^2 \rangle = \sum_g g^2 P_g. \tag{19–20}$$

We can ask one more type of question: If we pick one student at random, what are the odds that his or her grade lies between 15 and 19? The probability of finding that the student has a grade between 15 and 19 is the number of students with grades in that range divided by the total number of students. We write this as follows:

$$P(15 \le g \le 19) = \frac{1}{N} \sum_{g=15}^{19} N_g = \sum_{g=15}^{19} P_g.$$

For our example, this is equal to $13/100 = 0.13$. We will see that this type of question is especially relevant to quantities that vary continuously.

▲ **FIGURE 19–4** Bar graph of quiz grades in a class.

CONCEPTUAL EXAMPLE 19–5 A group of 100 totally unprepared students take a test of 20 true-or-false questions, and they all decide to answer the questions with random guesses. What are the odds of a student in the class getting a perfect score?

Answer A student with a perfect score has chosen the right answer with each of 20 guesses. Each guess is independent of a previous one (this is what we mean by random guesses), and thus the odds of choosing the right answer is $(1/2) \times (1/2) \times (1/2) \times \ldots$ with 20 factors. We know that $2^{10} = 1024$, so that this number is $(1/1024)^2$, or approximately 10^{-6}. One in a million students would have a perfect score; equivalently, the odds that any one student has a perfect score are one in a million. To find out the probability that there is one perfect score in the class, we remark that if a perfect score were to appear, it could be student 1, or student 2, or any of the students. As you likely know, in "or" situations, probabilities add, so the odds that there will be a perfect score at all is $100 \times 10^{-6} = 10^{-4}$.

Continuous Distributions

Some distributions do not involve discrete quantities, such as the grades on an exam, but instead involve continuous quantities, such as the heights of students. Much of the analysis of this case follows the discussion above of the discrete case, with one exception: We must recognize that the probability that any student will have a *particular* height x—say, $x = 1.73000055\ldots$ m, with an infinite number of digits—is zero. Instead, we must measure the probability $P(x)\,\Delta x$ that the outcome lies within some small interval Δx around the value x (Fig. 19–5). In our example, this could be the probability that a student has a height within an interval $\Delta x = 0.01$ m around $x = 1.73$ m. The normalization condition now expresses the fact that all students must have some height, so that the sum of all probabilities adds to 1:

$$\Sigma P(x)\,\Delta x = 1. \tag{19–21}$$

A sum appears here because—even if x takes on a continuous range of values—the intervals Δx have divided the region into a *finite* number of segments.

The concept we have described can be refined by making the interval smaller and smaller, thereby finding the probability that an outcome near a particular x-value occurs with more and more accuracy. Calculus notation is appropriate in this limit. The quantity Δx is written as dx, and the sum becomes an integral whose upper and lower limits are the upper and lower values of the variable in question, x_2 and x_1:

$$\int_{x_1}^{x_2} P(x)\,dx = 1. \tag{19–22}$$

This is the normalization condition in the case of continuous variables.

The averages of functions of x are found from formulas similar to those of the discrete case. For example,

$$\langle x^n \rangle = \int_{x_1}^{x_2} x^n P(x)\,dx. \tag{19–23}$$

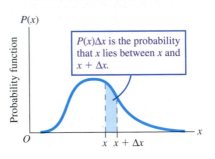

$P(x)$

$P(x)\Delta x$ is the probability that x lies between x and $x + \Delta x$.

Probability function

O $x \; x + \Delta x$ x

▲ **FIGURE 19–5** The probability $P(x)\,\Delta x$ associated with a continuous distribution is the area beneath the curve between x and $x + \Delta x$.

EXAMPLE 19–6 Suppose that all heights between 1.5 m and 1.8 m are equally likely among a group of students and that no students are outside this height range. Find the probability distribution and use it to calculate the average height.

Strategy Since sizes form a continuum, we must take the distribution to be continuous. The first step is to determine $P(x)$. where we use x to denote the height, within the distribution. The data tells us that $P(x) = 0$ for $x < 1.5$ m and for $x > 1.8$ m. In between, the probability of finding any given height is denoted by the constant P. Its value will be found using the normalization condition Eq. (19–22). The calculation of $\langle x \rangle$, the average height, is then a simple matter of working out an integral like that of Eq. (19–23).

Working It Out The normalization condition reads

$$1 = \int P(x)\,dx = P \int_{1.5\,m}^{1.8\,m} dx$$
$$= P \times (1.8\,m - 1.5\,m) = (0.3\,m)P.$$

We can solve this for P, $P = 3.33$ m^{-1}. The average height is then

$$\langle x \rangle = \int xP(x)\,dx = P \int_{1.5\,m}^{1.8\,m} x\,dx = P\left[\frac{x^2}{2}\right]_{1.5\,m}^{1.8\,m}$$
$$= (3.33\,m^{-1})(0.5)((1.8\,m)^2 - (1.5\,m)^2) = 1.65\,m.$$

As all heights are equally likely, this is a uniform distribution, so it is not surprising that the average lies right in the middle of the interval between 1.5 m and 1.8 m.

What Do You Think? Suppose the probability distribution $P(x)$ were still 0 for $x < 1.5$ m or > 1.8 m, but otherwise symmetrically U shaped, with the bottom (the least likely height) at 1.65 m. What would you expect the average to be in this case?

19–5 The Velocity Distribution of Gases

In this section, we extend our ideas about probability distributions to molecules in a gas. The velocity distribution function $F(\vec{v})$ is a probability distribution for the velocities of the gas molecules. Velocities form a continuum, so we use a probability function appropriate for continuous variables. The velocity distribution function for ideal gases was first described by James Clerk Maxwell in 1859, a time when the molecular model of matter was by no means universally accepted. As was true of the height distribution of Section 19–4, the probability of finding any particular velocity $\vec{v}$ is zero. Therefore we start with a number distribution $N(\vec{v})$ such that

$N(\vec{v}) \, d^3v$ = number of gas molecules with a velocity between $\vec{v}$ and $\vec{v} + d\vec{v}$.

What this means is that we describe the probability that a velocity will be somewhere in the neighborhood of $\vec{v}$. This "neighborhood" is a box of volume

$$d^3v = dv_x \, dv_y \, dv_z, \tag{19–24}$$

centered about the tip of the vector $\vec{v}$ (Fig. 19–6). The total number of molecules is N, so that

$$\int N(\vec{v}) \, d^3v = N. \tag{19–25}$$

The distribution $N(\vec{v})$ leads directly to a *probability* distribution $F(\vec{v})$:

$$F(\vec{v}) = \frac{1}{N} N(\vec{v}). \tag{19–26}$$

The meaning of this function is that

$F(\vec{v}) \, d^3v$ = probability that a gas molecule velocity is between $\vec{v}$ and $\vec{v} + d\vec{v}$.

From $F(\vec{v})$ we can calculate the averages for continuously varying quantities. For example, the average of the velocity squared, which is relevant to the internal energy and the temperature, is

$$\langle v^2 \rangle = \int v^2 F(\vec{v}) \, d^3v. \tag{19–27}$$

To determine the probability distribution $F(\vec{v})$, we must make some *physical* assumptions. The model of molecules darting about and occasionally colliding with other molecules suggests that a given molecule may have any velocity. This leads to a fundamental physical postulate: *Any way in which a gas's total energy and total momentum (which is zero) can be shared among the molecules is equally likely.* Once this postulate is accepted, we are in the same position as someone asking for the probability of getting any particular number between 100 and 600 when throwing 100 dice. What counts is the *number of ways* of getting that number. There is only one way of getting 100 or 600 (throwing 100 ones, or 100 sixes, respectively), but there are many ways of getting 400; thus the probability of getting 400 is much greater than the probability of getting 100 or 600. When this analysis is done for molecules whose individual kinetic energy is $\frac{1}{2}mv^2$ and whose average kinetic energy is $\frac{1}{2}m\langle v^2 \rangle = 3/2 \, kT$, the distribution function is found to be

$$F(\vec{v}) = \left(\frac{m}{2\pi kT} \right)^{3/2} e^{-mv^2/2kT}. \tag{19–28}$$

IDEAL GAS VELOCITY DISTRIBUTION

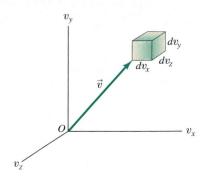

▲ **FIGURE 19–6** A molecule has a certain probability of having a velocity $\vec{v}$ within the "volume" $d\vec{v} = d^3v = dv_x \, dv_y \, dv_z$. Here we show the small box d^3v within which the velocity vector is located.

The prefactor in this equation ensures that this velocity distribution function is properly normalized, $\int F(\vec{v}) \, d^3v = 1$, meaning that the sum over all probabilities is 1. $F(\vec{v})$ is a function only of the magnitude of the velocity (that is, the speed), not the direction. This expresses the physically reasonable fact that there is no preferred direction in a container of gas in which the gas as a whole is not moving—that is, all directions are equally likely. It is possible to make a direct experimental verification of Eq. (19–28) by cutting a tiny hole in a container of gas and measuring the different speeds of the molecules that leak out.

The Average of the Velocity Squared

Here we look at an important concrete example of how F is used: Given F, we calculate $\langle v^2 \rangle$. We have

$$\langle v^2 \rangle = \int v^2 F(\vec{v}) \, d^3v = \left(\frac{m}{2\pi kT} \right)^{3/2} \int v^2 e^{-mv^2/2kT} d^3v. \qquad (19\text{–}29)$$

This integral appears to be a very tough one. It is multidimensional, and moreover it has some parameters in it that at first glance look hard to deal with. But we can extract the essential from this integral with very little work by using a technique known as **scaling**. This very useful technique allows us to extract the dependence of the integral on physical parameters of the problem. The idea is to change the integration variable to a dimensionless one, removing the dimensional dependence to a factor. In particular when there is an exponential function (or any transcendental function), the argument of the function is dimensionless. Accordingly, we change the integration variable in Eq. (19–29) from $\vec{v}$ to the dimensionless vector $\vec{u}$ by the definition

$$\vec{v} \equiv \vec{u} \sqrt{\frac{2kT}{m}}. \qquad (19\text{–}30)$$

This definition means that each component of the vector $\vec{v}$ is transformed; Eq. (19–30) stands for three equations. With this change the argument of the exponential function is simply $-u^2$, and u itself is dimensionless. This can be seen in another way: The integration variable v is not dimensionless. Both kT and mv^2 have the dimensions of energy, so v^2 has the dimensions of kT/m and v those of $(kT/m)^{1/2}$. By comparing this fact with Eq. (19–30), we see right away that u is indeed dimensionless.

In terms of our new variable, the volume element d^3v is

$$d^3v = \left(\frac{2kT}{m} \right)^{3/2} d^3u.$$

The factor $\sqrt{2kT/m}$ is cubed because d^3v stands for $dv_x \, dv_y \, dv_z$, and there is a factor for each variable. We also replace the factor v^2 in the integration of Eq. (19–29) with $u^2(2kT/m)$. Pulling everything together, we have

$$\langle v^2 \rangle = \left(\frac{m}{2\pi kT} \right)^{3/2} \left(\frac{2kT}{m} \right)^{5/2} \int u^2 e^{-u^2} d^3u = \frac{2kT}{m} \frac{1}{\pi \sqrt{\pi}} \int u^2 e^{-u^2} d^3u. \qquad (19\text{–}31)$$

The remaining integral on the right contains no quantities with dimension; it is a pure number. But even without knowing this number (which is indeed a hard integral), Eq. (19–31) shows that $\langle v^2 \rangle$ is a dimensionless constant times $2kT/m$. We anticipated that in our discussion of Eqs. (19–11) and (19–13) which combined with the ideal gas law led to

$$\langle v^2 \rangle = \frac{2U}{Nm} = \frac{3kT}{m}. \qquad (19\text{–}32)$$

Thus Eq. (19–31) is in accord with Eq. (19–32)—at least in terms of dimensions. To finish the evaluation, we note with the help of a table of integrals that:

$$\int u^2 e^{-u^2} d^3u = \frac{3}{2} \pi \sqrt{\pi}.$$

When this is inserted into the right side of Eq. (19–31), we find that the result of Eq. (19–32) is confirmed.

With Eq. (19–32) in hand, we can get a feel for molecular speeds from the rms speed v_{rms}:

$$v_{\text{rms}} = \sqrt{\langle v^2 \rangle} = \sqrt{\frac{3kT}{m}}. \qquad (19\text{–}33)$$

For helium gas, which has an atomic mass $m = 6.7 \times 10^{-27}$ kg, the rms speed is 556 m/s, 1362 m/s, and 2486 m/s for $T = 50$K, $T = 300$K, and $T = 1000$K, respectively. Contrast this with argon gas, which has a molecular mass 10 times greater than helium. The corresponding rms speeds are $\sqrt{10}$ times smaller: 176 m/s, 431 m/s, and 786 m/s.

One can also use the distribution F [Eq. (19–28)] to calculate the mean *speed* of the molecules. The expression will be similar to the rms speed, since dimensional analysis shows that it has to be of the form [energy/mass]$^{1/2}$, which must be (constant)$(kT/m)^{1/2}$. The fact that the speed is inversely proportional to the square root of the mass implies that lighter molecules (faster ones) escape in greater quantity through a tiny hole than heavier ones (slower ones). This is the practical basis for the separation of uranium isotopes.[†] The uranium isotope ^{238}U is more common, but the relatively rare isotope ^{235}U is more useful for nuclear reactors such as those used for generating electric power—it is more fissionable, and therefore more useful for the creation of chain reactions. These isotopes cannot be separated by chemical means because they have the same chemical properties. However, the fact that ^{238}U is a bit heavier than ^{235}U means that when they are mixed in a gas— uranium hexafluoride, UF_6, is used—the ^{238}U component moves, on average, at a lower speed. The UF_6 gas is allowed to escape through many tiny holes from one container at a given temperature into another container that is initially empty. The percentage of ^{235}U in the gas that has escaped is slightly higher than that in the original container. Thousands of repetitions of this process of isotope separation lead to significantly increased percentages of ^{235}U. This method was used during World War II in the Manhattan Project.

THINK ABOUT THIS . . .

WHY IS THERE SO LITTLE HYDROGEN IN EARTH'S ATMOSPHERE?

The most common element in the universe is hydrogen, yet there are virtually no H_2 molecules in Earth's atmosphere. It would seem reasonable that the material out of which the planets condensed contained a lot of hydrogen. Jupiter, for example, consists largely of hydrogen, as does the Sun. The explanation lies in the escape speed, 11.2 km/s for objects on Earth (see Section 12–3). This number can be compared to the rms speed of molecules in a gas at temperature T, $v_{rms} = \sqrt{3kT/m}$, where m is the molecular mass. With a T value of 290K, we get for air ($m = 29.2 \times 1.67 \times 10^{-27}$ kg, an average coming mainly from oxygen and nitrogen molecules) a value of $v_{rms} = 0.5$ km/s. For H_2 m is a factor of 14.6 smaller, so that $v_{rms} = 1.9$ km/s. Both numbers are small compared with 11.2 km/s, but since the number of molecules with some speed v is proportional to $e^{-v^2/2kT} = e^{-3/2(v/v_{rms})^2}$, there are many more hydrogen molecules at speeds >11.2 km/s than there are molecules of the other components of air moving with this speed. Thus the upper layers of the atmosphere will lose their faster (H_2) molecules much more rapidly than they lose their slower (O_2 or N_2) air molecules. It is a slow process, but ultimately all the hydrogen molecules get "filtered out."

19–6 The Maxwell–Boltzmann Distribution

An ideal gas may consist of molecules that have just one atom (monatomic) such as helium (He) or argon (Ar), or it may consist of molecules with two (diatomic) or more atoms, such as nitrogen (N_2), oxygen (O_2), or water (H_2O) (Fig. 19–7). Is there a difference between these types of ideal gases? The answer came in the last part of the 19th century from the work of Ludwig Boltzmann.

The velocity distribution of a monatomic ideal gas [Eq. (19–28)] has the general form

$$F(\vec{v}) = \frac{1}{Z}e^{-mv^2/2kT} = \frac{1}{Z}e^{-K/kT},$$
(19–34)

[†]Elements are characterized by their chemical properties, which depend on the number of electrons they carry, and hence on the number of (positively charged) protons in their nuclei. *Isotopes* of an element differ in the number of (electrically neutral) neutrons in the nucleus of the element, and thus differ in atomic weight. For uranium, the atomic weights 235 and 238 refer to the total number of protons plus neutrons.

Molecule	Bonding Schematic	Space-filling Schematic
Helium (He)		
Nitrogen (N_2)		
Carbon dioxide (CO_2)		
Water (H_2O)		
Sulfur trioxide (SO_3)		
Ammonia (NH_3)		
Methane (CH_4)		
Benzene (C_6H_6)		

▲ **FIGURE 19–7** Molecules come in different shapes, as shown here in two schematic representations.

where K is the kinetic energy ($K = mv^2/2$); the factor $1/Z$ takes into account the normalization. Boltzmann generalized the velocity distribution to a distribution that describes the probability that any one molecule has a given energy E. He discovered that in the original velocity distribution, Eq. (19–34), *the total energy of the molecule, E, should replace the kinetic energy of a point mass, K*:

$$F = \frac{1}{Z} e^{-E/kT}. \tag{19–35}$$

MAXWELL–BOLTZMANN DISTRIBUTION

This result was obtained under the assumption that *all* the variables describing molecular motion, such as velocity and angular velocity, have equal probability of taking on *any* value, subject to the constraint that energy must be conserved, very much in line with the assumptions that led to Eq. (19–28). This distribution function is known as the **Maxwell–Boltzmann distribution**. Equation (19–35) is one of the most frequently used formulas in the physics of the behavior of matter. This distribution function depends on all the dynamical variables that enter into the energy of a molecule. It is only for free, pointlike molecules that the energy depends on speed alone. The energy of a rotating molecule with rotational inertia also depends on the angular velocities of rotation about the various axes. Later in this section we will see how this affects the specific heat of gases.

The Energy Distribution for Diatomic Molecules

To see the effect of molecular structure on the thermal properties of a gas, let's consider a gas composed of diatomic molecules. To describe a symmetric diatomic molecule, such as O_2, we use a classical model in which the two atoms act as though they are connected by a rigid rod. The rotational inertia has the same value, I, about each of the two axes that pass through the center of the rod and are perpendicular to it (Fig. 19–8). (We can assume that the rotational inertia about the molecular axis is so small that we can ignore it.) The energy of the molecule is a sum of the overall translational kinetic energy from linear motion, or translation (Fig. 19–9a), and of the rotational motion of the atoms about the center of mass (Fig. 19–9b; see Chapters 9 and 10). The rotational energy is

$$E_{rot} = \frac{1}{2} I \omega_x^2 + \frac{1}{2} I \omega_y^2, \tag{19–36}$$

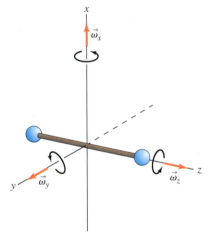

▲ **FIGURE 19–8** A dumbbell-like molecule can rotate about three axes. If the masses at the end of the rod are pointlike, the rotational inertia about the z-axis is zero. The rotational inertia about the x- and y-axes is I.

where ω_x and ω_y are the components of the angular velocity along the x- and y-axes, respectively (Fig. 19–8). The ω_z term is missing from Eq. (19–36) because the rotational inertia about the z-axis is zero. The probability distribution F now refers to the probability that the velocity of a molecule lies in a region d^3v about the velocity vector $\vec{v}$ *and* that the molecule's angular velocity lies in a region $d^2\omega = d\omega_x\, d\omega_y$ about the angular velocity vector ω whose components are ω_x and ω_y. The total energy of a molecule is now $E = K + E_{rot}$, and, from Eq. (19–35), the Maxwell–Boltzmann distribution is

$$F = \frac{1}{Z} e^{-(mv^2/2kT)-(I\omega_x^2/2kT)-(I\omega_y^2/2kT)}. \tag{19–37}$$

Since not only the molecular velocity but also the rotation speed can vary, the average energy $\langle E \rangle$ takes the form

$$\langle E \rangle = \tfrac{1}{2} m \langle v_x^2 \rangle + \tfrac{1}{2} m \langle v_y^2 \rangle + \tfrac{1}{2} m \langle v_z^2 \rangle + \tfrac{1}{2} I \langle \omega_x^2 \rangle + \tfrac{1}{2} I \langle \omega_y^2 \rangle. \tag{19–38}$$

▶ **FIGURE 19–9** Schematic diagram of (a) translation, (b) rotation, and (c) vibration of a diatomic molecule.

(a) Translation (b) Rotation (c) Vibration

The average is found by taking the integral of $E \times F$, and that integration is over d^3v, $d\omega_x$, and $d\omega_y$. The integrals are all of the form that lead to Eq. (19–31), and evaluation of each of the five terms in Eq. (19–38) is precisely $kT/2$. Thus for our rigid-rod diatomic molecule,

$$\langle E \rangle = \frac{5}{2}kT. \tag{19–39}$$

The total internal energy of the gas is then $U = N\langle E \rangle$, where N is the number of molecules. This result should be compared to the monatomic case, in which there are only the three terms of the kinetic energy of linear motion and $\langle E \rangle = \frac{3}{2}kT$ [Eq. (19–12)].

There is yet another aspect to diatomic molecules that we have not yet mentioned. If the diatomic molecules are not rigidly attached to each other, the bond between them acts like a spring with spring constant k (Fig. 19–9c). There are then two additional vibrational contributions to the energy, of the form $E_{vib} = (mV^2/2) + (kx^2/2)$. Here, V is the speed of the atoms of the molecule relative to the molecule's center of mass and x is the displacement of the vibrating atoms from an equilibrium position. Each of these terms also contributes $kT/2$ to the average energy and, for these molecules, $\langle E \rangle = \frac{7}{2}kT$.

We mention here in passing that the different average energies for molecules of different structure will be reflected in different values for the heat capacity. Thus these effects are directly translated to the macroscopic scale. We'll explore this in more detail below.

Equipartition

The discussion above shows that the total average energy per molecule depends on how many independent motions a molecule can have. If a molecule acts like a point mass, all it can do is move in the x-, y-, and z-directions. In this case, the energy has the three terms, proportional to v_x^2, v_y^2, and v_z^2, corresponding to linear motion in the x-, y-, and z-directions, respectively. If, in addition, the molecule is diatomic with a rotational inertia about axes x and y, we have two new terms in the energy—this time proportional to ω_x^2 and ω_y^2—making five terms in all. If vibration is possible, then there are two more terms, this time proportional to the squares of the relative speed of the atomic constituents and to their separation; thus there are in total seven such terms. Every term in the energy expression that is quadratic in an independent dynamical variable designates a **degree of freedom**. Generally, *the contribution of each degree of freedom to the average energy of a molecule is $kT/2$*. This result is called the **equipartition theorem**. If s is the number of degrees of freedom, then

$$\langle E \rangle = \frac{s}{2}kT. \tag{19–40}$$

EQUIPARTITION

Remember, $\langle E \rangle$ may contain contributions associated with translation, rotation, and vibration.

CONCEPTUAL EXAMPLE 19–7 Suppose you have a gas of rigid diatomic molecules constrained to move in just two spatial dimensions, for example between two plates spaced a molecular diameter apart. What would you expect the value of $\langle E \rangle$ to be?

Answer We may visualize this by thinking of the molecules as a collection of matches on a table forming the xy-plane, as in Fig. 19–10. The velocity of the center of mass can have only x- and y-components, and the integration that gives the average value of energy is over only two rather than the usual three translational degrees of freedom. Rotations can only occur when the nuclei rotate about their centers of mass in the xy-plane, that is, there are rotations only about a z-axis. This means there is only one rotational degree of freedom. The total number of degrees of freedom is three, just as for a monatomic gas. Therefore the average energy $\langle E \rangle$ is $3 \times kT/2$, the same result as for a three-dimensional monatomic gas.

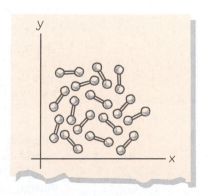

▲ **FIGURE 19–10** Diatomic molecules move in xy-plane.

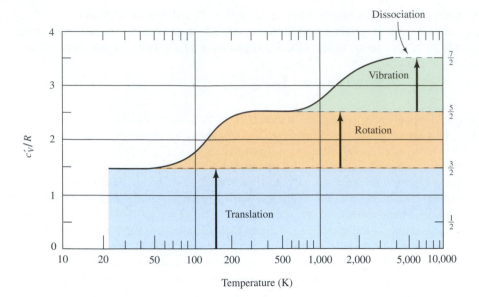

▶ **FIGURE 19–11** The constant-volume molar heat capacity divided by the gas constant, c_V'/R, for H_2 as a function of temperature.

A dilute gas—even one whose molecules have a complicated structure—continues to obey the ideal gas law. The internal energy is

$$U = N\langle E \rangle = \frac{s}{2}NkT. \qquad (19\text{–}41)$$

The internal energy continues to be linearly dependent on temperature, as for any ideal gas. The term by which the temperature is multiplied in Eq. (19–41) is the constant-volume heat capacity of the gas, C_V. The heat capacity therefore depends on the number of degrees of freedom of the molecules through the relation

$$C_V = \frac{s}{2}nR. \qquad (19\text{–}42)$$

Molecular hydrogen (H_2), for example, should have a *molar heat capacity, c'* or heat capacity per mole ($n = 1$), of $(7/2)R$ at constant volume: $(3/2)R$ is from the overall motion of the center of mass, $(2/2)R$ is from rotations about two axes, and $(2/2)R$ is from the kinetic and potential energies of the springlike bond that connects the two hydrogen atoms. How does this compare with experimental results? Figure 19–11 plots c_V'/R where c_V' is the constant-volume molar heat capacity, as a function of temperature. At lower temperatures, hydrogen gas behaves like a monatomic system, with a molar heat capacity of $(3/2)R$, which explains why we can treat H_2 at room temperature as a monatomic gas. The full factor of $(7/2)R$ comes in only at high temperatures. The steplike nature of the curve suggests that, somehow, the effects enter one at a time. The reason for this cannot be found in the realm of classical physics, but instead requires quantum mechanics. Quantum physics sets certain minimum temperatures for the excitation of the different degrees of freedom.

*19–7 Collisions and Transport Phenomena

Even in a dilute ideal gas, molecules follow a tortuous path in their container, colliding with one another and the walls, changing in direction and speed with each collision (Fig. 19–12). At STP an air molecule undergoes billions of collisions per second, but even so, the collisions are brief, and most of the time the molecule is free of the influence of other molecules. The *average* distance a molecule travels between collisions is a statistical quantity that can be calculated just as the average speed can be calculated.

It is by means of these collisions that molecules can carry physical properties through a gas. If there is local heating, that extra thermal energy can be carried to other parts of the gas through successive collisions. If some molecules with a particular odor or color are introduced in one location, they will spread through the gas through colli-

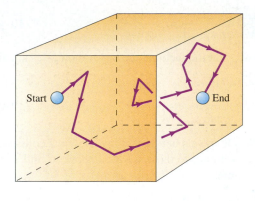

◀ **FIGURE 19–12**   A molecule follows a tortuous zigzag path because of its multiple collisions with other molecules.

(a)   (b)   (c)

◀ **FIGURE 19–13**   (a) Dye is placed with the help of a pipette at the bottom of a flask of water. (b) A combination of convection and diffusion has moved the dye into other regions of the flask. (c) Dye has diffused throughout the flask of water to make a uniform distribution. While stirring can speed the process, diffusion will make this happen if given enough time.

sions, even without wind currents (convection). The movement of such properties is called **transport** (of thermal energy, odor, and so forth), and, more particularly, the movement of molecules by random collisions is called **diffusion**. Figure 19–13 illustrates transport in a liquid. Ordinary thermodynamics has nothing to say about transport phenomena, but kinetic theory can explain them.

Collisions and Molecular Movement in a Gas

For a given gas density, the distance that a molecule travels before it has a collision depends on the size of the molecules. If molecules were infinitely small, they would never encounter each other; if they were very large, they would always be colliding. Molecules of diameter D collide when the path of the center of one molecule lies within an area πD^2 presented by the second molecule (Fig. 19–14). This area is the **collision cross section**, σ.

$$\sigma = \pi D^2. \tag{19–43}$$

Even if the molecules interact differently than billiard balls, there is some effective distance D that characterizes the collision, and the collision cross section still has the form of Eq. (19–43).

   Consider now a molecule that moves with speed v and sweeps out an area σ. If no other molecules are present, then in time t the molecule travels a distance $d = vt$, sweeping out a volume $V = \sigma d = \sigma vt$. However, our molecule is not alone: Within the volume V, there are $N = nV$ target molecules, where n is the number density of molecules (not the number of moles). The traveling molecule therefore suffers N collisions. Even though the path is bent with each collision, the volume V remains unchanged if the speed is not changed by the effect of multiple collisions, and, on average, it is not. Because the number of collisions in time t is N, there is on average a collision every t/N seconds. This is the **mean collision time**, τ. Since we are averaging over

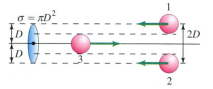

▲ **FIGURE 19–14**   If the projected centers of two molecules of diameter D that move together are less than or equal to a distance D apart, the molecules will collide. An area $\sigma = \pi D^2$ of one molecule, shown in blue, is available for collisions with another molecule.

many molecules, we replace v with some average value. It is reasonable to replace it with v_{rms} so that

$$\tau = \frac{t}{N} = \frac{t}{nV} = \frac{t}{n\sigma v_{rms}t} = \frac{1}{n\sigma v_{rms}}. \tag{19-44}$$

A more precise calculation takes into account the fact that the target molecules are also moving, and Eq. (19–44) is modified to

$$\tau = \frac{1}{\sqrt{2}n\sigma v_{rms}}. \tag{19-45}$$

The average distance a given molecule travels before it is involved in a collision is the **mean free path**, λ. Using arguments similar to those we used above to find τ, λ is the rms speed times the mean collision time:

$$\lambda = \tau v_{rms} = \frac{1}{\sqrt{2}n\sigma}. \tag{19-46}$$

It is reasonable that λ is inversely proportional to both the density and the collision cross section.

EXAMPLE 19–8 What is the mean free path in air at sea level when the temperature is 300K? By how much does the mean free path change when the temperature drops to 275K? Take $r \cong 10^{-10}$ m as a typical molecular radius, and treat the atmosphere as ideal.

Strategy To find the mean free path we need to assemble all the quantities that enter into the expression for it, Eq. (19–46). The collision cross section σ can be found with the given molecular radius using Eq. (19–43). The number density is obtained from the ideal gas law written in the form $pV = NkT$, from which we get the number of molecules per unit volume as $n = N/V = p/kT$. All that remains is to put in the numbers. For the second part of the problem, we notice that since $n \propto 1/T$, the mean free path $\lambda \propto T$. We can then get the mean free path at 275K by taking ratios.

Working It Out We write $\sigma = \pi(2r)^2$ and $n = p/kT$. We have $r = 10^{-10}$ m, $p = 1.0$ atm $= 1.0 \times 10^5$ Pa and $T = 300$K, so that

$$\lambda = \frac{1}{\sqrt{2}n\sigma} = \frac{1}{\sqrt{2}}\frac{kT}{p\pi(2r)^2}$$

$$= \frac{1}{\sqrt{2}}\frac{(1.38 \times 10^{-23} \text{ J/K})(300\text{K})}{(1.0 \times 10^5 \text{ N/m}^2)\pi(2 \times 10^{-10} \text{ m})^2} = 2.3 \times 10^{-7} \text{ m}.$$

For the second part of the problem (the mean free path at 275K), the linear dependence tells us that it is 90% of that at 300K:

$$\frac{\lambda(275\text{K})}{\lambda(300\text{K})} = \frac{275}{300} = 0.9.$$

What Do You Think? If you slowly increase the pressure by slowly moving a piston into a cylinder containing the gas, will the mean free path of the molecules increase or decrease?

The Random Walk and Diffusion

Molecules move through a gas by diffusion. How far does a molecule move, *on average*, from its initial position in a given amount of time? This problem is similar to a classic problem in mathematics called the **random walk**, known historically as the drunkard's walk. In this problem, a drunkard starts at a lamppost and takes steps that are equal in length but random in direction (Fig. 19–15). Compare Fig. 19–15 to Fig. 19–12, a typical molecular path in a gas. If the length of the drunkard's step is replaced by the mean free path, and if the time between steps is replaced by the mean collision time, then—aside from the fact that the molecule moves in three dimensions rather than two—the problems are very similar.

▲ **FIGURE 19–15** The random walk (a drunkard's walk around a lamppost).

We can find the average displacement of a molecule after N steps as follows. Let the successive displacements of the molecules be $\vec{L}_1, \vec{L}_2, \ldots \vec{L}_N$. These have random directions, but their magnitudes are all the same value, L. After N steps the net displacement of a molecule is $\vec{R}_N = \vec{L}_1 + \vec{L}_2 + \cdots + \vec{L}_N$. Squaring this quantity, we find that

$$R_N^2 = L_1^2 + L_2^2 + \cdots + L_N^2 + 2\vec{L}_1 \cdot \vec{L}_2 + 2\vec{L}_1 \cdot \vec{L}_3 + \cdots + 2\vec{L}_{N-1} \cdot \vec{L}_N.$$

We want the average value of this quantity. The directions of the vectors $\vec{L}_i$ are random, so all the dot products, such as $\vec{L}_1 \cdot \vec{L}_2$, have an average over time of zero, whereas the terms L_i^2 all equal L^2. We are left with

$$\langle R_N^2 \rangle = NL^2. \tag{19-47}$$

To connect this result to the properties of the gas, we replace L^2 by the mean free path squared. The time interval between steps is the mean collision time, τ, so after N steps, a time $t = N\tau$ has elapsed. N can thus be replaced by t/τ. Therefore, Eq. (19–47) states that after time t, a molecule will have moved on average a distance squared given by

$$\langle r^2 \rangle = \frac{t}{\tau}\lambda^2. \tag{19–48}$$

It is typical of random-walk problems that the displacement *squared* is linear in time, or equivalently that the displacement is proportional to the square root of the time.

This result (or the slightly modified result of a more detailed calculation) as it applies to molecules can be verified experimentally. Molecules that have been marked in one way or another (for example, by using some radioactive molecules) can be traced as they move. Even better, the actual motion of this random walk is observable when particles larger than molecules, such as smoke particles, are added to a gas. This motion is called **Brownian motion**. The observation of Brownian motion and its explanation in kinetic theory was a "smoking gun" that established once and for all the atomic picture of atoms.

THINK ABOUT THIS . . .

HOW DOES THE MOLECULAR PICTURE EXPLAIN THERMAL ENERGY CONDUCTION IN MATERIALS?

In Section 17–1 we discussed the various methods of thermal contact, which we subsequently learned involves the flow of thermal energy. We have learned here that thermal energy is the energy of motion of the constituents of matter and so we can now look at thermal conductivity in materials from a molecular point of view. It is useful to think about gases and solids separately.

Gases

In gases, thermal energy movement by conduction is a transport phenomenon of the type discussed in this section. Molecules that move more rapidly because they are in a region of higher temperature collide with molecules in a neighboring region, giving the adjacent molecules more kinetic energy and consequently more thermal energy. This model accounts satisfactorily for such quantities as the thermal conductivity κ (see Chapter 18).

Table 18–3 shows that air, like all gases, has a small thermal conductivity. So why is a house built of double walls with air between them poorly insulated? The reason is that convection, which involves large-scale movement ("wind currents"), is the main form of heat transport in fluids. Thermal energy is transported efficiently by convection. Fiberglass insulation is placed in walls so that air is trapped, unable to flow, or convect, easily.

Solids

It is a fact that metals conduct thermal energy better than nonmetals do. This is because in some sense metals contain a gas of electrons, and the thermal conduction mechanism available to gases is also available to metals. Within the atoms of metals, the outermost electrons that orbit the nucleus are attached very loosely (they are *weakly bound*) to the nucleus, and it takes very little energy to make them move almost freely through the material. The practically free electrons of metals are efficient in transferring energy from one end of a piece of metal to the other. The same electrons also transfer electric charge (Chapter 26), and thermal and electrical conductivity in metals are closely related. (The analogy between metals and gases only goes so far, and there is no analog to convection in metals.)

Nonmetals also conduct thermal energy, but less well than metals do. The outermost electrons of the atoms of nonmetals are attached tightly to their nucleus or to other atoms, and are therefore not readily available to conduct thermal energy. Both metals and nonmetals can move energy in the form of waves (Chapters 14 and 15), but purely elastic waves traverse a material without depositing energy along the way. Thus thermal energy cannot be distributed through material this way. What is required to do so is a set of impurities or defects in the medium, because when an elastic wave meets such obstacles it can indeed transfer some of its energy to them.

Summary

Molecules are sparse in gases under normal conditions. The kinetic theory of ideal gases is a statistical description in which molecules bounce randomly and relatively infrequently off each other and off the walls of their container. This theory accurately predicts many thermodynamic properties.

The internal energy, U, is given in this model by the number of molecules, N, times the average energy of a molecule:

$$U = N\langle K \rangle = N\left(\frac{1}{2}m\langle v^2 \rangle\right), \tag{19–2}$$

where $\langle v^2 \rangle$ is the average of the velocity squared. The pressure of a gas arises from multiple elastic collisions between the walls and the molecules. Pressure is related to internal energy by

$$pV = \frac{2}{3}U, \tag{19–11}$$

where V is the volume of gas. In the ideal gas law, pV is proportional to temperature, so temperature can be interpreted as a measure of the internal energy, or of the average energy of molecules:

$$kT = \frac{2}{3}\frac{U}{N} = \frac{2}{3}\langle K \rangle. \tag{19–13}$$

Probability distributions describe the probability that various outcomes will be represented in a large statistical sample. For gases, $F(\vec{v})\, d^3v$ describes the probability that a given molecule will have a velocity $\vec{v}$ in the range from $\vec{v}$ to $\vec{v} + d\vec{v}$. Averages are computed from the probability distribution by integration. For example, the average of v^2 is

$$\langle v^2 \rangle = \int v^2 F(\vec{v})\, d^3v. \tag{19–27}$$

The probability distribution $F(\vec{v})$ for the velocity of the molecules of a gas is

$$F(\vec{v}) = \left(\frac{m}{2\pi kT}\right)^{3/2} e^{-mv^2/2kT}. \tag{19–28}$$

Boltzmann generalized the velocity distribution to the cases in which molecules have internal energy as well as a center-of-mass kinetic energy, or in which the molecules are subject to external forces. The Maxwell–Boltzmann distribution is given by

$$F = \frac{1}{Z}e^{-E/kT}, \tag{19–35}$$

where E is the total energy of a molecule. Every variable that appears quadratically in the expression for the energy of a single molecule—such variables are called degrees of freedom—contributes $\frac{1}{2}kT$ to the average energy. If the number of degrees of freedom is labeled s, then the average energy of a molecule is

$$\langle E \rangle = \frac{s}{2}kT. \tag{19–40}$$

This result is the equipartition theorem. The internal energy is then $U = N\langle E \rangle$; in addition to the energy itself, the heat capacity, which is related to how fast U changes with temperature, is also determined, completing the link to thermodynamics.

Transport phenomena describe the motion of molecules, momentum, energy, and so forth through a gas as a result of multiple molecular collisions. Molecules have a collision cross section, σ, which describes the area available for collisions with another molecule. If the number of molecules per unit volume is n, then the mean time τ between collisions, called the mean collision time, is

$$\tau = \frac{1}{\sqrt{2}n\sigma v_{\text{rms}}}. \tag{19–45}$$

The average distance a given molecule travels before it suffers a collision is the mean free path, λ:

$$\lambda = \tau v_{\text{rms}} = \frac{1}{\sqrt{2}n\sigma}. \tag{19–46}$$

As a result of multiple collisions with other molecules, an individual molecule within a gas will undergo a random walk. In a time t, the molecule moves—diffuses—a distance that is proportional to the square root of t.

Understanding the Concepts

1. Consider an ideal gas in an insulated cylinder outfitted with a piston. The piston is moved slowly, so that the volume of the gas is decreased. What happens to the temperature of the gas? How do you explain your answer in terms of the motion of molecules?

2. On average, which will travel faster: the oxygen molecules or the nitrogen molecules in your room?

3. In Example 19–6 the calculated quantity P is 3.3 m^{-1}. Is this a probability? If so, how can it be greater than one?

4. In our discussion of pressure, we used the fact that the collisions of molecules with the walls are elastic. If there are diatomic molecules, they could be set into rotation by the collision, and thus their translational kinetic energy after the collision could be smaller than before the collision. How can these statements be reconciled? What is the effect on the pressure?

5. We made the connection between temperature and the average kinetic energy of a molecule without taking into account the large number of intermolecular collisions estimated in Section 19–1. Why do you think this approximation holds?

6. Why would more massive molecules in a gas at a given temperature have smaller average speeds?

7. In an extremely dilute gas, the rate of interatomic collisions is very small. As the gas becomes less dense, the rate of interatomic collisions becomes much less than the collision rate with the container's walls. Can we still make a connection between temperature and the average kinetic energy of a molecule? What are the possible problems in making the connection in this limit?

8. Very few hydrogen molecules are present in Earth's atmosphere today. Is this fact in conflict with the possibility of their abundance a long time ago? In explaining your answer, ignore the possibility that hydrogen is removed by chemical reactions.

9. Can we assign a temperature to a single molecule? Explain.

10. Why, from the point of view of kinetic theory, does the air near a hot stove become heated?

11. In Section 19–1 we stated that the typical distance a molecule travels before it suffers a collision in 1 mol of a gas at standard temperature and pressure is roughly 3×10^{-7} m. This will be much less than the size of a container of gas. How is it that the approximation that each molecule in the cylinder collides freely with the wall, without worrying about intermediate collisions between molecules, leads to a satisfactory result for the pressure, Eq. (19–11)?

12. Suppose you place a thin chip of paint of dimensions 1 cm by 1 cm in a gas. Do the collisions knock it to one side? Which side?

13. At standard temperature and pressure, a molecule of one mole of an ideal gas has a mean free path of about 3×10^{-7} m, some 100 times greater than the average intermolecular spacing. How is it possible for the mean free path to be greater than the average spacing?

14. You have a container of a dilute gas at a certain temperature and inject a small quantity of a different type of molecule into that container. What processes determine how long it takes the new molecules to come to thermal equilibrium with the original molecules?

15. If you were to consider helium gas and air at the same temperature, which gas would have a higher root-mean-square *momentum*?

16. Gravity keeps the atmosphere of Earth close to the surface. Why, then, is there no atmosphere on the Moon, which also exerts a gravitational force?

17. In Newtonian mechanics, energy is added to a system when work is done on it. In the language of classical mechanics, how is energy added to a gas in a container when a piston is pushed in and the container's volume is changed?

18. We have a container of helium gas and another of nitrogen gas. Both have molecules with the same average speed. What is the ratio of their temperatures?

19. Given our explanation of the origin of pressure, why is there no net force on a pane of glass placed in a container of gas, even though there may be a net force on a wall of the container?

20. Will the air just above the surface of hot water in a bowl be at the same temperature as the water? If so, what molecular mechanism is responsible for the equality of the two temperatures?

21. Why will a cup of hot water cool off much more rapidly when air is blown across its surface than when the air above the cup is stationary?

22. The smoke from the burning end of a cigarette rises; the flame of a candle points upward. Why?

23. In our discussion of the distribution of heights we claimed that in a group of persons all heights are equally likely. How would you check this premise? Can you know this with certainty, given that the number of people in the world is finite?

24. When a candle burns, carbon combines with the oxygen in air to form carbon dioxide, a compound used in fire extinguishers to put out fires. Why does the candle not snuff itself out? Could you burn a candle in a satellite in orbit?

25. Does temperature depend on the rotational kinetic energy of molecules?

Problems

19–1 A Microscopic View of Gases

1. (I) We can define a volume ratio R_V of a gas to be the ratio of the volume taken up by the molecules to the total volume of the container; similarly, we can define a spacing ratio R_L to be the ratio of the linear size of a molecule to the mean spacing between molecules. Compute R_V and R_L for a gas with molecules of diameter 2×10^{-10} m. Assume that the gas obeys the ideal gas law for the following values of temperature and pressure: (a) $T = 300$K, $p = 1.0$ atm; (b) $T = 5$K, $p = 1.0$ atm; (c) $T = 300$K, $p = 10^{-8}$ atm; (d) $T = 5$K, $p = 10^{-8}$ atm.

2. (I) For an ideal gas at 0°C and 1.0 atm, the ratio of the typical molecular radius (10^{-10} m) to the intermolecular spacing is about 1/20. What is this ratio in liquid water (H_2O, molecular weight 18 g/mol)?

3. (II) Assume that the carbon and hydrogen atoms that make up oil behave like spheres of diameter 10^{-8} cm, and that the fundamental molecular component of oil, CH_2, would take up a corresponding area of 10^{-15} cm^2 on a water surface. How many such components are there in 0.1 L of oil if the density of oil is 90 percent of the density of water? What would the area of a one-molecule-thick layer of this amount of oil on water be? (This type of experiment, said to have been performed by Benjamin Franklin, among others, is sometimes used to compute the size of molecules.)

19–2 Pressure and Molecular Motion

4. (I) A rubber ball, confined to move in the *x*-direction, bounces elastically back and forth between two walls. The ball, which has a mass of 85 g, moves at a speed of 11 m/s, and the spacing of the walls is such that the ball strikes the right-hand wall 20 times in a ten-second period. What is the average force on the right-hand wall due to the ball during that ten-second span?

5. (I) Grains of sand of mass 3×10^{-3} g each, fall from a height of 0.8 m on a sticky surface at a rate of 50 grains per second per cm^2. What pressure does this shower of sand exert on the surface, assuming that air resistance can be neglected?

6. (II) Use dimensional analysis to estimate the number of air molecules that strike the 12-in × 12-in screen of a television set during a 1-h program.

7. (II) One mol of a monatomic ideal gas is placed in a chamber under 5 atm pressure. The volume of the chamber is 5000 cm^3. (a) What is the internal energy of the gas? (b) What is the temperature of the gas? (c) Assuming that the mass of a molecule of the gas is 3.36×10^{-26} kg, what is the value of $\langle v^2 \rangle$ of a gas molecule? (d) What is the root-mean-square (rms) velocity?

8. (II) A cubic box of volume 0.080 m^3 contains helium gas at 0.85 atm pressure and 22°C. (a) How many total collisions do the molecules make with the walls per second? (b) What is the total internal energy of the gas? (c) If the temperature is doubled, what happens to the rate of collisions?

9. (II) Consider a gas constrained to move in a plane of area *A* rather than a volume in three dimensions. The pressure in that case is a force/unit length exerted at the edge. What is the relation between pressure and internal energy in this case?

10. (II) Consider a spherical satellite of mass *M* and radius *R*, moving with speed *v* through a dilute gas, the upper atmosphere. We shall take *v* to be much smaller than v_{rms} so that we can ignore the distribution in velocities of the molecules of the gas. Assuming that the gas molecules are stationary, and that the satellite is slowed down by collisions with the gas molecules, obtain an expression for the change of velocity of the satellite as a function of time in terms of *m*, the molecular mass, *n*, the number density of molecules in m^{-3}, and the characteristics of the satellite. [*Hint*: The area presented by the satellite to the stationary molecules is just the same as if the satellite were a disc of radius *R*, oriented perpendicular to the motion.]

11. (II) The apparatus shown in Fig. 19–16 is designed to demonstrate aspects of the kinetic theory of gases. It consists of a transparent cubical box, 20 cm on each side, containing 100 steel balls of diameter 5 mm. The density of steel is 7.8 g/cm^3. The bottom of the box vibrates so that the steel balls bounce around. The top of the box consists of a movable piston of mass 1 kg. What is the rms speed of the steel balls if the top of the box is in dynamic equilibrium with the "gas" of steel balls? Ignore gravity in your treatment of the motion of the steel balls. Is this approximation justified?

12. (II) When particles of mass 1.3×10^{-17} kg are suspended in a liquid at room temperature, they are observed to have an rms speed of 0.030 m/s. Use this to determine Avogadro's number. (You will need to look up the gas constant *R*.)

13. (II) The rms speed of galaxies in a large part of the visible universe is roughly 100 km/s. The number density of these galaxies is $3 \times 10^{-20}/ly^3$, and the average mass of a galaxy is 3×10^{41} kg. What is the pressure of a gas of such galaxies?

14. (II) Molecules are confined to move in a plane, for example, by putting them between two glass plates separated by a distance that is small compared to the mean free path. What is the relation between the pressure and the internal energy? What is the relation between internal energy and temperature for an ideal gas?

19–3 The Meaning of Temperature

15. (I) Consider the mole of helium gas with an internal energy of 3600 J that we discussed in Example 19–1. (a) What is the temperature of the gas? (b) If 1/1000 of this amount of gas had the same internal energy as the total gas, what would its temperature be? (c) What would the internal energy of 1/1000 of the original sample of gas be, at its original temperature?

16. (I) A gas is heated from 120K to 470K. What is the change in the rms speed?

17. (I) What is the average kinetic energy per molecule of a gas at room temperature, 293K? How fast would a baseball of mass 0.10 kg be moving to have this same kinetic energy?

18. (I) An ideal gas is contained in a vessel of volume 3.0×10^{-4} m^3 that is under a pressure of 1.2 atm. What is the internal energy of the gas?

19. (I) The temperature of hydrogen in a dilute plasma—a state of matter in which atoms are broken apart into electrons and nuclei—inside a nuclear fusion reactor needs to be on the order of 20 million K in order to initiate nuclear fusion. What is the rms speed of the hydrogen nuclei in the plasma?

20. (II) A mixture of nitrogen gas and sulfur dioxide gas is in equilibrium at 27°C. What is the rms speed of the sulfur dioxide molecules (SO_2, molecular weight 64 g/mol), and that of the nitrogen molecules (N_2, molecular weight 28 g/mol)?

21. (II) Calculate the rms speed at 300K of the principal components of air: O_2 molecules (molecular weight 32 g/mol), N_2 (molecular weight 28 g/mol), CO_2 (molecular weight 44 g/mol) and H_2 (molecular weight 2.0 g/mol).

22. (II) The rms speed of 1 mol of argon atoms (atomic weight 40 g/mol) in a box is 680 m/s. (a) What is the temperature inside the box? (b) What is the internal energy? (c) If the box has a volume of 10^{-3} m^3, what is the pressure? Treat the gas as ideal.

23. (II) Estimate v_{rms} for hydrogen atoms (a) on the surface of the Sun, where the temperature is 6000K; (b) on the surface of the Moon, where the temperature at one point is 150K.

▲ **FIGURE 19–16** Problem 11.

24. (II) A proton of kinetic energy 3×10^7 eV comes to a stop because of molecular collisions in a thin tube that contains 1.0×10^{-3} mol of oxygen at STP, as in Fig. 19–17. By how much is the temperature in the tube increased at equilibrium?

▲ FIGURE 19–17   Problem 24.

25. (II) A "gas" of water droplets 10^{-6} m in diameter is in equilibrium with air molecules at 300K. Given that the density of water is 1.0×10^3 kg/m³, what is the value of v_{rms} for the droplets?

26. (II) A relativistic gas is one in which a significant fraction of the constituents have speeds that are some finite fraction of the speed of light, $c \cong 3 \times 10^8$ m/s. Suppose that the rms speed of the constituents of a gas of atomic hydrogen is just 1% of c. What is the temperature? The energy that binds the electron and the proton together into a hydrogen atom is 2.18×10^{-18} J. Would this gas break down into a gas of electrons and protons? In the interior of stars, where temperatures comfortably exceed the temperature you have calculated, atoms cannot exist; they are broken into their components.

27. (II) If their kinetic energy greatly exceeds the average gravitational potential energy, galaxies in the sky may be viewed as an ideal gas. Given that the rms speed of galaxies is 100 km/s and the average mass is 3×10^{41} kg, what is the temperature of a gas of galaxies?

19–4   Probability Distributions

28. (I) A lottery that costs $1 to play has a jackpot of $10,000,000. The state that runs the lottery is obligated to tell the players that the probability of winning is one in 50,000,000. Is it a good idea from a financial point of view to play this lottery? Answer this question by calculating the ratio of your outlay to your revenue under the assumption that you play the game an unlimited number of times.

29. (I) Nine cars are measured at a given spot on a highway to have speeds v of 52.3, 54.5, 57.0, 57.2, 57.9, 63.6, 63.6, 68.1, and 82.2 mi/h. Calculate (a) the average speed, and (b) the rms speed.

30. (I) Take a deck of cards. Draw cards from that deck, one by one, without putting any cards back in the deck. What is the probability that the first card is a spade? What is the probability that the second one is also a spade? What is the probability that the third one is also a spade? (You may extend this procedure to calculate the probability that when you draw thirteen cards, they are all spades, by multiplying all the thirteen probabilities together).

31. (I) Four dice are tossed. What are the probabilities of getting (a) four sixes; (b) three sixes and a five?

32. (II) A cage of 50 Ping-Pong balls has air blowing through it to keep the Ping-Pong balls moving. Five Ping-Pong balls have the number 0 painted on them, five have the number 1, and so forth, finishing with five Ping-Pong balls painted with the number 9. The apparatus is used to choose the numbers for a lottery, and the balls are not returned to the cage after they are chosen. (a) The first drawing is for three balls. Your number is 186. Before any balls are drawn, what is the probability that you will win? (b) The first ball drawn is 1. What are your chances now? (c) The second ball drawn is 8. Now what is the probability that you will win?

33. (II) Consider the distribution of grades g discussed in Section 19–4. (a) Plot the distribution of the numbers D for the students, where D is defined as the deviation of the grade from the average. (D can be positive or negative.) (b) Calculate the average value of D^2. (c) Show that $\langle D \rangle = 0$ and $\langle D^2 \rangle = \langle g^2 \rangle - \langle g \rangle^2$.

34. (II) Find an expression for the probability of getting 13 spades in a bridge hand. (In bridge, the entire deck of 52 cards is dealt to four players.)

35. (II) In a true-or-false test, a correct answer is awarded $+1$ point and an incorrect answer is awarded -1 point. In a test involving 87 students, the following grade distribution is found (Fig. 19–18):

-50 to -30:	3 students;
-30 to -10:	18 students;
-10 to $+10$:	29 students;
$+10$ to $+30$:	22 students;
$+30$ to $+50$:	15 students.

(a) Calculate the average grade. (b) Calculate $\langle D^2 \rangle$, where D is defined in Problem 33. (c) Suppose that the distribution was such that every student has a grade between $+10$ and $+30$. With this information, what would the average be?

▲ FIGURE 19–18   Problem 35.

36. (II) Suppose that the distribution of grades (the number of test takers, N, with a grade x) in a national test is approximated by a continuous curve of the form (Fig. 19–19)

$$N(x) = 25x \qquad \text{for } 0 < x < 60$$
$$= 6000 - 75x \qquad \text{for } 60 < x < 80$$
$$= 0 \qquad \text{for } 80 < x < 100.$$

(a) What is the total number of students who took the test? (b) What is the average grade?

▲ FIGURE 19–19   Problem 36.

37. (III) A way to find the value of π (*Buffon's method*) is the following: Drop a needle of length L repeatedly onto a grid of parallel lines separated by a distance L (Fig. 19–20). Show that the probability of the needle falling on a line is $2/\pi$. [*Hint*: Take the lines as parallel to the x-axis. Calculate the average value of the component of the needle's length along the y-axis.]

▲ **FIGURE 19–20** Problem 37.

19–5 The Velocity Distribution of Gases

38. (I) What is the rms speed of a helium atom in the vapor above liquid helium at 20K? (The system is enclosed in a perfectly insulating bottle.)

39. (I) Oxygen atoms come in a variety of isotopes: ^{16}O, ^{17}O, and ^{18}O (with the superscript representing A). The latter two isotopes are present only in small quantities. A gas consists of diatomic molecules of oxygen (mostly $^{16}O_2$, with small admixtures of ^{16}O–^{17}O and ^{16}O–^{18}O). Calculate the ratios of the rms speeds for the latter two to that of the dominant (^{16}O–^{16}O) form. Does your answer depend on temperature?

40. (I) If the rms speed of molecules of gaseous H_2O is 200 m/s, what will be the rms speed of CO_2 molecules at the same temperature? Assume that both of these are an ideal gas.

41. (II) Calculate $\langle v \rangle$, the average speed for an ideal gas. In doing this calculation with the help of the velocity distribution $F(\vec{v})$, use the fact that for a spherically symmetric integrand, you may replace d^3v by $4\pi v^2\, dv$. You will need the integral

$$\int_0^\infty z^3 e^{-z^2}\, dz = \tfrac{1}{2}.$$

[*Hint*: This problem illustrates the use of dimensional analysis as an intermediate step.]

42. (II) Use the result of Problem 41 to find the numerical value of the average speed for an ideal gas of CO_2 at 550K, and compare it with v_{rms} at that temperature.

43. (III) Consider a gas being blown along at a velocity $\vec{u}$, so that its velocity distribution is given by

$$F(\vec{v}) = \frac{1}{Z} e^{-m(\vec{v}-\vec{u})^2/2kT}.$$

(a) Show that Z is the same as for the usual velocity distribution, as in Eq. (19–28). (b) Calculate $\langle \vec{v} \rangle$ and $\langle v^2 \rangle$ and interpret your results.

19–6 The Maxwell–Boltzmann Distribution

44. (I) The value of c'_p for carbon monoxide (CO) is 29.2 J/mol · K. What value do you expect for c'_V for the diatomic CO molecule?

45. (I) One mole of hydrogen gas is determined to have an internal energy of 1.87×10^4 J at 900K. How many degrees of freedom are there for the hydrogen molecule? Which degrees of freedom are likely to be available?

46. (I) A gas of N_2 molecules is in equilibrium at a temperature of 450K. Calculate the rms angular momentum $I\sqrt{\omega^2}$ for an N_2 molecule, given that the mass of a nitrogen atom is 2.33×10^{-26} kg and that the average separation of the two atoms in an N_2 molecule is approximately 0.70 nm (Fig. 19–21). (You may assume that the rotational—but not the vibrational—degrees of freedom participate in the equipartition theorem.)

▲ **FIGURE 19–21** Problem 46.

47. (II) The following data can be found in a table of thermodynamic properties of gases:

Argon (Ar): $c_p = 0.214$ cal/g · K; $c_V = 0.074$ cal/g · K

Oxygen (O_2): $c_p = 0.219$ cal/g · K; $c_V = 0.157$ cal/g · K

Water (H_2O): $c_p = 0.445$ cal/g · K; $c_V = 0.335$ cal/g · K

Carbon dioxide (CO_2): $c_p = 0.201$ cal/g · K;

$c_V = 0.156$ cal/g · K

What can you infer about the shape of the molecules from these data?

48. (II) Leaving out any contributions due to internal motion, what is the average molecular energy for a dilute gas of methane (see Fig. 19–7) at temperature T according to the Maxwell–Boltzmann distribution?

49. (II) A container has 1.3 mol of argon gas and 2.0 mol of nitrogen gas at room temperature. (a) What is the total internal energy of the gas? (b) What is the constant volume specific heat of the mixture?

50. (II) A gas of diatomic molecules is in equilibrium at a temperature of 350K. Assuming that the rotational inertia of the molecule [as used in Eq. (19–36)] is 2.4×10^{-40} kg · m², what is the rms angular speed of the molecules? What is the mean angular momentum of the molecules?

51. (III) Write down an integral for the probability that a molecule is found to have a velocity larger than 90 percent of v_{rms}. Show that the probability is independent of the temperature. Show that this is not the case if we ask for a velocity larger than some fixed number; for example, the escape velocity from a planet.

52. (III) A crystal can be thought of as a collection of N atoms arranged in a lattice, with all the atoms connected by springs (Fig. 19–22 see next page). Thus we can think of each atom as a mass at the end of three springs, aligned respectively along the x-, y-, and z-directions. Using this model of the lattice, count the number of degrees of freedom for each atom, then use the equipartition theorem to compute the molar heat capacity of the crystal at temperature T.

▲ FIGURE 19–22 Problem 52.

*19–7 Collisions and Transport Phenomena

53. (I) For a collision between two molecules of radii R_1 and R_2, respectively, show that the collision cross section is $\sigma = \pi(R_1 + R_2)^2$ (Fig. 19–23).

▲ FIGURE 19–23 Problem 53.

54. (I) Compute the pressure of a container of air molecules of diameter 2.5×10^{-10} m if the temperature is 273K and the mean free path is (a) 1 mm, (b) 1 μm. The gas should be treated as ideal.

55. (II) One can treat the flow of gas through a vacuum hose as if the gas were a fluid if the diameter of the hose is much larger than the mean free path. If, however, the mean free path is much larger than the diameter of the hose, the motion of the individual free molecules has to be considered. Consider a hose of diameter 1 cm. At what pressure is the mean free path of air at 300K shorter than 0.01 cm? Longer than 100 cm? Take the diameter of an air molecule to be 3×10^{-10} m.

56. (II) Using a reasonable value for the collision cross section or, more simply, a model of a gas in which the molecules are like tiny billiard balls, make an estimate of the density at which the mean free path becomes equal to the mean spacing between molecules.

57. (II) Ten tennis balls are rolling around randomly on a tennis court (Fig. 19–24). Estimate the mean free path for collisions.

▲ FIGURE 19–24 Problem 57.

58. (II) A student walks across a field that is 60 m × 40 m. Estimate the student's mean free path if there are (a) 10, (b) 100, (c) 300 other students on the field.

59. (II) A cubic box of sides 10 cm, contains helium atoms (diameter $\cong 1.0 \times 10^{-10}$ m) at a pressure of 3×10^4 Pa. If the mean free path in the box is 4×10^{-4} m, how many helium atoms are in the box? How many are there if the mean free path is 40 cm? What will the pressure be if the temperature is held fixed?

60. (III) According to one theory, planets are formed as a consequence of near collisions between stars. The number density of stars in our part of the galaxy is 10^{-51} m^{-3}. The velocity of the Sun relative to these stars can be taken to be 25 km/s, and the size of the stars is roughly 10^9 m. If we assume that a passage of two stars at a distance of 100 radii is close enough to set off the tidal forces that give rise to planet formation, what is the probability that the Sun would have acquired a planetary system in 10 billion yr?

General Problems

61. (I) Argon and methane (CH_4) gas are mixed together at 400K. What is the ratio of the rms speeds of the argon and methane atoms? What happens to this ratio if the temperature is halved?

62. (II) The density as well as the temperature of material in intergalactic space is rather uncertain. Suppose that the density and temperature were one hydrogen atom per cubic meter and 3K, respectively. (a) What is the pressure? (b) What is the mean free path if the diameter of a hydrogen atom is 10^{-10} m?

63. (II) The acceleration due to gravity on the surface of Mercury is 3.5 m/s^2, and Mercury's radius is 2.4×10^6 m. (a) What is the escape speed of a particle on Mercury? (b) Suppose that the atmosphere of Mercury were pure H_2 gas. What would the temperature be so that, if the molecules had the velocity distribution given by Eq. (19–28) for that temperature, the rms speed of the H_2 molecules matched the escape speed? Qualitatively, what is the effect on the temperature of the remaining gas? (c) If the temperature were less than the result of part (b), would there be a similar effect, and why or why not? (d) Suppose that Mercury's atmosphere has two or more molecular components. What happens to the composition of the atmosphere over time as a result of the effects discussed here?

64. (II) Some molecules (magnetic dipoles) carry a (vector) property $\vec{\mu}$ called the magnetic dipole moment. When such a dipole is placed in a magnetic field $\vec{B}$ it has an energy $-\vec{\mu} \cdot \vec{B}$. When a collection of N molecules with this property are placed in a box within which there is a magnetic field (meaning that at every point in the box the magnetic field has the magnitude and direction of the given $\vec{B}$), they each have an energy $-\vec{\mu} \cdot \vec{B}$. If the collection of molecules are at temperature T, what is the most likely orientation of $\vec{\mu}$? What is $P(30°)/P(60°)$, where $P(\theta)$ is the probability of finding one of them making an angle θ with the given magnetic field?

65. (II) In the quantum mechanical description of nature, electromagnetic radiation is carried in units called photons. These come with different frequencies. The equilibrium distribution of electromagnetic radiation in a cavity can be thought of as a collection of photons of different frequencies, appearing and disappearing as they are emitted or absorbed by the walls. For a given cavity volume and temperature, the probability of finding n photons of a given frequency f is $P_f(n) = x^n(1 - x)$, where $0 < x < 1$. (a) Is $P_f(n)$ normalized correctly? (b) What is $\langle n \rangle$, the average number of photons of frequency f? (c) What is the fractional spread $\dfrac{\sqrt{\langle n^2 \rangle - \langle n \rangle^2}}{\langle n \rangle}$? Does this ever go to zero?

66. (II) The surface of the Sun contains both hydrogen and helium atoms at 6000K. (a) What is the average kinetic energy of each type of atom? (b) What is the rms speed of each? (c) Which is most likely to escape from the Sun?

67. (II) The regions in which the beams of accelerated particles circulate in accelerators have high vacuums (10^{-9} torr) to allow the particles to circulate freely. (a) What are the mean free paths of air molecules inside the accelerator at room temperature? (b) By how much does the mean free path decrease if the pressure inside increases to 10^{-6} torr?

68. (II) Suppose that molecules were confined to move in a plane. Calculate the velocity distribution $F(\vec{v})$, assuming that it is of the general form $(1/Z)e^{-mv^2/2kT}$. In other words, calculate Z so that $\int_{\text{surface}} F(v)\, d^2v = 1$.

69. (II) Brownian motion is the random walk followed by individual molecules as well as by larger objects such as smoke particles. The Brownian motion of smoke particles in a room can be observed with the aid of a microscope. If the mass of the smoke particles is typically 3×10^{-15} kg, what is the rms speed of the particles at room temperature?

70. (II) Peter and Paul are standing in opposite corners of a large room, 14 m apart. The air in the room is still. Peter holds a small vial full of H_2S while Paul holds a vial containing Cl_2. They open their vials at the same time (Fig. 19–25). Paul smells H_2S after 3.5 min. When will Peter smell the Cl_2?

▲ **FIGURE 19–25** Problem 70.

71. (II) Calculate an expression for the pressure of a gas of massless particles, using the following facts: Massless particles always move with the speed of light, so that $v = c$; the energy of a massless particle, E, is related to its momentum magnitude P by $E = Pc$. Use this expression to show that for a gas of massless particles, $pV = U/3$. This result is applicable to the blackbody radiation discussed in Chapter 17.

72. (II) The relation $pV = 2U/3$ is correct only for a monatomic gas. Show that the more general relation is $pV = (\gamma - 1)U$, where $\gamma = c'_p/c'_V$.

73. (II) Consider a tank of volume 0.30 m^3 containing 2.5 mol of helium gas at 20°C. If we add 1.0 mol of O_2 at 0°C to the helium and let the whole system come to equilibrium, what will the equilibrium temperature be? What will the pressure be?

74. (II) Assume that an H_2O molecule has three rotational degrees of freedom in addition to its three linear degrees of freedom (Fig. 19–26). Calculate the molar heat capacity of its vapor. Compare this with the experimental value at low pressures of 0.48 cal/g·K. What role might vibrational motion play?

▲ **FIGURE 19–26** Problem 74.

75. (II) A vessel is pumped out to a high vacuum and then heated to get rid of water vapor. It is then filled with dry helium and pumped out again to a very high vacuum of 10^{-11} torr, so that it now contains only helium atoms at 300K. Helium atoms have a diameter of about 10^{-10} m. (a) What is the density of helium atoms? (b) the mean collision time? (c) the mean free path?

76. (II) The *law of partial pressures* states that the total pressure of a mixture of gases is the sum of the pressures that each gas would exert if it were present alone in the same amount as in the mixture. Use the kinetic theory to demonstrate this fact. Could this law be true if the gases were not ideal?

77. (III) The escape speed of a particle from Earth is 11 km/s. Assuming that the oxygen molecules at Earth's surface are at a temperature of 300K, how would you use the velocity distribution to calculate the probability that a particular molecule at the surface would escape Earth's gravitational attraction if the molecule were to suffer no further molecular collisions? Express this probability in the form of an integral, and manipulate the integral into a form that involves the ratio $v_{\text{rms}}/v_{\text{escape}}$, but do not attempt to perform the integration. [*Hint*: This problem is analogous to finding the probability that a given student in a class has a grade of 35 or more on a test.]

◄ The engines that power racing cars such as these are highly sophisticated, a world away from the first steam engines of the industrial revolution. Yet both types of engines are subject to fundamental limitations on their possible efficiency.

The Second Law of Thermodynamics

The first law of thermodynamics does not specify the direction of thermal-energy transfer. Nevertheless, we know from experience that the spontaneous transfer of thermal energy always proceeds from a hotter system to a cooler one. For example, if you put ice cubes in hot water, the ice cubes melt and the water cools off—nothing would be more surprising than to see the water becoming hotter and the ice cubes colder. This directionality in nature is enshrined in the second law of thermodynamics. The directionality of the spontaneous transfer of thermal energy can be understood from a molecular point of view by using the notions of probability developed in Chapter 19. These notions will allow us to develop the idea of order in a thermal system and to see that a spontaneous process proceeds from a more ordered to a less ordered state.

The second law has an important practical consequence: An engine, which is a device that takes thermal energy and uses it to do work, has a limit to how much of the thermal energy can be used in this way. (This limit is one that goes beyond the limits imposed by the first law of thermodynamics, which states that energy is conserved.) In this chapter we'll explore the meaning of the limit of how much thermal energy can be used to do work. What does this limit have to do with the direction of spontaneous processes? The answer to this question is provided by the concept of entropy, a new state variable for thermal systems, which quantifies the degree of order in thermal systems in thermodynamic terms.

20–1 Beyond Energy Conservation

The first law of thermodynamics extends the conservation of energy to thermal systems. We saw in Chapter 18 that we can construct engines, which are cyclic systems that do work on their surroundings—they may lift weights, or make vehicles go forward—by using thermal energy from thermal reservoirs. (These reservoirs may quite literally be heat baths, or they come in the form of hot gases that have been brought to high temperatures by burning fuel.) However, the first law cannot explain certain facts about thermal systems. For example, it is a fact that it is impossible to devise an engine that extracts all the thermal energy from a thermal reservoir and turns that energy into work. If we had such an engine, we could use it to power New York City—run its subway, light all its buildings, and provide heat in winter—while doing nothing more than cooling the Atlantic Ocean a little. Yet the existence of such an engine is completely consistent with the first law of thermodynamics.

There is a second set of facts about thermal systems that looks superficially different and that is also not explained by the first law. An example is the fact that when ice cubes are put into hot tea, the ice cubes melt and the tea cools. The first law of thermodynamics requires only that the heat flow into the ice cubes match the heat flow out of the hot tea, and heat flow could proceed in either direction. Nevertheless, lukewarm tea never *spontaneously* turns into hot tea with a couple of ice cubes floating in it. Another example is the observation that gas released into an evacuated chamber through a small opening spontaneously spreads throughout the available space—free expansion (Sections 18–1 and 18–7). We never see the gas spontaneously concentrating in a corner of the chamber, leaving most of the chamber evacuated.

These experimental phenomena can be summarized as follows:

1. *Not all the thermal energy in a thermal system is available to do work.*

2. *Thermal systems spontaneously change only in certain ways, and in particular, spontaneous heat flow always goes from a body at higher temperature to a body at lower temperature.*

These facts are expressions of the **second law of thermodynamics**. We will look in more detail at the second law and how it applies to point 1 in the next section, and how it applies to point 2 in Section 20–5. We can nevertheless make a few pertinent remarks that lean on the microscopic picture of thermal behavior, in which molecules undergo random motion. It is this chaos that lies behind the second law. The only way to get all the thermal energy out of a gas in a cylinder would be to get all the molecules moving in the same direction such that they can collide with and lift a weight together. But this is a very unlikely occurrence in a thermal system made of so many constituents. Similarly, the directionality of spontaneous processes has to do with probability—as we shall see, the directionality is determined by what configuration of molecules is the most probable one. Keep this very rough set of ideas in the back of your mind as you go through this chapter.

Engines and Refrigerators

In Sections 18–5 and 18–6 we described engines in terms of a set of thermal transformations that form a cycle. We emphasized that even though over one cycle the thermal system ends up where it starts, net work is done on the surroundings. To do this work, the engine has to get energy from someplace, and this comes in the form of thermal energy extracted from the equivalent of thermal reservoirs. An engine is thus a device that cyclically transforms thermal energy into mechanical energy. For example, in an automobile engine the thermal energy of burning fuel is converted into the energy of the automobile's motion. So is an automobile engine really like the abstract-sounding set of transformations we described in Chapter 18? The answer is yes. You can trace a set of transformations that describe the operation of an automobile engine on a p–V diagram—the engine does consist of cylinders in which gases expand against pistons and so forth. There is one difference between a real automobile engine and the systems

▲ **FIGURE 20–1** Abstract representation of an engine, which takes energy—here represented by a system at high temperature—to perform work. We shall learn in this chapter that it is not possible to convert thermal energy entirely into work, so there is some thermal energy left over, which is "dumped" at cooler temperatures.

we described: The automobile engine, even though it is well lubricated, has friction in it, and there is some irreversibility in its operation—some energy is lost this way. The engines we'll deal with here, at least to start, are idealized in that they are described by a set of reversible transformations, but below we will certainly address the issue of friction.

There is one feature of the cycles that we dealt with in Chapter 18 that we did not emphasize very much: There are parts of the cycle where there is positive heat flow *into* the system, and other parts of the cycle where there is positive heat flow *out of* the system. In every engine the conversion of thermal energy into mechanical energy is accompanied by the emission of exhaust gases, which carry off some thermal energy. This means that only part of the thermal energy of the burning fuel is converted into mechanical energy (Fig. 20–1).

We can restate the points above by recognizing that a useful engine exhibits two key features:

1. *An engine must work in cycles if it is to be useful.* We can understand this best by considering why a *noncyclic* operation is less useful. Consider an ideal gas confined to a cylinder with a movable piston that undergoes isothermal expansion. It is held at constant temperature through contact with a thermal reservoir; for example, it could literally be sitting in a large container of hot water. If there is a mass on top of the upward-oriented piston, the gas does work as it expands. To satisfy the first law, there must be a heat flow from the reservoir to the gas, and this means that thermal energy from the reservoir is converted into mechanical energy. But is this system an engine? It is not, because the process is not cyclic. The volume of the cylinder cannot increase indefinitely, because the pressure, which is inversely proportional to the volume, will decrease until at some point the expanding gas will no longer be able to push the piston. You will have an ineffectual container of gas that takes up a lot of room! To have a functioning engine, *the gas must be recompressed and the piston restored to its original position, ready to do work again.*

2. *A cyclic engine must include more than one thermal reservoir.* To understand this, consider the cylinder and piston again. If the piston were returned to its original position while the cylinder remained in thermal contact with the original thermal reservoir, then all the work that the gas did while it expanded would be needed to recompress the gas. In this scenario the original expansion is reversed, and nothing has been gained. In Chapter 18 we learned that the area enclosed by a closed path on a *p–V* diagram is the work done during the cycle. In this case the area enclosed is

▲ **FIGURE 20–2** A cycle in which a gas expands isothermally and is then compressed isothermally at the same temperature. The work done is zero.

zero, so the net work done by the system is also zero (Fig. 20–2). For finite work to be done over a cycle, thermal contact with the original thermal reservoir must be broken, and *temperatures other than the temperature of the original reservoir must come into play.* Here we would want a second large container of water, this time with cold water, so that the gas is compressed at a temperature that is lower than the temperature at which it expanded. If we do this, the pressure will be lower, and less work will be needed for the compression than was produced in the expansion. You will have done net work, and you will be back where you started, ready to repeat the operation. Note that in the process thermal energy is taken from the hot reservoir and dumped to the colder reservoir. We'll come back to this very important theme later in the chapter.

A young engineer, Sadi Carnot, gave an early but perfectly correct statement of the second law of thermodynamics in 1824. Working during the Industrial Revolution, he was interested in the very practical problem of improving the efficiency of steam engines. He found that there are limits to the possible efficiency of any engine. (It is interesting that his discovery of the second law was made even before the first law was understood!) Accordingly, we now look at engine efficiency.

Efficiency: Engines have an **efficiency**, which is a measure of what fraction of the heat flow from the hotter thermal reservoir (such as a burning fuel) is converted into work (Fig. 20–1). If the work done in one complete cycle is W, then we define the efficiency, η, as the ratio of the work done to the total positive heat flow supplied by the burning fuel, Q_h:

$$\eta \equiv \frac{W}{Q_h}. \tag{20–1}$$

DEFINITION OF ENGINE EFFICIENCY

According to the *first* law of thermodynamics, this efficiency can run from 0 to 1. When the efficiency is 1, or 100%, all the thermal energy taken from the thermal reservoir is converted to mechanical work; this is allowed by the conservation of energy (the first law). We will see, however, that an efficiency of 100% is impossible according to the second law. Figure 20–3 illustrates how the efficiency is calculated. This is an important quantity to be able to calculate, and we'll illustrate it by example in several places below.

Finally let us turn to refrigerators, which at first sight might seem to have nothing to do with engines. In fact, they are very closely related: A refrigerator is an engine run in reverse. Since the path on a *p–V* diagram of an idealized engine is reversible, we can indeed run it in reverse. Work will be done on the system by the surroundings, and thermal energy taken from a colder reservoir and released to a hotter reservoir. The motor in a refrigerator is even called a compressor; it compresses the gas of the idealized version

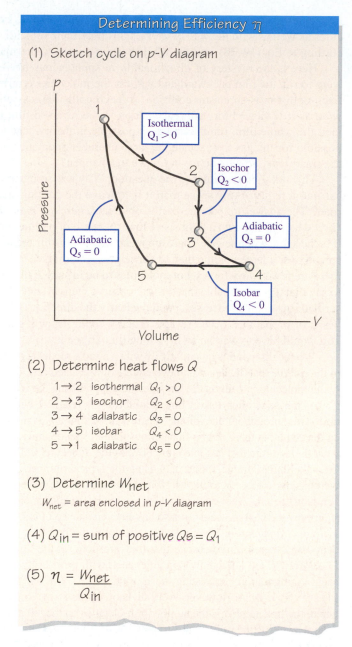

Determining Efficiency η

(1) Sketch cycle on p-V diagram

p

Pressure

1 | Isothermal $Q_1 > 0$

2 | Isochor $Q_2 < 0$

Adiabatic $Q_3 = 0$

Adiabatic $Q_5 = 0$

3

5 | 4

Isobar $Q_4 < 0$

Volume

V

(2) Determine heat flows Q

$1 \rightarrow 2$ isothermal $Q_1 > 0$
$2 \rightarrow 3$ isochor $Q_2 < 0$
$3 \rightarrow 4$ adiabatic $Q_3 = 0$
$4 \rightarrow 5$ isobar $Q_4 < 0$
$5 \rightarrow 1$ adiabatic $Q_5 = 0$

(3) Determine W_{net}

W_{net} = area enclosed in p-V diagram

(4) Q_{in} = sum of positive Qs = Q_1

(5) $\eta = \dfrac{W_{net}}{Q_{in}}$

◀ **FIGURE 20–3** An illustration of the technique for calculating efficiency.

of the refrigerator and thereby does work on the system. The colder reservoir is the ice-box—by removing thermal energy from it, you make it still colder or maintain its temperature when your little brother stands in front of the open door. Where is the hotter reservoir to which thermal energy is added? It is in the coils at the back of the refrigerator, and it ends up warming the air in the room in which the refrigerator sits.

Spontaneous Processes

The directionality of spontaneous processes is illustrated by the two examples we have mentioned—hot tea with ice cubes, and free expansion. Let's concentrate on free expansion, which we can discuss directly in terms of molecules. When a valve is opened and gas escapes from a small tank into a large, empty chamber, the gas spreads and fills the chamber until the density and temperature are uniform throughout that chamber. Now imagine a film of this process made on a scale in which individual molecules can be seen. The molecules move rapidly and collide frequently in the small tank. When the valve is opened, molecules that reach the opening escape. They enter a vast space, empty of molecules, and the first batch of molecules encounter nothing in their way. Their first collisions will be with the walls of the chamber. As more and more molecules escape into the large chamber, they will begin to collide with other molecules as well as with the walls. After a short time their velocities will be distributed according to Eq. (19–28), and the whole system will be in thermal equilibrium.

Here is the mystery of directionality in spontaneous free expansion. Suppose we were to run the film backward and witness the molecules returning to the smaller tank. Each collision is an elastic collision, so each collision is as convincing viewed backward as it is viewed forward; each action in the reversed film is perfectly possible. In fact, the whole film, run backward, is possible in theory, and so in theory it should be possible for the gas to evacuate the large chamber spontaneously and concentrate into the small tank. So why do we never see this happen?

This question is similar to the question, Why it is so much harder to parallel park a car into a tight parking space than it is to leave the parking space and get back on the road? The answer to that question is obvious: *There are many fewer ways of getting it right* when pulling into a tight space than there are of pulling out into a wide boulevard. It is the same with the gas. There are many more ways for the molecules of the gas to be spread over a large volume than to concentrate themselves into a small one. Saying that there are fewer ways for a set of molecules to be in state *B* than in state *A* is another way to say that it is improbable that we will observe them in state *B*. In this way, we have described irreversible processes, or equivalently the direction of spontaneous processes, in terms of probabilities. The reverse process could happen, but it is so improbable that you would have to wait an awfully long time to see it.

We can discuss this quantitatively with a simple model. Consider a box with a partition in the middle that divides it into two equal volumes labeled *L* (left) and *R* (right). If initially the molecules are all on the left side, then upon lifting the partition, molecules can move to the right, and after a short time equilibrium is reached, with the molecules distributed uniformly throughout the volume. There will be roughly as many molecules on the left as on the right, but there can be fluctuations that we can quantify. We could ask, for example, what the odds are that at some point 80% of the molecules will be on one side or the other. To make things simple, rather than using the typical number of order of $N_A = 6 \times 10^{23}$, we consider a gas consisting of 10 molecules. Any one molecule has an equal chance of being on the left (L) or on the right (R)—that is, the chance of being in, say, L is 1/2. Consider now the next molecule: it also has an even chance of being on the left. We now make a *physical assumption*: The molecules are not tied together in any way—they are independent—and therefore the chance that they are both on the left is $(1/2) \times (1/2)$. In fact *LL,LR,RL,RR* are all equally probable, each with probability $(1/2)^2$. With 10 molecules each configuration, consisting of a total of 10 L's and R's, has a probability of $(1/2)^{10} = 1/1024$. However, every division of the form $(10,0)$, $(9,1)$, $(8,2)$, $(7,3)\ldots$, where the first number is the number of molecules on the left and the second is the number of molecules on the right, is not equally likely. That is because the number of ways of getting them is different. Let us work out the number of ways of getting the configurations $(10,0)$, $(9,1)$, $(8,2)$, and so on. The number of ways in which we can have all 10 molecules on the left and none on the right is just one. It consists of the assignment $LLL \ldots L$ (10 terms). If we ask for the number of ways of having 9 on the left and 1 on the right, we can pick any one of the 10 molecules to go to the right, and therefore there are 10 *different* ways of having the division $(9,1)$. For the division of $(8,2)$ we have 10 ways of picking one molecule to go on the right, and then, since there are only 9 left, 9 ways of picking the second one. That would give us 10×9 ways. However, with this way of figuring things, we have overcounted. We have counted as different possibilities two different orders for choosing the two molecules: M_1 first and M_2 second, and the other way around. This is overcounting because we are interested only in the fact that there are 2 molecules on the right, and not in the order of their arrival. Thus we must divide by 2 for the 2 orders, and the number of ways of getting $(8,2)$ is not (10×9) but rather $(10 \times 9)/2 = 45$. For the order $(7,3)$, we can pick out the three molecules in sequence: there are 10 ways of picking the first, 9 ways of picking the second, and 8 ways of picking the third. Here again we have overcounting. If we label the molecules with 1,2,3, then they are counted as arriving in all the possible 6 sequences: (123), (132), (231), (213), (312), (321). We must therefore divide by 6, and we get $(10 \times 9 \times 8)/6 = 120$ for the $(7,3)$ division. It is not very hard to see what the whole picture yields [Fig. 20–4]:

▲ **FIGURE 20–4** The number of ways, and hence the relative probability, that in a collection of 10 molecules the configuration in a box is (# on left, # on right). For example, there are 120 ways to get 7 on the left and 3 on the right. We are assuming that the "choice" of whether a molecule goes left (or right) is independent of the other molecules.

Division	# of ways	Probability P
(10,0) or (0,10)	1	1/1024
(9,1) or (1,9)	10	10/1024
(8,2) or (2,8)	$(10 \times 9)/(2 \times 1) = 45$	45/1024
(7,3) or (3,7)	$(10 \times 9 \times 8)/(3 \times 2 \times 1) = 120$	120/1024
(6,4) or (4,6)	$(10 \times 9 \times 8 \times 7)/(4 \times 3 \times 2 \times 1) = 210$	210/1024
(5,5)	$(10 \times 9 \times 8 \times 7 \times 6)/(5 \times 4 \times 3 \times 2 \times 1) = 252$	252/1024

The probability results above show that with even as few as 10 molecules, the odds of equal division with a fluctuation of 1 [i.e. $(5, 5) + (6, 4) + (4, 6)$] compared with the outer extreme with a fluctuation of 1 [i.e. $(10, 0) + (9, 1) + (0, 10) + (1, 9)$] are in the ratio of $(252 + 210 + 210)$ to $(1 + 10 + 10 + 1) = 672/22 = 30.1$ to 1. There are many more ways of finding nearly equal numbers on both sides than there are ways of finding almost all the molecules on one side or the other. When we get to numbers as large as 10^{24} molecules, this effect is much bigger. In fact, the ratio of the odds becomes more like 10^{12} than 30.1. This means that once equilibrium is established, there is effectively no going back.

Our simple model suggests that the directionality of spontaneous processes is determined by what takes you to the state with the highest probability. There may be subsequent fluctuations around the most probable state, but they are relatively small when the numbers of molecules is high. The irreversibility of a spontaneous process such as free expansion is not guaranteed but it is an almost sure bet.

EXAMPLE 20–1 Three dice are thrown. Assuming honest dice, what is the probability of getting a total of 3? What is the probability of getting a total of 10?

Strategy When it lands, each die gives 6 possible outcomes, and each outcome is equally likely. Thus the probability of any one given outcome on one of the dice is 1/6. Since the three dice are totally independent, any single outcome has probability $(1/6)^3 = 1/216$. (Equivalently, there are 216 different outcomes for the set of three dice.) The difference between the three dice giving a total of three and a total of 10 lies in the number of ways of getting these totals with three dice. The simplest way of getting at these numbers is straight enumeration.

Working It Out There is only one way of getting a total of 3, and that is by the dice reading (111). This means that the probability of getting a total of 3 is 1/216.

Let's now enumerate ways of getting a total of 10: We list the triplets that can add up to 10 as follows: (136), (145), (226), (235), (244), (334). For an entry in this list in which all three digits differ, there are six different ways to get them. For example, if square brackets mean that we are keeping track of the order, we can get (136) as [136], [163], [316], [361], [613], or [631]. When two of the integers are the same, then there are only three ways of getting them—for example, (226), by the throws [226], [262], or [622]. Thus in the list of the number of ways to get 10 we have $6 + 6 + 3 + 6 + 3 + 3 = 27$ different throws. Thus the probability of getting 10 is 27/216, 27 times more probable than the probability of a 3.

What Do You Think? If you deal with the triplet (136), do you really have to list all six possibilities to convince yourself that there are six? *Answers to **What Do You Think?** questions are given in the back of the book.*

20–2 The Second Law of Thermodynamics

Here we'll give a precise formulation of the second law of thermodynamics, using some of the facts we gathered in the preceding section. The second law was discovered through the study of engines, and we shall first state the law in that context. There are in fact two equivalent formulations of the second law of thermodynamics associated with engines, and both are useful in understanding the conversion of thermal energy to mechanical energy. They are:

The Kelvin form. It is impossible to construct a cyclic engine that converts thermal energy from a body into an equivalent amount of mechanical work without a further change in its surroundings.

For example, there is no way to extract thermal energy from the ocean and use that energy to run an electrical generator without further effect on the surroundings of the system, such as a heating of the atmosphere. In the example discussed in Section 20–1 (a gas expanding against a piston without recompression), thermal energy from a thermal reservoir is indeed converted into work, but the engine is not cyclic. In an engine that takes thermal energy from a thermal reservoir cyclically and converts that energy to work, the additional effect is typically a heat flow from the engine into its lower-temperature surroundings, as in point 2 (page 574) of Section 20–1. This brings us to the second formulation of the second law of thermodynamics:

The Clausius form.† It is impossible to construct a cyclic engine whose only effect is to transfer thermal energy from a colder body to a hotter body.

In other words, there is no cyclic engine that can start with a pot of cold water, make it still colder by taking thermal energy from it, and use that energy to warm up a second pot of water *with no further effect*. In this case, the further effect is associated with the fact that the apparatus *must* take some mechanical energy from its surroundings (such as an electric motor could provide).

We will use mainly the Kelvin form, which was first stated by Lord Kelvin around 1850. Detailed analysis, however, shows that the two formulations of the second law of thermodynamics are entirely equivalent.

THINK ABOUT THIS . . .

HOW DO PERPETUAL MOTION MACHINES WORK?

The search for a *perpetual motion machine* has stimulated thought for centuries. The simplest version of such a machine is an engine that does more work than the energy that goes into it. (Analysis of a seemingly attractive example of such a machine can be found in Problem 16–67.) The discovery of the first law of thermodynamics, and the underlying law of conservation of energy, explained why no such machine (a "perpetual motion machine of the first kind") ever worked. Such knowledge has not stopped "inventors" from trying to violate this law, though the U.S. Patent Office no longer even considers granting patents on such contraptions. Inventors also attempt to construct machines that do not violate energy conservation but have the property that they take energy from a reservoir at a lower temperature and transfer it to one at a higher temperature, without requiring further energy input. Such a machine could take in thermal energy from the ocean and, for free, use it to heat a city on the shore. This is a very attractive idea. Unfortunately, it violates the second law of thermodynamics. It is this law that forbids the construction of this type of machine, also known as a "perpetual motion machine of the second kind."

We can construct wind-power generators that extract energy from the wind and convert it to useful power. The second law is followed in this case because not *all* of the energy in the air flux that hits a windmill can be converted to work in the form of the turning blades. In other words, a windmill does not have a 100% efficiency. Effectively the wind that flows in acts as a hot reservoir, and the outgoing air, which has been stirred up by its passage past the blades, acts as a cold reservoir. The role of such reservoirs in the operation of the second law is described in Section 20–3.

---

†Rudolf Clausius is best known for introducing the concept of entropy.

20–3 The Carnot Cycle

We argued earlier that thermal reservoirs at two (or more) different temperatures are necessary for the operation of a reversible thermodynamic cycle that performs work; that is, a reversible engine. A minimal version of such an engine—one that requires exactly two such reservoirs—at temperatures T_h and T_c, respectively, was invented by Sadi Carnot in 1824. The reversible cycle followed by this engine—known as a **Carnot cycle**—is of great importance in thermodynamics because it is the most efficient engine one can construct that operates between any two temperature extremes. While the Carnot cycle is not used for real engines in common use, some real engines at least approximately match it. By learning the efficiency of the Carnot cycle, we learn the best we can hope to accomplish in engine efficiency.

The Carnot cycle, demonstrated in Fig. 20–5 with a gas-filled cylinder and a piston, consists of four reversible steps. Figure 20–5a is a p–V diagram of these four steps, and Fig. 20–5b depicts the cycle schematically. Step I (AB in Fig. 20–5a) is an isothermal expansion: The expansion takes place with gas in the cylinder in thermal contact with a thermal reservoir—say, a kitchen hot plate—at temperature T_h. In step II (BC) the cylinder of gas has been insulated from thermal contact with its surroundings so that there is no heat flow ($\Delta Q = 0$; the transformation is adiabatic). The volume slowly expands further, while the pressure as well as the temperature of the gas decrease. The temperature of the gas is monitored throughout this adiabatic expansion. When the temperature has dropped to T_c (the temperature of a second, colder, thermal reservoir—for example, a bathtub full of cold water), it is placed in thermal contact with this second reservoir (step III). During this stage (CD), an isothermal compression of the gas at T_c

(a) **(b)**

▲ **FIGURE 20–5** (a) The Carnot cycle, which consists of two adiabatic and two isothermal transformations. (b) How the cycle might be realized.

takes place. The final pressure of step III is determined such that when the gas is removed from contact with the bathtub and again placed in thermal isolation, the adiabatic compression of step IV (DA) increases the temperature and pressure of the gas until it returns to the starting point, A.

In Fig. 20–5a the isothermal legs are drawn as hyperbolas, and the adiabatic legs follow the curve $p = $ (a constant) $V^{-\gamma}$ for $\gamma > 1$—as is appropriate for an ideal gas (see Chapter 18). However, the working fluid does not have to be an ideal gas.[†] *Any gas or other compressible fluid is suitable, as isothermal or adiabatic changes are possible for any substance. The crucial features of the Carnot cycle are independent of whether the working fluid is an ideal gas or not.*

The work W done *by* the gas in tracing out a complete Carnot cycle is the area enclosed by the curve in the p–V diagram. The cycle shown is clockwise, so W is positive. There is also heat flow associated with the Carnot cycle, although not during steps II and IV, which are by definition adiabatic. During step I, there is a positive heat flow Q_h from the hotter thermal reservoir into the thermal system. (For an ideal gas, $\Delta U = 0$ during an isothermal process, and then $Q_h = W_{A \to B}$, the positive work done by the gas in that expansion.) Similarly, there is a negative heat flow $-Q_c$ from the colder thermal reservoir to the system during step III. In other words, there is a positive heat flow Q_c from the thermal system to the colder reservoir. Thus the Carnot engine absorbs a heat flow Q_h during step I, gives up a heat flow Q_c during step III, and does a total amount of mechanical work W. (This is shown schematically in Fig. 20–1, without making precise the steps.) The first law of thermodynamics can be applied to the entire cycle. Because the engine returns to its original state, the internal energy is unchanged, a statement we could make even if the working fluid is not an ideal gas. Then, according to the first law,

$$0 = W - Q_h + Q_c. \tag{20–2}$$

Equation (20–2) allows us to find the efficiency of the Carnot cycle, $\eta_C = W/Q_h$:

$$\eta_C = \frac{Q_h - Q_c}{Q_h} = 1 - \frac{Q_c}{Q_h}. \tag{20–3}$$

When the working gas of the Carnot cycle is ideal, the efficiency can be calculated in terms of the temperatures of the thermal reservoirs:

$$\eta_C = 1 - \frac{T_c}{T_h}. \tag{20–4}$$

This result follows from an analysis (see below) of the Carnot cycle with an ideal working gas. The analysis makes use of the curves that describe both the adiabatic and isothermal curves for an ideal gas (see Chapter 18). Before we go on we should remark here that Eq. (20–4) is more general than it first appears to be: It will be seen to be the efficiency of the Carnot cycle with *any* working fluid. Anticipating this, we can say that for a Carnot engine to be efficient, the temperatures of the two reservoirs should be quite different from one another, and the temperature of the cold reservoir should be as low as possible. If T_c were to be zero, the efficiency would be 1 (100 percent), independent of T_h. This is the only situation that could give an efficiency of 1. Such a reservoir does not exist, and thus *a perfectly efficient Carnot engine is not possible*. Since we shall see that the Carnot engine is the most efficient engine operating between two temperature extremes, this means that we can never construct a perfectly efficient engine.

How to Find the Efficiency of the Ideal Gas Carnot Engine

Equation (20–4) is an important one. To derive it, we recall that for an ideal gas, the isothermal and adiabatic curves are

for an isothermal transformation: $pV = nRT = $ a constant,

for an adiabatic transformation: $pV^\gamma = $ a constant,

where the quantity $\gamma = C_p/C_V$ is greater than 1.

[†] The "working fluid" is the material that expands and contracts, allowing the engine to do work.

Refer now to Fig. 20–5a. For an ideal gas the thermal energy is a function of temperature only, so the thermal energy, U, is constant during the isothermal steps (I and III). For these steps, $\Delta U = 0$, and the first law of thermodynamics, $\Delta U = Q - W$, implies that

$$\text{for step I: } Q_h = W_{A \to B},$$

and

$$\text{for step III: } Q_c = -W_{C \to D}.$$

There is a minus sign in the second equation because we have defined Q_c as the heat flow *from* the thermal system to the cold reservoir, whereas in the first law of thermodynamics the heat flow is always *to* the thermal system.

The work done by the gas during the isothermal steps is

$$W_{A \to B} = \int_{V_A}^{V_B} p\,dV = \int_{V_A}^{V_B} \frac{nRT_h}{V}\,dV = nRT_h \ln\left(\frac{V_B}{V_A}\right);$$

$$W_{C \to D} = \int_{V_C}^{V_D} p\,dV = \int_{V_C}^{V_D} \frac{nRT_c}{V}\,dV = nRT_c \ln\left(\frac{V_D}{V_C}\right).$$

Using these results, we find the efficiency of the cycle:

$$\eta_C = 1 - \frac{Q_c}{Q_h} = 1 - \frac{-W_{C \to D}}{W_{A \to B}} = 1 + \frac{nRT_c \ \ln(V_D/V_C)}{nRT_h \ \ln(V_B/V_A)}$$

$$= 1 - \frac{T_c \ \ln(V_C/V_D)}{T_h \ \ln(V_B/V_A)}. \tag{20–5}$$

We have not yet used any information about the shape of the adiabatic curves in this expression. This information will relate the various volumes. Points B and C lie on the same adiabatic curve, so $p_B V_B^\gamma = p_C V_C^\gamma$. Similarly, for points A and D, $p_A V_A^\gamma = p_D V_D^\gamma$. The ratio of these two equations gives

$$\frac{p_B V_B^\gamma}{p_A V_A^\gamma} = \frac{p_C V_C^\gamma}{p_D V_D^\gamma}. \tag{20–6}$$

Next, we substitute for the ratio of the pressures on each side of this equation. Because A and B lie on an isotherm, $\dfrac{p_B}{p_A} = \dfrac{nRT_h/V_B}{nRT_h/V_A} = \dfrac{V_A}{V_B}$, and similarly, because C and D also lie on an isotherm, $\dfrac{p_C}{p_D} = \dfrac{nRT_c/V_C}{nRT_c/V_D} = \dfrac{V_D}{V_C}$. We insert these ratios into Eq. (20–6) to find that

$$\frac{V_B^{\gamma-1}}{V_A^{\gamma-1}} = \frac{V_C^{\gamma-1}}{V_D^{\gamma-1}}.$$

The same power, $\gamma - 1$, appears on both sides of this equation, so the arguments must be equal:

$$\frac{V_B}{V_A} = \frac{V_C}{V_D}.$$

This result considerably simplifies the expression for efficiency. The arguments of the logarithms in Eq. (20–5) are the same, so the logarithms cancel and we finish with Eq. (20–4).

EXAMPLE 20–2 In a Carnot engine with an ideal gas, calorimetry reveals that the heat flow Q_h from the hot thermal reservoir in one cycle equals 38 J, and the heat flow Q_c to the cold thermal reservoir is 28 J. The temperature of the cold reservoir is 290K. (a) What is the mechanical work done during one cycle? (b) What is the efficiency of the cycle? (c) What is the temperature of the hot reservoir?

Strategy For part (a) we find the mechanical work done during one cycle by finding the area enclosed by the cycle on a p–V diagram. This evaluation can in principle be done by integration, but only with great difficulty because all the parameters of the cycle are not yet known. A much simpler way to compute the work done is to use the first law of thermodynamics, which, for a closed cycle, states that the work done plus the heat flow out of a thermal system equals the heat flow into the system: $W + Q_c = Q_h$, from Eq. (20–2), so that $W = Q_h - Q_c$.

For part (b) we can use the expression for the efficiency in terms of the heat flows, given in Eq. (20–3). Once we have the efficiency, we can solve part (c): We express the efficiency in terms of the temperatures and solve this for the temperature of the hot reservoir, $T_h = \dfrac{T_c}{1 - \eta}$. Everything on the right hand side of this is known.

Working It Out (a) We have

$$W = Q_h - Q_c = 38\,\text{J} - 28\,\text{J} = 10\,\text{J}.$$

(b) Equation (20–3) expresses the efficiency directly in terms of the given heat flows:

$$\eta = 1 - \frac{Q_c}{Q_h} = 1 - \frac{28\,\text{J}}{38\,\text{J}} = 0.26.$$

(Remember, this the *most* efficient engine you can have that operates between these two temperature extremes.)

(c) We have

$$T_h = \frac{T_c}{1 - \eta} = \frac{290\text{K}}{1 - 0.26} = 3.9 \times 10^2\,\text{K}.$$

What Do You Think? What single parameter change will increase the efficiency of an ideal Carnot engine? If the efficiency cannot change, how could you make sure to get more work done by the gas in the engine?

CONCEPTUAL EXAMPLE 20–3 Describe the operation of a Carnot cycle run in reverse.

Answer Running a Carnot engine—or any reversible engine— in reverse means that the direction of the transformation in each step is reversed, so that the path runs counterclockwise on the p–V diagram. For example, step I in Figure 20–5a becomes an isothermal compression rather than an isothermal expansion. The work done during one cycle is then negative. Positive work must be supplied to the cycle from the exterior to make it run—while heat flow Q_c is taken from the cold reservoir and heat flow Q_h goes to the hot reservoir. We'll see in the subsection on heat pumps and refrigerators below how an engine run in reverse is useful to us.

The Importance of the Carnot Engine

The second law of thermodynamics implies two important results; the second of these explains why any discussion of engine efficiency must start with discussion of the Carnot engine.

1. *All Carnot cycles that operate between the same two temperatures have the same efficiency.* In particular, the efficiency of a Carnot cycle *does not depend on the use of an ideal gas.* The expression for the efficiency of a Carnot engine with an ideal gas, Eq. (20–4), is the expression for a Carnot engine with any working fluid.

2. *The Carnot engine is the most efficient engine possible that operates between any two given temperatures.* (We will demonstrate this in Section 20–4.) This fact is the reason why the Carnot cycle is so important. In a sense it provides us with a target for the best possible engine.

We can demonstrate the first result with a standard method: we show that we would violate the second law of thermodynamics if it were not true. Suppose that we have two Carnot engines, A and B, that operate between the same two temperatures. Further, suppose that somehow the efficiency $\eta_A = W_A/Q_{hA}$ of cycle A were 75 percent, whereas the efficiency $\eta_B = W_B/Q_{hB}$ of cycle B were 74 percent. Let's adjust the amount of working gas in the cycles so that $Q_{hA} = Q_{hB}$, say, 100 J. This does not change the efficiencies. The work done by the two engines, which is the product of efficiency and heat flows, would then be different; namely, $W_A = (75\%)(100\,\text{J}) = 75\,\text{J}$, whereas $W_B = (74\%)(100\,\text{J}) = 74\,\text{J}$. Cycle A does more work than cycle B by an amount $\Delta W = W_A - W_B = 1\,\text{J}$. We can now connect these two cycles in a way that violates the second law. Like any Carnot engine, engine B is reversible. We connect it to engine A as in Fig. 20–6 and run it in reverse. (In Conceptual Example 20–3 we described what is meant by running an engine in reverse.)

We can divide the 75 J of work done by engine A in Fig. 20–6 into two parts: We use 74 J to run engine B and 1 J is left over. An equivalent amount of heat flow goes into the hot reservoir as is taken out, so this reservoir plays no role. Thus the net effect of this combination engine is to employ only one reservoir to extract 1 J of work with each

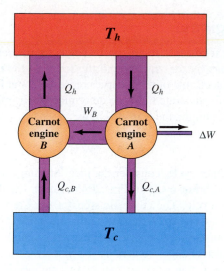

◀ **FIGURE 20–6** Carnot engines A and B are connected together, with B run in reverse. If engine A is more efficient than engine B, the net effect of the combination is to circumvent the Kelvin form of the second law of thermodynamics.

cycle. But such an arrangement violates the Kelvin form of the second law of thermodynamics. This result is quite general, and it is a remarkable one because Carnot engines can be constructed with many different working fluids.

We have shown that the efficiency of *any* Carnot cycle that operates between temperatures T_c and T_h is given by Eq. (20–4), whether or not the gas is ideal:

$$\eta_C = 1 - \frac{T_c}{T_h}. \tag{20–7}$$

EFFICIENCY OF ANY CARNOT CYCLE

We have found a universal expression for the efficiency of a Carnot cycle in terms of the temperatures of the thermal reservoirs. Note that the temperature scale here is the ideal-gas temperature scale—the Kelvin scale.

What happens if instead of the Carnot engine, with its idealized reversible cycle, we have an engine with some element of irreversibility? For example, there might be friction between the piston and the cylinder walls. We could reasonably expect that such an engine is less efficient than the Carnot engine that operates between the same two temperatures. We can, in fact, prove that this must be the case (see Problem 19).

CONCEPTUAL EXAMPLE 20–4 Consider a Carnot cycle with an ideal gas. Instead of sketching the steps in the cycle on a p–V diagram, sketch the steps on a U–V diagram, where U is the internal energy.

Answer For an ideal gas we have the relation $U(T) = C_V T$, with C_V independent of temperature, so that, aside from a scale change, plotting the steps on a U–V diagram is equivalent to plotting the steps on a T–V diagram (Fig. 20–7). The first step AB is a horizontal line at T_h extending from V_A to V_B. Similarly, step III is a horizontal line at T_c extending from V_C to V_D. Step II is an adiabatic expansion. This is described by the curve $pV^\gamma = $ constant. We are able to express p in terms of T and V since we are dealing with an ideal gas. Assuming one mol, we have $p = RT/V$. Thus the adiabatic curve is represented by $TV^{\gamma-1} = $ a constant. In fact, the constant is given by $T_h V_B^{\gamma-1}$. This allows us to determine V_C when the temperature is T_c. Step IV is the same kind of curve. Here $TV^{\gamma-1} = T_h V_A^{\gamma-1}$. This allows us to determine V_D at the lower temperature.

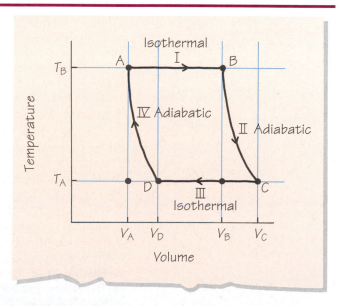

▲ **FIGURE 20–7** The Carnot cycle plotted on a T–V diagram.

20–4 Other Types of Engines

Any closed curve in the p–V diagram represents some sort of engine in the sense that the enclosed area represents work done by the engine, and the fact that the curve is closed implies that it can work in a cycle. Some of these curves represent, or at least approximate, real, functioning engines. To illustrate how one determines their efficiency, we consider a particularly interesting example, the Stirling engine.

The Stirling Engine

As an example of a reversible non-Carnot cycle, consider the ideal Stirling cycle (Fig. 20–8), and compare this cycle to the Carnot cycle in Fig. 20–5a. Steps I and III are isothermal steps, just as in the Carnot cycle. The adiabatic steps of the Carnot cycle, however, are replaced by changes in pressure and temperature that maintain a constant volume. No work is done during the constant-volume steps. However, because the temperature changes, the internal energy also changes so that, by the first law of thermodynamics, there is a heat flow to or from the working gas of the engine during the constant-volume steps. This heat flow occurs at decreasing temperatures during step II and at increasing temperatures during step IV. It is as though a series of ever colder thermal reservoirs enter into step II and ever hotter thermal reservoirs enter into step IV. Increasingly, hotter fuel must be used throughout step IV. Under these conditions, we note that the efficiency of an engine is really a measure of the work done for the total fuel burned; thus the denominator of the efficiency should be the entire positive heat flow to the thermal system, which we write as Q_{pos}.

In the subsection immediately below we work out the efficiency of the ideal Stirling engine, η_S, when the working fluid is 1 mol of an ideal gas. We find that

$$\eta_S = \frac{W}{Q_{pos}} = \frac{R(T_h - T_c)\ln(V_2/V_1)}{RT_h\ln(V_2/V_1) + C_V(T_h - T_c)}. \tag{20–8}$$

This result would be the corresponding Carnot efficiency if it were not for the term proportional to C_V in the denominator (due to step IV). Because this extra term is positive, the denominator of η_S is greater than the denominator of η_C, and thus *the efficiency of the Stirling cycle is less than that of the Carnot cycle.*

Stirling cycle

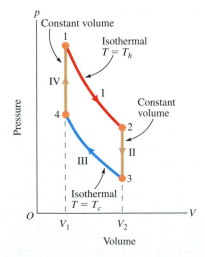

▲ **FIGURE 20–8** The ideal Stirling cycle. There is a positive heat flow to the engine in steps I and IV, and a heat flow from the engine in steps II and III. This cycle was described by Robert Stirling almost 20 years before Carnot discussed his own ideas. The engine Stirling patented in 1816 may be useful in outer space and under other conditions and has become the subject of active engineering work.

How to Get the Efficiency of the Stirling Engine

We assume that the working gas in a Stirling engine is ideal. We need the total positive heat flow $Q_{pos} = Q_I + Q_{IV}$ as well as W, the work done during one cycle (Fig. 20–8). During step I, the temperature is constant and $\Delta U = 0$. Thus, by the first law of thermodynamics, the heat flow to the engine is the work done during the transformation:

$$Q_I = W_I = \int_{V_1}^{V_2} p\,dV = RT_h \int_{V_1}^{V_2} \frac{dV}{V} = RT_h \ln\left(\frac{V_2}{V_1}\right).$$

According to the first law, the heat flow to the engine during step IV is the change in internal energy $U(T_h) - U(T_c)$ because no work is done by the engine during this portion of the cycle. Thus

$$Q_{IV} = U(T_h) - U(T_c) = C_V \times (T_h - T_c).$$

Combining these two equations, we see that the total heat flow to the engine is

$$Q_{pos} = Q_I + Q_{IV} = RT_h \ln\left(\frac{V_2}{V_1}\right) + C_V \times (T_h - T_c).$$

The total work done in the cycle is the area enclosed by the cycle on the p–V diagram:

$$W = W_I + W_{III} = RT_h \ln\left(\frac{V_2}{V_1}\right) + RT_c \ln\left(\frac{V_1}{V_2}\right) = RT_h \ln\left(\frac{V_2}{V_1}\right) - RT_c \ln\left(\frac{V_2}{V_1}\right)$$

$$= R(T_h - T_c) \ln\left(\frac{V_2}{V_1}\right).$$

The efficiency of this cycle, W/Q_{pos}, is then given by Eq. (20–8).

Other Engines and Their Efficiencies

In addition to the Stirling cycle there are other reversible cycles that represent idealized (friction ignored) versions of real engines. A short list would include the Diesel cycle (Fig. 20–9a), which as you know is used in many different settings; the Otto cycle (Fig. 20–9b), which is very close to an automobile engine; and the Brayton cycle (Fig. 20–9c). Even ignoring friction, these cycles are inherently less efficient than the Carnot cycle for given temperature extremes. With friction, which is always present in real engines, the efficiency is reduced still further. A more detailed treatment of these cycles and their efficiency can be found within the end-of-chapter problems.

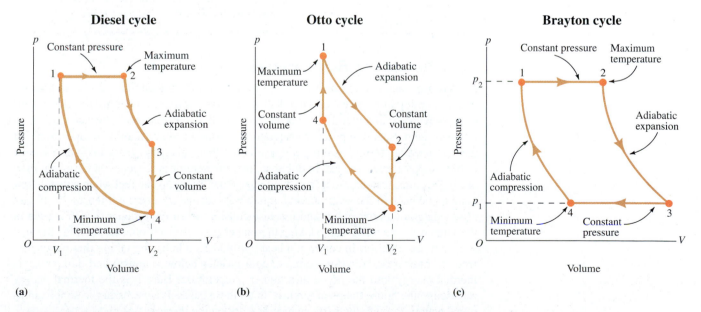

▲ **FIGURE 20–9** (a) The Diesel cycle, for the engine patented by Rudolf Diesel in 1892. (b) The Otto cycle, for the engine designed by Nikolaus Otto in 1876. (c) The Brayton cycle, named for George B. Brayton.

There is a way to decompose an arbitrary reversible cycle into a series of Carnot cycles operating at intermediate temperatures that allows one to prove that any non-Carnot cycle (even if it is reversible) that operates between two temperature extremes T_h and T_c is less efficient than the Carnot cycle that operates between those same temperatures. While we won't work this out in detail, we point out that a Carnot cycle is more efficient if the two temperatures are farther apart. An arbitrary reversible cycle is approximated by Carnot cycles operating at intermediate temperatures and therefore involves a decrease in efficiency.

THINK ABOUT THIS . . .

WHY IS IT SO HARD TO MAKE AUTOMOBILE ENGINES THAT DON'T POLLUTE?

From the end of the 1960s the U.S. government has been requesting that the efficiency of internal combustion engines be improved and air pollution reduced. Automobile engines of that period used an excess of fuel and the residue of unburnt hydrocarbons was a significant pollution source. The pollution problem was alleviated by adding air to the products of combustion as they left the engine's cylinder, thereby burning them. But, how is engine efficiency improved? The easiest way to accomplish this is to increase the maximum temperature at which the fuel is burned. But a higher temperature leads to a secondary reaction: the oxidation of the nitrogen in the added air, and the resulting oxides of nitrogen add very significantly to "acid rain" problems. Thus the government-mandated increase in efficiency resulted in a new pollution problem. The solution to this inherent conflict of efficiency and pollution is a complex one, involving recycled gases and catalysts, and it is not yet completely resolved. This area presents problems whose solution involves physics, chemistry and engineering. In the meantime, although we do have more efficient engines than in recent decades, some of us choose to drive oversized vehicles, and we are back in an era of vehicles that consume excessive amounts of fuel.

CONCEPTUAL EXAMPLE 20–5 The efficiency of a certain engine is half that of a Carnot engine operating between the same temperature extremes. Does that mean that there is an element of irreversibility in the working of the engine?

Answer An engine with an efficiency smaller than that of the Carnot cycle can differ from the Carnot cycle in two ways. First, the path traced out in a p–V diagram may not have the form of the Carnot cycle (two isothermal transformations and two adiabatic transformations), even though the path is reversible. These engines involve heat flows at temperatures between the two temperature extremes of the Carnot cycle and are in principle always less efficient than the Carnot cycle. This makes sense if we consider the fact that Carnot cycles themselves are more efficient when the two temperatures between which they work are larger. When one brings in intermediate temperatures, as for these "other" engines, the efficiency drops. Second, real engines generally involve some friction (that is, some irreversibility), and as you might expect this leads to less efficiency. But without knowing more details about the engine we cannot say which of these effects is the reason for the loss of efficiency, so we cannot definitely say whether or not there is irreversibility.

Heat Pumps and Refrigerators

Heat pumps and **refrigerators** are simply work-producing engines run in reverse. An engine running in reverse—counterclockwise on a p–V diagram—transfers thermal energy from colder to hotter thermal reservoirs. For this to occur, work must be performed *on* the system. Heat pumps and refrigerators are similar to one another in operation, but they are used differently. A refrigerator removes thermal energy from a colder reservoir (which keeps a freezer cold). In doing so, the refrigerator is responsible for a heat flow at a higher temperature into its surroundings, something you can feel when you put your hand at the back of the refrigerator, but that is of secondary interest. Figures 20–10 and 20–11 diagram a refrigerator and its close cousin, the air conditioner, which is used in the same way as a refrigerator, taking thermal energy from a room and dumping thermal energy to the exterior. In contrast, a heat pump heats a house by taking thermal energy from a colder reservoir (for example, cool ground below a house) and dumping this thermal energy into the house at a higher temperature. Here it is the thermal energy released to the house that is of interest. In all these cases, it is necessary to supply additional energy through the performance of work on the thermal system.

We can analyze heat pumps or refrigerators by the same method that we used to analyze heat engines. Because of the different ways these devices are used, their

(a)

(b)

▲ **FIGURE 20–10** (a) Diagram of how a refrigerator operates. A compressor, which is typically operated by electrical energy, does work on the working fluid (freon or an environmentally acceptable substitute) by compressing it, thereby raising its temperature. That thermal energy is released by thermal contact to the exterior of the refrigerator at the coils, and the working fluid is allowed to expand (at the expansion valve) to reduce its temperature. Again thermal contact with the interior of the refrigerator at the evaporator then cools the interior of the refrigerator, and the cycle repeats. (b) The condenser coils are normally on the back of the refrigerator and the compressor coils inside where we want it to be cold.

◀ **FIGURE 20–11** Diagram of a room air conditioner, whose operation follows the steps described for the refrigerator.

efficiencies are defined differently. For a heat pump, where the total positive heat flow rejected from the pump into a building, Q_{rej}, is of interest, we use the *coefficient of performance COP*, which is defined as

$$COP \equiv \frac{Q_{rej}}{W}, \tag{20–9}$$

where W is the total work that must be supplied in one cycle. For a refrigerator, where we are interested in the total positive heat flow absorbed from the interior of the refrigerator, Q_{abs}, the coefficient of performance K_{ref} is defined by

$$K_{ref} \equiv \frac{Q_{abs}}{W}, \tag{20–10}$$

where W is the work supplied to the refrigerator. For a heat pump, a larger coefficient of performance means that a given amount of work performed by the motor leads to a larger transfer of heat flow into the building. By the same token, for a refrigerator, a larger coefficient of performance corresponds to a larger heat flow from the interior of the refrigerator to its surroundings when the motor performs a given amount of work.

EXAMPLE 20–6 The Carnot cycle shown in Fig. 20–5a is run in reverse and used as a heat pump. Show that the *COP* is given by $T_h/(T_h - T_c)$.

Setting It Up Figure 20–12 is a schematic drawing of the operation, with W, Q_c, and Q_h all taken to be positive.

Strategy All formulas given previously for the operation of the Carnot cycle can be used, except that the signs must be changed because the cycle runs in reverse. A figure such as Fig. 20–12 helps us to figure out these signs. It is also necessary to interpret the total heat flow rejected, Q_{rej}, as the heat flow Q_h transferred to the thermal reservoir at T_h.

Working It Out By the first law of thermodynamics,

$$W = Q_h - Q_c,$$

so *COP* is given by

$$COP = \frac{Q_h}{Q_h - Q_c} = \frac{1}{1 - (Q_c/Q_h)},$$

according to Eq. (20–9). The ratio Q_c/Q_h is, in fact, the ratio of temperatures for any Carnot cycle—see Eq. (20–4) and the accompanying discussion. It is not important that the signs of the heat flows are reversed because we use only the ratio of these quantities. With $Q_c/Q_h = T_c/T_h$, we find

$$COP = \frac{1}{1 - (T_c/T_h)} = \frac{T_h}{T_h - T_c}. \qquad (20\text{–}11)$$

What Do You Think? The coefficient of performance, as defined in Eq. (20–10), for a Carnot cycle run as a refrigerator (see Problem 26) is given by

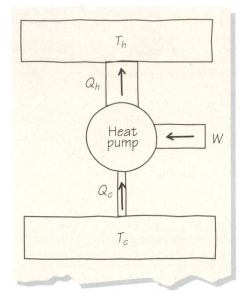

▲ **FIGURE 20–12** The Carnot cycle operated as a heat pump.

$$K_{ref} = \frac{T_c}{T_h - T_c}. \qquad (20\text{–}12)$$

If your refrigerator is operating in an igloo, where the air temperature is only a little above the temperature you want inside the refrigerator, do you need to (a) perform a lot of work, or (b) little work for a given heat flow?

The coefficient of performance of the Carnot heat pump, Eq. (20–11), is a number larger than 1 and it increases, or improves (you are using the energy to run the pump more efficiently), if the difference between the temperatures of the two thermal reservoirs is reduced: It is easier to transfer thermal energy from the cold ground to a warm house if the temperature difference is small. A heat pump used for heating a house in winter might operate between the temperatures $T_c = -5°C = 268K$ and $T_h = 45°C = 318K$. (Although 45°C is much hotter than the interior of a house, if the house is reasonably but not perfectly insulated, the temperature of the exhaust must be somewhat higher than the desired interior temperature.) The COP of an ideal Carnot cycle that operates between these temperatures is then $T_h/(T_h - T_c) = (318K)/(50K) = 6.36$. In contrast, a commercial heat pump that operates between the temperatures 45°C and −5°C has a COP that is only about half this value.

20–5 Entropy and the Second Law

In Section 20–1 we argued that the direction of spontaneous processes as well as the ability to extract thermal energy and use it to do work had something to do with the random nature of thermal energy and with the relative probability of certain states. Here we will find a way to measure the degree of disorder: the entropy. We will in fact start with a thermodynamic approach and use the Carnot cycle to identify entropy as a new thermodynamic variable. This variable is like internal energy in that it is a function of state. On a macroscopic level, we will see that the entropy is a measure of the energy unavailable to do useful work. By understanding entropy on a microscopic level, we will make the connection back to disorder and the role of probability.

Entropy as a Thermodynamic Variable

In a Carnot cycle, the efficiency, defined by $1 - (Q_c/Q_h)$, is given by Eq. (20–7): $\eta_C = 1 - (T_c/T_h)$. Thus Q_c/Q_h must equal T_c/T_h [see Eqs. (20–3) and (20–4)]. This relation can be expressed as

$$\frac{-Q_c}{T_c} + \frac{Q_h}{T_h} = 0. \tag{20–13}$$

Recall that Q_c is the heat flow from the Carnot engine to the cold thermal reservoir. The expression $-Q_c$ in the first term of Eq. (20–13), minus sign and all, is then the heat flow to the engine at the cold temperature, just as Q_h is the heat flow *to* the engine at the high temperature. The implication of all of this is that for the entire Carnot cycle, made of a series of steps on the p–V diagram,

$$\sum_i \frac{Q_i}{T_i} = 0, \tag{20–14}$$

where the index i labels the steps and where we always regard the heat flow as the heat flow *to* the engine. We will now show that this is a more general property of thermal systems going through reversible transformations, not just a particular property of the Carnot cycle. Our demonstration lacks in generality only in that it assumes that we are operating with an ideal gas. Consider the integral $\int_A^B \frac{dQ}{T}$ between any two points A and B on a p–V diagram. Then, working with one mole of gas,

$$\int_A^B \frac{dQ}{T} = \int_A^B \frac{dU + pdV}{T} = \int_A^B \frac{C_V\, dT}{T} + \int_A^B R\frac{dV}{V}$$
$$= (C_V \ln T_B + R \ln V_B) - (C_V \ln T_A + R \ln V_A). \tag{20–15}$$

In the first step we have used the first law of thermodynamics; to evaluate the temperature integral we have used the fact that C_V is independent of T. Equation (20–15) implies that the value of the integral depends only on the initial and final values of the variables, and not on the path along which the integral is evaluated. This property of the integral in fact holds even if the working fluid is not an ideal gas. This implies that there is a state function S which, like the internal energy U, depends only on the state of the gas, and not on how it reached that state. This state function is called the **entropy** of the system. It is precisely defined by

$$S(B) - S(A) = \int_A^B \frac{dQ}{T}. \tag{20–16}$$

This immediately leads to the generalization of (20–15),

$$\oint \frac{dQ}{T} = 0, \tag{20–17}$$

CLAUSIUS' THEOREM

for a reversible cycle. The circle placed on the integral is a reminder that the integral is taken over a cycle, i.e. a closed path. This result is known as **Clausius' theorem**. A mathematical way of expressing our result is to say that dS, defined by

$$dS = \frac{dQ}{T}, \tag{20–18}$$

is a perfect differential for any reversible transformation: its integral between two points is the difference of a function of each point, so that the integral over a closed path trip, for which the initial and final points are identical, vanishes. This statement also applies to dU but not to dQ.

EXAMPLE 20–7 Calculate the entropy change of 1.0 mol of an ideal gas that undergoes an isothermal transformation from an initial state of pressure 1.50 atm and volume 500 cm³ to a final state of pressure 0.90 atm. (Being able to find the entropy change in ideal gases will help us later to understand the connection between entropy and chaos on the microscopic scale.)

Setting It Up Figure 20–13 includes the isothermal curve for this transformation and labels the initial and final points i and f as well.

▲ **FIGURE 20–13** An isothermal change of state.

Strategy The specified path is one of constant temperature, so the integral of dQ/T along the specified path is $(1/T)$ times the integral of dQ along the path. By the first law we can write $dQ = dW + dU$, and since for an ideal gas thermal energy is a function of temperature only (that is, $dU = 0$ along this path), we have $dQ = dW = pdV$. For an ideal gas $pV = nRT$ so that $p = nRT/V$ and therefore $dS = \frac{1}{T}dQ = \frac{1}{T}\left(\frac{nRT}{V}dV\right) = nR\frac{dV}{V}$. This can be integrated between the limits. We are lacking the final volume V_f, but since the temperature is fixed, we can use $p_iV_i = p_fV_f$ to replace V_f with known quantities.

Working It Out The change in the entropy ΔS is

$$\Delta S = nR \int_{V_i}^{V_f} \frac{dV}{V} = nR \ln \frac{V_f}{V_i}.$$

Using $V_f/V_i = p_i/p_f$ this reads

$$\Delta S = nR \ln\frac{p_i}{p_f} = (1.0 \text{ mol})(8.3 \text{ J/mol} \cdot \text{K}) \ln\frac{1.5 \text{ atm}}{0.90 \text{ atm}} = 4.2 \text{ J/K}.$$

Note that we did not need the value of the initial volume of the gas to get the answer.

What Do You Think? What is the entropy change if the transformation from the initial state to the final state is accomplished in two steps: a first step at constant volume as the pressure drops from 1.50 atm to 0.90 atm, followed by an isobaric transformation with the volume increasing from 500 cm³ to the final volume of the example?

How Entropy Changes for Irreversible or Spontaneous Processes

Irreversible processes (processes with friction, for example) lead to increased inefficiency in any otherwise reversible engine. We'll now see what friction or other irreversibilities imply for heat flows to such a cycle. We start by looking again at efficiency, which for an arbitrary cycle is

$$\eta = \frac{W}{Q_{\text{abs}}} = 1 - \frac{Q_{\text{rej}}}{Q_{\text{abs}}},$$

where Q_{rej} is the total heat flow rejected by the cycle and Q_{abs} is the total heat flow absorbed. We have used the relation $W = Q_{\text{abs}} - Q_{\text{rej}}$ from the first law of thermodynamics. Each of these heat flows is positive.

The efficiency is reduced if $Q_{\text{rej}}/Q_{\text{abs}}$ gets bigger, and this can happen in two ways. First, reduce the denominator Q_{abs}, the heat flow *to* the system. Second, increase the numerator Q_{rej}, the heat flow *out* of the system. But an algebraic increase of Q_{rej} is the same as an algebraic decrease of $-Q_{\text{rej}}$, and $-Q_{\text{rej}}$ can be interpreted as heat flow *to* the system. Both ways of reducing the efficiency boil down to reducing the heat flow *to* the system. In other words, anything that causes the *algebraic sum* of the heat flows into the system to decrease will make the engine less efficient. But this is exactly the effect of irreversibilities that occur anywhere along the cycle. For example, in a real automobile engine, there is heat flow to the exhaust, but there is additional heat flow out (less heat flow in) due to the heating of the pistons and other parts of the engine. Broadly speaking, all irreversibilities cause the algebraic heat flow to the system to be reduced. More heat flow out is equivalent to less heat flow in. That means less energy to do work, and hence less efficiency.

What is the effect of a reduction in the heat flow to the system on Clausius' theorem, Eq. (20–17)? In that integral, dQ represents the differential flow to the system, so the integral will be be decreased. We can say that *for an irreversible cycle* (one with any element of irreversibility),

$$\oint \frac{dQ}{T} < 0. \qquad (20\text{–}19)$$

This result is known as **Clausius' inequality**.

Equations (20–17) and (20–19) can be combined into

$$\oint \frac{dQ}{T} \leq 0. \qquad (20\text{–}20)$$

CLAUSIUS' THEOREM AND INEQUALITY

The equality, Clausius' theorem [Eq. (20–17)], applies for a reversible path; the inequality applies when the transformations include any irreversibility.

Now we can use Equation (20–20) to learn about entropy changes in irreversible transformations. Consider the cycle shown in Fig. 20–14, in which leg $A \rightarrow B$ is irreversible. From Eq. (20–20),

$$0 > \oint \frac{dQ}{T} = \int_{\substack{A \\ \text{irrev}}}^{B} \frac{dQ}{T} + \int_{\substack{B \\ \text{rev}}}^{A} \frac{dQ}{T}.$$

The second term on the right-hand side of this equation is given by $S(A)-S(B)$ because it is taken over a reversible path. When we move this quantity to the left-hand side, we find that

$$S(B) - S(A) > \int_{\substack{A \\ \text{irrev}}}^{B} \frac{dQ}{T}. \qquad (20\text{–}21)$$

Thus *the difference in entropy between two points is greater than the integral of* dQ/T *over an irreversible change.*

How do we compute the increase in entropy when there is an irreversible transformation from thermodynamic state A to thermodynamic state B? We have seen that entropy is a function of state, so we can compute the change $S(B)-S(A)$ by any means we like; that is, *we can use a reversible path to find the change in entropy as long as that path connects the two given states.* We will illustrate this procedure in Section 20–6.

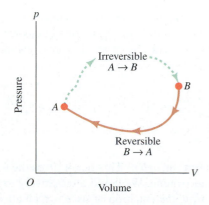

▲ **FIGURE 20–14** A cycle that contains an irreversible path from A to B and a reversible path from B back to A.

The Entropy of an Isolated System Never Decreases

In any reversible process in which there is no heat flow—an adiabatic process—the change in entropy will be zero. If there is no heat flow, but there are irreversible processes, then the entropy increases. That is because with no heat flow, the right-hand side of Eq. (20–21) is zero, and the equation reads $\Delta S > 0$. As an example, consider the free expansion of a gas from a small container into a bigger one, the entire system enclosed in Styrofoam. There is no heat flow into or out of the system, but the entropy increases. More generally a *spontaneous process in an isolated system increases the system's entropy*. Because the universe—our entire surroundings—is in contact with no other system, we say that irreversible processes increase the entropy of the universe. A consequence of this property of entropy is that the state of maximum entropy is a stable one. Once one has arrived at such a state, no further spontaneous changes can occur. (Or, more precisely, no more macroscopic changes occur—there will always be very small fluctuations about the equilibrium state, ones that are invisible on the macroscopic scale.) In other words, for an isolated thermal system *the state of maximum entropy is the state of stable equilibrium*. This is still another way to state the second law of thermodynamics. Below we are going to connect the entropy with the degree of disorder, with the direction of spontaneous processes determined by an increase in disorder. Thus what we have concluded here is that for an isolated thermal system the stable equilibrium state is the state of maximum disorder.

Sometimes you will hear people say that the existence of life is in contradiction with the second law, because living systems take thermal energy from the Sun and turn it into the order of a beautiful flower or a thinking human race that can organize parts of the entire planet. This reasoning is incorrect; a living being is not an isolated system. It is the whole system of Sun plus living creatures on Earth that is better approximated as isolated, and the entropy of this system is indeed increasing.

CONCEPTUAL EXAMPLE 20–8 Suppose you had never heard of engines, and only knew the second law of thermodynamics in the form $\Delta S \geq 0$. Is this enough to show that it is impossible to construct a device that transfers thermal energy from the ocean and converts it into an equivalent amount of work (which could be used, for example, to run the New York subways) without any other effect?

Answer The system in this case is the ocean together with wherever the thermal energy taken from it is used or dispersed. Let's denote the thermal energy taken from the ocean by Q. This implies that the reservoir, the ocean, absorbs thermal energy $-Q$. The entropy change of the ocean is therefore $\Delta S = -Q/T$. This is negative and if it were the only entropy change would violate the second law in the form $\Delta S \geq 0$. There must be entropy change somewhere else in the system, and hence there must be a heat flow somewhere else—the entire heat flow Q cannot be converted to work. The impossibility of taking heat flow and turning it into an equivalent amount of work is the Kelvin form of the second law (Section 20–2).

20–6 Entropy and Ideal Gases

When a gas is ideal, we can calculate its entropy change in reversible transformations. We demonstrated this when we used a combination of constant-volume and isothermal transformations to calculate the entropy of an ideal gas, resulting in Eq. (20–15). That result generalized to n moles is

$$S(B) - S(A) = C_V \ln\frac{T_B}{T_A} + nR \ln\frac{V_B}{V_A}. \tag{20–22}$$

ENTROPY OF AN IDEAL GAS

[We have used $\ln a - \ln b = \ln(a/b)$.] This result gives the entropy change in terms of the temperature and the volume. We could alternatively express it in terms of the pressure and volume by using the equation of state, which allows us to write $T_B/T_A = (p_B V_B)/(p_A V_A)$. If we additionally use $\ln(p_B V_B/p_A V_A) = \ln(p_B/p_A) + \ln(V_B/V_A)$, we find

$$S(B) - S(A) = C_V \ln\frac{p_B}{p_A} + (nR + C_V) \ln\frac{V_B}{V_A} = C_V \ln\frac{p_B}{p_A} + C_p \ln\frac{V_B}{V_A}, \qquad (20\text{–}23)$$

where we recall that $C_p = C_V + nR$ [Eq. (18–33)].

We used reversible paths to calculate the change in entropy. But because entropy is a state function, the same result obtains for *any* transformation, reversible or irreversible between the same states. In particular, we may use Eq. (20–22) to calculate the change of entropy in a free expansion of a gas at a fixed temperature, an irreversible process. Equations (20–22) and (20–23) can be used for any process whose end points can be put on any graph, be it a p–V plot, a T–V plot, or a p–T plot.

CONCEPTUAL EXAMPLE 20–9 (a) Plot the steps of a Carnot cycle on an S–T plot. (b) What is the significance of the area enclosed by the closed path thus formed in the S–T plane?

Answer (a) We shall follow the steps ABCD as in Fig. 20–5a. The S–T graph is shown as Fig. 20–15. The first step, AB, is an isothermal transformation. With our choice of vertical T axis, this is a horizontal line at $T = T_h$. The step BC is adiabatic. This means that there is no heat flow into the system and therefore the entropy does not change, and so the step is represented by a vertical line corresponding to a constant value of S—call it S_B. The next step is again isothermal so that the transformation CD is a horizontal line in the direction of decreasing S at the horizontal level of T_c. The last step, DA, is again *isentropic* (no change in entropy), so that it is represented by a vertical line at S_A.
(b) Although it is easy to find the area directly—it is just $(T_h - T_c) \times (S_B - S_A)$—the interpretation follows from an integral representation. The area under the upper horizontal line is equal to

$$\int_A^B T\,dS = \int_A^B T\frac{dQ}{T} = \int_A^B dQ.$$ The integral is taken along the T_h isotherm, and the result is the total heat flow from the higher temperature reservoir, that is Q_h. From this we must subtract the same

integral along the T_c isotherm, and by the same token, the result is Q_c. The total area is therefore $Q_h - Q_c = W$, the work done. This is the same result that one finds by calculating the enclosed area in the p–V plane.

▲ **FIGURE 20–15**   The Carnot cycle on an S–T plot. The area of the cycle is W, the work done.

EXAMPLE 20–10 You have 3.0 mol of an ideal gas in thermal contact with a thermal reservoir at $T = 300$K. The gas undergoes free expansion (no work is done) from 12 L to 24 L. What is the entropy change of the gas and of the universe? Repeat the exercise for an expansion from 12 L to 120 L.

Strategy The entropy change of the gas, being a state function, can be computed when the end states are known, as they are here, even if the transformation is irreversible. In this case, we can use Eq. (20–22) with fixed T. Because $\ln(1) = 0$, only the ratio of volumes is involved. What about the entropy change of the universe? To determine the entropy change in the rest of the surroundings, we must determine the heat flow to the thermal reservoir. The gas does no work in its expansion, so the heat flow out is equal to the change in internal energy. But this change is zero because the gas is ideal and because its internal energy depends only on the (unchanging) temperature. There is thus no heat flow to the thermal reservoir, and its entropy change, $\Delta Q/T$, is zero. The entropy change of the universe is therefore the entropy change of the gas. The fact that it is positive indicates that the transformation involved—in this case, free expansion—is irreversible.

Working It Out From Eq. (20–22) with $T_B = T_A$, we find $\Delta S = nR\ln(V_B/V_A)$, so that numerically $\Delta S = (3.0\text{ mol})(8.3\text{ J/mol}\cdot\text{K})\ln[(24\text{ L})/(12\text{ L})] = (25\text{ J/K})\ln 2 = 17\text{ J/K}$. The entropy change of the gas, which is also the entropy change of the universe, is positive. (The numerical result would have been the same had the question dealt with an isothermal leg of a Carnot cycle. In that case there would be a compensating $-\Delta Q/T$ contribution of entropy to the reservoir, so that there would be no net change to the entropy of the entire system of gas plus reservoir, as would be appropriate to a reversible transformation in a completely isolated system.)

When the expansion is from 12 L to 120 L, the entropy change is as just described, except that $\ln(2)$ is replaced by $\ln(10)$, and $\Delta S = 25\ln(10)\text{ J/K} = 58\text{ J/K}$. The natural logarithm changes fairly slowly with its argument, so, compared to the ratio of volume changes in the two cases, the difference in entropy changes is not very great.

What Do You Think? Suppose you had two containers, each of volume V. One contains one mole of CO_2 and the other contains one mole of N_2O. Suppose you now connect the two containers by a tube, open the stopcock on the tube, and let the gases mix. Do you think the entropy will increase or stay the same?

▲ **FIGURE 20–16** (a) When a plate breaks, it is more disordered—the entropy of the "plate system" increases. (b) In urban renewal, the demolition of an abandoned building raises its entropy very quickly!

20–7 The Meaning of Entropy

The entropy of a system and all its surroundings—the universe—increases when an irreversible process takes place. In effect, increased entropy corresponds to a lost heat flow, and hence a lost ability to do work. The loss of heat flow represents energy that can never be turned into work because, by the nature of irreversible processes, the process cannot be reversed to recapture this heat flow without making other changes. Of course, the energy is not really lost; energy is always conserved. But it has been turned into the random motion of the microscopic constituents of the thermal system and cannot be sufficiently well "organized" to allow all of it to go into, say, lifting a weight. By organization we mean order; if all the molecules were moving in precisely the same direction, for example, they would be well ordered. They might have the same total energy as a disorganized gas, but *all* of their energy could be used to lift a weight. Any degree of irreversibility reduces our ability to do work because it increases the disorder. Let's consider this idea more closely.

We began our discussion of spontaneous processes in Section 20–1 with two examples. In one, a gas expands into an empty space, its molecules move from a configuration that has a low probability (all the molecules are in one corner) to one that has the largest possible probability (the molecules are spread uniformly throughout the allowed volume). Initially, all the molecules are confined in a small box inside a larger one. This initial state is more *ordered* than the final state because the individual molecules are more localized. A loss of order is associated with a change of the system from a state of low probability to one of higher probability. Equation (20–22) applied at a fixed temperature illustrates how entropy increases for an ideal gas when the volume increases. From this perspective entropy is a measure of disorder, and the increase of the entropy in an irreversible process is the increase in disorder that accompanies such processes (Fig. 20–16).

We can illustrate this idea quite concretely by mixing two gases initially separated by a partition (Fig. 20–17). Each gas starts out with the same temperature and pressure. The partition is opened, and the gases in the two parts mix. Before the partition is opened, the degree of order is greater: All the molecules of each gas are on their respective sides. After the partition is opened, the disorder is greater: Any one molecule is somewhere in the greater volume, rather than being on one side or the other. It is more probable that each gas will spread separately throughout the greater volume than that it will stay on its respective side. Thus, we would expect the entropy to increase when the partition is opened, which is in line with the idea that the mixing of the two sides is an irreversible process. This increased entropy is the **entropy of mixing**, illustrated quantitatively in the example below.

▲ **FIGURE 20–17** (a) Two gases are initially unmixed. (b) A valve is opened, and each gas diffuses into the other's flask. (c) The final state is more disordered than the initial state. The system's entropy has increased.

EXAMPLE 20–11 A quantity of 0.20 mol of argon gas contained in a volume of 5.0 L mixes with 0.50 mol of neon gas contained in a volume of 12.5 L, making a total volume of 17.5 L. Both gases are at the same temperature and can be regarded as ideal. Each volume is thermally isolated. What is the change in entropy?

Setting It Up We draw the situation in Fig. 20–18, showing the situation before and after mixing.

Strategy The gases start at the same temperature, and they will be at that same temperature after expansion. Only their volumes will have changed. Each will then have an entropy change given by Eq. (20–22) with the temperature term absent, $\Delta S = nR \ln(V_f/V_i)$. This applies for the two gases separately, and to find the total entropy change we add the results.

Working It Out We have

$$\Delta S_{Ar} = (0.20 \text{ mol})(8.3 \text{ J/mol} \cdot \text{K}) \ln[(17.5 \text{ L})/(5.0 \text{ L})]$$
$$= (1.7 \text{ J/K}) \ln 3.5 = 2.1 \text{ J/K}$$

and

$$\Delta S_{Ne} = (0.50 \text{ mol})(8.3 \text{ J/mol} \cdot \text{K}) \ln[(17.5 \text{ L})/(12.5 \text{ L})]$$
$$= (4.2 \text{ J/K}) \ln 1.4 = 1.4 \text{ J/K}$$

The entropy of mixing is the sum of the increases of entropy of the two gases, so that

$$\Delta S = \Delta S_{Ar} + \Delta S_{Ne} = 3.5 \text{ J/K}.$$

What Do You Think? Is it possible to arrange things so that the two gases, originally in the separate containers described above, spontaneously pass into a common container whose volume is 3.0 L?

▲ **FIGURE 20–18** The neon and argon gases mix when the valve is opened. What happens to the entropy?

The second example of a spontaneous process described in Section 20–1 was the fact that ice cubes melt when they are placed in hot tea. We also interpret this case, and more generally the fact that systems with different temperatures tend to the same intermediate temperature when they are placed in thermal contact in terms of an increase in disorder. In this case, the more ordered system is the cooler one. On average, the velocity vectors of the molecules of steam at a lower temperature are confined to a smaller range than are those of steam at a higher temperature. The heated system is more disordered because, on average, the velocity vectors spread over a greater range. Equation (20–22), this time applied for changing temperature but constant volume, describes how entropy increases for an ideal gas when temperature changes. More advanced treatments of statistical physics quantify the idea that entropy is a measure of disorder. We have already described how the thermodynamic approach connects entropy change to the second law. Here we claim that entropy change is associated with an increase in disorder. We draw the conclusion that the second law follows from the interpretation of entropy as disorder. In fact, a more rigorous treatment than we can perform here confirms this: *All the consequences of the second law of thermodynamics follow from the treatment of entropy as a measure of disorder.*

Engines and Entropy

The principle that entropy never decreases is a way of formulating the second law of thermodynamics that emphasizes the directionality of thermal processes. The application of this formulation to engines is less obvious because the entropy in a reversible cycle does not change. Here (see also Conceptual Example 20–8) we show how the entropy formulation of the second law applies to engines and leads to our earlier formulations of the second law.

Consider a thermal reservoir at temperature T_h. Can we construct an engine that, in a single cycle, extracts thermal energy through a (positive) heat flow Q_h from that reservoir to lift a weight and does nothing else? Let's suppose that we could do so. There is no change in the entropy of the weight that has been lifted and there is no change in the engine after a full cycle. Thus, in the isolated system that consists of the weight, the engine, and the reservoir, the only entropy change is that of the reservoir, as given by $\Delta S = -Q_h/T_h$. This entropy change is negative and therefore violates the entropy formulation of the second law. We deduce that in order to satisfy the entropy formulation of the second law, there must be a second thermal reservoir (at temperature T_c) such that the net entropy change is positive. If this second reservoir is to have a positive entropy change, then there must be a heat flow Q_c into it. This gives rise to an entropy change $\Delta S' = Q_c/T_c$.

The isolated system, which now consists of the weight, the engine, and the two reservoirs, must satisfy the entropy formulation of the second law, namely $\Delta S + \Delta S' \geq 0$, or

$$-\frac{Q_h}{T_h} + \frac{Q_c}{T_c} \geq 0;$$

$$\frac{Q_c}{Q_h} \geq \frac{T_c}{T_h}.$$

Because the work done to lift the weight is $W = Q_h - Q_c$ and the efficiency of the engine is defined as $\eta \equiv W/Q_h = (Q_h - Q_c)/Q_h = 1 - (Q_c/Q_h)$, the entropy formulation of the second law leads to

$$\eta \leq 1 - \frac{T_c}{T_h}.$$

We are now on familiar ground. The equality applies for the reversible Carnot cycle, where there is no heat flow except at the two temperature extremes. In this case, ΔS_{tot} is zero, which requires that $(Q_c/Q_h) = (T_c/T_h)$, and the efficiency is $\eta = 1 - (T_c/T_h)$. For cycles with irreversibility, or for cycles such as the Stirling cycle in which there is heat flow at intermediate temperatures, the inequality sign holds. Thus, the entropy formulation of the second law implies that the Carnot engine is the most efficient type of engine.

We have now connected the two apparently dissimilar aspects of the second law of thermodynamics: the impossibility of constructing a perfectly efficient engine and the inevitable direction of spontaneous processes.

Order, Entropy, Probability, and the Arrow of Time

The title of this subsection carries all the catchwords that are associated with the second law of thermodynamics. Entropy as a state function was introduced through our study of engines, and the quantitative definition as given in Eq. (20–16) allowed us to learn how it changed in irreversible processes. There are other irreversible processes that do not involve thermodynamics. An example is a deck of cards arranged in order [4 aces, 4 kings, 4 queens, ... 4 threes and 4 twos]. We would call this an arrangement that has a great deal of *order* in it. Shuffle the deck. The order will be lost. Shuffle it over and over again, and different arrangements will appear, but order, as defined by the initial arrangement, will not reappear. We may say that the process of shuffling a deck leads to an *irreversible* loss of order. Can we apply the idea of entropy to such a process? One would expect that it should be possible, because the shuffling of the deck of cards bears some resemblance to what happens to molecules initially confined to a small box. When the lid of the small box is opened and the molecules are allowed to escape into a larger box that contains the small one, the system becomes much more disorderly. One would have a harder time finding a particular molecule in the larger box than in the confines of a smaller one. Furthermore, the process is irreversible: It is extremely improbable that all the molecules would go back into the small box. Common sense tells us that if the volume of the small box is 1/10 of the large one, then roughly 10% of the molecules might be found in the small one, but never 100% of them. What do the cards and the molecules have in common? In this context it is the notion of probability. We already saw in Chapter 19 that the kinetic theory of gases relies on probabilities through distributions which allow us to calculate the temperature as the average kinetic energy of molecules and relate pressure to the average momentum transfer to the walls of a container.

In fact, Boltzmann discovered a deep relationship between entropy and probability. We can give a hint of how this relationship works out by considering the very simple model of the small open box inside a large box, with molecules in the large box. If the small box has volume V_A and the large box containing it has volume V_B, the chance of a particular molecule being in the small box is V_A/V_B. A second molecule, independent of the first one, will have the same chance V_A/V_B of being found in the small box. This means that the chance of both of them being in the small box is $(V_A/V_B)^2$. With three molecules, the chance of all three being in the small box is $(V_A/V_B)^3$ and so on. Thus with n moles of gas, nN_A molecules, the probability of all of them being in the small box is $(V_A/V_B)^{nN_A}$. We are using the fact that probabilities for independent events multiply. But as we saw in Example 20–11, entropies add. How can we bring these facts together? If we have the relation $P_{12} = P_1 \times P_2$, then we can make this into an additive relation by taking logarithms of both sides: $\ln(P_{12}) = \ln(P_1) + \ln(P_2)$.

Apply this to the full set of molecules. The natural logarithm of $P = (V_A/V_B)^{nN_A}$ yields

$$\ln(P) = nN_A \ln(V_A/V_B).$$

Let us compare this with the change in entropy [Eq. (20–22)] in letting an ideal gas escape spontaneously (that is, without a change in temperature) from a volume V_A into a larger volume V_B. We found

$$S(B) - S(A) = nR \ln(V_B/V_A).$$

Taking into account that $R = kN_A$, where k is Boltzmann's constant and comparing the entropy calculation with the logarithm of the probability, we have the relation

$$\Delta S = k \ln(1/P) = -k \ln P. \qquad (20\text{–}24)$$

This relation between entropy and probability is in fact quite general, and it allows us to extend the notion of entropy to all processes in which irreversible changes occur.

The *arrow of time* is the name given to the universal appearance of directionality in processes in the world. This directionality is associated with the higher probability of a disordered state relative to an ordered state. Figure 20–16 dramatizes this directionality. A few boxes of TNT and an abandoned building represent a much more ordered state than the one that appears after the explosion. As we mentioned when we first discussed spontaneous processes, the indications of an arrow of time all around us is not in contradiction with the reversibility of the fundamental laws of nature at a microscopic level.

The elementary forces between the microscopic constituents of matter are, with one very tiny exception discussed in Chapter 45, indifferent to the direction of time, or *invariant under time reversal*. In other words, aside from the exception, this invariance applies to all the fundamental forces in nature, including the electromagnetic forces that govern the collisions of gas molecules. Nevertheless, we have seen that there is a preferred direction of time—an **arrow of time**—in which dropped eggs break rather than reassemble and leap up. Spontaneous processes, in which entropy increases, proceed in one direction and not the other.

We have learned that the reason is statistical. It is not against any physical laws for a broken egg to reassemble itself. It is just so unlikely that it would never happen in many lifetimes of the universe. The overwhelming majority of collisions between gas molecules lead to more disordered situations. The arrow of time expresses not a definite preferred direction of time in a fundamental sense but an overwhelming likelihood that out of a random sequence of microscopic interactions, less ordered situations will result. Irreversibility means only that the spontaneous reversal of the "irreversible" process is so unlikely as not to be worth considering.

Summary

The second law of thermodynamics limits the efficiency of cyclic engines and governs the direction of spontaneous processes in thermal systems. Cyclic engines do work in repeated cycles by extracting a heat flow from hotter external systems, converting part of that heat flow into work, and rejecting a heat flow to cooler thermal systems. The efficiency of such an engine is defined by

$$\eta \equiv \frac{W}{Q_h}, \qquad (20\text{--}1)$$

where W is the work done in one complete cycle and Q_h is the total positive heat flow the cycle extracts from its sources of fuel.

Two equivalent ways to state the second law of thermodynamics are the Kelvin form and the Clausius form. Each of these emphasizes the role of engines and thermodynamic processes. The Kelvin form states that it is impossible to convert thermal energy from a body at a given temperature into an equivalent amount of mechanical work with no other effect; the Clausius form states that it is impossible to transfer thermal energy from a colder body to a hotter body with no other effect.

In the Carnot cycle, which involves four reversible steps that occur at just two temperatures, a working fluid undergoes alternate adiabatic and isothermal transformations. It is the most efficient cycle possible that operates between two temperatures T_h and T_c. No matter what the nature of the working fluid, the efficiency of the Carnot engine is

$$\text{for any Carnot cycle: } \eta_C = 1 - \frac{T_c}{T_h}. \qquad (20\text{--}7)$$

The Carnot cycle that operates between two temperatures is more efficient than any other cycle that operates between the same temperature extremes. Any element of irreversibility also decreases the efficiency.

When the reversible cycles of work-producing engines are run in reverse, they act as refrigerators or heat pumps. Work is done on both of these machines while they extract a heat flow from a cold reservoir and release a heat flow to a hot reservoir.

The thermodynamic variable, entropy, S, measures the amount of disorder in a thermal system. Like thermal energy, it is a function of state, and only *changes* in entropy have physical significance. Entropy changes are path independent. For infinitesimal reversible transformations,

$$dS = \frac{dQ}{T}. \qquad (20\text{--}18)$$

For an ideal gas, entropy can be expressed as a function of T and V as

$$S(B) - S(A) = C_V \ln\frac{T_B}{T_A} + nR \ln\frac{V_B}{V_A}. \qquad (20\text{--}22)$$

Note that entropy decreases as temperature decreases.

The path independence of entropy changes is equivalent to the fact that, for a reversible cycle,

$$\oint \frac{dQ}{T} = 0, \qquad (20\text{--}17)$$

a result known as Clausius' theorem. More generally, we include the possibility of an element of irreversibility by modifying Clausius' theorem to Clausius' inequality,

$$\oint \frac{dQ}{T} \leq 0. \qquad (20\text{--}20)$$

The $<$ sign holds when there is irreversibility. If the change $A \rightarrow B$ is irreversible, the following relation holds:

$$S(B) - S(A) > \int_{\substack{A \\ \text{irrev}}}^{B} \frac{dQ}{T}. \qquad (20\text{--}21)$$

In particular, for transformations of thermally isolated systems, we find that $\Delta S \geq 0$.

A spontaneous process in an isolated system increases the system's entropy, and the state of maximum entropy is the state of stable equilibrium. The increase of entropy expresses the inevitable increase in the disorder of thermally isolated systems.

Understanding the Concepts

1. If the ocean is viewed as a reservoir at one temperature and the atmosphere as a reservoir at another temperature, when can thermal energy flow spontaneously from the ocean to the air?

2. When the brakes are applied, a car decelerates and then stops. A film of the process run in reverse obviously looks fake. Does the process in reverse violate the second law of thermodynamics? If so, in what way?

3. The historian Herman Daly has called the second law of thermodynamics "the law of random, ravage, and rust." Is he right?

4. Why is it not possible for an engine—even a Carnot engine—to have an efficiency of one?

5. In Russia the heat exhaust from nuclear power plants is sometimes used to heat local cities. Does this increase the efficiency of the "engine"? If so, how?

6. Suppose that a gas obeys the van der Waals equation of state. What is the efficiency of a Carnot engine that uses this gas as a working fluid in terms of the temperatures of the hot and cold reservoirs?

7. In what climates are heat pumps most likely to be used?

8. A professor spends a couple of hours straightening up his office—putting books away, filing papers, throwing old exam papers away, and so on. He has created a less probable configuration out of a more probable configuration. Has he violated the second law of thermodynamics?

9. You have a small bottle initially filled with a mysterious gas X in a tightly sealed air-filled room. Gas X is not a component of air. Both gases are initially at 1 atm of pressure. The bottle turns out to be permeable to gas X, but perfectly nonpermeable to any gas in air. What will happen?

10. Imagine a microscopic device at the boundary between two containers of a gas that could sense the arrival of a particularly fast molecule heading from container A toward container B and could open a tiny door to let that molecule through. The same device, which is known generically as a *Maxwell demon*, could sense the arrival of a particularly slow molecule heading from container B toward container A and could let the molecule pass through. How would such a device violate the second law of thermodynamics? Is such a device feasible?[†]

11. Give examples of things you have observed recently that exemplify the second law of thermodynamics. Discuss the entropy change associated with them.

12. Would heat pumps that use an underground water reservoir be more efficient than those that use the outside air as a thermal reservoir if the pumps are employed to (a) heat in the winter? (b) cool in the summer?

13. Are electric resistive heaters 100% efficient?

14. List and discuss some processes not already mentioned that will not violate the first law of thermodynamics but will violate the second law.

15. When gasoline is burned in the chamber of a piston, chemical energy is turned into mechanical energy that is used to propel an automobile. How is this process consistent with the second law of thermodynamics?

16. If you were to run a film of an egg that drops to the floor, you would have little trouble determining if the film were running backward or not. Yet if the film were in extreme closeup, so that you were observing the interaction of molecules in the dropped egg, you would be hard pressed to say if the film were running forward or backward. How are these two statements consistent?

17. Is it possible to cool a house by leaving a refrigerator door open? What would be the net effect if you were to leave the door open?

18. Turbocharged engines take hot exhaust gas and send it back to the engine to heat up the intake air passing into the engine. Is this a satisfactory way to increase the efficiency of an engine?

19. Is the entropy of a gas an extensive quantity (such as volume) or an intensive quantity (such as pressure or temperature)? (See Problem 17–3.)

20. Biological systems use energy (ultimately from the Sun) in organizing highly ordered organisms, such as humans. Does this mean that biological systems violate the second law of thermodynamics?

Problems

20–1 Beyond Energy Conservation

1. (I) An engine produces 3.8×10^5 J of mechanical energy accompanied by a total heat flow of 7.6×10^5 J into the engine. What is its efficiency?

2. (I) Suppose that you toss a pair of dice. Make a list of all the possible outcomes, and calculate the probability of getting totals of 2, 3, 4, ... 12, respectively.

3. (II) Six tokens, numbered one to six, are put into a hat. The tokens with even numbers are white; the ones with odd numbers are black. Suppose you draw four tokens at random, without putting them back into the hat. (a) List all the possible outcomes, without regard to order (for example, 1234 and 2134 do not count as separate outcomes). (b) What is the probability of drawing the tokens with 1, 2, 3, 4 on them? (c) What is the probability of drawing two white and two black tokens?

4. (II) N molecules of a gas are placed in the left half of a chamber and kept from entering the right half by a membrane. The membrane is then removed and the molecules distribute themselves uniformly throughout the whole volume (Fig. 20–19). What is the probability that, at some time during their subsequent random motion, they will all collect in the left half of the

▲ **FIGURE 20–19** Problem 4.

[†]For an interesting discussion of Maxwell demons, see Charles H. Bennett, "Demons, Engines, and the Second Law," *Scientific American*, Nov. 1987, pp. 108–116.

chamber, for the cases $N = 2, 100, 10^6,$ and 10^{23}? [*Hint*: This is equivalent to the probability of getting nothing but heads in 10^{23} coin tosses. To get an idea of this number in scientific notation, use the fact that $\frac{1}{2} \cong 10^{-0.3}$.]

5. (II) It is difficult to carry out experiments with just a few molecules. Here is an analogue to the problem of 10 molecules and on whether they are on the left or the right side of a container. Assume that you have 10 fair coins. Each coin gets tossed to produce a series of 10 heads or tails. Compare the probability that you end up with 10 heads (a maximally ordered array) with the probability of getting five heads and five tails in any order.

6. (II) You have a small 1-L bottle initially filled with a mysterious gas X in a tightly sealed air-filled room whose dimensions are 3 m × 3 m × 3 m. Gas X is not a component of air. Both gases are initially at 1 atm of pressure. The bottle turns out to be permeable to gas X but perfectly opaque to any gas in air. What will be the equilibrium pressure in the bottle, if any?

20–3 The Carnot Cycle

7. (I) What is the maximum efficiency of a steam engine whose boiler (the hottest-temperature reservoir in contact with the engine) is at 160°C and whose condenser (the coldest-temperature reservoir in contact with the engine) is at 35°C?

8. (I) A heat engine absorbs heat from a thermal reservoir at 550K, but the engine's maximum efficiency is only 38%. What must the temperature of the reservoir be to raise the efficiency to 41%?

9. (I) A heat engine receives heat from a thermal reservoir at 750K and transfers it to a thermal reservoir at 340 K. The efficiency of the engine is only 70% of the maximum possible. (a) What is the engine's actual efficiency? (b) How much work does the engine produce for each calorie of heat input?

10. (I) Temperature differences in the ocean have been proposed as a possible energy source in the tropics. Surface water at 29°C could act as a hot reservoir, and deep water at 3.0°C could serve as a cold reservoir. Ammonia gas could be a working fluid in a heat engine that runs between the two thermal reservoirs. What is the maximum efficiency of such an engine?

11. (II) A Carnot engine absorbs thermal energy from a reservoir at 450°C. Its efficiency is 0.55, and the work delivered by the engine is 5.0 kWh. (a) What is the temperature of the cold reservoir? (b) How much thermal energy flows into the system, and how much is rejected?

12. (II) What is the maximum efficiency of a heat engine that operates between 430°C and 125°C? If this engine generates 9800 J of mechanical energy, how many calories does it absorb from the hot reservoir, and how many calories does it transfer into the cold reservoir?

13. (II) A locomotive steam engine operating with a total train mass of 9.9×10^6 kg climbs a hill of height 1000 m in 10 minutes. If the steam engine produces steam at a temperature of 390K and exhausts the steam at 300K, and if the engine were at the theoretically best efficiency possible, what would be the rate, in mass per unit time, at which steam is exhausted? Real steam engines are much less efficient than the theoretical maximum.

14. (II) An athlete metabolizes energy at a rate such that his power output is 0.35 horsepower. (a) Assuming that the efficiency of his "engine" is 25%, how much thermal energy is dissipated? (b) If the dissipation is primarily through the evaporation of sweat, how much sweat is evaporated per minute?

15. (II) An inventor claims to have built an engine that takes in 3.0×10^8 J of thermal energy at 450K, rejects 1.4×10^8 J of thermal energy at 250K, and delivers 1.0×10^8 J of work in 1 h of cyclic operation. Is there anything wrong with this claim?

16. (II) A power plant generates 800 MW of electric power. At what rate does the plant generate waste heat if its efficiency is 28%? Assuming that the plant operates between 480°C and 80°C, what is the maximum efficiency possible?

17. (II) An ideal diatomic gas is used in a Carnot engine with thermal reservoirs at 77 K and 300 K. What is the ratio of the maximum volume to the minimum volume if the pressure drops by a factor of 2 during the isothermal expansion? [*Hint*: For a diatomic gas, $\gamma = 1.4$.]

18. (II) Prove that two reversible adiabatic paths cannot intersect. Do this by assuming that they do, constructing a cycle that consists of the two paths and another reversible transformation, and showing that such a cycle would violate the second law of thermodynamics (Fig. 20–20). (This result is simply shown for an ideal gas. Your proof should apply to a nonideal gas as well.)

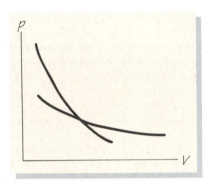

▲ **FIGURE 20–20** Problem 18.

19. (III) Any element of irreversibility makes a Carnot engine and, in fact, any engine, less efficient. Let engine A be a pseudo-Carnot engine, meaning that it has some irreversibility somewhere along the cycle; let engine B be an ideal Carnot engine—the two arranged in size so that they exhaust the same heat flow Q_c. If engine A is more efficient than engine B, then by combining the two engines such that A runs normally and B runs backward, as in Figure 20–6, show that you arrive at a contradiction of the Kelvin form of the second law of thermodynamics. Now that you have shown that engine A must be less efficient than engine B, why can the preceding argument not be reversed, running the Carnot cycle forward and the pseudo-Carnot cycle backward to get a contradiction of the second law?

20. (III) Show that if it is possible to build an engine that violates the Kelvin form of the second law of thermodynamics, it is also possible to build an engine that violates the Clausius form of the second law. [*Hint*: Consider a Carnot engine driven by a Kelvin-violating engine.]

20–4 Other Types of Engines

21. (I) An air conditioner is rated at 8000 Btu/h. Assuming maximum possible efficiency, an exhaust temperature of 38°C, and an interior temperature of 22°C, what is the electrical power consumption? (The work done on the air conditioner is supplied by electricity.)

22. (I) What is the coefficient of performance for an ideal Carnot heat pump operated between outside air at 40°C and a deep water well at 8°C? Would the coefficient of performance be

larger or smaller if your heat pump followed any other type of cycle?

23. (I) Two refrigerators can each be approximated as Carnot cycles run in reverse. Each refrigerator works with the same cold temperature reservoir $T_c = 273K$. Refrigerator 1 evacuates to a hot reservoir at temperature $T_h = 293K$, while refrigerator 2 evacuates to a hot reservoir at a temperature of $T_h = 303K$. Compare the coefficients of performance of these refrigerators. (Real refrigerators will have coefficients of performance substantially less than these.)

24. (I) A heat pump acts like a refrigerator that cools the outside of a house and gives rise to a heat flow to the inside of the house. If the outside temperature is $+3°C$ and the inside temperature is $21°C$, what is the maximum amount of British thermal units a pump can deliver per kilowatt-hour of electric energy supplied? (The work done on the pump has electric energy as its source.)

25. (I) Consider the diesel cycle in Fig. 20–9a. In which portions of the cycle is there heat flow into or out of the engine? What is the work done in the transformation $2 \rightarrow 3$ in terms of T_2 and T_3?

26. (II) An "ideal refrigerator," a Carnot engine that operates in reverse, has a compressor as its engine. It does work W to extract heat Q_c from a cold reservoir at temperature T_c and dumps heat Q_h into a hot reservoir at T_h. Use the known efficiency of a Carnot engine that operates between T_h and T_c and the conservation of energy for one cycle to show that the coefficient of performance $K_{ref} \equiv Q_c/W$ for the compressor is given by $K = T_c/(T_h - T_c)$. Note that K_{ref} is large when the cold and hot reservoirs have nearly the same temperature.

27. (II) A refrigerator with coefficient of performance $K_{ref} = 5.1$ (see Problem 26) gives rise to a heat flow out of the cooling compartment at a rate of 400 cal/min. What is the required power of the motor that operates this refrigerator?

28. (II) To compensate for a given energy loss when a refrigerator is opened too often, the refrigerator's compressor must do work. Take $T_c = 2.0°C$ and $T_h = 29°C$, and suppose that the refrigerator's coefficient of performance is 35% of its ideal value. What is the rate at which the compressor does work; that is, what is the power consumption for a given rate of heat loss dQ/dt?

29. (II) The specifications of a freezer claim that it can remove 80 cal/s heat flow from the compartment at $-20°C$ and release 94 cal/s into the room at $25°C$ while using 60 W of electrical energy to drive the compressor. Can you trust this statement?

30. (II) In order to make an ice cube from water, a refrigerator must extract 1400 cal of heat at the temperature of the freezer, $-12°C$. The room temperature is $30°C$. What minimum work must an ideal refrigerator do?

31. (II) You can buy a 5-lb bag of ice for about $2. Estimate the fraction of this price that pays for the electric energy needed to run the ice-making machine. Assume that the inside of the machine is at $-10°C$, that the outside temperature is $30°C$, that the initial temperature of the water is $20°C$, and that the coefficient of performance is 30% of the ideal. Electrical energy costs $0.10/kWh.

32. (II) The temperature inside a Carnot refrigerator placed in a room at $68°F$ is $38°F$. The heat flow from the refrigerator into the room is 0.95×10^8 J/h. How many watts are needed to operate the refrigerator?

33. (II) A heat engine operates on the cycle shown in Fig. 20–21, with heat absorbed or ejected on each leg as indicated. Its operating fluid is an ideal gas with heat capacity at constant volume C_V. (a) What are temperatures T_3 and T_4 at points 3 and 4, respectively, in terms of the indicated pressures and volumes?

▲ **FIGURE 20–21** Problem 33.

(b) What is the internal energy change along leg IV, ΔU_{IV}? Include the sign. (c) What is Q_{IV} in terms of ΔU_{IV} and the indicated pressures and volumes? (d) What is the efficiency of the engine in terms of the indicated pressures and volumes and C_V?

34. (II) Consider the ideal Stirling cycle shown in Fig. 20–8, working between a maximum temperature T_h and a minimum temperature T_c, and a minimum volume V_1 and a maximum volume V_2. Suppose that the working gas of the cycle is 0.1 mol of an ideal gas with $c'_V = 5R/2$. (a) What are the heat flows to the cycle during each leg? Be sure to give the sign. For which legs is the heat flow positive? (b) What work is done by the cycle during each leg? Again, be sure to include the sign. (c) If, in the definition of the efficiency of this cycle, $\eta = W/Q_{pos}$, where Q_{pos} is the total positive heat flow to the engine, what is the efficiency of the cycle when $T_h = 700K$ and $T_c = 400K$? $V_1 = 0.5$ L and $V_2 = 1.5$ L. Compare this efficiency to the efficiency of a Carnot cycle that operates between the same temperature extremes.

35. (II) The Otto cycle, represented in Figure 20–9b, runs between minimum and maximum volumes V_1 and V_2 and minimum and maximum temperatures T_1 and T_2, respectively. (a) Show that the efficiency is given by $\eta = 1 - (T_2 - T_3)/(T_1 - T_4)$. (b) Show that if the working fluid of the cycle is an ideal gas, the efficiency of this cycle can alternatively be written as $\eta = 1 - (V_1/V_2)^{\gamma-1}$, where $\gamma = C_p/C_V$. The efficiency of this cycle is thus independent of the temperatures between which it operates, depending instead only on γ and on geometry. The ratio V_2/V_1 is the compression ratio. A typical compression ratio is 8, and $\gamma = 1.4$, which gives a predicted efficiency of 56%.

36. (III) The Brayton cycle is represented by Fig. 20–9c. Show that when its working fluid is an ideal gas, the cycle's efficiency is given by the same formula as that for the Otto cycle in the previous problem. In this case, V_1 and V_2 are not the extreme volumes of the engine but are the volumes at the start and finish of the adiabatic expansion step, respectively.

37. (III) Find the efficiency of the diesel cycle illustrated in Fig. 20–9a in terms of $\gamma = C_p/C_V$ and the temperatures at points 1, 2, 3, and 4. Assume that it contains 1 mol of an ideal gas. (Rudolf Diesel tried to produce an engine based on the Carnot cycle. He was not able to produce such an engine, but he came close: A diesel engine nearly matches a Carnot engine in efficiency. Diesel's engine is one of the most efficient in use today.)

20–5 Entropy and the Second Law

38. (I) Consider a large, isolated thermal system in two parts. One part is a thermal reservoir at temperature T_1, labeled I, and the other

part is a thermal reservoir at temperature T_2, labeled II. Reservoir I is colder than reservoir II. These two reservoirs are briefly connected, and there is a heat flow Q from II to I. (a) What are the entropy changes for I and for II? (b) Compute the entropy change for the whole system, and show that it is positive.

39. (I) One end of a metal rod is in contact with a thermal reservoir at 1273K, and the other end is in contact with a reservoir at 293K. If the rate at which heat flow passes from the hot end to the cold end is 30 J/min, what is the rate of total entropy change? Is the process reversible?

40. (II) Find the entropy change for (a) 1 g of ice melting at 0°C; (b) 1 g of water evaporating at 100°C. You will need to look up the heats of melting and of vaporization.

41. (II) One kilogram of iron at 80°C is dropped into 0.5 L of water at 20°C. Given that the specific heat of water is 1 cal/g·K and that of iron is 0.107 cal/g·K, calculate the final equilibrium temperature of the system and the increase in entropy.

42. (II) One hundred and twenty grams of water at 100°C is changed reversibly into steam at 100°C. (a) Describe such a process. (b) What is the change in entropy of the water and of the universe? (c) How would part (b) change if the process were irreversible?

43. (II) A boater dumps 8.7 kg of ice from his cooler into the lake at the end of a fishing trip. The ice is at −3°C, and the lake is at 18.7°C. What is the change in entropy of the new lake–ice cube system after the ice cubes come to equilibrium with the lake?

44. (II) Calculate the entropy change of the universe when a 1-kg block of copper at a temperature of 200°C is put in a lake at a temperature of 4°C. The total heat capacity of this block of copper is 375 J/K.

45. (II) Calculate the change in entropy of the universe if 0.3 kg of water at 70°C is mixed with 0.2 kg of water at 15°C in a thermally insulated container. [*Hint:* Even though this process is irreversible, the change in entropy can be calculated by devising a reversible process that leads from the given initial state to the expected final state.]

46. (III) Thermodynamic tables often give the specific heat of materials in the form of a list of coefficients a, b, c, d in the expression $c_p = a + bT + cT^2 + d/T$. How can you use these data to find the entropy change of the material as it undergoes a transformation from T_i to T_f?

20–6 Entropy and Ideal Gases

47. (I) Five moles of an ideal gas is in thermal isolation and undergoes free expansion from 35 L to 100 L. What is the entropy change of the gas? of the universe?

48. (I) The heat capacity at constant volume of a sample of gas is determined to be 22 J/K. If this amount of gas is heated in a constant-volume container from 273K to 840K, what is the change in entropy?

49. (I) One mole of an ideal gas expands at constant pressure from an initial volume of 250 cm³ to a final volume of 650 cm³. What is the change in entropy, assuming that the molar specific heat is $c_V' = 3R/2$?

50. (I) One mole of ideal gas with molar specific heat $c_V' = 3R/2$ is heated from 273K to 350K at constant volume. What is the change in entropy?

51. (I) An ideal gas undergoes an isothermal compression. Does the entropy increase?

52. (II) Ten moles of a monatomic gas held in a tank at 17 atm escape from the tank isothermally at 27°C. What is the change in the entropy of the universe?

53. (II) One mole of a monatomic ideal gas at an initial pressure of 30 atm and a temperature of 600K is allowed to undergo a rapid (adiabatic) free expansion from a small vessel into a vessel of 50 times greater volume. Find the change in temperature and the increase in entropy.

54. (II) One mole of a monatomic ideal gas is carried in two steps from (p_1, V_1) to $(p_1, 5V_1)$ to $(0.3p_1, 5V_1)$ by reversible thermodynamic processes that are straight lines on a p–V diagram (Fig. 20–22). (a) What is the overall change in the gas's internal energy? (b) What is the overall work done by the gas? (c) How much net heat flow is added overall? (d) What is the net change in the entropy?

▲ **FIGURE 20–22** Problem 54.

55. (II) A gas obeys the well-known equation of state $pV = (\text{a constant})\,T$. The gas expands, doubling in volume. (a) Plot p versus V when the expansion is isobaric and when it is isothermal. (b) What work is done by the gas on its surroundings for both cases above? (c) What is the entropy change of the gas for both cases above?

56. (II) Show that in terms of temperature and pressure, the entropy of an ideal gas with constant heat capacities is $S(T, p) = C_p \ln(T/T_0) - nR \ln(p/p_0) + S(T_0, p_0)$.

57. (III) Dilute helium gas is taken through the following reversible cycle on a p–V diagram: AB is an isobaric compression; BC is an adiabatic expansion; CD is an isothermal expansion; DA is a constant-volume pressure increase (Fig. 20–23). Express the heat flow, work done on the system, change in internal energy, change in temperature, and change in entropy for each leg of the cycle in terms of the quantities T_A, T_B, T_C, and p_A. Pay attention to the signs.

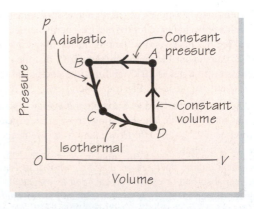

▲ **FIGURE 20–23** Problem 57.

20–7 The Meaning of Entropy

58. (I) A total of 12 molecules are in a large box, which we mentally divide into two halves, a left half and a right half. Any one molecule is equally likely to be in either of the two halves. What is the probability that all 12 molecules will be found in the left half? How does your answer change if there are 120 molecules?

59. (II) An ideal gas in a cylinder with a movable piston is placed in thermal contact with a thermal reservoir, and the gas is slowly compressed. How does the entropy of the gas in the cylinder change? How does the entropy of the reservoir change? Is the answer to these questions in conflict with the second law of thermodynamics? Explain your answer.

60. (II) A box of volume 3.0 L contains 1.8 L of He in one side and 1.2 L of O_2 in the other, separated by a partition. The temperature is 290K, and the pressure in both sides of the box is 1.2 atm. The partition is removed. (a) Describe the new equilibrium configuration. (b) What is the change in entropy?

61. (II) A thermally isolated container of volume $3V_0$ is divided by a partition. In one portion, of volume V_0, there are n mol of a monatomic ideal gas at temperature T_0. The other portion is empty. The partition is then broken, and the gas expands to fill the entire container (Fig. 20–24). (a) What is the final temperature of the gas? (b) The change in entropy of the gas? (c) The change in entropy of the universe? (d) Suppose that the gas is originally confined to volume V_0 by means of a piston instead of a partition. The piston is then slowly withdrawn until a final volume of $3V_0$ is obtained. What is the change in entropy of the gas?

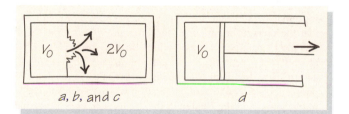

V_0 $2V_0$ V_0

$a, b,$ and c d

▲ FIGURE 20–24 Problem 61.

General Problems

62. (II) A nuclear power plant requires 1450 MW of cooling. If environmental concerns limit the temperature rise of the cooling water to 9°C, how much water flow is required to cool the heat generated by the plant? If the water is at an initial temperature of 287K, what is the entropy change per kilogram of water?

63. (II) Fifty grams of oxygen gas at 320K do 80 J of work while 40 cal of heat flow is absorbed by the gas. (a) What is the change in internal energy? (b) In the temperature of the gas? (c) In entropy, assuming that the change is isobaric?

64. (II) Calculate the entropy change when 1 kg of water is cooled from 100°C to 0°C. Calculate the entropy change when the water turns to ice at 0°C (Fig. 20–25). The latent heat of fusion is 80 cal/g.

65. (II) Sketch on an S–T diagram the following thermodynamic reversible transformations: (a) An adiabatic transformation at constant volume, involving a pressure change. (b) A rough picture of the Stirling cycle.

66. (II) You have a box of volume V divided by an imaginary partition (molecules pass right through it) into two parts, volumes V_1 and

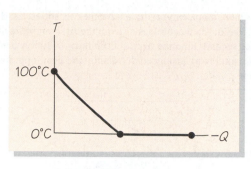

▲ FIGURE 20–25 Problem 64.

V_2. You start out with N_1 molecules of gas g_1 and N_2 molecules of gas g_2, all spread evenly through the box. What is the probability that you find all the g_1 molecules in V_1 and all the g_2 molecules in V_2? Work out the numbers for $V_1 = V/3$ and $V_2 = 2V/3$ while $N_1 = N_2 = 100$. Repeat, but for $N_1 = N_2 = 1000$.

67. (II) Two moles of helium at STP are compressed in an isothermal process to a pressure of 2 atm. (a) What is the final volume? (b) the final temperature? (c) the change in entropy?

68. (II) A Carnot engine for which the working fluid is 1 mol of an ideal gas with $c'_V = (5/2)R$ operates between two thermal reservoirs of temperatures $T_h = 550K$ and $T_c = 350K$. When the gas is in contact with the hot reservoir, its minimum volume is $V_1 = 0.70$ L, and when it is in contact with the cold reservoir, its maximum volume is $V_3 = 8.0$ L. (a) What is the efficiency of this engine? (b) For one cycle, what is the net energy change of the gas? (c) The net entropy change of the gas? (d) The net change in the entropy of the universe? (e) What is V_2, the maximum volume when the gas is in contact with the hot reservoir?

69. (II) Three moles of oxygen at STP undergoes an adiabatic compression to a final pressure of 3.0 atm. (a) What is the final volume? (b) The final temperature? (c) The change in entropy?

70. (II) One liter of nitrogen at STP is adiabatically compressed to a volume of 0.50 L when a membrane breaks and the nitrogen gas undergoes free expansion back to 1.0 L. (a) What is the final pressure? (b) The final temperature? (c) The change in entropy?

71. (II) A heat engine operates between the temperatures of 400K and 300K, doing 100 J of work. This work is used to run a refrigerator between the same temperatures. (a) Calculate the heat transfer between each reservoir and the engine or the refrigerator, assuming that both of them have ideal efficiency/coefficient of performance. (b) Repeat the calculation if the heat engine's efficiency is only 80% of the ideal. (c) Repeat the calculation for the case of an ideal heat engine but for a refrigerator whose coefficient of performance is 80% of the ideal.

72. (II) An inventor proposes that you invest in a newly developed engine. It purports to absorb 140,000 Btu at a temperature of 700°C, and reject 60,000 Btu at a temperature of 400°C. Should you invest?

73. (II) For one mole of a van der Waals gas, the energy change dE for a given temperature change dT and volume change dV takes the form

$$dE = C_V \, dT + \frac{a}{V^2} dV,$$

where in this problem we'll assume that C_V has no temperature dependence. Your gas undergoes free expansion from V_i, T_i to V_f, T_f. Does the temperature (a) increase, (b) decrease, or (c) remain unchanged? [*Hint*: In free expansion the energy is unchanged.

This energy has the form $E(T, V) = C_V \int_{T_0}^{T} dT + \int_{V_0}^{V} \frac{a}{V^2} dV$.]

74. (II) One Carnot engine drives another in series, as shown in Fig. 20–26. The heat released from the first engine is absorbed by the second. Find the overall efficiency, which is determined by the total work produced divided by the heat input to the first engine.

▲ **FIGURE 20–26** Problem 74.

75. (II) A coal-fired, 300-MW generating plant operates between 750K and 400K, with an efficiency 60% of the maximum efficiency possible. At what rate is waste heat flow produced? Suppose that water is used to carry off the waste heat flow, and it does so by being heated by 12°C. How much water must flow through the machinery per second?

76. (II) Consider the reversible thermodynamic cycle of the following form (Fig. 20–27): (i) One mole of a monatomic ideal gas at temperature T_a and pressure p_a is allowed to expand isobarically until the volume has increased by a factor r. (ii) The gas is then cooled so that its absolute temperature is reduced by a factor of 2, with the cooling performed at constant volume. (iii) The gas is then compressed isobarically to its original volume. (iv) Finally, the gas is heated to its starting temperature, T_a, at constant volume. (a) What is the mechanical work done by the engine? (b) The heat transfer in the four legs of the cycle? (c) The efficiency of the cycle?

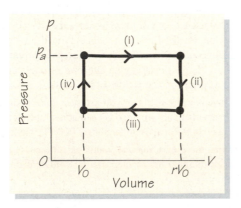

▲ **FIGURE 20–27** Problem 76.

77. (II) A heat engine operates in a reversible cycle with the following steps (Fig. 20–28): (i) An ideal gas characterized by $\gamma = 1.4$ expands adiabatically, so that it cools from an initial temperature of 330°C to 30°C; (ii) it is compressed isothermally until it reaches its initial volume; and (iii) it is then heated at constant volume until it reaches its initial temperature. Calculate the efficiency of the cycle and compare it with the Carnot efficiency that corresponds to these temperatures.

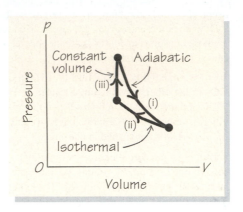

▲ **FIGURE 20–28** Problem 77.

78. (II) Suppose that there is steady flow of a mass M of gas through a turbine. Gas enters the turbine at an elevation z_1, pressure p_1, internal energy U_1, and speed v_1, and leaves at an elevation z_2, pressure p_2, internal energy U_1, and speed v_2. Thermal energy Q is supplied to the turbine. Work W is done by the turbine, and V_1 and V_2 are the respective volumes occupied by mass M on entering and leaving the turbine. (a) Use energy conservation to show that the change in enthalpy (see Problem 79) in terms of the specified quantities can be written in the form $H_2 - H_1 = Q - W + \frac{1}{2}mv_1^2 + Mgz_1 - \left(\frac{1}{2}mv_2^2 + Mgz_2\right)$. (b) Use this result to show that there is no change in enthalpy in the Joule experiment described in Section 18–7 (in which there is no change in elevation, and the change in the speed of the gas is so small that $v_2^2 - v_1^2$ can be neglected).

79. (II) A useful quantity called enthalpy, H, is defined by $H = U + pV$. Calculate the enthalpy of an ideal gas as a function of temperature.

80. (III) Three reversible paths, legs i, ii, and iii, connect points (p_2, V_2) and (p_1, V_1) in Fig. 20–29. (a) Calculate $\int_1^2 dW$ and $\int_1^2 dQ$ for the three paths. (b) Calculate $\int_1^2 dQ/T$ for the three paths.

▲ **FIGURE 20–29** Problem 80.

81. (III) The Brayton cycle is used to extract work from a high-temperature, gas-cooled nuclear reactor with helium (a monatomic ideal gas) as the working fluid. The helium enters a compressor, and its pressure is raised adiabatically. It is then heated at constant pressure to a high temperature in the reactor core. The energy is supplied by the fission of uranium or plutonium. The helium then enters a turbine, in which it is allowed to expand adiabatically. It then passes through a precooler, in which heat is rejected at constant pressure until it reaches the initial conditions. Suppose that helium enters the compressor at 30°C and a pressure of 20 atm. It is compressed to 50 atm and then heated to 1200°C in the reactor core (Fig. 20–30). (a) What are the temperatures at the end of compression and expansion? (b) What are the heat flows per kilogram of helium in each cycle? (c) How much net work is done per kilogram of helium in each cycle?

▲ **FIGURE 20–30** Problem 81.

The Système Internationale (SI) of Units

I-1 SOME SI BASE UNITS

Physical Quantity	Name of Unit	Symbol
length	meter	m
mass	kilogram	kg
time	second	s
electric current	ampere	A
thermodynamic temperature	kelvin	K
amount of substance	mole	mol

I-2 SOME SI DERIVED UNITS

Physical Quantity	Name of Unit	Symbol	SI Unit
frequency	hertz	Hz	s^{-1}
energy	joule	J	$kg \cdot m^2/s^2$
force	newton	N	$kg \cdot m/s^2$
pressure	pascal	Pa	$kg/m \cdot s^2$
power	watt	W	$kg \cdot m^2/s^3$
electric charge	coulomb	C	$A \cdot s$
electric potential	volt	V	$kg \cdot m^2/A \cdot s^3$
electric resistance	ohm	Ω	$kg \cdot m^2/A^2 \cdot s^3$
capacitance	farad	F	$A^2 \cdot s^4/kg \cdot m^2$
inductance	henry	H	$kg \cdot m^2/A^2 \cdot s^2$
magnetic flux	weber	Wb	$kg \cdot m^2/A \cdot s^2$
magnetic flux density	tesla	T	$kg/A \cdot s^2$

I-3 SI UNITS OF SOME OTHER PHYSICAL QUANTITIES

Physical Quantity	SI Unit
speed	m/s
acceleration	m/s^2
angular speed	rad/s
angular acceleration	rad/s^2
torque	$kg \cdot m^2/s^2$, or $N \cdot m$
heat flow	J, or $kg \cdot m^2/s^2$, or $N \cdot m$
entropy	J/K, or $kg \cdot m^2/K \cdot s^2$, or $N \cdot m/K$
thermal conductivity	$W/m \cdot K$

I-4 SOME CONVERSIONS OF NON-SI UNITS TO SI UNITS

Energy:
1 electron-volt (eV) = 1.6022×10^{-19} J
1 erg = 10^{-7} J
1 British thermal unit (BTU) = 1055 J
1 calorie (cal) = 4.185 J
1 kilowatt-hour (kWh) = 3.6×10^6 J

Mass:
1 gram (g) = 10^{-3} kg
1 atomic mass unit (u) = 931.5 MeV/c^2 = 1.661×10^{-27} kg
1 MeV/c^2 = 1.783×10^{-30} kg

Force:
1 dyne = 10^{-5} N
1 pound (lb or #) = 4.448 N

Length:
1 centimeter (cm) = 10^{-2} m
1 kilometer (km) = 10^3 m
1 fermi = 10^{-15} m
1 Angstrom (Å) = 10^{-10} m
1 inch (in or ″) = 0.0254 m
1 foot (ft) = 0.3048 m
1 mile (mi) = 1609.3 m
1 astronomical unit (AU) = 1.496×10^{11} m
1 light-year (ly) = 9.46×10^{15} m
1 parsec (ps) = 3.09×10^{16} m

Angle:
1 degree (°) = 1.745×10^{-2} rad
1 min (′) = 2.909×10^{-4} rad
1 second (″) = 4.848×10^{-6} rad

Volume:
1 liter (L) = 10^{-3} m^3

Power:
1 kilowatt (kW) = 10^3 W
1 horsepower (hp) = 745.7 W

Pressure:
1 bar = 10^5 Pa
1 atmosphere (atm) = 1.013×10^5 Pa
1 pound per square inch (lb/in^2) = 6.895×10^3 Pa

Time:
1 year (yr) = 3.156×10^7 s
1 day (d) = 8.640×10^4 s
1 hour (h) = 3600 s
1 minute (min) = 60 s

Speed:
1 mile per hour (mi/h) = 0.447 m/s

Magnetic field:
1 gauss = 10^{-4} T

APPENDIX ▌▌

Some Fundamental Physical Constants[†]

Constant	Symbol	Value	Error
speed of light in a vacuum	c	2.99792458×10^8 m/s	exact
gravitational constant	G	6.67259×10^{-11} m^3/kg·s^2	128
Avogadro's number	N_A	6.02214×10^{23} mol^{-1}	0.1
universal gas constant	R	8.31447 J/mol·K	8.4
Boltzmann's constant	k	1.38065×10^{-23} J/K	1.7
elementary charge	e	1.60218×10^{-19} C	0.004
permittivity of free space	ε_0	$8.85418781762 \times 10^{-12}$ C^2/N·m^2	exact
	$1/4\pi\varepsilon_0$	8.987552×10^9 kg·m^3·s^{-2}·C^{-2}	
permeability of free space	μ_0	$4\pi \times 10^{-7}$ T·m/A	exact
electron mass	m_e	9.10939×10^{-31} kg	0.1
proton mass	m_p	1.67262×10^{-27} kg	0.1
neutron mass	m_n	1.67493×10^{-27} kg	0.1
Planck's constant	h	6.62607×10^{-34} J·s	0.1
$h/2\pi$	$\hbar$	1.05457×10^{-34} J·s	0.1
		$= 6.58212 \times 10^{-22}$ MeV·s	0.1
	$\hbar c$	197.327 MeV·fm	0.3
electron charge-to-mass ratio	$-e/m_e$	-1.75882×10^{11} C/kg	0.1
proton-electron mass ratio	m_p/m_e	1836.15	0.15
molar volume of ideal gas at STP		22414.0 cm^3/mol	1.7
Bohr magneton	μ_B	9.27401×10^{-24} J/T	0.1
magnetic flux quantum	$\Phi_0 = h/2e$	2.06783×10^{-15} Wb	0.1
Bohr radius	a_0	0.529177×10^{-10} m	0.005
Rydberg constant	R_∞	1.09737×10^7 m^{-1}	0.00001

[†]P. J. Mohr and B. N. Taylor, "The 1998 CODATA Recommended Values of the Fundamental Physical Constants, Web Version 3.1," available at physics.nist.gov/constants (National Institute of Standards and Technology, Gaithersburg, MD 20899, 3 December 1999).

We have given values of the measured constants to six significant figures, even though they may be known to greater accuracy. The error, which expresses the uncertainty in the values of these constants, is in parts per million. Defined constants have no error, and we give their full definition; they are indicated by the notation "exact" in the error column.

APPENDIX III

Other Physical Quantities

III–1.1 SOME ASTRONOMICAL CONSTANTS

Constant	Symbol	Value
standard gravity at Earth's surface	g	9.80665 m/s^2
equatorial radius of Earth	R_e	$6.378 \times 10^6 \text{ m}$
mass of Earth	M_e	$5.976 \times 10^{24} \text{ kg}$
mass of Moon		$7.350 \times 10^{22} \text{ kg}$
		$= 0.0123 \, M_e$
mean radius of Moon's orbit around Earth		$3.844 \times 10^8 \text{ m}$
mass of Sun	$M_\odot$	$1.989 \times 10^{30} \text{ kg}$
radius of Sun	$R_\odot$	$6.96 \times 10^8 \text{ m}$
mean radius of Earth's orbit around Sun	AU	$1.496 \times 10^{11} \text{ m}$
period of Earth's orbit around Sun	yr	$3.156 \times 10^7 \text{ s}$
diameter of our galaxy		$7.5 \times 10^{20} \text{ m}$
mass of our galaxy		$2.7 \times 10^{41} \text{ kg}$
		$= (1.4 \times 10^{11}) \, M_\odot$
Hubble parameter	H	$2.5 \times 10^{-18} \text{ s}^{-1}$

III–1.2 PLANETARY DATA

Planet	Diameter (in km)	Relative[†]	Relative Mass[†]	Average Density (in g/cm³)	Period of Rotation	Surface Gravity[†] (in g)	Escape Speed (in km/s)	Semimajor Axis (AU)	Period of Solar Orbit	Average Orbital Speed (in km/s)
Mercury	4,800	0.38	0.05	5.4	58 d 15 h	0.38	4.3	0.387	87.96 d	47.8
Venus	12,100	0.95	0.82	5.2	243 d 4 h	0.90	10.3	0.723	224.7 d	35.0
Earth	12,750	1.00	1.00	5.5	23 h 56 min	1.00	11.2	1.000	365.26 d	29.8
Mars	6,800	0.53	0.11	3.9	24 h 37 min	0.38	5.0	1.524	687.0 d	24.1
									or 1.88 yr	
Jupiter	142,800	11.21	317.8	1.3	9 h 50 min	2.53	59.5	5.20	11.86 yr	13.1
Saturn	120,660	9.45	95.2	0.7	10 h 39 min	1.07	35.5	9.58	29.46 yr	9.7
Uranus	51,000	4.00	14.5	1.3	17 h	0.91	21.3	19.20	84.01 yr	6.8
Neptune	49,500	3.88	17.1	1.6	16 h	1.14	23.5	30.05	164.79 yr	5.4
Pluto	2,390	0.18	0.002	0.32	6 d 9 h 17 min	0.05	1.1	39.24	247.68 yr	4.7

[†]Relative to Earth.

III–2 ENERGY SUPPLY AND DEMAND[†]

[†]From the *Physics Vade Mecum*, Ed. Herbert L. Anderson, American Institute of Physics (New York, 1981); and U.S. Congress, Office of Technology Assessment, *Changing by Degrees: Steps to Reduce Green-house Gases*, OTA-O-482 (Washington, D.C.: U.S. Government Printing Office, February 1991).

III–2.1 FUEL RESOURCES (1980, ESTIMATED)

Resource	U.S. Resources	World Resources
coal (recoverable)	5×10^{21} J	2×10^{22} J
oil (not including oil shales)	10^{21} J	10^{22} J
natural gas	2×10^{21} J	10^{22} J
hydroelectric	10^{22} J/yr	6×10^{22} J/yr (North America)

III–2.2 ANNUAL USAGE OF RESOURCE (2001, PERCENTAGE OF TOTAL)
Source: www.energy.gov

Resource	U.S. Usage (total $= 1 \times 10^{20}$ J)	World Usage (total $= 4 \times 10^{20}$ J)
coal	23	24
oil	40	39
natural gas	24	23
nuclear	8	6
hydroelectric	2	7
biomass	3	1

III–2.3 ENERGY CONTENT OF FUELS

Fuel	Energy Content (in J/kg)
bread	10×10^{6}
glucose ($C_6H_{12}O_6$)	16×10^{6}
white pine wood	20×10^{6}
methyl alcohol (CH_4O)	23×10^{6}
anthracite coal	31×10^{6}
domestic heating oil	45×10^{6}
propane (C_3H_8)	50×10^{6}
natural gas (96% CH_4)	51×10^{6}
fission of U^{235}	5.8×10^{11}
perfect mass-energy conversion	9×10^{16}

III–2.4 SOLAR ENERGY OUTPUT

total radiated power from the Sun	4×10^{26} W
power per unit area at the top of Earth's atmosphere	1.4 kW/m^2
average power per unit area delivered to an average horizontal surface in the United States in 1 yr	0.2 W/m^2

III–2.5 ENERGY CONSUMPTION IN TRANSPORTATION

Mode	Energy Consumption (J/passenger · km)
bicycle	5×10^{4}
foot travel	1.5×10^{5}
automobile	1.9×10^{5}
intercity bus	6×10^{5}
intercity train	9×10^{5}
747 jet airplane	2.3×10^{6}
snowmobile	6×10^{6}

III–2.6 ENERGY CONSUMPTION OF ELECTRICAL APPLIANCE (*See* http://www.ianr.unl.edu/pubs/consumered/heg94.htm)

Appliance	Power (in W)	Energy Use per Year (in kWh)
window air conditioner	3750	3750
clock	2	17
dishwasher	1200	363
window fan	200	170
hair dryer	750	38
iron	1000	144
microwave oven	1450	190
radio	71	86
refrigerator-freezer	615	1830
stove	12,200	1175
color television	200	440
vacuum cleaner	630	46
washing machine	512	107

APPENDIX IV

Mathematics

IV–1 SOME MATHEMATICAL CONSTANTS†

Constant	Value
π	3.14159
e (Euler's constant)	2.71828
$\sqrt{2}$	1.41421
$1/\sqrt{2}$	0.707107
$\ln(10)$	2.30259
$\ln(2)$	0.693147
1 rad	57.2958°
1°	0.0174533 rad

†To six significant figures.

IV–2 SOLUTION OF QUADRATIC EQUATIONS

Quadratic equation:

$$ax^2 + bx + c = 0$$

Two solutions:

$$x = \frac{-b \pm \sqrt{b^2 - 4ac}}{2a}$$

IV–3 BINOMIAL THEOREM

$$(x + y)^n = \sum_{k=0}^{n} \binom{n}{k} x^{n-k} y^k$$

where

$$\binom{n}{k} = \frac{n!}{(n-k)!k!}$$

The factorial $m! \equiv 1 \cdot 2 \cdot 3 \ldots \cdot m$; $0! \equiv 1$. Some particular cases of the binomial theorem:

(1) $(x \pm y)^2 = x^2 \pm 2xy + y^2$;
(2) $(x \pm y)^3 = x^3 \pm 3x^2y + 3xy^2 \pm y^3$;
(3) $(x \pm y)^4 = x^4 \pm 4x^3y + 6x^2y^2 \pm 4xy^3 + y^4$.

IV–4 TRIGONOMETRY

1. For a right triangle with sides a, b, and c (the hypotenuse), where the angle opposite side a is θ_a (Figure A–1),

$$\text{sine of } \theta_a = \sin\theta_a = \frac{a}{c};$$

$$\text{cosine of } \theta_a = \cos\theta_a = \frac{b}{c};$$

$$\text{tangent of } \theta_a = \tan\theta_a = \frac{a}{b}.$$

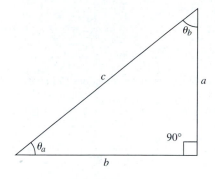

▲ FIGURE A–1

2. The cosine function is even, $\cos(-x) = \cos x$; the sine function is odd, $\sin(-x) = -\sin x$.

3. (1) $\tan\theta = \dfrac{\sin\theta}{\cos\theta}$

   (2) $\sec\theta = \dfrac{1}{\cos\theta}$

   (3) $\operatorname{cosec}\theta = \dfrac{1}{\sin\theta}$

   (4) $\cot\theta = \dfrac{1}{\tan\theta}$

4. (1) $\sin^2\theta + \cos^2\theta = 1$

   (2) $\sec^2\theta - \tan^2\theta = 1$

   (3) $\operatorname{cosec}^2\theta - \cot^2\theta = 1$

5. (1) $\sin(\theta_1 \pm \theta_2) = \sin\theta_1\cos\theta_2 \pm \cos\theta_1\sin\theta_2$

   (2) $\cos(\theta_1 \pm \theta_2) = \cos\theta_1\cos\theta_2 \mp \sin\theta_1\sin\theta_2$

   (3) $\sin\theta_1 \pm \sin\theta_2 = 2\sin\left(\dfrac{\theta_1 \pm \theta_2}{2}\right)\cos\left(\dfrac{\theta_1 \mp \theta_2}{2}\right)$

   (4) $\cos\theta_1 + \cos\theta_2 = 2\cos\left(\dfrac{\theta_1 + \theta_2}{2}\right)\cos\left(\dfrac{\theta_1 - \theta_2}{2}\right)$

   (5) $\cos\theta_1 - \cos\theta_2 = -2\sin\left(\dfrac{\theta_1 + \theta_2}{2}\right)\sin\left(\dfrac{\theta_1 - \theta_2}{2}\right)$

   (6) $\tan(\theta_1 + \theta_2) = \dfrac{\tan\theta_1 + \tan\theta_2}{1 - (\tan\theta_1)(\tan\theta_2)}$

   (7) $\cos\left(\theta \pm \dfrac{\pi}{2}\right) = \mp\sin\theta$

   (8) $\sin\left(\theta \pm \dfrac{\pi}{2}\right) = \pm\cos\theta$

   (9) $\sin\theta_1\sin\theta_2 = \dfrac{1}{2}\left[\cos(\theta_1 - \theta_2) - \cos(\theta_1 + \theta_2)\right]$

   (10) $\cos\theta_1\cos\theta_2 = \dfrac{1}{2}\left[\cos(\theta_1 - \theta_2) + \cos(\theta_1 + \theta_2)\right]$

   (11) $\sin\theta_1\cos\theta_2 = \dfrac{1}{2}\left[\sin(\theta_1 - \theta_2) + \sin(\theta_1 + \theta_2)\right]$

6. (1) $\sin(2\theta) = 2\sin\theta\cos\theta = \dfrac{2\tan\theta}{1+\tan^2\theta}$

(2) $\cos(2\theta) = \cos^2\theta - \sin^2\theta = 2\cos^2\theta - 1 = 1 - 2\sin^2\theta$

(3) $\tan(2\theta) = \dfrac{2\tan\theta}{1-\tan^2\theta}$

(4) $\sin\left(\dfrac{\theta}{2}\right) = \pm\sqrt{\dfrac{1-\cos\theta}{2}}$

(5) $\cos\left(\dfrac{\theta}{2}\right) = \pm\sqrt{\dfrac{1+\cos\theta}{2}}$

7. Expansions of trigonometric functions (θ in rad):

(1) $\sin\theta = \theta - \dfrac{\theta^3}{3!} + \dfrac{\theta^5}{5!} - \dfrac{\theta^7}{7!} + \cdots \qquad (\theta^2 < 1)$

(2) $\cos\theta = 1 - \dfrac{\theta^2}{2!} + \dfrac{\theta^4}{4!} - \dfrac{\theta^6}{6!} + \cdots \qquad (\theta^2 < 1)$

(3) $\tan\theta = \theta + \dfrac{1}{3}\theta^3 + \dfrac{2}{15}\theta^5 + \dfrac{17}{315}\theta^7 + \cdots \qquad \left(\theta^2 < \dfrac{\pi^2}{4}\right)$

IV–5   GEOMETRICAL FORMULAS

1. (circumference of a circle of radius r) $= 2\pi r$

2. (area of a circle of radius r) $= \pi r^2$

3. (area of a sphere of radius r) $= 4\pi r^2$

4. (volume of a sphere of radius r) $= \frac{4}{3}\pi r^3$

5. (area of a rectangle with sides of lengths L_1 and L_2) $= L_1 L_2$

6. For a right triangle with sides a, b, and c and angles θ_a and θ_b opposite the sides a and b, respectively (Fig. A–1):

(1) $a^2 + b^2 = c^2$ (the Pythagorean theorem)

(2) area $= \frac{1}{2}(\text{base})(\text{height}) = \frac{1}{2}ab$

7. For a triangle with sides a, b, and c opposite the angles θ_a, θ_b, and θ_c, respectively (Figure A–2):

(1) $\theta_a + \theta_b + \theta_c = 180° = \pi$ rad

(2) $a^2 = b^2 + c^2 - 2bc\cos\theta_a$

(3) $\dfrac{a}{\sin\theta_a} = \dfrac{b}{\sin\theta_b} = \dfrac{c}{\sin\theta_c}$

(4) $a = b\cos\theta_c + c\cos\theta_b$

(5) area $= \dfrac{1}{2}(\text{base})(\text{height}) = \dfrac{1}{2}ab\sin\theta_c = \dfrac{1}{2}a^2\dfrac{\sin\theta_b\sin\theta_c}{\sin\theta_a}$

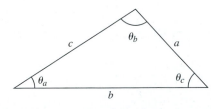

▲ **FIGURE A–2**

8. volume of a right cylinder of height h and radius $r = \pi r^2 h$

IV–6   SOME PROPERTIES OF ALGEBRAIC FUNCTIONS

1. General properties

(1) $a^x a^y = a^{x+y}$

(2) $a^0 = 1$

(3) $(ab)^x = a^x b^x$

2. Properties of exponential of x, $\exp(x)$ or e^x:

(1) $\exp(\ln x) = x$

(2) $\exp(x_1)\exp(x_2) = \exp(x_1 + x_2)$

(3) $\exp(0) = 1$

(4) expansion: $e^x = 1 + x + \dfrac{x^2}{2!} + \dfrac{x^3}{3!} + \cdots$

3. Properties of the natural logarithm of x, $\ln(x)$:

(1) $\ln(e^x) = x$

(2) $\ln(x_1 x_2) = \ln(x_1) + \ln(x_2)$

(3) $\ln(x_1/x_2) = \ln(x_1) - \ln(x_2)$

(4) $\ln(1) = 0$

(5) expansion:

$$\ln(1+x) = x - \dfrac{x^2}{2} + \dfrac{x^3}{3} - \dfrac{x^4}{4} + \cdots \quad (x^2 < 1)$$

IV–7   DERIVATIVES

In the following, b and p are constants, and u and v are functions of x:

1. $\dfrac{db}{dx} = 0$

2. $\dfrac{d}{dx}(bu) = b\dfrac{du}{dx}$

3. $\dfrac{d}{dx}(u+v) = \dfrac{du}{dx} + \dfrac{dv}{dx}$

4. $\dfrac{d}{dx}(uv) = v\dfrac{du}{dx} + u\dfrac{dv}{dx}$

5. $\dfrac{dx^p}{dx} = px^{p-1}$

6. Chain rule: If u is a function of y, and y is in turn a function of x, then $\dfrac{du}{dx} = \dfrac{du}{dy}\dfrac{dy}{dx}$.

7. $\dfrac{d}{dx}(\sin x) = \cos x$

8. $\dfrac{d}{dx}(\cos x) = -\sin x$

9. $\dfrac{d}{dx}(\tan x) = \dfrac{1}{\cos^2 x}$

10. $\dfrac{d}{dx}(e^{bx}) = be^{bx}$

11. $\dfrac{d}{dx}\ln(x) = \dfrac{1}{x}$

IV–8 TAYLOR EXPANSION

If $f(x)$ is well behaved near point $x = x_0$,

$$f(x) = f(x_0) + \frac{df}{dx}\bigg|_{x=x_0} (x - x_0) + \frac{1}{2!}\frac{d^2f}{dx^2}\bigg|_{x=x_0} (x - x_0)^2 + \cdots$$

$$(6) \quad \int_0^\infty \frac{b}{b^2 + x^2}\, dx = \left\{ \begin{array}{ll} \dfrac{\pi}{2} & (b > 0) \\[2mm] 0 & (b = 0) \\[2mm] -\dfrac{\pi}{2} & (b < 0) \end{array} \right\}$$

IV–9 INTEGRALS

In the following, b and p are constants, and u and v are functions of x:

1. $\displaystyle \int \frac{du}{dx}\, dx = u$

2. $\displaystyle \int_{x_1}^{x_2} \frac{du}{dx}\, dx = u(x_2) - u(x_1)$

3. $\displaystyle \int bu(x)\, dx = b\int u(x)\, dx$

4. $\displaystyle \int (u + v)\, dx = \int u\, dx + \int v\, dx$

5. $\displaystyle \int u\frac{dv}{dx}\, dx = uv - \int v\frac{du}{dx}\, dx$ (integration by parts)

6. If u is a function of y and y is in turn a function of x, then

$$\int u\, dy = \int u\frac{dy}{dx}\, dx$$

7. $\displaystyle \int x^p\, dx = \frac{x^{p+1}}{p + 1} \qquad (p \neq -1)$

8. $\displaystyle \int \frac{dx}{x} = \ln x$

9. $\displaystyle \int (\sin x)\, dx = -\cos x$

10. $\displaystyle \int (\cos x)\, dx = \sin x$

11. $\displaystyle \int e^{bx}\, dx = \frac{1}{b}e^{bx}$

12. $\displaystyle \int xe^{bx}\, dx = e^{bx}\left(\frac{x}{b} - \frac{1}{b^2}\right)$

13. Some definite integrals:

$(1) \quad \displaystyle \int_0^\infty x^n e^{-x}\, dx = n!$

$(2) \quad \displaystyle \int_0^\pi (\sin^2 x)\, dx = \int_0^\pi (\cos^2 x)\, dx = \frac{\pi}{2}$

$(3) \quad \displaystyle \int_0^\infty e^{-b^2 x^2}\, dx = \frac{\sqrt{\pi}}{2b} \qquad (b > 0)$

$(4) \quad \displaystyle \int_0^\infty xe^{-x^2}\, dx = \frac{1}{2}$

$(5) \quad \displaystyle \int_0^\infty x^2 e^{-x^2}\, dx = \frac{\sqrt{\pi}}{4}$

IV–10 SOME EXPANSIONS APPROPRIATE FOR $x^2 < 1$

1. The following expression is good for any n, positive or negative, integer or noninteger:

$$(1 + x)^n = 1 + nx + \frac{n(n - 1)}{2!}x^2 + \frac{n(n - 1)(n - 2)}{3!}x^3 + \cdots$$

2. $\displaystyle \sin x = x - \frac{x^3}{3!} + \frac{x^5}{5!} + \cdots$

3. $\displaystyle \cos x = 1 - \frac{x^2}{2!} + \frac{x^4}{4!} + \cdots$

4. $\displaystyle \tan x = x + \frac{x^3}{3} + \frac{2}{15}x^5 + \cdots$

5. $\displaystyle e^{ax} = 1 + ax + \frac{(ax)^2}{2!} + \frac{ax^3}{3!} + \cdots$

IV–11 SOME MATHEMATICAL NOTATION

1.	$=$	is equal to
2.	$\cong$	is approximately equal to
3.	$\propto$	is proportional to
4.	$\equiv$	is defined to be
5.	$\neq$	is unequal to
6.	$>$	is greater than
7.	$\geq$	is greater than or equal to
8.	$<$	is less than
9.	$\leq$	is less than or equal to
10.	Δx	the change in x
11.	$|x|$	the absolute value of x
12.	$O(N)$	on the order of the magnitude of N
13.	$\pm$	plus or minus
14.	$\mp$	minus or plus
15.	$\langle x \rangle$	average of x
16.	$\displaystyle\sum_{i=i_1}^{i_2} f_i$	the sum of all f_i over the integers i from a smallest integer i_1 to a largest integer i_2
17.	$\ln(x)$	natural logarithm of x
18.	$\log_{10}(x)$	logarithm to the base 10 of x
19.	$\displaystyle\int$	integral
20.	$\displaystyle\oint$	line integral around a loop

Periodic Table of the Elements

PERIODIC TABLE
Atomic Properties of the Elements

Adapted from:
U.S. DEPARTMENT OF COMMERCE
Technology Administration
National Institute of Standards and Technology
http://physics.nist.gov/PhysRefData/IonEnergy/periodic-table.pdf

For a description of the atomic data, visit physics.nist.gov/atomic

March 1999

Key:
- Atomic Number
- Symbol
- Name
- Atomic Weight[†]

Example:
58 Ce Cerium 140.116

Legend: □ Solids | □ Liquids | ■ Gases | ■ Artificially Prepared

Metals ← → Nonmetals

Group	Symbol	Name	Atomic Weight
IA			
1	H	Hydrogen	1.00794
3	Li	Lithium	6.941
11	Na	Sodium	22.98977
19	K	Potassium	39.0983
37	Rb	Rubidium	85.4678
55	Cs	Cesium	132.90545
87	Fr	Francium	(223)
IIA			
4	Be	Beryllium	9.01218
12	Mg	Magnesium	24.3050
20	Ca	Calcium	40.078
38	Sr	Strontium	87.62
56	Ba	Barium	137.327
88	Ra	Radium	(226)
IIIA			
21	Sc	Scandium	44.95591
39	Y	Yttrium	88.90585
57	La	Lanthanum	138.9055
89	Ac	Actinium	(227)
IVA			
22	Ti	Titanium	47.867
40	Zr	Zirconium	91.224
72	Hf	Hafnium	178.49
104	Rf	Rutherfordium	(261)
VA			
23	V	Vanadium	50.9415
41	Nb	Niobium	92.90638
73	Ta	Tantalum	180.9479
105	Db	Dubnium	(262)
VIA			
24	Cr	Chromium	51.9961
42	Mo	Molybdenum	95.94
74	W	Tungsten	183.84
106	Sg	Seaborgium	(263)
VIIA			
25	Mn	Manganese	54.93805
43	Tc	Technetium	(98)
75	Re	Rhenium	186.207
107	Bh	Bohrium	(264)
VIIIA			
26	Fe	Iron	55.845
44	Ru	Ruthenium	101.07
76	Os	Osmium	190.23
108	Hs	Hassium	(265)
27	Co	Cobalt	58.93320
45	Rh	Rhodium	102.90550
77	Ir	Iridium	192.217
109	Mt	Meitnerium	(268)
28	Ni	Nickel	58.6934
46	Pd	Palladium	106.42
78	Pt	Platinum	195.078
110	Uun	Ununnilium	(269)
IB			
29	Cu	Copper	63.546
47	Ag	Silver	107.8682
79	Au	Gold	196.96655
111	Uuu	Unununium	(272)
IIB			
30	Zn	Zinc	65.39
48	Cd	Cadmium	112.411
80	Hg	Mercury	200.59
112	Uub	Ununbium	
IIIB			
5	B	Boron	10.811
13	Al	Aluminum	26.98154
31	Ga	Gallium	69.723
49	In	Indium	114.818
81	Tl	Thallium	204.3833
IVB			
6	C	Carbon	12.0107
14	Si	Silicon	28.0855
32	Ge	Germanium	72.61
50	Sn	Tin	118.710
82	Pb	Lead	207.2
VB			
7	N	Nitrogen	14.00674
15	P	Phosphorus	30.97376
33	As	Arsenic	74.92160
51	Sb	Antimony	121.760
83	Bi	Bismuth	208.98038
VIB			
8	O	Oxygen	15.9994
16	S	Sulfur	32.066
34	Se	Selenium	78.96
52	Te	Tellurium	127.60
84	Po	Polonium	(209)
VIIB			
9	F	Fluorine	18.99840
17	Cl	Chlorine	35.4527
35	Br	Bromine	79.904
53	I	Iodine	126.90447
85	At	Astatine	(210)
VIII			
2	He	Helium	4.00260
10	Ne	Neon	20.1797
18	Ar	Argon	39.948
36	Kr	Krypton	83.80
54	Xe	Xenon	131.29
86	Rn	Radon	(222)

Lanthanides

Atomic Number	Symbol	Name	Atomic Weight
58	Ce	Cerium	140.116
59	Pr	Praseodymium	140.90765
60	Nd	Neodymium	144.24
61	Pm	Promethium	(145)
62	Sm	Samarium	150.36
63	Eu	Europium	151.964
64	Gd	Gadolinium	157.25
65	Tb	Terbium	158.92534
66	Dy	Dysprosium	162.50
67	Ho	Holmium	164.93032
68	Er	Erbium	167.26
69	Tm	Thulium	168.93421
70	Yb	Ytterbium	173.04
71	Lu	Lutetium	174.967

Actinides

Atomic Number	Symbol	Name	Atomic Weight
90	Th	Thorium	232.0381
91	Pa	Protactinium	231.03588
92	U	Uranium	238.0289
93	Np	Neptunium	(237)
94	Pu	Plutonium	(244)
95	Am	Americium	(243)
96	Cm	Curium	(247)
97	Bk	Berkelium	(247)
98	Cf	Californium	(251)
99	Es	Einsteinium	(252)
100	Fm	Fermium	(257)
101	Md	Mendelevium	(258)
102	No	Nobelium	(259)
103	Lr	Lawrencium	(262)

Period 1–7

[†]Based upon ^{12}C. ( ) indicates the mass number of the most stable isotope. For a description and the most accurate values and uncertainties, see J. Phys. Chem. Ref. Data, **26** (5), 1239 (1997).

Significant Dates in the Development of Physics

History can rarely be stated as a simple series of dates, and the history of science is no exception. Throughout the text we have alluded to important discoveries in physics. The list below is a personal choice and should be thought of as a guide. It oversimplifies some of the history, including stories that are covered more thoroughly in the text. Some of the dates are to be taken with a grain of salt, because discoveries are rarely made in a single identifiable moment. Our list includes some names (and discoveries) not mentioned in the text. Far more numerous are the names not listed, the names of those who built the experimental foundations, those who explored the false paths and cleared the way for those whose names we remember today, or those who verified the speculations that are now called laws.

1583	Galileo	Pendulum motion
1600	Gilbert	Study of magnets
1602	Galileo	Early statement of Newton's first law
1602	Galileo	Laws of falling bodies
1609	Kepler	First two laws of planetary motion
1619	Kepler	Third law of planetary motion
1620	Snell	Law of refraction
1648	Pascal	Atmospheric pressure
1650	Grimaldi	Diffraction of light
1661	Boyle	Chemical elements
1669	Newton	Light dispersion in prisms
1678	Huygens	Wave propagation
1687	Newton	Laws of motion; universal gravitation
1760	Black	Calorimetry
1785	Coulomb	Coulomb's law
1789	Lavoisier	Conservation of mass
1798	Cavendish	Measurement of G
1800	Volta	Electric battery
1801	Young	Interference of light
1801	Dalton	Laws of chemical combination
1802	Charles; Gay-Lussac	Ideal gases
1807	Dalton	Atomic theory
1812	Fourier	Decomposition of waves
1815	Fraunhofer	Discrete spectral lines
1819	Fresnel	Wave picture of light
1820	Oersted	Magnetic fields from currents
1820	Biot; Savart	Law of magnetic field produced by current
1824	Carnot	Second law of thermodynamics
1827	Ohm	Ohm's law
1827	Ampère	Ampère's law
1831	Faraday; Henry	Magnetic induction
1842	Joule	Mechanical equivalent of heat
1847	Helmholtz	Conservation of energy
1849	Fizeau	Direct measurement of the speed of light
1865	Maxwell	The laws of electricity and magnetism; light waves
1877	Boltzmann; Gibbs	Statistical mechanics
1879	Stefan	Blackbody radiation
1885	Osmond	Crystalline structure of metals
1887	Hertz	Electromagnetic waves
1887	Michelson and Morley	Constancy of the speed of light
1896	Becquerel	Radioactivity
1897	Thomson	Charge-to-mass ratio of the electron

1900	Planck	Quanta in blackbody radiation
1903	Rutherford; Soddy	Isotopes
1905	Einstein	Special relativity; quanta in photoelectric effect
1908	Kammerlingh Onnes	Superfluidity
1911	Kammerlingh Onnes	Superconductivity
1911	Rutherford	Nuclear structure of atom
1911	Millikan	Quantization of charge
1912	von Laue	X-ray diffraction in crystals
1912	Bragg and Bragg	Analysis of crystal structure with x-rays
1913	Bohr	Atomic structure
1916	Einstein	General relativity
1922	Compton	Scattering of x-rays
1923	Hubble	Discovery of galaxies
1924	de Broglie	Wave nature of particles
1925	Pauli	Exclusion principle
1925	Heisenberg	Formulation of quantum mechanics
1925	Goudsmit and Uhlenbeck	Electron spin
1926	Davisson and Germer; Thomson	Diffraction of electrons by crystals
1926	Schrödinger	Alternate formulation of quantum mechanics
1926	Born	Probabilistic interpretation of quantum theory
1927	Lemaitre	Big bang universe introduced
1927	Heisenberg	Uncertainty relations
1929	Hubble	Hubble's law
1930	Dirac	Antiparticles
1932	Anderson	The positron
1932	Lawrence and Livingston	The cyclotron
1932	Chadwick	The neutron
1934	Yukawa	Nuclear forces and the pi meson
1942	Fermi	Nuclear chain reaction
1948	Feynman; Schwinger; Tomonaga	Electromagnetism as a quantum theory
1954	Townes	The maser
1956	Reines and Cowan	Neutrinos observed
1957	Lee and Yang	Nonconservation of parity
1957	Bardeen, Cooper, and Schrieffer	Theory of superconductivity
1962	Josephson	Josephson junction
1964	Gell-Mann; Zweig	Quarks
1964	Penzias and Wilson	Background radiation of the universe
1967	Bell and Hewish	Neutron stars discovered (pulsars)
1967–1970	Glashow; Salam; Weinberg	Unification of electromagnetic and weak forces
1981	Binnig and Rohrer	Scanning tunneling electron microscope
1983	Taylor and Hulse	Gravitational radiation
1983	CERN	$W^{\pm}$ and Z^0 particles discovered
1986	Bednorz and Müller	High-temperature superconductors
1990	Davis and Koshiba	Neutrino oscillations and inference of neutrino mass

Answers to "What Do You Think?"

CHAPTER 1

1. Nothing is special about a year. One year refers to an astronomical accident in a sense, since if Earth traveled a slightly different orbit, the year would be different. The second is just as arbitrary. The importance of these or any units is that they establish a common standard with which we can all agree. A light-year is the same to anyone anywhere in the world.

2. Just because this answer is an exact one does not mean that you are obliged to use all the figures in application. It would in fact be an unusual situation for you to want to use the exact answer. You can always round off if fewer digits are called for.

3. The conversion between the millimeter and meter involves a definition, so no new uncertainty is introduced in the conversion.

4. If your calculator does not have enough digits, you might have to use a larger computer or you might have to be clever and use mathematical approximations. In fact, we could have used the approximation $\sin \theta \cong \theta - (\theta^3/3!) + \cdots$, correct if θ is small, to arrive at our numerical answer. But don't throw your calculator away; for almost all applications it will suffice.

6. No, you have a choice of any units you like with the right dimensions. For example, you could take the mass in slugs, the time in minutes, and the length in furlongs, in which case the units of G would be cubic furlongs per slug per minutes squared. But we would more normally use mks units.

7. If the radius of a sphere increases, so does the area of the sphere's surface. Thus there would be more area available for the same number of people, and the area available per person would increase.

8. The estimate l for the size in the example is in fact the cube root of the space taken up by each molecule. The only difference is that with the conditions stated the linear size of a molecule—the length of the side of a cube of the molecular volume—is $\frac{1}{10}l$. In other words, answer (a) is the correct one.

9. It would be unchanged. Displacements don't depend on the choice of origin. You can very easily check this explicitly in this case.

10. How the displacement is broken into components *can* differ according to the coordinate system. But the vector itself, that is its magnitude and direction in space, is independent of its description in a particular coordinate system.

CHAPTER 2

2. See Fig. 2–38.

3. You could certainly take the data in the table and plot it. The plot would contain no more and no less information than the table. When you draw a smooth line connecting the points of the

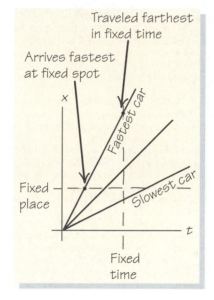

▲ **FIGURE 2–38** Discussion, What Do You Think? Example 2–2.

plot, then you are interpolating (guessing, with an assumption of smoothness) between the points. You could use the interpolated curve to find the velocity: The instantaneous velocity at a given time (or position) is given by the slope of the curve of position versus time at the given time. You would draw the straight line tangent to the curve at the given time, then just measure the slope of that straight line. Since the line is straight, you can measure its slope by examining any two points on it, just as we did in our discussion of Fig. 2–5. However, you should keep in mind that when you are taking the slope of a plot that is just a guess (an interpolation between points), you cannot *guarantee* that you have a precise value.

5. (1d) The ball first moves to the right, then to the left, and so on. When does any acceleration occur? Is it large? (2b) The two runners have different speeds, and the handoff occurs in an idealized fashion, with runners at their top speed. When is the acceleration of the baton? (3a) When the ball is first tossed, it has a large positive velocity; it slows down, passing through $v = 0$ at the top of its toss, at which point it falls with an increasing negative velocity. Since the acceleration is the slope of this curve and the slope is constant (and negative), the acceleration here is constant (and negative). (4c) The car is stopped, accelerates to a constant velocity, then has to stop again for another red light, and so on.

7. The last term in dv/dt is proportional to t^2 and therefore dominates the formula at large values of t in comparison to either the constant term

or the term linear in t. Since its coefficient is negative, the expression for the acceleration will eventually become negative. This will in turn eventually lead to a negative velocity, so that the rocket would eventually be moving faster and faster in the backward direction. This is a very peculiar situation, certainly not what the rocket designer had in mind; for that matter, with only a finite amount of fuel it is not possible for the acceleration to continue forever. You could surmise that the term in the acceleration proportional to t^2 is an approximate one meant to be used only for a brief period of time at the beginning, when it is not dominant.

8. Friction—here rolling friction—between the ball and the floor may be small, but it is not zero, and there is also air resistance. Often when we look at physical situations we simplify them by saying that we can neglect air resistance and friction, but these factors are important in many real situations involving motion. Here we have in fact exaggerated their effects. In a real bowling alley the degree to which the alley is level is what most influences the ball's motion.

10. See Fig. 2–39.

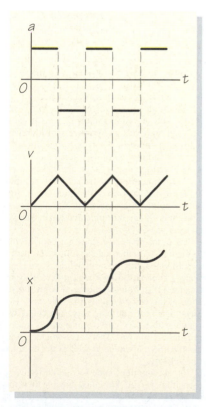

▲ **FIGURE 2–39** Discussion, What Do You Think? Example 2–10.

11. (b) The acceleration due to gravity is constant and downward with magnitude g, regardless of initial conditions, which here happens to be a positive (upward) initial velocity.

13. Under constant acceleration, the ball would fall four times as far in the first two time intervals as it did in the first time interval. (We are assuming that all the time intervals are the same.) This is a measurement that could easily be made on the photo in Fig. 2–23 and we encourage you to try it.

14. Refer to the sketches in Fig. 2–40. The acceleration is a constant, the same at the beginning of the motion, the end of the motion, and anywhere in between, including at the top of the trajectory (see Fig. 2–19).

▲ **FIGURE 2–40** Discussion, What Do You Think? Example 2–14.

CHAPTER 3

1. Both x and y are quadratic functions of time. However, in the range of time that comes into this problem, the quadratic term in x is not very important—you can see this in Fig. 3–3b, where x is close to linear in time. If x is linear in time, then a curve of y versus x will look very much like a curve of y versus t. The similarity of the y–x and y–t plots is just coincidence, and the x–t graph doesn't look at all like the x–y graph! Don't confuse the trajectory with a plot of a linear dimension versus time. The motion can also be given by means of the distance to the origin $r(t)$, and the angle that the line makes with the x-axis, $\theta(t)$.

2. The trajectory curve does not contain time information—whether the car followed a particular path while zipping or crawling along is not revealed. Thus you cannot read full velocity information off it. You can, however, know the direction of the velocity along the trajectory: It is always tangent to the trajectory. You could also imagine a trajectory curve along which successive times are marked. That curve would in principle give the magnitude as well as the direction of the velocity.

3. The acceleration is the second derivative of the position vector. We could have started this with the position vector as a function of time, as taking a double derivative is the same as taking two single derivatives, so we would have "passed

through" the calculation of Example 3–2 in any case.

5. The motion in the x- and y-directions are independent of each other. The horizontal travel distance is determined by a combination of things: how fast it moves horizontally (a constant given by the initial condition) and how much time it has to travel (also determined by initial conditions, but this time by the initial vertical component of velocity, zero in this problem). Of course, if the cliff were higher, the ball would be in flight for a longer time and would fall further out in the ocean.

6. The parabola describes the motion for a given acceleration and with the condition that the tee is at the top of the cliff and the velocity at the tee is $\vec{v}_0$. But if the cliff were not present, one could imagine that the golf ball was hit 57 m to the *left* of the tee and at sea level in a manner such as to have exactly the velocity $\vec{v}_0$ at a height of 52 m and 57 m to the right of the new tee-off point.

7. To see why the fact that the ground isn't level affects the θ_0-dependence of the maximum range, imagine as an example that there is a 6-ft fence a couple of feet in front of the tee-off point. (You can think of the fence as a model of the possible terrain.) Only by hitting the ball at a high enough angle could one hope to clear the fence and have the ball travel further.

9. The answer to this question is somewhat subjective, but the likely answer is the lower angle 36° because this trajectory has a shorter flight time. In addition, the target might conceivably more easily observe the water balloons heading toward them at the higher angle.

10. Air resistance or the presence of wind will affect the trajectory of both the rock and the coconut, as we'll discuss in more detail in Chapter 5.

11. Definitely not. If the automobile were rounding a curve with a larger radius of curvature, the acceleration would be less; as for the ramps onto an interstate highway, the car can negotiate a circular arc with a larger radius without danger. The extreme case is straight-line motion, for which 54 mph would pose no problem whatsoever. We'll see in Chapter 5 why too much acceleration can be dangerous, but you know from experience that it has something to do with the car skidding out.

12. The expression $a = \omega^2 r$ shows that the outermost parts of the top move with the largest acceleration. The angular speed is the same for all parts, so $a \propto r$.

13. (b) The rower would say that there is less current to fight, that he or she wouldn't have to row partly upstream just to stay even with the point across the river from the dock. With no current, the frames are identical; $u = 0$, so that $\cos \theta = u/v' = 0$, and the boat should be directed straight across. It would move across with speed v', rather than $v' \cos \theta < v'$. The crossing time is $(0.15 \text{ km})/(4 \text{ km/h}) = 0.04 \text{ h} \cong 2 \text{ min}$.

14. It is the solution to Example 3–13 that supplies the answer to this one. In effect, the flight path of the airplane must be longer (the hypotenuse of a right triangle) through the air in order to hold the north–south direction relative to the ground (a side leg of the right triangle). Thus the jet stream adds to the flight time even if is not directly opposed to the motion. You may want to think about the relative angles of flight direction and jet stream direction for which the jet stream can actually help.

CHAPTER 4

1. Not always. Following our rules for addition of vectors (tail to tip, as discussed in Chapter 1), the sum of the five forces for the fixed rods will end up as a single vector pointing somewhere in the three-dimensional space. We can get a zero net force if the two adjustable force vectors—the ropes—can add up to give a vector equal and opposite to this single vector. This is equivalent to asking whether three straight lines of fixed length can form a triangle, and this is only possible if the sum of the lengths of the adjustable force vectors is equal to or larger than the vector representing the sum of the fixed forces.

3. For a constant force, and therefore a constant acceleration, the final velocity is equal to $a \Delta t = F \Delta t/m$ and is therefore inversely proportional to the mass. If M is the mass of the box and m that of the bowling ball, the ratio of the final speed with one ball to the final speed with n balls is $(M + nm)/(M + m)$. The ratio of the speeds could still be used to count balls, but you would have to know the ratio, not necessarily an integer, of M to m. Notice that if $M \gg nm$, then the ratio is very nearly unity.

4. During the brief contact between the hand and the spaceship the astronaut exerts a force on the spaceship, and the spaceship exerts an equal and opposite force on the astronaut. They are both accelerated in opposite directions (inversely proportional to their masses) and move away from each other until a tether stops their separation by exerting forces on both of them. As the astronaut has less mass than the spaceship, she will have greater acceleration.

5. We can easily imagine that the passenger accelerates forward along with the sled. If he accelerates, there is a force on him. This force must be a contact force coming from the sled. But then, by the third law, the passenger must be exerting a force on the sled, and it must be backward. Why doesn't the sled accelerate backward as a result? That is because there is an additional force on the sled, one that comes from the person who is doing the pushing. The net force on the sled is forward.

7. Professor A would see her own tennis ball bounce on the floor and rise up vertically to her hand. She would conclude that the ball suffers no horizontal acceleration and that there is no horizontal force acting on the ball. Professor B would see that his ball does *not* bounce back into his hand. The reason is that the ball as it leaves the hand has the same instantaneous velocity as he does. While it falls, it maintains that velocity, whereas the professor and the train are accelerating. Thus, by the time the tennis ball comes back up, professor B has moved a little away from the ball. Professor B thinks that the same mysterious force—it has no visible source—that makes professor A accelerate backward also acts on the tennis ball.

8. The glue here ensures a sufficient amount of friction between the blocks. As long as these friction forces are large enough, nothing changes. If, however, the forces between the two blocks along the direction of motion are significantly reduced—for example, the blocks sliding past one another in the absence of friction—then there would be no way to transmit the pull from block 2 to block 1, and block 1 would be not be affected by the force acting on block 2—it would be left behind. As we shall see in the next chapter, friction has an "adjustable" quality up to a certain point.

11. During the contact time, the ball is indeed accelerated both horizontally and upward and acquires its initial velocity for its subsequent motion, during which only the force of gravity acts.

12. Look at Eq. (4–23): The x-position changes in time in a way that depends not only on the acceleration but also on v_{0x}. So we cannot make a rule about how far the nugget goes based on acceleration alone.

CHAPTER 5

2. Either fishermen are eternally optimistic about the size of fish they will catch or, more likely, fish can use their muscles and exert a force on the line much larger than their own weight, and the fishermen don't want the fish to be able to break the line.

3. Absolutely nothing would be different. The acceleration, Eq. (5–12), is independent of the mass of the sliding object. Everything placed on the ramp would slide down in the same way. Without friction neither the golf ball nor the car would roll. We are assuming the ramp will not collapse under the weight of a car.

4. If the ropes are not massless, we must include the gravitational force on them in the force equation. We would have to lift them! If the pulleys are not frictionless, then the tension throughout all the ropes is not the same, and it is no longer true that $T_1 = T_2 = T_3 = T_4$. The entire problem would become more complicated. All these effects reduce the ease with which the person could lift an object with the block and tackle, and for real machinery we wish to minimize these effects so that lifting is easier.

5. The masses would have to be equal. If $m_1 > m_2$, then m_1 will rest on the scale, and the scale would still have a nonzero reading.

6. The critical angle is determined by the *ratio* of the masses [Eq. (5–20)]. In part (a), the masses have the same ratio as in the example, and the system will remain motionless. (b) The mass m_1 will fall down pulling m_2 up the ramp. (c) The mass m_2 slides down the ramp and pulls m_1 up.

8. A force F_{hand} directed into the surface of the ramp would be added to the free-body diagram for the box. To balance this, we would have to increase the normal force from the ramp to the value $F_N = mg \cos\theta + F_{hand}$, and if the normal force increases, so does the maximum value of static friction and hence the angle at which slipping will start. This makes good sense; we know we can keep an object from sliding down a ramp by pushing down on it.

9. No, the y-component of the force will have a term mg down, which will complicate the last two equations in the example enough to lead to an angle that is not independent of F_{prof}. In effect, we have included a new dimensional quantity that has the same dimensions as F_{prof} and thus allows F_{prof} to enter through the dimensionless ratio mg/F_{prof}.

10. There is more weight on the front tires because the engine is in front. With more normal force pressing on the snow, the maximum static force is larger for the front tires than for the rear tires. This makes a significant difference in traction when μ_s is small and means that if the drive power is coming from the front, it can be larger than if it were at the back without the drive wheels slipping.

12. Gas efficiency for a given velocity would improve if the drag coefficient of the car is reduced, and for racing cars speed can be increased for a given engine power in this way. Drag coefficients for the boxy cars of years ago were as high as 0.7, whereas modern streamlined cars have C_D near 0.3. Gas mileage increases by 5 percent for every 10 percent improvement in drag coefficient. (Of course, the popularity of SUVs is evidence that gas efficiency may not be an important factor for some drivers.) Airplane wings have drag coefficients as low as 0.05.

13. Oil becomes "thicker," or more correctly more viscous (a word that describes the size of the coefficients in drag force—the higher the viscosity, the larger these coefficients), at low temperatures, and the marble in this example would move more slowly. Oil is a much better lubricant when it flows more easily, that is, at higher temperatures. But with today's multi-viscosity oils, car owners don't worry so much about warming up the engine before driving off.

14. From experience we know that the book will move toward the outside of the merry-go-round and fall off. It would move in a combination of radial and tangential directions because once static friction is replaced by kinetic friction (typically smaller than the largest value of static friction), the book will tend to continue motion in its current direction, which is initially tangential but as it moves out also attains a radial component.

15. To the inhabitants of the station, "up" is radially inward, opposite to the "force" that keeps them on the outer wall. Walking on the innermost wall would be like your walking on the ceiling, and walking on the side walls would be like walking on the side walls of your room! In both cases there would be no force to keep them accelerating centripetally toward the center of the wheel, and they couldn't walk on any surface except the outermost wall.

16. We can say that because static friction acts within a range, there will be some leeway for error that is not present when there is no static friction. There will be a range of speeds within which the car can negotiate the curve. See Problem 51, which treats this topic in more detail.

18. A skater will slow down when he or she hits the rough patch. The coefficient of friction of 0.10 in this example is small. A more realistic value, perhaps three or four times larger, would cause the speed to drop quickly enough to present a real danger: The skater might not be able to slow his or her upper body as quickly as his or her legs are slowed by the sand, and the skater may topple over! We are touching here on effects due to the finite size of the body that go beyond the motion of pointlike or perfectly rigid objects that are the main topics of our attention in these early chapters. Later in the book we'll come back to this point in much more detail.

CHAPTER 6

2. As we have not yet dealt with energy other than kinetic energy, we can only answer this question with our intuitive knowledge. We have the work–energy theorem, so we know that a change in kinetic energy implies that there is work done. The furniture mover has the ability to do work because his muscles have the ability to do work. The muscles are able to do work as they have an energy source they can call on—chemical energy from the metabolism of nutrition. The depletion of the chemical energy in the mover's body is measured by the kinetic energy acquired by the sofa.

3. For (i), the answer is (a): The work done is proportional to the force and the force of gravity is proportional to the mass, so that the work done by gravity is doubled. For (ii), the answer is (c): The relation between velocity, acceleration, and displacement is a kinematic one, which does not involve the mass of the object, so that the velocity is the same.

4. Initially the elevator (with its contents) is at rest. The downward force of gravity is exactly canceled by the upward force of the friction of the brakes and the tension of the cable. When the brakes are released and the tension of the cable is increased, there is a net force upward. The net work done is positive, giving rise to the kinetic energy of the elevator and contents. When the brakes grab and the tension decreases, the net work done is large and negative, canceling the kinetic energy possessed by the elevator before it stops.

5. The "pushing" force is unchanged, but the friction force that acts against it slowly increases. Since work done on the box depends on the *net* force, contributions to it become increasingly negative so that, by the theorem, the kinetic energy decreases until it reaches zero. The box comes to rest. At this point the static friction is just large enough to balance the external force and the box does not move.

6. We assume that when the box hits a downward slope, the push is still parallel to the floor. The force of friction is reduced because it is proportional to the normal force, which on the slope only involves the component of weight perpendicular to the slope. There is also an accelerating force due to the component of the force of gravity parallel to the floor. The result is that the net force is no longer zero, and the kinetic energy (and therefore the speed) of the crate increases as it goes down the slope.

7. The component of the force perpendicular to the ramp increases, so that the force of friction increases. The component of the force along the ramp decreases, so that the *net* force along the ramp decreases. There will be an angle at which the net force will go through zero and the crate will start slipping down.

9. Friction is certainly not a spring force. But in this case the friction force varies linearly with distance, so that the integration necessary to find the work done by it takes the same form as the integration necessary to find the work done by a spring force.

10. The circular motion itself implies that a centripetal force must also be acting. It is supplied by a tension in the rod that holds the ball. It does no work because the force is perpendicular to the motion; hence the scalar product that appears in the work is zero. Because it does no work, it cannot increase the speed; only a force with a tangential component can do that. Gravity also acts, but it is canceled by the contact force at the pivot that keeps the rod moving in the horizontal plane.

12. If the work done is independent of the path, then $W(A, B) = W(A, C) + W(C, B)$. This is equivalent to $\int_A^B F(x)\,dx = \int_A^C F(x)\,dx + \int_C^B F(x)\,dx$. This, however, is a general property of one-dimensional integrals (see the box Integration), so it is true for the spring force with the variable spring constant.

13. When an object slides on a rough surface, the motion is impeded by irregularities in the surfaces in contact. When these push past each other, energy is dissipated in bending them or in breaking them off (abrasion). In addition, bonds may be formed between atoms coming from the two different materials, and these bonds break when the object moves (see Section 5–2). In molasses, the object, in order to move, must push the fluid out of the way. Molasses is "hard to move" (liquids such as molasses are said to be viscous, and there are strong electrical bonds between neighboring molecules), while water, say, takes less energy to move it.

14. The horsepower is a rate, an energy per unit time. This can change from instant to instant, and it is therefore an "instantaneous" quantity. If the rate changes slowly, then an average rate over some interval of time may give a good approximate value for the instantaneous rate of delivery of energy.

CHAPTER 7

2. The shift $y \rightarrow y + y_0$ is just a shift in the coordinate system, nothing more. After you make this shift, you are measuring height y from the top of your building: $y = 1$ m in the new system is $y = 101$ m in the old system. If the expression for U in the old system is Mgy, its expression in the new system is $Mgy + mgy_0$. In the new system $U = 0$ at $y = -100$ m. This is a perfectly sensible result. You have not changed the *location* of the zero of U; that remains at sea level. But in the new coordinate system sea level is located 100 m below the building top, at $y = -100$ m.

3. One way to look at this is to note that the reverse process could be represented by a movie of the initial process *run backward*. Time does not enter into energy conservation equations, and for the square of the velocity (i.e., kinetic energy), it does not matter whether the movie runs backward or forward. We can conclude from this that the speed of the brick at roof level would have to be 15 m/s.

4. We have moved from one equilibrium situation to another, with no motion and hence no kinetic energy involved. The potential energy has, however, changed: the spring is compressed so that it has acquired a positive potential energy, while on the other hand the kilogram mass is now *lower*, so that the gravitational potential energy is lower. The total energy has changed because displacing the mass and compressing the spring involved work.

5. When the system is initially at rest, the spring has been extended to its equilibrium point, whose location is determined by the spring constant and *the sum of the masses attached to it*. This is not changed by the motion of the masses and so that motion will be the same. In terms of free-body diagrams we may enclose the pulley and the masses in a box, and then the relative positions of the masses and their motion are a matter that is internal to the box.

6. A sketch of the energy diagram shows that the $E = 0$ line just touches the top of the potential energy hill at $x = 0$. Our object may start at rest at the leftmost turning point, but in sliding down into the first valley it speeds up, reaching its maximum kinetic energy at $x = -2$. In climbing back up the other side of the first valley it loses kinetic energy, that is, it slows down, until at the top of the hill at $x = 0$ it has zero velocity. Will it go on to the right-side valley or will it slide back to the left-side valley? Actually it gets to the top and stays there since there are no perturbing forces that move it to one side or the other.

If you are inclined to worry how a solution of a perfectly well-defined $F = ma$ equation can have a solution that describes motion (the falling into the valley floor) and then at some later time describes a mass at rest, the answer turns out to be that it takes an infinitely long time for the mass to creep up to the very top of the hill. Remember, time information is something you cannot get from energy considerations; for that you need the full apparatus of Newton's second law.

7. No, we cannot tell. The speed would be the same in either case.

8. We see that as r becomes very large (approaches infinity), $U(r)$ goes to zero. The masses are very far apart, and since the force falls off with distance, it is reasonable that there be no potential energy where the force itself is zero. The total energy at large r ("at infinity") is just the kinetic energy.

9. Since the energy of the pendulum systematically decreases, its speed decreases. For example, at the bottom of the swing the energy is equal to the kinetic energy there, so the speed at the bottom of the swing will systematically decrease. The period of the clock, which is what marks the time, is the time for the pendulum to get back to its maximum elevation. After each swing both the speed and the distance that has to be traveled by the mass decrease, so it is conceivable that the period may be unchanged. This is indeed exactly what happens as long as the swing angle is small. The fact that the period is independent of the angle (for small starting angles) is what makes the pendulum a good clock. For more details, see Chapter 13.

CHAPTER 8

2. This is a matter of counting unknowns. The final state is formed of two balls each moving in the same plane. That means that the final state is specified with four pieces of information, the two velocity components of each ball. Originally we knew two pieces of information about the final state (the two components of the final velocity of the cue ball), and that was just enough to allow us to find the two components of the final velocity of the struck ball. Now we are replacing two pieces of information with one, the conservation of kinetic energy. This is no longer enough to allow us to find the final velocity of the struck ball.

3. We can suppose that the force applied to a ball has to do with your stroke and therefore is the same for either the strung racket or the solid bat. The strung racket is the choice to make because the ball remains in contact with it for a longer period as the strings "give." The impulse is then larger from the strung racket because of the larger Δt.

4. The more inelastic the better. In an elastic collision relatively more energy goes into the recoil of the car (as opposed to its distortion in collapse), the momentum change of the car is larger, and hence the impulsive force acting is greater. In other words, the acceleration of the occupants is greater in an elastic collision.

5. (b) Energy is lost as the boater and dog "give" in order to bring the dog to rest with respect to the boat—as they change their shape there is heating in the muscles and so forth. The collision was inelastic—the dog and boat have "coalesced" by means of friction between them. The answer cannot be (a) because all the motion is horizontal; it cannot be (c) because there is no friction between the boat and the ice.

6. We can often simplify a situation by using an approximation. In many situations a large quantity plus a small quantity can be closely approximated by the large quantity alone. Thus in the equation for v we can replace the denominator by m_{block}. But care must be taken—we certainly cannot ignore the mass of the bullet in the numerator; we would find $v = 0$ in that case!

7. With the initial object at rest, the momentum conservation equation reads $0 = m_{light}v_{light} + m_{heavy}v_{heavy}$, with solution $v_{light} = (m_{heavy}/m_{light})v_{heavy}$; the coefficient of v_{heavy} is greater than 1, so the answer is (b). None of this depends on details of the explosion.

8. With equal masses, the target ball moves off with the speed of the ball that struck it, while the projectile comes to rest in the place of the target ball. The struck ball then can rise to exactly the height from which the projectile started. The motion is symmetric and repetitive.

11. (b) The size of each of the orbs is taken care of by first recognizing that their center of mass lies at their respective centers, by symmetry. The center of mass of the entire system is then that of two *point* masses making a dumbbell. The center of mass is between the two ends of the dumbell, relatively closer to the more massive end. There is no reason that point could not lie within Earth's radius.

12. Certainly. The calculation of a center of mass is additive, and you can do addition in any combination you like. Try the suggested combination to show that this is the case.

13. Allowed angles in the laboratory frame are determined by the possible angles made by the vectors $\vec{p}' + \vec{p}/2$ and $\vec{p}' - \vec{p}/2$. To take a concrete example, suppose that nearly the maximum amount of energy were lost. This corresponds to the final two objects moving off arbitrarily slowly (i.e., very nearly sticking together—see Section 8–3). Then $\vec{p}'$ is very nearly zero, and the direction of $\vec{p}' \pm \vec{p}/2$ is that of $\vec{p}$ itself. In the laboratory frame the objects continue to move, nearly together, along the x-axis. They will not make a large angle with the direction of the projectile.

14. There are many objects that have an external center of mass. Some simple ones are a jackknife folded at a 90° angle—the center of mass will be somewhere within that angle—or a ring donut, which has the center of mass in the middle of the hole.

15. Symmetry places the center of mass at the center of the sheet, that is, $\left(\frac{1}{2}\text{ m}, \frac{1}{2}\text{ m}\right)$. If the mass density had not been constant, the calculation would have been more complex.

16. As we will learn in more detail in Chapter 12, the acceleration of gravity at the surface of the Moon is much smaller than its value at Earth's surface. It is possible to reach higher speeds far away from the Moon (and Earth) with a launch from the Moon, and some have proposed a Moon base just for this purpose.

CHAPTER 9

1. Not at all. The definition of the *average* acceleration depends only on the initial and final angular speeds, not on how they were reached.

2. This is a rotating rigid body: The angular acceleration is the same for all the points within it. We are of course repeating here what was used in the answer to the original question.

7. No. All the integrals would look the same. Of course, because the mass per unit area is just the total mass divided by the area, the correct total mass M would have to appear in the answer.

8. The rotational inertia "weights" points in an object according to their distance squared from the reference axis. Points in our object can be farther from the reference axis when the axis is at a corner than when it is at the center of mass.

10. True, as the equation for ω (the next-to-last equation of the example) shows; it contains the factor $F \times t$. This echoes the analogous result for linear velocity.

11. (c) Since the direction of both quantities is the same, the kinematic equation (9–34) becomes an equation for the magnitude of the angular velocity in terms of the magnitude of the angular acceleration. One doesn't know the time dependence of the change until the time dependence of α is known. In this case Eq. (9–33) shows that α is constant, and this will lead ω to change linearly with time.

13. Friction acts, and it is a nonconservative force, so energy is not conserved. This collision is analogous to the perfectly inelastic linear collision of Chapter 8, in which two objects collide and stick. That is in fact the most inelastic linear collision possible, and a similar remark holds for this "rotational" collision.

14. All points on the wheel have the same angular speed ω about the contact point. Because the top of the wheel is at a distance $2R$ from the point of contact, $v_{\text{top}} = 2R\omega = 2v$.

15. The ratio of the energy in rotation to the energy in linear motion is of the form $I\omega^2/(Mv^2)$. But for a fixed R the factor ω/v is fixed in rolling. Thus the percentage of the energy in rotation is greater for larger I given the same mass and radius.

16. The C factor for a thick cylinder lies in between its value for the solid case $\left(C = \frac{1}{2}\right)$ and the thin-walled case $(C = 1)$ From Eq. (9–52), the thick-walled cylinder will finish the race ahead of the thin-walled cylinder but behind the solid cylinder. The same equation reminds us that the results are independent of the radius.

CHAPTER 10

2. Crucially, there would no longer be a radial component to the angular momentum. This can be seen in our equation for L_{radial}: The angle θ would be $90°$, $d = R\sin\theta = R$, and hence $L_{\text{radial}} = 0$. When the rod lies in the plane of motion, its tension, which maintains the circular motion, is directed to the attachment point and hence exerts no torque.

3. It is unchanged. The instantaneous motion of the ball is the same, even if later the ball will go off as a projectile moving under only gravity. Since its motion looks the same at the moment the rope breaks, its angular momentum is the same.

4. As we had remarked in the "What do you think?" question for Example 10–2, the angular momentum is in this case purely vertical, and unchanging. One would then expect the torque to vanish, and it does because the radial component of tension (which is still present to maintain the circular motion) lies parallel to the radial vector; hence the cross product is zero.

5. (d) Answers (b) and (c) are closely related; as the moment arm is independent of distance

fallen, it translates into a moment arm independent of time. Note that this problem involves no rotations at all, yet there is a torque and hence there is an angular momentum. As we argued right at the beginning of Section 10–1, even a linearly moving point mass will generally have an angular momentum about some/any origin.

7. (c) The linear momentum transfer to the center of mass—the linear impulse—is independent of where the impulse is applied. This is not the case for the angular impulse.

9. (b) The angular momentum depends only on the mass of the comet, the speed at closest approach, and the distance of closest approach. The shape of the orbit is not a direct indication of the angular momentum.

11. The stool will stop rotating. Angular momentum is conserved, so two flips, bringing the wheel back to its original orientation, gets us back to the original situation, with the student and stool at rest.

12. Spool B has relatively more of its mass at larger radius and thus will have a larger rotational inertia than spool A. (The general form of I for a spool of mass M and radius R is CMR^2, where C is a geometrical factor, larger for spool B than for spool A.) Thus v is larger for spool A than for spool B.

13. The infinitesimal work done by the impulse is zero, because the scalar product of the angular impulse and the change in angle is zero. Thus the magnitude of the angular momentum is unchanged, although its direction is not.

CHAPTER 11

3. The mass density of the book must not be constant, because the book's center of mass must be somewhere directly over the table in equilibrium. This could be possible if, for example, the spine side were on the average denser than the side that opens.

4. The same fact that gravity acts on the center of mass tells us that if the map is placed with its center of mass on the point it will be in equilibrium; if not, it will rotate off the point.

5. (a) The coefficients of mg in the expressions for both F_A and F_B determine the answer to this question. If L decreases, then both F_B and F_A decrease.

6. If the crane is not fixed to the ground, the crane could pivot about the front tires and tip over! The torque that could make this happen would come from the suspended mass.

7. (b) As the window washer moves up the ladder the lever arm of the washer's weight with respect to the contact point with the ground increases. The torque due to the washer's weight about the contact point with the ground increases, until the torque about this point due to the normal force with the wall can no longer counteract the effect of the torque due to the weight. Note that (d) is in fact a consequence of (b).

8. Doing pull ups requires the lower arm to be rotated about the elbow to bring the body upwards. Thus the attachment distance x may be smaller for the person who cannot do pull ups as easily. Smaller x leads to a smaller torque about the elbow (xF_B).

9. The mass density is constant, so the volume V does not change. Therefore, $V = AL = A'(L + \Delta L)$ where A' is the new cross section. We see $A' = AL/(L + \Delta L)$; the new area will

be smaller by 1.2×10^{-3} or 0.1%. This is indeed a small correction, and the assumption that there was no change in the cross section was valid.

CHAPTER 12

2. (a) Look at the equation for the sun's mass M in this example. If, on the right hand side of the equation, we have a larger r, we must also have a larger T. This is independent of Mars' mass.

3. No. Once again a sketch is helpful. The centripetal force on the satellite always points to Earth's center, so that its orbit must be a circle in a plane that contains that point. The plane of a "great circle" passing though Washington makes an angle with the axis of Earth's rotation. This means that if at a given time the satellite is above Washington, a little later Washington will have moved along a circle at a fixed latitude, while the satellite will have moved to a different latitude and longitude.

4. The definition of a stable point can help us devise a test for this problem. We could imagine that $x = x_0 + \delta$, where $\delta \ll x_0$, and see whether the resulting (small) net force points back to x_0. In doing so one would want to use approximations based on the fact that $\delta/x_0 \ll 1$. This would be a test for stability along the line between Earth and the Moon. One could think about stability for motion perpendicular to this line in a similar way, with the vector aspect of the forces playing a more important role. The zero-point is in fact unstable: if, for example, the satellite is a little to the right of the zero-point, it tends to continue moving that way. See Problem 12–70.

You can also respond to the question by thinking about the potential energy as a function of the position on the line between Earth and the Moon. A graph of the potential energy would form a hill, and the zero-point is the top of the hill. The fact that you have a hill rather than a valley means that the equilibrium at the top of the hill is unstable; if the satellite goes a little off the top of the hill, it goes all the way off.

5. From Eq. (12–10), the escape speed depends on the ratio of M/R, which is a factor of 22 less for the Moon than for the Earth, so the escape speed from the Moon is significantly less than from Earth. This is why some people have proposed a moon base for initiation of longer voyages in space.

6. (d) The total energy is negative for a bound system, as in this example, and positive for an unbound system.

7. You cannot answer this without some algebraic thinking. Kepler's relation gives $R^3/T^2 =$ a constant. The orbital speed $v = 2\pi R/T$, and we can invert this to give $T = (2\pi R)/v$. This in turn implies that $R^3 \times v^2/(2\pi R)^2 =$ a constant, or $Rv^2 =$ a constant, or $v^2 \propto 1/R$, therefore as R increases the orbital speed *decreases*.

8. Keep in mind in thinking about orbital motions that all the conic sections described at the beginning of Section 12–2, and drawn in Fig. 12–4b, are possible. We can see from that figure that orbits can be elliptic ($E < 0$), parabolic ($E = 0$), or hyperbolic ($E > 0$). Information that a comet's orbit grazes the Sun is not enough to distinguish these.

9. (d) You can answer this one by thinking of the extreme case: If all the mass were concentrated right at the center, then the force would have the $1/r^2$ dependence characteristic of a point mass.

2. The acceleration is proportional to the displacement; it has maximum magnitude where the displacement has maximum magnitude, at $z = z_{\min}$ and $z_{\max}$. By the same reasoning the acceleration has minimum magnitude, namely 0, where $z = 0$.

4. (b) The time is still a quarter period, and the period does not depend on the speed at $t = 0$.

5. Simple harmonic motion is repetitive. The mass moves back out to $x = -X$, then returns to the origin, and the same maximum speed, a half period later. For this system, with an ideal spring, the pattern repeats forever. The times when $x = 0$, and hence the speed is a maximum, are at $t = T/4, 3T/4, 5T/4$, etc.

6. One would still expect simple harmonic motion. Although the springs may not be in a relaxed state in the stable equilibrium position, the restoring forces will always bring the mass toward this position as long as the spring is not stretched to the point that it no longer acts like a spring. As we saw earlier, any small motion about a stable equilibrium is harmonic, barring special circumstances.

7. The validity of an approximation such as the one we use for pendula depends on whether the first term of a mathematical expansion comes close enough to the exact answer. How close that is is a numerical question which can only be answered by first specifying a criterion for "close enough." In the case of the pendulum, the expansion is Eq. (13–32), and hence whether $\theta_{\max}^3/3! \ll \theta_{\max}$, or more usually whether the ratio $(\theta_{\max}^3/3!)/\theta_{\max} \ll 1$. For example, you may insist that $(\theta_{\max}^3/3!)/\theta_{\max} < 0.05$. In this example, $\theta_{\max} \cong (5.0 \text{ cm})/(2 \text{ m}) = 0.025 \cong 1.5°$, so that your criterion is satisfied.

9. (a) The monkey makes three things change: The total mass increases, the length increases, and the distribution of the mass over the new length has shifted to the monkey's end. The period is independent of the mass, so the first change is irrelevant. But the period increases with increased distance to the center of mass as well as with increased mass distribution toward the end (so that rotational inertia is increased.). Both the second and third changes increase the period.

10. At first sight we might say we can tell nothing whatsoever; the frequency of oscillation (and hence the period) is modified by the damping parameter b, but it depends on more than b alone. A little thought shows that there is something interesting we can say. You may know that sound is an oscillatory phenomenon, and the fact that one can continue to hear the gong even as its loudness decreases means that this system is not critically damped or overdamped. This establishes what you may already know: Sound involves oscillations that are much more rapid than the time scale associated with the damping of this gong.

CHAPTER 14

1. No. One of the inputs necessary to satisfy the wave equation is that the displacements can't be too large. If the string is infinitely long, then no matter how small a is, there will be x-values (positive and negative) for which the distortion is large. Of course, the function we gave could describe only a portion of a propagating distortion. No matter; this function displays the feature we are interested in here: It moves with speed v.

2. Greater. The relation $v = \lambda f$ shows this. If the system is really the same—for example, if water depth remains the same—v is unchanged. Then if λ decreases, f must increase to compensate.

4. The velocity will approach zero as either $T \to 0$, or $\mu \to \infty$, or both. The first condition of no tension implies that there is no force acting on any element of the wire. This implies no acceleration in the vertical direction and therefore no displacement of the wire from its equilibrium position. The second condition implies that the wire is more and more massive, and in order to get a transverse acceleration, the force (tension) would also have to get larger and larger. For a fixed T we get no acceleration in the limit.

6. Gravity causes the rope to sag, and the hand and attachment point will both have to supply a vertical component of force to allow for a vertical component of tension. We know from experience that if the rope is not too long, then the sag will be small, so that locally the rope will be straight and the conditions that went into our derivation of the wave equation will continue to be satisfied: The tension magnitude will be roughly the same, and the waves will be unchanged to first approximation.

8. We now have two separate situations. On the left we have a string whose length is 0.2 m, so that our expression for the wavelength changes because $L \to L/5$. Thus the wavelengths on the left side take the values $\lambda = \dfrac{2}{n_L}\dfrac{L}{5}$, that is, each allowed wavelength is shortened by a factor of 5. On the right side the length is changed from L to $4L/5$. Thus the wavelengths take the value $\lambda = \dfrac{2}{n_R}\dfrac{4L}{5}$—that is, they are also shortened, but only by a factor of $5/4$. Now we can certainly find values of n_L and n_R such that the wavelength has a common value. For example, $n_R = 4$, $n_L = 1$ have the same wavelength as the original string for $n = 5$ (which has a node at the newly clamped point). Nevertheless, these modes will not necessarily match the appearance of an unclamped string, because the amplitudes and the phases on either side of the clamping point must match.

9. Our expression for the relation between f and T shows that f is proportional to $T^{1/2}$. Thus an increase of 1% in f means that $T^{1/2}$ must increase by 1%, and therefore T must increase by 2%. (To see this, use the fact that $(1 + x)^{1/2} \cong 1 + \frac{1}{2}x$ for x small.) Piano strings differ from one another in mass density as well as tension. T is proportional to f^2, so if strings of different mass density were not used, the range of tensions in a piano would be enormous.

10. This is most easily answered by considering a long pipe with a rectangular cross section. Then it is clear that there can be standing waves across the height, as well as across the width of the pipe, since there is a wave equation for disturbances in those directions, and since the boundary conditions are the same as for the ends of the closed pipe. If the pipe is much longer than it is wide or high, then the wavelengths of the standing waves across the pipe are much smaller, and the frequencies much higher than for the waves along the pipe axis. Therefore they do not enter into the answer to this question. If the cross section is circular, then the standing waves in the directions perpendicular to the axis of the pipe look like Fig. 14–6. Again the distance between crests is much smaller than the wavelengths of

quite a large number of harmonics for waves along the pipe axis.

11. As we shall see in Chapter 16, blowing air across the opening of the bottle causes a change of pressure across the top of the bottle, which can set up a standing wave. This wave is analogous to the one-end-open organ pipe wave, so that the fundamental wavelength is roughly four times the length of the bottle. This is a fairly long wavelength, and therefore a fairly low frequency, or deeply pitched, wave. For a tube open at both ends, we have the situation shown in Fig. 14–27. These boundary conditions produce their own harmonic series.

13. Since the pitch does not change on reflection, we may view the process as the arrival of a sound wave at the cliff, and the subsequent emission of a sound wave with the same pitch by the cliff. The pitch as received by the cliff is raised from f_0 because the source is moving toward the receptor. As far as the reflected wave is concerned, we now have a moving observer dealing with a *raised* frequency of emission (but a stationary emitter). Thus the overall effect will again be a raised frequency. For a numerical evaluation

$$
\begin{aligned}
f_{\text{final}} &= \left(1 + \frac{v_{\text{car}}}{v}\right) f_{\text{refl}} \\
&= \left(1 + \frac{v_{\text{car}}}{v}\right) \frac{f_0}{1 - v_{\text{car}}/v} = \frac{v + v_{\text{car}}}{v - v_{\text{car}}} f_0.
\end{aligned}
$$

CHAPTER 15

1. We have set the wall separation so that the (fundamental) wavelength is fixed. The relation to this wavelength and the frequency with which the string vibrates is determined by the speed of waves on the string, i.e., by the tension and mass density of the string. There is no reason why this speed is the same as the wave speed in the medium. Thus there is no reason why the frequencies of the wave and the medium must match.

2. If we think of the vibrations of the prongs of a tuning fork as a harmonic oscillator, we recall that $\omega = \sqrt{k/m}$, where k is the spring constant, characteristic of the structure. Thus an increase in mass (such as the additional mass of the tape) reduces the frequency, and that is what happens when the tape is added.

3. In the first case, the different starting time would just reflect a different phase. *Nothing* is changed if we define a new time $t' = t - 5$ s and write the same equations that we wrote down with t replaced by t'. All we are doing is resetting the clock. In the second case we note that generically $\cos(kx - \omega t) = \cos \omega(t - x/v)$. Since the traveling waves move with the same velocity, a shift in the location where we observe the mutual maximum merely changes the origin of the t variable, and nothing really changes.

4. The motion is in a radial direction so that the angle θ does not change. The listener will still be at the first maximum. There are other effects: There is a Doppler shift, but this is very tiny since the shift is of magnitude $v/v_{\text{sound}} = 3/330 \cong 0.01$. The main effect is the loudness, since that falls off as $1/R^2$. This means a reduction of the sound intensity by a factor of 4.

5. Assuming constructive interference, the first maximum occurs on the center line with $\theta = 0$. The largest value of θ for an experimental setup as described is 90°, and this will be the limiting case of a "completely" spread-out pattern if that value of θ corresponds to $n = 1$. This occurs when d decreases to the size $d = \lambda$.

6. This wave is a standing wave in that at a given value of x the displacement varies between limits with a time dependence given by $\sin \omega t$. Since the wall was chosen to be at $x = 0$, the standing wave pattern holds for all values of $x < 0$. This pattern is identical to one in which a string is fixed at $x = 0$ and at $x = -2\pi/k$, for example, but its existence does not depend on having *two* ends fixed.

7. If we think in terms of a string attached to a second string, then the *continuity* of the strings—that is, the absence of a break at the connecting point (e.g. $x = 0$) provides a condition relating the incident wave, the reflected wave, and the transmitted wave at that point. Since this holds for all values of t, we will get another equation relating the three amplitudes. Unlike the energy condition we used here, which is quadratic in the amplitudes, this will be linear. We'll have enough information to solve for the amplitudes of the reflected and transmitted waves in terms of that of the incident wave.

CHAPTER 16

1. (a). The radius doubles, so the volume increases by a factor of 8. For density, which is mass per unit volume, the numerator (mass) doubles while the denominator (volume) goes up by a factor of 8; the density therefore drops by a factor of 4.

3. As in the example here, when the valve area is small it takes much less force to obtain a significant pressure across it, because $p = F/A$. This pressure is then transmitted everywhere within the fluid—in this case within the tire tube.

4. It is the difference in pressure between inside and outside the tube that gives rise to the force that pushes the mercury up inside the evacuated tube. If the top of the tube were open to the outside, the mercury would not rise at all. If there were, say, a third of an atmosphere above, then the net force up would be 2/3 of an atmosphere, and the height would be 2/3 of 760 mm.

5. The same reasoning that explains why the column of water (less dense) is higher than the column of amyl bromide also explains than the column of heptyl ether would be higher than the water column. A quick repeat of the calculation with heptyl ether would give $h_{he}/h_w = 1.23$.

7. (a) The density of salt water is greater than that of fresh water, and the final answer for the density would therefore be larger by a few percent according to the last equation in the example.

8. In contrast to the completely submerged sphere of Example 16–7, a floating object has a buoyant force that cancels the weight completely. The scale would read zero.

9. To reach a further distance, the water should come out of the nozzle at a higher speed—a projectile fired with a larger initial speed will go further. Thus a smaller nozzle diameter is required.

10. If the area changes, then the conservation of flux vA shows that the speed must change as well, increasing as the artery narrows. The full conservation law, Eq. (16–22), applies. A look at that equation shows that if both v and h increase as we go to the head, then p at the head will decrease even more than it did when only h increased. In turn, the pressure at the heart would have to be still higher to compensate. Narrowed arteries are dangerous, as blood flow to the body is reduced and the heart has to exert more effort.

11. An algebraic calculation as in the example would allow you to find the stream speed as a function of hole height. This could then be converted into an algebraic calculation of the horizontal distance carried by the stream as a function of hole height. One could then maximize this by setting the derivative of the horizontal distance with respect to height to zero. You can see that neither a hole near the top nor one near the bottom would give the farthest distance from the tank. For a hole near the top, the water will exit with a low speed, because the pressure is low. For a hole near the bottom, the interior pressure and hence the initial speed will be large, but the stream hits the ground very soon. Fig. 16–25b shows the trajectory of fluid for a variety of hole positions.

CHAPTER 17

2. Temperatures cannot be measured with this type of thermometer when the temperature and pressure are extremely low, because gases liquefy or freeze in this region of extreme cold. There are, however, other kinds of thermometers that we can use in this region, as we'll discuss later.

5. At the higher temperature the bridge would have expanded to fill up the 50-cm gap. At the lower temperature the bridge would have a 50-cm gap, enough to swallow small children if there were only a single joint. With many joints this is less of a problem! In any case, the joints tend to be closed in summer when it will be more pleasant to cross.

6. The inequality of the β-values states that the gasoline will shrink more than the container when the temperature goes down. You can safely fill to the brim on the hottest summer day, and when you look inside the container in winter, the container will not be full.

9. Air consists primarily of oxygen (molecular weight 32.0), nitrogen (molecular weight 28.0), and argon (molecular weight 39.9). The number 29.0 is a weighted average of the constituents' molecular weight according to their percentages in air: oxygen (21%), nitrogen (78%), and argon (1%). We have $M(\text{air}) = 0.21(32.0) + 0.78(28.0) + 0.01(39.9) = 29$.

11. None of them. The constant temperature line is shown on Fig. 17–13, while constant p is a horizontal line and constant V is a vertical line.

13. As it works its way to the surface, the radiation emitted from the Sun's interior is absorbed and reradiated at wavelengths characteristic of the successive temperatures of successive distances from the center.

CHAPTER 18

1. Because the temperature is constant, the transformation would be represented as a straight line parallel to the p axis as the pressure changes. The line would be at the fixed value of T.

3. One of the most common misunderstandings about thermodynamics is the incorrect assumption that heat and temperature are practically the same thing. Just because the temperature is unchanged does not mean there is no heat flow! In fact, there will generally be heat flow at constant temperature if other thermodynamic variables change.

4. (a) higher. The specific heat of aluminum is higher than that of iron (see Table 18–1). That means that for a given heat flow the temperature of the aluminum would change less, and that corresponds to a larger system temperature.

6. If the latent heat of fusion were much larger, then the freezing of lakes would be much rarer, since the air would have to extract a great deal more thermal energy from the lakes. Once a lake froze, melting would correspondingly be much more difficult. Ice cubes would not melt in drinks. (But they would still make them cold!) On the other hand, if the latent heat of fusion were very very small, the lakes would essentially have the same temperature as air, freezing quickly when air temperature fell below the freezing point, with the delay coming only from the time it takes water to conduct heat. Ice cubes would still have a cooling role, although a smaller one, and they would quickly melt and dilute your drink.

7. The rate of heat loss is proportional to the surface area, and a sphere has the smallest area for a given volume.

8. An examination of Eq. (18–12) shows that any term with a small R-value has a big effect. See the discussion following this example.

9. Yes. Kinetic energy will be converted to thermal energy when the containers crash on the road below. It does not matter how the containers receive the kinetic energy.

10. You are not reversing the entire process; in fact you are adding energy, whose source is your muscles, raising the weight and heating the water still further. We'll see in Chapter 20 that the impossibility of reversing the results of chaotic processes such as the churning of water is the basis of the second law of thermodynamics.

11. (b) You can see from the figure that the area enclosed is more than doubled. If you prefer an algebraic solution, the ratio of the new work to the old is $(2p_2 - p_1)/(p_2 - p_1) = (2p_2 - 2p_1)/(p_2 - p_1) + p_1/(p_2 - p_1) = 2 + p_1/(p_2 - p_1)$, a result larger than 2.

12. Because the cycle is closed, the internal energy over one cycle is unchanged, and the net heat flow in will always equal the work done. The heat flow in is positive if the path is clockwise and negative if it is counterclockwise.

13. The heat flow has to come from the thermal reservoir. The reservoir is, by definition, so large that you can take as much heat flow from it as you like without changing its temperature.

14. Since the temperature is constant, the internal energy does not change. Since the gas does work, thermal energy had to flow in, the same amount as the work done, so that positive thermal energy comes out of the reservoir.

CHAPTER 19

1. The internal energy enters in the expression for the collision rate in the form $(U)^{1/2}$; doubling the internal energy increases the collision rate by a factor of $2^{1/2}$.

3. The argon atoms are very few in number, so as soon as equilibrium sets in, they will be characterized by the same temperature as the helium atoms. Thus they have the same average kinetic energy as the helium atoms. Since the kinetic energy is of the form $m\langle v^2 \rangle/2$, and since argon atoms are much more massive than helium atoms, their rms speed will be smaller.

4. The b values measure the space occupied by a mole of molecules, and as the calculation above shows, the molecular radii of these substances must therefore be within a factor of $2^{1/3} = 1.3$ of each other. Positive a values decrease the pressure that the gas exerts on the surface of the containing vessel. This comes about because there is a force that slows down the molecules as they approach the container walls, so to speak, and therefore reduces the momentum transfer per wall collision and thus the pressure. This suggests that molecules exert an attractive force on each other.

6. The answer is "obviously" 1.65 m. If we were asked to justify this, our answer would be based on the symmetry of the U shape. An algebraic demonstration of this would be tedious but quite straightforward.

8. The dependence of the mean free path on the number density, and in particular $\lambda \propto 1/n = V/N$, is the relevant point here. The compression reduces the volume (the number of molecules does not change) and therefore the mean free path is reduced. Just how much the volume is reduced depends on whether the compression is adiabatic or not, but that is not our concern here. It makes sense that the mean free path is reduced: It is harder to get through a dense crowd than a dispersed one.

CHAPTER 20

1. We can count without writing out each case by saying that there are three ways to assign the first number, then two ways to assign the second, and only one choice for third number, making a total of 6. The traditional way to approach this is to think of three boxes and three integers, and count the ways in which they can be filled. For the first box: there are three integers, and any one of them can fill that box. When that is done there are two boxes left and two integers. There are two ways of filling the second box, and finally the third integer goes uniquely into the third box, so that the number of ways is $3 \times 2 \times 1 = 6$. This way of doing things would prove superior if you had to do this problem for four dice.

2. To increase the efficiency, decrease the ratio T_c/T_h. For the second part, probably the most sensible thing would be to scale the heat flow for both reservoirs. There are practical limits to how high we can make T_h, and to get to T_c below the temperature of available coolants, such as river water for power plants, is costly.

6. It is clear from Eq. (20–12) that for a small temperature difference the coefficient of performance of the refrigerator is large, and this means,

according to the definition in Eq. (20–10), that little work needs to be done. In the limiting case, when your kitchen is at 40° F, you can just unplug the refrigerator!

7. There is no need for a complicated calculation. We need only recall that the entropy is a state function, and therefore the change in entropy from an initial state to a final one is *independent of the path taken*. The result is unchanged as long as the initial and final states are the same.

10. The process is irreversible, since the gases, once mixed, will not spontaneously unmix. This means that $\Delta S > 0$.

11. A look at the expression for the entropy suggests that since the new volume, 3.0 L, is smaller than the original one, the entropy change would be negative, forbidden by the second law. But that assumes that all of each gas goes into the small container. In fact, such a process would not take place spontaneously. Some of the argon and some of the neon would pass into the common 3.0 L container, but much of the gas would stay where it was. The argon would increase its volume from 5.0 L to 8.0 L, and this would increase the entropy. The neon would be increasing its volume from 12.5 L to 15.5 L, and its entropy would also increase.

Answers to Understanding the Concepts Questions

CHAPTER 1

1. Yes; a vector $\vec{V}$ will have 4 components: $\vec{V} = (V_x, V_y, V_z, V_w)$.
3. Mine is 180 cm, or 1.80 m.
5. They are dimensionless.
7. Define $c = 299{,}792{,}458$ m/s, then 1 m $= c(1/299{,}792{,}458$ s$)$.
9. Yes.

11. No; if the angle between the two unit vectors are 60°.
13. About 99.6% accurate.
15. About 0.3% (assuming a weight of 150 lb and an uncertainty of about 0.5 lb).
17. The surface area of a mouse is large relative to its body size, and the energy radiated through the skin needs to be compensated with sufficient food intake.
19. No.
21. Three; infinite number of sets.

CHAPTER 2

1. In case the car in front of you suddenly stops, your car would travel further before you react, and it would take longer for it to stop after you apply the brake.
3. t is proportional to $g_x^{-1/2}$, and v is proportional to $g_x^{1/2}$.
5. No.
7. $h_{\text{Moon}}/h_{\text{Earth}} = g_{\text{Earth}}/g_{\text{Moon}}$, $h_{\text{Moon}} \approx 5$ m.
9. True.
11. 0; yes.
13. $a = $ constant and $v_0 = 0$.
15. False.
17. 3rd, 2nd, 1st; $v_1 = v_3 > v_2$.
19. Need to measure the diameter d of each wheel; number of rotations per unit time $= v/\pi d$.
21. A box sliding up or down a straight ramp, two unequal masses connected by a string hanging over a fixed pulley, a vehicle accelerating uniformly down a straight road, etc.
23.

$v = 0$ again at $t = 20$ s, when it returns to the point where it was initially dropped.

CHAPTER 3

1. A little higher.
3. A strong wind can alter the result significantly.
5. No.
7. R is proportional to v_0^2.
9. The ball will land ahead of the car.
11. No.
13. Tilt the umbrella forward at an angle θ, with $\tan \theta = $ running speed/raindrop speed.
15. (a) Outside the edge of the merry-go-round, (b) on yourself or next to you.
17. There is an initial eastward velocity.
19. Only (a) does.
21. No.

CHAPTER 4

1. A projectile motion with horizontal initial velocity.
3. Use a spring scale.
5. The reading is first greater than your weight, then becomes equal to it before getting less than it, and eventually becomes equal to it again.
7. False.
9. The tableware does not accelerate as fast as the tablecloth.
11. The direct cause is the normal force exerted by the person on the table.
13. Everything except gravitational and magnetic forces.
15. Gravity is balanced by the drag force of the air.
17. For example, pull the mass with a spring scale with a constant force F and measure the acceleration a.
19. The greater cross-sectional area of the rubber ball results in greater air resistance.
21. The friction force exerted by the road on the tires accelerates the vehicle; the engine forces the tires to turn, initiating the friction from the road.
23. She will land at a point further than half the radius away from the center of the merry-go-round.
25. Centrifugal force, the upward force that enables an astronaut to "float" inside a space shuttle, the backward force that keeps you at rest on the driver's seat as you accelerate your car forward.
27. Gravity speeds up the marble on its way down and slows it down on its way up, the bowl's contact force keeps it in a circular path, friction slows it down and eventually stops it.
29. It exceeds your weight.
31. Yes; the downward force from the branch.
33. Balance an object and check to see if $\Sigma \vec{F} = 0$.
35. The engine forces the driving tires to turn, which cause the road to exert a forward force of friction on the tires.
37. No.
39. Forces may not necessarily initiate motion.

CHAPTER 5

1. The tension in the wire must have a vertical component to balance the weight of the walker.
3. The mass of the string is not negligible, and/or the observer is in a non-inertial frame of reference.
5. The tension in the rope would vary; friction increases the tension necessary to move the load.
7. To increase the static friction from the road that serves as centripetal force.
9. No; yes.
11. Both help keep the boat from moving sideways; the keel also helps prevent the boat from overturning, while the centerboard is much lighter.
13. Yes in principle; friction reduces efficiency.
15. The weight of additional people causes the scull to sit lower in water, increasing the drag force of the water.
17. All of the statements are true.
19. Gravity serves as the centripetal force; no.
21. Yes.
23. No; the net force is never zero; tension in the string and the weight of the bob.
25. No.
27. Less static friction is available from the ground.
29. The lowest point, where the tension in the vine is the greatest.
31. The die will curve away sideways while descending the bowl's wall.
33. The first term dominates at low speeds ($v \ll b/c$), the second one dominates at high speeds ($v \gg b/c$).

CHAPTER 6

1. Use $(\vec{F}_{\text{drag}} + \vec{F}_g) \cdot dr = dK = d(mv^2/2)$ and integrate to find the speed of the baseball.
3. Yes; the force opposes the motion and reduces the kinetic energy of the object.
5. The centripetal force does zero work and does not change the speed (and the kinetic energy) of the object.
7. No.
9. The energy for the work done on the smokestack comes from the engine, not the man; the work done on the man by the deck equals the work he does on the smokestack.
11. No (if terminal speed has been reached).
13. Yes. For example, the friction of the road that accelerates a car from rest.
15. An amount equal in magnitude to the (negative) work of the drag forces attributed to your weight.
17. Friction from the wall does positive work on the participants while their speeds increase; no net work is done on them once they reach constant speed.
19. Measure the maximum height attained by the first acrobat and hence the kinetic energy imparted to him.
21. The one chopping the bone.
23. To the moving observer $\vec{F}$ and $\vec{v}$ are opposite in direction, so $\Delta K = W < 0$.

25. The mass m of the parachutist and the height h of the jump.

27. The net work done by the drag force is negative, rather than zero, for one complete round trip.

CHAPTER 7

1. No.

3. It means that U of the person at the bottom of the well is lower than that at sea level.

5. Loss of mechanical energy due to air friction as well as during the collision with the floor; a ball thrown down with a large initial speed has a significant amount of initial kinetic energy.

7. Yes; energy released from sugar and oxygen may be converted into the mechanical energy of motion, the heating of muscles and bones, and that in the chemical products that the body produces, etc.

9. No; if yes the spring would be able to oscillate forever (inside a vacuum tube).

11. Measure the final speed v of a falling object from height h (in a vacuum tube) to see if $mgh = mv^2/2$; observe the motion of satellites and planets.

13. Yes; air friction and the collision force from the floor.

15. The normal force is always perpendicular to the direction of motion (which is tangential) so it does no work; this can be verified from $E_i = E_f$.

17. Not necessarily.

19. If they were, then as the ball comes back to where it started its kinetic energy (and speed) would return to the same value as before.

21. None, although the energy of Earth as a whole is fairly constant, at least over time intervals that's much shorter than geological scales.

23. Back to the height reached after the first bounce.

CHAPTER 8

1. Yes (for the Sun-comet system).

3. Yes, the center of mass does stop; no, since $\vec{v}_{cm}$ can be zero before the collision.

5. No. It follows a parabolic path.

7. The body is curved like a horseshoe "draped" over the bar, so only part of the body is above the bar.

9. Yes; $x_{cm} = m_{point} R/(m_{circle} + m_{point})$.

11. Yes.

13. Yes; yes (albeit imperceptibly).

15. Short enough so that our ordinary senses cannot detect a finite time interval.

17. Assuming elastic collisions, the smaller ball can reach 9 times the initial height of the larger one.

19. Conservation of momentum of the gun-ammunition system.

21. It moves slightly forward.

23. As the parachute moves faster it collides with more air molecules (per unit time) moving at higher speeds towards it.

25. 45° (with no energy loss); less than 45° from the table (with energy loss).

CHAPTER 9

1. Down (with right-hand rule); up (with left-hand rule); yes, but one has to be consistent.

3. The angular momentum of the rotation stabilizes the motion.

5. The wheel moves downstream to an observer on the shore and backward relative to the boat.

7. Lower.

9. The one filled with water.

11. Shorten r by a factor of $1/\sqrt{2}$.

13. The force of the water against the side of the canoe can tip it over.

15. The angular speed, which cannot exceed a certain amount or the wheel will disintegrate.

17. To increase the torque applied by increasing the lever arm.

19. The longer wrench allows you to exert a greater torque by providing a longer lever arm.

21. The spring will be further stretched.

23. Two.

25. 1:1.

27. Not enough centripetal force available to support the circular motion.

CHAPTER 10

1. Yes.

3. The wheels have angular momentum so they tend to maintain their orientation.

5. To keep the center of mass as low as possible to increase stability.

7. Considerable torque is required to change the direction of the large spin angular momentum of the fan.

9. Little torque is exerted on the person using the bent leg technique; lifting weight with bent back requires you to counter the torque of the weight.

11. Acquire an initial rotation by using the contact force from the diving board to provide a torque about his/her center of mass.

13. The angular velocity is in the same direction as that of the torque from the cue about the center of mass; the spin angular momentum acquired equals the angular impulse delivered by the cue.

15. No. The friction from the ground exerts a torque on you.

17. Zero.

19. Yes.

21. The pilot must arrange his ailerons and rudder to introduce a force that will push the nose up.

23. The student will also turn, but by only half as much.

25. Slow down.

27. It is absorbed by Earth.

29. (b).

CHAPTER 11

1. Pushing against the slope results in a reactionary force from the slope that tends to push the climber away from it.

3. The force of gravity on the mass would produce a net torque about the point of suspension if the rope is not vertically aligned.

5. Stable equilibrium.

7. No; an object is approximately rigid if its deformation is considerably smaller than its dimensions.

9. True.

11. $\Sigma F_x = ma_x$, $\Sigma F_y = ma_y$, $\Sigma \tau_A = I_A \alpha$ (A is the contact point between the motorcycle and the curb); possible only if the curb extends below the center of the wheels.

13. Yes.

15. The one which makes an angle of 60° with the horizontal.

17. No.

19. No.

21. Unstable equilibrium.

23. It increases the tensile strength of bricks.

25. By heating up the diamond.

27. A structure (e.g., graphite) featuring layers of planes is more vulnerable to shear forces applied along the planes; yes.

CHAPTER 12

1. Drag forces in fluids, the tension in ropes or rods, the normal force between two objects in contact, etc.

3. Yes.

5. Launch the satellite near the equator, in the direction of Earth's rotation; launch it in the direction of the orbital motion of Earth around the Sun.

7. No.

9. A straight line passing through the Sun.

11. Yes.

13. As the Moon completes one revolution around Earth it also completes one rotation about its own axis.

15. Yes.

17. Consider Earth as a uniform sphere plus some local variation of density, and study the deviation of the g data from its uniform value to probe the density variation.

19. Assuming uniform density, g would be uniform on its surface if Earth were a perfect sphere.

21. (b) is correct.

23. The surface of Jupiter is not as rigid as that of Earth.

25. Zero.

27. Yes.

29. $g = 0$ inside a hollow Earth.

CHAPTER 13

1.

3. K_{max} increases by a factor of 4.

5. The thinner air on the mountain top provides less drag on the pendulum, while the g value there is lower—these two factors affect T in opposite directions.

7. Yes; the positive work done by the external driving force.

9. It should be shortened.

11. Increase.

13. T is decreased by a factor of $(3/2)^{1/2} \approx 1.22$; A is decreased by a factor of $(3/2)^{1/4} \approx 1.11$.

15. (a): Driven harmonic motion while wind is blowing; (b): driven harmonic motion; (c) driven harmonic motion while the child is pumping and damped while he or she is not; (d) and (e): damped harmonic motion.

17. Yes.

19. Yes; longer.

21. You are adjusting the driving frequency to match the natural frequency of the diving board.

23. No.

CHAPTER 14

1. Waves on a string is transverse, due to the elastic stretching of the string; sound waves are longitudinal, due to the compression and rarefaction of the air.

3. For more time, as no energy is transferred to the air to produce the sound wave.

5. (a) is correct.

7. For the same wavelength, the frequency of sound is higher in helium, as the speed of sound in helium is higher due to its lower density.

9. Sound waves cannot propagate in vacuum.

11. $h(x, t) = h_0 \sin(kx - \omega t + \phi_0)$.

13. Yes.

15. The wavelength must be decreasing, causing the frequency to increase.

17. Increase the tension in the wire.

19. No; v is more than doubled.

CHAPTER 15

1. Yes.

3. The pulses approach each other and superimpose as they meet, reinforcing each other if they are on the same side, canceling if they are on opposite sides, and forming a tilted pulse if they are perpendicular to each other. After passing through each other they move on, maintaining their integrity.

5. Stick the gum on one fork and listen to the beat again. If the beat frequency gets lower then this fork oscillates at a higher frequency than the other one.

7. f is unchanged, while λ decreases.

9. A single pulse is the superposition of many harmonic waves of different wavelengths, which travel at different speeds and cause the wave to spread into many ripples.

11. The energy of the wave on the string takes kinetic form rather than elastic potential form at the moment the two pulses cancel each other out, very much like the moment the string is flat in a standing wave.

CHAPTER 16

1. Hot air is less dense than cold air and the buoyant force of the surrounding cold air lifts the balloon.

3. The weight in water is less than that in air by that of the water displaced, from which we know the volume, and hence density, of the body. Fat content is then deduced from the density data.

5. It weighs more as the buoyant force is less than the weight of the compressed air inside.

7. Yes.

9. Its high density allows for a relatively short column to balance air pressure.

11. The head would implode if $p < 0$. The value of $\rho v^2/2$ is always positive; $\rho g h$ can be both positive or negative; p is usually positive but it can be assigned any value as long as $\Delta p = p_1 - p_2$ is unaffected.

13. $h_{max} = P_{air}/\rho_{water} g \approx 10.3$ m.

15. Yes if the sails are curved, which is not absolutely necessary but works better.

17. Styrofoam has greater volume and weighs less in air; estimate the volume and find the density from m/V.

19. The sides of the can are pushing in on the liquid.

21. The oil is closer to the center of the centrifuge—think of the centrifugal force as an effective gravity pointing away from the center of rotation.

23. The water pressure on one's chest and lungs prevents air from entering a long tube deep under the water surface.

25. It remains the same.

CHAPTER 17

1. $\approx 4 \times 10^{23}$ molecules.

3. The pressure increases as depth increases, and more air molecules are present in a given volume at a higher pressure.

5. An equilibrium is reached among all the three phases at the triple point, much like how ice and liquid water can coexist at 0°C.

7. Stand in front of a radiant heater and compare the heat it generates before and after it is turned on.

9. A very cold tire may have exceptionally low pressure, and boosting it to the nominal value while it is so cold could cause the pressure to be dangerously high when the tire is heated up by driving.

11. No.

13. The ball expands upon being heated so it cannot pass through the ring; heating the ring will work.

15. Measure the temperature of each substance to see if it has reached a common value.

17. As vast numbers of tiny molecules collide randomly with the wall of the container, there is no perceptible net force on the wall.

19. Nothing, except for the possible confusion it might cause.

21. Icebergs would sink to the bottom of the ocean. As enough ice builds up below the water surface, it could largely deplete the space occupied by liquid water, and that would be devastating to sea life.

CHAPTER 18

1. Work done on the air as it undergoes adiabatic compression causes its temperature to increase.

3. Yes.

5. (a) and (c) .

7.

9. To improve thermal insulation with the air trapped in between the layers.

11. An incredible amount of shaking would be required for any significant temperature increase, even if the container is thermally insulated.

13. Fluids with very large specific heat: as thermal reservoir to store and transfer heat and to limit temperature variation; fluids with very small specific heat: as lubricant in machinery.

15. $\Delta Q_{net} = c_1 m_1 \Delta T_1 - c_2 m_2 \Delta T_2 = 0$, so $\Delta T_1/\Delta T_2 = -c_2 m_2/c_1 m_1$.
If $c_1 = c_2 = c_{water}$ then $\Delta T_1/\Delta T_2 = -m_2/m_1$; and $\Delta T_{Hg}/\Delta T_{water} = -c_{water} m_{water}/c_{Hg} m_{Hg} = -m_{water}/[c_{Hg}m_{Hg}/c_{water}] = -m_{water}/m_{eff}$, i.e., 100 g of mercury behaves effectively like water of mass $c_{Hg}(100 \text{ g})/c_{water} = 3.3$ g.

17. It takes far more energy to completely separate water molecules from each other than to slightly "loosen them up".

19. $W_{net, 2} = 2W_{net, 1}$.

21. Not if $c(T)$ is known.

23. Friction turns 100% of the mechanical energy into heat, and a rapid process prevents thermal equilibrium from being established.

25. Yes. $c_{Hg} \ll c_{water}$ so $\Delta T_{Hg} \gg \Delta T_{water}$.

27. Keeping the engine idling greatly limits the airflow across the radiator surface that can help cool the engine, and winter air is typically drier and has lower specific heat.

CHAPTER 19

1. T increases; the gas molecules on average speed up due to the extra momentum imparted upon them by the inward movement of the piston.

3. It is the probability per a range of one meter.

5. The intermolecular collisions are largely elastic and leave the average molecular kinetic energy unchanged.

7. Thermal equilibrium can still be established between the wall and the gas molecules, yet relatively large fluctuations from the average value of molecular kinetic energy would make T less well defined.

9. Not quite meaningful as an indicator of average molecular kinetic energy, which is a statistical quantity involving a large number of molecules.

11. The intermolecular collisions are largely elastic which conserve momentum, so the same total momentum is carried by the molecules to the wall after the collisions.

13. The miniscule size of the molecules makes it harder for them to hit each other.

15. Air.

17. The inward moving piston collide with gas molecules to increase their average speed (and hence energy).

19. The pane of glass is bombarded on both sides with equal pressure.

21. The airflow helps more energetic water molecules to escape from the water surface

23. Measure the heights of a large group of people and divide the heights into small and equal-sized increments, then check the number of people whose heights fall within each increment. No absolute certainty.

25. The average molecular rotational kinetic energy is a function of temperature.

CHAPTER 20

1. Yes if the ocean temperature is higher than that of the air, as in winter times.

3. Only in the sense that it is associated with the "randomization" in spontaneous physical processes.

5. No.

7. In warm climates.

9. Gas X will permeate the bottle and diffuse throughout the room.

11. Cream poured into a cup of coffee diffuses and mixes up with the coffee, for example. Entropy is increased.

13. They can convert essentially 100% of the electrical energy into heat, but not to work.

15. Only part of the heat released from the chemical energy is converted into mechanical work.

17. No, the room would get even warmer.

19. Extensive.

Answers to Odd-Numbered Problems

CHAPTER 1

1. 3.0×10^3 green jelly beans.

3. 3.5600×10^4 cents

5. 10^7; 10^{14}.

7. 1.8×10^{10} atoms.

9. $1.18/\text{kg}$.

11. 32.2 ft/s^2.

13. 3.33 g/cm^3.

15. 34 mi/gal; 10 mi/gal.

17. 0.0402%; 8.49×10^{-6}%.

19. $0.2(1 \pm 30\%) \text{ m}^3$.

21. 5%.

23. $[MLT^{-1}]$.

25. $[ML^2T^{-1}]$.

27. No.

29. (a) $[L^{-1}]$, (b) $[L^5T^{-2}]$.

31. (a) $[M^{1/2}L^{3/2}T^{-1}]$, (b) $[ML^2T^{-2}]$.

33. 2 tons.

35. $3.5 \times 10^{-6} \text{ cm}^2$.

37. (a) 10^6 mechanics, (b) 10^6 mechanics, (c) 10^6 mechanics.

39. 1×10^{21} droplets.

41. 1×10^{-19}.

43. 9×10^6 automobiles.

45. 30 cm.

47. 1.2×10^{57} hydrogen atoms; 8.2×10^{56} hydrogen atoms;

49. ≈ 100 m.

51. Her position can be described by the clockwise angle ϕ from the north-south line drawn from the center to the starting point (south end of the lake). Her direction of travel will be tangent to the circle with a constant speed of 3 m/s at an angle ϕ clockwise to the south direction.

53. $(1.2\hat{i} + 8.2\hat{j})$ paces (8.3 paces, 82° north of east).

55. Take the origin where the catch was made, with the x-axis to the right and the y-axis upfield. Catch: $0\hat{i} + 0\hat{j}$; first turn: $0\hat{i} + 0\hat{j} + 15\hat{j} = 15\hat{j}$; second turn: $0\hat{i} + 15\hat{j} - 15\hat{i} = -15\hat{i} + 15\hat{j}$; third turn: $-15\hat{i} + 15\hat{j} + 10\hat{j} = -15\hat{i} + 25\hat{j}$; fourth turn: $-15\hat{i} + 25\hat{j} + 20\hat{i} = 5\hat{i} + 25\hat{j}$; fifth turn: $5\hat{i} + 25\hat{j} - 5\hat{i} = 25\hat{j}$; touchdown: $25\hat{j} + 65\hat{j} = 90\hat{j}$.

57. $V_x = +V \cos \alpha$.

59. $\overrightarrow{AB} = (7, 0) = 7$ at an angle of 0°; $\overrightarrow{BC} = (-7, 7) = 7\sqrt{2}$ at an angle of 135°; $\overrightarrow{CA} = (0, -7) = 7$ at an angle of 270°.

61. (a) $\vec{A} = -4\hat{i} + 2\hat{j}$, $\vec{B} = -\hat{i} + 4\hat{j}$, $\vec{C} = 2\hat{i} + 2\hat{j}$, $\mathbf{D} = 5\hat{i} - 3\hat{j}$, (b) $-11\hat{i} + 9\hat{j}$, 9.2

63. $A_x = A \sin(60° - \theta)/\sin 120°$; $A_y = A \sin \theta/\sin 120°$;

65. (a) 10^{-6} metric tons/g, (b) 10^{-6} m³/cm³.

67. 13 min.

69. 2×10^{27} molecules.

71. 3×10^{-8} cm.

73. 2×10^{41} kg/galaxy; 1.2×10^{68} H atoms.

77. 10^3 trucks/day; 2×10^3 trucks/day.

79. 1.1×10^8 km; 29 yr.

81. 10^{44} molecules.

83. (a) $\vec{v} = -v \sin \theta \, \hat{i} + v \cos \theta \, \hat{j}$, or $\vec{v} = +v \sin \theta \, \hat{i} - v \cos \theta \, \hat{j}$.

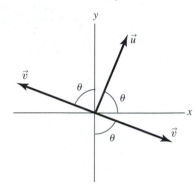

85. (a) $r \sin \theta \cos \phi$, (b) $r \cos \theta$, (c) $r \sin \theta \sin \phi$.

87. $t_0 = (\text{a constant})\ell(\lambda/t)^{1/2}$.

CHAPTER 2

1. +21 cm; 21 cm from the origin in the positive direction.

3. 252 m; 0;

5.

(a)

(b) 37 min and 15 km; 68 min and 53 km.

7.

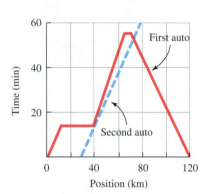

9. (a) $4.3\hat{j}$ mi; (b) $0.56\hat{j}$ mi/min.

11.

t (s)	v (m/s)	x (m)
0.0	0.00	0.00
0.5	0.75	0.19
1.5	1.75	1.44
2.5	8.75	6.69
3.5	21.75	21.94
4.5	39.75	52.69
5.5	62.75	103.94
6.5	90.75	180.69
7.5	122.75	287.44

13. 80 km.

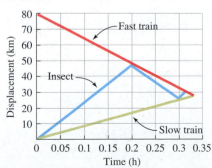

15. $1.4\hat{i}$ cm/s; $1.8\hat{i}$ cm/s; $3.5\hat{i}$ cm/s; $2.0\hat{i}$ cm/s.

17. 39 s.

19. $0.30g$.

21. 1.6 m/s².

23. $A\omega\cos(\omega t)$; $-A\omega^2\sin(\omega t)$.

25. The particle never gets farther from the origin than A; it oscillates back and forth through the origin. The magnitude of the velocity is maximum at the origin and zero at $x = \pm A$. The magnitude of the acceleration is maximum at $x = \pm A$ and zero at the origin.

27.

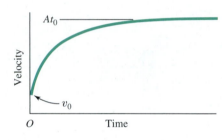

29. 2.02 ft/s² (0.61 m/s²).

31. 0.75 m/s.

33. (a) $8.0 - 0.5t$, easterly with t in s, v in m/s, (b) 34 m to the east.

35. 1×10^4 m/s²; 7×10^{-4} s.

37.
(a)

(b) -3.6 m/s².

39. 1.92 s.

41. (a) 2.96 m/s², (b) 97.2 s.

43. (a)

(b) 5940 ft (1.13 mi); (c) 54 s.

45. 3.65 s.

47. (a) 8.0×10^{-9} s; (b) 6.3×10^{14} m/s².

49. 4.0 m/s²; 18 m.

51. 54 m.

53. 9.0×10^5 m/s²; 6.7×10^{-4} s.

55. Not much time to say anything!

57. 1.1 s.

59. 3.5 s.

61. -22 m/s; 25 m; 20 m/s.

63. 28.4 m.

65. $-4e^{-0.5t}$; $(2 - v/2)$, with t in s, v in m/s, and a in m/s².

67. (a)

(b) 0.15 m/s, (c) 0.94 m/s²,

(d)

69. $(3v_f/\alpha)^{1/3}$.

71. 0.7 m; 3 m.

73. 12 m.

75. 1.6 m/s²; 15 m.

77. $x_0 + v_0 t + (1/2)a_0 t^2 + (1/6)Jt^3$.

79. (a) $-(14 \text{ m/s})\hat{j}$, (b) $(13 \text{ m/s})\hat{j}$, (c) $(1.4 \times 10^4 \text{ m/s}^2)\hat{j}$

81. 1.6 m; 0.08 s.

CHAPTER 3

1. $(15\hat{i} + 15\hat{j})$ km; $(30\hat{i} + 15\hat{j})$ km; $(30\hat{i} + 43\hat{j})$ km; 52 km, 55° N of E.

3. $\vec{r}_A = 0$; $\vec{r}_B = 25\hat{i}$ m; $\vec{r}_C = (5\hat{i} + 35\hat{j})$ m; $\vec{r}_D = 35\hat{j}$ m.

5. $(1.3\hat{i} + 2.7\hat{j})$ km; $(0.1\hat{i} + 4.9\hat{j})$ km.

7. $(3.60 \text{ m/s})t\hat{i} + (6.00 \text{ m/s})t\hat{j}$, $0 \leq t \leq 41.6$ s; $[-(141 \text{ m}) + (7.00 \text{ m/s})t]\hat{i} + (250 \text{ m})\hat{j}$, $41.6 \text{ s} \leq t \leq 77.3$ s.

9. $(2.0\hat{i} - 3.5\hat{j})$ m; 4.0 m; $-4.0\hat{j}$ m; $d = 4.0$ m; $4.0\hat{i}$ m; $d = 4.0$ m; $\theta(t) = -\pi t/T$.

11. 4.0 m/s^2, opposite to the direction of the velocities.

13. $(3.7 \text{ m/s}^2)\hat{j} - (2.4 \text{ m/s}^2)t\hat{k}$.

15. 30 km/h, 31.6° north of west, $(-4.3\hat{i} + 2.7\hat{j})$ km.

17. (a) $[(0.000225 \text{ m}^2/\text{s}) - (0.0009 \text{ m}^2/\text{s}^2)t]\hat{j}/$ $[(0.0169 \text{ m}^2) +$ $(0.00045 \text{ m}^2/\text{s})t - (0.0009 \text{ m}^2/\text{s}^2)t^2]^{1/2}$, (b) $0.017\hat{j}$ m/s; 0.

19. $\vec{v} = [-(4\pi/T)\sin(\pi t/T)\hat{i} - (4\pi/T)\cos(\pi t/T)\hat{j}]$ m/s; $\phi = \pi/2 - \pi t/T$.

21.

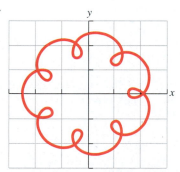

23. $[T^{-1}]$; $v_0 = 0$; $v \rightarrow -u$; $a_0 = -Bu$; $a \rightarrow 0$.

25. 12 m/s (horizontal); $h = 7.4$ m.

27. (a) 3.64 s, (b) 11.5 m, (c) -14.7 m/s (down); $+14.7$ m/s (up).

29. (a) 0.43 s, (b) 5.4 ft/s^2, to the left.

31. 5.2 m/s.

33. 187 m/s.

35. 25° below the horizontal.

39. 76°; 14°.

41. (a) 34 m/s, (b) 30 m, (c) 60 m.

43. 329 m.

45. (a) 1.4 s, (b) successful; 8.6 m above the bar.

47. 15° and 75°; 0° and 90°.

49. (a) 7.8×10^3 m/s, (b) 9.1 m/s^2 toward Earth's center.

51. 2.5 m/s^2.

53. 9.5 s.

55. 0.86 km.

57. (a) (0.15 m, -0.32 m), (b) $(-9.3 \times 10^2 \text{ m/s}^2)\hat{i}$, (c) $(-9.3 \times 10^2 \text{ m/s}^2)\hat{j}$.

59. 1.21R.

61. 15.8 km/h, 18° south of east.

63. 50 km/h.

65. (a) 15 ft/s, (b) 6.7 ft/s.

67. (a) 843 km/h, (b) 781 km/h, 26° W of S, (c) 3900 km, 26° W of S.

69. (a) 29.5 km/s, (b) 30.5 km/s, (c) 30.0 km/s.

71. (a) 45 m/s at 65°, (b) 84 m.

73. $(10\cos\theta + 6)t\hat{i} + (10\sin\theta)t\hat{j}$, with r in km and t in hr; 127°; 68 s.

75. The hammer will hit the deck.

77. (a) 4.4 m/s, (b) 0.88 m going up; 0.88 m coming down; 0, (c) 1.8 m.

79. $\vec{r} = [(5.0 \text{ m/s})t - (0.36 \text{ m})\sin\theta]\hat{i} + [(0.36 \text{ m})(1 - \cos\theta)]\hat{j}$; $\vec{v} = (5.0 \text{ m/s})[(1 - \cos\theta)\hat{i} + \sin\theta\,\hat{j}]$; $\vec{a} = (69 \text{ m/s}^2)(\sin\theta\,\hat{i} + \cos\theta\,\hat{j})$.

CHAPTER 4

1. (a) Force of gravity (toward Earth), (b) force of gravity (down), normal force from the ice (up); and a small friction force from the ice (opposite to the motion), (c) essentially none.

3. 600 N in the $-y$-direction.

5. Yes.

7. (a) 3.0×10^{-4} N, (b) 3.0×10^{-4} N.

9. (a) The acceleration opposite to the motion is due to a retarding force. (b) The observer sees the car (initially at rest) move backward with increasing speed until it reaches $\vec{v}_0$. She would say that this is due to a backward force from the wind, etc.

11.

13. -5.7×10^5 m/s^2; -1.1×10^3 N.

15. $(F/m)\hat{j}$; no change.

17. $(0.91 \text{ N})\hat{i} + (2.3 \text{ N})\hat{j}$.

19. 6.5×10^2 N.

21. $T = F$; $0.433F$.

23. 78 N.

25. 1.6×10^{-21} m/s^2.

27. (a) Earth, (b) the ice and Earth, (c) none.

29. (a) half, (b) twice.

31. (a) 0.14 m/s^2, (c) 1.2×10^4 N backward.

33. $(\hat{i} + \hat{j} + 2\hat{k})$ N.

35. Yes; yes; yes; 0.70 m/s^2; not real.

37. (a) Parallel to the window edge, (b) at an angle given by $\tan\theta = a'/g$, (c) parallel to the window edge.

39.

41.

43.

(a)

(b)

45.

(a)

(b)

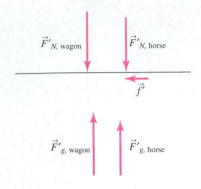

(c)

(d) a forward force from Earth on the horse's hooves.

47.

49.

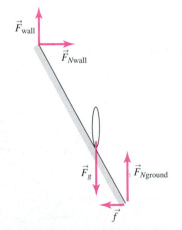

$\Sigma F_x = F_g \sin 21° - f = ma$ (x-direction);
$\Sigma F_y = F_N - F_g \cos 21° = 0$ (y-direction).

51. There is a net horizontal force from the wall.

53.

55. (a) $F_g = mg$ down; $F_N = F_g = mg$ up,
(b) mg down and $2mg$ up, (c) mg down.
57. $(1/M)(F_1^2 + F_2^2)^{1/2}$; $\tan^{-1}(F_2/F_1)$.
59. $(-1.1 \times 10^{-2}t^{-5/4})$ N, t in seconds.
61. (a) $c = (k/m)^{1/2}$,
(b) $A = 0$; $B = +v_0(m/k)^{1/2}$.
63. (a)

(b) 82.2 N up, (c) 1.55 m/s² up.
65. $T = (m_W g + m_B g)/4$;
$a = [4T - (m_W g + m_B g)]/(m_W + m_B)$.

67. (a) Backwards; 3.0×10^4 N,
(b) 1.6×10^4 N; 0.18 m/s², (c) 1.1×10^4 N
from the first car; 5.2×10^3 N from the third;
0.18 m/s².
69.
(a)

(b) same forces, but the upward force would be fictitious.
71. (a) $[A] = [ML^{-1}]$ with units of kg/m,
(b) $(Av^2 - mg)/m = (Av^2/m) - g$,
(c) $v_t = (mg/A)^{1/2}$.

73.

(a) force of gravity: mg (down); normal force of ground: F_N (up); friction force of ground: f (forward); wind resistance: F_w (backward),
(b) 0, (c) 0; $v_w + v$; 0, (d) $v + v_w$; 0; 0.
75. (a) 0.40 m/s²; 0.20 m/s²,
(b) 0.067 m/s²; 0.13 m/s², (c) 8.6 g.
77. $\phi = \theta$.

CHAPTER 5

1. 1.4×10^4 N.
3. 980 N.
5. 0.095 s.
7. (a) 24 N, (b) 7.8 N.
9. (a) 66 N, (b) 5.1 m/s² up.
11. (a) 2.5 m/s² forward,
(b) 3.5×10^3 N forward.
13. $M_{max} = 9.2$ kg; $M_{min} = 4.5$ kg; 0.
15. 9.8 mm/s²; 2.5 mm/s².
17. $a_1 = [(m_1 m_2 + m_1 m_3 - 4m_2 m_3)/(m_1 m_2 + m_1 m_3 + 4m_2 m_3)]g$;
$a_2 = [(m_1 m_2 - 3m_1 m_3 + 4m_2 m_3)/(m_1 m_2 + m_1 m_3 + 4m_2 m_3)]g$;
$a_3 = [(-3m_1 m_2 + m_1 m_3 + 4m_2 m_3)/(m_1 m_2 + m_1 m_3 + 4m_2 m_3)]g$;
$T_1 = [8m_1 m_2 m_3/(m_1 m_2 + m_1 m_3 + 4m_2 m_3)]g$;
$T_2 = [4m_1 m_2 m_3/T_2 = [4m_1 m_2 m_3/(m_1 m_2 + m_1 m_3 + 4m_2 m_3)]g$. If
$m_2 = m_3 \neq m_1$, then
$a_1 = [(m_1 - 2m_2)/(m_1 + 2m_2)]g$;
$a_2 = a_3 = [(2m_2 - m_1)/(m_1 + 2m_2)]g$;
$T_1 = [4m_1 m_2/(m_1 + 2m_2)]g$;
$T_2 = [2m_1 m_2/(m_1 + 2m_2)]g$.
19. 48.0 N; 96.0 N; 2.2 m/s² (up); -0.2 m/s² (down); -1.8 m/s² (down).
21. 0.12.
23. 0.61.
25. 3.6×10^2 N.
27. 0.40.
29. (a) 52 m, (b) 3.9 s.
31. (a) 1.8 s, (b) 2.6 N, (c) 0.65.
33. (a) $370/(\cos \theta + 0.75 \sin \theta)$ N,
(b) 37°; 2.9×10^2 N.
35. 0.077.
37. 9.5×10^2 N.
39. 0.060 m².
41. 0.57 m/s.
43. 0.67.
45. 5 m/s.
47. 1.3 m/s² toward the center.
49. 1.4×10^2 N toward the center.
51. $v_{max} = 32$ m/s, $v_{min} = 5.8$ m/s.
53. 0.87 m.
55. 0.26.
57. 1.0×10^3 m/s tangent to the orbit.

59. 4.43 m/s.

61. (a)

(b) 0.4 m/s.

63. 4.5 m.

65. 0.72 m.

67. 55 m/s.

69. (a) 21 N, (b) 18 N.

71. 1.6×10^{-12} kg/m^3.

73. 1.2×10^2 N.

75. (a) 0.5 N, (b) 1.2 N.

77. (a) $\Delta x_2 = -\Delta x_1/2$, (b) $a_1 = -1.78$ m/s^2 (down); $a_2 = +0.89$ m/s^2 (up), (c) 9.6 N.

79. -0.3 m/s^2, towards mass-1.

81. $\omega = (g/\mu_s R)^{1/2}$.

83. $\omega = [g/(\ell \cos \theta)]^{1/2}$.

85. 26 m.

87. 11 m/s.

89. 2.3 m.

CHAPTER 6

1. (a) 39 J, (b) 1.0 m/s; 89 m/s, (c) 0.40 m/s.

3. (a) The force of gravity = 98 N (down); the upward pull = 98 N; $F_{net} = 0$, (b) 0, (c) 98 J.

5. 1.0×10^4 J.

7. (a) 2.4×10^2 J, (b) friction force, (c) 0.

9. 730 J; -730 J.

11. -3.73 J.

13. 1.8×10^3 J.

15. 2.8×10^{11} J.

17. 2.5 m; $(0.916)^n(3.00$ m$)$.

19. 0.127 J; 0.159 J; 0.175 J.

21. $-mgR(1 - \cos \theta)$.

23. $+3$.

25. 0.89 m/s.

27. 2.53.

29. 3.9×10^3 J.

31. 6.1×10^3 J.

33.

$-\sin \theta \, \hat{i} + \cos \theta \, \hat{j}$; $\sin \theta \, \hat{i} - \cos \theta \, \hat{j}$.

35. 3.4.

37. -24 J.

39. $F_x L$.

41. 0.067 J.

43. $2g_1 - 4g_2$.

45. 8.6 m/s.

47. 8.1×10^{-2} J.

49. 14 J.

51. 2.1×10^3 J.

53. -15 J, -5 J; conservative.

55. (a) 6.3×10^{-2} J, (b) 0, (c) 5.9×10^{-2} J, (d) -5.9×10^{-2} J.

57. (a) 4 J; 6 J; 6 J; 4 J, (b) $+5$ m, (c) conservative.

59. Constant forces are included; not a function of position only.

61. (a) $0.16C$ J; 0, (b) $0.32C$ J.

63. 22 bulbs.

65. 0.7 kW.

67. 52 kW.

69. 5.0 m/s; 1.3 m/s.

71. 4.7×10^7 W.

73. (a) 16 m/s, (b) 0.5 s, (c) 12 s.

75. 5.8×10^{-12} J; 5.8×10^{-9} J.

77. 0.62; 0.31.

79. -0.2 J.

81. 0.034.

83. (a) 2.2×10^7 J, (b) -2.2×10^{-7} J.

85. (a) $mg = 26$ N down; $F_N = 22$ N perpendicular to plane (up); $f_k = 5.4$ N parallel to plane (down), (b) -18 J; 0; -7.0 J, (c) 4.4 m/s.

87. 5.3 J.

89. (a) 0, (b) -0.20 J, (c) 9.8 J.

91. $mgH(1 + \mu_k \cot \theta)$.

93. $W_g = mgL(\cos \theta_f - \cos \theta_i)$; $v = [2gL(1 - \cos \theta_i)]^{1/2}$; $W_T = 0$.

95. $0.01mK/x$.

CHAPTER 7

1. (a) 0.20 kJ, (b) 29 J; 2.3 kJ.

3. 25 m.

5. 1.2×10^{-2} J.

7. (a) $-8x$ J, with x in m, (b) $+18$ J, (c) 4.1 m/s.

9.

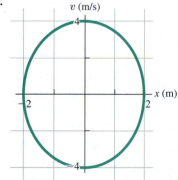

11. 0.541 m; 0.468 m.

13. 0.43 m; 0.48 m.

15. (a) 7.1 m/s, (b) -0.25 J, (c) 0.73.

17. $mg_0 y - (1/2)mg' y^2$.

19.

21. $E/(0.8$ N$)$, $-E/(2.5$ N$)$.

23. (a) To the right, (b) $(k/m)^{1/2}x$, (c) to the left; $-(k/m)^{1/2}x$.

25.

27. $x = 0$ is unstable and $x = \pm 0.24$ m is stable.

29. 503 m/s.

31. 130 m/s; same.

33. (b) $(2gH)^{1/2}$, (c) $2(gH)^{1/2}$.

37. -5.0×10^4 J.

39. (a) 1.4 m/s, (b) 1.1 m/s, (c) 8 cm from point a.

41. (a) $U(x) = \frac{1}{2}k\left(\sqrt{h^2 + x^2} - L\right)^2$.

(b) $U(x) = -\dfrac{k\left(\sqrt{h^2 + x^2} - L\right)x}{\sqrt{h^2 + x^2}}$.

43. (a) GMm/r^2 toward M, (b) $GMm/2r$, (c) $-GMm/2r$.

47. -6.2×10^4 J.

49. ≈ 20 cents.

51. 5.9×10^2 J.

53. (a) $(3 + 3x^2/2 - 0.2x^3/3)$ J, with x in m.

(b)

55. (a) Conservative, (b) not conservative.

57. $-A/3$; no.

59. -2 J; -2 J; $+2$ J.

61. (a) 7.7 m/s; 14.7 m/s (b) the skier leaves the surface.

63. (a) 4.4 m/s, (b) 3.7 m/s, (c) 5.9 N; 4.2 N.

65. (a) Conservative force, (b) $-C/r$, (c) 2.1×10^7 m/s.

67. (a) +1.13 J, (b) −0.083 J, (c) −0.30 J, (d) 0.70 m/s, (e) 8.8 cm, (f) 5.0 cm.

69. 1.1 m/s.

73. (a) $(1/2)mv^2 + (1/2)kr^2$, (b) $R(k/m)^{1/2}$, (c) $(2E/mr^2 - k/m)^{1/2}$.

75. $L[\sin \alpha + 2 \sin \alpha \cos^2 \alpha + 2(\cos^3 \alpha - \cos^6 \alpha)^{1/2}]$.

CHAPTER 8

1. (a) 1.2 kg·m/s, (b) 5.1 kg·m/s, (c) 7.0×10^2 kg·m/s, (d) 7.6×10^2 kg·m/s.

3. (a) 4.2×10^2 kg·m/s, (b) 6.0×10^6 kg·m/s, (c) 1.2×10^4 kg·m/s, (d) 3.3×10^{-22} kg·m/s, (e) 1.4×10^{-2} kg·m/s.

5. (a) −1.0 m/s, (b) 3.5 J.

7. (a) $(-1.9\hat{i} - 10.8\hat{j})$ kg·m/s, (b) $(-4.9\hat{i} - 6.8\hat{j})$ m/s, (c) $(-0.90\hat{i} - 2.9\hat{j} - 1.2\hat{k})$ m/s, (d) 24 J; 64 J; 19 J.

9. 40 m/s.

11. −12 kg·m/s, opposite to the original motion; 1.7×10^4 N, opposite to the original motion.

13. -1.2×10^3 kg·m/s, down; 1.3×10^4 N, up.

15. 2.2×10^2 kg·m/s, up.

17. 7.5×10^2 N, up.

19. (a) 5.7×10^2 N, (b) 1.4×10^2 J, (c) 0.011 s.

21. $0.60\sqrt{g(0.20 + 0.10N)}$ kg·m/s.

23. −1.3 kg·m/s, down; −0.65 kg·m/s, down.

25. −0.13 m/s, opposite to the direction of the first cart.

27. 0.74 m/s.

29. $m_1/m_2 = 3$.

31. 4×10^{-9}.

33. (a) $M/(m + M)$, (b) $[(m + M)/m]\sqrt{2gh}$.

35. (a) $v[m_2^2 + M(m_1 + m_2)]/M(m_2 + M)$ forward, (b) $v[m_1^2 + M(m_1 + m_2)]/M(m_1 + M)$ forward, (c) $v(m_1 + m_2)/M$ forward.

37. (a) −0.20 m/s, (b) inelastic; 64%.

39. 2.59 m/s.

41. 15°.

43. (a) $\sqrt{2gh}$, (b) $\sqrt{2gh}$, (c) $[(m - 3M)/(m + M)]\sqrt{2gh}$, up, (d) $[(m - 3M)/(m + M)]^2h$, (e) 9h.

45. 1.2 m/s 60° below the −x-axis, not elastic.

47. $(2.00 \text{ m/s})\hat{i} + (1.00 \text{ m/s})\hat{j}$; elastic.

49. 2.6 m/s.

51. 8.7 m/s, 20° above the horizontal.

53. The driver of the car was speeding!

55. 0.78 m from the heavier sphere.

57. $X = 0.38$ m; $Y = -0.66$ m.

59. 0.

61. $X = 0$; $Y = 1.0$ m from the bottom of the handle.

65. $4R/3\pi$ from the center of the arc, along the bisector.

67. $X = Y = 0.6$ m; $X = Y = 0.5$ m.

71. -3.3×10^3 kg/s.

73. 1% of mass must be discarded.

75. 2.5×10^3 m/s.

77. 5.0 m.

79. (a) Spit out seeds one at a time, (b) 6×10^{-3} m/s.

81. (a) 7.9 m/s (to the right), (b) 7.9 m/s (to the right), (c) −1.9 m/s (to the left); +1.1 m/s (to the right), (d) 2.1×10^4 N (to the right).

83. (a)

(b) 4.1 m/s recoil, (c) upward impulse provided by the ground.

85. (a) $[8M(M - m)/(M + m)^2]h$, (b) $-[4mM/(M + m)^2]h$.

87. 26 km/h.

89. $R/6$ from the center of the styrofoam sphere.

91. (a) 0.11 m/s, (b) 0, (c) −0.074 m, (d) 0.

CHAPTER 9

1. 0.68 rad/s².

3. −17 rad/s².

5. 7.3×10^{-5} rad/s from the South Pole to the North Pole; 2.0×10^{-7} rad/s perpendicular to the orbital plane.

7. (a) -2.62×10^2 rad/s², (b) 1.67×10^2 m.

9. (a) 0.10 rad/s², (b) 11 rev, (c) 2.1×10^2 m.

11. (a) $a = 1.5$ m/s²; $x = x_0 + (0.75 \text{ m/s}^2)t^2$, (b) 100 rad/s², (c) $\omega = (100 \text{ rad/s}^2)t$.

13. 1.35×10^{-4} m.

15. α is linearly proportional to t.

19. 9.

21. 2.8×10^{-3} kg·m².

23. 1.3×10^2 J; 4.0×10^2 J.

25. (a) 0.021 kg·m², (b) 0.083 kg·m².

27. $I_{\text{Earth}} = 9.8 \times 10^{37}$ kg·m²; $I_{\text{Neutron star}} = 1.0 \times 10^{39}$ kg·m² $\approx 10 \, I_{\text{Earth}}$.

29. $M(R_2^2 + R_1^2)/2$.

31. $(3/10)(\tan^2 \alpha)MH^2$.

33. $(8/15)\pi[(\rho_1 - \rho_2)R_1^5 + \rho_2R_2^5]$.

35. $(16/45)MR^2$.

37. 24 lb.

39. 98 kg.

41. $L(F_{1y} + F_{2y})/2$ perpendicular to the rod.

43. (a) 2.3 N·m, (b) change the 7-N force to 5.4 N.

45. 0.38 N·m.

47. (a) 2.13 kg·m², (b) 8.84 kg·m²/s, (c) 0.886 rev, (d) 18.3 J.

49. 3.8 rad/s up; rotate in the direction of the original rotation of the wheel.

51. -2.46×10^{-5}.

53. 2 s.

55. 0.12.

57. 13.4 m; 17.9 m.

59. 0.95 m.

61. 0.050 m/s².

63. 2.3 rev.

65. 5.1×10^6 m.

67. $(3v_0^2/4g) \sin \theta$.

69. m_1R_1/R_2; no.

71. (a) 0.72 kg·m²/s, (b) 3.8 J, (c) 7.6 J.

73. $3\mu MgR/(R + 2r)$.

75. $MgR^2/(R^2 + 2r^2)$.

CHAPTER 10

1. (a) 1.1×10^{10} kg·m²/s down, (b) 7.9×10^9 kg·m²/s down.

3. (a) 5×10^4 kg·m²/s down, (b) 5×10^4 kg·m²/s down.

5. 1.5 kg·m²/s north.

7. $-7.4t^2\hat{k}$ kg·m²/s (into the page).

9. $(mt^2/2)(bv\hat{i} - aw\hat{j} - av\hat{k})$.

11. $\vec{r} \times [m_2m_1/(m_1 + m_2)] \, d\vec{r}/dt = \vec{r} \times \mu \, d\vec{r}/dt$.

13. (a) $6.4 \times 10^{-2} \hat{k}$ kg·m²/s (along ω-direction), (b) same.

15. (a) $m\omega d^2$ along the axis of rotation, (b) $3m\omega d^2/4$ along the axis of rotation, (c) $m\omega d^2/2$ along the axis of rotation.

17. $(17\hat{i} - 19\hat{j} + 14\hat{k})$ N·m.

19. $140\hat{k}$ N·m (perpendicular to table).

21. $MgR \sin(\omega t)$.

23. $mvd \sin(2\theta)$ up; 0; impulsive force from the wall.

25. $-7\hat{i} + 9\hat{j} + 10\hat{k}$.

27. 1.1 rad/s.

29. 30 m/s.

31. 0.103 rad (0.016 rev).

33. 0.15 rad/s.

35. 5.3×10^2 N·m along the axis.

37. 409 rad/s.

39. 0.14 J.

41. $r_n = (n^2\hbar^2/mk)^{1/4}$, $n = 1, 2, \ldots$; $v_n = (n^2\hbar^2k/m^3)^{1/4}$, $n = 1, 2, \ldots$; $K_n = \sqrt{k/m} \, n\hbar/2$, $n = 1, 2, \ldots$.

43. No energy level 2.0 eV above the lowest state; possible excitation energies are 10.2 eV, 12.1 eV, 12.75 eV, ...

45. $\omega_p = \Delta\theta/\Delta t = Mg\ell/I\omega$ along the z-axis.

47. 2.8 rad/s.

49. 72 rad/s.

51. 0.62 rad/s; −33 J; friction.

53. 3.3 rad/s.

55. (a) 4 m/s, (b) 0.8 s.

57. h.

59. (a) 19.4 m/s, (b) 0.127, (c) 0.132.

61. 26 kg; 52 rad/s; 12 kg.

63. 4.0×10^3 J.

65. (a) K/L, (b) L^2/MK, (c) $2\pi L^3/MK^2$, (d) MK^3/L^4.

67. $K \approx 1 \times 10^{17}$ J; $L \approx 6 \times 10^{20}$ kg·m²/s.

69. (a) 2.4 m/s, (b) 2.4×10^2 rad/s.

CHAPTER 11

1. 40 kg.

3. 139 N down; 96 N down.

5. (2, 0, 1) m.

7. 1.7 m.

9. 0.8 m.

11. 14.2 N·m in xy-plane 27° from $-x$-axis ($\perp \vec{r}$).

13. (a) $3L/4$; $11L/12$.

15. 0.57 Mg.

17. 1.2×10^2 N; $F_{N_A} = 0$.

19. (a) 2.9 m.

21. (a) 2.9×10^2 N 13.5° above the horizontal, (b) $F_V = 3.5 \times 10^2$ N down; $F_H = 2.8 \times 10^2$ N.

23. (a) $\tan^{-1} \mu_s$, (b) $\tan^{-1}(\Delta/2L)$.

25. 0.58 Mg outward.

27. 1.2 kN.

29. 1.9×10^2 N right; 73 N left; 23 N left.

31. (a) $F_{N_A} = mg\ell_2/2\ell_1$ down; $F_{N_B} = mg(\ell_2 - \ell_1)/2\ell_1$ up, (b) seat will lose contact at B and turn clockwise, (c) 74 N; 25 N.

33. (a) 56 N, (b) $(164 \text{ N})\hat{i} + (46 \text{ N})\hat{j}$.

35. 353 kg.

37. (a) $F_{N_1} = 0$; $F_{N_3} = F_{N_2} = (M + m)g/2 - mgx/L$; $F_{N_4} = 2mgx/L$, (b) $F_{N_4} = 0$; $F_{N_1} = -2mgx/L$, $F_{N_3} = F_{N_2} = (M + m)g/2 + mgx/L$.

39. $MgR \sin \theta$.

41. 1.33×10^4 N.

43. (a) 0.255 nm.

45. 0.524; 0.740.

47. 2.97 cm.

49. $(\Delta L/L)_c = 0.014$; 14 mm.

51. 2.5×10^2 N/m^2.

53. $\Delta y \approx 10^{-13}$ m.

55. 0.12° below the horizontal.

57. 1.99928 cm.

61. 9.1×10^{-12} m^2/N.

63. 69 N.

65. (a) $(Mg/2) \cot \theta_0$, (b) $(3g \cos \theta_0)/2L$, (c) $\omega = \sqrt{(3g \sin \theta_0)/L}$.

67. 85 N.

69. 1.36×10^3 N.

71. (a) Near legs: $(1 + 2\mu_k)Mg/4$ up; $(1 + 2\mu_k)\mu_k Mg/4$ to the left; far legs: $(1 - 2\mu_k)Mg/4$ up; $(1 - 2\mu_k)\mu_k Mg/4$ to the left, (b) 0.

75. $\tan(\theta/2)$.

77. 6.88.

79. Top ball: mg, $0.577mg$, $1.155mg$; bottom ball: mg, $1.155mg$, $2.000mg$, $0.577mg$.

CHAPTER 12

1. 2.975×10^{-19} s^2/m^3.

3. (a) $[ML^4T^{-2}]$, (b) $\sqrt{mh}$, (c) $2\pi\sqrt{m/h}$ = a constant.

5. (a) 1.9×10^{-6} N, (b) must be the same.

7. 4.2×10^{-10} N; $\approx 10^{-7} W_{\text{fly}}$.

9. 6.6×10^6 m.

11. 1.51 h.

13. 9.56×10^6 m.

15. 4.9 m.

17. 0.25.

19. 6.18×10^5 m.

21. 26 m/s.

23. 1.6×10^{-3} rad/s.

25. 1.09×10^4 m/s.

27. (a) 1.85×10^8 m, (b) 7.99×10^8 m, (c) impossible.

29. 7.79 km/s tangent to the orbit; 1.52×10^{10} J; 2.56×10^{13} kg·m^2/s perpendicular to the orbit.

31. 1.74 h.

33. 1.49×10^{11} m.

35. 13 km/s.

37. 2.99 km/s.

39.

(a)

(b) 75.6 km/s perpendicular to the radius, (c) 4.76×10^7 km.

41. (a) 6.91×10^3 m/s, (b) 1.74×10^{13} kg·m^2/s, (c) 8.68×10^{12} kg·m^2/s, (d) the satellite crashes.

43. (a) 7.73×10^3 m/s, (b) 1.03×10^{14} kg·m^2/s, (c) -5.98×10^{10} J, (d) zero, (e) cannot be circular.

45. 0.996860; 0.996868.

47. $(2R^3/GM)^{1/2}$.

49. The tidal force on Jupiter is about 8% that on Earth.

51. 7.9×10^3 m/s.

53. 4.51×10^2 km from the Sun's center.

55. $\theta \approx 3 \times 10^{-10}$ rad; yes.

57. $4\pi(R^3/GM)^{1/2}$.

59. 3.5×10^{-15} m.

61. 0.40 rad/s.

63. (a) 1.66×10^3 m, (b) 2.8 m/s, (c) a speed of 1.76 m/s is needed to orbit the asteroid.

65. (a) Circular orbits are supported, (b) $T^2/r^{n+1} = 4\pi^2(m/k) = $ a constant.

67. $\Delta U \approx -4\pi fr$; $\Delta E = -2\pi fr$; $\Delta K \approx 2\pi fr$.

69. (a) 1.92 m/s relative to himself, which is 1.97 m/s relative to the ship, (b) 1.9 h.

71. (a) 1.68 m/s^2, (b) 1.95 m/s^2.

73. (a) Possible, (b) 1.83×10^7 s, (c) unstable.

CHAPTER 13

1. $\pi/2$ rad.

3.

5. 4.09 s.

7. -0.038 m.

9. 7.0137×10^{-6} m

11. $x = A \cos[(2.0 \text{ rad/s})t - 1.17 \text{ rad}]$, t in s.

13. (a) 1.6 rad/s; 4.0 s, (b) 1.3 m.

15. (a) 4.08 cm/s, (b) 6.41 cm/s; 23.6 cm/s^2.

17. (a) $x = (3.4 \text{ cm}) \sin[(3.0 \text{ rad/s})t + 2.1 \text{ rad}]$, (b) 1.8 s, (c) 0; 0.71 s; 1.1 s; 1.8 s,

(d)

19. R.

21. $x = R \cos(\omega t + \delta)$, with $R = 1.5 \times 10^{11}$ m and $\omega = 1.99 \times 10^{-7}$ rad/s.

25. 2.4 s; 15 cm/s.

27. 5.1 kg.

29. 0.18 N/m.

31. 1.5×10^2 N.

33. 1.2 s.

35. $x = (1.0 \text{ cm}) \sin[(71 \text{ rad/s})t + \pi/2]$.

37. 1.1 s.

39. 0.042 m/s.

41. 6.1 rad/s.

45. 2 m/s; 0.7 m; 0.3 J.

47. (a) $(\ell^3/R_E^3)M_E$, (b) $GM_E m\ell/R_E^3$ toward the center, (c) $-GM_E mx/R_E^3$ toward the center of the tunnel, (d) 1.4 h, (e) 1.4 h.

49. $F_{\text{net}} = -kx^3/2\ell^2$.

51. 9.79 m/s^2.

53. (a) 2.23 m, (b) 430 m.

55. (a) 1.7 s, (b) 0.05 m.

57. $E = mgL(1 - \cos \alpha) \approx mgL\alpha^2/2$ (for $|\alpha| \ll 1$).

59. 0.76 s.

61. $2\pi \sqrt{\dfrac{2(L^2 - 3Ly + 3y^2)}{3(L - 2y)g}}$.

63. (a) $MR^2/2 + M\ell^2$, (b) $-g\ell\theta = (R^2/2 + \ell^2)d^2\theta/dt^2$, (c) $2\pi \sqrt{\dfrac{R^2 + 3\ell^2}{2g\ell}}$, (d) $T \to \infty$; no torque.

65. (a) 2.7 s, (b) 1.0×10^{-4} J,
(c) 3.6×10^{-2} m/s.

67. 0.113; 8.4 min.

69. 0.095 kg/s.

71. $k \approx 6 \times 10^4$ kg/s^2; $b \approx 10^3$ kg/s.

73. $(2.0 + 0.00051)$s.

75. $x_{3.0} = -5.40$ cm; $x_{4.8} = -1.48$ cm;
$x_0 = -6.65$ cm.

79. 1.8 Hz.

81. (a) $2.2 \, \text{N} \cdot \text{s/m}$, (b) 0.25 s,
(c) $\Delta\omega = 8.0$ rad/s; $Q = 3.1$.

83. 0.36 m.

85. $\sqrt{2}\,\omega$.

87. 2.5 s.

89. (a) $0.553H$, (b) $0.347H$.

91. (a) 1.2 s, (b) $\pi\sqrt{\dfrac{2m}{k}}$, (c) $\pi\sqrt{\dfrac{2m}{k}}$,
(d) $2\pi\sqrt{\dfrac{m}{k}}$.

93. 9.0 s.

95. (a) $2A/e^2$,
(b) $U = -e^4/4A + (e^4/4A)x^2$; $e^4/(4A^3 m)^{1/2}$.

97. $\dfrac{1}{2\pi}\sqrt{\dfrac{g}{2R}}$.

99. $\dfrac{1}{2\pi}\sqrt{\dfrac{2k}{3m}}$.

CHAPTER 14

1.

3. 200 m/s.

5. 23 m^{-1}; 48 Hz.

7. 5.1×10^3 m/s; 3.5×10^3 m/s.

9. 7.0×10^4 MN/m^2.

11. 8.03×10^{11} dyne/cm^2.

17. Yes.

19. $Y = 3.2 \times 10^5$ MN/m^2;
$G = 8 \times 10^4$ MN/m^2; $L = 1.3 \times 10^4$ km.

21. $\mu\omega^2 R^2$.

23. 25 km/h.

25. $\ell(k/m)^{1/2}$.

27. 2.1×10^3 W; 5.3×10^2 W.

29. 8.9×10^{-10} m.

31. $\mu v \omega^2 z_0^2 \cos^2(kx + \omega t)$.

33. $P = \mu v \left(\dfrac{\omega}{k}\right)^2 \left[\dfrac{df}{d(x - vt)}\right]^2$.

35. 1.08 m; 0.54 m; 0.36 m;

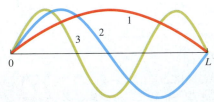

---

37. 13 N.

39. $m_2/m_1 = 1/9$.

41. 1.

43. 3.5 Hz, 22 rad/s, 0.28 s; 7.1 Hz, 44 rad/s,
0.14 s; 11 Hz, 67 rad/s, 0.094 s.

45. 8.49×10^2 N.

47. 1.28 kg/m^3; 0.085 kg/m^3.

49. 89 dB; 65 dB; 39 dB.

51. 6.3×10^{-2} W.

53. 1.97×10^9 N/m^2.

55. 343 m/s.

57. 147 Hz.

59. (a) 0.874 m; 378 Hz, (b) 372 Hz.

61. 1.5×10^7 m/s.

63. 1.88×10^3 Hz.

65. 440 Hz/$(1 + 0.060t)^{1/2}$.

67. 965 m/s (3.5×10^3 km/h).

69. 1.3 kHz.

71. $\lambda_n = 2.40/(2n - 1)$ m, where
$n = 1, 2, 3, \ldots$.

73. 3.2×10^6.

75. 400 Hz; 1.1 m.

77. (a) 484 Hz; 543 Hz, (b) D; B.

79. 3.5×10^2 m/s.

81.
$f_1 = \dfrac{1}{2L}\sqrt{\dfrac{T}{\mu} + \alpha\dfrac{\pi^2}{L^2}}$; $f_2 = \dfrac{1}{L}\sqrt{\dfrac{T}{\mu} + 4\alpha\dfrac{\pi^2}{L^2}}$;
the higher frequency.

CHAPTER 15

1. $A = \sqrt{2}$; $\tan\phi = 1$, or $\phi = \pi/4$.

3. $1.41z_0 \sin[kx - \omega t + (\pi/4)]$.

5. 3.6 cm.

7. $\delta = 2\pi/3$ and $-2\pi/3$.

9. $2z_0 \sin[kx + (\pi/4)]\cos[\omega t + (\pi/4)]$,
which represents a standing wave; x(nodes) $=$
$[n - (1/4)](\pi/k)$, where $n = 0, \pm 1, \pm 2, \ldots$.

11. (a) 11 Hz; 7.3 m/s;
(b) $z_\ell = 3\sin(kx + \omega t)$; $z_r = 3\sin(kx - \omega t)$.

15. 6 m.

17. 506 Hz.

19. 422 Hz.

21. -1.6%.

23. $z = 2A\cos(\omega t - kL)$.

25. 0.59 cm.

27. (a) 37°; (b) 17.5°; (c) 37°.

31. (a) -1 cm, 0, $+1$ cm; (b) (6 cm)
$\sin(\pi r - \pi vt)$, where $r = (y^2 + 4)^{1/2}$.

33. (d) Total energy is conserved.

35.

$z(x, t) = 0$ for $x < (vt - a)$ and
$x > vt$; $z(x, t) = k(x - vt + a)$ for
$(vt - a) < x < vt$.

---

37. $v_t = z_0\left[2(x - vt)\left(\dfrac{v}{\alpha^2}\right)\right]e^{-(x-vt)^2/\alpha^2}$;
$a_t = z_0\left[-\dfrac{2v^2}{\alpha^2} + 4(x - vt)^2\left(\dfrac{v^2}{\alpha^4}\right)\right]e^{-(x-vt)^2/\alpha^2}$;

39.

41. $-0.069A$; $1.07A$.

43. 6.6×10^{-3}.

45. $\dfrac{B}{A} = \dfrac{k_2 - k_1}{k_2 + k_1} = \dfrac{\sqrt{\mu_2} - \sqrt{\mu_1}}{\sqrt{\mu_2} + \sqrt{\mu_1}}$;
$\dfrac{C}{A} = \dfrac{2k_2}{k_2 + k_1} = \dfrac{2\sqrt{\mu_1}}{\sqrt{\mu_2} + \sqrt{\mu_1}}$;
$z_i = A\cos(k_1 x - \omega t)$;
$z_r = B\cos(k_1 x + \omega t + \pi)$;
$z_t = C\cos(k_2 x - \omega t)$.

47. 1 MHz.

49. (a) 400 Hz; 0.4 m; 160 m/s; (b) 7.7 cm.

51. $\Delta T/T = 9.1 \times 10^{-3}$.

53. $69\cos(4\pi t)$ Hz.

55. 90 Hz; 101 Hz; 111 Hz.

57. 6.0×10^{-7} m; 3.9×10^{-7} m.

59. (a) and (b) are linear.

61. z(envelope) $\simeq \sin[\Delta k(x/2) - \Delta\omega(t/2)$
$+ (\pi/4)]$; $f = (\omega_1 - \omega_2)/2\pi$;
$\lambda = 2\pi/(k_1 - k_2)$.

CHAPTER 16

1. 1.15×10^3 kg/m^3.

3. 50 L.

5. $\Delta d_1 = af/\sqrt{2}$; $\Delta d_2 = -af/\sqrt{2}$;

7. 2.1×10^5 N/m^2.

9. (a) 5.1×10^4 N, (b) 1.3×10^2 N (equivalent to lifting 13 kg).

11. 7.8×10^2 N/m^2; p will decrease.

13. 1.11×10^5 Pa; 2.02×10^5 Pa;
1.11×10^6 Pa; 1.01×10^8 Pa.

15. 1.02×10^5 Pa; 1.05×10^5 Pa.

17. $h_{\text{Venus}} = 0.838$ m; $h_{\text{Neptune}} = 0.619$ m.

19. 52 N; 2.2 mm.

21. Sink.

23. 8.4×10^2 kg/m^3.

25. 9.3 m.

27. 0.71 m^3.

29. 13 m^2.

31. 0.96×10^3 kg/m^3.

33. $\rho_1 < \rho_3 < \rho_2$;
$D/H = (\rho_3 - \rho_1)/(\rho_2 - \rho_1)$.

35. 8.1×10^2.

37.

(a)

(b)

39. 2.9 m/s; 0.23 kg/s.

41. 57 mi/h.

43. 0.81 cm.

45. 1.6×10^2 W (0.22 hp).

47. -4.7×10^2 Pa (gauge); 4.7×10^2 Pa.

49. 9×10^{-2} N; rise.

51. (a) 7.7 m/s; (b) 4.3 h.

53. (a) 2 m/s; (b) 2.2×10^5 Pa (2.2 atm).

55. $p_0 + (1/2)\rho v_0^2 + \rho g h_0 = p_1 + (1/2)\rho v_1^2 + \rho g h_0$;
$p_0 + (1/2)\rho v_0^2 + \rho g h_0 = p_2 + (1/2)\rho v_2^2 + \rho g h_0$; $v_0 D^2 = v_1 d_1^2 + v_2 d_2^2$.

57. $y = y_0 + (\omega^2/2g)r^2$ (which describes a parabola).

59. $a = 1, b = 1$.

61. 5.4 cm.

63. 3.9 atm; less air.

65. (a) 27 m/s; (b) $\approx$200 times the rate of flow of the hose; (c) 1.9×10^{-3} m^2;
(d) 3.65×10^5 Pa (gauge).

67. Yes.

69. (a) 6.0×10^2 kg/m^3, (b) -0.39×10^3 Δ N (up), (c) 0.39×10^3Δ N (down), (d) 1.44 Hz.

71. (a) 1.8×10^5 Pa, (b) 4.8×10^5 Pa,
(c) 7.0×10^{-4} m/s, (d) 13 m/s; (e) 0.37 L/s.

73. $[2g(h_1 - h_0)]^{1/2}$.

75. 8.6 min.

CHAPTER 17

1. (b), (e), (f) are in thermal equilibrium.

3. The extensive variable are (a) and (d), and the intensive variable are (b) and (c).

5. (a) $t_F = -40°F; t_C = -40°C$, (b) none,
(c) $t_F = +575°F, T = +575$K.

7. 95.9°F to 108.5°F.

9. $\Delta p = 0.29$ Pa.

11. (a) 457°C; 854°F, (b) −196°C; −321°F,
(c) 297K; 24°C, (d) 2.2°C; 276K.

13. $T = 1.25t_R + 273$; $t_F = 2.25t_R + 32$.

15. 672°R; 7.6°R; 10,800°R; 492°R; 430°R; 447°R.

17. 52°C (126°F).

19. 0.5 mm.

21. 8.0×10^{-4} K^{-1}.

23. 7.0 cm^3.

25. (a) $\Delta T = 104°C$, (b) 57×10^{-6} K^{-1},
(c) 5.9×10^{-3}.

27. 1.3×10^{-4} cm^2.

29. (a) steel,
(b) 0.99971 Hz $\leq f_{aluminum} \leq$ 1.00029 Hz;
0.99979 Hz $\leq f_{copper} \leq$ 1.00021 Hz;
0.99987 Hz $\leq f_{steel} \leq$ 1.00013 Hz.

31. 2.1%.

33. 1.4×10^3 cm^3.

35. 7.4×10^4 m^3.

37. 0.47 kg/m^3.

39. $p_2/p_1 = 0.562$.

41. 1.30×10^5 Pa; 0.28×10^5 Pa; 1.58×10^5 Pa.

43. 3.1×10^{-10} m.

45. (a) 6.1×10^4 Pa; 308K, (b) 3.5×10^{-3} m^3,
(c) 0.77 atm.

47. (a) $pV = NKT$,
(b) $P(V - bN) = N(kT - aN/V)$,
(c)

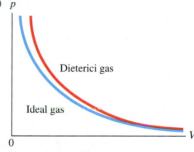

49. 22.4×10^{-3} m^3; 1.29 kg/m^3.

51. 6 kg/m^3.

53. 2.42×10^9 molecules.

55. (a) 71 atm, (b) 71 atm.

57. 40.5 cm^3; 142 cm^3;

59. 1.4 kg/m^3; 0.74 m^3/kg.

61. (a) $p_2 = 10^5$ Pa; $V_2 = 600$ cm^3; $T_2 = 568$K,
(b) $p_3 = 1.30 \times 10^5$ Pa; $V_3 = 600$ cm^3;
$T_3 = 762$K,
(c) $p_4 = 1.30 \times 10^5$ Pa; $V_4 = 300$ cm^3;
$T_4 = 381$K.

63. 4.0×10^{-2} m^3.

65. 2.56×10^3 J; 2.15×10^3 J; the energy of the gas is $\approx 10^4$ the energy of the ball.

67. 373.15K.

69. (b) 421K (148°C).

71. $E_{3200°C} = 1.84 \times 10^4 E_{25°C}$.

73. (a) 6.25×10^{10} Hz, (b) 5.83×10^{12} Hz,
(c) 1.67×10^{13} Hz, (d) 6.25×10^{13} Hz.

75. 9.4×10^{-6} m.

77. 4.47×10^{26} W.

79. 2.05×10^{17} W; almost the same.

81. 5.67×10^{-8} W/m$^2 \cdot$ K^4.

83. 37 kg.

85. $a = (T_1Z_2 - T_2Z_1)/(Z_1 - Z_2)$;
$b = (T_1 - T_2)/(Z_1 - Z_2)$; not necessarily.

87. 0.1 m^3.

89. 26 m/s.

91. 0.754×10^{-23} J/K'.

CHAPTER 18

1. Reversible transformations: (c), (e) and (h); irreversible transformations: (a), (b), (d), (f), and (g).

(a)   **(b)**

(c)   **(d)**

(e)

3. 5×10^2 cal/s.

5. $c'_p pV/R$.

7. (a) 262.8 g of water and 37.2 g of ice at 0°C,
(c) 300 g of water at 7.6°C.

9. (a) 0.22 cal/g $\cdot$ K; 5.9 cal/mol $\cdot$ K,
(b) 0.103 cal/g $\cdot$ K; 5.8 cal/mol $\cdot$ K,
(c) 0.04 cal/g $\cdot$ K; 8 cal/mol $\cdot$ K.

11. 308K.

13. 167K.

15. 7.6%.

17. 287K (14°C).

19. 7.4 kW.

21. Pine: 144 Btu/h; Fiberglas: 62 Btu/h; $R_{eff} = 16 \text{ ft}^2 \cdot \text{h} \cdot °F/\text{Btu}$.

23. (a) $\kappa_{eff} = \kappa_1\kappa_2(L_1 + L_2)/(\kappa_2 L_1 + \kappa_1 L_2)$, (b) $(\kappa_1 L_2 T_1 + \kappa_2 L_1 T_2)/(\kappa_1 L_2 + \kappa_2 L_1)$.

29. $\dfrac{1}{L}\dfrac{dQ}{dt} = \dfrac{2\pi\kappa(T_2 - T_1)}{\ln(1 + \alpha/R_1)}$.

31. 6.7°C.

33. 33 times.

35. 1 cal = 4.14 J.

37. (a) 11.9°C, (b) 10.1°C, (c) ΔT of the water is less than in part (a), (d) net negative temperature change.

39. 1.0×10^6 J.

41. 3.0×10^5 J.

43. 0.012 m³; 0.0034 m³; -1.6×10^3 J.

45. (a) $[ML^{-4}T^{-2}]$, (b) $(V_2 - V_1)[p_0 - (V_2 + V_1)\beta/2]$, (c) same as part (b).

47. 4.1×10^3 J.

49. 633 cal.

51. (a) 2.7×10^4 cal, (b) -3.5×10^4 cal; liberated.

53. (a) $U_A = (c_V'/R)p_A V_A$; $U_B = (c_V'/R)p_B V_B$; $U_C = (c_V'/R)p_B V_A$, (b) $(1/2)(p_A - p_B)(V_B - V_A)$.

55. (a) 2.8×10^2 K; 6.1×10^2 K, (b) 1.2×10^3 J, (c) 2.3×10^2 J, (d) 1.4×10^3 J (into the gas).

57. -0.72K (cooled).

59. 5.0×10^4 J.

63. 20°C.

65. (a) No, (b) no, (c) no.

67. $(p_0V_0 - p_1V_1)/(\gamma - 1)$.

69. $\Delta p_{adiabatic}/\Delta p_{isothermal} = 1.24$.

71. $C_V/C_p = 0.683$; $W_{on} = 1.43p_1V_1$; $\Delta U = 1.43p_1V_1$.

73. 5.4 atm; 210°C.

75. 61.7 cm³; 151 cm³.

79. (a)

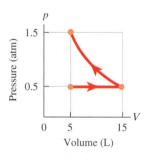

(b) 1.5 atm, (c) -3.2×10^2 J.

81. (a) $+7.1 \times 10^3$ J, (b) $+2.5 \times 10^3$ J.

83. $\text{slope}_{adiabatic}/\text{slope}_{isothermal} = \gamma$.

85. 255K.

87. 3.85×10^3 J.

89. $dm/dt = 20 \times 10^3$ kg/s.

91. (a) 10 atm, (b) -5.7×10^3 J, (c) -5.7×10^3 J (from the gas).

93. (a) $C_p = 3R/2 + RV(pV^2 - a)/(pV^3 - aV + 2ab)$, (b) $C_V = 3R/2$,

(c) $\gamma = 1 + 2V(pV^2 - a)/[3(pV^3 - aV + 2ab)]$.

95. 32°F.

97. $+2.48 \times 10^3$ J.

99. (a) 7.18×10^5 Pa (7.1 atm), (b) 5.9×10^{-2} m³, (c) 2.0×10^4 J, (d) 137K, (e) $+2.0 \times 10^4$ J, (f) $+1.2 \times 10^4$ J.

101. 0.90 W.

CHAPTER 19

1. (a) 1.0×10^{-4}; 5.8×10^{-2}, (b) 6.1×10^{-3}; 0.23, (c) 1.0×10^{-12}; 1.2×10^{-4}, (d) 6.1×10^{-11}; 4.9×10^{-4}.

3. 4×10^{24} components; 4×10^5 m².

5. 6 Pa.

7. (a) 3.8×10^3 J, (b) 304K, (c) 3.8×10^5 m²/s², (d) 6.1×10^2 m/s.

9. $PA = U$.

11. 11 m/s.

13. 4×10^{-17} Pa.

15. (a) 290K, (b) 2.90×10^5 K, (c) 3.6 J.

17. (a) 6.1×10^{-21} J, (b) 3.5×10^{-10} m/s.

19. 7.0×10^5 m/s.

21. 484 m/s; 517 m/s; 412 m/s; 1934 m/s.

23. (a) 1.2×10^4 m/s, (b) 1.9×10^3 m/s.

25. 4.9×10^{-3} m/s.

27. 7.2×10^{73} K.

29. (a) 61.8 mi/h, (b) 62.4 mi/h.

31. (a) 1/1296, (b) 1/324.

33. (a)

(b) 124.

35. (a) $+6.4$, (b) 474, (c) $+20$.

39. 0.984; 0.970; independent of the temperature.

41. $v_{av} = 2(2kT/\pi m)^{1/2}$.

43. $\langle \vec{v} \rangle = \vec{u}$; $\langle \vec{v}^2 \rangle = v_{rms,\,0}^2 + u^2$.

45. $s = 5$.

47. monatomic gas; dumbbell shape; atoms not linear; linear arrangement.

49. (a) 17 kJ, (b) 0.45 kJ/kg·K.

55. 1×10^{-3} atm; 1×10^{-7} atm.

57. ≈ 100 m.

59. 6×10^{19} atoms; 6×10^{16} atoms; 30 Pa.

61. 0.63; no change.

63. (a) $v_{esc} = 4.1 \times 10^3$ m/s, (b) 1350 K; v_{rms} and T both decrease, (c) similar effect, (d) lighter component will make up a smaller fraction of the atmosphere.

65. (a) Yes, (b) $x/(1 - x)$, (c) $x^{-1/2}$; no.

67. (a) 7.0×10^5 m, (b) The mean free path decreases by a factor of 1000 to 7.0×10^2 m.

69. 2×10^{-3} m/s.

71. $p = nE/3$.

73. (a) 285K, (b) 2.8×10^4 Pa.

75. (a) 3.2×10^{11} atoms/m³, (b) 6×10^4 s, (c) 7×10^7 m.

77. $P = \dfrac{4}{\pi} \displaystyle\int_{u_0}^{\infty} u^2 e^{-u^2}\, du$, where $u_0 = \sqrt{3/2}\,v_{escape}/v_{rms}$.

CHAPTER 20

1. 50%.

3. (a) The possible outcomes are

1 2 3 4	1 2 3 5	1 2 3 6	1 2 4 5	1 2 4 6
1 2 5 6	1 3 4 5	1 3 4 6	1 3 5 6	1 4 5 6
2 3 4 5	2 3 4 6	2 3 5 6	2 4 5 6	3 4 5 6

(b) 1/15, (c) 3/5.

5. 252 times more likely.

7. 29%.

9. (a) 38%, (b) 0.38 cal.

11. (a) 325K (52°C), (b) 9.1 kWh; 4.1 kWh.

13. 270 kg/s.

15. Violation of the first law of thermodynamics.

17. 60.

19. Engine A is not reversible.

21. 127 W.

23. 13.7; 9.1.

25. $C_V(T_2 - T_3)$.

27. 5.5 W (0.0074 hp).

29. The statement cannot be trusted.

31. ≈ 1 cent.

33. (a) $T_3 = p_L V_U/nR$; $T_4 = p_L V_L/nR$, (b) $\Delta U_{IV} = -C_V p_L(V_U - V_L)/nR$, (c) $Q_{IV} = +[1 + (C_V/nR)]p_L(V_U - V_L)$, (d) $\eta = nR(p_U - p_L)(V_U - V_L)/[C_V(p_U V_U - p_L V_L) + nRp_U(V_U - V_L)]$.

37. $1 - [(T_3 - T_4)/\gamma(T_2 - T_1)]$.

39. 7.9×10^{-2} J/K·min; irreversible.

41. 31°C; $+10$ J/K.

43. $+0.75$ kJ/K.

45. $+8$ J/K.

47. $\Delta S_{gas} = +44$ J/K; $\Delta S_{universe} = +44$ J/K.

49. 20 J/K.

51. The entropy of the gas decreases.

53. $\Delta T = 0$; $\Delta S = +32.5$ J/K.

55. (a)

(b) $W_{isobaric} = p_1V_1$; $W_{isothermal} = 0.693p_1V_1$, (c) $\Delta S_{isobaric} = 0.693nc_p'$; $\Delta S_{isothermal} = 0.693nR$.

57. Process $A \rightarrow B$ (isobaric):
$Q = (nR(T_B - T_A)$, $W = nR(T_B - T_A)$,
$\Delta U = (3/2)nR(T_B - T_A)$, $\Delta T = T_B - T_A$,
$\Delta S = (5/2)nR \ln(T_B/T_A)$;
Process $B \rightarrow C$ (adiabatic): $Q = 0$,
$\Delta U = (3/2)nR(T_C - T_B)$,
$W = (3/2)nR(T_B - T_C)$,
$\Delta T = T_C - T_B$, $\Delta S = 0$;
Process $C \rightarrow D$ (isothermal): $\Delta T = 0$,
$\Delta U = 0$, $W = (3/2nRT_C \ln(T_A^{2/3}T_C/T_B^{5/3})$,
$Q = (3/2)nRT_C \ln(T_A^{2/3}T_C/T_B^{5/3})$,
$\Delta S = (3/2)nR \ln(T_A^{2/3}T_C/T_B^{5/3})$;
Process $D \rightarrow A$ (constant volume):
$Q = (3/2)nR(T_A - T_C)$, $W = 0$,
$\Delta U = (3/2)nR(T_A - T_C)$, $\Delta T = T_A - T_C$,
$\Delta S = (3/2)nR \ln(T_A/T_C)$.

59. $\Delta S_{\text{gas}} = Q/T < 0$ (decreases);
$\Delta S_{\text{reservoir}} = -Q/T > 0$ (increases); no conflict because this is a reversible process.

61. (a) $T = T_0$, (b) $+nR \ln 3$, (c) $+nR \ln 3$,
(d) 0.

63. (a) $+87$ J, (b) 2.7K, (c) $+0.38$ J/K.

65.
(a) T

Adiabatic process

 S

(b) T

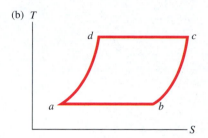

 S

67. (a) 2.25×10^{-2} m^3, (b) 273K, (c) -11.5 J/K.
69. (a) 3.07×10^{-2} m^3, (b) 373K, (c) 0.

71. (a) $Q_{h,\text{ engine}} = 400$ J; $Q_{c,\text{ engine}} = 300$ J;
$Q_{c,\text{ refrigerator}} = 300$ J; $Q_{h,\text{ refrigerator}} = 400$ J;
$Q_{h,\text{ net}} = 0$; $Q_{c,\text{ net}} = 0$,
(b) $Q_{h,\text{ engine}} = 500$ J; $Q_{c,\text{ engine}} = 400$ J;
$Q_{c,\text{ refrigerator}} = 300$ J; $Q_{h,\text{ refrigerator}} = 400$ J;
$Q_{h,\text{ net}} = -100$ J (from the reservoir);
$Q_{c,\text{ net}} = +100$ J (to the reservoir),
(c) $Q_{h,\text{ engine}} = 400$ J; $Q_{c,\text{ engine}} = 300$ J;
$Q_{c,\text{ refrigerator}} = 240$ J; $Q_{h,\text{ refrigerator}} = 340$ J;
$Q_{h,\text{ net}} = -60$ J (from the reservoir);
$Q_{c,\text{ net}} = +60$ J (to the reservoir).

73. T decreases.

75. $dQ_h/dt = 1070$ MW; $dQ_c/dt = 770$ MW;
$dm/dt = 1.5 \times 10^4$ kg/s.

77. 0.30; 60% of the Carnot efficiency.

79. nc_pT.

81. (a) $T_1 = 437$K $(164°$C);
$T_3 = 1020$K $(747°$C),
(b) $Q_{\text{abs}}/m = 5.38 \times 10^6$ J/kg;
$Q_{\text{rej}}/m = -3.73 \times 10^6$ J/kg;
$Q_{\text{net}}/m = 1.65 \times 10^6$ J/kg (absorbed);
(c) 1.65×10^6 J/kg.

Photo Credits

Index

Note: Italics indicate a definition or primary entry for multiple entries, where applicable.

Absolute temperature, 494–97
Absolute zero, 495
Acceleration, *36–48*, 64–65, 78, 87
  angular, 249–50, 261
  average, *36–37*
  of center of mass, 228
  centripetal, *75–76*, 136
  constant:
    motion with, 40–41, 66–68
    special case of, 51
  dimensions of, 36
  due to gravity, 12, *44–46*
    experimental tests of, 45
  instantaneous, *37–39*, 64–65, 76
  magnitude, 76
  nonconstant, motion with, 40–41
  radial, 250
  in relative motion, 77–79
  in simple harmonic motion,
    369–70
  for simple pendulum, 378–79
  tangential, 140
  in uniform circular motion, 75–77
  units of, A–1, 33
Accelerators, 174, 237
Accelerometer, 39
Accuracy, 9–11
Action and reaction, law of equal, 98
Adams, John, 356
Addition, vectors, 17–18
Adiabatic calorimetry, 520
Adiabatic transformations,
    530–31, 534
  of ideal gas, 535–37
Air bag, 97
Air pump, 468
Airstream speed, 481
Algebraic functions, properties of,
    A–6
Allowed states, *302*
Alpher, Ralph, 506
Ampere, A–1
Amplitude, *367*, 376, 388, *404*
  in damped harmonic motion,
    383–84
  in driven harmonic motion, 385–86
  in resonance, 386–87
  in simple harmonic motion,
    367–68, 372, *375*
  of waves, 404
Angle, A–1
  elevation, 69
  measured in radians, 67, 367
  in pendulum motion, 378–79
  plane polar coordinates, 74
  in rotational motion, 247
  of vector, 19
Angstrom (Å), 8
Angular acceleration, *249–50*, 261
  constant, 249
  units of, 249, A–1
Angular displacement, 247
Angular frequency, *367*
  of damped harmonic motion,
    384–85

fundamental, 452
  of precession, 305
  in resonance, 385
  under spring force, 372–73
  of waves, 404, 406–407, 413,
    416, 428
Angular impulse, *291–93*
Angular momentum, *265*
  and central forces, 293–94, 305
  conservation of, 265–67,
    293–98
    central forces, 293–95
    nonrigid objects, 296–98
  generalization of, 280–85
  independence of time, 266
  of point mass, 280–81
  quantization of, 301–302
  reference point, role of, 289
  transferring in collisions, 267
  as vector product, 284–85
Angular motion, work and energy in,
    299–301
Angular speed, 74, 76
  units of, A–1
Angular variables, 251
Angular velocity, 247–49
  instantaneous, *248*
  of rolling, 267–70, 272
  units of, 248
Antinodes, 415, 420
Aphelion, 340, 348
Apogee, 348, 363
Apollo 13, 73
Arc length, *74*
Arch, 317
Archimedes, 472
Archimedes' principle, 472–73
Aristotle, 88–89
Arrow of time, 598–99
Associative addition, 18
Astronomical constants, A–3
Astronomical objects, 3
Astronomical unit (AU), 8, A–1
Atmosphere, 410, 429, 464
Atmosphere (unit), A–1
Atmospheric temperature with
    height, 537
Atomic mass weight (u), A–1
Atomic weight, *499*
Atoms, 3, 549
  energy gap in, 302
  energy levels of, 302
  nuclear structure of, 237
  probing by collisions, 237
Atwood machine, 126, 144
Automobile engines, 588
Average speed, 32
Average velocity, *32–33*
Averages, 549, 554–57
Avogadro's number, *499*, 549,
    A–2
Axes, Cartesian or Euclidean, 21,
    61, 103
Axis of rotation, 248, 250, 252,
    256–57, 262–63, 265, 272

Babylonians, 2–3
Background radiation of the universe,
    discovery/measurement
    of, 506
Background temperature of the uni-
    verse, 503–506
Ballistic pendulum, 241–42
Bar, A–1
Barometer, 469
Barrier penetration energy, *195*
Barriers, potential energy, 201
Base units, SI, A–1
Beam, 318–19, 325
Beat frequency, calculation, 440–42
Beats, *439–42*
  in sound, 441–42
Bernoulli, Daniel, 477
Bernoulli's equation, 476–77
  applications of, 478–80
    Bernoulli effect, *478*
    flow from a tank, 479–80
    fluid motion with constant
      speed, 479
Big Bang model, 496, 506
Binomial theorem, A–5
Black holes, 359
Blackbody, 504
Black-body radiation, *503–506*
Block and tackle, 125
Bohr magnetron, A–2
Bohr, Niels, 301
Bohr radius, A–2
Boltzmann distribution, *560*
Boltzmann, Ludwig, 500, 504, 548,
    559, 566, 599
Boltzmann's constant, *500*, 504, 507,
    516, 549, 552
Bonds, 325
Boundary conditions, 417, 419–21
Boyle, Robert, 499
Boyle temperature, 512
Boyle's law, *499–500*, 502
Braginsky, Vladimir Borisovitch,
    356
Brahe, Tycho, 339–40
Brayton cycle, 587
Brayton, George B., 587
British engineering system, 7, 96
British thermal unit (BTU), 518, A–1
Broadening in interference, 446
Brownian motion, 565
BTU (British thermal unit), 518, A–1
Buffon's method for pi, 570
Bulk modulus, 24, *327*, 328, 432
Bundle, *475*
Buoyancy, 469–71
Buoyant force, 470
Burnout velocity, 244

Caloric, 521
Calorie (unit) (cal), 518, 520–521,
    538, A–1
  numerical definition of, 527
Calorimeter, 521, 533, 540
Calorimetry, 520–21, 538, 617

Carnot cycle, 545, 581–88, 591, 598,
    600
Carnot engine:
  finding the efficiency of, 582–84
  ideal gas, finding the efficiency of,
    582–84
  importance of, 584–85
Carnot, Sadi, 576, 581
Cartesian axes, 21, 61, 103
Cavendish experiment, 343, 357
Cavendish, Henry, 342, 357
Celsius temperature scale, 496–97
Center of gravity, 316
Center of mass, 210, *225–34*, 238
  acceleration of, 228
  and angular momentum, 288
  of continuous mass distribution,
    230–33
  finding, 233–34
    holes, dealing with, 234
    subsystems, 233
    symmetry, 233
  torque on, 277
Center of mass motion:
  in absence of external forces, 227
  in presence of external forces,
    228–29
Center-of-mass frame, 230–33
Center-of-mass frame of two collid-
    ing objects, 225
Centimeter (cm), 7, A–1
Central dogmas of physics, 277
Central forces, *293*
  and angular momentum,
    293–95, 305
  potential energy for, 197, 201
Central value of measurement, 9–10
Centrifugal force, *141*
Centripetal acceleration, *75–76*, 136
Centripetal force, 136–37
cgs system, 7
Change of phase, 521
Charles, Jacques, 512
Chemical energy, 151
Cherenkov, Pavel, 427
Cherenkov radiation, *427*
Circle, *341*
Circular motion, 136–41, 370–71
  with changing speed, 139–40
  and noninertial frames, 140–41
Circular orbits, 346–47
Circulation, 474–75, 480–82, 492
Clausius form of second law of
    thermodynamics, 580
Clausius, Rudolf, 580
Clausius's inequality, 593
Clausius's theorem, 591, 593, 600
Coefficient:
  drag, 133
  of friction, *129–31*
  of performance (COP), 589–90,
    602–603
  of thermal expansion, *497*
  of viscosity, 482
  of volume expansion, *497*

Coherence, 436–37
Collision cross section, 563–64
Collisions, 212–16, 239
  classification of, 214–15
  elastic, 215
    two-body collisions in one
      dimension, 220–23
  energy considerations in, 215–16
  impulsive forces, 212–14
  inelastic, 216
  perfectly inelastic, 216–19
    energy loss in, 217–18
  between pulses, 447–48
  and transport phenomena, 562–65
Comets, 347
Common forces, 105–107, 119–27
  constant forces, 124–26
  gravity, 119–21
  normal force, 123
  tension, 122–23
Commutative addition, 17
Component vectors, *19–21*
Compressibility, 214, 335, 464
Compression, *418*, 464
Compressional strain, 325
Compressor, 576–77, 589, 603
Condition of no torque, 315–16
Conduction:
  thermal contact by, 492–93
  transport of thermal energy
    by, 565
Conductivity, thermal, *524*, 538, 541
Conic pendulum, 148, 389
Conic sections, 341, 345, 348
Conservation of angular momentum,
  265–67, 293–98
Conservation of energy, 186
  and allowed motion, 191–93
  applications of, 186–89
    gravity, 187–88
    spring force, 188–89
  collisions, 212–16
  explosions, 219–20
  as a general principle, 198
  and nonconservative forces,
    198–200
  principle of, *152*
  as an underpinning of physics,
    200–201
Conservation of energy, consequences of, 450
Conservation of flux, *475*
Conservation of mass, *475*
Conservation of momentum, 210–12
  in collisions, 209, 210–11
  in different inertial frames,
    229–30
  for many-body systems, 212
  principle of, *211*
  in supernovas, 297
Conservative forces, 171–72,
  175, 183
Constant acceleration:
  motion with, 40–41, 66–68
  special case of, 51
Constant angular acceleration, 249
Constant forces, 124–26, 136,
  373–74
  in more than one dimension,
    159–62
Constant velocity, 89, 99
Constant-pressure transformation,
  530, 534
Constant-volume transformation,
  530, 534
Construction cranes, 322

Constructive interference, *436*, 456
Contact forces, *91*
Continuity, equation of, *475–76*
Continuous distributions, 556
Continuous mass distribution, center
  of mass of, 230–33
Continuous objects, in two and three
  dimensions, 232–33
Convection:
  thermal contact by, 492–93
  transport of thermal energy
    by, 565
Conversion of units, 8–9, A–1
Coordinates, 19
  plane polar, *74*
COP (coefficient of performance),
  589–90, 602–603
Copernicus, Nicolaus, 339
Coriolis force, 149
Coulomb (C), A–1
Crab supernova, 2–4
Crane, construction, 319
Crest, wave, 400, 405
Critical strains, 328
Critical stresses, 328
Critically damped system, *384*
Cross product, 283–84
Crystal, 323–24, 463, 497, 570, 618
  liquid, 463–64, 493
Cyclic transformations, 528–30
Cycloid, 267
Cylinder, rotational inertia of,
  271–72

da Vinci, Leonardo, 129
Damped harmonic motion, 383–85
  critically damped system, 384
  overdamped system, 384
  underdamped system, 384
Damping coefficient, *383*
Damping, critical, *384*
Damping factor, *383*
Damping parameter, *383*
Dark matter, 351–52
Day (d), A–1
Deceleration, *37*
Decibel scale, 422
Decibels (dB), 422
Decomposition, 453–54
Defect, 565
Definite integrals, *49*
Degree, A–1
Degree of freedom, *561–62*
Density, 464–65
  of air, 134, 147
  of Earth, 120
  energy, 413–14, 456, 504–507
  mass, 7, 15, 24, 26, 27, 133, 145,
    147, 231–33, 238, 465
  of water, 7, 26, 173
Derivatives, 34–35, 38–40, 50–52,
  369, 401, A–1, A–6
  partial, 200, 401
Derived units, 8, A–1
Destructive interference, *436*, 456
*Dialogue Concerning Two New
  Sciences* (Galileo), 14
Diatomic molecules, energy distribu-
  tion of, 560–61
Dicke, Robert, 356
Diesel cycle, 587
Diesel engine, 545, 603
Diesel, Rudolf, 587, 603
Differential equation, 110
Diffusion, 563
  and the random walk, 564–65

Dimensional analysis, 12–14, 22
Dimensions, 12
  matching, 12–13
Discrete frequencies, 302, 426
Disorder and entropy, 573, 580,
  590–92
Dispersion, *428*
Displacement, *16*, 28–31, 51, 153
  angular, 247
  as an integral of velocity over time,
    48–49
  net, 30
Distance, 4
Distant shout vs. nearby shout, 413
Distribution functions:
  continuous, 556
  for position, 554
  for velocity, 554–55, 557
dm notation, 252
Domesday Book, 181
Doppler, Christian, 423
Doppler effect, 423–26
  moving medium, 425–26
  moving observer, 424–25
  moving source, 423–24
  moving source and observer, 425
Doppler radar, 426
Doppler shift, for sound, 426
Dot product, *160*
Drag coefficient, 133
Drag forces, 114, 116, 119,
  133–35
  better approximation to, 135
  drag coefficient, *133*
  terminal speed, *133–34*
Driven harmonic motion, 385–88
  equations of motion for, 385–86
  resonance:
    properties of, 386–87
    and uncertainty, 387–88
Driving frequency, 386, 388
Dumbbell, rotational inertia of, 254
Dynamics:
  of rolling, 269–72
  of rotations, 288
  in wave motion, 400–401
Dyne, 96, A–1

Earth, mass of, 120
Eccentricity, 348
Efficiency, engines, 576–77
Egyptians, building of the
  Pyramids, 159
Einstein, Albert, 7, 141, 339, 357
Ejection seats, 97
Elastic collisions, 215
  in two and three dimensions,
    223–25
  two-body collisions in one dimen-
    sion, 220–23
Elastic modulus, *326*
Electrical appliances, energy
  consumption of, A–4
Electrical energy, 151
Electricity, 141
Electromagnetic force, 141
Electromagnetic radiation, *503*
Electromagnetism, 141
Electron charge-to-mass ratio, A–2
Electron mass, A–2
Electron-volt (e-V), A–1
Electroweak force, *141–42*
Elementary charge, A–2
Elevation angle, 69–70
Ellipse, 340, *341–42*, 348, 355
Elliptical orbits, 346–47

Energy, *151–52*, *186*, A–1,
  See also Kinetic energy;
    Potential energy
  in angular motion, 299–301
  colloquial meaning, 153–54
  of an extended object in motion,
    299–300
  quantization of, 302
  and simple harmonic motion,
    374–77
  of a simple pendulum, 380–81
  society's production/use of, 200
  and standing waves, 417–18
  supply and demand, A–4
  of systems, 189–90
  thermal, *500*
  total, 215
  total mechanical, *186*
  transport, 413
  units of, 154–55
  in waves, 412–14
Energy conservation, *See* Conservation of energy
Energy consumption, A–4
Energy content of fuels, A–4
Energy density, 413–14, 456,
  504–507
Energy diagrams, *191–93*
  equilibrium points, 193
Energy distribution, of diatomic molecules, 560–61
Energy gaps, 302
Energy levels, *302*
  in atoms, quantized, 302
Engines, 574–76
  automobiles, 588
  Brayton, 587
  Carnot, 582–84, 587
  diesel, 545, 587, 603
  efficiency of, 576–77
  and entropy, 597–98
  jet, 220
  Otto, 587
  Stirling, 586–87
Enthalpy, 606
Entropy, 573, 580, 590–92, 594, 598
  and direction of spontaneous
    processes, 592–93
  and engines, 597–98
  and ideal gases, 594–95
  of an isolated system, 594
  meaning of, 596–99
  microscopic interpretation of, 591
  of mixing, 596
  path dependence of, 600
  and the second law, 590–94
  as state function, 591
  as a thermodynamic variable, 591
  units of, A–1
Envelope, 384
Eötvös experiments, 120, 356
Eötvös, Loránd von, 356
Epicycles, 339
Equal action and reaction, law of, *98*
Equation of continuity, *475–76*
Equation of state, *500*
Equation of state of gases, 499–503
  real gases compared to ideal gases,
    502–503
  thermodynamic variables, *499*
    changing, 501
  van der Waals equation of
    state, 503
Equations of motion, *109*
  for rockets, 235–36
  for rotations, 269–70

Equilibrium:
  stable, *193*
  thermal, *492–93*
  unstable, *193*
Equilibrium points, 193
Equipartition theorem, *561–62*
Equipotential, *196*
Equivalence principle, *356–57*
  black holes, 359
  gravitational lenses, 358–59
  light falls under the influence of
    gravity, 357–58
  precession of planetary orbits, 359
  predictions of, 357–59
Erg, 154, A–1
Escape speed, 346
Estimates, 14–15
Euclidean axes, 21
Euler's constant, A–6
Explosions, 219–20
Extended force diagrams, 316
Extended object in motion, energy of,
  299–300
Extended objects:
  and gravitation, 349–55
    dark matter, 351–52
    spherically symmetric object,
      gravitational force due to,
      349–51
    tidal forces, 352–55
Extensive variables, 508
External forces, 105

Fahrenheit temperature scale, 497
Farad (F), A–1
Fermi, A–1
Fictitious forces, *100–101*, 353
  in circular motion, 141
Filter, 387
First harmonic, 415
First law of motion, *89–90*
First law of thermodynamics, *See*
  Thermodynamics, first law of
Flight time of projectile, 69
Flow:
  of fluids, 462
  from tank, 479–80
Flow from a tank, 479–80
Fluid motion with constant speed,
  479
Fluid speed in a closed pipe,
  measuring, 478
Fluids:
  Archimedes' principle, 472–73
  Bernoulli's equation, 476–77
    applications of, 478–80
  buoyancy, 469–71
  circulation, *474*
  conservation of flux, *475*
  convection of, *474*
  density, 464–65
  equation of continuity, *475–76*
  hydrostatics, *464*
  irrotational flow, *474*
  laminar flow, *474*
  lift, 480–81
  in motion, 474–75
  pressure, 465–66
    in a fluid at rest, 466–69
  properties of, 462–90
  real, 481–83
    turbulence, 482–83
    viscosity, 482
  rotational flow, *474*
  steady flow, *474*
  streamline, *474*

turbulent flow, *474*
viscosity, *474*
Flux, *475*
  conservation of, *475*
Flywheels, 252
Foot (ft), 4, A–1
Force diagrams, 89, 199
  extended, 316
Force law, *109*
  inverse-square, 340–45
Force(s), *88*, 91, A–1
  centripetal, 136–37
  circular motion, 136–41
    with changing speed, 139–40
    and noninertial frames, 140–41
  common, 105–107
  conservative, 171–72
  constant, 124–6, 136
  contact, *91*
  drag, 133–35
  electroweak, *141–42*
  external, 105
  fictitious, *99*
  fundamental, 141–42
  of gravity, *91*
  identifying, 103–109
  internal, 105
  net, 105
  nonconservative, 171–72
  normal, *91*, 106, 123
  nuclear, *141*
  and solids, 323–328
  spring, 170
  strong, *141–42*
  of universal gravitation, *141*
  variable:
    in magnitude and direction,
      166–67
    in one dimension, 163–65
Foundations, constructing, 214
Fourier analysis, 454
Fourier decomposition, 452–54
Fourier analysis and the ear, 454
  of pulses, 452
  and uncertainty principle,
    454–56
  of triangular waves, 453–54
Fourier expansion, 452
Fourier's theorem, *452*, 457
Frames of reference, 77
  geocentric, *339*
  heliocentric, 339
  inertial, 90, 92, 99
  noninertial, 99–103, 106, 111, 140
Franklin, Benjamin, 351
Free expansion, 516, 533
Free fall, 73
Free-body diagrams, *103*, 262–64
  identifying, 103–109
Freely falling objects, 44–45
Frequency, 75, 249, *368*, *404*
  angular, *See* Angular frequency
  beat, calculation of, 440–42
  fundamental, 416
  natural, *383–84*
  of simple harmonic motion,
    368, 373
  and uniform circular motion, *75*
Friction, *127–33*, 169–70
  coefficient of, *129–31*
  forward-directing/backward-
    directing, 132
  kinetic, *128–29*
  as nonconservative force, 171
  origins of, 133
  quantitative properties of, 129–31

sliding, *128–29*
static, *128–29*, 132–33, 137–38
Fuel resources, A–4
Fundamental angular frequency, 452
Fundamental forces, 141–42
Fundamental frequency, 416
  harmonics of, 452
Fundamental mode, 415
Fundamental physical constants, A–2
Fundamental physical quantities, 4–7
  length, 5
  mass, 6
  time, 6
Fundamental wavelength, 416
Fundy National Park (New
  Brunswick, Canada), 353
Fusion, latent heat of, 521–22
Fusion, nuclear, 182

Galileo Galilei, 2, 14, 44–45, 57, 89,
  214, 617
Gamow, George, 506
Gas, 463
  ideal, *See* Ideal gases
Gas constant, A–2
  universal, 500, 507, 516
Gas pressure, 410
Gauge pressure, 486
Gauss, A–1
Gaussian function, 459–60
Gaussian pulse, 460
Gay-Lussac, Joseph, 512
General Conference on Weights and
  Measures, 5
General theory of relativity, 7, 339,
  357, *504*
Generalization:
  of angular momentum, 280–85
  of torque, 285–87
Geocentric reference frame, *339*
Geometrical formulas, A–6
Geosynchronous orbit, 362
Gibbs, Josiah-Willard, 548
Gilbert, William, 2
Glashow, Sheldon L., 141
Global Positioning System (GPS),
  6–7, 294, 339
Grain, 323
Gram (g), 7–8, A–1
Gravitation, 338–65
  Einstein's theory of, 357–59
  equality of inertial and gravitation-
    al masses, 356
  equivalence principle, *357*
    black holes, 359
    gravitational lenses, 358–59
    light falls under the influence of
      gravity, 357–58
    precession of planetary
      orbits, 359
    predictions of, 357–59
  and extended objects, 349–55
    dark matter, 351–52
    spherically symmetric object,
      gravitational force due to,
      349–51
    tidal forces, 352–55
  Newton's inverse-square law,
    340–45, 357
  orbits, types of, 346–48
  planetary motion, early observa-
    tions of, 339–40
    Copernican picture, 339
    Kepler's laws, 340
  planets, 345–49
  satellites, 345–49

Gravitational constant, A–2
Gravitational lens, 359
Gravitational lensing, 358–59
Gravitational mass, 356
Gravitational potential energy, *193*
Gravity, 119–21, 169, 187–88
  acceleration due to, 44–45
  and extended objects, 264–65
  force of, *91*
  and rigid bodies, 316–17
  rocket motion in the presence
    of, 236
  specific, *464*
Gyroscopes, 303

Halley's comet, 347
Harmonic driving force, *387*
Harmonic series, 416
Harmonic waves, 411
Harmonics, 416–17
  *n*th harmonic, 416, 419
Harmonics of the fundamental fre-
  quency, 452
Hearing, 421–22
  and Fourier analysis, 454
Heat, *See also* Heat flow
  mechanical equivalent of, 526–27
  specific, 518
Heat bath, 492, *492*
Heat capacity, 518–19
  constant-pressure, 530, 534
  constant-volume, 530, 534
  molar, 518
  units of, 519
Heat flow:
  in materials, 523–26
  path dependence of, 519–20
  and phase changes, 521–22
  units of, A–1
Heat pump, 584, 588–90
Heisenberg uncertainty relation, *237*
Heisenberg, Werner, 237
Heliocentric reference frame, *339*
Helmholtz, Hermann von, 201
Henry (H), A–1
Herman, Robert, 506
Hertz, Heinrich, 368
Hertz (Hz), *368*, A–1
Hooke, Robert, 326
Hooke's law, *164*, 188, 371, 388
Horsepower (hp), *173*
Hubble, Edwin, 426
Huygens, Christian, 214
Hydrodynamics, 474
Hydrostatics, *464*
Hyperbola, 341, 534
Hyperbolic orbit, 346

Ice point, 497
Ideal-gas Carnot engine, finding the
  efficiency of, 582–84
Ideal gas law, *500*
Ideal gases, 491–514, *495*
  and absolute temperature, 494–97
  adiabatic transformations of,
    535–36
  entropy of, 594–95
  isothermal transformations of,
    501, 534–35
  temperature of, 552
Impulse, *212*
  angular, *291–93*
Impulsive forces, *212–14*
Inch (in), 7, 9–10, A–1
Incident waves, interference of, 439
Incoherence, *437*

Incompressibility of fluids, *464*
Indefinite integrals, *49*
Inelastic collisions, 216
  perfectly inelastic, 216–19
Inertia, 92
  law of, 89
  moment of, *252*
  rotational, *See* Rotational inertia
Inertial frames, *90*, 92, 99
Inertial mass, 92, 120, 261, 356
Infrasonic, 422
Initial conditions, *368*
Instantaneous acceleration, 37–39, 51
Instantaneous angular velocity, *248*
Instantaneous power, *173*
Instantaneous speed, 31
Instantaneous velocity, 33–35, 51, 63
Integrals, A–7
  definite, 49
  indefinite, 49
  line, 167
  scaling technique for, 558
Integrand, 49, *49–50*
Integration, 48–49, 164
  as inverse of differentiation, 50
Intensity, of sound, *422*
Intensive variables, 508
Interatomic bonds, 325
Interference, *436*, 456
  of incident and reflected
    waves, 439
  spatial, *See* Spatial interference
    phenomena
  standing waves through, 437–39
Interference patterns, *443*, 446
Interference phenomena, *436*
Internal energy, 474, 500, 518
  change of, in thermal
    transformation, 532
  as a function of state, 539
  of ideal gas, 532–35
  and pressure in gas, 532
  and temperature, 530–31
Internal forces, 105
  role in the torque on a system, 288
International System of
    measurements, 4
Inverse-square force law, 340–45
Irreversible loss of order, 598
Irreversible processes, 516–17
  and loss of order, 596
Irrotational flow, *474*
Isobaric transformation, *501*
Isochore, 530, 596
Isotherm, 517, 519, 535
Isothermal transformations, *501*
  of an ideal gas, 534–35
Isotope separation, 559
Isotopes, 559
Isotropic material, 497

Jeans, James, 504
Jet engines, 220
Joule experiment:
  on energy of ideal gas, 532–33
  on mechanical equivalent of heat,
    526–27, 531
Joule (J), 154, 173, 175, 306, A–1
Joule, James Prescott, 201, 526, 618
Jupiter, 57, 339

Kelvin form of second law, of ther-
    modynamics, 580, 600
Kelvin (K), A–1
Kelvin, Lord (Sir William
    Thompson), 495

Kelvin temperature scale, 495–96,
    507, 585
Kepler, Johannes, 340
Kepler's laws, 340
  second law, *295*
Kilo-, 8
Kilocalorie (kg), 5, A–1
Kilogram (kg), 4–7, A–1
Kilometer (km), A–1
Kilowatt-hour (kWh), *173*, A–1
Kinematics, *28*, 51, 87, 238
  of rolling, 267–68
  of rotations about an axis, 246
Kinetic energy, 151, *152*–60, 175,
    183, 239
  expressed in terms of
    momentum, 210
  relativistic, 174
  of rolling, 268–69
  rotational, 251–53
  at very high speeds, 174
  and work, 151–82
Kinetic friction, *128–29*
Kinetic theory of gases, 548
Kirchoff, Gustav Robert, 504

Laminar flow, *474*
Latent heat:
  fusion of: 521–22
  of vaporization, 521–22
Lattice, 463
Lattice structure, *323*
Lavosier, 521, 526
Law of equal action and
    reaction, 98
Law of inertia, 89
Law of partial pressures, 572
Le Verrier, Urbain, 356
Length, 5, A–1
  units of, A–1
Leonardo da Vinci, 129
Lever arm, *259*
Lifetime, of damped harmonic
    motion, 384
Lift, 480–81
Light waves, 443
Light-year (ly), 8, A–1
Line integrals, 167
Linear momentum, 210
Linear motion, *259*
  collecting parallels between rota-
    tional motion and,
    300–301
  parallels between rotational motion
    and, 266–67
Linear wave equation, 447
Linearity, *436*
Liquid crystal, 463–64, 493
Liquids, 464
  shear modulus, 327
  thermally isolated, *493*
Liter (L), A–1
Local velocity, 474
Logarithmic scale, *422*
Longitudinal stresses, 325
Longitudinal waves, *399*,
    407–09, 418
Loudness, *421*

Mach number, 427
Machine, simple, 159
Magnetic field, A–1
Magnetism, 141
Magnitude, vectors, *16*, 33
Major heat capacity, 562
Manhattan Project, 559

Manometer, *465*
Mass, 4, 6, 121, A–1
  center of, 210
  conservation of, *475*
  gravitational, 356
  inertial, *92*
  reduced, 308
  units of, 4–5, A–1
Mass densities, 465
Mass density, 7, 15, 24, 26, 145,
    231–33, 238, 465
Mass density function, 252
Mass energy, 151
Mathematics, A–5 to A–7
  algebraic functions, properties
    of, A–6
  binomial theorem, A–5
  constants, A–5
  derivatives, A–6
  expansions, A–7
  geometrical formulas, A–6
  integrals, A–7
  notation, A–7
  quadratic equations, A–5
  Taylor expansion, A–7
  trigonometry, A–5 to A–6
Matter, states of, 462–64
  gases, 463
  liquids, 464
  solids, 463
Maxima, location of, 444–46
Maximum entropy, state of, 594
Maximum height in projectile
    motion, 71
Maxwell demon, 601
Maxwell, James Clerk, 548, 557
Maxwell–Boltzmann distribution,
    559–62, 566
Mayer, Julius Robert, 201, 542
Mean collision time, 563
Mean free path, 564
Mean life, of damped harmonic
    motion, 384
Measurement, uncertainty in,
    9–10
Mechanical equivalent of heat,
    526–27
Mechanical waves, 401, 409, 429
Mechanically isolated system, 520
Mechanics, 28
Meter (m), 4–6, A–1
Meters per second (m/s), 8
Metric (mks) system, 5
Metric system, 5
Micrometer (μm), 8
Micron, 8
Mile (mi), A–1
Mile per hour (mi/h), A–1
Minima, location of, 444–46
Minute (min), A–1
Mixing, entropy of, 596
mks system, 5
Modes of vibration, 415
Molar heat capacity, 518
Mole (mol), 499, A–1
Molecular weight, *499*
Molecules, 549
Moment arm, 259, 282, 287, 314
Moment of inertia, 252
Momentum, *210*
  angular, 265
    conservation of, 265–67,
    293–98
    generalization of, 280–85
    transferring in collisions, 267
  conservation of, 210–12

in different inertial frames,
    229–30
  for a system of many objects,
    212
  linear, 210
Momentum transfer, 237
Monomolecular layer, 25
Motion, 87
  acceleration, 64–65
  angular, work and energy in,
    299–301
  with constant acceleration, 40–41,
    66–68
  damped harmonic, 383–85
  energy of an extended object in,
    300–301
  equations of, *109*
  finding, 109–10
  first law of, 89–90
  fluids in, 474–75
  graphing, 34–36
  linear, 259, 266–67
    parallels between rotational
      motion and, 300–301
  oscillary, 366–96
  periodic, 192, 366
  projectile, 68–73
    trajectory, 69–71
  relative, 77–80
  rhythmic, 366
  rocket, 234–36
  rotational, parallels between linear
    motion and, 300–301
  second law of, 92–97, 103,
    119–20, 151, 239, 261–62,
    351–52, 401
  simple harmonic, *366*, 367–70
    and energy, 374–77
    kinematics of, 367–68
    properties of, 368–69
    relations among position,
      velocity, and acceleration
      in, 369–70
    and springs, 371–74
  straight-line, 28–59
  third law of, 97–99, 209,
    211, 342
  trajectories, representing, 65–66
  in two or three dimensions,
    60–86, 195–97
    central forces, 197
    potential energy for projectile
      motion, 196–97
  uniform circular, 74–77, 388
  velocity, 62–64
Mount Everest, 355
Musical instruments, 376, 385, 421,
    435, 438

Nanometer, 8
Nanometer (nm), 8
Narrowing in interference, 446
NASA, 73
National Institute of Standards and
    Technology, 6
Natural frequency, *383–84*
Neptune, discovery of, 356
Net displacement, 30
Net force, *88*, 105
Net work, 153–54, 156
Neutral equilibrium, *193*
Neutrino, 174
Neutron mass, A–2
Neutron star, 275, 297–98, 353, 465
Newton, Isaac, 2–3, 87, 301,
    338–45, 353, 357, 359–60

Newton's inverse-square law, 340–45
  gravitational force, potential energy associated with, 344
  superposition principle, 344–45, 356
  universal gravitation, 341–43
Newton's law of universal gravitation, *341*
Newton's laws, 87–118, 159
  applications of, 119–50
  common forces, 105–107, 119–27
    constant forces, 124–26
    gravity, 119–21
    normal force, 123
    tension, 122–23
  drag forces, 133–35
  external forces, 105
  finding the motion, 109–10
  first law of motion, *89–90*
  force, *88*, 91
  forces:
    and circular motion, 136–41
    identifying, 103–109
    friction, *127–33*
      coefficient of, *129–31*
      forward-directing/backward-directing, 132
      kinetic, *128–29*
      quantitative properties of, 129–31
      sliding, *128–29*
      static, *128–29*
  fundamental forces, 141–42
  internal forces, 105
  noninertial frames of reference, 99–103
  second law of motion, 92–97, 103, 109, 119–20, 151, 239, 261–62, 351–52, 401
  third law of motion, 97–99, 209, 211, 342
Newtons (N), 96, A–1
Nodes, 415
Noise levels, 422
Noise-canceling headphones, 447
Noncircular orbits, properties of, 348
Nonconservative forces, 171–72, 198–200
  friction as, 171
  sliding friction as, 171
Nonharmonic period waves, 452
Noninertial forces, *100*
Noninertial frames of reference, *99–103*
  and circular motion, 140–41
Nonrigid bodies:
  angular momentum, 296–98
    astrophysical example, 297–98
Normal force, *91*, 106, 123
Normalization condition, *555*
Nozzles, 476
*n*th harmonic, 416, 419
Nuclear energy, 151
Nuclear force, *141*
Nuclear submarines, 294
Nuclei, 549
Null vector, *18*
Nutation, *302*

Observation, 2
Ohm, A–1
Optical pyrometer, *503*
Orbits, *339*
  closed, 341
  geosynchronous, 362
  noncircular, properties of, 348

parabolic, 346
  types of, 346–48
Order, 598
Order-of-magnitude calculation, 14
Orders of magnitude:
  for length, 5
  for mass, 6
  for time, 6
Oscillary motion, 366–96
  and circular motion, 370–71
  damped harmonic motion, 383–85
  driven harmonic motion, 385–88
    equations of motion for, 385–86
    resonance, and uncertainty, 387–88
    resonance, properties of, 386–87
  physical pendulum, 381–82
  simple harmonic motion, 367–70
    kinematics of, 367–68
    properties of, 368–69
    relations among position, velocity, and acceleration in, 369–70
    and springs, 371–74
  simple pendulum, 378–81, 388
    accuracy of pendulum clock, 379
    energy of, 380–81
Otto cycle, 587
Otto, Nikolaus, 587
Overdamping, *384*

Parabola, *341*
Parabolic motion of projectiles, 69
Parabolic orbit, 346
Parallel connection, of thermal resistance, 525
Parallel-axis theorem, 257–59
Parsec (pc), 8, A–1
Partial derivatives, 401
Partial pressures, law of, 572
Pascal, Blaise, 466
Pascal (Pa), 466, A–1
Pascal's principle, *467*
Path dependence, 169–72
  of entropy, 600
  friction, 169–70
  gravity, 169
  of heat flow, 519–20
  spring force, 170
Peak, resonant, 387
Peebles, James, 506
Pendulum:
  ballistic, 241–42
  conical, 148, 389
  period, 13
  physical, 381–82
  simple, 378–81, 388
Penzias, Arno, 506
Percentage uncertainty, 10
Perfectly inelastic collisions, 216–19, 230
  energy loss in, 217–18
Perigee, 348
Perihelion, 348
Period, 13, *368*, 388, *404*
  of a motion, *248*
  and uniform circular motion, *74*
Periodic motion, 192–93, 366
Periodic table of elements, A–8
Periodic waves, 403–405
Permeability of free space, A–2
Permittivity of free space, A–2
Perpetual motion machines, 489, 580
Phase, *367*, 464
  of simple harmonic motion, 367

Phase, change of, 521
Phase changes, 521
  and heat flow, 521–22
Phase diagrams, *501*
Phase difference, 371
Phase plot, 203
Physical pendulum, 381–82
Physical quantities, 4–7, A–3 to A–4
  SI units, A–1
Physical world, progress in understanding, 2–3
Physics:
  classical, 3
  modern, 3
  quantum, *See* Quantum physics
  revolutionary developments in (1900–1925), 3
  significant dates in development of, A–9 to A–10
  and society, 1
  statistical, 548–49
Pitch, *421*
Planck formula, *504–506*
Planck, Max, 504
Planck's constant, 301, *504–505*, A–2
Plane polar coordinates, 74
Planetary data, A–3
Planetary motion:
  Copernican picture, 339
  Kepler's laws, 340
Planets, 345–49, A–3
  orbits, types of, 346–48
Plasma, 464
Plumb bob, 44
Poiseuille flow, *482*
Poiseuille, Jean, 482
Poisson, Simeon, 326
Poisson's ratio, 326–27
Polar coordinates, 247
Position distribution function, *554*
Position vectors, *19*, 30
Potential energy, 151, 172, 183–97, 201
  conservation of energy, 186
    applications of, 186–89
  and conservative forces, 184–85
  for projectile motion, 196–97
Potential energy barriers, 201
Potential energy well, 376
Pound (lb), 6, A–1
Power, *172–74*, A–1
  horsepower (hp), *173*
  instantaneous, *173*
  kilowatt-hour (kWh), *173*
  watt (W), *173*
  in waves, 412–14
Powers of 10, unit prefixes for, 7
Precession, 302–305
  angular frequency of, 305
  torque on a spinning top, 304–305
Pressure, 465–66, A–1
  in a fluid at rest, 466–69
  gas, 410
  gauge, 486
  and internal energy, 532
  law of partial, 572
  and molecular motion, 550–52
  origin of, 550–51
  STP (standard temperature and pressure), 510, 562
Priestley, Joseph, 351
Primary dimensions, 12
Principle of conservation of momentum, 211
Principle of superposition, *344–45*, 356, 438–39

Principle of the conservation of energy, *152*
Probability distributions, 554–55, 566, *See also* Distribution functions
Problem-solving techniques, 32, 107
Projectile motion, 68–73
  potential energy for, 196–97
  trajectory, 69–71
Proportions, 14
Proton mass, A–2
Proton-electron mass ratio, A–2
Pseudoforce, *100*
Ptolemaic description of planetary motion, 339
Ptolemy, Claudius, 339
Pulse frequency, 458, 461
Pulses, 446–52
  change in velocity with change in medium, 451
  collisions between, 447–48
  conservation of energy, consequences of, 450
  reflection, 448–49
  transmitted, 449–52
  and uncertainty principle, 454–56
  width, 455
Pyramids, 159
Pyrometer, optical, *503*
Pythagoras's theorem, 19, 21

Q factor, *384*
Quadratic equations, solution of, A–5
Quantization:
  of angular momentum, 301–302
  of energy, 302
Quantum mechanics, 237, 266, 301–302
Quantum physics, 3, 505, 562
Quantum theory, *504*
Quarks, *237*
Quasars, 358

Radial acceleration, 250
Radians, *74*, 247, A–1
Radians per second (rad/s), 248
Radiation, 493
  black-body, *504–506*
  Cherenkov, *503*
  electromagnetic, *503*
  emitted, 504
  and thermal equilibrium, 503
Random walk, 564–65
Range, 69–70
Rankine temperature scale, 509
Rankine, William, 509
Rarefaction, *418*
Rayleigh, Lord, 504–505
Reaumur temperature scale, 509
Reciprocal relation, *455–56*
Red-shift, 433
Reduced mass, 308
Reference frames,
  See Frames of reference
Reference point:
  choosing, 290–91
  role of, 289
Reflected pulses, 448–49
  amplitudes of, 450
  uses for, 451
Reflected waves, 457
  interference of, 439
Refrigerators, 588–90
Reinforced concrete, 328
Relative motion, *77–80*
  and reference frames, 91–92

Relative velocity, in elastic
    collisions, 221
Relativity, general theory of, 7, 339,
    357, *504*
Relativity principle, *97*
Resonance, *385*, 421
    properties of, 386–87
    and uncertainty, 387–88
Resources, annual usage of, A–4
Restoring force, 366–67, 371, 373,
    377, 388, 398–99
    and wave speed, 409–10
Resultant vector, *17*
Reversible processes, 516–17
Reynolds number, *482–83*
Rhythmic motion, 366
Right-hand plane, 248
Right-hand rule:
    for axes, 21
    in rotations, 248, 272, 282, 299
    for torque, 261, 270, 273, 285, 304
    for vector product, 282,
        283–84, 305
Rigid bodies, *246*
    gravity, 316–17
    motions of, 247
    rotating, 246–79
        acceleration of a point in,
            250–51
        angular acceleration, 249–50
        angular velocity, 247–49
        rolling, 267–72
        rotational inertia, 253–59
        torque, 259–65
        rotational kinetic energy, 251–53
        statics for, 313–16
Ripple tank, *443*
rms speed, 550
Rocket motion, 234–36
    in the presence of gravity, 236
    thrust, *235*
Rockets, 99
    burnout velocity, 244
Rolling, *267–72*
    dynamics of, 269–72
    energy in, 268–69
    kinematics of, 267–68
Rolling cylinder, 267
Root-mean-square (rms) speed, 550
Rotating rigid bodies, acceleration of
    a point in, 250–51
Rotation axis, 247
Rotational flow, *474*
Rotational inertia, *252*, 253–59
    evaluation of, 253–59
        dumbbell, 254–55
        parallel-axis theorem, 257–59
        rod, 256
        solid cylinder, 255–56
        thin cylinder, 255
    of simple solids, 253
Rotational kinetic energy, 251–53
Rotational motion:
    collecting parallels between linear
        motion and, 300–301
    dynamical equation, 260–62
Rotations:
    about fixed axis, 247, 252
    dynamics of, 288–93
    equations of motion for, 269–70
    of rigid bodies, 246–79
    work-energy theorem for, 299
Rumford, Count (Benjamin Thomp-
    son), 521, 526
Rutherford, Ernest, 237
Rydberg constant, A–2

Salam, Abdus, 141
Satellites, 339, 345–49
    orbits, types of, 346–48
Scalar multiplication, vectors, *18*
Scalar product, *160–161*
Scalars, *15*
Scales, 3–4
Scaling, 558
    laws, 22
    technique for integrals, 558
Scanning tunneling microscopy
    (STM), 195
Scientific method, 2
Scientific notation, 4–9
    distance, 4
    length, 5
    mass, 4, 6
    time, 4, 6
    unit prefixes, 7–8
Sears Tower (Chicago), 352
Second law of motion, 92–97, 103,
    109, 119–20, 151, 239,
    261–62, 351–52, 401
    and force, 93–95
    and observers in different inertial
        frames, 96–97
Second law of thermodynamics,
    See Thermodynamics, second
    law of
Second (s), 4, 6–7, A–1
Seismic waves, 411
Seismograph, 411
Semi-major axis, 348
Semi-minor axis, 348
Series connection:
    of springs, 391
    thermal, 540–42, 586
Shear force, 410–11, 464
Shear modulus, *327*
Shear strain, *326*
Shear stress, *326*
Shear waves, in solids, 410–11
Shock waves, *426–28*
SI (Système International), 5,
    600, A–1
    base units, A–1
    conversion to non-SI units, A–1
    derived units, A–1
Significant figures, 10–11
Simple cubic lattice, *323*
Simple harmonic motion, *366*,
    367–70
    amplitude of, 372
    energy of, 374–77
    frequency of, 368, 373
    kinematics of, 367–68
    period of, 367–68
    phase of, 367
    properties of, 368–69
    relations among position, velocity,
        and acceleration in, 369–70
    and springs, 371–74
    and uniform circular motion, 368
    velocity of, 369–70
Simple machine, 159
Simple pendulum, 139, 199,
    378–81, 388
    accuracy of pendulum clock, 379
    energy of, 380–81
Siphon mechanism, 484
Skyscrapers, sway of, reducing, 386
Sliding friction, *128–29*
    as nonconservative force, 171
Slug, 7
Solar energy output, A–4
Solids, 463

crystal, *323–24*
    defects, 323–24
    and forces, 323–28
    lattice structure, 323
    shear waves in, 410–11
    strains, *325–28*
    strength of, 327–28
    stresses, *325–28*
    tensile strengths of, 328
Sonic, 422
Sound:
    beats in, 441–42
    cancellation of, 435
    Doppler shift for, 426
    speed of, 409–10, 419
        in solids, 409
Sound waves, *407*, 409, 418–22, 443
    hearing sounds, 421–22
    nature of, 418–19
    standing, 419–21
Spatial interference phenomena,
    443–46
    maxima/minima, locations of,
        444–46
Special relativity, 3
Specific gravity, *464*
Specific heat, 518
Speed, 31, 41, A–1
    airstream, 481
    angular, *74*, 76
    average, 32
    root-mean-square (rms), 550
    traveling waves, 409–10
    units of, A–1
Speed of light in a vacuum, A–2
Spherically symmetric object, gravi-
    tational force due, 349–51
Spoilers, 481
Spontaneous processes, 577–79,
    594, 599–600
    direction of, 592–93
Spring constant, *164*, 326, 371
Spring force, 170, 188–89, *See also*
    Hooke's law
    path dependence, 170
Springs, and simple harmonic mo-
    tion, 371–74
Stable equilibrium, *193*, 194–95,
    204, 314, *594*
Standard temperature and pressure
    (STP), 510, 562
Standing sound waves, 419–21
Standing waves, *399*, *414*,
    414–18, 456
    and energy, 417–18
    frequencies of, 416
    harmonics, 416–17
    modes, 415
    through interference, 437–39
    in two and three dimensions, 417
State of gases, equation of,
    499–503
State of maximum entropy, 594
States, 464
States of matter, 462–64
    gases, 463
    liquids, 464
    solids, 463
Static equilibrium, 130, 320,
    333, 336
Static friction, *128–29*, 132–33,
    137–38
Statics, 313–37
    applications of, 318–23
    underdetermined systems,
        322–23

for rigid bodies, 313–16
Statistical mechanics, *491–92*
Statistical physics, 548–49
Steady flow, *474*
Stefan, Josef, 504
Stefan-Boltzman formula,
    504–505, 513
Stirling cycle, 587
Stirling engines, 586–87
    finding the efficiency of, 587
Stirling, Robert, 586
Stoke's law, 489
STP (standard temperature and pres-
    sure), 510, 562
Straight-line motion, 28–59
    acceleration, 36–48
    freely falling objects, 44–45
    integration/motion in one dimen-
        sion, 48–51
Strains:
    compressional, 325
    critical, 328
    relationship between stresses and,
        326–28
    shear, *326*
    volume, *326*
Streamlines, *474*
Streamtube, *475*
Stresses:
    critical, 328
    longitudinal, 325
    relationship between strains and,
        326–28
    shear, *326*
    volume, *326*
Strong force, *141–42*
Subtraction, vectors, 18
Sun:
    energy output of, A–4
    temperature of, 496
Superposition, and interference of
    waves, 435–61
Superposition principle, 344–45,
    356, 438–39
    beats, *439–42*
    coherence, *436–37*
    interference, *436*
    pulses, 446–52
    spatial interference phenomena,
        443–46
Système International (SI), *See* SI
    (Système International)
Systems, energy of, 189–90

Tangential acceleration, 140
Tangential forces, 139–40
Taylor expansion, 376, A–7
Temperature(s):
    background radiation of the uni-
        verse, discovery/measure-
        ment of, 506
    background temperature of the
        universe, 503–506
        blackbody radiation, *503*–506
    Celsius scale, *496–97*
    equation of state of gases,
        499–503
    Fahrenheit scale, 497
    gradient, *524*
    and ideal gases, 491–514
        absolute temperature, 494–97
    Kelvin scale, *495–96*
    meaning of, 552–54
    occurring in nature, 496
    Rankine scale, 509
    Reaumur scale, 509

Temperature(s) (cont.)
and thermal equilibrium, 492–93
thermal expansion, 497–99
thermometers, 493–94
Tensile strength, 327–29
Tension, 122, 124, 126, 325
Terminal speed, 133–34
Terminal velocity, 118
Tesla (T), A–1
Thermal conductivity, 524, 538, 541
SI unit, A–1
Thermal contact, 492
Thermal energy, 151, 198, 500,
515–16, 518, 523, 525–27,
530–31, 535, See also Internal energy
Thermal equilibrium, 492–93
and radiation, 503
Thermal expansion, 497–99
coefficient of, 497
Thermal insulator, 524–25
Thermal isolation, 520, 530, 543
Thermal process, 501
Thermal reservoir, 492
Thermal resistance, 524–26
in parallel, 525–26
in series, 525
Thermal systems, 492–93, 499–501,
507, 515
changes in, 516–17
Thermal transformation, 501,
516, 534
adiabatic, 535–36
cyclic, 528–30
irreversible, 516–17
isobaric, 501
isothermal, 534–35
reversible, 516–17
Thermally isolated liquids, 493
Thermodynamic variable, 499–500,
516, 518–19
entropy as, 591–92
extensive, 508
of a gas, changing, 501–502
intensive, 508
Thermodynamics, 491
first law of, 515, 531–32, 533, 535
in closed cycles, 532
second law of, 573–607, 580
Clausius form, 580
second law of motion, Kelvin
form, 580
zeroth law of, 492
Thermometers, 493–94
Third law of motion, 97–99, 209,
211, 342
Thompson, Benjamin (Count Rumford), 521, 526
Thomson, Sir William (Lord Kelvin),
495
Thrust, 235
Tidal forces, 352–55
Tides, 353, 355, 360, 361, 363
Time, 4, 6, A–1
units of, 4, A–1
Time reversal, invariance under, 599
Top, motion of, 302–303
Torque, 259–65
on center of mass, 277
condition of zero, 315–16
generalization of, 285–87
and internal forces, 288
and interparticle forces, 288
and physical pendulum, 381–82
right-hand rule for, 261, 270, 273,
285, 304

rotational motion, dynamical equation for, 260–62
on a spinning top, 304–305
units of, A–1
Torsion constant, 394
Torsion pendulum, 393–94
Total mechanical energy, 186
Track, 428
Trajectory, 60, 69–71
flight time, 70–71
maximum height, 71
range, 69–70
representing, 65–66
Transformations, See Adiabatic transformations; Constant-pres-
sure transformation;
Constant-volume transformation; Cyclic transformations;
Isobaric transformation;
Isothermal transformation;
Thermal transformation
Transmitted pulses, 449–52
amplitudes of, 450
uses for, 451
Transport phenomena, 548
and collisions, 562–65
Transportation, energy consumption
in, A–4
Transverse motion, 399, 447
Transverse waves, 399, 405–407
Traveling waves, 405–12
arbitrary waveforms, propagation
of, 411–12
harmonic, 405
longitudinal waves, 399, 407–409
shear waves in solids, 410–11
speed of, 406–407
speed of sound in air, 410
transverse waves, 399, 405–407
wave speed, 409–10
Tribology, 133
Triple point, 495
Tunneling, 195
Turbulence, 482–83
Turbulent flow, 474
Turning points, 192

Ultrasonic, 422
Uncertainty, 9–10
percentage, 10
Uncertainty principle, 387–88
and pulses, 454–56
Uncertainty relations, 455–56, See
also Heisenberg uncertainty
relation
Underdamped motion, 384
Underdamping, 384
Underdetermined systems,
322–23
Unification, 141–42
Uniform circular motion, 74–77,
136, 167–68, 388
acceleration in, 75–77
frequency of, 75
period of, 74
and simple harmonic motion, 368
and work, 167–68
Unit prefixes, 7–8
Unit vectors, 18
Units, 3–8, A–1, See also SI
(Système International)
British engineering system, 7, 96
cgs system, 7
conversions, 8–9, A–1
mks system, 5
prefixes for, 7–8

Universal gas constant, 500, 500,
507, 516, A–2
Universal gravitation, 141, 341–43
gravitational constant, 342–43,
See also Gravitation; Gravity
Universe, background temperature of,
503–506
Unstable equilibrium, 193
Upward buoyant force, 470

van der Waals equation of state, 503
van der Waals gas, 512, 546
interpretation of, 553–54
van der Waals, Johannes D., 503
Vaporization, 521, 604
latent heat of, 522, 543
Variable forces, in one dimension,
163–65
Vector equations, 21
Vector product, 283–84
Vectors, 15–18, 88
addition, 17–18
components of, 19
displacement, 16
magnitude, 16, 33
null, 18
position, 19, 30
resultant, 17
scalar multiplication, 18
subtraction, 18
in three-dimensional space, 21
unit, 18
vector equations, 21
Velocity, 31–36, 62–64, 87
angular, 247–49
average, 32–33
burnout, 244
constant, 89, 99
instantaneous, 33–35, 51, 63
instantaneous angular, 248
as an integral of acceleration over
time, 49
linear dependence on time, 40
relative, 221
root-mean-square, 550
of simple harmonic motion,
369–70
terminal, 118
units of, 33
zero, 89
Velocity distribution, 554
of gases, 557–59
average of the velocity squared,
558–59
Velocity distribution function, 554,
557, 560
Venturi flowmeter, 478
Virial coefficient, 514
Virial expansion, 514
Viscosity, 474, 482
coefficient of, 482
Volt (V), A–1
Volume, A–1
Volume expansion, coefficient
of, 497
Volume strain, 326
Volume stress, 326
Von Guericke, Otto, 113

Wake, 428
Wallis, John, 214
Water waves, 443
Watt, James, 173
Watt (W), 8, 173, A–1
Wave equation, 401–403
Wave fronts, 400

Wave number, 404, 440–41
of harmonic standing waves, 428
of traveling waves, 406
Waveform, 406
Wavelength, 403–407, 415, 428
beats, 439
fundamental, 416
of harmonic standing waves,
415–16
of traveling waves, 405–407
Waves, 397–434
amplitude of, 404
angular frequency of, 404,
406–407, 413, 416, 428
crest, 400, 405
Doppler effect, 423–26
Doppler shift, 426
energy in, 412–14
Fourier decomposition of, 452–54
harmonic, 411, 435
incident, interference of, 439
light, 443
longitudinal, 399, 407–409, 418
mechanical, 401, 409, 429
periodic, 403–405
periodic waves, 403–405
power delivered by, 412–14
reflected, 457
interference of, 439
seismic, 411
shock, 426–28
sound, 407, 409, 418–22, 443
hearing sounds, 421–22
nature of, 418–19
standing, 419–21
standing, 399, 414–18, 456
and energy, 417–18
frequencies of, 416
harmonics, 416–17
superposition, See Superposition
transverse, 399, 405–407
traveling, 405–12
arbitrary waveforms, propagation of, 411–12
longitudinal waves, 399,
407–409
shear waves in solids, 410–11
speed of sound in air, 410
transverse waves, 399, 405–407
wave speed, 409–10
triangular, Fourier decomposition
of, 453–54
types of, 398–401
wave equation, 401–403
wavelength of, See Wavelength
Weak force, 141
Weber (Wb), A–1
Weight, 7, 91, 121, See also Gravity,
force of
Weinberg, Steven, 141
Weisskopf, Victor, 355
Width at half-maximum, 387
Width, pulses, 455
Wien, Wilhelm, 504
Wilson, Robert, 506
Work, 152, 183
in angular motion, 298–300
colloquial meaning, 153–54
done by a spring, 164–66
done by conservative forces,
171, 183
done by individual forces, 156–57
done by thermal systems, 528–31
in transformations, 530
done by torque, 299
friction for, 169–70

Work (cont.)
 by individual forces, 156–57
 as integral, 167, 170
 and kinetic energy, 151–82
 net, 153
 path dependence of, 169–70
 relationship between energy
  and, 152
 as scalar product, 160–61

sign of, 153
for spring force, 170
and uniform circular motion,
 167–68
units of, 154–56
for variable forces,
 163–65
Work-energy theorem, *153*, 169,
 172, 175, 201

 for rotations, 299
Wren, Christopher, 214

Year (yr), A–1
Young, Thomas, 326
Young's modulus, *326–27*, 408
Yukawa potential energy
 function, 208

Zeno's paradox, 53
Zero, 4, 10, 12
 absolute, 495
 density, 502
 gravity, 91
 slope, 34
 velocity, 34, 63, 89
Zeroth law of thermodynamics, *492*